The chemistry of **amides**

THE CHEMISTRY OF FUNCTIONAL GROUPS

A series of advanced treatises under the general editorship of
Professor Saul Patai

The chemistry of alkenes (published in 2 volumes)
The chemistry of amides (published)
The chemistry of the carbonyl group (published in 2 volumes)
The chemistry of the ether linkage (published)
The chemistry of the amino group (published)
The chemistry of the nitro and nitroso groups (published in 2 parts)
The chemistry of carboxylic acids and esters (published)
The chemistry of the carbon–nitrogen double bond (published)

The chemistry of
amides

Edited by

JACOB ZABICKY

The Weizmann Institute of Science,
Rehovoth, Israel

1970

INTERSCIENCE PUBLISHERS
a division of John Wiley & Sons
LONDON — NEW YORK — SYDNEY — TORONTO

Library of Congress Catalogue Card No. 76–116520
ISBN 0 471 98049 8

Made and printed in Great Britain by
William Clowes and Sons Limited, London and Beccles

To **Judith**

A woman of valour who can find? . . .
Proverbs 31 : 10–29

Foreword

The amide function as considered in the present volume is based on

the trivalent group $-\overset{\overset{\displaystyle O}{\|}}{C}-N\big<$. On attaching various radicals to

the three free valences of this group, one may obtain many classes of compounds. In order to keep the volume within a reasonable size, some restrictions were imposed as to the type of radical that might be attached at each site. On the carbonyl side of the group only bonds to H or C atoms were allowed, while on the amino moiety of the group N could also be attached. These limitations leave compounds such as carbamates, ureas and semicarbazides, outside the scope of the book, but include in it amides, lactams, imides, diacylamines, triacylamines and hydrazides. The thiono analogues of these compounds are also discussed.

Studies on the peptidic carboxamido group are usually biologically oriented and extensively reviewed in other publications. Authors were asked to avoid or abbreviate discussions on compounds containing this link, unless they afford good illustrations of features present in other types of amides too.

The nomenclature rules of the International Union of Pure and Applied Chemistry [*Pure Appl. Chem.* **11**, Nos. 1–2 (1965)] and Chemical Abstracts [*Chem. Abstr.* **66**, Introduction to Subject Index (1967)] are not always adequate to designate all the intermediates and compounds mentioned in the text and are sometimes in conflict with widely accepted usage. No attempt was therefore made to attain a uniform nomenclature throughout the book, however any confusion that might arise from this fact should be dispelled by the profusion of formulae that accompany the text.

It is regretted that chapters on analysis and mass spectrometry of amides failed to materialize.

I wish to thank Prof. Saul Patai for his helpful advice and my wife for her patience and encouragement.

Rehovoth, March 1970   Jacob Zabicky

vii

The Chemistry of the Functional Groups
Preface to the series

The series 'The Chemistry of the Functional Groups' is planned to cover in each volume all aspects of the chemistry of one of the important functional groups in organic chemistry. The emphasis is laid on the functional group treated and on the effects which it exerts on the chemical and physical properties, primarily in the immediate vicinity of the group in question, and secondarily on the behaviour of the whole molecule. For instance, the volume *The Chemistry of the Ether Linkage* deals with reactions in which the C—O—C group is involved, as well as with the effects of the C—O—C group on the reactions of alkyl or aryl groups connected to the ether oxygen. It is the purpose of the volume to give a complete coverage of all properties and reactions of ethers in as far as these depend on the presence of the ether group, but the primary subject matter is not the whole molecule, but the C—O—C functional group.

A further restriction in the treatment of the various functional groups in these volumes is that material included in easily and generally available secondary or tertiary sources, such as Chemical Reviews, Quarterly Reviews, Organic Reactions, various 'Advances' and 'Progress' series as well as textbooks (i.e. in books which are usually found in the chemical libraries of universities and research institutes) should not, as a rule, be repeated in detail, unless it is necessary for the balanced treatment of the subject. Therefore each of the authors is asked *not* to give an encyclopaedic coverage of his subject, but to concentrate on the most important recent developments and mainly on material that has not been adequately covered by reviews or other secondary sources by the time of writing of the chapter, and to address himself to a reader who is assumed to be at a fairly advanced post-graduate level.

With these restrictions, it is realized that no plan can be devised for a volume that would give a *complete* coverage of the subject with *no* overlap between chapters, while at the same time preserving the readability of the text. The Editor set himself the goal of attaining *reasonable* coverage with *moderate* overlap, with a minimum of cross-references between the chapters of each volume. In this manner, sufficient freedom is given to each author to produce readable quasimonographic chapters.

The general plan of each volume includes the following main sections:

(a) An introductory chapter dealing with the general and theoretical aspects of the group.

(b) One or more chapters dealing with the formation of the functional group in question, either from groups present in the molecule, or by introducing the new group directly or indirectly.

(c) Chapters describing the characterization and characteristics of the functional groups, i.e. a chapter dealing with qualitative and quantitative methods of determination including chemical and physical methods, ultraviolet, infrared, nuclear magnetic resonance, and mass spectra; a chapter dealing with activating and directive effects exerted by the group and/or a chapter on the basicity, acidity or complex-forming ability of the group (if applicable).

(d) Chapters on the reactions, transformations and rearrangements which the functional group can undergo, either alone or in conjunction with other reagents.

(e) Special topics which do not fit any of the above sections, such as photochemistry, radiation chemistry, biochemical formations and reactions. Depending on the nature of each functional group treated, these special topics may include short monographs on related functional groups on which no separate volume is planned (e.g. a chapter on 'Thioketones' is included in the volume *The Chemistry of the Carbonyl Group*, and a chapter on 'Ketenes' is included in the volume *The Chemistry of the Alkenes*). In other cases, certain compounds, though containing only the functional group of the title, may have special features so as to be best treated in a separate chapter as e.g. 'Polyethers' in *The Chemistry of the Ether Linkage*, or 'Tetraaminoethylenes' in *The Chemistry of the Amino Group*.

This plan entails that the breadth, depth and thought-provoking nature of each chapter will differ with the views and inclinations of the author and the presentation will necessarily be somewhat uneven. Moreover, a serious problem is caused by authors who deliver their manuscript late or not at all. In order to overcome this problem at least to some extent, it was decided to publish certain volumes in several parts, without giving consideration to the originally planned logical order of the chapters.   If after the appearance of the originally planned parts of a volume, it is found that either owing to non-delivery of chapters, or to new developments in the subject, sufficient material has accumulated for publication of an additional part, this will be done as soon as possible.

The overall plan of the volumes in the series 'The Chemistry of the Functional Groups' includes the titles listed below:

The Chemistry of Alkenes (published in two volumes)
The Chemistry of the Carbonyl Group (published in two volumes)
The Chemistry of the Ether Linkage (published)
The Chemistry of the Amino Group (published)
The Chemistry of the Nitro and Nitroso Group (published in two parts)
The Chemistry of Carboxylic Acids and Esters (published)
The Chemistry of the Carbon–Nitrogen Double Bond (published)
The Chemistry of the Cyano Group (in press)
The Chemistry of the Amides (published)
The Chemistry of the Carbon–Halogen Bond
The Chemistry of the Hydroxyl Group (in press)
The Chemistry of the Carbon–Carbon Triple Bond
The Chemistry of the Azido Group (in preparation)
The Chemistry of Imidoates and Amidines
The Chemistry of the Thiol Group
The Chemistry of the Hydrazo, Azo and Azoxy Groups
The Chemistry of the Carbonyl Halides (in preparation)
The Chemistry of the SO, SO_2, —SO_2H and —SO_3H Groups
The Chemistry of the —OCN, —NCO and —SCN Groups
The Chemistry of the —PO_3H_2 and Related Groups

Advice or criticism regarding the plan and execution of this series will be welcomed by the editor.

The publication of this series would never have started, let alone continued, without the support of many persons.   First and foremost among these is Dr. Arnold Weissberger, whose reassurance and trust encouraged me to tackle this task, and who continues to help and advise me.   The efficient and patient cooperation of several staff-members of the Publisher also rendered me invaluable aid (but unfortunately their code of ethics does not allow me to thank them by name).   Many of my friends and colleagues in Jerusalem helped me in the solution of various major and minor matters and my thanks are due especially to Prof. Y. Liwschitz, Dr. Z. Rappoport and Dr. J. Zabicky. Carrying out such a long-range project would be quite impossible without the non-professional but none the less essential participation and partnership of my wife.

The Hebrew University,   SAUL PATAI
Jerusalem, ISRAEL

Contributing authors

Harold Basch — Ford Motor Company, Dearborn, Michigan, U.S.A.

A. L. J. Beckwith — University of Adelaide, Australia.

Joseph F. Bieron — Canisius College, Buffalo, New York, U.S.A.

F. A. Bovey — Bell Telephone Laboratories, Murray Hill, N.J., U.S.A.

R. U. Byerrum — Michigan State University, East Lansing, Michigan, U.S.A.

B. C. Challis — Imperial College of Science and Technology, London, England.

Judith Challis — Imperial College of Science and Technology, London, England.

Frank J. Dinan — Canisius College, Buffalo, New York, U.S.A.

R. B. Homer — University of East Anglia, Norwich, England.

C. D. Johnson — University of East Anglia, Norwich, England.

Hans Paulsen — University of Hamburg, Germany.

J. E. Reimann — Michigan State University, East Lansing, Michigan, U.S.A.

K. J. Reubke — University of Hamburg, Germany.

M. B. Robin — Bell Telephone Laboratories, Murray Hill, N.J., U.S.A.

Oscar Rosado — University of Puerto Rico, Mayaguez, Puerto Rico.

Ionel Rosenthal — Weizmann Institute of Science, Rehovoth, Israel.

J. A. Shafer — University of Michigan, Ann Arbor, Michigan, U.S.A.

Dieter Stoye — University of Hamburg, Germany.

J. Voss — University of Hamburg, Germany.

W. Walter — University of Hamburg, Germany.

Owen H. Wheeler — Puerto Rico Nuclear Center, Mayaguez, Puerto Rico.

Contents

CHAPTER **1**

Molecular and electronic structure of the amide group

M. B. ROBIN, F. A. BOVEY, AND HAROLD BASCH*

Bell Telephone Laboratories, Incorporated, Murray Hill,
New Jersey, U.S.A.

* Present Address: *Scientific Laboratories, Ford Motor Company, Dearborn, Michigan,*
U.S.A.

I. INTRODUCTION

Our present knowledge of the molecular structure of the amide group
(1) is founded largely upon the results of x-ray and electron diffraction

$$
\begin{array}{c}
O \qquad\quad R_{(\alpha)} \\
\diagdown\!\!\!\diagdown \quad\diagup \\
C\!\!-\!\!N \\
\diagup \qquad\diagdown \\
R' \qquad\quad R_{(\beta)} \\
(1)
\end{array}
$$

analysis, while that of the electronic structure is derived both from
diverse spectroscopic experiments (u.v.–i.r.–n.m.r.–microwave) and
the results of various semi-empirical and *ab initio* quantum mechanical
calculations.   As might be expected, problems still exist in regard to
both the molecular and electronic structures of amides, but sufficient
theoretical and experimental data are now at hand to allow more or
less detailed descriptions of not only the ground state of the amide
group but certain excited states as well.

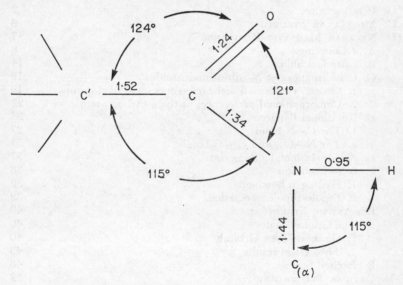

FIGURE 1.   Mean values of the distances (Å) and angles within the amide group
in a crystalline environment.

Much of the recent interest in the amide group is no doubt related
to the fact that it is the repeating unit in the biologically important
polypeptide macromolecules.   Since the amide groups within the polypeptide molecule appear to be only weakly interacting, they retain

their character to a large extent as isolated amide groups, and for this reason the study of isolated, i.e. monomeric amides, has been intense. However, because the polypeptide aspects of the amide structure have been adequately reviewed recently [1-3], we shall largely ignore them here, except where experiments on polypeptides have provided pertinent data which otherwise has not been available for the monomeric amides. In our discussion of the amide group, we shall also largely ignore intermolecular effects such as solvent shifts, dimer formation, hydrogen bonding, complexation, etc. Instead, we present first, some of the pertinent molecular structural data, and then go on to descriptions of the n.m.r. spectra, theoretical electronic structure, and finally, electronic and photoelectron spectra of various amides. We present this discussion in a book dedicated almost totally to the organic chemistry of the amide group in the hope that organic chemists may find in it some physical-chemical information either of value in explaining the results of their experiments, or so stimulating as to encourage new experiments.

II. MOLECULAR STRUCTURE

X-ray structure analyses of a variety of crystalline amides show a fairly constant geometry for the amide group (Table 1), the mean values of which are given in Figure 1. The heavier atoms of the primary amide group (C',C,O,N) are essentially coplanar, however, the evidence would seem to indicate that the protons of the NH_2 group are not in the plane of the heavier atoms. In many of the crystals listed in Table 1, the amides are found as centrosymmetric dimer pairs (**2**).

(**2**)

The results of structure determinations on monomeric amides in the gas phase are distinctly different from those in the crystalline state, the gaseous amides having the $C{=}O$ distance reduced to 1·19–1·21 Å and a concomitant opening of the $C{-}N$ distance to 1·36–1·37 Å. The angles however appear to be sensibly the same, with $C'\hat{C}N$ being 113–117°, and the other two about 120–125° each. A plot of the $C{=}O$ vs. $C{-}N$ distances of several amides in both the gas and crystalline phases suggests a reciprocal dependence, the shortest $C{=}O$ distances having

TABLE 1. Structural parameters for various simple amides (cf. Figure 1).

Compound	State	C=O	C—N	C—H or C—C'	N—H or N—C(α)	OĈN	C'ĈO	Ĉ'CN	HN̂H	CN̂C(α)	Ref.
$HCONH_2$	crystal	1·255	1·300			121·5°					4
$HCONH_2$	poly-crystalline		1·376								5
$HCONH_2$	gas	1·193		1·102	1·036 / 1·014[a] / 1·002[b]	123·8°			118·98°		6
CH_3CONH_2	crystal	1·28	1·38	1·51		122°	129°	109°			7
CH_3CONH_2	gas	1·21	1·36	1·53		125°	122°				8
$C'H_3CONHC_{(α)}H_3$	gas	1·230	1·293	1·550	1·44					117°	8
$C'H_3CONHC_{(α)}H_3$	crystal	1·24	1·31	1·48	1·457	123°	120°	117°		121°	9
$C_6H_5CONH_2$	crystal	1·28	1·31	1·47		122°	122°	116°			10
$CH_3(CH_2)_8CONH_2$	crystal	1·23		1·49		114·4°	129°	116·3°			11
$CH_3(CH_2)_{12}CONH_2$	crystal	1·243	1·26			116·3°	130·4°	113·1°			12
$H_2NCOCONH_2$-trans	crystal		1·315	1·542		125·7°	119·5°	114·8°			13
$H_2NCOCONH_2$-cis	poly-crystalline				1·039						5
$H_2NCO(CH_2)_2CONH_2$	crystal	1·238	1·333	1·512 / 1·501		122°	122·38°	115·60°			14
$H_2NCO(CH_2)_3CONH_2$	crystal	1·22	1·34	1·52	0·93 / 0·90	122°	121°	117°	111°		15
$H_2NCO(CH_2)_6CONH_2$	crystal	1·25	1·32	1·52	0·92 / 1·25	123°	120·5°	116·5°	118°		16
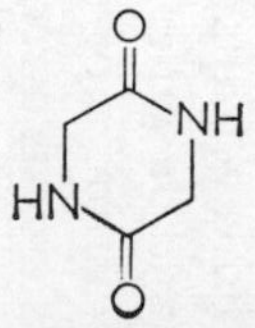 (cyclic diamide structure)	crystal	1·239	1·325	1·499	0·86	122·6°	118·5°	118·9°		126·0°	17
(cyclic diamide structure)	crystal	1·22	1·38	1·44	0·98	121°	121·5°	117·5°		121·5°	18

[a] Trans.   [b] Cis.

the longest C—N (Figure 2). However, the axes in both directions only encompass 0·1 Å, and the points for some crystalline amides (acetamide, decanamide, tetradecanamide) are so far off the line as to be off the graph. The significance of this reciprocal bond-length dependence will be discussed below. It is also to be noticed that there

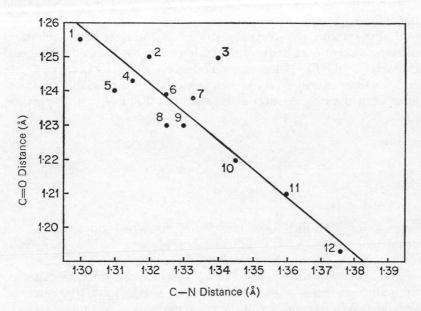

FIGURE 2. C=O versus C—N distances in (1) formamide (crystal), (2) suberamide, (3) aureomycin, (4) *trans*-oxamide, (5) benzamide, (6) diketopiperazine, (7) succinamide, (8) N,N,N',N'-tetramethyl-α,α'-dibromosuccinamide (*meso*-form), (9) chloroacetamide (α-form), (10) acetamide (gas), and (11) formamide (gas).

is a slight shortening of the C'—C bond in amides of about 0·02–0·06 Å below that of ethane, which can be ascribed to the smaller radius of the —C= atom in amides[19].

In theoretical studies of the amide group, it has always been assumed that all the atoms depicted in Figure 1 were coplanar, allowing maximal conjugation of the π electrons in the C=O bond and the 'non-bonding' pair of the —$\dot{N}R_2$ group. Actually, the gas-phase microwave absorption spectrum of formamide suggests this is not quite correct, for the hydrogens of the NH_2 group are 0·15 Å out of the NCO plane[6]. In diketopiperazine, it is also claimed that the amino proton is 0·02 Å out of the NCO plane[17]. These situations appear to be much like that of aniline, where simple π-electron arguments would predict sp^2 hybridization at the nitrogen atom and a planar, conjugated structure, whereas the hybridization at the nitrogen, in fact, is found to lie between sp^2 and sp^3, resulting in a non-planar structure[20]. Curiously, in crystalline benzamide[10], the molecule is grossly non-planar, with the amide portion turned 26° out of the plane of the phenyl ring.

It appears from the structural studies that in spite of possible non-planarity, there does exist appreciable conjugation and double-bond character in the C—N linkage. Thus the reciprocity between C—N and C=O bond lengths could well be explained by variable contributions from the two valence-bond structures **3** and **4**, with increasing

amounts of **4** leading to shorter C—N distances and longer C=O distances. In fact, the C—N distance of, say, 1·376 Å in gaseous formamide, is intermediate between the single bond C—N distance of 1·47 Å and the double bond C=N distance of 1·24 Å, as expected from such an explanation. However, it would seem that such an argument is superficial, for in the same molecule, one finds the C=O bond distance (1·19 Å) to be just that expected for a *pure* C=O double bond, the intermediate length being approximately 1.31 Å. Thus there appears to be something going on in the σ-electron systems of amides which affects the bond lengths, and which does not allow a simple π-electron explanation of the reciprocal relationship shown in Figure 2, if indeed, such a relationship exists at all.

A variable contribution of polar structure **4** to the ground states of different amides as suggested by the C—N bond lengths might reasonably be expected to express itself as a variation in the dipole moments of these molecules. Meighan and Cole[21] infer 'the presence of considerable difficulties' in the measurement of amide dipole moments in solution, but they present gas-phase values which are remarkably constant from molecule to molecule. They report at 110°C: N-methylformamide, 3·82 D; N,N-dimethylformamide, 3·80 D; acetamide, 3·75 D; N-methylacetamide, 3·71 D; N,N-dimethylacetamide, 3·80 D; and N-methylpropionamide, 3·59 D. The microwave determination of the dipole moment of gaseous formamide gave $3 \cdot 7$ D[22]. From these values, it seems that the amide group itself can be assigned a moment of $3 \cdot 75$ D, with variations in the amount of structure **4** and σ-bond moments due to various alkyl groups having a net effect of only $\pm 0 \cdot 1$ D.

Another interesting feature of the molecular structure of amides is the possible geometric isomerization about the C—N and C—C′ bonds of this group. This feature of the structure of amides, together with the thermodynamic aspects of isomerization are best studied using the techniques of n.m.r. spectroscopy as described in the following section.

III. NUCLEAR MAGNETIC RESONANCE

Amides have been intensively studied by n.m.r. Though most of these studies have dealt with rotational isomerization about the C—N bond, chemical shifts, J couplings, proton exchange, and association have received considerable attention as well. A general, overriding motivation for such studies is their relevance to polypeptides and proteins. This aspect, however, will not be emphasized here.

Although formamide has been shown by microwave studies to have a slightly pyramidal conformation about the nitrogen atom, it is permissible to regard the

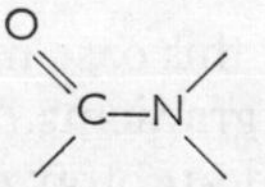

$$\underset{\diagup}{\overset{\text{O}}{\parallel}}\text{C}-\overset{\diagup}{\text{N}}$$

bond system as planar for most amides, at least so far as n.m.r. interpretations are concerned. The pyramidal conformation, if present, will invert very rapidly[6], and so will appear effectively planar.

Rotation about the C—N bond is slow, the barrier being of the order of 20 kcal/mole. This is attributed[23] to partial double-bond character derived from the structure **5**. As a result, the environments

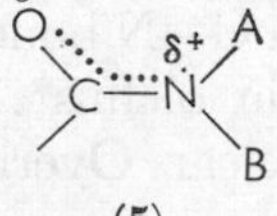

$$\overset{\delta-}{\text{O}}\cdots\overset{\diagup}{\underset{\diagdown}{\text{C}}}=\overset{\delta+}{\text{N}}\overset{\text{A}}{\underset{\text{B}}{}}$$

(5)

of groups A and B are not averaged, and they can usually be observed separately by n.m.r. even at temperatures well above room temperature. These features will be described further in section III.D.

A. J Couplings

The ^{14}N nucleus has a spin of 1 and consequently an electric quadrupole moment. It therefore tends to couple strongly to the motions of the molecular framework via the electric-field gradients in the molecule. Scalar coupling of ^{14}N to protons directly bonded to it varies from 30 to 70 c.p.s., depending upon the hybridization of the bond. For amides, $J_{^{14}N-^{1}H}$ is usually about 65 c.p.s. The $^{14}N—^{1}H$ coupling is most clearly observed in molecules such as the ammonium ion, the proton spectrum of which is a $1:1:1$ triplet with a spacing of 52 c.p.s. Here, the molecular environment is highly symmetrical and the interaction of the nuclear quadrupole with the molecular electric-field gradients is very weak, and would in fact be zero if it were not for vibrational perturbations of the tetrahedral symmetry. The coupling of the ^{14}N quadrupole to the tumbling of the molecule in less symmetrical environments, such as the amide group, tends to remove the $^{14}N—^{1}H$ coupling and if sufficiently effective will collapse the triplet completely. An example is provided by N-acetyl-L-valine (6)[24], the NH spectrum of which is a sharp doublet (from coupling

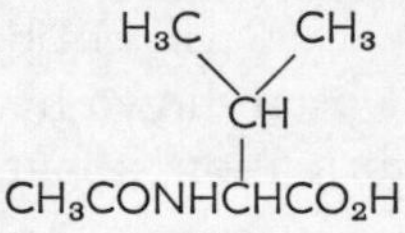

$$H_3C \diagdown \diagup CH_3$$
$$CH$$
$$|$$
$$CH_3CONHCHCO_2H$$

(6)

to the α-proton); probably in this case molecular asymmetry produces unusually large electric-field gradients.

More commonly, the amide-proton resonance appears as a broad singlet, 10–100 c.p.s. in width. Figure 3a shows the normal spectrum of formamide; the spectrum of the amide protons is broad and featureless. Upon irradiation at the resonance of ^{14}N, the 'decoupling' of the nitrogen and directly bonded protons can be made complete and separate resonances for each proton can be seen (Figure 3b). The assignments and couplings are shown in the figure and caption. The relative chemical shifts indicated are based on the assumption that *trans* vicinal couplings across a C—N bond with double-bond character will be larger than *cis*, as in olefins*. (The assignment could in principle be proved by nuclear Overhauser effect measurements; see below.)

* The *cis–trans* nomenclature has been recently changed as explained in section I.C.2 of Chapter 8. The new designation is not used in the present chapter.

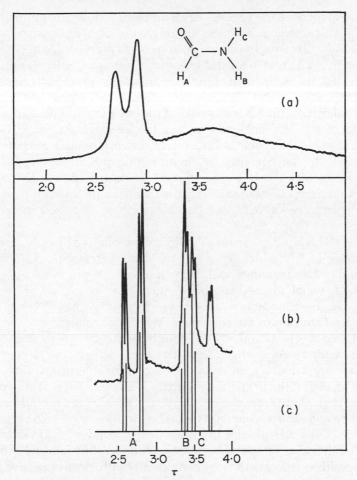

FIGURE 3. The 60 MHz n.m.r. spectrum of neat formamide, previous to, (a), and upon decoupling of the ^{14}N nucleus by irradiation, (b). From the calculated spectrum, (c), the following parameters are determined: $\tau_A = 2.70 \ \tau$; $\tau_B = 3.35 \ \tau$; $\tau_C = 3.56 \ \tau$; $J_{AB} = 2.1$ c.p.s.; $J_{AC} = 13.3$ c.p.s.; $J_{BC} = 2.1$ c.p.s.

The 'decoupling' effects of molecular motion in amides are, as one might expect, temperature dependent. Thus, Roberts[25] found that the broad singlet resonance of the NH_2 group of formamide begins to assume a triplet character above about 50°c; for acetamide and N-methylacetamide the transition was in the neighbourhood of 200°; for N-methylformamide it remained a singlet up to 250°, the highest temperature employed. Conversely, it was found[26] that on adding glycerol to increase the viscosity of an aqueous formamide solution, the NH_2 peak became a doublet, although not so well resolved as in Figure 3b. Likewise, the amide protons of polyacrylamide appear as a doublet[26]. The reason for this behaviour is that as molecular motion is increased, the component of the local magnetic noise spectrum at the resonant frequency of ^{14}N is decreased, and consequently the spin-lattice relaxation time is increased. Under these conditions, the spin lifetime of the ^{14}N nucleus becomes long enough for the ^{14}N—^1H coupling to be observable. Conversely, when motion is restricted by increasing the solvent viscosity or attaching the amide group to a larger molecule, the ^{14}N spin lifetime is shortened. It should be noted that these effects are entirely distinct from the marked temperature dependence of N,N-dialkylamide spectra, discussed in section III.D.

Since ^{14}N couplings are usually difficult to observe directly, use has been made of ^{15}N, which has a spin of $\frac{1}{2}$ and therefore no quadrupole moment. ^{15}N couplings will bear a constant proportion to ^{14}N couplings, equal to the ratio of magnetogyric ratios, γ^{15}/γ^{14} or ca. 1·41. Because the magnetic moment of ^{15}N is negative, ^{15}N couplings are opposite in sign to the corresponding ^{14}N couplings.

In Table 2, ^1H–^1H and ^{15}N–^1H couplings are given for a number of simple amides. As we have seen, the *trans* vicinal ^1H–^1H coupling (J_{13} in entry 1 and J_{12} in entry 7a) is comparable in magnitude to that in an olefin; the *cis* coupling (entries 1, 3b, and 7b) is considerably smaller than that in an olefin. Both are positive, as are all known vicinal proton–proton couplings on carbon frameworks. The geminal coupling of the NH protons in unsubstituted amides (J_{23} in entries 1 and 2) is likewise comparable in magnitude to that in vinyl groups, but is positive, whereas in vinyl compounds with electronegative substituents such as oxygen on the β carbon, this coupling is negative. The four-bond couplings to methyl groups (J_{13} in entries 3b and 4) are likewise comparable in magnitude and sign (negative) to the corresponding olefinic couplings except that the transoid coupling is slightly larger than the cisoid. This is the reverse of observations made in olefins. Thus, there is a fairly close similarity between amides and olefins, at least as far as n.m.r. parameters are concerned.

Nitrogen couplings to protons fall off rapidly with the number of intervening bonds, being always negative for ^{15}N (and positive for ^{14}N) across one bond. All ^{15}N–proton couplings are negative in the HCONH bond system. Two-bond ^{15}N couplings to protons on tetrahedral carbon are weaker and are positive (see $J_{^{15}N-^1H_{(1)}}$ vs. $J_{^{15}N-^1H_{(3)}}$ in

TABLE 2. *J* couplings in amides (c.p.s.).

Entry no.	Compound	Solvent	$J_{^1H{-}^1H}{}^{a}$	$J_{^{15}N{-}^1H}$	Ref.
1	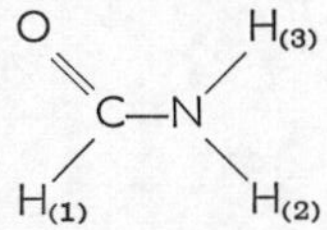 formamide: O=C(H$_{(1)}$)—N(H$_{(2)}$)(H$_{(3)}$)	neat	$^1H_{(1)}$—$^1H_{(2)}$, +2·1; $^1H_{(1)}$—$^1H_{(3)}$, +13·3; $^1H_{(2)}$—$^1H_{(3)}$, +2·1		27
2	N-methylformamide: O=C(H$_{(3)}$)—N(H$_{(2)}$)—CH$_{3(1)}$	H$_2$O	$^1H_{(1)}$—$^1H_{(2)}$, <0·2; $^1H_{(1)}$—$^1H_{(3)}$, <0·2; $^1H_{(2)}$—$^1H_{(3)}$, 2·4	^{15}N—$^1H_{(1)}$, −19·0; ^{15}N—$^1H_{(2)}$, −91·9; ^{15}N—$^1H_{(3)}$, −95·4	28
		H$_2$O			29
		H$_2$O		^{15}N—$^1H_{(2)}$, −89; ^{15}N—$^1H_{(3)}$, −89	30
3a	*cis*: O=C(H$_{(1)}$)—N(H$_{(2)}$)—CH$_{3(3)}$	neat	$^1H_{(2)}$—$^1H_{(3)}$, 4·5		31
		H$_2$SO$_4$	$^1H_{(1)}$—$^1H_{(3)}$, 0·9		
3b	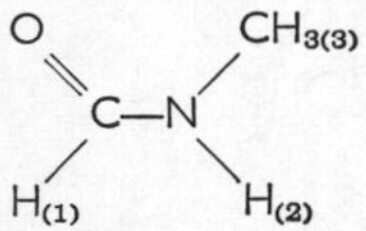 *trans*: O=C(H$_{(1)}$)—N(CH$_{3(3)}$)(H$_{(2)}$)	neat	$^1H_{(1)}$—$^1H_{(2)}$, +1·8; $^1H_{(1)}$—$^1H_{(3)}$, −0·9; $^1H_{(2)}$—$^1H_{(3)}$, +5·0		31, 32
		H$_2$O	$^1H_{(1)}$—$^1H_{(2)}$, 2·3; $^1H_{(1)}$—$^1H_{(3)}$, 0·8; $^1H_{(2)}$—$^1H_{(3)}$, 5·0	^{15}N—$^1H_{(1)}$, −14·1; ^{15}N—$^1H_{(2)}$, −93·6; ^{15}N—$^1H_{(3)}$, +1·4	28, 30–32
		H$_2$SO$_4$	$^1H_{(1)}$—$^1H_{(2)}$, 5·0; $^1H_{(1)}$—$^1H_{(3)}$, 1·3; $^1H_{(2)}$—$^1H_{(3)}$, 5·6		31

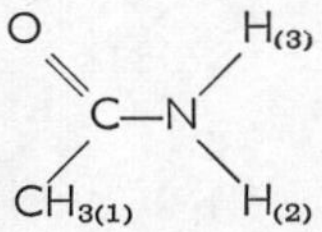

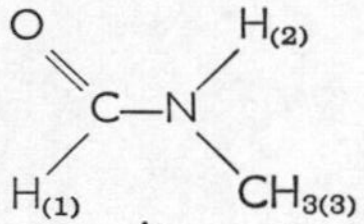

TABLE 2 (*Cont.*)

Entry no.	Compound	Solvent	$J_{^1H-^1H}{}^a$	$J_{^{15}N-^1H}$	Ref.
4	(CH$_{3(3)}$, CH$_{3(2)}$, N, O=C, H$_{(1)}$)	all solvents except H$_2$SO$_4$ H$_2$SO$_4$	${}^1H_{(1)}$—${}^1H_{(2)}$, −0·4 ${}^1H_{(1)}$—${}^1H_{(3)}$, −0·7 ${}^1H_{(2)}$—${}^1H_{(3)}$, <0·1 ${}^1H_{(1)}$—${}^1H_{(2)}$, 1·2 ${}^1H_{(2)}$—${}^1H_{(3)}$, 1·7		32–35 35
		neat		${}^{15}N$—${}^1H_{(1)}$, −15·6 ${}^{15}N$—${}^1H_{(2)}$, +1·1 ${}^{15}N$—${}^1H_{(3)}$, +1·2	30
5	(CH$_{3(3)}$, H$_{(2)}$, CH$_{3(3)}$, CH$_{3(2)}$, N, N, O=C, CH$_{3(1)}$, O=C, CH$_{3(1)}$)	neat H$_2$O	${}^1H_{(2)}$—${}^1H_{(3)}$, 4·7 ${}^1H_{(1)}$—${}^1H_{(3)}$, 0·4		29 36
6	(H$_{(2)}$, N, O=C, H$_{(1)}$)	HCONH$_2$ CCl$_4$	${}^1H_{(1)}$—${}^1H_{(2)}$, ca. 0 ${}^1H_{(1)}$—${}^1H_{(3)}$, 0·4 ${}^1H_{(2)}$—${}^1H_{(3)}$, ca. 0		34 37
7a	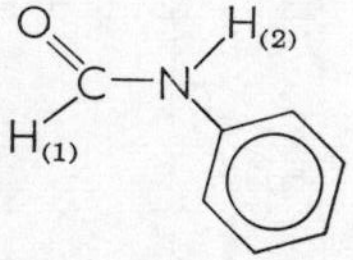 (H$_{(2)}$, N, O=C, H$_{(1)}$; phenyl) *cis*	CDCl$_3$	${}^1H_{(1)}$—${}^1H_{(2)}$, 11·0	${}^{15}N$—${}^1H_{(1)}$, −15·0 ${}^{15}N$—${}^1H_{(2)}$, −88·0	38
7b	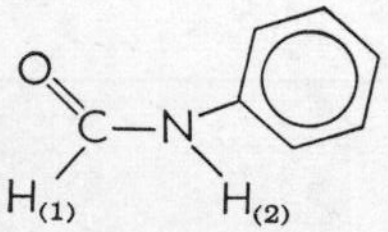 (phenyl; H$_{(2)}$, N, O=C, H$_{(1)}$) *trans*	CDCl$_3$	${}^1H_{(1)}$—${}^1H_{(2)}$, 2·0	${}^{15}N$—${}^1H_{(1)}$, −16·3 ${}^{15}N$—${}^1H_{(2)}$, −91·2	

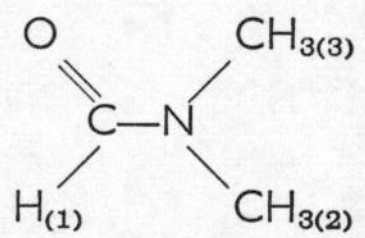

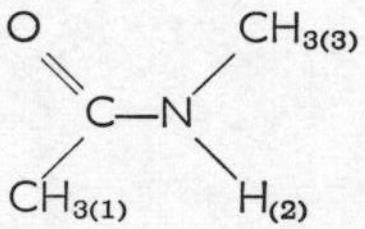

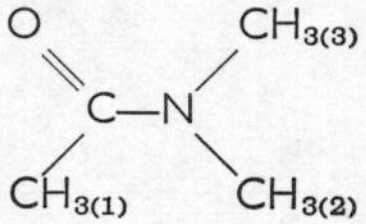

8	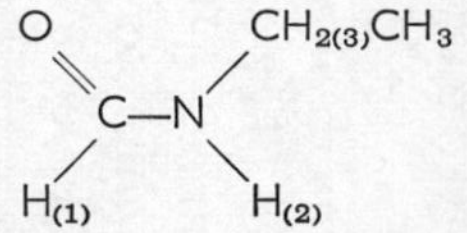	neat	$^1H_{(1)}$ — $^1H_{(3)}$, $^1H_{(2)}$ — $^1H_{(3)}$,	0·7 6·0	31
9	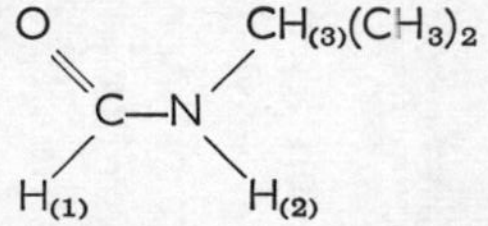	neat	$^1H_{(1)}$ — $^1H_{(3)}$, $^1H_{(2)}$ — $^1H_{(3)}$,	1·0 7·7	31

a Where no sign is given, it has not been determined, but is assumed to be positive.

entry 3b). There are indications that directly bonded N—H couplings increase with the fractional s-character of the nitrogen bonding orbital, in a manner reminiscent of the well-established dependence of ^{13}C—^1H couplings. The magnitudes of the ^{15}N—H couplings in amides are consistent with an approximately sp^2 bond hybridization, as might be expected. There are also a few scattered ^{13}C—^1H coupling measurements for formamides; Muller[39] and Malinowski[40] observed a value of 192 c.p.s. for the formyl coupling in dimethylformamide. From relationship (1)[39], where ρ_{CH} is the percentage of s character in the

$$\rho_{CH} = 0.20 \, J_{^{13}C-^1H} \qquad (1)$$

bonding C orbital, we find a value of 38% for this bond, again consistent with sp^2 hybridization at the carbon atom.

The vicinal coupling $J_{N\alpha}$ between the NH proton and the α-carbon proton in monoalkyl-substituted amides is of particular importance because of its relationship to the conformations of poly(α-amino acid) chains. There is evidence that it is dependent upon the H—N—$C_{(\alpha)}$—H dihedral angle. We shall discuss this question further in section III.C.

B. Chemical Shifts

As we have seen, substituent groups on the amide nitrogen atom can be distinguished because amides are planar or nearly so and rotation about the C—N bond is slow (sections III.A and III.D). In formamide, the more shielded of the two NH protons is taken to be that which is *cis* to the carbonyl. The spectrum of acetamide (Figure 4) is fundamentally similar, except that here quadrupolar relaxation is evidently more effective and the NH protons appear as a doublet without irradiation of the nitrogen. (The coupling of the NH protons to each other is weak and can be resolved only upon irradiation of ^{14}N; cf. Table 2, entry 2.)

In N,N-dialkylamides such as dimethylformamide and dimethylacetamide, the N-methyl groups are likewise in non-equivalent environments and give separate resonances. In Figure 5 is shown the spectrum of dimethylformamide in chloroform. The more shielded methyl protons are more strongly coupled to the formyl proton. In this instance, the assignment of the methyl protons is not based on assumptions concerning the relative magnitudes of the couplings, for it has been demonstrated by nuclear Overhauser experiments that the methyl group *cis* to the carbonyl is the more shielded. Before discussing this further, let us consider briefly some features of the shielding anisotropy and other shielding effects of the carbonyl group, since this undoubtedly makes a major contribution to the differentiation in chemical shift of the methyl groups.

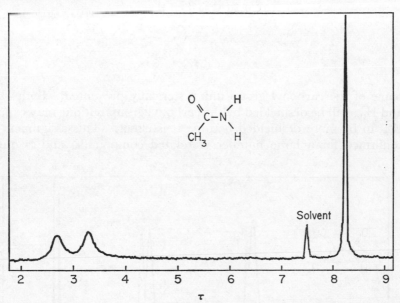

FIGURE 4.   The 100 MHz n.m.r. spectrum of a 10% (wt /vol) solution of acetamide in DMSO-d_6 at 25°c.   A residual resonance due to DMSO-d_5 appears at ca. 7·5 τ.   The doublet near 3 τ is the resonance of the NH$_2$ group; the methyl group appears at 8·22 τ.

There is ample qualitative evidence for shielding and deshielding effects by carbonyl groups.   These effects are thought to arise primarily from the magnetic anisotropy of the C=O double bond, with a contribution from electric-field effects arising from the strong electric dipole moment of the C=O group, particularly in amides[41-44].

Jackman[45] has suggested that the diamagnetism of the carbonyl group is as shown in Figure 6.   In the volume marked '+', increased shielding will be experienced by a neighbouring proton, while in the '−' volume, deshielding will be observed.   Both effects decrease toward zero as the conical nodal surface is approached.   Alternatively, Pople[46] has suggested that shielding effects arise from a paramagnetism centred on the carbon atom and in the plane of the bonds.   The very strong deshielding of aldehydic protons which appear at 0-1 τ, is in

accord with this picture; it may be significant that the formyl proton
of formaldehyde is markedly more deshielded than that of formamide
(Figure 3). Olefinic protons in α,β-unsaturated esters (**7**) lie in the

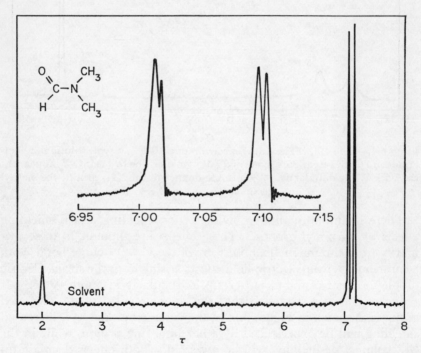

plane of the carbonyl group unless sterically prevented.   Both $H_{(a)}$
and $H_{(b)}$ will be deshielded by the carbonyl group but one may expect
$H_{(a)}$ to be more deshielded because it is closer.   This assignment is
confirmed for a large number of related compounds, and is quite

FIGURE 5.   The 100 MHz spectrum of a 10% (wt/vol) solution of N,N-di-
methylformamide in $CDCl_3$ at 25°.   Residual $CHCl_3$ is responsible for the peak
at ca. 2·7 τ. The formyl proton appears at 2 τ, and the methyl resonances give a
doublet near 7 τ.   As the 20-fold expanded methyl spectrum shows, each methyl
peak is itself a doublet resulting from coupling to the formyl proton.

secure when $R' = H$ because of the very well-established relative magnitudes of *cis*- and *trans*-olefinic couplings (section III.A). It is, however, the reverse of the assignment in formaldehyde.

In *N,N*-dimethylformamide (**8**), one would expect $CH_{3(b)}$ to be

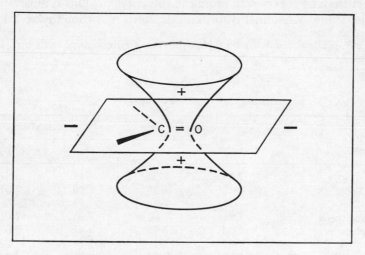

(**8**)

more shielded than $CH_{3(a)}$ if the representation in Figure 6 is reasonably correct, and if diamagnetic anisotropy is the dominant influence. In fact, as we have indicated, it has been shown by nuclear Overhauser effect measurements[47] that $CH_{3(a)}$ is more shielded than $CH_{3(b)}$, confirming earlier assumptions. An observable positive nuclear Overhauser effect, i.e. an increase in the resonance signal of one proton

FIGURE 6. The diamagnetic shielding, $(+)$, and deshielding, $(-)$, regions of the carbonyl group.

upon irradiation of another proton or group of protons to which it is coupled, requires that (*i*) the direct (anisotropic) dipole–dipole coupling between them be substantially larger than the electron-mediated (scalar) coupling, and (*ii*) the spin-lattice relaxation of the observed proton arise primarily from the proximity and motion of the proton(s) to be irradiated. These conditions may be expected to be fulfilled for N,N-dimethylformamide when the formyl proton is observed as the methyl protons are irradiated, since the scalar coupling is weak (Table 2, entry 4) and the formyl proton has no other intramolecular neighbours.   From the structural viewpoint, the most important factor is that the dipole–dipole coupling depends upon r^{-6}, r being the internuclear distance, or its average when one proton group is, as here, a rotating methyl group.   This very strong dependence enables one to distinguish $CH_{3(a)}$ from $CH_{3(b)}$, since $(r_{(a)}/r_{(b)})^6$ is ~ 6. Upon irradiation of the less shielded methyl protons, Anet and Bourn [47] found an 18% increase in the formyl peak intensity, whereas upon irradiation of the more shielded methyl protons no appreciable effect was observed.   However, it is by no means assured that analogous assignments can be assumed for all N,N-dialkylamides, and it will be of interest to extend these measurements to other compounds.

It has been found that the relative shieldings of alkyl groups in N,N-dialkylamides are highly dependent upon the solvent and in fact may be reversed in aromatic solvents [33].   This is illustrated in Table 3 for N,N-dimethylformamide and N,N-dimethylacetamide.

TABLE 3.   The dependence on solvent of N-methyl chemical shifts in

amides

$$ \underset{R}{\overset{O}{\diagdown}} C-N \underset{CH_{3(b)}}{\overset{CH_{3(a)}}{\diagup}} . $$

	R = H		R = Me	
Solvent	$\tau_{(a)}$	$\tau_{(b)}$	$\tau_{(a)}$	$\tau_{(b)}$
Neat	7·18	7·00	7·15	6·97
CDCl$_3$	7·12	7·03	7·10	7·01
DMSO	7·11	7·01		
Nitrobenzene	7·06	7·04	7·01	7·01
Anisole	7·48	7·75	7·31	7·62
Quinoline	7·27	7·71	7·11	7·62
Benzene	7·58	8·08	7·35	7·83
Thiophene	7·50	7·86	7·30	7·68

In the latter, the N-methyl couplings are not resolved, and the assignment was based on the greater breadth of the $CH_{3(a)}$ resonance.   This behaviour is probably due to a geometrically specific association of the aromatic ring of the solvent with the amide.   Hatton and Richards [33] have suggested that the preferred arrangement (9) has the nitrogen, with its fractional positive charge, over the centre of the benzene ring and the carbonyl as far away from the centre as possible, the amide and benzene planes remaining approximately parallel. This conformation places both methyls in the shielding region of the ring[48-50], but $CH_{3(b)}$ will be more strongly shielded than $CH_{3(a)}$.

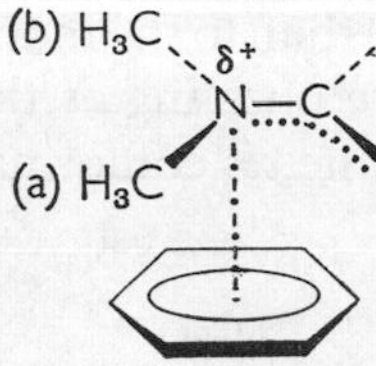

(9)

There is evidence[34] that in thioamides the relative shieldings of the N-methyl protons are reversed. The origin of this effect is not clear. Whatever it may be, one would expect aromatic solvents to enhance the separation of peaks, and this in fact is observed[34].

C. Conformations of N-Substituted Amides

1. Cis-trans conformational preferences of N-alkyl groups

In N-monosubstituted amides, the *trans* conformation (11) has been shown to be strongly preferred over the *cis* (10) by a variety of methods, including dipole-moment measurements and infrared, Raman, and ultraviolet spectroscopy (for a detailed bibliography see reference 31). By these means, however, it has not been possible, in general, to determine whether a minor proportion of the *cis* conformer exists in equilibrium with the *trans*. La Planche and Rogers[31] have observed the n.m.r. spectra of a number of N-monosubstituted amides (10) and (11)

in which $R_{(1)}$ was Me, Et, i-Pr and t-Bu, and $R_{(2)}$ was H, Me, Et and i-Pr. The neat compounds were observed at 35°c except for N-ethylisobutyramide ($R_{(1)}$ = Et; $R_{(2)}$ = i-Pr), N-isopropylisobutyramide ($R_{(1)}$ and $R_{(2)}$ = i-Pr), and N-t-butylacetamide ($R_{(1)}$ = t-Bu; $R_{(2)}$ = Me), which are solids under these conditions and were observed in carbon tetrachloride solution. Only when $R_{(2)}$ was H could an N-substituent resonance corresponding to the *cis* isomer (10) be detected;

its fraction increased from 0·08 for N-methylformamide ($R_{(1)}$ = Me; cf. Table 2, entry 3) to 0·12 when $R_{(1)}$ = Et or i-Pr, and to 0·18 when $R_{(1)}$ = t-Bu. This is to be expected as a result of steric interference between the N-substituent and the carbonyl oxygen atom. It is noteworthy, however, that even the bulky t-butyl group strongly prefers to be *cis* to the oxygen rather than to the much less sterically-demanding hydrogen. This is the more puzzling as the hydrogen-bonded dimer (**12**) would seem to require the *cis* conformation.

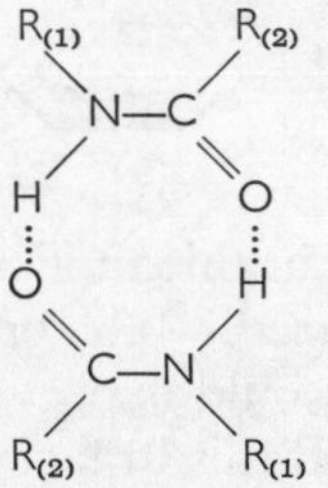

(**12**)

Bourn and coworkers[38] have found that the *cis* conformation is much more favoured in formanilide ($R_{(1)}$ = C_6H_5; $R_{(2)}$ = H) than in any of the N-alkylformamides just discussed. In 1·5 mol % deuterochloroform solution at 35°c, its fraction is 0·45; it increases to 0·73 at 52·5 mol %, as would be expected if dimer formation promoted the *cis* conformation (cf. Table 2, entry 7).

Observations of conformational preferences in unsymmetrically substituted N,N-dialkylamides can be more rationally explained in terms of steric competition. It is found that in formamides[51,52] the bulkier substituent tends to be *cis* to the formyl proton, but *trans* to the methyl group in acetamides. Thus, for formamides having one N-methyl group, La Planche and Rogers[52] find the following fractions of the preferred *cis* conformer (**13**): $R_{(1)}$ = Et, 0·60; n-Bu, 0·61; cyclo-

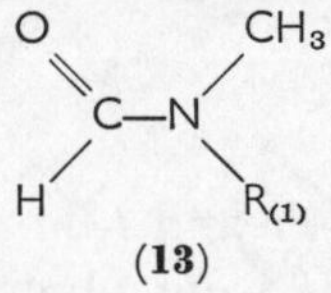

(**13**)

C_6H_{11}, 0·66; i-Pr, 0·67; t-Bu, 0·89. For acetamides, the preference for **14** was as follows: $R_{(1)}$ = Et, 0·51 (or no preference within experimental error); n-Bu, 0·53 (probably the same comment applies); cyclo-C_6H_{11}, 0·55; i-Pr, 0·58; t-Bu, indeterminate, only one t-butyl proton peak being observed.

For amides of the structure **15** it is found[53] that the rotamer in which $R_{(1)}$ = Me, and $R_{(2)}$ = benzyl, is present to the extent of 0·73 in quinoline and carbon tetrachloride at about 40°c. When $R_{(2)}$ = cyclohexyl, this rotamer fraction is 0·70 in carbon tetrachloride. The

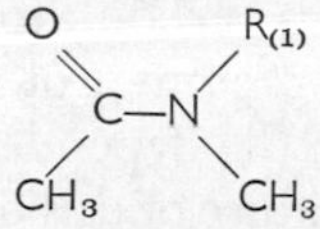

(14)

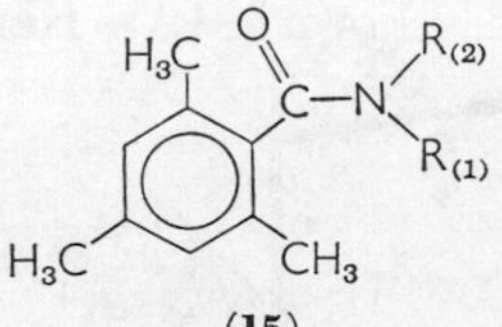

(15)

conformer in which $R_{(1)}$ = i-Pr and $R_{(2)}$ = benzyl is present to the extent of 0·78 in quinoline.

The infrared analysis by Miyazawa[54] of the vapour- and condensed-phase spectra of *N*-methylformamide suggests the presence of a small amount of *cis* isomer along with the *trans*, while in *N*-methylacetamide only *trans* isomer was observed[55]. Though these observations parallel the n.m.r. results quoted above, Jones thinks that the vapour of *N*-methylformamide is predominantly *cis*[56]. Crystallographic studies of acetanilide (**16**)[57], *N*-methylacetamide[9], and acetylglycine (**17**)[58]

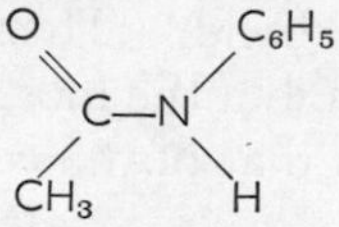

(16)

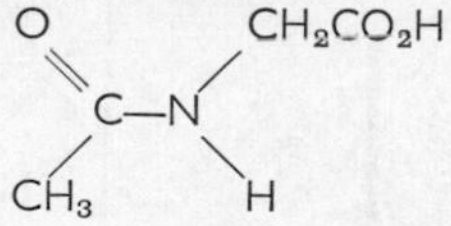

(17)

also showed these molecules to have the *trans* conformation. In general, it appears that the *trans* conformation of an *N*-alkylamide is the more stable one, except when steric repulsion is overpoweringly unfavourable for the *trans* conformation, and in the case of the small-ring lactams where the conformation is necessarily *cis*, but in which it also becomes *trans* if the ring is sufficiently large[59].

2. Conformational preferences at the α-carbon atom

As we have indicated, the conformation at the α-carbon atom in monosubstituted amides is of particular significance in relation to the conformations of poly-(α-amino acid) chains. Knowledge of conformational preferences at this carbon atom depends primarily upon measurements of the vicinal proton–proton coupling in the fragment **18**, $J_{N\alpha}$, and a knowledge of the dependence of $J_{N\alpha}$ on this dihedral angle. Unfortunately, the nature of this dependence has not yet been determined. We shall define the dihedral angle in the way customary among organic chemists, as shown in **18**. Here the $C_{(\alpha)}$—H bond is in

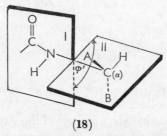

(18)

the plane II, defined by N—$C_{(\alpha)}$—H and groups A and B are found above and below this plane. We shall designate this angle φ' to distinguish it from the somewhat different definition of this angle now adopted by peptide chemists[60], and designated φ.

In order to make use of even the scanty data available, it is helpful first to decide upon the form of the potential-energy function for rotation about the N—$C_{(\alpha)}$ bond. There is by no means universal agreement as to this function, but the most plausible assumption seems to be that it is analogous to that of the conformers of propene[61] and

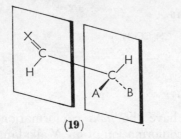

(19)

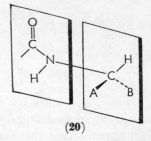

(20)

acetaldehyde[62] as deduced from microwave studies. In these molecules, the preferred conformer is that in which one of the substituents on the tetrahedral carbon atom eclipses the vinyl or carbonyl group, as in **19**. The analogous conformation of an N-substituted amide (**20**) with A and B other than H, corresponds to $\varphi' = 180°$. When A = B, we may suppose that, assuming a three-fold potential barrier, there will be in addition to this *trans* conformer, two equally populated mirror-image *gauche* conformers, corresponding to $\varphi' = \pm 60°$. When A ≠ B, as in a polypeptide chain, these *gauche* conformers will in general be unequally populated. It should be noted that the minima usually assumed in polypeptide conformational energy calculations differ from these by $\pm 60°$, and that furthermore, φ' for the α-helix is ca. 120°. This is comparatively unimportant, however, as the conformational preferences of polypeptide chains are determined primarily by other factors, chiefly side-chain steric interactions.

Examining in Table 2 the series N-methylformamide (entry 3b), N-ethylformamide (entry 8) and N-isopropylformamide (entry 9), we observe a marked increase in J_{23}, i.e. $J_{N\alpha}$, as the bulkiness of the alkyl group increases. Closely analogous behaviour has been observed by Bothner-By and coworkers for a series of alkylethylenes[63]. The most probable explanation seems to be that $J_{N\alpha}$ has a Karplus-like dependence on φ', i.e. is at a maximum when $\varphi' \sim 180°$, and presumably a minimum near 90° with relatively small values near 60°. However, more data are needed to establish this conclusion.

Hammaker and Gugler[64] have suggested that for N,N-diethyl-acetamide the conformation 21 is preferred, while for N,N-diisopropyl-amides, the methyl groups are turned away from the N atom as in 22.

(21)   (22)

Conformation 21 appears to be consistent with steric requirements, but 22 is decidedly unlikely. It has been further suggested[64] that they are supported by the chemical shifts of the CH_2 and CH protons compared to those of the methyl groups, but in view of the uncertain rationale of chemical shifts of N-alkyl protons, this cannot be considered strong support. These questions are considered in section III.D.2.

D. Rotational Barriers

We have alluded at the beginning of section III to the high barrier to rotation about the C—N bond in amides, enabling one to observe separate signals for otherwise equivalent N-substituents. Restricted rotation about this bond was first demonstrated by Phillips[65], who observed two N-methyl peaks in the spectra of N,N-dimethylform-amide and N,N-dimethylacetamide. Although the period of rotation at room temperature is relatively short on the ordinary time scale of chemical reactions, ca. 0·1 sec, it is long on the n.m.r. time scale. In Figure 7 is shown the spectrum of neat N,N-dimethylformamide at various temperatures. At 35°c, the spectrum shows two relatively narrow peaks, similar to those in Figure 5. As the temperature is raised further, these begin to broaden and at 118°, they coalesce to a single peak which becomes increasingly narrow as the temperature is further increased. It is now well known that this temperature dependence is a consequence of the lifetime τ of the methyl protons in each state becoming shorter as the temperature is increased and the molecules surmount the rotational barrier at an increasing rate. Peak coalescence occurs when $\tau < \sqrt{2}/[2\pi(\nu_A - \nu_B)]$, where ν_A and ν_B are the resonant frequencies of two like N-substituents. It may be crudely thought of as an uncertainty broadening, although this does not suffice to explain the line narrowing which occurs above the coales-cence temperature. A full explanation can only be given in terms of the modified Bloch equations expressing the behaviour of the macro-scopic nuclear magnetic moment of the system as a function of the rate of exchange of nuclei between the two methyl sites[66-69].

It is not within the scope of this article to describe in detail the application of n.m.r. to the study of chemical rate processes. Reviews will be found in references 70–73. From careful measurements of the line shape and application of the Bloch equations, rate constants may be determined as a function of temperature and from these the activation entropy $\Delta S^{\ddagger}$ and activation enthalpy $\Delta H^{\ddagger}$ (in the Eyring formulation) may be obtained from an Arrhenius plot in the usual way.

In its simplest form, the line-shape function[66,67] describing the spectra of Figure 7 takes account only of broadening arising from the kinetic exchange process, but not from spin–spin relaxation, charac-terized by the relaxation time T_2, nor is account taken of scalar coupling of the nuclei to each other or to other nuclei—the formyl proton in the present case. Neglect of such factors can lead to serious errors, particularly in the estimation of activation enthalpies. Their effect is to produce overestimates of rates below peak coalescence and underestimates above coalescence, thus tipping the Arrhenius plot toward lower values of $\Delta H^{\ddagger}$. This no doubt accounts for the reports

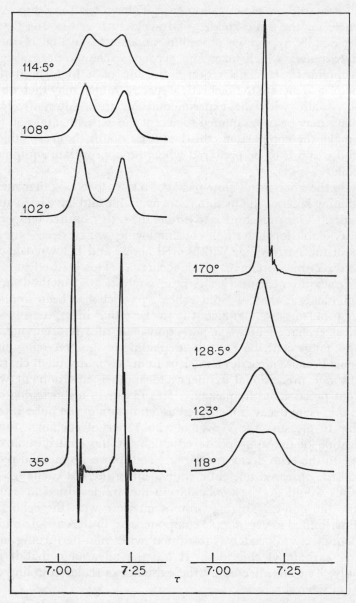

FIGURE 7.   The n.m.r. spectrum of neat N,N-dimethylformamide at various temperatures.

by different authors of activation energies from less than 8 to over 20 kcal/mole for the same amide. To avoid such errors, the best and most generally applicable procedure appears to be that of the *total line-shape analysis* which takes into account exchange narrowing, T_2, and coupling[74-77]. This requires the use of a high-speed digital computer to generate the theoretical spectra, which may then be compared visually with the experimental spectra. Alternatively, the computer may be programmed to accept the experimental points and itself make the comparison; this last is no doubt the least subjective procedure and is to be preferred when the appropriate equipment is available.

Beside the analysis of 'slow-passage' n.m.r. spectra, other methods of obtaining kinetic data by n.m.r. are available and have been applied to amides. One procedure is to extend the rather limited temperature range available for such studies by employing 'wiggle decay' measurements of the narrow line widths well above and below coalescence, where direct measurements are inaccurate. This is in effect a more precise measurement of T_2. A more sophisticated method for such measurements is that of 'spin echos'[78], which has been applied to kinetic problems[79,80]. Suffice it to say here that in a system in which chemical exchange between non-equivalent sites is occurring, each such exchange contributes to the dephasing of the precessing nuclear moments because it carries a nucleus from an environment characterized by one precessional frequency to another environment with a different precessional frequency. This results in a shortening of T_2. If the spin-echo decay is measured as a function of the pulse rate, it is possible to measure the τ over a wide range of exchange rates by application of the appropriate equation for the contribution of exchange to the echo decay rate[79,80]. The spin-echo method requires that all the observed nuclei be equivalent in the fast exchange limit, and thus would not be applicable to the amides listed in Table 2, because the non-exchanging protons interfere with the echo signal. It is thus limited to very simple compounds or, in the case of amides, to compounds appropriately substituted with non-perturbing nuclei, such as carbamoyl chlorides. It has the advantage that it is not seriously, if at all, affected by the existence of scalar coupling of the nuclei of interest.

A further method which has been applied to amides, is to isolate one of the two conformers by crystallization[81,82] or complexation[83] at a temperature at which equilibration is slow, and then to observe the approach to equilibrium by monitoring the intensities of the n.m.r.

peaks of the two forms; this is not inherently an n.m.r. method, although this is by far the most convenient way of carrying it out.

I. The C—N bond

Most of the significant reported data are summarized in Table 4. The amides are listed in order of complexity of the group attached to the carbonyl, and within each such class, in order of complexity of the groups attached to nitrogen. Kinetic data are expressed in terms of the Eyring formulation, on the assumption (not necessarily valid) that the transmission coefficient is unity, and therefore that $\Delta S^{\ddagger}$ is given by $2 \cdot 303R(\log A - \log kT/h)$, A being the frequency factor.

In some respects, these data are discouraging. The disagreement in the reported magnitude of $\Delta H^{\ddagger}$ and in the magnitude and *sign* of $\Delta S^{\ddagger}$ for the same compound makes it difficult to draw any conclusions concerning the effects even of major structural variations, much less those of solvent and concentration. For N,N-dimethylformamide, values of $\Delta H^{\ddagger}$ vary from 7 to 26·3 kcal/mole and of $\Delta S^{\ddagger}$ from -56 to 14·3 e.u. Similar discrepancies can be seen for N,N-dimethylacetamide and many of the other amides. The data for N,N-dimethylcarbamoyl chloride are particularly instructive. As Neuman and coworkers[77] point out, the reported values of $\Delta H^{\ddagger}$ and $\Delta S^{\ddagger}$ vary widely for this compound, but nearly all the calculated values of $\Delta G^{\ddagger}$, determined near the coalescence temperature, are within the range $16 \cdot 6 \pm 0 \cdot 2$ kcal/mole. Much of the other data shows the same trend. We have already indicated that the principal reason for this state of affairs is that the rate constants near coalescence can be determined fairly accurately, while at other temperatures the errors of the measurements are such as to tip the Arrhenius plot towards lower values of $\Delta H^{\ddagger}$ and consequently more negative values of $\Delta S^{\ddagger}$. It is notable that the most recent and most careful measurements yield the highest values of $\Delta H^{\ddagger}$ and less negative or even positive values of $\Delta S^{\ddagger}$. Strongly negative entropies of activation for internal rotation in amides are difficult to understand, as, on the basis of the resonance formulation (**3** and **4**), the transition state should be less polar and less solvated than the ground state. However, this assumes that only the π-electron density changes on going through the transition state.

The effect of solvent on the barrier height has been studied by several authors (Table 4). From the above considerations, one would expect that polar solvents would preferentially stabilize the ground state and thus increase the barrier height. Dipole–dipole association might also be expected to stabilize the ground state and thus increase

TABLE 4. Activation enthalpies and entropies for rotation about the central C—N bond of amides.

Entry no.	Compound	State	Method[a]	$\Delta H^{\ddagger\,b}$ (kcal/mole)	$\Delta S^{\ddagger}$ (e.u.)	Ref.
1	HCONH$_2$ (^{15}N-labelled)	10 mol % in acetone	l.a.	18 ± 3		28
2a	HCON(CH$_3$)$_2$	neat	l.a.	7 ± 3		68
2b		neat	l.a.	8.9 ± 1.5	−56 ± 6	35
2c		in 100% H$_2$SO$_4$	l.a.	12.0 ± 1.5	−30	35
2d		neat	l.a.	18.3 ± 0.7	−22	84
2e		neat	l.a.	15.9 ± 2.0	−9.7	85
2f		0.0858 mol frac. in acetone-d_6	l.a.	16.8 ± 2.0		85
2g		0.105 mol frac. in CFCl$_3$	l.a.	11.3 ± 2.0		85
2h		0.0633 mol frac. in HMDS[c]	l.a.	9.4 ± 1.0		85
2i		in hexachlorobutadiene	l.a.	17.6	−11.4	53
2j		neat	l.a.	18.0 ± 0.9	−5.1 ± 0.6	86
2k		12 mol % in formamide	l.a.	25.6 ± 2.6	12.0 ± 1.5	86
2l		neat	l.a.	26.3	14.3	87
2m		neat	c.l.a.	19.9 ± 0.2	−0.3	76
3	HCON(C$_2$H$_5$)$_2$	neat	l.a.	19.6 ± 0.5^d		64
4a	HCON(CH$_3$)(CH$_2$C$_6$H$_5$)	neat	c.l.a.	23.2^e	5.6	88
4b		neat	equil.^{f}	19.5	−6.5	88
5	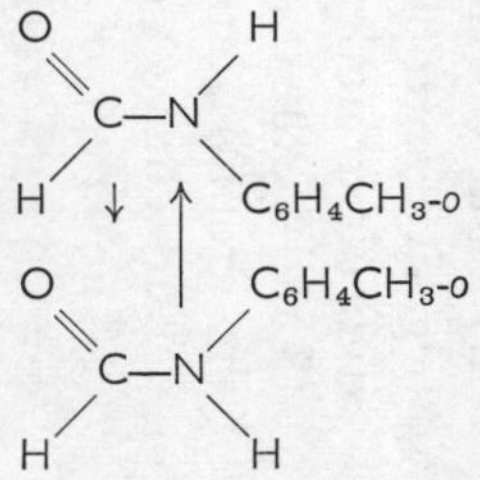 (structure)	CDCl$_3$	equil.^{g}	ca. 15.0	ca. 14.7	82

Entry 5 structure:

$$O=C \begin{array}{c} H \\ \\ \cdots N \end{array} \begin{array}{c} C_6H_4CH_3\text{-}o \\ \\ \end{array} \rightleftharpoons \begin{array}{c} C_6H_4CH_3\text{-}o \\ \\ N \cdots \end{array} C=O \begin{array}{c} H \\ \\ \\ H \end{array}$$

6a	CH$_3$CON(CH$_3$)$_2$	neat	l.a.	12 ± 2	−22 ± 4	68
6b		neat	l.a.	10·6 ± 0·4	−23·5	84
6c		neat	l.a.	23·3	14·3	87
6d		0·10 mol frac. in formamide	l.a.	24·0 ± 0·8	16·1	86
6e		in hexachlorobutadiene	l.a.	9·9	−25·2	53
7a	CF$_3$CON(CH$_3$)$_2$	neat	l.a.	9·3 ± 0·6	−28 ± 4	84
7b		neat	s.e.[h] short τ limit			89
			sinh^{-1}F fit	19·9 ± 1·4	4·22	90
8a	CCl$_3$CON(CH$_3$)$_2$	neat	s.e.	17·8 ± 2·7	−2·33	90
			l.a.[i]	13·9 ± 0·6	−1·8	80
8b		neat	p.s.	9·6 ± 1·7	−15·1 ± 1·3	80
8c		neat	p.v.r	11·9 ± 1·2	−8·7 ± 0·9	80

(continued)

[a] l.a.: Slow-passage line-shape analysis.
c.l.a.: Complete slow-passage line-shape analysis (see text).
equil.: Equilibrium measurements.
s.e.: Spin echo.
p.s.: Peak separation. See note i.
p.v.r.: Peak-to-valley ratio. See note i.
l.n.: Line narrowing. See note i.
[b] Many authors express the temperature dependence of rate as activation energy, ΔE or E_a; this is converted to $\Delta H^{\ddagger}$ by the relationship $\Delta H^{\ddagger} = \Delta E - RT$ except where RT (ca. 0·7 kcal/mole) is substantially smaller than the reported experimental uncertainty.
[c] Hexamethyldisiloxane, (CH$_3$)$_3$SiOSi(CH$_3$)$_3$.
[d] $\Delta G^{\ddagger}$ at 298°K; $\Delta H^{\ddagger}$ and $\Delta S^{\ddagger}$ not reported.
[e] Averaged value of both barriers from observation of benzyl- and formyl-proton resonances.
[f] By measurement of return to equilibrium of mixture in which the rotamer with the benzyl group cis to carbonyl is favoured by crystallization of uranyl nitrate adduct.
[g] For trans to cis.
[h] See text and references therein for details of the two measurements indicated.
[i] These entries refer to specific methods of line-shape analysis; 'peak separation' refers to the observation that just below coalescence the components of the doublet appear to move together; 'peak-to-valley' ratio refers to the measurement of the ratio of either doublet peak height to the intensity midway between; 'line narrowing' refers to the line above coalescence.

TABLE 4 (*Cont.*)

Entry no.	Compound	State	Method[a]	$\Delta H^{\ddagger}$[b] (kcal/mole)	$\Delta S^{\ddagger}$ (e.u.)	Ref.
8d		neat	l.n.	13·4 ± 0·5	−4·1 ± 0·3	80
8e		neat	p.v.r.	9·9 ± 0·3	−17·5 ± 0·2	84
9	$CH_3CON(C_2H_5)_2$	neat	l.a.	16·9 ± 0·2[a]		64
10	$CH_3CON(n\text{-}Pr)_2$	neat	l.a.	16·3 ± 0·4[a]		64
11	$CH_3CON(i\text{-}Pr)_2$	neat	l.a.	15·0 ± 0·2[a]		64
12a	$CH_3CON(CH_2C_6H_5)_2$	38·1 mol % in CH_2Br_2	l.a.	6·6 ± 0·5	−31·0 ± 0·3	91
12b		38·3 mol % in CCl_4	l.a.	5·7 ± 0·6	−34·0 ± 0·4	91
13a	$C_2H_5CON(CH_3)_2$	neat	l.a.	9·2 ± 0·7	−25·8 ± 0·5	84, 91
13b		84·7 mol % in CH_2Br_2	l.a.	9·4 ± 0·8	−23·0 ± 0·5	91
		68·9 mol % in CH_2Br_2	l.a.	9·5 ± 0·7	−22·5 ± 0·5	91
		58·0 mol % in CH_2Br_2	l.a.	9·5 ± 0·6	−22·5 ± 0·4	91
		40·6 mol % in CH_2Br_2	l.a.	9·2 ± 0·7	−23·8 ± 0·5	91
		22·2 mol % in CH_2Br_2	l.a.	8·4 ± 0·2	−27·0 ± 0·1	91
		10·2 mol % in CH_2Br_2	l.a.	6·0 ± 0·7	−34·4 ± 0·5	91
		69·3 mol % in CCl_4	l.a.	7·9 ± 0·6	−32·2 ± 0·4	91
		39·9 mol % in CCl_4	l.a.	6·8 ± 0·4	−29·0 ± 0·3	91
		11·1 mol % in CCl_4	l.a.	5·6 ± 0·4	−33·0 ± 0·3	91
14	$H_2C{=}CHCON(CH_3)_2$	neat	l.a.	6·1 ± 0·1	−31·6 ± 0·5	84
15	$C_6H_5CON(CH_3)_2$	36·3 mol % in CH_2Br_2	l.a.	7·0 ± 0·5	−25·0 ± 0·4	84
16	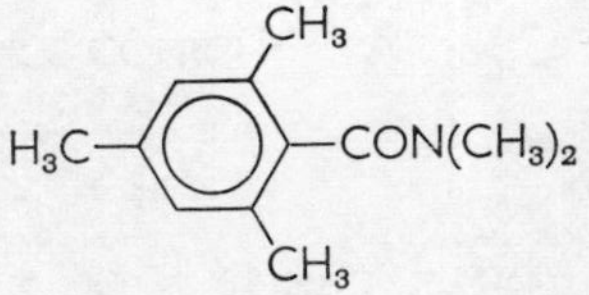	hexachlorobutadiene; 1:1 α-chloronaphthalene/ benzotrichloride	l.a.	22·5 ± 0·3[f]	0 ± 5	53, 81

17a	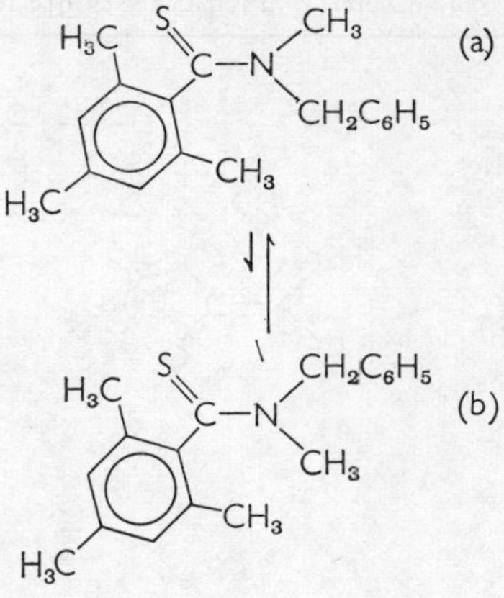 (a)	1:1 α-chloronaphthalene/benzotrichloride	equil.[k]	$22\cdot9 \pm 2\cdot7$ $(a) \rightarrow (b)$	-1 ± 9	53, 81
17b	(b)	1:1 α-chloronaphthalene/benzotrichloride	equil.[k]	$22\cdot2 \pm 2\cdot2$ $(b) \rightarrow (a)$	-1 ± 7	53, 81
18a	ClCON(CH₃)₂	neat	l.a.,[l] p.v.r.	$6\cdot6 \pm 0\cdot5$	$-31\cdot0 \pm 0\cdot3$	84, 91
18b		90·0 mol % in CH_2Br_2	l.a.	$7\cdot5 \pm 0\cdot3$	$-28\cdot6 \pm 0\cdot9$	90
		63·4 mol % in CH_2Br_2	l.a.	$8\cdot4 \pm 0\cdot6$	$-25\cdot6 \pm 1\cdot8$	90
		40·9 mol % in CH_2Br_2	l.a.	$7\cdot9 \pm 0\cdot6$	$-27\cdot1 \pm 1\cdot8$	90
		10·7 mol % in CH_2Br_2	l.a.	$6\cdot5 \pm 0\cdot6$	$-31\cdot3 \pm 1\cdot8$	90
18c		71·3 mol % in CCl_4	l.a.	$6\cdot2 \pm 0\cdot5$	$-32\cdot2 \pm 1\cdot8$	90
		40·1 mol % in CCl_4	l.a.	$5\cdot9 \pm 0\cdot5$	$-28\cdot6 \pm 1\cdot4$	90
		11·0 mol % in CCl_4	l.a.	$6\cdot1 \pm 0\cdot2$	$-31\cdot6 \pm 0\cdot9$	90
		1·2 mol % in CCl_4	c.l.a.	$17\cdot7 \pm 0\cdot9$	$1\cdot0 \pm 0\cdot9$	77
18d		neat	s.e.	$13\cdot3 \pm 0\cdot9$	$-9\cdot2 \pm 0\cdot6$	80
18e		neat	l.a.; p.s.	$7\cdot9 \pm 1\cdot7$	$-27\cdot6 \pm 1\cdot2$	80
18f		neat	p.v.r.	$6\cdot6 \pm 0\cdot5$	$-31\cdot3 \pm 0\cdot9$	80
18g		neat	c.l.a.	$16\cdot9 \pm 0\cdot5$	$0\cdot3 \pm 0\cdot4$	77
19	H₂C=CHCON(CH₃)₂	36·3 mol % in CH_2Br_2	l.a.	$7\cdot7 \pm 0\cdot5$	$-28\cdot1 \pm 0\cdot5$	84

[l] $\Delta G^{\ddagger}$ at 433°K.

[k] By measurement of return to equilibrium of rotamer (a) and observed equilibrium rotamer ratio.

(continued)

TABLE 4 (Cont.)

Entry no.	Compound	State	Method[a]	$\Delta H^{\ddagger b}$ (kcal/mole)	$\Delta S^{\ddagger}$ (e.u.)	Ref.
20a	HCSN(CH$_3$)$_2$	neat	l.a.	27.1 ± 1.1	3.3 ± 0.4	91
20b		40% in o-dichlorobenzene	l.a.	35.4 ± 1.7	28.6 + 7.0	91
21a	HCSN(i-Pr)$_2$	neat	l.a.	31.0 ± 2.8	9.7 ± 3.1	91
21b		40% in o-dichlorobenzene	l.a.	23.4 ± 5.6	−0.36 ± 4.6	91
22	CH$_3$CSN(CH$_3$)$_2$	9.0 mol % in formamide	l.a.	42.9 ± 5.6		86
23	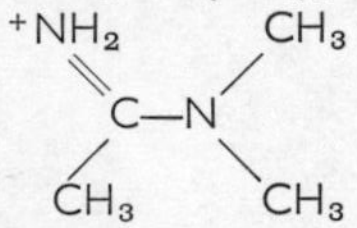 (iminium structure)	5.0 mol % in formamide	l.a.	18.9 ± 1.0	−0.7 ± 0.5	86
24a	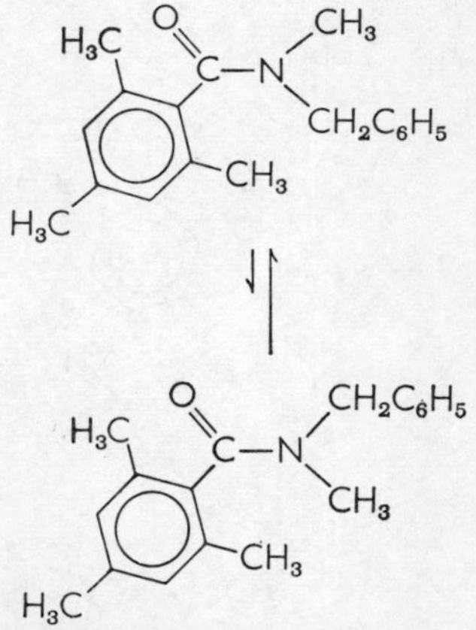 (structure a)	in CDCl$_3$	equil.[k]	27.3 (a) → (b)		92
24b	(structure b)			26.8 (b) → (a)		
25a	ClCSN(CH$_3$)$_2$	2.1 M in CCl$_4$	c.l.a.	ca. 19[l]		77
25b		0.062 M in CCl$_4$	c.l.a.	ca. 19[m]		77

[l] $\Delta G^{\ddagger}$ near 316°K; somewhat uncertain due to effects of association.
[m] $\Delta G^{\ddagger}$ near 290°K; somewhat uncertain due to effects of association.

the barrier at higher concentrations in non-polar solvents. Woodbrey and Rogers[90] have reported that the activation energy for N,N-dimethylcarbamoyl chloride increased with concentration in carbon tetrachloride, while in the more polar solvent CH_2Br_2 it passed through a maximum (Table 4, entry 18b). They rationalized this result in terms of dimer formation and solvation, but the claimed effects were probably well within experimental error. It is notable again that the activation free energy was 16.4 ± 0.2 kcal/mole and independent of solvent and concentration. For N,N-dimethylpropionamide, there was an indication of a lower activation free energy in carbon tetra-chloride than in CH_2Br_2 (Table 4, entry 13b), but the effect was slight. Again, Whittaker and Siegel[85] have reported apparent effects of solvent on the activation energy for isomerization in N,N-dimethylformamide, but the behaviour of the methyl-proton chemical shifts with temperature complicated the kinetic analysis to such an extent as to give slight credence to their conclusion.

The replacement of one methyl group by a benzyl group in N,N-dimethylformamide appears to raise the barrier somewhat as judged by line-shape analysis but not by equilibration measurements (Table 4, entries 4a, 4b). Larger and more branched N-alkyl groups have no marked effect. A mesityl group on the carbonyl, however, does appear to increase the barrier measurably (Table 4, entries 16, 17). The substitution of sulphur for oxygen unmistakeably increases the barrier, the effect being more marked for acetamides (compare entries 6d and 22) than for carbamoyl chlorides (compare entries in 18b, 25a, 25b) and formamides (compare entries in 2 'neat' to 20a and 21a).

Neuman and Young[86] have shown that in the series of N,N-dimethylformamides in which the carbonyl oxygen is replaced by S, NH, and NH_2^+, the barrier height increases with ^{13}C—^1H coupling

(23)

at the formyl proton. The correlation is not impressive, however, as the change in $J_{^{13}C-^1H}$ is small.

The actual isolation of pure crystalline isomers of amides is reported by Siddall[93], who obtained two isomers of 23 and studied the rates of their interconversion in sym-tetrachloroethane solution. The isomerization proceeds with an activation energy of 25 kcal/mole from either direction, but is thought to involve N—Ar rotation, rather than OC—N rotation.

The non-equivalence of the two amino protons of acetamide due to hindered rotation about the C—N bond has also been demonstrated using e.p.r. spectroscopy[94]. In these experiments, acetamide reacts with ·OH radicals in a flowing system. Analysis of the e.p.r. spectrum of the ·CH_2CONH_2 radical so produced, shows that the hyperfine coupling constants to the two amino protons are not equal (1·96 ± 1 G and 2·53 ± 1 G). Similar treatment of formamide leads to the radical HCONH.

Since the systematic errors in the determinations of the activation energy for C—N bond rotation all act to make it appear too small, it is now felt that the higher values of 18–20 kcal/mole are more nearly correct. This can be compared to 65 kcal/mole measured for isomerization about the full double bond of dideuteroethylene[95]. It is interesting to note that a barrier to C—N rotation of 20 kcal/mole is very nearly equal to the thermochemically[96] (21–22 kcal/mole) and theoretically [23,97] (21, 23·6 kcal/mole) determined resonance energies of the amide group. As the situation now stands, the errors in the activation energy are so large, and the measurement parameters so non-uniform, that comparison of barriers in different compounds must be looked upon with caution.

2. The N—C$_{(\alpha)}$ and C'—C bonds

Siddall and Prohaska[98] observed that in amides of structure **24** (the conformational preferences of which have already been discussed

$$\begin{array}{c} O \qquad\quad R \\ \backslash\backslash \qquad / \\ C—N \\ / \qquad \backslash \\ H_3C \qquad CH_2C_6H_5 \end{array}$$

(**24**)

in section III.C.1), the benzyl protons were non-equivalent, and therefore appeared as an AB quartet, when R was o-tolyl, but as a singlet when R was phenyl or 2,6-dimethylphenyl. They interpreted this result as indicating 'slow inversion' at the nitrogen atom, but this explanation was justifiably condemned by Shvo and co-workers[99] since, as we have seen in the introduction to section III such inversion would be very rapid, if indeed the nitrogen atom is appreciably pyramidal, which is by no means certain. A much more

plausible explanation[99] is that in the conformation **25** rotation of the
o-tolyl group is sterically restricted. This would make the benzyl
protons non-equivalent. If two *ortho* methyls are present, rotation is
no doubt still slower, but the conformer is symmetrical. If none are
present, the conformer is also symmetrical, but rotation may also be
less restricted. As Shvo and coworkers point out, restricted rotation

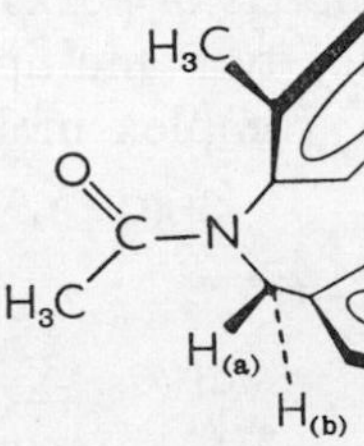

(**25**)

around benzene–nitrogen bonds has long been known[100]. For the
acetamide illustrated (**25**) they calculated from the AB quartet
collapse, a $\Delta G^{\ddagger}$ for rotation, near coalescence at 135°c, of 20·0
kcal/mole. For the cyclic amide (**26**) a $\Delta G^{\ddagger}$ of 17·3 kcal/mole was

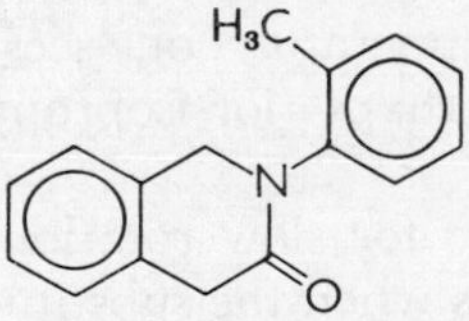

(**26**)

reported. Slow rotation in a number of open-chain amides with
aromatic substituents on nitrogen has been reported in a number of
papers by Siddall's group[101–108]. The preferred conformations of
rotating N-aryl rings in these compounds are not known, but may be
similar to that of biphenyl, i.e. with the amide and phenyl groups
neither coplanar nor orthogonal (both of which conformations probably represent energy maxima), but at intermediate angles, with two
energy minima and two maxima per 180° rotation[99] (see also the case
of benzamide in section II).

There is also evidence of substantial barriers to rotation about
N—$C_{(\alpha)}$ bonds in some *N,N*-dialkylamides. Whittaker and Siegel[109]
observed an unusual temperature dependence of the chemical shifts

of the alkyl protons in N,N-diisopropylacetamide and N,N-diiso-propylformamide. Siddall and Stewart[110] reexamined N,N-di-isopropylacetamide and extended their observations to N,N-diiso-propylisobutyramide, N,N-di-3-amylacetamide and N,N-di-3-amyliso-butyramide. It was observed that at temperatures of ca. $-10°$ to $-40°$c, one of the two methine proton septets broadened markedly, the other remaining unaffected. At still lower temperatures, this resonance reappeared as *two* subsets of peaks for the larger molecules (for N,N-diisopropylacetamide, two multiplets could not be distinctly seen). In other, more complex molecules, such as **27** both

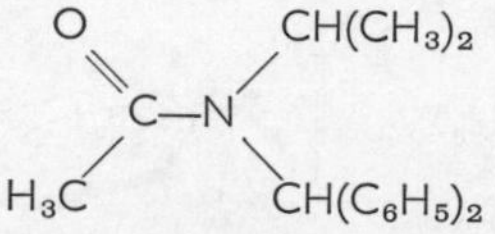

(27)

methine multiplets underwent these changes but in different tempera-ture ranges. A rather complicated explanation is needed for such results, and a completely satisfactory one is not yet at hand. Each compound has its own peculiar behaviour and conformational preferences. There seems little reason to doubt, however, that the alkyl groups interlock upon rotation, each impeding the rotation of the other. Barriers of the order of 10 kcal/mole are thus developed. Below the coalescence temperature, one sees the spectra of at least two preferred conformers, perhaps (for isopropyl groups) something like **28** and **29**.

There is also evidence for slow rotation about the aryl-carbonyl bond in aromatic amides when the substituents at both nitrogen and

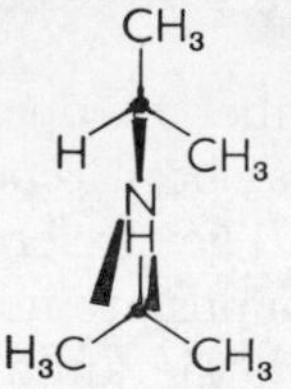

(28) 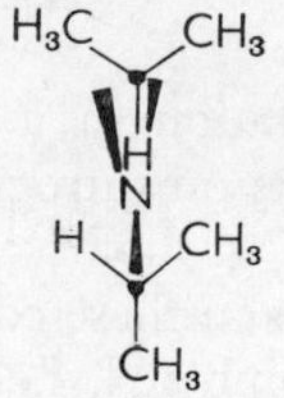   (29)

carbonyl are sufficiently bulky[104,105]. In the spectrum of **30** doubling of the signals of most of the protons indicates the usual *cis* and *trans* conformers about the OC—N bond. It is further observed that, even well above room temperature, the methoxyl groups give four signals, indicating slow rotation about the benzene-carbonyl bond, probably

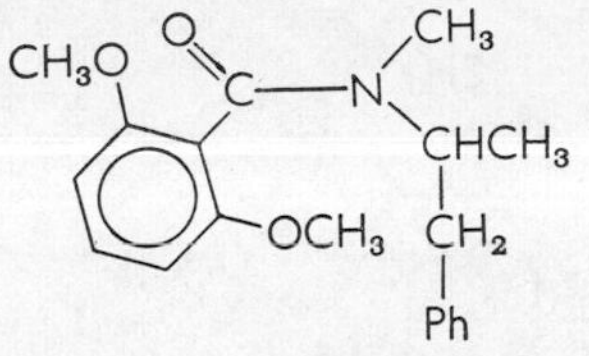

(30)

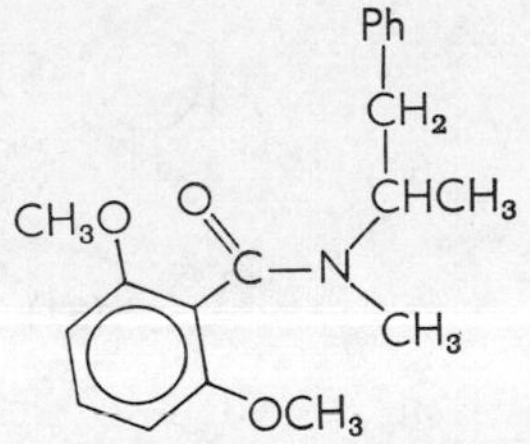

(31)

with a barrier of about 20 kcal/mole, approximately that for rotation about the OC—N bond. Inspection of molecular models suggest that these rotations may be synchronized, as the benzene ring attached to the carbonyl group clearly cannot rotate in conformer **30**, but should be able to in conformer **31**.

E. Amide–Iminol Tautomerism

Yet another geometrical isomerization in amides has been proposed by Potapov and coworkers[111], who presented optical rotatory dispersion evidence that N-benzoyl-α-phenylethylamine and its derivatives in benzene solution exist as the amide (**32**), whereas in methanol, the predominant form has the iminol structure (**33**). However, later n.m.r. and u.v. work on these materials[112,113] claimed that only the amide form (**32**) is present in these solutions, and that the ORD solvent effects may instead be consequences of *cis–trans* isomerization. Certain

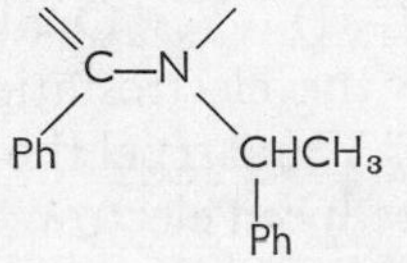

(32)

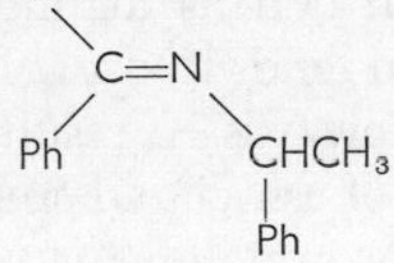

(33)

of these amides may be obtained in two crystal forms which may prove to be the *cis* and *trans* isomers. In the isomeric systems **34–35** and **36–37** the equilibria at 130°c were also found to lie completely towards the right, with no traces of **34** or **36** detectable in the equilibrium mixtures[114].

The most recent claim for the iminol form of an amide is that of Brown and coworkers[115] who studied the n.m.r. and i.r. spectra of the

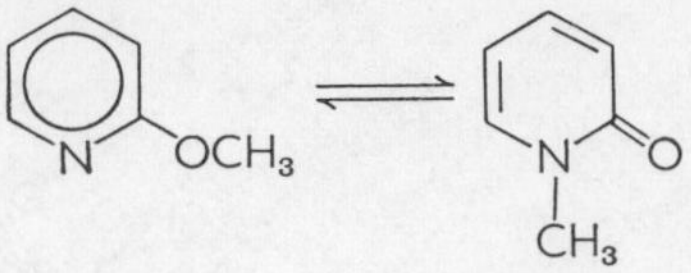

(34)   (35)

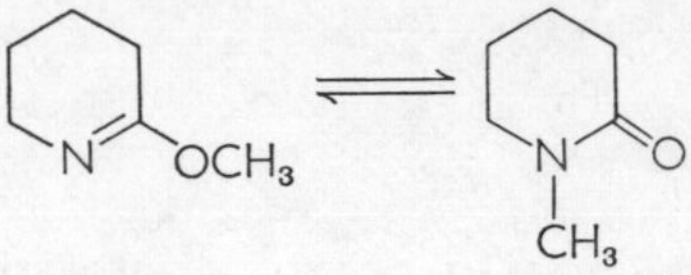

(36)   (37)

bis-trimethylacetamide complex of $PtCl_2$, and found the ligand to have the iminol form with Pt—NH=C coordination.

F. Self-association

We append this section to point out briefly how association effects are of consequence in the interpretation of n.m.r. spectra of amides.

I. Hydrogen bonding

Many n.m.r. studies of self-association of molecules capable of hydrogen bonding have been carried out. These depend upon marked deshielding of the protons involved in the hydrogen bond. The dominant contribution to this effect appears to be a distortion of the electronic structure of the X—H bond by the presence of the electron donor Y, usually depicted as X—$\overset{\delta+}{H}$ ··· $\overset{\delta-}{Y}$. Qualitatively, this may be understood as arising because the electrostatic field of the hydrogen bond tends to draw H towards Y and repel the X—H bonding electrons towards X, resulting in a reduced electron density about H. Dilution of associated species in an inert solvent such as carbon tetrachloride or cyclohexane results in an upfield shift of the bonding proton, and from the shape of the dilution curve, equilibrium constants can be derived provided the equilibria involved are not too complex, and provided the chemical shift of the unassociated proton can be established. For relatively weakly associating molecules such as alcohols and amines, fairly satisfactory results can be obtained as dimer or trimer equilibria strongly predominate. But for strongly associated systems such as mono-N-alkylamides and carboxylic acids, the monomer chemical shift is difficult to establish, since very high dilutions would be required; the broad signals of NH protons aggravate this problem.   In addition, the equilibria are complex with large polymeric species tending to predominate.

There appear to be no studies of this kind on unsubstituted amides, such as formamide and acetamide, mainly for the good reason that they are not appreciably soluble in inert solvents.   La Planche and coworkers[116] have attempted to deal with the association of N-methyl-, N-isopropyl-, and N-t-butylacetamide in both inert (carbon tetrachloride and cyclohexane) and hydrogen-bonding solvents (dioxane, chloroform, diethylketone, and dimethylsulphoxide).   They analysed their results in terms of a monomer–dimer and a generalized equilibrium among higher aggregates, and also included association with hydrogen-bonding solvents.   They observed that association in inert solvents was very strong, decreasing appreciably with increasing bulk of the N-alkyl substituent.   The association was markedly reduced in chloroform because the aggregates were broken up by hydrogen bonding between chloroform and the amide carbonyl group.   Dioxane and diethylketone are about as effective as chloroform in this regard, but dimethylsulphoxide is much more successful, although some association persists even in relatively dilute solutions.

2. Dipole–dipole association

In N,N-dialkylamides, hydrogen-bonded self-association is not possible (except, perhaps, that involving the formyl protons in formamides, for which there is no conclusive evidence).   Neuman and coworkers[117] have observed small but definite shifts, of the order of 0·1 p.p.m., of the NCH_3 protons of N,N-dimethylacetamide, N,N-dimethylformamide, N,N-dimethylthioacetamide, and N,N-dimethylthioformamide upon dilution in carbon tetrachloride.   It was found that the cis-methyl peaks moved upfield and the trans-methyl peaks moved downfield by the same amount.   This result was explained in terms of dipole–dipole association to form dimers.   The conformation of the dimer is not clear, since the most obvious head-to-tail, parallel-plane model predicts chemical-shift trends just opposite to those observed.   The association was in all cases weak, the calculated equilibrium constants being slightly greater than unity only for N,N-dimethylthioformamide.

Rather different conclusions are indicated by Pines and Rabinovitz[76], who, on the basis of evidence not yet given in detail, report dipole–dipole interactions of the order of 6 kcal/mole for N,N-dimethylformamide in solution.   They suggest that these interactions could markedly influence the measurement of the OC—N rotational barrier, in a manner already mentioned (section III.D).

IV. ELECTRONIC STRUCTURE

A. The Ground State

Virtually all of the preliminary quantum chemistry required for a discussion of the electronic structure of amides has already been set out in detail in Coulson and Stewart's[118] contribution to this series of books.   Our discussion will differ principally from theirs, in that we will consider calculations performed in a Gaussian-Type Orbital (GTO) basis rather than in a Slater-Type Orbital (STO) basis, and our calculations explicitly consider *all* electrons, σ and π, rather than just π. Moreover, since our principal interest in amides has been electronic–spectroscopic, we shall place a heavy emphasis on that here.   However, first we shall describe the results of ground-state calculations, and then go on to the excited states (section IV.B) and the optical properties of amides (section V).

I. Gaussian-Type Orbitals

Since the earliest days of quantum chemistry the electronic states of molecules have been discussed and described within the framework of some sort of Molecular-Orbital (MO) theory.   For very practical reasons it was immediately found necessary in actual calculations to introduce an arbitrary, although usually very reasonable, partitioning of the electrons in a molecule into what might be termed chromophore or valence-shell electrons—those that are considered to be involved in ordinary physical and chemical phenomena, and non-chromophore or core electrons—those that remain unaffected in chemical or photo-chemical processes.   In any event, even with the very limited number of electrons under consideration, or perhaps more likely because of it, in order to get reasonably good agreement with experiment it was found necessary to further tamper with the basic theory and make use of empirical parameters.   Thus we have the Hückel, extended Hückel, and Pariser–Parr–Pople theories in organic chemistry and the Wolfsberg–Helmholz ligand-field theory in inorganic chemistry. The obvious trend nowadays, however, as reflected in the recent development by Pople's group of the family of CNDO methods[119], has been towards greater rigour, fewer parameters, and a more complex and sophisticated calculation.

Omitting a lengthy discussion of these semi-empirical theories, their importance and usefulness, it is sufficient to say that at the present time it is possible to carry this tendency towards greater rigour to its logical conclusion, and to perform all-electron, all-integrals, non-empirical electronic-structure calculations on moderately large-size molecules within the framework of the Roothaan SCF–MO method. This fact is not very widely known or understood. Such complete calculations are performed using a Gaussian-Type Orbital (GTO) basis set.

Analytic forms for the Gaussian orbitals are given in equations (2), with representative examples of s-, p-, and d-type orbitals. They are really quite similar to Slater orbitals except for the r^2 dependence in the exponential part.

$$
\begin{aligned}
\text{GTO} &= Nx^l y^m z^n e^{-\alpha r^2} \\
l + m &+ n \leq 2 \\
s &\sim e^{-\alpha_i r^2} \\
p &\sim xe^{-\alpha_j r^2} \\
d &\sim xye^{-\alpha_k r^2}
\end{aligned}
\tag{2}
$$

The exponential factors α_i, α_j, etc., are determined from atomic calculations. One disadvantage of using GTO's is that it is not immediately obvious in all cases how the wave function can be related to the more common Atomic-Orbital (AO) functions of well-defined principle quantum number. Thus the lowest s-type function on an atom can most likely be considered safely as $1s$, but the next higher s-type orbital may be a linear combination of $1s$, $2s$, and $3s$ AO's; one can only say that it is the next higher s-type function, but cannot analyse it further into a combination of STO's.

Individual Gaussian Orbitals are known to be poorer representations of AO's than are single STO's. This situation is remedied by using fixed linear combinations of GTO's (called a contracted basis) as the basis functions. The proper linear combinations needed to form the GTF's are easily obtained either by directly fitting the Hartree–Fock AO's with GTO's or by doing an analytic atomic SCF calculation in a GTO basis. It turns out that on the average, only 2–3 GTO's per STO are required for quantitatively comparable results. This is demonstrated in Table 5, where the total energies and dipole moments computed for carbon monoxide in various bases are compared. If each AO of the basis is represented by one STO, then the basis is referred to as Single Zeta, whereas a Double-Zeta basis employs two STO's per AO. It is seen from this table that the GTO basis used for our calculations[122,123] gives results intermediate between those of Best-Atom Double-Zeta and Best-Molecule Double-Zeta STO calculations. Since the STO Double-Zeta basis is composed of four s-type and two p-type functions and the corresponding Gaussian basis uses ten s-type and five p-type functions, the ratio of only 2–3 GTO's per STO for quantitatively comparable results is demonstrated. It has recently been shown that for molecules such as formamide, the Double-Zeta GTO basis gives essentially the s-p limit wave function, and that only the addition of d- and f-type orbitals to the basis are needed to reach the Hartree–Fock limit[124].

TABLE 5. Calculations on the ground state of carbon monoxide[a].

Basis	Total energy (a.u.)	Dipole moment (D)
Best-Atom Single-Zeta[b,c]	−112·3261	0·593 (C⁻O⁺)
Best-Molecule Single-Zeta[b,c]	−112·3927	0·389 (C⁻O⁺)
Best-Atom Double-Zeta[b,c]	−112·6755	0·603 (C⁺O⁻)
Gaussian-Type Orbitals[d]	−112·6762	0·416 (C⁺O⁻)
Best-Molecule Double-Zeta[b,e]	−112·7015	0·393 (C⁺O⁻)
Hartree–Fock Limit[b,e]	−112·7860	0·274 (C⁺O⁻)
Experimental[e]	−113·377	0·112 (C⁻O⁺)

[a] Computed at the equilibrium internuclear distance of 2·132 a.u.
[b] Slater-Type Orbitals basis.
[c] H. Basch, unpublished calculations.
[d] Exponents and fixed coefficients taken from work of Whitten[120] and Huzinaga[121].
[e] W. H. Huo, *J. Chem. Phys.*, **43**, 624 (1965).

Of course, the use of a large GTO-basis set (compared to STO's) means that one has that-many-times-more to the fourth power number of integrals to compute. But, multicentre integrals over GTO's are evaluated using simple analytic formulas which are easily coded in simple FORTRAN language without recourse to complex numerical integration techniques, numerous difficultly convergent expansions, or sophisticated programming structure. Thus the speed with which GTO integrals can be computed more than compensates for the handicap of the larger basis and the resultant need to compute the greater number of integrals. We dwell upon the discussion of the use of GTO functions in molecular calculations because their use is not widely appreciated, as is that of STO's, and because the best and most reliable calculations on the electronic structure of the amide group have been carried out using such GTO-basis sets.

There are, of course, a great many earlier calculations in the literature on the electronic structure of the amide group, all of them in a Single-Zeta STO basis, or worse. While our Double-Zeta GTO calculations are still incomplete in that they do not include d- or f-type functions or allow for any of the correlations of electronic motions which carry one beyond Hartree–Fock, they do avoid many of the other objections one can raise to the earlier calculations. Consequently, we shall devote most of our attention to the results of the GTO calculations, which, in fact, encompass and elaborate on the approximations and results of the earlier STO calculations.

2. Theoretical results

Assuming the non-planar geometry of Costain and Dowling[6] (Figure 8), and the essentially Double-Zeta basis of reference 122, Roothaan's SCF–MO procedure leads to the molecular-orbital scheme of Figure 9, for the ground state of formamide. The corresponding

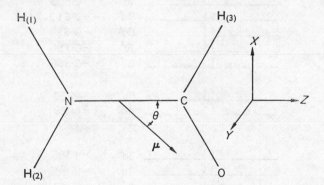

FIGURE 8.   Numbering of the atoms and coordinate system used for formamide GTO calculations. The coordinate system shown on the right is centred at the midpoint of the C—N bond, and defines the X, Y, and Z directions for the atomic p orbitals on all centres. $H_{(1)}$ and $H_{(2)}$ have a negative Y coordinate; N,C,O and $H_{(3)}$ are in the $Y = 0$ plane; μ is the dipole moment.

molecular-orbital wave functions have been analysed to yield the orbital populations and charge densities given in Tables 6 and 7. We have temporarily presumed a planar structure for assigning symmetry species to the various orbitals; those labelled a′ are 'σ' and those labelled a″ are 'π'. Robb and Csizmadia[125] have presented three-dimensional diagrams which illustrate the patterns of nodes and antinodes in each of the occupied and unoccupied MO's of formamide.

The highest filled orbital in formamide is computed to be 2a″, a π MO consisting of almost equal amounts of $2p$ π AO's on oxygen and nitrogen, with a node through the intervening carbon atom.   Because there is very little overlap between the π orbitals on oxygen and nitrogen, 2a″ is appropriately described as non-bonding[126], even

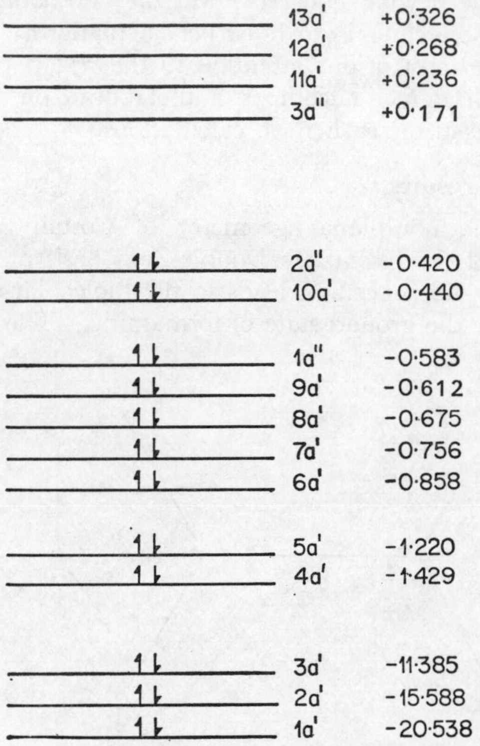

	13a′	+0·326
	12a′	+0·268
	11a′	+0·236
	3a″	+0·171
↑↓	2a″	−0·420
↑↓	10a′	−0·440
↑↓	1a″	−0·583
↑↓	9a′	−0·612
↑↓	8a′	−0·675
↑↓	7a′	−0·756
↑↓	6a′	−0·858
↑↓	5a′	−1·220
↑↓	4a′	−1·429
↑↓	3a′	−11·385
↑↓	2a′	−15·588
↑↓	1a′	−20·538

FIGURE 9.   Orbital symmetries, energies (a.u.) and occupancies computed for the ground state of formamide in the non-planar geometry[6]. The level spacings are not drawn to scale.

though of π type.   The lower occupied π orbital 1a″ is strongly bonding, having all AO's in phase, with maximum density on the nitrogen atom.   It is the pair of electrons in this π orbital which are responsible for the rotational barrier about the C—N bond of amides.

The highest filled σ orbital 10a′ is largely oxygen-centred and corresponds to the conventional 'non-bonding oxygen lone pair' of ketonic substances.   The fact that neither the 10a′ or 2a″ MO's contain any appreciable electron density on adjacent atoms explains their near degeneracy, even though the former appears localized and the latter delocalized. One sees, however, from the population analysis that 10a′ does have a non-zero density on $H_{(3)}$, and somewhat smaller densities on the C and N atoms as well. Thus it is seen that in the ground state, the conventional sigma lone pair is in part delocalized

TABLE 6.   Orbital population analysis of the formamide ground state.

	8a′	9a′	10a′	1a″	2a″	3a″[a]
$H_{(1)}$	0·0187	0·0791	0·0158	0	0	0
$H_{(2)}$	0·2111	0·0573	0·0147	0	0	0
$H_{(3)}$	0·1683	0·1331	0·2238	0	0	0
C_s	0·0916	0·0190	0·0151	0	0	0
N_s	0·0007	0·0000	0·0056	0	0	0
O_s	0·0876	0·2390	0·0000	0	0	0
C_x	0·1068	0·3661	0·0229	0	0	0
N_x	0·2942	0·1329	0·0009	0	0	0
O_x	0·3296	0·4867	0·4859	0	0	0
C_y	0	0	0	0·6607	0·0344	1·4475
N_y	0	0	0	0·7621	1·0439	0·1340
O_y	0	0	0	0·5772	0·9217	0·4185
C_z	0·4106	0·0391	0·0277	0	0	0
N_z	0·2678	0·0094	0·1101	0	0	0
O_z	0·0145	0·4369	1·0773	0	0	0

[a] This orbital is not occupied in the ground state of formamide.

TABLE 7.   Charge densities in the ground state of formamide.

Atom	Net charge	π-Electron density
H^1	+0·357	0
H^2	+0·368	0
H^3	+0·152	0
C	+0·258	0·695
N	−0·758	1·806
O	−0·377	1·499

over the adjacent framework. As will be seen in section IV.B this is not necessarily true in the excited states of formamide.

Application of a Mulliken population analysis to the occupied π orbitals of formamide shows that the nitrogen atom has only 1·8 π

electrons, the other 0·2 being in the C=O group which itself is strongly imbalanced towards oxygen. The paucity of π-electron density on nitrogen is just that expected from a consideration of the valence-bond structures **3** and **4**. Various experiments and calculations place the relative weights of **3** and **4** in the ground-state wave

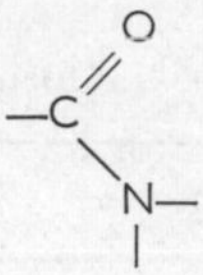

(3) 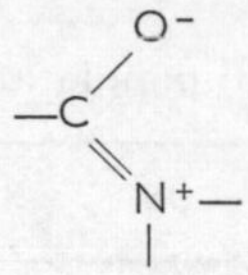   (4)

function of amides at from 2:1 to 4:1. It can also readily be seen from Table 7, that such π-electron-only considerations can be grossly misleading, for although the nitrogen atom is π-electron poor (net π charge of $+0·19$), it seems to be *overall* electron rich (net charge of $-0·76$). In fact, the populations show a back-donation effect, with charge leaving nitrogen via the π system, and more than compensating this, a flow of charge to nitrogen from the protons via the σ system. A similar effect is reported for all-electron calculations on formaldehyde[127]. Because the carbon atom loses 0·3 electron to oxygen via the π system and has an overall charge of $+0·3$, it is seen to be essentially electroneutral in its σ system. The same conclusion applies to the oxygen of formamide. The total net charge on the NH_2 and CHO fragments of formamide are calculated to be $-0·033$ and $+0·033$, respectively.

The π-electron deficiency at carbon suggests that the ground state of formamide must include the structure **38** as an ingredient at least as

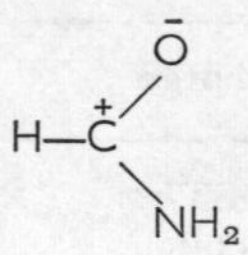

(38)

important as structure **4** (even more important according to the MO calculations), and that valence-bond structures which we dare not even pictorialize, but which transfer charge from the protons to nitrogen, are also necessary to give correctly the *overall* charge distribution.

The dipole moment of formamide can be calculated from the electron distribution in the Costain–Dowling geometry to be 4·39 D with an angle $\theta = 42·6°$ (Figure 8). Kurland and Wilson[22] report a dipole moment of 3·7 $\pm$ 0·06 D at an angle of 39·6° for gaseous formamide in

its ground state. Surprisingly, a value of 4·2 D has also been calculated for the dipole moment of the amide group, considering the π
electrons only[128]. We see, however, from the strong electronic
polarization of the σ system of formamide, that comparison of the
measured moment with the moment calculated in the π-electron
approximation is meaningless. The apparent agreement arises from
the fact that though the σ system has suffered a large reorganization of charge on forming the molecule from its atoms, the centres of
gravity of the positive and negative charges in the σ system are acci-
dentally very nearly coincident, thus having very little effect on the
dipole moment.

Realizing that the 10a' and 2a'' MO's were quite nearly degenerate in
formamide, Hunt and Simpson[126] long ago raised the question as to
which orbital was involved in the lowest ionization potential (10·2 ev)
of this molecule. In MO calculations of the SCF type, the ionization
potentials are usually taken as the negatives of the calculated orbital
energies (Koopmans' theorem). Using this approach, we calculate
that ionizations from the non-bonding σ (10a') and π (2a'') orbitals of
formamide require 11·99 and 11·42 ev, respectively. This approximation, however, presumes a certain three-fold vertical nature to the
process: (*i*) the nuclear geometry does not change on ionization, (*ii*)
the occupied molecular orbitals of the system do not reorganize on
ionization, and (*iii*) the correlation energy error does not change on
ionization. The approximation (*i*) would appear to be a valid one,
whereas (*ii*) and (*iii*) can be shown to be poor approximations.

The effect of electronic reorganization, (*ii*), can be evaluated by
calculating the ionization potentials as the differences between the
computed total energy of the neutral molecule and the *recomputed*
total energies of the positive ions of interest. When applied to formamide, this technique gives the ionization potential from the σ orbital
10a' as 8·80 ev, and that from 2a'' π as 9·7 ev[122], just the reverse order
predicted using Koopmans' theorem. Even though this technique
does account for 1·5–3·0 ev of reorganization energy, the error in-
curred in approximation (*iii*) still leaves the theoretical question of the
ground-state symmetry of the formamide positive ion unanswered.

The correlation error (*iii*) can be treated simply in the following
way. Using a Mulliken population analysis, we first reduce the MO
occupations of the neutral molecule and the positive ions of interest,
to AO populations. Then using Nesbet's tables of AO correlation
energies[129], we can add atomic contributions to get the final molecular
quantities. When performed in this way, the reorganization and

correlation corrections give an ionization potential of 11·06 ev from the 10a′ σ orbital, and 12·9 ev from the 2a″ π orbital. Other work, quoted in section V.A, also supports the n orbital ionization potential as lower than that from π, in amides. A discussion of the higher ionization potentials of formamide is deferred to section V.C.

It has recently been shown[130] that the total energy of formamide as calculated in the Double-Zeta basis can also be used to calculate the heat of the reaction: $CO + NH_3 \rightarrow HCONH_2$, for this is simply equal to the difference of the total energy of formamide and the sum of those of carbon monoxide and ammonia. After correction for zero-point energies, the calculated value of 9·4 kcal/mole compares nicely with the observed value of 12·6 kcal/mole.

Other quantities amenable to calculation with the ground-state wave function are the components of the electric-field gradient tensor at the quadrupolar nitrogen nucleus, and the molecular quadrupole moment. The first of these has been measured in formamide by Kurland and Wilson[22], and in the principal axis system has the values, $q_{aa} = +1·90$, $q_{bb} = +1·70$, and $q_{cc} = -3·60$ mc/sec, which compare favourably with the calculated values[131*], of $q_{aa} = +1·99$, $q_{bb} = +1·67$, and $q_{cc} = -3·66$ mc/sec. The quadrupole moment of formamide has not been measured as yet, but in the principal axis system, it has computed components of $Q_{\alpha\alpha} = 4·419$, $Q_{\beta\beta} = -2·232$, and $Q_{\gamma\gamma} = -2·186$, all in units of 10^{-26} e.s.u. cm^2.

B. Excited States

In order to interpret optical absorption spectra and various excited-state properties, one needs a method of accurately calculating excited-state (open-shell) wave functions. However, the calculation of open-shell states in the Hartree–Fock scf approximation is a difficult problem which has not yet been solved in a general way. Ideally, electronic excited-state properties should be calculated from the appropriate scf solution for the electronic state in question. For example, excitation energies would be computed by subtracting the total scf energies of the excited and ground states. This procedure, however, even if fully implemented, would not be totally satisfactory since it neglects the change in correlation energy due to the different

* Calculated assuming a ^{14}N nuclear quadrupole moment of $1·470 \times 10^{-26}$ cm^2, which is an experimental lower limit to this quantity. Presuming the upper limit of $1·604 \times 10^{-26}$ cm^2, leads to equally good agreement between calculated and experimental quadrupole coupling constants.

number of electron pairs in the closed-shell ground state and the open-shell excited state.

Short of directly calculating upper-state wave functions, the most common method of constructing excited states uses the virtual MO's (11a', 3a'', etc.) which are obtained as a by-product of the solution of the ground-state SCF equations. In this procedure, excited-state configurations are constructed by promoting electrons from the higher occupied MO's to the lower unoccupied (virtual) MO's which are assumed to be good representations of the excited-state terminating MO's; the excitation energies are then calculated using well-known formulas[132]. In addition to ignoring changes in correlation energy, the virtual-orbital approach also ignores any electronic rearrangement which may occur among the unexcited electrons when the optical electron changes orbit.

If there were some reason for believing that the virtual-orbital MO's obtained from the ground-state calculation corresponded to those that one would get by doing the excited-state calculation directly, then the virtual-orbital method would be a useful procedure for simply obtaining excited-state wave functions, subject still to correlation deficiencies. In fact, our calculations on formamide excited states show that for certain situations the virtual-orbital approximation is acceptable and in others, totally unacceptable. An extreme example of how misleading the virtual-orbital MO's can be in describing excited states is furnished by the Rydberg orbital calculations on formamide.

In the usual type of electronic excitation, termed valence-shell excitation, an electron is excited from an MO composed of a certain set of AO's (say, $2s$ and $2p_\sigma$) with fixed phases, to another composed usually of the same AO's, but having different phases. In a Rydberg excitation the same electron is promoted to an orbital composed, instead, of AO's having a principal quantum number higher than any of those occupied in the ground-state configuration (say, $3s$ or $4p$). Moreover, since terminating orbitals can be constructed with an ever-increasing principal quantum number, one usually talks in terms of families of Rydberg states, called series, all of which have the same azimuthal quantum number, but different principal quantum numbers.

For a complete and proper interpretation of the electronic absorption spectrum of amides, one needs to calculate both valence-shell and Rydberg excited states. This is accomplished by putting both valence shell ($1s$, $2s$, $2p$) and Rydberg ($3s$, $3p$) AO functions into the basis. We have done this, and an orbital population analysis of the resulting MO's is shown in Table 8, in which the populations of the highest

TABLE 8.   Orbital population analysis for formamide.

	Ground state				$\pi \to \pi^*$ triplet		$\pi \to \sigma^*$ triplet		$n \to \pi^*$ triplet		$n \to \sigma^*$ triplet	
	$n(10a')$	$\sigma^*(11a')$	$\pi(2a'')$	$\pi^*(3a'')$	$(2a'')$	$(3a'')$	$(2a'')$	$(11a')$	$(10a')$	$(3a'')$	$(10a')$	$(11a')$
BADZ + 3p												
H$_{(1)}$	0·009	0·000	0·0	0·0	0·0	0·0	0·0	0·008	0·003	0·0	0·002	0·022
H$_{(2)}$	0·007	−0·010	0·0	0·0	0·0	0·0	0·0	0·031	0·001	0·0	0·001	−0·001
H$_{(3)}$	0·110	0·025	0·0	0·0	0·0	0·0	0·0	0·069	0·030	0·0	0·030	0·115
C valence shell	0·033	−0·066	0·016	−0·003	0·081	0·666	0·011	−0·049	0·004	0·849	0·007	−0·071
C 3p	0·001	1·321	0·000	1·740	0·005	0·004	0·000	0·751	0·002	0·024	0·002	0·951
N valence shell	0·058	0·000	0·519	0·003	0·210	0·124	0·135	−0·019	0·019	0·079	0·020	−0·020
N 3p	0·000	−0·040	0·005	−0·355	0·001	0·007	0·000	0·344	0·000	−0·008	0·001	−0·124
O valence shell	0·781	0·004	0·458	0·006	0·711	0·199	0·856	0·001	0·945	0·074	0·943	−0·003
O 3p	0·000	−0·235	0·003	−0·391	−0·008	0·000	−0·002	−0·136	−0·004	−0·018	−0·007	−0·134
BADZ												
H$_{(1)}$	0·008	0·822	0·0	0·0	0·0	0·0			0·003	0·0		
H$_{(2)}$	0·007	0·223	0·0	0·0	0·0	0·0			0·002	0·0		
H$_{(3)}$	0·112	0·031	0·0	0·0	0·0	0·0			0·030	0·0		
C valence shell	0·033	0·130	0·017	0·724	0·099	0·662			0·002	0·846		
N valence shell	0·058	−0·213	0·522	0·067	0·188	0·125			0·019	0·080		
O valence shell	0·782	0·007	0·461	0·209	0·713	0·213			0·944	0·074		

occupied and lowest unoccupied MO's in formamide are presented for two different calculations; a strictly valence-shell basis (BADZ) and a valence shell + Rydberg $3p$ (BADZ + $3p$) basis.   In the ground state, the $\sigma^*(11a')$ and $\pi^*(3a'')$ MO's are both unoccupied.   First it should be noted that the orbital populations in the $n(10a')$ and $\pi(2a'')$ MO's are almost identical for the two bases; as expected, the Rydberg-basis functions do not contribute to the ground-state wave function. However, the population analysis of the $\sigma^*(11a')$ and $\pi^*(3a'')$ virtual orbitals shows that in the BADZ + $3p$ basis both of the MO's are computed to be almost 100% Rydberg.   Thus, in a basis containing both valence-shell and Rydberg orbitals, the first $n \rightarrow \pi^*$ and $\pi \rightarrow \pi^*$ excitations in formamide are predicted to be of the Rydberg type, contrary to the experimental evidence supplied by the spectra of formamide in condensed phases (section V.A).

In the same table are shown the results of SCF calculations performed directly on the upper-state triplets corresponding to the singlet states of interest.   The triplet-state SCF wave functions are readily obtainable, and if we make the presumption that triplet and singlet configurations differ only in the spin parts of their wave functions, then the space wave functions of the triplet (Table 8) will be the same as that of the singlet.   The table clearly shows that when the terminating MO is directly involved in the SCF procedure, a clear-cut separation is obtained between valence-shell and Rydberg excited states.   Thus, when calculated in this way, the 3a'' MO in both $n \rightarrow \pi^*$ and $\pi \rightarrow \pi^*$ excitations are valence shell, in *both* BADZ and BADZ + $3p$ basis set calculations, as observed experimentally.   Note also that the composition of the 3a'' MO is significantly different in the ground, $n \rightarrow \pi^*$, and $\pi \rightarrow \pi^*$ states.   Similar reorganization effects are observed for the n orbital, 10a'.   In the ground state, this orbital is 78% on oxygen, with the remaining 22% distributed among the σ orbitals of the other atoms.   However, in the $n \rightarrow \pi^*$ excited state, the electron remaining in the n orbital is now 94% on oxygen.   Apparently the positive hole formed on oxygen in the $n \rightarrow \pi^*$ upper state pulls the remaining n electron back onto oxygen.   The reorganization is even more dramatic for the 2a'' orbital in the 2a'' $\rightarrow$ 3a'' transition.

To sort out these somewhat confusing computational results the following procedure was used in the amide calculations.   The triplet-state calculations were used to determine the nature of the excited state in question (Rydberg or valence shell) and the virtual-orbital excited state in the corresponding appropriate basis used to calculate properties.   It should be noted that under the assumption that the space part of the wave functions is the same for corresponding singlets and triplets, then the energy of the singlet (E_S) is related to that of the triplet (E_T) by equation (3), where K is the familiar exchange integral

$$E_S = E_T + 2K \tag{3}$$

over the two singly occupied MO's. This has been called the indirect SCF method and has been applied to formamide calculations.

Another quantity of concern to us is the oscillator strength connecting the ground state to each of the excited states. There are two alternative equations, (4) and (5), for computing the oscillator strength of a transition of energy ΔE (atomic units). If the Ψ's were exact,

$$f(\boldsymbol{r}) = \tfrac{4}{3}\langle\Psi_0|\boldsymbol{r}|\Psi_1\rangle^2 \times \Delta E \tag{4}$$

$$f(\boldsymbol{\nabla}) = \tfrac{4}{3}\langle\Psi_0|\boldsymbol{\nabla}|\Psi_1\rangle^2/\Delta E \tag{5}$$

TABLE 9. Virtual-orbital excited states of formamide.

	E_T(ev)	E_S(ev)	$f(\boldsymbol{\nabla})$	$f(\boldsymbol{r})$
$n(10a') \rightarrow \pi^*(3a'')$	6·36	6·89	0·011	0·006
$\pi(2a'') \rightarrow \pi^*(3a'')$	6·06	10·50	0·233	0·760
$n'(9a') \rightarrow \pi^*(3a'')$	10·65	11·19	$<10^{-5}$	6×10^{-5}
$\pi(2a'') \rightarrow \sigma^*(11a')$	10·35	10·93	0·001	0·001
$n(10a') \rightarrow \sigma^*(11a')$	13·12	13·63	0·089	0·209
$\pi(1a'') \rightarrow \pi^*(3a'')$	9·58	13·98	0·095	0·281

TABLE 10. Singlet Rydberg excited states of formamide.

	Virtual-orbital energy (ev)	$f(\boldsymbol{r})$	$f(\boldsymbol{\nabla})$	Indirect SCF energy (ev)	$f(\boldsymbol{\nabla})$
$n \rightarrow 3s$	9·76	0·009	0·003	5·83	0·033
	10·21	0·014	0·004		
	10·21	0·024	0·024		
$n \rightarrow 3p_\sigma$	9·76	0·055	0·036	6·39	0·063
	10·08	0·030	0·026		
	11·43	0·008	0·010		
$\pi \rightarrow 3s$	8·89	0·002	0·002	6·53	0·033
	9·27	0·009	0·011		
	9·76	0·009	0·001		
$\pi \rightarrow 3p_\sigma$	9·12	0·002	0·002	7·26	0·0003
	9·23	0·032	0·022		
	10·42	0·004	0·004		

then the oscillator strengths calculated using dipole length matrix elements (equation 4) and dipole velocity matrix elements (equation 5) would be equal.   However, since the Ψ's are inexact, we report both $f(r)$ and $f(\nabla)$ for the various transitions of formamide, but prefer the latter.

The calculated optical spectrum of formamide is assembled in Tables 9 and 10, and will be discussed along with the observed spectrum in section V.A.

Excited-state dipole moments can also be calculated with the wave functions derived using the indirect scf technique.   Such calculations show that the dipole-moment directions in the $n \rightarrow \pi^*$ and $\pi \rightarrow \pi^*$ upper states are very nearly coincident with that of the ground state, but that the moments are depressed by about 2.5 and 1.5 D, respectively [22].

V. OPTICAL PROPERTIES

A. Absorption Spectra

It has been shown[122,123] that the electronic spectra of amide-containing molecules is very characteristic in the region 40,000–80,000 cm^{-1} (2500–1250 Å, 5–10 ev), the spectrum of formamide showing an obvious resemblance to those of, say, 39 and 40 as in

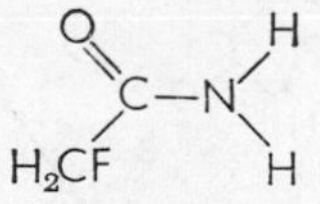

(39) 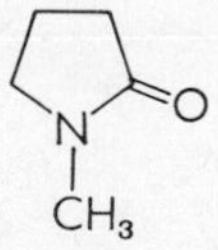   (40)

Figure 10.   In the region beyond 70,000 cm^{-1}, the resemblance may be only superficial, however, since it is in this region that the various alkyl groups have their first absorptions.   Labelling the amide bands sequentially as W, R_1, V_1, R_2, and Q, we feel that the first four bands retain their spectroscopic individuality in all simple amides, and that Q is similarly present but may be overlapped with alkyl-group absorption.

Of the five bands of the amide group, the W band is least controversial, with all investigators agreeing to a singlet–singlet $n \rightarrow \pi^*$ (10a' $\rightarrow$ 3a'') assignment.   The presence of this band in amides was only barely hinted at for a long time, until Litman and Schellman[133] pointed out that by going from the usual hydroxylic solvents to non-polar hydrocarbons, the W band is red shifted, whereas the adjacent absorption is blue shifted, thus uncovering the W-band profile. The $n \to \pi^*$ frequency in simple amides and lactams is near 45,000 cm^{-1}, and as is appropriate for such excitations, their molar extinction co-

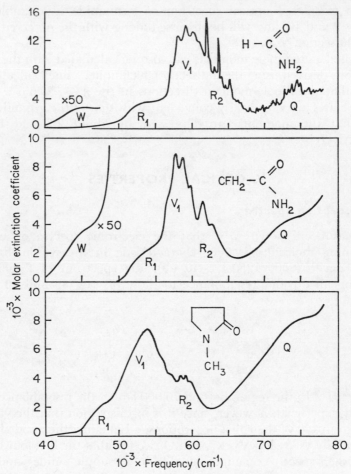

FIGURE 10. The electronic spectra of various amides in the gas phase. [Reproduced, by permission, from ref. 123.]

efficients when corrected for overlapping absorption are less than 100 (the extinction coefficient for the $n \to \pi^*$ band of formamide is 64). It is our experience that the $n \to \pi^*$ bands of amides in acetonitrile solvent are almost always smooth and featureless, the only structure of which we are aware being a few 1200–1500 cm^{-1} intervals in the $n \rightarrow \pi^*$ band of N,N'-dimethyloxamide (**41**).

Two interesting exceptions to these simple generalizations should be recognized. First, in the oxamides, the π^* orbital is delocalized over both amide groups, thus shifting the $n \rightarrow \pi^*$ excitation to 36,000 cm^{-1}.

$$\underset{\textbf{(41)}}{\overset{\displaystyle \begin{array}{c} \text{O} \qquad \text{NHCH}_3 \\ \diagdown \,\diagup \\ \text{C} \\ | \\ \text{C} \\ \diagup \,\diagdown \\ \text{O} \qquad \text{NHCH}_3 \end{array}}{}}$$

Similar shifts to lower frequencies may be expected whenever a π-electron system is placed α to the amide group. Second, as can be seen from Figure 10, the $n \rightarrow \pi^*$ excitation is not evident in the absorption spectra of tertiary amides[134]. However, it is readily confirmed by circular dichroism spectra that the band is merely covered in these compounds by the stronger absorption to the blue (cf. section V.B).

Overall, the GTO calculations would appear to be doing a more than adequate job on the $n \rightarrow \pi^*$ excitation, for the predicted excitation energy of 6·42 ev is in agreement with the 5·65 ev observed, and the $n \rightarrow \pi^*$ oscillator strength is observed to be about 0·002–0·004[122] in various amides, while 0·007 is calculated. The GTO calculations also predict that the $n \rightarrow \pi^*$ transition has associated with it a magnetic transition moment, unmeasured as yet, but calculated to be 0·7736 Bohr magnetons.

In addition to the frequency, the molar extinction coefficient and the possible change in spin multiplicity of an electronic transition, the electric dipole polarization direction is another measurable quantity which characterizes the excitation. The polarization direction of an electronic transition is that direction of the **E** vector of incident polarized light in a molecule-fixed coordinate system for which the light is absorbed maximally at the absorption frequency of interest. Since the absorption intensity varies as the square of the cosine of the angle between the polarization direction and the direction of the incident **E** vector, an out-of-plane polarized transition (in a planar molecule) will have no absorption intensity for in-plane polarized light, and vice versa. Peterson and Simpson[135] attempted to measure this direction of maximum absorption using single crystals of myristamide and light polarized along one or the other of the principal directions of the crystal. While they did not see a specific $n \rightarrow \pi^*$ maximum in their experiments (Figure 11), they found a ratio of absorption intensity for the two principal directions at about 45,000 cm^{-1} which suggested to them that the polarization direction was largely in-plane for the $n \rightarrow \pi^*$ band of this amide.   Since simple group theoretical analysis shows that the $n \rightarrow \pi^*$ band of amides should be out-of-plane polarized, they concluded that the problem was being complicated by the interaction of nuclear vibrations with the electronic motions.

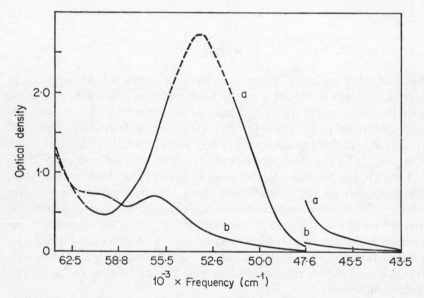

FIGURE 11. Absorption spectrum of the myristamide crystal with light polarized along the crystallographic a and b directions. [Reproduced, by permission, from ref. 135.]

The R_1 band is a relative newcomer to the overall picture of amide spectra, having first been reported in 1966–1967[122,136,137]. It is very unlike almost all other transitions studied by spectroscopists, for though it is quite evident in gas-phase spectra, it does not appear at all in *any* dense or condensed phase.   The suggestion has been made that this peculiar behaviour arises whenever the optical electron is excited into a Rydberg orbital, i.e. one in which the optical electron orbit is very large compared with those remaining in the core.   Such Rydberg orbitals can be visualized as linear combinations of AO's having principal quantum numbers 3 and higher, as well as azimuthal quantum numbers, $0(s)$, $1(p)$, $2(d)$, etc.   Thus, for example, the first

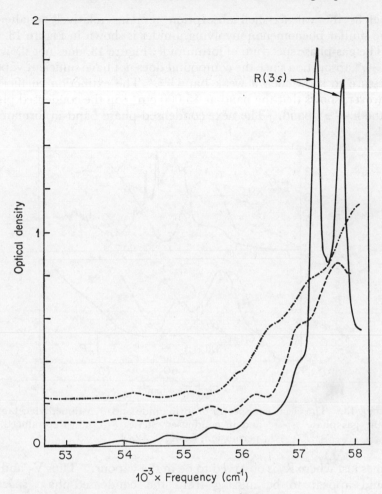

FIGURE 12.   The electronic spectrum of ethylene as a gas at low pressure (——),
with 150 atm of N_2 gas added (– – –), and as a polycrystalline film at 24°K
(–·–).   [Reproduced, by permission, from ref. 138.]

Rydberg excitation in ethylene[138] involves the excitation of an electron from the $2p$ π-bonding orbital into an MO composed of $3s$ AO's on
the carbon atoms.   This excitation is readily observed in the gas-
phase spectrum of ethylene, Figure 12, as the prominent feature
marked $R(3s)$ poised on the edge of the valence-shell (non-Rydberg)
$\pi \rightarrow \pi^*$ excitation.   However, on adding 150 atm of N_2 gas or by
forming a polycrystalline film at 24°K, $R(3s)$ is seen to disappear

3*

whereas the valence-shell absorption remains relatively unaltered. The similar phenomenon involving amides is shown in Figure 13.

The gas-phase spectrum of formamide, Figure 13, does not show an $n \rightarrow \pi^*$ absorption since the compound does not have sufficient vapour pressure to show such a weak band[122]. The extinction coefficient, however, shows that the band at 45,000 cm^{-1} in the condensed phase is the $n \rightarrow \pi^*$ band. The next condensed-phase band in formamide

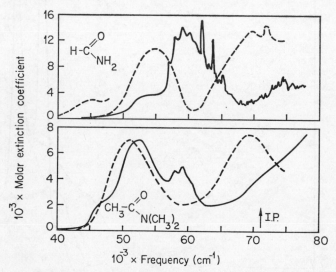

FIGURE 13. The electronic spectra of formamide and N,N-dimethylacetamide in the gas phase (——) and in condensed phases (---). [Reproduced, by permission, from ref. 123.]

comes just where R_1 is observed to be in the vapour. Thus V_1 and R_2 would appear to be missing from the condensed-phase spectra. However, in dimethylacetamide, where condensed-phase intensities were measured accurately, it was found that the band remaining in the 45,000–60,000 cm^{-1} region has an extinction coefficient compatible only with its being V_1. In this way it was concluded that both R_1 and R_2 are absent in condensed-phase amide spectra, with V_1 being more or less red shifted into the R_1 region. Room-temperature experiments with a large number of other amides as solutes in trimethyl phosphate or hexafluoroacetone hemihydrate also showed only one band in the 50,000–60,000 cm^{-1} region, where three are clearly indicated in the gas phase. Thus is it concluded that R_1 and R_2 are Rydberg bands.

Peterson and Simpson[135] found that the R_2 band of formamide could be fitted as the first term of a Rydberg series, all terms of which are given by equation (6), where E_R is the excitation energy measured

$$E_R = \text{I.P.} - \frac{109,720}{(n-\delta)^2} \tag{6}$$

downward from the ionization potential I.P. ($82,566 \text{ cm}^{-1}$), n is an integer running upwards from 3, and δ is the quantum defect, equal to 0·639 for this series of formamide. Application of equation (6) to R_1 of formamide suggests that if it also has $n = 3$ and the same ionization potential, then it must have $\delta = 1·03$. Now it has been found experimentally[139] that the value of δ in such Rydberg series is often characteristic of the symmetry type of the Rydberg orbital. Thus excitations to s, p, or d Rydberg orbitals have quantum defects of approximately 1·0, 0·6, or 0·1, respectively. Thus is it clear that R_1 is the first member of an s Rydberg series and R_2 the first member of a p series. Using these arguments in reverse, the ionization potential of any amide can be estimated by adding $109,720/(3-1)^2$ or $109,720/(3-0·6)^2$ to its observed R_1 or R_2 gas-phase absorption frequencies.

Kaya and Nagakura[136] have also observed the R_1 bands of several amides in the gas phase, but chose to assign them as valence-shell excitations in hydrogen-bonded dimers. This argument, however, would fail to explain the clear presence of the R_1 bands in tertiary amides such as dimethylacetamide and 1-methyl-2-pyrrolidone (**40**), Figures 10 and 13, which cannot form hydrogen bonds (see however, the possibility of dipole–dipole coupling, section III.F.2).

Inasmuch as Rydberg excitations are quite sharp normally, with more or less clearly defined vibrational structure, it is rather surprising to note that the R_1 Rydberg band in the dozen or so amides in which it has been observed is always completely structureless. The usual explanation for structureless absorption is that the upper state is not bound with respect to certain motions of the nuclei. If this is the case for the R_1 bands of amides, then one could expect irradiation of amides in the R_1 band to lead to extensive photochemistry. Moreover, the photochemistry would be very different in condensed phases where there is no R_1 absorption. A second possibility, however, is that the R_1 state is relaxed so quickly to the manifold of states below it that the component levels are broadened due to their very short lifetime.

The assignment of the R_1 and R_2 bands as Rydbergs is supported as well by the indirect SCF calculations. Consideration of GTO bases containing both $3s$ and $3p$ Rydberg AO's on the C, N, and O atoms, leads to the prediction that the lowest energy Rydberg excitation (5·83 ev) is $n \rightarrow 3s$, and that of the Rydberg excitations terminating at $3p_\sigma$, the lowest one (6·39 ev) is again $n \rightarrow 3p$ (Table 10). These results predict automatically that the lowest ionization potential of formamide involves the n orbital 10a′ and not the 2a″ π orbital.

The V_1 band (53,000 − 59,000 cm^{-1}) is the most prominent of those in the amides[140], and corresponds to the singlet–singlet excitation $\pi \rightarrow \pi^*$ (2a″ $\rightarrow$ 3a″) in the MO scheme. A second roughly equivalent and very useful approach to the assignment of the V_1 band considers the amide molecule as composed of an amino part —N̈R$_2$ and a keto part $>$C$=$O in the ground state, while the upper state has the charge-transfer configuration —N̈R$_2^+$; $>$C$=$O$^-$ [97,141,142]. Because the energy of such a charge transfer will depend directly on the ionization potential of the —N̈R$_2$ group, the frequencies of the V_1 bands of a number of amides can be readily correlated with the ionization potentials of their amino parts (Figure 14)[142]. Our GTO calculations lend but little support to these simple ideas about charge distribution in the V_1 excited state. Thus the population analyses show that the 3a″ π^* orbital in the π, π^* configuration is almost 90% within the carbonyl group, as presumed in the charge-transfer model, but that in

TABLE 11. π-Electron densities computed from the ground and $\pi \rightarrow \pi^*$ excited states of formamide.

	$(2a'')^2(3a'')^0$	$(2a'')^1(3a'')^1$
C	0·034	0·761
N	1·044	0·313
O	0·922	0·926

the ground state, the originating 2a″ π orbital is not at all localized on the nitrogen atom, but rather is equally distributed between nitrogen and oxygen. Moreover, the 2a″ orbital reorganizes appreciably in the V_1 excited state, so that the π-electron density changes on excitation as shown in Table 11.

Thus it is calculated that the net effect of the $\pi \rightarrow \pi^*$ transition on 2a″ and 3a″ *only*, is the transfer of 0·7 electron from N to C, the net π-electron density on O remaining fixed.

A peculiar intensity effect has been found for the V_1 band of amides. The molar extinction coefficient of the V_1 band of formamide is 15,000, which is almost twice that of the other alkylated amides (approximately 8,000)[140]. When translated into oscillator strength, the V_1 band of formamide amounts to 0·37, which is still appreciably

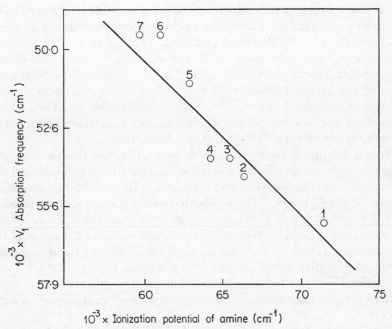

FIGURE 14. Correlation of the ionization potential of $NHR_{(1)}R_{(2)}$ and the $N \rightarrow V_1$ excitation energy in $CH_3CONR_{(1)}R_{(2)}$, where (1) $R_{(1)} = R_{(2)} = H$; (2) $R_{(1)} = H$, $R_{(2)} = Me$; (3) $R_{(1)} = H$, $R_{(2)} = Et$; (4) $R_{(1)} = H$, $R_{(2)} = n\text{-}Bu$; (5) $R_{(1)} = R_{(2)} = Me$; (6) $R_{(1)} = R_{(2)} = Et$; (7) $R_{(1)} = R_{(2)} = n\text{-}Pr$. [Reproduced, by permission, from ref. 142.]

larger than those of the alkylated amides (0·23–0·27). A similar effect is also found for the V_1 bands of formic acid relative to other carboxylic acids.

The polarization of the amide V_1 band has been deduced by Peterson and Simpson[135] from the spectrum of myristamide single crystals in polarized light. Their data are shown in Figure 11. Using the absorption ratios evident in this figure and the known orientation of the amide groups in the crystal, it was concluded that the V_1 excitation is in-plane polarized with $\theta = 17\cdot9°$. This is very nearly the direction one would predict if the excitation involved the transfer of an electron

from the —$\ddot{\text{N}}$H$_2$ group to the centre of the $\rangle$C$=$O group.

Even with our limited experience, it has become quite clear that the *ab initio* calculation of the $\pi \to \pi^*$ excitation energy in whatever planar system, is going to be too high by about 2 ev, even after indirect SCF or limited configuration interaction. Formamide has proved to be no exception, with the calculated excitation energy coming at 10·41 ev, more than 3 ev higher than observed. In accord with this energy discrepancy, the calculated oscillator strength (0·236) also differs appreciably from the observed value of 0·37. Peterson and Simpson's $\pi \to \pi^*$ polarization direction of $\theta = 17·9 \pm 10°$ is reproduced about as well as one could expect ($\theta = 30·8°$) considering that the $\pi \to \pi^*$ properties are not too well calculated, and that the experimental number is for myristamide crystal, not formamide gas.

It is a feature of electronic spectroscopy that the deeper one goes into the vacuum ultraviolet, the more excitations are possible and the more difficult it becomes to sort them out. In amides, this confusion begins at the Q band. The condensed-phase spectra show a band at the Q position, but it is many times more intense than the gas-phase Q band, and whatever its origin, its presence obscures the result of the condensed-phase Rydberg/valence-shell test. The Q band of formamide had earlier been assigned as a second valence-shell $\pi \to \pi^*$ excitation[97,135,143], but this is only one of a large number of possibilities for this excitation, which include $n \to \sigma^*$ and $\pi \to \sigma^*$ assignments.

The indirect SCF calculations, while indirect for the singlet excited states, are directly applicable to the triplet states of formamide. To our knowledge, no report of the experimental determination of the triplet energies in simple amides has appeared in the literature, and for this reason it may be pertinent to remark that the GTO calculations place the $n \to \pi^*$ and $\pi \to \pi^*$ triplet states at 3·87 and 4·40 ev, respectively.

B. Circular Dichroism Spectra

The virtually complete lack of circular dichroism (CD) spectra of optically active monomeric amides is puzzling in view of the frantic effort expended on the CD spectra of polypeptides. Only in the last few years have the first dribbles of CD data for amides appeared in the literature. The problem stems in part from the fact that one has to penetrate to about 45,000 cm^{-1} (220 mμ region) to reach the first band of amides, but this has been feasible with commercial instruments for many years.

Litman and Schellman[133] studied the optical rotatory dispersion (ORD) spectrum of the optically active lactam **42** and observed a

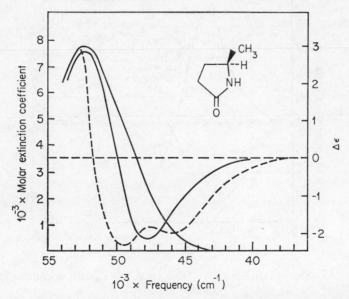

Cotton effect in dioxane solution at 43,000 cm^{-1}. This was the first observation of the Cotton effect at the $n \to \pi^*$ band of a simple amide,* though it had been seen earlier in the CD spectra of helical polypeptides[144]. A more complete CD spectrum of a similar amide (**43**) is reported by Urry[145], (Figure 15). In water solution, amide **43**

FIGURE 15. The absorption and CD spectra in water of γ-valerolactam (——), and the CD spectrum in cyclohexane, (---). [Reproduced, by permission, from ref. 145.]

* Actually the NH$_2$ group itself has an absorption at about 43,000 cm^{-1}, which however is a Rydberg excitation and, as solution spectra show, does not appear in condensed phases.

shows a negative $n \rightarrow \pi^*$ band at 47,700 cm^{-1} and a positive $\pi \rightarrow \pi^*$ band at 52,700 cm^{-1}. In cyclohexane solution, the 47,700 cm^{-1} band is split into two, at 45,800 cm^{-1} and 49,500^{-}cm^1, which is most probably due to an association phenomenon.   As can be seen from this

figure, the $n \rightarrow \pi^*$ band is completely obscured in absorption, but is quite evident in the CD spectrum.

D-Lupanine perchlorate (44) is another amide in which the $n \rightarrow \pi^*$ band is completely covered by the stronger $\pi \rightarrow \pi^*$ band in absorption, but is most conspicuous in the CD spectrum (Figure 16).

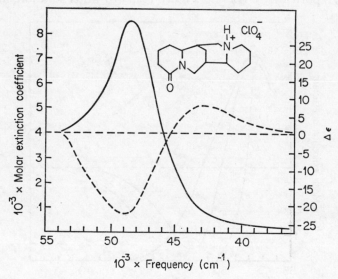

FIGURE 16.   The absorption (———) and CD (– – –) spectra of D-lupanine perchlorate in water.

The absorption and CD spectra of the lactam of aminolauronic acid (45) in acetonitrile solution (Figure 17) show $n \rightarrow \pi^*$ and $\pi \rightarrow \pi^*$ bands clearly at 45,200 cm^{-1} and 51,500 cm^{-1}, respectively[123].   The vapour-phase absorption and CD of this lactam have also been recorded, and interestingly, they both show the presence of Rydberg absorption between the $n \to \pi^*$ and $\pi \to \pi^*$ bands, which of course does not appear in the solution spectra.

It is still too early in the CD game to get very much fundamental information from spectra such as those in Figures 15–17, for it still has not even been settled as to whether the $n \to \pi^*$ rotation of amides follows a quadrant or an octant rule. However, one can say first, that

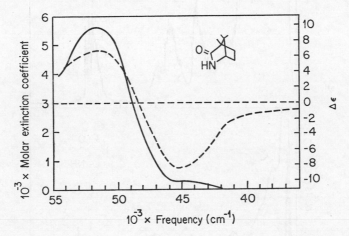

FIGURE 17. The absorption (——) and CD (– – –) spectra of amide (45) in acetonitrile solution. [Reproduced, by permission, from ref. 123.]

since the rotatory strengths of the $n \to \pi^*$ and $\pi \to \pi^*$ bands of amides are not in the same ratio as their oscillator strengths, $n \to \pi^*$ bands which are effectively invisible in ordinary absorption spectra may be readily observed in the CD spectra, and second, that Rydberg excitations are wiped out in condensed-phase CD spectra, just as in condensed-phase absorption spectra.

C. Photoelectron Spectra

Another type of experiment which is potentially of great value in understanding the electronic structures of small molecules is photoelectron spectroscopy. In this type of experiment, a monochromatic photon beam of energy 21·23 ev impinges on the gaseous molecule of interest, resulting in photoionization. The electrons so produced then travel with kinetic energies equal to 21·23 ev diminished by the ionization potentials of the molecular orbitals from which they came. Kinetic-energy analysis of the photoejected electron spray then yields all the ionization potentials of the molecule up to 21 ev.

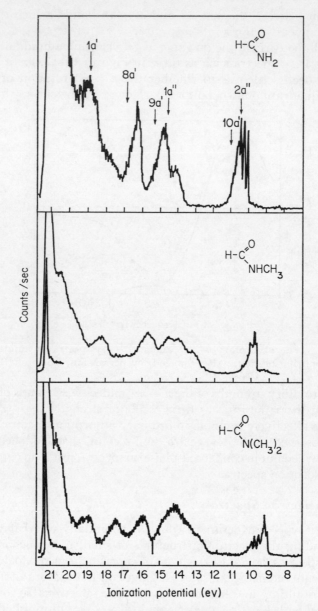

FIGURE 18. Photoelectron spectra of the various methylated formamides in the gas phase.

By Koopmans' theorem, the quantized energy decrements of the photoelectrons are simply the negatives of the computed orbital energies of Figure 9. However, our experience has been that the Koopmans' theorem values are uniformly too high by 8%, so that the purely empirical adjustment of the results of Double Zeta calculations by this amount leads to good agreement with experiment. The photoelectron spectra of various formamides[131] are shown in Figure 18, along with the theoretical energy levels for formamide.

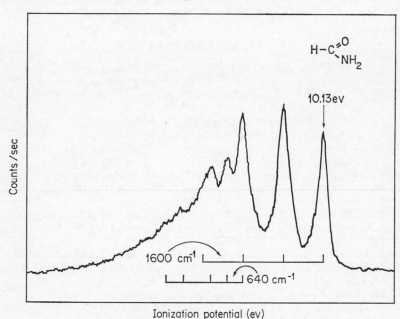

FIGURE 19.   Detail of the first two overlapping bands in the photoelectron spectrum of formamide (cf. Figure 18).

According to the predictions, the first two photoelectron peaks correspond to ionization processes originating in the n (10a') and π (2a'') orbitals. Under higher resolution, (Figure 19), the 10–11 ev region in formamide definitely does appear as two bands, one with a vibrational spacing of 1600 cm^{-1} (C=O stretch in the $^2A'$ positive ion), and one with a spacing of 640 cm^{-1} in the $^2A''$ positive ion. The identical, interleaved vibrational pattern is found as well in the $3p$ Rydberg bands of the optical spectrum of formamide (Figure 10). The next four ionization processes in formamide are identified as indicated by the remaining four arrows in Figure 18.

Like the CD measurements, it is not yet clear just how important photoelectron spectroscopy will prove to be in the study of the electronic structure of amides. Up to the moment, it has confirmed in a striking way the very near degeneracy of the highest occupied n and π orbitals of simple amides and in general supports the orbital energies calculated for the ground state. Additionally, the photoelectron spectra of N-methylformamide and N,N-dimethylformamide strongly suggest that in these compounds the n and π levels are reversed, with π slightly above n.

VI. REFERENCES

1. J. A. Schellman and C. Schellman, in *The Proteins*, Vol. II, 2nd ed., (Ed. H. Neurath), Academic Press, New York, 1964, p. 1.
2. W. F. Harrington, R. Josephs, and D. M. Segal, *Ann. Rev. Biochem.*, **35**, Pt. II, 599 (1966).
3. D. W. Urry, *Ann. Rev. Phys. Chem.*, **19**, 477 (1968).
4. J. Ladell and B. Post, *Acta Cryst.*, **7**, 559 (1954).
5. W. G. Moulton and R. A. Krumhout, *J. Chem. Phys.*, **25**, 34 (1956).
6. C. C. Costain and J. M. Dowling, *J. Chem. Phys.*, **32**, 158 (1960).
7. F. Senti and D. Harker, *J. Am. Chem. Soc.*, **62**, 2008 (1940).
8. M. Kimura and N. Aoki, *Bull. Chem. Soc. Japan*, **26**, 429 (1953).
9. J. L. Katz, *Dissertation Abstr.*, **17**, 2039 (1957).
10. B. R. Penfold and J. C. B. White, *Acta Cryst.*, **12**, 130 (1959).
11. J. R. Brathoude and E. C. Lingafelter, *Acta Cryst.*, **11**, 729 (1958).
12. J. D. Turner and E. C. Lingafelter, *Acta Cryst.*, **8**, 551 (1955).
13. E. M. Ayerst and J. R. C. Duke, *Acta Cryst.*, **7**, 588 (1954).
14. D. R. Davies and R. A. Pasternak, *Acta Cryst.*, **9**, 334 (1956).
15. M. Hospital and J. Housty, *Acta Cryst.*, **21**, 413 (1966).
16. M. Hospital and J. Housty, *Acta Cryst.*, **20**, 368 (1966).
17. R. Degeilh and R. E. Marsh, *Acta Cryst.*, **12**, 1007 (1959).
18. B. K. Vainshtein, *Zh. Fiz. Khim.*, **29**, 327 (1956).
19. C. A. Coulson, Victor Henri Memorial Volume, *Contribution a l'etude de la Structure Moleculaire*, Desoer, Liege, 1948, p. 15.
20. D. G. Lister and J. K. Tyler, *Chem. Commun.*, 152 (1966).
21. R. M. Meighan and R. H. Cole, *J. Phys. Chem.*, **68**, 503 (1964).
22. R. J. Kurland and E. B. Wilson, Jr., *J. Chem. Phys.*, **27**, 585 (1957).
23. L. Pauling, *The Nature of the Chemical Bond*, Cornell University Press, Ithaca, New York, 1948.
24. G. V. D. Tiers and F. A. Bovey, *J. Phys. Chem.*, **63**, 302 (1959).
25. J. D. Roberts, *J. Am. Chem. Soc.*, **78**, 4495 (1956).
26. F. A. Bovey and G. V. D. Tiers, *J. Polymer Sci.*, A, **1**, 849 (1963).
27. J. B. Stothers and T. G. Hill, private communication; see also, L. H. Piette, J. D. Ray, and R. A. Ogg, *J. Mol. Spectry.*, **2**, 66 (1958).
28. B. Sunners, L. H. Piette, and W. G. Schneider, *Can. J. Chem.*, **38**, 681 (1960); see also ref. 29.
29. H. Kamei, *Bull. Chem. Soc. Japan*, **38**, 1212 (1965).

30. G. Binsch, J. B. Lambert, B. W. Roberts, and J. D. Roberts, *J. Am. Chem. Soc.*, **86**, 5564 (1964). This paper also reports some longer-range $^{15}N-^1H$ coupling constants.

31. L. A. La Planche and M. T. Rogers, *J. Am. Chem. Soc.*, **86**, 337 (1964).

32. A. J. R. Bourn and E. W. Randall, *Mol. Phys.*, **8**, 567 (1964); A. J. R. Bourn, D. G. Gillies, and E. W. Randall, in *Nuclear Magnetic Resonance in Chemistry* (Ed. B. Pesce) Academic Press, New York, 1965, p. 277.

33. J. V. Hatton and R. E. Richards, *Mol. Phys.*, **5**, 139 (1962).

34. R. C. Neuman, Jr. and L. B. Young, *J. Phys. Chem.*, **69**, 1777 (1965).

35. G. Fraenkel and C. Franconi, *J. Am. Chem. Soc.*, **82**, 4478 (1960).

36. A. Berger, A. Loewenstein, and S. Meiboom, *J. Am. Chem. Soc.*, **81**, 62 (1959).

37. D. G. De Kowalewski and V. J. Kowalewski, *Arkiv Kemi*, **16**, 373 (1960).

38. A. J. R. Bourn, D. G. Gillies, and E. W. Randall, *Tetrahedron*, **20**, 1811 (1964).

39. N. Muller, *J. Chem. Phys.*, **36**, 359 (1962).

40. E. Malinowski, *J. Am. Chem. Soc.*, **84**, 2649 (1962).

41. J. W. ApSimon, W. G. Craig, P. V. Demarco, D. W. Mathieson, A. K. G. Nasser, L. Saunders, and W. B. Whalley, *Chem. Commun.* 754 (1966).

42. H. Sternlicht and D. Wilson, *Biochemistry*, **6**, 288 (1967).

43. H. Paulsen and K. Todt, *Angew. Chem. Intern. Ed. Engl.*, **5**, 899 (1966).

44. T. H. Siddall and W. E. Stewart, *J. Mol. Spectry.*, **24**, 290 (1967).

45. L. M. Jackman, *Applications of Nuclear Magnetic Resonance Spectroscopy in Organic Chemistry*, Pergamon Press, London, 1959, p. 123.

46. J. A. Pople, *Proc. Roy. Soc. (London)*, *Ser. A*, **239**, 541, 550 (1957).

47. F. A. L. Anet and A. J. R. Bourn, *J. Am. Chem. Soc.*, **87**, 5250 (1965).

48. J. A. Pople, *J. Chem. Phys.*, **24**, 111 (1956).

49. J. S. Waugh and R. W. Fessenden, *J. Am. Chem. Soc.*, **79**, 846 (1957); **80**, 6697 (1958).

50. C. E. Johnson, Jr. and F. A. Bovey, *J. Chem. Phys.*, **29**, 1012 (1958).

51. C. Franconi, *Z. Elektrochem.*, **65**, 645 (1961); *Scienza Technica, n.s.*, **4**, 183 (1960).

52. L. A. La Planche and M. T. Rogers, *J. Am. Chem. Soc.*, **85**, 3728 (1963).

53. A. Mannschreck, *Tetrahedron Letters*, 1341 (1965).

54. T. Miyazawa, *J. Mol. Spectry.*, **4**, 155 (1960).

55. S. Mizushima, T. Simanouti, S. Nagakura, K. Kuratani, M. Tsuboi, H. Baba, and O. Fujioka, *J. Am. Chem. Soc.*, **72**, 3490 (1950).

56. R. L. Jones, *J. Mol. Spectry.*, **2**, 581 (1958).

57. C. J. Brown and D. E. C. Corbridge, *Nature*, **162**, 72 (1948).

58. G. B. Carpenter and J. Donohue, *J. Am. Chem. Soc.*, **72**, 2315 (1950).

59. S. Mizushima and T. Shimanouchi, *Advan. Enzymol.*, **23**, 1 (1961).

60. J. T. Edsall, P. J. Flory, J. C. Kendrew, A. M. Liquori, G. Nemethy, G. Ramachandran, and H. A. Scheraga, *Biopolymers*, **4**, 121 (1966).

61. D. R. Herschbach and L. C. Krisher, *J. Chem. Phys.*, **28**, 728 (1958).

62. R. W. Kilb, C. C. Lin, and E. B. Wilson, *J. Chem. Phys.*, **26**, 1695 (1957).

63. A. A. Bothner-By, C. Naar-Colin, and H. Günther, *J. Am. Chem. Soc.*, **84**, 2748 (1962).

64. R. M. Hammaker and B. A. Gugler, *J. Mol. Spectry.*, **17**, 356 (1965).

65. W. D. Phillips, *J. Chem. Phys.*, **23**, 1363 (1955).

70   M. B. Robin, F. A. Bovey, and Harold Basch

66. H. S. Gutowsky, D. W. McCall, and C. P. Slichter, *J. Chem. Phys.*, **21**, 279 (1953).
67. H. S. Gutowsky and A. Saika, *J. Chem. Phys.*, **21**, 1688 (1953).
68. H. S. Gutowsky and C. H. Holm, *J. Chem. Phys.*, **25**, 1228 (1956).
69. H. M. McConnell, *J. Chem. Phys.*, **28**, 430 (1958).
70. C. S. Johnson, Jr., in *Advances in Magnetic Resonance*, Vol. I (Ed. J. S. Waugh), Academic Press, New York, 1965.
71. J. W. Emsley, J. Feeney, and L. H. Sutcliffe, *High Resolution Nuclear Magnetic Resonance Spectroscopy*, Pergamon Press, London, 1965, Chap. 9.
72. F. A. Bovey, *High Resolution Nuclear Magnetic Resonance Spectroscopy*, Academic Press, New York, 1968, Chap. VII.
73. J. A. Pople, W. G. Schneider, and H. J. Bernstein, *High Resolution Nuclear Magnetic Resonance*, McGraw-Hill Book Co., New York, 1959, Chap. 10 and 13.
74. J. Jonas, A. Allerhand, and H. S. Gutowsky, *J. Chem. Phys.*, **42**, 396 (1965).
75. A. Allerhand, H. S. Gutowsky, J. Jonas, and R. A. Meinzer, *J. Am. Chem. Soc.*, **88**, 3185 (1966).
76. A. Pines and M. Rabinovitz, *Tetrahedron Letters*, 3529 (1968).
77. R. C. Neuman, Jr., D. N. Roark, and V. Jonas, *J. Am. Chem. Soc.*, **89**, 3412 (1967).
78. E. L. Hahn, *Phys. Rev.*, **80**, 580 (1950).
79. Z. Luz and S. Meiboom, *J. Chem. Phys.*, **39**, 366 (1964).
80. A. Allerhand and H. S. Gutowsky, *J. Chem. Phys.*, **41**, 2115 (1964).
81. A. Mannschreck, A. Mattheus, and G. Rissmann, *J. Mol. Spectry.*, **23**, 15 (1967).
82. T. H. Siddall, III, W. E. Stewart, and A. L. Marston, *J. Phys. Chem.*, **72**, 2135 (1968).
83. H. S. Gutowsky, J. Jonas, and T. H. Siddall, III, *J. Am. Chem. Soc.*, **89**, 4300 (1967).
84. M. T. Rogers and J. C. Woodbrey, *J. Phys. Chem.*, **66**, 540 (1962).
85. A. G. Whittaker and S. Siegel, *J. Chem. Phys.*, **42**, 3320 (1965).
86. R. C. Neuman, Jr. and L. B. Young, *J. Phys. Chem.*, **69**, 2570 (1965).
87. C. W. Fryer, F. Conti, and C. Franconi, *Ric. Sci. Rend. A*, **8**, 788 (1965).
88. H. S. Gutowsky, J. Jonas, and T. H. Siddall, III, *J. Am. Chem. Soc.*, **89**, 4300 (1967).
89. K. H. Abramson, P. T. Inglefield, E. Krakower, and L. W. Reeves, *Can. J. Chem.*, **44**, 1685 (1966).
90. J. C. Woodbrey and M. T. Rogers, *J. Am. Chem. Soc.*, **84**, 13 (1962).
91. A. Loewenstein, A. Melera, P. Rigny, and W. Walter, *J. Phys. Chem.*, **68**, 1597 (1964).
92. A. Mannschreck, *Angew. Chem.*, **4**, 985 (1965).
93. T. H. Siddall, *Tetrahedron Letters*, 4515 (1965).
94. P. Smith and P. B. Wood, *Can. J. Chem.*, **44**, 3085 (1966).
95. J. E. Douglas, B. S. Rabinovitch, and F. S. Looney, *J. Chem. Phys.*, **23**, 315 (1955).
96. J. Tanaka, *J. Chem. Soc. Japan*, **78**, 1636 (1958).
97. S. Nagakura, *Mol. Phys.*, **3**, 105 (1960).
98. T. H. Siddall and C. A. Prohaska, *Nature*, **208**, 582 (1965).

99. Y. Shvo, E. C. Taylor, K. Mislow, and M. Raban, *J. Am. Chem. Soc.*, **89**, 4910 (1967).
100. R. Adams, *Record Chem. Progr.* (*Kresge-Hooker Sci. Lib.*), **10**, 91 (1949).
101. T. H. Siddall and C. A. Prohaska, *J. Am. Chem. Soc.*, **88**, 1172 (1966).
102. T. H. Siddall and M. L. Good, *Naturwissenschaften*, **53**, 502 (1966).
103. T. H. Siddall, *Tetrahedron Letters*, 2027 (1966).
104. T. H. Siddall and R. H. Garner, *Tetrahedron Letters*, 3513 (1966).
105. T. H. Siddall and R. H. Garner, *Can. J. Chem.*, **44**, 2387 (1966).
106. T. H. Siddall, *J. Phys. Chem.*, **70**, 2050 (1966).
107. T. H. Siddall, *J. Phys. Chem.*, **70**, 2249 (1966).
108. T. H. Siddall, *J. Org. Chem.*, **31**, 3719 (1966).
109. A. G. Whittaker and S. Siegel, *J. Chem. Phys.*, **43**, 1575 (1965).
110. T. H. Siddall and W. E. Stewart, *J. Chem. Phys.*, **48**, 2928 (1968).
111. V. M. Potapov, V. M. Dem'yanovich, and A. P. Terent'ev, *Zh. Obshch. Khim.*, **31**, 3046 (1961); and earlier work.
112. L. Skulski, G. C. Palmer, and M. Calvin, *Tetrahedron Letters*, 1773 (1963).
113. L. Skulski, *Bull. Acad. Polon. Sci.*, **12**, 299 (1964).
114. P. Beak, J. Bonham, and J. T. Lee, Jr., *J. Am. Chem. Soc.*, **90**, 1569 (1968).
115. D. B. Brown, R. D. Burbank, and M. B. Robin, *J. Am. Chem. Soc.*, **91**, 2895 (1969).
116. L. A. La Planche, H. B. Thompson, and M. T. Rogers, *J. Phys. Chem.*, **69**, 1482 (1965).
117. R. C. Neuman, Jr., W. Snider, and V. Jonas, *J. Phys. Chem.*, **72**, 2469 (1968).
118. C. A. Coulson and E. T. Stewart, 'Wave Mechanics and the Alkene Bond,' in *The Chemistry of Alkenes* (Ed. S. Patai), John Wiley and Sons, London, 1964, Chap. 1, pp. 1–147.
119. J. A. Pople and G. A. Segal, *J. Chem. Phys.*, **44**, 3289 (1966).
120. J. L. Whitten, *J. Chem. Phys.*, **44**, 359 (1966).
121. S. Huzinaga, *J. Chem. Phys.*, **42**, 1293 (1965).
122. H. Basch, M. B. Robin, and N. A. Kuebler, *J. Chem. Phys.*, **47**, 1201 (1967).
123. H. Basch, M. B. Robin, and N. A. Kuebler, *J. Chem. Phys.*, **49**, 5007 (1968).
124. D. Neumann and J. Moskowitz, unpublished calculations on H_2O, H_2CO, and C_2H_4.
125. M. A. Robb and I. G. Csizmadia, *Theor. Chim. Acta* (*Berlin*), **10**, 269 (1968).
126. H. D. Hunt and W. T. Simpson, *J. Am. Chem. Soc.*, **75**, 4540 (1953).
127. D. B. Cook and R. McWeeny, *Chem. Phys. Letters*, **1**, 588 (1968).
128. S. Yamosa, *Biopolymers Symp.*, **1**, 1 (1964).
129. R. K. Nesbet, private communication.
130. L. C. Snyder and H. Basch *J. Am. Chem. Soc.* **91**, 2189 (1969).
131. C. R. Brundle, D. W. Turner, M. B. Robin and H. Basch, *Chem. Phys. Letters*, **3**, 292 (1969).
132. C. C. J. Roothaan, *Rev. Mod. Phys.*, **23**, 69 (1951).
133. B. J. Litman and J. A. Schellman, *J. Phys. Chem.*, **69**, 978 (1965).
134. See also, M. L. Good, T. H. Siddall, III, and R. N. Wilhite, *Spectrochim. Acta*, **23A**, 1161 (1967).
135. D. L. Peterson and W. T. Simpson, *J. Am. Chem. Soc.*, **79**, 2375 (1957).
136. K. Kaya and S. Nagakura, *Theor. Chim. Acta* (*Berlin*), **7**, 117, 124 (1967).
137. W. Rhodes and D. G. Barnes, *J. Chem. Phys.* **48**, 817 (1968).

138. M. B. Robin, H. Basch, and N. A. Kuebler, *J. Chem. Phys.*, **48**, 5037 (1968).
139. S. R. La Paglia, *J. Mol. Spectry.*, **10**, 240 (1963).
140. E. B. Nielson and J. A. Schellman, *J. Phys. Chem.*, **71**, 2297 (1967).
141. E. J. Rosa and W. T. Simpson, in *Physical Processes in Radiation Biology*, (Eds. L. Augenstein, P. Mason, and B. Rosenberg), Academic Press, New York, 1964, p. 43.
142. J. A. Schellman and E. B. Nielsen, *J. Phys. Chem.*, **71**, 3914 (1967).
143. M. Suard, G. Berthier, and B. Pullman, *Biochim. Biophys. Acta*, **52**, 254 (1961).
144. W. B. Gratzer, G. M. Holzwarth, and P. Doty, *Proc. Natl. Acad. Sci. U.S.*, **47**, 1785 (1961).
145. D. W. Urry, *J. Phys. Chem.*, **72**, 3035 (1968).

CHAPTER 2

Synthesis of amides

A. L. J. Beckwith

University of Adelaide, Australia

73

I. INTRODUCTION

In this chapter we shall be concerned with the synthesis of compounds containing the amide function, including simple amides, diacyl- and triacylamines, lactams, and imides. Some limitation of the area surveyed has been necessary in order to keep the size of the review within reasonable bounds. Accordingly, discussion of methods of preparation of carbamates and ureas has been precluded except in those cases where such compounds are intermediates in the formation of amides; nor in general do we consider the preparation of amides by reactions involving modification of molecules which already contain the carbamyl group. However, the principle that only those transformations involving introduction of a new amide function will be reviewed has been relaxed in section VI, where, because of their particular significance in lactam synthesis, some aspects of the alkylation of amides are discussed.

Also, for reasons of space it has not been possible to devote to methods for the preparation of lactams such close scrutiny as has been accorded acyclic amides. However, this deficiency is partly rectified by the availability of reviews of the chemistry of α- and β-lactams[1,2], the former very recent, whilst discussions of synthetic routes to higher lactams are to be found in most texts on heterocyclic chemistry.

Each section and subsection includes, where possible, some consideration of mechanistic aspects, as well as a survey of the scope and limitations of the particular reaction under examination. In the belief that those readers who consult this chapter will include some who are confronted with practical problems in amide synthesis, I have endeavoured to select from the vast range available a number of specific recent examples of preparations of amides which proceed in high yield and which illustrate well the best experimental procedures. Further information on practical methods is available in texts; those by Hickinbottom[3], Wagner and Zook[4], and Fieser and Fieser[5] are especially useful. Excellent discussions of the general chemistry of amides, including synthesis, are to be found in the new edition of Sidgwick[6] and in Smith's recent monograph[7].

II. ACYLATION OF AMINES AND AMIDES

A. General Considerations

This section is concerned with formation of amides, imides and lactams via acylation at nitrogen in ammonia, amines or amides. Apart from a few special cases (e.g. acylation with ketenes) the reactions are of the general form of equation (1) and thus represent examples of

$$R^1COX + R^2R^3NH \longrightarrow R^1CONR^2R^3 + HX \qquad (1)$$

nucleophilic substitution at the carbonyl carbon atom. For such processes three distinct mechanisms (2–4) may be formulated (where Y^- represents the nucleophile).

$$RCOX \rightleftharpoons RCO^+ + X^- \xrightarrow{Y^-} RCOY \qquad (2)$$

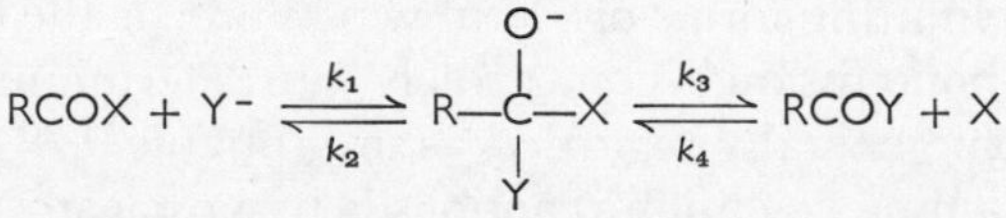

$$RCOX + Y^- \underset{k_2}{\overset{k_1}{\rightleftharpoons}} R-\overset{\overset{\displaystyle O^-}{|}}{\underset{\underset{\displaystyle Y}{|}}{C}}-X \underset{k_4}{\overset{k_3}{\rightleftharpoons}} RCOY + X^- \qquad (3)$$

$$RCOX + Y^- \rightleftharpoons \left[R-\overset{\overset{\displaystyle O}{\|}}{\underset{\underset{\displaystyle X}{\vdots}}{C}}\cdots Y \right]^- \longrightarrow RCOY + X^- \qquad (4)$$

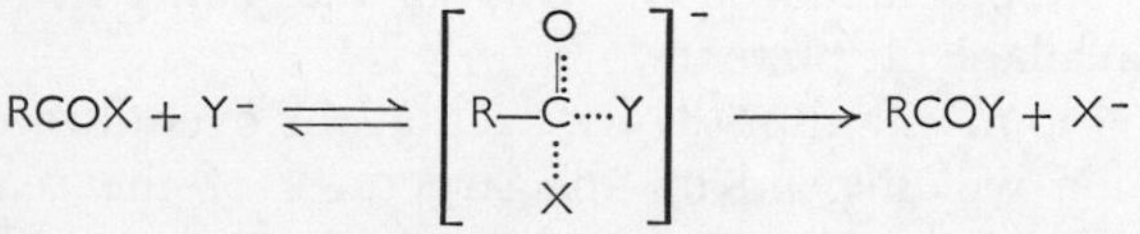

Mechanism (2) is most likely to be observed in solvents of high polarity when X forms a very stable anion (i.e. when HX is a strong acid) and when Y^- is weakly nucleophilic. In view of the relatively high nucleophilicity of nitrogen in most amines and amides mechanism (2) is improbable for the majority of acylation reactions of such compounds. However, in some special cases (e.g. acylation with acyl tetrafluoroborates)[8] acylium-ion intermediates are involved.

The addition–elimination mechanism (3) is generally considered to apply to most acylation reactions of amines and amides[9–12] although substantial evidence for its existence has been amassed only in the case of aminolysis of esters (section II.D). When applied to amine acylation, mechanism (3) must include a step involving loss of a proton from nitrogen and then is most simply represented as equation (5).

$$R^1COX + R^2R^3NH \rightleftharpoons R-\overset{\overset{\displaystyle O^-}{|}}{\underset{\underset{\displaystyle +N-H}{|}}{C}}-X \overset{-H^+}{\rightleftharpoons} R^1-\overset{\overset{\displaystyle O^-}{|}}{\underset{\underset{\displaystyle N}{|}}{C}}-X \rightleftharpoons R^1CONR^2R^3 + X^-$$

$$\text{(5)}$$

Although this formulation provides a useful generalization of predictive value for estimating the probable effects of changes in reactant structure and experimental conditions it undoubtedly presents an over-simplified view. In particular the nature and position in the reaction sequence of proton-transfer steps and the factors affecting them, such as catalysis by acids and bases, are imperfectly understood and present complex problems requiring further investigation. In general, careful consideration of such subtleties of mechanism lies outside the scope of the present discussion.

Mechanism (4) involves synchronous bond breaking and bond making. It covers a wide range of mechanistic behaviour depending on the relative importance of the two processes, and mechanisms (2) and (3) are seen as limiting forms of (4) in which one or the other of bond formation and bond fission becomes solely rate determining. Mechanism (4) has been generally regarded as improbable[11,13] but its occurrence in some amine acylation reactions is by no means inconceivable and cannot be precluded on the basis of the scanty mechanistic information available at present.

Returning to mechanism (3) we see that the overall rate of the forward reaction will depend on the structures of the reactants[11,14]. Thus, increase in the electron-attracting power of R in RCOX will stabilize the intermediate complex hence increasing k_1 which is the most significant factor in the rate expression, and enhancing the overall rate of acylation. Conversely, electron-donating groups R, particularly those which stabilize RCOX by resonance interaction with the carbonyl groups, will decrease the rate of the forward reaction.

Similar generalizations can be made concerning the nature of X. In broad terms increase in the electron-withdrawing character of X will increase k_1 and k_3 and decrease k_4 thus enhancing the rate of acylation, whilst electron release by X, particularly through conjugation with the carbonyl group will diminish the rate. Such considerations are in accord with the observed approximate order of reactivity of the main classes of acylating agents and its parallelism with the order of acidity of HX.

$$RCOR < RCONR_2 < RCO_2R < (RCO)_2O < RCOHal < RCOBF_4$$

Within each class of acylating agent more subtle relationships between structure and reactivity can be detected. They are discussed in the appropriate sections of this chapter. However, the profound effect of conjugative release in X is worthy of special note since it accounts for the particular effectiveness of reagents of the general type RCO—A—B=D such as nitrophenyl esters, adducts of acids with carbodiimide, and acyl azides.

For reactions proceeding by mechanism (3) (or mechanism 4) the rate of the forward reaction should depend on the nucleophilicity of Y^-. With amines and amides a reasonable parallelism is observed between nucleophilic power, as approximately represented by base strength, and ease of acylation (e.g. alkylamines > arylamines > amides). An important practical consequence of this relationship, nicely illustrated for example by the Haller–Bauer reaction (section II.F) or by alkoxide-catalysed aminolysis of esters (section II.D), is that the conjugate bases, RNH^- or $RCONH^-$, of amines and amides, being more powerful nucleophiles undergo acylation by reagents which are either inert towards, or react very slowly with, the parent compounds.

Finally, it is noteworthy that intramolecular acylation of nitrogen in suitably constituted derivatives of amino acids (i.e. those leading to 5- or 6-membered rings) occurs more readily than analogous intermolecular reactions, presumably because of the more favourable entropy term in the rate expression. Consequently it is frequently possible to prepare pyrrolidones, piperidones and related compounds under experimental conditions much milder than those generally employed for amide formation.

B. Acyl Halides

Ammonia, and most primary and secondary amines are readily acylated by treatment with acyl halides (equation 6). The reaction

$$R^1COHal + R^2R^3NH \longrightarrow R^1CONR^2R^3 + HHal \qquad (6)$$

often proceeds with vigour; presumably this is why it has so infrequently been chosen for mechanistic investigation. Nevertheless, the rather scanty information available[15–18] accords with prediction (section II.A) in that acetyl chloride is more reactive than its higher homologues (ascribed both to the increased $+I$ effect and to the greater steric interactions of higher alkyl groups), that the reactivity of acyl halides is enhanced by electron-attracting substituents, and that crotonyl and benzoyl chlorides, in which there is conjugative stabilization of the carbonyl group, are less reactive than saturated acyl chlorides. Some typical reactivity series are:

$$CH_3COCl > CH_3CH_2COCl > CH_3CH_2CH_2COCl > (CH_3)_2CHCOCl$$
$$ClCH_2COCl > PhCH_2COCl > CH_3COCl$$

The order of ease of displacement of the various halogens is $I > Br > Cl > F$. Apparently, as in nucleophilic displacement of halogen from saturated carbon, the effect of C—Hal bond strength outweighs that of electronegativity[12]. Finally it should be noted that although most discussions of mechanism have assumed direct attack of amine on RCOHal, there is a possibility, at least under some experimental conditions[19], of prior participation of oxygen-containing solvents in the mechanism through formation of oxo-oxonium salts (equation 7).

$$R^1COHal + R^2—O—R^3 \rightleftharpoons R^1CO—\overset{+}{O}R^2R^3Hal^- \qquad (7)$$

For preparative purposes acyl chlorides and bromides are usually employed rather than the less readily available fluorides and iodides, but formyl fluoride is used for formylation[20,21], and in other special cases (e.g. preparation of acetoacetamides[22]) acyl fluorides offer advantages. Acyl tetrafluoroborates, hexafluoroantimonates, and similar oxocarbonium salts are highly efficient N-acylating agents[8,21,23]. Methods recently developed for the preparation of acyl chlorides[24] and bromides[25] under very mild conditions will undoubtedly extend the application of these reagents in amide synthesis.

Acyl halides react with ammonia and with amines under a wide range of experimental conditions[26] and the choice of the best procedure depends on the nature and availability of the starting materials. Acylation of ammonia and the lower alkylamines is often conducted by adding the halide to a cold, stirred, aqueous solution of the base[27-29], a method which has the advantages of technical simplicity and efficiency, although yields usually diminish as the homologous series is ascended. The lower yields encountered when amides are prepared from long-chain acyl halides probably arise from the difficulty of ensuring intimate contact between the hydrophobic acyl halide and the water-soluble amine, and also because the insoluble product tends to form a protective film around unreacted halide. These difficulties can be avoided by shaking a solution of the acyl halide in a suitable inert solvent with the aqueous amine. Ether is often employed, and the product amide is then obtained by evaporation of the organic layer[30,31]. Tetrahydrofuran has been used in the preparation of adamantane-1-carboxamide[32]. Frequently, when a solvent immiscible with water is employed, the product, being insoluble in either phase, precipitates at the interface. Examples of this procedure, which often affords excellent yields include the preparation of aromatic amides[33] (e.g. **1**) and steroid amides[30,34] (e.g. **2**).

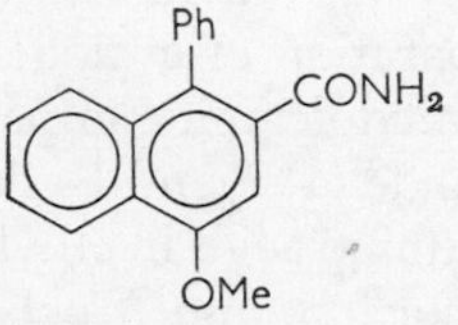

(**1**) 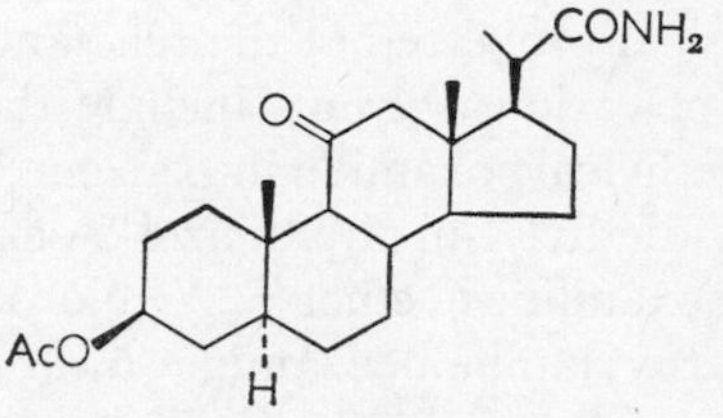   (**2**)

Aqueous ammonia is not a suitable reagent for the preparation of primary carboxamides which, because of low molecular weight or the presence of hydrophilic functions, have high water solubility. In such cases it is usual to pass gaseous ammonia into, or over*, a solution of the acyl halide in a suitable organic solvent. Philbrook[35] claims that the reaction in benzene gives consistently higher yields of fatty acid amides than other methods. Similarly, treatment of 3-methoxy-2-methylacryloyl chloride in ethylene chloride with gaseous ammonia affords the amide **3** in excellent yield whereas aqueous ammonia gives a very poor yield[36].

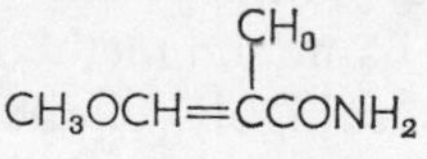

(**3**) 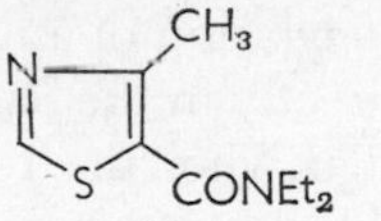   (**4**) 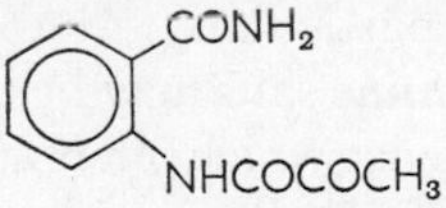   (**5**)

The reactions of lower acyl halides with ammonia are frequently inconveniently vigorous. A milder method consists of treating the acyl chloride with ammonium acetate in acetone[37]. The reaction,

* Considerable experimental difficulties can arise by blocking of the inlet tube with ammonium chloride when ammonia is passed *into* solutions of acyl halides in organic solvents. Passing ammonia *over* the surface of a stirred solution is a better procedure.

which is believed to involve free ammonia formed by dissociation of
the ammonium salt, proceeds in good yield and has been applied to a
wide range of representative compounds.   Ammonium carbonate in
water has similarly been used for mild ammonolysis of highly reactive
halides[38].

Organic solvents which offer the convenience of a homogeneous
reaction mixture have been very widely used in the acylation of
alkyl- and arylamines.   Almost the whole range of common organic
liquids, including methanol and ethanol, has been employed; the
precise choice for any particular reaction obviously depends on the
physical properties of the reactants.   Representative examples of the
use of various solvents include the preparation of cyclohexane-N,N-
dimethylcarboxamide in benzene[39], N-butyloleamide in petroleum[28],
N,N-diethyl-4-methylthiazol-5-carboxamide (4)[40] and 1-adamantane-
carboxanilide in ether[41], N-chloroacetylamino acids in ethyl acetate[42],
2-pyruvylaminobenzamide (5) in chloroform[43], and N-octyltrichloro-
acetamide in ethylene dichloride[44].

In this procedure the use of an excess of amine is necessary since
part of it is consumed in reaction with hydrogen halide liberated during
the acylation.   The resultant amine salts usually precipitate and are
removed by filtration or dissolution in water.   The maximum possible
yield of amide from amine is thus 50%.   When a higher conversion
is desirable inorganic or tertiary amine bases are added to the reaction
mixture.   In the widely used Schotten–Baumann method the bases
most frequently employed are aqueous sodium hydroxide and potassium hydroxide.   The procedure is technically simple; typically the
acid halide is slowly added to a vigorously stirred suspension or solution of the amine in aqueous caustic.   The method has been very
widely used for the preparation of aromatic amides and anilides, and
is particularly suitable for acylation of amino acids and peptides[45,46].
Amine salts may be used directly in the Schotten–Baumann method, a
device which offers advantages when the free bases are highly volatile[47];
[15]N-labelled amides may thus be conveniently prepared with minimal
loss of ammonia from acyl halides and ammonium nitrate or other
ammonium salts[48].   Aqueous sodium carbonate and sodium bicarbonate are also suitable inorganic bases for the Schotten–Baumann
method[49], as is magnesium oxide in dioxan–water[50].

When aqueous bases are employed hydrolysis of the acyl halide to
the acid competes, albeit inefficiently, with amide formation, and for
this reason an excess of the acylating agent is generally used.   This disadvantage may be overcome by using an inorganic base in an organic

solvent, e.g. calcium oxide[51] or sodium carbonate[52] in benzene, or sodium carbonate in acetone[53].

Organic bases may also be used to consume hydrogen halide liberated during the acylation reaction. Pyridine, dimethylaniline, triethylamine, and tertiary alkylamines in general are suitable reagents and they may be used either in an organic solvent or neat. Examples illustrative of such methods include benzoylation of *o*-nitroaniline in dimethylaniline solution[54], and acylation of aziridine in ether–triethylamine[55].

Both inorganic bases and tertiary amines appear to exert a catalytic effect on acylation reactions, particularly when the substrate is a weakly basic amine such as diphenylamine or nitroaniline. In some reactions the catalytic effect of tertiary amines, like that of inorganic bases, may be ascribed to their role as proton acceptors (equation 5), but in others, particularly acylation of weakly nucleophilic substrates, it may be reasonably attributed to formation of acylammonium salts of high acylating power (equation 8). Under carefully controlled conditions acylammonium compounds can be isolated.

$$R^1COHal + R_3^2N \rightleftharpoons R^1CO\overset{+}{N}R_3^2Hal^- \qquad (8)$$

Known examples include acetyl-, *p*-nitrobenzoyl-, and furoylpyridinium chlorides[56], benzoyltrimethyl- and benzoyltriethylammonium hexachloroantimonates[57], acyl trialkylammonium chlorides[58], and dimethylacryloyltrimethylammonium chloride (**6**)[59]. They are readily hydrolysed by moisture, and on heating ketenes are usually produced (equation 9).

$$\begin{array}{ccc} Me & & Me \\ \diagdown & & \diagdown \\ C=CHCO\overset{+}{N}Me_3 \; Cl^- & \longrightarrow & C=C=C=O + Me_3NH^+Cl^- \\ \diagup & & \diagup \\ Me & & Me \end{array} \qquad (9)$$

(**6**)   (**7**)

However, in certain cases, notably aryl-substituted tertiary amines, reaction with acyl halides leads to amides via fragmentation (equation 10)[60].

$$\begin{array}{c} ArNCH_2NR_2 + PhCOCl \longrightarrow ArNCOPh + ClCH_2NR_2 \\ | \qquad\qquad\qquad\qquad\qquad\qquad | \\ Me \qquad\qquad\qquad\qquad\qquad\quad Me \end{array} \qquad (10)$$

The formation of amides by acylation of amines or ammonia with acyl halides is applicable to a wide range of structural types. Some typical examples illustrative of the scope of the reaction, experimental conditions and yields, are given in equations (11–19).

4+c.o.a.

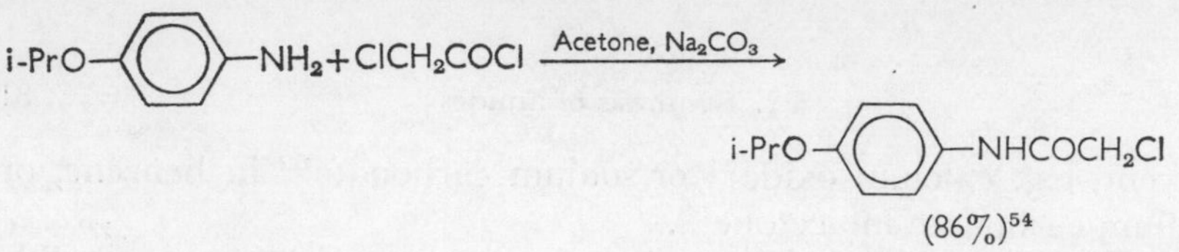

$$\text{i-PrO}-\langle\text{C}_6\text{H}_4\rangle-\text{NH}_2 + \text{ClCH}_2\text{COCl} \xrightarrow{\text{Acetone, Na}_2\text{CO}_3}$$

$$\text{i-PrO}-\langle\text{C}_6\text{H}_4\rangle-\text{NHCOCH}_2\text{Cl} \quad (11)$$
$$(86\%)^{54}$$

$$\text{Cl(CH}_2)_4\text{COCl} + \text{EtNH}_2 \xrightarrow{\text{H}_2\text{O, cold}} \text{Cl(CH}_2)_4\text{CONHEt} \quad (12)$$
$$(71\%)^{61}$$

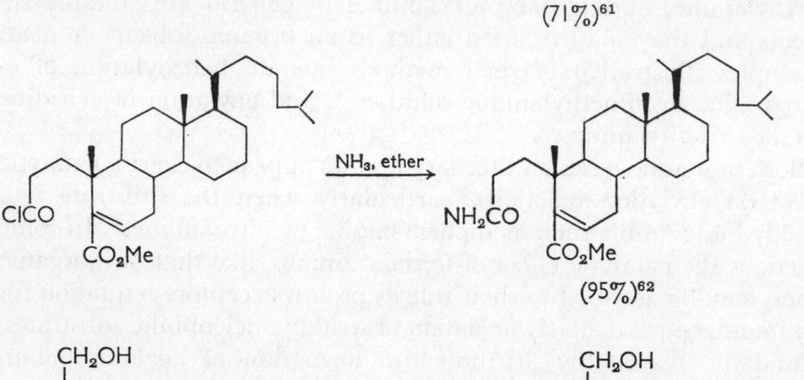

$$(13)$$
$$(95\%)^{62}$$

$$+ \text{CH}_3(\text{CH}_2)_{14}\text{COCl} \xrightarrow[-10°]{\text{NaOH, H}_2\text{O}}$$

$$(14)$$
$$(100\%)^{63}$$

$$\text{CH}_2{=}\text{CHCH}_2\text{NH}_2 + \text{HCOF} \xrightarrow{\text{Ether, 0°}} \text{CH}_2{=}\text{CHCH}_2\text{NHCHO} \quad (15)$$
$$(79\%)^{20}$$

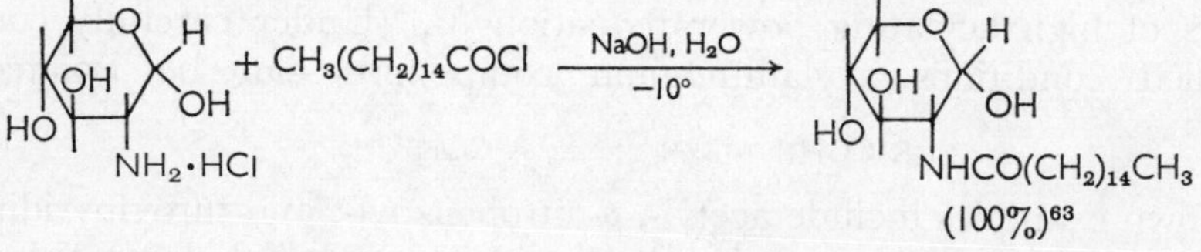

$$\text{Ph}\text{-}\langle\text{aziridine}\rangle\text{NH} + \text{ClCO}-\langle\text{C}_6\text{H}_4\rangle-\text{NO}_2 \xrightarrow{\text{Et}_3\text{N, ether}} \text{Ph}\text{-}\langle\text{aziridine}\rangle\text{NCO}-\langle\text{C}_6\text{H}_4\rangle-\text{NO}_2$$
$$(89\%)^{64} \quad (16)$$

$$\xrightarrow{\text{NH}_4\text{OAc, acetone}}$$

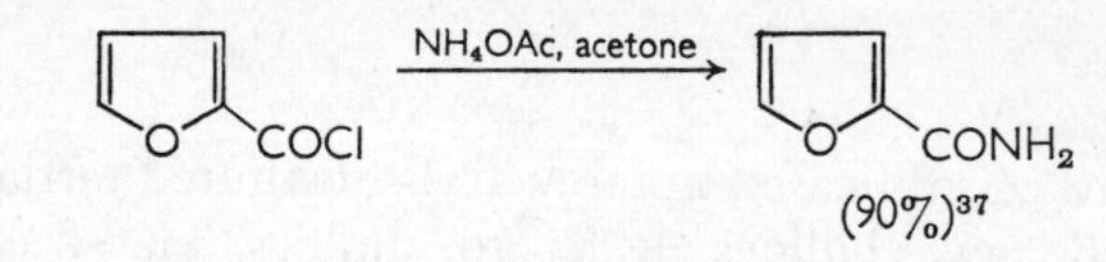

$$(17)$$
$$(90\%)^{37}$$

$$+ \text{CH}_3\text{NH}_2 \xrightarrow[25°]{\text{PhCH}_3, \text{H}_2\text{O}}$$

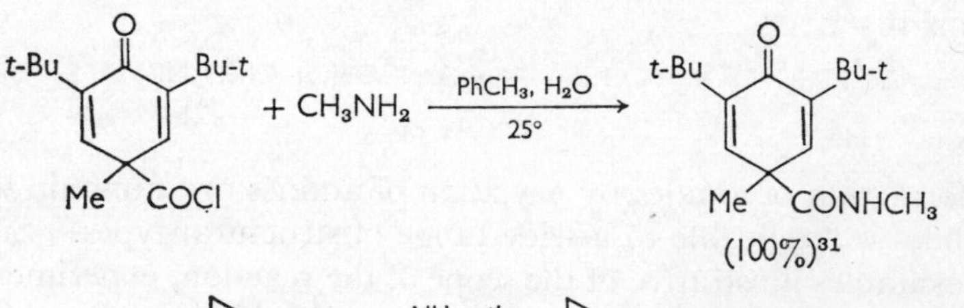

$$(18)$$
$$(100\%)^{31}$$

$$\langle\text{cyclopropyl}\rangle\text{-COCl} \xrightarrow[0°]{\text{NH}_3, \text{ether}} \langle\text{cyclopropyl}\rangle\text{-CONH}_2 \quad (19)$$
$$(91\%)^{65}$$

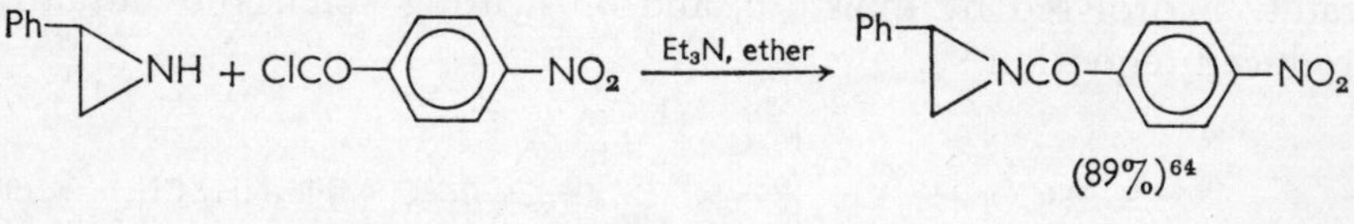

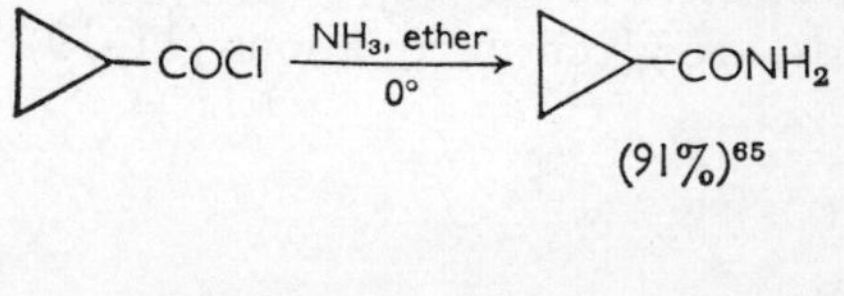

Generally the course of the reaction is unaffected by the presence of other potentially reactive functions in either the acid chloride or amine. Unsaturated amines and acyl halides react normally, and the unstable polyacetylenic amides found in nature can be prepared without difficulty[66]. The acylation of aziridines proceeds readily[55,64,67] and is of considerable interest because the products may be transformed into aldehydes[55] or N-acyl-β-haloalkylamines (equation 20)[67].

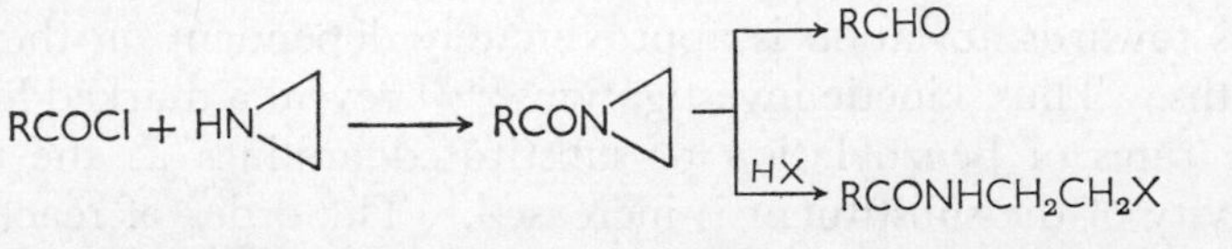

$$RCOCl + HN\diagdown \longrightarrow RCON\diagdown \begin{array}{l} \longrightarrow RCHO \\ \\ \xrightarrow{HX} RCONHCH_2CH_2X \end{array} \qquad (20)$$

The ease of reaction of amines with the —COHal function is illustrated by the successful application of the reaction to acyl halides containing other groups sensitive to aminolysis, e.g. alkyl halide[53,61], benzyl halide[68], and ester[30,62,69]. Similarly the high nucleophilic power of the amino function allows selective N-acylation of amino alcohols[63] and amino phenols[70].

Polyfunctional amines and acyl halides usually react in the expected manner. Examples include the preparation of adipamide[71], methylsuccinamide[72], biphenyl-3,3',5,5'-tetracarboxamide[73], and the diarylterephthalamide 8[74] from the appropriate acyl halides (equation 21).

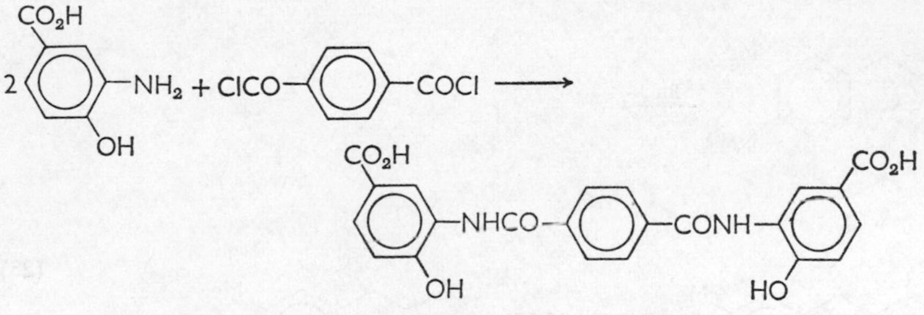

$$2 \underset{OH}{\overset{CO_2H}{\diagdown}} -NH_2 + ClCO-\bigcirc-COCl \longrightarrow$$

$$\underset{OH}{\overset{CO_2H}{\diagdown}} -NHCO-\bigcirc-CONH-\underset{HO}{\overset{CO_2H}{\diagdown}} \qquad (21)$$

(8)

Reactions of diacyl halides with diamines are of great technical importance for the preparation of polymers.

Unlike their higher homologues acyl halides derived from 1,2-dicarboxylic acids often undergo side-reactions when treated with amines. Thus ammonolysis of phthaloyl chloride or of phenylsuccinoyl chloride affords the cyano acids (equation 22)[75]. For this reason, and because of the difficulty of preparing the required acyl

halides, 1,2-dicarboxamides are generally prepared from esters, or
imides. Chlorides derived from half acid esters of 1,2-dicarboxylic
acids must also be used with caution since rearrangements can occur
during their preparation[76].

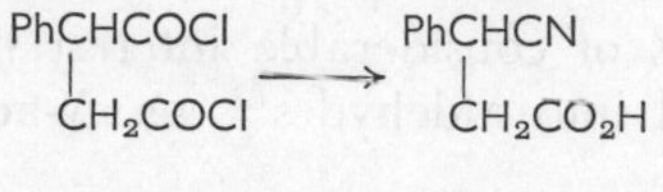

$$
\begin{array}{cc}
\underset{|}{\text{PhCHCOCl}} & \underset{|}{\text{PhCHCN}} \\
\text{CH}_2\text{COCl} & \text{CH}_2\text{CO}_2\text{H}
\end{array} \qquad (22)
$$

In accord with prediction (section II.A) the relative reactivity of
amines towards acylation is approximately dependent on their basic
strengths. Thus, kinetic investigations[18,77] reveal a marked lowering
in the rates of benzoylation of substituted anilines as the electronegativity of the substituent is increased. The order of reactivity of
substituted anilines, $RC_6H_4NH_2$, is

$$R = p\text{-CH}_3 > m\text{-CH}_3 > H > p\text{-Cl} > m\text{-Cl} > m\text{-NO}_2 > p\text{-NO}_2$$

However, the preparation of amides even from very weakly basic
amines rarely presents difficulty provided forcing conditions, such as
heating of the reactants in dimethylaniline solvent, are employed.
Acylation is also sensitive to steric hindrance. Thus direct reaction of
2,2-dimethylbutanoyl chloride with the highly hindered amine **9**
appears to be impossible. However acylation is successfully accomplished by prior addition of butyllithium to the amine[78]; presumably
the reactive intermediate is the amide ion **10**.

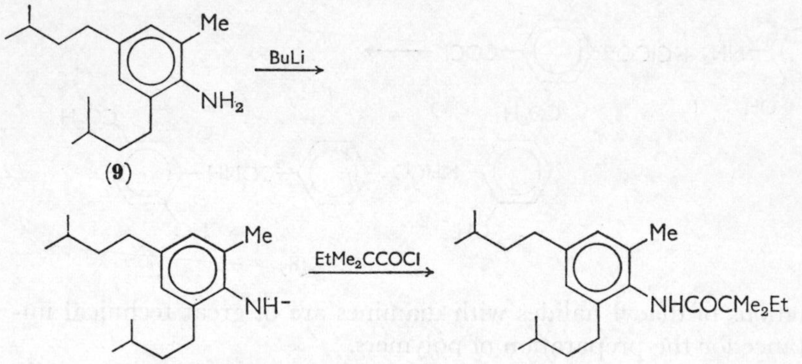

Reactions of imines with acyl halides (equation 24) have not been
extensively studied. It appears that the products are generally acyl-
α-haloalkylamines[79–81].

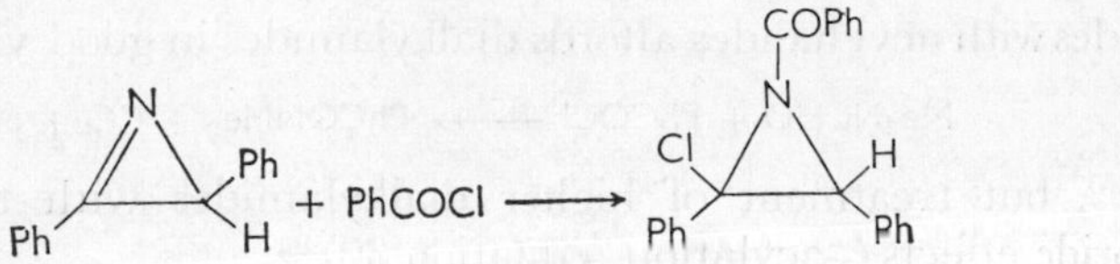

$$(24)$$

The reaction probably involves addition of halide to an acyl-iminium ion (equation 25).

$$\text{RCOHal} + \overset{|}{\underset{|}{N}}\!\!=\!\!\overset{|}{C} \longrightarrow \text{RCO}\!-\!\overset{|}{\underset{+}{N}}\!\!=\!\!\overset{|}{\underset{|}{C}}\ \text{Hal}^- \longrightarrow \text{RCON}\!-\!\overset{|}{\underset{|}{C}}\text{Hal} \quad (25)$$

The well-known formation of Reissert's compounds[82] by acylation of pyridine and related molecules may be similarly rationalized. However, O-alkyllactims (e.g. 11) when treated with acylating agents undergo dealkylation affording N-acyllactams (equation 26)[83].

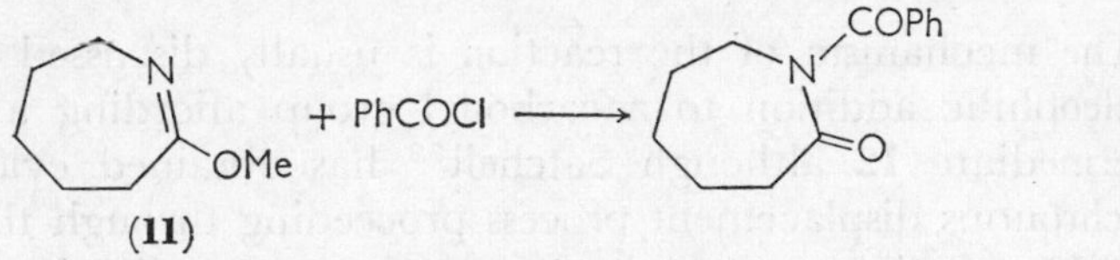

$$(26)$$

(11)

When acylation of an amine is conducted with an excess of acyl halide, di- or triacylamines are sometimes formed (equation 27)[84].

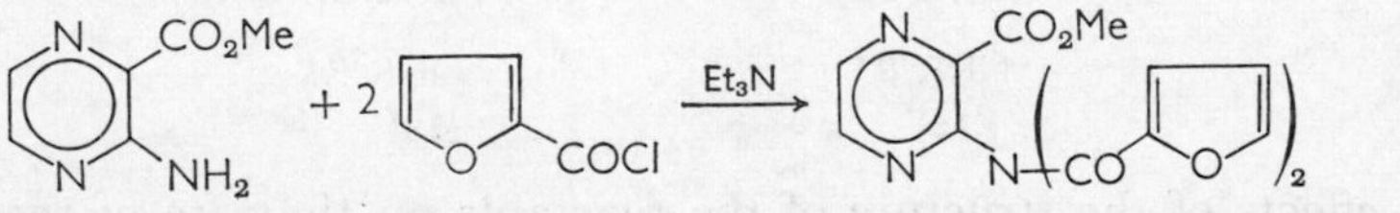

$$(27)$$

A new method for proceeding directly to *sym*-triacylamines involves treatment of lithium nitride with acyl halides[85]. Presumably the reaction is initiated by attack on the acylating agent of Li_2N^- or a similar nucleophilic species.

Amides and lactams are acylated by treatment with acyl halides and related reagents. A mild method[86] for the preparation of tri-acylamines involves consecutive addition of pyridine and mono-acylamine to a solution of acyl chloride in chloroform below 0°. N-Acylpyrrolidones[87] and -azetidinones[88] are readily formed by conventional methods. Another useful procedure[87,89] for the formation of diacylamines consists of addition of acyl halide to the ion formed by reaction of a suitable base with the monoacyl compound (equation 28)[89]. Finally, it is noteworthy that acylation of N,N-dialkylform-

$$\text{PhNHCOCH}_3 + \text{EtMgBr} \longrightarrow \text{PhN̄COCH}_3 \xrightarrow{\text{RCOCl}} \text{Ph}\overset{\text{COR}}{\underset{|}{N}}\text{COCH}_3 \quad (28)$$

amides with acyl halides affords dialkylamides in good yield (equation

$$Me_2NCHO + PhCOCl \longrightarrow PhCONMe_2 + CO + HCl \tag{29}$$

29)[90], but treatment of higher dialkylamides with trichloroacetyl chloride effects C-acylation (equation 30)[91].

$$CCl_3COCl + PhCH_2CONEt_2 \longrightarrow PhCHCONEt_2 \tag{30}$$
$$\underset{\displaystyle COCCl_3}{|}$$

C. Anhydrides

Carboxylic acid anhydrides, although generally less reactive than acyl halides, are useful reagents for acylation of amines and amides (equation 31).

$$R^1CO_2COR^2 + R^3R^4NH \longrightarrow R^1CONR^3R^4 + R^2CO_2H \tag{31}$$

The mechanism of the reaction is usually discussed in terms of nucleophilic addition to a carbonyl group affording a tetrahedral intermediate **12** although Satchell[92] has obtained evidence for a synchronous displacement process proceeding through the transition state **13**. Both hypotheses lead to the same generalizations concerning

$$
\begin{array}{cc}
\overset{\displaystyle -O}{\underset{\displaystyle |}{|}} & \overset{\displaystyle O}{\|} \\
R^1-\overset{|}{C}-OCOR^2 & R^1-C\cdots OCOR^2 \\
\underset{\displaystyle +NHR^3R^4}{|} & \vdots \\
 & NHR^3R^4 \\
(\mathbf{12}) & (\mathbf{13})
\end{array}
$$

the effects of the structure of the reactants on the rate of reaction. Increase in the electron-attracting power of R in the anhydride $(RCO)_2O$ will increase the reaction rate by enhancing the electrophilic character of the carbonyl carbon atom and by stabilizing the leaving group, RCO_2^-. Hence anhydrides containing strongly electronegative substituents, e.g. trifluoroacetic anhydride, are highly effective acylating agents. Conversely, increase in the electron-attracting power of the groups R^3 and R^4 in the amine, by lowering its nucleophilicity, will decrease the rate of acylation. For example p-methoxyaniline reacts some fifty times more rapidly than m-chloroaniline with benzoic anhydride in dioxan–water[92]. Very weakly basic amines, such as the nitroanilines and diarylamines react very slowly with most anhydrides and special conditions are needed for the efficient preparation of their acyl derivatives. Acylation with anhydrides appears to be catalysed by acid[15]; in the absence of an excess of added acid the process is usually autocatalytic[93].

Unsymmetrical carboxylic anhydrides offer two possible sites for attack by amines leading to formation of two different acylated products (equation 32).   The course of such reactions is controlled by

$$R^1C\overset{O}{\underset{\|}{}}-O-\overset{O}{\underset{\|}{C}}R^2 + R_2^3NH \longrightarrow \begin{array}{l} \longrightarrow R^1CONR_2^3 + R^2CO_2H \\ \text{or} \\ \longrightarrow R^2CONR_2^3 + R^1CO_2H \end{array} \qquad (32)$$

the steric and electronic effects of R^1 and R^2.   Steric effects are readily predicted; attack by the amine will occur preferentially at that carbonyl group adjacent to the less bulky substituent.   Electronic effects are more complex.   If R^2 is more electron attracting than R^1 we should expect (*i*) that the initial rate of addition of R_2^3NH will be greater at the carbonyl group adjacent to R^2, and (*ii*) that $R^2CO_2^-$ will be a more effective leaving group than $R^1CO_2^-$.   Thus it is clear that the eventual outcome of the reaction will depend upon which is the more important, the effect of substituents on carbonyl electrophilicity or on leaving-group stability.   In terms of the synchronous displacement mechanism (equation 4) if bond formation in the transition state **13** is more important than bond breaking, then aminolysis will occur at the carbonyl group adjacent to R^2.

For the addition–elimination mechanism (equation 3) proceeding through a tetrahedral intermediate (**12**) the overall rate constant is $k = k_1/(k_2/k_3 + 1)$.   If k_1 is more sensitive to change in the electron-attracting power of substituents than is the function $(k_2/k_3 + 1)$ then reaction will occur adjacent to the more powerfully electron-attracting group.   Both approaches indicate that the course of these reactions is likely to be affected by the nature of the amine and the reaction conditions.   In many cases the interplay of electronic and steric effects leads to the formation of both possible products.   However, attack of amines on carbonic carboxylic (**14**) and carbamic carboxylic (**15**) anhydrides usually proceeds selectively at the acyl carbonyl

$$R^1C\overset{O}{\underset{\|}{}}-O-\overset{O}{\underset{\|}{C}}OR^2 \qquad\qquad R^1C\overset{O}{\underset{\|}{}}-O-\overset{O}{\underset{\|}{C}}NR_2^2$$
$$\text{(14)} \qquad\qquad\qquad\qquad \text{(15)}$$

groups, R^1CO—, because the electrophilicity of the alternative positions is lowered by mesomeric release from the adjacent O or N atoms.   Finally, in discussing mechanism we should note that mixed anhydrides of carboxylic acids with such other acids as sulphuric, sulphonic and phosphoric acids (i.e. **16**, **17**, **18**) in accord with the concepts adumbrated above undergo selective attack at the carbonyl group and are effective reagents for *N*-acylation.

$$\underset{(16)}{R\overset{\overset{\displaystyle O}{\|}}{C}-OSO_3^-} \qquad \underset{(17)}{R^1\overset{\overset{\displaystyle O}{\|}}{C}-OSO_2R^2} \qquad \underset{(18)}{R^1\overset{\overset{\displaystyle O}{\|}}{C}-OPO(OR^2)_2}$$

For preparative work the anhydrides most widely used for acylation of amines and amides are the easily obtainable symmetrical compounds, i.e. the lower aliphatic carboxylic anhydrides, benzoic anhydrides, and cyclic anhydrides. Recently developed methods[94] for the simple preparation of carboxylic acid anhydrides widen the scope of the reaction.

A wide variety of experimental procedures is available for acylation of amines by anhydrides (equation 31). Frequently, as in the acetylation of benzylamine[95] or imidazole[96] the two reagents are mixed without solvent and heated if necessary. Inert solvents such as ether, acetone, toluene and petroleum are often employed; acetic acid is a particularly useful solvent for acetylation with acetic anhydride. Pyridine and tertiary amine bases catalyse acylation by anhydrides and often provide convenient solvents. It seems clear that in many cases, as for example in the acylation of weakly nucleophilic amines, the effective acylating agent is the acylammonium ion (equation 33)[97].

$$R^1CO_2COR^2 + R_3^3N \rightleftharpoons R^1CO\overset{+}{N}R_3^3 \ R^2CO_2^- \tag{33}$$

A useful method developed by Chattaway[98] for the acetylation of aromatic amino acids and amino phenols involves addition of acetic anhydride to a solution or suspension of the amine in ice-cold aqueous caustic soda. As in the Schotten–Baumann method hydrolysis of the acylating agent is usually unimportant.

For acylation of weakly basic amines, e.g *o*-nitro-*N*-methylaniline[99], sulphuric acid is an effective catalyst. It functions, presumably, by protonation of the anhydride thus facilitating attack by the amine.

Under suitable experimental conditions selective partial acylation of diamines, amino alcohols and amino phenols can be accomplished. Typical examples illustrative of methods employed are given in equations (34–37).

Acetylenic amines react normally with acetic anhydride yielding the expected acetylenic acetylamines. The products, however, are sensitive to acid and are rapidly converted to the keto amides unless care is taken to keep the reaction mixture alkaline (equation 38)[103].

$$+ (CH_3CO)_2O \xrightarrow{\text{Acetone}}{\text{r.t.}} \quad (34)$$

$$(92\%)^{100}$$

$$+ (PhCO)_2O \xrightarrow{\text{MeOH, reflux}} \quad (35)$$

$$(50–70\%)^{101}$$

$$+ (PrCO)_2O \xrightarrow{\text{MeO}^-,\text{ MeOH}} \quad (36)$$

$$(100\%)^{63}$$

$$+ (CH_3CO)_2O \xrightarrow{\text{Pyridine}}{\text{r.t.}} \quad (37)$$

$$(85\%)^{102}$$

$$CH{\equiv}C{-}\underset{Me}{\overset{Me}{C}}{-}NHMe \xrightarrow{Ac_2O} CH{\equiv}C{-}\underset{Me}{\overset{Me}{C}}{-}\underset{COCH_3}{N}Me \xrightarrow{H^+,\,H_2O} CH_3{-}\overset{O}{\overset{\|}{C}}{-}\underset{Me}{\overset{Me}{C}}{-}\underset{COCH_3}{N}Me$$

$$(38)$$

As expected on mechanistic grounds trifluoroacetic anhydride is a vigorous acylating agent, and readily forms trifluoroacetyl derivatives of aromatic and aliphatic amines[104]. It can be used in a variety of organic solvents including ether, chloroform, carbon tetrachloride and trifluoroacetic acid. It reacts readily with α-amino acids[105] and is employed for protecting amino groups in the synthesis of peptides[106,107] and other complex molecules[108].

Acetylation of α-amino acids with acetic anhydride also proceeds

4*

readily under mild conditions and affords excellent yields of α-acetamido acids (e.g. preparation of acetylglycine[109]), but when the same reactants are heated in the presence of pyridine the Dakin–West reaction occurs leading to the formation of α-acylamino ketones (e.g. preparation of acetamidoacetone (equation 39)[110]). The Dakin–West reaction proceeds through cyclization and further acetylation of

$$
\begin{array}{c}
CH_2CO_2H \\
| \\
NH_2
\end{array} + Ac_2O \longrightarrow
\begin{cases}
\xrightarrow[r.t.]{H_2O}
\begin{array}{c}
CH_2CO_2H \\
| \\
NHAc \\
(89\text{–}92\%)
\end{array} \\
\\
\xrightarrow{\text{reflux in pyridine}}
\begin{array}{c}
CH_2COCH_3 \\
| \\
NHAc \\
(70\text{–}78\%)
\end{array}
\end{cases}
\tag{39}
$$

the initially formed acetamido acid[46,111] and it can be applied to a variety of α-acylamino compounds (equation 40).

$$
\underset{(19)}{\overset{\displaystyle CH_3 \quad CH_3}{CH_2{=}CCONHCHCO_2H}} \xrightarrow{Ac_2O, \text{ pyridine}} \underset{(20)}{\overset{\displaystyle CH_3 \quad CH_3}{CH_2{=}CCONHCHCOCH_3}}
\tag{40}
$$

The formation of amides by acylation of amines prepared *in situ* by reduction of suitable precursors is a well-known reaction. Aromatic nitro compounds, for example, readily afford acetanilides when treated with reducing agents in acetic anhydride. Under suitable conditions selective acetylation can be achieved. Thus o-hydroxyacetanilide is obtained from o-nitrophenol in acetic acid–acetic anhydride by reduction with stannous chloride or by hydrogenation[112]. Similarly, reductive acetylation with zinc and acetic anhydride in acetic acid, of the oximino compound 21, affords a convenient synthesis of acetamidomalonic ester (22)[113]. Acetylamines are

$$
\underset{(21)}{HON{=}C{\overset{\displaystyle CO_2Et}{\underset{\displaystyle CO_2Et}{\Big\langle}}}} \xrightarrow[AcOH]{Zn/Ac_2O} \underset{(22)}{AcNHCH{\overset{\displaystyle CO_2Et}{\underset{\displaystyle CO_2Et}{\Big\langle}}}}
\tag{41}
$$

obtained in excellent yield by catalytic hydrogenation of nitriles in acetic anhydride (equation 42)[114].

$$
NC(CH_2)_4CN + H_2 \xrightarrow[NaOAc]{Raney \ Ni/Ac_2O} AcNH(CH_2)_6NHAc
\tag{42}
$$
$$
(100\%)
$$

Acyclic unsymmetrical anhydrides of simple carboxylic acids offer two different points of attack for nucleophilic reagents. With amines mixtures of both possible amides are frequently obtained, and the course of such reactions is often altered dramatically by small changes in experimental conditions. Thus aniline when treated with acetic chloroacetic anhydride in benzene yields mainly the chloroacetyl derivative, but in aqueous acetone the same reaction affords a mixture of which acetanilide is the major component. Similar results were obtained with other mixed carboxylic anhydrides leading Emery and Gold[115] to suggest that in non-polar media reaction occurs predominantly, although not always exclusively, at the carbonyl group adjacent to the more powerfully electron-attracting substituent. As expected on theoretical grounds the course of the reaction also depends on the strength of the nucleophile. Aniline when treated with trifluoroacetic acetic anhydride affords both the acetyl and trifluoroacetyl derivatives, whereas the weakly basic acetanilide undergoes solely trifluoroacetylation[116]. Surprisingly, treatment of glucosamine with acetic butyric anhydride in methanol affords only the N-butyryl derivative[63].

Acetic formic anhydride, which is readily prepared by warming acetic anhydride with formic acid, reacts with amines selectively at the formyl carbonyl group and thus provides an excellent method for the preparation of formamides[117]. Amino acids are smoothly formylated with acetic formic anhydride[118], and the reagent has found considerable use for protecting amino groups in synthesis of peptides[106,107] and other natural products, e.g. the prostoglandins[119]. Mixed anhydrides of lower fatty acids with α-amino acids react with amines preferentially in the amino acid moiety[120]. The reaction is most selective when a mixed anhydride of an α-amino acid with a sterically hindered acid, e.g. isovaleric or diphenylacetic, is used, and the reaction finds application in the preparation of peptides[107,121].

Mixed anhydrides of carboxylic acids with carbonic acids and carbamic acids (14, 15) undergo aminolysis selectively at the acyl carbonyl group. Reactions of the latter are discussed in section V.B.2. The former find extensive use in peptide synthesis[107,121,122] and have also been used for the preparation of other amides. Thus bicyclo[2,2,2]octane-1-carboxylic acid (23) reacts with ethyl chloroformate in chloroform containing triethylamine affording in situ the mixed anhydride, which when treated with ammonia gives the amide 24 in 82% yield (equation 43)[123]. Other examples include the preparation of amides of penicillin[69], amino sugar nucleosides[124], ricinoleic acid[125], and other hydroxy acids and fatty acids[126].

$$\text{(23)} \quad \text{—CO}_2\text{H} + \text{ClCO}_2\text{Et} \longrightarrow$$

$$\text{—COOCO}_2\text{Et} \xrightarrow{\text{NH}_3} \text{—CONH}_2 + \text{CO}_2 + \text{EtOH} \quad (43)$$

$$\text{(24)}$$

Mixed anhydrides of carboxylic acids with inorganic acids can be used for the preparation of amides. For example, addition of sulphur trioxide in the form of its crystalline dimethylformamide complex to an alkali metal salt of an acid gives an acyl sulphate which readily undergoes aminolysis at the acyl group (equation 44)[127].

$$\text{R}^1\text{CO}_2^- + \text{SO}_3 \longrightarrow \text{R}^1\text{CO}_2\text{SO}_3^- \xrightarrow{\text{R}^2_2\text{NH}} \text{R}^1\text{CONR}^2_2 + \text{HSO}_4^- \quad (44)$$

Carboxylic sulphonic anhydrides may be prepared by reaction of sulphonic anhydrides with carboxylic acids[128], by treatment of silver carboxylates with arenesulphonyl chlorides[129], or, most conveniently, by mixing an arenesulphonyl chloride with a carboxylic acid in pyridine[130]. They react readily with amines giving high yields of amides[129,130]. Thus addition of aniline to a solution of benzoic acid and benzenesulphonyl chloride in pyridine affords benzanilide in 94% yield (equation 45)[130]. Cyclic sulphonic carboxylic anhydrides

$$\text{PhCO}_2\text{H} + \text{PhSO}_2\text{Cl} \longrightarrow \text{PhCO}_2\text{SO}_2\text{Ph} \xrightarrow{\text{PhNH}_2} \text{PhCONHPh} + \text{PhSO}_3\text{H} \quad (45)$$

react similarly (equation 46), and o-sulphobenzoic anhydride (**25**) has been recommended as a reagent for the estimation of amino groups[131].

$$\text{(25)} \quad \begin{array}{c} \text{CO} \\ \text{O} \\ \text{SO}_2 \end{array} + \text{RNH}_2 \longrightarrow \begin{array}{c} \text{CONHR} \\ \text{SO}_3\text{H} \end{array} \quad (46)$$

Acyl phosphates and their esters react readily with amines yielding amides (equations 47 and 48)[132]. The reaction has found synthetic application mainly in the preparation of peptides[107,121]. It is probable that the convenient formation of amides by interaction of carboxylic acids and amines in polyphosphoric acid[133] proceeds through a similar type of intermediate.

$$\text{R}^1\text{CO}_2\text{PO}_3^{2-} + \text{R}^2\text{NH}_2 \longrightarrow \text{R}^1\text{CONHR}^2 + \text{HOPO}_3^{2-} \quad (47)$$

$$\text{R}^1\text{CO}_2\text{PO}(\text{OR}^3)_2 + \text{R}^2\text{NH}_2 \longrightarrow \text{R}^1\text{CONHR}^2 + \text{HOPO}(\text{OR}^3)_2 \quad (48)$$

Cyclic carboxylic anhydrides react readily with ammonia or amines affording half acid amides. Thus phthalic anhydride, on shaking with aqueous methylamine, affords N-methylphthalamic acid in 80% yield (equation 49)[134]. Other examples, illustrative of the methods

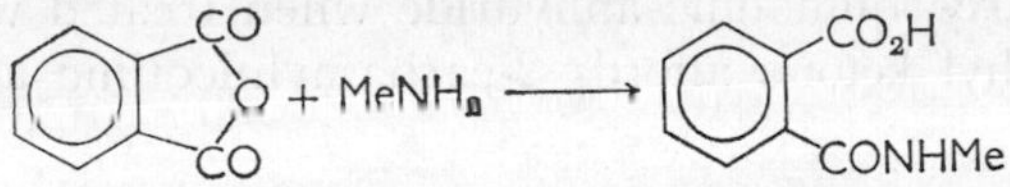

$$\text{(structure)} + \text{MeNH}_2 \longrightarrow \text{(structure)} \qquad (49)$$

used and the scope of the reaction include the preparation of phthalanilic acid by treatment of phthalic anhydride with aniline in chloroform[135], N-2-pyridylsuccinamic acid from succinic anhydride and 2-aminopyridine in benzene[136], and the monoanilide of $\alpha,\alpha,\alpha',\alpha'$-tetramethyladipic acid from the anhydride and aniline in benzene[137]. Phthalamic acids derived from α-amino acids are readily prepared in high yield by addition of an aqueous solution of the amino acid and triethylamine to phthalic anhydride in tetrahydrofuran[138]. The formation of phthalamic acids by treatment with phthalic anhydride in benzene has been recommended for the characterization of amines[139]. m-Aminophenol when heated with succinic anhydride is acylated preferentially at the amino group[140]. In all such reactions careful control of conditions is necessary to avoid cyclization of the amic acid to the imide (equation 50).

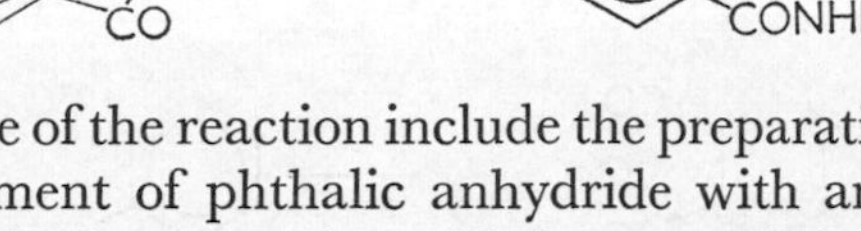

$$\text{(structure)} \longrightarrow \text{(structure)} \qquad (50)$$

Unsaturated cyclic anhydrides react normally. Thus maleic anhydride is readily converted in high yield to maleanilic acid by treatment with aniline in ether[141], whilst maleamic acid is prepared by passing ammonia over a solution of the anhydride in xylene or dioxan[142].

Like their acyclic analogues, cyclic anhydrides containing substituents of strong electron-attracting or -donating character when treated with amines in non-polar media undergo reaction selectively at the more electron-deficient carbonyl group. Thus 3-nitrophthalic anhydride in carbon tetrachloride or benzene reacts readily with amines to yield exclusively 3-nitro-2-phthalamic acids (equation 51)[143]. The reaction is useful for the identification of amines and for

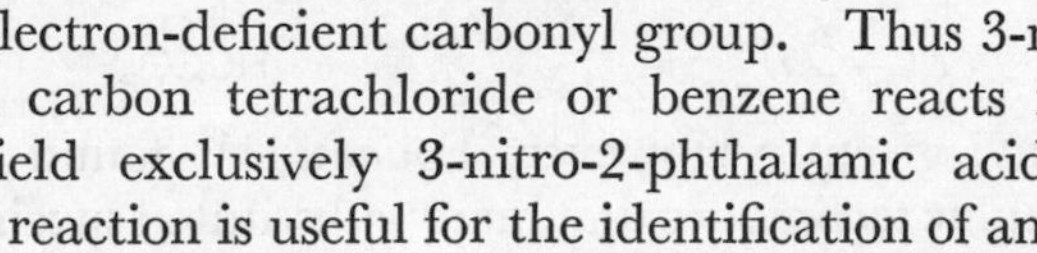

$$\text{(structure)} + \text{R}^1\text{R}^2\text{NH} \longrightarrow \text{(structure)} \qquad (51)$$

their separation since only the products from primary amines undergo cyclization on heating to yield neutral N-substituted 3-nitrophthalimides. Similarly, quinolinic anhydride when treated with ammonia in methyl ethyl ketone affords 2-carbamylnicotinic acid (equation 52) [144].

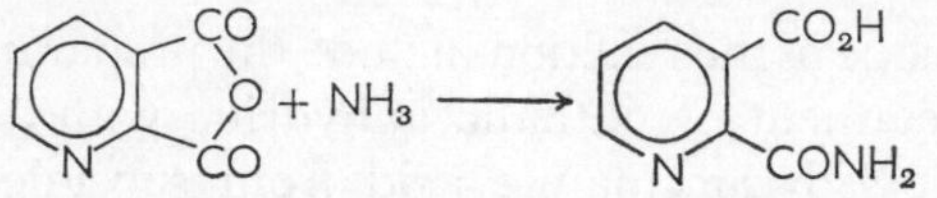

$$\text{(52)}$$

3-Methoxyphthalic anhydride affords a further example of the influence of polar substituents since it reacts with aniline preferentially at the 1-position (equation 53) [145].

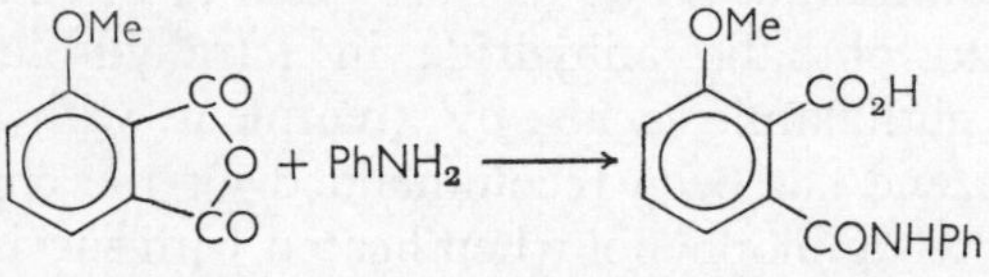

$$\text{(53)}$$

The directive effects of alkyl substituents on aminolyses of cyclic anhydrides are more ambiguous. Treatment of methylsuccinic anhydride with methylaniline affords β-carboxy-N-methylbutyranilide in good yield (equation 54) [146]. A similar specificity of attack of

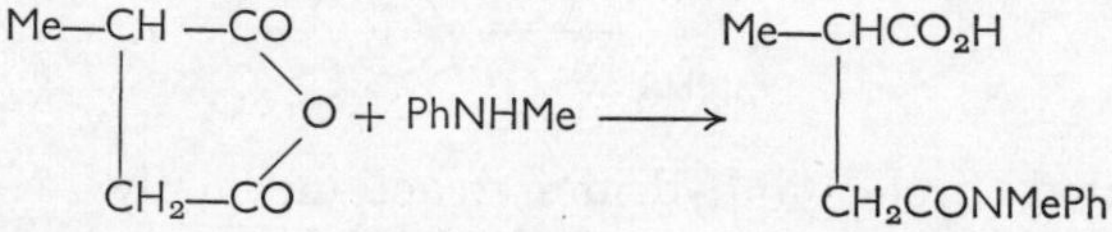

$$\text{(54)}$$

various amines at the β-carbonyl group in alkyl- and aryl-substituted succinic anhydrides and in itaconic anhydride has been noted [147], but methylmaleic anhydride undergoes aminolysis at the α-position (equation 55) [148], presumably because of conjugative stabilization of

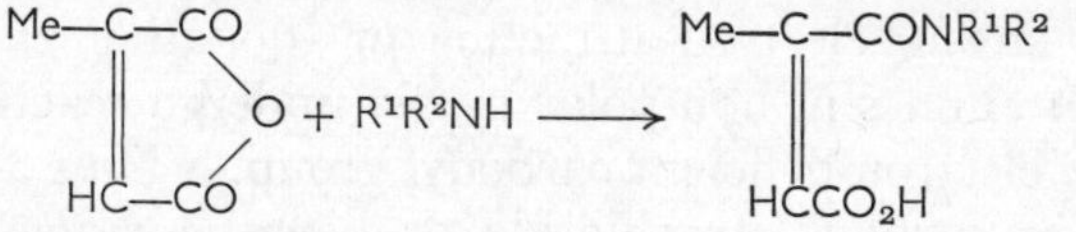

$$\text{(55)}$$

the β-carbonyl group. However, Foucaud [149] found by the use of chromatographic techniques that many substituted succinic anhydrides give mixtures of both possible products when treated with ammonia in ether. He interprets his results on the basis of the steric and inductive effects of the substituents.

Finally, it is interesting to note that isatoic anhydride (26) and similar cyclic carboxylic carbamic anhydrides do not always react, as expected on theoretical grounds, at the carbonyl group remote from the nitrogen atom[150]. However, it has become apparent from a

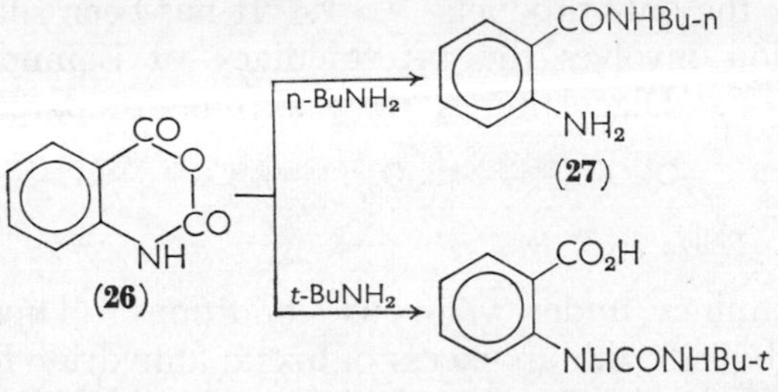

$$(56)$$

recent kinetic study that only the normal amide products (**27**) arise by direct attack of the amine at a carbonyl group; ureides (**28**) are formed via an initial rearrangement of the anhydride to an isocyanate[151].

Tertiary amines react reversibly with carboxylic anhydrides yielding acylammonium salts, but in most cases no further reaction ensues and the amine is recovered after addition of water to the mixture. However, benzylic tertiary amines undergo fragmentation with the formation of benzyl esters and acyl derivatives of secondary amines[152]. Similarly, treatment of hexahydroindolopyrrocoline (**29**) with cold acetic anhydride affords the cleavage product **30** in high yield[153].

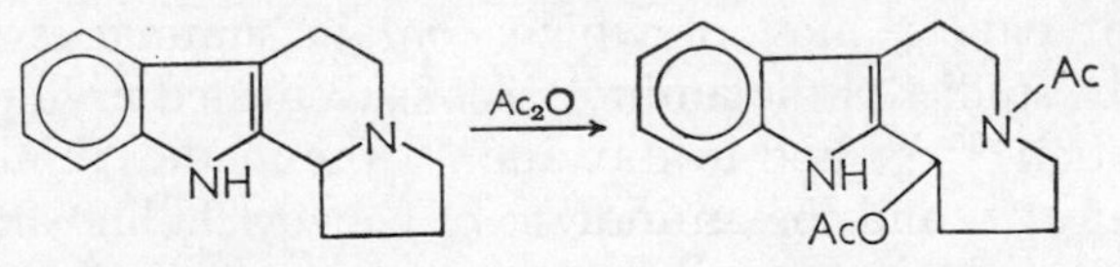

$$(57)$$

The reaction has proved useful in alkaloid synthesis[154].

Imines react with anhydrides yielding α-acyloxyamides (equation 58). The reaction is thought to involve initial N-acylation followed

$$PhCH{=}NPh + Ac_2O \longrightarrow \underset{\underset{OAc\ Ac}{|\quad|}}{PhCH{-}NPh}$$

$$(31) \hspace{4cm} (32)$$

$$(58)$$

successively, or accompanied simultaneously, by an internal C-acyloxylation[155].

Acylation of amides with carboxylic anhydrides is frequently used as a route to diacylamines (equation 59)[156]. The reaction, however,

$$R^1CONHR^2 + (R^3CO)_2O \longrightarrow \underset{\underset{R^2}{|}}{R^1CONCOR^3} + R^3CO_2H$$

$$(59)$$

is not always as straightforward as might be expected, for nitriles are often produced from primary carboxamides and under vigorous conditions may be the only products[157,158]. It has been suggested that nitrile formation involves the intermediacy of isoimidinium salts (equation 60)[157]. Diacylamines may also be prepared by direct

$$R^1C\!\!-\!\!OCOR^2 \longrightarrow R^1C\!\!\equiv\!\!N + R^2CO_2H + H^+ \tag{60}$$
$$\underset{+NH_2}{\|}$$

acylation of amines under vigorous conditions. Thus, aromatic amines when boiled with an excess of acetic anhydride for $\frac{1}{2}$ to 1 hr afford diacetyl derivatives in good yield (equation 61)[159]. Treatment

$$ArNH_2 + 2Ac_2O \longrightarrow ArNAc_2 + 2HOAc \tag{61}$$

of dimethylformamide with carboxylic anhydrides gives dimethylamides (equation 62)[90].

$$Me_2NCHO + (RCO)_2O \longrightarrow RCONMe_2 + RCO_2H + CO \tag{62}$$

D. Esters

Although aminolysis of esters is probably less frequently used for the preparation of amides than acylation of amines with acyl halides or anhydrides the reaction has been the subject of intensive investigation. Recent papers, most of which contain summaries of earlier work, describe studies of the aminolysis of substituted aryl acetates in aqueous solution[160–162] and in dioxane[163], the methoxyaminolysis of phenyl acetates[164], and the aminolysis of benzoylcholine and related compounds[165].

Qualitatively the reaction appears to be a nucleophilic substitution at a carbonyl carbon atom proceeding through a tetrahedral intermediate (equation 63), but there are many subtleties of mechanism

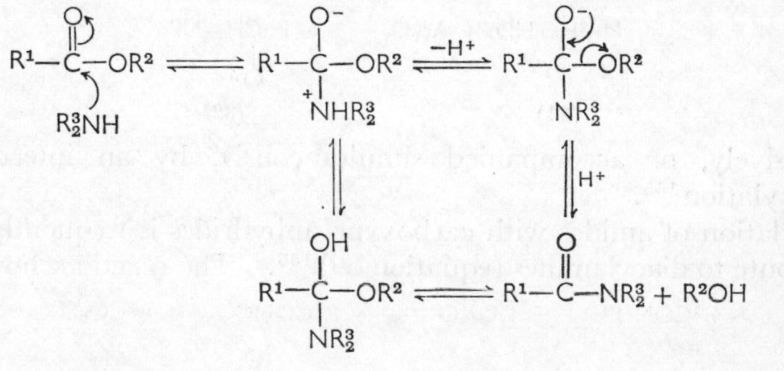

$$\begin{array}{ccccccc}
& O & & O^- & & O^- \\
& \| & & | & & | \\
R^1\!\!-\!\!C\!\!-\!\!OR^2 & \rightleftharpoons & R^1\!\!-\!\!C\!\!-\!\!OR^2 & \xrightarrow{-H^+} & R^1\!\!-\!\!C\!\!-\!\!OR^2 \\
| & & +| & & | \\
R_2^3NH & & NHR_2^3 & & NR_2^3 \\
\end{array}$$

$$\tag{63}$$

$$\begin{array}{ccc}
OH & & O \\
| & & \| \\
R^1\!\!-\!\!C\!\!-\!\!OR^2 & \rightleftharpoons & R^1\!\!-\!\!C\!\!-\!\!NR_2^3 + R^2OH \\
| & & \\
NR_2^3 & &
\end{array}$$

concerned with the effects of catalysts, media, and reactant structure on proton-transfer steps which lie beyond our present discussions. Suffice it to say that aminolysis of esters may be subject in certain circumstances to acid catalysis, base catalysis or both. In the former, incipient protonation of the carbonyl oxygen in the transition state **33** is thought to favour attack by the nucleophile, whilst in the latter the nucleophilicity of the amine is enhanced by incipient amide ion formation (transition state **34**).

$$
\begin{array}{cc}
\text{O}\cdots\cdots\text{H}\cdots\cdots\text{A}^{\delta-} & \text{O}^{\delta-} \\
\| & \| \\
\text{R}-\text{C}-\text{OR} & \text{R}-\text{C}-\text{OR} \\
| & | \\
{}^{\delta+}\dot{\text{N}}\text{HR}_2 & \dot{\text{N}}\text{R}_2 \\
& | \\
& \text{H}\cdots\text{B}^{\delta+} \\
(33) & (34)
\end{array}
$$

However, even the simplest mechanistic concepts allow one to make predictions concerning the gross effects of reactant structure on reaction rate; namely that (*i*) electron-attracting substituents R^1 in the acyl function will enhance reactivity by decreasing electron density at the carbonyl carbon atom, (*ii*) electron-attracting substituents R^2 in the alkoxy group will enhance the stability and hence the ease of displacement of R^2O^-, and (*iii*) the rate of reaction will depend on the nucleophilicity of the amine. In general these predictions are confirmed by experiment (see below). Also kinetic investigations reveal the sensitivity of aminolysis to steric factors; in particular some cyclic amines such as azetidine in which steric hindrance is minimized by constraint of the CÑC bond angle show unexpectedly high reactivity[160].

Reaction conditions employed for the acylation of ammonia and of amines with esters vary widely according to the nature of the substrates. Ammonia is quite an effective nucleophile and reacts with many esters, particularly those containing electron-attracting substituents, in aqueous media. Representative examples of the preparation of primary carboxamides by treatment of esters with concentrated ammonia solution include cyanoacetamide from ethyl cyanoacetate[166], fumaramide from diethyl fumarate[167], nicotinamide from ethyl nicotinate[168], malondiamide from diethyl malonate[169], and the monoamide of malonic acid from the potassium salt of the monomethyl ester[170]. Successful preparations of mono- and dichloroacetamide by ammonolysis of the appropriate ethyl esters at 0° give evidence both of the activating effect of the halo substituents and of the selectivity of attack at the carbonyl function[171].

Ammonia in alcoholic solution is a useful reagent for ammonolysis of esters which are too insoluble or insufficiently reactive to undergo attack in water[172]. Liquid ammonia has also been employed. Sometimes, as in the preparation of 4-methoxynicotinamide 1-oxide[173] the reaction proceeds efficiently at the boiling point of the reagent, but in others, such as preparation of mandelamide[174] or lactamide[175] it is necessary to conduct the reaction in a pressure vessel at room temperature.

Yet another procedure for the preparation of primary carboxamides from esters is the reaction with ammonium salts. Thus, in the synthesis of compounds in the tetracycline series the amide **36** was prepared by fusing the ester **35** with ammonium formate under nitrogen[176].

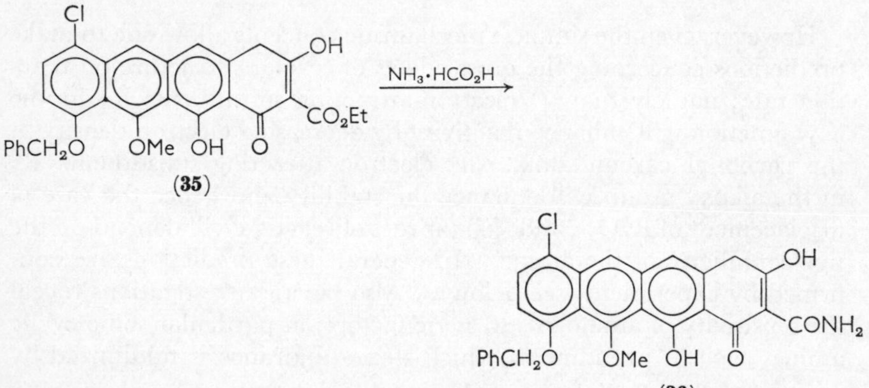

(35)

(36)

Acylation of amines with esters has been conducted under a wide variety of conditions. Lower fatty esters react with simple amines at room temperature, albeit rather slowly. For example, N-methylheptamide is obtained by allowing heptanoic ester to stand with aqueous methylamine for several days[29]. However, more vigorous conditions are necessary for the acylation of anilines and higher alkylamines. Representative examples of methods used include the preparation of benzoylacetanilide from ethyl benzoylacetate and aniline in a continuous reactor at 135°[177], of the monoamide of ethylene diamine and piperidylacetic acid from the amine and ethyl ester in refluxing ethanol[178], and of a series of ethanolamides by heating ethanolamine with ethyl esters of the homologous fatty acids at 160°[179]. Salicyl-o-toluidide[180] is prepared by heating phenyl salicylate with o-toluidine

at 183–202° in 1,2,4-trichlorobenzene solvent or in α-methylnaphtha-
lene at 230°.

Strongly basic, highly nucleophilic, amines readily undertake
aminolysis of esters[181]. Cyclohexylamine reacts exothermically with
ethyl formate at 0° yielding N-cyclohexylformamide[181], and benzylamine in tetrahydrofuran has been recommended for the cleavage of
active esters[182]. For example treatment of carbobenzoxyglycine 4-
phenylazophenyl ester with benzylamine in tetrahydrofuran affords
the benzylamide of carbobenzoxyglycine[182]. Alkoxides have been
used to promote aminolysis of esters. Thus addition of small amounts
of sodium methoxide to the reaction mixture greatly enhances the rate
of ammonolysis of methyl phenylacetate with methanolic ammonia[183],
and makes possible the easy preparation of the diethanolamide of
lauric acid[184]. Some aminolyses which are inconveniently slow even
in the presence of catalytic amounts of alkoxide proceed readily when
one molar equivalent of the base is added. The method is particularly
useful for the preparation of secondary amides and anilides. It is
conducted by heating the ester, amine, and sodium methoxide in
benzene under reflux[185]. The mechanism is believed to involve
generation of the highly nucleophilic amide ion, R^1NH^- (equation
64), attack of R^1NH^- on the ester proceeding through the usual

$$R^1NH_2 + MeO^- \rightleftharpoons R^1NH^- + MeOH \tag{64}$$

tetrahedral intermediate (equation 65), and stabilization of the pro-

$$R^2CO_2Me + R^1NH^- \rightleftharpoons R^2{-}\underset{\underset{NHR^1}{|}}{\overset{\overset{O^-}{|}}{C}}{-}OMe \rightleftharpoons R^2{-}\overset{\overset{O}{\parallel}}{C}{-}NHR^1 + MeO^- \tag{65}$$

duct amide by formation of an acylamide ion $R\overset{-}{CONR^1}$ (equation 66)

$$R^2\overset{\overset{O}{\parallel}}{C}NHR^1 + MeO^- \rightleftharpoons R\overset{\overset{O}{\parallel}}{C}\overset{-}{N}R^1 + MeOH \tag{66}$$

which survives until working-up of the reaction mixture.

Other procedures which take advantage of the high nucleophilicity
of amide ions include the formation of amides under very mild conditions by treating esters in ethereal solution with a lithium aluminium
amide complex (prepared by passing dry ammonia into ethereal
lithium aluminium hydride)[186], and the preparation of anilides from
hindered esters and sodium anilide in toluene[187].

The Bodroux reaction[188], in which amides are formed by reaction
of an ester with the magnesium amide obtained by interaction of an

amine with a Grignard reagent is mechanistically related.   The yields of amides from simple esters are often poor[189], but the reaction proceeds with much greater efficiency when the substrate contains an ester group adjacent to some function which is able to coordinate with magnesium[190].

Esters having the general formula RCOOA=B in which nucleophilic attack on the carbonyl group is aided by conjugation, readily undergo aminolysis and are of great utility for the preparation of amides, and particularly for peptide synthesis.   A wide selection of such 'activated intermediates' is now available and a full account of their formation and reactions is beyond the scope of this review. Examples illustrative of the types of compound used and their application to the preparation of amides are given in equations (67–72).

nitrophenyl esters[191]

$$p\text{-}NO_2C_6H_4OCHO + NH_2(CH_2)_3\underset{\underset{NH_2}{|}}{C}HCO_2H \xrightarrow[\text{r.t.}]{\text{Tetrahydrofuran}} HCONH(CH_2)_3\underset{\underset{NH_2}{|}}{C}HCO_2H \quad (67)$$

$$(52\%)$$

vinyl esters[192]

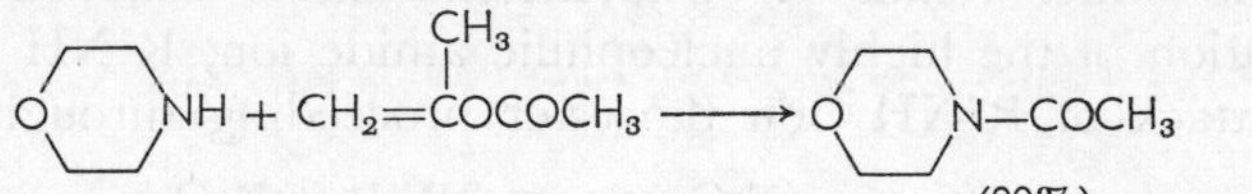

$$(90\%) \quad (68)$$

acid–carbodiimide adducts[193]

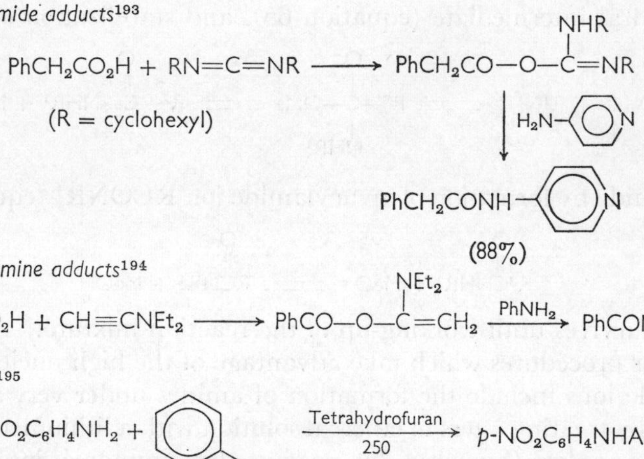

$$(88\%) \quad (69)$$

acid–alkynylamine adducts[194]

$$PhCO_2H + CH\equiv CNEt_2 \longrightarrow PhCO-O-\underset{\underset{CH_2}{\|}}{\overset{\overset{NEt_2}{|}}{C}} \xrightarrow{PhNH_2} PhCONHPh \quad (70)$$

pyridyl esters[195]

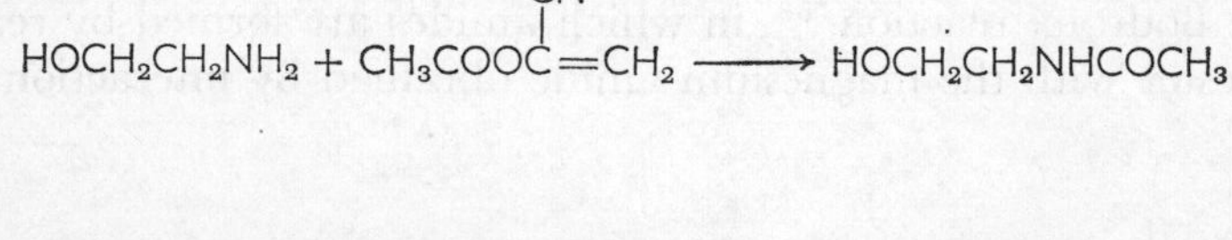

$$(82\%) \quad (71)$$

α-cyanovinyl esters[196]

$$HOCH_2CH_2NH_2 + CH_3COO\underset{\underset{CH_2}{\|}}{\overset{\overset{CN}{|}}{C}} \longrightarrow HOCH_2CH_2NHCOCH_3 \quad (72)$$

Other compounds of similar type whose use has been restricted mainly to peptide synthesis[107,121] include azlactones[197], acetylenic esters[198], and adducts of carboxylic acids with cyanamide[199], ketenimines[200], ethoxyacetylene[201] and isoxazolium salts[202].

Finally in this group of highly active acylating agents we should include the hydroxylamine esters. 1-Benzoyloxypiperidine, for example, readily undergoes aminolysis under mild conditions (equation 73)[203]. The high reactivity of this reagent is thought to be due to the

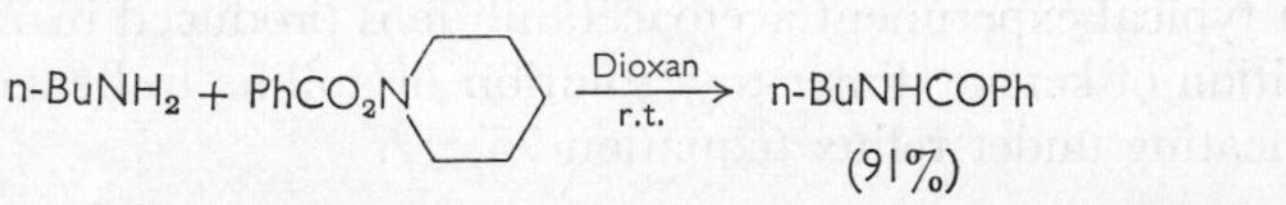

$$\text{n-BuNH}_2 + \text{PhCO}_2\text{N} \overset{\text{Dioxan}}{\underset{\text{r.t.}}{\longrightarrow}} \text{n-BuNHCOPh} \qquad (73)$$
$$(91\%)$$

inductive effect of the nitrogen which facilitates nucleophilic attack at the carbonyl group and aids expulsion of the leaving group. In accord with this view, acids which protonate the nitrogen catalyse acylations. Other reagents of similar type, but in which the inductive effect of the nitrogen is reinforced by adjacent carbonyl groups are esters of N-hydroxyphthalimide (37) and N-hydroxysuccinimide (38)[204]. Both have proved exceedingly useful in peptide synthesis[107,121].

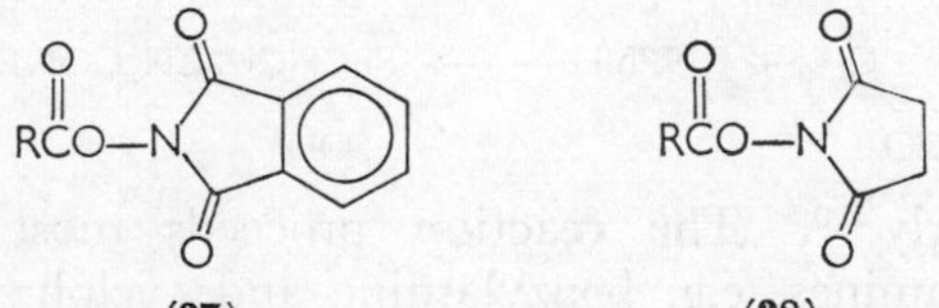

(37)　　　　　　　　　　(38)

Aminolysis of acyclic esters proceeds by nucleophilic displacement at the carbonyl group, i.e. by acyl–oxygen fission. The same is usually true of their cyclic analogues, the lactones, but in propiolactone, presumably because of the effect of ring stain, alkyl–oxygen fission is sometimes observed[205,206]. Thus, reaction of propiolactone with ammonia in water affords the expected β-hydroxypropionamide, but aniline gives N-phenyl-β-aminopropionic acid (equation 74)[206].

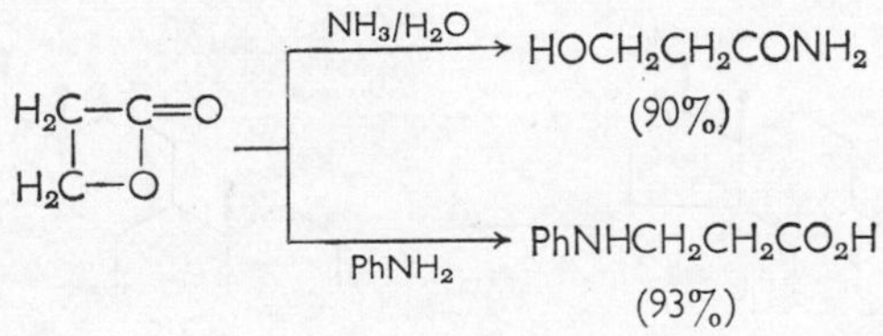

$$
\begin{array}{c}
\text{H}_2\text{C}-\text{C}=\text{O} \\
| \quad\quad | \\
\text{H}_2\text{C}-\text{O}
\end{array}
\left\{
\begin{array}{l}
\xrightarrow{\text{NH}_3/\text{H}_2\text{O}} \text{HOCH}_2\text{CH}_2\text{CONH}_2 \\
\quad\quad\quad (90\%) \\
\\
\xrightarrow[\text{PhNH}_2]{} \text{PhNHCH}_2\text{CH}_2\text{CO}_2\text{H} \\
\quad\quad\quad (93\%)
\end{array}
\right.
\qquad (74)
$$

In many cases aminolysis of propiolactone proceeds simultaneously by both routes giving a mixture of products. Also, the course of the reaction is sensitive to the experimental conditions; use of acetonitrile solvent enhances the yield of β-alanine derivatives.

By comparison, diketene, which is a cyclic analogue of an enol ester, and in which acyl–oxygen fission is aided by mesomeric release, reacts readily and preferentially at the carbonyl group forming amides of β-keto acids[207]. The reaction provides an extremely convenient and efficient procedure for the preparation of acetoacetamides. In a typical experiment acetoacetanilide is produced in 75% yield by addition of ketene dimer to a solution of aniline in benzene followed by heating under reflux (equation 75)[208].

$$\begin{array}{c} CH_2\text{---}C\text{=}O \\ | \qquad | \\ CH_2\text{=}C\text{----}O \end{array} + PhNH_2 \longrightarrow CH_3\overset{O}{\overset{\|}{C}}CH_2\overset{O}{\overset{\|}{C}}NHPh \qquad (75)$$

Amines usually react with γ- and δ-lactones under mild conditions by acyl–oxygen fission affording γ- and δ-hydroxy amides. Thus, γ-phenylbutyrolactone reacts cleanly with primary and secondary amines to form the appropriate N-substituted derivatives of γ-hydroxy-γ-phenylbutyramide (equation 76)[209]. Phenylvalerolactone

$$\begin{array}{c} PhCH\text{----}CH_2 \\ | \qquad\quad | \\ O \qquad CH_2 \\ \diagdown\, / \\ CO \end{array} + R^1R^2NH \longrightarrow PhCH CH_2CH_2CONR^1R^2 \qquad (76)$$
$$\ \underset{OH}{|}$$

behaves similarly[210]. The reaction proceeds most readily with strongly basic amines, e.g. benzylamine and cyclohexylamine, and has been used to characterize γ- and δ-lactones[30]. Under more vigorous conditions, treatment of lactones with amines produces amino acids and amino amides, possibly via intermediate lactams.

Like ketene dimer, γ- and δ-lactones containing unsaturation α to the oxygen atom readily undergo aminolysis with the formation of keto amides. For example, the steroid lactone **39** when treated at room temperature with ammonia in benzene is rapidly converted into the amide (equation 77)[211]. The 17-acetoxy group is unaffected.

$$(77)$$

(39)

Lactones derived from α,β-unsaturated acids show more complex behaviour on aminolysis (equation 78). Three reaction pathways are conceivable: (*i*) alkyl–oxygen fission, (*ii*) acyl–oxygen fission, (*iii*) Michael addition. In practice either or both of the routes (*ii*)

$$
\begin{array}{l}
\quad (i) \\
\xrightarrow{\quad} RNHCH_2CH{=}CHCO_2H \\
(ii) \\
\xrightarrow{\quad} HOCH_2CH{=}CHCONHR \qquad (78) \\
(iii) \\
\xrightarrow{\quad} RNH{-}
\end{array}
$$

and (*iii*) are followed, and the nature of the products isolated depends on experimental conditions and reactant structure[209,212].

Intramolecular aminolysis of ester groups in suitably constituted amino esters affords a convenient and widely used method of lactam synthesis. The reaction occurs most readily when it leads to the formation of 5- or 6-membered lactam rings. Thus heating of ethyl 5-amino-2,2-diethylpentanoate gives 3,3-diethylpiperidone (equation 79)[213]. Interestingly, this amino ester was prepared by treatment of

$$
H_2NCH_2CH_2CH_2CEt_2CO_2Et \longrightarrow \qquad (79)
$$

the appropriate bromo ester with ammonia. Apparently steric hindrance of the ester carbonyl group greatly retards ester ammonolysis which normally proceeds more readily than nucleophilic displacement of halogen. Eschenmoser's[214] classical work on the synthesis of corrins contains some interesting examples of lactam formation via intramolecular aminolysis including the formation of the bicyclic lactams **41** and **42** by appropriate treatment of the aziridine **40**.

Amino esters required as intermediates in lactam synthesis frequently are prepared *in situ* and converted directly to the final products without isolation. Some of the methods used are treatment of bromo esters with ammonia or amines[210], reduction of nitro esters (equation 81)[215] or reduction of nitrile esters (equation 82)[216]. The route via nitro esters has proved specially useful for the synthesis of steroidal lactams[217].

Seven-membered lactam rings are readily formed by intramolecular aminolysis[215,218], but interference from intermolecular amide formation becomes increasingly serious as the homologous series is ascended,

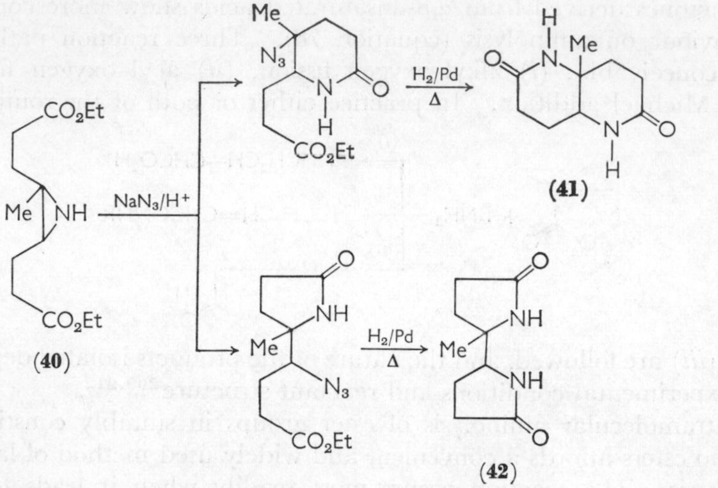

(80)

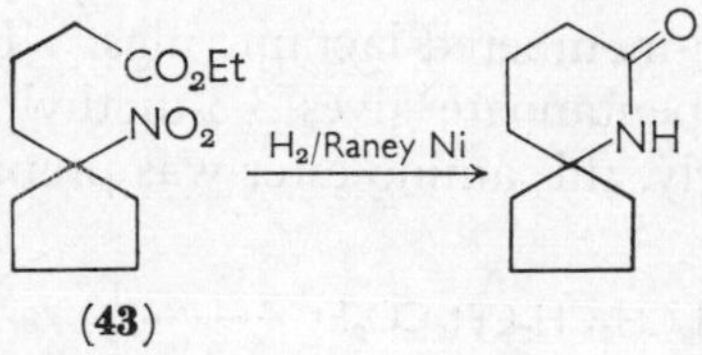

(81)

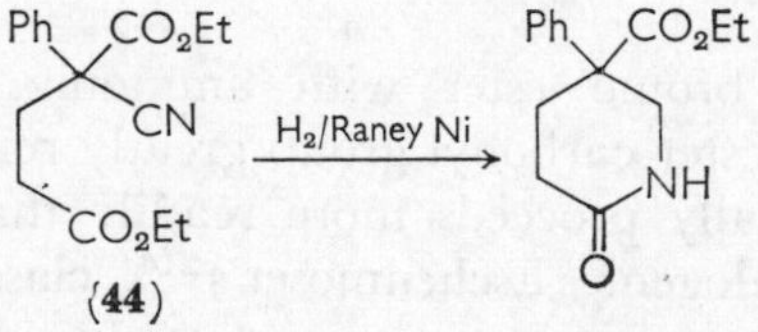

(82)

and the higher lactams are best prepared by the Beckmann or Schmidt reactions (section IV).

Intramolecular aminolysis leading to β-lactams from β-amino esters does not proceed efficiently under normal conditions, but can be accomplished by use of the Bodroux reaction. Thus the parent compound, azetidinone, was first prepared, albeit in low yield, by treatment of ethyl β-aminopropionate with ethereal ethylmagnesium bromide, which reagent has also been used for preparation of substituted compounds[219]. Apparently direct attack of the Grignard reagent on the ester function competes with magnesium amide formation, since use of the sterically hindered reagent prepared from bromomesitylene dramatically improves the yield[220]. Structural and synthetic studies on penicillin and related compounds provide many examples of azetidinone formation by cyclization of β-amino acid derivatives. A recent elegant example from Woodward's synthesis of cephalosporin C is the formation of **46** from the amino ester **45** by treatment with triisobutylaluminium[221].

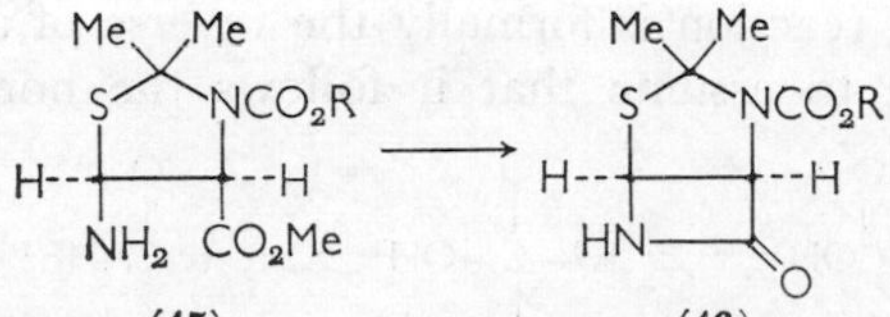

$$\text{(83)}$$

(45)   (46)

Acylation of amides with esters has occasionally been employed as a method of amide and imide preparation. Intermolecularly, the reaction occurs readily only if active esters are used. For example amides react with isopropenyl esters affording diacylamines[222]. Strong bases powerfully catalyse the reaction of amides with esters, but the final products are then not diacylamines but amides formed by transamidation; thus methyl esters when treated with formamide or N-methylformamide give good yields of the appropriate acylamines (equation 84)[223]. The reaction is believed to involve intermediate

$$RCO_2Me + MeNHCHO \longrightarrow RCONHMe + MeOCHO \qquad (84)$$

formation of the strongly nucleophilic formamide ion, $H\overline{C}ONCH_3$. In a similar reaction alkyl acrylates are converted to acrylamides by heating with fatty acid amides and lithium hydroxide[224]. Intramolecular acylation of amides occurs more readily and has been used for example for the preparation of diphenylsuccinimide[225] and of 4,6-dihydroxyimidazo-4,5,c-pyridines (equation 85)[226].

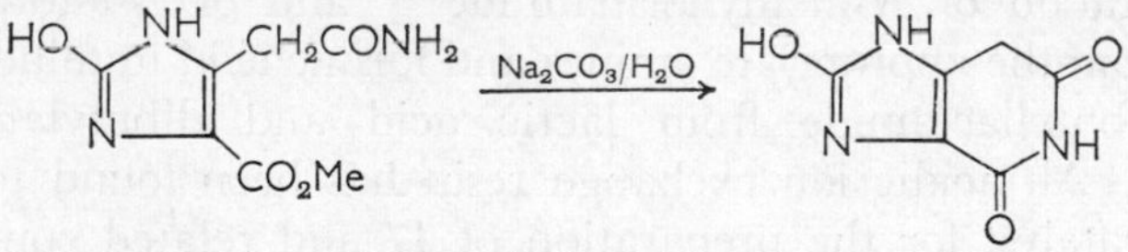

$$\text{(85)}$$

However, most investigations of intramolecular acylation of amides have been concerned with details of mechanism rather than synthetic application[227].

E. Carboxylic Acids

The standard procedure for the preparation of acetamide involves strong heating of ammonium acetate[228]; it provides a simple example of a general method (86) for the synthesis of amides. The mechanism

$$R^1NH_2 + R^2CO_2H \longrightarrow R^1NHCOR^2 + H_2O \qquad (86)$$

of the reaction has not yet been completely clarified, but it undoubtedly involves the free amine and acid in equilibrium with the
salt. Since the reaction is formally the reverse of amide hydrolysis
it is reasonable to assume that it follows the normal route (87),

$$R^1NH_2 + R^2\overset{O}{\overset{\|}{C}}OH \rightleftharpoons R^2\underset{\underset{+NH_2R^1}{|}}{\overset{O^-}{\overset{|}{C}}}OH \rightleftharpoons R^2\overset{O}{\overset{\|}{C}}NHR^1 + OH^- + H^+ \qquad (87)$$

proceeding through a tetrahedral intermediate with displacement of
OH⁻. There is evidence in support of this mechanism for reactions
of amines with monocarboxylic acids in aqueous solution[229], but with
dicarboxylic acids the reaction appears to proceed by initial formation
of anhydrides[230].

A good general procedure involves heating a mixture of acid and
amine at about 200°[231]. Examples of its use include the preparation
of benzanilide[232], substituted N-β-(phenylethyl)phenylacetamides[231],
and N-phenyloleiamide[233]. An interesting adaption of the method
involves formation of piperazinediones by heating of α-amino acids
(equation 88)[234,235]. Variations of the procedure include the use of

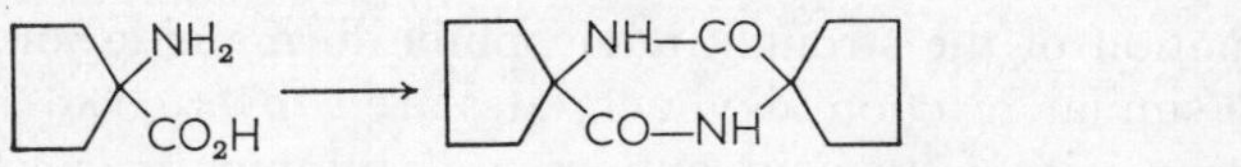

$$(88)$$

silica gel, which apparently acts as a catalyst[236], and of high-boiling
hydrocarbon solvents which allow azeotropic removal of water from
the reaction mixture[237]. Examples of the latter technique include
the preparation of N-methylformanilide[237] and of N-o-tolylforma-
mide[238] from the appropriate amines and formic acid in toluene, and
of N,N-dibutyllactamide from lactic acid and dibutylamine in
xylene[239]. An acidic ion exchange resin has been found to be an
excellent catalyst for the preparation of 47 and related compounds
in xylene (equation 89)[240].

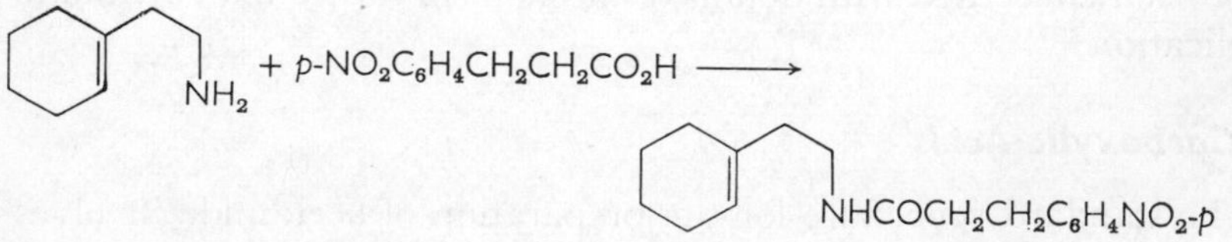

$$\qquad + p\text{-}NO_2C_6H_4CH_2CH_2CO_2H \longrightarrow$$

NHCOCH₂CH₂C₆H₄NO₂-p   (89)

(47)

Intramolecular amide formation takes place readily in suitably constituted amino acids and is a commonly used reaction for the preparation of piperidones and pyrrolidones (equation 90). Fre-

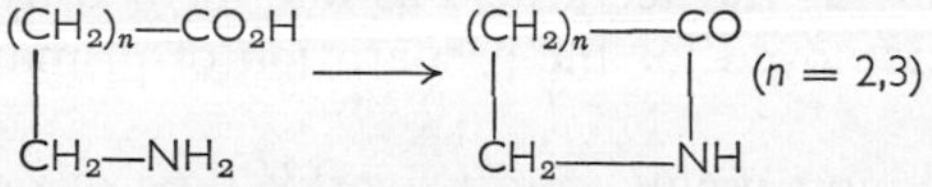

$$\begin{array}{c}(CH_2)_n\text{---}CO_2H \\ | \\ CH_2\text{---}NH_2 \end{array} \longrightarrow \begin{array}{c}(CH_2)_n\text{------}CO \\ | \qquad | \\ CH_2\text{------}NH \end{array} \quad (n = 2,3) \tag{90}$$

quently, the amino acid required for cyclization is prepared *in situ*. Suitable methods include the reduction of imino acids and nitro acids[241,242], reductive amination of keto acids[243], and aminolysis of lactones[244]. Amines, when treated with γ- and δ-keto acids give unsaturated lactams, presumably via cyclization of an intermediate imino acid. The reaction has found considerable application in the preparation of aza steroids[245], e.g. 4-aza-5-cholesten-3-one (equation 91)[246].

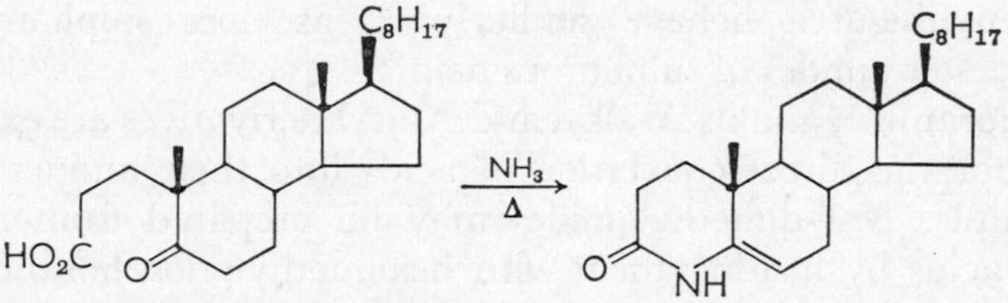

$$\tag{91}$$

Amides react with carboxylic acids but usually diacylamines are formed only when the reaction takes place intramolecularly so that the products are cyclic imides. The mechanism of the reaction has been recently discussed[247]. Glutarimide is conveniently made by heating the monoamide of glutaric acid[248] but for preparative purposes the required amido acids are usually prepared *in situ* by reaction of amines with 1,2- or 1,3-dicarboxylic acids or their anhydrides. Thus succinimide[249] is prepared by heating ammonium succinate, and phthalimide[250] by heating phthalic anhydride with ammonium carbonate. α-Amino acids react with phthalic anhydride in toluene yielding phthaloyl derivatives[251]. Alternatively, the phthalamic acids obtained from the same reactants in dioxan can be cyclized by addition of triethylamine and further heating of the reaction mixture[138]. The reaction provides a useful method of protecting the amino group for peptide synthesis.

Treatment of simple amides with acids usually results in a transacylation reaction (92). When one acid is much lower boiling than

$$R^1NHCOR^2 + R^3CO_2H \rightleftharpoons R^1NHCOR^3 + R^2CO_2H \tag{92}$$

the other it is possible to use the reaction for preparative purposes[252-254]. For example *N*-methylbenzanilide is produced in good yield by heating a mixture of benzoic acid and *N*-methylacet-anilide and removing the acetic acid as it is formed by distillation[253]. Formamide is employed for the preparation of primary amides from acids[254].

However, the most useful amides for this type of reaction are urea and related compounds[255-257]. Thus, heptanoic acid when heated with urea at 140–180° gives heptamide in good yield (equation 93)[256]. The reaction, which is of wide application, is believed to involve the intermediacy of carboxylic carbamic anhydrides[257], and is thus closely related to methods of amide formation involving acylation of isocyanates and of carbamyl halides (section V). Thiourea and

$$RCOOH + NH_2\!-\!CONH_2 \longrightarrow RCO\!-\!O\!-\!CONH_2 \longrightarrow RCONH_2 + CO_2$$

$$(93)$$

sym-diphenylthiourea behave similarly[258], as does sulphamide[259] and monoester amides of sulphurous acid[260].

Phosphoramide, and its *N*-alkyl and *N*-aryl derivatives are excellent reagents for the direct conversion of acids into their amides[261,262]. For example, *N*,*N*-dimethylamides may be prepared from a wide range of acids by heating them with hexamethylphosphoramide[262]. Other phosphoramides for use in this type of reaction are readily prepared *in situ* from amines and suitable phosphoryl halides[261]. Thus heating of a carboxylic acid with an amine and phosphoryl chloride in benzene affords the appropriate amide in excellent yield[263]. Amides of diphenylphosphinic acid are yet another class of reagent capable of bringing about direct amidation of acids[264]. Finally, in this group of 'activated amine' derivatives incorporating phosphorus mention must be made of phosphazo compounds which are readily prepared *in situ* from an amine and phosphorus trichloride, and which yield the appropriate amide when treated with a carboxylic acid (equation 94)[265]. The method has been mainly used for peptide

$$R^1NH_2 + PCl_3 \longrightarrow R^1N\!=\!P\!-\!NHR^1 \xrightarrow{R^2CO_2H} R^2CONHR^1 + (PHO_2)_x \quad (94)$$

synthesis[266], but can be applied generally to the preparation of amides from amines and carboxylic acids by warming them in benzene with phosphorus trichloride[267].

Acids can also be converted directly to amides in good yield by treatment with tris-dialkylaminoboranes[268]. The reaction, which is exothermic and rapid, is conducted by mixing the two reagents in

benzene.   The mechanism is thought to involve intermediate formation of an acyloxyborane derivative (equation 95).   Possibly the

$$R^1CO_2^- + B \overset{\overset{\displaystyle NR_2^2}{|}}{\underset{\displaystyle NR_2^2}{|}} \overset{+}{N}R_2^2H \longrightarrow$$

$$R^1 - \overset{\overset{\displaystyle O}{\|}}{C} - O - B \overset{NR_2^2}{\underset{NR_2^2}{}} \longrightarrow R^1CONR_2^2 + R_2^2NH + BONR_2^2 \quad (95)$$

$$HNR_2^2$$

reductive acylation of Schiff bases with a carboxylic acid and trimethylamine borane[269] is mechanistically related.

Some other methods for the conversion of carboxylic acids into
amides involve the intermediate formation in the reaction mixture of
mixed anhydrides with carbonic acids, carbamic acids and inorganic
acids, or of active esters, and are discussed in sections II.C, II.D, V.A,
and V.B.

F. Aldehydes and Ketones

Ammonia, and amines, readily undertake nucleophilic addition
to the carbonyl groups in aldehydes and ketones affording, initially,
tetrahedral zwitterionic intermediates **48** of the same general type as
those involved in reactions of amines with carboxylic acid derivatives.
However, because of the low stability of most carbanions, there is
usually little tendency for addition to be followed by C—R bond
scission, and the intermediate is stabilized by proton transfer yielding
the carbinolamine, and eventually (in the case of primary amines)
the Schiff base (equation 96).

$$R^1 - \overset{\overset{\displaystyle O}{\|}}{C} - R^2 \rightleftharpoons R^1 - \overset{\overset{\displaystyle O^-}{|}}{\underset{\displaystyle \overset{+}{N}H_2R^3}{C}} - R^2 \rightleftharpoons R^1 - \overset{\overset{\displaystyle OH}{|}}{\underset{\displaystyle NHR^3}{C}} - R^2 \longrightarrow R^1 - \overset{C}{\underset{\displaystyle NR^3}{\|}} - R^2 \quad (96)$$

$$NH_2R^3$$
$$(48)$$

Formation of amides by acylation of amines with aldehydes or
ketones becomes practicable when one of the alkyl groups attached
to the carbonyl carbon contains substituents which, by stabilizing the
related carbanion, allow it to function as a leaving group.   Thus
trihalomethyl ketones and aldehydes, when treated with amine
undergo addition–elimination according to the general mechanism

previously discussed, with formation of amides and haloform (equation 97).

$$
\underset{\underset{R^2NH_2}{\displaystyle\diagup}}{\overset{\overset{\displaystyle O}{\parallel}}{R^1\!-\!C\!-\!CX_3}} \;\Longleftrightarrow\; \underset{\overset{+}{N}H_2R^2}{\overset{\overset{\displaystyle O^-}{\mid}}{R^1\!-\!C\!-\!CX_3}} \;\longrightarrow\; \overset{\overset{\displaystyle O}{\parallel}}{R^1\!-\!C\!-\!NHR^2} + \underbrace{X_3C^- + H^+}_{\displaystyle X_3CH} \quad (97)
$$

This reaction, which is mechanistically closely related to the final step in the haloform reaction, is of considerable value for the preparation of formamides under mild conditions. The method involves slow addition of one molecular equivalent of chloral to a cold solution of the amine in chloroform; it is applicable to both primary and secondary amines and the yields are usually excellent[270]. It has recently been employed for the preparation of N-methyl-N-(β-chloroethyl)formamide and related compounds (equation 98)[271].

$$
\underset{\displaystyle ClCH_2CH_2\overset{\displaystyle Me}{\overset{\mid}{N}}H} + CCl_3CHO \longrightarrow \underset{\displaystyle ClCH_2CH_2\overset{\displaystyle Me}{\overset{\mid}{N}}CHO} + CHCl_3 \quad (98)
$$

Ketones containing the trichloromethyl group also function as acylating agents. For example, amines when treated with hexachloroacetone in hexane are smoothly and efficiently converted into their trichloroacetyl derivatives (equation 99)[272]. The reaction has

$$
R^1R^2NH + CCl_3COCCl_3 \longrightarrow R^1R^2NCOCCl_3 + CHCl_3 \quad (99)
$$

been applied to a wide range of aromatic and aliphatic amines and usually gives yields in the range 80–90%.

Unsymmetrical trihalomethyl ketones are less frequently used as acylating agents but are sometimes valuable in special circumstances. Thus in the synthesis of chloroamphenicol the amino group in the amino alcohol **49** was selectively acylated by treatment with pentachloroacetone in dioxan (equation 100)[273].

$$
\underset{\textbf{(49)}}{p\text{-}NO_2C_6H_4\overset{\displaystyle OH}{\overset{\mid}{C}}H\!-\!\overset{\displaystyle NH_2}{\overset{\mid}{C}}HCH_2OH} \;\xrightarrow{CCl_3COCHCl_2}\; \underset{(100)}{p\text{-}NO_2C_6H_4\overset{\displaystyle OH}{\overset{\mid}{C}}H\!-\!\overset{\displaystyle NHCOCHCl_2}{\overset{\mid}{C}}HCH_2OH}
$$

Another method of somewhat more general application for the formation of amides from ketones is the Haller–Bauer reaction[274], which involves heating a non-enolizable ketone with sodium amide in benzene, toluene or similar aprotic solvent. The mechanism of the reaction probably involves addition of amide ion, NH_2^-, to the carbonyl centre, followed by elimination of a carbanion (equation 101).

$$NH_2^- + R-\overset{\overset{\displaystyle O}{\|}}{C}-R \rightleftharpoons R-\overset{\overset{\displaystyle O^-}{|}}{\underset{\underset{\displaystyle NH_2}{|}}{C}}-R \rightleftharpoons R-\overset{\overset{\displaystyle O}{\|}}{C}-NH_2 + R^-$$

$$\text{(50)} \qquad\qquad\qquad \text{(51)} \qquad\qquad \text{(52)}$$

$$\longrightarrow R-\overset{\overset{\displaystyle O}{\|}}{C}-NH^- + RH \quad (101)$$

$$\text{(53)}$$

Undoubtedly the success of the method is due to (*i*) the high nucleophilicity of NH_2^- which forces the equilibrium **50** ⇌ **51** far to the right, (*ii*) the absence from the reaction mixture of any acid sufficiently strong to stabilize the intermediate **51** by protonation of oxygen and (*iii*) stabilization of the product amide as its ion **53**. The reaction has considerable merit as a means of preparing the amides of tertiary carboxylic acids, compounds which are rarely easily accessible by other routes. The customary starting materials are alkyl phenyl ketones which are rapidly prepared by alkylation of acetophenone and similar compounds. The preparation of 2,2,4-trimethylpentanamide provides a typical example of the reaction sequence generally employed (equations 102 and 103)[275].

$$Me_2CHCOCl \xrightarrow[AlCl_3]{PhH} Me_2CHCOPh \xrightarrow[Me_2CHCH_2Br]{NaNH_2} Me_2CHCH_2\overset{\overset{\displaystyle Me}{|}}{\underset{\underset{\displaystyle Me}{|}}{C}}COPh$$

$$(102)$$

$$Me_2CHCH_2\overset{\overset{\displaystyle Me}{|}}{\underset{\underset{\displaystyle Me}{|}}{C}}COPh \xrightarrow[toluene]{NaNH_2} Me_2CHCH_2\overset{\overset{\displaystyle Me}{|}}{\underset{\underset{\displaystyle Me}{|}}{C}}CONH_2 + PhH \qquad (103)$$

Usually as in this example, cleavage of the ketone proceeds in that direction which affords the alkanecarboxamide rather than that leading to the aromatic amide. However, alkyl phenyl ketones containing large, highly branched alkyl groups give low yields of alkanamide and benzamide is also obtained.

The reaction is applicable to a wide range of alkyl aryl ketones. An interesting example is that of *l*-1-benzoyl-1-methyl-2,2-diphenyl-cyclopropane, which undergoes cleavage by expulsion of the cyclopropyl group and which proceeds with retention of optical activity (equation 104) [276].

$$Ph-\overset{\overset{O}{\|}}{\underset{\underset{Ph}{|}}{C}}\overset{Me}{\underset{Ph}{\diagup}} \xrightarrow{NaNH_2} PhCONH_2 + \overset{H}{\underset{Ph}{\underset{|}{Me}}}\diagup Ph \qquad (104)$$

$$[\alpha]_D = -32.5° \qquad\qquad\qquad [\alpha]_D = +127.6°$$

On the basis of these results it was suggested that the Haller–Bauer reaction could not involve the intermediacy of free carbanions and that cleavage and proton transfer must occur synchronously (equation 105) [276]. However, the force of the argument was considerably

$$Ph-\overset{\overset{O^{(-)}}{|}}{\underset{\underset{H}{N}\underset{H}{}}{C}}\overset{Me}{\underset{Ph}{\diagup}}\diagdown Ph \longrightarrow Ph-\overset{\overset{O}{\|}}{C}-NH^- + \overset{Me}{\underset{Ph}{\underset{|}{H}}}\diagup Ph \qquad (105)$$

weakened by the subsequent demonstration that cyclopropyl carbanions can maintain configurational stability [277].

Fission of diaryl ketones occurs under Haller–Bauer conditions. The reaction has little value for amide synthesis but gives further information concerning the reaction mechanism. As expected, cleavage occurs in that direction which affords the more stable aryl anion. Thus, p-methoxybenzophenone yields mainly the amide of anisic acid (equation 106) [278].

$$p\text{-MeOC}_6\text{H}_4\text{COPh} \xrightarrow{NaNH_2} p\text{-MeOC}_6\text{H}_4\text{CONH}_2 + \text{PhCONH}_2 \qquad (106)$$
$$\qquad\qquad\qquad\qquad (72\%) \qquad\qquad (28\%)$$

o-Chlorobenzophenone reacts with sodium amide with exceptional facility yielding benzamide and aniline (equation 107) [279]. The formation of the latter product is considered to arise through the intermediacy of benzyne and thus provides additional evidence for the involvement of aryl anions in the reaction mechanism.

The Haller–Bauer procedure can also be used for the preparation of amides from non-enolizable aliphatic and alicyclic ketones, but the

$$Ph-\overset{\overset{O}{\|}}{C}-\overset{Cl}{\diagdown} +NH_2^- \longrightarrow PhCONH_2 + {}^-\overset{Cl}{\diagdown} \longrightarrow \diagdown\diagdown \xrightarrow{NH_3} \overset{NH_2}{\diagdown}$$

$$(107)$$

means of preparing amides in good yield under mild conditions when direct aminolysis of the starting material is impracticable.

The preparation of lysergamide from ergotamine and other ergot alkaloids provides a good example of the use of the azide route to amides. The hydrazide obtained by treating the alkaloid with hydrazine[286], is treated successively at 0° with aqueous nitrous acid and bicarbonate, and the resultant azide is extracted with ether and saturated with gaseous ammonia[287]. Other amides may be similarly prepared[286,287]. Recent examples include the preparation of macrocyclic diamides[288], and of acyl derivatives of amino acids[289].

The azide method for forming amide linkages finds its most important application in peptide synthesis[107] where it offers the special advantage of proceeding without racemization[290]. Its uses have included the preparation of penicillamine dipeptides[291], and of N-acyl dipeptides from melphalan[292].

Benzoyl azide reacts with imines yielding, after hydrolysis of the initial product, N-acylbenzamides (equation 111). The reaction, however, is not a simple acylation. It proceeds with evolution of nitrogen and affords initially an imino-amide[293]. Undoubtedly, a preliminary dipolar addition is involved.

$$PhCON_3 + Me_2CHCH{=}N{-}R \xrightarrow{-N_2} \underset{\substack{\| \\ O}}{Ph\overset{O}{C}}{-}NH{-}\underset{\substack{\| \\ NR}}{\overset{NR}{C}}CHMe_2$$

$$\downarrow H_2O$$

$$\underset{\substack{\| \\ O}}{Ph\overset{O}{C}}{-}NH{-}\underset{\substack{\| \\ O}}{\overset{O}{C}}CHMe_2 \tag{111}$$

The most serious disability to the azide method of amide preparation is the concurrent occurrence of the Curtius rearrangement (112) leading, under the usual reaction conditions, to the formation of ureas[282]. On mechanistic grounds we should expect this side-reac-

$$RCON_3 \longrightarrow RN{=}C{=}O \longrightarrow RNHCONHR \tag{112}$$

tion to be most serious when the attacking amine is weakly nucleophilic or highly sterically hindered, when the group R through the operation of electronic or steric factors diminishes the reactivity of the adjacent carbonyl function, and when R has a high migratory aptitude. Occasionally it becomes so important that it takes precedence over aminolysis even under the mildest possible conditions[294]. Curtius rearrangement occurs so readily in azides of aromatic acids synthetic potential of the reaction appears to be rather limited. A full account of the mechanism, scope and limitations of the Haller–Bauer reaction and related synthetic processes is available[264].

G. Acyl Azides

Acyl azides undergo nucleophilic attack by amines yielding amides (equation 108). The reaction almost certainly follows the general mechanism previously adumbrated (section II.A). Since the azide

$$R^1{-}\overset{\overset{\displaystyle O}{\|}}{C}{-}N_3 + R_2^2NH \longrightarrow R^1{-}\underset{\underset{\displaystyle N_3}{|}}{\overset{\overset{\displaystyle O^-}{|}}{C}}{-}\overset{+}{N}HR_2^2 \longrightarrow R^1{-}\overset{\overset{\displaystyle O}{\|}}{C}{-}NR_2^2 + HN_3 \quad (108)$$

group offers little steric hindrance towards attack at the adjacent carbonyl carbon, and since the azide ion, being a relatively weak base, is a good leaving group, we should expect the reaction to proceed readily, and this appears to be the case. Both towards hydrolysis and aminolysis acyl azides seem to show reactivity comparable with that of related carboxylic anhydrides.

Two routes are commonly employed for the preparation of acyl azides. The first, involving reaction of sodium azide with acyl halide in aqueous acetone[280], or in water with pyridine catalyst[281] (equation 109), is convenient but offers little advantage for the preparation of amides which can normally be obtained more directly by aminolysis of the halide. The second method consists of treating an acyl

$$RCOCl + N_3^- \longrightarrow RCON_3 + Cl^- \quad (109)$$

hydrazide with nitrous acid generated from sodium nitrite or nitrite ester in a suitable solvent[282], or preferably, since undesirable side-reactions are thereby diminished, from nitrosyl chloride or t-butyl nitrite with hydrogen chloride in ether or tetrahydrofuran[283]. Because of the ease with which they undergo hydrolysis, azides prepared in aqueous medium are usually extracted into an organic solvent before further reaction; ether[284] or ethyl acetate are suitable.

Hydrazine, being a highly active nucleophile reacts more readily than amines with esters and amides[282–285], and the route represented in equation (110), although circuitous at first sight, often offers a

$$\begin{array}{c} R^1CO_2R^2 \\ \\ R^1CONHR^2 \end{array} \xrightarrow{NH_2NH_2} R^1CONHNH_2 \xrightarrow{HNO_2} R^1CON_3 \xrightarrow{R_2^3NH} R^1CONR_2^3 \quad (110)$$

5+c.o.a.

that their reaction with most amines to give ureas has been recommended for purposes of characterization[295]. However, benzylamine, presumably because of its high nucleophilicity, smoothly undertakes aminolysis of aroyl azides; the relative reactivities of a series of azides is compatible with the effects of substituents on the electron density at the carbonyl function[296].

Another important side-reaction—formation of primary amides—occurs when azides are prepared by treatment of hydrazides with nitrous acid. It is believed to involve elimination of nitrous oxide from a nitroso hydrazide (equation 113)[283,297].

$$RCONHNH_2 \xrightarrow{HNO_2} (RCONHNH-N=O \rightleftharpoons RCONH-N=N-OH) \longrightarrow$$
$$RCONH_2 + N_2O \quad (113)$$

H. Ketenes

Amines react readily with ketene and substituted ketenes yielding their N-acyl derivatives (equation 114)[298]. Although formally an

$$R^1R^2C=C=O + R^3NH_2 \longrightarrow R^1R^2CHCONHR^3 \quad (114)$$

addition process the reaction is believed to be mechanistically related to other N-acylation reactions in that it involves initial nucleophilic attack of the amine on the carbonyl group (equation 115). The

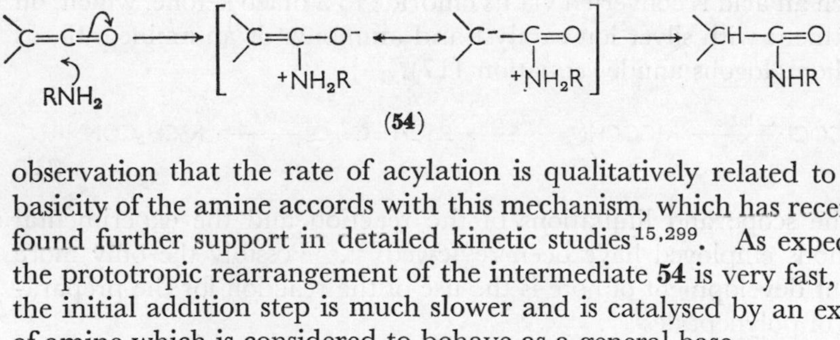

$$(54) \qquad\qquad\qquad\qquad (115)$$

observation that the rate of acylation is qualitatively related to the basicity of the amine accords with this mechanism, which has recently found further support in detailed kinetic studies[15,299]. As expected the prototropic rearrangement of the intermediate **54** is very fast, but the initial addition step is much slower and is catalysed by an excess of amine which is considered to behave as a general base.

The ketene most widely used in preparative work is ketene itself which acetylates amines rapidly and in high yield. In a typical experiment acetanilide was obtained quantitatively by passing ketene vapour into aniline dissolved in ether[300]. The reaction can be conducted also in aqueous or alcoholic solution and affords exclusively N-acylated products from amino phenols or amino alcohols[301]. Cysteine, however, affords the N,S-diacetyl derivative[302]. Acyl-ketenes have been employed for the preparation of amides of β-ketocarboxylic acids[303].

Because of its high reactivity ketene is of great utility for the acetylation of amides and imides. Typical examples include the formation of N-acetylbenzamide by passing ketene into a suspension of benzamide in benzene containing a catalytic amount of sulphuric acid[304], and the preparation of N-acetylsuccinimide from succinimide in carbon tetrachloride[305].

Ketenes react with imines by an addition process affording β-lactams. For example, addition of ketene to the imine **55**, obtained by treating aldehydes with sulphuryl isocyanate halide, affords **56** which is readily hydrolysed to an azetidinone (equation 116)[306].

$$
\begin{array}{ccccc}
\text{RCHO} & & \text{RCH} & \text{RCH—CH}_2 & \text{RCH—CH}_2 \\
+ & \xrightarrow{\hspace{1cm}} & \underset{\text{N}}{\overset{\|}{}} \xrightarrow{\text{CH}_2=\text{C}=\text{O}} & \underset{\text{N—C}=\text{O}}{|\quad\quad|} \xrightarrow{\text{H}_2\text{O}} & \underset{\text{NH—C}=\text{O}}{|\quad\quad|} \\
\text{NCO} & & | & | & \\
| & & \text{SO}_2\text{X} & \text{SO}_2\text{X} & \\
\text{SO}_2\text{X} & & & &
\end{array}
$$

$$\text{(55)} \qquad\qquad\qquad \text{(56)} \tag{116}$$

Other examples of the formation of β-lactams are summarized in Lacey's review[298].

An important method for the preparation of amides which probably involves the intermediacy of ketenes is the Arndt–Eistert reaction in which an acid is converted via its chloride to a diazo ketone, which, on treatment with silver ion catalyst and ammonia or an amine, affords the homologous amide (equation 117).

$$R^1\text{COCl} \xrightarrow{\text{CH}_2\text{N}_2} R^1\text{COCHN}_2 \xrightarrow{\text{Ag}^+} R^1\text{CH}=\text{C}=\text{O} \xrightarrow{\text{R}^2\text{NH}_2} R^1\text{CH}_2\text{CONHR}^2 \tag{117}$$

The scope and limitations of the reaction and the experimental methods employed have been reviewed[307]. Possibly the only more recent development of note is the use of the reaction for the preparation of polypeptides[308].

I. Miscellaneous Acylating Agents

1. Amides and imides

When an amine or its salt is heated with a primary amide, exchange (118) occurs.

$$R^1\text{NH}_2 + R^2\text{CONH}_2 \longrightarrow R^2\text{CONHR}^1 + \text{NH}_3 \tag{118}$$

The reaction is of wide applicability[309] and is particularly useful for the preparation of N-formyl compounds which are obtained in high yield by heating an amine hydrochloride with formamide at 60–70° for several minutes. An improved procedure for formylation involves treating an amine with N,N-dimethylformamide and sodium methoxide[310]. The reaction has recently been employed for the preparation of formamido tetrazoles[311].

Aminolysis of cyclic imides occurs particularly readily and the reaction has been extensively employed for the preparation of mixed diamides of 1,2-dicarboxylic acids. Succinimide, for example, when shaken with aqueous methylamine gives N-methylsuccinamide (equation 119)[312]. The reaction is reversible: heating of such diamides neat or in acid or base affords imides[313]. Dichloromaleimide

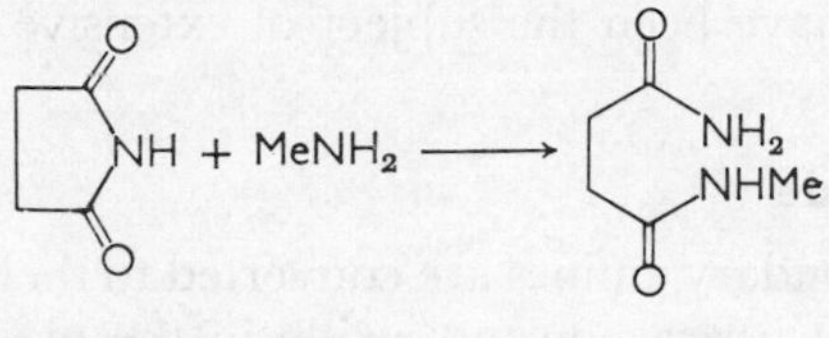

$$\text{(equation 119)}$$

and related compounds show particularly high reactivity towards amines because of the electronic effects of the halo substituents[314]. N-Carboethoxyphthalimide is also highly reactive and is useful for the phthaloylation of amino acids[315].

As expected on mechanistic grounds, N-acylimidazoles (**57**)[316] and N-acylpyrazoles (**58**)[317] are particularly effective acylating agents and both have found extensive application in peptide synthesis[107,121].

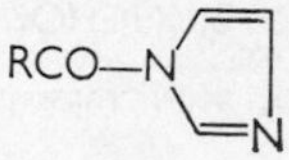

$$RCO-N \qquad\qquad RCO-N$$

(**57**) 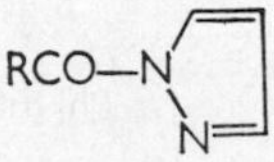   (**58**)

The acylimidazoles are conveniently made by a transamidation reaction of carbonyl-1,1′-diimidazole with carboxylic acids (equation 120)[316,318]. Acetylation of pyrrole with N-acetylimidazole affords

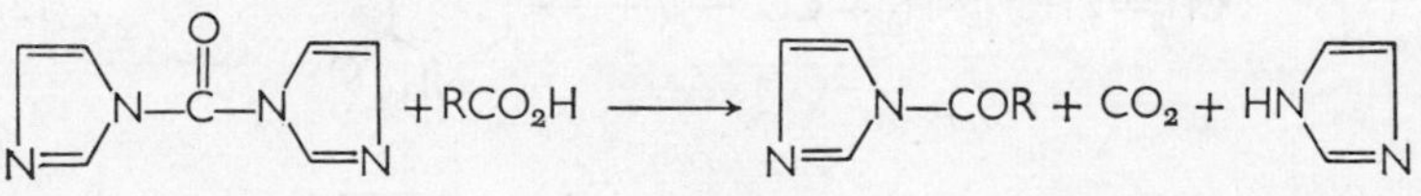

$$\text{(equation 120)} \quad + RCO_2H \longrightarrow \quad N-COR + CO_2 + HN$$

N-acetylpyrrole which is difficult to prepare by conventional methods[319].

2. Thioacids and thiolesters

Thioacids, thiolesters, and thiolcarboxylic anhydrides each react with amines yielding acyl derivatives (equations 121–123). In

$$R^1COSH + R^2NH_2 \longrightarrow R^1CONHR^2 + H_2S \tag{121}$$

$$R^1COSR^3 + R^2NH_2 \longrightarrow R^1CONHR^2 + R^3SH \tag{122}$$

$$R^1COSCOR^1 + R^2NH_2 \longrightarrow R^1CONHR^2 + R^1COSH \tag{123}$$

general such acylations occur more readily than those with the oxygen analogues: however, thioacid derivatives usually offer little advantage over other reagents for the preparation of amides except in the field of peptide synthesis[121]. Acylation reactions of thioacids and their derivatives are of considerable theoretical and biochemical interest because of their possible relevance to the mode of action of acetyl coenzyme A, and have been the subject of extensive mechanistic investigation[92,320].

3. Carbon monoxide

Primary and secondary amines are converted to their formyl derivatives by treatment with carbon monoxide in the presence of sodium methoxide, cobalt octacarbonyl or various other metallic salts (equation 124)[321–324]. Tertiary alkylamines when similarly treated

$$R^1R^2NH + CO \longrightarrow R^1R^2NCHO \tag{124}$$

suffer loss of one of the substituents giving dialkylformamides (equation 125), but aniline derivatives undergo a carbonyl insertion reaction (equation 126)[321]. Suitably constituted alkenylamines

$$Bu_3N + CO \longrightarrow Bu_2NCHO \tag{125}$$

$$PhNEt_2 + CO \longrightarrow PhNCOEt \atop | \atop Et \tag{126}$$

react with carbon monoxide at both the amino and olefinic functions to yield lactams (equation 127)[323], whilst acrylamides undergo a similar reaction giving cyclic imides (equation 128)[324].

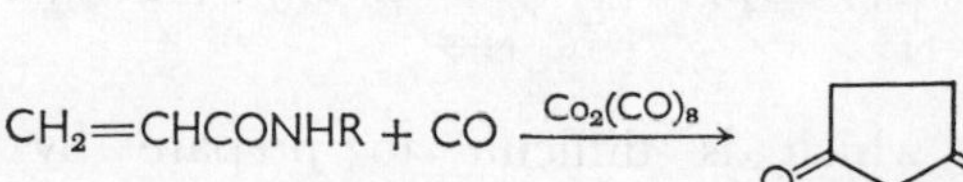

$$CH_2{=}CHCH_2NH_2 + CO \xrightarrow{\text{Co}_2(CO)_8} \tag{127}$$

$$CH_2{=}CHCONHR + CO \xrightarrow{\text{Co}_2(CO)_8} \tag{128}$$

4. Hydrazides

Amines are acylated when treated with an acylhydrazine in the presence of a suitable oxidizing agent such as iodine or N-bromo-succinimide. The reaction, which is useful in peptide synthesis, probably involves intermediate formation of an acyldiazonium salt (equation 129)[325].

$$R^1CONHNH_2 \xrightarrow{I_2} R^1CON_2^+ \xrightarrow{R^2NH_2} R^1CONHR^2 + N_2 + H^+ \qquad (129)$$

III. PREPARATION OF AMIDES FROM NITRILES

A. Hydration

Hydrolysis of nitriles to carboxylic acids involves intermediate formation of amides (equation 130). The relative rates of the two

$$RCN \xrightarrow{H_2O} RCONH_2 \xrightarrow{H_2O} RCO_2H + NH_3 \qquad (130)$$

steps vary according to the structure of the substrate and the experimental conditions. Hydration of nitriles can be of value for the preparation of amides only when it is much faster than subsequent hydrolysis of the initial product.

The hydration reaction is subject both to acid and base catalysis. The mechanism for the base-catalysed reaction involves initial addition of hydroxide ion to the $C\equiv N$ group (equation 131), whilst the acid-catalysed reaction proceeds through the protonated nitrile (equation 132).

$$RC\equiv N + OH^- \longrightarrow RC\underset{OH}{\overset{N^-}{\diagdown}} \xrightarrow{H_2O} RC\overset{O}{\overset{\|}{-}}NH_2 + OH^- \qquad (131)$$

$$RC\equiv N + H_3O^+ \longrightarrow [R\overset{+}{C}=NH \leftrightarrow RC\equiv \overset{+}{N}H] \xrightarrow{H_2O}$$

$$RC\underset{\overset{+}{O}H_2}{\overset{NH}{\diagdown}} \xrightarrow{H_2O} R-\overset{O}{\overset{\|}{C}}-NH_2 + H_3O^+ \qquad (132)$$

The most widely used procedure for the hydration of nitriles involves treating the substrate with mineral acid. Strong sulphuric acid is a particularly useful reagent for the preparation of aromatic amides and amides of highly sterically hindered aliphatic acids. Examples of its use include the preparation of tributylacetamide[326] in

80% sulphuric acid at 100°, and of diisopropylacetamide[327] in 96% acid at 140–150°.

The success of these syntheses illustrates the extreme resistance to acid-catalysed hydrolysis characteristic of hindered amides, and indicates that hydration of the nitrile group is less susceptible to steric factors than is the subsequent hydrolysis of the amide. However, in straight-chain compounds hydrolysis of the amide often competes effectively with the hydration step, and the reaction is suitable as a preparative method only under very carefully controlled experimental conditions.

Sulphuric acid catalysed hydration of nitriles sometimes proceeds with remarkable specificity. Thus hydration of the nitrile **59** in 97% sulphuric acid proceeds without disruption of the ester groups (equation 133)[238], whilst the imide and lactone functions in **60** survive similar treatment (equation 134). Hydration of **61** in 96% sulphuric

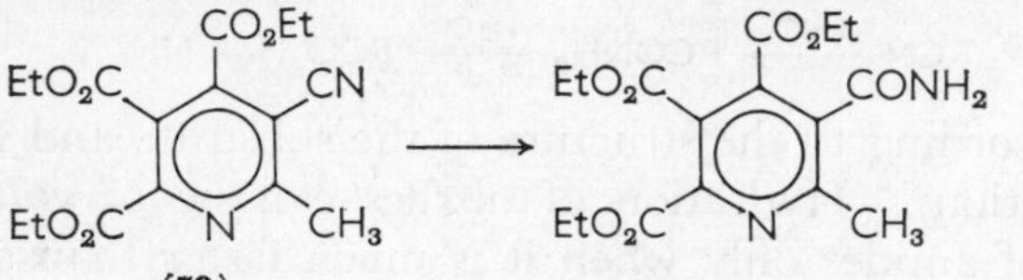

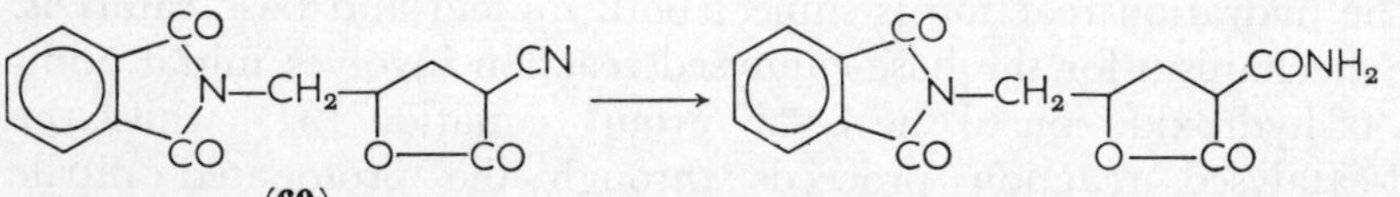

$$(133)$$

$$(134)$$

acid proceeds without simultaneous hydrolysis of the ester, but in 75% acid, hydrolysis of both amide and ester functions occurs followed by decarboxylation to yield **62** (equation 135)[330].

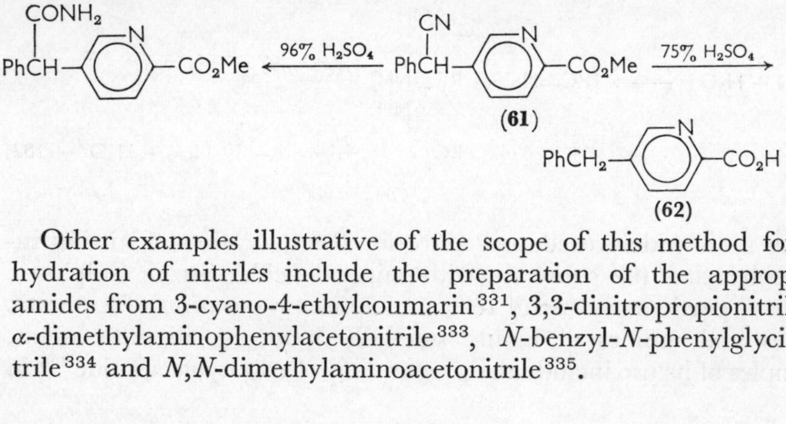

$$(135)$$

Other examples illustrative of the scope of this method for the hydration of nitriles include the preparation of the appropriate amides from 3-cyano-4-ethylcoumarin[331], 3,3-dinitropropionitrile[332], α-dimethylaminophenylacetonitrile[333], N-benzyl-N-phenylglycinonitrile[334] and N,N-dimethylaminoacetonitrile[335].

α-Keto nitriles can be hydrated in strong acid to yield α-keto amides (equation 136). Cyanohydrins often undergo simultaneous loss of the

$$RCOCN \longrightarrow RCOCONH_2 \qquad (136)$$

alcoholic hydroxyl giving α,β-unsaturated amides. Thus, treatment of acetone cyanohydrin with strong sulphuric acid affords a convenient preparation of methylacrylamide (equation 137)[336].

$$CH_3-\underset{\underset{OH}{|}}{\overset{\overset{CH_3}{|}}{C}}-CN \xrightarrow{H_2SO_4} CH_2{=}\underset{}{\overset{\overset{CH_3}{|}}{C}}-CONH_2 \qquad (137)$$

Hydrochloric acid is a suitable reagent for hydration of nitriles. Phenylacetamide is formed in excellent yield by stirring benzyl cyanide with the concentrated acid at 40–50° and the same procedure is applicable to a wide range of substituted arylacetonitriles[337]. The unsaturated cyanohydrin **63** is converted to the hydroxy amide **64** by treatment with a mixture of aqueous hydrochloric and sulphuric acids (equation 138)[338], and p-chlorobenzyl cyanide (**65**) when mixed with concentrated nitric and sulphuric acids undergoes concomitant hydration and nitration (equation 139)[339].

$$PhCH{=}CH\underset{\underset{(63)}{}}{\overset{\overset{OH}{|}}{C}}HCN \longrightarrow PhCH{=}CH\underset{\underset{(64)}{}}{\overset{\overset{OH}{|}}{C}}HCONH_2 \qquad (138)$$

$$Cl{-}\langle\!\!\!\bigcirc\!\!\!\rangle{-}CH_2CN \xrightarrow{HNO_3/H_2SO_4} Cl{-}\langle\!\!\!\bigcirc\!\!\!\rangle{-}CH_2CONH_2 \qquad (139)$$

(65)

Polyphosphoric acid has been recommended for the preparation of amides from nitriles. Benzonitrile when heated at 110° for 1 hour with polyphosphoric acid affords benzamide in 96% yield, and other aryl and benzyl cyanides are hydrated under similar conditions[340]. However the reaction is not successful when applied to cyanomesitylene, presumably because of steric hindrance. Other transformations illustrative of the utility of polyphosphoric acid are the formation of α-hydroxybutyramide from acetone cyanohydrin, ethyl malonamate from ethyl cyanoacetate[340], and β-keto amides from β-keto nitriles[341].

Another useful reagent for accomplishing hydration of nitriles is boron trifluoride. Excellent yields of amide are obtained by passing

5*

boron trifluoride into a solution of the nitrile in acetic acid containing a small amount of water[342]. When an anhydrous boron trifluoride–acetic acid complex is used, some nitriles give amides, but from others mixtures of amide and acid are obtained[342]. Hydration of β-keto nitriles to β-keto amides can be accomplished with boron trifluoride in aqueous acetic acid[341].

The method of choice for the preparation of N-acetyl-α-phenylace-toacetamide consists of treating benzyl cyanide with boron trifluoride in acetic anhydride (equation 140)[343]. Phenylacetamide is probably

$$\overset{\overset{\displaystyle COCH_3}{\displaystyle |}}{} $$

$$PhCH_2CN \longrightarrow PhCH_2CONH_2 \longrightarrow PhCHCONHCOCH_3 \qquad (140)$$

an intermediate in the reaction. Boron trifluoride in acetic anhydride converts β-keto nitriles into N-acetyl-β-ketoamides[344].

Base-catalysed hydration of nitriles has been less frequently used for the preparation of amides; in many compounds the reaction proceeds to the acid by further hydrolysis of the amide. Examples of its successful use include the conversion of 7,12-dicyanobenz[k]fluoranthene, which is inert to acids, into the appropriate diamide by heating with potassium hydroxide in ethoxyethanol[345], and formation of 2,3,6,7-tetramethylnaphthalene-1,4-dicarboxamide by similar treatment of the dinitrile[346].

A convenient and reliable method for the preparation of amides involves treatment of nitriles with alkaline hydrogen peroxide. For example, o-toluamide is formed in 90% yield when o-tolunitrile is warmed in ethanol with sodium hydroxide and 30% aqueous hydrogen peroxide[347], and veratramide may be similarly prepared from veratronitrile[348]. Other examples illustrative of the method include the preparation of benzocyclobutane-1-carboxamide[349] and 3-nitrobiphenyl-4-carboxamide[350]. Unlike acid- and base-catalysed hydration reactions this procedure is applicable to the preparation of simple aliphatic amides[351].

Although the reaction was first described in 1885 its mechanism was not closely scrutinized until 1953 when Wiberg showed that the reaction rate exhibits first-order dependence on the concentrations of nitrile, H_2O_2 and OH^-, that benzonitrile oxide is not an intermediate, and that the oxygen evolved is derived from hydrogen peroxide[352]. A mechanism consistent with these observations involves initial nucleophilic addition of hydroperoxide ion to the $C \equiv N$ group followed by hydride transfer from a second molecule of peroxide (equation 141). When an olefin is added to the reaction mixture it

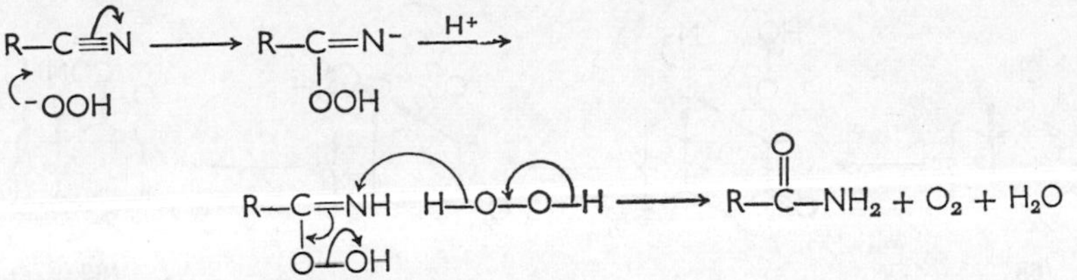

$$R-C\equiv N \longrightarrow R-C=N^- \xrightarrow{H^+}$$
$$\overset{|}{C}\text{-OOH} \qquad\qquad \overset{|}{O}OH$$

$$R-C=NH \quad H-O-O-H \longrightarrow R-\overset{\displaystyle O}{\overset{\|}{C}}-NH_2 + O_2 + H_2O \qquad (141)$$
$$\overset{|}{O}-OH$$

interacts with the intermediate peroxyimine yielding an epoxide in good yield (equation 142)[353].

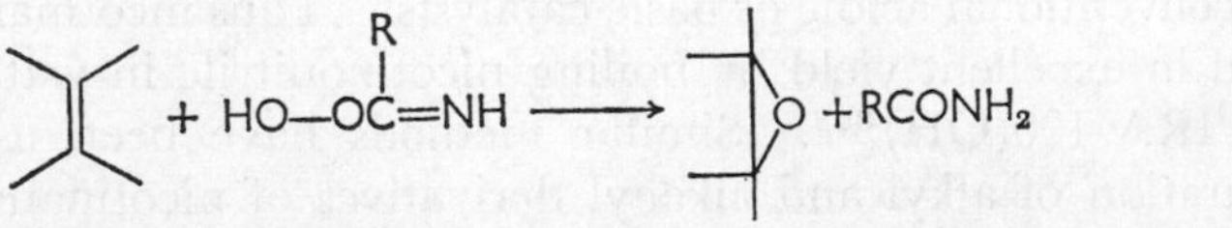

$$\underset{}{\bigtimes} + HO-O\overset{R}{\overset{|}{C}}=NH \longrightarrow \underset{}{\bigtriangleup}O + RCONH_2 \qquad (142)$$

When α,β-unsaturated nitriles are treated with hydrogen peroxide, intramolecular epoxidation of the olefinic linkage occurs affording epoxy amides, e.g. acrylonitrile gives glycidamide (equation 143)[354].

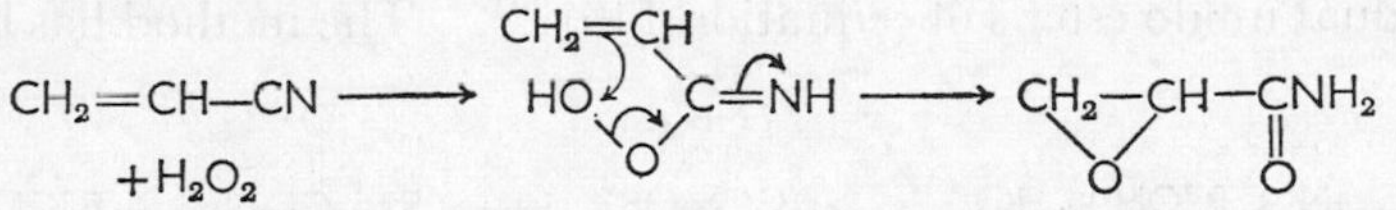

$$CH_2=CH-CN \longrightarrow HO \overset{CH_2=CH}{\underset{O}{\bigvee}} C=NH \longrightarrow CH_2-CH-CNH_2 \qquad (143)$$
$$+H_2O_2 \qquad\qquad\qquad\qquad\qquad\qquad \overset{\diagdown}{O}\quad\overset{\|}{O}$$

In work towards the synthesis of lysergic acid the nitrile **66** was similarly converted into the epoxy amide **67** in quantitative yield (equation 144)[355].

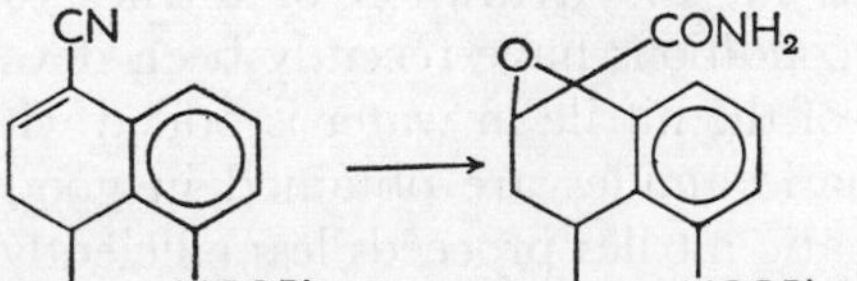

$$(144)$$

(66)   **(67)**

An interesting example of intramolecular attack of the hydroperoxy group on a cyano function was uncovered by Barton in his synthesis of β-amyrin when he found that a compound containing the system **68** is converted into the epoxy amide **69** in high yield by treatment with alkaline hydrogen peroxide (equation 145)[356].

Hydration of nitriles may be accomplished in the absence of added acid or base but it is normally necessary to heat the reaction mixture strongly in an autoclave[357]. However, the use of ion exchange resins sometimes allows formation of amides under relatively mild conditions

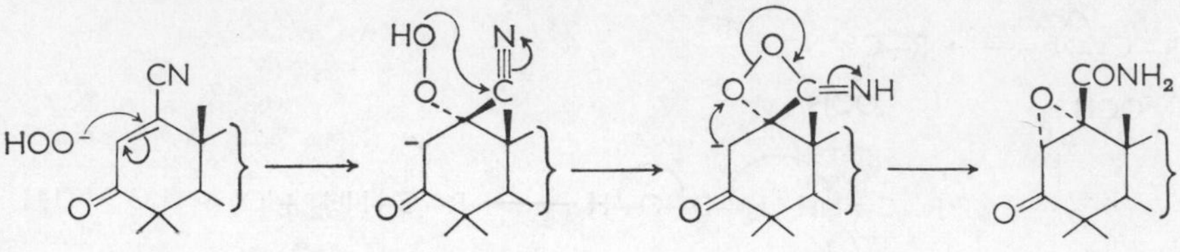

(68)   (69)

(145)

from nitriles which undergo complete hydrolysis to acids when treated with conventional acidic or basic catalysts. Thus nicotinamide is obtained in excellent yield by boiling nicotinonitrile in water with the resin IRA-400(OH)[358]. Similar methods have been used for the preparation of alkyl and alkenyl derivatives of nicotinamide[359] and for the partial hydrolysis of dinitriles to cyano amides[360].

An indirect procedure for the hydration of nitriles (the Pinner reaction) involves treating a nitrile with an alcoholic solution of hydrogen chloride, removing the solvent by evaporation and heating the residual imido ester salt (equation 146)[361]. The method has been

$$R^1C{\equiv}N + R^2OH + HCl \longrightarrow R^1\overset{\overset{+}{N}H_2Cl^-}{\underset{}{C}}-OR^2 \longrightarrow R^1-\overset{O}{\underset{}{C}}-NH_2 + R^2Cl$$

(146)

recommended as a convenient synthesis for α-hydroxy[362] and α-amino amides[363].

Efficient methods for the hydration of nitriles to amides under essentially neutral conditions have recently been developed. In one of them a mixture of the nitrile in water is boiled with a zinc–nickel catalyst[364]. Aromatic amides are obtained in good yield, but the reaction with aliphatic nitriles proceeds less efficiently. A very mild procedure involves shaking of a solution of the nitrile in dichloromethane with manganese dioxide at room temperature[365]. The method is applicable to alkyl and aryl cyanides, to acetone cyanohydrin, and to α,β-unsaturated nitriles (equation 147). The mechanisms of these reactions have not yet been elucidated. Perhaps it is

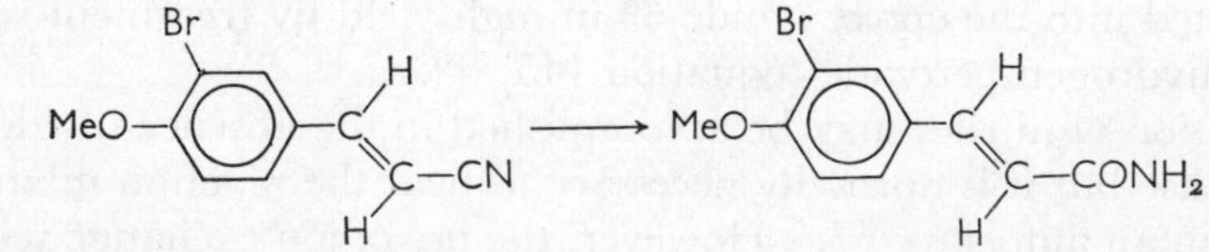

(147)

(90%)

relevant that hydration of 2-cyano-1,10-phenanthroline to the amide is powerfully catalysed by Cu^{2+}, Ni^{2+} and other metal ions[366].

B. Alkylative Hydration—The Ritter Reaction

Although the interaction of olefins with hydrogen cyanide in the presence of a strong acid (HCl–$AlCl_3$) to yield formamides was first described in 1930[367] the synthetic potential of this general type of transformation was not realized until 1948 when Ritter[368] showed that treatment of alkenes with nitriles and concentrated sulphuric acid affords N-substituted amides in good yield.   In a typical example of the method N-t-butylacetamide was prepared in 85% yield by passing isobutene into an acetic acid solution of acetonitrile and sulphuric acid, and pouring the mixture into water (equation 148).   Ritter

$$CH_3\!-\!\underset{\underset{CH_2}{\|}}{\overset{\overset{CH_3}{|}}{C}} + CH_3CN \xrightarrow[\text{(2) H}_2\text{O}]{\text{(1) H}_2\text{SO}_4} CH_3\!-\!\underset{\underset{CH_3}{|}}{\overset{\overset{CH_3}{|}}{C}}\!-\!NHCOCH_3 \qquad (148)$$

also showed in his early work that tertiary alcohols could be used in place of olefins[369], that the method was applicable to dinitriles[370], and that the reaction with hydrogen cyanide, which yields initially readily hydrolysed formyl compounds, provides a convenient route to tertiary alkylamines[369].

Mechanistically, the reaction is closely related to acid-catalysed hydration of nitriles, in that it is initiated by attack of an electrophilic species—in this case a carbonium ion formed by protonation of an olefin or dehydration of an alcohol—on the weakly basic cyanide nitrogen atom yielding a nitrilium salt which readily undergoes hydration on addition of water (equation 149). Nitrilium salts

$$R^1OH \xrightarrow{\text{H}^+} \overset{+}{R^1} + \big(N\!\equiv\!C\!-\!R^2 \longrightarrow [R^1\!-\!\overset{+}{N}\!\equiv\!C\!-\!R^2 \longleftrightarrow$$
$$\text{or alkene}$$
$$R^1\!-\!N\!=\!\overset{+}{C}\!-\!R^2] \xrightarrow{\text{H}_2\text{O}} R^1NHCOR^2 + H^+ \qquad (149)$$

cannot be isolated from the reaction mixture under normal Ritter conditions but the mechanism has been supported by a recent kinetic study of the reaction of isobutylene with acrylonitrile[371] and by investigations of the chemistry of nitrilium salts prepared in other ways[372,373].

The Ritter reaction is applicable to a very wide range of substrates*.

* See sections IV.A and IV.B for a discussion of the intervention of the Ritter mechanism in Beckmann and Schmidt rearrangements.

Alcohols and olefins which afford tertiary carbonium ions on treatment with strong acid react particularly readily giving high yields of amides, and other compounds capable of giving stabilized carbonium ions (e.g. benzyl alcohol)[374] are also suitable substrates. The reaction is applicable to unsaturated nitriles[375], to halohydrins and halo alkenes[376], to long-chain nitriles[368], to nitrilo esters[368], to cycloalkanols[377] and to compounds containing other reactive functions. Examples (150–156) illustrate the scope of the method. The reaction works well with heterocyclic alcohols[382], and has recently been used

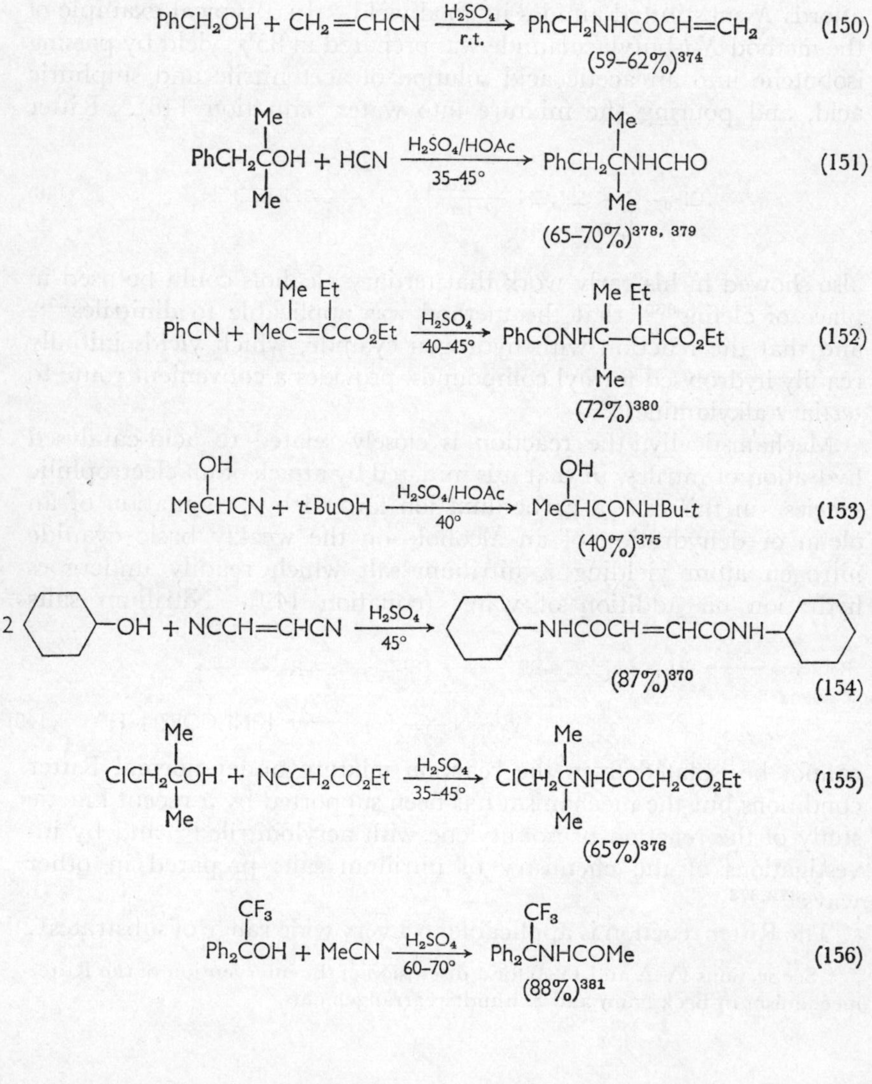

$$PhCH_2OH + CH_2=CHCN \xrightarrow[\text{r.t.}]{H_2SO_4} PhCH_2NHCOCH=CH_2 \qquad (150)$$
$$(59\text{–}62\%)^{374}$$

$$\underset{\underset{Me}{|}}{\overset{\overset{Me}{|}}{PhCH_2COH}} + HCN \xrightarrow[35\text{–}45°]{H_2SO_4/HOAc} \underset{\underset{Me}{|}}{\overset{\overset{Me}{|}}{PhCH_2CNHCHO}} \qquad (151)$$
$$(65\text{–}70\%)^{378, \ 379}$$

$$PhCN + \underset{\underset{Me}{|}}{\overset{\overset{Me \ Et}{| \ |}}{MeC=CCO_2Et}} \xrightarrow[40\text{–}45°]{H_2SO_4} \underset{\underset{Me}{|}}{\overset{\overset{Me \ Et}{| \ |}}{PhCONHC-CHCO_2Et}} \qquad (152)$$
$$(72\%)^{380}$$

$$\overset{\overset{OH}{|}}{MeCHCN} + t\text{-BuOH} \xrightarrow[40°]{H_2SO_4/HOAc} \overset{\overset{OH}{|}}{MeCHCONHBu\text{-}t} \qquad (153)$$
$$(40\%)^{375}$$

$$2 \langle \rangle\text{-OH} + NCCH=CHCN \xrightarrow[45°]{H_2SO_4} \langle \rangle\text{-NHCOCH=CHCONH-}\langle \rangle$$
$$(87\%)^{370} \qquad (154)$$

$$\underset{\underset{Me}{|}}{\overset{\overset{Me}{|}}{ClCH_2COH}} + NCCH_2CO_2Et \xrightarrow[35\text{–}45°]{H_2SO_4} \underset{\underset{Me}{|}}{\overset{\overset{Me}{|}}{ClCH_2CNHCOCH_2CO_2Et}} \qquad (155)$$
$$(65\%)^{376}$$

$$\overset{\overset{CF_3}{|}}{Ph_2COH} + MeCN \xrightarrow[60\text{–}70°]{H_2SO_4} \overset{\overset{CF_3}{|}}{Ph_2CNHCOMe} \qquad (156)$$
$$(88\%)^{381}$$

to prepare polyamides by condensation of dinitriles with a diester in sulphuric acid (equation 157)[383]. However, α-amino nitriles when

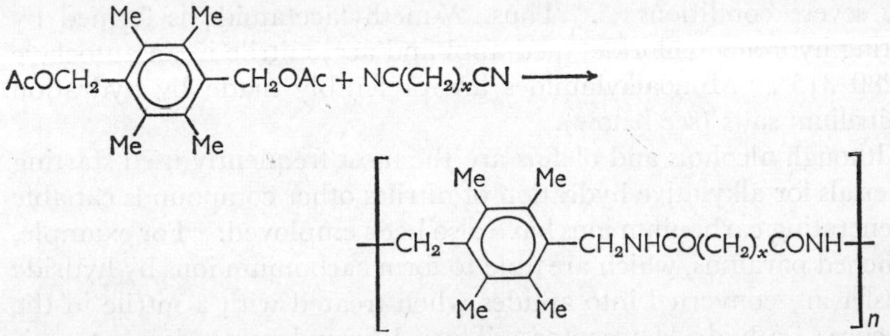

$$\text{AcOCH}_2\text{-}\underset{\underset{\text{Me}\quad\text{Me}}{}}{\overset{\overset{\text{Me}\quad\text{Me}}{}}{\bigcirc}}\text{-CH}_2\text{OAc} + \text{NC(CH}_2)_x\text{CN} \longrightarrow$$

$$\left[\text{-CH}_2\text{-}\underset{\underset{\text{Me}\quad\text{Me}}{}}{\overset{\overset{\text{Me}\quad\text{Me}}{}}{\bigcirc}}\text{-CH}_2\text{NHCO(CH}_2)_x\text{CONH-}\right]_n \quad (157)$$

treated with t-butanol in sulphuric acid undergo replacement of the amino group; the products obtained in fair yield are α-hydroxy amides (equation 158)[384]. The products obtained from the Ritter

$$\text{R}^1\text{CHCN} + t\text{-BuOH} \xrightarrow{\text{H}_2\text{SO}_4} \text{R}^1\text{CHCONHBu-}t \quad (158)$$
$$\underset{\text{NR}_2^2}{|} \qquad\qquad\qquad \underset{\text{OH}}{|}$$

reaction with halohydrins and with allyl halides are useful compounds for the preparation of oxazoline derivatives[376,385]. This and other applications of the Ritter reaction to heterocyclic synthesis have recently been reviewed[386].

Alcohols and olefins which afford secondary carbonium ions on treatment with acid, undertake the Ritter reaction less readily than tertiary compounds. For example cyclohexene when treated under normal Ritter conditions with acrylonitrile and sulphuric acid gives only cyclohexyl acetate but the required amide was formed in good yield when neat sulphuric acid was used (equation 159)[371]. The

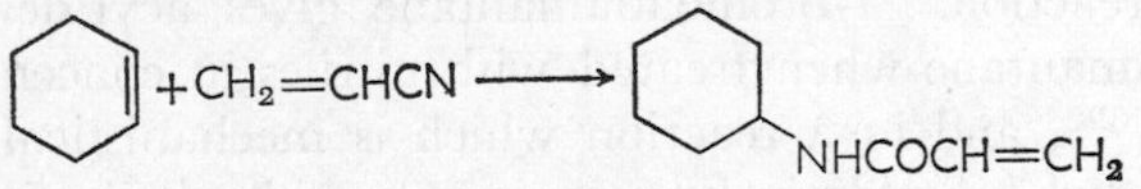

$$\bigcirc + \text{CH}_2\text{=CHCN} \longrightarrow \underset{\text{NHCOCH=CH}_2}{\bigcirc} \quad (159)$$

reactions of propylene with various nitriles have been studied. The yield of amide from different nitriles increases in the order MeCN < PhCN < CH_2=CHCN thus reflecting the effect of mesomeric release from the substituent on the basicity of the nitrogen atom. Formation of the acrylamide by treatment of cis-6-octadecenoic acid with acrylonitrile in sulphuric acid is another example of the application of the Ritter reaction to 1,2-disubstituted olefins[387]. Improved methods for the reaction with 1-alkenes have recently been described[388].

The Ritter reaction can be applied to primary alcohols only under very severe conditions[389]. Thus, N-methylacetamide is formed by heating hydrogen chloride, methanol and acetonitrile in an autoclave at 280–315°. Monoalkylamides are preferably made by hydration of nitrilium salts (see below).

Although alcohols and olefins are the most frequently used starting materials for alkylative hydration of nitriles other compounds capable of generating carbonium ions have also been employed. For example, branched paraffins, which are able to form carbonium ions by hydride transfer are converted into amides when treated with a nitrile in the presence of a hydride acceptor. Thus, 1-formylaminoadamantane is obtained in good yield when t-butanol and hydrogen cyanide are added to an emulsion of adamantane, hexane and sulphuric acid (equation 160)[390].

$$\text{(adamantane–H)} + Me_3C^+ \longrightarrow \text{(adamantyl}^+\text{)} \xrightarrow[\text{(2) H}_2\text{O}]{\text{(1) HCN}} \text{(adamantyl–NHCHO)} \qquad (160)$$

Tertiary carboxylic acids undergo the Ritter reaction when treated with a nitrile and concentrated sulphuric acid (equation 161). The reaction involves decarboxylation generating a carbonium ion, which then interacts with the nitrile in the usual way. Yields are usually good[391].

$$R_3^1CCO_2H + H^+ \xrightarrow{-H_2O, -CO} R_3^1C^+ \xrightarrow[\text{(2) H}_2\text{O}]{\text{(1) R}^2\text{CN}} R_3^1CNHCOR^2 \qquad (161)$$

Alkyl halides have been used for the preparation of amides by the Ritter reaction. 1-Bromoadamantane gives acyl derivatives of 1-aminoadamantane when treated with nitriles in concentrated sulphuric acid[392], and in a reaction which is mechanistically related, diphenylmethyl bromide in benzene reacts with nitriles in the presence of silver sulphate (equations 162)[393].

$$Ph_2CHBr + Ag^+ \longrightarrow Ph_2\overset{+}{C}H + AgBr \qquad (162a)$$

$$Ph_2\overset{+}{C}H \xrightarrow[\text{(2) H}_2\text{O}]{\text{(1) RCN}} Ph_2CHNHCOR \qquad (162b)$$

Reactions of halides with nitriles in the presence of Lewis acids were earlier investigated by Cannon and his coworkers who studied the effect of aluminium chloride on ethereal solutions of nitriles and cyclo-hexyl and cyclopentyl halides. They concluded that the method was less useful than the normal Ritter reaction[394]. Other methods for generating the required intermediate carbonium ions include de-composition of arenediazonium salts[395] and anodic oxidation of hydrocarbons[396].

Sometimes the carbonium ion initially generated under Ritter conditions undergoes rearrangement either by alkyl or hydride shifts, before it interacts with the nitrile[397]. The phenomenon assumes practical importance in the preparation of derivatives of 9-amino-decalin from the readily available decahydro-β-naphthol (equation 163)[398].

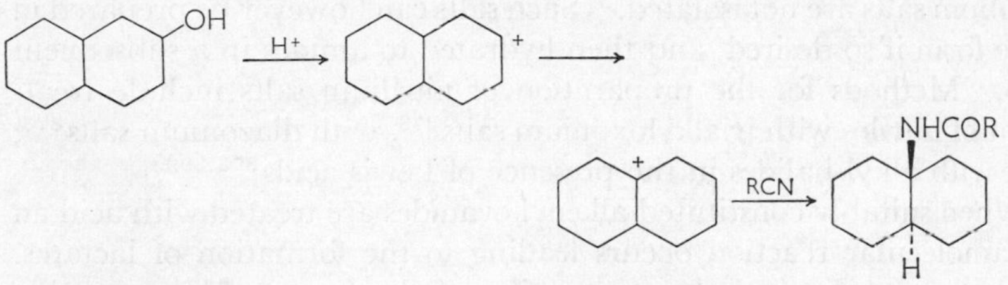

$$(163)$$

This transformation is obviously closely related mechanistically to the formation of decalin-9-carboxylic acid by treatment of 2-decalol with formic acid in sulphuric acid[399]. However, preferential formation of the kinetically favoured *trans*- or the thermodynamically favoured *cis*-carboxylic product, which can be achieved by appropriate regulation of the experimental conditions, appears not to have been accomplished in the case of amide formation.

A reaction which is mechanistically closely related to the Ritter reaction involves the synthesis of *N*-(2-chloroalkyl)amides by passing chlorine into a mixture of olefin and nitrile and pouring the resultant solution into water[400]. Bromoalkylamides are similarly made. The reaction proceeds via the halonium ion (equation 164).

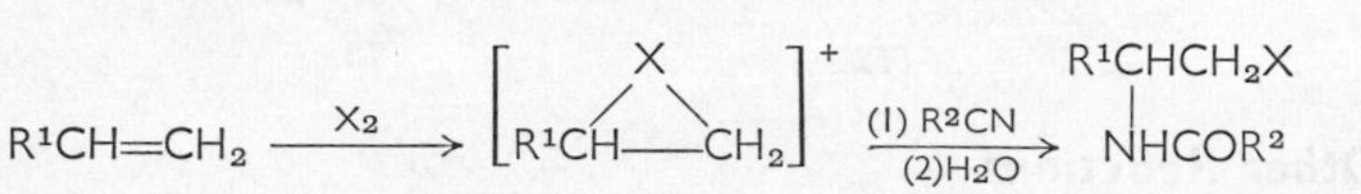

$$(164)$$

This method has recently been improved by Hassner and his co-workers who reasoned that competition of halide ion with nitrile for reaction with the intermediate halonium ion decreases the yield of product[401]. Accordingly, they conducted their reactions in the presence of silver perchlorate to remove halide ion and then obtained enhanced yields of β-haloalkylamides (equation 165). As expected on the basis of a reaction mechanism involving an intermediate cyclic bromonium ion the reaction affords *trans* products. Iodine

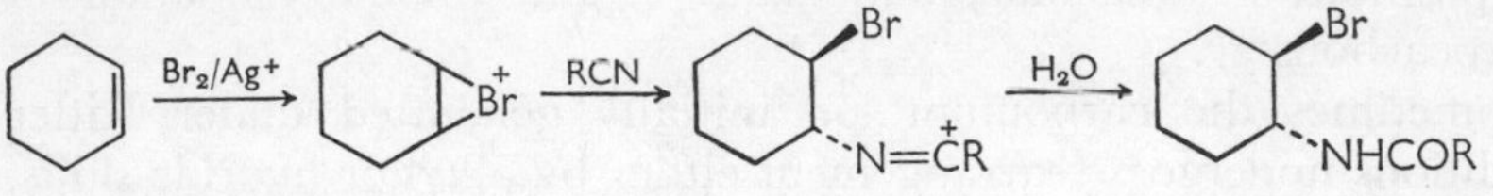

$$\tag{165}$$

monofluoride has been recommended for the preparation of N-(2-iodoalkyl)amides by a similar route, but even iodine alone is effective [402].

In the usual procedure for the Ritter reaction the intermediate nitrilium salts are not isolated. Such salts can however be prepared in pure form if so desired, and then hydrated to amides in a subsequent step. Methods for the preparation of nitrilium salts include treatment of nitriles with trialkyloxonium salts [373], with diazonium salts [373], and with alkyl halides in the presence of Lewis acids [373,399].

When suitably constituted alkenyl cyanides are treated with acid an intramolecular reaction occurs leading to the formation of lactams. Recent examples include preparation of the lactam **71** by heating 3-cyano-4-stilbazole (**70**) with polyphosphoric acid [403] and the formation of the lactam **73** from *cis-cis*-5-cyano-3-methylsorbic acid (**72**) [404].

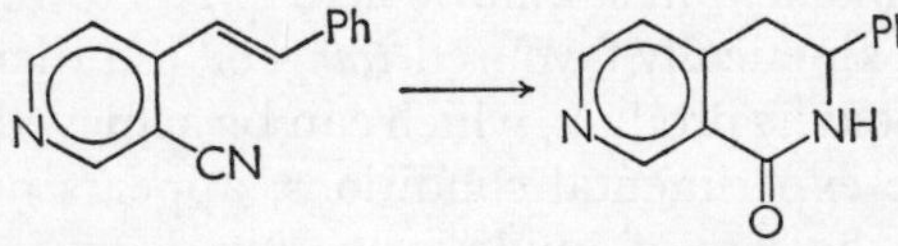

$$\tag{166}$$

(**70**)   (**71**)

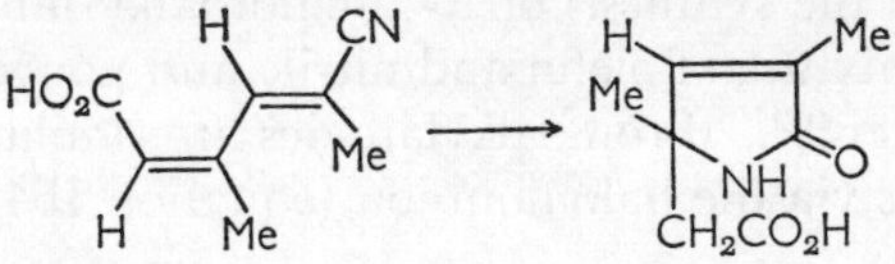

$$\tag{167}$$

(**72**)   (**73**)

C. Other Reactions

I. Acylation

Nitriles, when strongly heated with carboxylic acids or their anhydrides, yield amides and imides. The reaction is believed to involve a complex system of equilibria (168) [157,405].

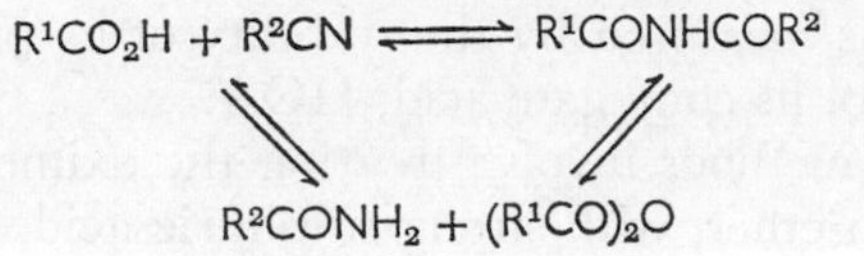

$$R^1CO_2H + R^2CN \rightleftharpoons R^1CONHCOR^2$$

$$R^2CONH_2 + (R^1CO)_2O$$

$$(168)$$

2. Aminolysis

Some nitriles react with aqueous ammonia or amines to yield amides, presumably via intermediate formation of amidines (equation 169). The reaction has been used for the preparation of nicotinamide

$$R^1CN + R^2R^3NH \rightleftharpoons R^1C\overset{\displaystyle NH}{\underset{\displaystyle NR^2R^3}{\Big\langle}} \xrightarrow{H_2O} R^1CONR^2R^3 \qquad (169)$$

from nicotinonitrile[406] and of a series of N,N'-disubstituted amides from succinonitrile and similar dinitriles[407].

IV. REARRANGEMENT REACTIONS

A. The Beckmann Reaction

In 1887 Beckmann described the first examples of the reaction which now bears his name; the rearrangement of oximes to amides when treated successively with an acid, or acylating agent, and water (equation 170). After many years of considerable speculation and

$$R_2C{=}NOH \xrightarrow[\text{(2) } H_2O]{\text{(1) } H^+} RCONHR \qquad (170)$$

controversy it is now agreed[408,409] that the reaction is initiated by conversion of the oximino hydroxyl into a suitable leaving group (OA) by protonation, coordination with a Lewis acid, or acylation. Heterolysis of the N—O bond with synchronous migration of the *anti* substituent via a quasi three-membered transition state affords an iminium salt, hydration of which gives the product (equation 171). The rapidity with which the rearrangement occurs is expected to depend on the migratory aptitude of R^1 and on the efficiency of OA as a

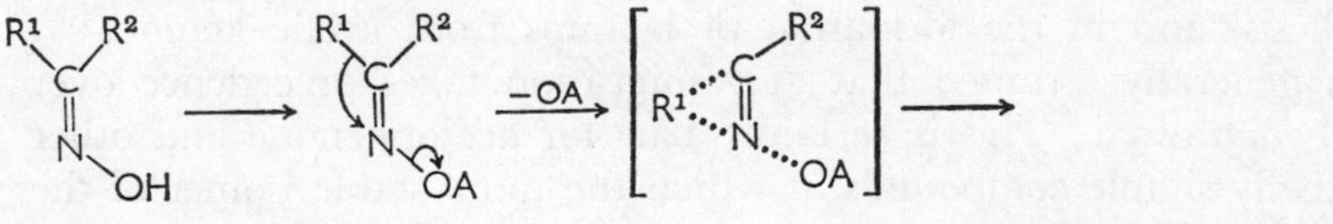

$$R^2C{\equiv}\overset{+}{N}R^1 \xrightarrow{H_2O} R^2CONHR^1 \qquad (171)$$

leaving group as measured by the stability of its product ion, OA^-, or the strength of its conjugate acid, HOA.

The classical methods involve treating the oxime with phosphorus pentachloride in ether, with strong sulphuric acid, or with hydrogen chloride in acetic acid–acetic anhydride. Newer procedures include the use of polyphosphoric acid[410-412], p-acetamidobenzenesulphonyl chloride[413], p-toluenesulphonyl chloride[414] and similar arene-sulphonyl halides, phosphorus oxychloride[415], boron trifluoride[416], iodine pentafluoride[417], thionyl chloride[418], trifluoroacetic anhydride[419] and formic acid[420].

The migratory aptitude of R^1 should be enhanced by electron-donating substituents. Kinetic measurements on p-substituted aceto-phenone oxime picrates reveal a linear Hammett relationship with a negative regression constant[421]. An earlier claim[422] that the migration step shows a reverse ^{14}C isotope effect has been discounted[423].

The Beckmann rearrangement normally proceeds with retention of configuration at the point of attachment of the migrating group[424], and there is usually a *trans* relationship between the leaving group and R^1. Nevertheless there are numerous reports of the formation of mixtures of amides from stereochemically pure oximes. Undoubtedly, in many instances such mixtures arise because *syn–anti* isomerization of the oxime occurs more rapidly than its rearrangement. Purely aliphatic ketoximes undergo isomerization particularly readily, especially in the presence of protic acids. Phosphorous pentachloride in ether is a much better reagent for the rearrangement of such compounds. Passage of oxime p-toluenesulphonates over alumina has been claimed to give high yields of amides with negligible prior isomerization of the starting material[425].

The iminium ion ($RC{\equiv}\overset{+}{N}R$) produced by migration may be intercepted by nucleophiles other than water. Variations on the normal Beckmann procedure include the formation of imidate esters by reactions with alcohols, of imidoyl azides (which cyclize to tetrazoles) with azide ion, and of α-amino nitriles with cyanide ion.

The Beckmann reaction finds its most important applications in the rearrangement of alkyl aryl ketones, obtainable by Friedel–Crafts synthesis, and in the formation of lactams from cyclic ketones[409]. It is generally assumed that aryl migration takes precedence over alkyl migration. This is certainly true for acetophenone and other relatively simple compounds in which the more stable isomer of the oxime has the bulky aryl group in the *anti* configuration, but substrates containing branched-alkyl substituents may show migration of the alkyl group. Thus pivalophenone oxime, when treated with non-protic catalysts, gives *N-t*-butylbenzamide (**75**) in good yield thus indicating the *anti* arrangement of the OH and *t*-butyl substituents (equation 172)[426]. However, with hydrogen chloride in acetic acid the anilide **76** is formed; it appears that under these conditions an equilibrium mixture of the two isomeric oximes (**74a** ⇌ **74b**) is rapidly generated and the overall course of the reaction then reflects the greater migratory aptitude of the phenyl group.

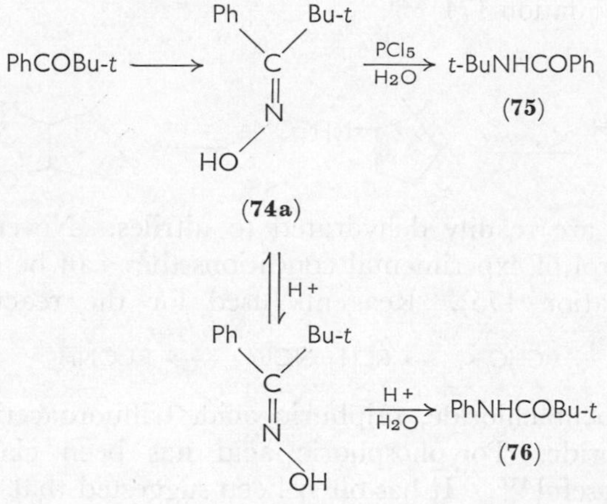

$$PhCOBu\text{-}t \longrightarrow \underset{(\textbf{74a})}{\underset{HO}{\overset{Ph\quad Bu\text{-}t}{\underset{\parallel}{\underset{N}{C}}}}} \xrightarrow[\text{H}_2\text{O}]{\text{PCl}_5} \underset{(\textbf{75})}{t\text{-BuNHCOPh}}$$

$$\Big\Updownarrow \text{H}^+$$

$$\underset{(\textbf{74b})}{\underset{\dot{O}H}{\overset{Ph\quad Bu\text{-}t}{\underset{\parallel}{\underset{N}{C}}}}} \xrightarrow[\text{H}_2\text{O}]{\text{H}^+} \underset{(\textbf{76})}{PhNHCOBu\text{-}t} \tag{172}$$

Alkyl aryl ketones containing a wide range of other functional groups may be safely subjected to the Beckmann procedure[409]. A recent example[427] consists of the rearrangement of oximes derived from aryl-substituted Mannich bases (**77**) and vinyl ketones (**78**).

$$\underset{(\textbf{77})}{ArCOCH_2CH_2NMe_2} \qquad \underset{(\textbf{78})}{ArCOCH{=}CH_2}$$

A novel variation of the Beckmann method for the preparation of anilides involves treatment of an arene with a hydroxamic acid in polyphosphoric acid[428]. The reaction is thought to involve intermediate formation of a ketoxime which then undergoes rearrangement in the usual way (equation 173). Suitably constituted aryl-sub-

$$ArH + RCONHOH \longrightarrow Ar\overset{\overset{\displaystyle NOH}{\parallel}}{C}R \longrightarrow ArNHCOR \tag{173}$$

stituted hydroxamic acids undergo intramolecular reaction yielding lactams[428].

When applied to dialkyl ketones the Beckmann reaction usually affords mixtures of amides in which the major constituent is that compound formed by migration of the bulkier group. A variety of catalysts have been used. Trifluoroacetic anhydride has been recommended for rearrangement of water-soluble compounds of low molecular weight[419], whilst phosphorus oxychloride–pyridine gives exceptionally good yields in side-chain degradation of steroids[415]. α,β-Unsaturated ketoximes give the expected products under mild conditions but vigorous treatment causes cyclization to 2-isooxazoline derivatives (equation 174)[429].

$$\text{(174)}$$

Aldoximes are readily dehydrated to nitriles. Nevertheless with suitable control of experimental conditions they can be converted to amides (equation 175). Reagents used for the reaction include

$$RCHO \longrightarrow RCH{=}NOH \longrightarrow RCONH_2 \qquad (175)$$

phosphorus pentachloride, sulphuric acid, trifluoroacetic acid, and boron trifluoride. Polyphosphoric acid has been claimed to be particularly useful[410]. It has often been suggested that apparent rearrangement of aldoximes involves formation of nitriles followed by acid-catalysed hydration. Although such a mechanism is conceivable under the acidic conditions often employed, it seems unlikely to apply to a new and useful method involving treatment of aldoximes with nickel acetate in toluene[430].

Cyclic ketoximes rearrange to lactams under Beckmann conditions. The reaction is applicable to rings of all sizes and to polycyclic and heterocyclic systems; rearrangement of cyclohexanone oxime to caprolactam has been extensively studied because of its commercial importance in polyamide manufacture. Recently it has found extensive use in the synthesis of aza steroids[413,431].

The scope and limitations of the Beckmann method for lactam preparation are well covered in Donaruma and Heldt's review[409]. Interesting recent examples of its application are given in equations (176–178)[432,433].

The dioxime of cyclodecane-1,6-dione (79) behaves normally when treated with sulphur trioxide in sulphur dioxide, and gives the expected lactam (80), but if the reaction is conducted with thionyl chloride a

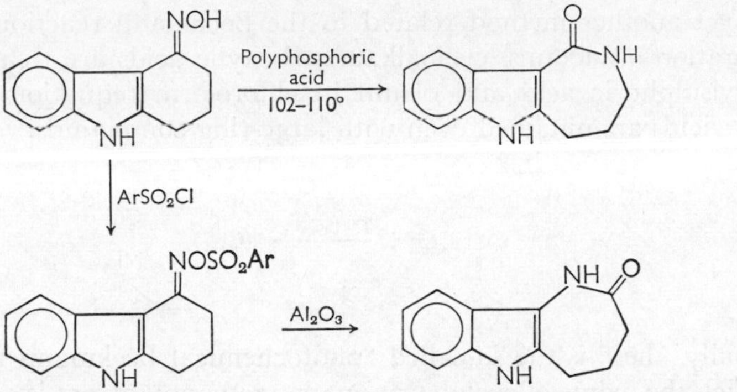

(176)

(177)

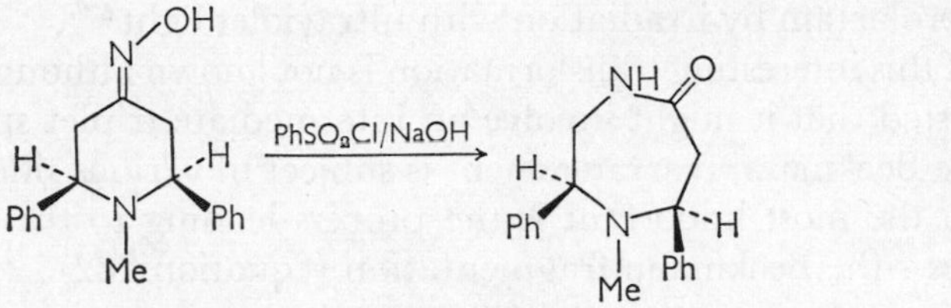

(178)

curious transannular rearrangement occurs giving rise to valero-lactam (equation 179b)[434].

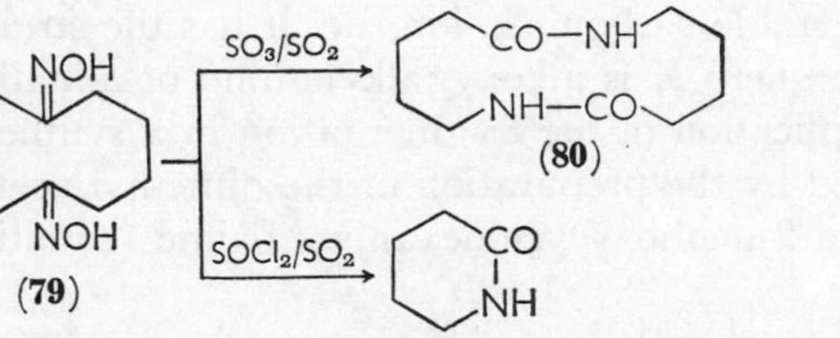

(179a)

(80)

(79)

(179b)

An interesting variation in the preparation of lactams from cyclic ketoximes involves treatment of the oxime with triphenylphosphine and halogen in benzene[435]. The reaction mechanism is believed to involve intermediate formation of halonitroso alkanes and phosphonium salts (equation 180).

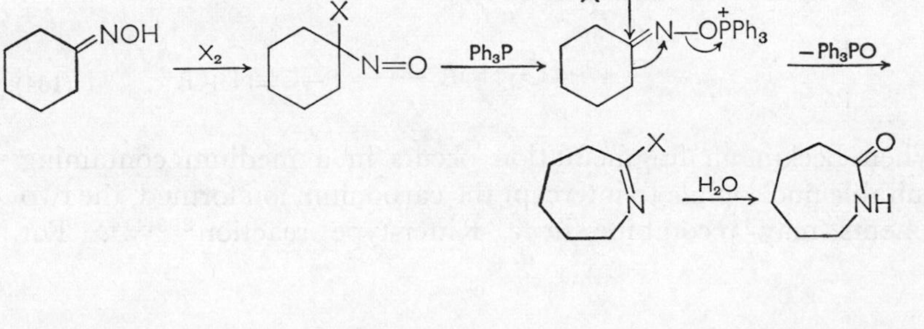

(180)

In yet another method related to the Beckmann reaction for the preparation of lactams, cycloalkanecarboxylic acids are treated with nitrosylsulphuric acid and oleum in chloroform (equation 181)[436]. Good yields are obtained even with large-ring compounds.

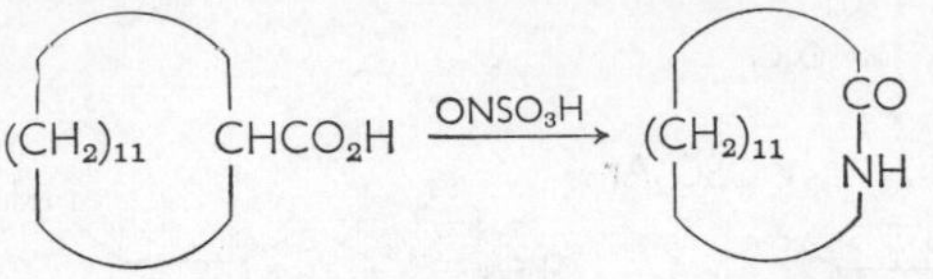

$$(CH_2)_{11} \quad CHCO_2H \xrightarrow{ONSO_3H} (CH_2)_{11} \quad \overset{CO}{\underset{NH}{|}} \tag{181}$$

Finally, there is the so-called 'photochemical Beckmann reaction' in which the oxime of cyclohexanone in methanol solvent is converted to caprolactam by irradiation with ultraviolet light[437]. The mechanism of this interesting transformation is not known although it has been suggested that it might involve an intermediate triplet species.

The Beckmann rearrangement is subject to various side-reactions of which the most important is the process leading to the formation of nitriles—the Beckmann fragmentation (equation 182).

$$R_2C=NOA \longrightarrow R^+ + RC\equiv N + OA^- \tag{182}$$

In general, this type of disruption of the oxime is most likely to occur when one of the groups is so constituted as to readily form a relatively stable carbonium ion, i.e. it has the structure —CHAr$_2$ or —CR$_2$X, where X is alkoxy, alkylamino or alkylthio. An example of the application of the fragmentation in a synthetically useful way is provided by the preparation of the dimethyl acetal of 5-cyanopentanal from 2-methoxycyclohexanone oxime (equation 183)[438]. The

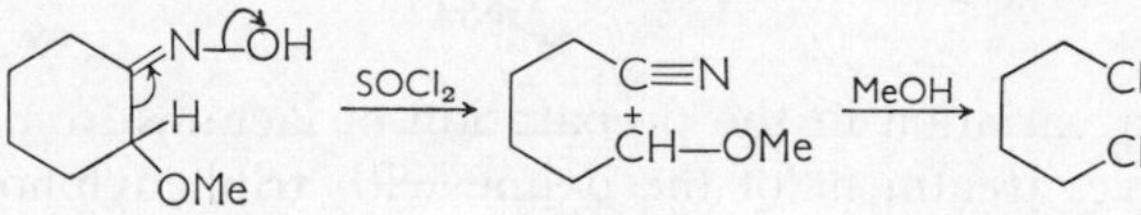

$$\xrightarrow{SOCl_2} \qquad \xrightarrow{MeOH} \qquad \tag{183}$$

mechanism and scope of the reaction have recently come under close scrutiny[439] and there is now some evidence that rearrangement often proceeds in the normal way and that the iminium ion produced then undergoes heterolysis (equation 184).

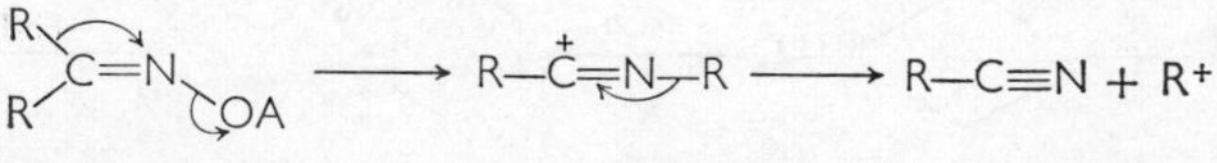

$$\underset{R'}{\overset{R}{>}}C=N\underset{OA}{\overset{}{\diagdown}} \longrightarrow R\overset{+}{-}C\equiv\overset{+}{N}\text{—}R \longrightarrow R-C\equiv N + R^+ \tag{184}$$

When Beckmann fragmentation occurs in a medium containing no suitable nucleophile to intercept the carbonium ion formed, the two fragments may recombine in a Ritter-type reaction[398,414]. For example the oxime of 9-acetyl-*cis*-decalin (**81**) when treated with *p*-toluenesulphonyl chloride in pyridine undergoes the normal Beckmann rearrangement with retention of configuration, but in sulphuric or polyphosphoric acid it affords 9-acetylamino-*trans*-decalin (**82**) whose formation is formulated as proceeding through the more stable *trans*-iminium ion (equation 185).

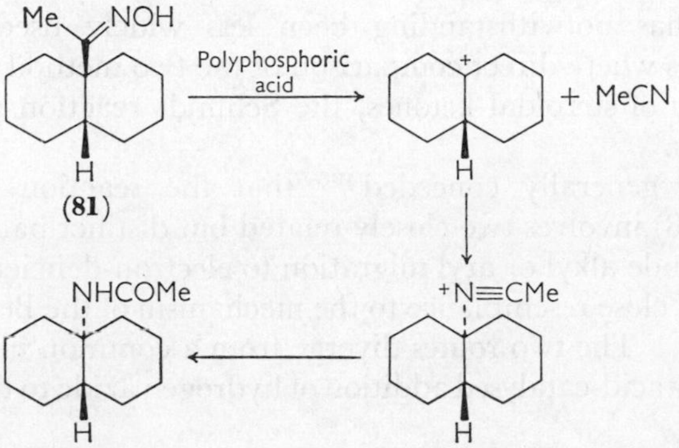

$$(185)$$

As expected, when a mixture of two oximes is subjected to treatment with polyphosphoric acid a mixture of all four possible amides is obtained thus indicating the intermediacy of 'free' carbonium ions[398].

Other interesting departures from the normal course of the Beckmann reaction include intramolecular aromatic substitution by the intermediate iminium ion[440] or insertion into an adjacent C—H bond (equation 186)[441].

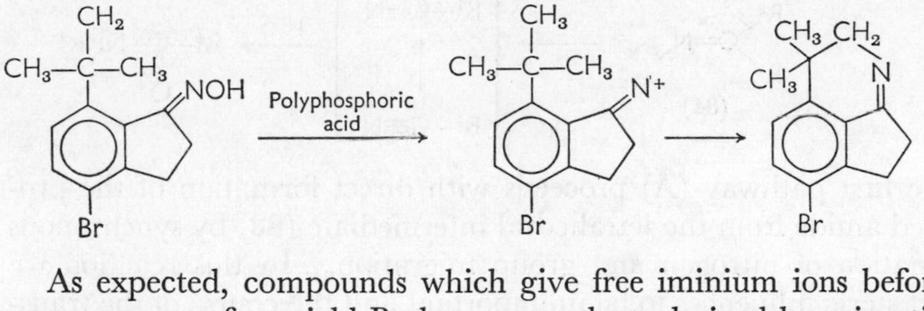

$$(186)$$

As expected, compounds which give free iminium ions before rearrangement often yield Beckmann products derived by migration of the *syn* substituent[441].

B. The Schmidt Reaction

The formation of amides by treatment of ketones, and occasionally aldehydes or carboxylic acids, with hydrazoic acid is one of the many variations of the Schmidt reaction (equation 187).   Like the Beckmann

$$R^1COR^2 + HN_3 \longrightarrow R^1CONHR^2 + N_2 \qquad (187)$$

rearrangement it finds its widest synthetic application in the preparation of amides from alkyl aryl ketones, and of lactams from cyclic ketones.   By comparison with the Beckmann reaction it offers the convenience of a one-step procedure leading directly from ketone to amide.   It has notwithstanding been less widely used, although in many cases where direct comparison of the two methods is possible, e.g. reactions of steroidal ketones, the Schmidt reaction gives better overall yields.

It is now generally conceded[408] that the reaction mechanism (equation 188) involves two closely related but distinct pathways both of which include alkyl or aryl migration to electron-deficient nitrogen, and so bear a close resemblance to the mechanism of the Beckmann rearrangement.   The two routes diverge from a common intermediate, **83**, formed by acid-catalysed addition of hydrogen azide to the carbonyl group.

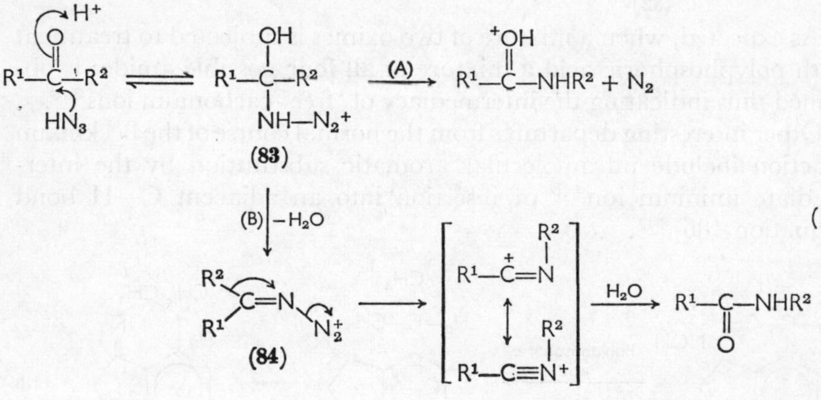

The first pathway (A) proceeds with direct formation of the protonated amide from the tetrahedral intermediate (**83**) by synchronous elimination of nitrogen and group migration.   In this reaction we expect steric influences to be unimportant and the course of the transformation to be controlled by the relative migratory aptitudes of R^1 and R^2.

The second pathway (B) involves an intermediate, **84**, analogous to that in the Beckmann rearrangement, which will lose nitrogen with synchronous migration of the group in the *anti* position.   In an unsymmetrical ketone the operation of this mechanism should generally lead to that product formed by migration of the bulkier substituent. However, under some circumstances dehydration of the tetrahedral intermediate is rapidly reversible, in which case equilibrium is established between the two configurations of **84** and the eventual course of the reaction may then be determined as in mechanism (A) by the relative migratory aptitudes of the two substituents.

As in the Beckmann rearrangement intermediate iminium ions may be intercepted by suitable nucleophiles. Thus the Schmidt reaction sometimes affords imidate esters when conducted in the presence of alcohols (equation 189) or tetrazoles when an excess of hydrazoic acid is employed (equation 190).

$$R^1C \equiv \overset{+}{N}R^1 + R^2OH \longrightarrow R^1\overset{\overset{\displaystyle OR^2}{|}}{C} = NR^1 + H^+ \tag{189}$$

$$RC \equiv \overset{+}{N}R + HN_3 \longrightarrow \quad \quad \longrightarrow \quad \quad + H^+ \tag{190}$$

Both pathways (A) and (B) are initiated by acid-catalysed addition of hydrazoic acid to the carbonyl group and the overall rate of reaction, therefore, should reflect the extent of the equilibrium involving **83**. In fact the observed order of reactivity is as expected: dialkyl ketones > alkyl aryl ketones > diaryl ketones. A practical consequence is that dialkyl ketones undergo the Schmidt reaction in concentrated hydrochloric acid whereas diaryl ketones are inert under these conditions[442].

In view of Wolff's review[443] of the scope and limitations of the method and Smith's more recent discussion[408] it will only be necessary here to touch briefly on its more important aspects.

Alkyl aryl ketones are readily converted into amides under Schmidt conditions. Hydrazoic acid in sulphuric acid has been widely used as a reagent but Conley's results suggest that sodium azide in polyphosphoric acid is preferable (equation 191)[444]:

$$PhCOMe + NaN_3 \xrightarrow{\text{Polyphosphoric acid}} \underset{(98\%)}{PhNHCOMe} \tag{191}$$

This reaction provides an illustration of the general rule that aryl methyl ketones yield products formed by migration of the aryl group. Since acetyl arenes are often readily obtainable they sometimes provide the most convenient point of access to arylamines. Examples include the preparation of 2-aminophenanthrene (equation 192)[445].

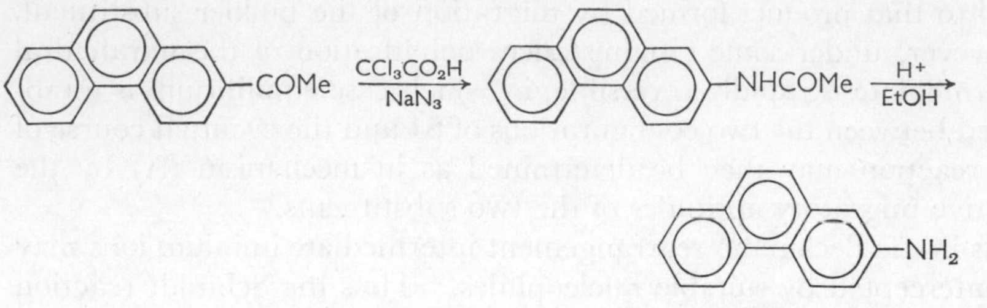

$$\text{(88\%)} \qquad (192)$$

11-Acetylaminofluoroanthene may be similarly obtained from the appropriate acetyl compound[446].

When the reaction is applied to alkyl aryl ketones containing groups more bulky than methyl, mixtures of amides are likely to be produced. Thus, the amount of product formed by alkyl migration in PhCOR increases in the order Me < Et < i-Pr; the last compound affords almost equal quantities of isobutyranilide and N-isopropylbenzamide[447].

Diaryl ketones are converted into the appropriate amides by treatment with hydrazoic acid in sulphuric acid or polyphosphoric acid. Substituents at *meta* and *para* positions appear to have little effect on the course of the reaction and unsymmetrical benzophenones usually give a mixture of products. *o*-Substituted benzophenones give anomalous results in that the products obtained are often those formed by migration of the less bulky substituent. The behaviour of such compounds has been rationalized in terms of participation of the substituent in iminium-ion stabilization through ring formation, and of the effect of substituent enhancement of conjugation on the configuration of the intermediate iminodiazonium ion (**84**)[448].

Dialkyl ketones undergo the Schmidt reaction with facility. As expected, unsymmetrical ketones yield as major products the compounds arising from migration of the more bulky substituent but the reaction is rarely completely specific. Since dialkyl ketones rearrange under relatively mild conditions it is possible to apply the reaction to substrates containing other reactive functions. For example, substituted acetoacetic esters are converted by treatment with hydrazoic acid into α-acetylamino esters. The yields are excellent and the reaction provides a convenient route to α-amino acids (equation 193)[449].

$$CH_3COCR^1R^2CO_2Et \xrightarrow{HN_3} CH_3CONHCR^1R^2CO_2Et \xrightarrow[H_2O]{H_2SO_4} NH_2CR^1R^2CO_2H$$

$$(193)$$

Similar reactions have been applied to γ- and δ-keto esters[450].

An excellent illustration of the specificity and selectivity of the Schmidt reaction when applied to a complex substrate is the formation in 90% yield of the acetylamino compound **86** by treatment of warfarin (**85**) in chloroform with sodium azide and sulphuric acid[451].

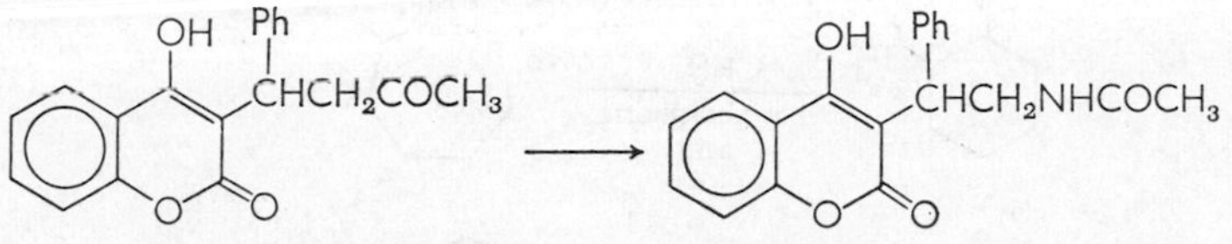

(**85**) (**86**) (194)

Some α,β-unsaturated ketones are anomalous in that they do not yield amides when treated with hydrazoic acid, even under very mild conditions. The products are usually α-dicarbonyl compounds which are thought to arise via Michael addition of azide to the conjugated system (equation 195)[452]. Benzalacetone, however, reacts normally

$$CH_2{=}CHCOCH_3 \xrightarrow{HN_3} \underset{\displaystyle N_3}{CH_2}CH_2COCH_3 \longrightarrow CH_3\underset{\displaystyle \| \atop NH}{C}COCH_3 \longrightarrow CH_3COCOCH_3 \qquad (195)$$

yielding *N*-methylcinnanamide[444].

The Schmidt reaction has been widely employed for the preparation of lactams from cyclic ketones. Yields of products are frequently very good and the reaction is applicable both to small, e.g. cyclobutanone[453], and to large ring compounds, e.g. cyclohexadecanone[454]. The reaction with fluorenone affords a convenient preparation of phenanthridone[444] but asymmetrically substituted fluorenones often afford mixtures of lactams[455]. Benzolactams are formed by treatment of tetralones and homologous compounds with hydrazoic acid in acetic acid[456]. Selective reaction of cyclic ketones containing other reactive functions can be achieved (equations 196–198)[444,457,458].

The Schmidt reaction of *cis*-8-methylhydrindan-1-one, like that of many other cyclic ketones, yields a mixture of both possible isomeric amides together with tetrazoles and fission products (equation 199)[459]. The course of the reaction has been rationalized on the basis of concurrent operation of both possible mechanistic pathways described above (equation 188). Reactions of amino ketones of the general type **87** have been studied by Schmid and his coworkers[460] who have shown that electrostatic repulsion between the protonated nitrogen function and the positive centre of the iminodiazonium intermediate controls the direction of migration. Similar effects are probably reponsible

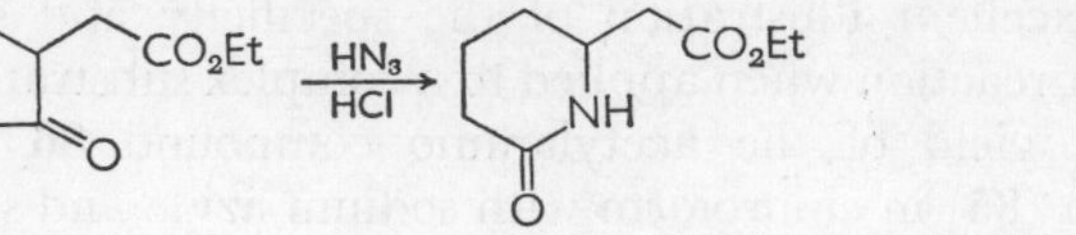

$$\text{(196)}$$

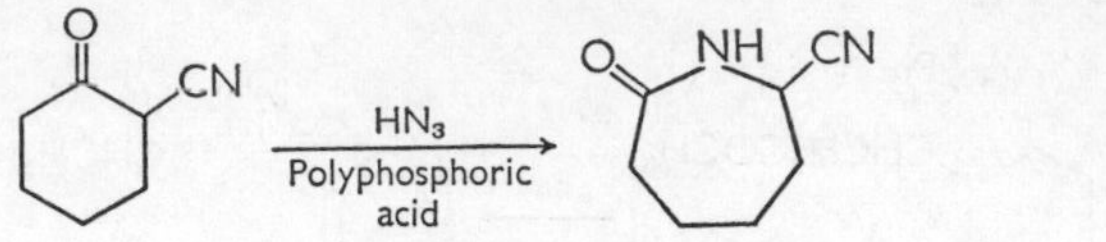

$$\text{(197)}$$

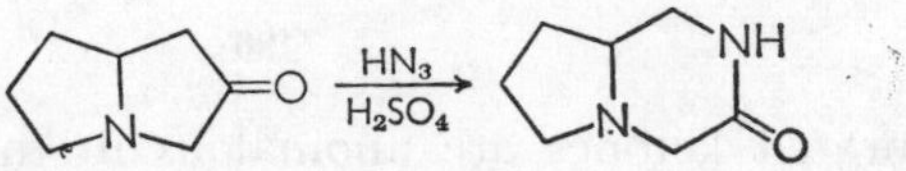

$$\text{(198)}$$

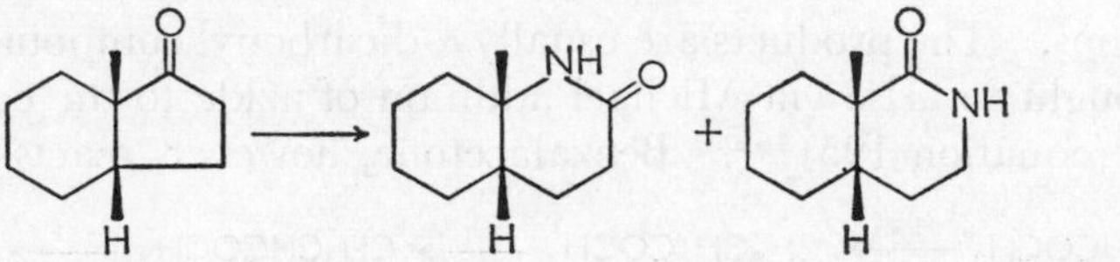

$$\text{(199)}$$

for the anomolous rearrangement of the phenothiazone derivative **88** (equation 200)[461].

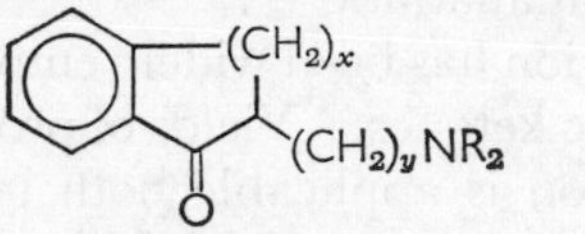

(87)

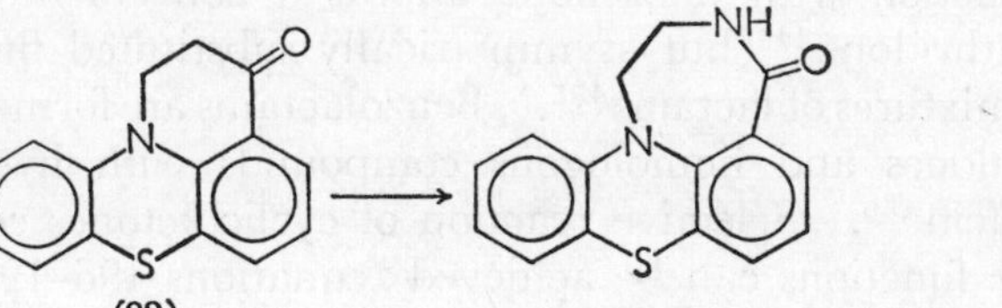

(88)

$$\text{(200)}$$

Polycylic ketones often give good yields of lactams when subjected to the Schmidt reaction, and the method has been widely used for the introduction of the aza group into steroids and triterpenes[411,462]. Uyeo and his colleagues[463] have made extensive use of the Schmidt reaction for the elaboration of seven-membered lactam rings in alkaloid synthesis. Among their many interesting results was the observation that compound **89** ($R = H$) gives predominantly **90** under Schmidt conditions whereas **89** ($R = OMe$) gives mainly **91** (equation 201).

Reactions of hydrazoic acid with quinones are complex and can

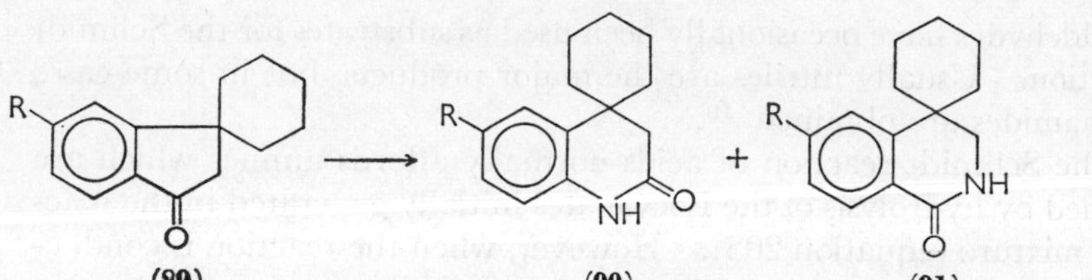

$$
\text{(89)} \longrightarrow \text{(90)} + \text{(91)} \tag{201}
$$

give rise either to ring-expanded or ring-contracted products. Thus anthraquinone, and its 1- and 2-amino derivatives undergo the normal reaction at one carbonyl group (equation 202) [464], but 2-hydroxy-1,4-

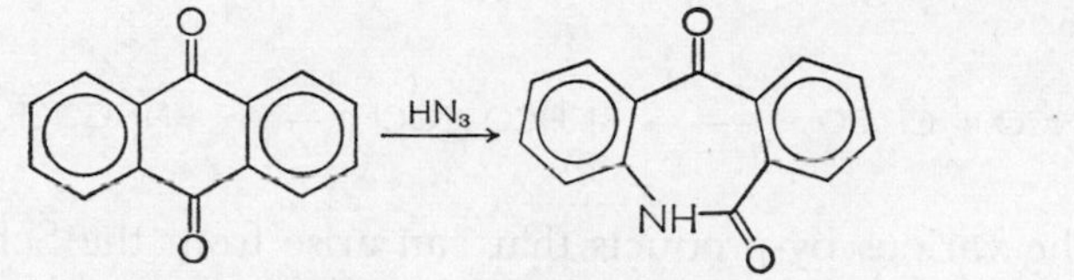

$$
\xrightarrow{\text{HN}_3} \tag{202}
$$

naphthaquinone affords the ring-contracted compound (equation 203). The mechanism of the reaction has been discussed [465].

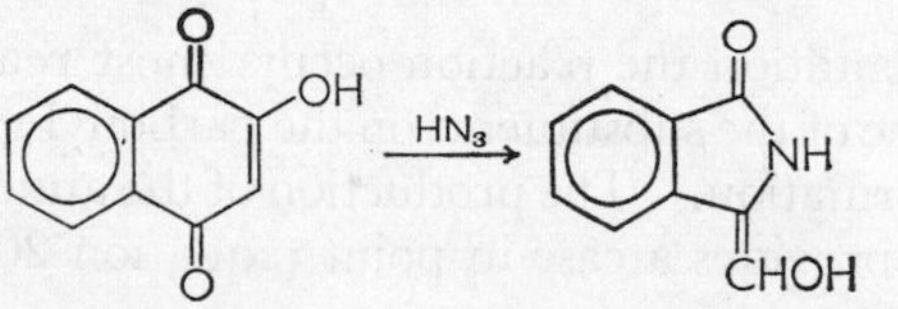

$$
\xrightarrow{\text{HN}_3} \tag{203}
$$

The Schmidt reaction with substituted benzoquinones affords a useful route to azepine derivatives not easily accessible by other methods (equation 204) [466,467]. Initially the structures of the products were

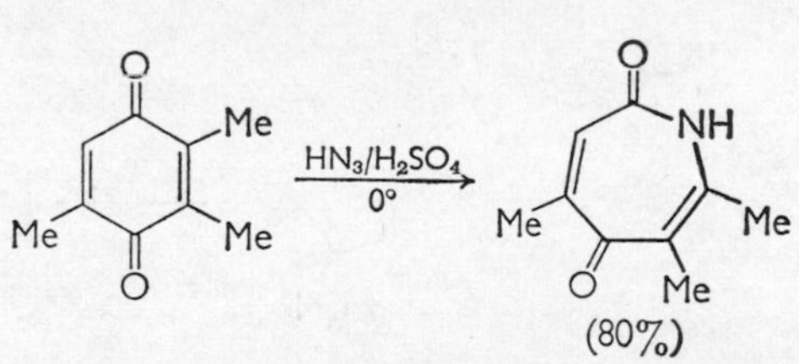

$$
\xrightarrow[0°]{\text{HN}_3/\text{H}_2\text{SO}_4} \tag{204}
$$

(80%)

incorrectly assigned [466], but on reexamination [467] it became clear that in accord with general mechanistic principles their formation was primarily under steric control and involved preferential attack at the less hindered carbonyl function followed by migration of the more bulky substituent.

Aldehydes have occasionally been used as substrates for the Schmidt reaction. Usually nitriles are the major products, but in some cases formamides are obtained[449].

The Schmidt reaction of acids normally affords amines which are formed by hydrolysis of the isocyanates initially generated in the reaction mixture (equation 205). However, when the reaction is conduc-

$$RCO_2H \xrightarrow{HN_3} RCONHN_2^+ \xrightarrow[-H^+]{-N_2} RNCO \xrightarrow{H_2O} RNH_2 + CO_2 \qquad (205)$$

ted with aromatic acids in a cold mixture of trifluoroacetic acid and its anhydride, isocyanates are obtained in good yield[468], whilst the same reactants at higher temperature give trifluoroacetamides (equation 206)[469].

$$RNCO + CF_3CO_2H \longrightarrow RNHCO_2COCF_3 \longrightarrow RNHCOCF_3 + CO_2 \qquad (206)$$

Of the various by-products that can arise from the Schmidt reaction undoubtedly the most important are those formed by fragmentation of the intermediate iminodiazonium ion (equation 207). As with

$$R_2C=N-N_2^+ \longrightarrow RCN + N_2 + R^+ \qquad (207)$$

Beckmann fragmentation the reaction occurs most readily whenever the structure of one of the substituents on the carbonyl group is such as to favour cation formation. The production of the amide-acid **92** from camphorquinone provides a case in point (equation 208)[470].

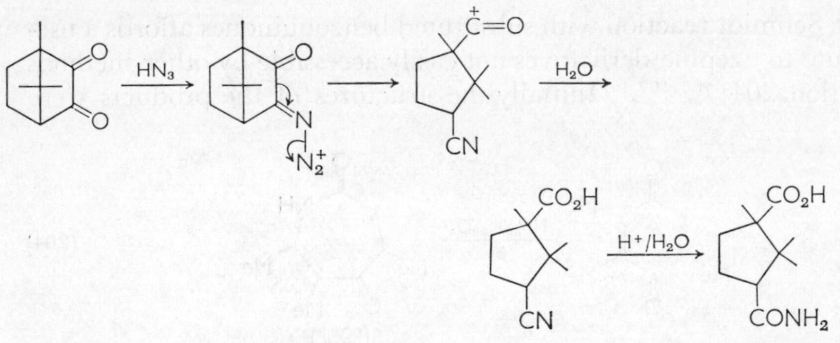

$$(92)$$

Recently it has become apparent that the formation of some of the by-products obtained from the Schmidt reaction involves the intermediacy of free iminium ions (equation 209)[441]. The evidence in

$$R_2C=N-N_2^+ \longrightarrow R_2C=N^+ + N_2 \qquad (209)$$

support of such intermediates has been briefly outlined in the discussion of the Beckmann reaction.   From studies of this nature it is clear that the principle of migration of the *anti* substituent in oximes and iminodiazonium ions must be used with caution in assigning the structures of products and the stereochemistry of starting materials.

C. The Willgerodt Reaction

The Willgerodt reaction[471] in its original form involves formation of an amide by heating a ketone, usually an alkyl aryl ketone, with ammonium polysulphide in a sealed tube.   The characteristic features of the reaction, namely reduction of the carbonyl group to methylene, and oxidation of the terminal methyl group in the alkyl chain, are well illustrated in the preparation of γ-phenylbutyramide from butyrophenone (equation 210)[472].   A number of experimental variations

$$PhCOCH_2CH_2CH_3 \xrightarrow{(NH_4)_2S_x} PhCH_2CH_2CH_2CONH_2 \qquad (210)$$

have been devised for the reaction which finds its most important application in the preparation of ω-aryl-alkylcarboxamides from alkyl aryl ketones readily obtained by the Friedel–Crafts reaction.

Despite considerable investigation[473] the mechanism of the reaction has not yet been completely elucidated.   It is known that skeletal rearrangement does not occur[474], that the reaction proceeds via loss of hydrogen and its subsequent partial replacement at the position β to the carbonyl group[475], that apparent migration of the carbonyl group cannot proceed beyond a quaternary position, and that such migration takes place preferentially along the shorter of the two alkyl chains[476].   Possible intermediates in the reaction include imines and α-mercapto ketones and other sulphur-containing compounds[473]. The suggestion that migration proceeds via acetylenic or olefinic bonds[477,478] was not supported by a study of some unsaturated ketones which appeared to react without change in the position of unsaturation[479].   However, the question has recently been reopened by the demonstration that the structures of some of the products from the earlier work were incorrectly assigned[480].

The scope and limitations of the Willgerodt reaction have been discussed[471].   Several variations of the original experimental procedure are now available.   For example it is often advantageous to add an organic solvent such as dioxan[481] or pyridine[482] to the aqueous reagent which is usually made by adding 10% by weight of sulphur to a solution of hydrogen sulphide in concentrated ammonia.   Thus,

6+c.o.a.

δ-phenylvaleramide is obtained in 29% yield[482] by heating butyl phenyl ketone with ammonium polysulphide, sulphur, aqueous ammonia and pyridine at 165°, and 3-acetylpyrene is converted into 3-pyrenylacetamide in 92% yield by heating at 160° with ammonium polysulphide in dioxan[483].

The Willgerodt reaction is applicable to a wide range of substituted alkyl aryl ketones. Illustrative examples include preparations of m-methoxyphenylacetamide (equation 211)[482], p-hydroxyphenylacetamide (equation 212)[484], and 2-dibenzofurylacetamide (equation 213)[485].

$$m\text{-MeOC}_6\text{H}_4\text{COCH}_3 \xrightarrow[200°]{\text{NH}_3/\text{S}} m\text{-MeOC}_6\text{H}_4\text{CH}_2\text{CONH}_2 \quad\quad (211)$$
$$(53\%)$$

$$p\text{-HOC}_6\text{H}_4\text{COCH}_3 \xrightarrow{(\text{NH}_4)_2\text{S}_x} p\text{-HOC}_6\text{H}_4\text{CH}_2\text{CONH}_2 \quad\quad (212)$$
$$(67\%)$$

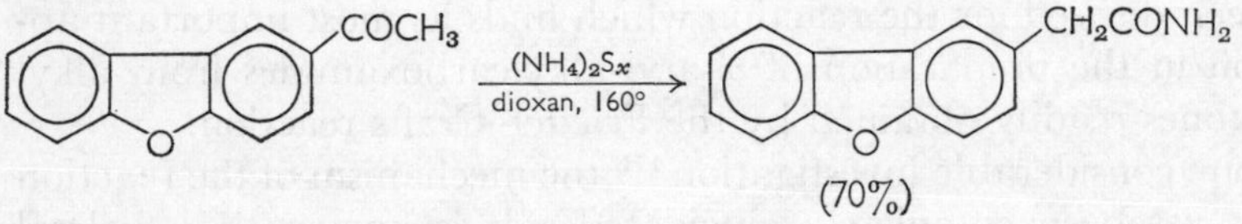

$$(213)$$
$$(70\%)$$

The Kindler modification of the Willgerodt procedure in which a ketone is heated with sulphur and an amine, usually morpholine[486], is widely employed for the preparation of arylacetic acids. The intermediate, however, is not an amide but the thioamide (equation 214).

$$\text{ArCOCH}_3 + S + \text{NHR}_2 \longrightarrow \text{ArCH}_2\overset{\overset{\displaystyle S}{\|}}{\text{C}}\text{NR}_2 \quad\quad (214)$$

Dialkyl ketones have been used as starting materials in the Willgerodt reaction. 2-Heptanone for example yields heptamide when heated with ammonium polysulphide in pyridine[487], and other straight-chain ketones behave similarly. With branched-chain ketones there appears to be a preference for migration of the carbonyl function along the less branched substituent. Thus pinacolone gives t-butylacetamide (equation 215) and isobutyl methyl ketone gives isocaproamide[487].

$$\text{Me}_3\text{CCOCH}_3 \xrightarrow{(\text{NH}_4)_2\text{S}_x} \text{Me}_3\text{CCH}_2\text{CONH}_2 \quad\quad (215)$$
$$(58\%)$$

Transformations closely related to the Willgerodt reaction and conducted under similar experimental conditions include the formation of primary amides from olefins (equation 216)[488], acetylenes (equation 217)[489], alcohols (equation 218)[489], thiols (equation 219)[478], aldehydes (equation 220)[487], and α-halo acids (equation 221)[490].

$$Me_2C{=}CH_2 \longrightarrow Me_2CHCONH_2 \qquad\qquad (216)$$
$$(70\%)$$

$$CH_3(CH_2)_4C{\equiv}CH \longrightarrow CH_3(CH_2)_5CONH_2 \qquad (217)$$
$$(35\%)$$

$$PhCMe_2OH \longrightarrow PhCHMeCONH_2 \qquad\qquad (218)$$
$$(38\%)$$

$$PhCH_2CH_2SH \longrightarrow PhCH_2CONH_2 \qquad\qquad (219)$$
$$(95\%)$$

$$CH_3(CH_2)_5CHO \longrightarrow CH_3(CH_2)_5CONH_2 \qquad\quad (220)$$
$$(50\%)$$

$$CH_3(CH_2)_3\underset{\underset{Br}{|}}{C}HCO_2H \longrightarrow CH_3(CH_2)_3CONH_2 \qquad (221)$$
$$(77\%)$$

D. Miscellaneous Rearrangement Reactions

I. Diazotization of hydrazones and semicarbazones

Diazotization of hydrazones and of semicarbazones with sodium nitrite in sulphuric acid[491] or polyphosphoric acid[441] gives amides, which are formed by rearrangement reactions closely related in mechanism to the Beckmann and Schmidt reactions (equation 222).

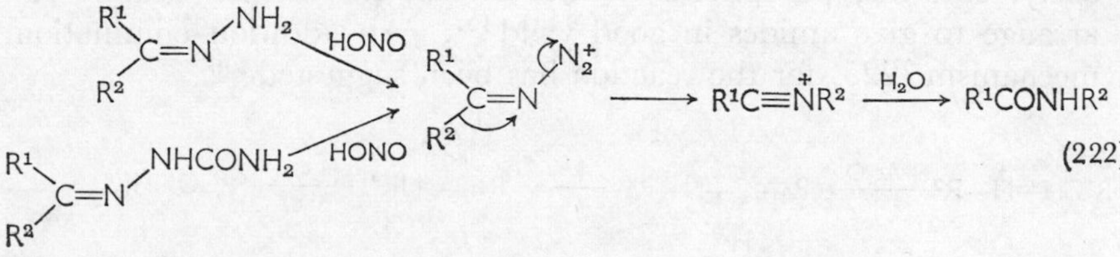

$$\begin{array}{c}
R^1 \\ \diagdown \\ R^2 \end{array}\!\!C{=}N{-}NH_2 \xrightarrow{HONO} \begin{array}{c} R^1 \\ \diagdown \\ R^2 \end{array}\!\!C{=}N{-}\overset{+}{N}{\equiv}N \longrightarrow R^1C{\equiv}\overset{+}{N}R^2 \xrightarrow{H_2O} R^1CONHR^2$$

$$\begin{array}{c} R^1 \\ \diagdown \\ R^2 \end{array}\!\!C{=}N{-}NHCONH_2 \xrightarrow{HONO}$$

$$(222)$$

These transformations give good results with diaryl and alkyl aryl ketones containing a variety of substituents[491], with indanones, and with tetralones[441]. Caprolactam is formed by treating cyclohexanone semicarbazone with sodium nitrite in aqueous acid[492], or by adding acid to an aqueous solution of nitrocyclohexane, hydrazine and sodium nitrite[493].

2. The Chapman rearrangement

When strongly heated, aryl N-arylbenzimidates undergo rearrangement via a 1,3-migration of an aryl group from oxygen to nitrogen leading to the formation of N,N-diarylamides (equation 223). The

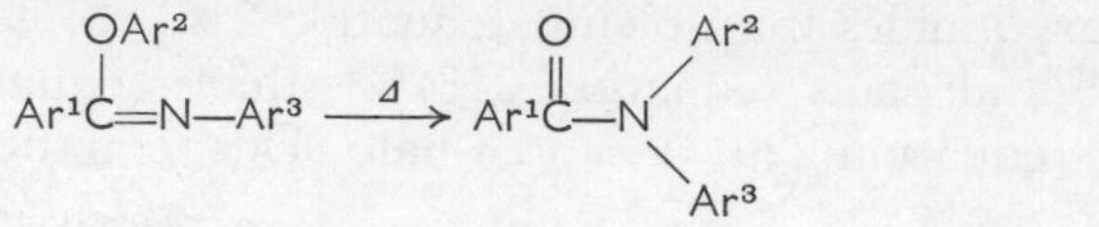

$$\overset{\overset{\displaystyle OAr^2}{|}}{Ar^1C}=N-Ar^3 \xrightarrow{\Delta} Ar^1\overset{\overset{\displaystyle O}{\|}}{C}-N\overset{\nwarrow Ar^2}{\underset{\searrow Ar^3}{}} \qquad (223)$$

scope and limitations of the reaction, which provides a useful route to unsymmetrical diarylamines and their aroyl derivatives, have been fully discussed in a recent review[494].

Oxazolines, being cyclic imidates, might be expected to undergo Chapman rearrangement on heating to yield acylaziridines. In fact, the products of vigorous pyrolysis are N-allylamides, but it is possible that these are derived from intermediate aziridines by further rearrangement (equation 224)[495].

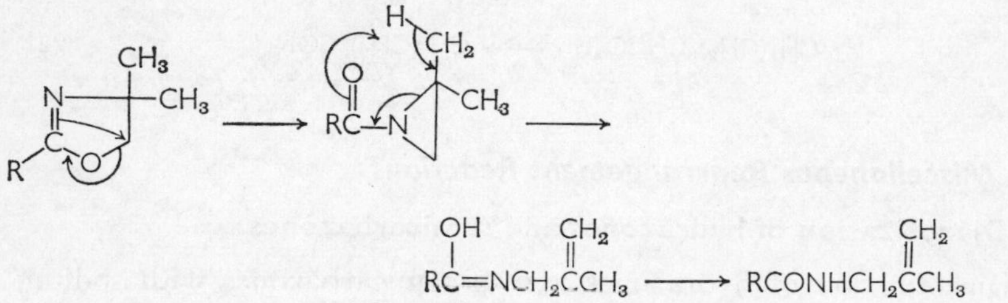

$$\text{(structure)} \longrightarrow \text{(structure)} \longrightarrow$$

$$\underset{\displaystyle RC=NCH_2CCH_3}{\overset{\displaystyle OH \qquad CH_2}{\overset{|}{}\overset{\|}{}}} \longrightarrow \underset{\displaystyle RCONHCH_2CCH_3}{\overset{\displaystyle CH_2}{\overset{\|}{}}} \qquad (224)$$

3. Rearrangements of nitrones[496] and of oxaziridines

Aldonitrones when heated with such reagents as acetic anhydride, acetyl chloride, phosphorus halides or sulphur dioxide readily rearrange to give amides in good yield[497]. An addition–elimination mechanism (225) for the reaction has been suggested[498].

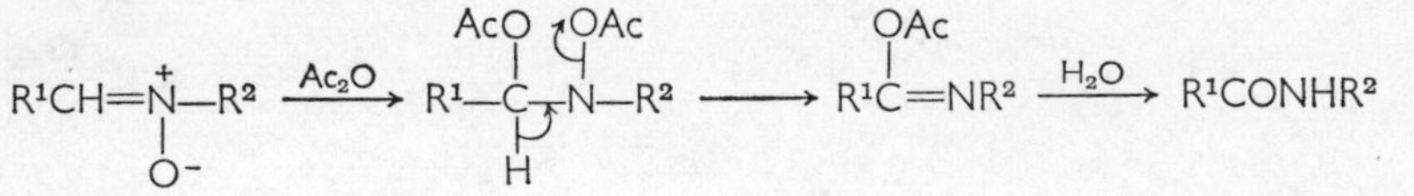

$$\underset{\displaystyle \overset{|}{O^-}}{R^1CH}\overset{+}{=}N-R^2 \xrightarrow{Ac_2O} R^1\underset{\displaystyle H}{\overset{\displaystyle AcO \; \; OAc}{\overset{|\;\;|}{C}}}N-R^2 \longrightarrow R^1\overset{\overset{\displaystyle OAc}{|}}{C}=NR^2 \xrightarrow{H_2O} R^1CONHR^2 \qquad (225)$$

Under basic conditions C-benzoyl-N-arylnitrones undergo rearrangement with migration of the benzoyl group (equation 226).

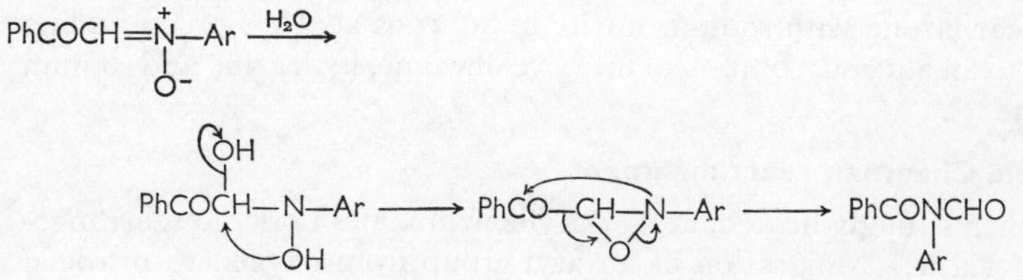

$$\underset{\displaystyle \overset{|}{O^-}}{PhCOCH}\overset{+}{=}N-Ar \xrightarrow{H_2O}$$

$$\underset{\displaystyle \overset{|}{OH}}{PhCOCH}-\overset{\overset{\displaystyle OH}{|}}{N}-Ar \longrightarrow PhCO-CH-N-Ar \longrightarrow \underset{\displaystyle Ar}{\overset{|}{PhCONCHO}} \qquad (226)$$

The reaction is thought to proceed via an oxaziridine intermediate; similar rearrangements of oxaziridine derivatives have been described[499,500]. Photolysis of nitrones also affords oxaziridines which are readily converted to amides by heating (equation 227).

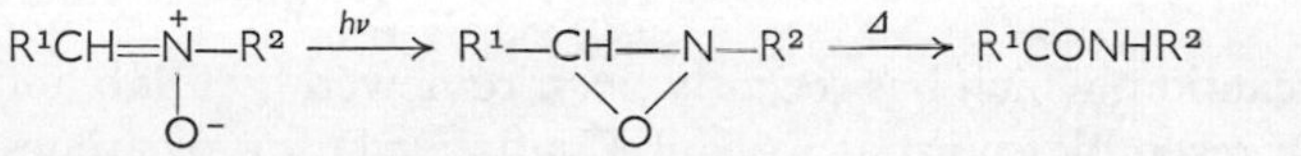

$$R^1CH{=}\overset{+}{\underset{O^-}{N}}{-}R^2 \xrightarrow{h\nu} R^1{-}CH\underset{O}{\overset{}{-}}N{-}R^2 \xrightarrow{\Delta} R^1CONHR^2 \qquad (227)$$

4. Rearrangement of α-amino thiolesters

Heating of dialkylaminomethyl thiolesters affords N,N-dialkylamides in good yield, and polythioformaldehyde (equation 228)[501].

$$R^1R^2NCH_2SCOR^3 \longrightarrow R^1R^2NHCOR^3 + (CH_2S)_x \qquad (228)$$

5. Reaction of chloramine with 2,6-dialkylphenols

Treatment of the sodium salts of 2,6-dialkylphenols with ethereal chloramine affords derivatives of azepinone[502]. The mechanism of the reaction involves rearrangement of an intermediate aziridine (equation 229).

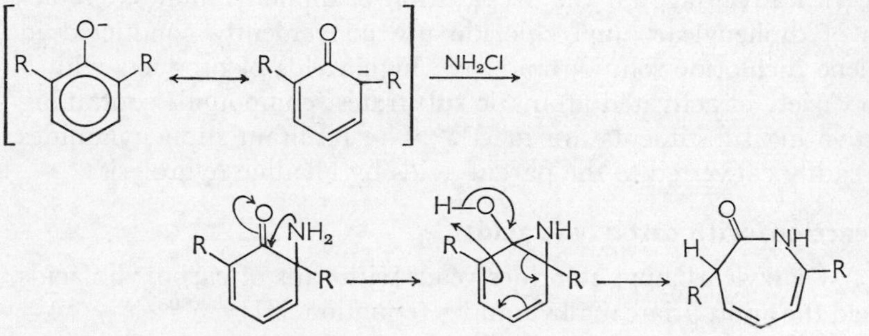

$$(229)$$

V. REACTIONS OF CARBAMYL HALIDES, ISOCYANATES AND ISOCYANIDES

A. Carbamyl Halides

I. Reactions with arenes

Carbamyl chloride, NH_2COCl, and its mono- and dialkyl and aryl derivatives undertake electrophilic attack on aromatic compounds in the presence of Lewis acids to give arenecarboxamides and related compounds (equation 230).

$$R^1R^2NCOCl + AlCl_3 \longrightarrow \left\{ \begin{array}{c} \overset{O}{\overset{\|}{R^1R^2N-C\cdots Cl\bar{A}lCl_3}} \\ \text{or} \quad {}^+ \\ {}^+O-\bar{A}lCl_3 \\ \overset{\|}{R^1R^2N-C-Cl} \end{array} \right\} \xrightarrow{\text{ArH}} ArCONR^1R^2 + HCl$$

$$(230)$$

The reaction, which has recently been reviewed by Olah and Olah[503], closely resembles normal Friedel–Crafts acylation of arenes in that it involves generation of an electrophilic species by interaction of the halide with Lewis acid.

The scope of the reaction was originally investigated by Gattermann who worked mainly with carbamyl chloride (the Gattermann amide synthesis) but also used the N-methyl and N-ethyl derivatives. Later development was due to Hopff[504] who showed that carbamyl chloride when treated with Lewis acid halides forms stable complexes which can be stored for long periods and which react with aromatic compounds to give excellent yields of amides. Reactions of N-methyl-N-phenylcarbamyl chloride with aromatic compounds were investigated by Weygand[505], and recently Wilshire[506] has shown that N,N-diphenylcarbamyl chloride, which is commercially available, can be used with advantage for the preparation of diphenylamides. Reactions of diphenylcarbamyl chloride are conveniently conducted in ethylene dichloride solution and give high yields of products with a wide variety of activated aromatic substrates (compounds containing deactivating substituents are inert). The resultant diphenylamides are readily converted to the parent acids by alkaline hydrolysis.

2. Reactions with carboxylic acids

N,N-Dialkylcarbamyl chlorides react with salts of carboxylic acids to yield the appropriate dialkylamides (equation 231)[507,508].

$$R^1R^2NCOCl + R^3CO_2^-M^+ \longrightarrow R^1R^2NCOR^3 + CO_2 + MCl \qquad (231)$$

The reaction, which has been applied to the preparation of a wide range of dimethyl- and diethylacylamines, including those derived from dicarboxylic acids, undoubtedly proceeds via an intermediate carbamic carboxylic anhydride (equation 232) (see section V.B.2).

$$\overset{O}{\overset{\|}{R^1R^2N-C-Cl}} + {}^-OCOR^3 \longrightarrow \overset{O}{\overset{\|}{R^1R^2N-C-O-\overset{O}{\overset{\|}{C}}-R^3}} \longrightarrow$$

$$\overset{O}{\overset{\|}{R^1R^2N-C-R^3}} + CO_2 \quad (232)$$

The usual experimental procedure involves heating the ammonium, sodium or potassium salt of the acid with the carbamyl chloride in the absence of solvent. Alternatively the chloride and the free acid are mixed in pyridine.

B. Isocyanates

I. Alkylation and arylation

Isocyanates, like ketenes, with which they are isoelectronic, are highly susceptible to nucleophilic attack at the electron-deficient carbonyl carbon atom. The addition process is usually completed by subsequent protonation of the initial adduct (equation 233).

$$RN{=}C{=}O + Y^- \longrightarrow \left[R\bar{N}{-}\underset{Y}{C}{=}O \longleftrightarrow RN{=}\underset{Y}{C}{-}\bar{O} \right] \xrightarrow{H^+} RNHCOY$$

$$(233)$$

Such reactions are of importance for the preparation of amides when Y^- is a carbon nucleophile as, for example, when isocyanates are treated with Grignard reagents (equation 234). The yields of pro-

$$R^1NCO + R^2MgX \longrightarrow R^1N{-}\underset{MgX}{COR^2} \xrightarrow{H_2O} R^1NHCOR^2$$

$$(234)$$

ducts from this type of reaction are often excellent; addition of phenyl-magnesium bromide to 2-furyl isocyanate affords N-furylbenzamide in 80% yield (equation 235)[509]. However, the utility of the reaction

$$(235)$$

for synthesis is restricted by the limited availability of isocyanates.

Other types of carbanionic reagent will undertake similar reactions. Metal derivatives of malonic ester and other active methylene compounds have occasionally been employed. Thus phenyl isocyanate when treated with ethyl nitroacetate and potassium carbonate affords the amido ester **93**[510], whilst the amido acid **94** is produced by the

$$PhNCO + O_2NCH_2CO_2Et \xrightarrow{K_2CO_3} PhNHCOCHCO_2Et$$

$$\underset{NO_2}{}$$

$$(236)$$

(93)

reaction of phenyl isocyanate with the di(chloromagnesium) derivative of phenylacetic acid (equation 237)[511].

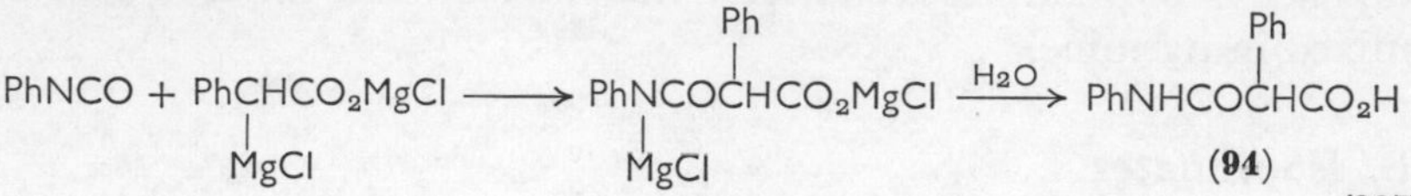

$$PhNCO + \underset{\underset{MgCl}{|}}{PhCHCO_2MgCl} \longrightarrow \underset{\underset{MgCl}{|}}{PhNCO\overset{\overset{Ph}{|}}{C}HCO_2MgCl} \xrightarrow{H_2O} PhNHCO\overset{\overset{Ph}{|}}{C}HCO_2H$$

$$(94)$$

$$(237)$$

Isocyanates are alkylated by olefins but the reaction usually occurs readily only when relatively nucleophilic species such as enamines and enols are employed. Thus the enamine **95** derived from morpholine and cyclopentanone on treatment with *o*-nitrophenyl isocyanate in chloroform affords the amide **96** which is readily hydrolysed to the β-keto amide **97** by dilute acid[512]. The reaction has been

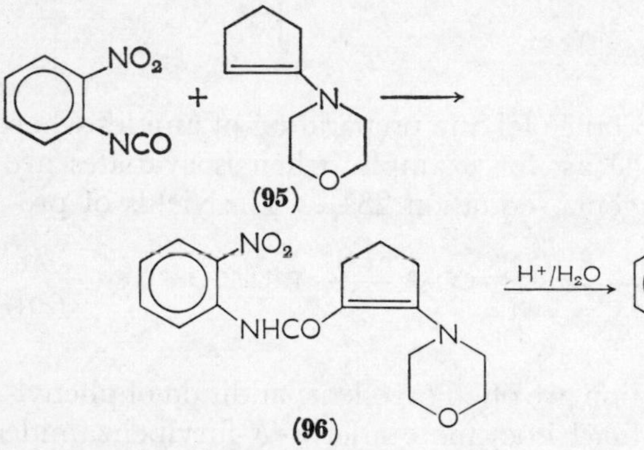

$$(238)$$

used for the elaboration of heterocyclic systems[513]. When treated with excess of simple isocyanates enamines undergo diaddition (equation 239)[514], whilst β,β-disubstituted enamines derived from

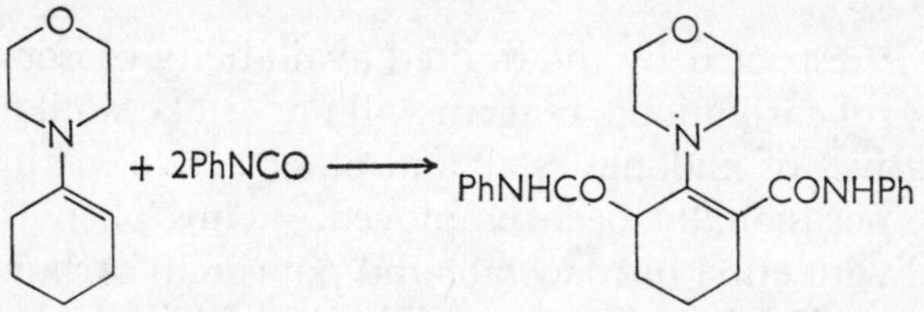

$$(239)$$

aldehydes give β-amino-β-lactams[515], hydrolysis of which affords β-formylamides (equation 240).

Intramolecular alkylation of isocyanates with enols provides an elegant route for the elaboration of lactams. Thus a key step in a

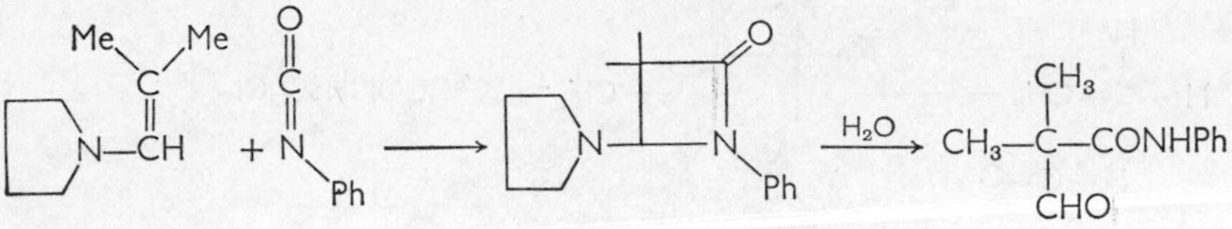

$$(240)$$

synthesis of the alkaloid atisine[516] consists of treating the isocyanate **98**, generated *in situ* from the appropriate azide, with *p*-toluene-sulphonic acid in benzene when the lactam **99** is formed in good yield (equation 241).

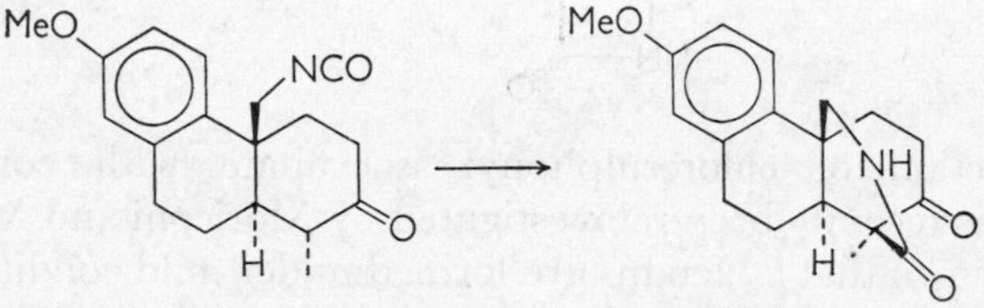

$$(241)$$

In stereochemically favourable situations intramolecular reaction between an isocyanate function and a non-activated olefinic bond can occur[517]. Thus *cis*-2-vinylcyclopropyl isocyanate spontaneously undergoes conversion into an azepinone derivative in a reaction which is formally analogous to the Cope rearrangement (equation 242).

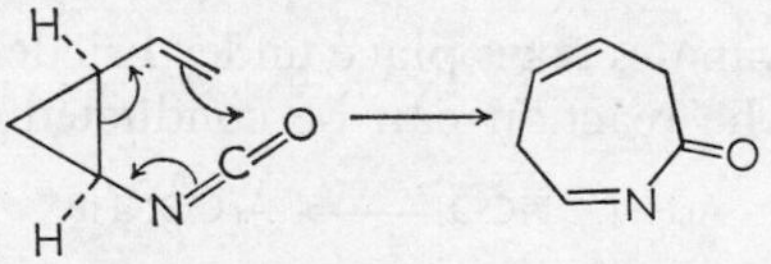

$$(242)$$

Chlorosulphonyl isocyanate (**100**)[518] in which direct attachment of the strongly electron-attracting SO_2Cl group to the isocyanate function greatly enhances the electrophilic character of the carbonyl carbon atom is sufficiently reactive to undertake intermolecular addition to non-activated double bonds[519,520]. Thus with isobutene it affords a readily separable mixture of a sulphamyl chloride and a β-lactam derivative which is readily converted into the parent compound by hydrolysis or mild reduction (equation 243)[521].

The reaction provides an excellent method for the preparation of β-lactams from a variety of substrates including allenes[522] and poly-cyclic olefins[523]. The mechanism was originally thought to involve a dipolar intermediate, but the recent observation that the reaction proceeds by stereospecific *cis* addition now casts doubt on this interpretation[524].

6*

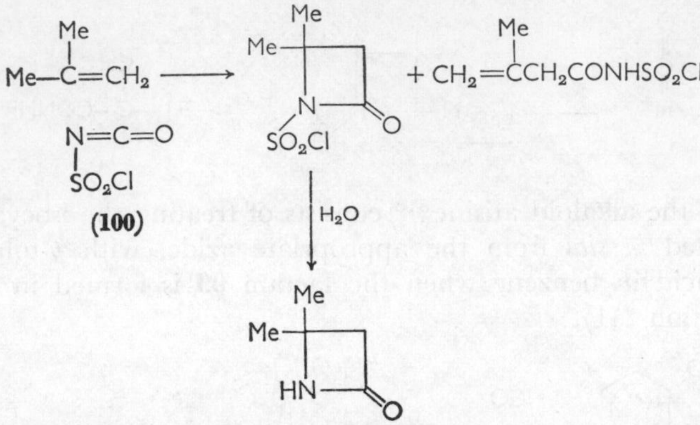

$$
\underset{\underset{\underset{SO_2Cl}{|}}{N=C=O}}{\overset{\overset{Me}{|}}{Me-C=CH_2}} \longrightarrow \underset{\underset{SO_2Cl}{|}}{Me-\overset{Me}{\underset{}{\Box}}{N}}{\overset{}{\underset{}{O}}} \quad + CH_2=\overset{\overset{Me}{|}}{C}CH_2CONHSO_2Cl
$$

$$
\textbf{(100)}
$$

$$
\downarrow H_2O \tag{243}
$$

$$
\underset{HN-}{Me-\overset{Me}{\underset{}{\Box}}{}}{\overset{}{\underset{}{O}}}
$$

The reaction of chlorosulphonyl isocyanate with conjugated dienes[525] has recently been reinvestigated by Moriconi and Meyer[526] who have shown that β-lactams are formed under mild conditions but readily undergo rearrangement on warming (equation 244).

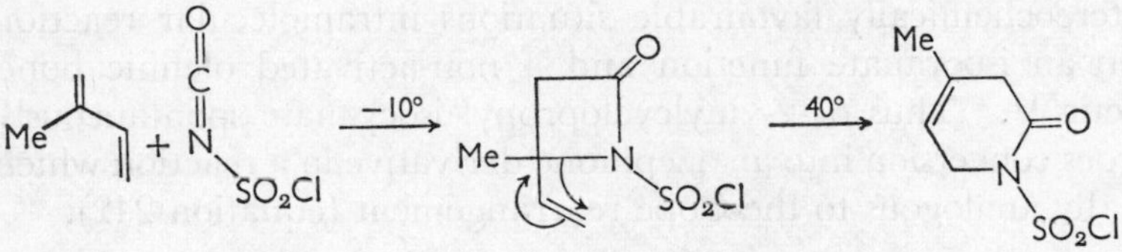

$$
\underset{Me}{\overset{}{\diagdown}}{=} + \underset{\underset{SO_2Cl}{|}}{\overset{\overset{O}{\|}}{\underset{}{C}}{N}} \xrightarrow{-10^\circ} Me-\overset{}{\underset{}{\Box}}{\underset{SO_2Cl}{\overset{}{\underset{}{N}}}}{\overset{}{\underset{}{O}}} \xrightarrow{40^\circ} \underset{\underset{SO_2Cl}{|}}{\overset{\overset{Me}{}}{\underset{}{N}}}{\overset{}{\underset{}{O}}} \tag{244}
$$

Arylation of isocyanates takes place under Friedel–Crafts conditions (equation 245). The reaction can be conducted in the presence of

$$
ArH + RNCO \longrightarrow ArCONHR \tag{245}
$$

Lewis acids, hydrogen chloride or polyphosphoric acid. Illustrative examples include the preparation of N-4-bromophenylferrocene-carboxamide by the reaction of ferrocene with 4-bromophenyl isocyanate and aluminium chloride in methylene chloride[527], and the cyclization of 2-biphenylyl isocyanate to phenanthridone in poly-phosphoric acid[528]. The reaction proceeds particularly readily with chlorosulphonyl isocyanate affording N-arylsulphamyl chlorides which are converted to primary amides by mild hydrolysis (equation 246)[519].

$$
ArH + ClSO_2NCO \longrightarrow ArCONHSO_2Cl \xrightarrow{H_2O} ArCONH_2 \tag{246}
$$

Finally, mention should be made of a reaction which although formally similar to isocyanate alkylation is mechanistically un-related since it probably involves free-radical intermediates. It con-

sists of the direct reduction of isocyanates to formamides by treating
them with triphenyltin hydride (equation 247)[529].

$$ArNCO + 2Ph_3SnH \longrightarrow ArNHCHO + Ph_3SnSnPh_3 \qquad (247)$$

2. Reactions with carboxylic acids

Isocyanates, when warmed with carboxylic acids react with the
evolution of carbon dioxide and the formation of acylamines (equation 248)[530]. The mechanism undoubtedly involves the intermediacy
of carbamic carboxylic anhydrides (**101**) generated by nucleophilic
addition of carboxylate ion to the electropositive carbon centre in the
isocyanate function[531].

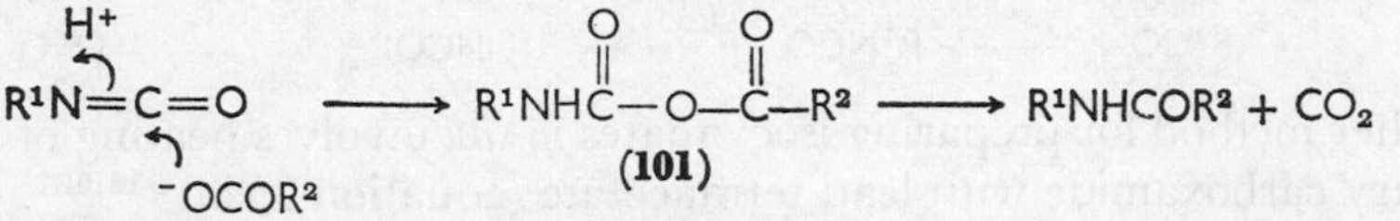

$$R^1N{=}C{=}O \xrightarrow{H^+} R^1NH\overset{\overset{O}{\|}}{C}{-}O{-}\overset{\overset{O}{\|}}{C}{-}R^2 \longrightarrow R^1NHCOR^2 + CO_2 \qquad (248)$$
$$^-OCOR^2 \qquad \qquad \textbf{(101)}$$

It has been shown that the carbon dioxide liberated during the
reaction is derived from the isocyanate[532], and it has been suggested
that its formation involves intramolecular acylation within **101**.
However, in view of recent studies on similar cyclic compounds[533]
it now seems more probable that decomposition of the intermediate
101 proceeds by a chain mechanism involving free amine (equations
249a and 249b). Selective attack of the amine at the carbonyl group

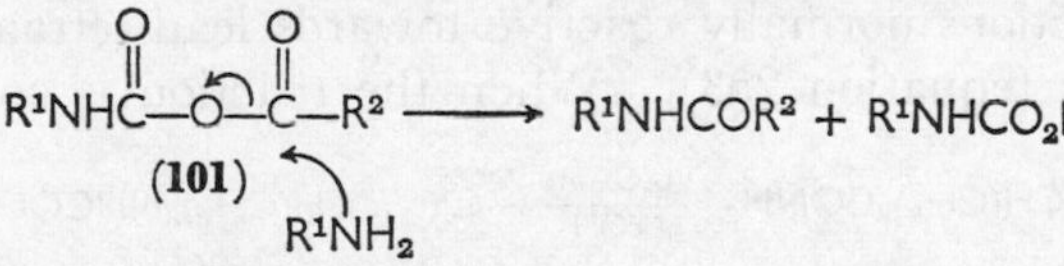

$$R^1NH\overset{\overset{O}{\|}}{C}{-}O{-}\overset{\overset{O}{\|}}{C}{-}R^2 \longrightarrow R^1NHCOR^2 + R^1NHCO_2H \qquad (249a)$$
$$\textbf{(101)} \qquad R^1NH_2$$

$$R^1NHCO_2H \longrightarrow R^1NH_2 + CO_2 \qquad (249b)$$

remote from the nitrogen atom is in accord with mechanistic principles
outlined previously (section II.C). Nevertheless, the formation of
dialkylureas as by-products[53] indicates that reaction at the carbamyl
group is not completely precluded.

Isocyanates, as such, are infrequently used for the preparation of
amides by this method. One illustrative example is the preparation
of the diamide **103** in 76% yield from the diisocyanate **102** by heating
it with acetic acid in chlorobenzene (equation 250)[534].

More frequently, isocyanates are prepared *in situ*. Thus the formation of acylamines by decomposition of acyl azides in carboxylic acid

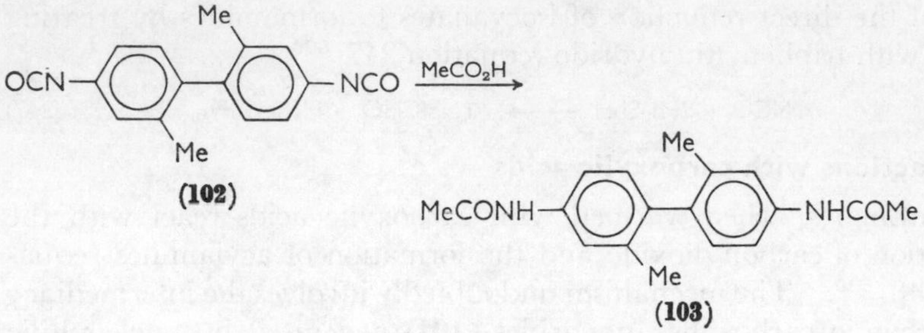

(102)

(103)

$$\text{(250)}$$

solvents is a well-known variation of the Curtius reaction (equation 251)[535].

$$R^1CON_3 \longrightarrow R^1NCO \xrightarrow{R^2CO_2H} R^1NHCOR^2 \tag{251}$$

Another method for preparing isocyanates *in situ* involves heating of a primary carboxamide with lead tetraacetate (equation 252a)[536,537]. When the reaction is conducted in hot benzene the isocyanate reacts with acetic acid liberated in the first step to give acetylamines in reasonable yield (equation 252b)[536]. Acetic acid is a somewhat

$$RCONH_2 + Pb(OAc)_4 \longrightarrow RNCO + 2HOAc + Pb(OAc)_2 \tag{252a}$$

$$RNCO + HOAc \longrightarrow RNHAc + CO_2 \tag{252b}$$

better solvent and allows the preparation in good yield of acetyl-amines derived from a wide range of amides, including some containing other functions normally reactive towards lead tetraacetate, e.g. olefinic bonds (equation 253). When the reaction is conducted in

$$CH_2{=}CH(CH_2)_8CONH_2 \xrightarrow[\text{AcOH}]{Pb(OAc)_4} CH_2{=}CH(CH_2)_8NHCOCH_3 \tag{253}$$
$$(59\%)$$

propionic acid solvent propionamides are formed (equation 254)[536].

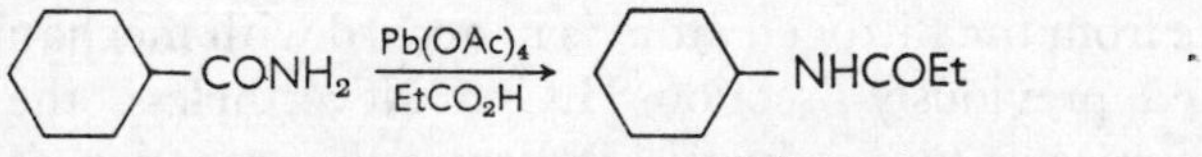

$$(69\%)$$

When an isocyanate group is generated adjacent to a carboxyl group within the same molecule cyclic carboxylic carbamic anhydrides are formed (equation 255)[538]. Those derived from aromatic compounds (e.g. **104**) are reasonably stable, but simple monocyclic compounds such as Leuchs anhydrides (**105**) readily decompose affording polypeptides (equation 256)[539]. The mechanism of this interesting

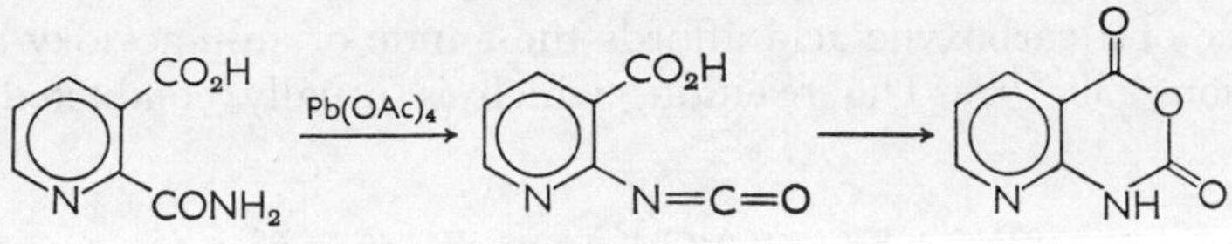

$$(255)$$

$$(256)$$

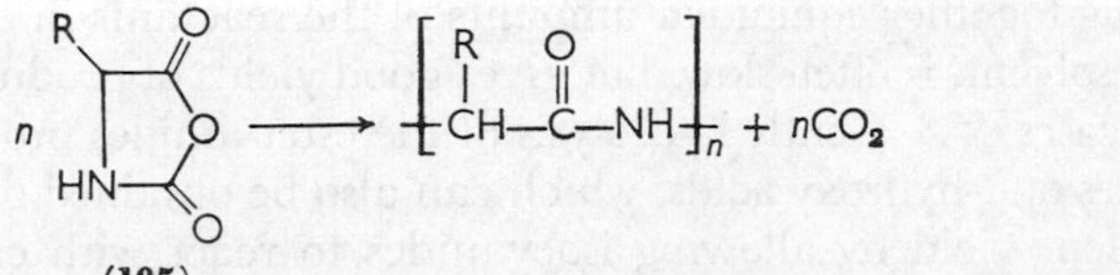

reaction, which is of utility in peptide synthesis, has recently come under careful scrutiny[533] and it now appears that the 'normal' reaction, i.e. that conducted in the absence of strong base, involves attack of free amine on the carbonyl at position 5 (equation 257).

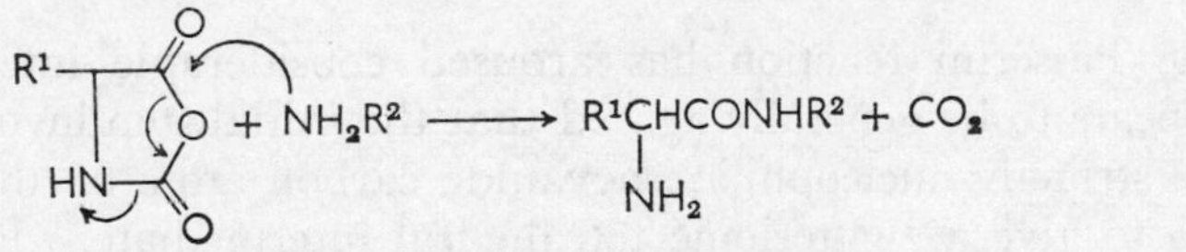

$$(257)$$

The homologous six-membered cyclic anhydrides decompose similarly yielding polypeptides derived from β-amino acids[540].

Carbamic carboxylic anhydrides can be prepared by other methods not involving isocyanates. They include reaction of carbamyl halides with carboxylate salts (section V.A.2), and the action of phosgene on amino acids[541].

Finally, we should note that chlorosulphonyl isocyanate reacts particularly readily with carboxylic acids yielding carbamyl sulphonyl chlorides via mixed anhydride intermediates (equation 258a)[542]. Gentle hydrolysis of the products affords primary amides (equation 258b).

$$RCO_2H + OCNSO_2Cl \longrightarrow RCOOCONHSO_2Cl \longrightarrow RCONHSO_2Cl + CO_2$$

$$(258a)$$

$$RCONHSO_2Cl \xrightarrow{H_2O} RCONH_2 \qquad (258b)$$

C. Isocyanides

Isocyanides undergo a number of reactions leading to amides and their derivatives. Possibly the best known is the Passerini reaction in which treatment of an isocyanide with an aldehyde or ketone in the presence of a carboxylic acid affords the amide of an α-acyloxy acid (equation 259)[543]. The reaction, which is usually conducted by

$$R^1NC + R^2_2CO + AcOH \longrightarrow R^1NHCO\overset{\displaystyle OAc}{\underset{|}{C}}R^2_2 \tag{259}$$

mixing together equimolar amounts of the reactants in ether or other inert solvent, is often slow, but gives good yields of products in favourable cases[544]. Gentle hydrolysis of the ester amides produced affords amides of α-hydroxy acids, which can also be obtained directly and in excellent yield by allowing isocyanides to react with carbonyl compounds in aqueous mineral acid[545] (equation 260). The mechanism

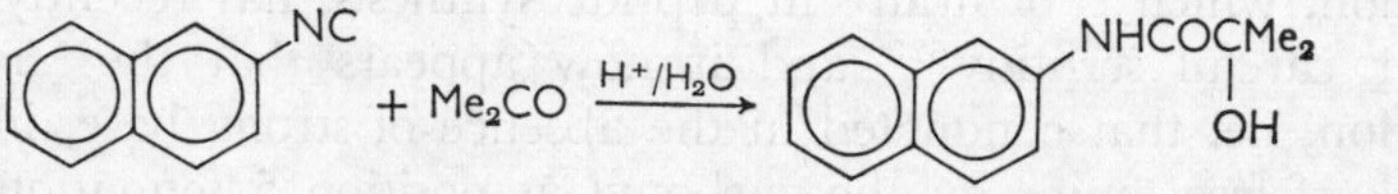

$$(84\%)$$

of the Passerini reaction has aroused considerable interest[543–549]. It appears to be generally agreed that the initial step involves attack of the strongly nucleophilic isocyanide carbon atom on the carbonyl group to give a zwitterionic tetrahedral intermediate. Possibly the next step in the reaction sequence consists of protonation of the oxygen atom and addition of carboxylate ion affording an imidoyl anhydride (equation 261a) which is then converted into the final product by acyl migration. However, very recent work[547] suggests that the true intermediate is an imino-oxirane species (106).

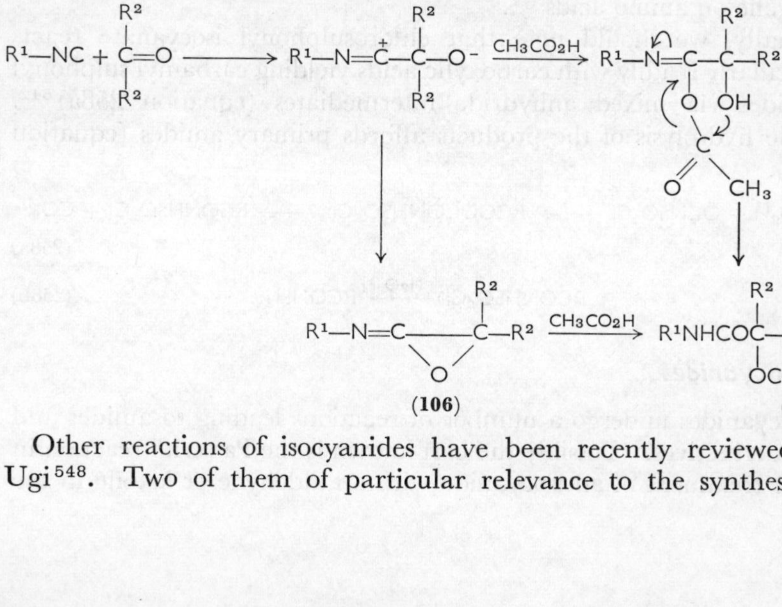

Other reactions of isocyanides have been recently reviewed by Ugi[548]. Two of them of particular relevance to the synthesis of amides are (i) the formation of α-acylaminocarboxylic acid amides and α-amino acid amides by treatment of isocyanides with ammonia, or an amine, a carbonyl compound and a carboxylic acid (equations 262 and 263) [548,549], and (ii) the preparation of β-lactam derivatives by

$$R^1NC + R^2_2CO + R^3NH_2 \longrightarrow R^1NHCOCR^2_2 \qquad (262)$$
$$\underset{NHR^3}{|}$$

$$R^1NC + R^2_2CO + NH_3 + AcOH \longrightarrow R^1NHCOCR^2_2 \qquad (263)$$
$$\underset{NHAc}{|}$$

the reaction of isocyanides with carbonyl compounds and β-amino acids (equation 264) [548].

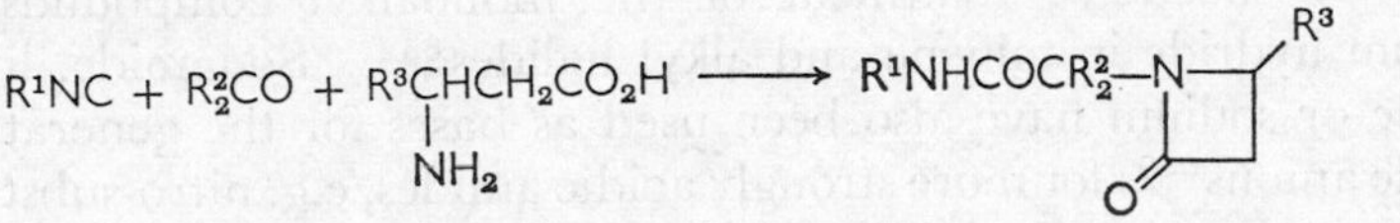

$$R^1NC + R^2_2CO + R^3CHCH_2CO_2H \longrightarrow R^1NHCOCR^2_2{-}N \qquad (264)$$
$$\underset{NH_2}{|}$$

VI. ALKYLATION OF AMIDES, IMIDES AND LACTAMS

Amides present three possible sites for alkylation—the oxygen and nitrogen centres in the amide function and the carbon atom at the α-position. Examples of all three types of reaction are known. However, amides are feeble nucleophiles and intermolecular alkylation under neutral conditions takes place slowly and requires active alkylating agents such as trialkyloxonium salts [550] or dialkyl sulphates [551]. Under these circumstances alkylation, like protonation of amides, occurs predominantly at oxygen affording imidates (equation 265). Presumably the course of the reaction reflects the greater thermodynamic stability of the oxonium ion **107**, due to the delocalization of charge, as compared with its ammonium isomer **108**.

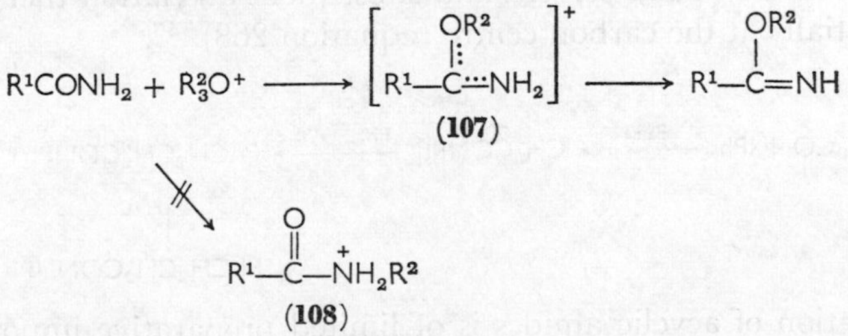

$$R^1CONH_2 + R^2_3O^+ \longrightarrow \left[\underset{(107)}{R^1{-}C{\cdots}NH_2}\overset{OR^2}{\overset{||}{}}\right]^+ \longrightarrow R^1{-}\overset{OR^2}{\underset{}{C}}{=}NH$$

$$(265)$$

$$R^1{-}\overset{O}{\overset{||}{C}}{-}\overset{+}{N}H_2R^2$$
$$(108)$$

On the other hand, alkylation of the anions generated from amides by treatment with a suitable strong base leads to N-alkylated products (equation 266). In this case both of the possible products of alkylation at O or N are neutral molecules and the course of the reaction is controlled by the greater nucleophilicity of the nitrogen centre.

$$R^1CONH_2 + B^- \longrightarrow \left[R^1\!\!-\!\!\overset{\overset{\displaystyle O}{\|}}{C}\cdots NH \right]^- \xrightarrow{R^2X} RCONHR^2$$

$$\xrightarrow{} \underset{\substack{|\\ R\!\!-\!\!C=\!NH}}{OR^2}$$

(266)

Thus, a number of N,N-dialkylamides have been prepared in 60–90% yield by successive treatment of the monoalkyl compounds with sodium hydride in toluene and alkyl halides[552]. Sodamide, lithium amide or sodium have also been used as bases for the generation of amide anions[553]: for more strongly acidic amides, e.g. nitro-substituted anilides, potassium hydroxide in acetone is adequate[554].

When N,N-dialkylamides are treated with a strong base and an alkyl halide, alkylation occurs at the α-position via carbanion intermediates (equation 267). In a recent survey of the reaction[555] best

$$RCH_2CONR^1_2 + B^- \longrightarrow R\bar{C}HCONR^1_2 \xrightarrow{R^2X} \underset{\substack{|\\R^2}}{RCHCONR^1_2}$$ (267)

results were obtained when sodamide in toluene or benzene was employed.

In unsubstituted or N-monoalkylamides C-alkylation competes effectively with N-alkylation only when there is some special structural feature of the molecule which enhances the acidity of the α-hydrogens, as, for example, in acetoacetamide. However, some secondary amides are converted into their C,N-dianions when treated with butyllithium in a non-polar solvent and subsequent alkylation then occurs preferentially at the carbon centre (equation 268)[556].

$$CH_3CONHPh \xrightarrow{BuLi} \overset{\overset{\displaystyle Li}{|}}{CH_2}\overset{\overset{\displaystyle Li}{|}}{CONPh} \xrightarrow{PhCH_2Cl} PhCH_2CH_2\overset{\overset{\displaystyle Li}{|}}{CONPh}$$

$$\downarrow H_2O$$

$$PhCH_2CH_2CONHPh$$ (268)

Alkylation of acyclic amides is of limited preparative importance; the reaction when applied to lactams is of much greater utility since the parent unsubstituted compounds are often readily available via the

Schmidt or Beckmann reactions.   Once again the position of alkylation depends on the reagents used and the experimental conditions. An interesting illustration is provided by reactions of the enamide **109** which was studied by Eschenmoser's group during the course of their work on the synthesis of corrins[214].   When treated with trimethyloxonium fluoroborate, **109** yields exclusively the O-methyl compound, methylation of the potassium salt affords the N-methyl compound, whilst methylation of the silver salt takes place at the methine group.   Methylation of caprolactam with dimethyl sulphate

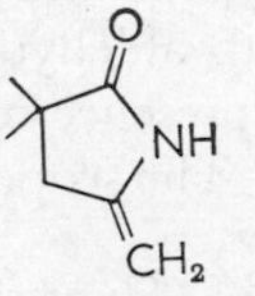

(109)

in benzene affords initially the expected O-methyl compound, but on further heating of this product with an excess of reagent, N-methyl-caprolactam is formed[557].

N-Alkylation is conveniently achieved by treating a lactam successively with sodium hydride and an alkyl halide in benzene.   The method has been recently used for the methylation of large-ring lactams[558] and for the preparation of the interesting allenamide **111** by spontaneous rearrangement under the reaction conditions of the initial acetylenic product **110**[559].

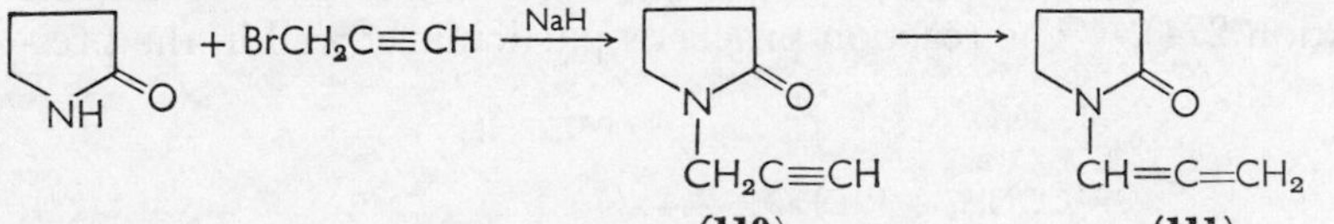

$$+ BrCH_2C\equiv CH \xrightarrow{NaH} \qquad\longrightarrow \qquad (269)$$

(110) **(111)**

N-Substituted lactams are readily alkylated at the α-position.   For example N-methylpyrrolidone is converted into the 3-ethyl derivative by successive treatment with sodamide in liquid ammonia and ethyl bromide[560].   When an excess of reagents is used dialkylation occurs (equation 270).

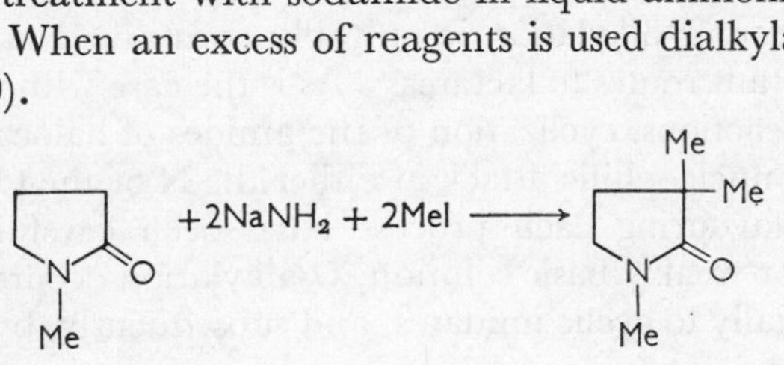

$$+2NaNH_2 + 2MeI \longrightarrow \qquad (270)$$

The metal salts of imides undergo N-alkylation when heated with alkyl halides. Probably the most widely used variation of this reaction is the Gabriel synthesis of amines via alkylation of phthalimide (equation 271). Excellent yields are obtained when the alkylation

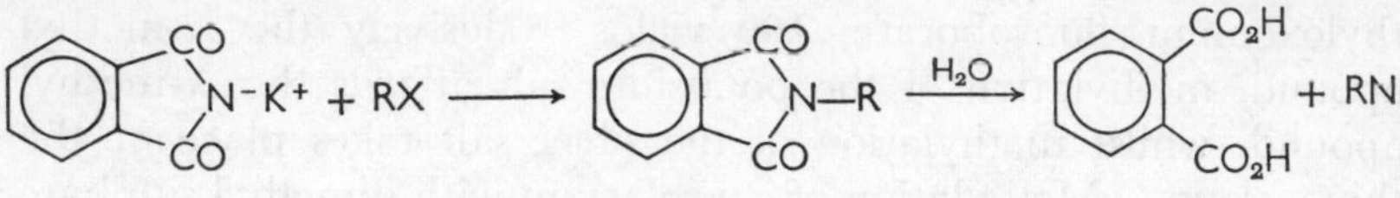

$$\text{(phthalimide)} N^-K^+ + RX \longrightarrow \text{(phthalimide)} N\!-\!R \xrightarrow{H_2O} \text{(benzene)} \begin{matrix} CO_2H \\ CO_2H \end{matrix} + RNH_2 \tag{271}$$

step is conducted in dimethylformamide solution[561].

Amides, when treated with aldehydes, yield hydroxyalkyl derivatives (equation 272); in the presence of alcohols N-alkoxyalkylamides are formed (equation 273). The scope and limitations of the reaction

$$R^1CONH_2 + R^2CHO \longrightarrow R^1CONHCHR^2 \underset{OH}{\mid} \tag{272}$$

$$R^1CONH_2 + R^2CHO + R^3OH \longrightarrow R^1CONHCHR^2 \underset{OR^3}{\mid} \tag{273}$$

have been reviewed[562]. The method has been recently employed for the preparation of N-(alkoxymethyl)acrylamides[563], and the extent and nature of the reaction when applied to N-substituted amides has been investigated by n.m.r. techniques[564]. When aldehydes other than formaldehyde or chloral are employed the reaction usually progresses beyond the monohydroxyalkyl stage and affords bis-amides (equation 274). The reaction proceeds particularly well in the pres-

$$2R^1CONH_2 + R^2CHO \longrightarrow \begin{matrix} R^1CONH \\ \diagdown \\ CHR^2 \\ \diagup \\ R^1CONH \end{matrix} \tag{274}$$

ence of perchloric acid[565]. Similar products are obtained by treating amides with acetals in sulphuric acid[566].

Intramolecular alkylation in suitably constituted amides provides some important routes to lactams. As is the case with intermolecular alkylation reactions, cyclization of the amides of halocarboxylic acids can involve nucleophilic attack at either the N or the O centre. The conditions favouring each process have been carefully studied[567]. In neutral or weakly basic solution, O-alkylation occurs preferentially leading initially to cyclic imidates, and subsequently, by hydrolysis, to lactones (equation 275)[30,568]. Under strongly basic conditions lac-

$$
\underset{\substack{\text{CONH}_2\\(CH_2)_n\ X\\C}}{} \longrightarrow \underset{\substack{\overset{\displaystyle NH}{\|}\\ \overset{\displaystyle C}{}\\(CH_2)_n\ O\\C}}{} \xrightarrow{\text{H}_2\text{O}} \underset{\substack{\overset{\displaystyle O}{\|}\\ \overset{\displaystyle C}{}\\(CH_2)_n\ O\\C}}{} \tag{275}
$$

tams are formed via N-alkylation in the amide anion. The method has been widely used for the preparation of pyrrolidones, azetidinones and other lactams. Recent examples include the preparation of δ-lactams by heating N-alkyl-5-chlorovaleramides with sodium ethoxide in ethanol (equation 276)[569], and of a series of azetidinones, pyrrolidones

$$
\underset{\text{Cl}}{\text{CONHR}} \longrightarrow \underset{\substack{\text{N}\\R}}{\text{C}}\text{O} \tag{276}
$$

and piperidones by treating N-aryl-ω-haloamides with such strong bases as sodamide in liquid ammonia and sodium hydride in dimethyl sulphoxide[570]. By careful use of potassium t-butoxide or metallic sodium as base even α-lactams can be prepared by intramolecular alkylation[571]. Recent examples include the formation of the unusually stable lactams containing t-butyl or adamantanyl substituents (equation 277)[572,573].

$$
\underset{\substack{|\\Br}}{\text{RCHCONHR}} \longrightarrow \text{RCH}\underset{\substack{|\\NR}}{\overset{C=O}{|}} \tag{277}
$$

It has been suggested that α-lactams may be intermediates in the formation of cyclic imino esters from α-halo lactams (equation 278)[574],

$$
\underset{\substack{\text{HN}\qquad\text{CH}-\text{X}\\(CH_2)_n}}{\overset{\overset{\displaystyle O}{\|}\\C} } \xrightarrow{\text{OR}^-} \underset{\substack{\text{N}\qquad\text{CH}\\(CH_2)_n}}{\overset{\overset{\displaystyle O}{\|}\\C}} \xrightarrow{\text{OR}^-} \underset{\substack{\text{NH}-\text{CH}-\text{CO}_2\text{R}\\(CH_2)_n}}{} \tag{278}
$$

and in the amination of α-chlorodiphenylacetamide[575], but an alternative route has recently been proposed for the latter reaction[576].

Intramolecular N-alkylation leading to lactam formation can also be accomplished by treatment of a suitable unsaturated amide with polyphosphoric acid (equation 279)[577,578]. Preparation of the lactam **112** provides a recent illustration of the use of the reaction[578]. The

(279)

(112)

reaction, probably involving acid-induced electrophilic attack on the amide function, presumably does not follow the O-alkylation pathway observed in intramolecular reactions because of the constraints of ring size. Some aryl-substituted unsaturated amides undergo photochemical conversion into β-lactams[579].

The formation of lactams from acylic precursors can also be achieved via intramolecular alkylation at the α-carbon atom. The preparation of azetidinones by consecutive treatment of Schiff bases with cyanoacetyl chloride and triethylamine involves such a process (equation 280)[580]. Another interesting example of this general type

$$\text{ArCH}{=}\text{NR} + \text{ClCOCH}_2\text{CN} \longrightarrow \underset{\underset{\text{NCCH}_2\overset{|}{\text{CO}}}{|}}{\overset{\overset{\text{Cl}}{|}}{\text{ArCH}}}{-}\text{NR} \overset{\text{Et}_3\text{N}}{\longrightarrow} \underset{\underset{\text{NCCH}{-}\overset{|}{\text{CO}}}{|}}{\overset{|}{\text{ArCH}}}{-}\text{NR} \quad (280)$$

of transformation is the preparation of an α-lactam by treatment of N-t-butyl-N-chlorophenylacetamide with potassium butoxide (equation 281)[581]. A rather unusual nucleophilic displacement of chlorine from nitrogen appears to be implicated. The formation of β-lactams

$$\underset{\overset{|}{\text{Cl}}}{\text{PhCH}_2\text{CONBu-}t} \overset{t\text{-BuO}^-}{\longrightarrow} \text{PhCH}\overset{\overset{\text{O}}{\parallel}}{\underset{\underset{\text{Cl}}{}}{{}_{\text{C}}}}\text{NBu-}t \longrightarrow \text{Ph}{-}\underset{}{\text{CH}}\overset{\overset{\text{O}}{\parallel}}{\underset{}{{}_{\text{C}}}}\text{N}{-}\text{Bu-}t \quad (281)$$

by intramolecular alkylation at an activated methylene group α to the amino nitrogen has been extensively explored by Bose and his colleagues[582].

In a few cases alkylation of amides has been achieved via a free-radical mechanism[583–586]. Thus, photolysis of mixtures of terminal olefins with formamide gives aliphatic amides in good yield by a radical chain process (equations 282)[584]. A similar reaction of formamide with acetylenes leads to the formation of 2:2 adducts[585], whilst

2-pyrrolidone when photolysed with olefins undergoes alkylation at the 3- and 5-positions[586].

$$RCH{=}CH_2 + \overset{\bullet}{C}ONH_2 \longrightarrow R\overset{\bullet}{C}HCH_2CONH_2 \qquad (282a)$$

$$R\overset{\bullet}{C}HCH_2CONH_2 + HCONH_2 \longrightarrow RCH_2CH_2CONH_2 + {}^\bullet CONH_2 \quad (282b)$$

Finally, it is convenient here to consider photolysis of acyl azides which, although not formally an alkylation reaction of amides, can lead in favourable circumstances to the formation of an N—C bond[587,588]. The reaction, which probably involves the intermediacy of acylnitrenes, affords lactams by insertion at an unactivated C—H position and has been used for this purpose in natural-product synthesis (equation 283)[588].

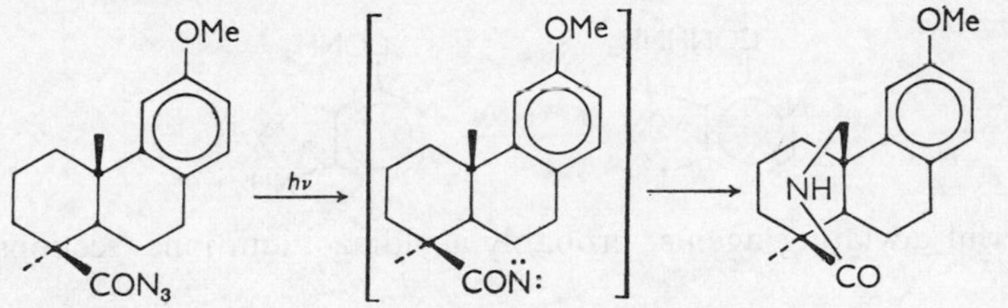

(283)

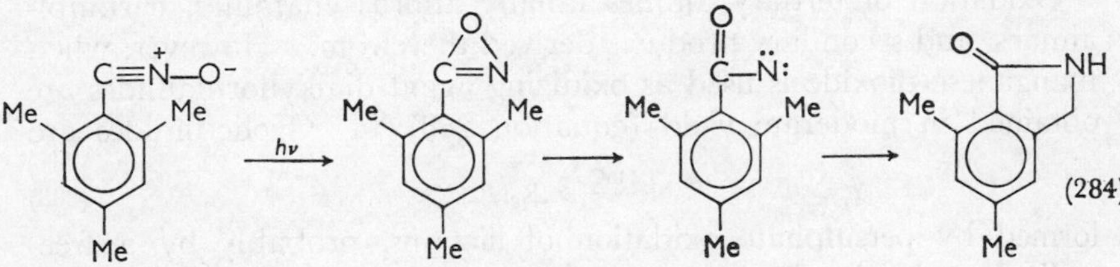

(284)

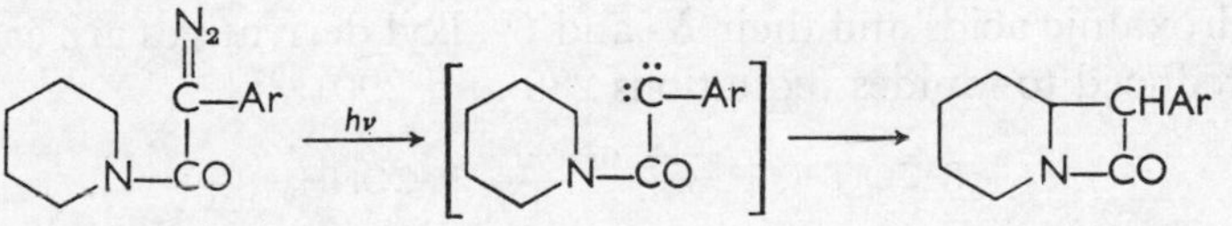

(285)

Photolysis of nitrile oxides can also be used for the generation of nitrene intermediates (equation 284)[589], whilst α-diazo amides undergo lactamization in a formally similar manner via carbenes (equation 285)[590].

VII. OXIDATION AND REDUCTION REACTIONS

Among the oxidation and reduction reactions that have occasionally been employed for the preparation of amides those which utilize acylhydrazines are probably the most important.   Reduction of hydrazides is readily achieved by heating them with an excess of Raney nickel or with Raney nickel and hydrazine in alcohol[591–593].   Hydrazine because of its high nucleophilicity often attacks esters and N-substituted amides that are inert towards ammonolysis and pathway (286) thus

$$\begin{array}{c} R^1CO_2R^2 \\ \text{or} \\ R^1CONR^2_2 \end{array} \xrightarrow{NH_2NH_2} R^1CONHNH_2 \xrightarrow{Ni} R^1CONH_2 \qquad (286)$$

provides a convenient and mild route to primary amides from such substrates[592].   Diacylhydrazines are similarly reduced by Raney nickel[593].   Somewhat surprisingly the same transformation of hydrazide to amide can be accomplished by ferricyanide oxidation (equation 287)[594].   Oxidation of hydrazides with other reagents generates

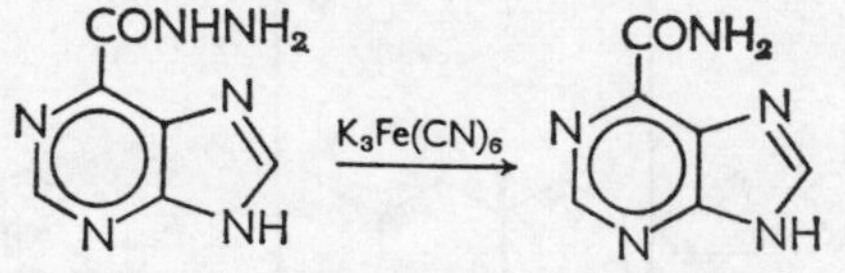

$$\qquad (287)$$

powerful acylating agents, probably acyldiazonium ions (sections II. I.4).

Oxidation of tertiary amines usually affords enamines, carbinolamines, and secondary products derived therefrom.   However, when manganese dioxide is used as oxidizing agent dialkylformamides are obtained in moderate yield (equation 288)[595].   Cyclic imides are

$$R_3N \xrightarrow{MnO_2} R_2NCHO \qquad (288)$$

formed by persulphate oxidation of lactams, probably by a free-radical mechanism[596].

Hydroxamic acids and their N- and O-alkyl derivatives are catalytically reduced to amides (equations 289 and 290)[597].

$$R^1CONHOR^2 \xrightarrow{H_2,\ Catalyst} R^1CONH_2 \qquad (289)$$

$$R^1CONR^2 \xrightarrow{H_2,\ Catalyst} R^1CONHR^2 \qquad (290)$$
$$\quad\ \ | \\ \quad\ \ OH$$

Oxidation of hydroxamic acid by a variety of agents including periodate, bromine, iodine and N-bromosuccinimide in the presence of amines affords amides (equation 291)[598].   Experimental evidence[598] supports the earlier suggestion[599] that an intermediate nitroso compound is formed which behaves as a powerful acylating agent.

$$R^1CONHOH \longrightarrow R^1CONO \xrightarrow{R^2_2NH} R^1CONR^2_2 + [HNO] \qquad (291)$$

Oxidative amination of aldehydes, and benzylic and allylic alcohols is readily accomplished with nickel peroxide and ammonia in ether (equation 292)[600]. Similar transformations using ammonium poly-

$$\text{RCHO(or RCH}_2\text{OH)} \xrightarrow{\text{NiO}_2/\text{NH}_3} \text{RCONH}_2 \tag{292}$$

sulphide have been previously mentioned (section IV.C).

VIII. ACKNOWLEDGMENT

The Author gratefully acknowledges the kind hospitality of Professor R. O. C. Norman and the Staff of the Chemistry Department, University of York, where this chapter was written during tenure of a Visiting Professorship in Chemistry.

IX. REFERENCES

1. I. Lengyel and J. C. Sheehan, *Angew. Chem. Intern. Ed. Engl.*, **7**, 25 (1968).
2. J. C. Sheehan and E. J. Corey, *Org. Reactions*, **9**, 388 (1957).
3. W. J. Hickinbottom, *Reactions of Organic Compounds*, 3rd ed., Longmans, Green and Co., New York, 1957.
4. R. B. Wagner and H. D. Zook, *Synthetic Organic Chemistry*, John Wiley and Sons, London, 1953.
5. L. F. Fieser and M. Fieser, *Reagents for Organic Synthesis*, John Wiley and Sons, New York, 1967.
6. N. V. Sidgwick, *The Organic Chemistry of Nitrogen*, 3rd ed. (revised and re-written by I. T. Millar and H. D. Springall), Clarendon Press, Oxford, 1966.
7. P. A. S. Smith, *The Chemistry of Open Chain Nitrogen Compounds*, Vols. I and II, W. A. Benjamin Inc., New York, 1965.
8. G. A. Olah, S. J. Kuhn, W. S. Tolgyesi, and E. B. Baker, *J. Am. Chem. Soc.*, **84**, 2733 (1962).
9. E. S. Gould, *Mechanism and Structure in Organic Chemistry*, Holt, Rinehart, and Winston, New York, 1959.
10. J. Hine, *Physical Organic Chemistry*, 2nd ed., McGraw-Hill, Book Co., New York, 1962.
11. M. L. Bender, *Chem. Rev.*, **60**, 53 (1960).
12. C. K. Ingold, *Structure and Mechanism in Organic Chemistry*, G. Bell and Sons, London, 1953.
13. J. F. Bunnett in *Theoretical Organic Chemistry*, The Kekule Symposium, Butterworths, London, 1959, p. 144.
14. D. P. N. Satchell, *Quart. Rev. (London)*, **17**, 160 (1963).
15. P. J. Lillford and D. P. N. Satchell, *J. Chem. Soc. (B)*, 360 (1967).
16. D. P. N. Satchell, *J. Chem. Soc.*, 1752 (1960).
17. H. S. Venkataraman and Sir C. Hinshelwood, *J. Chem. Soc.*, 4977 (1960).
18. A. N. Bose and Sir C. Hinshelwood, *J. Chem. Soc.*, 4085 (1958); F. J. Stubbs and Sir C. Hinshelwood, *J. Chem. Soc.*, S 71 (1949).

19. E. Ronwin and C. B. Warren, *J. Org. Chem.*, **29**, 2276 (1964); E. Ronwin and D. E. Horn, *J. Org. Chem.*, **30**, 2821 (1965).

20. G. A. Olah and S. J. Kuhn, *J. Am. Chem. Soc.*, **82**, 2380 (1960).

21. G. A. Olah and S. J. Kuhn, *Chem. Ber.*, **89**, 2211 (1956).

22. G. A. Olah and S. J. Kuhn, *J. Org. Chem.*, **26**, 225 (1961).

23. G. A. Olah, W. S. Tolgyesi, S. J. Kuhn, M. E. Moffatt, I. J. Bastien, and E. B. Baker, *J. Am. Chem. Soc.*, **85**, 1328 (1963).

24. J. B. Lee, *J. Am. Chem. Soc.*, **88**, 3440 (1966).

25. H. J. Bestmann and L. Mott, *Ann. Chem.*, **693**, 132 (1966).

26. N. O. V. Sonntag, *Chem. Rev.*, **52**, 237 (1953).

27. R. E. Kent and S. M. McElvain, *Org. Syn.*, Coll. Vol. III, John Wiley and Sons, New York, 1955, p. 490.

28. E. T. Roe, J. T. Scanlan, and D. Swern, *J. Am. Chem. Soc.*, **71**, 2215 (1949).

29. G. F. D'Alelio and E. E. Reid, *J. Am. Chem. Soc.*, **59**, 109 (1937).

30. D. H. R. Barton, A. L. J. Beckwith, and A. Goosen, *J. Chem. Soc.*, 181 (1965).

31. H. A. Stansbury and R. F. Cantrell, *J. Org. Chem.*, **32**, 824 (1967).

32. H. Stetter, J. Mayer, M. Schwarz, and K. Wulff, *Chem. Ber.*, **93**, 226 (1960).

33. C. F. Koelsch, *J. Org. Chem.*, **26**, 1003 (1961).

34. B. Acott, A. L. J. Beckwith, and A. Hassanali, *Australian J. Chem.*, **21**, 185, 197 (1968).

35. G. E. Philbrook, *J. Org. Chem.*, **19**, 623 (1954).

36. K. C. Murdock and R. B. Angier, *J. Org. Chem.*, **27**, 3317 (1962).

37. P. A. Finan and G. A. Fothergill, *J. Chem. Soc.*, 2824 (1962).

38. E. Fischer and A. Dilthey, *Chem. Ber.*, **35**, 855 (1902).

39. A. C. Cope and E. Ciganek, *Org. Syn.*, Coll. Vol. IV, John Wiley and Sons, New York, 1963, p. 339.

40. E. R. H. Jones, F. A. Robinson, and M. N. Strachan, *J. Chem. Soc.*, 87 (1946).

41. H. Stetter and E. Rauscher, *Chem. Ber.*, **93**, 1161 (1960).

42. E. Ronwin, *J. Org. Chem.*, **18**, 127 (1953).

43. P. J. Suter and W. B. Turner, *J. Chem. Soc.* (*C*), 2240 (1967).

44. A. D. Swensen and W. E. Weaver, *J. Am. Chem. Soc.*, **70**, 4060 (1948).

45. R. E. Steiger, *J. Org. Chem.*, **9**, 396 (1944); H. E. Carter, R. F. Frank, and H. W. Johnston, *Org. Syn.*, Coll. Vol. III, John Wiley and Sons, New York, 1955, p. 167; A. W. Ingersoll and S. H. Babcock, *Org. Syn.*, Coll. Vol. II, John Wiley and Sons, New York, 1943, p. 328.

46. Y. Iwakura, F. Toda, and H. Suzuki, *J. Org. Chem.*, **32**, 440 (1967).

47. S. M. McElvain and C. L. Stevens, *J. Am. Chem. Soc.*, **69**, 2668 (1947).

48. B. A. Geller and L. S. Samosvat, *Zh. Obshch. Khim.*, **30**, 1594 (1960).

49. R. W. Holley and A. D. Holley, *J. Am. Chem. Soc.*, **74**, 3069 (1952).

50. J. C. Sheehan and V. S. Frank, *J. Am. Chem. Soc.*, **71**, 1856 (1949).

51. W. Reeve and W. M. Eareckson, *J. Am. Chem. Soc.*, **72**, 5195 (1950).

52. J. H. Billman and E. E. Parker, *J. Am. Chem. Soc.*, **66**, 538 (1944).

53. J. Büchi, G. Lauener, L. Ragaz, H. Böniger, and R. Lieberherr, *Helv. Chim. Acta*, **34**, 278 (1951).

54. P. Ruggli and J. Rohner, *Helv. Chim. Acta.*, **25**, 1533 (1942).

55. H. C. Brown and A. Tsukamoto, *J. Am. Chem. Soc.*, **83**, 2016 (1961).

56. V. Prey, *Chem. Ber.*, **75**, 537 (1942); H. Adkins and Q. E. Thompson, *J. Am. Chem. Soc.*, **71**, 2242 (1949).
57. F. Klages and E. Zange, *Ann. Chem.*, **607**, 35 (1957).
58. D. Cook, *Can. J. Chem.*, **40**, 2362 (1962).
59. G. B. Payne, *J. Org. Chem.*, **31**, 718 (1966).
60. H. Böhme, *Angew. Chem. Intern. Ed. Engl.*, **5**, 849 (1966).
61. H. Wamhoff and F. Korte, *Chem. Ber.*, **100**, 2122 (1967).
62. C. W. Shoppee, R. W. Killick, and G. Kruger, *J. Chem. Soc.*, 2275 (1962).
63. Y. Inouye, K. Onodera, S. Kitaoka, and S. Hirano, *J. Am. Chem. Soc.*, **78**, 4722 (1956).
64. H. W. Heine and M. S. Kaplan, *J. Org. Chem.*, **32**, 3069 (1967).
65. M. J. Schlatter, *J. Am. Chem. Soc.*, **63**, 1733 (1941).
66. L. Crombie and M. Manzoor-i-Khuda, *J. Chem. Soc.*, 4984 (1963).
67. G. R. Pettit, S. K. Gupta, and P. A. Whitehouse, *J. Med. Chem.*, **10**, 692 (1967).
68. O. Ferno and T. Linderot, *Swedish Pat.*, 137,579 (1952); *Chem. Abstr.*, **48**, 1444, (1954).
69. K.-W. Glombitza, *Ann. Chem.*, **673**, 166 (1964).
70. H. E. Fierz-David and W. Kuster, *Helv. Chim. Acta*, **22**, 82 (1939).
71. J. von Braun and G. Lemke, *Chem. Ber.*, **55**, 3526 (1922).
72. G. F. Morrell, *J. Chem. Soc.*, **105**, 1737 (1914).
73. H. Burton and J. Kenner, *J. Chem. Soc.*, **123**, 1045 (1923).
74. J. Preston, W. de Winter, and W. L. Hofferbert, *J. Heterocyclic Chem.*, **5**, 269 (1968).
75. S. Widequist, *Arkiv. Kemi*, **7**, 117 (1954).
76. B. H. Chase and D. H. Hey, *J. Chem. Soc.*, 553 (1952); S. Ställberg-Stenhagen, *J. Am. Chem. Soc.*, **69**, 2568 (1947); J. Cason, *J. Am. Chem. Soc.*, **69**, 1548 (1947).
77. L. M. Litvinenko, *Izv. Akad. Nauk SSSR., Otd. Khim. Nauk*, 1737 (1962); L. M. Litvinenko and N. M. Oleinick, *Zh. Obshch Khim.*, **32**, 2290 (1962); *Chem. Abstr.*, **58**, 8869 (1963).
78. G. Casnati, M. R. Langella, F. Piozzi, A. Ricca, and A. Umani-Ronchi, *Gazz. Chim. Ital.*, **94**, 1221 (1964).
79. H. Leuchs and A. Schlotzer, *Chem. Ber.*, **67**, 1572 (1934).
80. H. Böhme, S. Ebel, and K. Hartke, *Chem. Ber.*, **98**, 1463 (1965); H. Böhme and K. Hartke, *Chem. Ber.*, **96**, 600 (1963).
81. F. W. Fowler and A. Hassner, *J. Am. Chem. Soc.*, **90**, 2875 (1968).
82. J. Weinstock and V. Boekelheide, *Org. Syn.*, Coll. Vol. IV, John Wiley and Sons, New York, 1963, p. 641; F. D. Popp, W. Blount, and P. Melvin, *J. Org. Chem.*, **26**, 4930 (1961).
83. B. Stoll and W. Griehl, *Helv. Chim. Acta*, **48**, 1805 (1965).
84. F. Dallacker and G. Steiner, *Ann. Chem.*, **660**, 98 (1962).
85. F. P. Baldwin, E. J. Blanchard, and P. E. Koenig, *J. Org. Chem.*, **30**, 671 (1965).
86. Q. E. Thompson, *J. Am. Chem. Soc.*, **73**, 5841 (1951).
87. M. F. Shostakovskii, M. G. Zelenskaya, F. P. Sidel'kovskaya, and B. V. Lopatin, *Izv. Akad. Nauk SSSR Otd. Khim. Nauk.*, 505 (1962); *Chem. Abstr.*, **57**, 16532 (1962).
88. P. Schlack, *Ger. Pat.*, 1,186,065 (1965); *Chem. Abstr.*, **62**, 10382 (1965).

89. K. Heyns and W. Pyrus, *Chem. Ber.*, **88**, 678 (1955).
90. G. M. Coppinger, *J. Am. Chem. Soc.*, **76**, 1372 (1954).
91. A. J. Speziale, L. R. Smith, and J. E. Fedder, *J. Org. Chem.*, **30**, 4303 (1965).
92. J. Hipkin and D. P. N. Satchell, *J. Chem. Soc.* (*B*), 345 (1966).
93. M-H. Loucheux and A. Banderet, *Bull. Soc. Chim. France*, 2242 (1961).
94. P. Rambacher and S. Mäke, *Angew. Chem. Intern. Ed. Engl.*, **7**, 465 (1968); R. K. Smalley and H. Suschitzky, *J. Chem. Soc.*, 755, (1964); T. Mukaiyama, I. Kuwajima, and Z. Suzuki, *J. Org. Chem.*, **28**, 2024 (1963)
95. K. Heyns and W. v. Bebenburg, *Chem. Ber.*, **86**, 278 (1953).
96. G. S. Reddy, L. Mandell, and J. H. Goldstein, *J. Chem. Soc.*, 1414 (1963).
97. A. R. Butler and V. Gold, *J. Chem. Soc.*, 4362 (1961); V. Gold and E. G. Jefferson *J. Chem. Soc.*, 1409 (1953).
98. F. D. Chattaway, *J. Chem. Soc.*, 2495 (1931).
99. C. H. Roeder and A. R. Day, *J. Org. Chem.*, **6**, 25 (1941).
100. G. D. Parkes and A. C. Farthing, *J. Chem. Soc.*, 1275 (1948).
101. K. A. Watanabe and J. J. Fox, *Angew. Chem. Intern. Ed. Engl.*, **5**, 579 (1966).
102. A. L. J. Beckwith and W. B. Gara, unpublished work; cf. L. C. Raiford and C. E. Greider, *J. Am. Chem. Soc.*, **46**, 430 (1924).
103. N. R. Easton and R. D. Dillard, *J. Org. Chem.*, **28**, 2465 (1963).
104. E. Sawicki and F. E. Roy, *J. Am. Chem. Soc.*, **75**, 2266 (1953); E. J. Bourne, S. H. Henry, C. E. M. Tatlow, and J. C. Tatlow, *J. Chem. Soc.*, 4014 (1952).
105. F. Weygand and R. Geiger, *Chem. Ber.*, **89**, 647 (1956).
106. R. A. Boissonnas, *Advan. Org. Chem.*, **3**, 159 (1963).
107. E. Schroeder and K. Luebke, *The Peptides*, Vol. I, Academic Press, New York, 1965.
108. J. F. W. McMomie, *Advan. Org. Chem.*, **3**, 191 (1963).
109. R. M. Herbst and D. Shemin, *Org. Syn.*, Coll. Vol. II, John Wiley and Sons, New York, 1943, p. 11.
110. J. D. Hepworth, *Org. Syn.*, **45**, 1 (1965); see also R. H. Wiley and O. H. Borum, *Org. Syn.*, Coll. Vol. IV, John Wiley and Sons, New York, 1963, p. 5.
111. W. Steglich and G. Höfle, *Tetrahedron Letters*, 1619 (1968).
112. T. E. de Kiewiet and H. Stephen, *J. Chem. Soc.*, 82 (1931); M. Friefelder, *J. Org. Chem.*, **27**, 1092 (1962).
113. A. J. Zambito and E. G. Howe, *Org. Syn.*, **40**, 21 (1960); (see correction in *Org. Syn.*, **45**, 32 (1965)).
114. F. E. Gould, G. S. Johnson, and A. F. Ferris, *J. Org. Chem.*, **25**, 1658 (1960).
115. A. R. Emery and V. Gold, *J. Chem. Soc.*, 1443, 1447, 1455 (1950).
116. E. J. Bourne, S. H. Henry, C. E. M. Tatlow, and J. C. Tatlow, *J. Chem. Soc.*, 4014 (1952).
117. G. R. Clemo and G. A. Swan, *J. Chem. Soc.*, 603 (1945); C. W. Huffman, *J. Org. Chem.*, **23**, 727 (1958).
118. J. C. Sheehan and D-D. H. Yang, *J. Am. Chem. Soc.*, **80**, 1154 (1958); K. Hofman, E. Stutz, G. Spühler, H. Yajima, and E. T. Schwartz, *J. Am. Chem. Soc.*, **82**, 3727 (1960).
119. E. J. Corey, N. H. Andersen, R. M. Carlson, J. Paust, E. Vedejs, I. Vlattas, and R. E. K. Winter, *J. Am. Chem. Soc.*, **90**, 3245 (1968).

120. J. R. Vaughan and R. L. Osato, *J. Am. Chem. Soc.*, **73**, 5553 (1951); G. Losse and E. Demuth, *Chem. Ber.*, **94**, 1762 (1961).
121. N. F. Albertson, *Org. Reactions*, **12**, 157 (1962).
122. G. W. Anderson, F. M. Callahan, and J. E. Zimmerman, *J. Am. Chem. Soc.*, **89**, 5012 (1967).
123. H. P. Fischer and C. A. Grob, *Helv. Chim. Acta.*, **47**, 564 (1964).
124. H. A. Friedman, K. A. Watanabe, and J. J. Fox, *J. Org. Chem.*, **32**, 3775 (1967).
125. T. H. Applewhite, J. S. Nelson, and L. A. Goldblatt, *J. Am. Oil Chemists Soc.*, **40**, 101 (1963).
126. T. H. Applewhite and J. S. Binder, *J. Am. Oil Chemists Soc.*, **44**, 423 (1967).
127. G. W. Kenner and R. J. Stedman, *J. Chem. Soc.*, 2069 (1952).
128. L. Field and P. H. Settlage, *J. Am. Chem. Soc.*, **76**, 1222 (1954).
129. C. G. Overberger and E. Sarlo, *J. Am. Chem. Soc.*, **85**, 2446 (1963).
130. J. H. Brewster and C. J. Ciotti, *J. Am. Chem. Soc.*, **77**, 6214 (1955).
131. V. Iyer and N. K. Mathur, *Anal. Chim. Acta*, **33**, 554 (1965).
132. A. W. D. Avison, *J. Chem. Soc.*, 732 (1955); G. D. Sabato and W. P. Jencks, *J. Am. Chem. Soc.*, **83**, 4393 (1961).
133. H. R. Snyder and C. T. Elston, *J. Am. Chem. Soc.*, **76**, 3039 (1954).
134. F. S. Spring and J. C. Woods, *J. Chem. Soc.*, 625 (1945).
135. M. L. Sherrill, F. L. Schaeffer, and E. P. Shoyer, *J. Am. Chem. Soc.*, **50**, 474 (1928).
136. L. Schmid and H. Mann, *Monatsh. Chem.*, **85**, 864 (1954).
137. E. H. Farmer and J. Kracovski, *J. Chem. Soc.*, 680 (1927).
138. E. Hoffmann and H. Schiff-Shenhav, *J. Org. Chem.*, **27**, 4686 (1962).
139. N. V. Subba Rao and C. V. Ratnam, *J. Sci. Ind. Res.*, (*India*), **21B**, 45 (1962).
140. R. Chiron and Y. Graff, *Bull. Soc. Chim. France*, 1904 (1967).
141. M. P. Cava, A. A. Deana, K. Muth, and M. J. Mitchell, *Org. Syn.*, **41**, 93 (1961).
142. C. K. Sauers and R. J. Cotter, *J. Org. Chem.*, **26**, 6, (1961).
143. J. W. Alexander and S. M. McElvain, *J. Am. Chem. Soc.*, **60**, 2285 (1938); W. Flitsch, *Chem. Ber.*, **94**, 2494 (1961).
144. F. G. Mann and J. A. Reid, *J. Chem. Soc.*, 2057 (1952).
145. V. V. Kiselev and R. A. Konovalova, *Zh. Obshch. Khim.*, **22**, 2233 (1952); *Chem. Abstr.*, **48**, 691 (1954).
146. S. Akaboshi and M. Suzuki, *Pharm. Bull.* (*Japan*), **3**, 65 (1955); *Chem. Abstr.*, **50**, 1638 (1956).
147. M. Naps and I. B. Johns, *J. Am. Chem. Soc.*, **62**, 2450 (1940); A. Zilkha and U. Golik, *J. Org. Chem.*, **28**, 2007 (1963).
148. T. V. Sheremeteva and V. A. Gusinskaya, *Bull. Acad. Sci. USSR, Div. Chem. Science*, 657 (1966).
149. A. Foucaud, *Bull. Soc. Sci. Bretagne*, **35**, 88 (1960); *Chem. Abstr.*, **55**, 3516 (1961).
150. R. P. Staiger and E. B. Miller, *J. Org. Chem.*, **24**, 1214 (1959); R. P. Staiger and E. C. Wagner, *J. Org. Chem.*, **18**, 1427 (1953); **13**, 347 (1948).
151. J. F. Bunnett and M. B. Naff, *J. Am. Chem. Soc.*, **88**, 4001 (1966).
152. L. J. Dolby and S.-i. Sakai, *J. Am. Chem. Soc.*, **86**, 1890 (1964).

153. G. H. Foster, J. Harley-Mason, and W. R. Waterfield, *Chem. Commun.*, 21 (1967).
154. J. Harley-Mason and Atta-ur-Rahman, *Chem. Commun.*, 1048 (1967).
155. A. W. Burgstahler, *J. Am. Chem. Soc.*, **73**, 3021 (1951).
156. J. B. Polya and T. M. Spotswood, *Rec. Trav. Chim.*, **67**, 927 (1948).
157. D. Davidson and H. Skovronek, *J. Am. Chem. Soc.*, **80**, 376 (1958).
158. R. H. Wiley and W. B. Guerrant, *J. Am. Chem. Soc.*, **71**, 981 (1949).
159. S. A. Abbas and W. J. Hickinbottom, *J. Chem. Soc.* (*C*), 1305 (1966).
160. T. C. Bruice, A. Donzel, R. W. Huffman, and A. R. Butler, *J. Am. Chem. Soc.*, **89**, 2106 (1967).
161. L. R. Fedor, T. C. Bruice, K. L. Kirk, and J. Meinwald, *J. Am. Chem. Soc.*, **88**, 108 (1966).
162. W. P. Jencks and M. Gilchrist, *J. Am. Chem. Soc.*, **88**, 104 (1966).
163. A. Sami A. S. Shawali, and S. S. Biechler, *J. Am. Chem. Soc.*, **89**, 3020, (1967).
164. L. do Amaral, K. Koehler, D. Bartenbach, T. Pletcher, and E. H. Cordes, *J. Am. Chem. Soc.*, **89**, 3537 (1967).
165. S.-H. Chu and H. G. Mautner, *J. Org. Chem.*, **31**, 308 (1966).
166. B. B. Corson, R. W. Scott, and C. E. Vose, *Org. Syn.*, Coll. Vol. I, 2nd ed., John Wiley and Sons, New York, 1941, p. 179.
167. D. T. Mowry and J. M. Butler, *Org. Syn.*, Coll. Vol. IV, John Wiley and Sons, New York, 1963, p. 486.
168. F. B. La Forge, *J. Am. Chem. Soc.*, **50**, 2477 (1928).
169. W. Rohrs and S. Lang, *J. Prakt. Chem.*, **158**, 109 (1941).
170. A. Galat, *J. Am. Chem. Soc.*, **70**, 2596 (1948).
171. W. A. Jacobs and M. Heidelberger, *Org. Syn.*, Coll. Vol. I, 2nd ed., John Wiley and Sons, New York, 1941, p. 153; E. L. d'Ouville and R. Connor, *J. Am. Chem. Soc.*, **60**, 33 (1938).
172. K. Butler, D. R. Lawrance, and M. Stacey, *J. Chem. Soc.*, 740 (1958).
173. E. C. Taylor and A. J. Crovetti, *J. Am. Chem. Soc.*, **78**, 214 (1956).
174. E. Mosettig and J. W. Krueger, *J. Org. Chem.*, **3**, 317 (1939).
175. J. Kleinberg and L. F. Audrieth, *Org. Syn.*, Coll. Vol. III, John Wiley and Sons, New York, 1955, p. 576.
176. J. H. Boothe, A. S. Kende, T. L. Fields, and R. G. Wilkinson, *J. Am. Chem. Soc.*, **81**, 1006 (1959).
177. C. F. H. Allen and W. J. Humphlett, *Org. Syn.*, Coll. Vol. IV, John Wiley and Sons, New York, 1963, p. 80.
178. H. Baganz and L. Domaschke, *Arch. Pharm.*, **295**, 758 (1962).
179. G. F. D'Alelio and E. E. Reid, *J. Am. Chem. Soc.*, **59**, 111 (1937).
180. C. F. H. Allen and J. Van Allan, *Org. Syn.*, Coll. Vol. III, John Wiley and Sons, New York, 1955, p. 765.
181. B. C. McKusick and M. E. Hermes, *Org. Syn.*, **41**, 14 (1961); G. Moffat, M. V. Newton, and G. J. Papenmeier, *J. Org. Chem.*, **27**, 4058 (1962).
182. A. Barth and G. Losse, *Z. Naturforsch*, **19B**, 264 (1964).
183. R. L. Betts and L. P. Hammett, *J. Am. Chem. Soc.*, **59**, 1568 (1937).
184. J. A. Monick, *J. Am. Oil Chemists Soc.*, **39**, 213 (1962).
185. R. J. De Feoand and P. D. Strickler, *J. Org. Chem.*, **28**, 2915 (1963).
186. J. Petit and R. Poisson, *Compt. Rend.*, **247**, 1628 (1958).
187. E. S. Stern, *Chem. Ind.* (*London*), 277 (1956).

188. F. Bodroux, *Bull. Soc. Chim. France*, **33**, 831 (1905).
189. H. L. Bassett and C. R. Thomas, *J. Chem. Soc.*, 1188 (1954).
190. R. P. Houghton and C. S. Williams, *Tetrahedron Letters*, 3929 (1967).
191. K. Okawa and S. Hase, *Bull. Chem. Soc. Japan*, **36**, 754 (1963).
192. W. P. Utermohlen, *U.S. Pat.*, 2,472,633 (1949); *Chem. Abstr.*, **43**, 6652 (1949).
193. A. Buzas, F. Canac, C. Egnell, and P. Freon, *Compt. Rend.*, *Ser. C.*, **262**, 658 (1966).
194. H. G. Viehe, *Angew Chem. Intern. Ed. Engl.*, **6**, 767 (1967); H. G. Viehe, R. Fuks, and M. Reinstein, *Angew. Chem. Intern. Ed. Engl.*, **3**, 581 (1964).
195. Y. Ueno, T. Takaya, and E. Imoto, *Bull. Chem. Soc. Japan*, **37**, 864 (1964).
196. J. Pokorný, *Chem. Zvesti*, **18**, 218 (1964).
197. H. E. Carter, *Org. Reactions*, **3**, 198 (1946).
198. M. Bodánszky, *Chem. Ind.* (*London*), 524 (1957).
199. G. Losse and H. Weddige, *Ann. Chem.*, **636**, 144 (1960).
200. C. L. Stevens and M. E. Munk, *J. Am. Chem. Soc.*, **80**, 4065, 4069 (1958).
201. J. F. Arens, *Rec. Trav. Chim.*, **74**, 769 (1955).
202. R. B. Woodward, R. A. Olofson, and H. Mayer, *J. Am. Chem. Soc.*, **83**, 1010 (1961).
203. B. O. Handford, J. H. Jones, G. T. Young, and T. F. N. Johnson, *J. Chem. Soc.*, 6814 (1965).
204. G. H. L. Nefkens and G. I. Tesser, *J. Am. Chem. Soc.*, **83**, 1263 (1961); G. W. Anderson, J. E. Zimmerman, and F. M. Callahan, *J. Am. Chem. Soc.*, **86**, 1839 (1964).
205. C. D. Hurd and S. Hayao, *J. Am. Chem. Soc.*, **74**, 5889 (1952).
206. T. L. Gresham, J. E. Jansen, F. W. Shaver, R. E. Bankert, and F. T. Fiedorek, *J. Am. Chem. Soc.*, **73**, 3168 (1951).
207. C. E. Kaslow and N. B. Sommer, *J. Am. Chem. Soc.*, **68**, 644 (1946); K. Schank, *Chem. Ber.*, **100**, 2292 (1967).
208. J. W. Williams and J. A. Krynitsky, *Org. Syn.*, Coll. Vol. III, John Wiley and Sons, New York, 1955, p. 10.
209. N. H. Cromwell and K. E. Cook, *J. Am. Chem. Soc.*, **80**, 4573 (1958).
210. A. Burger and A. Hofstetter, *J. Org. Chem.*, **24**, 1290 (1959).
211. M. Uskokovic and M. Gut, *Helv. Chim. Acta*, **42**, 2258 (1959).
212. J. B. Jones and J. M. Young, *Can. J. Chem.*, **44**, 1059 (1966).
213. J. A. Baker and J. F. Harper, *J. Chem. Soc.* (*C*), 2148 (1967).
214. A. Eschenmoser, *Pure Appl. Chem.*, **7**, 297 (1963); E. Bertele, H. Boos, J. D. Dunitz, F. Elsinger, A. Eschenmoser, I. Felner, H. P. Gribi, H. Gschwend, E. F. Meyer, M. Pesaro, and R. Scheffold, *Angew. Chem.*, *Intern. Ed. Engl.*, **3**, 490 (1964).
215. R. K. Hill, *J. Org. Chem.*, **22**, 830 (1957).
216. D. C. Bishop and J. F. Cavalla, *J. Chem. Soc.* (*C*), 802 (1966); J. W. Lynn, *J. Org. Chem.*, **21**, 578 (1956).
217. A. A. Patchett, F. Hoffman, F. F. Giarrusso, H. Schwam, and G. E. Arth, *J. Org. Chem.*, **27**, 3822 (1967).
218. W. E. Coyne and J. W. Cusic, *J. Med. Chem.*, **10**, 541 (1967); K-H. Wünsch, A. Ehlers, and H. Beyer, *Z. Chem.*, **7**, 185 (1967).
219. R. W. Holley and A. D. Holley, *J. Am. Chem. Soc.*, **71**, 2129 (1949); G. Cignarella, G. F. Cristiani, and E. Testa, *Ann. Chem.*, **661**, 181 (1963).

220. S. Searles and R. E. Wann, *Chem. Ind. (London)*, 2097 (1964).
221. R. B. Woodward, K. Heusler, J. Gosteli, P. Naegeli, W. Oppolzer, R. Ramage, S. Ranganathan, and H. Vorbruggen, *J. Am. Chem. Soc.*, **88**, 852 (1966).
222. E. S. Rothman, S. Serota, and D. Swern, *J. Org. Chem.*, **29**, 646 (1964).
223. E. A. Allred and M. D. Hurwitz, *J. Org. Chem.*, **30**, 2376 (1965).
224. N. Jochum, K. Riefstahl, and A. Tilly, *Ger. Pat.*, 1,164,397 (1964); *Chem. Abstr.*, **60**, 15742 (1964).
225. F. Salmon-Legagneur, *Bull. Soc. Chim. France*, 580 (1952).
226. R. K. Robins, J. K. Horner, C. V. Greco, C. W. Noell, and C. G. Beames, *J. Org. Chem.*, **28**, 3041 (1963).
227. See *inter alia*: R. M. Topping and D. E. Tutt, *Chem. Commun.*, 698 (1966); J. H. Shafer and H. Morawetz, *J. Org. Chem.*, **28**, 1890 (1963); M. T. Behme and E. H. Cordes. *J. Org. Chem.*, **29**, 1255 (1964).
228. G. H. Coleman and A. M. Alvarado, *Org. Syn.*, Coll. Vol. 1, 2nd ed., John Wiley and Sons, New York, 1941, p. 3.
229. H. Morawetz and P. S. Otaki, *J. Am. Chem. Soc.*, **85**, 463 (1963).
230. T. Higuchi, T. Miki, A. C. Shah, and A. K. Herd, *J. Am. Chem. Soc.*, **85**, 3655 (1963).
231. E. R. Shepard, H. D. Porter, J. F. Noth, and C. K. Simmans, *J. Org. Chem.*, **17**, 568 (1952).
232. C. N. Webb, *Org. Syn.*, Coll. Vol. 1, 2nd ed., John Wiley and Sons, New York, 1941, p. 82.
233. E. T. Roe, J. T. Scanlan, and D. Swern, *J. Am. Chem. Soc.*, **71**, 2219 (1949).
234. S. M. McElvain and E. H. Pryde, *J. Am. Chem. Soc.*, **71**, 326 (1949).
235. D. V. Nightingale and J. E. Johnson, *J. Heterocyclic Chem.*, **4**, 102 (1967).
236. Y. I. Leitman and M. S. Pevzner, *Zh. Prikl. Khim.*, **36**, 632 (1963); *Chem. Abstr.*, **59**, 7422 (1963).
237. L. F. Fieser and J. E. Jones, *Org. Syn.*, Coll. Vol. III, John Wiley and Sons, New York, 1955, p. 590.
238. B. C. McKusich and O. W. Webster, *Org. Syn.*, **41**, 102 (1961).
239. M. L. Fein and E. M. Filachione, *J. Am. Chem. Soc.*, **75**, 2097 (1953).
240. M. Walter, H. Besendorf, and O. Schnider, *Helv. Chim. Acta*, **44**, 1546 (1961).
241. E. C. Taylor, A. McKillop, and R. E. Ross, *J. Am. Chem. Soc.*, **87**, 1984, 1990 (1965).
242. R. L. Frank, W. R. Schmitz, and B. Zeidman, *Org. Syn.*, Coll. Vol. III, John Wiley and Sons, New York, 1955, p. 328.
243. R. B. Moffett, *J. Org. Chem.*, **14**, 862 (1949).
244. F. B. Zienty and G. W. Steahly, *J. Am. Chem. Soc.*, **69**, 715 (1947).
245. See *inter alia* R. B. Woodward, F. Sondheimer, D. Taub, K. Heusler, and W. M. McLamore, *J. Am. Chem. Soc.*, **74**, 4223 (1952); N. J. Doorenbos and C. L. Huang, *J. Org. Chem.*, **26**, 4548 (1961).
246. N. J. Doorenbos, C. L. Huang, C. R. Tamorria, and M. T. Wu, *J. Org. Chem.*, **26**, 2546 (1961).
247. J. Brown, S. C. K. Su, and J. A. Shafer, *J. Am. Chem. Soc.*, **88**, 4468 (1966).
248. G. Paris, L. Berlinguet, and R. Gaudry, *Org. Syn.*, Coll. Vol. IV, John Wiley and Sons, New York, 1963, p. 496.

249. H. T. Clarke and L. D. Behr, *Org. Syn.*, Coll. Vol. II, John Wiley and Sons, New York, 1943, p. 562.
250. W. A. Noyes and P. K. Porter, *Org. Syn.*, Coll. Vol. I, 2nd ed., John Wiley and Sons, New York, 1941, p. 457.
251. A. K. Bose, *Org. Syn.*, **40**, 82 (1960).
252. M. Michman, S. Patai, and I. Shenfeld, *J. Chem. Soc.* (*C*), 1337 (1967).
253. R. N. Ring, J. G. Sharefkin, and D. Davidson, *J. Org. Chem.*, **27**, 2428 (1962).
254. S. Sugasawa and II. Shigehara, *J. Pharm. Soc. Japan*, **62**, 532 (1942); *Chem. Abstr.*, **45**, 2861 (1951); R. G. Jargue, M. de la Morena Calvet, and F. Marquez Archilla. *Anales Real. Soc. Espan. Fis. Quim.*, (*Madrid*), *Ser. B*, **54**, 233, (1958).
255. E. Cherbuliez and F. Landolt, *Helv. Chim. Acta*, **29**, 1438 (1946).
256. J. L. Guthrie and N. Rabjohn, *Org. Syn.*, Coll. Vol. IV, John Wiley and Sons, New York, 1963, p. 513.
257. A.-U. Rahman, *Rec. Trav. Chim.*, **75**, 164 (1956).
258. A.-U. Rahman, M. A. Medrano, and O. P. Mittal, *Rec. Trav. Chim.*, **79**, 188 (1960); A.-U. Rahman, M. A. Medrano, and B. E. Jeanneret, *J. Org. Chem.*, **27**, 3315 (1962).
259. A. V. Kirsanov and Y. M. Zolotov. *J. Gen. Chem. USSR* (*Engl. Transl.*), **12**, 675 (1949).
260. H. G. O. Becker and K. F. Funk, *J. Prakt. Chem.*, **14**, 55 (1961).
261. S. Goldschmidt and F. Obermeier, *Ann. Chem.*, **588**, 24 (1954).
262. R. D. Youssefyeh and R. W. Murray, *Chem. Ind.* (*London*), 1531 (1966).
263. J. Klosa, *J. Prakt. Chem.*, **19**, 45 (1963).
264. I. N. Zhmurova and Y. Voitsekhovskaya, *Zh. Obshch. Khim.*, **29**, 2083 (1959); *Chem. Abstr.*, **54**, 8681 (1960).
265. H. Schönenberger, J. Holzheu-Eckardt, and E. Bamann, *Arzneimittel-Forsch*, **14**, 324 (1964).
266. W. Grassmann and E. Wünsch, *Chem. Ber.*, **91**, 449 (1958).
267. A. K. Bose, *J. Indian Chem. Soc.*, **31**, 108 (1954).
268. P. Nelson and A. Pelter, *J. Chem. Soc.*, 5142 (1965).
269. J. H. Billman and J. W. McDowell, *J. Org. Chem.*, **27**, 2640 (1962).
270. F. F. Blicke and C. J. Lu, *J. Am. Chem. Soc.*, **74**, 3933, (1952).
271. D. G. Gehring, W. A. Mosher, and G. S. Reddy, *J. Org. Chem.*, **31**, 3436 (1966).
272. B. Sukornick, *Org. Syn.*, **40**, 103 (1960).
273. J. Kollonitsch, Λ. Hajós, V. Gábor, and M. Kraut, *Acta Chim. Acad. Sci. Hung.*, **5**, 13 (1954).
274. K. E. Hamlin and A. W. Weston, *Org. Reactions*, **9**, 1 (1957).
275. A. L. J. Beckwith, *J. Chem. Soc.*, 2248 (1962).
276. H. M. Walborsky and F. J. Impastato, *Chem. Ind.* (*London*), 1690 (1958).
277. H. M. Walborsky and F. J. Impastato, *J. Am. Chem. Soc.*, **81**, 5835 (1959); H. M. Walborsky, A. A. Youssef, and J. M. Motes, *J. Am. Chem. Soc.*, **84**, 2465 (1962); D. E. Applequist and A. H. Peterson, *J. Am. Chem. Soc.*, **83**, 862 (1961).
278. T. R. Lea and R. Robinson, *J. Chem. Soc.*, 2351 (1926).
279. J. F. Bunnett and B. F. Hrutfiord, *J. Org. Chem.*, **27**, 4152 (1962).
280. C. F. H. Allen and A. Bell, *Org. Syn.*, Coll. Vol. III, John Wiley and Sons, New York, 1955, p. 846.

281. R. K. Smalley and H. Suschitzky, *J. Chem. Soc.*, 755 (1964).
282. P. A. S. Smith, *Org. Reactions*, **3**, 337 (1946).
283. J. Honzl and J. Rudinger, *Collection Czech. Chem. Commun.*, **26**, 2333 (1961).
284. S. Sugasawa and H. Tomisawa, *Pharm. Bull. (Japan)*, **3**, 32, (1955).
285. For a recent example of a typical procedure for hydrazide formation see: P. A. S. Smith, *Org. Syn.*, Coll. Vol. IV, John Wiley and Sons, New York, 1963, p. 819.
286. A. Stoll and A. Hoffman, *Helv. Chim. Acta.*, **26**, 922, 944 (1943).
287. L. Bernardi and O. Goffredo, *Gazz. Chim. Ital.*, **94**, 947 (1964).
288. H. Zahn and J. Kunde, *Chem. Ber.*, **94**, 2470 (1961).
289. F. Weygand and K. Hunger, *Chem. Ber.*, **95**, 1 (1962); M. Brenner and P. Zimmerman, *Helv. Chim. Acta*, **40**, 1933 (1957).
290. N. A. Smart, G. T. Young, and M. W. Williams, *J. Chem. Soc.*, 3902 (1960).
291. W. Baker and W. D. Ollis, *J. Chem. Soc.*, 556 (1951).
292. F. Bergel, J. M. Johnson, and R. Wade, *J. Chem. Soc.*, 3802 (1962).
293. R. D. Burpitt and V. W. Goodlett, *J. Org. Chem.*, **30**, 4308 (1965).
294. J. W. Hinman, E. L. Caron, and H. N. Christensen, *J. Am. Chem. Soc.*, **72**, 1620 (1950).
295. P. T. Sah and S. C. Chen, *J. Chinese Chem. Soc.*, **14**, 74 (1946); *Chem. Abstr.*, **43**, 7447 (1949); K. J. Karrman, *Svensk Kem. Tidskr.*, **60**, 61 (1948).
296. T. Shingaki, *Nippon Kagaku Zasshi*, **80**, 55 (1959); *Chem. Abstr.*, **55**, 4408 (1961).
297. J. Rudinger, *Pure Appl. Chem.*, **7**, 335 (1963).
298. R. N. Lacey, in *The Chemistry of Alkenes* (Ed. S. Patai), Interscience Publishers, London, 1964, p. 1175.
299. J. M. Briody and D. P. N. Satchell, *Tetrahedron*, **22**, 2649 (1966).
300. C. D. Hurd, *Org. Syn.*, Coll. Vol. I, 2nd ed., John Wiley and Sons, New York, 1941, p. 330.
301. G. Quadbeck, *Angew. Chem.*, **68**, 369 (1956); M. Bergmann and F. Stern, *Chem. Ber.*, **63**, 437 (1930); *Ger. Pat.*, 453,577 (1927).
302. A. Neuberger, *Biochem. J.*, **32**, 1452 (1938).
303. H. Bredereck, R. Gompper, and K. Klemm, *Chem. Ber.*, **92**, 1456 (1959).
304. R. E. Dunbar and G. C. White, *J. Org. Chem.*, **23**, 915 (1958).
305. R. E. Dunbar and W. M. Swenson, *J. Org. Chem.*, **23**, 1793 (1958).
306. R. Graf., *Chem. Ann.*, **661**, 111 (1963).
307. W. E. Bachmann and W. S. Struve, *Org. Reactions*, **1**, 38 (1942).
308. D. Fleš and A. Markovac-Prpić, *Croat. Chem. Acta*, **29**, 79 (1957); **28**, 73 (1956); *Chem. Abstr.*, **52**, 11744 (1958); **51**, 1841 (1957).
309. A. Galat and G. Elion, *J. Am. Chem. Soc.*, **65**, 1566 (1943).
310. G. R. Pettit and E. G. Thomas, *J. Org. Chem.*, **24**, 895 (1959); G. R. Pettit, M. V. Kalnins, T. M. H. Liu, E. G. Thomas, and K. Parent, *J. Org. Chem.*, **26**, 2563 (1961).
311. F. Einberg, *J. Org. Chem.*, **32**, 3687 (1967).
312. F. S. Spring and J. C. Woods, *J. Chem. Soc.*, **625**, (1945).
313. A. K. Bose, F. Greer, J. S. Gots, and C. S. Price, *J. Org. Chem.*, **24**, 1309 (1959).
314. R. Oda, Y. Hayashi, and T. Takai, *Tetrahedron*, **24**, 4051 (1968).
315. G. H. L. Nefkens, G. I. Tesser, and R. J. F. Nivard, *Rec. Trav. Chim.*, **79**, 688 (1960).

316. G. W. Anderson and R. Paul, *J. Am. Chem. Soc.*, **80**, 4423 (1958); R. Paul and G. W. Anderson, *J. Am. Chem. Soc.*, **82**, 4596 (1960).

317. H. C. Beyerman and W. M. Van Den Brink, *Rec. Trav. Chim.*, **80**, 1372 (1961).

318. H. A. Staab, M. Luking, and F. H. Dürr, *Chem. Ber.*, **95**, 1275 (1962).

319. J. S. Reddy, *Chem. Ind. (London)*, 1426 (1965).

320. see *inter alia*, J. Hipkin and D. P. N. Satchell, *J. Chem. Soc. (B)*, 365 (1967); M. J. Gregory and T. C. Bruice, *J. Am. Chem. Soc.*, **89**, 2121 (1967); K. A. Connors and M. L. Bender, *J. Org. Chem.*, **26**, 2498 (1961); R. K. Chaturvedi, A. E. MacMahon, and G. L. Shmir, *J. Am. Chem. Soc.*, **89**, 6984 (1967).

321. W. Reppe, *Ann. Chem.*, **582**, 14 (1953).

322. H. W. Sternberg, I. Wender, R. A. Friedel, and M. Orchin, *J. Am. Chem. Soc.*, **75**, 3148 (1953); A. Schiffers and F. Glaser, *Chem. Ztg.*, **85**, 435 (1961).

323. J. Falbe and F. Korte, *Chem. Ber.*, **98**, 1928 (1965).

324. J. Falbe and F. Korte, *Chem. Ber.*, **95**, 2680 (1962).

325. Y. Wolman, P. M. Gallop, A. Patchornik, and A. Berger, *J. Am. Chem. Soc.*, **84**, 1889 (1962).

326. H. Sperber, D. Papa, and E. Schwenk, *J. Am. Chem. Soc.*, **70**, 3092 (1948).

327. S. Sarel and M. S. Newman, *J. Am. Chem. Soc.*, **78**, 5416 (1956).

328. R. G. Jones, *J. Am. Chem. Soc.*, **73**, 5610 (1951).

329. R. Gaudry and C. Godin, *J. Am. Chem. Soc.*, **76**, 139 (1954).

330. K. Scholz and L. Panizzon, *Helv. Chim. Acta*, **37**, 1605 (1954).

331. C. Wiener, C. H. Schroeder, and K. P. Link, *J. Am. Chem. Soc.*, **79**, 5301 (1957).

332. A. Mulford and L. W. Kissinger, *J. Org. Chem.*, **30**, 945 (1965).

333. G. F. Morris and C. R. Hauser, *J. Org. Chem.*, **26**, 4741 (1961).

334. R. A. Turner and C. Djerassi, *J. Am. Chem. Soc.*, **72**, 3081 (1950).

335. R. A. Turner, *J. Am. Chem. Soc.*, **68**, 1607 (1946).

336. R. H. Wiley and W. E. Waddey, *Org. Syn.*, Coll. Vol. III, John Wiley and Sons, New York, 1955, p. 560.

337. H. Wenner, *Org. Syn.*, Coll. Vol. IV, John Wiley and Sons, New York, 1963, p. 760.

338. F. Nerdel and H. Rachel, *Chem. Ber.*, **89**, 671 (1956).

339. R. I. Meltzer, R. J. Stanaback, S. Farber, and W. B. Lutz, *J. Org. Chem.*, **26**, 1418 (1961).

340. H. R. Snyder and C. T. Elston, *J. Am. Chem. Soc.*, **76**, 3039 (1954).

341. C. R. Hauser and C. J. Eby, *J. Am. Chem. Soc.*, **79**, 725 (1957).

342. C. R. Hauser and D. S. Hoffenberg, *J. Org. Chem.*, **20**, 1448 (1955).

343. J. F. Wolfe, C. J. Eby, and C. R. Hauser, *J. Org. Chem.*, **30**, 55 (1965).

344. J. F. Wolfe and C. L. Mao, *J. Org. Chem.*, **31**, 3069 (1966).

345. M. Orchin and L. Reggel, *J. Am. Chem. Soc.*, **73**, 436 (1951).

346. W. L. Mosby, *J. Am. Chem. Soc.*, **75**, 3600 (1953).

347. C. R. Noller, *Org. Syn.*, Coll. Vol. II, John Wiley and Sons, New York, 1943, p. 586.

348. J. S. Buck and W. S. Ide, *Org. Syn.*, Coll. Vol. II, John Wiley and Sons, New York, 1943, p. 44.

349. M. P. Cava, R. L. Little, and D. R. Napier, *J. Am. Chem. Soc.*, **80**, 2257 (1958).

350. C. M. Atkinson and C. J. Sharpe, *J. Chem. Soc.*, 2858 (1959).

178 A. L. J. Beckwith

351. L. McMaster and C. R. Noller, *J. Indian Chem. Soc.*, **12**, 653 (1935).
352. K. B. Wiberg, *J. Am. Chem. Soc.*, **75**, 3961 (1953).
353. G. B. Payne, D. H. Denning, and P. H. Williams, *J. Org. Chem.*, **26**, 659 (1961); G. B. Payne, *Tetrahedron*, **18**, 763 (1962).
354. G. B. Payne and P. H. Williams, *J. Org. Chem.*, **26**, 651 (1961).
355. E. C. Kornfeld, E. J. Fornefeld, G. B. Kline, M. J. Mann, D. E. Morrison, R. G. Joncs, and R. B. Woodward, *J. Am. Chem. Soc.*, **78**, 3087 (1956).
356. D. H. R. Barton, E. F. Lier, and J. F. McGhie, *J. Chem. Soc. (C)*, 1031 (1968).
357. C. Sannie and H. Lapin, *Bull. Soc. Chim. France*, 369 (1952).
358. A. Galat, *J. Am. Chem. Soc.*, **70**, 3945 (1948).
359. J. M. Bobbitt and D. A. Scola, *J. Org. Chem.*, **25**, 560 (1960); J. M. Bobbitt and R. E. Doolittle, *J. Org. Chem.*, **29**, 2298 (1964).
360. C. Berther, *Chem. Ber.*, **92**, 2616 (1959).
361. S. M. McElvain and B. E. Tate, *J. Am. Chem. Soc.*, **73**, 2233 (1951).
362. H. E. Johnson and D. G. Crosby, *J. Org. Chem.*, **28**, 3255 (1963).
363. H. E. Johnson and D. G. Crosby, *J. Org. Chem.*, **27**, 798 (1962).
364. K.-i. Watanabe, *Bull. Chem. Soc. Japan*, **37**, 1325 (1964); K.-i. Watanabe and K. Sakai, *Bull. Chem. Soc. Japan*, **39**, 8 (1966).
365. M. J. Cook, E. J. Forbes, and G. M. Khan, *Chem. Commun.*, 121 (1966).
366. R. Breslow, R. Fairweather, and J. Keana, *J. Am. Chem. Soc.*, **89**, 2135 (1967).
367. H. Wieland and E. Dorrer, *Chem. Ber.*, **63B**, 404 (1930).
368. J. J. Ritter and P. P. Minieri, *J. Am. Chem. Soc.*, **70**, 4045 (1948).
369. J. J. Ritter and J. Kalish, *J. Am. Chem. Soc.*, **70**, 4048 (1948).
370. F. R. Benson and J. J. Ritter, *J. Am. Chem. Soc.*, **71**, 4128 (1949).
371. G. Glikmans, B. Torck, M. Hellin, and F. Coussemant, *Bull. Soc. Chim. France*, 1383 (1966).
372. F. Klages and W. Grill, *Ann. Chem.*, **594**, 21(1955); F. Klages, R. Ruhnau, and W. Hauser, *Ann. Chem.*, **626**, 60 (1959).
373. H. Meerwein, *Angew. Chem.*, **67**, 379 (1955); H. Meerwein, P. Laasch, R. Mersch and J. Spille, *Chem. Ber.*, **89**, 209 (1956).
374. C. L. Parris, *Org. Syn.*, **42**, 16 (1962).
375. H. Plaut and J. J. Ritter, *J. Am. Chem. Soc.*, **73**, 4076 (1951).
376. R. M. Lusskin and J. J. Ritter, *J. Am. Chem. Soc.*, **72**, 5577 (1950).
377. H. J. Barber and E. Lunt, *J. Chem. Soc.*, 1187 (1960).
378. J. J. Ritter and J. Kalish, *Org. Syn.*, **44**, 44 (1964).
379. B. V. Shetty, *J. Org. Chem.*, **26**, 3002 (1961).
380. L. W. Hartzel and J. J. Ritter, *J. Am. Chem. Soc.*, **71**, 4130 (1949).
381. A. Kaluszyner, S. Blum, and E. D. Bergmann, *J. Org. Chem.*, **28**, 3588 (1963).
382. V. A. Zagorevskii and K. I. Lopatina, *Zh. Organ. Khim.*, **1**, 366 (1965); *Chem. Abstr.*, **62**, 16190 (1965).
383. F. L. Ramp, *J. Polymer Sci.*, *A*, **3**, 1877 (1965).
384. D. Giraud-Clenet and J. Anatol, *Compt. Rend.*, *Ser. C.*. **262**, 224 (1966).
385. S. Julia and C. Papartoniou, *Compt. Rend.*, *Ser. C.*, **260**, 1440 (1965).
386. F. Johnson and R. Madroñero, *Advan. Heterocyclic Chem.*, **6**, 95 (1966).
387. R. L. Holmes, J. P. Moreau, and G. Sumrell, *J. Am. Oil Chemists Soc.*, **42**, 922 (1965).

388. T. Clarke, T. Devine, and D. W. Dicker, *J. Am. Oil Chemists Soc.*, **41**, 78 (1964).

389. K. Hamamoto and M. Yoshioka, *Nippon Kagaku Zasshi*, **80**, 326 (1959); *Chem. Abstr.*, **55**, 4349 (1961).

390. W. Haaf, *Chem. Ber.*, **97**, 3234 (1964).

391. W. Haaf, *Chem. Ber.*, **96**, 3359 (1963).

392. H. Stetter, J. Mayer, M. Schwarz, and K. Wulff, *Chem. Ber.*, **93**, 226 (1960); T. Sasaki, S. Eguchi, and T. Toru, *Bull. Chem. Soc. Japan*, **41**, 236 (1968).

393. J. Cast and T. S. Stevens, *J. Chem. Soc.*, 4180 (1953).

394. G. W. Cannon, K. K. Grebber, and Y.-K. Hsu, *J. Org. Chem.*, **18**, 516 (1953).

395. W. E. Hanby and W. A. Waters, *J. Chem. Soc.*, 1792 (1939); L. G. Makarova and A. N. Nesmeyanov, *Izv. Akad. Nauk SSSR, Otd. Khim. Nauk.*, 1019 (1954); *Chem. Abstr.*, **50**, 241 (1956).

396. L. Eberson and K. Nyberg, *Tetrahedron Letters*, 2389 (1966); V. D. Parker and B. E. Burgert, *Tetrahedron Letters*, 2411 (1968).

397. A. Laurent, E. Laurent-Dieuzeide, and P. Mison, *Bull. Soc. Chim. France*, 945 (1965); R. Jacquier and H. Cristol, *Bull. Soc. Chim. France*, 556 (1954).

398. R. K. Hill, R. T. Conley, and O. T. Chortyk, *J. Am. Chem. Soc.*, **87**, 5646 (1965).

399. P. D. Bartlett, R. E. Pincock, J. H. Rolston, W. G. Schindel, and L. A. Singer, *J. Am. Chem. Soc.*, **87**, 2590 (1965).

400. T. L. Cairns, P. J. Graham, P. L. Barrick, and R. S. Schreiber, *J. Org. Chem.*, **17**, 751 (1952).

401. A. Hassner, L. A. Levy, and R. Gault, *Tetrahedron Letters*, 3119 (1966).

402. R. H. Andreatta and A. V. Robertson, *Australian J. Chem.*, **19**, 161 (1966).

403. J. M. Bobbitt and R. E. Doolittle, *J. Org. Chem.*, **29**, 2298 (1964).

404. A. T. Balaban, T. H. Crawford, and R. H. Wiley, *J. Org. Chem.*, **30**, 879 (1965).

405. W. S. Durrell, J. A. Young, and R. D. Dresdner, *J. Org. Chem.*, **28**, 830 (1963).

406. C. F. Krewson and J. P. Couch, *J. Am. Chem. Soc.*, **65**, 2256 (1943).

407. L. E. Exner, M. J. Hurwitz, and P. L. de Benneville, *J. Am. Chem. Soc.*, **77**, 1103 (1955).

408. P. A. S. Smith, in *Molecular Rearrangements*, Vol. I, (Ed. P. de Mayo), Interscience Publishers, New York, 1963, p. 457.

409. I. G. Donaruma and W. Z. Heldt, *Org. Reactions*, **11**, 1 (1960).

410. E. C. Horning and V. L. Stromberg, *J. Am. Chem. Soc.*, **74**, 2680, 5151 (1952); R. A. Barnes and M. T. Beachem, *J. Am. Chem. Soc.*, **77**, 5388 (1955).

411. N. J. Doorenbos and M. T. Wu, *J. Org. Chem.*, **26**, 2548 (1961).

412. D. E. Pearson and R. M. Stone, *J. Am. Chem. Soc.*, **83**, 1715 (1961).

413. St. Kaufmann, *J. Am. Chem. Soc.*, **73**, 1779 (1951); H. Heusser, J. Wohlfahrt, M. Müller, and R. Anliker, *Helv. Chim. Acta.*, **38**, 1399 (1955); G. Rosenkranz, O. Mancera, F. Sondheimer, and C. Djerassi, *J. Org. Chem.*, **21**, 520 (1956).

414. R. K. Hill and O. T. Chortyk, *J. Am. Chem. Soc.*, **84**, 1064 (1962).

415. J. Schmidt-Thomé, *Chem. Ber.*, **88**, 895 (1955); *Ann. Chem.*, **603**, 43 (1956); H. Dannerberg and T. Köhler, *Chem. Ber.*, **97**, 140 (1964).

416. C. R. Hauser and D. S. Hoffenberg, *J. Org. Chem.*, **20**, 1482, 1496 (1955).
417. T. E. Stevens, *J. Org. Chem.*, **26**, 2531 (1961).
418. H. Stephen and B. Staskun, *J. Chem. Soc.*, 980 (1956).
419. W. D. Emmons, *J. Am. Chem. Soc.*, **79**, 6522 (1957).
420. T. van Es, *J. Chem. Soc.*, 3881 (1965).
421. R. Huisgen, J. Witte, H. Walz, and W. Jira, *Ann. Chem.*, **604**, 191 (1957).
422. Y. Yukawa and M. Kawakami, *Chem. Ind. (London)*, 1401 (1961).
423. I. T. Glover and V. F. Raaen, *J. Org. Chem.*, **31**, 1987 (1966).
424. J. Kenyon and D. P. Young, *J. Chem. Soc.*, 263 (1941); A. Campbell and J. Kenyon, *J. Chem. Soc.*, 25 (1946).
425. J. Cymerman-Craig and A. R. Naik, *J. Am. Chem. Soc.*, **84**, 3410 (1962).
426. R. F. Brown, N. M. van Gulick, and G. H. Schmid, *J. Am. Chem. Soc.*, **77**, 1094 (1955).
427. F. L. Scott, R. J. MacConaill, and J. C. Riordan, *J. Chem. Soc. (C)*, 44 (1967).
428. F. W. Wassmundt and S. J. Padegimas, *J. Am. Chem. Soc.*, **89**, 7131 (1967).
429. J. Wiemann and P. Ham, *Bull. Soc. Chim. France*, 1005 (1961).
430. L. Field, P. B. Hughmark, S. H. Shumaker, and W. S. Marshall, *J. Am. Chem. Soc.*, **83**, 1983 (1961).
431. see *inter alia.* H. R. Nace and A. C. Watterson, *J. Org. Chem.*, **31**, 2109 (1966); C. W. Shoppee, R. E. Lack, and B. C. Newman, *J. Chem. Soc.*, 3388 (1964); J. T. Edward and P. F. Morand, *Can. J. Chem.*, **38**, 1316 (1960).
432. J. B. Hester, *J. Org. Chem.*, **32**, 3804 (1967).
433. G. G. Lyle and E. T. Pelosi, *J. Am. Chem. Soc.*, **88**, 5276 (1966).
434. N. Tokura, R. Tada, and K. Suzuki, *Bull. Chem. Soc. Japan*, **32**, 654 (1959).
435. M. Ohno and I. Sakai, *Tetrahedron Letters*, 4541 (1965).
436. W. Ziegenbein and W. Lang, *Angew Chem.*, **74**, 943 (1962); H. Metzger and L. Beer, *Z. Naturforsch.*, **18B**, 986 (1963).
437. R. T. Taylor, M. Douek and G. Just, *Tetrahedron Letters*, 4143 (1966).
438. M. Ohno and I. Terasawa, *J. Am. Chem. Soc.*, **88**, 5684 (1966).
439. see *inter alia* A. Hassner and E. G. Nash, *Tetrahedron Letters*, 525 (1965); M. P. Cava, E. J. Glamkowski, and P. M. Weinstraub, *J. Org. Chem.*, **31**, 2755 (1966); K.-i. Morita and Z. Suzuki, *J. Org. Chem.*, **31**, 233 (1966); W. Eisele, C. A. Grob, E. Renk, and H. von Tschammer, *Helv. Chim. Acta.*, **51**, 816 (1968).
440. P. T. Lansbury and R. P. Spitz, *J. Org. Chem.*, **32**, 2623 (1967); P. A. S. Smith, *J. Am. Chem. Soc.*, **76**, 431 (1954).
441. P. T. Lansbury and N. R. Mancuso, *J. Am. Chem. Soc.*, **88**, 1205 (1966).
442. P. A. S. Smith, *J. Am. Chem. Soc.*, **70**, 320 (1948).
443. H. Wolff, *Org. Reactions.* **3**, 307 (1946).
444. R. T. Conley, *J. Org. Chem.*, **23**, 1330 (1958).
445. J. R. Dice and P. A. S. Smith, *J. Org. Chem.*, **14**, 179 (1949).
446. N. Campbell, W. K. Leadill, and J. F. K. Wilshire, *J. Chem. Soc.*, 1404 (1951).
447. P. A. S. Smith and J. P. Horwitz, *J. Am. Chem. Soc.*, **72**, 3718 (1950).
448. P. A. S. Smith and E. P. Antoniades, *Tetrahedron*, **9**, 210 (1960).
449. K. F. Schmidt, *Chem. Ber.*, **57B**, 704 (1924).
450. W. Pritzkow and K. Dietzsch, *Chem. Ber.*, **93**, 1733 (1960).

451. C. Wiener, C. H. Schroeder, B. D. West, and K. P. Link, *J. Org. Chem.*, **27**, 3086 (1962).
452. A. J. Davies, A. S. R. Donald, and R. E. Marks, *J. Chem. Soc. (C)*, 2109 (1967).
453. J. Jaz and J. P. Davreux, *Tetrahedron Letters*, 277 (1966).
454. L. Ruzicka, M. W. Goldberg, M. Hürbin, and H. A. Boeckenoogen, *Helv. Chim. Acta*, **16**, 1323 (1933).
455. C. L. Arcus, M. M. Coombs, and J. V. Evans, *J. Chem. Soc.*, 1498 (1956).
456. R. Huisgen, I. Ugi, H. Brade, and E. Rauenbusch, *Ann. Chem.*, **586**, 30 (1954).
457. L. Birkofer and I. Storch, *Chem. Ber.*, **86**, 749 (1953).
458. L. A. Paquette and M. K. Scott, *J. Org. Chem.*, **33**, 2379 (1968).
459. G. Di Maio and V. Permutti, *Tetrahedron*, **22**, 2059 (1966).
460. H. J. Schmid, A. Hunger, and K. Hoffmann, *Helv. Chim. Acta.*, **39**, 607 (1956).
461. T. Ichii, *J. Pharm. Soc. Japan*, **82**, 999 (1962).
462. see *inter alia*, B. Stevenson, *J. Org. Chem.*, **28**, 188 (1963); H. Singh, V. V. Parashar, and S. Padmanabham, *J. Sci. Ind. Res. (India)*, **25**, 200 (1966).
463. S. Uyeo, *Pure Appl. Chem.*, **7**, 269 (1963).
464. G. Caronna and S. Palazzo, *Gazz. Chim. Ital.*, **83**, 315 (1953).
465. H. W. Moore and H. R. Shelden, *J. Org. Chem.*, **32**, 3603 (1967).
466. D. Misiti, H. W. Moore, and K. Folkers, *Tetrahedron*, **22**, 1201 (1966).
467. R. W. Rickards and R. M. Smith, *Tetrahedron Letters*, 2361 (1966); G. R. Bedford, G. Jones, and B. R. Webster, *Tetrahedron Letters*, 2367 (1966).
468. K. G. Rutherford and M. S. Newman, *J. Am. Chem. Soc.*, **79**, 213 (1957).
469. K. G. Rutherford, S. Y.-S. Ing, and R. J. Thibert, *Can. J. Chem.*, **43**, 541 (1965).
470. K. N. Carter, *J. Org. Chem.*, **31**, 4257 (1966).
471. for reviews see M. Carmack and M. A. Spielman, *Org. Reactions*, **3**, 83 (1946); R. Wegler, E. Kühle, and W. Schäfer, *Newer Methods of Preparative Organic Chemistry*, Vol. III, (Ed. W. Foerst), Academic Press, New York, 1964, Chap. 1, pp. 1–46.
472. C. Willgerodt and F. H. Merk, *J. Prakt. Chem.*, **80**, 192 (1909).
473. F. Asinger, W. Schäfer, K. Halcour, A. Saus, and H. Triem, *Angew. Chem., Intern. Ed. Engl.*, **3**, 19 (1964), and references cited therein.
474. E. V. Brown, E. Cerwonka, and R. C. Anderson, *J. Am. Chem. Soc.*, **73**, 3735 (1951).
475. E. Cerwonka, R. C. Anderson, and E. V. Brown, *J. Am. Chem. Soc.*, **75**, 30 (1953).
476. E. Cerwonka, R. C. Anderson, and E. V. Brown, *J. Am. Chem. Soc.*, **75**, 28 (1953).
477. M. Carmack and D. F. DeTar, *J. Am. Chem. Soc.*, **68**, 2029 (1946).
478. J. A. King and F. H. McMillan, *J. Am. Chem. Soc.*, **68**, 632 (1946).
479. D. Nightingale and R. A. Carpenter, *J. Am. Chem. Soc.*, **71**, 3560 (1949); G. A. R. Kon, *J. Chem. Soc.*, 224 (1948).
480. T. Bacchetti, A. Alemagna, and B. Danieli, *Tetrahedron Letters*, 2001 (1965).
481. L. F. Fieser and G. W. Kilmer, *J. Am. Chem. Soc.*, **62**, 1354 (1940).
482. D. F. DeTar and M. Carmack, *J. Am. Chem. Soc.*, **68**, 2025 (1946).
483. W. E. Bachmann and M. Carmack, *J. Am. Chem. Soc.*, **63**, 2494 (1941).

484. A. C. Ott, L. A. Mattano, and G. H. Coleman, *J. Am. Chem. Soc.*, **68**, 2633 (1946).
485. H. Gilman and S. Avakian, *J. Am. Chem. Soc.*, **68**, 2104 (1946).
486. E. Schwenk and D. Papa, *J. Org. Chem.*, **11**, 798 (1946).
487. L. Cavalieri, D. B. Pattison, and M. Carmack, *J. Am. Chem. Soc.*, **67**, 1783 (1945).
488. M. A. Naylor and A. W. Anderson, *J. Am. Chem. Soc.*, **75**, 5392 (1945).
489. D. B. Pattison and M. Carmack, *J. Am. Chem. Soc.*, **68**, 2033 (1946).
490. H. Feichtinger, *Chem. Ber.*, **95**, 2238 (1962).
491. D. E. Pearson, K. W. Carter, and C. M. Greer, *J. Am. Chem. Soc.*, **75**, 5905 (1953).
492. L. G. Donamura, *U.S. Pat.*, 2,777,841 (1957); *Chem. Abstr.*, **51**, 10565 (1957).
493. L. G. Donamura, *U.S. Pat.*, 2,763,644 (1956); *Chem. Abstr.*, **51**, 5822 (1957).
494. J. W. Schulenberg and S. Archer, *Org. Reactions*, **14**, 1 (1965).
495. H. L. Wehrmeister, *J. Org. Chem.*, **30**, 664 (1965).
496. For a review of nitrone chemistry see G. R. Delpierre and M. Lamchen, *Quart. Rev. (London)*, **19**, 329 (1965).
497. O. L. Brady and F. P. Dunn, *J. Chem. Soc.*, 2411 (1926).
498. F. Kröhnke, *Ann. Chem.*, **604**, 203 (1957).
499. A. Padwa, *Tetrahedron Letters*, 2001 (1964).
500. J. S. Splitter and M. Calvin, *J. Org. Chem.*, **30**, 3427 (1965).
501. S. Searles, S. Nukina, and E. R. Magnuson, *J. Org. Chem.*, **30**, 1920 (1965).
502. L. A. Paquette, *J. Am. Chem. Soc.*, **84**, 4987 (1962); **85**, 3288 (1963); *Org. Syn.*, **44**, 41 (1964).
503. G. A. Olah and J. A. Olah in *Friedel–Crafts and Related Reactions*, Vol. III, (Ed. G. A. Olah), Interscience Publishers, New York, 1964, pp. 1262–1267.
504. H. Hopff and H. Ohlinger, *Angew. Chem.*, **61**, 183 (1949).
505. F. Weygand and R. Mitgau, *Chem. Ber.*, **88**, 301 (1955).
506. J. F. K. Wilshire, *Australian J. Chem.*, **20**, 575 (1967).
507. E. Stein and O. Bayer, *Ger. Pat.*, 875,807 (1953); *Chem. Abstr.*, **52**, 10183 (1958).
508. J. K. Lawson and J. A. T. Croom, *J. Org. Chem.*, **28**, 232 (1963).
509. H. M. Singleton and W. R. Edwards, *J. Am. Chem. Soc.*, **60**, 540 (1938).
510. R. N. Boyd and R. Leshin, *J. Am. Chem. Soc.*, **75**, 2762 (1953).
511. F. F. Blicke and H. Zinnes, *J. Am. Chem. Soc.*, **77**, 4849 (1955).
512. R. Fusco and S. Rossi, *Gazz. Chim. Ital.*, **94**, 3 (1964).
513. W. Ried and W. Käppeler, *Ann. Chem.*, **673**, 132 (1964).
514. G. A. Berchtold, *J. Org. Chem.*, **26**, 3043 (1961); R. Fusco, G. Bianchetti and S. Rossi, *Gazz. Chim. Ital.*, **92**, 825 (1962).
515. G. Opitz and J. Koch, *Angew. Chem. Intern. Ed. Engl.*, **2**, 152 (1963).
516. R. W. Guthrie, Z. Valenta, and K. Wiesner, *Tetrahedron Letters*, 4645 (1966).
517. E. Vogel, R. Erb, G. Lenz, and A. A. Bothner-By, *Ann. Chem.*, **682**, 1 (1965); O. Schindler, R. Blaser, and F. Hunziker, *Helv. Chim. Acta.*, **49**, 985 (1966).
518. For reviews of the chemistry of chlorosulphonyl isocyanate see R. Graf, *Angew. Chem. Intern. Ed. Engl.*, **7**, 172 (1968); H. Ulrich, *Chem. Rev.*, **65**, 369 (1965).

519. R. Graf, *Ann. Chem.*, **661**, 111 (1963).
520. E. J. Moriconi and P. H. Mazzocchi, *J. Org. Chem.*, **31**, 1372 (1966).
521. R. Graf, *Org. Syn.*, **46**, 51 (1966).
522. E. J. Moriconi and J. F. Kelly, *J. Am. Chem. Soc.*, **88**, 3657 (1966).
523. E. J. Moriconi and W. C. Crawford, *J. Org. Chem.*, **33**, 370 (1968).
524. E. J. Moriconi and J. F. Kelly, *Tetrahedron Letters*, 1435 (1968).
525. H. Hoffmann and H. J. Diehr, *Tetrahedron Letters*, 1875 (1963).
526. E. J. Moriconi and W. C. Meyer, *Tetrahedron Letters*, 3823 (1968).
527. M. Rausch, P. Shaw, D. Mayo, and A. M. Lovelace, *J. Org. Chem.*, **23**, 505 (1958).
528. E. C. Taylor and N. W. Kalenda, *J. Am. Chem. Soc.*, **76**, 1699 (1954); for a more recent example see J. Schmutz, F. Künzle, F. Hunziker, and A. Bürki, *Helv. Chim. Acta.*, **48**, 336 (1965).
529. D. H. Lorenz and E. I. Becker, *J. Org. Chem.*, **28**, 1707 (1963).
530. J. H. Saunders and R. J. Slocombe, *Chem. Rev.*, **43**, 203 (1948); R. G. Arnold, J. A. Nelson, and J. J. Verbanc, *Chem. Rev.*, **57**, 47 (1957).
531. C. Naegeli and A. Tyabji, *Helv. Chim. Acta*, **17**, 931 (1934); **18**, 142 (1935).
532. A. J. Fry, *J. Am. Chem. Soc.*, **75**, 2686 (1953).
533. M. Goodman and J. Hutchison, *J. Am. Chem. Soc.*, **88**, 3627 (1966); E. Peggion, M. Terbojevich, A. Cosani, and C. Colombini, *J. Am. Chem. Soc.*, **88**, 3630 (1966); N. H. Grant, D. E. Clark and H. E. Alburn, *J. Am. Chem. Soc.*, **88**, 4071 (1966).
534. T. Lieser and G. Nischk, *Ann. Chem.*, **569**, 59 (1950).
535. P. A. S. Smith, *Org. Reactions*, **3**, 337 (1946).
536. B. Acott, A. L. J. Beckwith, and A. Hassanali, *Australian J. Chem.*, **21**, 185 (1968); B. Acott and A. L. J. Beckwith, *Chem. Commun.*, 161 (1965).
537. H. E. Baumgarten and A. Staklis, *J. Am. Chem. Soc.*, **87**, 1141 (1965).
538. A. L. J. Beckwith and R. J. Hickman, *J. Chem. Soc.*, (*C*), 2756 (1968).
539. M. Szwarc, *Advan. Polymer Sci.*, **4**, 1 (1965).
540. L. Birkofer and R. Modic, *Ann. Chem.*, **604**, 56 (1957); S. A. Lepetit, *Belg. Pat.*, 622,901 (1963), *Chem. Abstr.*, **59**, 7534 (1963); H. E. Winberg, *U.S. Pat.*, 2,600,596 (1962); *Chem. Abstr.*, **47**, 7536 (1953).
541. E. R. Blout and R. H. Karlson, *J. Am. Chem. Soc.*, **78**, 941 (1956).
542. R. Graf, *Ger. Pat.*, 931,225 (1955); *Chem. Abstr.*, **50**, 7861 (1956).
543. M. Passerini, *Gazz. Chim. Ital.*, **61**, 964 (1931), and preceding papers.
544. R. H. Baker and L. E. Linn, *J. Am. Chem. Soc.*, **70**, 3721 (1948).
545. I. Hagedorn and U. Eholzer, *Chem. Ber.*, **98**, 936 (1965).
546. R. H. Baker and D. Stanonis, *J. Am. Chem. Soc.*, **73**, 699 (1951).
547. T. Saegusa, N. Taka-ishi, and H. Fujii, *Tetrahedron*, **24**, 3795 (1968).
548. I. Ugi, *Angew. Chem. Intern. Ed. Engl.*, **1**, 8 (1962).
549. J. W. McFarland, *J. Org. Chem.*, **28**, 2179 (1963).
550. H. Muxfeldt and W. Rogalski, *J. Am. Chem. Soc.*, **87**, 933 (1965); S. Petersen and E. Tietze, *Ann. Chem.*, **623**, 166 (1959).
551. R. Roger and D. G. Neilson, *Chem. Rev.*, **61**, 179 (1961).
552. W. S. Fones, *J. Org. Chem.*, **14**, 1099 (1949).
553. I. A. Kaye, C. L. Parris, and N. Weiner, *J. Am. Chem. Soc.*, **75**, 744 (1953).
554. I. J. Pachter and M. C. Kloetzcl, *J. Am. Chem. Soc.*, **74**, 1321 (1952).
555. H. L. Needles and R. E. Whitfield, *J. Org. Chem.*, **31**, 989 (1966).
556. R. L. Gay and C. R. Hauser, *J. Am. Chem. Soc.*, **89**, 1647 (1967).

557. R. E. Benson and T. L. Cairns, *J. Am. Chem. Soc.*, **70**, 2115 (1948).
558. R. M. Moriarty, *J. Org. Chem.*, **29**, 2749 (1964).
559. W. B. Dickinson and P. C. Lang, *Tetrahedron Letters*, 3035 (1967).
560. P. G. Gassman and B. L. Fox, *J. Org. Chem.*, **31**, 982 (1966).
561. J. C. Sheehan and W. A. Bolhofer, *J. Am. Chem. Soc.*, **72**, 2786 (1950).
562. H. E. Zaugg and W. B. Martin, *Org. Reactions*, **14**, 52 (1965).
563. R. Dowbenko, R. M. Christenson, and A. N. Salem, *J. Org. Chem.*, **28**, 3458 (1965).
564. J. P. Chupp and A. J. Speziale, *J. Org. Chem.*, **28**, 2592 (1965).
565. N. Yanaihara and M. Saito, *Chem. Pharm. Bull.*, (*Tokyo*), **15**, 128 (1967).
566. H. Böhme and G. Berg, *Chem. Ber.*, **99**, 2127 (1966).
567. H. E. Zaugg, R. J. Michaels, A. D. Schaefer, A. M. Wenthe, and W. H. Washburn, *Tetrahedron*, **22**, 1257 (1966).
568. C. J. M. Stirling, *J. Chem. Soc.*, 255 (1960).
569. H. Wamhoff and F. Korte, *Chem. Ber.*, **100**, 2122 (1967).
570. M. S. Manhas and S. J. Jeng, *J. Org. Chem.*, **32**, 1246 (1967).
571. see *inter alia* H. E. Baumgarten, J. F. Fuerholzer, R. D. Clark, and R. D. Thompson, *J. Am. Chem. Soc.*, **85**, 3303 (1963); J. C. Sheehan and I. Lengyel, *J. Am. Chem. Soc.*, **86**, 1356 (1964).
572. J. C. Sheehan and J. H. Beeson, *J. Am. Chem. Soc.*, **89**, 362 (1967).
573. E. R. Talaty, A. E. Dupuy, and A. E. Cancienne, *J. Heterocyclic Chem.*, **4**, 657 (1967); K. Bott, *Tetrahedron Letters*, 3323 (1968).
574. H. T. Nagasawa and J. A. Elberling, *Tetrahedron Letters*, 5393 (1966).
575. S. Sarel, A. Taube, and E. Breuer, *Chem. Ind.* (*London*), 1095 (1967).
576. K. Nagarojon and C. L. Kulkarni, *Tetrahedron Letters*, 2717 (1968).
577. R. K. Hill, *J. Org. Chem.*, **22**, 830 (1957).
578. R. R. Wittekind, C. Weissman, S. Farber, and R. I. Meltzer, *J. Heterocyclic Chem.*, **4**, 143 (1967).
579. O. L. Chapman and W. R. Adams, *J. Am. Chem. Soc.*, **89**, 4243 (1967).
580. H. Böhme, S. Ebel, and K. Hartke, *Chem. Ber.*, **98**, 1463 (1965).
581. H. E. Baumgarten, *J. Am. Chem. Soc.*, **84**, 4975 (1962).
582. A. K. Bose, M. S. Manhas, and R. M. Ramer, *Tetrahedron*, **21**, 449 (1965), and references cited therein.
583. A. Rieche, E. Schmitz, and E. Gründemann, *Angew. Chem.*, **73**, 621 (1961).
584. D. Elad and J. Rokach, *J. Org. Chem.*, **29**, 1855 (1964).
585. G. Friedman and A. Komem, *Tetrahedron Letters*, 3357 (1968).
586. J. Sinnreich and D. Elad, *Tetrahedron*, **24**, 4509 (1968).
587. G. T. Tissue, S. Linke, and W. Lwowski, *J. Am. Chem. Soc.*, **89**, 6303 (1967); I. Brown and O. E. Edwards, *Can. J. Chem.*, **45**, 2599 (1967); R. F. C. Brown, *Australian J. Chem.*, **17**, 47 (1964); W. L. Meyer and A. S. Levinson, *J. Org. Chem.*, **28**, 2859 (1963).
588. J. W. ApSimon and O. E. Edwards, *Can. J. Chem.*, **40**, 896 (1962).
589. G. Just and W. Zehetner, *Tetrahedron Letters*, 3389 (1967).
590. E. J. Corey and A. M. Felix, *J. Am. Chem. Soc.*, **87**, 2518 (1965).
591. C. Ainsworth, *J. Am. Chem. Soc.*, **76**, 5774 (1954); F. P. Robinson and R. K. Brown, *Can. J. Chem.*, **39**, 1171 (1961).
592. E. Shaw, *J. Am. Chem. Soc.*, **81**, 6021 (1959).
593. R. L. Hinman, *J. Org. Chem.*, **22**, 148, (1957).
594. A. Giner-Sorolla and A. Bendich, *J. Am. Chem. Soc.*, **80**, 3932 (1958).

595. H. B. Henbest and M. J. W. Stratford, *Chem. Ind. (London)*, 1170 (1961).
596. H. L. Needles and R. E. Whitfield, *J. Org. Chem.*, **31**, 341 (1966).
597. R. M. Gipson, F. H. Pettit, G. G. Skinner, and W. Shive, *J. Org. Chem.*, **28**, 1425 (1963); M. Masaki and M. Ohtake, *Bull. Chem. Soc. Japan*, **38**, 1802 (1965); D. V. Nightingale and J. E. Johnson, *J. Heterocyclic Chem.*, **4**, 102 (1967).
598. B. Sklarz and A. F. Al-Sayyab, *J. Chem. Soc.*, 1318 (1964); A. D. Ward, personal communication.
599. A. L. J. Beckwith and G. W. Evans, *J. Chem. Soc.*, 130 (1962).
600. K. Nakagawa, H. Onoue, and K. Minami, *Chem. Commun.*, 17 (1966).

CHAPTER 3

Acid–base and complexing properties of amides

R. B. HOMER and C. D. JOHNSON

School of Chemical Sciences, University of East Anglia, Norwich, England

I. INTRODUCTION

The amides are relatively weak bases, and the elucidation of their protonation behaviour and related phenomena such as hydrogen bonding and Lewis acid complexation has aroused a good deal of attention. Emphasis has been placed on such interactions since the amide function is the basic structural unit of peptides and proteins.

The question of the site of protonation is of primary importance; this has resulted in considerable controversy, although the dispute has now been largely settled in favour of the oxygen site. The resolution of this problem is a good example of the application of physical techniques to the determination of fine details of chemical structure.

The accurate determination of thermodynamic pK_a values is also of considerable consequence. Since the amides are weak bases, their protonation behaviour must be observed in moderately or very concentrated acid, which entails the consideration of relevant acidity functions and their underlying theory. Many amides are however sufficiently basic to allow determination of their pK_a values by titration of their solutions in very weak bases such as nitromethane or acetic acid.

Closely linked to the question of protonation is hydrogen bonding. The biological activity of proteins must be correlated very intimately with the inter- and intramolecular hydrogen bonding in which such structures participate, and therefore the examination of such bonding in the parent amide group is of fundamental significance. The formation of such bonds is subsequently discussed, and then other phenomena relating to complexation of the amide function to electron-pair acceptors other than the hydrogen ion, which by definition may be termed Lewis acids.

Attention is also given to amides functioning as weak acids, which again necessitates the consideration of acidity functions, this time applicable to strongly alkaline solutions.

Only where relevant is the discussion extended to thioamides, ureas and thioureas, pyridones, and carbamic acid and its esters, and any other structures which are analogous to simple amides in that they contain a nitrogen atom directly linked to a carbonyl or thiocarbonyl group.

II. THE SITE OF PROTONATION

Two potential sites of protonation, N and O, exist in amides, giving rise to cationic structures 2 and 3.

The amino group is inherently much more basic than the carbonyl group, suggesting on this simple basis 2 as the most likely structure for a protonated amide. However, in structure 3 there will be important contributions from canonical forms 4 and (especially) 5 to the resonance hybrid, resulting in sharing of the positive charge between oxygen, carbon, and, presumably most important, nitrogen.

The available evidence, which we now discuss, is overwhelmingly in favour of a large predominance of the O-protonated form (**3**) over the N-protonated form (**2**) in all cases to be considered. Similarly, S-protonation is favoured in thioamide structures. This evidence has

$$
\begin{array}{ccc}
\underset{\text{(1)}}{R^1\overset{\overset{\displaystyle O}{\|}}{C}{-}NR^2R^3} & \underset{-H^+}{\overset{+H^+}{\rightleftharpoons}} & \underset{\text{(2)}}{R^1\overset{\overset{\displaystyle O}{\|}}{C}{-}\overset{+}{N}HR^2R^3}
\end{array}
$$

$$
\left[
\begin{array}{ccc}
\underset{\text{(4)}}{R^1\overset{\overset{\displaystyle OH}{|}}{\overset{+}{C}}{-}NR^2R^3} & \longleftrightarrow & \underset{\text{(3)}}{R^1\overset{\overset{\displaystyle \overset{+}{O}{-}H}{\|}}{C}{-}NR^2R^3} & \longleftrightarrow & \underset{\text{(5)}}{R^1\overset{\overset{\displaystyle OH}{|}}{C}{=}\overset{+}{N}R^2R^3}
\end{array}
\right]
$$

already been critically reviewed[1], and we cite therefore the most conclusive of the earlier pieces of information, together with recent relevant data.

The most convincing evidence comes from n.m.r. studies in concentrated acids. The groups of Fraenkel[2] and Berger[3] have studied the n.m.r. spectra of various N-methylamides in sulphuric and deuteriosulphuric acid, and other acid mixtures. The former workers also showed, from cryoscopic measurements, that the amides are all monoprotonated in 100% sulphuric acid. Other workers have also shown that amides are protonated in trifluoroacetic acid[4]. The spectrum of N-methylacetamide shows a doublet methyl peak due to spin–spin coupling with the proton bound to nitrogen. On acidification the doublet collapses to a single peak due to rapid N—H proton exchange, but in strong acid, containing dioxan to slow down the rate of exchange, a doublet (rather than the triplet expected for the —$\overset{+}{N}$H$_2$CH$_3$ group) reappears, suggesting that O- rather than N-protonation has occurred.

The n.m.r. spectra of N,N-dimethylformamide and N,N-dimethylacetamide show doublets for the methyl protons in neutral solution, due to restricted rotation about the C—N bond, resulting in different environments for the methyl protons (**6**). This doublet remains in

$$
\underset{\text{(6)}}{\overset{\displaystyle O^-}{\underset{\displaystyle}{\overset{|}{\underset{+}{C}}}}{=}N\overset{\displaystyle CH_3}{\underset{\displaystyle CH_3}{}}}
\qquad\qquad
\underset{\text{(7)}}{\overset{\displaystyle OH}{\underset{\displaystyle}{\overset{|}{\underset{+}{C}}}}{=}N\overset{\displaystyle CH_3}{\underset{\displaystyle CH_3}{}}}
$$

both sulphuric and deuteriosulphuric acid, the former indicating that

O-protonation has occurred retaining the partial double-bond character of the carbon–nitrogen bond (**7**), and the latter that the doublet is not due to splitting of the methyl-group protons by a proton on nitrogen.

Spinner[5] suggested that rotation about the C—N linkage will be restricted even if N-protonation occurs, giving rise to *trans* (**8**) and *gauche* (**9**) forms in protonated N-methylamides and *trans/gauche* (**10**) and *gauche/gauche* (**11**) forms in protonated N,N-dimethylamides. The *trans* form (**8**) will be energetically preferred, and thus the observed doublet is said to be due to the signals from the different protons H_a and H_b. Such a system would, however, give rise to a methyl-

group doublet of relative areas 1:2, and not the 1:1 ratio actually found. This explanation also presupposes no spin–spin coupling between the methyl-group protons and the two protons on nitrogen. Similarly, the *trans/gauche* form (**10**) in the dimethyl case is considered to be more abundant than the *gauche/gauche* form (**11**), and gives rise to the observed doublet due to the non-equivalent methyl groups. Again, however, the criticism of the lack of spin–spin coupling with the proton on nitrogen applies. Moreover, Fraenkel is reported[1] to have measured the spectrum of N-methylacetamide in both its neutral and protonated forms at 40 and 60 mc/s, and shown that the coupling constant $J_{\text{NH,CH}_3}$ is 3·8 c.p.s. in all cases. The observed doublet is therefore due to a spin–spin interaction and not a chemical shift.

More recently, Gillespie and Birchall[6,7] have studied the n.m.r. spectra of formamide, acetamide and benzamide, and their N-methyl and N,N-dimethyl derivatives, in fluorosulphuric acid, an acid in which proton exchange is slower than for sulphuric acid. At low temperatures ($\sim -90°$) a peak appears which from its area and the fact it is a singlet must be due to a proton on the carbonyl oxygen. These workers confirm the splitting of the amino-group protons into a doublet, due to restricted rotation about the C—N bond. Typical data for acetamide are shown in Figure 1. Other experiments[7] showed S-protonation in the case of thioamides and thioureas, although sulphonamides apparently protonated on nitrogen. There was also evidence for diprotonation of ureas and thioureas.

Evidence from infrared data has also been accumulated which suggests *O*-protonation in these compounds. Interpretation of such data is complicated by the mixed character of the absorption bands. Thus, calculations on *N*-methylacetamide[8] have shown that the amide I band arises predominantly but not completely (> 80%) from the

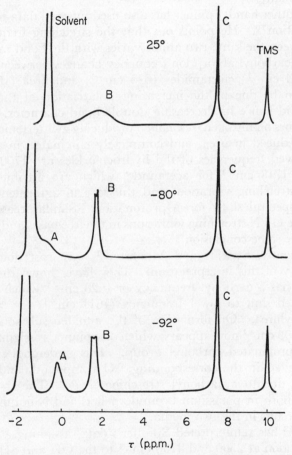

FIGURE 1. N.m.r. spectra of acetamide in fluorosulphuric acid. Tetramethylsilane (TMS) used as external standard. A, proton on carbonyl group, relative area 1·07; B, protons on nitrogen, relative area 2·10; C, methyl protons, relative area 3·00.
[Reproduced by permission of the National Research Council of Canada from the *Canadian Journal of Chemistry*, Vol. 41, p. 150 (1963).]

carbonyl stretching mode, and the amide II band is compounded of
N—H in-plane bending ($\sim 60\%$) and C—N stretching modes
($\sim 40\%$).

Gompper and Altreuther[9] have prepared fixed forms of cations
5 and 2, $C_6H_5C(OCH_3)\!\!=\!\!\overset{+}{N}(C_2H_5)_2$ and $C_6H_5CO\!-\!\overset{+}{N}(C_2H_5)_3$ respectively, and shown that the infrared spectrum of the former re-
sembles the spectrum of protonated benzamide very closely, but not
that of the latter.

On the other hand, Spinner has also used infrared data to argue for
N-protonation[10]. He points out that the stretching frequency of a
bond between the same two atoms varies with the bond multiplicity.
Thus, the carbonyl absorption frequency changes between formaldehyde (1744 cm^{-1}), acetamide (1675 cm^{-1}) and urea (1627 cm^{-1})
are said to be due to the increasing polarization of the carbonyl
linkage, and hence its decreasing double-bond character. If protonation occurs on nitrogen, resonance producing zwitterionic canonical
forms is reduced in urea, and completely eliminated in acetamide.
The observed frequencies of the hydrochlorides are 1700 cm^{-1} for
urea and 1718 cm^{-1} for acetamide, which are attributed to the
carbonyl stretching vibration, and thus are in agreement with the
expected spectral shifts for N-protonation. Similar reasoning, applied to the C—N stretching vibrations in the free bases and their salts,
leads to the same conclusion.

Subsequently, Stewart and Muenster[11] have cast doubt on the
authenticity of this interpretation. They have shown that dicyclohexylurea has a carbonyl frequency of 1628 cm^{-1} which undergoes
the expected shift to lower frequency (1611 cm^{-1}) in ^{18}O-labelled
dicyclohexylurea. On formation of the p-toluenesulphonate salt, a
band at 1699 cm^{-1} now appears, which on Spinner's assignment is due
to the unprotonated carbonyl group. This undergoes no isotopic
shift, however, in the corresponding ^{18}O compound, and therefore
cannot be due to a carbonyl stretching mode. The spectra were
measured both in potassium bromide pellets and Nujol mull, giving
identical results in the two media.

Janssen[12] has reinterpreted Spinner's data, assigning a broad absorption region at 2500 and 2100 cm^{-1} to the OH and SH bands respectively, of the hydrochlorides of urea and acetamide and their
thio analogues, and explaining the apparent upward displacement
of the carbonyl band on protonation by reassignment of this frequency to $>\!\!C\!\!=\!\!\overset{+}{N}\!\!<$ in the protonated form.

The infrared spectra of the hydrochlorides of N,N-dimethylacet-amide and N,N-di-n-butylacetamide in the form of Nujol mulls have also been interpreted[13] on the assumption that the proton goes to oxygen, but later workers[14] postulate N-protonation for the 1:1 salt of N-methylacetamide with hydrochloric acid, on the basis of the Raman spectra of the solid. Surprisingly, however, these workers, give no discussion of the significance of this result in the light of other investigations. They also propose a hydrogen-bonded structure **12** for the 2:1 salt.

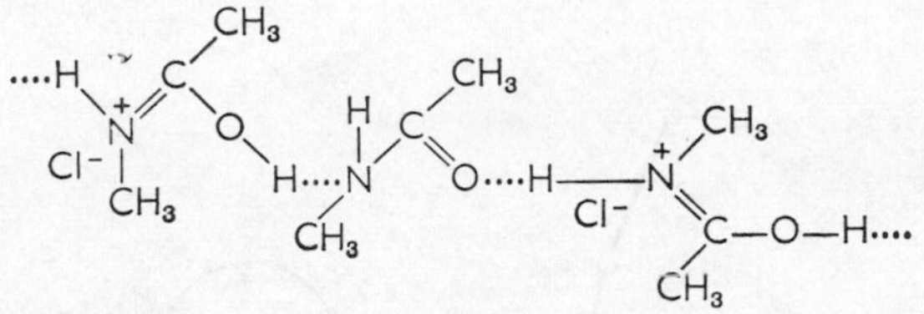

(12)

Janssen has also argued convincingly[12] for S-protonation in the case of thioacetamide, thiourea, and other thioamides, basing his conclusions on the differences between the ultraviolet spectra in ethanol, and aqueous sulphuric acid solutions of sufficiently high acidity to produce the conjugate acids. (The amides themselves show no absorption above 220 mμ.) Thus, the spectrum of **13**, it is reasoned, should be similar to that of **14** (λ_{max} 327, 266 mμ), since the inductive effects of the CH_3 and $\overset{+}{N}H_3$ groups will have little influence on the ultraviolet spectra. It is found, however, that there are no bands above 220 mμ

(13)   (14)

in the spectrum of the conjugate acid of thiourea, which is thus incompatible with N-protonation.

However, other workers[15] have found that the ultraviolet spectra of protonated benzamides resemble very closely those of unprotonated acetophenones, suggesting on the above argument N-protonation, and indicating that the absorption characteristics of protonated benzamides are commensurate with an N- or O-protonated structure (this point is further discussed in section III.D).

Next, we turn to basicity studies for a final vindication of O-protonation. Huisgen[16] argues for this on the basis of the larger effects of

N-substituents on the pK_a values of amines compared with amides. This argument is not very convincing, since a good deal of positive charge will reside on the nitrogen of the O-protonated amides (**5**), but a second argument provides more persuasive evidence[17]. Figure 2 shows the variation of pK_a with ring size of cyclic amides and amines. The variation for the amines is explained in terms of steric interactions in the protonated form due to the additional hydrogen atom

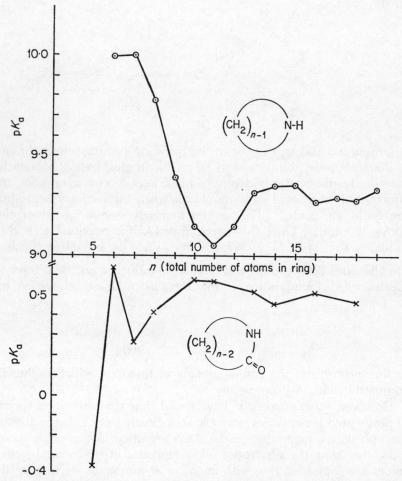

FIGURE 2. The variation of pK_a values with ring size in cyclic amines and lactams.

[Reproduced, by permission, from *Chem. Ber.*, **90**, 1437 (1957).]

attached directly to the ring, and its attendant solvent shell.   The very different variation for the lactam series thus suggests protonation at an alternative site removed from the ring, namely oxygen.

Since protonation on oxygen in amides is considered to arise from the resonance stabilization of the cation as in $3 \leftrightarrow 5$, it is of interest to note the basicity of 2,2-dimethylquinuclidone (15)[18].   In this molecule no overlap can occur between the nitrogen sp^3 orbital containing the lone-pair electrons, and the p orbitals of oxygen and carbon (as in structure 5 for the usual form of the protonated amide molecule). The compound thus apparently protonates on nitrogen, yielding a

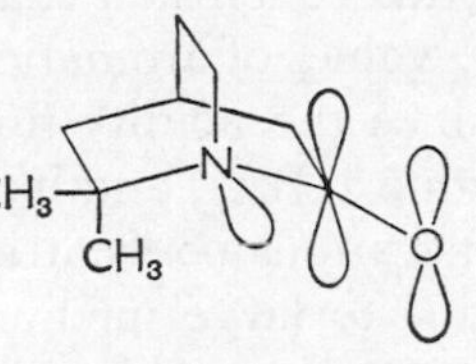

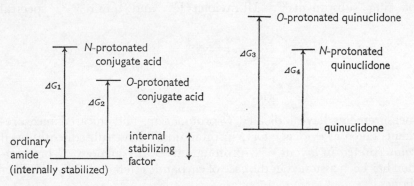

(15)

pK_a value of 5·33, far higher than for normal amides (see section III), and is very rapidly hydrolysed in conditions under which acetamide, for example, would be stable.   If resonance stabilization is important enough to cause protonation on oxygen rather than on the intrinsically more basic nitrogen in general for amides, why does removal of this form of stabilization as in 2,2-dimethylquinuclidone then lead to a base-strengthening effect?   The answer may lie in a consideration of stabilizing effects on both base and conjugate acid, as in the following diagram:

		O-protonated quinuclidone
N-protonated conjugate acid	ΔG_3	
ΔG_1		N-protonated quinuclidone
	O-protonated conjugate acid	ΔG_4
	ΔG_2	
ordinary amide (internally stabilized)	internal stabilizing factor	quinuclidone

for which $\Delta G_4 < \Delta G_2 < \Delta G_1 \leqslant \Delta G_3$

(We would like to thank Professor K. Yates both for bringing our attention to this anomaly, and also for supplying the possible explanation as given above.)

Further evidence for the site of protonation in aromatic amides arising from correlation of pK_a values with Hammett σ constants is discussed (in section III.D). However, on the basis of a linear correlation between pK_a values and the carbonyl stretching frequency of various aromatic aldehydes and ketones, the protonation of benzamide has been considered to occur on nitrogen[19], since the carbonyl stretching frequency for this compound is widely different from that predicted by its pK_a and the straight line defined by the other carbonyl compounds. However, this conclusion may be in error for several reasons. Firstly, the pK_a values of aromatic carbonyl compounds are not accurately known, since the acidity function which they follow may not be H_0[20] (see section III.A), and interpretation of their ultraviolet spectral changes in various concentrations of acid for pK_a determination is difficult due to large medium as well as protonation effects[20,21]. Secondly, the 1675 cm^{-1} band of benzamide is not solely due to the carbonyl stretching frequency, but contains a contribution from an N—H mode[8]. Thirdly, extra resonance stabilization is possible for protonated amides, compared to benzaldehyde, acetophenone etc., which increases their basicity over that predicted by the $pK_a - \nu_{C=O}$ correlation, since $\nu_{C=O}$ refers only to the free base[22].

It can be seen that, overall, there is a tremendous weight of evidence in favour of O-protonation of amides. A similar picture for 2- and 4-pyridones, which may be considered as vinylogous amides, has been established. These compounds exist predominantly in the oxo tautomeric form (16), although K_T can be altered markedly by the presence of ring substituents[23]. Ultraviolet[24,25] and n.m.r.[26-29] spectral

(16)   (17)

measurements have indicated O-protonation, and infrared measurements taken as evidence for N-protonation[30] have subsequently been reinterpreted in favour of O-protonation[31]. By analogy, O-protonation has been assumed in the case of carbamic esters[32] and pyrimidinediones[33].

Moodie [*Chem. Commun.*, 1362 (1968)] has now presented evidence for N-protonation of ethyl N,N-diisopropylcarbamate on the basis of its n.m.r. spectrum in sulphuric acid, and the fact that protonation follows an acidity function other than H_A.

III. pK_a VALUES OF AMIDES FUNCTIONING AS BASES

The pK_a value of a relatively strong base, falling well within the limits of the pH range, is a thermodynamic quantity capable of accurate experimental determination[34]. On the other hand, determination of pK_a values for weak bases is a much more arbitrary process, and there are many sources of error.

One very widely used method for the determination of the pK_a values of amides in aqueous sulphuric, perchloric or hydrochloric acid involves the measurement of accurate ionization ratios at known acidities, and subsequent use of appropriate acidity functions. For the direct evaluation of such ratios for amides, the use of ultraviolet spectroscopy is widespread, but n.m.r., including fluorine as well as proton resonance[32,35,36], and Raman[37] spectral techniques have been employed.

One frequently encountered drawback to the accurate estimation of ionization ratios by ultraviolet or n.m.r. spectroscopy is a medium effect. In this, the absorption maximum of a given base or its conjugate acid suffers a variation both in wavelength and intensity, with changing acid concentration, other than that due to protonation. Various methods for compensation of this effect have been described[21,38]; one very successful method was indeed first described for amides[39]. However, such an effect appears to be small and often negligible in the case of amides in general[32,40,41], and we therefore consider it no further.

Extensive compilations of aqueous amide pK_a values based on the H_0 scale are given in Arnett's review[38] (which also includes basicity data in other media), and by Yates and Stevens[42], who have considered the conversion of such values to the original H_A scale.

Titration techniques in non-aqueous solvents also constitute an important method. Hall[43] has shown that the pK_a values in acetic acid of a whole series of organic bases parallel their pK_a values in water, and a good deal of work for amides has been carried out in this organic solvent, as well as in formic acid and nitromethane. This method is particularly significant in the case of aliphatic amides which lack suitable ultraviolet spectra, and thus cannot readily be estimated in aqueous acid.

There appears to be little information concerning the application of other methods[38], cryoscopy, differential solubility, and conductivity, to this class of compounds.

A. The H_A Acidity Function and Determination of pK_a Values in Strong Aqueous Acid

The most straightforward general method for determining pK_a values comes from the determination of conjugate acid to free base ratios at varying hydrogen ion concentrations of aqueous acid. This ratio is termed an indicator ratio, and given the symbol I. Use of equation (1) then gives the pK_a value directly[34].

$$\text{pH} = pK_a - \log \frac{[BH^+]}{[B]} = pK_a - \log I \qquad (1)$$

However, for the weakly basic amides, the region of observable protonation (say 5–95%) occurs in strong acid, i.e. in an acidity region where the stoichiometric concentration of hydrogen ions and free base and conjugate acid molecules is no longer accurately equatable to the activities of these species.

On introducing activity coefficients equation (1) then becomes

$$pK_a = \log I - \log \frac{a_{H^+} f_B}{f_{BH^+}} \qquad (2)$$

Introducing the symbol H for the final term of equation (2), we obtain

$$H = -\log \frac{a_{H^+} f_B}{f_{BH^+}} = pK_a - \log I \qquad (3)$$

and H can be considered as a quantitative measure of the ability of the acid to transfer a proton to the base B.

We may write, for two bases A and B, using equation (2),

$$pK_A - pK_B = \log \frac{[AH^+]}{[A]} - \log \frac{[BH^+]}{[B]} + \log \frac{f_{AH^+} f_B}{f_A f_{BH^+}} \qquad (4)$$

If it can be shown that $\log \dfrac{[AH^+]}{[A]} - \log \dfrac{[BH^+]}{[B]}$ is constant for various pairs of bases of necessarily similar pK_a values over the whole acidity range, by demonstrating the parallelism of plots of $\log I$ vs. % acid from dilute solution to concentrated acid, then

$$\log \frac{f_{AH^+}}{f_A} = \log \frac{f_{BH^+}}{f_B} \qquad (5)$$

at any given acidity, and the bases can be said to follow the same acidity function H. By anchoring such a set of bases with an initial base to which equation (1) applies, the pK_a values of the whole series of bases may be found, and the values of H appropriate to their ionization behaviour can be evaluated.

In his pioneer studies of acidity functions, Hammett[44] derived an acidity scale incorporating aromatic amine and carbonyl indicators, which he called H_0, for aqueous sulphuric and perchloric acid solutions. The log I values were calculated using a colorimetric technique. Subsequently, H_0 values have been calculated for a whole variety of acid systems[21,45]. The Hammett acidity scale H_0 has since been reevaluated using entirely primary aniline indicators for both aqueous sulphuric acid[46,47] and perchloric acid solutions[60], using the spectrophotometric technique. The ionization of many other types of base appears to be reasonably accurately described by this scale incorporated in equation (3).

It should be noted that some compounds do not adhere to the same acidity scale over the complete acid region[38]. Thus benzoic acid is a Hammett base at high acidities, but deviates significantly from H_0 as the acidity is reduced. Fortunately, such cases appear to be few in number, and there is no report of any amide behaving in this fashion.

Certain other bases, however, depart markedly from equation (3) (writing H_0 for H), and follow instead equation (6). Such compounds

$$H_0 = H_0 \text{ (half protonation)} - n \log I \qquad (6)$$

include the olefins[48], indoles[49], tertiary anilines[50] and carbinols[51] (in all of which cases $n < 1$), and, most important from the point of view of this review, amides[41], where $n > 1$.

For all of these groups of compounds, new acidity scales have been estimated in aqueous sulphuric acid (olefins, H_R'; indoles, H_I; tertiary amines, H_0'''; carbinols, H_R; amides, H_A), by employing the overlapping indicator technique, and measuring ionization ratios spectrophotometrically. These scales yield approximately linear plots against H_0[52] (Figure 3), which intersect approximately at a common point, $H_0 = 0$.

The amide H_A scale in aqueous sulphuric acid is of particular interest because it is the only distinct function which involves an extent of proton uptake with increasing acidity smaller than H_0. Initial work on the accurate determination of indicator ratios for amides using the ultraviolet technique[39,40,53,54] demonstrated, by use of equations of the form of (6), that amides were not adhering to the H_0

acidity function. This was followed by the establishment of the H_A scale using substituted benzamides[41], at 25°. The ionization curves for these indicators are given in Figure 4, together with that for the second pK_a of phenazine 5,10-dioxide, an indicator which has been used to extend the scale to 93% sulphuric acid[55]. The resultant pK_a

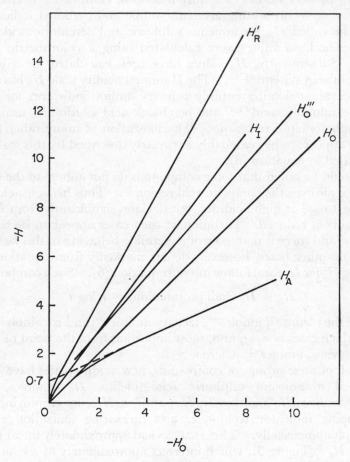

FIGURE 3.   The linear dependence of H on H_0.
[Reproduced, by permission, from *J. Am. Chem. Soc.*, **89**, 2686 (1967).]

values are shown in Table 1, and the H_A scale in Table 2 and Figure 5 where it is compared with H_0. Figure 4 illustrates the very good degree of parallelism achieved by the indicators used in the scale. The H_A function thus established shows that amides protonate more

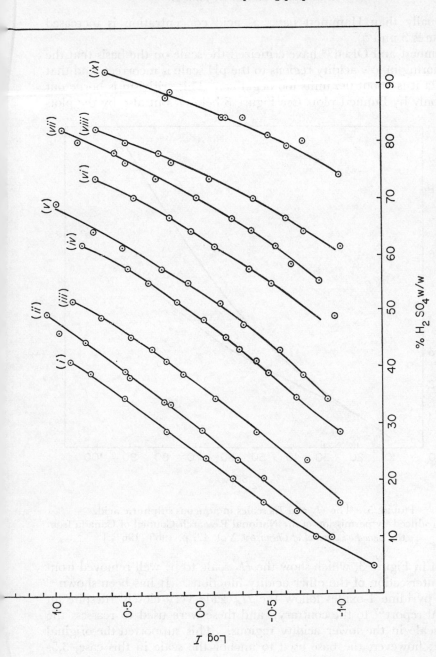

FIGURE 4. Ionization ratio curves for amide indicators in aqueous sulphuric acid. (i) Pyrrole-2-carboxamide. (ii) 4-Methoxybenzamide. (iii) 3,4,5-Trimethoxybenzamide. (iv) 3-Nitrobenzamide. (v) 4-Methyl-3,5-dinitro-benzamide. (vi) 2,3,6-Trichlorobenzamide. (vii) 2,4-Dichloro-3,5-dinitrobenzamide. (viii) 2,4,6-Trinitrobenzamide. (ix) 5-Hydroxyphenazinium 10-oxide.

[Reproduced by permission of the National Research Council of Canada from the Canadian Journal of Chemistry, Vol. 42, p. 1963 (1964).]

gradually than Hammett bases as acid concentration is increased (Figures 3 and 5).

Bunnett and Olsen[56] have criticized the scale on the basis that the anchoring in low acidity regions to the pH scale is incorrect, and that in fact it is about 0·3 units too negative. This criticism is borne out not only by Bunnett plots (see Figure 8 below), but also by the plots

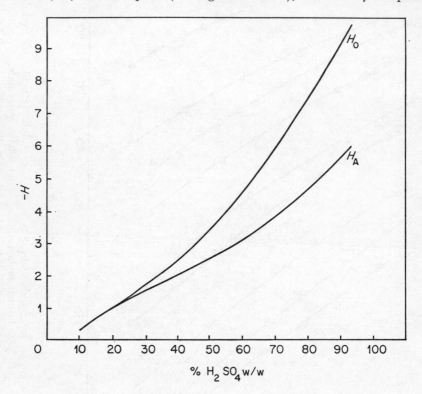

FIGURE 5.   The H_A and H_0 scales in aqueous sulphuric acid.
[Reproduced by permission of the National Research Council of Canada from the *Canadian Journal of Chemistry*, Vol. 42, p. 1965 (1964).]

given in Figure 3, which show the H_A scale to be well removed from the intersection of the other acidity functions. It has been shown[55] that pyridine 1-oxides follow the H_A scale very closely (despite an initial report[57] to the contrary), and these were used to reassess the H_A scale in the lower acidity regions. This supported the original scale; however, the base used to anchor the scale in this case, 3,5-dimethyl-4-nitropyridine 1-oxide, appears to be a Hammett base.

Data for several other N-oxides of pK_a approximately zero or less show deviation from Hammett base behaviour in these low acidity regions. Details of these bases are given in Table 3, and reanchoring of the H_A scale is indicated in Figures 6 and 7.   This yields pK_a values of -0.15

TABLE 1.   pK_a values of the H_A-scale indicators[41].

Compound	pK_a
Pyrrole-2-carboxamide	-1.23
4-Methoxybenzamide	-1.44
3,4,5-Trimethoxybenzamide	-1.82
3-Nitrobenzamide	-2.42
4-Methyl-3,5-dinitrobenzamide	-2.69
2,3,6-Trichlorobenzamide	-3.30
2,4-Dichloro-3,5-dinitrobenzamide	-3.73
2,4,6-Trinitrobenzamide	-4.08
5-Hydroxyphenazinium 10-oxide	-5.12

TABLE 2.   The H_A scale in aqueous sulphuric acid[41,55], at 25°.

% H_2SO_4 w/w	$-H_A$	% H_2SO_4 w/w	$-H_A$
15	0.69	70	3.74
20	0.97	75	4.13
25	1.25	80	4.56
30	1.50	82	4.74
35	1.74	84	4.91
40	2.00	86	5.12
45	2.25	88	5.34
50	2.50	90	5.57
55	2.78	92	5.79
60	3.06	93	5.90
65	3.38		

for 5-nitroquinoline 1-oxide and -1.03 for pyrrole-2-carboxamide, compared with -1.23 given by Yates, Stevens, and Katritzky[41] for this latter compound.   These results thus disclose the amide scale as presented in Table 2 to be 0.20 units too negative.

Yates and Riordan[58] have established the H_A scale in aqueous hydrochloric acid, at 25°.   The indicators used are given in Table 4, together with their pK_a values which show reasonable agreement with

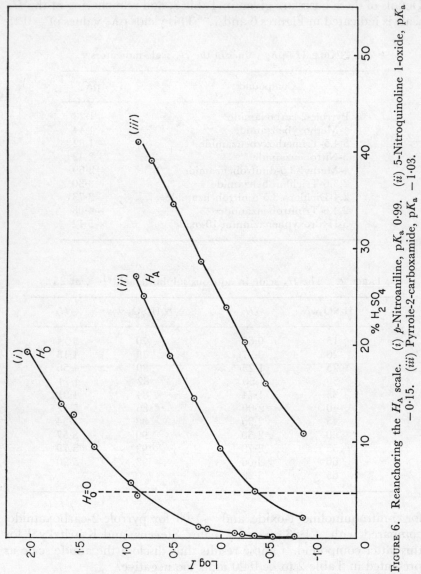

FIGURE 6.   Reanchoring the H_A scale.   (i) p-Nitroaniline, pK_a 0.99.   (ii) 5-Nitroquinoline 1-oxide, pK_a −0.15.   (iii) Pyrrole-2-carboxamide, pK_a −1.03.

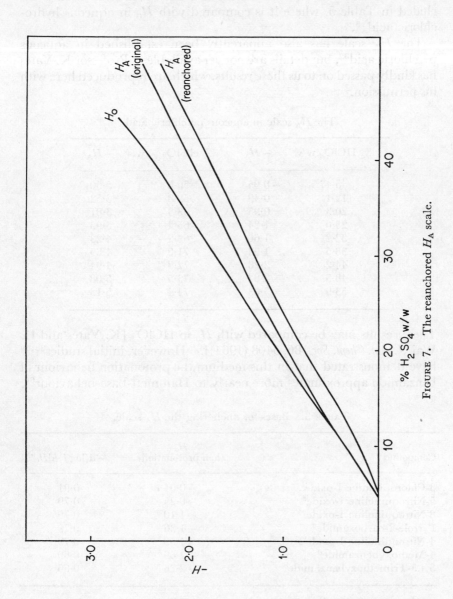

FIGURE 7. The reanchored H_A scale.

the values found in aqueous sulphuric acid.   The scale itself is included in Table 5, where it is compared with H_0 in aqueous hydrochloric acid[45].

The H_A scale has also apparently been established in aqueous perchloric acid[32], but details are not yet available.   Professor K. Yates has kindly passed on to us these results, which are reproduced here with his permission.

The H_A scale in aqueous perchloric acid.

% HClO$_4$ w/w	$-H_A$	% HClO$_4$ w/w	$-H_A$
5·47	−0·05	56·8	3·30
12·1	0·49	59·8	3·52
20·3	0·95	63·3	3·81
25·0	1·24	65·4	3·93
32·2	1·60	68·2	4·23
38·5	1·93	71·0	4·65
44·2	2·28	72·9	4·94
48·5	2·62	73·5	5·08
53·9	3·03	74·4	5·13

These results may be compared with H_0 in HClO$_4$ [K. Yates and H. Wai, *J. Am. Chem. Soc.*, **86**, 5408 (1964)].   However, initial studies[32,59] have demonstrated that in this medium the protonation behaviour of benzamide approximates more nearly to Hammett-base behaviour[60].

TABLE 3.   Bases for anchoring the H_A scale.

Compound	H_0 (half protonation)	$-\mathrm{d}(\log I)/\mathrm{d}H_0{}^a$
3-Chloropyridine 1-oxide[b]	−0·04	0·91
5-Nitroquinoline 1-oxide[c]	−0·25	0·79
8-Nitroquinoline 1-oxide[c]	−1·10	0·79
Pyrrole-2-carboxamide[d]	−1·36	0·79
4-Nitroquinoline 1-oxide[c]	−1·53	0·79
4-Methoxybenzamide[d]	−1·69	0·69
3,4,5-Trimethoxybenzamide[d]	−2·26	0·66

    [a] Or $-m$ in equation (9).
    [b] N. Shakir, *Ph.D. Thesis*, University of East Anglia, 1966.
    [c] J. T. Gleghorn, R. B. Moodie, E. A. Qureshi, and K. Schofield, *J. Chem. Soc. (B)*, 316 (1968);
J. T. Gleghorn, *Ph.D. Thesis*, University of Exeter, 1966.
    [d] Reference 41.

In general, the evaluation of the pK_a value of a weak base from a knowledge of the variation of log I with acidity involves plotting log I against the various acidity functions H until a line of approximately unit slope has been obtained. Equation (3) then applies, and the pK_a value may be read off at the point where log $I = 0$.

If the compound appears not to follow any known acidity function,

TABLE 4. pK_a values of H_A-scale indicators from measurements in aqueous hydrochloric acid[58].

Compound	pK_a
2-Nitroaniline	-0.31
Pyrrole-2-carboxamide	-1.23
4-Methoxybenzamide	-1.46
3,4,5-Trimethoxybenzamide	-1.86
3-Nitrobenzamide	-2.25
4-Methyl-3,5-dinitrobenzamide	-2.77
2,3,6-Trichlorobenzamide	-3.10

TABLE 5.   The H_0 and H_A scales in aqueous hydrochloric acid[58].

% HCl w/w	$-H_A$	$-H_0$
3	0.13	0.13
4	0.31	0.29
5	0.47	0.43
6	0.59	0.57
7	0.71	0.69
8	0.82	0.81
10	1.03	1.01
12	1.23	1.23
14	1.43	1.44
16	1.63	1.64
18	1.82	1.87
20	2.01	2.11
22	2.21	2.35
24	2.42	2.60
26	2.61	2.87
28	2.81	3.12
30	3.02	3.39
32	3.22	3.67
34	3.44	3.95
35	3.56	4.11

an approximation to the thermodynamic pK_a may be estimated by Bunnett and Olsen's equation[56]

$$\log I - \log [H^+] = (\phi - 1)(H_0 + \log [H^+]) + pK_a \qquad (7)$$

where $(\phi - 1)$ is a slope parameter. The validity of this equation is demonstrated by the linearity of plots of $(H + \log [H^+])$ vs. $(H_0 + \log [H^+])$ shown in Figure 8, the former sum of terms being equivalent to $(-\log I + \log [H^+])$ for a base following the H acidity function of zero pK_a.

An alternative method[52] is the use of equation (8)

$$mH_0 - mH_0(\text{half protonation}) = -\log I \qquad (8)$$

where $m = 1/n$, and n is found according to equation (6), the validity of which follows from Figure 3, and thus

$$\log I = -mH_0 + pK_a \qquad (9)$$

Although in general the use of both equations (7) and (9) leads to practically equivalent values of pK_a[20], Bunnett and Olsen's equation does give better agreement specifically with pK_a values of amides using the adjusted H_A scale. This is illustrated in Table 6, and also in Figures

TABLE 6. pK_a values of amide indicators.

Amide	pK_a^a	pK_a^b	pK_a^c
Pyrrole-2-carboxamide	$-1 \cdot 03$	$-1 \cdot 06$	$-1 \cdot 07$
4-Methoxybenzamide	$-1 \cdot 24$	$-1 \cdot 17$	$-1 \cdot 15$
3,4,5-Trimethoxybenzamide	$-1 \cdot 62$	$-1 \cdot 53$	$-1 \cdot 47$
3-Nitrobenzamide	$-2 \cdot 22$	$-2 \cdot 03$	$-1 \cdot 86$
4-Methyl-3,5-dinitrobenzamide	$-2 \cdot 49$	$-2 \cdot 25$	$-2 \cdot 01$
2,3,6-Trichlorobenzamide	$-3 \cdot 10$	$-3 \cdot 00$	$-2 \cdot 70$
2,4-Dichloro-3,5-dinitrobenzamide	$-3 \cdot 53$	$-3 \cdot 16$	$-2 \cdot 79$
2,4,6-Trinitrobenzamide	$-3 \cdot 88$	$-3 \cdot 87$	$-3 \cdot 55$
5-Hydroxyphenazinium 10-oxide	$-4 \cdot 92$	$-5 \cdot 12$	$-4 \cdot 95$

^{a} (Value from Table 1) $+ 0 \cdot 2$.
^{b} Calculated by Bunnett and Olsen's method (equation 8).
^{c} mH_0(half protonation).

3 and 8. In the latter figure, the H_A correlation line intersects the abscissa much closer ($0 \cdot 3$ units) to the origin than in the former ($0 \cdot 7$ units). Table 7, essentially an extension of that due to Yates and Stevens[42], gives protonation data for a series of primary, secondary

and tertiary amides. The pK_a values quoted have been derived in general by two methods. One involves the conversion of the H_0 (half protonation) value to the adjusted H_A scale, and the other direct use of equation (9). Unfortunately, Bunnett and Olsen's treatment could not be evaluated for any of these amides, as the necessary

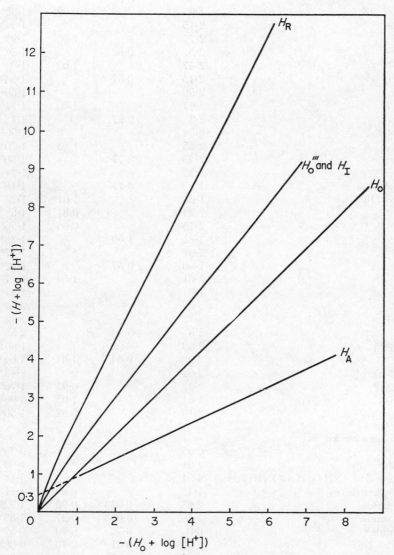

FIGURE 8. Linearity of plots of $-(H + \log [H^+])$ against $-(H_0 + \log [H^+])$.

8+c.o.a.

Table 7. pK_a values and protonation data for amides[a].

Amide	$-H_0$(half protonation)	$-\dfrac{d \log I}{d H_0}$ (m)	$-\dfrac{d \log I}{d H_A}$	$-pK_a^{b,c}$
1. Substituted benzamides				
Benzamide[d]	2·16		0·99	1·54[b]
Benzamide[e]	2·1	0·66		1·5[b], 1·4[c]
Benzamide[g] (18°)	1·85		0·98	1·38[b]
Benzamide[i]	2·15			1·54[b]
m-Bromo[d]	2·75		1·00	1·93[b]
m-Bromo[e]	2·5	0·52		1·8[b], 1·3[c]
p-Bromo[d,f]	2·47		1·03	1·82[b]
p-Bromo[e]	2·15	0·62		1·54[b], 1·33[c]
m-Chloro[d]	2·59			1·89[b]
p-Chloro[d]	2·47		1·00	1·77[b]
p-Chloro[e]	2·3	0·77		1·7[b], 1·8[c]
p-Fluoro[d]	2·24			1·60[b]
m-Methoxy[d]	2·35		1·03	1·70[b]
m-Methoxy[e]	2·45	0·75		1·80[b], 1·84[c]
p-Methoxy[d,h]	1·80			1·34[b]
p-Methoxy[e]	1·55	0·77		1·18[b], 1·19[c]
p-Methoxy[g] (18°)	1·60		1·01	1·22[b]
o-Methyl[i]	2·24		0·81	1·6[b]
m-Methyl[d]	2·15		0·95	1·56[b]
m-Methyl[e]	1·85	1·00		1·38[b], 1·85[c]
p-Methyl[d]	2·01			1·47[b]
p-Methyl[e]	1·90	0·67		1·42[b], 1·27[c]
p-Methyl[f]	2·01		1·04	1·47[b]
p-Methyl[i]	1·81			1·34[b]
m-Nitro[d,h]	3·07			2·10[b]
p-Nitro[d]	3·23			2·22[b]
p-Nitro[e]	3·8	0·39		2·5[b], 1·48[c]
p-Nitro[g] (18°)	2·58			1·86[b]
p-Phenyl[j]	2·22	0·645	1·01	1·64[b], 1·43[c]
4-Bromo-3-nitro[f]	3·67		0·95	2·43[b]
2,6-Dimethyl[i]	2·42		0·95	1·76[b]
3,5-Dinitro[f]	4·61		1·05	3·06[b]
2,4,6-Trimethyl[i]	2·10		0·92	1·50[b]
2. Other primary amides				
Acetamide[k]	0·55			0·35[l]
2-Furamide[d]	2·61			1·90[b]
Glycinamide .H$^+$ $(NH_3^+CH_2CONH_2)$[k]	4·91			3·1[b]
3-Hydroxyphenylurea[f]	1·45		0·99	1·11[b]
1-Naphthamide[j]	2·63	0·638	0·98	1·92[b], 1·68[c]
2-Naphthamide[j]	2·56	0·600	0·97	1·88[b], 1·54[c]
Propionamide[k]	0·8	0·70	0·70	0·57[c]
Butyramide[m]	1·20	0·61	0·78	0·88[b], 0·73[c]
Ethyl carbamate[m,n]	4·39	0·57	1·10	2·83[b], 2·50[c]

TABLE 7. *(Cont.)*

Amide	$-H_0$ half protonation	$-\dfrac{d \log I}{d H_0}$ (m)	$-\dfrac{d \log I}{d H_A}$	$-pK_a{}^{b,c}$
3. *Secondary amides*				
N-Acetylglycine[k]	3·10			2·1[b]
N-Butyrylglycine[m]	2·42	0·53	0·89	1·73[b], 1·28[c]
N-Ethoxycarbonylglycine[m]	5·88	0·56	1·09	3·57[b], 3·29[c]
Ethyl N-methylcarbamate[m]	4·18	0·74	1·40	2·72[b], 3·09[c]
N-n-Butylacetamide[k]	0·41			0·29[l]
N-i-Butylacetamide[k]	0.62			0·42[l]
N-Methylacetamide[k]	1·25			1·0[b]
N-Methylbenzamide[d]	2·13		1·02	1·5[b]
N-Methylbenzamide[e]	2·0	0·53		1·47[b], 1·06[c]
N-Methyl-1-naphthamide[j]	2·17	0·749	1·11	1·55[b], 1·62[c]
N-Methyl-2-naphthamide[j]	2·13	0·741	1·11	1·54[b], 1·58[c]
N-Methyl-p-phenylbenzamide[j]	2·17	0·715	1·07	1·56[b], 1·55[c]
N-Ethylbenzamide[o]	2·33	0·69	0·95	1·72[b]
N-Ethyl-m-bromobenzamide[o]	3·22	0·51	0·99	2·20[b]
N-Ethyl-m-chlorobenzamide[o]	2·94	0·52	0·91	2·08[b]
N-Ethyl-p-chlorobenzamide[o]	2·81	0·52	0·92	2·01[b]
N-Ethyl-p-methoxybenzamide[o]	1·88	0·73	1·01	1·40[b]
N-Ethyl-m-methylbenzamide[o]	2·77	0·66	0·90	1·68[b]
N-Ethyl-p-methylbenzamide[o]	2·12	0·66	0·90	1·57[b]
N-Ethyl-m-nitrobenzamide[o]	3·49	0·41	0·79	2·34[b]
N-Trifluoroethylbenzamide[o]	5·00	0·54	0·96	3·13[b]
N-Trifluoroethyl-m-bromobenzamide[o]	5·61	0·58	1·11	3·44[b]
N-Trifluoroethyl-m-chlorobenzamide[o]	5·56	0·49	0·95	3·42[b]
N-Trifluoroethyl-p-chlorobenzamide[o]	5·15	0·51	0·98	3·21[b]
N-Trifluoroethyl-p-methoxybenzamide[o]	3·78	0·61	1·17	2·50[b]
N-Trifluoroethyl-m-methylbenzamide[o]	4·72	0·46	0·88	3·00[b]
N-Trifluoroethyl-p-methylbenzamide[o]	4·35	0·55	1·05	2·80[b]
N-Trifluoroethyl-m-nitrobenzamide[o]	5·95	0·55	1·06	3·63[b]
4. *Tertiary amides and thioamides*				
N,N-Dimethylbenzamide[d]	1·62		0·90	1·2[b]
N,N-Dimethyl-1-naphthamide[j]	1·91	0·634	0·90	1·37[b], 1·21[c]
N,N-Dimethyl-2-naphthamide[j]	1·69	0·578	0·77	1·3[b], 0·98[c]
N,N-Dimethyl-p-phenylbenzamide[j]	1·30			1·00[b]
Ethyl N,N-dimethylcarbamate[m]	4·65	0·67	1·30	3·16[b], 3·11[c]
N,N-Dimethylthioformamide[p]	2·54	1·35		3·43[c]
N,N-Dimethylthioacetamide[p]	1·53	1·45		2·22[c]
N,N-Dimethylthiopivalamide[p] [(CH₃)₃CCSN(CH₃)₂]	1·51	1·50		2·26[c]

[a] Temperature 25° unless stated otherwise. [b] H_A (half protonation) + 0·2.
[c] $m H_0$(half protonation). [d] Reference 15. [e] Reference 39. [f] Reference 42.
[g] J. T. Edward and S. C. R. Meacock, *J. Chem. Soc.*, 2000 (1957).
[h] Also used as an indicator for the H_A scale, see Table 1. [i] Reference 53. [j] Reference 40.
[k] Reference 54. [l] Calculated from Figure 6. [m] Reference 32.
[n] Methyl carbamate stated to yield similar results, but no values given.
[o] Personal communication from Dr. R. B. Moodie. [p] Reference 61.

$\%H_2SO_4$ vs. log I tabulations are not given in the literature. Yates and Stevens' table is the source of the d log I/dH_A values.

In some cases, agreement between the two methods is reasonable; where there is a large difference, this is due either to the protonation behaviour of the compound deviating widely from H_A, or to poor agreement among different workers of either the H_0 (half protonation) value, or the d log I/dH_0 and d log I/dH_A values, which should obviously bear a constant relationship to one another.

All of the primary amides except o-methyl- and p-nitrobenzamide follow H_A quite closely. Yates and Stevens[42] attribute the apparent deviation of the latter to inaccuracies in the ultraviolet measurements; the former is less easy to account for in the light of the conformity of the other o-methyl-substituted benzamides, 2,6-dimethyl- and 2,4,6-trimethylbenzamide. Propionamide and butyramide, the former having been cited as not following H_A[42], are especially interesting. These are stronger bases than most of the others given in Table 7, and in fact stronger than all of the others for which gradient data (d log I/dH_0 or of d log I/dH_A) are given. The values of such gradients for these two compounds are smaller than for the N-oxides of similar basicity used for anchoring the scale. Anchoring the scale with these two (and a good case could be made for this, since it seems reasonable to define the H_A scale in its entirety by amides alone) would make it even less negative than by the 0·2 units suggested, and it seems very likely that the scale would then pass through or much closer to the origin of Figures 3 and 8.

Certain secondary and tertiary amides also deviate from H_A, but no overall trend can be detected, and many others follow H_A closely. In fact, judged within the context of experimental inaccuracies, discrepancies between the data of one set of workers and another, and the vagaries of acidity-function theory, the amides as a whole, irrespective of substituent type, number and position, seem to follow H_A satisfactorily.

As the last portion of Table 7 shows, and as Janssen[61] has observed, however, thiones do not follow H_A, plots of log I against H_0 having slopes of unity or greater[61,62].

There is also some doubt about the acidity function obeyed by the pyridones. 2-Pyridones appear to follow H_A[63], as do 4-pyridones, judging from kinetic evidence[64]. However, the situation is complicated by the fact that electron-withdrawing substituents may drastically alter the position of equilibrium **16** $\rightleftharpoons$ **17** (end of section II), tending to favour the hydroxy pyridine tautomer.

B. The H_0 and H_A Acidity Functions and Related Hydration Theories

The difference between the Hammett function, H_0, and H_A (equation 10 where B refers to a Hammett base, and A to an amide base), lies in the medium dependence of the activity coefficient term.

$$H_0 - H_A = \log\frac{f_{BH^+}f_A}{f_B f_{AH^+}} \tag{10}$$

Yates has recently determined[65] the medium dependence of some aromatic amide activity coefficients by solubility measurements, referring the ionic activity coefficients to the tetraethylammonium ion (TEA$^+$) through use of the pentacyanopropenide anion as Boyd[66] had done for some aniline and carbinol indicators. Interpretative difficulties arise because it is not possible to measure f_B and f_{BH^+}/f_{TEA^+} for the same compound over the same range of acid concentration. The activity coefficients of several neutral amides, f_A, showed a similar medium dependence decreasing as acid concentration increased above 20% sulphuric acid. The size of this decrease is directly proportional to the number of nitro groups attached to the aromatic nucleus, and this is taken to reflect strong hydrogen bonding between such groups and the acidic solvent. The benzamidonium ion activity coefficient measured as f_{AH^+}/f_{TEA^+} increases rapidly with acid concentration. Figure 9 shows typical acidity dependence of the various terms involved, whence it can be seen that the order of importance of the individual terms in the expression for H_A (11 and 12),

$$-H_A = \log f_{H^+} + \log[H^+] + \log f_A - \log f_{AH^+} \tag{11}$$

and thus

$$-H_A = \log\frac{f_{H^+}}{f_{TEA^+}} + \log[H^+] + \log f_A - \log\frac{f_{AH^+}}{f_{TEA^+}} \tag{12}$$

is $f_{H^+} > f_{AH^+} \gg [H^+] > f_A$, for the acidity range 5–40% acid, and at higher acid concentrations $f_{H^+} \gg f_{AH^+} \gg f_A > [H^+]$.

Comparison with Boyd's data on anilines showed that the difference between H_0 and H_A arises because $f_{AH^+} > f_{BH^+}$ and $f_A < f_B$, where B represents an aniline base. This may be interpreted as indicating that the protonated amide has greater hydration requirements than the corresponding anilinium ions, and for the neutral molecules that the amide is less strongly hydrated than the aniline.

Other attempts have been made to explain the deviations from H_0 in the ionization behaviour of amides in terms of the differential hydration of the cation and free base, as Taft[67] had done for the N-substituted anilines, by considering the equilibrium (13) for amides,

$$AH^+(H_2O)_m + (a + p - m)H_2O \rightleftharpoons A(H_2O)_a + H^+(H_2O)_p \qquad (13)$$

and for a Hammett indicator

$$BH^+(H_2O)_n + (b + p - n)H_2O \rightleftharpoons B(H_2O)_b + H^+(H_2O)_p \qquad (14)$$

Thence

$$\frac{h_A}{h_0} = (f_{BH^+}f_A/f_B f_{AH^+})a_{H_2O}^{(m-a-(n-b))} \qquad (15)$$

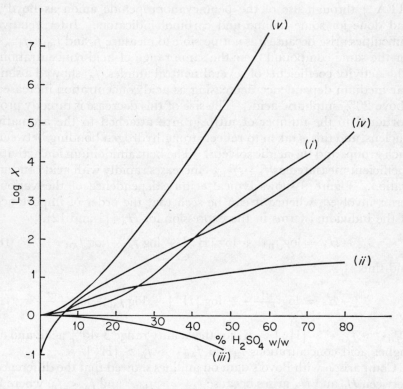

FIGURE 9.   Concentration dependence of individual terms ($\log X$) contributing to the H_A acidity function.   (i) Log (f_{AH^+}/f_{TEA^+}) for benzamidionium; (ii) log $[H^+]$; (iii) log f_A for 2,4-dinitrobenzamide; (iv) $-H_A$; (v) log (f_{H^+}/f_{TEA^+}). [Reproduced by permission of the National Research Council of Canada from the *Canadian Journal of Chemistry*, Vol. 44, p. 2402 (1966).]

where the activity coefficients now refer to hydrated species; if the activity coefficient term is assumed to be invariant with acidity, expression (16) results

$$\frac{d(H_0 - H_A)}{d \log a_{H_2O}} = (m - a) - (n - b) \qquad (16)$$

Edward and Wang[54] investigated this expression for propionamide ionization in aqueous sulphuric acid, by plotting it in its alternative form

$$\log I + H_0 = -h \log a_{H_2O} + pK_a \qquad (17)$$

but obtained a curve with slope $(-h)$ increasing as acidity increased. Less basic amides which protonate at acidities where the supply of water may be considered restricted, above 35% sulphuric acid (1 mole H_2SO_4 to 10 moles of water), must compete for available water with the $H_4O_9{}^+$ ion which is itself capable of forming relatively stable hydrogen bonds with a further six water molecules[68]. Using a similar expression to Edward and Wang's (16), but incorporating H_R' rather than H_0, since hydration is presumably unimportant in the protonation equilibria of olefins, Homer and Moodie[40] obtained linear plots for several aromatic amides in the region above 40% sulphuric acid. Assuming the free amide is unhydrated $(a = 0)$, values of m for primary amides were about 5 and for N-methylamides about 4. Deno[21] considers that the interaction forces between solvent and solute are in general similar for various solute species, but in one case vary enormously, that of hydrogen bonding of solvent to free base, and most important, to charged conjugate acid. Thus hydration as in 18

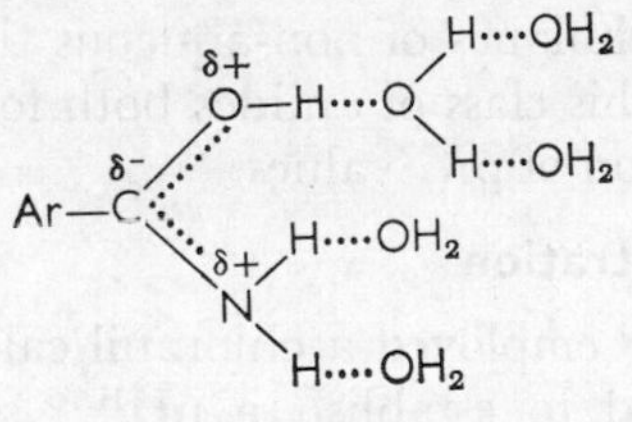

(18)

was proposed for the amide conjugate acid[40], a similar structure being arrived at by Yates[41] in applying the same arguments to the amides used in establishing the H_A scale. N-Methylation removes one of the hydrogen-bonded water molecules.

In this connexion it is interesting to recall that thioamides protonate a good deal more sharply than the amides themselves (Table 7).

This could either be due to increased hydration requirements of the thioamide free base as suggested by Edward and Stollar[62], or to less hydration of the conjugate acid. It is well known that sulphur compounds are not strongly hydrogen bonded (cf. H_2O and H_2S), and this difference between oxygen and sulphur may also be true when each bears a positive charge, though in this case we are dealing with hydrogen-bond formation between the hydrogen atoms on the protonated amide or thioamide and the water molecules of the solvent.

N,N-Dimethylamides did not yield a linear plot from equation (16) and values of m could not be obtained[40]. This is possibly due to their greater basicity preventing accurate measurement of log I in the region above 40% sulphuric acid. However, tertiary amide ionizations apparently follow H_A as closely as primary or secondary amides (Table 7), indicating that other factors besides hydration must be important in determining the medium effect on activity coefficients. In fact, recent work by Arnett[50,69], utilizing an equation of the form of (17) and applying it to various acidic media, has demonstrated that the difference in acidity functions cannot, in general, be explained completely in terms of simple hydration theory.

The low basicity of 2,6-dimethylbenzamide (see Table 7) has been attributed in part to steric hindrance to solvation in the conjugate acid[53].

C. Determination of pK$_a$ Values in Non-aqueous Media

It is difficult (although not impossible[54]) to measure the pK_a values of aliphatic amides by ultraviolet spectroscopy due to the absence of any strong absorption bands at a convenient wavelength. Consequently the techniques of non-aqueous titration have been extensively applied to this class of amides both for quantitative estimation and determination of pK_a values.

I. Potentiometric titration

Hall and Conant[70] employed a chloranil/calomel electrode system in glacial acetic acid to establish a pH^{HAc} scale in this medium. pK_a^{HAc} was then defined as the pH^{HAc} at which a 0.2 M solution of the base in acetic acid was half-neutralized by a 2 M solution of sulphuric acid in acetic acid. The presence of liquid-junction potentials makes the absolute magnitude of such a scale somewhat uncertain, and the high concentration of base is a disadvantage. Hall[43] subsequently used perchloric acid in acetic acid as titrant and reduced the base concentration to 0.05 M. Data were obtained for a large number of anilines and some amides. The basicities of the former compounds were related to their pK_a values in water by equation (18) which has been used by subsequent workers[16,17,71] to determine the pK_a of

$$pK_a^{H_2O} = pK_a^{HAc} + 1.7 \qquad (18)$$

many aliphatic amides and lactams. This equation probably overestimates amide basicity in water by about 0.5 pK_a units and only values of pK^{HAc}, which are useful measures of relative basicity, are given in Table 8. There are insufficient data available to allow us to compare meaningfully pK_a values in aqueous solution on the one hand

TABLE 8. pK_a^{HAc} values of some amides[a] $R^1CONR^2R^3$.

R^1	R^2	R^3	pK_a^{HAc}
H	H	H	-2.18^b
Me	H	H	-1.59^b, -1.59^c, -1.64^d
Et	H	H	-1.58^c
n-Pr	H	H	-1.65^c
n-Pentyl	H	H	-1.75^c
n-Hexyl	H	H	-1.76^c
i-Pr	H	H	-1.97^c
i-Bu	H	H	-1.69^c
t-Bu	H	H	-2.18^c
i-Pentyl	H	H	-1.72^c
Ph	H	H	-2.70^d
H	Me	H	-1.74^b
H	n-Bu	H	-1.67^b
H	PhCH$_2$	H	-2.03^b
H	(Ph)$_2$CH	H	-2.33^b
H	Ph	H	3.49^d
Me	Me	H	-0.90^b
Me	Et	H	-0.91^b
Me	n-Bu	H	-0.86^b
Me	Cyclohexyl	H	-0.93^b
Me	PhCH$_2$	H	-1.39^b
Me	Ph	H	-2.59^d
n-Bu	Me	H	-0.91^b
t-Bu	PhCH$_2$	H	-2.13
H	Me	Me	-1.71^b
Me	Me	Ph	-2.17^d
Me	n-Pr	Ph	-2.27^d

[a] Measured by potentiometric titration with perchloric acid at 20°.
[b] Reference 16.
[c] Reference 71.
[d] Reference 43.

8*

and those in acetic acid or any other organic solvent on the other. Little data in acetic acid exist for the protonation of substituted benzamides whose pK_a values in aqueous sulphuric acid are well established; for aliphatic amides, where much work has been done in organic solvents, only the pK_a of propionamide is known with any confidence in aqueous solution. As noted above, the latter is due to experimental difficulties. However, determination of pK_a values for the benzamides in organic solvents seems a very relevant and feasible investigation.

Streuli[72] has determined basicities in nitromethane by potentiometric titration with perchloric acid employing the glass and calomel electrodes. A linear relation was found between e.m.f. at half neutralization and $pK_a^{H_2O}$ for a large number of amines but again there were insufficient data to determine whether amides followed the same relation. Adelman[73] has employed the same technique for N-substituted amides. Acetic anhydride has also been used as a solvent[74].

2. Indicator methods

Hammett[75] showed that it was possible to determine the pK_a value of a weak base by using an indicator to measure the change in H_0 when the base is added to an acidic solution in formic acid. Lemaire and Lucas[76] developed this method for perchloric–acetic acid media and it has been applied to the determination of amide basicity[17,76,77] yielding pK_a values differing by a constant from those obtained by potentiometric titration[17] for a series of lactams (the pK_a values determined by the indicator method are plotted in Figure 2).

In glacial acetic acid solutions the dissociation constant of ion pairs is very low and equilibria such as equation (19), involving the base A

$$A + HX \xrightleftharpoons{K_i} AH^+ . X^- \xrightleftharpoons{K_d} AH^+ + X^- \qquad (19)$$

and indicator B with the titrating acid must be considered[78]. Recent work due to Grunwald and Ceska[79] on *meta*- and *para*-substituted anilines has demonstrated that K_i is a better measure of basicity than the overall equilibrium constant, in acetic acid. In fact measurements give $K_{ex} = (K_i^{BHX}/K_i^{AHX})$, and Higuchi[80] has evaluated K_{ex} for a large number of amides using perchloric acid as titrant and Sudan III, an azo dye ($K_i^{BHX} = 700$), as indicator. In addition to the inherent basicity of the amide the ion-pair formation constant depends on the acid used and possibly on the indicator employed (compare the results obtained with Sudan III[80] and 2,4-dinitro-N,N-diethylaniline[81] as indicators for acetamide and N-methylacetamide in Table 10).

Table 9 compares the basicity in acetic acid with that in water for those few compounds for which both sets of data are available. A moderate correlation exists for the aromatic amides expressed by equation (20). However, propionamide is considerably more basic

$$pK_a = \log K_i^{AHX} - 4 \cdot 8 \tag{20}$$

than predicted which suggests that the aliphatic amides may follow a different relation. More data on aqueous pK_a values are clearly needed before this can be established.

TABLE 9.   A comparison of basicity in acetic acid and aqueous solutions.

Amide	Log K_i^{AHX}	$-pK_a^c$	$\log K_i - pK_a$
Benzamide	2·92[a]	1·55	4·47
N-Methylbenzamide	3·28[a]	1·50	4·78
N,N-Dimethyl-1-naphthamide	3·74[a]	1·20	4·94
1-Naphthamide	2·65[a]	1·92	4·57
2-Naphthamide	2·92[a]	1·88	4·80
N,N-Dimethyl-2-naphthamide	3·77[a]	1·30	5·07
Propionamide	3·52[b]	0·57	4·09

[a] Reference 80.
[b] Reference 81.
[c] Table 7.

The protonation of some aromatic amides in anhydrous sulphuric–acetic acid and hydrobromic–acetic acid mixtures has been studied spectrophotometrically together with the effect of added salts (tetraethylammonium bromide and tetraethylammonium hydrogen sulphate)[82]. Plots of log I vs. the logarithm of stoichiometric molar concentration of acid (log $M_{H_2SO_4}$) were linear, but the slopes depended on the degree of N-methylation as had been observed in aqueous acid[40]. A measure of the basicity in acetic acid solution was obtained from the H_0 values at half protonation, derived from data due to Hall and Spengeman[83]. The agreement between the relative basicities in water and acetic acid was poor.

D. The Correlation of pK_a Values and Structure

Substituents adjacent to the amide function will affect the basicity of the site by reason of inductive, resonance and steric interactions.

The nature of these effects has in general been extensively elucidated in organic systems in a qualitative manner, and considerable progress, by means of Hammett-type parameters and MO calculations, has been made towards their quantitative assessment.

The extensive work on the basicity of aliphatic amides in acetic acid discussed in the previous section provides a large body of data for the assessment of structural effects though the range of substituents is not large, being almost entirely confined to alkyl groups. Inspection of the potentiometric data in Table 8 and especially the indicator data in Table 10 shows that N-substitution is base strengthening as expected from the inductive effect. A second N-alkyl group has less effect on the basicity than the first, as is found in amines[80]. For mono-N-substituted acetamides log K_i^{AHX} correlates very approximately with Taft's σ^* values giving $\rho^* = -1 \cdot 2$. Alkyl substitution in the acyl group produces the predicted base strengthening on going from

TABLE 10. Substituent effects on basicity of aliphatic amides[a] $R^1CONR^2R^3$.

R^1	R^2	R^3	Log K_i^{AHX}
H	H	H	$3 \cdot 13^b$
H	H	Me	$3 \cdot 74^b$
H	Me	Me	$3 \cdot 70^b$
Me[d]	H	H	$3 \cdot 85^b$, $3 \cdot 66^c$
Me[d]	H	Me	$4 \cdot 49^b$, $4 \cdot 37^c$
Me	Me	Me	$4 \cdot 81^b$
Me[d]	H	Et	$4 \cdot 42^c$
Me[d]	H	i-Pr	$4 \cdot 49^c$
Me[d]	H	t-Bu	$4 \cdot 60^c$
Et	H	H	$3 \cdot 52^c$
i-Pr	H	H	$3 \cdot 37^c$
t-Bu	H	H	$3 \cdot 23^c$
CH_2Cl	H	H	$1 \cdot 59^b$
$CHCl_2$	H	H	$< -1 \cdot 6^b$

[a] Measured by indicator methods with perchloric acid in acetic acid.
[b] Sudan III indicator[80].
[c] 2,4-Dinitro-N,N-diethylaniline indicator[81].
[d] Correlate approximately with σ^* (see text).

formamides to acetamides, but further alkyl substitution affects the basicity in the opposite way to that demanded by the relative inductive effects, although the chloroacetamides behave in the manner demanded by their negative inductive effect (Table 10). This may reflect the importance of steric factors on solvation of the conjugate acid and on ion-pair formation. Hyperconjugation could also explain this phenomenon as Martin and Reese suggest[81].

Adelman[73] has reported an attempt to correlate the carbonyl stretching frequencies (measured in isooctane) and the basicity (measured potentiometrically in nitromethane) of some N,N-di-substituted aliphatic amides with σ^* values. For an amide of general formula $XCONY_2$ it was found that the sum of inductive and reson-ance contributions for $-NY_2$ $(2\sigma_Y^*)$ fortuitously equalled that of X through consideration of the effect of changes in X and Y on the carbonyl stretching frequency. A good linear plot of $\nu_{C=O}$ vs. $\sum\sigma^*$ was obtained. However, the plot of pK_a vs. $\sum\sigma^*$ showed deviations from linearity for propionamides whose base strengths were lower than predicted from the ρ^* value of 1·24. This is a similar anomaly to that observed above, and was ascribed by Adelman to steric interference with ion-pair formation in the conjugate acid, since he found that deviations occurred with lower homologues when the size of the com-plexing acid was increased in the series perchloric acid < phenol < iodine, chloroform.

The basicities of a series of *meta*- and *para*-substituted benzamides, measured in sulphuric acid by spectrophotometric means[15], show better linearity on plotting against $\sigma(r = 0·988)$ than σ^+ $(r = 0·958)$. This is taken by the authors to indicate N-protonation, since O-protonation would give rise to enhanced resonance in the protonated form with electron-donating substituents and thus correlate with σ^+. In analogous cases, acetophenones[84], benzaldehydes[85], and benzoic acids[86] the pK_a values correlate well with σ^+. However the authors themselves question the validity of their interpretation and suggest that the apparent lack of conjugation can be reconciled with O-protonation if it is supposed that the protonated amide group is not coplanar with the ring. Some recent and very relevant data on nuclear substituted N-ethyl- and N-trifluoroethylbenzamides have been provided by Dr. R. B. Moodie (whom we thank for permission to re-produce them here) and are included in Table 7. The pK_a values of the former compounds correlate with σ rather than σ^+ but the pK_a values of the latter correlate with σ^+. Here the strongly electron-withdrawing properties of the CH_2CF_3 group apparently reduce the conjugation of the nitrogen lone pair with the carbonyl group. The result is that the amide grouping becomes coplanar with the ring allowing the stabilizing carbonyl–aromatic resonance and thus corre-lation of the pK_a values with σ^+. A suggestion[40] that the non-planar structure was stabilized by hydrogen bonding of the localized positive charge to water molecules has been disproved by recent work[87], which showed that the dissociation constants of boron trifluoride complexes with substituted benzamides in tetrahydrofuran, where hydrogen bonding is impossible, also correlate with σ rather than σ^+ (see section V). The pK_a values reported by Edwards and coworkers[15] were measured on the assumption that they followed the H_0 acidity scale. This does not rule out the validity of the pK_a vs. σ correlation however[20],

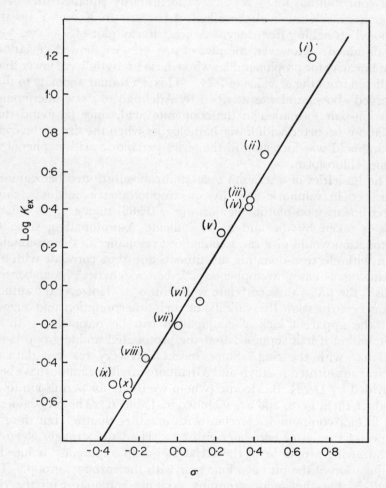

FIGURE 10. σ–ρ plot for basicities of substituted acetanilides.   The substituents are as follows: (*i*) 2,4,6-tribromo; (*ii*) 4-carboxy; (*iii*) 3-chloro; (*iv*) 3-carboxy; (*v*) 4-chloro; (*vi*) 3-methoxy; (*vii*) unsubstituted; (*viii*) 4-methyl; (*ix*) 2,6-dimethyl; (*x*) 4-methoxy.

since conversion to approximate thermodynamic pK_a values will involve multiplication of each H_0 (half protonation) value by the constant m (equation 9), which in this case is 0·6. Thus the ρ value becomes 0·78(= 1·30 × 0·6).

Yates and Scott[53] have determined the effect of o-methyl substituents on the basicities of benzamides and other weak bases. Assuming that the electronic effect of a substituent in the *ortho* position is the same as that in the *para*, and that substituent effects are additive, they attribute the low basicity of 2-methylbenzamide to steric inhibition of resonance through lack of coplanarity of the amide function and ring system. However, this cannot account entirely for the low basicity of the 2,6-dimethyl derivative and here steric hindrance to solvation is thought to occur.

In the series of ring-substituted acetanilides indicator measurements using Sudan III in acetic acid[80,89] show that $\log K_{ex}$ correlates linearly with σ (Figure 10), giving $\rho = 1·66$ ($r = 0·987$). Again, steric effects may account for the low basicity found for the 2,4,6-tribromo and 2,6-dimethyl derivatives, although in these cases σ_p values have been taken for the *ortho* substituents (an assumption which is probably correct for resonance effects but grossly inaccurate for inductive and steric interactions).

IV. HYDROGEN BONDING

In the same way as the lone pairs of electrons on the oxygen and nitrogen atoms of the amide function may be utilized by direct bonding to a proton to form a cation on treatment with a strong acid, so may a weakly acid molecule complex with the amide function by a hydrogen bond. An amide molecule may in fact function in the dual role of both proton acceptor and donor resulting in dimerization or oligomerization. Similar interactions in polypeptides are vital in stabilizing the secondary structures, notably the α-helix. A proton donor molecule may be the solvent, e.g. chloroform, or may be present together with the amide in an inert solvent, e.g. phenol in carbon tetrachloride.

The extent of hydrogen bonding may be defined by the association constant K_{as} for the equilibrium (21). The designation of O rather

$$HX + \begin{array}{c} O \\ \| \\ -C-N \\ \diagdown \end{array} \xrightarrow{\;K_{as}\;} \begin{array}{c} HX \\ \vdots \\ O \\ \| \\ -C-N \\ \diagdown \end{array} \qquad (21)$$

than N as the acceptor site of hydrogen bonding in the amide follows logically from consideration of O as the site of direct protonation, and evidence supporting this deduction comes from the infrared work of Bellamy and Pace[90]. These workers have studied hydrogen bonding between pyrrole and various carbonyl compounds, in which O is indisputably the basic site, such as esters, aldehydes and lactones in the solvent carbon tetrachloride. Plotting $\nu_{C=O}$ vs. ν_{N-H} (pyrrole) gave a smooth curve on which the amides also fitted.

In general one would expect a correlation between the basic strength of a compound and its ability to accept hydrogen bonds. Arnett[38] has collected together the pK_a values of some 42 bases with values ranging from -12 to $+13$ and plotted them against the shifts of the ν_{COD} peak in O-deuteromethanol solutions relative to the peak for the pure solvent. The magnitude of this shift is proportional to the strength of hydrogen bonding and the resultant graph shows reasonable linearity. N,N-Dimethylacetamide is the only amide included in this compilation.

A. Self-association

The most common example of hydrogen bonding is association in concentrated solutions where each amide molecule donates and accepts a hydrogen bond. This is readily demonstrated by the shifts to lower frequency of the N—H stretching and amide I bands in the infrared spectra of amides, on going from the vapour to the condensed phase[91]. As in protonation and metal complex formation the low-frequency shift of the amide I band is evidence for the carbonyl oxygen atom

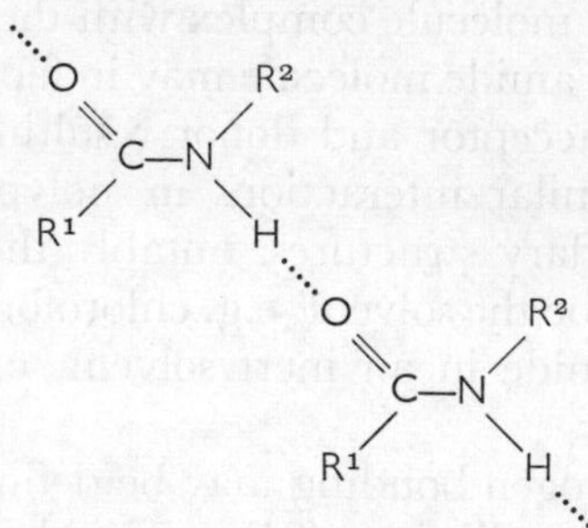

(19)

being the hydrogen acceptor. Two structures are possible, the linear polymer (19), formed for example by N-methylacetamide, and the cyclic dimer (20) formed by lactams and some other amides. The two classes may usually be distinguished by consideration of ΔH for the association. The linear polymer involving one hydrogen bond per

$$
\begin{array}{c}
\text{R}^2 \\
| \\
\text{N}-\text{H}\cdots\text{O} \\
\text{R}^1-\text{C} \qquad\qquad \text{C}-\text{R}^1 \\
\text{O}\cdots\text{H}-\text{N} \\
| \\
\text{R}^2
\end{array}
$$

(20)

dimer unit has, typically $\Delta H = -3\cdot6$ kcal/mole, whereas ΔH for cyclic dimer formation is twice this value (see Table 11 for examples).

Cyclic dimer formation requires the *cis* conformation of the N—H and carbonyl groups about the C—N bond whereas the preponderant form, except of course for lactams, is *trans*. LaPlanche and Rogers[88] have shown by n.m.r. that it requires very bulky *N*-substituents to produce a significant proportion of the *cis* form. *N-t*-Butylformamide is only 18% *cis* at room temperature in benzene solution. On protonation however, and presumably also on hydrogen-bond complexation, the proportion of *cis* form in equilibrium increases considerably. Davies and Thomas[93] have shown through energy considerations that the type of association is not necessarily determined by the predominant conformation of the free amide. *N*-Methyltrichloro-acetamide and *N*-methylacetamide are probably both *trans* but whereas the latter associates to form a linear polymer, the former gives a cyclic dimer. *N*-Methylacetamide (all-*trans*, K_{as} 5·4 l/mole) associates more strongly than *N*-phenylurethane (95% *cis*, K_{as} 1·5 l/mole) (Table 11) but other factors must be at work as all-*cis* lactams associate to a much larger extent. Acetamide itself is reported to form cyclic trimers[94] whereas trichloroacetamide dimerizes[93].

Infrared spectroscopy is most generally employed in studying amide association. Measurements are conveniently made at the first overtone of the N—H absorption[92]. Isopiestic measurements have been used[93], as have cryoscopic and ebullioscopic methods[94]. A detailed discussion is given by Pimental and McClellan[95].

Association constants are collected in Table 11. Inspection reveals that such constants are very sensitive to the solvent. ΔH and ΔS values are generally much more negative in carbon tetrachloride than in benzene or chloroform. Chloroform is known to hydrogen bond to amides[73] and competes with dimerization. It has also been suggested

TABLE 11. Self-association of amides and lactams.

Amide	Solvent	K_{as} (l/mole) (T if not 25°)	$-\Delta H$ (kcal/mole)	$-\Delta S$ (e.u.)	Ref.
N-Methylformamide	C_6H_6	19	3·5	1·3	a
Formanilide	C_6H_6	0·45	3·8	10·6	b
Acetamide	C_6H_6	3·6 (30°)			c
	$CHCl_3$	3·3	2·2		d
N-Methylacetamide	C_6H_6	6·1	3·6	3·7	a
	CCl_4	4·7	4·2	11	e
		5·4	4·7		f
	H_2O	0·005	0·0	10	e
	Dioxan	0·52	0·8	4	e
	trans-CHCl=CHCl	1·52	3·3	10	g
	cis-CHCl=CHCl	1·00	1·5	5·0	g
N-Propylacetamide	C_6H_6	4·5 (22°)			a
Trichloroacetamide	C_6H_6	8·3	7·2	20·6	a
N-Methyl-trichloroacetamide	C_6H_6	0·75	7·2	24·4	a
Propionamide	C_6H_6	2·4 (30°)			c
	CCl_4	60·5	6·8		d
	$CHCl_3$	2·4	1·8		d
n-Butyramide	C_6H_6	5·5			c
	CCl_4	55	6·3		d
	$CHCl_3$	2·3	1·8		d
Isobutyramide	C_6H_6	8·2 (30°)			c
Benzamide	C_6H_6	3·6 (30°)			c
		36	9	23	b
		6·7 (80°)			h
m-Chlorobenzamide	C_6H_6	25·8 (30°)			c
p-Chlorobenzamide	C_6H_6	14·2 (30°)			c
m-Bromobenzamide	C_6H_6	19·6 (30°)			c
p-Bromobenzamide	C_6H_6	5·3 (30°)			c
N-Methylbenzamide	C_6H_6	0·5	3·6	13·6	a
γ-Butyrolactam	CCl_4	288	7·0		i
ε-Caprolactam	CCl_4	106	5·5		j
		78·8	6·8	14·2	g
	$CHCl_3$	1·55			g
	CH_2ClCH_2Cl	1·43			g
	Dioxan	0·48			g
	Tetrahydrofuran	1·96			g
	$MeOCH_2CH_2OMe$	0·70			g
	trans-CHCl=CHCl	11·6	6·8	18·3	g
	cis-CHCl=CHCl	2·17	5·3	16·2	g
δ-Valerolactam	CCl_4	270	10·3	23·5	k
N-Phenylurethane	CCl_4	1·5			f

a. Reference 93.　　　　b. W. Scheele and A. Hartmann, Kolloid-Z. **131**, 126 (1953).
c. M. E. Hobbs and W. W. Bates, J. Am. Chem. Soc., **74**, 746 (1952).
d. J. Fruwert, D. Dombrowski, and G. Geiseler, Z. Physik. Chem. (Leipzig), **227**, 349 (1964).
e. Reference 92.　　　f. Reference 109.　　　g. Reference 99.
h. K. L. Wolf and G. Metzger, Ann. Chem., **563**, 157 (1949).
i. H. E. Affsprung, S. D. Christian and J. D. Worley, Spectrochim. Acta., **20**, 1415 (1964).
j. R. C. Lord and T. J. Porro, Z. Electrochem., **64**, 672 (1960).
k. M. Tsuboi, Bull. Chem. Soc. Japan, **24**, 75 (1951).

that benzene can complex amides[96]; evidence to support this is given in section IV.C.3.

Woodbrey and Rogers[97] consider that an observed increase in the energy barrier to rotation about the C—N bond of N,N-disubstituted amides with increasing concentration, deduced from variable-temperature n.m.r. measurements, might result from association through dipolar interactions as shown in **21**. A similar suggestion has been made by Hatton and Richards[98].

$$
\begin{array}{c}
\overset{|\delta+}{-N}\diagdown\overset{/}{C} \\
\delta-O\diagup\diagup O\delta- \\
\overset{}{C}-\overset{}{N}{-} \\
\diagup\quad|\delta+
\end{array}
$$

(21)

This mechanism of association, additional or alternative to hydrogen bonding, is supported by an interesting study, by Franzen and Stephens[99] of the effect of changing dielectric constant in mixtures of *cis*- and *trans*-dichloroethylene on the association of N-methylacetamide and ϵ-caprolactam. The association constant decreased as the dielectric constant was increased, as expected if dipolar interactions were important, but the authors also point out that amide–solvent interactions could explain the result.

It is difficult to predict the electronic effect of substituents on dimerization as the donor and acceptor role of each molecule will be oppositely affected by the substituents present. This will lead to partial cancelling of such effects, a conclusion borne out by the generally rather small spread of values of K_{as} in Table 11. Steric effects have been investigated by Lumley Jones[100], who showed that a considerable reduction in the extent of dimerization resulted from the substitution of methyl groups for hydrogen atoms in N-methylacetamide.

B. Amide–Phenol Complexes

A considerable amount of work has been done on the measurement of association constants between phenol and amides, chiefly N,N-disubstituted aliphatic amides. Techniques used include those based on the shift of the hydroxyl stretching frequency in the infrared spectra[101], changes in the ultraviolet spectrum of phenol upon complexation[102], and quenching of the fluorescence of phenol and tyrosine in their complexes with acetamides[103]. The association constants and thermodynamic data, measured in carbon tetrachloride solution and corrected where necessary to 25°, are presented in Table 12.

In the series of N,N-dimethyl aliphatic amides, $RCON(CH_3)_2$, the enthalpy change for the formation of addition compounds with phenol correlated linearly with σ^* for R, but log K_{as} did not[102], the equilibrium constant for N,N-dimethylpropionamide being less and that for N,N-dimethyltrichloroacetamide being much greater than expected. It was suggested that this was due to an unfavourable entropy term arising from one of the rotamers of N,N-dimethylpropionamide (**22**, X = Me, R = H) being unfavourable for complex formation through steric interference between the X group and the hydrogen-bonded oxygen atom in the complex.

(**22**)   (**23**)

In the case of trichloroacetamide (R = X = Cl) the rotamers are identical and stabilization of the complex is postulated through dispersion forces or dipole–dipole interaction between a chlorine atom and the coordinating phenol. It is pertinent at this point to recall Adelman's work[73] (section III.D) that the order of increasing steric requirements of acids complexing with amides, as judged by the breakdown of linearity in log K vs. σ^* plots, is perchloric acid < phenol < iodine, chloroform. With chloroform all amides with acyl groups larger than formyl showed approximately the same complexing ability as estimated by the shift in their carbonyl stretching frequencies. He proposed that steric strain in amide complexes could be partially relieved by twisting around the C—N bond.

Some doubt has been thrown on the steric-hindrance hypothesis by the work of Schmulbach and Hart[104] who showed that N-methyllactams had considerably greater association constants with phenol than N,N-dimethylpropionamide. Ring size influences the phenol association constant of lactams in the same way as basicity[17], i.e. six > seven ~ five-membered rings. Phenols substituted with bulky groups adjacent to the hydroxyl function show markedly reduced association constants with N-methyl- and N,N-dimethylacetamides[105].

TABLE 12.   The association of phenol with amides and
lactams in carbon tetrachloride solution, at 25°C.

Amide or lactam	K_{as} (l/mole)	$-\Delta H$ (kcal/ mole)	$-\Delta S$ (e.u.)	Ref.
N,N-Dimethylformamide	64	6.1	12.1	a
	67.3	5.4	9.9	b
N-Methylacetamide	120	4		c
N,N-Dimethylacetamide	134	6.4	11.7	a
N,N-Diethylacetamide	136	5.1	7.4	b
N,N-Dicyclohexylacetamide	150	5.6	8.9	b
N-Acetylpiperidine	146	5.1	7.7	b
N,N-Diphenylacetamide	50.7	5.2	9.7	b
N,N-Dimethylchloroacetamide	38	4.7	8.5	a
N,N-Diethylchloroacetamide	40	4.6	8.3	b
N,N-Dicyclohexylchloroacetamide	45.7	5.1	10.0	b
N-(Chloroacetyl)piperidine	41.5	4.8	8.8	b
N,N-Diphenylchloroacetamide	19.0	3.5	5.9	b
N,N-Dimethyltrichloroacetamide	32	3.8	5.5	a
N,N-Dimethyltrifluoroacetamide		3.6		d
N,N-Dimethylpropionamide	107	6.4	12.1	a
N,N-Diethylpropionamide	114	5.6	9.4	b
N,N-Dicyclohexylpropionamide	90.1	5.2	8.5	b
N-Propionylpiperidine	105	5.3	8.9	b
N,N-Diphenylpropionamide	38.3	4.9	8.9	b
N,N-Diethyl-n-butyramide	105	5.4	8.9	b
N,N-Dicyclohexyl-n-butyramide	95	5.8	10.5	b
N-n-Butyrylpiperidine	109	5.8	10.6	b
N,N-Diphenyl-n-butyramide	40	4.5	7.8	b
N,N-Dimethylbenzamide		5.2		d
N,N-Diethylbenzamide	82.9	5.5	9.7	b
N,N-Dicyclohexylbenzamide	89.4	5.9	11.0	b
N-Benzoylpiperidine	40.1	4.2	6.9	b
N,N-Diphenylbenzamide	31.1	3.9	6.3	b
N,N-Diethyl-p-nitrobenzamide	39.8	5.9	12.5	b
N,N-Dicyclohexyl-p-nitrobenzamide	36.3	4.2	7.0	b
N-p-Nitrobenzoylpiperidine	40.1	4.2	6.9	b
N-Methyl-2-pyridone	170	6.0	10.7	b
N-Methyl-γ-butyrolactam	137	5.9	10.0	e
	150	6.0	10.0	f
δ-Valerolactam	126	5.3	8.2	e
N-Methyl-δ-valerolactam	186	5.6	8.4	f
ϵ-Caprolactam	131	5.2	7.8	e
	154	5.7	9.0	f
N-Methyl-ϵ-caprolactam	149	6.4	11.5	f

a. Reference 102.   b. Reference 105.   c. Reference 101.
d. R. L. Middaugh, R. S. Drago, and R. J. Niedzielski, *J. Am. Chem. Soc.*, **86,** 388 (1964).
e. T. Gramstad and W. J. Fuglevik, *Spectrochim. Acta*, **21,** 343 (1965).
f. Reference 104.

C. Other Hydrogen-bonded and π-Complex Systems

I. Amides as hydrogen-bond acceptors

Recent work has extended the range of complexes in which amides act as hydrogen-bond acceptors to include amines[106,107] and thiols[108]. Equilibrium constants were measured by infrared spectroscopy in cyclohexane (C_6H_{12}) solution[107] and n.m.r. in chloroform solution[106]. In the latter paper the authors describe the corrections necessary to take into account the hydrogen bonding between the amides and chloroform and derive the appropriate equilibrium constants. The data, corrected to 25°, are collected in Table 13. The association constants with amines are an order of magnitude smaller than those with phenol, which is in accord with the greater acidity of the latter, but it is notable that thiophenol is a poor donor.

2. Amides as hydrogen-bond donors

Amides functioning in this capacity are exhibiting their acidic character (section VI), and we have already observed that when self-associating in solution or in proteins, amides are acting as both donors and acceptors. In view of the biological importance of these interactions it is surprising that such little work has been done on the ability of amides to donate hydrogen bonds to suitable acceptors. Some data, due to Bhaskar and Rao[109], on the equilibrium constants between N-methylacetamide and some oxygen, sulphur and nitrogen acceptors are included in Table 13. That the basicity of the acceptor alone does not determine the equilibrium constant is evident from the similarity of K_{as} for pyridine and benzophenone, which differ widely in pK_a. Further work is required to elucidate the factors responsible.

3. Amide interactions with π-electron systems

Studies, by n.m.r., of the solvent effect on the barrier to internal rotation in amides, suggested two types of solvent effect[97,98,110]; firstly an electrostatic effect in which the planar ground state is stabilized more by polar solvents than the less polar transition state[97], in which the orbital overlap necessary for the participation of the zwitterionic canonical form in the resonance hybrid is diminished or completely absent, and secondly a specific interaction of the amide with the solvent. In an extensive investigation covering thirty-one solvents Hatton and Richards[98] proposed complex formation between dimethylformamide and aromatic solvents to explain the differential high-field shifting of the two methyl group resonances in the n.m.r.

TABLE 13.　The association of amides with hydrogen-bond donors (D) and acceptors (A), at 25°.

Amide	Hydrogen-bond partner		Solvent	K_{as} (l/mole)	$-\Delta H$ (kcal/mole)	$-\Delta S$ (e.u.)	Ref.
	Compound	Type					
N,N-Dimethyl-acetamide	Aniline	D	C_6H_{12}	0·33 (1:1)	4·4		a
				9·3 (1:2)	4·85	11·8	a
			$CHCl_3$	7·8	3·3	6·9	b
	N-Methylaniline	D	C_6H_{12}	7·36	5·16	13·3	a
	Solvent	D	$CHCl_3$	0·90	1·1	3·7	b
	Thiophenol	D	CCl_4	0·24	1·8		c
	Methanol	D	CCl_4	5·5	3·72	9·1	d
	Ethanol	D	CCl_4	3·5	3·88	10·5	d
	i-Propanol	D	CCl_4	2·74	2·4		e
	t-Butanol	D	CCl_4	2·9	3·92	11·0	d
N-Methylacet-amide	Solvent	D	$CHCl_3$	6·1	−0·4	0·2	b
	Aniline	D	$CHCl_3$	6·6	1·5	1·2	b
	t-Butylamine	D	$CHCl_3$	4·5	1·8	3·0	b
	Thiophenol	D	CCl_4	0·14	0·9		f
	i-Propanol	D	CCl_4	4·93	4·1		e
	Pyridine	A	CCl_4	1·8	4·6		g
	Benzophenone	A	CCl_4	1·9	2·9		g
	Ethylenetrithio-carbonate	A	CCl_4	0·65	1·0		g
	Methyl ethyl sulphide	A	CCl_4	0·30	1·4		g

a. Reference 107.
b. Reference 105.
c. R. Mathur, E. D. Becker, R. B. Bradley, and N. C. Li, *J. Phys. Chem.*, **67**, 2192 (1963).
d. E. D. Becker, *Spectrochim. Acta.*, **17**, 436 (1961).
e. F. Takahashi and N. C. Li, *J. Phys. Chem.*, **68**, 2136 (1964).
f. Reference 108.
g. Reference 109.

spectrum measured in these solvents. Introduction of electron-attracting groups in the solvent reduced the effect, but amino groups, especially in 1-naphthylamine, produced very pronounced shifts, probably due to additional interaction through hydrogen bonding. Structures in which a methyl group was over the centre of the aromatic system were proposed (**24** and **25**).

Moriarty[110] suggested a structure similar to **24** for the N-cyclohexyl-N-methylacetamide–benzene complex, but found that in pyridine

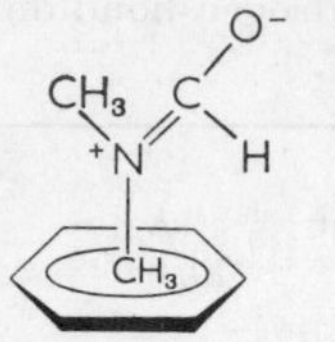

(24) 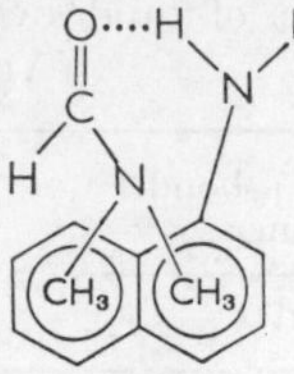   **(25)**

the N-methyl resonance underwent a low-field shift in contrast to the behaviour in benzene. For the pyridine complex, structure **26** was proposed in which the ring system is perpendicular to the amide.

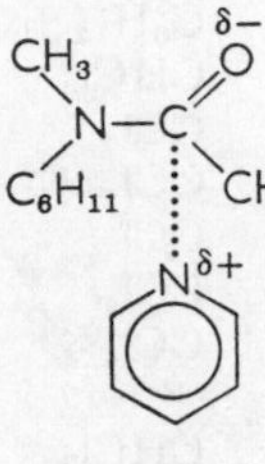

(26)

An equilibrium constant for the association between N-methyl-acetamide and benzene of 0·25 l/mole has been found recently by infrared spectroscopy[109].

V. COMPLEXES WITH OTHER LEWIS ACIDS

Gerrard and coworkers[111] have examined the infrared and n.m.r. spectra of eleven amide complexes with boron trichloride and tri-bromide and titanium tetrachloride, and once more the picture is of coordination at oxygen rather than nitrogen. Values of ν_{N-H} for such complexes are very similar in both methylene chloride solution and Nujol mull, whereas the equivalent bands in the free amides show large differences (Table 14). These shifts for the free amides are explained in terms of intermolecular hydrogen bonding, which thus must be absent in the complexes. For a complex of type **27**, one would

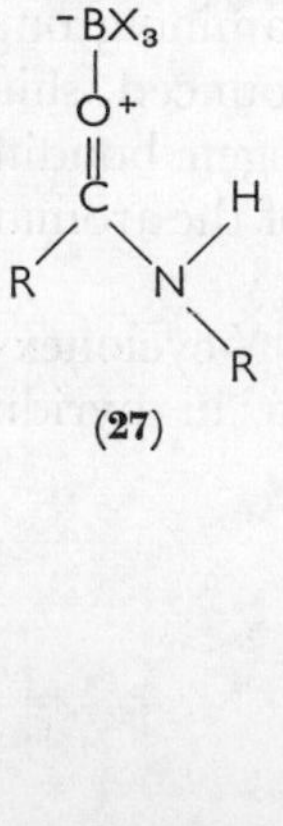

(27) 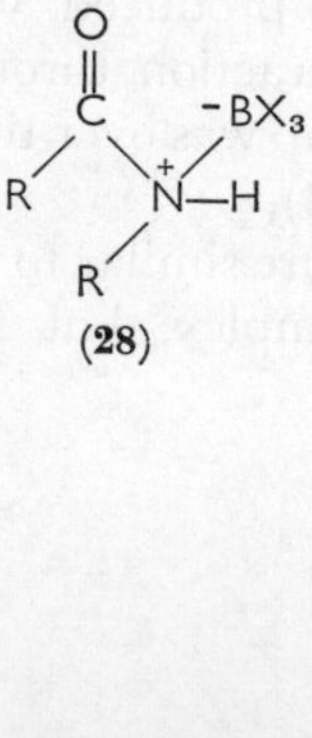   **(28)** $(RCO^+)(RNH—BX_3^-)$

(29)

TABLE 14. Infrared measurements for BCl_3 complexes of acetamide, N-methylacetamide, and N,N-dimethylacetamide[111].

Compound	Vapour phase	Free Amide CH_2Cl_2	Nujol	BCl_3 Complex CH_2Cl_2	Nujol
		1. N—H stretching frequencies (cm^{-1})			
CH_3CONH_2		3559	3333	3448	3436
		3436	3175	3367	3367
				3289	3289
$CH_3CONHMe$	3500	3460	3300	3367	3367
		2. The amide I band (cm^{-1})			
CH_3CONH_2		1685	1685	1661	1658
$CH_3CONHMe$	1718	1669	1653	1650	1661
CH_3CONMe_2		1634	1634	1633	1645
		3. The amide II band (cm^{-1})			
CH_3CONH_2		1595	1626	1550	1548
$CH_3CONHMe$	1487	1528	1567	1538	1536
		4. The amide III band (cm^{-1})			
$CH_3CONHMe$	1247	1266	1299	1323	1332

expect such association between molecules to be eliminated, but not for structure **28**, where apart from steric effects, associative tendencies between molecules would be enhanced. Structure **29** was rejected in the absence of carbonyl absorption characteristics of the acetylium ion. The authors state that the data on the amide I, II, and III bands, confirmed this conclusion, but here the evidence is much less clear cut.

For the primary amides, two N—H stretching modes would be expected, and were found. However, the corresponding complexes gave three bands, explicable in terms of geometrical isomerization, arising either from restricted rotation about the C—N bond (**30**), or

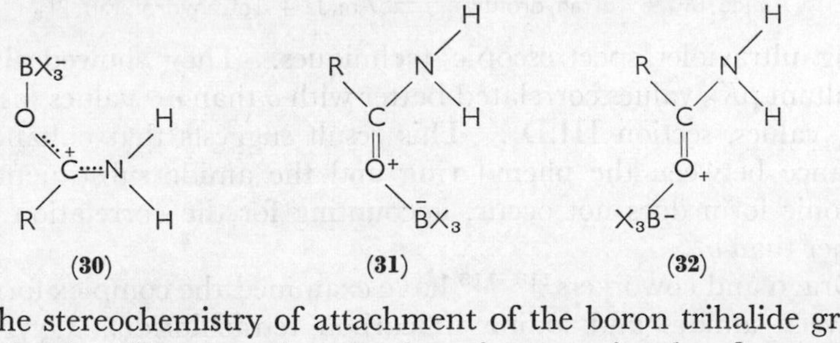

(30)   (31)   (32)

from the stereochemistry of attachment of the boron trihalide group (**31** and **32**). However, these explanations require that four, rather than three, bands should be observed. It was also observed that the amide I band, due predominantly to the carbonyl group, was at lower frequency in the complexes than the free amides, but the magnitude varied with the structure of the amide.   The direction and magnitude of the shifts, for the primary and secondary amides were in favour of structure **30**, but the tertiary amides did not fit into this scheme. Table 14 gives selected data illustrating the details of all these points.

The n.m.r. spectrum of the N,N-dimethylformamide–boron tri-chloride complex in methylene chloride[111] showed two distinct sets of methyl protons, ascribable to structure **33**.   On the other hand,

$$CH_3 \qquad O\text{—}\bar{B}Cl_3$$
$$N\overset{+}{\cdots}C$$
$$CH_3 \qquad H$$

(**33**)

N,N-dimethylacetamide showed only a single line for the NMe_2 group, and only a slightly separated asymmetric doublet for the complex.   This is explained in terms of prevention of coplanarity of the molecule, although other workers[98] have obtained a doublet for the methyl groups in the spectrum of N,N-dimethylacetamide.

Infrared and n.m.r. spectra of the boron trifluoride complex of ϵ-caprolactam and N-methyl-ϵ-caprolactam were compared with the spectra of the boron trifluoride complexes of cyclohexanone and piperidine, with the conclusion that the oxygen lone pairs are those involved in coordination[112].

An interesting study, with relevance to the discussion of correlations between pK_a values and substituent effects in benzamides, has been made by Ellul and Moodie[87] on the boron trifluoride complexes of substituted benzamides.   These workers investigated equilibrium (22)

$$\text{Amide.BF}_3 + \text{Tetrahydrofuran} \overset{K_D}{\rightleftharpoons} \text{Amide} + \text{Tetrahydrofuran.BF}_3 \qquad (22)$$

using ultraviolet spectroscopic techniques.   They showed that the resultant pK_D values correlated better with σ than σ^+ values (as do the pK_a values, section III.D).   This result suggests that enhanced resonance between the phenyl ring and the amide substituent in its cationic form does not occur, accounting for the correlation with σ rather than σ^+.

Drago and coworkers[113–115] have examined the complex formation between amides and iodine in carbon tetrachloride.   Oxygen coordination is indicated by the shift in the infrared of the carbonyl stretching mode to higher frequencies on formation of the adduct.

The thermodynamic data for this interaction are given in Table 15. ΔH correlates linearly with σ^*, but log K does not, the propionamide falling below and the trichloroacetamide very much above the correlation line.   This is explained by steric interactions leading to an unfavourable entropy term in the former case, and stabilizing interactions between the chloro substituent and iodine in the latter (as for the phenol complex, section IV.B).   Drago has also reported an interesting study[116] on the enthalpies and equilibrium constants for the N,N-dimethylacetamide–iodine adduct ($DMA.I_2$) in various solvents.   The

TABLE 15.   Thermodynamic data for formation of iodine–amide adducts in carbon tetrachloride, at 25°[113–115].

Compound	K (l/mole)	$-\Delta H$ (kcal/ mole)	$-\Delta S$ (e.u.)
N,N-Dimethylformamide	2·9	3·7	10·4
N,N-Dimethylacetamide	6·8	4·0	9·3
N,N-Dimethylpropionamide	3·9	4·0	10·7
N,N-Dimethylchloroacetamide	1·3	3·3	10·5
N,N-Dimethyltrichloroacetamide	0·3	2·5	10·7
N,N-Dimethylbenzamide	3·91	4·0	10·7

breakdown of the overall enthalpy term ΔH_{obs} in terms of the contributing factors according to equation (23) has been considered, in which the terms are given by the enthalpy cycle (24).

$$\Delta H_{obs} = \Delta H_{CCl_4} + \Delta H_B + \Delta H_A + \Delta H_C \qquad (23)$$

$$DMA\ (CCl_4) + I_2\ (CCl_4) \xrightarrow{\Delta H_{CCl_4}} DMA.I_2\ (CCl_4)$$

$$\Bigg\uparrow \Delta H_B \qquad \Bigg\uparrow \Delta H_A \qquad \Bigg\downarrow \Delta H_C \qquad (24)$$

$$DMA\ (solvated) + I_2\ (solvated) \xrightarrow{\Delta H_{obs}} DMA.I_2\ (solvated)$$

ΔH_A and ΔH_B are usually endothermic and ΔH_C exothermic.   For benzene, ΔH_{obs} is $-3\cdot3$ kcal/mole at 25°, compared with ΔH_{CCl_4} of $-4\cdot0$.   Therefore, $\Delta H_A + \Delta H_B \neq -\Delta H_C$.   However, studies of the temperature variation of the benzene–iodine equilibrium yielded a value of ΔH_A, whence it was determined that the difference in enthalpy of solvation of the base, ΔH_B, and the complex, ΔH_C, is the same in benzene as in chloroform.

For methylene chloride, ΔH_{obs} is $-2\cdot6$ kcal/mole at 25°.   A hydrogen-bonding enthalpy of $-2\cdot2$ kcal/mole for the adduct DMA–

CH_2Cl_2 in carbon tetrachloride was adduced. An enthalpy of $-1 \cdot 1$ kcal/mole was attributed to non-specific solvation of the $DMA-I_2$ complex in methylene chloride.

An n.m.r. study[117] of the N,N-dimethylpropionamide–iodine complex in carbon tetrachloride showed that the $N(CH_3)_2$ doublet was broadened by the addition of iodine. This was regarded as evidence that some N-coordination takes place.

A large number of metal complexes involving various aliphatic amides, mainly di-N-substituted, have been described. The infrared spectra of the complexes show a low frequency shift of the amide I band compared with the free amide, which has been generally interpreted as indicating that the carbonyl group is the donor. Transition metal ions seem able to achieve their maximum coordination number with amides, but mixed complexes with some halide ligands or water are known.

Bull and coworkers[118] prepared and characterized dimethylacetamide complexes of fourteen metals, e.g. $[Cr(DMA)_6](ClO_4)_3 \cdot H_2O$, $[Zn(DMA)_2Br_2]$, by electrolytic conductance and magnetic susceptibility data. Rollinson and White[119] report the absorption spectra of Cr^{III} complexes with a variety of amides and lactams. The latter have been extensively studied as ligands[120], especially ϵ-caprolactam[121], γ-butyrolactam[122] and N-methyl-γ-butyrolactam[122,123], and they have been shown to form hexacoordinated octahedral complexes.

N-Methyl-γ-butyrolactam also complexes with a wide variety of non-transition metals[124]. Titanium chloride complexes of formamide and N,N-dimethylformamide have been studied by infrared[125] and magnetic susceptibility[126] measurements, and a range of complexes characterized. Mercuric chloride has been shown to complex several tertiary amides affording a useful method of purifying N-formyl compounds[127].

Drago and coworkers[128] have prepared both octahedral $[Co(DMA)_6](ClO_4)_2$ and tetrahedral $[Co(DMA)_4](ClO_4)_2$ complexes of Co^{II}. Spectroscopic criteria show that the latter complex is somewhat distorted, and it is suggested that weak nitrogen coordination is occurring in addition to that of the carbonyl oxygen. Another possibility is extensive ion pairing, and the result of an x-ray crystallographic investigation is awaited.

Bull and Ziegler[129] have shown that the diamide N,N,N',N'-tetramethylmalonamide behaves as a bidentate ligand forming octahedral complexes, e.g. $[Cr(L)_3](ClO_4)_3$.

TABLE 16.   Nickel(II) complexes of amides $[\text{Ni}(\text{Amide})_6](\text{ClO}_4)_2$.

Ligand	Solvent for spectral measurement	Dq (cm^{-1})
1. *Amides*		
N-Methylformamide	N-Methylformamide	838
N,N-Dimethylformamide	N,N-Dimethylformamide	850
N,N-Diethylformamide	N,N-Diethylformamide	840
Acetamide	Acetone	824
N-Methylacetamide	N-Methylacetamide	752
N,N-Dimethylacetamide	Methylene chloride	758
N,N-Dimethylacetamide	N,N-Dimethylacetamide	769
N,N-Dimethylbutyramide	N,N-Dimethylbutyramide	749
2. *Lactams*		
γ-Butyrolactam	γ-Butyrolactam	810
N-Methylbutyrolactam	N-Methylbutyrolactam	780
δ-Valerolactam	3·0 M δ-Valerolactam in methylene chloride	833
N-Methyl-δ-valerolactam	N-Methyl-δ-valerolactam	759
ε-Caprolactam	4·3 M ε-Caprolactam in methylene chloride	834
N-Methyl-ε-caprolactam	N-Methyl-ε-caprolactam	749

Drago and coworkers have examined the octahedral complexes of nickel(II) and chromium(III) with a series of amides[130] and lactams[120], relating the ligand structure to ligand-field splittings, Dq, as estimated from the appropriate absorption band in the near infrared (see Table 16 for the nickel complexes). Drago points out that the lack of correlation between basicity as measured by the frequency of the O—H stretching vibration in the complex of the amides with phenol, and the Dq values, may well be due to steric factors. When R^1 and R^3 are both alkyl groups the *trans* structure for the parent amide is favoured and in the complex **34** steric repulsions between R^1 and R^3 arise, and

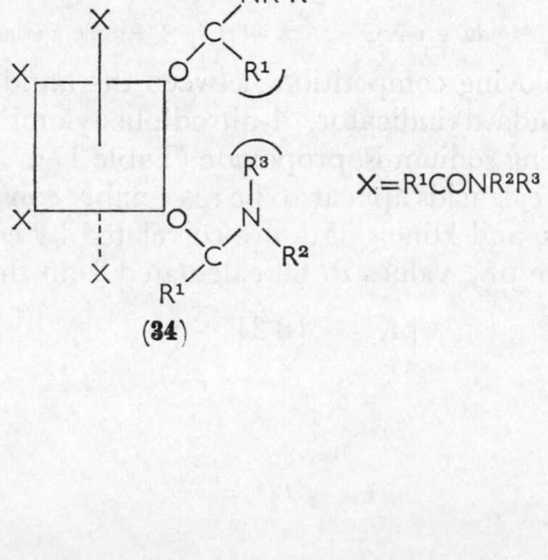

$X = R^1 CONR^2 R^3$

(**34**)

Dq is lowered.   When R^1 is hydrogen, this steric interaction is largely absent, and the resultant strong interaction between the metal ion and the ligand produces a higher Dq value.   It is suggested that steric factors may also be important in determining the magnitude of Dq in the lactam complexes.

VI. AMIDES AS ACIDS

In contrast to the coordination of the amide function with a proton, a proton may be lost from the amide NH_2 or NH group, enabling the function to act in an acidic capacity.   Although such a process is facilitated by the neighbouring carbonyl group (equation 25), the

$$\text{(25)}$$

amides are still only relatively weak acids.   Such a process therefore only occurs to any significant extent in strongly basic media.   Again, as for the determination of pK_a values of amides as bases, the quantitative expression of their acidic behaviour requires the establishment of acidity functions appropriate in this case for strongly basic media. The setting up of such scales for weak acids in general has bcen reviewed[131,132], and it appears that a single function, H_-, will suffice for a whole range of different indicator acid structures, a result in marked contrast to the effect of base structure in setting up acidity scales in strong acid (section III).   However, much less work has been done on this than for strongly acidic media, and, specifically, the amides appear to have been the subject of only a few investigations.

The most extensive data are those of Hine and Hine[133], who report work on a whole series of weak bases including twelve amides.   Values of K_e for equilibrium (26) were found by a spectrophotometric

$$\text{Amide} + \text{i-PrO}^- \xrightleftharpoons{K_e} \text{i-PrOH} + \text{Amide anion} \qquad \text{(26)}$$

technique, employing competition between the amide anion and the anion of a standard indicator, 4-nitrodiphenylamine, in isopropyl alcohol containing sodium isopropoxide (Table 17).   The pK_a values of a variety of weak acids appear to be reasonably constant in a variety of basic systems, and Hine's data are correlated by equation (27)[131]. This enables the pK_a values to be calculated from the data in Hine's

$$pK_a = 18\cdot31 + pK_e \qquad \text{(27)}$$

study and these are also given in Table 17.   In view of the almost total lack of comparable data it is difficult to assess the absolute values. There is rather poor agreement with Edward and Wang's work[134] in which the pK_a of thioacetamide (13·4) was measured spectrophotometrically in aqueous sodium hydroxide solutions.   The same technique was used to establish the pK_a of trifluoroacetanilide (9·54)[135]

TABLE 17.   Acidity constants for some amides.

Amide	pK_e^a	pK_a^b
Benzamide	> 0·7	> 19·0
N-Methylbenzamide	> 0·7	> 19·0
Acetanilide	− 0·72	17·59
Phenylacetanilide	− 1·00	17·31
Formamide	− 1·11	17·20
Phenoxyacetamide	− 1·11	17·20
p-Bromobenzamide	− 1·18	17·13
Benzanilide	− 1·78	16·53
p-Nitrobenzamide	− 2·46	15·85
p-Bromobenzanilide	− 2·58	15·73
Formanilide	− 2·75	15·56
Thioacetamide	− 3·6	14·7

[a] Reference 133.
[b] Obtained from pK_e values by equation (27).

which is in good agreement with the value of 9·51 obtained potentiometrically[136].   These results invalidate the previous estimate of 11·9 from hydrolysis data[137].   A value of 9·98 has been determined potentiometrically for the pK_a of trichloroacetanilide[136].

An early attempt to measure the acidic dissociation constants through conductivity measurements in alkaline solutions yielded a reasonable value for acetamide ($pK_a = 15·1$), but that for benzamide ($pK_a = 14–15$) probably overestimates its acidity[138] (see Table 17).

Berger, Lowenstein, and Meiboom[3] have shown that the N-methyl doublet in the n.m.r. spectrum of N-methylacetamide collapses to a singlet at pH 13, due to rapid exchange of the NH proton with the solvent.

Platinum blue, a diamagnetic polymeric complex, probably containing platinum–platinum bonds, of empirical formula $Pt(CH_3CONH)_2 \cdot H_2O$[139], appears to involve the acetamide anion structure, and related compounds have been prepared[140].

It may be seen that investigations of the behaviour of the amide function as an acid are sparse.   It is to be expected that many of the omissions, particularly in the field of acidity-function behaviour will be remedied shortly.   The exploration of an H_--type scale generated by the function, using standard overlap techniques, the assessment of its degree of correlation with the H_- scale for nitrogen and carbon acids, and structure–acidity correlations, particularly involving comparison with the pK_a values of amides as bases, are all questions deserving attention.

VII. REFERENCES

1. A. R. Katritzky and R. A. Y. Jones, *Chem. Ind. (London)*, 722 (1961).
2. G. Fraenkel and C. Niemann, *Proc. Natl. Acad. Sci. U.S.*, **44**, 688 (1958); G. Fraenkel and C. Franconi, *J. Am. Chem. Soc.*, **82**, 4478 (1960).
3. A. Berger, A. Lowenstein, and S. Meiboom, *J. Am. Chem. Soc.*, **81**, 62 (1959).
4. I. M. Klotz, S. F. Russo, S. Hanlon, and M. A. Stake, *J. Am. Chem. Soc.*, **86**, 4774 (1964).
5. E. Spinner, *J. Phys. Chem.*, **64**, 275 (1960).
6. R. J. Gillespie and T. Birchall, *Can. J. Chem.*, **41**, 148 (1963).
7. T. Birchall and R. J. Gillespie, *Can. J. Chem.*, **41**, 2642 (1963).
8. T. Miyazawa, T. Shimanouchi, and S. Mizushima, *J. Chem. Phys.*, **29**, 611 (1958).
9. R. Gompper and P. Altreuther, *Z. Anal. Chem.*, **170**, 205 (1959).
10. E. Spinner, *Spectrochim. Acta*, **15**, 95 (1959).
11. R. Stewart and L. J. Muenster, *Can. J. Chem.*, **39**, 401 (1961).
12. M. J. Janssen, *Spectrochim. Acta*, **17**, 475 (1961).
13. W. D. Kumler, *J. Am. Chem. Soc.*, **83**, 4983 (1961).
14. O. D. Bonner, K. W. Bunzl, and G. B. Woolsey, *Spectrochim. Acta*, **22**, 1125 (1966).
15. J. T. Edward, H. S. Chang, K. Yates, and R. Stewart, *Can. J. Chem.*, **38**, 1518 (1960).
16. R. Huisgen and H. Brade, *Chem. Ber.*, **90**, 1432 (1957).
17. R. Huisgen, H. Brade, H. Walz, and I. Glogger, *Chem. Ber.*, **90**, 1437 (1957).
18. H. Pracejus, *Chem. Ber.*, **92**, 988 (1959).
19. M. Liler, *Spectrochim. Acta*, **23A**, 139 (1967).
20. C. C. Greig and C. D. Johnson, *J. Am. Chem. Soc.*, **90**, 6453 (1968).
21. N. C. Deno in *Survey of Progress in Chemistry*, Vol. I (Ed. A. F. Scott), Interscience Publishers, New York, 1963, p. 155.
22. J. E. Leffler and E. Grunwald in *Rates and Equilibria of Organic Reactions*, John Wiley and Sons, New York, 1963, p. 257.
23. A. R. Katritzky and J. M. Lagowski in *Advances in Heterocyclic Chemistry*, Vol. I. (Ed. A. R. Katritzky), Academic Press, New York, 1963, p. 339; A. Gordon, A. R. Katritzky, and S. K. Roy, *J. Chem. Soc. (B)*, 556 (1968).
24. V. I. Bliznyukov and V. M. Reznikov, *Zh. Obshch. Khim.*, **25**, 1781 (1955).
25. S. F. Mason, *J. Chem. Soc.*, 1253 (1959).

26. R. A. Y. Jones, A. R. Katritzky, and J. M. Lagowski, *Chem. Ind.* (*London*), 870 (1960).
27. A. R. Katritzky and R. A. Y. Jones, *Proc. Chem. Soc.*, 313 (1960).
28. P. J. Van der Haak and Th. J. de Boer, *Rec. Trav. Chim.*, **83**, 186 (1964).
29. A. R. Katritzky and R. E. Reavill, *J. Chem. Soc.*, 753 (1963).
30. E. Spinner, *J. Chem. Soc.*, 1226 (1960).
31. E. Spinner and J. C. B. White, *J. Chem. Soc.* (*B*), 996 (1966).
32. V. C. Armstrong and R. B. Moodie, *J. Chem. Soc.* (*B*), 275 (1968).
33. A. R. Katritzky and A. J. Waring, *J. Chem. Soc.*, 1540 (1962).
34. A. Albert and E. P. Serjeant, *Ionisation Constants of Acids and Bases*, Methuen and Co., London,, 1962.
35. J. T. Edward, J. B. Leane, and I. C. Wang, *Can. J. Chem.*, **40**, 1521 (1962).
36. R. W. Taft and P. L. Levins, *Anal. Chem.*, **34**, 436 (1962).
37. N. C. Deno and M. J. Wisotsky, *J. Am. Chem. Soc.*, **85**, 1735 (1963).
38. E. M. Arnett in *Progress in Physical Organic Chemistry*, Vol. I (Eds. S. G. Cohen, A. Streitwieser, and R. W. Taft), Interscience Publishers, New York, 1963 p. 223.
39. A. R. Katritzky, A. J. Waring, and K. Yates, *Tetrahedron*, **19**, 465 (1963).
40. R. B. Homer and R. B. Moodie, *J. Chem. Soc.*, 4377 (1963).
41. K. Yates, J. B. Stevens, and A. R. Katritzky, *Can. J. Chem.*, **42**, 1957 (1964).
42. K. Yates and J. B. Stevens, *Can. J. Chem.*, **43**, 529 (1965).
43. N. F. Hall, *J. Am. Chem. Soc.*, **52**, 5115 (1930).
44. L. P. Hammett and A. J. Deyrup, *J. Am. Chem. Soc.*, **54**, 2721 (1932).
45. M. A. Paul and F. A. Long, *Chem. Rev.*, **57**, 1 (1957).
46. M. J. Jorgenson and D. R. Hartter, *J. Am. Chem. Soc.*, **85**, 878 (1963).
47. R. S. Ryabova, I. M. Medvetskaya, and M. I. Vinnik, *Russ. J. Phys. Chem.*, **40**, 182 (1966).
48. N. C. Deno, P. T. Groves, and G. Saines, *J. Am. Chem. Soc.*, **81**, 5790 (1959).
49. R. L. Hinman and J. Lang, *J. Am. Chem. Soc.*, **86**, 3796 (1964).
50. E. M. Arnett and G. W. Mach, *J. Am. Chem. Soc.*, **86**, 2671 (1964).
51. N. C. Deno, J. J. Jaruzelski, and A. Schriesheim, *J. Am. Chem. Soc.*, **77**, 3044 (1955).
52. K. Yates and R. A. McClelland, *J. Am. Chem. Soc.*, **89**, 2686 (1967).
53. K. Yates and B. F. Scott, *Can. J. Chem.*, **41**, 2320 (1963).
54. J. T. Edward and I. C. Wang, *Can. J. Chem.*, **40**, 966 (1962).
55. C. D. Johnson, A. R. Katritzky, and N. Shakir, *J. Chem. Soc.* (*B*), 1235 (1967).
56. J. F. Bunnett and F. P. Olsen, *Can. J. Chem.*, **44**, 1899 (1966).
57. C. D. Johnson, A. R. Katritzky, B. J. Ridgewell, N. Shakir, and A. M. White, *Tetrahedron*, **21**, 1055 (1965)
58. K. Yates and J. C. Riordan, *Can. J. Chem.*, **43**, 2328 (1965).
59. K. Yates and H. Wai, *Can. J. Chem.*, **43**, 2131 (1965); R. B. Moodie, P. D. Wale, and T. J. Whaite, *J. Chem. Soc.*, 4273 (1963).
60. K. Yates and H. Wai, *J. Am. Chem. Soc.*, **86**, 5408 (1964).
61. M. J. Janssen, *Rec. Trav. Chim.*, **81**, 650 (1962); **82**, 1197 (1963).
62. J. T. Edward and H. Stollar, *Can. J. Chem.*, **41**, 721 (1963).
63. A. Gordon, *Ph.D. Thesis*, University of East Anglia, 1967.
64. P. Bellingham, C. D. Johnson, and A. R. Katritzky, *J. Chem. Soc.* (*B*), 1226 (1967).

65. L. M. Sweeting and K. Yates, *Can. J. Chem.*, **44**, 2395 (1966).
66. R. H. Boyd, *J. Am. Chem. Soc.*, **85**, 1555 (1963).
67. R. W. Taft, *J. Am. Chem. Soc.*, **82**, 2965 (1960).
68. R. Grahn, *Arkiv. Fysik*, **21**, 13 (1962).
69. E. M. Arnett and G. W. Mach, *J. Am. Chem. Soc.*, **88**, 1177 (1966).
70. N. F. Hall and J. B. Conant, *J. Am. Chem. Soc.*, **49**, 3047 (1927).
71. K. Heyns, H. F. Grutzmacher, and A. Roggenbuck, *Chem. Ber.*, **93**, 1488 (1960).
72. C. A. Streuli, *Anal. Chem.*, **31**, 1652 (1959).
73. R. L. Adelman, *J. Org. Chem.*, **29**, 1837 (1964).
74. C. A. Streuli, *Anal. Chem.*, **30**, 997 (1958).
75. L. P. Hammett and A. J. Deyrup, *J. Am. Chem. Soc.*, **54**, 4239 (1932).
76. H. Lemaire and H. J. Lucas, *J. Am. Chem. Soc.*, **73**, 5198 (1951).
77. R. Huisgen, H. Brade, and E. Rauenbusch, *Ann. Chem.*, **586**, 30 (1954).
78. I. M. Kolthoff and S. Bruckenstein, *J. Am. Chem. Soc.*, **78**, 2974 (1956).
79. E. Grunwald and G. W. Ceska, *J. Am. Chem. Soc.*, **89**, 1377 (1967).
80. T. Higuchi, C. H. Barnstein, H. Ghassemi, and W. E. Perez, *Anal. Chem.*, **34**, 400 (1962).
81. R. J. L. Martin and I. H. Reese, *J. Chem. Soc.*, 4697 (1960).
82. R. B. Homer and R. B. Moodie, *J. Chem. Soc.*, 812 (1965).
83. N. F. Hall and W. F. Spengeman, *J. Am. Chem. Soc.*, **62**, 2483 (1940).
84. R. Stewart and K. Yates, *J. Am. Chem. Soc.*, **80**, 6355 (1958).
85. K. Yates and R. Stewart, *Can. J. Chem.*, **37**, 664 (1959).
86. K. Yates and R. Stewart, *J. Am. Chem. Soc.*, **82**, 4059 (1960).
87. B. M. J. Ellul and R. B. Moodie, *J. Chem. Soc.* (*B*), 253 (1967).
88. L. A. LaPlanche and M. T. Rogers, *J. Am. Chem. Soc.*, **86**, 337 (1964).
89. J. H. Wang and L. Parker, *Proc. Natl. Acad. Sci. U.S.*, **58**, 2451 (1967).
90. L. J. Bellamy and R. J. Pace, *Spectrochim. Acta*, **19**, 1831 (1963).
91. R. D. McLachlan and R. A. Nyquist, *Spectrochim. Acta*, **20**, 1397 (1964).
92. I. M. Klotz and J. S. Franzen, *J. Am. Chem. Soc.*, **84**, 3461 (1962).
93. M. Davies and D. K. Thomas, *J. Phys. Chem.*, **60**, 767 (1956).
94. M. Davies and H. E. Hallam, *Trans. Faraday Soc.*, **47**, 1170 (1951).
95. G. C. Pimental and A. L. McClellan in *The Hydrogen Bond*, W. H. Freeman and Co., San Francisco, California, 1960.
96. W. Klemperer, M. W. Cronyn, A. H. Maki, and G. C. Pimental, *J. Am. Chem. Soc.*, **76**, 5846 (1954).
97. J. C. Woodbrey and M. T. Rogers, *J. Am. Chem. Soc.*, **84**, 13 (1962).
98. J. V. Hatton and R. E. Richards, *Mol. Phys.*, **3**, 253 (1960); **5**, 139 (1962).
99. J. S. Franzen and R. E. Stephens, *Biochemistry*, **2**, 1321 (1963).
100. R. Lumley Jones, *Spectrochim. Acta*, **22**, 1555 (1966).
101. T. Gramstad and W. J. Fuglevik, *Acta. Chem. Scand.*, **16**, 1369 (1962).
102. M. D. Joesten and R. S. Drago, *J. Am. Chem. Soc.*, **84**, 2696 (1962).
103. R. W. Cowgill, *Biochem. Biophys. Acta*, **109**, 536 (1965).
104. C. D. Schmulbach and D. M. Hart, *J. Org. Chem.*, **29**, 3122 (1964).
105. F. Takahashi and N. C. Li, *J. Phys. Chem.*, **69**, 1622 (1965).
106. F. Takahashi and N. C. Li, *J. Phys. Chem.*, **69**, 2950 (1965).
107. J. H. Lady and K. B. Whetsel, *J. Phys. Chem.*, **71**, 1421 (1967).
108. R. Mathur, S. M. Wang, and N. C. Li, *J. Phys. Chem.*, **68**, 2140 (1964).
109. K. R. Bhaskar and C. N. R. Rao, *Biochem. Biophys. Acta*, **136**, 561 (1967).

110. R. M. Moriarty, *J. Org. Chem.*, **28**, 1296 (1963).
111. W. Gerrard, M. F. Lappert, and J. W. Wallis, *J. Chem. Soc.*, 2141 (1960);
     W. Gerrard, M. F. Lappert, H. Pyszora, and J. W. Wallis, *J. Chem. Soc.*,
     2144 (1960).
112. E. F. J. Duynstee, W. van Raayen, J. Smidt, and Th.A. Veerkamp, *Rec.
     Trav. Chim.*, **80**, 1323 (1961).
113. R. S. Drago, R. L. Carlson, N. J. Rose, and D. A. Wenz, *J. Am. Chem.
     Soc.*, **83**, 3572 (1961).
114. R. S. Drago, D. A. Wenz, and R. L. Carlson, *J. Am. Chem. Soc.*, **84**, 1106
     (1962).
115. R. L. Carlson and R. S. Drago, *J. Am. Chem. Soc.*, **84**, 2320 (1962).
116. R. S. Drago, T. F. Bolles, and R. J. Niedzielski, *J. Am. Chem. Soc.*, **88**,
     2717 (1966).
117. R. S. Drago and D. Bafus, *J. Phys. Chem.*, **65**, 1066 (1961).
118. W. E. Bull, S. K. Madan, and J. E. Willis, *Inorg. Chem.*, **2**, 303 (1963).
119. C. L. Rollinson and R. C. White. *Inorg. Chem.*, **1**, 281 (1962).
120. J. H. Bright, R. S. Drago, D. M. Hart, and S. K. Madan, *Inorg. Chem.*, **4**,
     18 (1965).
121. S. K. Madan and H. H. Denk, *J. Inorg. Nucl. Chem.*, **27**, 1049 (1965).
122. S. K. Madan and J. A. Sturr, *J. Inorg. Nucl. Chem.*, **29**, 1669 (1967).
123. R. J. Niedzielski and G. Znider, *Can. J. Chem.*, **43**, 2618 (1965).
124. S. K. Madan, *Inorg. Chem.*, **6**, 421 (1967).
125. J. Archambault and R. Rivet, *Can. J. Chem.*, **36**, 1461 (1958).
126. P. Ehrlich and W. Siebert, *Z. Anorg. Allgem. Chem.*, **303**, 96 (1960).
127. I. Baxter and G. A. Swan, *J. Chem. Soc.*, 3011 (1965).
128. B. B. Wayland, R. J. Fitzgerald, and R. S. Drago, *J. Am. Chem. Soc.*, **88**,
     4600 (1966).
129. W. E. Bull and R. C. Ziegler, *Inorg. Chem.*, **5**, 689 (1966).
130. R. S. Drago, D. W. Meek, M. D. Joesten, and L. LaRoche, *Inorg. Chem.*,
     2, 124 (1963).
131. K. Bowden, *Chem. Rev.*, **66**, 119 (1966).
132. C. H. Rochester, *Quart. Rev. (London)*, **20**, 511 (1966)
133. J. Hine and M. Hine, *J. Am. Chem. Soc.*, **74**, 5266 (1952).
134. J. T. Edward and I. C. Wang, *Can. J. Chem.*, **40**, 399 (1962).
135. P. M. Mader, *J. Am. Chem. Soc.*, **87**, 3191 (1965).
136. S. O. Eriksson and C. Holst, *Acta Chem. Scand.*, **20**, 1892 (1966).
137. S. S. Biechler and R. W. Taft, *J. Am. Chem. Soc.*, **79**, 4927 (1957).
138. G. E. K. Branch and J. O. Clayton, *J. Am. Chem. Soc.*, **50**, 1680 (1928).
139. R. D. Gillard and G. Wilkinson, *J. Chem. Soc.*, 2835 (1964).
140. I. I. Chemyaev and L. A. Nazarova, *Izvest. Sektora Platiny Drug. Blagarodn.
     Metal. Inst. Obshch. i Neorgan. Khim. Akad. Nauk SSSR*, **26**, 101 (1951).

CHAPTER 4

Rearrangement and elimination of the amido group

JOSEPH F. BIERON and FRANK J. DINAN

Canisius College, Buffalo, New York, U.S.A.

I. REARRANGEMENT OF AMIDES

In any discussion involving rearrangement reactions of amides, the main topic invariably is the Hofmann degradation of amides to amines (equation 1).

$$R-\overset{\overset{\displaystyle O}{\|}}{C}-NH_2 \xrightarrow[OH^-]{Br_2} R-NH_2 \tag{1}$$

As a consequence, this reaction is reviewed in numerous publications[1] and will not be discussed in this chapter. Instead, it was decided that a review of lesser known rearrangement and dehydration reactions of amides and some of their N-substituted derivatives would be undertaken.

A. Aziridine Derivatives

A wide variety of rearrangement reactions involving amide nitrogen atoms incorporated into aziridine rings have been reported in recent years. Whereas these reactions proceed variously by pyrolysis, nucleophilic and electrophilic attack, the primary driving force seems to be the strain introduced by the incorporation of the amide nitrogen atom into the three-membered ring.

I. Pyrolytic rearrangement to oxazolines

The first pyrolytic rearrangement of a 1-aroylaziridine was reported by Gabriel and Stelzner[2] who observed 1-benzoylaziridine (**1**) to undergo rearrangement to afford 2-phenyl-2-oxazoline (**2**) in good yield at 250°.

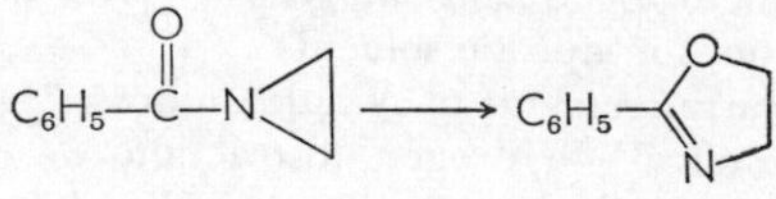

        (**1**)   (**2**)

This rearrangement was subsequently observed to take place at a significantly lower temperature, 125°, when a reduced-pressure distillation of **1** was attempted[3].

It should be noted that this pyrolytic rearrangement of aziridine to oxazoline normally occurs only when no substituents which would allow the formation of a six-membered cyclic transition state are present on the aziridine ring. The presence of such substituents normally results in isomerization of the aziridine to an unsaturated amide[4] (equation 2).

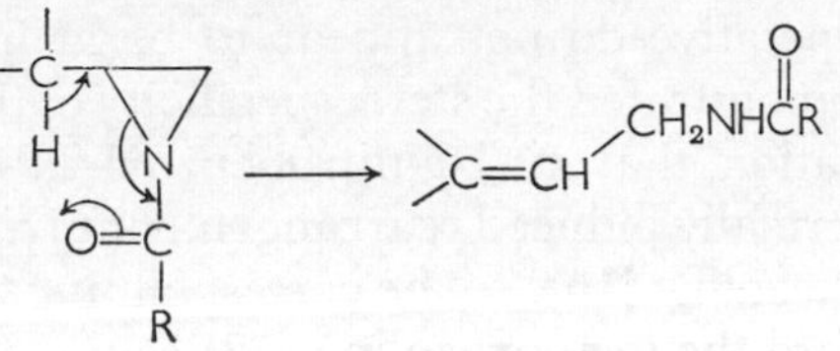

$$\text{(2)}$$

Winternitz and coworkers[5] reported an apparent exception to this generalization with the observation that *N*-benzoylcyclohexeneimine (**3**) rearranged to form an isomeric *trans*-oxazoline (**4**). A subsequent investigation of this reaction revealed that the product was not the oxazoline (**4**) but an unsaturated benzamide (**5**), the formation of which would be predicted via a six-membered cyclic transition state[6].

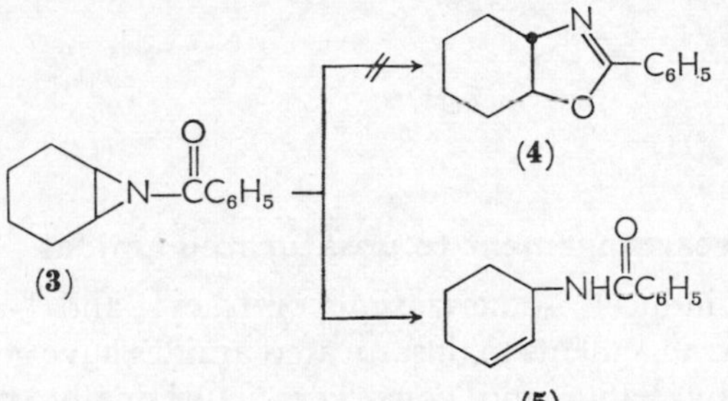

It has been observed, however, that when the substituents on the aziridine ring are part of a five-membered ring, rearrangement to the corresponding oxazoline is apparently preferred over isomerization to the unsaturated amide. Fanta and Walsh[7] have reported that the pyrolysis of 6-benzoyl-3-oxa-6-azabicyclo[3.1.0]hexane (**6**) afforded the oxazoline (**7**) rather than the anticipated unsaturated amide (**8**).

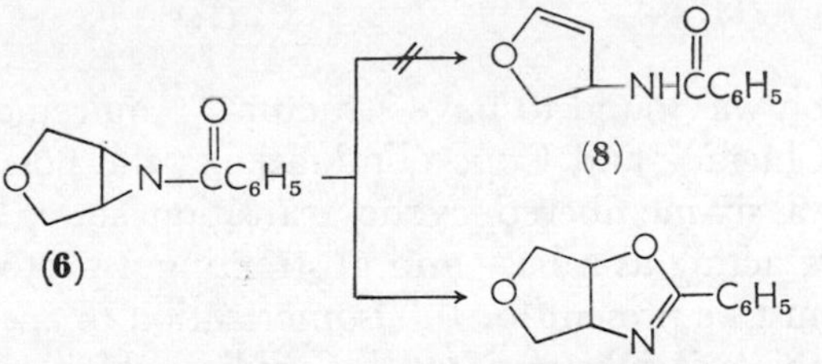

The rearrangement of 1-aroylaziridines to 2-aryl-2-oxazolines must proceed through a four-membered transition state which can involve either concerted bond making and breaking, or the formation of a tight ion pair. However, data which would determine which of these mechanisms is operative do not appear to be at hand.   Heine and Kaplan[8] have demonstrated the stereospecificity of this rearrangement with the observation that *cis*-1-*p*-nitrobenzoyl-2,3-diphenylaziridine (**9**) undergoes thermally induced rearrangement to *cis*-2-*p*-nitrophenyl-4,5-diphenyl-2-oxazoline (**10**).   The corresponding *trans*-aziridine (**11**) was shown to afford the *trans*-oxazoline (**12**) as the principal product.

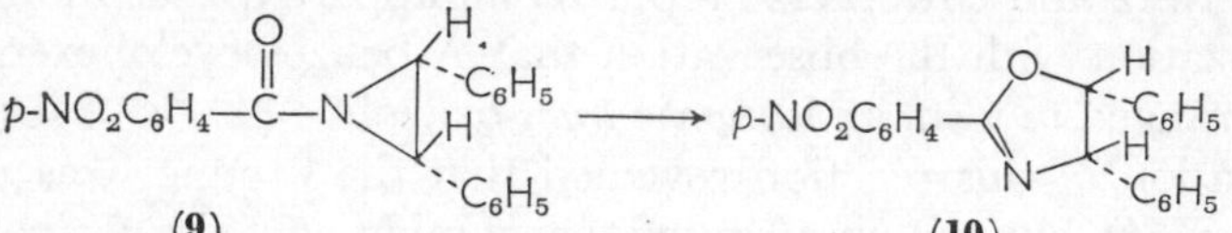

(**9**)   (**10**)

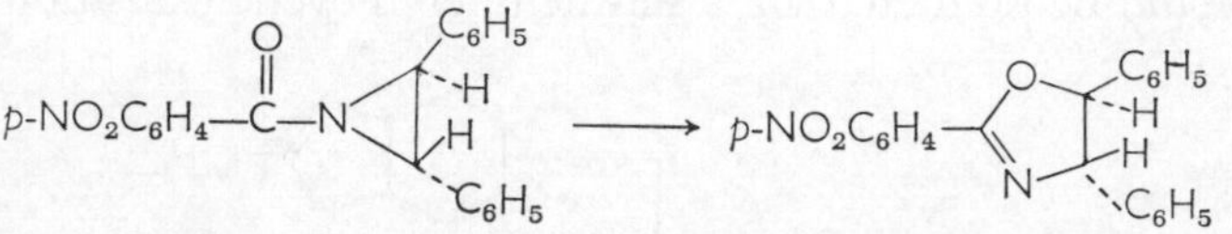

(**11**)   (**12**)

2. Pyrolytic rearrangement to unsaturated amides

Thermally induced isomerizations of 1-acyl- and 1-aroylaziridines involving rearrangements to unsaturated amides have geen extensively investigated by Fanta and coworkers.   Originally, pyrolysis of 1-acetyl-2,2-dimethylaziridine (**13**) was found to afford *N*-(β-methylallyl)acetamide (**14**) in good yield[9].

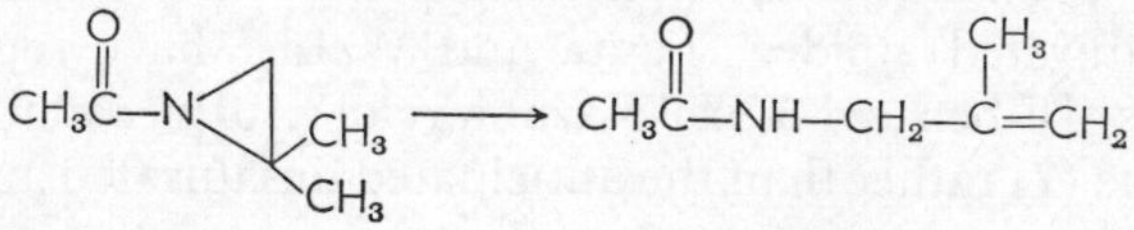

(**13**)   (**14**)

This reaction was found to have structural requirements similar to those of the Chugaev and Cope eliminations; the possibility for the formation of a six-membered cyclic transition state, in which the amide oxygen, acting as a base and abstracting a proton from the 2-substituent, must be present for the isomerization to occur.

This mechanism is substantiated by studies which clearly demonstrate the importance of the basicity of the amide oxygen.   Thus, the more weakly basic *N*-*p*-nitrobenzoyl derivative of cycloocteneimine (**15**) requires a significantly higher temperature to undergo isomerization to the corresponding unsaturated amide[10] (**16**) than does the corresponding benzoyl derivative (**15a**)[11].   The effect of basicity was

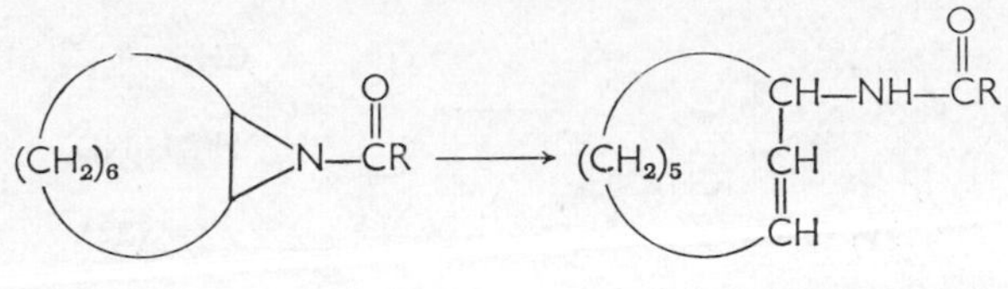

(15) R=p-NO$_2$C$_6$H$_4$
(15a) R=C$_6$H$_5$

(16) R=p-NO$_2$C$_6$H$_4$
(16a) R=C$_6$H$_5$

also demonstrated in the cyclohexeneimine series by the observation that the more basic N-benzoylcyclohexeneimine (3) undergoes pyrolytic isomerization to form the unsaturated amide (5) whereas pyrolysis of the less basic N-p-nitrobenzoyl derivative of cyclohexeneimine (17) results in rearrangement to the corresponding 2-oxazoline (18).

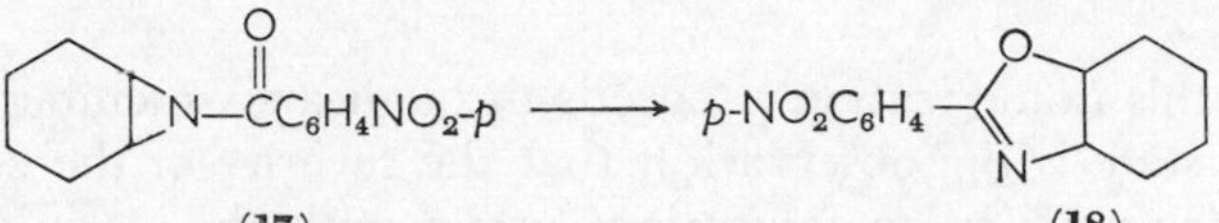

(17)   (18)

The kinetic behaviour of isomerization reactions has been studied by Fanta and Kathan[12] who investigated the pyrolysis of 2,2-dimethyl-1-p-nitrobenzoylaziridine (19) to N-(β-methylallyl)-p-nitrobenzamide (20). This reaction was found to be first order over a range of tem-

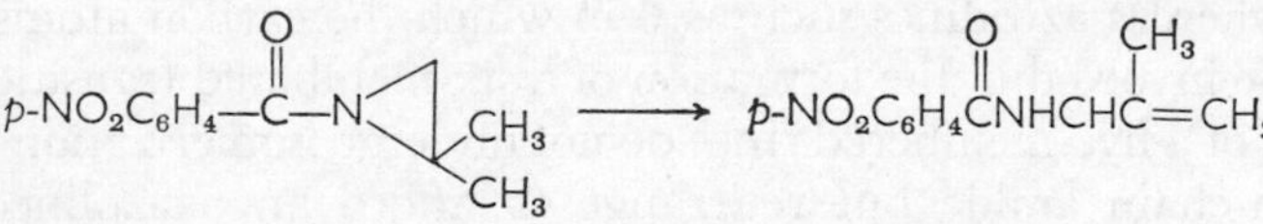

(19)   (20)

peratures. A large, negative entropy of activation was observed which supports the idea that this reaction proceeds through a highly ordered transition state, expected for equation (2).

Stereochemical data also support the concept of an ordered transition state and show the isomerization to open-chain amides to be a stereospecific, cis, intramolecular reaction similar to the Cope rearrangement of amine oxides[13]. This view is supported by the observation that 2-benzyl-1-p-nitrobenzoylaziridine (21) affords N-($trans$-cinamyl)-p-nitrobenzamide (22) as the exclusive product of isomerization.

This reaction was proposed to proceed through conformation 23 which would lead to the formation of the $trans$ product via cis elimina-

9*

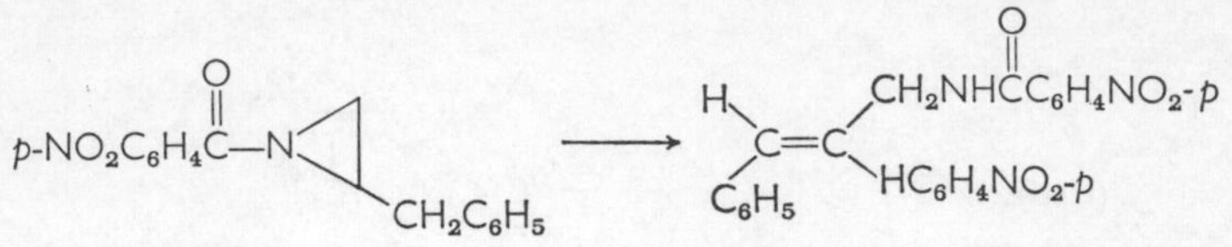

(21)   **(22)**

tion. Reaction through the less sterically favourable conformation **24** would lead to the formation of the *cis*-alkene. Also supporting the

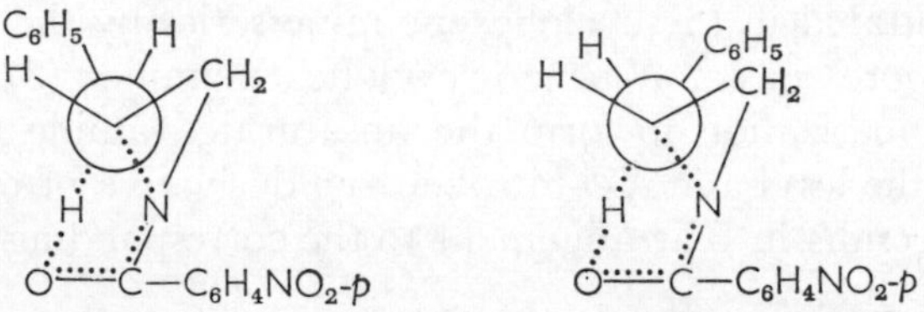

(23)   **(24)**

idea that this isomerization proceeds through a six-membered cyclic transition state is the observation that the severity of the conditions required for reaction to take place grows with the steric strain involved in forming the requisite six-membered ring. When two of the carbon atoms involved in the formation of the six-membered transition state are part of an eight-membered ring as in **15a** isomerization takes place at temperatures below 80°c[11]. With a six-membered ring, as in **3**, temperatures above 200°c are required to induce isomerization[6]; whereas aziridines such as **6** in which the carbon atoms which would be involved in the formation of a six-membered transition state are part of a five-membered ring, do not undergo isomerization to form an open-chain amide but rearrange to afford an oxazoline as the reaction product[6].

3. Nucleophile-catalysed rearrangements

Treatment of 1-aroylaziridines with nucleophiles such as iodide ion in acetone results in the formation of 2-aryl-2-oxazolines in high yield[14]. This reaction is generally assumed to involve a nucleophilic attack on the aziridine ring by iodide ion followed by ring opening and a second nucleophilic attack by the amide oxygen to displace iodide ion with formation of the oxazoline ring (equation 3).

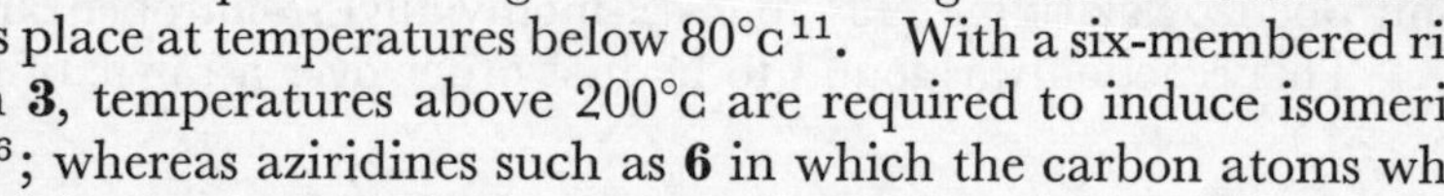

(3)

This mechanism is supported by the stereospecificity which has been shown to be characteristic of 2-alkyl-1-aroylaziridine reactions. Treatment of 2,2-dimethyl-1-(*p*-nitrobenzoyl)aziridine (**25**) with iodide ion in acetone results in the selective formation of 4,4-dimethyl-2-(*p*-nitrophenyl)-2-oxazoline (**26**) as the principal product of rearrangement[14]. This product would result from initial iodide ion

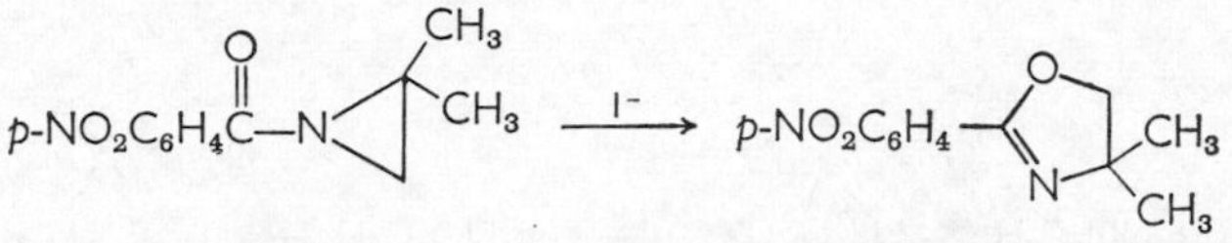

(**25**)   (**26**)

attack at the 3-position of the aziridine ring which is the electronically and sterically favoured site for nucleophilic attack. None of the isomeric 5,5-dimethyl-2-(*p*-nitrophenyl)-2-oxazoline which would result from iodide ion attack at the 2-ring position of **25** was formed.

Heine and Kaplan have demonstrated the importance of electronic effects in determining the direction of ring opening in these rearrangements[8]. Treatment of 1-(*p*-nitrobenzoyl)-2-phenylaziridine (**27**) with iodide ion resulted in the formation of 1-(*p*-nitrophenyl)-5-phenyl-2-oxazoline (**28**) in 89% yield. This result was interpreted as

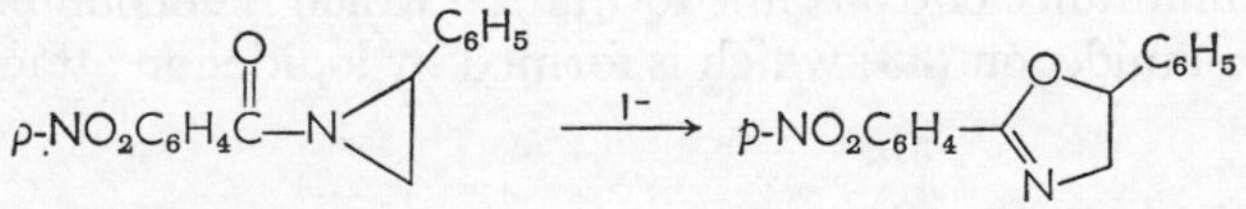

(**27**)   (**28**)

demonstrating the dominance of electronic effects over steric effects since **28** was presumed to have been formed via iodide ion attack at the electronically favoured, but sterically unfavourable 2-position of the aziridine ring.

Heine, King and Portland[15] investigated the stereochemistry of iodide-ion-catalysed rearrangement of 2,3-dialkyl-1-aroylaziridines to oxazolines and found that both *trans*-2,3-dimethyl-1-(*p*-nitrobenzoyl)-aziridine (**29**) and the corresponding *trans*-diphenyl compound (**29a**) rearranged to form the *trans*-oxazolines **30** and **30a**. Similarly, the corresponding *cis*-dimethylaziridine **31** rearranged to the *cis*-oxazoline **32**.

These results were interpreted as being consistent with a two-step nucleophilic attack in which a double inversion of configuration results in overall retention. The first inversion corresponds to attack by iodide ion which results in ring opening and the formation of the

$$p\text{-NO}_2\text{C}_6\text{H}_4\overset{\overset{\text{O}}{\|}}{\text{C}}-\text{N} \underset{\text{R}}{\overset{\text{R}}{<}} \xrightarrow{\text{I}^-} p\text{-NO}_2\text{C}_6\text{H}_4-\overset{\text{O}}{\underset{\text{N}}{<}}\overset{\text{R}}{\underset{\text{R}}{<}}$$

(29) R=CH₃;	(30) R=CH₃;
(29a) R=C₆H₅	(30a) R=C₆H₅

(29) R=CH_3;   (30) R=CH_3;
(29a) R=C_6H_5 (30a) R=C_6H_5

$$p\text{-NO}_2\text{C}_6\text{H}_4\overset{\overset{\text{O}}{\|}}{\text{C}}-\text{N}\underset{|}{\overset{\text{R}}{<}}\text{R} \xrightarrow{\text{I}^-} p\text{-NO}_2\text{C}_6\text{H}_4-\overset{\text{O}}{\underset{\text{N}}{<}}\underset{|}{\overset{\text{R}}{<}}\text{R}$$

(31) R=CH_3;   (32) R=CH_3;
(31a) R=C_6H_5 (32a) R=C_6H_5

postulated iodo-amide ion intermediate; the second inversion results from a nucleophilic attack by the amide oxygen which displaces the iodide ion converting the iodo-amide intermediate to the oxazoline of the same geometry as the starting aziridine.

In apparent conflict with these observations, the *cis*-diphenylaziridine (31a) rearranged to the *trans*-oxazoline (30a) rather than the anticipated *cis* isomer (32a). This result, however, is apparently due to steric hindrance of oxazoline formation which is encountered in the *threo*-iodo-amide ion (33) which is formed by iodide ion attack on 31a.

31a

$\downarrow$ I$^-$

$$32a \xleftarrow{\;\;/\!/\;\;} \underset{(33)}{\text{Ar}-\overset{\overset{\text{O}}{\|}}{\text{C}}\cdots\text{N}\begin{matrix} & \text{H} \\ \text{H}-\text{C}-\text{C}\text{-}\text{C}_6\text{H}_5 \\ | \quad\quad | \\ \text{C}_6\text{H}_5 \quad \text{I} \end{matrix}} \xrightarrow{\text{I}^-} \underset{(34)}{\text{Ar}\overset{\overset{\text{O}}{\|}}{\text{C}}\cdots\text{N}\begin{matrix} & \text{C}_6\text{H}_5 \\ \text{H}\text{-}\text{C}-\text{C}\text{-}\text{H} \\ | \quad\quad | \\ \text{C}_6\text{H}_5 \quad \text{I} \end{matrix}} \xrightarrow{\;\;} 30a$$

Formation of 30a was postulated to be preceded by the conversion of 33 to the *erythro* configuration 34 via a Finkelstein reaction, followed by ring closure.

4. Acid-catalysed rearrangements

Rearrangements of 1-aroylaziridines to 2-aryl-2-oxazolines have been shown to be catalysed by a variety of acids. Heine and Proctor[16] reported the AlCl$_3$-catalysed rearrangement of N-p-ethoxybenzoyl-aziridine to 2-p-ethoxyphenyl-2-oxazoline in 97% yield in refluxing heptane, with similar reactions having been catalysed with concentrated sulphuric acid.   For example, rearrangement of **25** takes place at room temperature in concentrated H$_2$SO$_4$ to afford 5,5-dimethyl-2-p-nitrophenyl-2-oxazoline (**35**) in 97% yield.

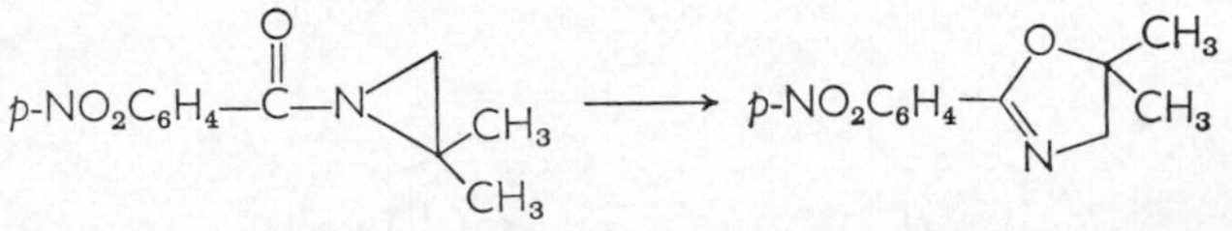

p-NO$_2$C$_6$H$_4$—C—N⟨CH$_3$/CH$_3$ ⟶ p-NO$_2$C$_6$H$_4$—oxazoline with CH$_3$, CH$_3$

(**25**)   (**35**)

It will be recalled that when rearrangement of **25** is catalysed by iodide ion, the corresponding, 4,4-dimethyl-2-oxazoline (**26**) is the principal reaction product[14].

Observations such as this indicate the acid-catalysed rearrangement to proceed through a carbonium ion intermediate.   Protonation of the amide results in ring opening occurring in the direction which affords the more stable carbonium ion.   Nucleophilic attack by the amide oxygen in a subsequent step results in formation of the oxazoline ring (equation 4).

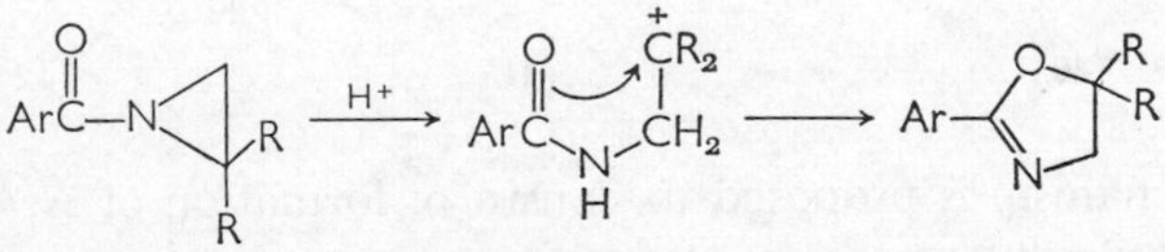

$$\text{ArC—N} \xrightarrow{\text{H}^+} \text{ArC}_{\text{N,H}}\text{CH}_2 \longrightarrow \text{Ar—oxazoline} \tag{4}$$

In accord with this mechanism, it has been observed that both the *cis* and *trans* forms of the 2,3-diphenylaziridine, **31a** and **29a**, undergo acid-catalysed rearrangement to give the *trans*-diphenyloxazoline **30a** as the principal product[15].   These reactions presumably proceed through a common carbonium ion intermediate which undergoes ring closure to the sterically more favourable *trans*-oxazoline.

B. Quinoline Synthesis from Acetanilide Rearrangement

Ardasher and coworkers[17] have reported the rearrangement of anilides to quinolines in low yield under fairly drastic reaction conditions.   For example, N-ethylformanilide (**36**) with zinc chloride at approximately 200°c gave an 11% yield of quinoline (**37**).   Similarly, n-propylformanilide (**38**) gave a 5% yield of 3-methylquinoline (**39**) [17].

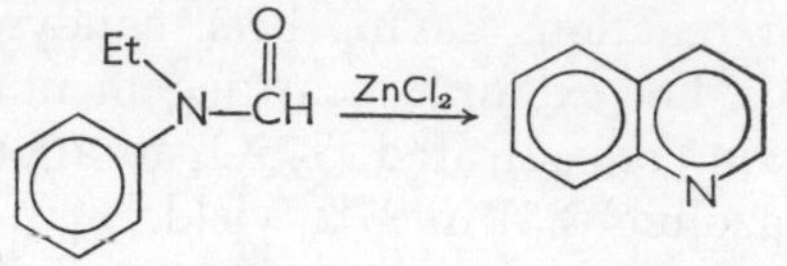

(**36**)   (**37**)

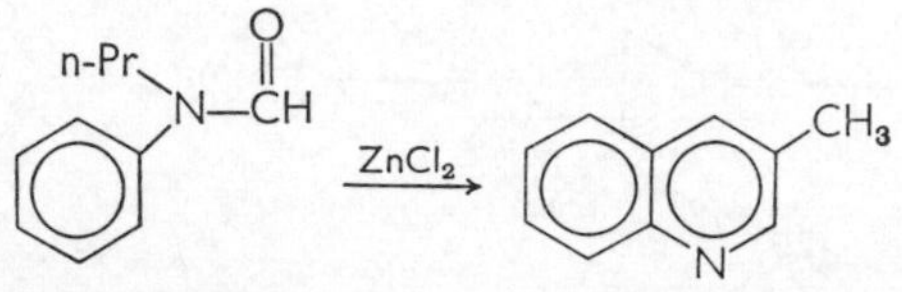

(**38**)   (**39**)

Rearrangement of acetanilide (**40**) with $ZnCl_2$ at 220°c resulted in a 5% yield of flavaniline (**41**), and 10% yield of p-aminoacetophenone (**42**) [18].

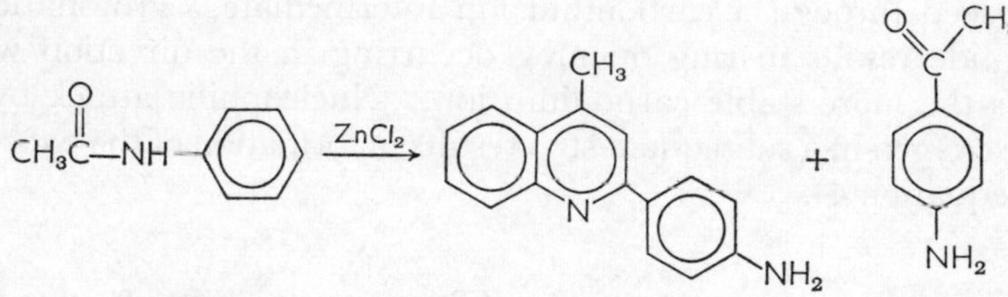

(**40**)   (**41**)   (**42**)

The mechanism is proposed to consist of formation of *N,N*-diaryl-acylamidines as intermediates and subsequent conversion to a mixture of *o*- and *p*-anils of amino ketones [19]. This postulation is based on the fact that rearrangement of acetanilide hydrochloride (**43**) in a sealed tube at 200°c yielded *N,N'*-diphenylacetamidine (**44**) which on reaction with zinc chloride at 290°c for 4 hours resulted in the production of flavaniline (**41**).

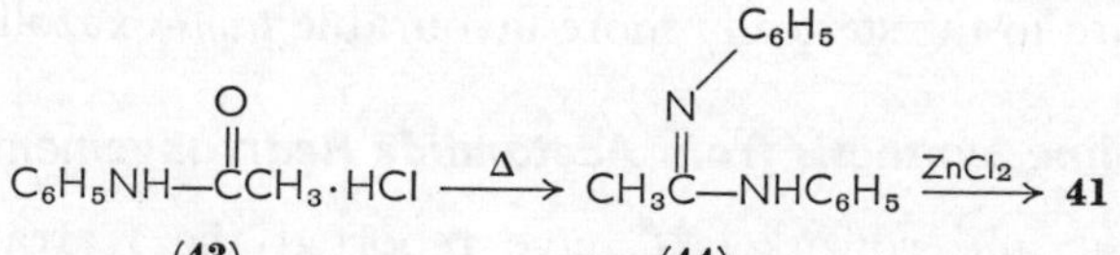

(**43**)   (**44**)

Obviously, the mechanism for this reaction involves a multi-step process and could best be considered as unknown at this time.

II. REARRANGEMENT OF N-SUBSTITUTED AMIDES

A. Rearrangement of N-Nitrosoamides

In general, amides can be converted to N-nitroso derivatives by a number of reagents, the preferable one appears to be nitrogen tetroxide in an acetate buffer solution[20] (equation 5).

$$RNHCR^1 + N_2O_4 \xrightarrow{OAc^-} RNCR^1 + HOAc + NO_3^-$$

(5)

The transformations of nitrosoamides can be conveniently classified depending on the reaction conditions. A basic medium, especially where R is a methyl group, is commonly used in the preparation of diazoalkanes[21]. The reaction probably proceeds through a rearrangement similar to that observed under neutral conditions (equation 6a) and the presence of base allows the formation of the diazo derivative and its isolation (equation 6b).

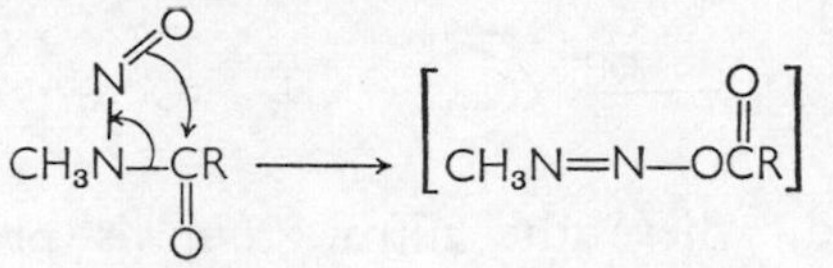

$$CH_3N-CR \longrightarrow \left[CH_3N=N-OCR \right]$$

(6a)

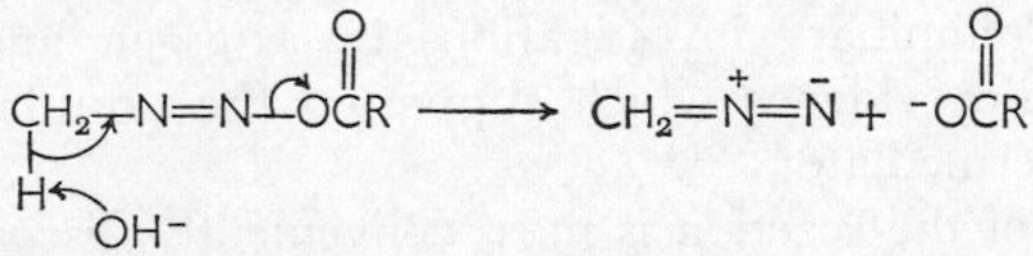

$$CH_2-N=N-OCR \longrightarrow CH_2=\overset{+}{N}=\overset{-}{N} + {}^-OCR$$

(6b)

This type of reaction has been reviewed elsewhere and will not be discussed further[22].

A second type of reaction involving N-nitrosoamides is their thermal rearrangement in a variety of solvents. Extensive research has been carried out to elucidate the mechanism of this reaction chiefly by White and Huisgen, with the use of elegant techniques.

In general, the reaction can be represented as yielding a mixture of ester, olefin, and carboxylic acid (equation 7). The relative amount of each product is influenced by a number of factors, including the nature of R and R^1, the solvent, and the temperature.

An exception to this general reaction route is where R is aromatic. In this instance, it has been demonstrated[23] that a free-radical inter-

mediate is involved in the main reaction pathway. Decomposition
of *N*-nitrosoacetanilide in methanol yields benzene as the main

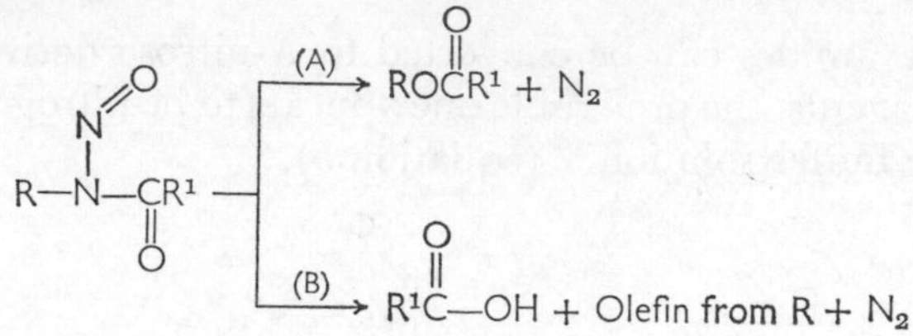

$$(7)$$

product. The intermediate postulated in the reaction was a diazo
ester (equation 8). In nitrobenzene a mixture of nitrobiphenyl

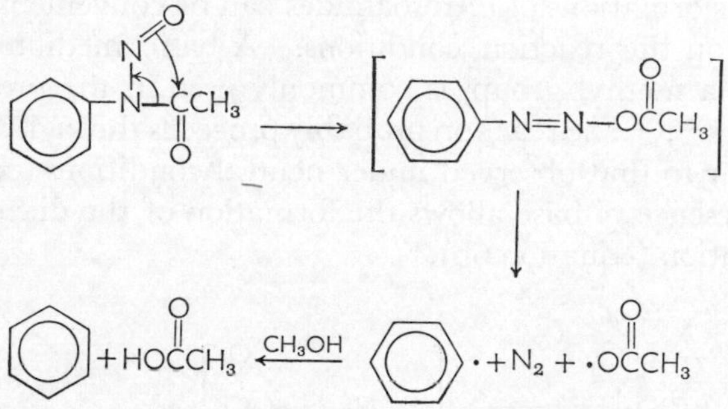

$$(8)$$

derivatives is obtained, where the nitro group is predominantly
ortho–para directing, further supporting the free-radical nature of the
reaction[24]. Preliminary investigations by Huisgen and coworkers
provided kinetic evidence which supported the postulation for the
diazo-ester intermediate[25-27].

The course of the reaction is most conveniently discussed by consideration of the *N*-nitrosoamides as derivatives of either primary or
secondary carbinamines. The R^1 group in equation (7) has little
effect on the course of the reaction but the nature of the R group
exerts a large influence. *N*-Nitrosoamides where R is primary
yield predominantly ester products[28]. *N*-(n-Butyl)-*N*-nitroso-3,5-
dinitrobenzamide (**45**) produced approximately an 80% yield of n-
butyl 3,5-dinitrobenzoate (**46**). Highest yields of ester are obtained
at the lowest temperature at which the reaction will proceed.

Investigations by Streitwieser[29] indicate that the intermediate
diazo ester produced in the reaction decomposes to a diazoalkane
intermediate. The decomposition of optically active *N*-(1-butyl-1-*d*)-
N-nitrosoacetamide (**47**) produced optically inactive esters with the

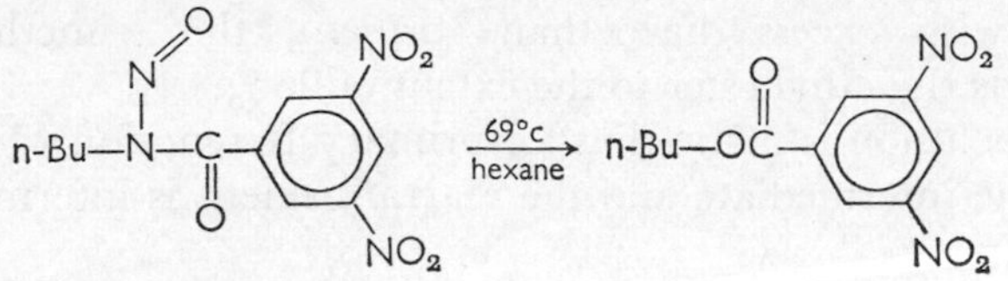

(45)   (46)

following deuterium distribution: 22% **48**, 56% **49** and 22% **50**. In another experiment employing unlabelled nitrosoamide, in the presence of *O*-deuterioacetic acid, some **49** was formed. In order to explain the disproportionation of the deuterium label and the loss of optical activity, the formation of a diazoalkane by α-elimination was postulated.

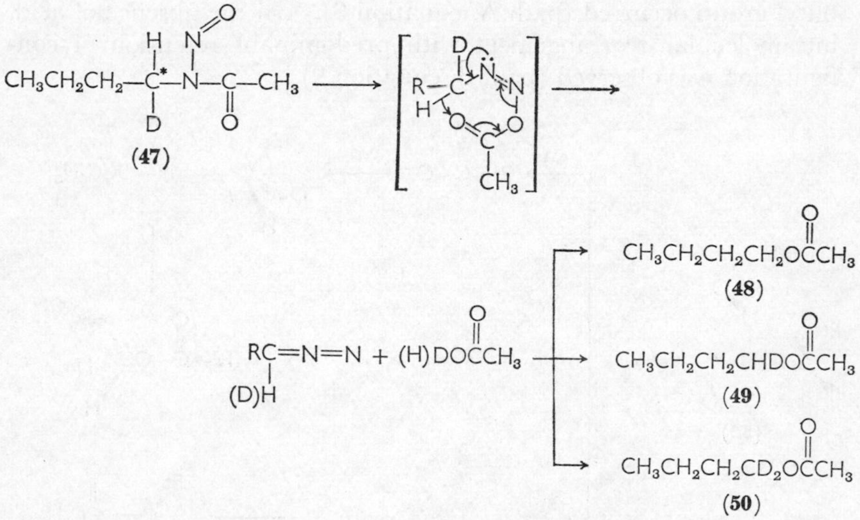

(47)

$$CH_3CH_2CH_2CH_2OCCH_3$$

(48)

$$RC{=}N{=}N + (H)DOCCH_3 \longrightarrow$$
$$(D)H$$

$$CH_3CH_2CH_2CHDOCCH_3$$

(49)

$$CH_3CH_2CH_2CD_2OCCH_3$$

(50)

The existence of a diazoalkane intermediate was confirmed by intercepting the intermediate carboxylic acid with diazoethane[30]. When *N*-(n-butyl)-*N*-nitrosotrimethylacetamide (**51**) was decomposed in

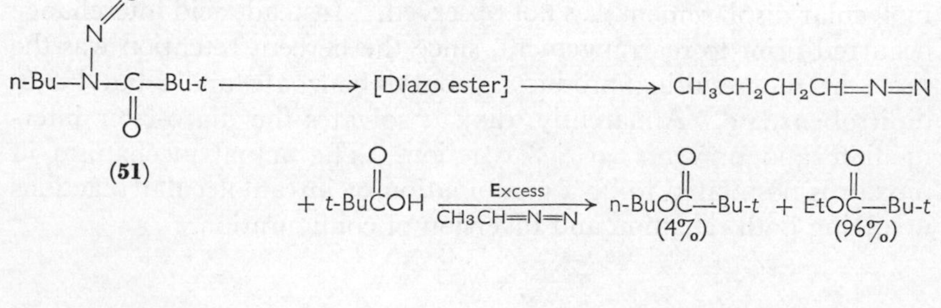

(51)

$$+\ t\text{-BuCOH} \xrightarrow[CH_3CH=N=N]{Excess} n\text{-BuOC—Bu-}t + EtOC—Bu-}t$$
$$(4\%) \qquad (96\%)$$

pentane, with excess diazoethane present, the trimethyl acetate appeared as the ethyl ester to the extent of 96%.

The conclusion is made that for primary R reaction (7) involves a diazoalkane intermediate and the rearrangement is intermolecular in nature.

When R is secondary a greater proportion of olefin and carboxylic acid compared to ester is produced (path B, equation 7). Preliminary evidence indicates that the rearrangement is intramolecular[31], since (+)-N-(s-butyl)-N-nitrosobenzamide (52) decomposed to s-butyl benzoate with retention of configuration (path B, equation 9). The details of the mechanism were expanded by observing the reaction in different solvents. In a non-polar solvent, pentane, with acetic acid added, a bimolecular rearrangement with inversion of the (+)-N-s-butyl group occurred (path A, equation 9). In the absence of acid, intramolecular rearrangement with predominant retention of configuration was observed (path B, equation 9).

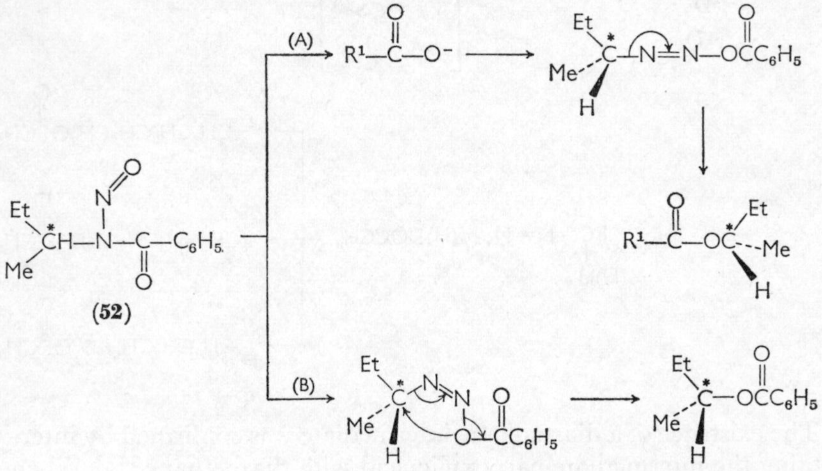

With dioxan as the solvent and 3,5-dinitrobenzoic acid added to compete with the benzoic acid produced on olefin formation, bimolecular displacement was not observed. Instead, acid interchange occurred prior to rearrangement, since the percent retention was the same in the resulting products, s-butyl benzoate and s-butyl 3,5-dinitrobenzoate. Apparently, dioxan solvates the diazo-ester intermediate and prevents an S_N2 reaction. The actual mechanism in dioxan is postulated to be a combination of intramolecular reactions involving both retention and inversion of configuration.

In acetic acid, the results are similar to that in dioxan concerning configurational changes but isomerization of the butyl group occurred much more extensively in acetic acid. It is concluded that in this solvent, there must be some charge separation in the intermediate (**53**) involved.

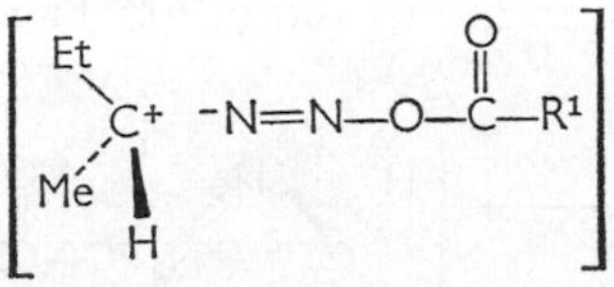

$$\left[\begin{array}{c} \overset{Et}{\underset{Me}{\diagdown}}\overset{}{\underset{|}{C^+}} \quad {}^-N{=}N{-}O{-}\overset{\overset{O}{\|}}{C}{-}R^1 \\ \quad\quad H \end{array}\right]$$

(**53**)

Finally, the mechanism for production of olefin is presented as involving an intramolecular rearrangement as in **54** (since reaction is unaffected by added acid), but not completely concerted (since polarity of solvent influences olefin distribution).

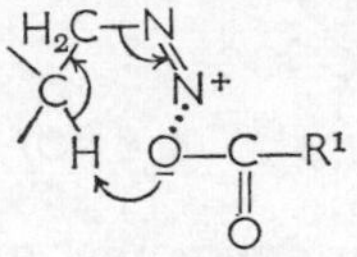

(**54**)

In order to study the decomposition mechanism of N-alkyl-N-nitrosoamides having a secondary alkyl group, a tracer study using ^{18}O was carried out[32]. Optically pure $(+)$-N-nitroso-N-(1-phenylethyl)-2-naphthamide (**55**) was decomposed in acetic acid to yield 1-phenylethyl 2-naphthoate. Analysis of the optical activity

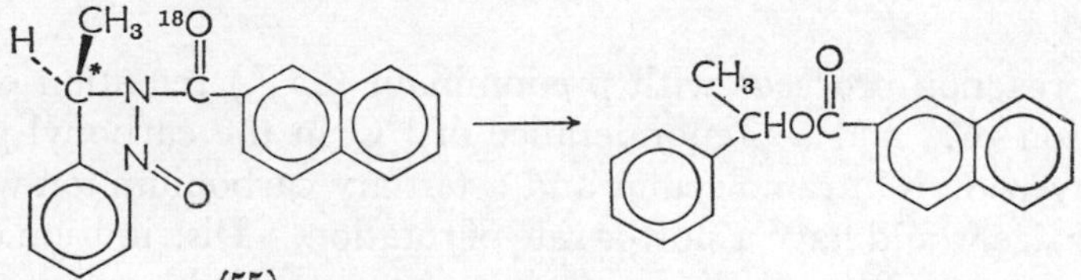

(**55**)

of the ester indicated 81% of the product corresponded to retention of configuration. Esters resulting from both retention of configuration and racemization contained 69% of the ^{18}O in the carbonyl oxygen. Based on this data, a mechanism (10) was proposed which involved

$$\left[\begin{array}{c} \overset{18O}{\|} \\ R^*N{=}N{-}O{-}\overset{}{C}R^1 \end{array}\right] \longrightarrow \left[\begin{array}{c} \overset{18O}{\|} \\ R^*N{=}N^+{-}O{-}\overset{}{C}R^1 \end{array}\right] \rightleftharpoons \left[\begin{array}{c} \overset{O}{\|} \\ R^*N{=}N^+{-}{}^{18}O{-}\overset{}{C}R^1 \end{array}\right]$$

(10)

decomposition of the diazo ester to an ion pair in the first step. Scrambling of the ^{18}O at this ion-pair stage is supported by the fact that a long-lived diazonium ion such as that derived from apocamphyl-amine produces extensive equilibration of ^{18}O. Loss of nitrogen is accompanied by both inversion of the carbonium ion and further scrambling of the oxygen atoms (equation 11).

$$
\left[R^*N{=}N{-}O\overset{O}{\overset{\|}{C}}R^1 \right] \longrightarrow \left[\underset{C_6H_5}{\overset{H}{\underset{|}{\overset{\diagdown}{C^+}}}}\;CH_3 \quad N_2 \;\; \overset{^{18}O}{\underset{}{\overset{\|}{O{-}CR^1}}} \right]
$$

$$
\overset{20\%}{\diagup \diagdown}
$$

H $\underset{C_6H_5}{\overset{H_3C}{\underset{|}{\overset{\diagdown}{C}}}}\overset{O}{\overset{\|}{-}O{C}R^1}$ (one ^{18}O) $\underset{C_6H_5}{\overset{H_3C}{\underset{|}{\overset{\diagdown}{C}}}}\overset{H}{\;}\overset{O}{\overset{\|}{-}O{-}CR^1}$ (one ^{18}O)

(11)

Ruled out by this data are two other mechanisms which could have been postulated; that of a concerted mechanism involving an $S_N i$ process and any mechanism which would involve a long-lived carbonium or diazonium ion.

Concerning the rearrangement of N-alkyl-N-nitrosoamides with a tertiary alkyl group, the nitrosobenzamide (**56**) derived from optically active 2-phenyl-2-butylamine, containing ^{18}O in the carbonyl oxygen was decomposed to yield a mixture of ester, carboxylic acid and olefin[33].

The reaction proceeds with predominant (95%) retention of configuration (**57**) and a preponderance of ^{18}O in the carbonyl group. Any inversion is intramolecular and a tertiary carbonium ion with its greater size would have a slower rate of rotation. Distribution of ^{18}O is independent of the N-alkyl group and consistent with a low activation energy for the loss of nitrogen from the diazo-ester intermediate. In comparison, a secondary carbonium ion such as that in equation (11) is a smaller carbonium ion and has only 75% retention of configuration since it rotates with a greater facility.

Recent studies have been concerned with further elucidating the nature of the ion-pair intermediate[34]. It has been shown that the ion pair (**59**) generated from the reaction of diphenyldiazomethane with benzoic acid in ethanol[35], is different from that obtained by thermal

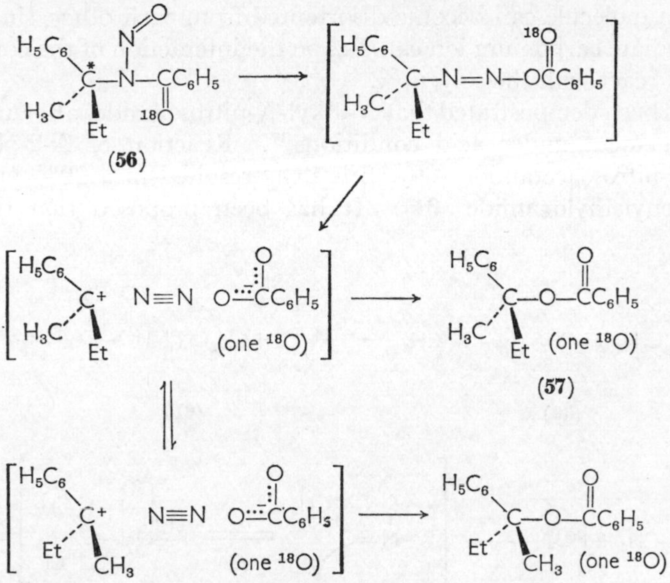

$$H_5C_6 \overset{N=O}{\underset{Et}{\overset{|}{\underset{|}{C}}}} \overset{*}{\underset{H_3C}{}}-N-\overset{O}{\underset{^{18}O}{\overset{||}{C}}}C_6H_5 \longrightarrow \left[H_5C_6 \overset{^{18}O}{\underset{H_3C}{\overset{||}{C}}}-N=N-O\overset{||}{C}C_6H_5 \right]$$

(56)

$$\left[\overset{H_5C_6}{\underset{H_3C}{}}C^+ \quad N\equiv N \quad O\overset{\cdots||}{\underset{}{C}}C_6H_5 \atop \text{(one }^{18}O) \right] \longrightarrow \overset{H_5C_6}{\underset{H_3C}{\overset{}{}}}\overset{O}{\underset{Et}{C}}-O-\overset{||}{C}C_6H_5 \quad \text{(one }^{18}O)$$

(57)

$$\left[\overset{H_5C_6}{\underset{Et}{}}C^+ \quad N\equiv N \quad O\overset{\cdots||}{\underset{}{C}}C_6H_5 \atop \text{(one }^{18}O) \right] \longrightarrow \overset{H_5C_6}{\underset{CH_3}{\overset{}{}}}\overset{O}{\underset{}{C}}-O-\overset{||}{C}C_6H_5 \quad \text{(one }^{18}O)$$

decomposition of *N*-benzhydryl-*N*-nitrosobenzamide (**58**) in ethanol solution. This was based on the different ratio of ester to (ester +

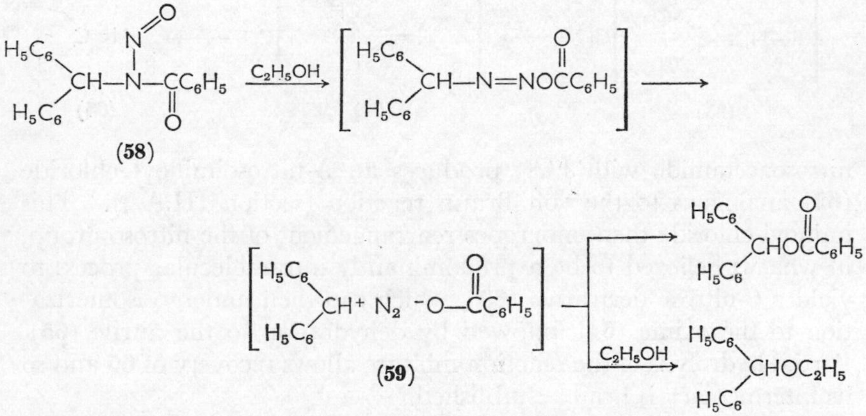

$$\overset{H_5C_6}{\underset{H_5C_6}{}}CH-\overset{N=O}{\underset{O}{\overset{|}{N}}}-\overset{||}{C}C_6H_5 \xrightarrow{C_2H_5OH} \left[\overset{H_5C_6}{\underset{H_5C_6}{}}CH-N=NO\overset{O}{\overset{||}{C}}C_6H_5 \right] \longrightarrow$$

(58)

$$\left[\overset{H_5C_6}{\underset{H_5C_6}{}}CH^+ \ N_2 \ ^-O-\overset{O}{\overset{||}{C}}C_6H_5 \right] \begin{array}{l} \longrightarrow \overset{H_5C_6}{\underset{H_5C_6}{}}CHO\overset{O}{\overset{||}{C}}C_6H_5 \\[2em] \xrightarrow{C_2H_5OH} \overset{H_5C_6}{\underset{H_5C_6}{}}CHOC_2H_5 \end{array}$$

(59)

ether), obtained in each reaction. The failure to incorporate deuterium into the ester products when the reaction was conducted in *O*-deuterioacetic acid, indicates that under the reaction conditions diphenyldiazomethane is not an intermediate in the nitrosoamide decomposition.

It has also been observed that the ions in the ion pair separated by a

nitrogen molecule can become disoriented from each other, since the
intermediate carbonium ion can escape the interaction of the counter-
ion and react with the solvent[36].

It has been demonstrated that N-alkyl-N-nitrosoamides also undergo
rearrangement under acid conditions[37]. Reaction of N-2-phenyl-
ethyl-N-nitrosoacetamide (**60**) with PCl_5 resulted in a 90% yield of
N-2-phenylethyloxamide (**61**). It has been proposed that the N-

$$C_6H_5CH_2CH_2-\overset{\overset{\displaystyle N}{\|}}{\underset{\underset{\displaystyle O}{\|}}{N}}-\overset{}{\underset{}{C}}-CH_3 \xrightarrow{PCl_5} C_6H_5CH_2CH_2NH\overset{\overset{\displaystyle O}{\|}}{C}-\overset{\overset{\displaystyle O}{\|}}{C}-NH_2$$

(**60**)   (**61**)

$$R-\overset{\overset{\displaystyle N}{\|}}{\underset{\underset{\displaystyle O}{\|}}{N}}-C-CH_3 + PCl_5 \longrightarrow \left[R-\overset{+}{N}=C\overset{CH_3}{\underset{Cl}{}} \rightleftharpoons R-\overset{}{NH}-C\overset{CH_2}{\underset{Cl}{}} \right] \longrightarrow$$

(**62**)

$$\left[R-\overset{+}{NH}=C\overset{CH_2}{\underset{Cl}{}} \right] Cl^- \longrightarrow \left[\overset{H}{\underset{R}{}}\overset{+}{N}=C\overset{CH}{\underset{Cl}{}} \right] Cl^- \longrightarrow \overset{}{\underset{R}{N}}=C\overset{C\equiv N}{\underset{Cl}{}}$$

(**63**)   (**64**)   (**65**)

nitrosoacetamide with PCl_5 produces an N-nitrosoiminoyl chloride
(**62**) analogous to the von Braun reaction (section III.A.2). This
iminoyl chloride then undergoes rearrangement of the nitroso group,
in what is believed to be a predominantly intramolecular process to
yield a C-nitroso derivative (**63**), which can then undergo isomerization to the oxime (**64**) followed by dehydration to the nitrile (**65**).
Partial hydrolysis of the reaction mixture allows recovery of **65** and so
its intermediacy is firmly established.

B. Rearrangement of N-Nitroamides

In a reaction analogous to the decomposition of N-nitrosoamides,
the N-nitro derivatives of amides are known to rearrange to give
similar products[38]. It is postulated that reaction proceeds through a
diazoxy-ester intermediate to yield either an ester (path A, equation

12) or a carboxylic acid plus an alkene in addition to nitrous oxide (path B). The course of the reaction depends mainly on the nature of

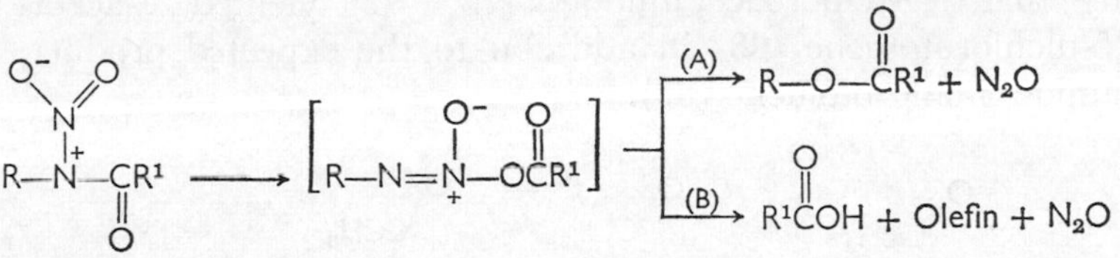

$$\begin{matrix} O^- \\ | \\ N \\ \end{matrix} \qquad R-N \overset{+}{-}CR^1 \longrightarrow \left[R-N \overset{+}{=} N-OCR^1 \right] \xrightarrow[\text{(B)}]{\text{(A)}} \begin{matrix} R-O-CR^1 + N_2O \\ R^1COH + \text{Olefin} + N_2O \end{matrix} \tag{12}$$

the R group, with ester formation predominating when R is primary and alkene formation favoured when R is secondary.

C. Rearrangement of N-Haloamides

I. The Orton rearrangement

The reaction of an *N*-haloacetanilide, also referred to as the Orton rearrangement, can be generalized as in equation (13).

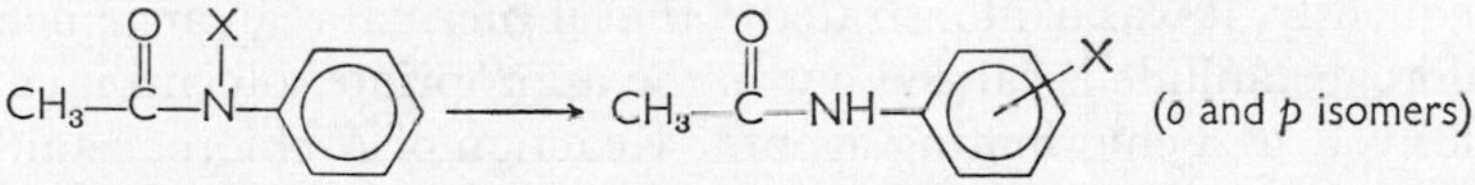

$$CH_3-\overset{O}{\underset{}{C}}-\overset{X}{\underset{}{N}}-\bigcirc \longrightarrow CH_3-\overset{O}{\underset{}{C}}-NH-\bigcirc^X \quad (o \text{ and } p \text{ isomers}) \tag{13}$$

Most of the reported work involves the *N*-chloro derivatives of acetanilides and the reaction can take place under three different sets of experimental conditions, namely:

a) a free-radical process promoted by either heat or light,
b) in protic solvents with specific halogen acid catalysis, and
c) in aprotic solvents with carboxylic acid catalysis.

The mechanism of this reaction under each set of conditions has been investigated.

a. Free-radical mechanism. If *N*-chloroacetanilide (**66**) is reacted in CCl_4 with a peroxide catalyst in the dark, smooth isomerization to a mixture of *o*- and *p*-chloroacetanilide is observed[39].

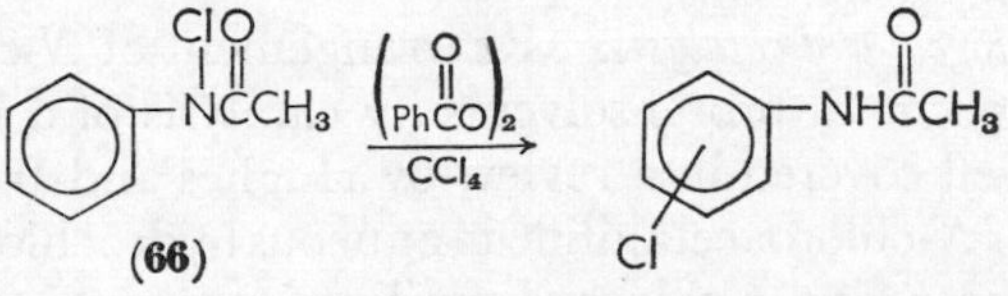

$$\bigcirc-\overset{Cl}{\underset{}{N}}\overset{O}{\underset{}{C}}CH_3 \xrightarrow[CCl_4]{\left(\overset{O}{\underset{}{PhCO}}\right)_2} \bigcirc-NH\overset{O}{\underset{}{C}}CH_3 \tag{14}$$
$$\quad\quad (\mathbf{66}) \qquad\qquad\qquad Cl$$

Preliminary investigation concerned itself with the question of whether the rearrangement was intra- or intermolecular in nature, that is, whether the chlorine atom was ever removed from the reacting acetanilide molecule. Insight into this question was gained by observing that *p*-*N*-chloroacetamidotoluene (**67**) yielded 4-acetamido-3,5-dichlorotoluene (**68**) in addition to the expected product, 4-acetoamido-3-chlorotoluene (**69**).

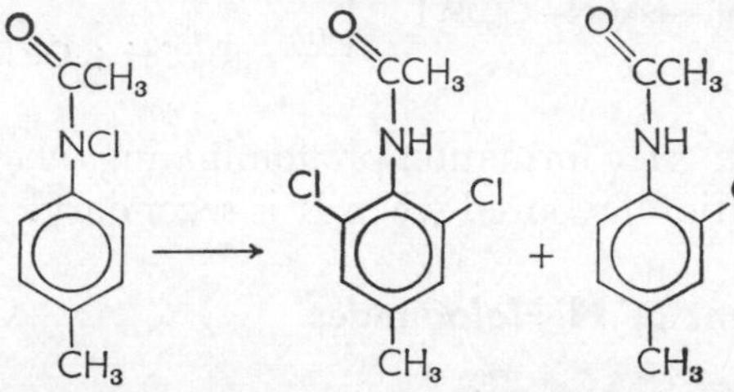

This result suggested that rearrangement was occurring through an intermolecular pathway.

Subsequently, it was demonstrated[40] that if thermal rearrangement of *N*-chloroacetanilide is carried out in the appropriate solvent, it can act exclusively as a chlorinating agent. Reaction of *N*-chloroacetanilide with *o*-nitroaniline in tetrachloroethane gave 4-chloro-2-nitroaniline as the only chlorinated product (equation 15). The

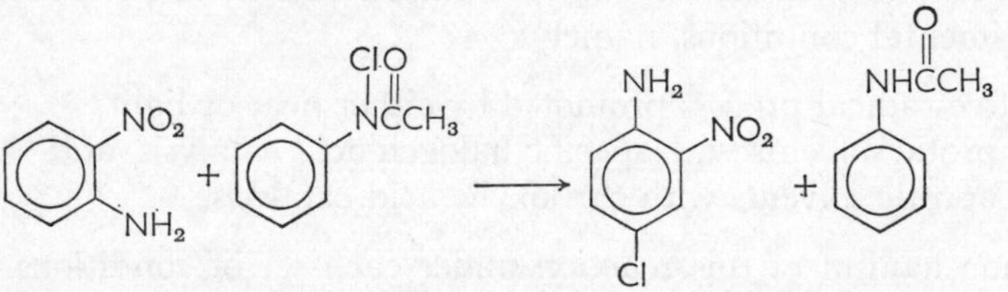

mechanism for reaction (15), therefore can be formulated as the sequence of equations (16)–(18).

If another moiety is present which would form a more stable radical, chlorine abstraction could be effected by this radical as in the case of nitroaniline (equation 15).

b. Specific halogen acid catalysis. Rearrangement of *N*-chloroacetanilides can also occur in protic solvents by catalysis of a halogen acid. This work is well covered in a review by Hughes and Ingold[41]. Rearrangement of *N*-chloroacetanilide in aqueous hydrochloric acid yields a mixture of the *o*- and *p*-chloroacetanilide isomers, similar to results obtained under free-radical conditions. The postulated mechanism in this instance involves the production of free chlorine and is therefore

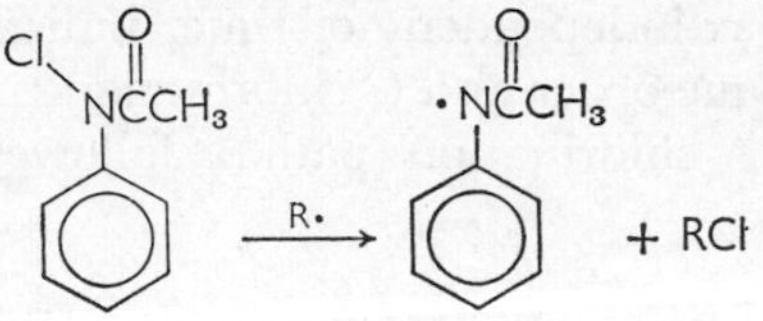

$$\text{(16)}$$

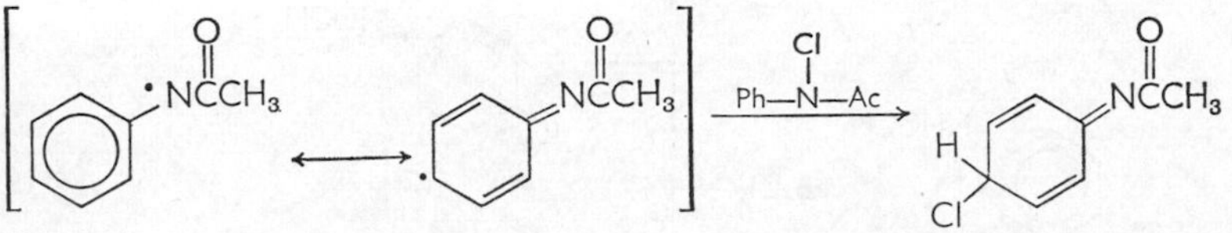

$$\text{(17)}$$

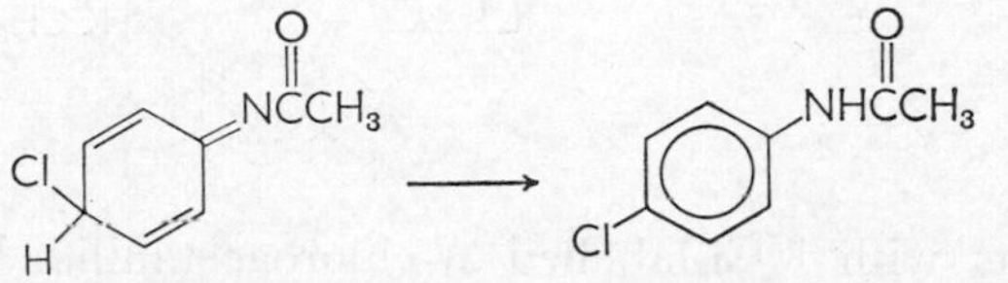

$$\text{(18)}$$

an intermolecular reaction.   The working hypothesis for the reaction
was formulated as an initial reversible acidolysis (equation 19) and
subsequent aromatic substitution by elemental chlorine (equation 20).

$$C_6H_5\overset{Cl}{\underset{}{N}}\overset{O}{\underset{}{C}}CH_3 + HCl \rightleftharpoons C_6H_5NH\overset{O}{\underset{}{C}}CH_3 + Cl_2 \qquad (19)$$

$$C_6H_5NH\overset{O}{\underset{}{C}}CH_3 + Cl_2 \longrightarrow ClC_6H_4NH\overset{O}{\underset{}{C}}CH_3 \ (o \ \text{and} \ p \ \text{isomers}) \qquad (20)$$

Evidence lending further support to this mechanism included the
following:
1) Orton[42] observed that elemental chlorine was evolved from the
reaction and if a more reactive substrate towards electrophilic substitution was introduced, it was preferentially chlorinated.   In this
manner N-chloro-2,4-dichloroacetanilide (**70**) chlorinated anisole.

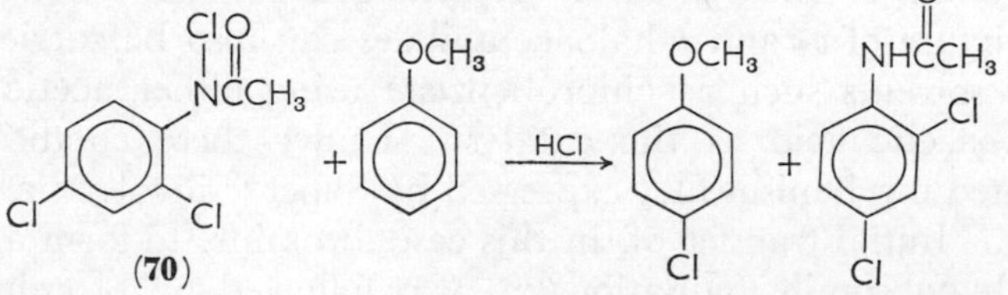

2) A series of rate studies involving substituted acetanilides and free
chlorine[43] (equation 21) demonstrated that the ratio of nitrogen to

carbon chlorination was independent of time, indicating that both reactions were of the same order and C-chlorination (path B) was not dependent on previous N-chlorination (path A) followed by rearrangement (path C).

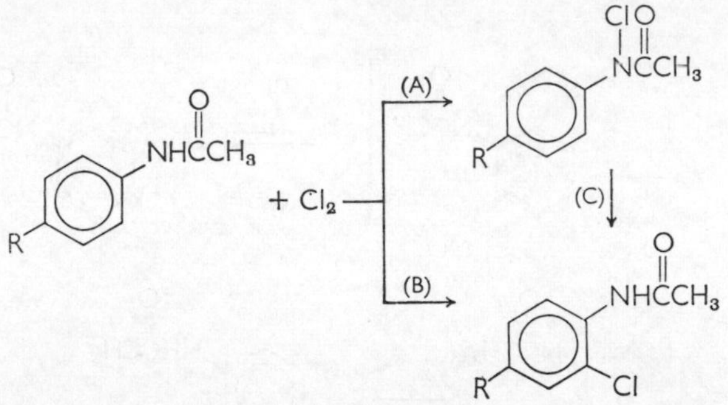

(21)

3) Experiments with ^{35}Cl-labelled N-chloroacetanilide[44] showed a dilution by the inorganic chloride present in the reaction mixture, again demonstrating an intermolecular transfer of chlorine.

Kinetic studies of the reaction indicate the reaction to be third order; first order with respect to the chloroamide and second order with respect to hydrochloric acid. Hughes and Ingold[41] consequently proposed a mechanism which can be classified as an S_N2-type process, involving attack of chloride ion on the protonated chloroamide as the first step in the reaction (equation 22).

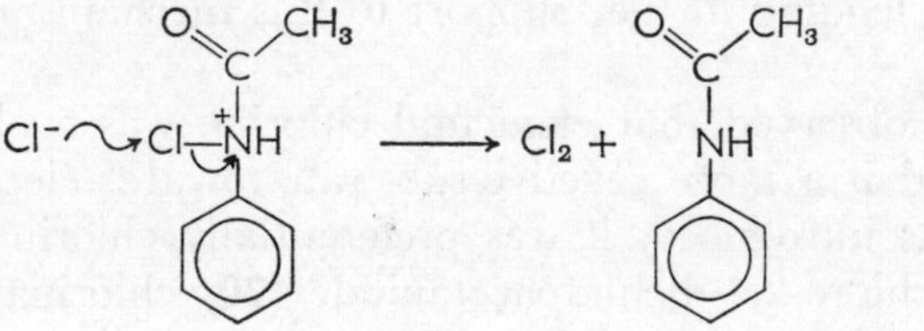

(22)

c. Reactions in aprotic solvents. Rearrangement of N-haloacetanilide to a mixture of o- and p-haloacetanilides can also be carried out in aprotic solvents such as chlorobenzene using either acetic acid or trichloroacetic acid as the catalyst. Under these conditions, the postulated mechanism first expressed by Soper[45] involved a two-step process. Initial transfer of, in this case, bromine, to form a reactive acetyl hypobromite (equation 23) was followed by bromination of either acetanilide or some other reactive substrate such as anisole, which can be introduced into the reaction medium (equation 24).

Previous work by Bell[46] had established the fact that the rate-determining step involves proton transfer. Subsequently, Dewar[47] introduced the possibility of a π complex and the two mechanisms can be summarized as depicted in equation (25).

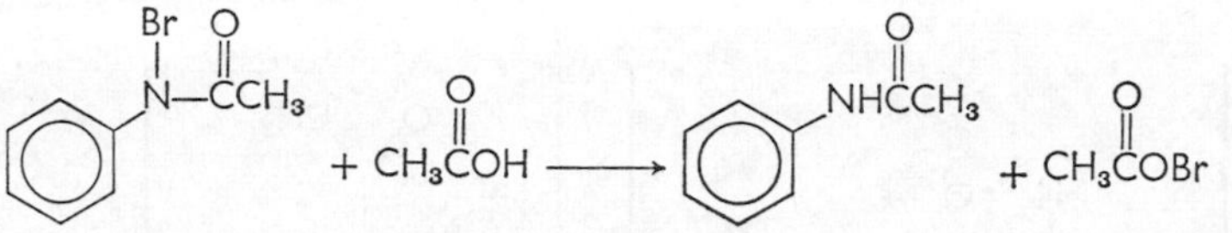

$$\text{(23)}$$

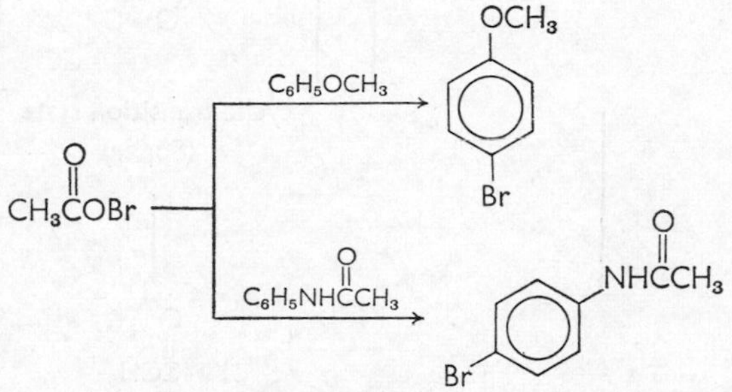

$$\text{(24)}$$

Further work in this area indicated that these relatively simple mechanisms cannot explain all the experimental facts[48]. Experiments using ^{14}C as a tracer element indicated that a rapid equilibrium is established prior to any subsequent rearrangement (equation 26), and therefore, it is not possible to determine whether the rearrangement is intra- or intermolecular.

Dewar and Couzens[48] also ruled out the Soper mechanism which involves the intermediacy of an acetyl hypobromite. When acetyl hypobromite and acetanilide are reacted together under experimental conditions similar to those employed in the rearrangement of N-bromoacetanilide C-bromination of the aromatic ring occurs very rapidly. Since C-bromination by the hypobromite is extremely fast as compared to the lifetime of N-bromoacetanilide in the radioactive exchange reactions (26), hypobromite intermediacy can be ruled out and the bromine transfer must be direct. If one includes the equilibrium reaction, then the Soper mechanism (equations 27–29) requires the carbon bromination step (29) to be a slow reaction, in contradiction with the results just described.

In summary, the Orton rearrangement in aprotic solvents appears to be a complex reaction and not easily described by simple mechanistic

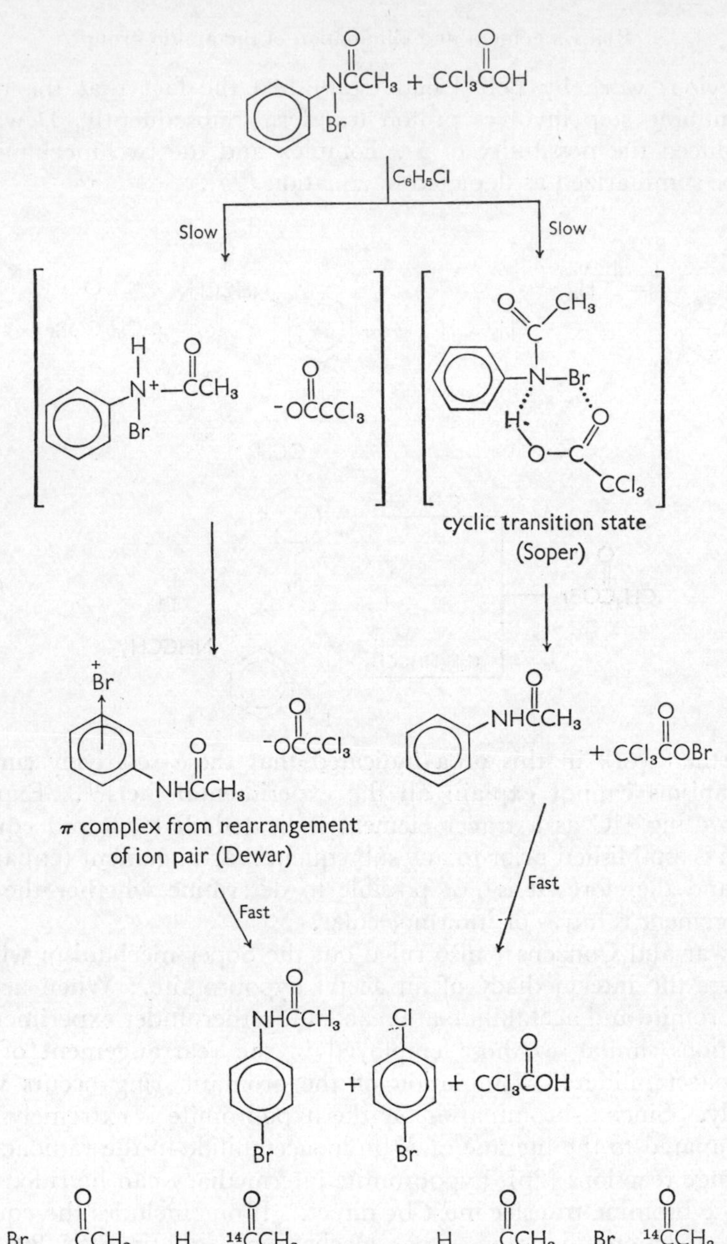

$$(25)$$

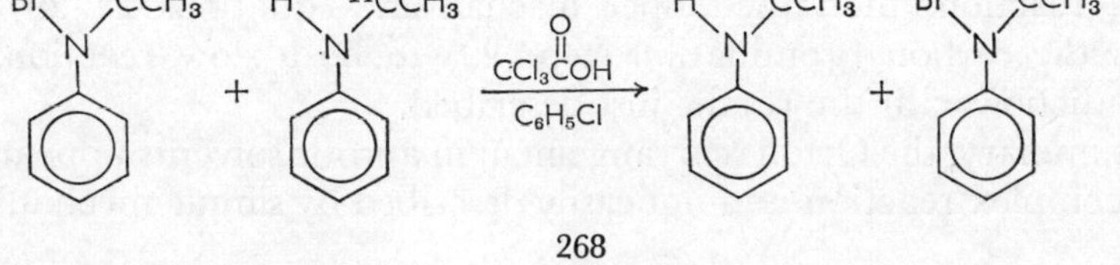

cyclic transition state
(Soper)

π complex from rearrangement
of ion pair (Dewar)

$$(26)$$

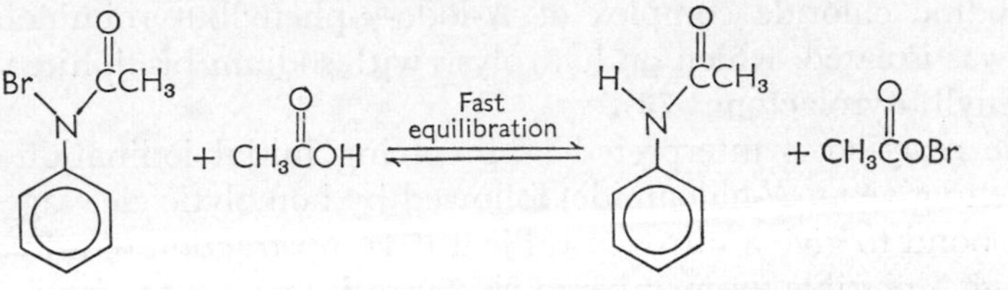

$$\text{(27)}$$

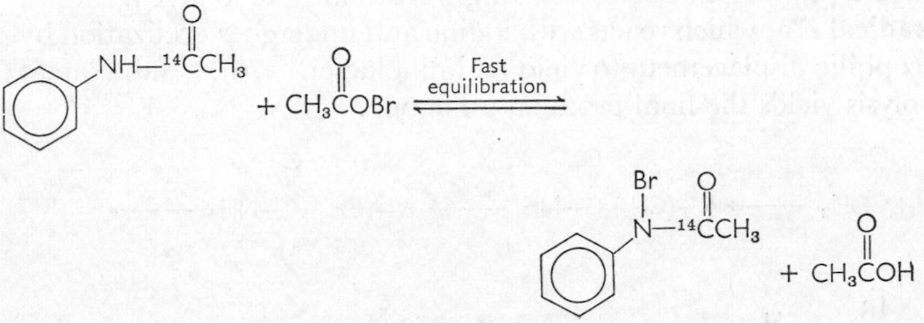

$$\text{(28)}$$

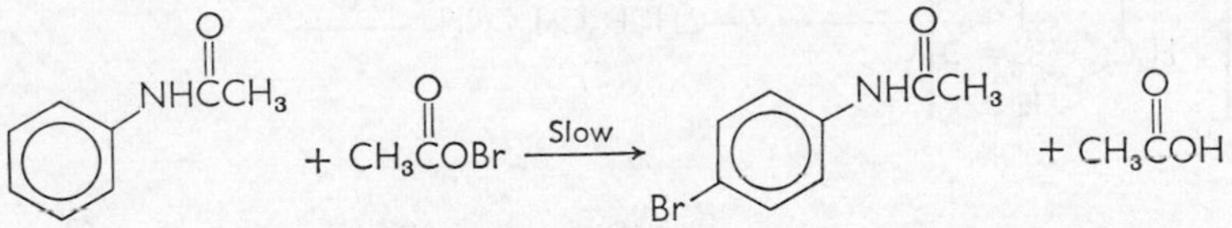

$$\text{(29)}$$

pathways. Perhaps this should be expected when polar molecules undergo reaction in non-polar media, a condition which promotes the formation of molecular aggregates and leads to complications in the study of reaction kinetics.

2. Other rearrangements of N-haloamides

a. Free-radical γ-hydrogen abstraction. As stated previously, *N*-halo-acetanilides undergo the Orton rearrangement on exposure to heat or u.v. radiation. Recently, it has been demonstrated[49] that if the structure of the *N*-haloamide is altered, different courses of reaction are possible. Specifically, it was found that if γ-phenylbutyramide (**71**) was photolysed in the presence of *t*-butyl hypochlorite and iodine,

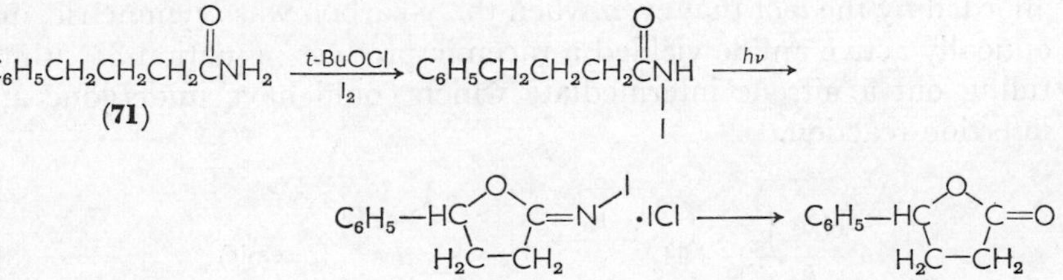

$$C_6H_5CH_2CH_2CH_2\overset{O}{\overset{\|}{C}}NH_2 \xrightarrow[I_2]{t\text{-BuOCl}} C_6H_5CH_2CH_2CH_2\overset{O}{\overset{\|}{C}}NH \xrightarrow{h\nu}$$

(**71**)

(**72**)   (**73**)

the iodine chloride complex of N-iodo-γ-phenylbutyroiminolactone (**72**) was isolated, which on hydrolysis with sodium bisulphite yielded γ-phenylbutyrolactone (**73**).

The reaction is interpreted as involving initial iodination of the amide to give an N-haloamide, followed by homolytic cleavage of the N—I bond to give a nitrogen radical (**74**), abstraction of a hydrogen through a possible six-membered ring transition state to give a carbon free radical (**75**) which reacts with iodine and undergoes cyclization by nucleophilic displacement to yield an iminolactone (**76**). Subsequent hydrolysis yields the final product, a lactone.

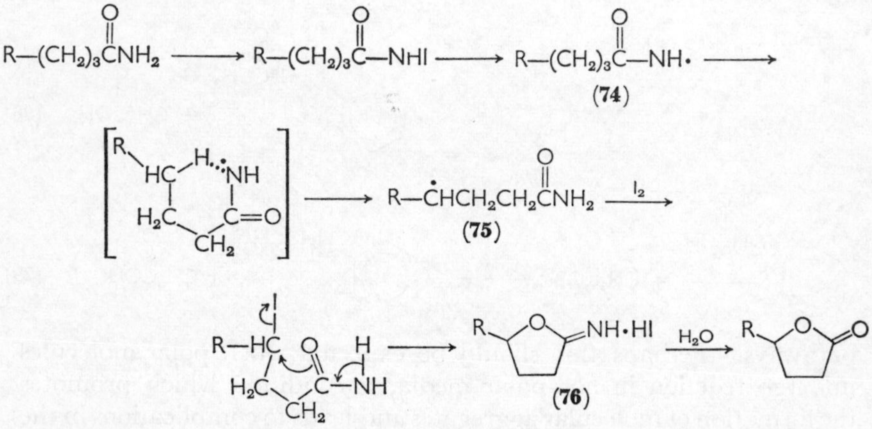

The yield of the reaction is limited to a maximum of 50% because free iodine is also liberated by the reaction of the iminolactone and the N-iodoamide (equation 30).

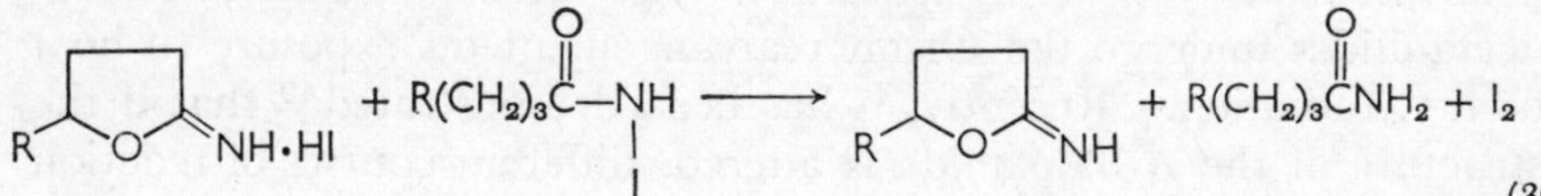

$$(30)$$

Evidence for formation of the free-radical intermediate **75** was afforded by the fact that even when the γ-carbon was asymmetric, an optically active amide yielded a racemic product (equation 31), thus ruling out a nitrene intermediate which could have undergone an insertion reaction.

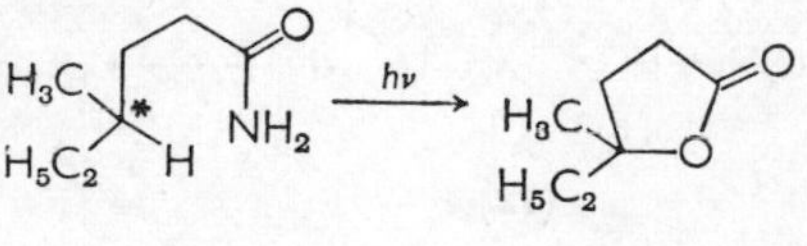

$$(31)$$

Similar results have been obtained from N-chloroamides[50].

Specific reactivity of the N-haloamides has been noted if the amide is substituted with a t-butyl group[51]. Reaction of N-bromo-N-t-butylpentanamide (**77**) in benzene solution under photolytic conditions produced 2-t-butylimino-5-methyltetrahydrofuran hydrobromide (**78**) as the main product. Stability of **78** towards hydrolysis of the

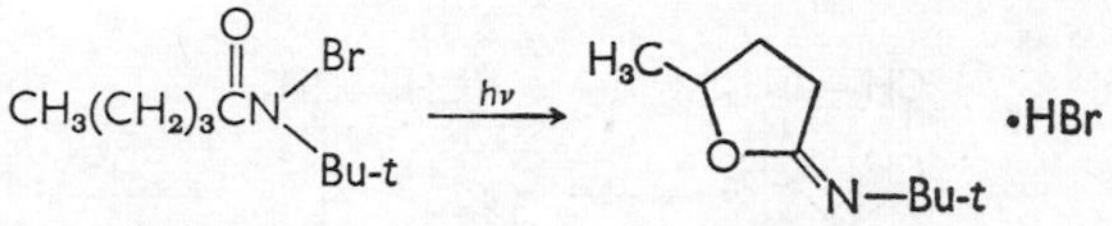

imino group is attributed to the bulkiness of the t-butyl group. Support for the radical mechanism proposed above by Barton is provided by the N-chloro derivative of N-t-butylpentanamide (**79**), which was rearranged under the same conditions to the 4-chloro isomer (**80**).

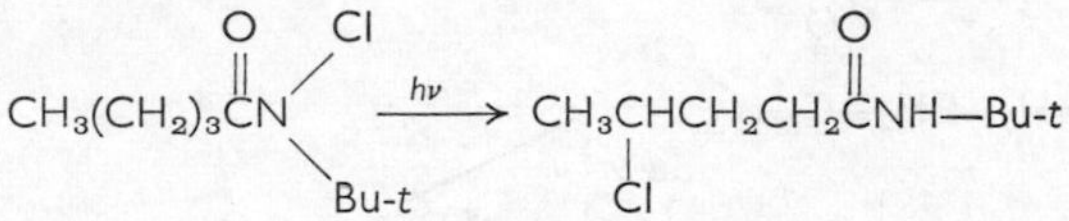

In all reactions investigated, benzene was the solvent of choice. Investigation of some of the by-products formed indicated that the N-chloroamide can act as a chlorinating agent by intermolecular reaction, similar to results concerning the Orton rearrangement of N-haloacetanilides.

It appears that the photolytic rearrangement of N-haloamides that fulfils the requirement of an abstractable hydrogen in a γ-position parallels the Hofmann–Loffer rearrangement of N-haloamines[52] (equation 32).

Apparently, in the N-haloamides, the halogen–nitrogen bond is weak enough to undergo homolytic cleavage and does not have to be activated by protonation, as in this case with haloamines.

b. Formation of α-lactams. Reactions of N-haloamides under basic conditions will in most instances yield amines, which are products of the Hofmann rearrangement. However, in certain instances, variations in reaction pathways are observed which are caused by some alteration of the reactants or reactant medium.

In non-aqueous media, it is possible to isolate α-lactams from the reaction of a base and an N-haloamide. The formation and reactions

$$
\begin{array}{c}
\underset{\underset{\underset{R}{|}}{\overset{\underset{Cl}{\diagdown}}{N}}}{\overset{\overset{CH_2-CH_2}{|}}{\underset{CH_3\ \ CH_2}{}}}
\ \xrightarrow{\ H^+\ }\
\underset{\underset{Cl\ \ \ R}{\overset{+}{HN}}}{\overset{CH_2-CH_2}{\underset{CH_3\ \ CH_2}{}}}
\ \xrightarrow{\ h\nu\ }\
\underset{\underset{Cl\cdot}{}\ \ \overset{+}{N}H-R}{\overset{CH_2-CH_2}{\underset{CH_3\ \ CH_2}{}}}
\ \longrightarrow\
\underset{\underset{R}{|}\ \overset{+}{N}H_2}{\overset{CH_2-CH_2}{\underset{\cdot CH_2\ \ CH_2}{}}}
\end{array}
$$

$$\Bigg\downarrow Cl_2 \qquad (32)$$

$$
\underset{\underset{R}{|}}{\overset{CH_2-CH_2}{\underset{N}{\overset{|\ \ \ \ \ |}{CH_2\ \ CH_2}}}}
\ \longleftarrow\
\underset{\underset{\underset{H}{|}}{\overset{}{:N-R}}}{\overset{CH_2-CH_2}{\underset{Cl\ \ \ CH_2\ \ CH_2}{}}}
\ \xleftarrow{\ OH^-\ }\
\underset{\underset{R}{|}}{\overset{CH_2-CH_2}{\underset{Cl\ \ \ \overset{+}{N}H_2}{\overset{|\ \ \ \ \ |}{CH_2\ \ CH_2}}}}
$$

of these cyclic derivatives, referred to as aziridinones has been reviewed recently[53]. A representative example[54] is the cyclization of N-t-butyl-N-chlorophenylacetamide by reaction with potassium t-butoxide to yield 1-t-butyl-3-phenyl-2-aziridinone (equation 33).

$$
\begin{array}{c}
C_6H_5-CH_2-\overset{\overset{\displaystyle O}{\|}}{C}-\underset{\underset{Bu\text{-}t}{}}{N}\!\!\diagup^{Cl} \\[2mm]
C_6H_5-\underset{\underset{Cl}{|}}{CH}-\overset{\overset{\displaystyle O}{\|}}{C}-NH-Bu\text{-}t
\end{array}
\ \xrightarrow{\ t\text{-}BuO^-\ }\
\underset{\underset{H_5C_6}{}}{\overset{\overset{\displaystyle O}{\|}}{\underset{CH-N-Bu\text{-}t}{C}}}
\qquad (33)
$$

The same lactam was formed under similar reaction conditions[55] by starting with N-t-butyl-α-chlorophenylacetamide.

The isolation of α-lactams substantiated earlier postulations concerning the intermediacy of an aziridone in a number of rearrangements, such as those observed by Sarel and his coworkers. For instance[56], reaction of α-chloro-α,α-diphenylacetanilide (81) with

$$
\underset{\underset{H_5C_6}{}}{\overset{H_5C_6}{\underset{C}{\overset{\overset{\displaystyle Cl}{|}}{C}}}}\!\!-\!\overset{\overset{\displaystyle O}{\|}}{C}-NH-C_6H_5
\ \xrightarrow{\ NH_2^-\ }\
\underset{\underset{H_5C_6}{}}{\overset{H_5C_6}{C}}\!\!\diagup^{\overset{\displaystyle Cl}{|}}_{\underset{\underset{\underset{O}{\|}}{C}}{N-C_6H_5}}
\ \longrightarrow
$$

(81)

$$
\left[\underset{\underset{H_5C_6}{}}{\overset{H_5C_6}{\underset{C}{C}}}\!\!\diagdown_{\underset{C_6H_5}{N}}^{\overset{\overset{\displaystyle O}{\|}}{C}}\ \ NH_2^- \right]
\ \longrightarrow\
\underset{\underset{H_5C_6}{}}{\overset{H_5C_6}{\underset{C-N}{C}}}\!\!\diagdown_{C_6H_5}^{\overset{\displaystyle O^-}{C}}\ NH_2
\ \xrightarrow{\ NH_3\ }\
\underset{\underset{H_5C_6}{}}{\overset{H_5C_6}{CH}}\!-\!\underset{\underset{C_6H_5}{|}}{N}\!-\!\overset{\overset{\displaystyle O}{\|}}{C}-NH_2
$$

(82)

sodium amide in liquid ammonia yielded the benzhydrylurea **82**. The reaction can be visualized as proceeding through an aziridinone intermediate and subsequent ring opening by attack with amide ion.

c. *Rearrangement of* α,N-*dihaloamides.*   Rearrangement of α,N-dihaloamides in aqueous base yields alkylidene halides and sodium isocyanate as products (equation 34). The reaction is thought to proceed through intramolecular rearrangement of the conjugate base **83**.

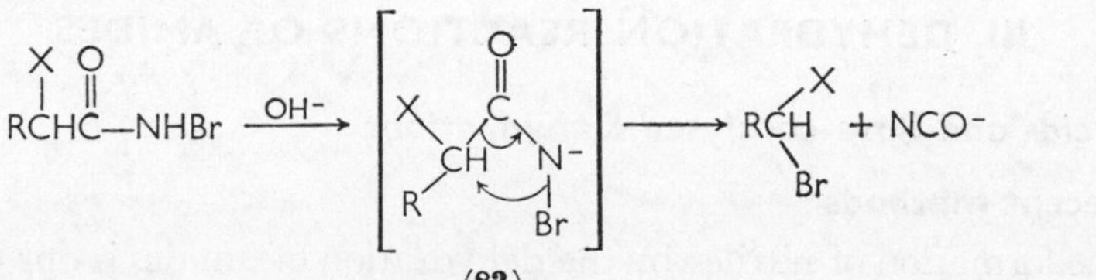

$$RCHC\text{—}NHBr \xrightarrow{OH^-} \left[\begin{array}{c} \text{(83)} \end{array} \right] \longrightarrow RCH + NCO^- \tag{34}$$

Support for this mechanism appears in the fact that rearrangement of optically active D-α-chlorohydrocinnamamide yielded a product indicative of retention of configuration[57]. Tracer studies employing [82]Br also supported an intramolecular pathway[58]. Reaction of α-chloro-N-bromoisobutyramide in base with an excess of [82]Br$^-$ ion present yielded 2-bromo-2-chloropropane with no incorporation of [82]Br. This was cited as exclusive evidence for pathway (A) over pathway (B) in equation (35).

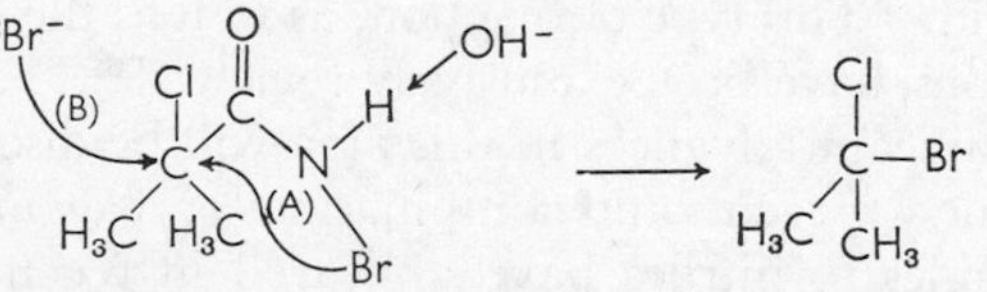

$$\tag{35}$$

A similar reaction has been noted[59] for the rearrangement of α-nitroacetamide to nitrodibromomethane in the presence of hypobromite ions (equation 36).

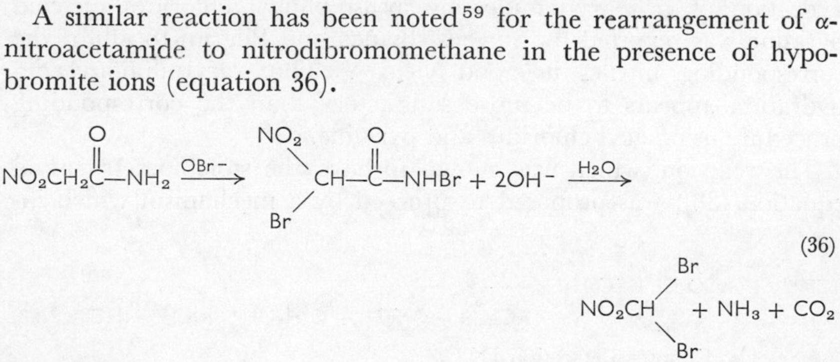

$$NO_2CH_2C\text{—}NH_2 \xrightarrow{OBr^-} CH\text{—}C\text{—}NHBr + 2OH^- \xrightarrow{H_2O}$$

$$\tag{36}$$

$$NO_2CH \begin{array}{c} Br \\ \\ Br \end{array} + NH_3 + CO_2$$

10+c.o.a.

An exception has been noted for this mechanistic pathway if the α-halogen is fluorine[60]. N-Bromoheptafluorobutyramide did not react unless an excess of base was present. It exhibited complex reaction kinetics, and in the presence of oxygen yielded the heptafluorobutyrate (equation 37).

$$C_3F_7\overset{O}{\overset{\|}{C}}NHBr + O_2 + OH^- \longrightarrow C_3F_7\overset{O}{\overset{\|}{C}}O^- + NO_2 + Br^- + H_2O \qquad (37)$$

III. DEHYDRATION REACTIONS OF AMIDES

A. Acid- and Base-catalysed Dehydrations

I. Recent methods*

The formation of nitriles by the dehydration of amides is one of the oldest known reactions in organic chemistry, dating back to the first synthesis of a nitrile by Wohler and Leibig's dehydration of benz-amide to benzonitrile[61]. This reaction occurred during an attempt to distill the amide over barium oxide and is typical of one type of amide dehydration reaction in which the amide is heated together with a catalytic agent such as alumina, silica, etc. This type of procedure is of relatively little use in the laboratory where it has been supplanted by the use of chemical dehydrating agents.

Laboratory-scale amide dehydrations are most commonly effected by heating the amide together with the halide or anhydride of a mineral acid. This type of reaction, as well as the catalytic dehydration reactions, have been extensively reviewed[62,63] and therefore only recent, significant advances in this area will be discussed.

Several new reagents and combinations of reagents for the dehydration of amides to nitriles have appeared in recent years. The dehydration of primary amides by arylsulphonyl chlorides in cold pyridine was reported by Stevens, Bianco and Pilgrim to afford the corresponding nitriles in good yield[64]. Thus, arylsulphonyl dehydration appears to be more satisfactory than the corresponding procedure using acyl chlorides and pyridine.

The reaction, which was shown to have the stoichiometry as in equation (38), was proposed to proceed by a mechanism which in-

$$R\overset{O}{\overset{\|}{C}}NH_2 + ArSO_2Cl + 2C_5H_5N \longrightarrow$$
$$RC\equiv N + ArSO_3^- + C_5H_5NH^+ + C_5H_5\overset{+}{N}HCl^- \qquad (38)$$

* See also Chapter 13, section IX.C.

volves oxygen, rather than nitrogen sulphonation (equation 39). This proposal is largely based on the observation that compounds of

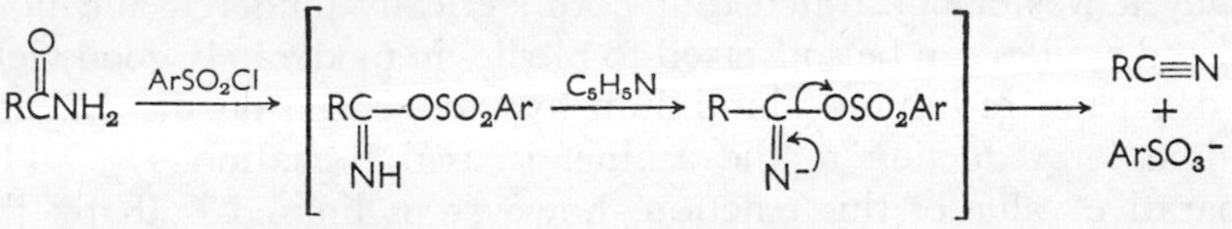

$$\underset{\substack{\| \\ O}}{RCNH_2} \xrightarrow{ArSO_2Cl} \left[\underset{\substack{\| \\ NH}}{RC-OSO_2Ar} \xrightarrow{C_5H_5N} \underset{\substack{\| \\ N^-}}{R-C\!\!\curvearrowright\!\!OSO_2Ar} \right] \longrightarrow \begin{array}{c} RC\!\equiv\!N \quad (39) \\ + \\ ArSO_3^- \end{array}$$

the type $RCONHSO_2Ar$ which would result from nitrogen sulphonation have been prepared from acyl halides and sulphonamides[65] and are stable under the reaction conditions.

Substituted and unsubstituted amides have been converted to nitriles by treatment with silanes[66].

The use of the complex formed between dimethylformamide (DMF) and thionyl chloride has been shown to be effective in cases where more conventional methods for amide dehydration have failed[67]. For example, pyromellitonitrile (**85**) has been obtained in good yield by adding $SOCl_2$ to a stirred suspension of **84** in DMF at 0°. Under the same conditions, benzamide and phthalamide were

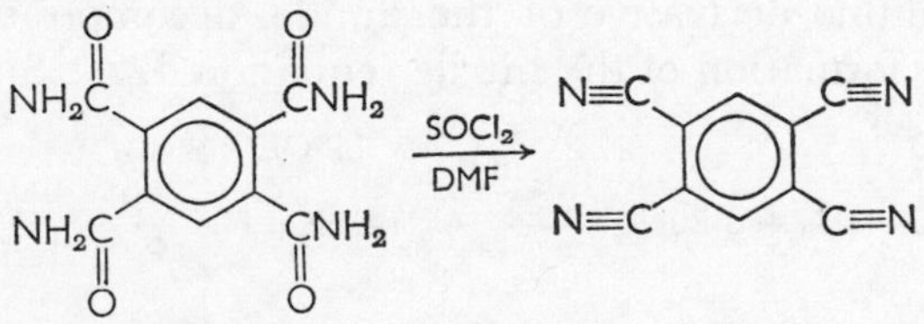

(**84**)   (**85**)

quantitatively converted to benzonitrile and phthalonitrile. This reagent also brought about the cyclodehydration of dibenzoyl-hydrazide to a diphenyloxadiazole in 45% yield at 0° (equation 40).

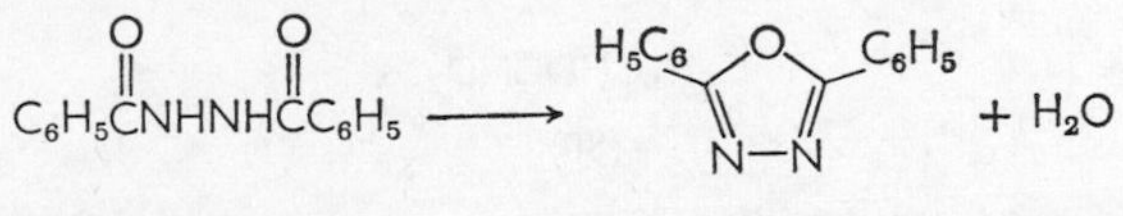

The conversion of amides to nitriles using basic reagents is much less common than procedures which involve acidic reagents or combinations of acidic reagents and amines. However, basic reagents are useful alternatives to the acidic methods for molecules which contain groups prone to attack by acids, and several basic dehydration reagents have been reported.

Newman and Fukunaga[68] established that the widely used lithium aluminium hydride reduction of unsubstituted amides to primary amines actually proceeds through a nitrile intermediate. Additionally it was demonstrated that both sterically hindered and non-hindered amides can be converted to nitriles in moderately good yield if a deficiency of lithium aluminium hydride, based on the amount required for reduction to the amine, is used (equation 41). The preparative value of this reaction, however, is limited both by the necessity of using a deficiency of the basic reagent, and to cases in which the rate of conversion of amide to nitrile (k_1) is significantly greater than the rate at which the nitrile is converted to the corresponding primary amine (k_2).

$$R-\overset{\overset{\displaystyle O}{\|}}{C}-NH_2 \xrightarrow[k_1]{LiAlH_4} R-C\equiv N \xrightarrow[k_2]{LiAlH_4} R-CH_2NH_2 \qquad (41)$$

The strongly basic reagent n-butyllithium in ether–hexane, or tetrahydrofuran–hexane solution has been shown to be an effective reagent for the conversion of phenyl-, diphenyl-[69] and monoalkyl-phenylacetamides[70] to the corresponding nitriles. Treatment of the amides with three equivalents of n-butyllithium results in the formation of a trilithio derivative of the amide, decomposition of which results in the formation of the nitrile (equation 42). Support for the

$$C_6H_5CH_2\overset{\overset{\displaystyle O}{\|}}{C}-NH_2 + 3\,BuLi \longrightarrow C_6H_5\overset{\overset{\displaystyle Li}{|}}{C}H\overset{\overset{\displaystyle OLi}{|}}{C}=NLi \xrightarrow{H^+} C_6H_5CH_2C\equiv N$$
$$(42)$$

trilithio intermediate was obtained from the observation that treatment of phenylacetamide with three moles of n-butyllithium followed by hydrolysis with D_2O resulted in the incorporation of three deuterium atoms into the molecule and the formation of **86**.

$$C_6H_5CHD\overset{\overset{\displaystyle O}{\|}}{C}ND_2$$
$$(86)$$

Cram and Haberfield have demonstrated that optically active amides can be converted to nitriles without significant loss of optical purity by a variety of acidic reagents[71]. Phosphorus pentoxide had previously been shown to be an effective reagent for this purpose[72]. Thus, optically pure (+)-2-phenylbutyramide was converted to (−)-2-phenylbutyronitrile of high optical purity by treatment of the amide with P_2O_5. Similarly, optically active 2-methyl-3-phenyl-propionamide was converted to the corresponding nitrile by this reagent with very little isomerization[73].

An investigation of *cis-* and *trans*-4-*t*-butylcyclohexanecarboxamides disclosed that these compounds can be dehydrated to the corresponding nitriles by phosphorus pentoxide and thionyl chloride without the occurrence of any geometric isomerization[74]. Phosphorus oxychloride was found to be a less effective reagent for this purpose. However, in all cases more than 97% retention of geometric configuration was observed.

2. von Braun reaction*

The reaction of an *N*-alkylbenzamide with phosphorus pentahalide to yield benzonitrile and an alkyl halide (equation 43) was first discovered by von Pechmann[75], but bears the name of von Braun because of his extensive research in this area.

$$C_6H_5\overset{\overset{O}{\|}}{C}\text{—NHR} + PX_5 \longrightarrow C_6H_5C{\equiv}N + RX + HX + POX_3 \qquad (43)$$

The early research of von Braun has recently been reviewed, including a complete bibliography of his published work[76]. In spite of the apparent general synthetic usefulness of the reaction for converting amines to halides, little work has been carried out since the initial investigations of von Braun to determine either the mechanism or scope of the reaction.

Leonard and Nommensen have made a number of experimental observations in an attempt to elucidate the mechanism[77]. von Braun[78] obtained a 78% yield of 1,5-dibromopentane from the reaction of *N*-benzoylpiperidine with phosphorus pentabromide (equation 44). Similarly *N*-benzoyl-2,6-dimethylpiperidine (87) gave a 19%

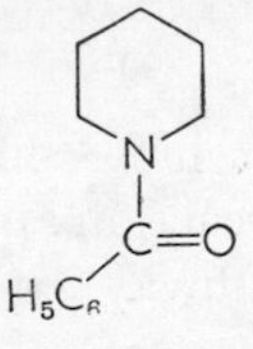

$$\underset{\underset{\underset{H_5C_R}{|}}{\underset{\overset{|}{C=O}}{N}}}{\bigcirc} \xrightarrow{\text{PBr}_5} Br(CH_2)_5Br + C_6H_5C{\equiv}N + POBr_3 \qquad (44)$$

yield of 2,6-dibromoheptane while *N*-benzoyl-2,2,6,6-tetramethyl-piperidine (88) gave no reaction. Obviously, steric bulk at the α_N-carbon decreases the reaction rate and favours a bimolecular nucleophilic substitution mechanism. Supporting this view is the fact that

* See also Chapter 13, section IX.B.1.

N-benzoyl-$(+)$-s-butylamine reacts to yield $(-)$-s-butyl bromide, corresponding to inversion of configuration in the substitution reaction.

In an attempt to both extend the synthetic utility of the reaction and elaborate on the proposed mechanism of the reaction, Vaughan and

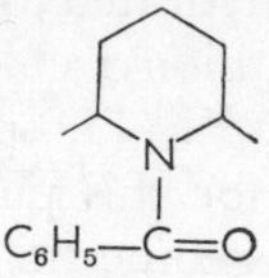

(87) 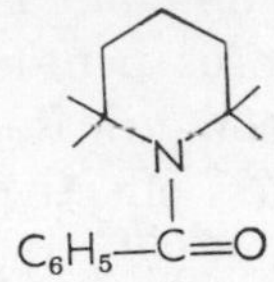　　　　　　　　　　(88)

Carlson[79] substituted thionyl chloride for phosphorus pentachloride in the reaction. The reaction was limited to a variety of secondary amides (equation 45).

$$RCNHR^1 + SOCl_2 \longrightarrow R^1Cl + RC{\equiv}N + SO_2 + HCl \qquad (45)$$

Best results were obtained when the α_N-carbon afforded a stable carbonium ion. For example, the reaction of N-benzhydrylbenzamide with thionyl chloride gave an 80% yield of benzhydryl chloride (equation 46).

$$C_6H_5C{-}NH{-}CH(C_6H_5)_2 \xrightarrow{SOCl_2} C_6H_5C{\equiv}N + (C_6H_5)_2CHCl \qquad (46)$$

The reaction is visualized as proceeding through the attack of $SOCl_2$ or PCl_5 yielding an intermediate structure **89** which can eliminate either SO_2 or $POCl_3$ to form an ion pair which can collapse to the iminoyl chloride (**90**). Reaction of benzanilide with thionyl

$$RC{-}NHR^1 \xrightarrow[Z = SO \text{ or } PCl_3]{ZCl_2} RC{=}NR^1 \longrightarrow RC{=}NR^1 + ZO$$

(89)　　　　　　　　　(90)

chloride afforded N-phenylbenzimidoyl chloride which could be isolated and characterized. Two possible modes of reaction appear to be open for reaction of the iminoyl chloride. One pathway (equation 47)

$$\underset{RC{=}N}{\overset{Cl \quad R^1}{|\quad|}} \longrightarrow RC^+{=}N \longrightarrow RC{\equiv}N + R^1Cl \qquad (47)$$

is based on the data of Leonard and Nommensen[77] involving inversion of optically active R^1, and is an S_N2 process.

An alternative mechanistic pathway (equations 48 and 49) is supported by the fact that in the series where the N-alkyl group was benzyl, p-methoxybenzyl, α-methylbenzyl and benzhydryl, the yield

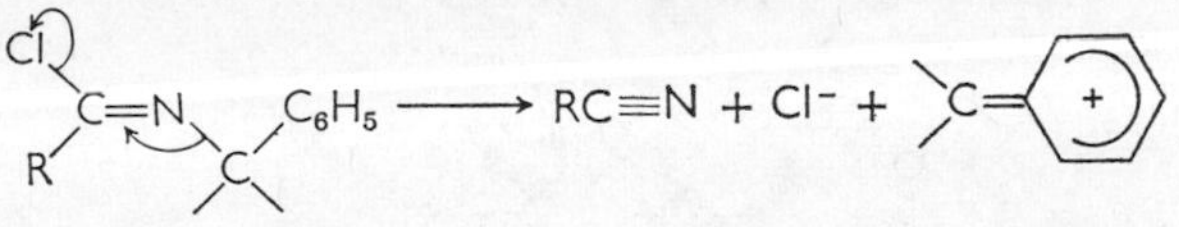

$$\text{(48)}$$

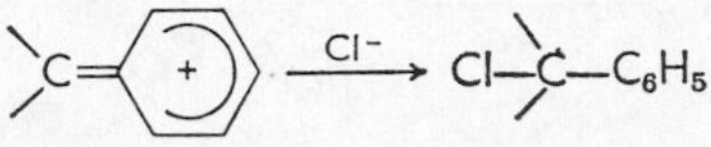

$$\text{(49)}$$

of chloride was the highest in the case of benzhydryl. Furthermore, reaction of optically active $(-)$-N-(α-methylbenzyl)acetamide produced a chloride which had lost most of its optical activity. These results are interpreted as supporting a fragmentation process involving an S_N1-type mechanism.

B. Amide Pyrolysis

The formation of nitriles via the pyrolytic dehydration of amides is a relatively little used laboratory procedure. The reaction usually produces ammonia as well as other by-products. Dehydration to nitriles is normally brought about more cleanly by the chemical dehydration procedures described in section III.A. The pyrolytic reaction, however, has both historical and industrial significance and has been adequately covered in several review articles [80,81].

More recent evidence indicates the pyrolytic dehydration of primary amides to proceed through an imide or isoimide [82]. Partial pyrolysis of propionamide led, for example, to the formation of a mixture of propionimide and propionitrile (equation 50). The imide was

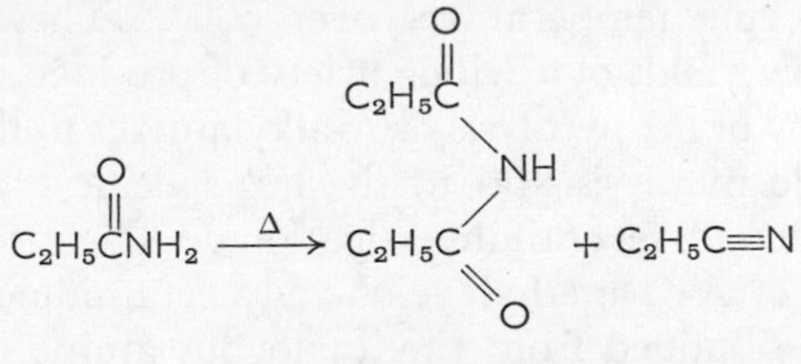

$$\text{(50)}$$

readily isolated and, on further heating afforded additional nitrile. Imide formation is postulated to take place via a bimolecular deamination to isoimide (equation 51), further decomposition of which resulted in the formation of imide, nitrile and carboxylic acid (equation 52).

Pyrolysis of variously substituted amides can form a number of products other than simple nitriles. Whereas the pyrolysis of unsubstituted amides leads primarily to the formation of nitriles, pyrolysis of N-alkyl- and N-methyl-N-alkylamides has been found to be a

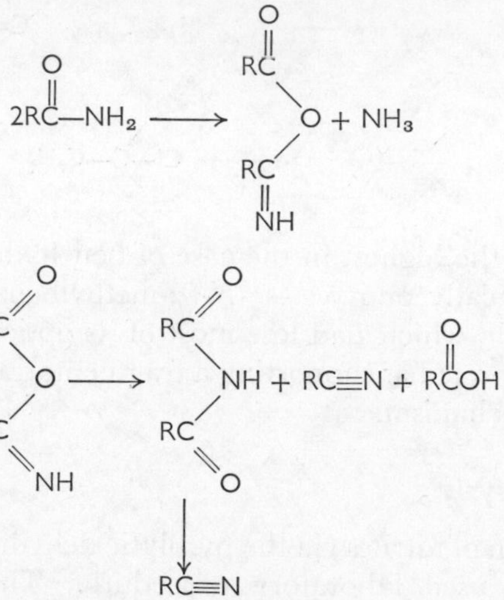

$$2RC(=O)-NH_2 \longrightarrow \quad \begin{array}{c} RC(=O) \\ \diagdown \\ O \\ \diagup \\ RC(=NH) \end{array} + NH_3 \qquad (51)$$

$$\begin{array}{c} RC(=O) \\ \diagdown \\ O \\ \diagup \\ RC(=NH) \end{array} \longrightarrow \begin{array}{c} RC(=O) \\ \diagdown \\ NH \\ \diagup \\ RC(=O) \end{array} + RC{\equiv}N + RC(=O)OH \qquad (52)$$

$$\downarrow$$
$$RC{\equiv}N$$

synthetically useful method for the formation of olefins. This elimination reaction is at least in appearance similar to both the Cope pyrolytic elimination of amine oxides, and to the pyrolysis of esters. The reaction is synthetically useful only for amides which undergo reaction at temperatures below 500°. This essentially limits the use of this reaction to amides which are substituted with secondary or tertiary N-alkyl groups[83]. Pyrolysis of compounds of this type has been found to take place at significantly lower temperatures than pyrolysis of N-alkylamides in which the alkyl group is primary, generally requiring temperatures over 600°. These reactions normally afford low yields of a wide variety of products.

The similarity of the pyrolysis of N-alkylamides to the highly stereospecific N-oxide pyrolysis and to the less specific ester pyrolysis has been the subject of several investigations. The products obtained from pyrolysis of N-(1-methylcyclohexyl)acetamide (**91**) were compared to those obtained from the analogous amine oxide **92** and 1-methylcyclohexyl acetate (**93**)[84].

Pyrolysis of the amide (**91**) was found to give a mixture of olefins (**94, 95**) and acetamide as the principal reaction products, with only small amounts of acetonitrile formed. Analysis of the olefin mixture demonstrated that the more thermodynamically stable isomer **94** was predominating, approximately in a 4:1 ratio.

These results compared closely with those obtained on analysis of the olefin mixture which resulted from pyrolysis of the ester (**93**); 76% *endo* isomer (**94**) and 24 per cent *exo* isomer (**95**), thus indicating a

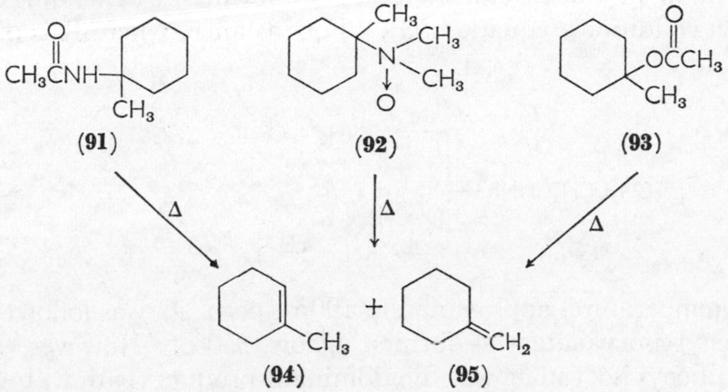

(91) (92) (93)

(94) (95)

probable mechanistic and steric similarity between *N*-alkylamide and ester pyrolysis.

In marked contrast to this similarity, pyrolysis of the amine oxide (**92**) resulted in the formation of 97% of the *exo*cyclic olefin(**95**) and only 3% of *endo*cyclic olefin (**94**)[85].

The results of pyrolysis of **91** appear to be in agreement with an earlier postulation[86] of a six-membered ring transition state (**96**).

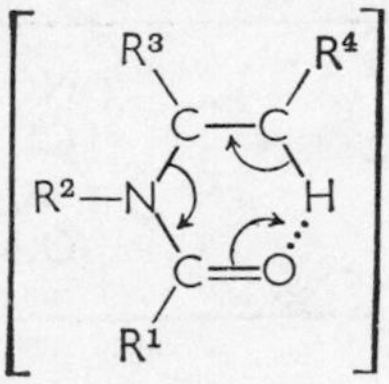

(96)

The pyrolytic reaction of *N*-alkylamides appears to be of little synthetic value because the lack of stereospecificity results in a mixture of products. For example[86], *N*-(1,3-dimethyl)butylacetamide (**97**)

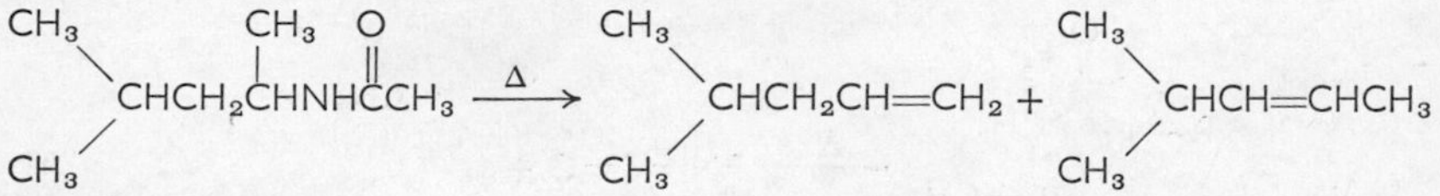

(97) (98) (99)

10*

was pyrolysed at 590°C to give an 18% yield of olefins which consisted of a mixture of 4-methyl-1-pentene (**98**) and both the *cis* and *trans* isomers of 4-methyl-2-pentene (**99**).

A more useful synthetic reaction appears to be the pyrolytic decomposition of acetoacetamides[87]. Two routes are open for decomposition, producing either ketenes or isocyanates, depending on the reaction conditions (equation 53). For instance, where R is phenyl,

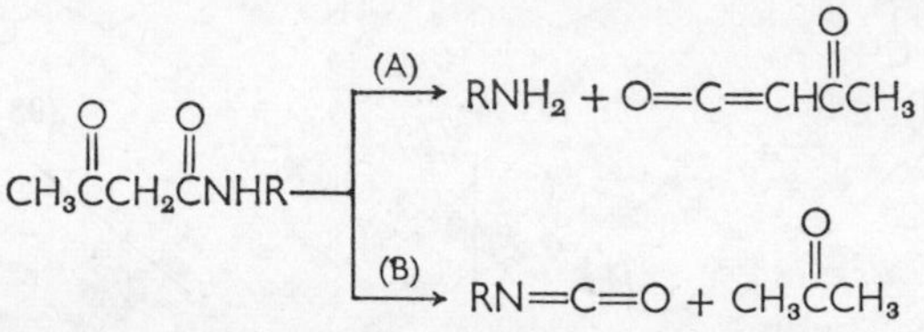

$$
\begin{array}{c}
\underset{\substack{\parallel \\ CH_3CCH_2CNHR}}{O \quad O} \xrightarrow{\begin{array}{c}(A)\end{array}} RNH_2 + O{=}C{=}CHCCH_3 \\[2mm]
\xrightarrow{(B)} RN{=}C{=}O + CH_3CCH_3
\end{array}
\tag{53}
$$

at low temperature, approximately 190°C, path (A) was followed and no phenyl isocyanate was formed below 350°C. However, above 500°C, phenyl isocyanate was the dominant product (path B) together with acetone. When the aromatic ring was substituted with electron-releasing groups, formation of isocyanate was facilitated while electron-withdrawing groups inhibited its formation, as shown in Table 1.

TABLE 1. Substituent effects on isocyanate formation by pyrolysis of substituted acetoacetanilides.

$CH_3\overset{O}{\overset{\parallel}{C}}C_{(2)}H_2\overset{O}{\overset{\parallel}{C}}_{(1)}NH$—⟨◯⟩—X	X	Isocyanate (%)
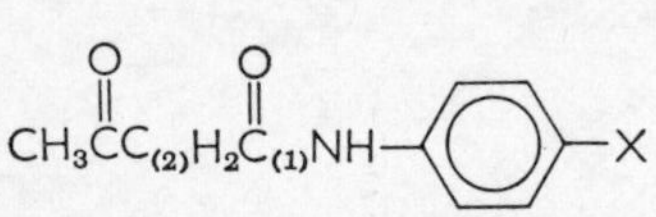	NO_2	0
	CH_3	50
	OCH_3	30
	$COOC_2H_5$	0

The data of Table 1 were interpreted as indicating that the reaction is dependent on the ease of $C_{(2)}$—$C_{(1)}$ bond cleavage to give $CH_3COCH_2^-$ and $RNHCO^+$ fragments, with the anion then abstracting the NH hydrogen of the cation to yield products (equation 54).

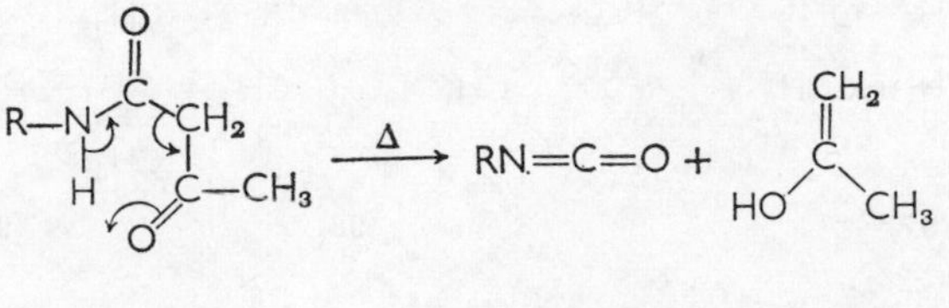

$$
\underset{H}{\overset{\displaystyle R{-}N}{\big|}}\overset{O}{\overset{\parallel}{C}}\Big\langle \begin{array}{c} CH_2 \\ C{-}CH_3 \\ \parallel \\ O \end{array} \xrightarrow{\Delta} RN{=}C{=}O + \underset{HO}{\overset{\displaystyle CH_2}{\overset{\parallel}{\underset{\displaystyle}{C}}}}\underset{CH_3}{}
\tag{54}
$$

Based on these results, the pyrolytic decomposition of trihaloacet-amides was attempted in anticipation of producing chloroform and isocyanate[88]. Instead, decomposition of 2,2,2-trichloroacetanilide at 520°c gave a 61% yield of benzonitrile and very little phenyl isocyanate (equation 55).

$$\underset{\substack{\|\\ \text{O}}}{\text{Cl}_3\text{CCNHC}_6\text{H}_5} \xrightarrow{\Delta} C_6H_5C\equiv N + C_6H_5N{=}C{=}O + Cl_2 + HCl + COCl_2 \tag{55}$$

In this case, it appears that there are two reaction pathways after the initial bond cleavage (equation 56). Path (A) of equation (57),

$$\underset{\substack{\|\\ \text{O}}}{\text{RNHCCCl}_3} \longrightarrow \underset{(100)}{\underset{\substack{\|\\ \text{O}}}{\text{RNHC}^+}} + CCl_3^- \tag{56}$$

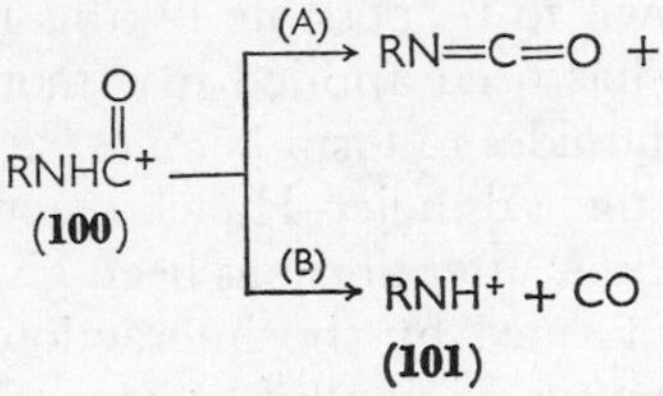

$$\underset{(100)}{\underset{\substack{\|\\ \text{O}}}{\text{RNHC}^+}} \begin{cases} \xrightarrow{(A)} RN{=}C{=}O + H^+ \\ \xrightarrow{(B)} \underset{(101)}{RNH^+} + CO \end{cases} \tag{57}$$

$$RNH^{+\cdot} + CCl_3^- \longrightarrow \underset{(102)}{R\text{—}NH\text{—}CCl_3} \longrightarrow \underset{(103)}{R\text{—}NC} + Cl_2 + HCl$$

the anticipated route, does not occur to any great extent in this case. Therefore, cation **100** must decompose to the ion **101** which then combines with the trichloromethyl anion to yield **102**. Elimination of chlorine and hydrogen chloride would yield an isonitrile (**103**) which has been demonstrated to rearrange under these conditions to the nitrile. Once more, as in the case with acetoacetamides, electron-withdrawing groups substituted on the aromatic ring R greatly decrease the yield of nitrile.

IV. ELIMINATION OF NR₂ GROUPS*

The Vilsmeier–Haack reaction[89] involves the acylation of activated aromatic rings, using a complex formed between a substituted amide and phosphorus oxychloride as the acylating agent. The overall reaction involves the elimination of the NR₂ group from the amide

* See also section IX.A.3 of Chapter 13.

and the attachment of the acyl portion of the molecule to the ring. Typically, dimethylaniline, when treated with $POCl_3$ and dimethylformamide (DMF) undergoes acylation to afford p-dimethylaminobenzaldehyde in good yield[90] (equation 58).

$$\begin{array}{c}\text{H}_3\text{C} \quad \text{CH}_3 \\ \text{N} \\ \bigcirc \\ \end{array} + \text{HCN}\begin{array}{c}\text{O} \\ \|\\ \end{array}\begin{array}{c}\text{CH}_3 \\ \text{CH}_3 \end{array} + POCl_3 \longrightarrow \begin{array}{c}\text{H}_3\text{C} \quad \text{CH}_3 \\ \text{N} \\ \bigcirc \\ \text{CH}{=}\text{O} \end{array} \qquad (58)$$

The reaction is not limited to the formylation of aromatic rings and has been applied to a wide variety of hydrocarbons[91]; oxygen-[92], nitrogen-[93], and sulphur-containing[94] heterocyclic rings; phenols[95] and steroids[96]. More recently, the formylation of unsaturated hydrocarbons has been shown to be possible by this reaction[97,98]. Additionally, the reaction has been applied in a more limited number of cases with higher acylamides to form ketones[93].

The mechanism of the Vilsmeier–Haack reaction has been studied extensively and particular attention has been given to the structure of the reactive complex formed by the interaction between DMF and $POCl_3$. This complex was originally[89] proposed to have a completely covalent structure. However, subsequent investigations have clearly shown the complex to be ionic and the structures 104 and 105 are most frequently postulated.

$$\left[\begin{array}{c}\text{H}_3\text{C} \quad\quad \text{OPOCl}_2 \\ \text{N}{\cdots}\text{C} \\ \text{H}_3\text{C} \quad\overset{+}{}\quad \text{H}\end{array}\right]\text{Cl}^- \qquad \left[\begin{array}{c}\text{H}_3\text{C} \quad\quad \text{Cl} \\ \text{N}{\cdots}\text{C} \\ \text{H}_3\text{C} \quad\overset{+}{}\quad \text{H}\end{array}\right]\text{PO}_2\text{Cl}_2^-$$

$$\quad\quad (104) \quad\quad\quad\quad\quad\quad (105)$$

Structure 104 has been proposed by a number of authors[93,99,100] and the infrared absorption spectrum of the complex has been interpreted as supporting this structure[101]. However, the more recently acquired evidence mentioned below tends to support structure 105 which was originally proposed by Lorenz and Wizinger[102].

Arnold studied the complex which results from the action of phosgene on DMF (equation 59)[105]. He observed that CO_2 was liberated in this reaction and a complex having structure 106 was formed.

By analogy, structure 105 was proposed for the DMF–$POCl_3$ complex. More recently, a proton magnetic resonance study of the DMF–

$POCl_3$ complex was conducted by Martin and Martin[103]. This investigation failed to disclose the presence of the substantial P—H coupling which should be observed for the H—C—O—P bond in **104**[104]. Additionally, comparison of the p.m.r. spectrum obtained

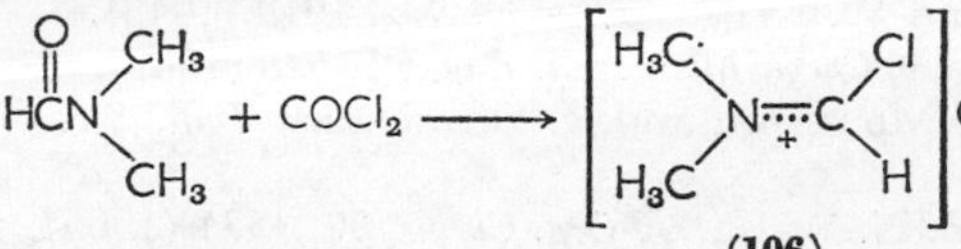

$$HCN\begin{matrix}CH_3\\\\CH_3\end{matrix} + COCl_2 \longrightarrow \left[\begin{matrix}H_3C\\\\H_3C\end{matrix}N\overset{+}{\cdots}C\begin{matrix}Cl\\\\H\end{matrix}\right]Cl^- + CO_2 \qquad (59)$$
$$(106)$$

from the DMF–$POCl_3$ complex with the unambiguous structure **106**, established their similarity and further supported the view that structure **105** best represents this complex.

In a typical reaction, the complex is formed and then the compound to be formylated is added in a separate step. Electrophilic attack generally occurs only at activated ring positions and frequently results in the formation of an isolatable intermediate which, upon alkaline hydrolysis, can be converted to the corresponding aldehyde. The intermediate species formed in the formylation of indole was isolated as the perchlorate salt, investigated spectroscopically and assigned structure **107**[99]. Hydrolysis of this salt afforded 3-formylindole (**108**) in good yield.

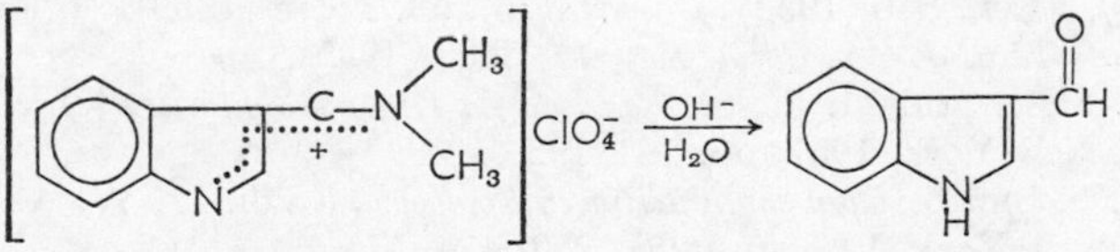

(107)   (108)

Similarly, Jutz and Muller[98] treated *dl*-camphene with a DMF–$POCl_3$ complex and obtained the cation **109** which, on alkaline hydrolysis, gave the unsaturated aldehyde **110**.

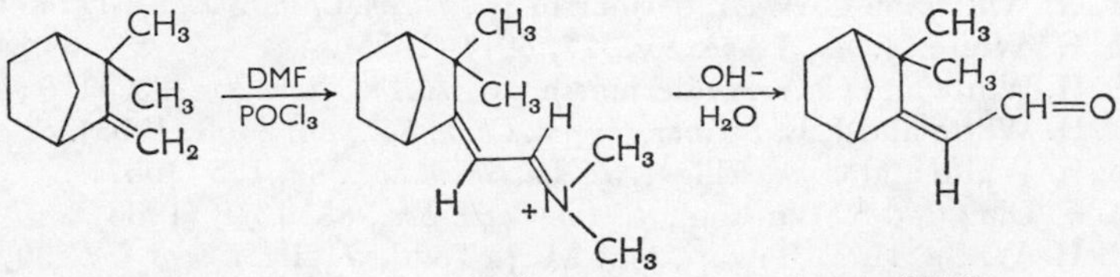

(109)   (110)

V. REFERENCES

1. E. S. Wallis and J. L. Lane, *Organic Reactions*, Vol. III, 1946, Chap. 7; P. A. S. Smith, in *Molecular Rearrangements* (Ed. P. DeMayo), Interscience Div., John Wiley and Sons, New York, 1963, Chap. 8.
2. S. Gabriel and R. Stelzner, *Chem. Ber.*, **28**, 2929 (1895).
3. A. A. Goldberg and W. Kelley, *J. Chem. Soc.*, 1919 (1948).
4. H. W. Heine, *Angew. Chem. Intern. Ed. Engl.*, **1**, 528 (1962).
5. F. Winternitz, M. Mousseron and R. Dennilauler, *Bull. Soc. Chim. France*, 382 (1956).
6. P. E. Fanta and E. N. Walsh, *J. Org. Chem.*, **30**, 3574 (1965).
7. P. E. Fanta and E. N. Walsh, *J. Org. Chem.*, **31**, 59 (1966).
8. H. W. Heine and M. S. Kaplan, *J. Org. Chem.*, **32**, 3069 (1967).
9. P. E. Fanta and A. S. Deutsch, *J. Org. Chem.*, **23**, 72 (1958).
10. D. V. Kashelikar and P. E. Fanta, *J. Am. Chem. Soc.*, **82**, 4927 (1960).
11. P. E. Fanta, L. J. Pandya, W. R. Groskopf, and H. J. Su, *J. Org. Chem.*, **28**, 413 (1963).
12. P. E. Fanta and M. Kathan, *J. Heterocyclic Chem.*, **1**, 293 (1964).
13. D. V. Kashelikar and P. E. Fanta, *J. Am. Chem. Soc.*, **82**, 4930 (1960).
14. H. W. Heine, M. E. Fetter, and E. M. Nicholson, *J. Am. Chem. Soc.*, **81**, 2202 (1959).
15. H. W. Heine, D. C. King, and L. A. Portland, *J. Org. Chem.*, **31**, 2662 (1966).
16. H. W. Heine and Z. Proctor, *J. Org. Chem.*, **23**, 1554 (1958).
17. B. I. Ardasher and V. I. Minkin, *Zh. Obshch. Khim.*, **28**, 1578 (1958); *Chem. Abstr.*, **53**, 1348c (1959).
18. B. I. Ardasher and V. I. Minkin, *Zh. Obshch. Khim.*, **27**, 1261 (1957); *Chem. Abstr.*, **52**, 2856c (1958).
19. B. I. Ardasher, V. I. Minkin, and M. B. Minkin, *Nauchn. Dokl. Vysshei Shkoly. Khim.*, 526 (1958); *Chem. Abstr.*, **53**, 3227e (1959).
20. E. H. White, *J. Am. Chem. Soc.*, **77**, 6008 (1955).
21. L. F. Fieser and M. Fieser, *Reagents for Organic Synthesis*, John Wiley and Sons, New York, 1967, pp. 191–193.
22. P. A. S. Smith, *Open-chain Nitrogen Compounds*, Vol. II, W. A. Benjamin, Inc., New York, 1966, pp. 257, 258, 474, 475.
23. D. F. DeTar, *J. Am. Chem. Soc.*, **73**, 1446 (1951).
24. D. F. DeTar and H. J. Scheifele, Jr., *J. Am. Chem. Soc.*, **73**, 1442 (1951).
25. R. Huisgen, *Angew Chem.*, **62**, 369 (1950).
26. R. Huisgen and L. Krause, *Ann. Chem.*, **574**, 157 (1951).
27. R. Huisgen, *Ann. Chem.*, **574**, 171 (1951).
28. E. H. White, *J. Am. Chem. Soc.*, **77**, 6011 (1955).
29. A. Streitwieser, Jr. and W. D. Schaeffer, *J. Am. Chem. Soc.*, **79**, 2893 (1957).
30. E. H. White and C. A. Aufdermarsh, Jr., *J. Am. Chem. Soc.*, **83**, 1174 (1961).
31. E. H. White, *J. Am. Chem. Soc.*, **77**, 6014 (1955).
32. E. H. White and C. A. Aufdermarsh, Jr., *J. Am. Chem. Soc.*, **83**, 1179 (1961).
33. E. H. White and J. E. Stuber, *J. Am. Chem. Soc.*, **85**, 2168 (1963).
34. E. H. White and C. A. Elliger, *J. Am. Chem. Soc.*, **89**, 165 (1967).
35. A. F. Diaz and S. Winstein, *J. Am. Chem. Soc.*, **88**, 1318 (1966).
36. E. H. White, H. P. Tiwari, and M. J. Todd, *J. Am. Chem. Soc.*, **90**, 4734 (1968).

37. M. Murakami, K. Akagi, and K. Takahoshi, *J. Am. Chem. Soc.*, **83**, 2002 (1961).
38. E. H. White and D. W. Grisley, Jr., *J. Am. Chem. Soc.*, **83**, 1191 (1961).
39. K. N. Ayad, C. C. Beard, R. F. Garwood, and W. J. Hickinbottom, *J. Chem. Soc.*, 2981 (1957).
40. C. C. Beard, J. R. B. Boocock, and W. J. Hickinbottom, *J. Chem. Soc.*, 520 (1960).
41. E. D. Hughes and C. K. Ingold, *Quart. Rev. (London)*, **6**, 34 (1952).
42. K. J. P. Orton and W. J. Jones, *Proc. Chem. Soc. (London)*, **25**, 196, 233, 305 (1909).
43. K. J. P. Orton, F. G. Soper, and G. Williams, *J. Chem. Soc.*, 998 (1928).
44. A. R. Olson, C. W. Porter, F. A. L. Bong, and R. S. Halford, *J. Am. Chem. Soc.*, **58**, 2467 (1936).
45. G. C. Israel, A. W. N. Tuck, and F. G. Soper, *J. Chem. Soc.*, 547 (1945).
46. R. P. Bell, *J. Chem. Soc.*, 1154 (1936).
47. M. J. S. Dewar, *Electronic Theory of Organic Chemistry*, Oxford University Press, Oxford, 1949, p. 168.
48. M. J. S. Dewar, *Theoretical Organic Chemistry*, Kekulé Symposium, Butterworths Scientific Publications, 1958.
49. D. H. R. Barton, A. L. J. Beckwith, and A. Goosen, *J. Chem. Soc.*, 181, (1965).
50. A. L. J. Beckwith and J. E. Goodrich, *Australian J. Chem.*, **18**, 747 (1965).
51. R. S. Neale, N. L. Marcus, and R. G. Schepers, *J. Am. Chem. Soc.*, **88**, 3051 (1966).
52. M. E. Wolff, *Chem. Rev.*, **63**, 55 (1963).
53. I. Lengyel and J. C. Sheehan, *Angew. Chem. Intern. Ed. Engl.*, **7**, 25 (1968).
54. H. E. Baumgarten, *J. Am. Chem. Soc.*, **84**, 4975 (1962).
55. H. E. Baumgarten, J. J. Fuerholzer, R. D. Clark, and R. D. Thompson, *J. Am. Chem. Soc.*, **85**, 3303 (1963).
56. S. Sarel, F. D'Angeli, J. T. Klug, and A. Taube, *Israel J. Chem.*, **2**, 167 (1964).
57. C. L. Stevens, H. Dittmer, and J. Kovacs, *J. Am. Chem. Soc.*, **85**, 3394 (1963).
58. C. L. Stevens, M. E. Munk, A. B. Ash, and R. D. Ellicott, *J. Am. Chem. Soc.*, **85**, 3390 (1963).
59. S. K. Brownstein, *J. Org. Chem.*, **23**, 113 (1958).
60. W. P. Judd and B. E. Swedlund, *Chem. Commun.*, 43 (1966).
61. F. Wohler and J. Leibig, *Ann. Chem.*, **3**, 249 (1832).
62. D. T. Mowrey, *Chem. Rev.*, **42**, 189 (1948).
63. R. Wagner and H. Zook, *Synthetic Organic Chemistry*, John Wiley and Sons, New York, 1953, pp. 596–8.
64. C. R. Stevens, E. J. Bianco, and F. J. Pilgrim, *J. Am. Chem. Soc.*, **77**, 1701 (1955).
65. Q. E. Thompson, *J. Am. Chem. Soc.*, **73**, 5841 (1951).
66. R. Calas, E. Frainnet, and A. Bazouin, *Compt. Rend.*, **254**, 2357 (1962).
67. J. C. Thurman, *Chem. Ind. (London)*, 752 (1964).
68. M. S. Newman and T. Fukunaga, *J. Am. Chem. Soc.*, **82**, 693 (1960).
69. E. M. Kaiser, R. L. Vaulx, and C. R. Hauser, *J. Org. Chem.*, **32**, 3640 (1967).

70. E. M. Kaiser and C. R. Hauser, *J. Org. Chem.*, **31**, 3873 (1966).
71. D. J. Cram and P. Haberfield, *J. Am. Chem. Soc.*, **83**, 2354 (1961).
72. J. Kenyon and W. A. Ross, *J. Chem. Soc.*, 3407 (1951).
73. D. J. Cram and P. Haberfield, *J. Am. Chem. Soc.*, **83**, 2363 (1961).
74. B. Rickborn and F. R. Jensen, *J. Org. Chem.*, **27**, 4608 (1962).
75. H. von Pechmann, *Chem. Ber.*, **33**, 611 (1900).
76. P. Kurtz, *Chem. Ber.*, **99**, June 1966.
77. N. J. Leonard and E. W. Nommensen, *J. Am. Chem. Soc.*, **71**, 2808 (1949).
78. J. von Braun, *Chem. Ber.*, **37**, 3210 (1904).
79. W. R. Vaughan and R. D. Carlson, *J. Am. Chem. Soc.*, **84**, 769 (1962).
80. A. C. Cope and E. R. Trumbull, *Organic Reactions*, Vol. XI, 1960, Chap. 5, p. 371 ff.
81. R. B. Wagner and H. D. Zook, *Synthetic Organic Chemistry*, John Wiley and Sons, New York, 1953.
82. D. Davidson and M. Karten, *J. Am. Chem. Soc.*, **78**, 1066 (1956).
83. H. E. Baumgarten, F. A. Bower, R. A. Setterquist, and R. E. Allen, *J. Am. Chem. Soc.*, **80**, 4588 (1958).
84. W. J. Bailey and W. F. Hale, *J. Am. Chem. Soc.*, **81**, 651 (1959).
85. A. C. Cope, C. L. Baumgardner, and E. E. Schweizer, *J. Am. Chem. Soc.*, **79**, 4729 (1957).
86. W. J. Bailey and C. N. Bird, *J. Org. Chem.*, **23**, 996 (1958).
87. T. Mukaiyama, M. Tokizaua, H. Nohira, and H. Takei, *J. Org. Chem.*, **26**, 4381 (1961).
88. T. Mukaiyama, M. Tokizaua, and H. Takei, *J. Org. Chem.*, **27**, 803 (1962).
89. A. Vilsmeier and A. Haack, *Chem. Ber.*, **60**, 119 (1927).
90. H. Bassard and H. Zollinger, *Helv. Chim. Acta*, **42**, 1659 (1959).
91. L. F. Fieser, J. L. Hartwell, and E. Jones, *J. Am. Chem. Soc.*, **60**, 2542 (1938).
92. R. Royer, E. Bisagni, A. M. L. Jeantet, and J. P. Marquet, *Bull. Soc. Chim. France*, 2607 (1965).
93. W. C. Anthony, *J. Org. Chem.*, **25**, 2049 (1960).
94. E. Campaigne and W. L. Archer, *J. Am. Chem. Soc.*, **75**, 989 (1953).
95. J. H. Wood and R. W. Bost, *J. Am. Chem. Soc.*, **59**, 1721 (1937).
96. R. Sciaky, V. Pallini, and A. Consonni, *Gazz. Chim. Ital.*, **96**, 1284 (1966).
97. C. Jutz, W. Muller, and E. Muller, *Chem. Ber.*, **99**, 2479 (1966).
98. C. Jutz and W. Muller, *Chem. Ber.*, **100**, 1536 (1967).
99. G. F. Smith, *J. Chem. Soc.*, 842 (1954).
100. C. Jutz, *Chem. Ber.*, **91**, 850 (1958).
101. H. Bredereck, R. Gumpper, K. Klemm, and H. Rem, *Chem. Ber.*, **92**, 1456 (1959).
102. H. Lorenz and R. Wizinger, *Helv. Chim. Acta*, **28**, 600 (1945).
103. G. Martin and M. Martin, *Bull. Soc. Chim., France*, 1637 (1963).
104. G. O. Dudek, *J. Chem. Phys.*, **33**, 624 (1960).
105. Z. Arnold, *Collection Czech. Chem. Commun.*, **24**, 4048 (1959).

CHAPTER **5**

Photochemistry of the amido group

IONEL ROSENTHAL

Department of Chemistry, The Weizmann Institute of Science, Rehovoth, Israel

I. INTRODUCTION

This chapter is concerned with the photochemistry of simple amides. Photochemical reactions of this class of compounds have not been extensively investigated and the aim of the present review is to assemble and correlate the results that have been reported.

A knowledge of the behaviour of simple amides towards light irradiation is a necessary first step in understanding the photochemical reactions undergone by more complicated systems, such as those derived from amino acids or purine and pyrimidine bases in a great variety of biologically important substances. The photochemical transformations undergone by proteins and nucleic acids have been extensively reviewed previously[1].

Further impetus in the photochemical study of simple amides has been provided by the interest shown in the effects of ultraviolet and visible light on polyamide plastics, for such reactions are of practical significance.

Although the light-absorption process in amides is centred exclusively around the carboxamido group, subsequent chemical transformations do not necessarily take place in this part of the molecule and carbon–carbon and carbon–hydrogen bond cleavage may frequently occur at adjacent atoms.

In particular the reactivity of the carbon–hydrogen bonds adjoining the amide chromophore has been reported. Such bond cleavage gives rise to free radicals which subsequently undergo further chemical reactions.

Some few examples of the photochemical reactions of amides, induced by the presence of photoinitiators, such as ketones, have also been reported. These reactions were successfully used for synthetic purposes.

Although much work has been conducted on model systems for complex biochemically significant compounds, there are relatively few publications concerned with the photochemical reactions of amides themselves, and there exists an obvious need for further investigations in this promising field.

II. LIGHT ABSORPTION OF AMIDES

The main absorption bands of saturated amides, lactams and imides lie below 200 nm (1 nm = 10 Å), the end absorption going as high as 260 nm[2–19]. The maxima are reported to be around 170–190 nm ($\epsilon \sim 10,000$) and 130–160 nm. These high-intensity bands are attributed to $N \rightarrow V_1$, Rydberg ($2p$, $3s$), and $N \rightarrow V_2$ electronic transitions[14,16]. The location and intensities of these maxima depend on the substituents attached to the carboxamido group. Thus, an absorption band which appears in formamide at 172 nm appears in dimethylformamide at 197 nm[14], and in N-methylsuccinimide at 204 nm[17]. In general, the presence of alkyl substituents at C_α in both amides and imides results in a progressive decrease in intensity of the 191 nm band[12] (ca. 175 nm in the amides). Further the effect of an N-methyl group on the ultraviolet spectra of amides is that the maximum is shifted to a longer wavelength and the absorption is intensified[17]. This has been explained as deriving from the electron-repelling effect of the methyl group which facilitates the electronic transitions. Absorption bands of amides above 200 nm have also been reported. It is claimed that the long-wavelength tail of the $N \rightarrow V_1$ absorption may be interpreted as containing a separate electronic transition near 210 nm with ϵ about 100 and designated as

$n \rightarrow \pi$ transition[6,10,15,16]. A long-wavelength absorption band in amides at 280 nm has also been reported[20,21], whose origin however is under dispute. This absorption band has been regarded as due to dissociated amide molecules, while an alternative suggestion[22], assigns it to the $RC(=O)N^+HR^1R^2$ structure, or to impurities.

Recently it has been reported[71,72] that α-lactams possess a distinct absorption in the ultraviolet $[\lambda_{max}^{n\text{-hexane}}$ 250 nm, $\epsilon \sim 10^2$ 1/mol cm] which exhibits a hypsochromic shift on changing the solvent from n-hexane to ethanol, a characteristic of $n \rightarrow \pi^*$ excitations.

III. PHOTOCHEMICAL REACTIONS OF AMIDES

A. Photolysis of Amides and Lactams

The photolysis of amides and lactams in gas phase and in solution has been reported by a number of groups[23-28]. Light of wavelengths $\lambda < 250$ nm was usually employed. The gaseous products in the photolysis of acetamide in the vapour phase[23] were CO, C_2H_6, CH_4, NH_3, CH_3CN and H_2O, together with traces of CO_2, N_2 and H_2. The experimental results obtained may be explained by Scheme 1

$$CH_3CONH_2 \xrightarrow{h\nu} \dot{C}H_3 + \dot{C}ONH_2$$
$$CH_3CONH_2 \xrightarrow{h\nu} CH_3CN + H_2O$$
Primary processes

$$\dot{C}H_3 + CH_3CONH_2 \longrightarrow CH_4 + \dot{C}H_2CONH_2$$

$$\dot{C}H_3 + \dot{C}ONH_2 \longrightarrow CH_4 + HNCO$$

$$\dot{C}H_3 + \dot{C}H_3 \longrightarrow C_2H_6$$

$$\dot{C}ONH_2 \longrightarrow CO + \dot{N}H_2$$

$$\dot{R} + \dot{C}ONH_2 \longrightarrow RCONH_2$$

$$H_2\dot{N} + CH_3CONH_2 \longrightarrow NH_3 + \dot{C}H_2CONH_2$$

$$\dot{N}H_2 + \dot{R}^1 \longrightarrow R^1NH_2$$

SCHEME 1.

where R and R^1 are radicals, not necessarily different, which are present in the system. Photolysis of acetamide in aqueous solution has been reported[25] to yield among other products, acetic acid (equation 1).

$$CH_3CONH_2 \xrightarrow[H_2O]{h\nu} CH_3COOH + NH_3 + CO_2 + CO + CH_4 + N_2 \qquad (1)$$

Irradiation of aliphatic amides in organic solvents such as dioxan or hexane[24] led to carbon monoxide, hydrogen, amines and unsaturated hydrocarbons. Two different modes of decomposition can be advanced to account for these products (Scheme 2).

$$CH_3CH_2CH_2CONH_2 \xrightarrow{h\nu} CH_3CH_2CH_2NH_2 + CO \qquad \text{Type I}$$

$$CH_3CH_2CH_2CONH_2 \xrightarrow{h\nu} CH_2{=}CH_2 + CH_3CONH_2 \quad \text{Type II}$$

$$CH_3CH_2CH_2NH_2 \xrightarrow{h\nu} CH_3CH_2CH_2\dot{N}H + \dot{H} \quad \text{etc.}$$

$$CH_3CONH_2 \xrightarrow{h\nu} CH_3NH_2 + CO$$

$$CH_3NH_2 \xrightarrow{h\nu} CH_3\dot{N}H + \dot{H}$$

SCHEME 2.

Both carbon–carbon and carbon–nitrogen bonds are cleaved in these reactions. It is proposed that either the C=O group or the $CONH_2$ group as a whole, rather than the isolated NH_2 group, controls the decomposition of amides[24]. The u.v. degradation products of N-alkylamides in the absence of oxygen[26] were CO, H_2O, hydrocarbons, carboxylic acid and N-alkylamides with chain lengths differing from that of the parent compound; in addition there was some evidence for the formation of primary amines. A reaction mechanism involving homolytic dissociation of both NH—CH_2 and NH—CO bonds can explain the products in a way similar to the previous cases. The results of photolysis of gelatin[24], show that the photochemical degradation of proteins occurs at least partially in a manner similar to amide photolysis.

Recently, it has been shown by e.s.r. spectroscopy[73], that the primary free-radical producing step in the photolysis of formamide is the formation of $\dot{H}$ and $\dot{C}ONH_2$, whereas the photolysis of acetamide produces $\dot{C}H_3$ and $\dot{C}ONH_2$. Substitution of a methyl group on the nitrogen atom in these compounds does not change the primary step (C—H or C—C bond scission for N-methylformamide and N-methylacetamide respectively).

The photoinduced decomposition of α-lactams has been reported[71,72] to yield carbon monoxide and the corresponding imine (equation 2). This photochemical reaction is in contrast with the

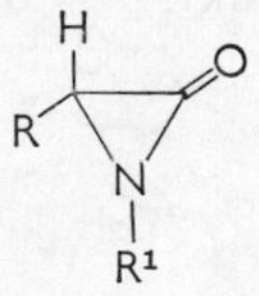

$$\underset{\substack{| \\ R^1}}{\overset{\text{H}}{\underset{N}{\overset{O}{R}}}} \xrightarrow[\lambda > 200 \text{ nm}]{h\nu, \text{ pentane}} RCH{=}NR^1 + CO \qquad (2)$$

thermal decomposition of these compounds[74] which gives a carbonyl compound and an isocyanide (equation 3).

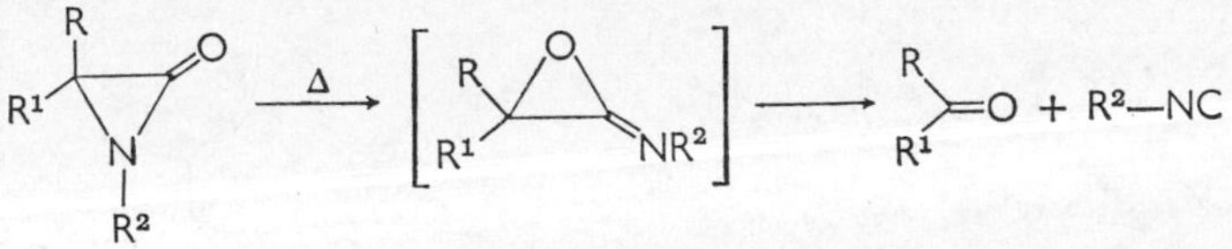

$$ (3) $$

It has been shown[78] that the β-lactams may react in a similar way.

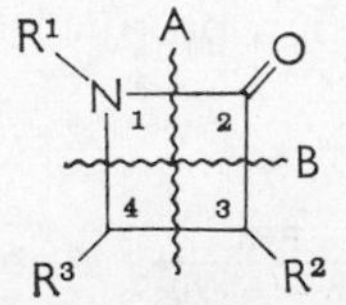

Thus a β-lactam which is not substituted in position 4 undergoes photolysis according to pattern A like α-lactams. The presence of a substituent in that position, leads to fragmentation according to both patterns A and B; the contribution of the splitting according to pattern B increases with electron-attracting character of the R^3 substituent.

The splitting of the CO—NH linkage has also been reported in the case of amides containing supplementary chromophoric groups where the light is absorbed by the latter. For instance, it was shown[34] that stearic anilide, $C_{17}H_{35}CONHC_6H_5$, when irradiated as a monolayer with light of wavelength 235–240 and 248 nm was decomposed into stearic acid and aniline. In this experiment the benzene ring functioned as the light-absorbing group. Moreover, N-benzylstearamide and N-(β-phenylethyl)stearamide which contain respectively one and two CH_2 groups, interposed between the chromophore and the keto–imino linkage, undergo photolysis in a similar way[36,37]. From these results it was hoped[36,37] that ultraviolet light might prove to be a useful tool for selective cleavage of protein molecule at CONH groups adjacent to the side-chain-bearing chromophoric groups.

Photolysis of sulphanilamide in water led to the liberation of unidentified acids and ammonia[35].

Several instances in which reaction occurred at centres adjacent to a carboxamido group following absorption of ultraviolet light by such a group are known. Thus the photolysis of N,N-dimethacrylyl-methacrylamide in ether solution has been reported to give a photo-isomer in 61% yield[29] (reaction 4).

The photolysis of maleimide in an aromatic hydrocarbon solution,

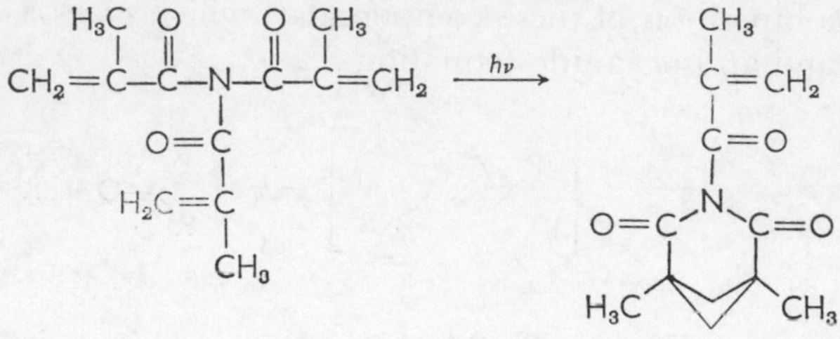

$$\text{(4)}$$

in the presence of a triplet sensitizer like acetophenone, has been reported to lead to an addition product[32] (reaction 5).

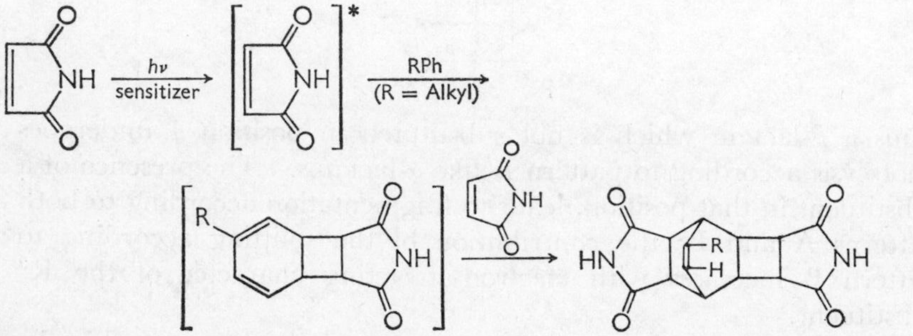

$$\text{(5)}$$

In these cases reaction occurred at a double bond conjugated with the carbonyl position of the carboxamido group. Such types of reactions are of great interest, since such systems are chemically similar to the biologically important pyrimidine bases[1].

Irradiation of α,β-unsaturated amides in degassed solution gave, in each case, the *cis*- and *trans*-β-lactams in the yields indicated, together

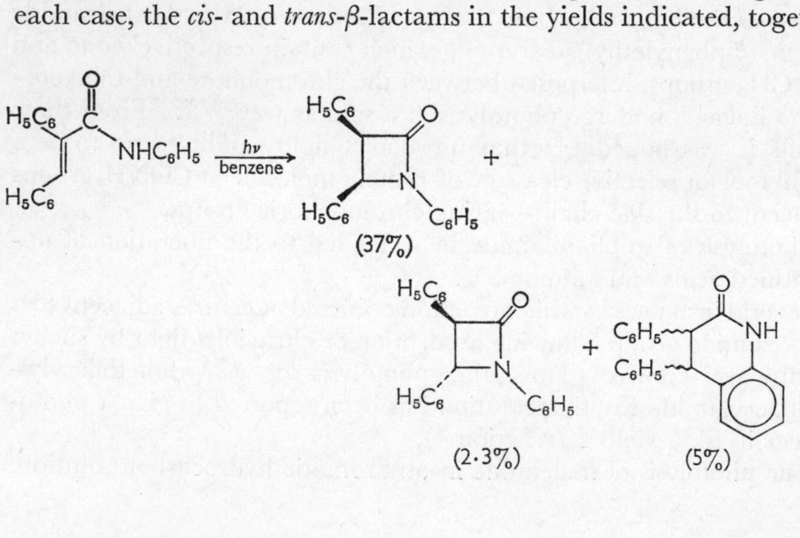

(37%)

(2·3%)   (5%)

$$\text{(6)}$$

with other products[30,31]. Thus, irradiation of *cis*-α-phenylcinnamanilide in benzene gave *trans*-1,3,4-triphenyl-2-azetidinone (2·3%), *cis*-1,3,4-triphenyl-2-azetidinone (37%) and 3,4-diphenyl-3,4-dihydrocarbostyril (5%), according to equation (6). Irradiation, of *cis*-α-phenylcinnamamide in degassed benzene for 70 hr gave a complex mixture from which it proved possible to isolate *trans*-stilbene (2·5%), *cis*-3,4-diphenyl-2-azetidinone (13%), *trans*-3,4-diphenyl-2-azetidinone (3%) and an unidentified product (reaction 7).

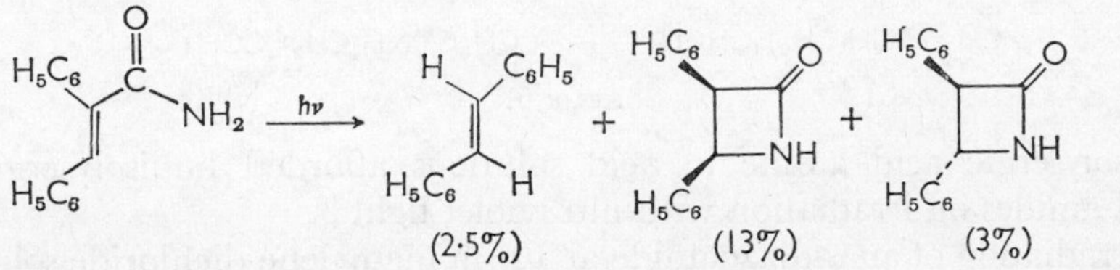

$$(7)$$

The photolysis of nitroso amides has been reported[33,75,76]; thus, *N*-methyl-*N*-nitrosoacetamide irradiated in methanol solution yielded *N*-methylacetamide, nitrous oxide and formaldehyde. In isopropyl alcohol, the products were *N*-methylacetamide, nitrous oxide and acetone. The mechanism described in Scheme 3 was proposed.

$$CH_3CON(N{=}O)CH_3 \xrightarrow{h\nu} NO + CH_3CO\overset{\cdot}{N}CH_3$$

$$CH_3CO\overset{\cdot}{N}CH_3 + CH_3OH \longrightarrow CH_3CONHCH_3 + \overset{\cdot}{C}H_2OH$$

$$\overset{\cdot}{C}H_2OH + NO \longrightarrow CH_2O + HNO$$

$$2HNO \longrightarrow H_2O + N_2O$$

SCHEME 3.

When cyclohexene was used as solvent, the products were *N*-methylacetamide and cyclohexenone oxime, according to Scheme 4.

The same reaction has been carried out with higher *N*-alkyl-*N*-nitrosoacetamides when the abstraction of a hydrogen atom by an

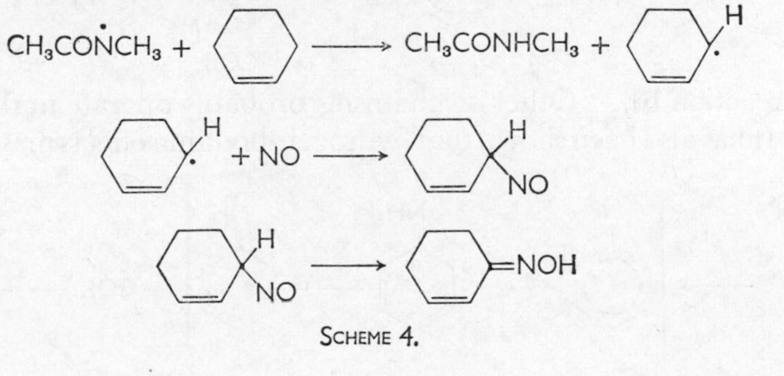

SCHEME 4.

amide radical can also take place intramolecularly, as exemplified by N-nitroso-N-pentylacetamide (Scheme 5).

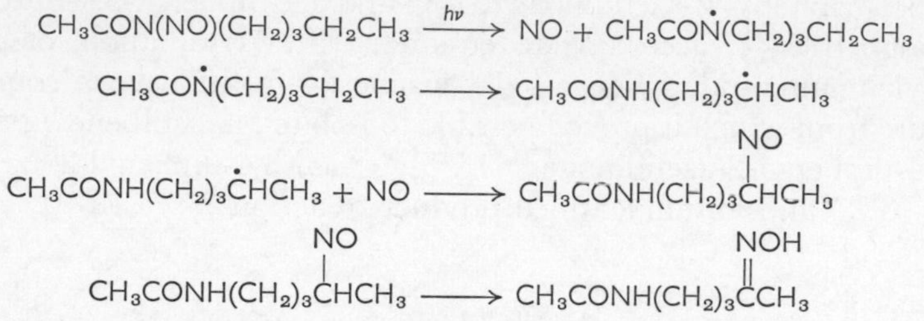

$$CH_3CON(NO)(CH_2)_3CH_2CH_3 \xrightarrow{h\nu} NO + CH_3C\overset{\centerdot}{O}N(CH_2)_3CH_2CH_3$$

$$CH_3C\overset{\centerdot}{O}N(CH_2)_3CH_2CH_3 \longrightarrow CH_3CONH(CH_2)_3\overset{\centerdot}{C}HCH_3$$

$$CH_3CONH(CH_2)_3\overset{\centerdot}{C}HCH_3 + NO \longrightarrow CH_3CONH(CH_2)_3\overset{\overset{\displaystyle NO}{|}}{C}HCH_3$$

$$CH_3CONH(CH_2)_3\overset{\overset{\displaystyle NO}{|}}{C}HCH_3 \longrightarrow CH_3CONH(CH_2)_3\overset{\overset{\displaystyle NOH}{\|}}{C}CH_3$$

Scheme 5.

Isolysergic acid amide in acid solutions afforded lumisolysergic acid amides on irradiation with ultraviolet light[38].

Irradiation of an α-diazoamide at 10° in methylene dichloride solution has been reported to yield a β-lactam[39]. For example reaction (8) where the final product is methyl 6-phenylpenicillanate, such reaction appearing to offer an approach to penicillin synthesis.

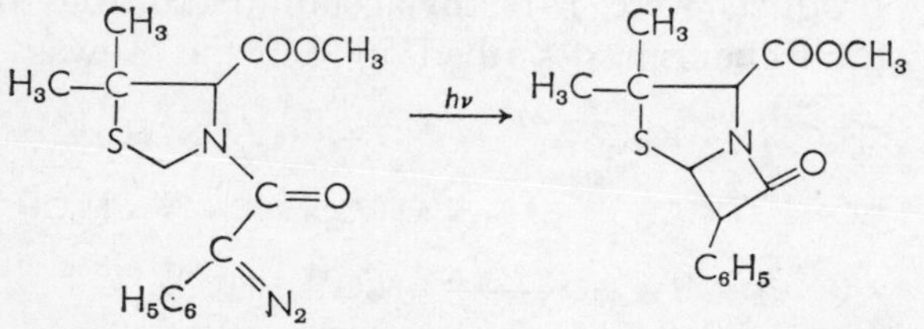

$$(8)$$

The photolysis of acetanilide and benzanilide led to a mixture of aminoacetophenones and -benzophenones, respectively[40,41] (reaction 9). A free-radical mechanism has been proposed for this rearrange-

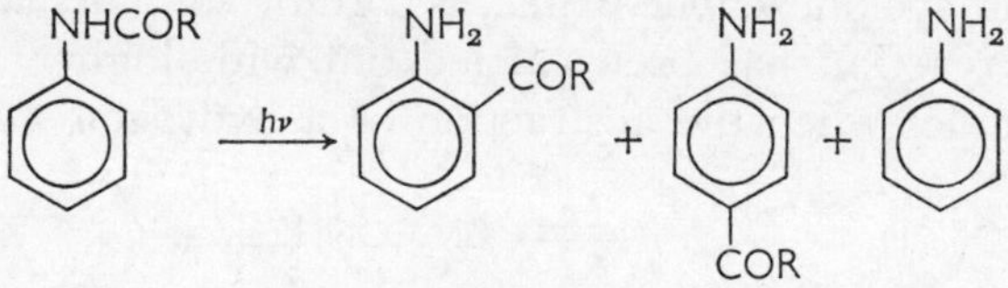

$$(9)$$

ment (reaction 10). Other mechanisms probably operate in this reaction. It has also been shown that hydrogen-bonding effects may govern

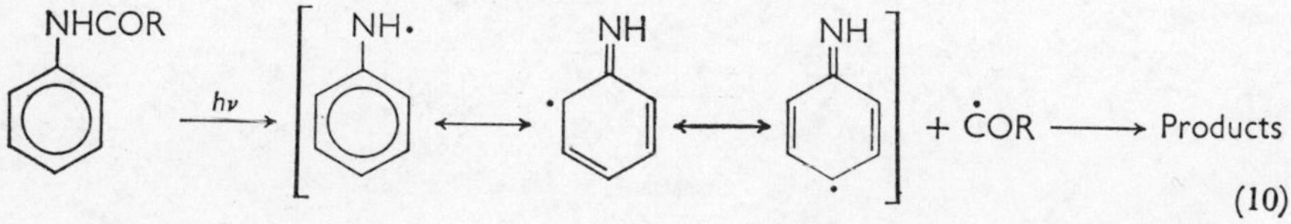

$$(10)$$

this rearrangement[42]. When salicylanilide was irradiated in methanol, the rearrangement took place only to a very small extent (4%
conversion) to yield 2-amino-2'-hydroxybenzophenone and 4-amino-2'
hydroxybenzophenone (reaction 11). The inactivity of salicylanilide

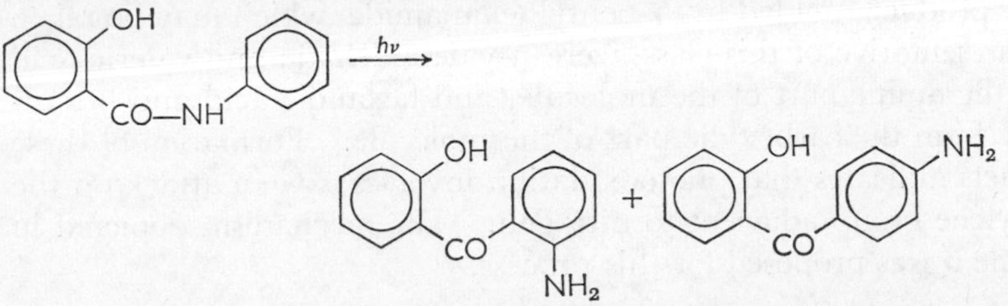

was suggested as being due to a fast decay of the excited state in the
form of a tautomeric shift via hydrogen bonding, involving a six-
membered ring (reaction 12). The O-methyl ether of salicylanilide

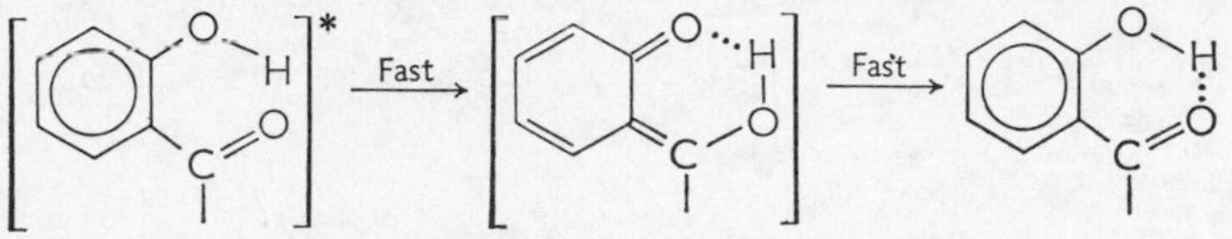

in which the intramolecular hydrogen bond to the carbonyl group is
absent, was converted to products in up to 24% yield by irradiation,
but the yields of defined products were very low, probably owing to
side-reactions of the intermediates, leading to tarry material.

The photoanilide rearrangement has also been extended to N-
aryllactams[77] (equation 13).

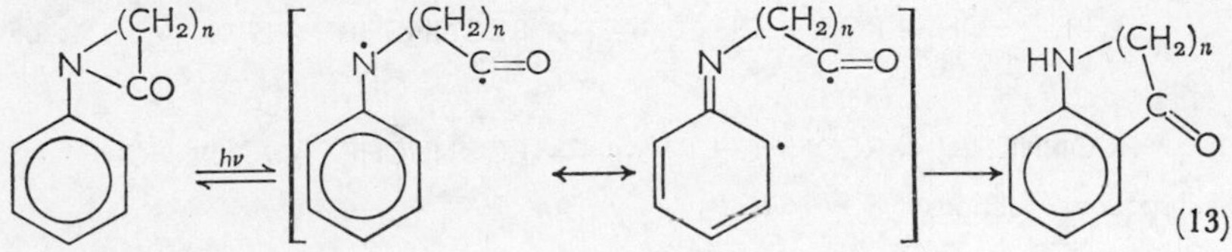

$(n = 5, 6, 11)$

Ultraviolet irradiation of benzanilide, the anilide of o-iodobenzoic
acid, or the benzoyl derivative of o-iodoaniline led to phenanthridone
in various yields[43] (reaction 14).

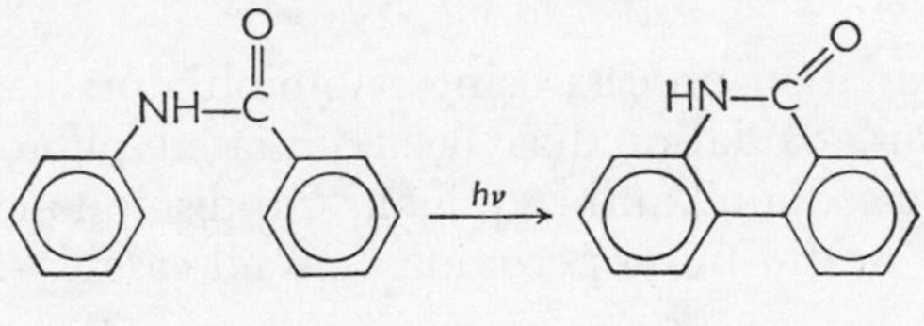

B. Photooxidation of Amides

There are a number of reports in the literature on the photooxidation of amides. The photooxidation of N-alkylamides has been reported to yield aldehydes, acids, and amides[44]. For example, the major products yielded by N-pentylhexanamide, which may be taken as representative of this class, were n-valeraldehyde and valeric acid from the amine part of the molecule, and hexanoic acid and hexanamide from the carboxylic part of the molecule. Formation of these products indicates that photooxidation involves oxygen attack on the methylene group adjacent to nitrogen. The mechanism depicted in Scheme 6 was proposed for this process.

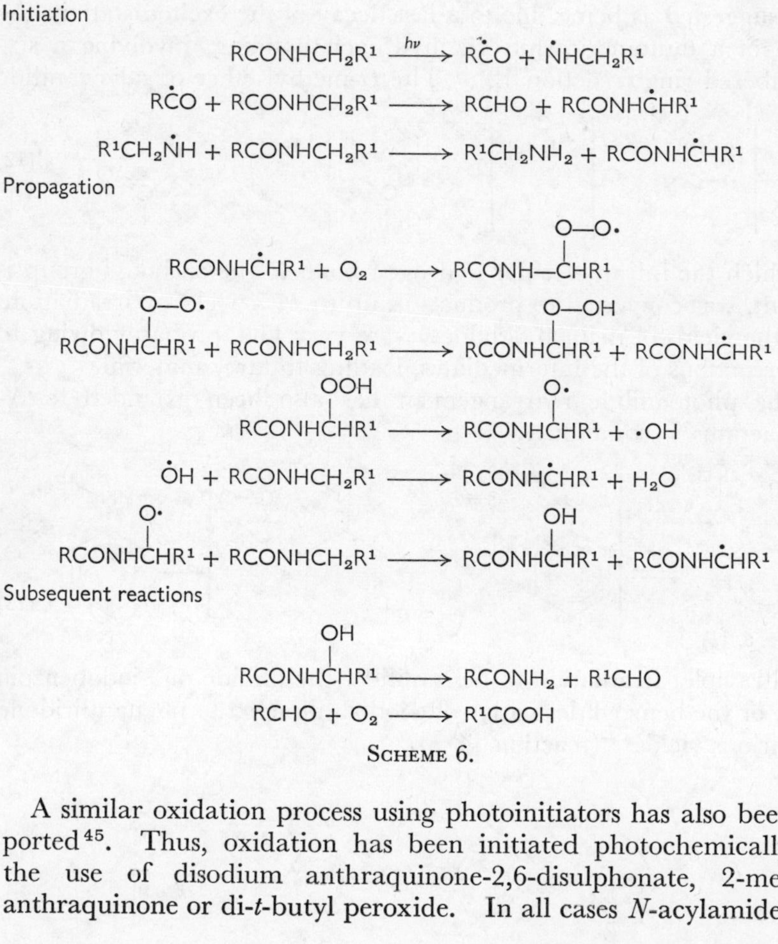

Initiation

$$RCONHCH_2R^1 \xrightarrow{h\nu} R\overset{\cdot}{C}O + \overset{\cdot}{N}HCH_2R^1$$

$$R\overset{\cdot}{C}O + RCONHCH_2R^1 \longrightarrow RCHO + RCONH\overset{\cdot}{C}HR^1$$

$$R^1CH_2\overset{\cdot}{N}H + RCONHCH_2R^1 \longrightarrow R^1CH_2NH_2 + RCONH\overset{\cdot}{C}HR^1$$

Propagation

$$RCONH\overset{\cdot}{C}HR^1 + O_2 \longrightarrow RCONH\overset{\underset{\displaystyle |}{O-O\cdot}}{-}CHR^1$$

$$RCONH\overset{\underset{\displaystyle |}{O-O\cdot}}{C}HR^1 + RCONHCH_2R^1 \longrightarrow RCONH\overset{\underset{\displaystyle |}{O-OH}}{C}HR^1 + RCONH\overset{\cdot}{C}HR^1$$

$$RCONH\overset{\underset{\displaystyle |}{OOH}}{C}HR^1 \longrightarrow RCONH\overset{\underset{\displaystyle |}{O\cdot}}{C}HR^1 + \cdot OH$$

$$\overset{\cdot}{O}H + RCONHCH_2R^1 \longrightarrow RCONH\overset{\cdot}{C}HR^1 + H_2O$$

$$RCONH\overset{\underset{\displaystyle |}{O\cdot}}{C}HR^1 + RCONHCH_2R^1 \longrightarrow RCONH\overset{\underset{\displaystyle |}{OH}}{C}HR^1 + RCONH\overset{\cdot}{C}HR^1$$

Subsequent reactions

$$RCONH\overset{\underset{\displaystyle |}{OH}}{C}HR^1 \longrightarrow RCONH_2 + R^1CHO$$

$$RCHO + O_2 \longrightarrow R^1COOH$$

SCHEME 6.

A similar oxidation process using photoinitiators has also been reported[45]. Thus, oxidation has been initiated photochemically by the use of disodium anthraquinone-2,6-disulphonate, 2-methylanthraquinone or di-t-butyl peroxide. In all cases N-acylamides are the major products, a fact not previously recorded. The reaction courses (15) were noted.

$$RCONHCH_2R^1 \begin{cases} \longrightarrow RCONHCOR^1 \\ \longrightarrow RCONHCHO \\ \longrightarrow RCONH_2 + R^1CHO \end{cases} \quad (15)$$

Similar results have also been reported by other groups[26]. Such photooxidation reactions have potential significance due to the light they may shed on the photochemical degradation of the chemically similar polyamide material used in nylon manufacture.

C. Photoamidation

Amides and lactams have been shown to undergo light-induced addition reactions to unsaturated systems. The photoaddition reactions of formamide to olefins, acetylenes and aromatic systems have been extensively studied[46-54,57]. These reactions, which are usually initiated photochemically by acetone, acetophenone or benzophenone, involve the addition of formamide to the double bond yielding higher amides. For example, in the case of terminal olefins the reaction is formulated as in (16)[46].

$$RCH{=}CH_2 + H{-}CONH_2 \xrightarrow[\text{ketone}]{h\nu} RCH_2CH_2CONH_2 \quad (16)$$

This addition reaction has been studied with a variety of olefins and the high yields obtained (up to 90%) may lead to its use for synthetic purposes.

Photoamidation has been applied to non-terminal olefins[47] and to reactive double bonds such as those of norbornene[48] or α,β-unsaturated esters[49,79]. The same reaction with terminal acetylenes, leads to 2:2 adducts as the major products[53] (reaction 17), whereas non-terminal

$$RC{\equiv}CH + H{-}CONH_2 \xrightarrow[\text{ketone}]{h\nu} \begin{array}{c} CONH_2 \\ | \\ RCH{-}CH{-}CONH_2 \\ | \\ CH{=}CHR \end{array} \quad (17)$$

isolated acetylenes were found to yield 1:2 adducts, under similar reaction conditions[50] (reaction 18).

$$RO_2CC{\equiv}CCO_2R + 2HCONH_2 \xrightarrow[R=CH_3, C_2H_5]{h\nu} \begin{array}{c} RO_2CCHCONH_2 \\ | \\ RO_2CCHCONH_2 \end{array} \quad (18)$$

With aromatic compounds substitution takes place at the aromatic nucleus as well as at the side-chain[51] (reactions 19, 20).

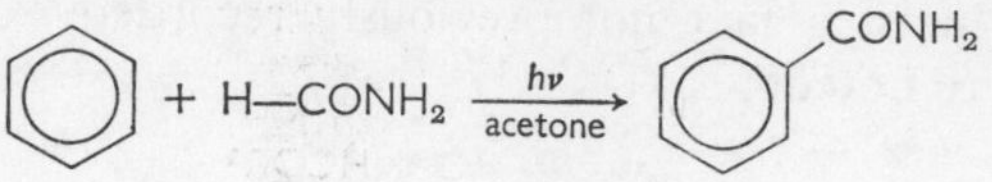

$$(19)$$

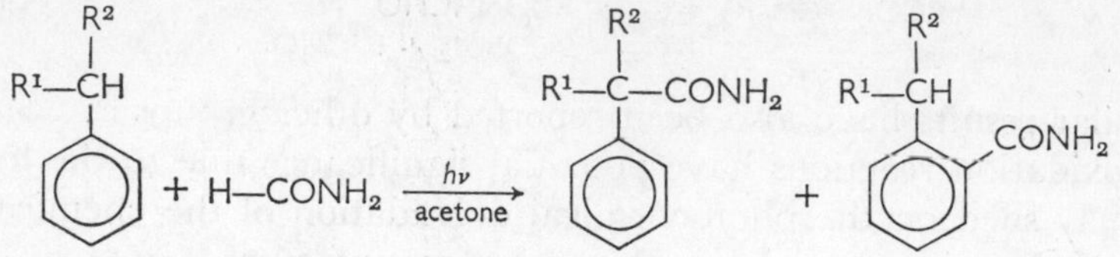

$$(20)$$

The reaction with isolated dienes leads to 1:1 adducts[54] (reactions 21, 22).

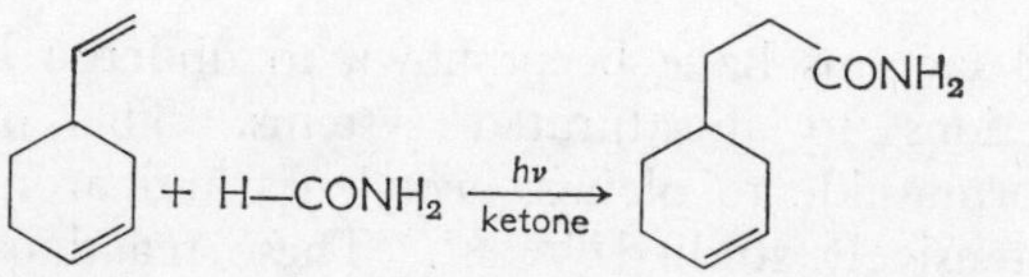

$$(21)$$

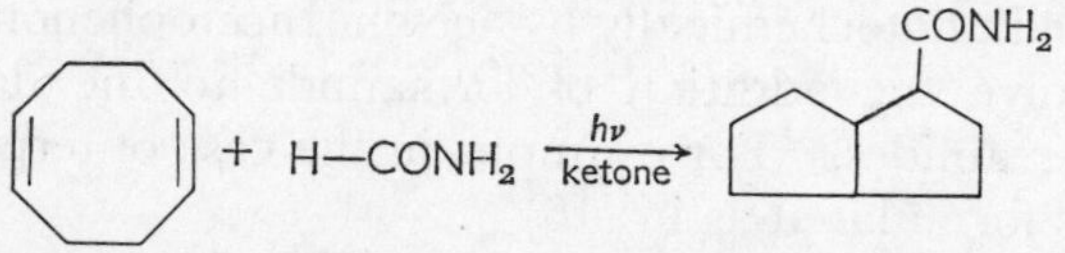

$$(22)$$

The photoamidation reactions of carbon–carbon unsaturated compounds are described as free-radical reactions involving a carbamoyl radical $\cdot CONH_2$, which is formed on hydrogen atom abstraction from formamide by the excited ketone molecule (Scheme 7). This

$$H\text{—}CONH_2 \xrightarrow{h\nu} \dot{C}ONH_2$$

$$H\text{—}CONH_2 \xrightarrow[\text{ketone}]{h\nu} \dot{C}ONH_2$$

$$RCH{=}CH_2 + \dot{C}ONH_2 \longrightarrow R\dot{C}HCH_2CONH_2$$

$$R\dot{C}HCH_2CONH_2 + H\text{—}CONH_2 \longrightarrow RCH_2CH_2CONH_2 + \dot{C}ONH_2$$

$$R\dot{C}HCH_2CONH_2 + RCH{=}CH_2 \longrightarrow \underset{\underset{CH_2\dot{C}HR}{|}}{RCHCH_2CONH_2}$$

$$\underset{\underset{CH_2\dot{C}HR}{|}}{RCHCH_2CONH_2} + H\text{—}CONH_2 \longrightarrow \underset{\underset{CH_2CH_2R}{|}}{RCHCH_2CONH_2} + \dot{C}ONH_2$$

SCHEME 7.

mechanism has been confirmed by direct examination of the radicals using paramagnetic resonance techniques[55].

The lower-limit values of quantum yields for the photochemical addition of formamide to 1-hexene were reported to be 0·06–0·08[57]. In spite of these values it was suggested that side-reaction hindered the proposed chain mechanism.

The addition of formamide to olefins and dienes could be also induced by γ-rays and electron radiation[57–59].

In the absence of olefin the excited ketone was found to interact with formamide[56] affording equal molecular amounts of $NH_2COCONH_2$ and α-hydroxyisobutyramide, together with small amounts of cyanuric acid and $H_2NCOOCMe_2CONH_2$. Acetone is proposed to yield biradicals on light absorption. These subsequently react with formamide, and the observed reaction products can thus be accounted for according to Scheme 8.

Application of photoamidation to more complicated unsaturated systems has also been reported. Thus acetone-induced addition of formamide to longifolen[62] affords two longifolen-ω-carboxamides in 11% and 15% yields respectively.

The reaction has been extended also to carbon–nitrogen double bonds. Thus, on irradiation of iminium salts in formamide with light of $\lambda \sim 250$ nm, in the absence of oxygen and in the presence of benzophcnone as sensitizer, primary photochemical addition products are formed which, however, undergo further reaction[60]. For example, cyclohexylidenepyrrolidinium perchlorate undergoes reaction (23).

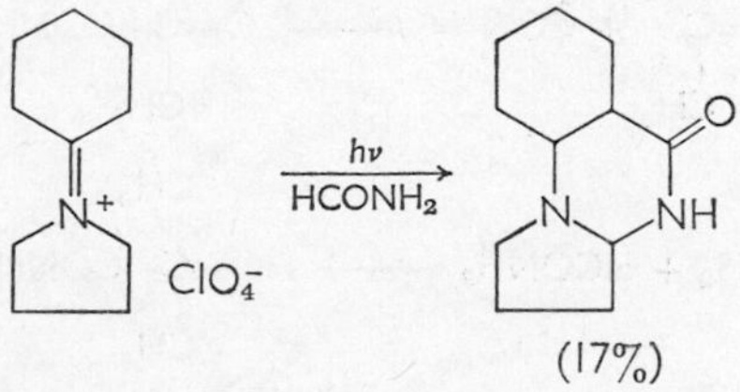

$$(23)$$

$$(17\%)$$

N-Methylacetamide has been found to react with olefins under ultraviolet irradiation in the presence of acetone to give substitution products, among which the one resulting from substitution at the N-methyl group predominates[61] (reaction 24).

$$RCH{=}CH_2 + CH_3CONHCH_3 \xrightarrow[\text{acetone}]{h\nu}$$

$$CH_3CONHCH_2(CH_2)_2R + R(CH_2)_2CH_2CONHCH_3$$

major product   minor product

$$(24)$$

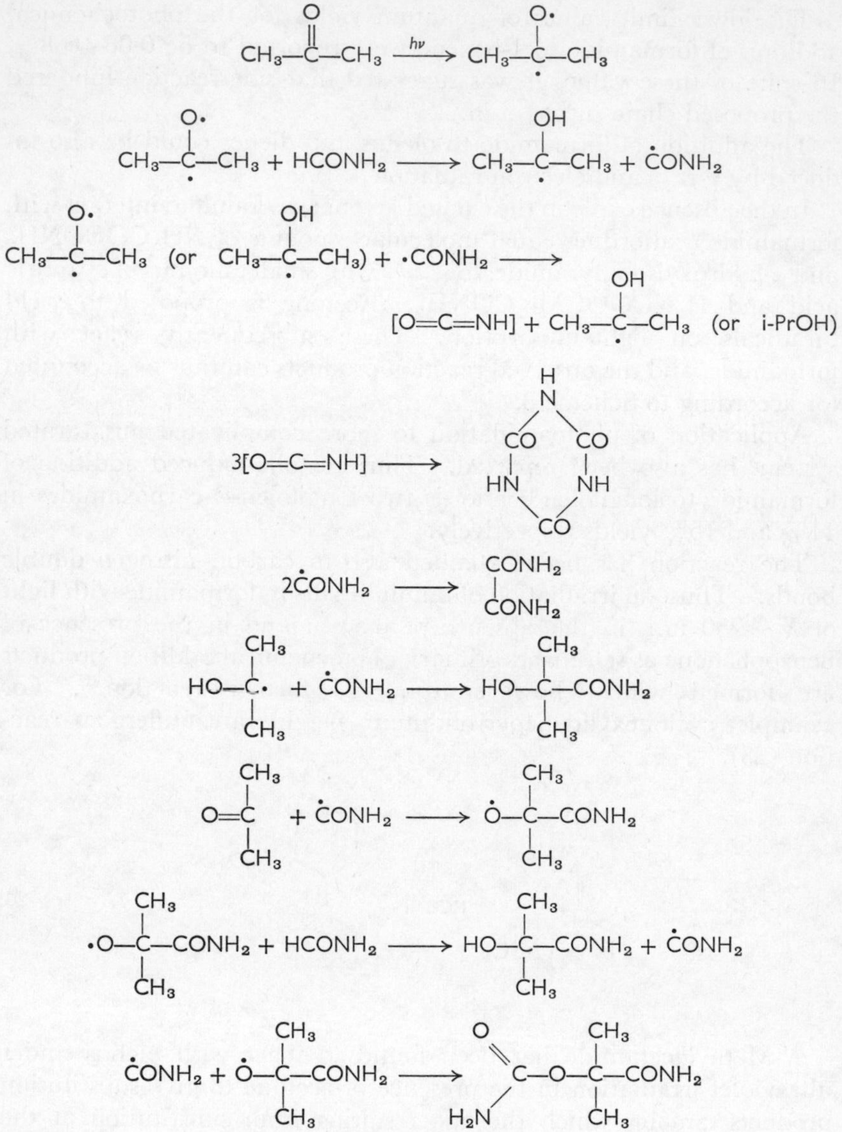

$$CH_3-\overset{\overset{\displaystyle O}{\|}}{C}-CH_3 \xrightarrow{h\nu} CH_3-\overset{\overset{\displaystyle O\cdot}{|}}{C}-CH_3$$

$$CH_3-\overset{\overset{\displaystyle O\cdot}{|}}{\underset{\cdot}{C}}-CH_3 + HCONH_2 \longrightarrow CH_3-\overset{\overset{\displaystyle OH}{|}}{\underset{\cdot}{C}}-CH_3 + \cdot CONH_2$$

$$CH_3-\overset{\overset{\displaystyle O\cdot}{|}}{\underset{\cdot}{C}}-CH_3 \ (or \ CH_3-\overset{\overset{\displaystyle OH}{|}}{\underset{\cdot}{C}}-CH_3) + \cdot CONH_2 \longrightarrow$$

$$[O{=}C{=}NH] + CH_3-\overset{\overset{\displaystyle OH}{|}}{\underset{\cdot}{C}}-CH_3 \ (or \ i\text{-}PrOH)$$

$$3[O{=}C{=}NH] \longrightarrow \begin{matrix} & H & \\ & N & \\ CO & & CO \\ HN & & NH \\ & CO & \end{matrix}$$

$$2\dot{C}ONH_2 \longrightarrow \begin{matrix} CONH_2 \\ | \\ CONH_2 \end{matrix}$$

$$HO-\overset{\overset{\displaystyle CH_3}{|}}{\underset{\underset{\displaystyle CH_3}{|}}{C}}\cdot + \dot{C}ONH_2 \longrightarrow HO-\overset{\overset{\displaystyle CH_3}{|}}{\underset{\underset{\displaystyle CH_3}{|}}{C}}-CONH_2$$

$$O{=}\overset{\overset{\displaystyle CH_3}{|}}{\underset{\underset{\displaystyle CH_3}{|}}{C}} + \dot{C}ONH_2 \longrightarrow \dot{O}-\overset{\overset{\displaystyle CH_3}{|}}{\underset{\underset{\displaystyle CH_3}{|}}{C}}-CONH_2$$

$$\cdot O-\overset{\overset{\displaystyle CH_3}{|}}{\underset{\underset{\displaystyle CH_3}{|}}{C}}-CONH_2 + HCONH_2 \longrightarrow HO-\overset{\overset{\displaystyle CH_3}{|}}{\underset{\underset{\displaystyle CH_3}{|}}{C}}-CONH_2 + \dot{C}ONH_2$$

$$\dot{C}ONH_2 + \dot{O}-\overset{\overset{\displaystyle CH_3}{|}}{\underset{\underset{\displaystyle CH_3}{|}}{C}}-CONH_2 \longrightarrow \underset{H_2N}{\overset{O}{\diagdown}}C-O-\overset{\overset{\displaystyle CH_3}{|}}{\underset{\underset{\displaystyle CH_3}{|}}{C}}-CONH_2$$

SCHEME 8.

The photoalkylation of 2-pyrrolidone led to a 1:2 mixture of 3-alkyl-2-pyrrolidone and 5-alkyl-2-pyrrolidone in yields up to 60% [52] (reaction 25). The presence of a ketone for initiating this reaction is vital, since in its absence no 1:1 adducts could be detected. This

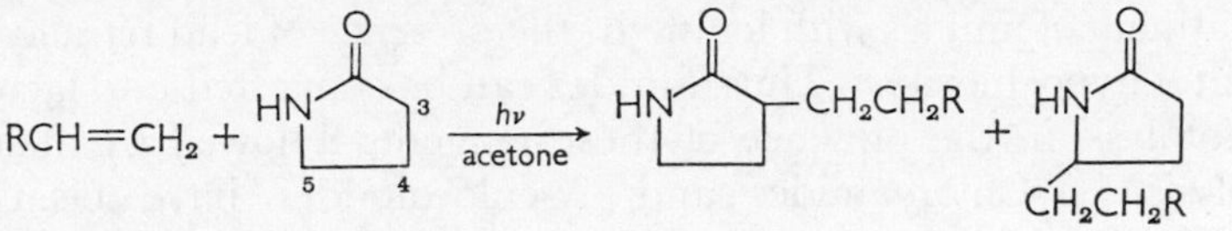

$$RCH{=}CH_2 + \quad \xrightarrow[\text{acetone}]{h\nu} \quad + \quad \tag{25}$$

reaction is described as a free-radical chain reaction initiated by the abstraction of a hydrogen atom from 2-pyrrolidone by an excited ketone molecule, according to equation (26).

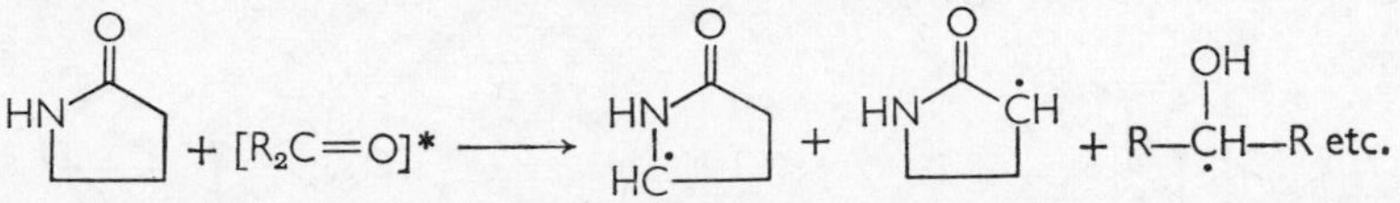

$$+ [R_2C{=}O]^* \longrightarrow \quad + \quad + R{-}\underset{|}{CH}{-}R \text{ etc.} \tag{26}$$

D. Photochemical Reactions of Halo Amides

Several photochemical reactions of halo amides have been described in the literature[63-70]. The N-halogen bond in these compounds is fairly labile and is split during the photochemical processes. The free radicals produced in these reactions react further to give a variety of products. The use of N-bromosuccinimide is common in organic chemistry and in some cases the fission of the N—Br bond is achieved by irradiation. The reactions of N-bromosuccinimide have been reviewed and will not be dealt with here[70].

Synthesis of γ-lactones by irradiation of N-iodoamides has been reported[67,68]. The mechanism involves photolysis of the N—I bond followed by intramolecular hydrogen transfer and coupling of the resultant radical with iodine. Hydrolysis of the γ-iodoamide so produced would give γ-lactone via an intermediate γ-iminolactone, as shown in reaction (27). Iodoamides can in fact be generated *in situ*

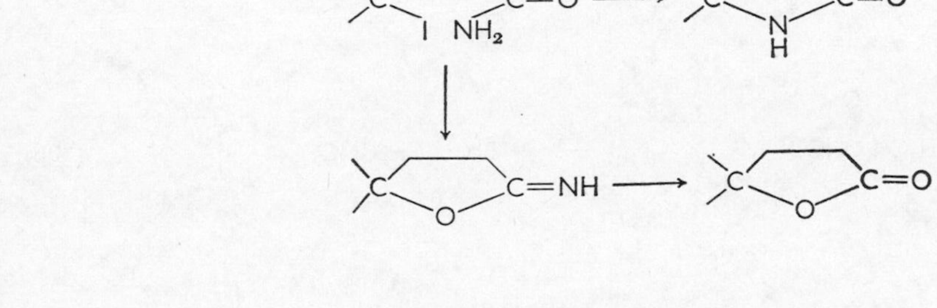

(27)

by treatment of amides with iodine in the presence of lead tetraacetate or t-butyl hypochlorite. Thus, amides can be converted into lactones by photolysis in the presence of these reagents followed by alkaline hydrolysis. N-Chloroamides and N-chloroimides have been reported[66] to yield the same photochemical products.

These reactions have been applied to steroidal amides[67–69]. Thus 3β-acetoxy-11-oxo-5α-pregnane-20-carboxamide yielded the corresponding lactone, according to equation (28).

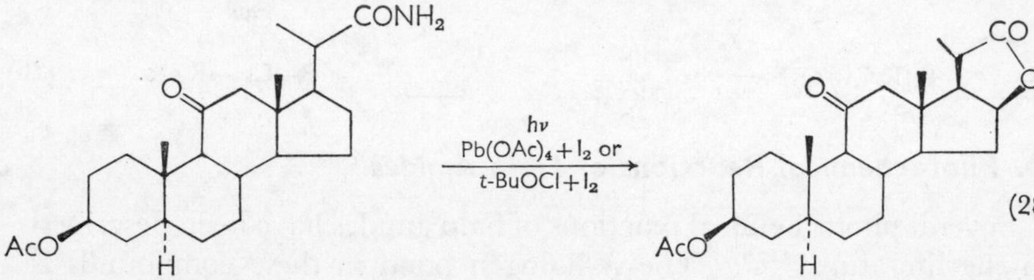

$$\text{(28)}$$

Some aliphatic N-bromoamides and N-chloroamides have been found to rearrange photochemically in a different way leading to the isomeric 4-halo amides[64]. The reaction is most efficient with N-t-butyl derivatives and takes place readily in benzene or carbon tetrachloride; the products are cyclized by heating to the corresponding iminolactone salt in yields up to 71% (equation 29).

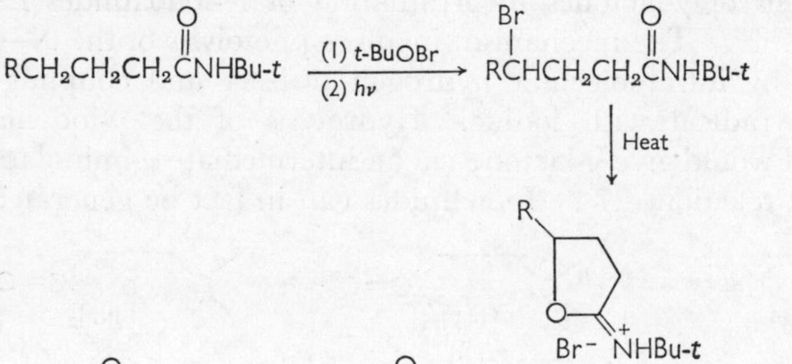

$$\text{(29)}$$

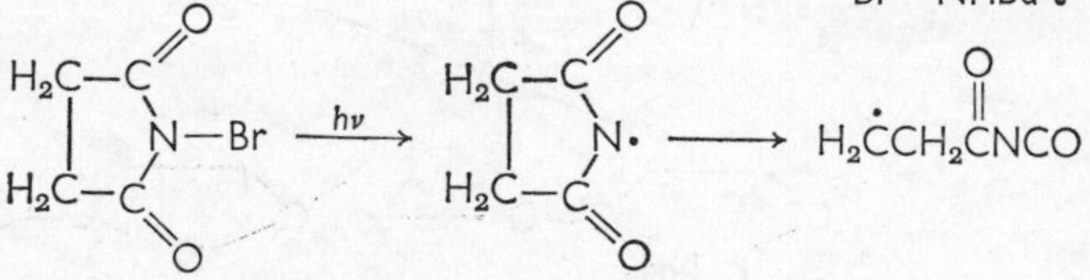

$$\text{(30a)}$$

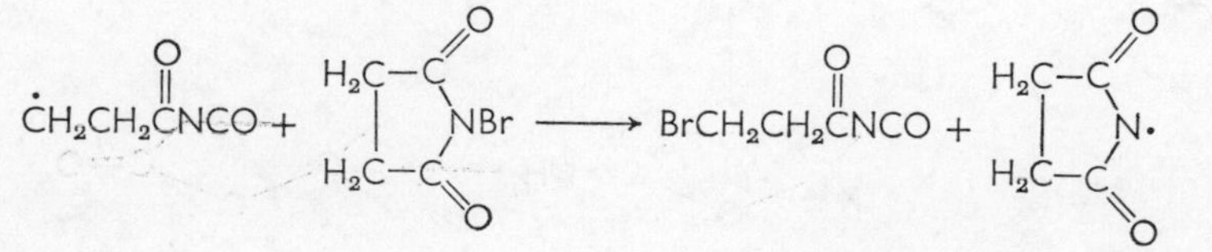

$$\text{(30b)}$$

The isomerization of N-bromosuccinimide when irradiated in carbon tetrachloride[65] leads to β-bromopropionyl isocyanate. A possible mechanism would involve initial cleavage of the N—Br bond followed by ring opening of the resulting succinimide radical, according to equations (30).

N-Bromoacetanilide undergoes halogen migration to the aromatic nucleus on ultraviolet irradiation[63] leading to p-bromoacetanilide as the major product (78%). The detailed mechanism of the reaction is given in Scheme 9.

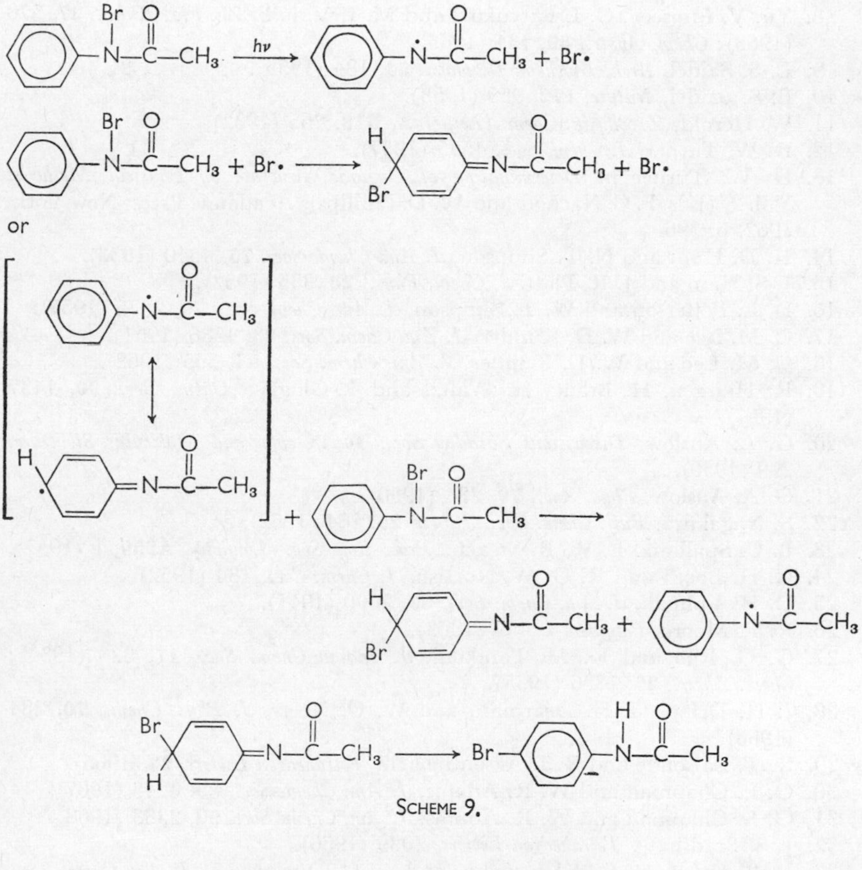

SCHEME 9.

IV. ACKNOWLEDGMENT

The author wishes to express his appreciation to Prof. D. Elad for most valuable discussions and comments.

11+c.o.a.

V. REFERENCES

1. A. D. McLaren and D. Shugar, *Photochemistry of Proteins and Nucleic Acids*, Pergamon Press, Oxford, London, Edinburgh, New York, Paris, Frankfurt, 1964.
2. J. Bielecki and V. Henri, *Compt. Rend.*, **156**, 1860 (1913).
3. H. Levy and B. Arends, *Z. Physik. Chem.* (*Leipzig*), **B17**, 177 (1932).
4. L. J. Saidel, *J. Am. Chem. Soc.*, **77**, 3892 (1955).
5. G. Oster and E. H. Immergut, *J. Am. Chem. Soc.*, **76**, 1393 (1954).
6. C. N. R. Rao, *Ultraviolet and Visible Spectroscopy*, Butterworths, London, 1967, p. 193.
7. P. Ramat-Lucas, *Bull. Soc. Chim. France*, **9**, 850 (1942).
8. Yu. V. Moiseev, G. I. Batyukov, and M. I. Vinnik, *Zh. Fiz. Khim.*, **37**, 570 (1963); *Chem. Abstr.*, **59**, 134 (1963).
9. L. S. Saidel, *Arch. Biochem. Biophys.*, **54**, 184 (1955).
10. L. S. Saidel, *Nature*, **172**, 955 (1953).
11. W. Herold, *Z. Physik. Chem.* (*Leipzig.*), **B18**, 265 (1932).
12. D. W. Turner, *J. Chem. Soc.*, 4555 (1957).
13. D. W. Turner in *Determination of Organic Structure by Physical Methods*, Vol. 2, (Eds. F. C. Nachod and W. D. Phillips), Academic Press, New York, 1962, p. 396.
14. H. D. Hunt and N. T. Simpson, *J. Am. Chem. Soc.*, **75**, 4540 (1953).
15. J. S. Ham and J. R. Platt, *J. Chem. Phys.*, **20**, 335 (1952).
16. D. L. Petterson and W. T. Simpson, *J. Am. Chem. Soc.*, **79**, 2375 (1957).
17. C. M. Lee and W. D. Kumler, *J. Am. Chem. Soc.*, **83**, 4586 (1961).
18. C. M. Lee and W. D. Kumler, *J. Am. Chem. Soc.*, **84**, 565 (1962).
19. R. Huisgen, H. Brade, H. Waltz, and I. Glogger, *Chem. Ber.*, **90**, 1437 (1957).
20. G. A. Anslow, *Discussions Faraday Soc.*, *Spectroscopy and Molecular Structure*, 299 (1950).
21. G. A. Anslow, *Phys. Rev.*, **79**, 234 (1950).
22. S. Nagakura, *Bull. Chem. Soc.*, *Japan*, **25**, 164 (1952).
23. B. C. Spall and E. W. R. Steacie, *Proc. Roy. Soc.* (*London*), **A239**, 1 (1957).
24. G. H. Booth and R. G. W. Norrish, *J. Chem. Soc.*, 188 (1952).
25. D. H. Volman, *J. Am. Chem. Soc.*, **63**, 2000 (1941).
26. R. F. Moore, *Polymer*, **4**, 493 (1963).
27. G. G. Rao and K. M. Pandalai, *J. Indian Chem. Soc.*, **11**, 623 (1934); *Chem. Abstr.*, **29**, 1326 (1935).
28. J. H. Delap, H. H. Dearman, and W. C. Neely, *J. Phys. Chem.*, **70**, 284 (1966).
29. R. T. LaLonde and R. I. Aksentijevich, *Tetrahedron Letters*, **23** (1965).
30. O. L. Chapman and W. R. Adams, *J. Am. Chem. Soc.*, **89**, 4243 (1967).
31. O. L. Chapman and W. R. Adams, *J. Am. Chem. Soc.*, **90**, 2333 (1968).
32. J. S. Bradshaw, *Tetrahedron Letters*, 2039 (1966).
33. L. P. Kuhn, G. G. Kleinspehn, and A. C. Duckworth, *J. Am. Chem. Soc.*, **89**, 3858 (1967).
34. E. K. Rideal and J. S. Mitchell, *Proc. Roy. Soc.* (*London*), **159A**, 206 (1937); *Chem. Abstr.*, **31**, 7760 (1937).
35. S. M. Rosenthal and H. Bauer, *Science*, **91**, 509 (1940).

36. D. C. Carpenter, *Science*, **89**, 251 (1939).
37. D. C. Carpenter, *J. Am. Chem. Soc.*, **62**, 289 (1940).
38. H. Hellberg, *Acta Chem. Scand.*, **16**, 1363 (1962).
39. E. J. Corey and A. M. Felix, *J. Am. Chem. Soc.*, **87**, 2518 (1965).
40. D. Elad, *Tetrahedron Letters*, 873 (1963).
41. D. Elad, D. V. Rao, and V. I. Stenberg, *J. Org. Chem.*, **30**, 3252 (1965).
42. D. V. Rao and V. Lambreti, *J. Org. Chem.*, **32**, 2896 (1967).
43. B. S. Thyagarajan, N. Kharasch, H. B. Lewis, and W. Wolf, *Chem. Commun.*, 614 (1967).
44. W. H. Sharkey and W. E. Mochel, *J. Am. Chem. Soc.*, **81**, 3000 (1959).
45. M. V. Lock and F. B. Sagar, *Proc. Chem. Soc. (London)*, 358 (1960).
46. D. Elad and J. Rokah, *J. Org. Chem.*, **29**, 1855 (1964).
47. D. Elad and J. Rokah, *J. Org. Chem.*, **30**, 3361 (1965).
48. D. Elad and J. Rokah, *J. Chem. Soc.*, 800 (1965).
49. J. Rokah and D. Elad, *J. Org. Chem.*, **31**, 4210 (1966).
50. D. Elad, *Proc. Chem. Soc.*, 225 (1962).
51. D. Elad, *Tetrahedron Letters*, 77 (1963); D. Elad and G. Friedman, unpublished results.
52. J. Sinreich and D. Elad, *Tetrahedron*, **24**, 4509 (1968).
53. G. Friedman and A. Komem, *Tetrahedron Letters*, 3357 (1968).
54. D. Elad, *Fortsch. Chem. Forsch.*, **7**, 528 (1967).
55. R. Livingston and H. Zeldes, *J. Chem. Phys.*, **47**, 4173 (1967).
56. H. Grossman, *Z. Naturforsch.*, **20b**, 209 (1965).
57. D. P. Gush, N. S. Marans, F. Wessells, W. D. Addy, and S. J. Olfky, *J. Org. Chem.*, **31**, 3829 (1966).
58. J. Rokah, C. H. Krauch, and D. Elad, *Tetrahedron Letters*, 3253 (1966).
59. C. H. Krauch, J. Rokah, and D. Elad, *Tetrahedron Letters*, 5099 (1967).
60. W. Dörscheln, H. Tiefenthaler, H. Göth, P. Cerutti, and H. Schmid, *Helv. Chim. Acta*, **50**, 1759 (1967).
61. D. Elad and J. Sinreich, unpublished results.
62. M. Fisch and G. Ourisson, *Bull. Soc. Chim. France*, 1325 (1966).
63. D. D. Tanner and E. Protz, *Can. J. Chem.*, **44**, 1555 (1966).
64. R. S. Neale, N. L. Marcus, and R. G. Schepers, *J. Am. Chem. Soc.*, **88**, 3051 (1966).
65. J. C. Martin and P. D. Bartlett, *J. Am. Chem. Soc.*, **79**, 2533 (1957).
66. R. C. Petterson and A. Wambsgans, *J. Am. Chem. Soc.*, **86**, 1648 (1964).
67. D. H. R. Barton and A. J. L. Beckwith, *Proc. Chem. Soc. (London)*, 335 (1963).
68. D. H. R. Barton, A. J. L. Beckwith, and A. Goosen, *J. Chem. Soc.*, 181 (1965).
69. *Neth. Appl.*, 6,407,607, Jan. 4, 1965; *Chem. Abstr.*, **62**, 16341 (1965).
70. A. Schönberg, G. O. Schenck, and O. A. Neumüller, *Preparative Organic Photochemistry*, Springer-Verlag, Berlin, Heidelberg, New York, 1968, p. 356.
71. J. C. Sheenan and M. M. Nafissi-V, *J. Am. Chem. Soc.*, **91**, 1176 (1969).
72. E. R. Talaty, A. E. Dupuy, Jr., and T. H. Golson, *Chem. Commun.*, 49 (1969).
73. S. R. Bosco, A. Cirillo, and R. B. Timmons, *J. Am. Chem. Soc.*, **91**, 3140 (1969).
74. J. C. Sheenan and J. H. Beeson, *J. Am. Chem. Soc.*, **89**, 362 (1967).

75. O. E. Edwards and R. S. Rosich, *Can. J. Chem.*, **45**, 1287 (1967).
76. Y. L. Chow and J. N. S. Tam, *Chem. Commun.*, 747 (1969).
77. M. Fischer, *Tetrahedron Letters*, 4295 (1968).
78. M. Fischer, *Chem. Ber.*, **101**, 2669 (1968).
79. M. Itoh, M. Tokuda, K. Kihara, and A. Suzuki, *Tetrahedron*, **24**, 6591 (1968).

CHAPTER 6

Radiation chemistry of amides

OWEN H. WHEELER

Puerto Rico Nuclear Centre, Mayaguez, Puerto Rico

I. INTRODUCTION

Gamma and x-rays are short-wavelength, high-energy, forms of electromagnetic radiation. The amount of radiation absorbed (dI) in passing through a small thickness of absorber (dx) is given by

$$-dI = I\mu dx$$

where I is the intensity of the incident radiation, and μ is the linear absorption coefficient for the material of the absorber. On integration this gives the exponential relation

$$I = I_0 e^{-\mu x}$$

The absorption coefficient is the sum of the individual absorption coefficients of the atoms composing the material. The absorption coefficients for hydrogen and carbon for 1·24 mev γ rays, corresponding to a wavelength of 0·01 Å, are 0·117 and 0·059, respectively. The values for oxygen and nitrogen are similar to those of carbon. The radiation is thus absorbed by all the atoms in the material. This is in

309

contrast to the absorption of ultraviolet and visible light, where the radiation is absorbed by certain chromophoric groups (see Chapter 5).

The absorption of gamma radiation occurs through three physical processes. The photoelectric effect, in which an atomic electron is ejected, occurs on the absorption of low-energy γ rays. Medium-energy γ rays produce Compton scattering in which an electron and a γ ray of lesser energy are emitted from an atom. At energies of 1·02 MEV and above, the absorption of gamma energy gives rise to the production of an electron–positron pair. The positron is subsequently annihilated by combination with another electron, and the annihilation energy is emitted as two γ rays of 0·51 MEV. The commonly used sources for γ irradiations are cobalt-60. This radioisotope emits γ's of 1·17 and 1·33 MEV, which give rise mainly to Compton scattering in an absorbing material[1,2].

The absorption of electrons by material results from interaction with the atomic electrons. These atomic electrons are either excited, or ejected from the atom, leaving a cation. The ejected secondary electrons can themselves produce excitation of another atom.

The absorption of neutrons occurs through an entirely different process involving interaction with the nuclei of the atoms of the absorber. Fast and moderately slow (epithermal) neutrons undergo elastic scattering, in which part of their energy is imparted to the nuclei. The energy transfer is greater for light elements, such as hydrogen. Lower-energy neutrons undergo inelastic scattering, in which the incident nucleus is excited, and subsequently emits its excitation energy in the form of gamma radiation. In certain cases the nucleus may capture a neutron forming a new isotope as a result of the nuclear reaction. Hydrogen produces deuterium in the reaction $^1H(n,\gamma)^2H$. The cross-section for this reaction is 0·3 barn (1 barn = 10^{-24} cm^2) and the emitted γ ray has an energy of 2·2 MEV. Nitrogen-14 affords carbon-14 through the reaction $^{14}N(n,p)^{14}C$. The cross-section is 1·5 barn, and the proton has an energy of 0·66 MEV. The carbon-14 formed is a β-emitter, with a half-life of 5,760 years. Oxygen and carbon have only very low cross-sections for neutron capture ($^{16}O < 0·00002$, and ^{12}C 0·003 barn, respectively).

The absorption of ionizing radiation generally leads to the formation of excited molecules (equation 1) or of cations (equation 2). The secondary electrons emitted in the latter process can, in turn excite

$$M \longrightarrow M^* \tag{1}$$

$$M \longrightarrow M^+ + e^- \tag{2}$$

or ionize other molecules. However, the majority of the secondary electrons react with other cations to afford excited molecules. The energy of the excited molecules is largely lost in fragmentation processes, which proceed by homogeneous breakage of a bond, and result in the formation of two free radicals (or atoms). In the case of amides, the present evidence (see below) is that bond rupture does not occur at the amide group, but elsewhere in the amide molecule.

The radicals formed on irradiating solid amides are usually stable and can be detected by their electron paramagnetic (spin) resonance (see section II)[3]. The radical yields are expressed as G values, which are the number of radicals formed per 100 ev of energy absorbed. In dilute solution, the radicals initially formed are from the solvent molecules, and these radicals then attack the amides present in the solution (see section III). The majority of the studies on amides have been on derivatives of amino acids and on peptides in relation to the radiolysis of proteins, and these topics are discussed separately (see section IV). The reactions resulting from neutron bombardment involve the excited radioactive atoms formed in the nuclear transformations and are the so called hot-atom reactions. These reactions are different in nature from the radical processes caused by ionizing radiation, and are not usually considered a branch of radiation chemistry. However, free radicals probably play a role in the recombination processes which follow the nuclear recoil, and the hot-atom chemistry of amides is considered in a separate section (see section V).

II. SOLID AMIDES

The e.s.r. spectra of x-ray-irradiated acetamide showed the presence of the $\dot{C}H_2CONH_2$ radical[4], and N,N-dideuteroacetamide (CH_3COND_2) gave an identical spectrum[5]. The spectra of propionamide and acetanilide indicated the presence of $CH_3\dot{C}HCONH_2$ and $\dot{C}H_2CONHC_6H_5$ radicals, respectively. The spectra of radicals from butyramide and formamide at 77°K were also reported. The spectrum from thioacetamide was entirely different from that given by acetamide[4].

The e.s.r. spectra of acetamide irradiated with 1 mev electrons also indicated the presence of the $\dot{C}H_2CONH_2$ radical[6]. Propionamide similarly formed the $CH_3\dot{C}HCONH_2$ radical, whereas isobutyramide apparently afforded both $(CH_3)_2\dot{C}CONH_2$ and $CH_3\dot{C}HCONH_2$

radicals.   However, on aging or warming, only the latter radical disappeared[7].   Trimethylacetamide, which has no free hydrogen on the β-carbon atom gave a $(CH_3)_3\dot{C}$ radical.   X-ray-irradiated single crystals of monochloroacetamide gave an e.s.r. spectrum due to the $\dot{C}HClCONH_2$ radical[8].   When caproamide (1) was irradiated at $-80°$, the crystals developed a colour which was stable at room temperature in the absence of air, and had an absorption maximum at 3850 Å[9].   The radical formed was that resulting from the loss of an α-hydrogen atom.   The N-deuterated amide gave the same radical. Irradiation of caprolactam (2) gave a complex e.s.r. spectrum, which could not be interpreted[9], while irradiated single crystals of succinamide and N,N,N',N'-tetradeuterosuccinamide showed e.s.r. spectra of the radical 3[10].

$$CH_3(CH_2)_4CONH_2 \qquad\qquad (CH_2)_5 \begin{array}{c} C=O \\ | \\ NH \end{array}$$

(1)   (2)

$$NH_2COCH_2\dot{C}HCONH_2$$

(3)

A number of studies have been reported on N-alkylamides, and radicals were detected in irradiated crystals of N-methyl-, N,N-dimethyl-, N,N-diethyl-, N,N-di-n-propyl, N-isopropyl- and N,N-diisopropylacetamide, N-methyl- and N,N-dimethylpropionamide, N,N-dimethyl- and N,N-diethylacrylamide, N-ethyl- and N,N-diethylformamide, N,N-dimethyl- and N,N-diethyl-n-butyramide, N,N-methyl- and N-ethylisobutyramide and N,N-diethylchloroacetamide[6].   In all these cases the radicals were formed by loss of a hydrogen atom from carbon, and the loss occurred from an N-alkyl group in preference to the acyl group[7].   When crystals of N-ethylpropionamide were irradiated with 2 Mev electrons, the e.s.r. spectra showed the presence of the $CH_3CH_2CONH\dot{C}HCH_3$ radical, and the same type of radical was formed on irradiating N-ethylbutyramide and N-ethylhexanamide. N-n-Propylpropionamide, N-n-propylbutyramide and N-n-hexylpropionamide gave a $CONH\dot{C}H$ type radical.   The radical yield $(G_\dot{R})$ was 4.7 ± 1.4 for N-n-propylpropionamide. N-t-Amylpropionamide and N-t-amylbutyramide gave a radical on the β-carbon atom attached to the nitrogen atom $(CONHC(CH_3)_2\dot{C}H_2)$.   However, N-neopentylpropionamide gave rise to two radicals (**4** and **5**). The evidence was that these radicals were formed directly and not by initial formation of a radical at some other site followed by migration along the chain. This was because N-(n-propyl-2,2-d_2)-propionamide gave a radical spectra which showed

$$CH_3CH_2CONH\dot{C}HC(CH_3)_3 \qquad CH_3\dot{C}HCONHCH_2C(CH_3)_3$$

(**4**)   (**5**)

interaction with only one hydrogen atom indicating that the two deuterium atoms on the carbon atom were still in their original position in the radical (**6**)[11].

$$CH_3CH_2CONH\dot{C}HCD_2CH_3 \qquad \begin{array}{c} RCONHCHR^1 \\ | \\ OOH \end{array}$$

(**6**)   (**7**)

Another study of the e.s.r. spectra of irradiated crystals of N-n-butylpropionamide and N-n-butylbutyramide has indicated the presence of $RCONH\dot{C}HR^1$ radicals. The spectra were resolved into 7 lines at $-196°$, which passed into a 4-line spectra at -130 to $-90°$. The yields of hydrogen, carbon monoxide and methane were independent of the presence of oxygen. However, oxygen reduced the yields of propane and propylene, producing carbon monoxide. The liquid products were organic peroxide, hydrogen peroxide, aldehydes and alcohols. n-Butyraldehyde was the principal aldehyde and was probably formed via the hydroperoxide **7**[12].

The e.s.r. spectra of gamma-irradiated perfluoroacetamide indicated the presence of a $\dot{C}F_2CONH_2$ radical[13,14], and a study of the fluorine hyperfine spectra of irradiated single crystals[15] showed that in this radical, which was stable in air, the CF_2CON partial structure was planar. Difluoroacetamide gave equal amounts of $\dot{C}HFCONH_2$ and $\dot{C}F_2CONH_2$ radicals, of which the latter were more thermally stable. Monofluoroacetamide, only afforded the $\dot{C}HFCONH_2$ radical[14]. Irradiated crystals of pentafluoropropionamide showed the presence of the $CF_3\dot{C}FCONH_2$ radical stable at $300°K$[13-16]. The free rotation of the trifluoromethyl group was stopped by cooling to $77°K$. At room temperature the radical was attacked by oxygen to produce a $RO\dot{O}$ type radical, together with another radical which was probably $CF_3CF_2CO\dot{N}H$, and was formed by rupture of an N—H bond[14]. Perfluoro-n-butyramide similarly gave a $CF_3CF_2\dot{C}FCONH_2$ radical[13].

11*

Gamma-irradiated monomeric acrylamide and methacrylamide gave an e.s.r. signal which decayed on heating[17,18]. The spectrum of the radical from acrylamide was also changed on exposure to sulphur dioxide, and that from methacrylamide was destroyed on exposing the crystal to nitric oxide[19]. Other studies of the e.s.r. spectrum of irradiated acrylamide suggested that radical polymerization involved the $^+CH_2\dot{C}HCONH_2$ radical-ion formed in the following sequence of reactions[20,21].

$$CH_2{=}CHCONH_2 \xrightarrow{\gamma} [C_3H_5NO]^{\ddot{+}} + e^-$$

$$[C_3H_5NO]^{\ddot{+}} \longrightarrow \overset{\pm}{C}H_2\dot{C}HCONH_2$$

This radical-ion apparently scavenged hydrogen below $-125°$ to give the same radical formed in the radiation of propionamide. On warming from $-125°$ to $-30°$ polymerization began[17,21]. The e.s.r. spectra of irradiated acrylamide, N,N'-methylenebisacrylamide (8), N-methylolacrylamide (9) and N-t-butylacrylamide have also been measured[22].

$$(CH_2{=}CHCONH)_2CH_2 \qquad\qquad CH_2{=}CHCONHCH_2OH$$

$$\text{(8)} \qquad\qquad\qquad\qquad\qquad\qquad \text{(9)}$$

Measurements of the yields of hydrogen, carbon monoxide and methane formed in the radiolysis of solid solutions of acrylamide and propionamide indicated energy-transfer reactions and scavenging of hydrogen atoms[23]. The radiation-induced polymerization of N-vinylsuccinimide has been studied in the liquid[24] and solid state[25].

The infrared spectra of gamma-irradiated urea inclusion complexes, and of mixtures of urea with various compounds have been measured[26], but no general conclusion could be drawn.

N-Acetylamino acids serve as convenient models for the peptide linkage. The spectra of irradiated N-acetylglycine showed the presence of the $CH_3CON\dot{H}CHCO_2H$ radical, and the deuterated compound 10 gave an identical doublet, which was also due to the

$$CH_3CONDCH_2CO_2D$$

$$\text{(10)}$$

$-\dot{C}H-$ radical[27]. Acetylglycine, glycylglycine, diglycylglycine and triglycylglycine showed similar radical spectra[28]. The spectra of γ-irradiated single crystals of glycylglycine indicated the presence of the radical 11 on the carbon atom, since there was coupling from the single hydrogen atom remaining on the carbon[5] and hyperfine splitting from the ^{14}N atom[29,30]. The doublet of the glycylglycine radical was broad at room temperature, but could be resolved into a four-line spectrum at liquid nitrogen temperature[31]. Glycylglycine hydrochloride formed a similar radical **12**[32].

$$\overset{+}{N}H_3CH_2CONH\overset{\cdot}{C}HCO_2^- \qquad Cl^-\overset{+}{N}H_3CH_2CONH\overset{\cdot}{C}HCO_2H$$

$$(11) \qquad\qquad\qquad (12)$$

The zero-field electron spin resonance spectra of irradiated acetylglycine and glycylglycine showed the presence of four groups of lines, due to two different but similar radicals[33]. These resulted from the removal of one or the other of the non-equivalent hydrogen atoms on the α-methylene group. In the case of glycylglycine, the radical must be formed from the group nearest to the carboxylic acid end of the molecule, since the spectrum was very similar to that of acetylglycine. The spectra of N-deuteroglycylglycine was also similar to that of the undeuterated compound[5,33]. Nitrogen-15 labelled acetylglycine gave a similar spectrum to the normal compound, indicating that there was no interaction between the nitrogen nucleus and the unpaired electron in the radical. The zero-field spectrum of irradiated glycylglycine hydrochloride also indicated the presence of the radical on the α-carbon atom, at the acid end of the molecule[33].

The free radical yield ($G_R^{\cdot}$) for x-ray-irradiated crystals of glycylglycine was 7·0. Mixtures of glycylglycine with L-tyrosine and L-tryptophane showed a decrease in the G value. Furthermore, glycyl-L-tyrosine and glycyl-L-tryptophane had $G_R^{\cdot}$ values of 3·1 and 3·3, respectively, indicating radiation protection by the tyrosine and tryptophane units[34]. The e.s.r. spectra of polycrystalline glycylglycine, glycylleucine, glycylvaline, glycylleucylvaline and glycylglycylleucylglycine have been measured[35,36]. The radical yields $G_R^{\cdot}$ of a number of di- and tripeptides at $-196°$ were reported to be: glycylglycine (2·5), glycylproline (**13**) (3·3), α-alanyl-α-alanine (0·5), β-alanyl-β-alanine (0·4), glycylglycylleucine (**14**) (1·6), glycylleucylglycine (0·7) and glutathione (**15**) (4·4)[37].

When N-acetylglycine and N-acetylalanine were irradiated, the e.s.r. spectra showed the presence of radicals (**16** and **17**, respectively), which were stable to oxygen[38]. On dissolving the irradiated crystals of acetylglycine in water in the absence of oxygen, bis(acetylamino)-succinic acid (**18**) (resulting from the dimerization of **16**) was formed (G 0·60), together with ammonia (G 2·0). If the crystals were irradiated *in vacuo* and then dissolved in water in the presence of oxygen,

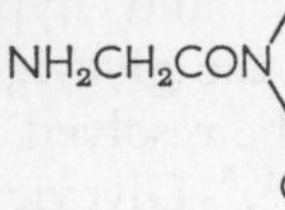

$$NH_2CH_2CON\underset{CO_2H}{\overset{\big|}{\big<}}$$

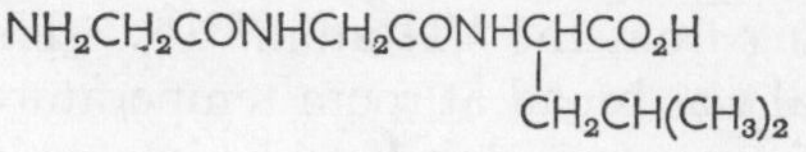

$$NH_2CH_2CONHCH_2CONHCHCO_2H$$
$$\overset{|}{CH_2CH(CH_3)_2}$$

(13)   (14)

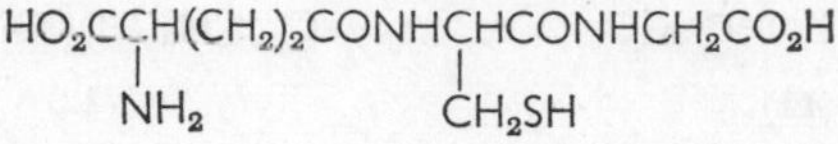

$$HO_2CCH(CH_2)_2CONHCHCONHCH_2CO_2H$$
$$\overset{|}{NH_2}\qquad\qquad\overset{|}{CH_2SH}$$

(15)

$$CH_3CONH\overset{\bullet}{C}HCO_2H$$ $$CH_3CONH\overset{\bullet}{C}(CH_3)CO_2H$$

(16)   (17)

less bis(acetylamino)succinic acid (G 0·12) and more ammonia (G 3·2) were formed, since the radical **16** was scavenged by oxygen. Irradiation in oxygen and dissolution in the presence of oxygen further decreased the yield of **18**, while increasing G_{NH_3}. The solution of irradiated acetylglycine contained glyoxylic acid ($G_{total\ carbonyl}$ 0·65)

$$CH_3CONHCHCO_2H$$
$$\overset{|}{CH_3CONHCHCO_2H}$$

(18)

and direct nitrogen–carbon bond cleavage must have occurred (equation 3). However, since the G_{NH_3} was higher than the $G_{carbonyl}$, cleavage following equation (4) must also have taken place.

$$CH_3CONHCH_2CO_2H \longrightarrow CH_3CHO + NH{=}CHCO_2H \qquad (3)$$

$$CH_3CONHCH_2CO_2H \longrightarrow CH_3CONH + \overset{\bullet}{C}H_2CO_2H \qquad (4)$$

The carbobenzoxy derivatives (**19**) of eleven amino acids were irradiated in the solid state, as were carbobenzoxyserylalanylalanine ethyl ester, carbobenzoxyprolylglycylphenylalanine ethyl ester and carbobenzoxyprolylglycylphenylalanine methyl ester[39]. Carbobenzoxyserylalanylalanine ethyl ester afforded a new unidentified amino acid on hydrolysis subsequent to irradiation, while proline and hydroxyproline were destroyed on irradiating their respective derivatives. The G_{CO_2} values were also measured and found to be 0·10–0·30, indicating that protection of the amino group rendered the amino acid more resistant to radiation. Typical values were G_{CO_2} 0·25 for carbobenzoxyglycine, and 0·14 for the derivatives of phenylalanine and arginine, with 0·18 for those of glutamic acid and tryptophane. The radical yields were from $0·6 \times 10^{-6}$ to $2·1 \times 10^{-6}$, except for carbobenzoxyglycine and carbobenzoxyaspartic acid, which had G_R^* values of 10^{-5} and 4.9×10^{-5}, respectively. Carbobenzoxytryptophane (20) formed stable radicals, which could not be identified[40]. The chemical destruction of the compounds was determined by dissolving the crystals and then hydrolysing the derivatives. The G_{-M} values (for destruction of the original carbobenzoxy compound) were 2–11, being respectively 3.8 and 11.0 for the derivatives of glycine and tryptophane (20)[40].

PhCH₂OCONHR

(19)

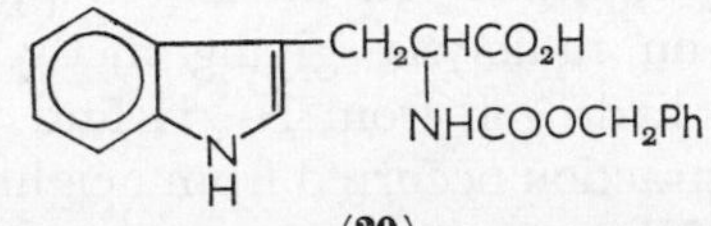

CH_2CHCO_2H
$NHCOOCH_2Ph$

(20)

Crystalline peptides have been irradiated with x-rays to doses of 0.5–150 Mrad, and then dissolved and hydrolysed[41]. Glycylglycyl-glycine afforded glycylglycylmethylamine, formed by decarboxylation of the C-terminal amino acid, and also some glyoxylic acid. γ-Glutamylalanine (21) formed glutamic and pyruvic acid, and ammonia, presumably via the α,β-unsaturated compound 22. Gly-

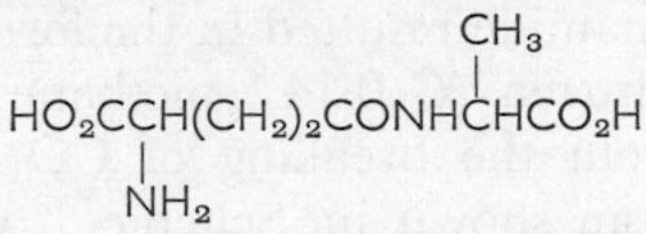

$$HO_2CCH(CH_2)_2CONHCHCO_2H$$

with CH₃ and NH₂ substituents

(21)

$$HO_2CCH(CH_2)_2CON=CCO_2H$$

with CH₃ and NH₂ substituents

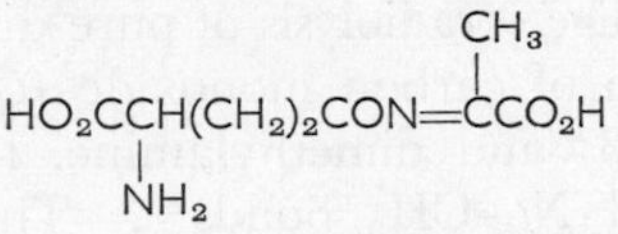

(22)

cylvalylalanine also gave glycylvaline and ethylamine, after irradiation and hydrolysis, whereas glutathione (15) was converted to γ-aminobutyrocysteinylglycine (23) and γ-glutamylcysteinylmethylamine (24).

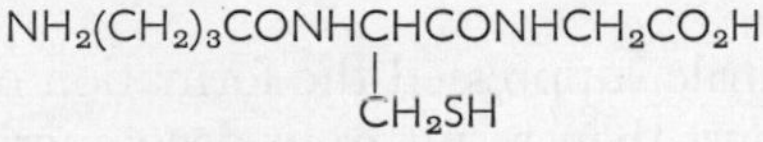

$$NH_2(CH_2)_3CONHCHCONHCH_2CO_2H$$

with CH₂SH substituent

(23)

$$HO_2CCH(CH_2)_2CONHCHCONHCH_3$$

with NH₂ and CH₂SH substituents

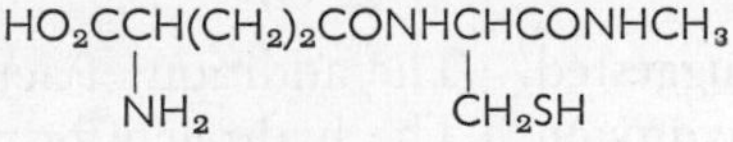

(24)

Recently, the radiolysis of acetamide in the stable rhombohedral crystal form, in the metastable orthorhombic form and in the molten state has been studied[42]. The more stable crystal form was less stable to radiolysis and formed more acetonitrile (G 1.0–1.2). This may be due to stronger hydrogen bonding in the crystal, resulting in the formation of the enol form (equation 5). The principal reaction, however,

$$CH_3CONH_2 \longrightarrow CH_3C(=NH)OH \longrightarrow CH_3CN + H_2O \qquad (5)$$

was breakage of carbon–hydrogen bonds, giving $\dot{C}H_2CONH_2$ radicals, which afforded succinamide (**25**) on melting the crystals. Some malonamide (**26**) was also formed, and must have arisen from

$$CH_2CONH_2 \qquad\qquad CH_2(CONH_2)_2$$
$$|$$
$$CH_2CONH_2$$

$$\text{(25)} \qquad\qquad\qquad\qquad \text{(26)}$$

$\dot{C}ONH_2$ radicals. In the liquid phase carbon–carbon bonds were broken on radiolysis, giving methyl radicals. These radicals abstracted hydrogen from N—H and C—H bonds forming methane. The abstraction occurred from neighbouring bonds in the crystal and CH_3COND_2 gave nearly equal amounts of methane and deuteromethane (CH_3D). However, in the liquid state, abstraction took place from weaker C—H bonds and the ratio $CH_4 : CH_3D$ became 10.

III. AMIDES IN SOLUTION

A. Non-aqueous Solution

The γ radiolysis of pure dimethylformamide resulted in the formation of carbon monoxide (G 2·6), hydrogen (G 0·14), methane (G 0·93) and dimethylamine, resulting from the breaking of CO –N and N—CH_3 bonds[43]. The mechanism shown in Scheme 1 was

$$HCON(CH_3)_2 \longrightarrow H\dot{C}O + \overset{+}{N}(CH_3)_2$$

$$HCON(CH_3)_2 \longrightarrow HCO\dot{N}CH_3 + \dot{C}H_3$$

$$HCO\dot{N}CH_3 \longrightarrow \dot{H} + CH_3NCO \quad \text{or} \quad \dot{C}H_3 + HNCO$$

SCHEME 1.

suggested. The addition of acrylonitrile suppressed the formation of hydrogen. The hydrogen formed must then result from decomposition of the $H\dot{C}O$ radical at nearly thermal energies and cannot result from direct abstraction from dimethylformamide (equation 6).

$$HCON(CH_3)_2 \longrightarrow \dot{H} + \dot{C}ON(CH_3)_2 \qquad\qquad (6)$$

Acrylonitrile also depressed the formation of carbon monoxide by 30%, and this amount must have been formed through secondary reactions of thermalized $H\dot{C}O$ radicals. Dissolved ferric chloride had no effect on the carbon monoxide yield and $G_{(-FeCl_3)}$ was 12·4 ± 1·0, which probably represented the true radical yield[43]. The poly-merization of acrylonitrile in dimethylformamide has also been studied[44].

Acrylamide has been used as a scavenger of the precursors of mole-cular hydrogen, in studies on the yields of the radical and molecular products formed in the radiolysis of both ordinary and heavy water[45]. The unsaturated amide served as a convenient water-soluble reactive olefin. The yield of deuterium gas from x-ray-irradiated heavy water has been measured with added acetamide, and glycylglycine[46]. The G_{D_2} value in pure heavy water was 0.49 ± 0.02, and decreased to 0.40 ± 0.02 in the presence of the amides. Thus the deuterium atoms formed on radiolysis cannot abstract a second deuterium atom from either the $COND_2$ or $COND$ groups, which were formed by exchange with the heavy water.

When a 4% solution of 1-octene in a 1 : 1 mixture of formamide and t-butanol was irradiated with γ rays or electrons, a 53% yield of non-amide was formed[47]. 1-Heptene, 1-dodecene and cyclohexene also gave the corresponding higher homologous amides in yields of 50–78%, with G values of 3 to 3.5. Norbornene formed exo-norbornanecarbox-amide in 80% yield, and the G value was 2.5. The mechanism sug-gested involved the formation of formamide radicals (equation 7), which then added onto the double bond of the olefin (equation 8).

$$HCONH_2 \longrightarrow \dot{H} + \dot{C}ONH_2 \tag{7}$$
$$(27)$$

$$RCH{=}CH_2 + \dot{C}ONH_2 \longrightarrow R\dot{C}HCH_2CONH_2 \tag{8}$$
$$(28)$$

$$R\dot{C}HCH_2CONH_2 + HCONH_2 \longrightarrow RCH_2CH_2CONH_2 + \dot{C}ONH_2 \tag{9}$$

Evidence for the mechanism was that oxamide (29) was a biproduct of electron irradiation, and can arise through dimerization of the formamide radical obtained as shown (equation 9). Alkylsuccin-amides, 30, were also isolated, resulting from combination of the intermediate carboxamide radical 28 with a formamide radical 27. Other amides have also been reported to add to the double bond of olefins[48].

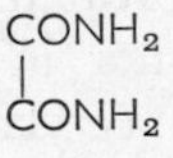

$$\begin{array}{cc} CONH_2 & RCHCH_2CONH_2 \\ | & | \\ CONH_2 & CONH_2 \\ (29) & (30) \end{array}$$

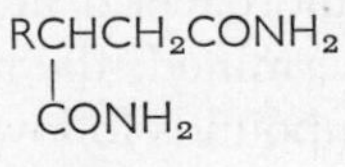

The formation of hydrogen chloride during the γ radiolysis of 5 mole % solutions of amides in carbon tetrachloride has been studied[49]. Other products were chloroform, tetrachloroethylene and hexachloroethane. The amides used were formamide and acetamide, and their N-methyl, and N-phenyl derivatives. G_{HCl} was 8·46 to 23·29 and the N-methylated compounds generally gave higher G values, up to 106 for N,N-dimethylacetamide. The reaction sequence suggested involved N-chloromethylamides (**31**) formed by hydrogen abstraction from the methyl group of the amide, by trichloromethyl radicals (Scheme 2).

$$CCl_4 \longrightarrow \dot{C}l + \dot{C}Cl_3$$

$$RCONR^1CH_3 + \dot{C}Cl_3 \longrightarrow RCONR^1\dot{C}H_2 + CHCl_3$$

$$RCONR^1\dot{C}H_2 + CCl_4 \longrightarrow RCONR^1CH_2Cl + \dot{C}Cl_3$$

$$(31)$$

SCHEME 2.

B. Aqueous Solution

The majority of the ionizing radiation absorbed by a dilute aqueous solution is absorbed by the water molecules. In the absence of oxygen, the solvated electrons (e_{aq}^-) and the hydrogen atoms cause reduction of any organic molecules present. However, in an oxygenated solution the solvated electrons or the hydrogen atoms react with oxygen to form hydroperoxyl ($H\dot{O}_2$) radicals (equations 10 and 11). These are both oxidizing species[50].

$$e_{aq}^- + O_2 \longrightarrow \dot{O}_2^-$$

$$\dot{O}_2^- + H^+ \longrightarrow H\dot{O}_2 \qquad (10)$$

$$\dot{H} + O_2 \longrightarrow H\dot{O}_2 \qquad (11)$$

The rates of reaction of hydrated electrons with N-acylamino compounds have been measured by competition with chloroacetic acid. The effect of the acylamino compounds at varying concentration in reducing the G value for the radiolysis of chloroacetic acid, leads to a value for the ratio of the rate constants. Since the rate constant for the reaction of chloroacetic acid with solvated electrons has been independently determined, the rate constants (in units l/mole sec) for the acylamino compounds follow. For N-acetylalanine, the rate constants were $2·3 \times 10^8$ and $5·7 \times 10^6$ at pH 3 and 7, respectively.

The rate constant for N-ethylacetamide was $5\cdot7 \times 10^6$ at pH 3[51]. Similar rate constants have been determined directly by pulsed radiolysis[52–54]. N-Acetylglycine showed a rate constant of 2×10^7 at pH $5\cdot95$, and N-acetylalanine gave a value of 10^7 pH $8\cdot6 - 9\cdot0$. These values were lower than those of the corresponding free amino acids. However, the acylamino acids exist in solution as negative ions ($CH_3CONHRCO_2^-$). The decrease in reactivity was due to the decrease in the collision frequency because of the repulsion by the negative charge, and also to the absence of the attractive force of the protonated amino group. The rate constants for reaction with acetamide at pH $10\cdot9$ were $1\cdot7 \times 10^7$[55], for acylamide (pH 7) $3\cdot3 \times 10^{10}$[56], formamide (pH 11) $4\cdot2 \times 10^7$[55] and succinimide (pH $8\cdot0$) $7\cdot2 \times 10^9$[57]. The rate constant for reaction of hydrogen atoms with acetanilide at pH 7 has been determined to be $6\cdot7 \times 10^8$ by using isopropanol as scavenger[58]. Similarly the rate constants for the reaction of hydroxyl radicals at pH 9 with acetamide, acetanilide and benzamide were found to be $7\cdot8 \times 10^6$[59], $3\cdot0 \times 10^9$[60] and $2\cdot6 \times 10^9$[60], respectively, by employing p-nitrosodimethylaniline as scavenger.

While most of the studies of the radiolysis products of amide compounds in aqueous solutions have been with polypeptides[61], some studies have been carried out using N-acetylamino acids as models for the peptide linkage. In oxygen-free solution the major product formed in the radiolysis of N-acetylglycine was identified as α,α'-diaminosuccinic acid (G $1\cdot6$) after hydrolysing the solution. This compound was formed, as its diacetyl derivative, through the dimerization of two radicals. The other products were ammonia (G $0\cdot90$), glyoxylic acid together with other carbonyl compounds (G $0\cdot65$) and aspartic acid (**32**) (G $0\cdot13$)[38]. Reductive cleavage of the nitrogen–

$$HO_2CCH_2CH(NH_2)CO_2H$$

(**32**)

carbon bond (equation 12) was not important, since the yield of aspartic acid, formed by attack of the $\dot{C}H_2CO_2H$ radical on acetylglycine, was low and only very small amounts of acetic acid (from $CH_3\dot{C}O$ or $\dot{C}H_2CO_2H$), glycine (equation 13), and succinic acid

$$CH_3CONHCH_2CO_2H + \dot{H} \longrightarrow CH_3CONH_2 + \dot{C}H_2CO_2H \qquad (12)$$

$$CH_3CONHCH_2CO_2H + \dot{H} \longrightarrow CH_3\dot{C}O + NH_2CH_2CO_2H \qquad (13)$$

(from the dimerization of $\dot{C}H_2CO_2H$) were formed. In oxygenated solution, the intermediate radicals reacted with oxygen to form peroxy radicals, which break down according to Scheme 3[62]:

$$CH_3CONHCH_2CO_2H + \dot{O}H \longrightarrow CH_3CONH\dot{C}HCO_2H + H_2O$$

$$CH_3CONH\dot{C}HCO_2H + O_2 \longrightarrow CH_3CONHCH(\dot{O}_2)CO_2H$$

$$CH_3CONH\dot{C}HCO_2H + O_2 \longrightarrow CH_3CON{=}CHCO_2H + H\dot{O}_2 \longrightarrow$$
$$CH_3CONH_2 + CHOCO_2H$$

$$\text{or} \quad CH_3CON{=}CHCO_2H + H_2O \longrightarrow CH_3CONHCH(OH)CO_2H \longrightarrow$$
$$CH_3CONH_2 + CHOCO_2H$$

SCHEME 3.

The G_{NH_3} for acetylglycine and acetylalanine was about 3 in oxygenated 0·1 M solution[63]. This corresponded to G_{OH}[64] and suggested that all the hydroxyl radicals were being scavenged. The yield of carbonyl compounds, however, was 0·8. When ferric ion was used instead of oxygen to scavenge radicals, the G_{NH_3} value for acetylalanine increased from 0·7 to a limiting value of 3·3, and equal amounts of pyruvic acid were formed. Acetylglycine similarly gave equal

$$CH_3CONH\dot{C}HCO_2H + Fe^{III} + H_2O \longrightarrow CH_3CONHCH(OH)CO_2H + Fe^{II} + H^+$$
$$\tag{14}$$

amounts of ammonia and glyoxylic acid. The scavenging reaction involved the reduction of the radical to the hydroxy amide (equation 14). At high Fe^{III}:peptide ratios the following relation existed

$$-G_{amide} = G_{NH_3} = G_{RCOCO_2H} \approx 3.2 \approx G_{OH} + G_{H_2O_2}$$

since the ferric ion scavenged the solvated electrons and hydrogen atoms. At lower ratios the hydrogen atoms were not scavenged and

$$-G_{amide} = G_{NH_3} = G_{RCOCO_2H} \approx 4.0 \approx G_{OH} + G_{H_2O_2} + G_{\dot{H}}$$

The low yield of carbonyl compounds in the presence of oxygen again indicated that the peroxy radical underwent complex reactions[63]. At concentrations of acetylalanine above 0·1 M, the G_{NH_3} value increased, reaching a value of 3 in 2–3 M oxygen-free solution[65]. The increase in ammonia yield was not accompanied by a corresponding increase in the yield of carbonyl compounds, which remained at $G \sim 0·7$. Formate ion did not reduce the G_{NH_3} value, which cannot arise from reactions of $\dot{O}H$. The major organic product found was propionic acid, formed with G 1·6 in 2 M solution. However, no propionic acid was detected in 0·5 M acetylalanine solution. Propionic acid was not formed through a direct reaction with hydrated electrons (equation 15), since the addition of chloroacetate did not reduce G_{NH_3}. However, naphthalene sulphonic acid, which is an effective quencher of exicted states, decreased G_{NH_3} to 1·0 at only 0·025 M concentration. Apparently excitation of acetylalanine was caused by low-energy electrons, and the excited species that reacted with another molecule (equation 16).

$$e_{aq}^- + CH_3CONHCH(CH_3)CO_2H \longrightarrow (CH_3CONHCH(CH_3)CO_2H)^{\bar{\cdot}} \longrightarrow$$
$$CH_3CONH^- + \dot{C}H(CH_3)CO_2H \quad (15)$$

$$CH_3CONHCH(CH_3)CO_2H^* + CH_3CONHCH(CH_3)CO_2H \longrightarrow$$
$$CH_3CONH\dot{C}(CH_3)CO_2H + CH_3CONH_2 + \dot{C}H(CH_3)CO_2H \quad (16)$$

An early study of the effect of x-ray on the ultraviolet light absorption of N-acetyl-N-methyl-α-aminobutyric acid and N-benzoyl-α-aminoisobutyric acid[66] had shown that the absorption was displaced to longer wavelength. Ammonia was liberated, indicating that hydrolysis had occurred. In more recent studies[67], the irradiation of alanine anhydride (33) in oxygen-free solution did not alter the ultraviolet spectra. However, in the presence of oxygen a maximum at 3200 Å developed, and this was accelerated by adding either acid or base. The absorption was shown to be due to 2-hydroxy-3,6-dimethylpyrazine (34), which was formed with a G value of 0·4[68].

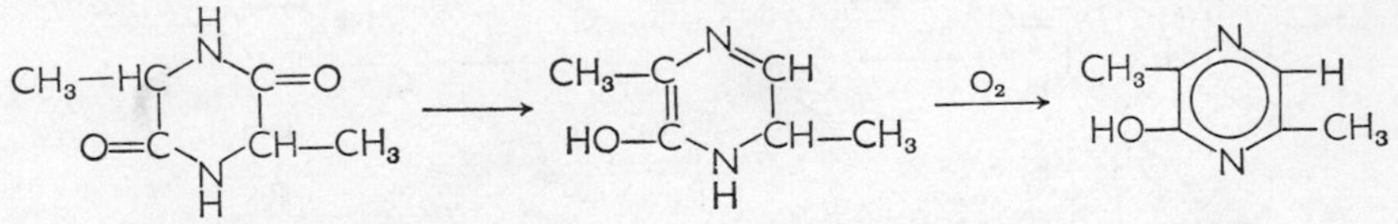

(33)   (34)

Further studies have shown that glycine anhydride and the mixed glycine–serine (35a) and alanine–serine (35b) anhydrides, on irradiation in the solid state and dissolution in oxygen-free water, showed no

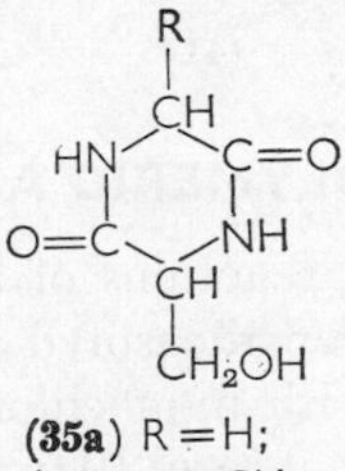

(35a) R = H;
(35b) R = CH$_3$

ultraviolet absorption. However, the addition of 0·02 M sodium hydroxide resulted in rearrangement of the dehydropeptides (e.g. **36**) with the formation of the dihydroxypyrazines (**37a,b**) with maxima at

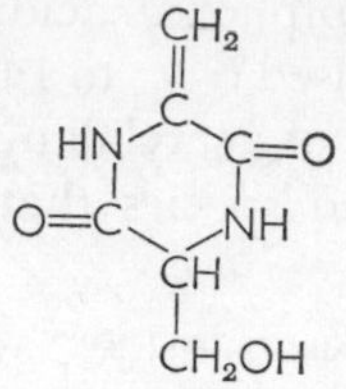

(**36**)

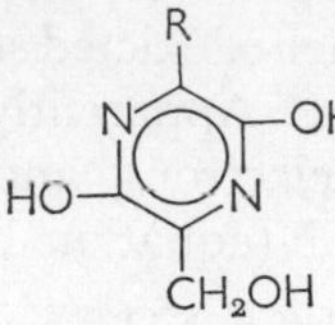

(**37a**) R=H;
(**37b**) R=CH$_3$

3700 Å.   The anhydrides of phenylalanine–glycine (**38a**) and phenylalanylalanine (**38b**) under the same conditions formed 3-benzylidene-2, 5-diketopiperazines (**39**), which existed in the enol forms (**40**) in basic solution, and showed a maximum at 3250 Å[69].

Nicotinamide (niacinamide) (**41**) was reported to be destroyed on radiolysis in aqueous solution[70]. N-Alkylnicotinamides were reduced in air-free solution[71], and nicotinamide N-methyl chloride[72] and N-propyl iodide[73] gave dihydro compounds, which were not the 1,4-dihydro derivatives, but appeared to be dihydro dimers[74].

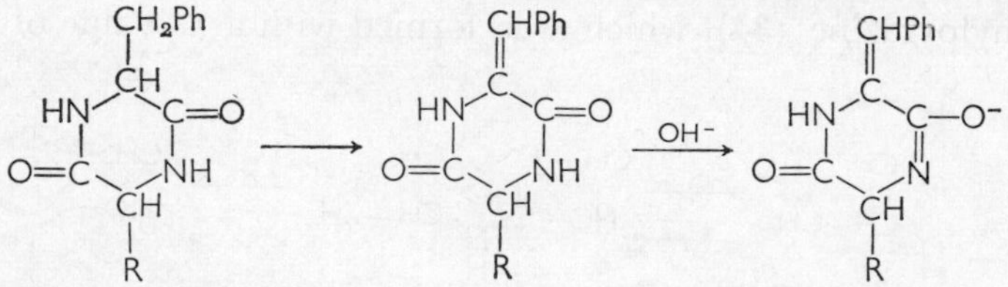

(**38a**) R=H;
(**38b**) R=CH$_3$

(**39**)

(**40**)

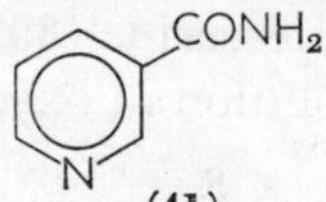

(**41**)

IV. PEPTIDES AND POLYMERIC AMIDES IN SOLUTION

The rate constants for the reactions of a number of peptides with hydrated electrons have been measured in pulsed radiolysis experiments, or by competition with monochloroacetic acid (see Table 1). The peptides generally react at rates 10 times higher than those of their

TABLE 1.   Rate constants for reaction of peptides with
hydrated electrons.

Compound	pH	Method[a]	$10^{-8}k_2$ (l/mole sec)	Ref.
Glycylglycine		pr	3.4	75
	3	ca	8·9	52,76
	6·38	pr	2·5	54,75
	6·7	ca	1·0	76
	7	ca	1·9	51
	11·75	pr	0·5	54,75
Glycylalanine	6·22	pr	2·9	75
Glycylleucine	5·9	pr	1·5	75
	6·46	pr	2·8	54
	8·74	pr	0·7	54
	8·94	pr	0·65	54
Glycylvaline	5·97	pr	2·6	75
Glycylasparagine	5·33	pr	5·4	54
	11·41	pr	0·8	54
Glycylproline	6·66	pr	11	54
Glycylphenylalanine	6·7	pr	1·6	75
Glycyltyrosine	6·13	pr	4·1	54
Glycyltryptophane	6·37	pr	4·5	54
Alanylglycine	6·22	pr	2·1	75
Alanylleucine	6·46	pr	1·3	54
Alanylalanine	6·27	pr	1·3	54
Leucylglycine	6·09	pr	1·1	54
Leucylalanine	6·1	pr	1·5	54
Leucylleucine	5·97	pr	0·9	54
Phenylalanylphenylalanine	5·66	pr	4·5	54
Histidinylhistidine	5·51	pr	79·2	54
	6·83	pr	24·3	54
	7·3	pr	13	75
	8·37	pr	2·85	75
	11·0	pr	0·51	75
Glycylglycylglycine	3	ca	30	76
	6·0	pr	9	54,75
	6·7	ca	7·2	76
	6·7	pr	9	52,54
	7·0	ca	7·2	51
	11·1	pr	0·9	54,75
Leucylglycylglycine	6·0	pr	2·0	54
	9·5	pr	0·5	54
Glutathione		pr	32	75
Glutathione oxidized		pr	46	75

[a] pr = pulsed radiolysis; ca = competition with chloroacetic acid.

constituent amino acids. The protonated amino group had a larger rate constant when the proton was less tightly bound and was the reactive site for hydrated electrons. The higher reactivities of the protonated peptides can be related to their lower pK_a's as compared to the amino acids[52], and a linear relation generally existed between the pK_a and the logarithm of the rate constant[54]. The rate of reaction of hydrated electrons with the peptide bond itself was less than 2×10^7 l/mole sec. Peptides with side-chain amino acids, such as glycylaspartylglycyltyrosine, were more reactive. The rate of reaction of histidinylhistidine varied greatly with the pH, and the changes in rate could be correlated with the various states of protonation of the imidazole groups[54]. The rate constants for reaction of peptides with hydroxyl radicals have also been determined by competition with thiocyanate ion (see Table 2).

TABLE 2. Rate constants for reaction of peptides with hydroxyl radicals.

Compound	pH	$10^{-8}k_2$ (l/mole sec)	Ref.
Glycylglycine	2	0·95	77
	6	1·3	77
	7	1·6	78
Glycylalanine	2	1·1	77
	6	2·1	77
Glycylleucine	2	15	77
Glycylisoleucine	2	14	77
Glycylvaline	2	7·0	77
Glycylserine	2	3·4	77
Glycylmethionine	2	0·65	77
	5	1·3	77
Glycylphenylalanine	2	5·1	77
Glycyltryrosine	2	56	77
Glycylproline	2	8·4	77
Histidinylhistidine	6	54	77
	7	31	79
Glycylglycylglycine	2	0·88	77
	3	1·5	77
	6	2·0	77
	8·6	1·1	77
Triglycylglycine	2	1·4	77
	2·5	2·1	77
	6	2·7	77
	7·8	7·1	77
	9·6	18	77

The simplest dipeptide is glycylglycine, and in early studies[80] it was noted that this compound formed about 50% more ammonia on radiolysis than glycine, whereas diglycylglycine gave slightly less than the amino acid. Glycylalanine and glycylserine afforded 12 and 18% ammonia, respectively[81], and glycylleucine formed glycine (G 0·97), but no leucine on radiolysis in aqueous solution[82]. The G_{-M} value for glycylglycine has been variously reported to be 4·02 at 0·2 Mrad[82], 2·0 at zero dose[83], and 2.79 at 10 Mrad[84,85]. The G_{-M} value has also been reported to decrease with increasing dose[82,86]. The G_{NH_3} value for glycylglycine increased with the concentration of solute and approached a limiting value of 3·0, when all the reactive water species ($\dot{O}H$, $\dot{H}$ and e^-_{aq}) were scavenged[76]. The addition of formate ion, which preferentially scavenges hydrogen atoms, decreased the G_{NH_3} value to 2·5. Since the $G_{e^-_{aq}}$ in water was 2·8, the difference (of 0·3) may be due to the conversion of solvated electrons to hydrogen atoms by the glycylglycine. When chloroacetic acid was added to an oxygen-free solution the G_{NH_3} value decreased to 0·5. Since chloroacetic acid scavenges hydroxyl radicals, the principal reaction liberating ammonia from glycylglycine was due to these radicals.

In early studies of the effect of x-rays on the ultraviolet spectra of peptides and derivatives in solution[66], the following compounds were investigated: glycylleucine, acetylalanyl-α-aminoisobutyric acid and acetylglycylleucylanilide in water, and acetylphenylalanylglycine, acetylphenylalanylalanine, acetylphenylalanylvaline, acetylphenyl-alanylaminoisobutyrylamide, acetylphenylalanylalanylanilide, and acetylglycylleucylanilide in ethanol. The absorption curves of the acetyl dipeptides were displaced to longer wavelength, although the effect was smaller for the anilides than for the compounds with a free carboxyl group. All the dipeptides liberated ammonia, suggesting that the compounds hydrolysed to simple amino acids, which were then dehydrogenated to $NH{=}CR_2$-type intermediates.

More recent work has included the analysis of the products formed on radiolysis[87,88]. The G_{-M} values in oxygen-free solution of leucyl-glycine and glycylleucine were 2·4 and 2·9, respectively[87]. Leucine and glycine individually had G_{-M} values of 2·1 and 1·57, whereas when a mixture was radiolysed the leucine was destroyed preferentially (G_{-Leu} 3·0, G_{-Gly} 0·9). Leucylglycylglycine showed G_{-M} 2·2, with G_{NH_3} 1·25 and G_{CO_2} 0·15. Traces of carboxyl compounds and volatile acids were also formed (each G 0·1). However, the main products were monoamino- and diaminodicarboxylic acids (G 0·7 and 0·55, respectively). The monoaminodicarboxylic acids were identified as

2-amino-3-methyladipic acid (**44**) formed by reaction of radical **42**, derived from leucine, with the glycyl radical **43**, and 2-amino-4-

$$\underset{\text{(42)}}{\underset{|}{HO_2CCH_2CH_2\overset{CH_3}{\overset{|}{\overset{\bullet}{C}H}}}}\underset{NH_2}{\overset{|}{}} + \underset{\text{(43)}}{\overset{\bullet}{C}HCO_2H} \longrightarrow \underset{\text{(44)}}{HO_2CCH_2CH_2\overset{CH_3}{\overset{|}{CH}}CHCO_2H\underset{NH_2}{\overset{|}{}}}$$

methylpimelic acid (**47**) formed by reaction of the radical (**45**) from leucine with the radical **46** from glycine. The diaminodicarboxylic

$$\underset{\text{(45)}}{\overset{CH_3}{\overset{|}{\overset{\bullet}{C}H_2CHCH_2CHCO_2H}}}\underset{NH_2}{\overset{|}{}} + \underset{\text{(46)}}{\overset{\bullet}{C}H_2CO_2H} \longrightarrow \underset{\text{(47)}}{HO_2CCH_2CH_2\overset{CH_3}{\overset{|}{CH}}CH_2CHCO_2H\underset{NH_2}{\overset{|}{}}}$$

acids were diaminosuccinic acid (dimer of **43**), 2,5-diamino-3,4-dimethyladipic acid (**48**) and 2,9-diamino-4,7-dimethylsebacic acid (**49**) (dimer of **45**).

$$\underset{\text{(48)}}{HO_2CCHCH—CHCHCO_2H}$$

$$\underset{\text{(49)}}{HO_2CCHCH_2CHCH_2CH_2CHCH_2CHCO_2H}$$

An analysis of the products formed in the radiolysis of dipeptides derived from glycine, alanine and leucine[88] has shown that the principal reactions in oxygen-free solution resulted from recombination of these radicals. Recombination of the acyl radical (e.g. **46**) with the original peptide gave a new peptide linked by a monoaminodicarboxylic acid. Two radicals (such as **43**) also recombined to form a peptide linked by a diaminodicarboxylic acid, and this accounted for 40% of the products from leucylglycine. Reductive deamination of the *N*-terminal amino acid gave ammonia and a radical on the end carbon atom, which formed an acyl peptide, and amounted for 10% of the reactions. Decarboxylation of the *C*-terminal amino acid also occurred. In the presence of oxygen oxidative deamination afforded an imino compound, which hydrolysed to ammonia and an aldoacyl or ketoacyl peptide (equation 17). Direct cleavage of the peptide also occurred giving an amide and a keto acid (equation 18). Methyl

$$NH_2CHR^1COR \xrightarrow{-2H} NH{=}CR^1COR \xrightarrow{+H_2O} NH_3 + O{=}CR^1COR \tag{17}$$

$$RCONHCHR^1CO_2H \longrightarrow RCON{=}CR^1CO_2H \longrightarrow RCONH_2 + OCR^1CO_2H \tag{18}$$

and methylene groups in the side-chain were also hydroxylated. Thus, 3-, 4- and 5-hydroxyleucine were found in a radiolysed solution of leucylglycine after hydrolysis. The scission of the carbon chain also occurred, since γ-methylglutamylglycine (50), aspartylglycine (51) and homoserylglycine (52) were formed. Terminal decarboxylation was a minor reaction in the presence of oxygen[88].

$$\underset{\substack{\mid\\ NH_2}}{HO_2C\overset{\displaystyle CH_3}{\overset{\displaystyle\mid}{C}}HCHCH_2CHCONHCH_2CO_2H} \qquad \underset{\substack{\mid\\ NH_2}}{HO_2CCH_2CHCONHCH_2CO_2H}$$

(50)   (51)

$$\underset{\substack{\mid\\ NH_2}}{HOCH_2CH_2CHCONHCH_2CO_2H}$$

(52)

Glycylglycyltyrosine was found to afford 3,4-dihydroxyphenylalanine (dopa) on hydrolysis following radiolysis. However, glycylleucyltyrosine and leucyltyrosine did not form dopa[89]. The radiolysis of 0·01 M solutions of glycylmethionine and alanylmethioninc to a dose of 5·1 Mrad destroyed one half of the methionine. However, a dose of 7·3 Mrad was required to destroy one half of the glycine or alanine moiety[90]. Sulphur compounds and other amino acids also protected glycylglycine from radiolysis[84].

The radiolysis of frozen 1M aqueous solutions of glycylglycine, α-alanyl-α-alanine and β-alanyl-β-alanine resulted in the formation of radicals[91], and these had the same structure as those formed in the solid state[92]. The radiolysis of glycylglycylglycine has also been studied[93]. The G_{-M} value for synthetic poly-α-L-glutamic acid has been found to be 3·5 in oxygenated solution[94]. On hydrolysis, amide groups (G 2·3) were formed. The glutamic acid liberated was found to have racemized (G 0·45), and the interpretation given was that the initial radical formed, recombined with a hydrogen atom. The monolayer properties of poly-D,L-alanine and poly-L-tyrosine were affected by irradiation, since there was a reduction in their areas. However, this was apparently due to crosslinking and not to attack on the peptide bond[95].

V. HOT-ATOM REACTIONS

The majority of recoil studies involving carbon-14 and organic compounds have employed amines or other nitrogen-containing compounds as external sources of nitrogen-14 for the $^{14}N(n,p)^{14}C$ reaction. The amide group, however, provides a convenient internal source for the recoil reaction. The carbon-14 recoil atom chemistry of several amides has been studied. Only the carbon-14 products can be analysed, although other non-radioactive compounds must also be formed.

Acetamide was activated and after hydrolysis, 6·4 to 8·1% of the total activity was in the form of acetic acid. This acid had 62% of its activity in the methyl group, although only half the activity was expected, on the basis of equal 'reentry' into the parent molecule. Another product was propionic acid formed in a radiochemical yield* of 4·8 to 6·5%. The propionic acid had 52% of its activity in the methyl group, and 24% in both the methylene and carboxyl groups. The major reaction was then insertion of a carbon-14 atom into a carbon–hydrogen bond of the original acetamide. A trace of labelled acetone (0·13%) was also formed. This had 40% of its activity in each methyl group. A direct replacement of nitrogen for carbon-14 ('knock-on' process) would have resulted in a 50% distribution. Labelled acetonitrile was detected as a product of radiation damage prior to activation[96].

A more detailed analysis of the products from acetamide[97] has given the following radiochemical yields: formaldehyde (1·4), acetamide (3·6), acetic acid (2·0), propionamide (5·0), propionic acid (1·0), acetone (0·5), diacetamide (53) (3·0), acetylacetone (54) (4·2), malonamide (6·1) and succinamide (7·1%).

$$CH_3CONHCOCH_3 \qquad\qquad CH_3COCH_2COCH_3$$
$$(53) \qquad\qquad\qquad\qquad (54)$$

In other recoil atom studies acetamide was activated with epithermal neutrons in a reactor, producing carbon-14 compounds, and also activated with rapid neutrons from an accelerator target, giving carbon-11 labelled compounds, via the reaction $^{12}C(n,2n)^{11}C$. The samples suffered radiation damage. However, the true recoil yields in the absence of radiation damage could be calculated by extrapolating the data. Labelled acetamides formed by replacement amounted to 0·5% of the total radioactive products, N-methylacetamide and propionamide, formed by insertion, amount to less than 5% each, and 2% acetone was also formed by replacement of nitrogen by carbon. The other radioactive products were gaseous. The re-

---

*The radiochemical yield is the yield of a labelled compound based on the total yield of radioactive products.

placement of the carbon atoms of the carbonyl and methyl groups was in the ratio of $65:35$[98].

The activation of propionamide afforded the labelled parent compound ($3 \cdot 3\%$ radiochemical yield), n-butyramide ($2 \cdot 1\%$), isobutyramide ($1 \cdot 7\%$) and a trace of labelled ethyl methyl ketone. The individual products were not degraded in order to determine the extent of labelling in each position. However, the four products were formed in the ratio $4 \cdot 2:2 \cdot 1:2 \cdot 7:1 \cdot 0$. Propionamide and n-butyramide again resulted from reentry and insertion into a carbon–hydrogen bond of the methyl group of propionamide. Isobutyramide also resulted from insertion into a carbon–hydrogen bond of the methylene group, while ethyl methyl ketone was formed by a 'knock-on' reaction of carbon-14 on nitrogen. The gamma radiolysis of solid propionamide also afforded n-butyramide and isobutyramide in the ratio $2 \cdot 0:3 \cdot 1$[99].

Another study showed that the yields of reentry and synthesis for propionamide and malonamide depended on the number of carbon hydrogen atoms involved. Oxamide afforded ^{14}C-labelled oxamide and $\frac{1}{14}$ as much labelled malonamide, formed by insertion of a $^{14}CH_2$ group[100].

Benzamide gave labelled benzoic acid ($3 \cdot 8\%$), after hydrolysis[101], with 87% of the activity in the ring compared to the theoretical amount of $85 \cdot 7\%$ ($\frac{6}{7}$) based on a random 'knock-on' process. Another study indicated $4 \cdot 1\%$ benzamide and $0 \cdot 7\%$ acetophenone[102]. Benzanilide ($1 \cdot 5\%$) has also been reported as a labelled product[103]. Nicotinamide afforded benzoic acid ($0 \cdot 4\%$), as well as the labelled parent compound ($3 \cdot 4\%$)[104].

VI. REFERENCES

1. A. J. Swallow, *Radiation Chemistry of Organic Compounds*, Pergamon Press, New York, 1960.
2. J. W. T. Spinks and R. J. Woods, *An Introduction to Radiation Chemistry*, John Wiley and Sons, New York, 1964.
3. P. B. Ayscough, *Electron Spin Resonance in Chemistry*, Barnes and Noble, New York, 1968.
4. C. F. Luck and W. J. Gordy, *J. Am. Chem. Soc.*, **78**, 3240 (1956).
5. I. Miyagawa and W. Gordy, *J. Am. Chem. Soc.*, **83**, 1036 (1961).
6. M. T. Rogers, S. J. K. Bolte, and P. S. Rao, *J. Am. Chem. Soc.*, **87**, 1875 (1965).
7. S. J. K. Bolte, *Thesis*, Michigan State University, 1963.
8. R. P. Kohin, *Thesis*, University Maryland, 1962.
9. R. G. Bennett, R. L. McCarthy, B. Nolin, and J. Zimmerman, *J. Chem. Phys.*, **29**, 249 (1958).

10. M. Kashiwagi, *J. Chem. Phys.*, **44**, 2823 (1966).
11. E. J. Burrell, *J. Am. Chem. Soc.*, **83**, 574 (1961).
12. A. S. Fomenko, T. M. Abramova, E. P. Dav'eva, and A. A. Galina, *Khim. Vys. Energ.*, **1**, 314 (1967).
13. R. H. Schramm and G. B. Cox, *Am. Phys. Soc. Mtg.*, New York, 1962.
14. M. Iwasaki, K. Toriyama, and B. Eda, *J. Chem. Phys.*, **42**, 63 (1965).
15. R. J. Lontz and W. Gordy, *Paramagnetic Resonance*, Vol. II, Academic Press, New York, 1963, p. 795.
16. R. J. Lontz, *J. Chem. Phys.*, **45**, 1339 (1966).
17. A. J. Restaino, R. B. Mesrobian, H. Morawetz, D. S. Ballantine, G. J. Dienes, and P. J. Metz, *J. Am. Chem. Soc.*, **78**, 2939 (1956).
18. H. Ueda and Z. Kuri, *J. Polymer Sci.*, **61**, 333 (1962).
19. H. Ueda, Z. Kuri, and S. Shida, *J. Polymer Sci.*, **56**, 251 (1962).
20. H. Morawetz and T. A. Fadner, *Makromol. Chem.*, **34**, 162 (1959).
21. T. A. Fadner and H. Morawetz, *J. Phys. Chem.*, **45**, 475 (1966).
22. G. Adler, D. Ballantine, and B. Baysal, *J. Polymer Sci.*, **48**, 195 (1960).
23. G. Adler, D. Ballantine, R. Ranganthan, and T. Davis, *J. Phys. Chem.*, **68**, 2184 (1964).
24. G. Hardy, J. Varga, G. Nagy, and A. Helez, *Magy. Kem. Folyoirat*, **71**, 313 (1965).
25. G. Hardy, J. Varga, G. Nagy, F. Cser, and J. Eroe, *Magy. Kem. Folyoirat*, **73**, 55 (1967).
26. W. Seaman, *J. Phys. Chem.*, **65**, 2029 (1961).
27. I. Miyagawa, Y. Kurita, and W. Gordy, *J. Chem. Phys.*, **33**, 1599 (1960).
28. D. K. Ghosh and D. H. Whiffen, *Mol. Phys.*, **2**, 285 (1959).
29. M. Katayama and W. Gordy, *J. Chem. Phys.*, **35**, 117 (1961).
30. W. C. Lin and C. A. McDowell, *Mol. Phys.*, **4**, 333 (1961).
31. H. G. Freund, *Am. Phys. Soc. Mtg.*, Washington, D.C., 1964.
32. H. C. Box, H. G. Freund, and K. T. Lilga, *J. Chem. Phys.*, **38**, 2100 (1963).
33. R. S. Mangiaracina, *Radiation Res.*, **26**, 343 (1965).
34. J. Depireux and A. Mueller, *Proc. Colloq. AMPERE*, **12**, 2625 (1963).
35. B. Sanaev, K. G. Yanova, V. A. Sharpatyi, A. P. Ibragimov, D. D. Margolin, and B. V. Maslov, *Zh. Fiz. Khim.*, **39**, 2510 (1965).
36. W. Gordy, *Report* PB-181506, Duke Univ., Durham, N.C., 1966, p. 85.
37. V. A. Sharpatyi, K. G. Ianova, A. V. Tuichiev, and A. P. Ibragimov, *Dokl. Akad. Nauk SSSR*, **157**, 660 (1964).
38. W. M. Garrison and B. M. Weeks, *Radiation Res.*, **17**, 341 (1962).
39. R. Badiello and A. Breccia, *Intern. J. Appl. Radiation Isotopes*, **15**, 763 (1964).
40. R. Badiello, A. Breccia, and R. Budini, *Intern. J. Appl. Radiation Isotopes*, **17**, 61 (1966).
41. B. Rajewsky and K. Dose, *Z. Naturforsch.*, **12b**, 384 (1957).
42. K. Narayana Rao and A. O. Allen, *J. Phys. Chem.*, **72**, 2181 (1968).
43. N. Colebourne, E. Collinson, and F. S. Dainton, *Trans. Faraday Soc.*, **59**, 886 (1963).
44. N. Colebourne, E. Collinson, D. J. Currie, and F. S. Dainton, *Trans. Faraday Soc.*, **59**, 1357 (1963).
45. D. A. Armstrong, E. Collinson, and F. S. Dainton, *Trans. Faraday Soc.*, **55**, 1375 (1959).
46. P. Riesz and B. E. Burr, *Radiation Res.*, **16**, 661 (1962).

47. J. Rokach, C. H. Krauch, and D. Elad, *Tetrahedron Letters*, **28**, 3253 (1966).
48. K. Okaka and K. Isawa, *Kagaku Kojo*, **7**, 22 (1963).
49. R. H. Wiley, R. L. S. Patterson, G. F. Chesnut, and E. Grünhut, *Radiation Res.*, **22**, 253 (1964).
50. A. J. Swallow, *Photochem. Photobiol.*, **7**, 682 (1968).
51. R. L. S. Willix and W. M. Garrison, *Report* UCRL-16580, Lawrence Rad. Lab., U. Cal., Berkeley, 1965, p. 207.
52. R. Braams, in *Pulse Radiolysis* (Ed. M. Ebert), Academic Press, New York, 1965, p. 171.
53. R. Braams, *Radiation Res.*, **27**, 319 (1966).
54. R. Braams, *Radiation Res.*, **31**, 8 (1967).
55. M. Anbar and P. Neta, *Intern. J. Appl. Radiation Isotopes*, **18**, 493 (1967).
56. K. Chambers, E. Collinson, F. S. Dainton, and W. Seddon, *Chem. Commun.*, 498 (1966).
57. A. Szutka, J. K. Thomas, S. Gordon, and E. J. Hart, *J. Phys. Chem.*, **69**, 289 (1965).
58. M. Anbar, D. Meyerstein, and P. Neta, *Nature*, **209**, 1348 (1966).
59. M. Anbar, D. Meyerstein, and P. Neta, *J. Chem. Soc. (B)*, 742 (1966).
60. M. Anbar, D. Meyerstein, and P. Neta, *J. Phys. Chem.*, **70**, 2660 (1966).
61. O. H. Wheeler, *Photochem. Photobiol.*, **7**, 675 (1968).
62. B. M. Weeks, S. Cole, and W. M. Garrison, *Report* UCRL-10185, Lawrence Rad. Lab., U. Cal., Berkeley, 1962.
63. H. L. Atkins, W. Bennett-Corniea, and W. M. Garrison, *J. Phys. Chem.*, **71**, 772 (1967).
64. E. Hayon, *Trans. Faraday Soc.*, **61**, 723 (1965).
65. M. A. J. Rodgers and W. M. Garrison, *J. Phys. Chem.*, **72**, 758 (1968).
66. A. J. Allen, R. E. Steiger, M. A. Magill, and R. G. Franklin, *Biochem. J.*, **31**, 195 (1937).
67. M. Kland-English and W. M. Garrison, *Nature*, **189**, 302 (1961).
68. M. Kland-English and W. M. Garrison, *Nature*, **197**, 895 (1963).
69. M. Kland-English and W. M. Garrison, *Report* UCRL-16580, Lawrence Rad. Lab., U. Cal., Berkeley, 1965, p. 210.
70. M. Sjöstedt and L. E. Ericson, *Acta Chem. Scand.*, **16**, 1989 (1962).
71. G. Stein, *J. Chim. Phys.*, **52**, 634 (1955).
72. G. Stein and A. J. Swallow, *Nature*, **173**, 937 (1954).
73. G. Stein and G. Stiassny, *Nature*, **176**, 734 (1956).
74. G. Stein and A. J. Swallow, *J. Chem. Soc.*, 306 (1958).
75. J. V. Davies, M. Ebert, and A. J. Swallow, in *Pulse Radiolysis* (Ed. M. Ebert), Academic Press, New York, 1965, p. 165.
76. R. L. S. Willix and W. M. Garrison, *Radiation Res.*, **32**, 452 (1967).
77. G. Scholes, P. Shaw, R. L. Wilson, and M. Ebert, in *Pulse Radiolysis* (Ed. M. Ebert), Academic Press, New York, 1965, p. 151.
78. G. E. Adams, J. W. Boag, J. Currant, and B. D. Michael, in *Pulse Radiolysis* (Ed. M. Ebert), Academic Press, New York, 1965, p. 131.
79. D. R. Kalkwarf, *Hanford Radiol. Science Res. Dev., Ann. Rept.*, HW-81746 (1963).
80. W. M. Dale, J. V. Davies, and C. W. Gilbert, *Biochem. J.*, **45**, 93 (1949).
81. E. S. G. Barron, *Ann. N.Y. Acad. Sci.*, **59**, 574 (1959).
82. H. Hatano, *J. Radiation Res. (Japan)*, **1**, 38 (1960).

83. M. S. Kang, *Kisul Yon Guso Pogo*, **4**, 46 (1965).
84. O. H. Wheeler, M. Santos, R. A. Ribot, and M. Castro, *Radiation Res.*, **36**, 601 (1968).
85. O. H. Wheeler and D. Julian, unpublished results.
86. O. H. Wheeler and G. Infante, unpublished results.
87. J. Liebster and J. Kopoldova, *Nature*, **203**, 636 (1964).
88. J. Liebster and J. Kopoldova, *Radiation Res.*, **27**, 162 (1966).
89. G. L. Fletcher and S. Okada, *Radiation Res.*, **15**, 349 (1961).
90. F. Shimazu, U. S. Kumta, and A. L. Tappel, *Radiation Res.*, **22**, 276 (1964).
91. V. A. Sharpatyi, K. G. Yanova, A. V. Topchiev, and A. P. Ibragimov, *Zh. Fiz. Khim.*, **39**, 232 (1965).
92. B. Sanaev, K. G. Yanova, V. A. Sharpatyi, A. P. Ibragimov, D. D. Morgolin, and B. V. Maslov, *Zh. Fiz. Khim.*, **39**, 2510 (1965).
93. C. A. Leone and M.-S. Kang, *Radioisotopes (Tokyo)*, **14**, 303 (1965).
94. E. M. Southern and D. N. Rhodes, *Radiation Preservation of Foods*, Am. Chem. Soc., Monograph, Washington, (1966).
95. K. S. Korgaonkar and S. V. Joshi, *Radiation Res.*, **35**, 213 (1968).
96. A. P. Wolf, C. S. Redvanly, and R. C. Anderson, *J. Am. Chem. Soc.*, **79**, 3717 (1957).
97. T. W. Lapp and R. W. Kiser, *J. Phys. Chem.*, **66**, 1730 (1962).
98. B. Diehn, *Thesis*, Univ. of Kansas, (1964).
99. E. Tachikawa and G. Tsuchihashi, *Bull. Chem. Soc. Japan*, **34**, 770 (1961).
100. E. Tachikawa and G. Tsuchihashi, *Radioisotopes*, **10**, 420 (1961).
101. A. P. Wolf, *Chemical Effects of Nuclear Transformations*, Vol. II, International Atomic Energy Agency, Vienna, 1961, p. 3.
102. M. Zifferero and D. Sordelli, *Recerca Sci.*, **26**, 1194 (1956).
103. F. Cacace, L. Cieri, and M. Zifferero, *Ann. Chim. (Rome)*, **47**, 892 (1957).
104. R. C. Anderson, E. Penna-Franca, and A. P. Wolf, *Brookhaven National Lab. Quart. Progress Rept.*, BNL 326 (S-24), (1954).

CHAPTER 7

Chemistry of imidic compounds

OWEN H. WHEELER

Puerto Rico Nuclear Center, Mayaguez, Puerto Rico
and
OSCAR ROSADO

University of Puerto Rico at Mayaguez, Puerto Rico

I. INTRODUCTION

The imide group is considered here as an amino group flanked by two carbonyl groups (**1**). This functional group occurs in acyclic di-acylamines (**2**), such as diacetamide (**2**, $R^1 = R^2 = CH_3$), or as

$$-CONHCO-$$

(1)

$$R^1CONHCOR^2$$

(2)

mixed imides, such as *N*-acetylpropionamide (**2**, $R^1 = C_2H_5$, $R^2 = CH_3$). The imide group is also present in the 4-, 5- and 6-membered ring compounds, malonimide (2,4-azetidinedione, **3**), succinimide (2,5-pyrrolidinedione, **4**) and glutarimide (2,6-piperidinedione, **5**).

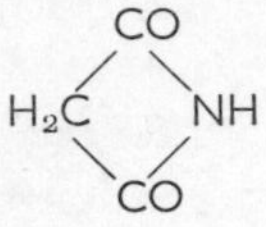

(3) 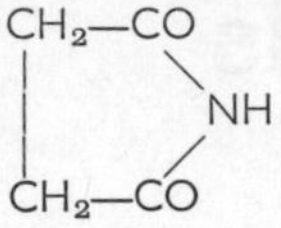   (4) 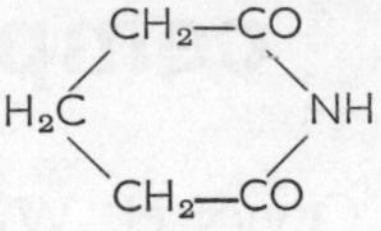   (5)

The unsaturated imide, maleimide (2,5-pyrroledione, **6**) and the aromatic imide, phthalimide (1,3-isoindoledione, **7**) are important compounds. Other imides are named as derivatives of the dicar-

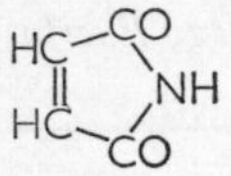

(6) 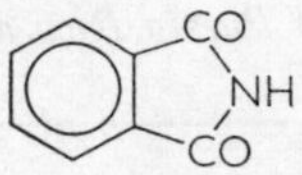   (7)

boxylic acids by adding *imide* to the root name instead of *ic*. Thus compound **8** is cyclohexane-1,2-dicarboximide (hexahydrophthalimide) and compound **9** is naphthalene 2,3-dicarboximide (2,3-benz[*f*]isoindole-1,3-dione). More complex compounds are named

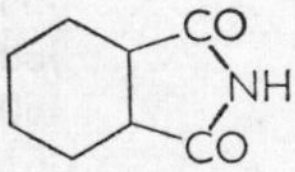

(8) 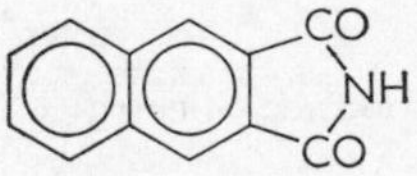   (9) 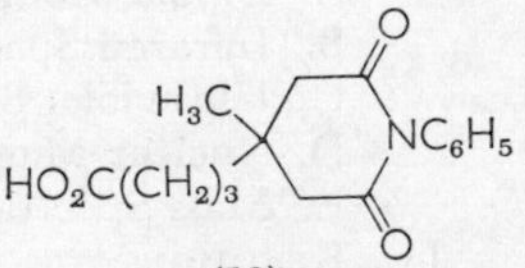   (10)

systematically from the corresponding ring system. Thus, compound **10** is 4-(4-methyl-2,6-dioxo-1-phenyl-4-piperid-4-yl)butyric acid.

The presence of other functional groups can modify the properties of the imidic group. The chemistry of hydantoin (**11**) and isocyanuric

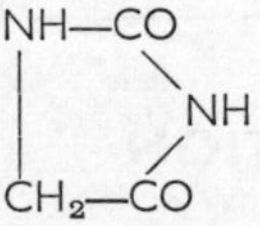

(11) 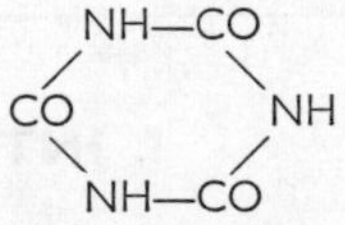   (12)

acid (**12**) is vastly different from that of the imides and is not considered in this chapter. The carbodiimides (**13**) are not imidic compounds. The reactions of the isoimides, such as *N*-alkylphthalisoimide

$$RN{=}C{=}NR$$
(**13**)

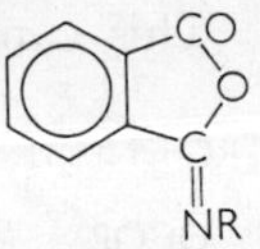

(**14**)

(**14**) are only considered in relation to those of the corresponding imides.

II. SYNTHESIS

A. Synthesis from Amides and Amidic Acids

The diamides of some dicarboxylic acids can be converted to the corresponding imides by heating. Thus maleic acid diamide (maleamide) gave maleimide[1]. Diglycollamide (**15**) also afforded diglycollimide (**16**) when heated. Maleamide was also converted

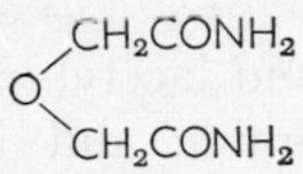

to maleimide by heating with zinc chloride[2], and treatment of succinamide with sulphur monochloride gave succinimide[3]. Phthalamide (**17**) lost ammonia on melting at 222° to form phthalimide[4].

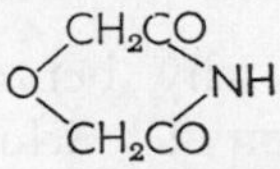

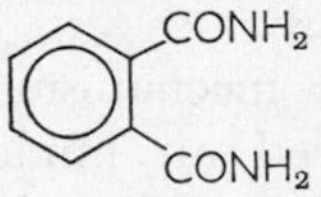

Diacetamide was formed by treating acetamide with acetyl chloride in benzene[5]. Propionamide when heated under reflux with triphenylmethyl chloride gave dipropionamide (**18**). Heating isobutyramide with triphenylmethyl carbinol also afforded diisobutyramide (**19**) and *N*-(triphenylmethyl)isobutyramide (**20**)[6].

(EtCO)₂NH	(i-PrCO)₂NH	i-PrCONHCPh₃
(**18**)	(**19**)	(**20**)

Amides react with the acid anhydrides in two ways to form either imides or nitriles. The uncatalysed reaction produces nitriles, while strong acid catalysis produces the imides[7]. Thus, the uncatalysed reaction of propionamide with propionic anhydride resulted in a high yield of propionitrile[8].

Since dipropionamide did not form propionitrile under the same

12+c.o.a.

conditions, it follows that the imide cannot be considered an intermediate in the formation of propionitrile. The primary product of
the attack of an anhydride on an amide may be the isoimidinium
carboxylate (21). This can decompose into a nitrile or rearrange

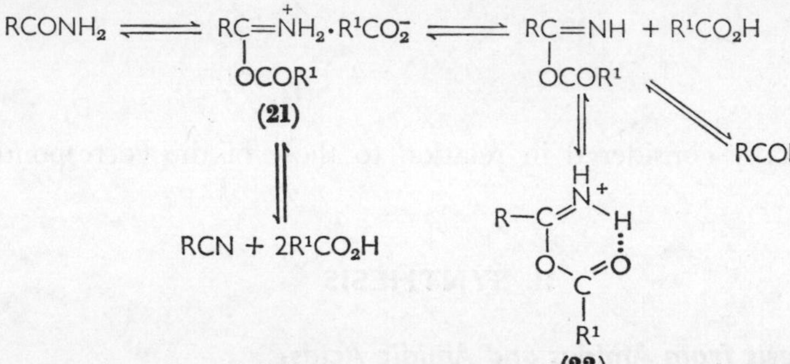

$$RCONH_2 \rightleftharpoons \underset{\underset{OCOR^1}{|}}{RC}{=}\overset{+}{N}H_2 \cdot R^1CO_2^- \rightleftharpoons \underset{\underset{OCOR^1}{|}}{RC}{=}NH + R^1CO_2H$$

$$(21) \qquad\qquad\qquad RCONHCOR^1$$

$$RCN + 2R^1CO_2H$$

(22)

into an imide, but the rate of decomposition is much greater than that
of the uncatalysed rearrangement. The decomposition reaction is of
the six-centre type (22) and not subject to acid catalysis, while the rearrangement is an intramolecular acylation susceptible to acid
catalysis.

This mechanism explains why benzamide and acetyl chloride
produced an imide (equation 1), while acctamide and benzoyl
chloride gave a nitrile (equation 2). The inductive effects of the

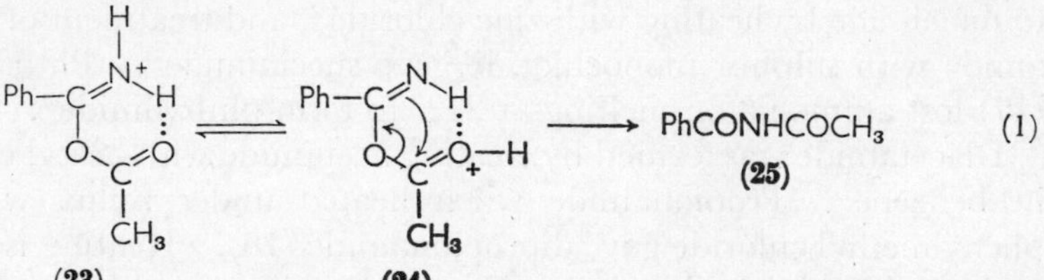

$$\longrightarrow PhCONHCOCH_3 \qquad (1)$$

$$(25)$$

(23)   (24)

$$\longrightarrow CH_3CN + PhCO_2H + H^+ \qquad (2)$$

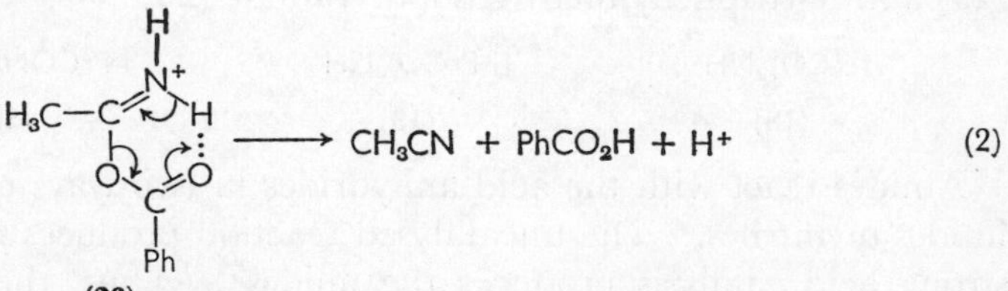

(26)

phenyl and methyl groups in the intermediate 23 formed from benzamide, promoted proton transfer to the tautomeric form 24, which

readily rearranged to the imide (**25**) by attack of the activated carbonyl group on the tertiary nitrogen atom[8]. However, in the intermediate from acetamide (**26**), the interchange of the groups results in
inhibiting the proton transfer sufficiently to allow the six-centre
reaction to predominate, and this then breaks down into the nitrile
and carboxylic acid.

The catalysed reaction has been used to prepare dipropionamide
in good yield from propionic anhydride, propionamide, and sulphuric
acid[9]. Such catalysts as hydrogen chloride, acyl chloride, or sulphuric acid are helpful with acid anhydrides; and pyridine has been
used for acyl halides.

Benzamide reacted with acetic anhydride to produce N-acetyl-

$$PhCONH_2 + (MeCO)_2O \longrightarrow PhCONHCOMe$$

(**25**)

benzamide (**25**). Similarly a 52% yield of N-benzoyl-N-phenyl-
acetamide (**27**) was obtained when a mixture of phenylacetic
anhydride and benzamide was heated in the presence of sulphuric
acid. When two equivalents of phenylacetic anhydride were used,
bis-phenylacetimide (**28**) was produced in 60% yield. The same compound was obtained in 96% yield when benzamide was heated with an
excess of phenylacetyl chloride. This behaviour was explained by a
mechanism involving the formation of phenylacetamide (**29**) from
the protonated form of **27**[9]. The fact that the imide **28** was formed

$$PhCONH_2 + PhCH_2COCl \longrightarrow PhCH_2CONHCOPh \xrightarrow{PhCH_2COCl} \begin{array}{c} PhCH_2CO \\ \diagdown \\ NH \\ \diagup \\ PhCH_2CO \end{array}$$

(**27**)   (**28**)

from either phenylacetic anhydride or phenylacetyl chloride is evi-

$$PhCONH_2 \underset{\xleftarrow{}}{\overset{PhCH_2\overset{+}{C}=O}{\rightleftharpoons}} PhCO\overset{+}{N}H_2COCH_2Ph \rightleftharpoons Ph\overset{+}{C}=O + PhCH_2CONH_2$$

(**29**)

$$\xrightarrow{PhCH_2\overset{+}{C}=O} PhCH_2CONHCOCH_2Ph$$

(**28**)

dence for the phenylacetyl carbonium ion acting as intermediate, since
it is common to both the anhydride and chloride. The greater
reactivity of the phenylacetyl carbonium ion over the benzyl

carbonium ion explains why the reaction produced the symmetric imide **28**.

In pyridine, the reaction of imides with acid chlorides yields a mixture of the triimide, diimide and a nitrile.　Since the isoimide (**31**) is a stronger base than the original amide (**30**), it follows that it is

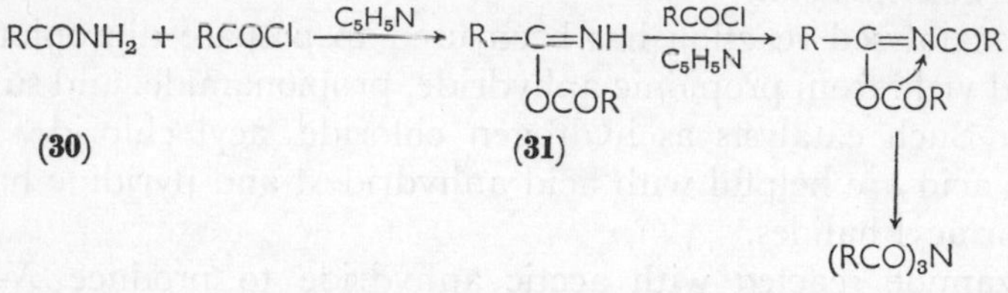

$$RCONH_2 + RCOCl \xrightarrow{C_5H_5N} R-\underset{\underset{OCOR}{|}}{C}=NH \xrightarrow[C_5H_5N]{RCOCl} R-\underset{\underset{OCOR}{|}}{C}=NCOR$$

$$(\textbf{30}) \qquad\qquad\qquad (\textbf{31})$$

$$\downarrow$$

$$(RCO)_3N$$

acylated in preference[9].　N-Acetylpyrrolidone (**32**, $n = 3$), N-acetyl-caprolactam (**32**, $n = 5$) and similar compounds were formed directly from the cyclic lactam and acetic anhydride[10].　Succinimide, phthalimide, tetrahydrophthalimide and naphthalimide have been N-acetylated using ketene in the presence of acetic acid in carbon tetrachloride[11].

The reaction of thioacetamide with acetyl chloride in acetonitrile in the presence of pyridine produced N-acetylthioacetamide (**33**), and

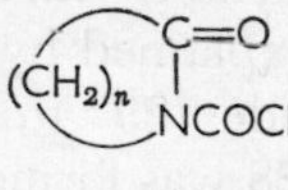

$$(\textbf{32}) \qquad\qquad\qquad (\textbf{33}) \qquad\qquad\qquad (\textbf{34})$$

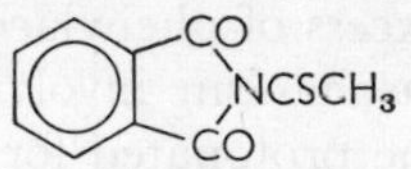

phthaloyl chloride reacted with thioacetamide under similar conditions to produce N-thioacetylphthalimide (**34**)[12].

The amide and carboxylic groups in amidic acids react internally to form cyclic imides.　Thus, glutarimide can be synthesized by distilling glutaramic acid[13].　Phthalamic acid (**35**, R = H) also cyclized to phthalimide when heated to 155°[14], and phthalanilic acid

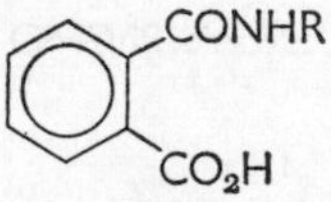

$$(\textbf{35}) \qquad\qquad\qquad (\textbf{36})$$

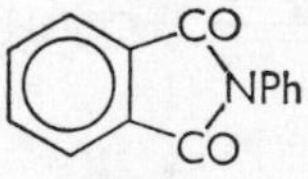

(**35**, R = Ph) gave phthalanil (**36**) when melted at 170°[15].　However, treatment of phthalanilic acid with acetyl chloride produced 3-phenyliminophthalide (as-phthalanil, **37**)[16].　This compound rearranged to phthalanil (**36**) when shaken with concentrated potassium

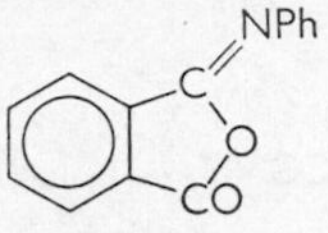

$$\text{(37)} \qquad\qquad \text{(38)} \qquad\qquad \text{(39)}$$

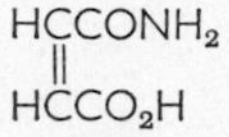

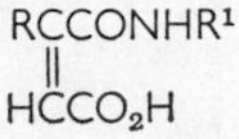

carbonate[17]. Maleamic acid (38) cyclized to maleimide[18]; N-alkyl-maleamic acids (39, R = H) and N-alkylcitraconamic acids (39, R = CH$_3$) ring-closed to the corresponding substituted maleimides[19]. N-(p-Anisyl)maleamic acid, was converted to N-(p-anisyl)maleimide when heated with phosphorus pentoxide in toluene[20]. Similarly, N-phenylmaleamic acid gave N-phenylmaleimide when heated with phosphorus pentoxide or with sodium acetate and acetic anhydride. N-α-Naphthylmaleamic acid also formed the corresponding male-imide, but N-(p-carboxyphenyl)maleamic acid (40) and N-carbeth-oxymaleamic acid (41) could not be cyclized[21]. Oxanilic acid (42)

$$\text{HCO}_2\text{CH}{=}\text{CHCONHC}_6\text{H}_4\text{CO}_2\text{H-}p \qquad\qquad \text{HCO}_2\text{CH}{=}\text{CHCONHCO}_2\text{C}_2\text{H}_5$$

$$\text{(40)} \qquad\qquad\qquad\qquad\qquad \text{(41)}$$

has been reported to give oxanil (43, R = C$_6$H$_5$) when treated with

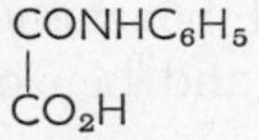

$$\text{(42)} \qquad\qquad \text{(43)} \qquad\qquad \text{(44)}$$

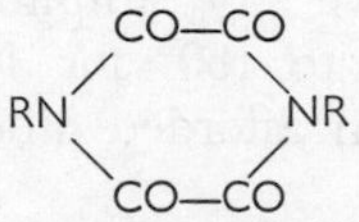

thionyl chloride[23]. However, the product was shown to be 1,4-diphenyl-2,3,5,6-tetraketopiperazine (44, R = C$_6$H$_5$)[24]. Attempts to form oximide (43, R = H) also resulted in the formation of tetra-ketopiperazine (44, R = H)[25].

3,3-Diaryl- and dialkyl-2,4-azetidinediones have been synthesized in a series of reactions in which the disubstituted cyanoacetic esters (45) were partially hydrolysed with concentrated sulphuric acid to the ester monoamides. These were then hydrolysed with alcoholic potassium hydroxide to the malonic monoamides (46), which were cyclized by treating with thionyl chloride in pyridine[26]. Ethyl cyclohexyl-n-propylcyanoacetate (45, R = C$_6$H$_{11}$; R^1 = n-Pr; R^2 = Et) was similarly converted to 3-cyclohexyl-3-n-propylmalonimide[27]. The hydrolysis of 3-carbethoxy-2-phenylsuccinamic acid (47) with sodium hydroxide gave α-carboxy-α'-phenylsuccinimide (48)[28].

Succinamic acid azide (49) afforded succinimide when treated with

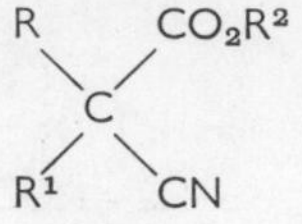

R   CO$_2$R^2
  \   /
   C
  /   \
R^1 CN

(45)

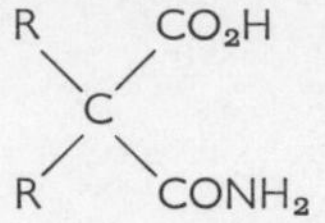

R   CO$_2$H
  \   /
   C
  /   \
R   CONH$_2$

(46)

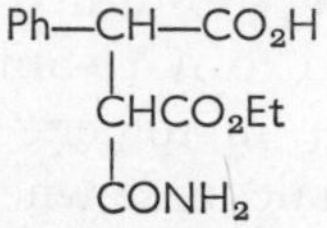

Ph—CH—CO$_2$H
      |
    CHCO$_2$Et
      |
    CONH$_2$

(47)

HO$_2$C
     \
      CH—CO
     /   \
                NH
     \   /
      CH—CO
     /
   Ph

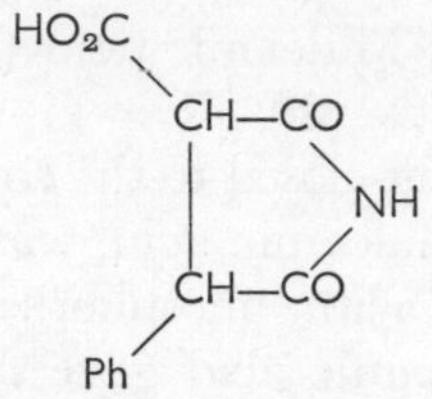

(48)

0·2 N sodium hydroxide[29].   The hydrochloride of 5-cyanovaleramide

NH$_2$COCH$_2$CH$_2$CON$_3$

(49)

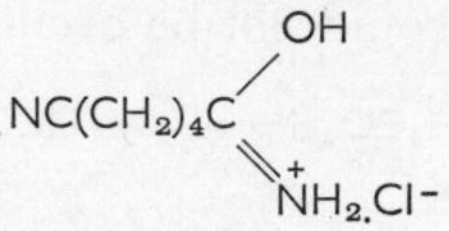

                OH
               /
NC(CH$_2$)$_4$C
               \\
                NH$_2$.Cl$^-$

(50)

(NC(CH$_2$)$_4$CO)$_2$NH

(51)

(50) gave 25% adipimide and 23% bis(5-cyanovaler)amide **(51)** on heating to 180° for 1 hr. The hydrochloride of benzamide (**52**, R = Ph) afforded dibenzamide (**53**), benzonitrile and benzoic acid

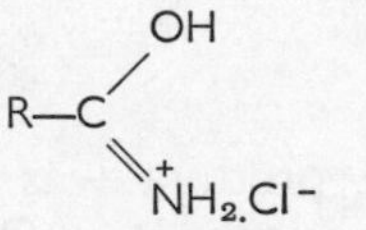

             OH
            /
    R—C
            \\
             NH$_2$.Cl$^-$

(52)

(PhCO)$_2$NH

(53)

on heating, and acetamide hydrochloride (**52**, R = CH$_3$) formed di-acetamide, acetyl chloride and acetic acid[30].

The rate of cyclization of the N-arylmaleamic acids to imides in the solid phase below their melting points has been studied.   N-(p-Aminophenyl)maleamic acid was 7% cyclized in 199 min at 214°. The N-p-chlorophenyl compound was 96% cyclized in 40 min at 143°, and N-(p-tolyl)maleamic acid gave 76% imide at 160° and 92% at 205° min[31].

A number of cyclic amino acids (**54**, n = 2–4, **55**, **56**, and **57**) could not be cyclized to bicyclic amides or imides, because of the strain which would be introduced in the intermediates leading to the bicyclic compounds (Bredt's rule)[32].

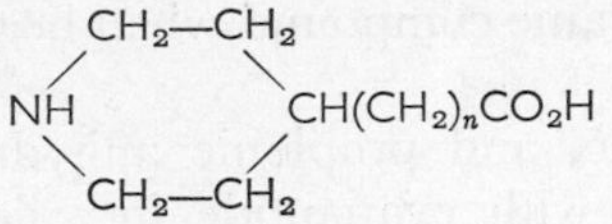

(54)

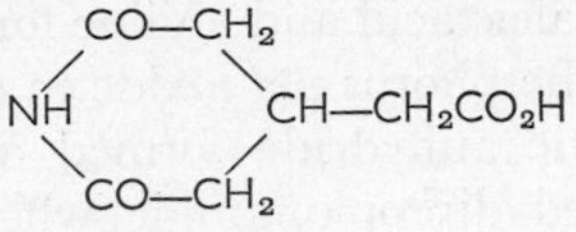

(55)

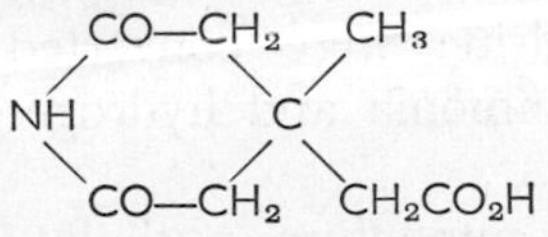

(56)

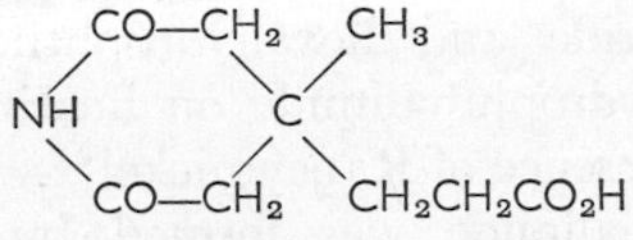

(57)

B. Synthesis from Carboxylic Acids and Derivatives

The reaction of an acid or acid derivative with amines or ammonia has been extensively used in the preparation of imides.

A comparison has been made of the yields of N-substituted succinimides and glutarimides formed by heating primary amines with succinic or glutaric acid, either at 230–260° for 8–10 hr in a sealed tube, or at 125–175° for 3–5 hr distilling off the water, or by azeotropic distillation of water with p-cymene[33]. The latter method gave better yields. Glutaric acid formed some glutaric acid diamide, although no succinic diamide was formed.

When succinic anhydride was heated with ammonia, succinimide as well as succinamic acid and succinamide were formed[34]. Glutarimide was obtained when glutaric acid and ammonium hydroxide were refluxed and then heated to 170–180°[35]. *cis*-Cyclohexane-1,3-dicarboximide was also prepared by distilling a mixture of the corresponding dicarboxylic acid and ammonium hydroxide[36]. Cyclopentane-1,2-dicarboximide was similarly prepared from the corresponding anhydride[37]. N-(n-Dodecyl)-*cis*-cyclohexane-1,2-dicarboximide resulted from heating the anhydride with n-dodecylamine in xylene[38]. N-(2-Hydroxyethyl)camphorimide (58) was prepared by heating camphoric anhydride with ethanolamine. This compound was converted to the N-vinylimide (59) by heating in a sealed tube to 500°[39]. Aniline maleate formed N-phenylmaleimide on heating[1],

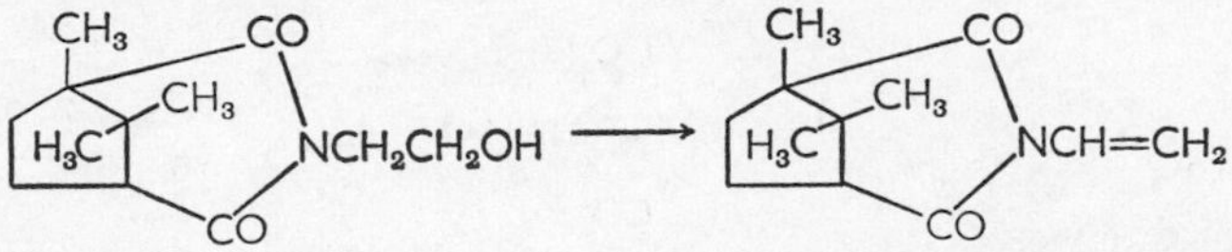

(58)   (59)

and maleic acid and aniline formed the same compound when heated with phosphorus pentoxide in dioxan[22].

Acetic anhydride formed diacetamide, and propionic anhydride afforded dipropionamide when heated with cyanamide in xylene. A 2:1 mixture of acetic and propionic anhydrides gave 19·5% diacetamide and 22·5% dipropionamide[40]. Tetrahydrophthalic anhydride and hexahydrophthalic anhydride were converted to hexahydrophthalimide on heating with ammonia and hydrogen in the presence of Raney nickel[41].

Phthalimide was formed by heating ammonium phthalate or phthalic anhydride and gaseous or aqueous ammonia to 300°[42]. Heating phthalic acid or phthalic anhydride with ammonium carbonate or urea also gave phthalimide[43]. Phthalic anhydride also reacted with hydrazine in ethanol to form N-aminophthalimide (60, R = H). However, on heating with an excess of hydrazine or on

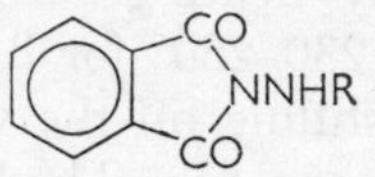

(60) 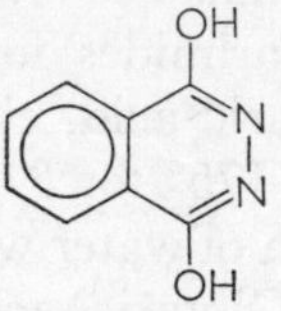   (61)

treatment with acid or alkali this compound transformed to 1,4-dihydroxyphthalazine (61)[44]. Phthalic anhydride and phenylhydrazine yielded N-anilinophthalimide (60, R = C$_6$H$_5$)[45]. 4-Amino-N-arylphthalimides (63) have been prepared by reacting the dimethyl p-aminophthalate (62) with anilines[46].

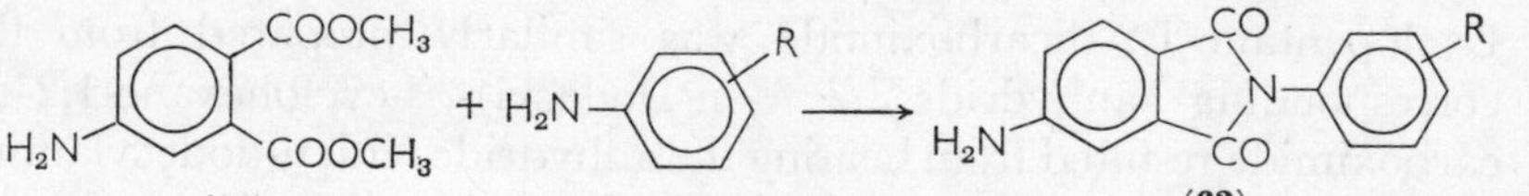

(62)   + H$_2$N—⟨ ⟩—R ⟶   (63)

Disubstituted malonic acids were converted to the substituted N-aminomalonimides (64) by reaction with hydrazine or substituted hydrazines[47]. Thus, 1-amino-3,3-diethyl-2,4-azetidinedione (64,

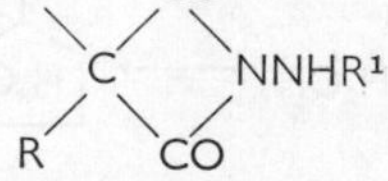

(64)

$R = Et$; $R^1 = H$) was synthesized by reacting diethylmalonyl chloride with acetone hydrazone in methylene chloride in the presence of triethylamine, followed by treatment with cold hydrogen chloride in ethanol[48].

Phthalic anhydride reacted with amino azides to form N-azido-alkylphthalimides[49]. In addition, azidoacetamide (**65**, R = H) con-

$$RCH(N_3)CONH_2 + N_3CH_2COCl \longrightarrow RCH(N_3)CONHCOCH_2N_3$$
$$\quad\quad (65) \quad\quad\quad\quad (66) \quad\quad\quad\quad\quad\quad\quad (67)$$

densed with azidoacetyl chloride (**66**) in boiling xylene to form bis-(azidoacetyl)imide (**67**, R = H). The azidoimide was converted to diglycylimide (**68**, R = H) on reaction with hydrogen bromide in

$$RCH(NH_2)CONHCOCH_2NH_2 \quad\quad\quad RCH(NH_2)CONHCH_2CO_2H$$
$$\quad\quad\quad (68) \quad\quad\quad\quad\quad\quad\quad\quad\quad\quad (69)$$

acetic acid and acetone.   Diglycylimide rearranged on treatment with a trace of triethylamine in 85% ethanol to glycylglycine (**69**, R = H). Other azidoamides (**65**, R = Me, i-Pr etc.) were subjected to the same series of reactions, and the asymmetric imides (**68**) rearranged to a mixture of dipeptides[50].

α-N-Phthalimidoglutarimide (thalidomide, **71**) has been synthesized by first reacting phthalimide with glutamic acid to form α-(N-phthalimido)glutaric acid (**70**). This compound was converted successfully to the anhydride and the glutaric acid amide, which was cyclized to **71** by heating to $200°$[51].   The 3- and 4-hydroxy-phthalimide derivatives, which are metabolites of thalidomide were prepared in an analogous fashion[52].

A general route for the preparation of imides is the reaction of a carboxylic acid with a nitrile (equation 3)[53].   This reaction, pre-

$$RCN + RCO_2H \longrightarrow (RCO)_2NH \quad\quad\quad\quad\quad (3)$$
$$ArCO_2H + RCN \longrightarrow RCO_2H + ArCN \quad\quad\quad (4)$$

viously considered purely thermal, is acid catalysed.   Aromatic carboxylic acids give apparent metathesis (equation 4), while acid anhydrides form triamides (**72**).   Cyclic imides can be synthesized using this reaction, and succinimide was formed from β-cyanopropionic

12*

$$RCN + (RCO)_2O \longrightarrow (RCO)_3N$$

$$(72)$$

acid by internal addition[54,55]. A mixture of benzyl cyanide and phenylacetic acid formed 17% bis-phenylacetamide and 6·7% phenylacetic anhydride on heating. Amyl cyanide and hexanoic acid similarly afforded 9·4% dihexanoamide and 2% hexanoic anhydride[8].

Homophthalimide (74) has been prepared by the action of acids on the dinitrile, o-cyanobenzyl cyanide (73)[56]. The reaction must

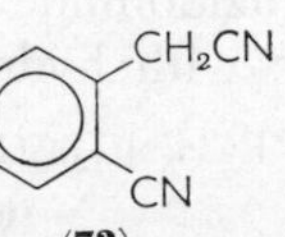

(73) 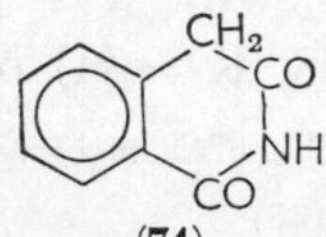   (74)

involve a partial hydration of one of the cyanide groups[57]. 1,2-

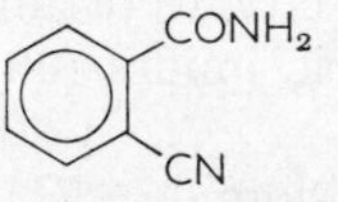

(75) 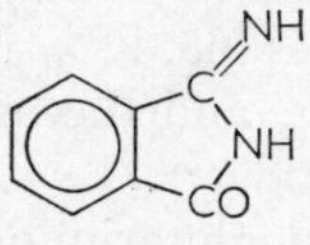   (76)

Dicyanoethane gave succinimide on heating with dilute sulphuric acid at 153–170° for 6 hr, and 1,3-dicyanopropane formed glutarimide by heating for 5–10 hr at 180–200°[58]. o-Cyanobenzamide (75), however, formed imidophthalimide (76) on heating[58,59].

The reaction of a thiocarbonyl group with a nitrile is a general method for preparing cyclic imides with a sulphur atom in the ring[60–62]. Thus, 2,4-thiazolidenedione (77) was synthesized by treating thiocyanoacetic acid with sulphuric acid[60]. β-Thiocyanopropionic acid was also converted into 1,3-thiazane-2,4-dione (78) using small amounts of thionyl chloride, phosphorus oxychloride or aluminium trichloride[54]. This same compound was prepared by treating β-iodopropionic acid with xanthogenamide (79) and acetic anhydride[63].

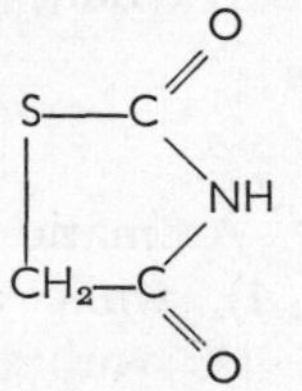

(77) 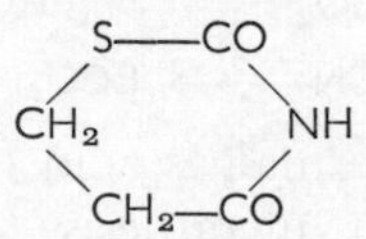   (78) 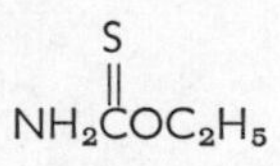   (79)

C. Synthesis from Isocyanates

The reaction of an isocyanate with an acid anhydride has been used for the preparation of imides. Phthalic anhydride was found to react with phenyl isocyanate giving N-phenylphthalimide (**80**) in 71% yield. Acetic anhydride also reacted with phenyl isocyanate to yield N-phenyldiacetamide[64]. To explain the formation of imides, it is

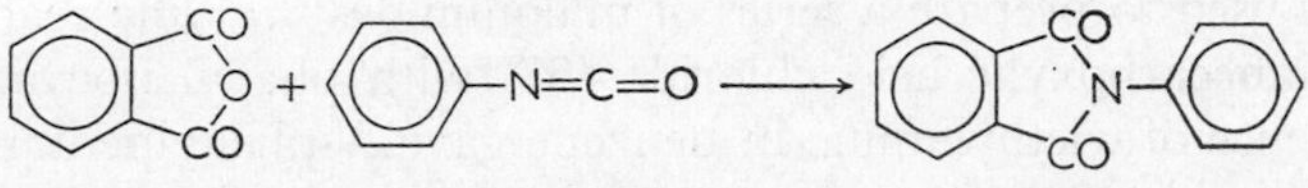

(**80**)

assumed that there is an initial addition of the anhydride to the isocyanate. The resulting mixed anhydride **81** is unstable and

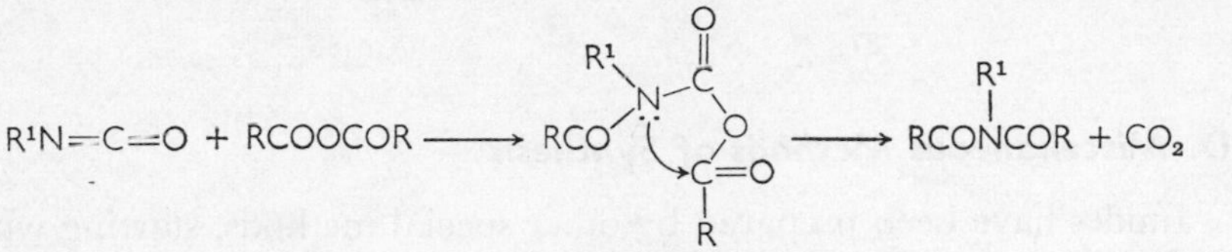

$$R^1N{=}C{=}O + RCOOCOR \longrightarrow RCO \longrightarrow RCONCOR + CO_2$$

(**81**)

loses carbon dioxide[9] originating from the isocyanate carbonyl group. Evidence for this came from the fact that carbon oxysulphide (COS) was evolved in the reaction of isothiocyanates (RNCS), and the reaction of phenyl isocyanate and acetic-1-^{14}C acid liberated unlabelled carbon dioxide (equation 5)[65].

$$PhNCO + CH_3{}^{14}CO_2H \longrightarrow PhNHCOO^{14}COCH_3 \longrightarrow PhNH^{14}COCH_3 + CO_2$$

(5)

Monothiophthalimides (**83**) and monothiohomophthalimides have been synthesized by the cyclization of acyl isothiocyanates (**82**) with aluminium trichloride[66]. Acylureas can also be prepared by the reaction of isocyanates with imides (equation 6), and the substituents

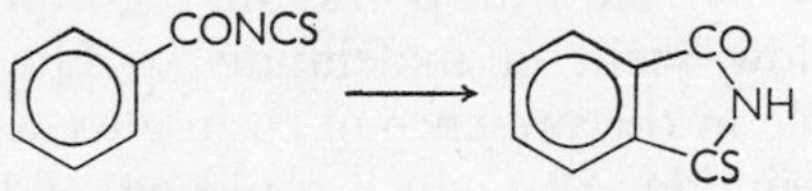

(**82**)   (**83**)

can be alkyl or aryl groups[67,68].

$$RNCO + R^1CONH_2 \longrightarrow RNHCONHCOR^1$$

(6)

When diphenylketene (**84**) was reacted with methyl isocyanate (**85**), the product was *N*-methyldiphenylmalonimide (**86**), and this reac-

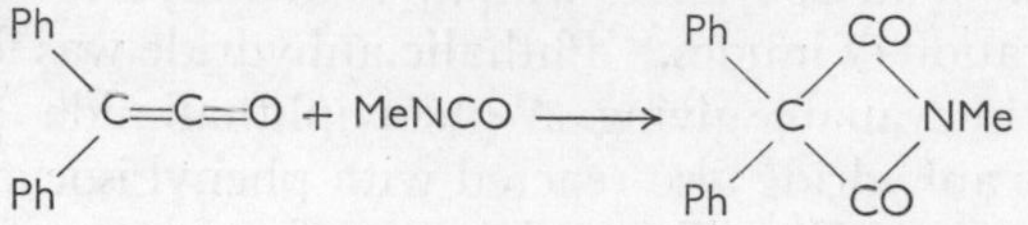

$$\underset{\text{(84)}}{\overset{\text{Ph}}{\underset{\text{Ph}}{\diagdown}}\text{C}{=}\text{C}{=}\text{O}} + \underset{\text{(85)}}{\text{MeNCO}} \longrightarrow \underset{\text{(86)}}{\overset{\text{Ph}}{\underset{\text{Ph}}{\diagdown}}\text{C}\overset{\text{CO}}{\underset{\text{CO}}{\diagup}}\text{NMe}}$$

tion was used to prepare a series of malonimides[69]. The reaction of cyclohexanecarboxylic acid chloride (**87**) with phenyl isocyanate in the presence of triethylamine in benzene gave *N*-phenylpentamethyl-enemalonimide (**88**), from the ketene formed *in situ*[70].

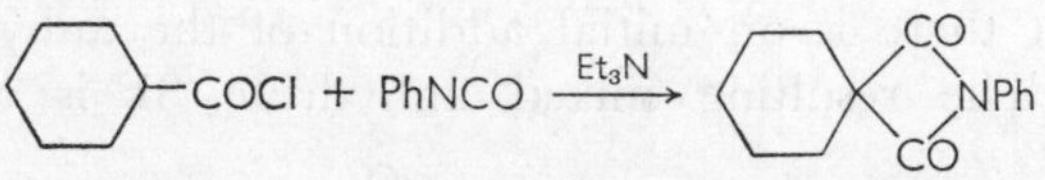

D. Miscellaneous Methods of Synthesis

Imides have been prepared by other special methods, starting with hydrocarbons, or from nitrogen-containing compounds.

Phthalimide was formed when *o*-xylene was passed with air and ammonia over a catalyst of vanadium pentoxide or alumina[71]. However, *o*-tolunitrile and phthalonitrile were also formed[72]. A tin vanadate catalyst has also been used[73]. The use of a 16% molybdenum trioxide and 2% vanadium pentoxide on alumina catalyst resulted in 95·8% phthalimide and 2·5% phthalonitrile[74]. A phosphorus pentoxide–vanadium pentoxide catalyst afforded largely phthalonitrile with little phthalimide[75]. Substituents *ortho* to the methyl groups in *o*-xylene inhibited the formation of phthalonitriles, but did not reduce the amounts of phthalimides[76]. 1-Nitronaphthalene has been oxidized in the vapour phase at 340–350° over vanadium pentoxide to a mixture of phthalimide and phthalic anhydride[77].

The oxidation of pyrrole with potassium dichromate and sulphuric acid afforded a low yield of maleimide[78]. However, the photo-oxidation of pyrrole in the presence of eosin gave a good yield (32%) of the dihydromaleimide (**89**), which could be oxidized to maleimide with manganese dioxide. *N*-Methylpyrrole similarly gave *N*-methylmaleimide[79]. 2-Formylpyrrole (**90**) was also oxidized to succinimide with hydrogen peroxide in the presence of pyridine[80].

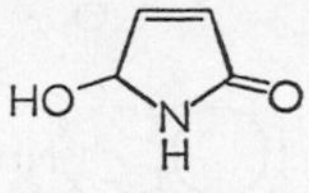

(89) 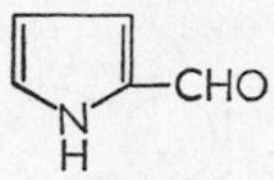　　　　　　(90)

The reaction of acrylamide with carbon monoxide at 160–180° and 100–300 atm using Raney nickel, cobalt salts or cobalt carbonyl as catalyst gave an 81% yield of succinimide. Cyclohexene-1-carboxamide similarly formed cyclohexane-1,2-dicarboximide, and crotonamide (91) yielded a mixture of 68% α-methylsuccinimide (92) and 19% glutarimide[81,82]. The catalytic reduction of adiponitrile over a nickel–magnesium oxide catalyst at 80–120° and 105–120 atm afforded 26% adipimide (93) and 41% adipamide[83].

$$CH_3CH=CHCONH_2$$

(91)　　　　　　　　(92) 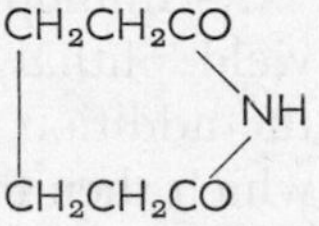　　　　　　　　　(93)

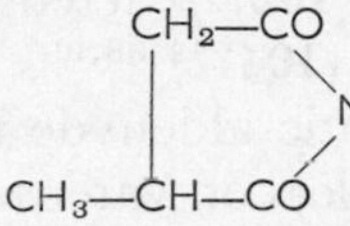

Lactones react with ammonium dithiocarbamate (95) to form imides. Thus, β-isovalerolactone (94) was converted to ammonium β-dithiocarbamylisovalerate (96) which on hydrolysis yielded 4-keto-6,6-dimethyl-2-thiono-1,3-thiazone (97)[84]. Five- and six-

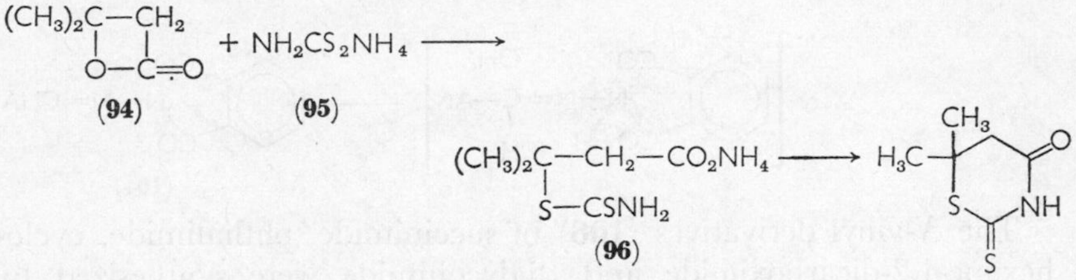

(94)　　　　　(95)

(96)

(97)

membered lactams (98, n = 2,3) reacted with aqueous potassium persulphate yielding imides (99, n = 2,3) as primary products[85].

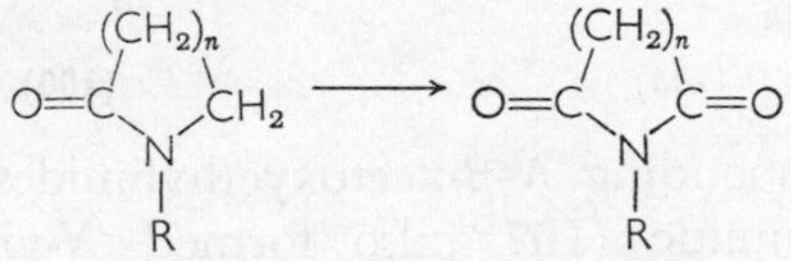

(98)　　　　　　(99)

Phthalonitrile and succinonitrile added ammonia to yield imidines (100, 101), which on hydrolysis formed imides[86,87].

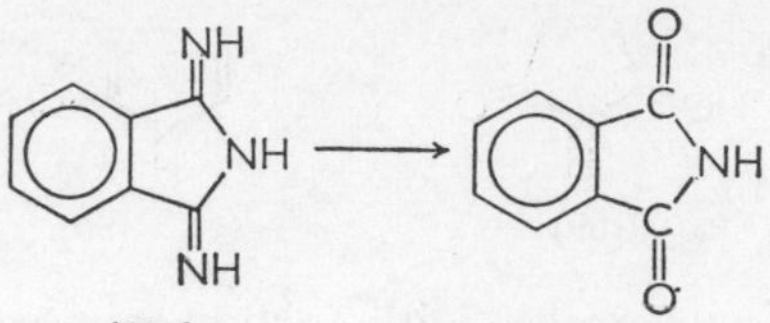

(100)

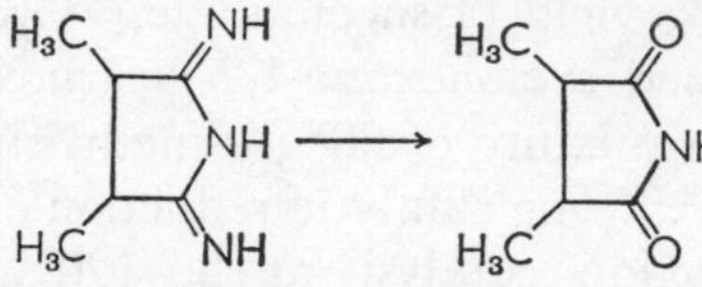

(101)

N-Aminophthalimides (**102**) reacted with aromatic aldehydes to yield phthalylhydrazides (**104**)[44,88,89]. The primary reaction was the addition of the aromatic aldehyde to yield an intermediate **103** which then cyclized with loss of water to the phthalylhydrazide **104**.

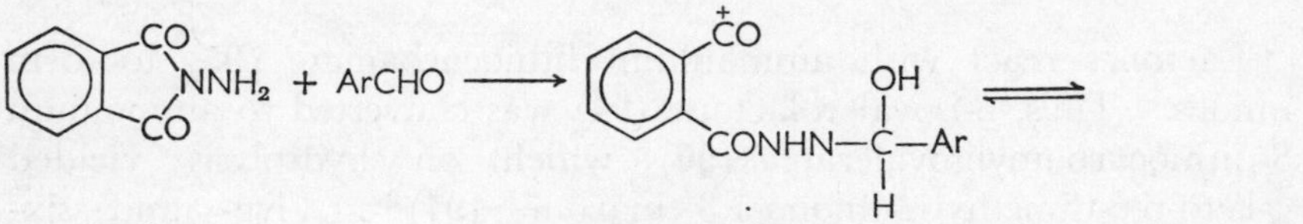

(102)   (103)

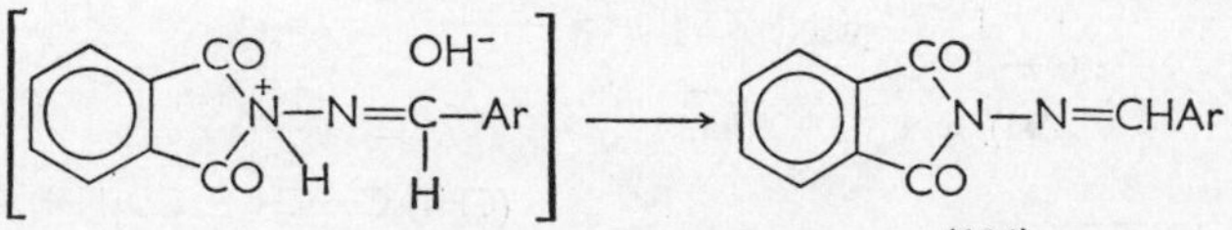

(104)

The *N*-vinyl derivatives (**106**) of succinimide, phthalimide, cyclohexane-1,2-dicarboximide and diglycollimide were synthesized by

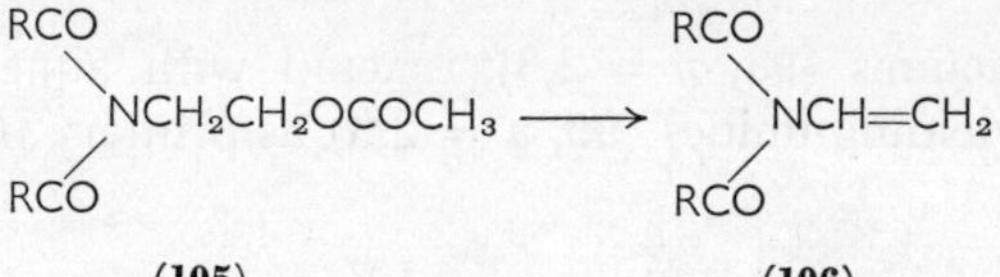

(105)   (106)

heating the corresponding *N*-β-acetoxyethylimides (**105**)[90-92]. *N*-α-Butoxyethylsuccinimide (**107**) also formed *N*-vinylsuccinimide on heating with sodium bisulphate or sulphuric acid[93].

1-Aza-7-oximinocycloheptan-2-one (**108**) was transformed to adipimide on treatment with thionyl chloride in ether[94]. Citraconyl-

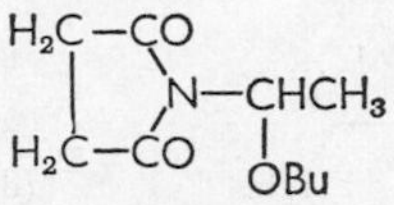

(107)

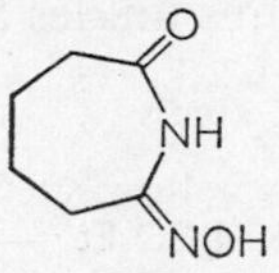

(108)

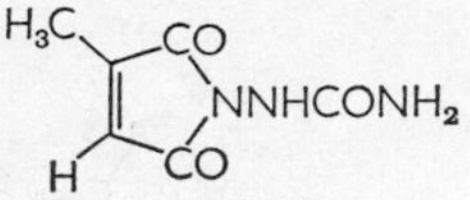

(109)

semicarbazide (109) formed 15% citraconimide on treatment with sodium nitrite in acetic acid[95].

Cycloheximide (actidione) is an antibiotic produced by strains of *Streptomyces griseus*[96]. The structure has been shown to be β-[2-(3,5-dimethyl-2-oxocyclohexyl)-2-hydroxyethyl]glutarimide (110)[97] with the indicated stereochemistry[98,99]. The compound has been syn-

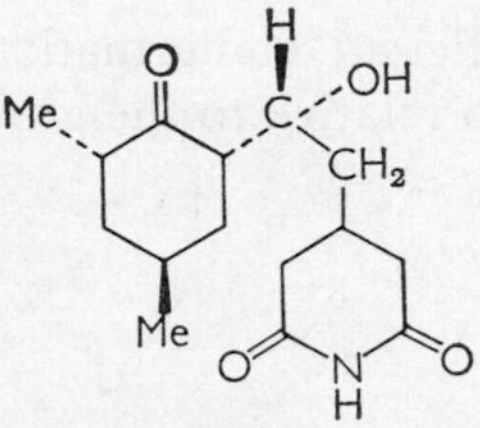

(110)

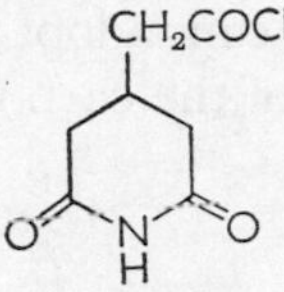

(111)

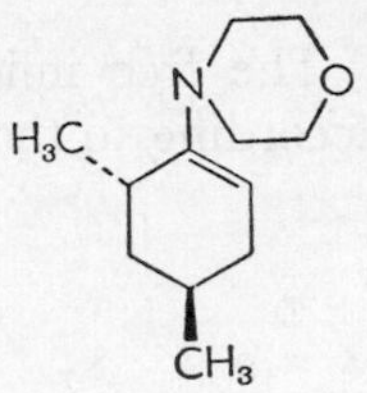

(112)

thesized in a series of reactions, which involved a key step of the addition of β-glutarimidylacetyl chloride (111) to the double bond of *N*-(1-cyclohexenyl-*trans*-4,6-dimethyl)morpholine (112)[100].

Dialkyl ketones and cyclic ketones react with cyanoacetic ester and ammonia to give Guareschi's imides (113) (equation 7). Aryl

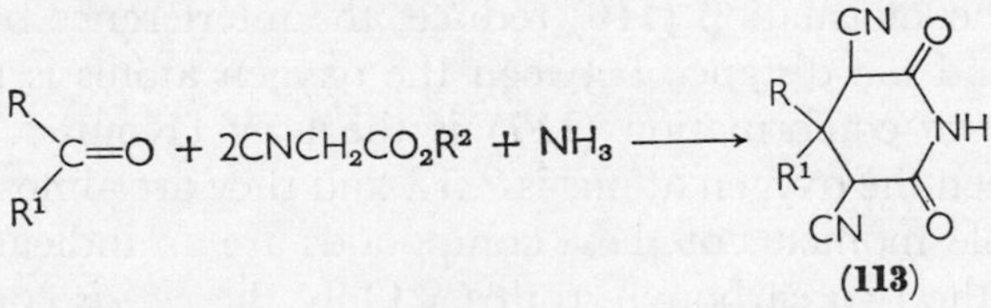

$$\underset{R^1}{\overset{R}{\diagdown}}C{=}O + 2CNCH_2CO_2R^2 + NH_3 \longrightarrow \qquad\qquad (7)$$

(113)

alkyl ketones can react in two stages by first forming an arylidene-cyanoacetic ester (114) employing sodium acetate in acetic acid.

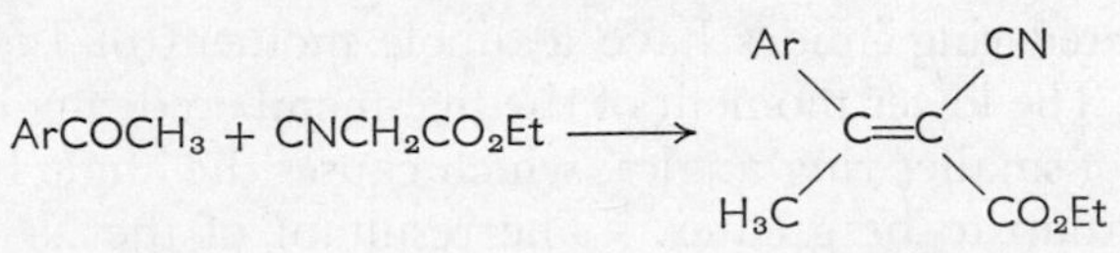

$$ArCOCH_3 + CNCH_2CO_2Et \longrightarrow$$

(114)

The arylidenecyanoacetic esters are then reacted with a second equivalent of cyanoacetic ester (equation 8) in the presence of sodium ethoxide[101,102]. The Guareschi's imides are important intermediates

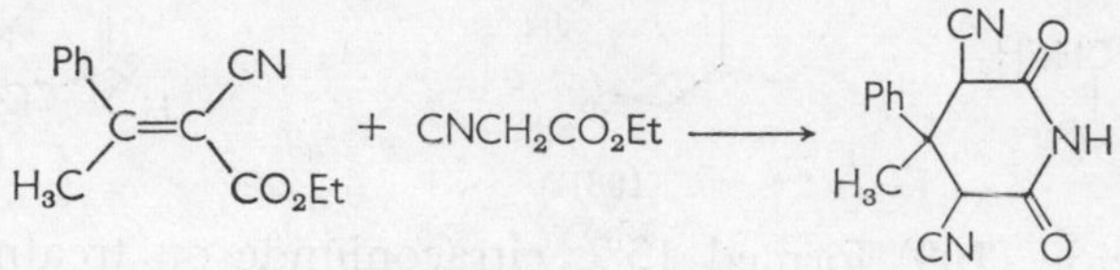

$$Ph\underset{H_3C}{\overset{Ph}{>}}C=C\underset{CO_2Et}{\overset{CN}{<}} + \text{CNCH}_2\text{CO}_2\text{Et} \longrightarrow \qquad (8)$$

for the synthesis of cyclic compounds and pharmaceuticals.

III. PHYSICAL PROPERTIES

A. Dipole Moments

The free imide group can adopt three different conformations, according to the position of the carbonyl group relative to the group

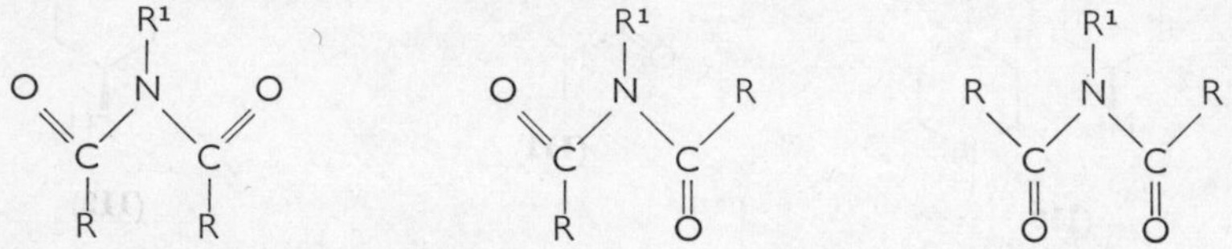

(115)   (116)   (117)

on nitrogen. However, because of resonance, the free imide group is essentially planar[103]. The *cis–cis* conformation (115) is the favoured conformation when R is small since the carbonyl groups are farther apart (5·6 Å) and electrostatic repulsions are less. When R is large, the *cis–trans* conformation (116) reduces the interference between the R groups and the distance between the oxygen atoms is then 4·8 Å. The *trans–trans* conformation (117) is the least favoured. The distance between the oxygen atoms is 2·5 Å and they are almost touching.

The dipole moments of these compounds are an indication of the position of the two carbonyl groups. Only the *cis–cis* conformation (115) is possible in the five- and six-membered cyclic imides. The six-membered ring imides have a dipole moment of 2·6–2·9 D while the five-membered ring imides have a dipole moment of 1·5–2·2 D (see Table 1). The lower moment of the five-membered ring imides is an indication of smaller ring angles, which causes the angle between the carbonyl group to be greater. The resultant of the N^+—O^- contribution is then opposed to the resultant of the carbonyl dipole[103].

In the *cis–cis* conformation, low moments are expected since the resultant of the carbonyl and the N—H dipoles are subtracted from

TABLE 1. Dipole moments of cyclic imides[a].

Compound	μ (D)
Cyclohexane-1,3-dicarboximide	2·89
N-Methylcyclohexane-1,3-dicarboximide	2·88
Glutarimide	2·58
N-Methylglutarimide	2·70
Bemegride (Megimide, A)	2·92
Glutethimide (B)	2·83
Aminoglutethimide (C)	3·64
Succinimide	1·47
N-Methylsuccinimide	1·61
Cyclohexane-1,2-dicarboximide	1·74
3,6-Methano-1,2-cyclohexanedicarboximide	2·24

[a] In dioxan at 30° (ref. 103).

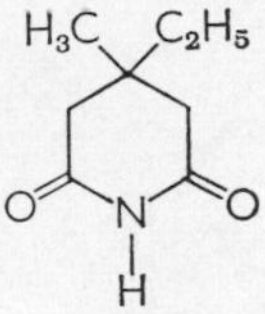

(A) 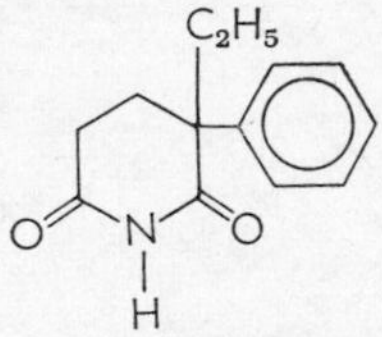   (B) 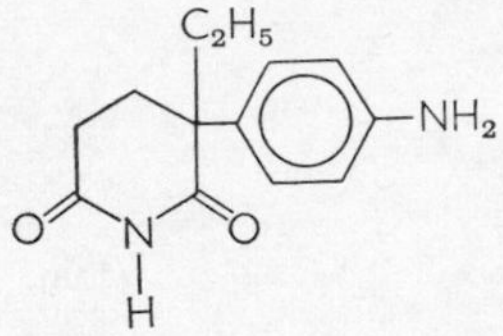   (C)

each other. Furthermore, the additional contribution to the moment from the usual imide resonance is not present, since the N^+—O^- dipoles are at 180° to each other.

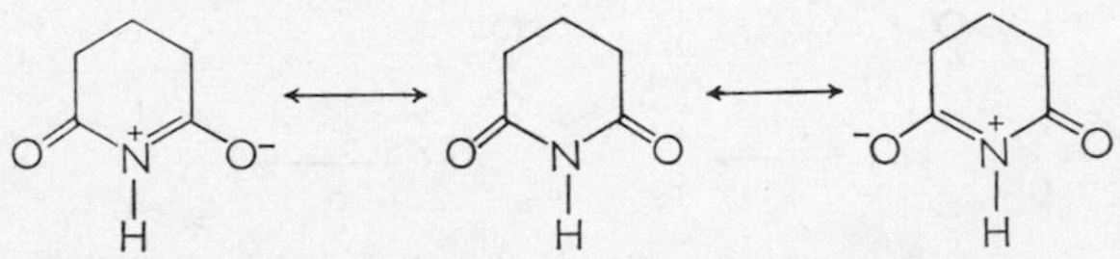

N-Acetyllactams (118, R = CH₃) are in the *cis–trans* conformation and have moments of 3·0–3·2 D. The moments of six-membered ring N-acetyllactams are slightly higher than those of acetylated five-membered ring lactams, and these in turn are larger than those of unacetylated five-membered ring lactams. Ring size affects the moment of N-acetyllactams by varying the angle between the carbonyls group and by changing the amounts of *s* character in the exocyclic bond. However N-benzoyllactams (118, R = C₆H₅) have higher dipole moments than N-acetyllactams (Table 2). This is probably due to the increased conjugation of the benzene ring, and

TABLE 2.   Dipole moments of substituted imides.

Compound	μ (D)	Ref.
N-Benzoylpyrrolidine	2·69	104
N-Benzoylpiperidine	3·07	104
N-Benzoylcaprolactam	3·47	104
Phthalimide	2·91	105
	2·14	106
N-Methylphthalimide	2·24	106
1,8-Naphthalimide	4·73	105

the larger ring size benzoyl lactam (N-benzoylcaprolactam) has a higher dipole moment due to the greater ease of conjugation with the phenyl group.

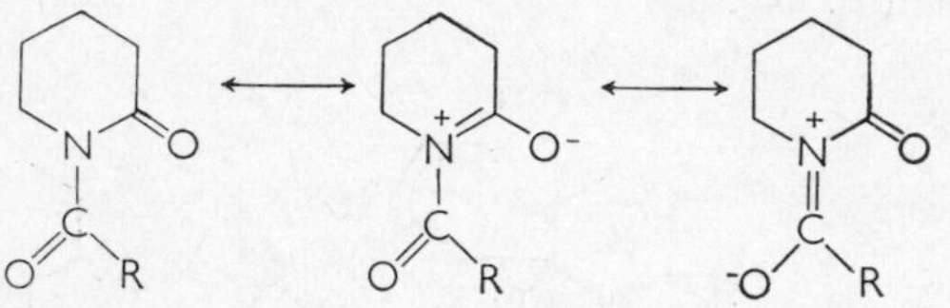

(118)

Dipole moments of aromatic imides such as phthalimide (119) and naphthalimide are higher than the non-aromatic imides, succinimide and glutarimide (Table 2), and this is due to the increased resonance in the aromatic compound. 1,8-Naphthalimide (120) has an additional resonance form which causes its moment to be higher than that

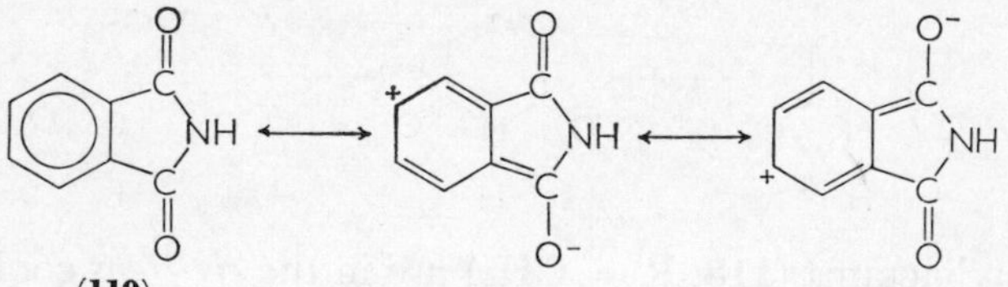

(119)

of phthalimide (119)[105]. The dipole moments of substituted phthalimides have been reported[106].

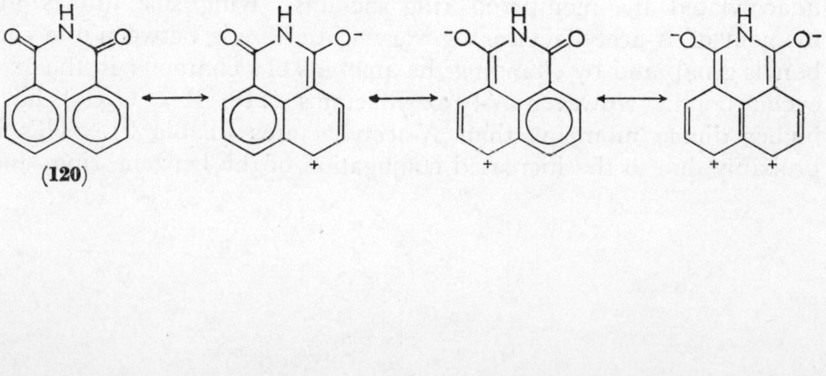

(120)

N-Methyldiacetamide exists in the *cis–trans* conformation. Diacet-amide, while having the moment expected for the *cis–trans* conformation in dioxan, has a much lower moment in benzene and heptane, suggesting hydrogen bonding in a cyclic dimer (**121**)[107]. The low

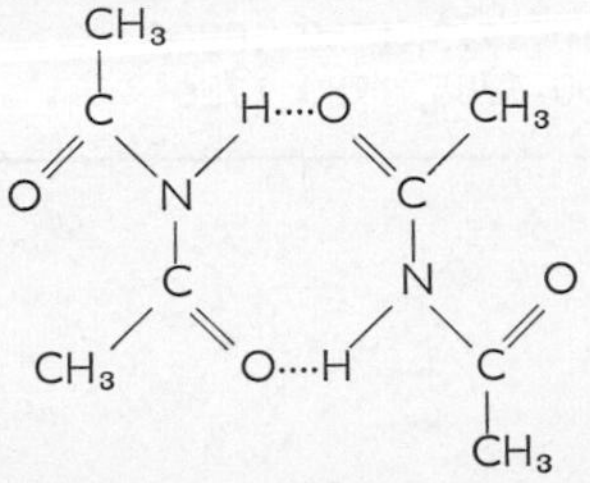

(**121**)

value of the dipole moment for *N*-methyldiformamide (**122**) is explained as due to hydrogen bonding between the formyl hydrogen and the carbonyl group (**122**)[107].

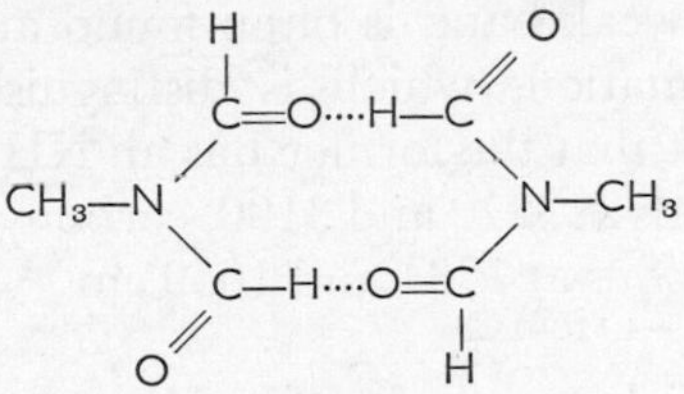

(**122**)

B. Infrared Spectra

Imides give rise to two bands due to the vibrational coupling of the two carbonyl groups. These bonds absorb at 1700–1650 cm^{-1} and at 1790–1710 cm^{-1}[108]. The vibrational coupling between the two carbonyl groups depends on the nature of the substituent on the nitrogen atom, suggesting an electronic and not a purely mechanical origin. Conjugation shifts the imide bands to lower frequencies while in cyclic compounds in which the carbonyl group is part of the ring, ring strain causes a shift to higher frequencies.

In the solid state, imides have a strong band at 1740–1670 cm^{-1} [109] and a bonded NH band near 3250 cm^{-1} [109-113]. Diacylamines (**123**) show a *trans–trans* conformation in the crystal form, while in non-polar solvent such as carbon tetrachloride their configuration inverts to the *cis–trans*[114]. The *trans–trans* form is the one in which the carbonyl groups are parallel and *trans–trans* relative to NH[115] (see Table 3).

TABLE 3.   Infrared spectra of imides[a].

Type	C=O stretching	N—H stretching[b]
Aliphatic	1714, 1690[c]	∼3421
5-Membered ring	1810–1775, 1750–1680[d]	∼3426
6-Membered ring	1800–1785, 1780–1720[e]	∼3386

[a] In cm^{-1}, for CHCl$_3$ solns.
[b] Refs. 109–113.
[c] Ref. 108.
[b] Refs. 116, 117.
[e] Ref. 116.

In non-cyclic imides the C—N stretching vibration gives rise to bands at 1507–1053 cm^{-1} and 1236–1167 cm^{-1} similar to those in monosubstituted amides. The *trans–trans* configuration of acyclic imides is characterized by bands at 3280–3200, 1737–1733, 1505–1503, 1230–1167, and 739–732 cm^{-1}, the latter band being ascribed to NH wagging[110,111]. A weak band is often found at 1695–1690 cm^{-1} in the *trans–cis* conformation, which is distinguished from the *trans–trans* form by the fact that the former has an NH band at 3245 cm^{-1}, accompanied by bands at 3270 and 3190, carbonyl bands at 1700 cm^{-1} with weaker companions at 1734 and 1659 cm^{-1} and the NH wagging band at 836–816 cm^{-1}[110–112].

Cyclic imides which are part of a six-membered ring, such as glutarimides, show carbonyl stretching bands near 1800 and 1700 cm^{-1} while five-membered cyclic imides have bands at higher frequencies (1800 and 1770 cm^{-1}). These compounds can only exist in the *cis–cis* conformation[116]. Five-membered cyclic imides such as phthalimides, have a C=O band at 1790–1735 and 1745–1680 cm^{-1}[116,117]. N-Substituted phthalimides show a doublet at 1790–1778 and 1747–1721, N-substituted succinimides at 1780–1769 and 1728–1705 and N-substituted maleimides, 1780–1770 and 1737–1711 cm^{-1}[118]. The lower-frequency band is always the more intense in five-membered ring imides. The increased absorbance of the five-membered cyclic imides has been explained in terms of the hybridization of the carbon in the carbonyl group. Contraction of the ring gives more *p* character to the ring carbons, which confer more *s* (triple-bond character) to the exocyclic bond. The increased strength of the carbonyl bond will be reflected in a higher force constant and hence in an increased absorption. The NH stretching vibrations in cyclic imides increase with decreasing ring size. The cyclic imides do not have the 1505 cm^{-1}

C—N band present in non-cyclic imides. The corresponding semi-cyclic compounds, *N*-acyl- and *N*-benzoyllactams, which are in the *cis–trans* form, absorb at a higher frequency than the *cis–cis* cyclic compounds[119].

The two crystalline forms of diacetamide and *N*-deuterodiacetamide have different infrared spectra and appear to be conformational isomers. The less stable form has a *trans–trans* planar conformation[111]. Higher diacylamines (**123**, R = Me, Et or Pr; R[1] = Et or Pr) also have a *trans–trans* conformation in the solid state[114].

$$RCONHCOR^1$$

(**123**)

The doublet of *N*-substituted naphthalimides (at 1720–1700 and 1680–1600 cm^{-1}) is shifted by the mesomeric and inductive effects of the substituents. A comparison of the carbonyl frequencies of *N*-substituted phthalimides, pyromellitic diimide (**124**) and naphthalene-1,4,5,8-tetracarboxylic acid diimide (**125**) shows that the six-membered ring imides absorbed at a lower frequency than the five-membered ring compounds[120].

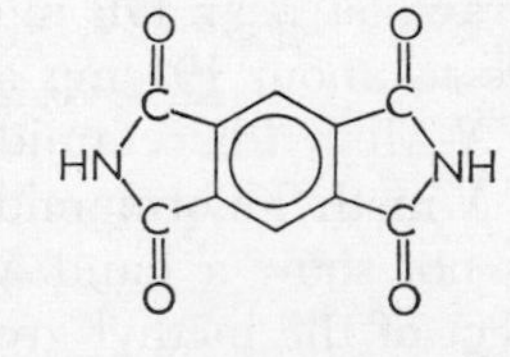

(**124**) 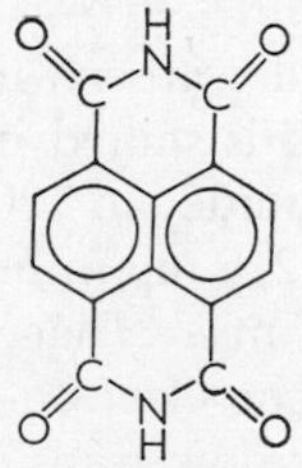   (**125**)

N-Arylmaleisoimides (**126**) and *N*-arylphthalisoimides (**127**) can be distinguished from imides by comparing the extinction coefficient of the carbonyl absorption. The value of the molecules extinction coefficient is of the order of 200–400 for the isoimides, and 850–1300 for imides[121].

Alkyl isocyanurates (alkyl isocyanate trimers, **128a**) have a strong

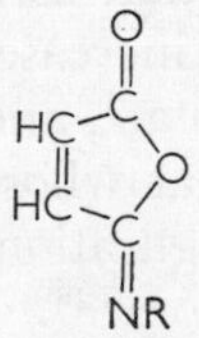

(**126**) 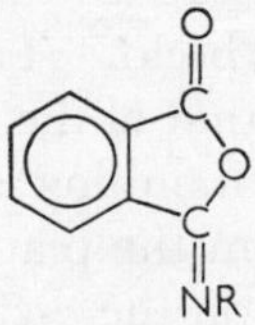   (**127**) 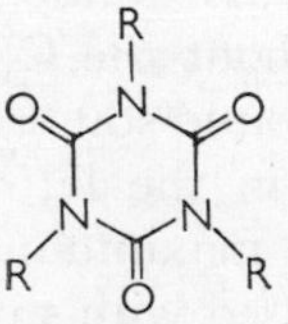   (**128a**) 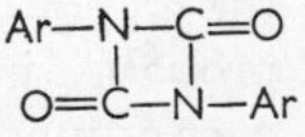   (**128b**)

carbonyl band at 1700–1680 cm^{-1} with a weaker shoulder near 1755 cm^{-1}. The aryl analogues shows carbonyl absorption at a higher frequency (1715–1710 cm^{-1}). Aromatic isocyanate dimers (**128b**) show a carbonyl band at 1785–1775 cm^{-1}[122].

The i.r. absorption spectra of N-benzoyloxyhomophthalimides and N-hydroxyhomophthalimides showed they are represented by the structure **129**, rather than as tautomers **130** and **131**[123]. The

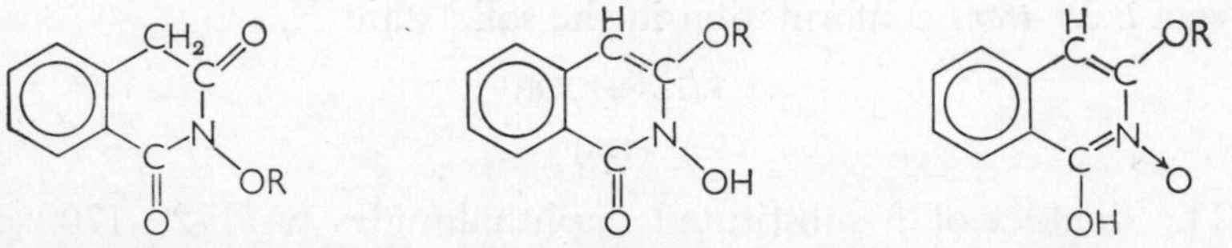

(129)   (130)   (131)

colourless and yellow crystalline forms of N-hydroxyphthalimide have identical infrared spectra and the yellow colour is due to an impurity. The i.r. spectra confirms the structure as an N-hydroxy compound[124].

C. Ultraviolet Spectra

Simple imides absorb in the far ultraviolet near 178 mμ (ϵ 8000) and this band is shifted in succinimide to about 191 mμ and in N-methylsuccinimide to 206 mμ[107]. N-Methyldiacetamide has a maximum at 216 mμ in hexane, while N-methyldiformamide absorbs at about 207 mμ. Diacetamide does not show a band above 200 mμ[107,125]. The electron-repelling effect of the methyl group facilitates the electronic transition and the maximum is shifted to longer wavelength, and there is a higher extinction coefficient. The substitution of formyl for acetyl causes the wavelength to decrease[107]. The approximate doubling of the molar absorptivity in imides compared to amides is explained by the presence of two C(=O)N linkages in the former[87].

Glutarimides show a shift to longer wavelength of about 7 mμ as compared to succinimides[87]. This bathochromic shift is attributed to non-planarity of the imide group on the $n \rightarrow \pi^*$ transition as a result of twisting about the C(=O)—N bond. This twist increases the energy of the ground state relative to that of the excited state, and probably involves, in the latter, an antibonding π^* orbital whose energy function is a maximum at 90° from the planar configuration. The intensity increased with substitution on the α-carbon atom[126].

N-Hydroxysuccinimide and N-hydroxyglutarimide have absorption maxima of high intensity at 215 mμ. This absorption is probably due to the presence of two chromophoric interactions. It was suggested that this absorption was due to the presence of the hydroxy N-oxide (**132**) in equilibrium with the N-hydroxyimide form (**133**) in solution[123]. A similar explanation has been given for the absorption

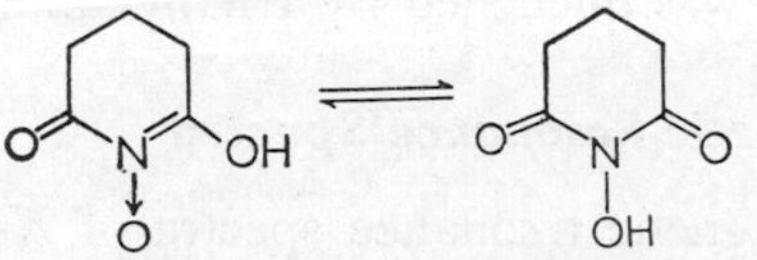

(**132**)   (**133**)

of N-benzyloxyimides, such as N-benzyloxy-α,β-dimethylglutaconimide (**134**), at about 280 mμ. The ultraviolet spectra indicate that these compounds are ionized in ethanol giving an equilibrium mixture[127].

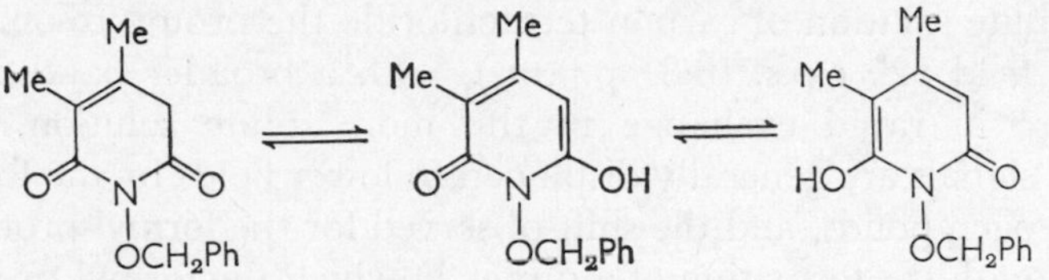

(**134**)

The spectrum of N-benzyloxysuccinimide (**135**) was similar to those of N-hydroxysuccinimide and N-pentyloxysuccinimide and confirmed the structure of the compound[127]. N-Acetyllactams showed absorp-

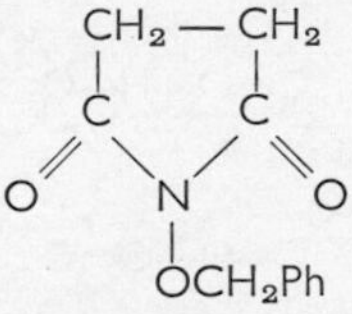

(**135**)

tion bands at 216–219 mμ ($\epsilon = 9,000 - 11,000$), while N-benzoyl-lactams had bands at 228–232 mμ, due to the chromophore $C(=O)NRC(=O)$[104]. A second band, which varied from 280 to 268 mμ present in the spectra of the five-, six- and seven-membered N-benzoyllactams was probably due to resonance between the ring carbonyl and the free carbonyl group. This resonance was shown in the extinction coefficient which increased as the amount of resonance increased[104].

The high-resolution ultraviolet spectra of phthalimide has been reported[128]. The spectra of hydroxy- and methoxyphthalimides and

N-phenylphthalimides[129], and of N-arylsuccinimides[130] and N-aryl-phthalimides[131-133] have been measured.

The fluorescence spectra of 4-aminophthalimide[134], 3-amino-phthalimide and 3-amino-N-methylphthalimide[135], and of other phthalimide derivatives[136] have been studied. The fluorescence of 3-aminophthalimide was quenched by triethylammonium iodide[137].

D. Nuclear Magnetic Resonance Spectra

The nuclear magnetic resonance spectra of N-methyldiacetamide show two peaks of area 2:1. The larger peak, 138 c.p.s. from TMS ($\delta = 2.30$ p.p.m.), corresponds to the six acetyl hydrogens while the other peak, 189 c.p.s. from TMS ($\delta = 3.15$ p.p.m.), is the absorption of the N-methyl hydrogens[107]. The formyl proton in liquid N-methyldiformamide appears 546 c.p.s. from TMS ($\delta = 9.1$ p.p.m.). In a dilute solution of carbon tetrachloride the proton resonates at a higher field 529 c.p.s. (8.81 p.p.m.), with a broader band due to a decrease in rapid exchange in the more dilute solution. N.m.r. proton signals are generally displaced to lower fields by the formation of hydrogen bonds, and the shift observed for the formyl protons is in accord with the formation of a dimer in which the formyl protons are hydrogen bonded to the carbonyl group (**136**). The presence of the electron donor C=O group draws the proton away from its binding electron and reduces the electron density immediately around it.

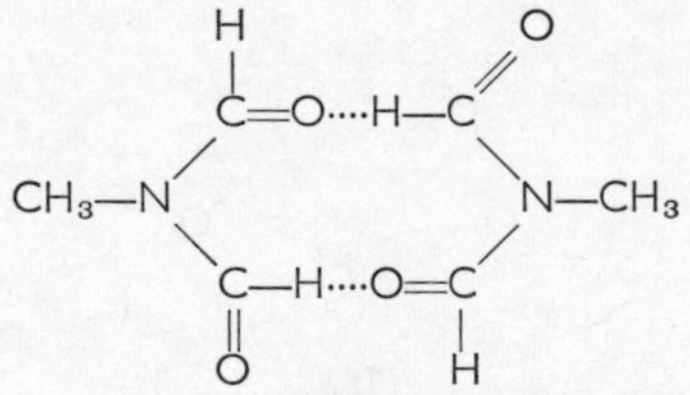

(**136**)

The n.m.r. spectra of N-methylsuccinimide and N-methyldithio-glutarimide show that the N-methyl-group protons of the five-membered ring compound resonate at a higher field than those of the six-membered compound. This may be due to the increase in bond angle in the five-membered ring, which causes the N-methyl group to deviate considerably more from the plane of the thiocarbonyl group[136]. The n.m.r. spectra of N-substituted succinimides and maleimides show that the protons on the carbon atoms resonate at 20 c.p.s. towards higher field than in the corresponding anhydrides[127]. N-Substituted alkyl groups do not affect the chemical shifts of imides, but the N-substituted alkyl groups lower the chemical shifts of the ring protons by 19 c.p.s.[137].

The ^{14}N resonance of succinimide (199 p.p.m.) is shifted from that of ammonia (376 p.p.m.), because of the increased electronegativity of the nitrogen[138].

E. Mass Spectra

The mass spectrometry studies of imides are very recent. Whereas the most abundant ions in the spectra of five-membered lactams comes from cleavage alpha to nitrogen, in alkylated succinimides the most abundant ions involve a double hydrogen transfer from the alkyl chain. N-n-Butylsuccinimide showed relatively intense molecular ions probably represented by 137 or 138[139]. Both N-alkylsuccinimides

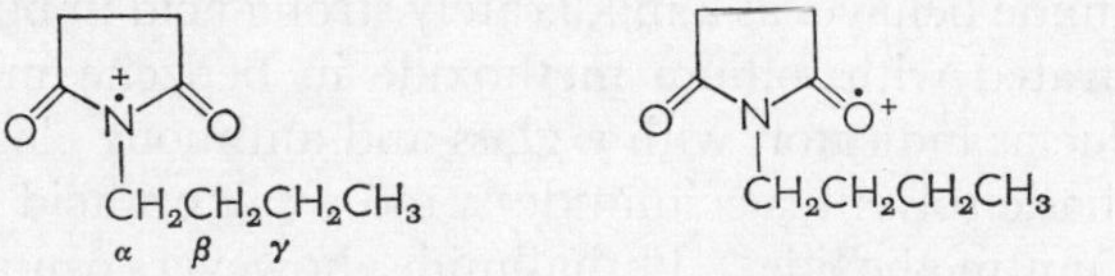

(137)   (138)

and N-alkyl-2-pyrrolidines (where alkyl = n-propyl, n-butyl) showed α- and β-cleavage, the n-butyl analogue showing γ-cleavage of the alkyl chain. In N-n-propylsuccinimide loss of a methyl group occurred to an extent of 60% by β-cleavage. The remaining ion yield arises from expulsion of the α-carbon of the alkyl chain with its attached hydrogen and this must involve transfer of an ethyl radical in the molecular ion (139) to give the charged species 140 and 141[139].

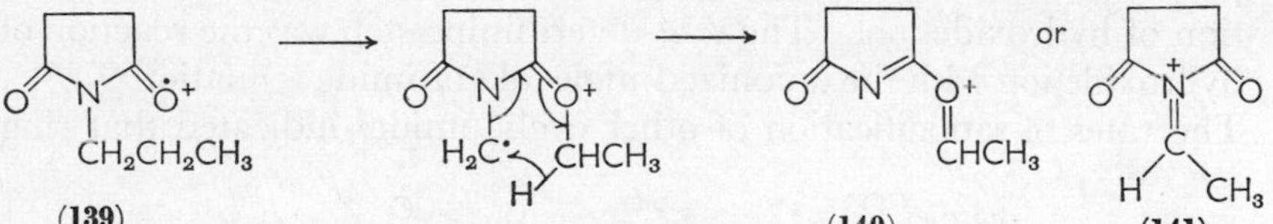

(139)   (140)   (141)

The base peak in the spectrum of N-n-butylsuccinimide occurred at $m/e = 100$ ($\Delta M - 55$), corresponding to the loss of the alkyl chain with transfer of two hydrogen atoms to the charged entity (143) principally from the β- and γ-carbons of the alkyl chain. The process involved intramolecular abstraction of a hydrogen atom from the γ-carbon of the side-chain by oxygen in the resonance form of the imide (142) and

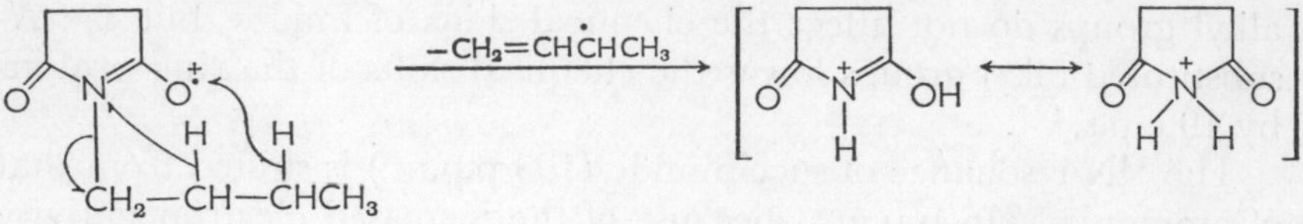

(142) (143)

concomitant hydrogen transfer from the β-carbon atom to nitrogen with synchronous nitrogen–α-carbon bond rupture[139].

Diacetamide underwent skeletal rearrangement on electron impact[140].

IV. REACTIONS

A. Hydrolysis and Exchange

Succinimide behaves as a moderately strong acid in butylamine and can be titrated with sodium methoxide in benzene–methanol using thymol blue as indicator, with a glass and antimony electrode system. In a less basic solvent succinimide is too weak an acid to be titrated using sodium methoxide. Phthalimide, however, cannot be titrated under similar conditions[141]. Succinimide was hydrolysed by aqueous alkali to succinamic acid, which on prolonged hydrolysis formed succinic acid[142]. Succinimide on heating with solid sodium or potassium hydroxide was converted directly to succinate and ammonia[143]. Phthalimides are less readily hydrolysed, and attempts to determine the saponification equivalents of N-alkylnaphthalimides were unsuccessful, since the imide linkage resisted hydrolysis[144].

The rates of saponification of diacetamide and succinamide were proportional to the fraction of ionized imide and not to the concentration of hydroxide ion. The rate-determining step was the reaction of hydroxide ion with the unionized molecule of imide (equation 9)[144a]. The rates of saponification of other cyclic imides indicated that ring

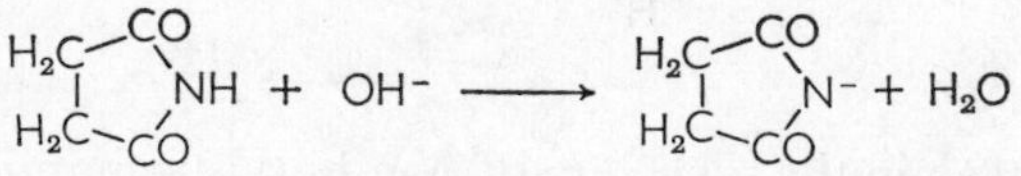

$$\begin{matrix} H_2C-CO \\ | \quad\quad NH \\ H_2C-CO \end{matrix} + OH^- \longrightarrow \begin{matrix} H_2C-CO \\ | \quad\quad N^- \\ H_2C-CO \end{matrix} + H_2O \quad\quad (9)$$

opening was not the rate-determining step. However, ring strain and electronic effects in the ring affected the rate of hydrolysis[145]. The rates of hydrolysis of N-methyl-[145], N-n-butyl- and N-phenyl-[146] diacetamide have also been measured. The rates of alkaline hydrolysis of succinanil (N-phenylsuccinimide) and methyl-substituted succinanils followed those of the corresponding anhydrides. The order found was unsubstituted > monomethyl > *meso*-dimethyl > 2,2-dimethyl > *dl*-dimethyl > trimethyl > tetramethyl. The rates covered a range of 83 fold. The tetramethyl compound existed in equilibrium as the major species with the corresponding succinamic acid at pH 8[147]. The α-alkyl-α-phenylsuccinimides gave a mixture of amidic acids on hydrolysis[148].

The acid hydrolysis of phthalimide (145) to phthalamic acid (146) has been studied spectroscopically at 80–100°, as an intermediate step in the hydrolysis of phthalamide (144)[149,150]. Phthalimide and *o*-carboxyphthalimide (147) showed a normal acid-catalysed hydrolysis

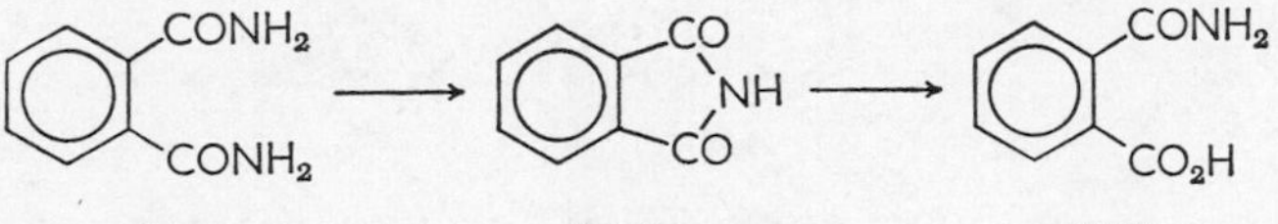

(144)   (145)   (146)

below pH 1. The hydrolysis of phthalimide above pH 3 followed simple base catalysis, as did the behaviour of 147 at pH 5. However, the rate of hydrolysis of *o*-carboxyphthalimide increased in the pH range 1–4. This behaviour was interpreted as due to an intramolecular general acid catalysis (equation 10) in which perpendicular attack on the carbonyl carbon atom was hindered[151].

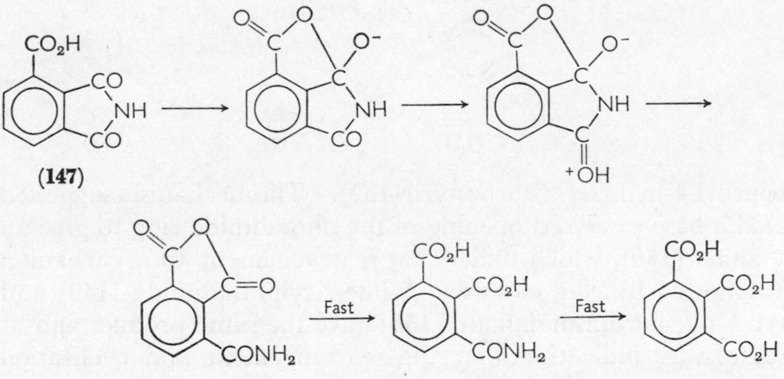

(147)

(10)

The hydrolysis of *N*-phenylphthalimide showed no ionic strength effect in 30% ethanol at pH 9–11[152]. The addition of adenine increased the rate of hydrolysis of phthalimide in sodium carbonate solution at pH 10–10·5, and this was found to be due to the formation of a complex[153]. The rate of hydrolysis of thalidomide (71) in 0·001 and 0·002 M sodium hydroxide was twice as fast as that of phthalimide and N-n-butylphthalimide, and this increased rate was due to the glutarimide part of the molecule [154]. Cycloheximide (110) underwent an acid-catalysed dehydration to anhydrocycloheximide, which was then rehydrated stereospecifically to form an isomer of cycloheximide. The hydrolysis of the imide occurred simultaneously with dehydration, and the rate-determining step was hydrolysis to the acid amide, which existed in equilibrium with the imide and dicarboxylic acid [155].

N-(ω-Aminoalkyl)phthalimides (148, R = H, CH_3 or C_2H_5; n = 2 or 3) underwent very ready hydrolysis, and dissolved in base at room

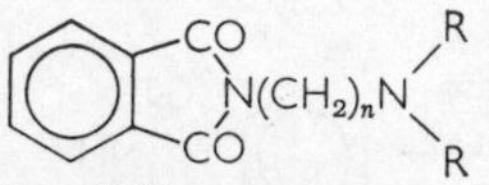

(148)

temperature to afford the salt of the N-substituted phthalamic acid [156].

N-Phenacylphthalimide (149) was rearranged by sodium methoxide

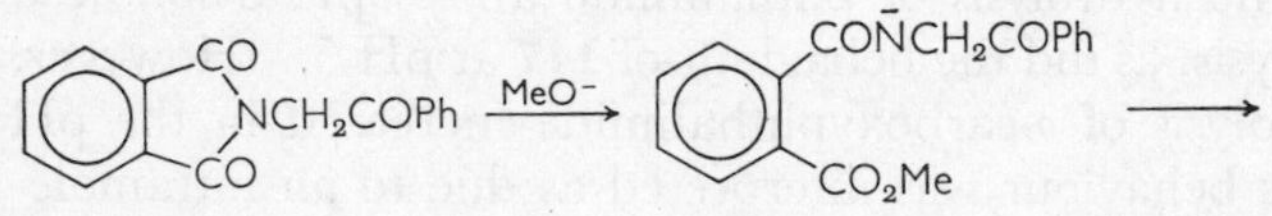

(149) (150)

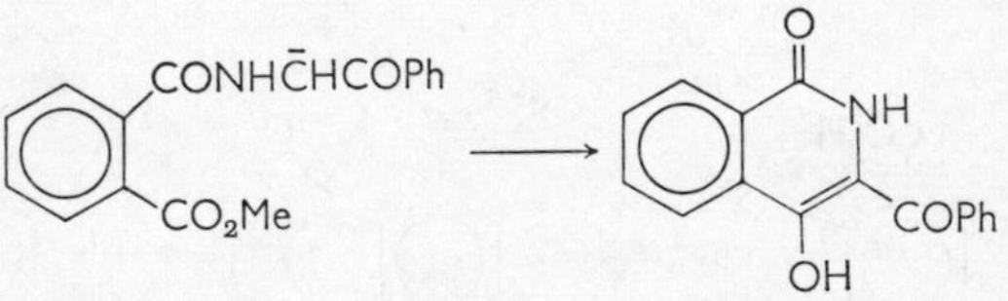

(151) (152)

to 3-benzoyl-4-hydroxyisocarbostyril (152). The mechanism suggested involved a base-catalysed opening of the phthalimide ring to give an imide anion (150), which underwent rearrangement to a carbanion (151), followed by ring closure. N-Phenacylphthalimide (149) and methyl N-phenacylphthalamate (153) gave the same product and at the same rate, indicating that the rearrangement and cyclization were the slow steps. The rate of reaction was slower using t-butoxide

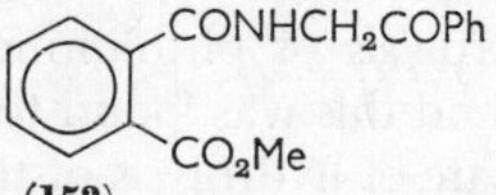

(153)

ion. The rates for the methoxide-catalysed reaction of various phenacyl compounds (p-methyl and p-methoxy) followed a linear Hammett relation, with $\rho = 1 \cdot 98$[157].

The acetyl group in N-alkyldiacetamides exchanged with the corresponding group in acetic anhydride in the presence of pyridine, and the exchange has been studied using ^{14}C-labelled anhydride[158]. No exchange occurred with N-aryldiacetamides[159]. The ease of trans-acetylation for the different alkyl groups studied was i-Pr > i-Bu > n-Bu > n-Pr > Et. This corresponded to the differences in nucleophilicity of the nitrogen atom. The diacetylamine and acetic anhydride were assumed to dissociate (equations 11 and 12). The

$$RN(COCH_3)_2 \rightleftharpoons R\bar{N}COCH_3 + CH_3\overset{+}{C}O \tag{11}$$

$$(CH_3C^*O)_2O \rightleftharpoons CH_3C^*O_2^- + CH_3\overset{+}{C}{}^*O \tag{12}$$

$$R\bar{N}COCH_3 + CH_3\overset{+}{C}{}^*O \rightleftharpoons RN(C^*OCH_3)_2 \tag{13}$$

fragments so-formed then recombined giving a labelled diacetamide (equation 13). The rate of this recombination of the N-acetyl anion and acetyl cation depended on the nature of the N-alkyl group. The rate of exchange was less in the absence of pyridine[160].

B. Reduction

The reduction of imides has been accomplished by many different methods the most common being reduction with lithium aluminium hydride or sodium borohydride. The reduction of N-alkylsuccinimides (**154**) to the corresponding N-alkylpyrrolidines (**155**) by LiAlH$_4$ provides a convenient laboratory method for their preparation relatively free from pyrrolidones[161]. The reduction of phthalimides

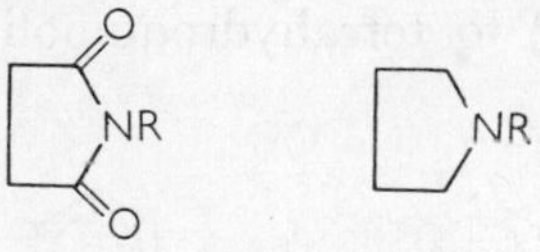

(154)   (155)

has also been employed to obtain isoindole derivatives[162]. N-Alkyl-cis-Δ^4-tetrahydrophthalimide (**156a**) gave the hexahydroisoindole derivative (**156b**) in good yield[163,164].

Cyclopentane-1,2-dicarboximide (**157a**) and 1-3-dicarboximide (**158a**) have been reduced by lithium aluminium hydride to 3-azabicyclo[3.3.0]octane (**157b**)[37,165] and 3-azabicyclo[3.2.1]octane (**158b**)[165] respectively.

(156a)   (156b)   (157a)   (157b)

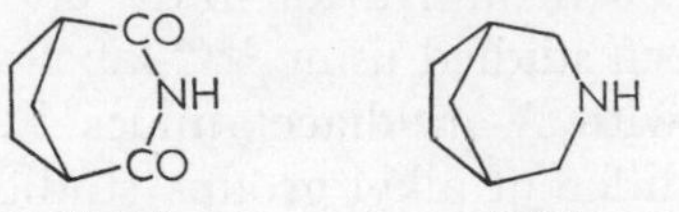

(158a)   (158b)

Hexahydroisophthalimide similarly afforded 3-azabicyclo[3.3.1]-nonane[166]. Reduction of N-methylglutarimide with sodium aluminium hydride gave a 62% yield of 1-methylpiperidine, and 3-ethyl-3-methylglutarimide was reduced in 28% yield to 4-ethyl-4-methylpiperidine[167]. Bicyclic imides with nitrogen as the bridge atom (159a) can be reduced with $LiAlH_4$ to the corresponding amines (159b)[168]. α,α-Disubstituted succinimides and glutarimides can be

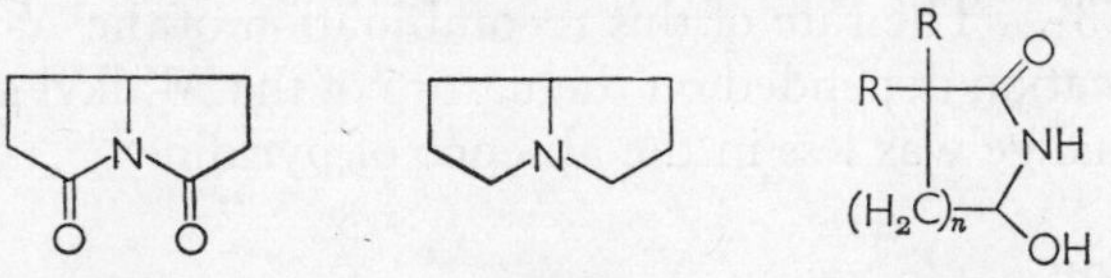

(159a)   (159b)   (160)

selectively reduced, primarily at the one group far from the substituents. Hydroxy lactams (160, $n = 1,2$) were formed as the initial reduction product[169]. Similarly, phthalimides can be reduced selectively with sodium borohydride, but phthalide (162) and o-hydroxymethyl-benzamides (163) often appear as by-products[170]. Lithium aluminium hydride reduced both reactive groups in monothio-homophthalimide (164) to tetrahydroquinoline (165)[163]. However,

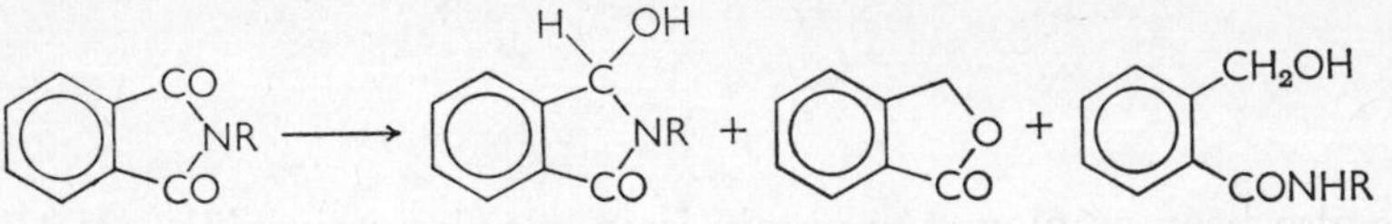

(161)   (162)   (163)

Guareschi imides (166) could not be reduced by lithium aluminium hydride[171].

Succinimide on reduction with sodium and alcohol gave pyrrolidine, but remained unchanged when shaken with hydrogen over platinum oxide. However, under this condition N-acylphthalimides

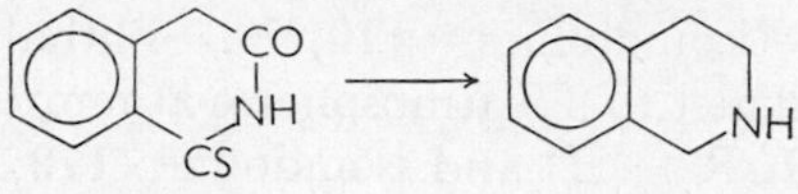

(164) → **(165)**

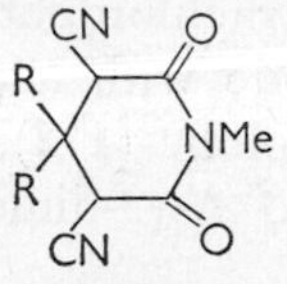

(166)

reduced incompletely to N-phthalimidylcarbinols[172]. Isoindolones were also produced by the reduction of phthalimide with tin and hydrochloric acid; reduction with zinc and sodium hydroxide afforded phthalide (**162**) and 3-hydroxyisoindolone (**161**, R = H)[173]. The distillation of succinimide, glutarimide and homophthalimide with zinc dust produced pyrrole, pyridine and isoquinoline, respectively[174]. Catalytic hydrogenation of glutarimide gave piperidine[174]. One of the carbonyls of phthalimide could be reduced over copper chromite to yield 1-isoindolinone. The use of Raney nickel afforded hexahydrophthalimide (cyclohexane-1,2-dicarboximide)[175].

Imides of dibasic acids have been reduced electrolytically in acid solution to cyclic lactones and cyclic amines. Generally a lead cathode was used but cadmium and amalgamated zinc have been employed in some cases[176,177]. The product from the reduction of only one carbonyl group of succinimide may be isolated in good yield, while electroreduction of both carbonyl groups gave poor yields of the corresponding cyclic amine. Succinimide in 50% sulphuric acid yielded pyrrolidone in moderate amounts with only a trace of pyrrolidine[176]. Pyrrolidone is difficult to prepare by other methods.

The imide of camphoric acid (**167**) underwent a similar reduction, the products being camphidene (**168**) and camphidone (**169**)[176]. The electrolytic reduction of phthalimides has been used to prepare

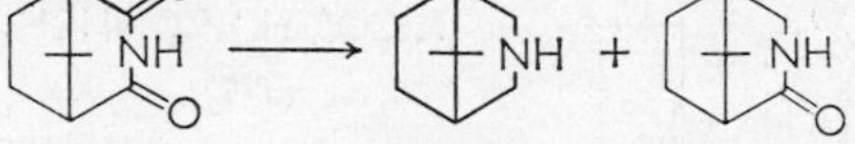

(167) ⟶ **(168)** + **(169)**

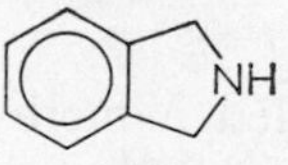

(170)

dihydroisoindoles (isoindolines, **170**)[178]. Phthalimide was electrolytically reduced at 1 to 8·5 atmospheres at temperatures below 49° to isoindolone (**161**, R = H) and isoindoline (**170**), at a lead or zinc amalgam electrode. The yield increased at higher pressure[179]. *N*-(2-Dimethylaminoethyl)tetrachlorophthalimide (**171**) can be reduced at a palladium cathode with a potential of −0·68 v to the corresponding hydroxyisoindolone (**172**), which was further reduced at a potential of −1·19 v to *N*-(2-dimethylaminoethyl)tetrachloroisoindoline (**173**)[180].

(**171**) (**172**) (**173**)

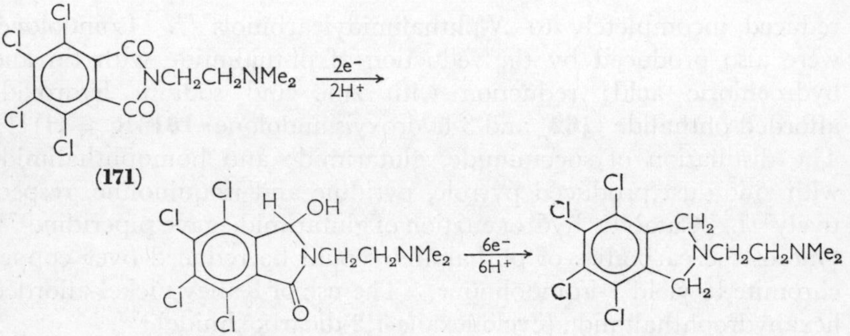

C. Reactions with Grignard Reagents

N-Phenylsuccinimide (**174**) reacted with ethylmagnesium bromide to form 2-ethyl-2-hydroxy-1-phenyl-5-pyrrolidone (**175**)[181]. *N*-Methylglutarimide (**176**) and phenylmagnesium bromide or benzyl-

(**174**) (**175**)

(**176**) (**177**) (**178**)

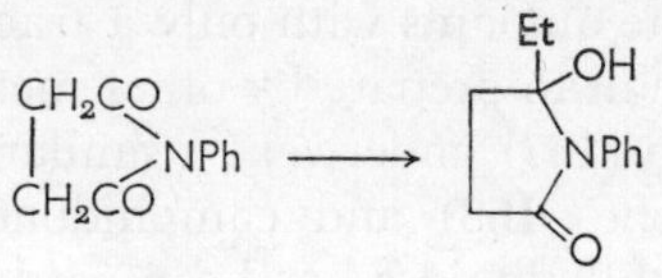

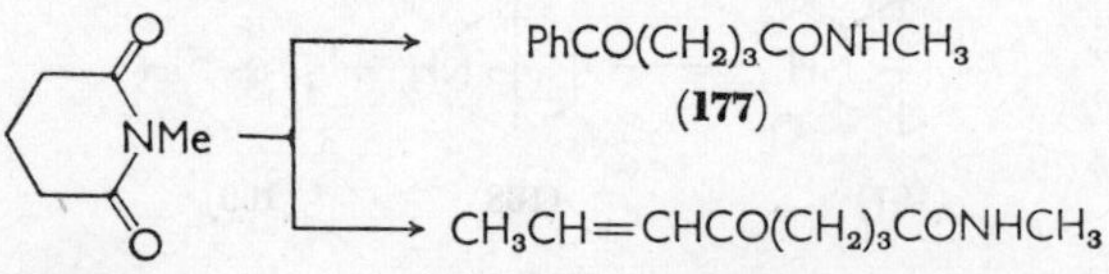

magnesium bromide afforded *N*-methyl-5-oxo-5-phenylpentanamide (**177**) and *N*-methyl-5-oxo-6-phenylhexanamide, respectively[182].

The reaction of **176** with allylmagnesium bromide gave *trans-N*-methyl-5-oxo-Δ^6-octenamide (**178**)[183]. *N*-Arylmaleimides (**179**) ring-opened with Grignard reagents to give β-aroyl-*N*-arylacrylamides (**180**)[184].

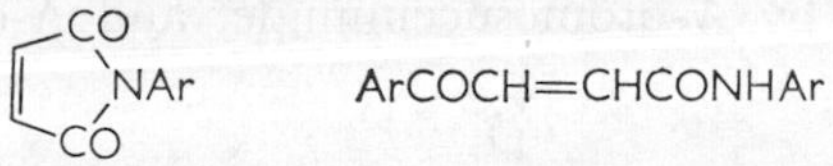

$$\text{ArCOCH}=\text{CHCONHAr}$$

(**179**)   (**180**)

N-Arylphthalimides[185] (**181**, R = Aryl) and *N*-ethylphthalimide[186] (**181**, R = Et) reacted with alkyl Grignard reagents, similarly to **174**, to form the corresponding compounds **182**.

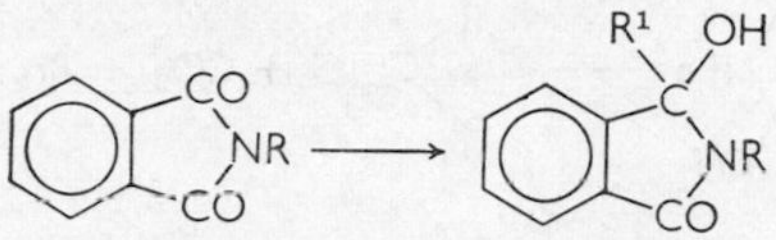

(**181**)   (**182**)

D. N-Haloimides

The acidic hydrogen atom on the nitrogen atom of imides can be replaced by chlorine, bromine or iodine[187]. *N*-Bromosuccinimide and *N*-bromophthalimide are prepared by adding bromine to an ice-cold solution of the imide in sodium hydroxide, and *N*-chlorophthalimide by passing chlorine into a suspension of phthalimide in water[188]. *N*-Bromotetramethylsuccinimide was obtained by the addition of bromine to tetramethylsuccinimide in sodium bicarbonate solution[189], and *N*-bromotetrafluorosuccinimide from tetrafluorosuccinimide and bromine in trifluoroacetic anhydride[189] or trifluoroacetic acid[190] in the presence of silver oxide. *N*-Bromoglutarimide and *N*-iodosuccinimide were prepared by reacting the silver salt of the imide with the appropriate halogen[35].

N-Bromoimides, particularly *N*-bromosuccinimide, are used to introduce a bromine atom into a position adjacent to a double bond or a benzene ring[191]. This is the Wohl–Ziegler reaction and proceeds by a radical-chain mechanism[192–195]. An allylic hydroxyl group may also be introduced and this resulted from hydrolysis of the allylic bromide[196]. In aqueous medium *N*-bromosuccinimide forms hypobromous acid which can convert an olefin to a bromohydrin[196]. Reaction with *N*-bromosuccinimide in the presence of pyridine or quinoline can result in dehydrogenation via elimination from an

13 + C.O.A.

intermediate bromide[196]. An important reaction of N-bromo- and N-chlorosuccinimide in pyridine and t-butanol is the oxidation of carbinols to ketones[196]. The reaction apparently involves the halogenation of the α-carbon atom (equation 14), since benzyl ether is oxidized to benzaldehyde by N-bromosuccinimide, and N-chlorosuccinimide

$$\underset{\substack{|\\ |}}{\overset{\substack{H\\ |}}{-C-OH}} \longrightarrow \underset{\substack{|}}{\overset{\substack{X\\ |}}{-C \sbond OH}} \longrightarrow C=O + HX \qquad (14)$$

converts benzaldehyde to benzoyl chloride[197]. In an allied reaction, α-hydroxy acids (**183**) were oxidized to ketones and CO_2 by two

$$\underset{(\textbf{183})}{\underset{R^1}{\overset{R}{\underset{\diagdown}{\diagup}}}\underset{CO_2H}{\overset{OH}{\underset{\diagdown}{C}}}} + 2 \underset{O}{\overset{O}{\bigsqcup}}NBr \longrightarrow \underset{R^1}{\overset{R}{\underset{\diagdown}{\diagup}}}C=O + CO_2 + Br_2 + 2 \underset{O}{\overset{O}{\bigsqcup}}NH \qquad (15)$$

equivalents of N-bromosuccinimide (equation 15)[196]. Tertiary amines underwent cleavage of a carbon–nitrogen bond forming secondary amines, N—CH_2 bonds being cleaved preferentially[196]. Alanine gave 50% acetaldehyde with N-bromosuccinimide and 25–35% with N-bromophthalimide. Glycine was 25–40% cleaved with either reagent[198]. These reactions of bromination and oxidation with N-bromosuccinimide have been covered in review articles[188,196], and only the more important recent developments will be mentioned here.

N-Bromosuccinimide, N-bromotetrafluorosuccinimide and N-bromotetramethylsuccinimide showed identical selectivity to substituted toluenes, suggesting that the rate-controlling step was the abstraction of a hydrogen atom by a bromine atom[189]. The relative selectivities for bromination by N-bromosuccinimide in carbon tetrachloride were for toluene = 1, secondary aliphatic hydrogens < 0·01 and allylic hydrogens = 50–100. The rates of bromination of toluenes followed the σ^+ constants of the substituents, with a Hammett reaction constant ρ of $-1·38$[199]. In the case of α-substituted toluenes the rate depended on the relative capacity of the substituent to stabilize the transition state of the attack of a bromine atom on hydrogen, by electron release through resonance[200–202]. The yield of benzaldehyde from alkyl benzyl ethers was insensitive to the alkyl group and probably proceeded via a benzyl ether radical (equation 16). N-Bromo- and

$$PhCH_2OR \longrightarrow PhCHOR \longrightarrow PhCH(Br)OR \longrightarrow PhCHO + RBr \qquad (16)$$

N-chlorosuccinimide in sulphuric acid solution caused ionic halo-genations of benzene derivatives, giving 50–95% yields of monohalo-genated products. Toluene formed o-, m- and p-bromotoluene in the proportions 67:2:31 and similar proportions of chlorotoluenes. The proportions of bromochlorobenzenes from chlorobenzene were *ortho* 39·4, *meta* 1·5, and *para* 59·1% [204]. N-Chlorosuccinimide showed the relative rate of chlorination of cyclohexane to that of toluene to be $(3·9 \pm 0·3):1$ [205].

Arylalkyl hydrocarbons underwent photobromination with N-bromosuccinimide in methylene chloride. The relative rates of photobromination were the same as those for photobromination by bromine itself, and the reaction must involve a hydrogen abstraction by bromine atoms [206]. The photochemical reaction of N-bromo-succinimide with $(+)$-1-bromo-2-methylbutane gave $(-)$-1,2-di-bromo-2-methylbutane in a chain reaction initiated by bromine atoms (equation 17) [207]. The illumination of solutions of ketone and

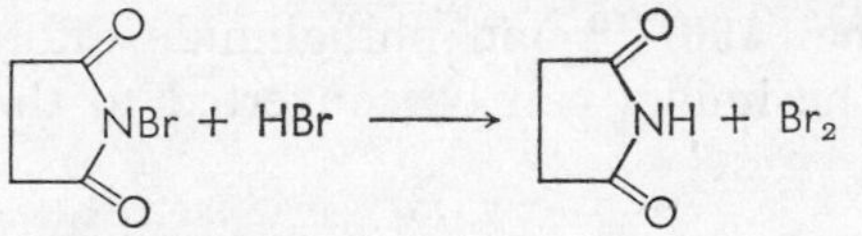

$$\dot{Br} + RH \longrightarrow HBr + \dot{R}$$

$$\dot{R} + Br_2 \longrightarrow RBr + \dot{Br} \qquad (17)$$

N-bromosuccinimide in carbon tetrachloride produced α-bromo-ketones. Thus, diethyl ketone formed 1-bromoethyl ethyl ketone (**184**), which was further brominated to bis(1-bromoethyl) ketone (**185**) [208].

$$\text{C}_2\text{H}_5\text{COCHBrCH}_3 \qquad\qquad \text{CH}_3\text{CHBrCOCHBrCH}_3$$
$$\textbf{(184)} \qquad\qquad\qquad\qquad \textbf{(185)}$$

N-Bromoimides were generally less reactive than N-bromoamides in the addition of bromine to styrene. N-Bromophthalimide and N-bromosuccinimide did not form dibromostyrene. N-Bromoglutar-imide afforded a low yield of dibromide on prolonged reaction, and formed 66% 3-bromocyclohexene on reaction with cyclohexene [35]. N-Bromosuccinimide was more reactive to olefins than N-bromo-phthalimide. 2,3-Dimethyl-1,3-butadiene (**186**) was converted to

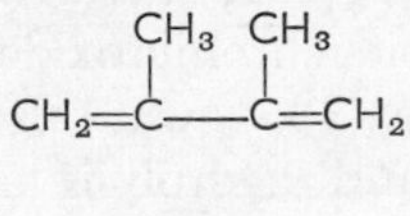

$$\textbf{(186)} \qquad\qquad\qquad \textbf{(187)} \qquad\qquad\qquad \textbf{(188)}$$

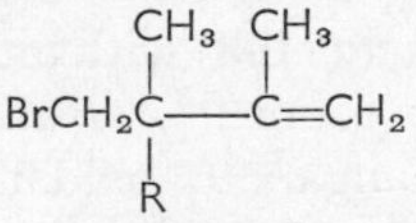

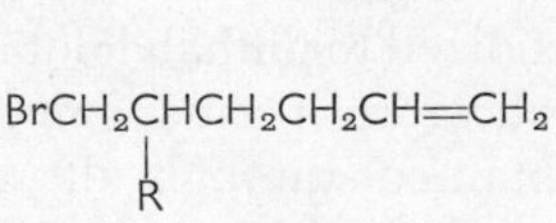

1-bromo-2-succimido- (**187**, R = $(CH_2CO)_2N$) or 2-phthalimido-
(**187**, R = $C_6H_4(CO)_2N$)-2,3-dimethyl-3-butene, and biallyl formed
1-bromo-2-succimido- (**188**, R = $(CH_2CO)_2N$) and 2-phthalimido-

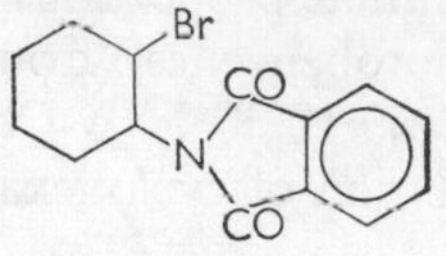

(189)

(**188**, R = $C_6H_4(CO)_2N$-)-5-hexene. These reactions all proceeded
by 1,2-addition[209]. Cyclohexene reacted with N-bromophthalimide
to form *trans*-2-bromo-1-phthalimidocyclohexane (**189**) with 5% of the
cis isomer[210].

E. Miscellaneous Reactions

The action of hypohalite on succinimide (the Hofmann reaction)
produced β-alanine (**190**)[210] and phthalimide afforded anthranilic
acid (**191**)[212]. The imides can be converted to the corresponding

$$NH_2(CH_2)_2CO_2H$$

(190)   **(191)**

acid amides and free acids by electrolytic oxidation. Thus the elec-
trolysis of a solution of the potassium salt of succinimide yielded
ammonia at the cathode and succinic acid at the anode (equation
18)[213]. However, imides are generally resistant to oxidation, and

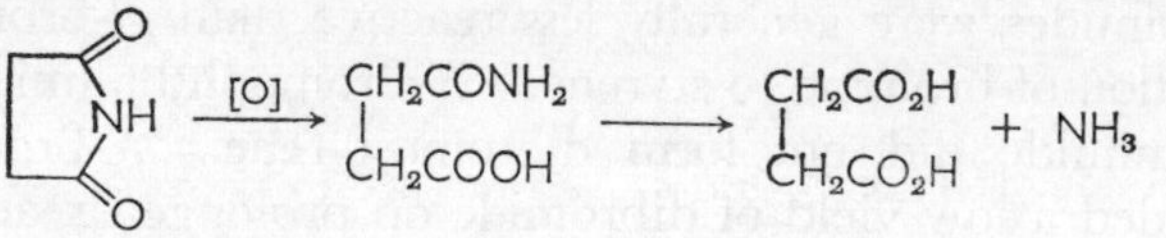

can be used as protecting agents for easily oxidized groups. For
example, glutamic acid was synthesized by first reacting phthalic
anhydride with 3-aminocyclopentene. The product, **192**, was then
oxidized to phthalylglutamic acid (**193**) with nitric acid, chromic oxide,
potassium permanganate or ozone[214]. β-Alanine (**195**) was also
obtained through the permanganate oxidation and hydrolysis of
β-phthalimidopropionaldehyde (**194**)[215].

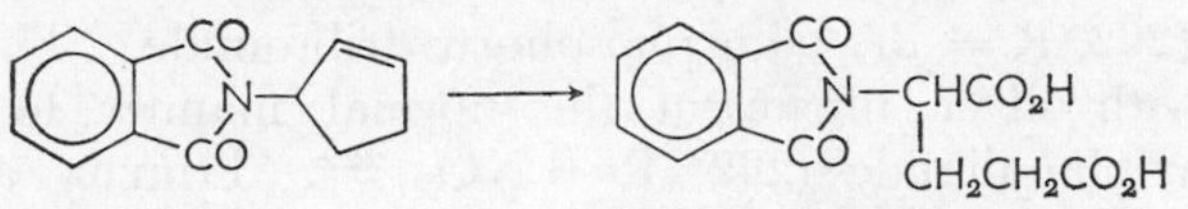

(192) (193)

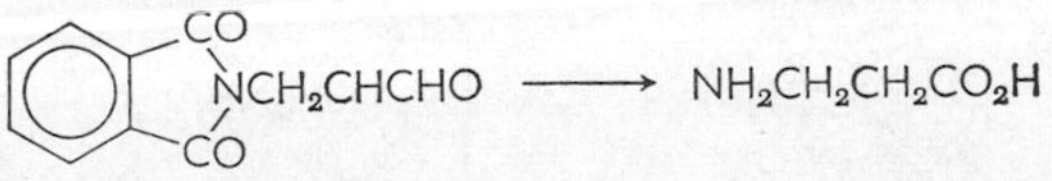

(194) (195)

Glutarimide reacted with phosphorus pentachloride to form 2,3,6-trichloropyridine (**196**) (53%), 2,6-dichloropyridine (38%) and 2,3,5,6-tetrachloropyridine (9%)[216]. The phosphorus oxychloride formed in the reaction caused by-products, and no chlorinated pyridines resulted when phosphorus oxychloride was used as solvent[217]. However, homophthalimide was converted by phosphorus oxychloride to 1,3-dichloroisoquinoline (**197**)[218]. The potassium salt of phthalimide formed N,N-thiobiphthalimide (**198**) on reaction with sulphur monochloride[219].

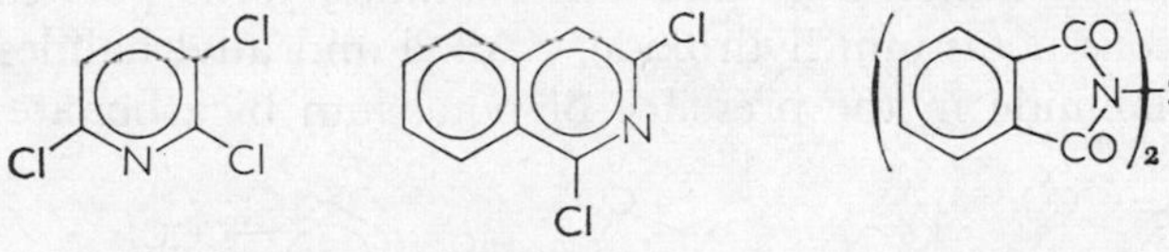

(196) (197) (198)

Diacetamide on heating with ethylene carbonate gave N-(2-hydroxyethyl)diacetamide (**199**), and phthalimide similarly formed N-(2-hydroxyethyl)phthalimide[220].

$$(CH_3CO)_2NCH_2CH_2OH$$

(199)

N,N-Diacylanilines (**200**) rearranged in the presence of acid to give 4-acylaminophenyl alkyl ketones (**201**). The reaction involved acylium ions[221].

$$PhN(COAr)_2$$

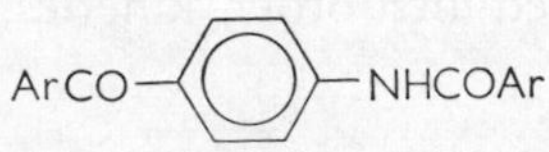

(200) (201)

Phthalimide reacted with formaldehyde to form N-hydroxymethylphthalimide (**202**, R = OH), which could be converted to the bromide (**202**, R = Br) with phosphorus tribromide. The bromide reacted with silver nitrite in the normal manner to form *N*-nitromethylphthalimide (**202**, R = NO$_2$)[222]. Primary arylamines condensed with formaldehyde and succinimide to form *N*-(arylamino-

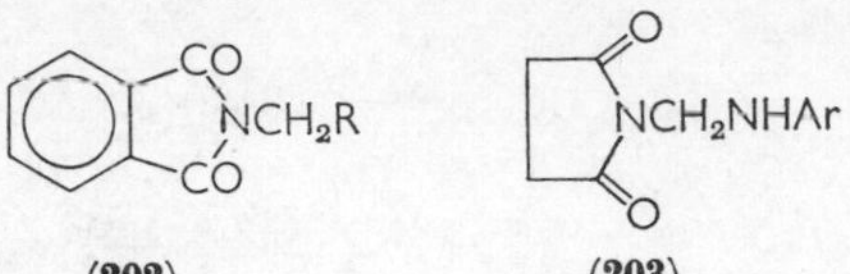

(**202**)   (**203**)

methyl)succinimides (**203**). The reaction may proceed via an *N*-(hydroxymethyl)succinimide or *N*-(hydroxymethyl)arylamine[223].

The NH group in an imide is acidic, to a much higher degree than in an amide. This is because the anion can be stabilized by resonance, and the second acyl group provides a larger orbital for electron de-

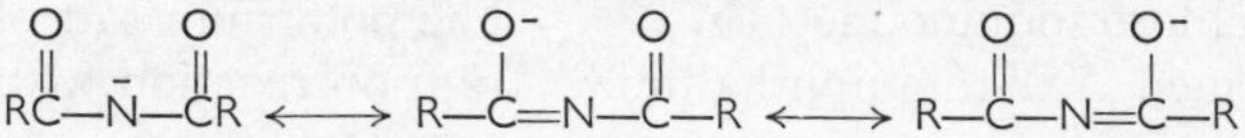

localization. Phthalimide, and other imides, form potassium salts with aqueous potassium hydroxide. Alkyl and allyl halides reacted with phthalimide in the presence of potassium bicarbonate to form

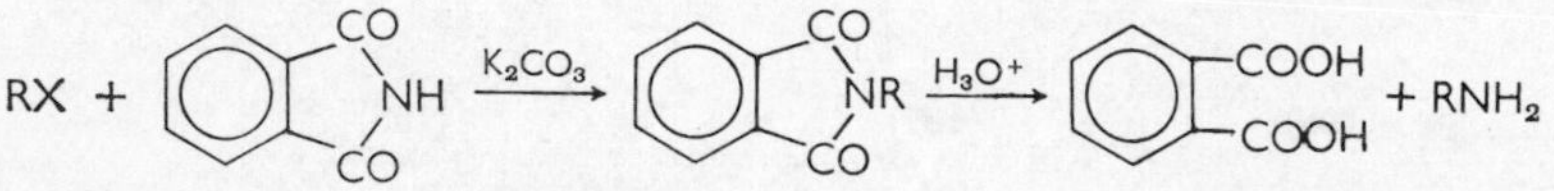

N-alkyl- and *N*-allylphthalimides, which were hydrolysed to the corresponding primary amines (Gabriel synthesis, equation 19)[224,225]. The hydrolysis of the *N*-alkylphthalimide was conveniently carried out using hydrazine[224].

The photolysis of succinimide vapour resulted in a mixture of products formed in four reactions; equations (20) and (21) accounted each for 40% of the products and equations (22) and (23) each for 10% of the products. The reactions probably proceeded through a common initial ring-opening step. The pyrolysis of succinimide vapour followed first-order kinetics, with equations (22) and (23)

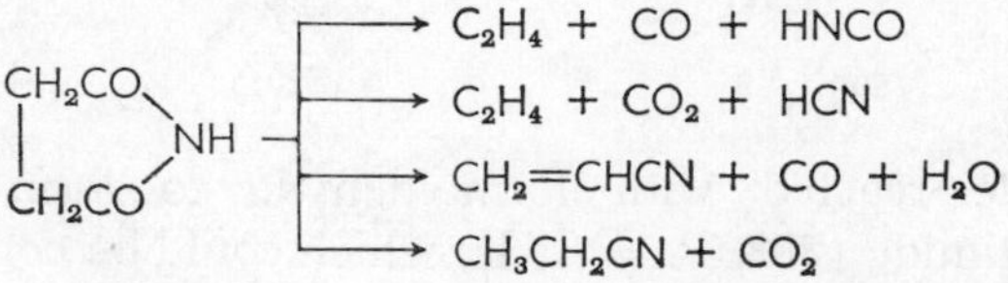

$$C_2H_4 + CO + HNCO \qquad (20)$$
$$C_2H_4 + CO_2 + HCN \qquad (21)$$
$$CH_2{=}CHCN + CO + H_2O \qquad (22)$$
$$CH_3CH_2CN + CO_2 \qquad (23)$$

predominating. The pyrolysis and photolysis of *N*-methylsuccinimide followed equation (20) giving methyl isocyanate[226].

Thiodiglycollimide (**204**) formed 1,4-thiazine (**205**) on passing over alumina on pumice at 450°[227].

(**204**)   (**205**)

V. REFERENCES

1. R. Anschtüz and Q. Wirtz, *Ann. Chem.*, **239**, 154 (1887).
2. I. J. Rinkes, *Rec. Trav. Chim.*, **48**, 960 (1929).
3. P. Hope and L. A. Wiles, *J. Chem. Soc.*, 4583 (1965).
4. F. M. Rowe, E. Levin, A. C. Burns, J. S. H. Davies, and W. Tepper, *J. Chem. Soc.*, 690 (1926).
5. A. F. Nagornyi, N. F. Semenova, T. I. Komarenko, and G. A. Lyubchenko, *Metody Polucheniya Khim. Reactivov i Preparatov, Glos. Kom. Sov. Min. SSSR po Khim.*, 24 (1964).
6. H. Brcdereck, R. Gompper, and D. Bitzer, *Chem. Ber.*, **92**, 1139 (1959).
7. D. E. Cadwallader and J. P. La Rocca, *J. Am. Pharm. Assoc.*, **45**, 480 (1956).
8. D. Davidson and H. Skovronek, *J. Am. Chem. Soc.*, **80**, 376 (1958).
9. C. D. Hurd and A. G. Prapas, *J. Org. Chem.*, **24**, 388 (1959).
10. H. K. Hall, M. K. Brandt, and R. M. Mason, *J. Am. Chem. Soc.*, **80**, 6420 (1958).
11. R. E. Dunbar and W. M. Swenson, *J. Org. Chem.*, **23**, 1793 (1958).
12. J. Goerdeler and K. Stadelbauer, *Chem. Ber.*, **98**, 1556 (1965).
13. G. Pans, L. Berlinguet, and R. Gaudry, *Org. Syn.*, **37**, 47 (1957).
14. E. Chapman and H. Stephens, *J. Chem. Soc.*, **127**, 1793 (1925).
15. M. L. Sherrill and F. L. Schaeffer, *J. Am. Chem. Soc.*, **50**, 475 (1928).
16. M. L. Sherrill, F. L. Schaeffer, and E. P. Shoyer, *J. Am. Chem. Soc.*, **50**, 474 (1928).
17. P. Pummerer and G. Dorfmuller, *Chem. Ber.*, **45**, 292 (1912).
18. A. Boucherle, G. Carraz, A. M. Revol, and J. Dodu, *Bull. Soc. Chim. France*, 500 (1960).
19. N. B. Mehta, A. P. Phillips, F. Lu-Lin, and R. E. Brooks, *J. Org. Chem.*, **25**, 1012 (1960).
20. W. R. Roderick, *J. Am. Chem. Soc.*, **79**, 1710 (1957).
21. B. Matkovics, L. Ferenczi, and G. Selmeczi, *Acta Univ. Szeged., Acta Phys. Chem.*, **4**, 134 (1958).
22. M. Z. Barakat, S. K. Shebab, and M. M. El-Sadr, *J. Chem. Soc.*, 4133 (1957).
23. W. H. Warren and R. A. Briggs, *Chem. Ber.*, **64**, 26 (1931).
24. D. Buckley and H. B. Henbest, *J. Chem. Soc.*, 1888 (1956).

25. A. T. de Mouilpied and A. Rule, *J. Chem. Soc.*, **91**, 176 (1907).
26. E. Testa, L. Fontanella, G. Cristiani, and L. Mariani, *Helv. Chim. Acta*, **42**, 2370 (1959).
27. E. Testa and L. Fontanella, *Ann. Chem.*, **660**, 118 (1962).
28. A. Foucaud and M. Duclos, *Compt. Rend.*, **256**, 4033 (1963).
29. R. A. Clement, *J. Org. Chem.*, **27**, 1904 (1962).
30. A. E. Kulikova and E. N. Zilberman, *Zh. Obshch. Khim.*, **30**, 596 (1960).
31. M. K. Gluzman, *Zh. Obshch. Khim.*, **28**, 2987 (1958).
32. R. Lukes and M. Ferles, *Collection Czech. Chem. Commun.*, **24**, 1297 (1959).
33. G. B. Hoey and C. T. Lester, *J. Am. Chem. Soc.*, **73**, 4473 (1951).
34. A. V. Kirsanov and Y. M. Zolotov, *Zh. Obshch. Khim.*, **20**, 1145 (1950).
35. R. E. Buckles and W. J. Probst, *J. Org. Chem.*, **22**, 1728 (1957).
36. H. K. Hall, *J. Am. Chem. Soc.*, **80**, 6412 (1958).
37. L. M. Rice and C. H. Grogan, *J. Org. Chem.*, **24**, 7 (1959).
38. J. R. Geigy, *Netherlands Pat.*, 6,505,972; Nov. 15, 1965; *Chem. Abstr.*, **64**, 14007 (1966).
39. S. Kuroda and K. Nishimune, *J. Pharm. Soc. Japan*, **64**, 157 (1944).
40. L. E. Kretov and A. P. Momsenko, *Zh. Obshch. Khim.*, **33**, 397 (1963).
41. A. Schulz and O. Stichnoth, *Ger. Pat.*, 1,086,704, Aug. 11, 1960; *Chem. Abstr.*, **55**, 16490 (1961).
42. W. A. Noyes and P. K. Potter, *Org. Syn.*, Col. Vol. I., 446 (1941).
43. W. Herzog, *Z. Anorg. Allgem. Chem.*, **32**, 301 (1919).
44. H. D. K. Drew and H. H. Hatt, *J. Chem. Soc.*, 16 (1937).
45. F. D. Chattaway and W. Tesh, *J. Chem. Soc.*, **117**, 711 (1920).
46. A. Arcasis and A. Salerno, *Boll. Sedute Accad. Gicinia Sci. Nat. Catania*, **4**, 195 (1957).
47. E. Jucker, *Angew. Chem.*, **71**, 321 (1959).
48. E. Jucker, A. Ebnöether, E. Rissi, A. Vogel, and R. Steiner, *Swiss Pat.* 367,174 (1963).
49. M. Aeberli and H. Erlenmeyer, *Helv. Chim. Acta*, **31**, 470 (1948).
50. T. Wieland and H. Urbach, *Ann. Chem.*, **613**, 84 (1958).
51. S. Sabiniewicz, *Przemysl Chem.*, **44**, 253 (1965).
52. M. Menard, L. Erichomovitch, M. Le Brooy, and F. L. Chubb, *Can. J. Chem.*, **41**, 1722 (1963).
53. F. C. Whitmore, *Organic Chemistry*, D. Van Nostrand Co., New York, N.Y.. 1937, p. 501.
54. C. M. Hendry, *J. Am. Chem. Soc.*, **80**, 973 (1958).
55. S. Gabriel and T. Posner, *Chem. Ber.*, **27**, 2492 (1894).
56. A. W. Day and S. Gabriel, *Chem. Ber.*, **23**, 2478 (1890).
57. M. T. Bogert and D. C. Eccles, *J. Am. Chem. Soc.*, **24**, 20 (1902).
58. O. Allendorff, *Chem. Ber.*, **24**, 2346 (1891).
59. O. Allendorff, *Chem. Ber.*, **24**, 3264 (1891).
60. W. Davies and J. A. McLaren, *J. Chem. Soc.*, 2595 (1951).
61. H. L. Wheeler and B. Barner, *J. Am. Chem. Soc.*, **22**, 80 (1900).
62. P. Klason, *Chem. Ber.*, **10**, 1349 (1877).
63. N. A. Langlet, *Chem. Ber.*, **24**, 3851 (1951).
64. K. Kato and S. Wade, *Nippon Kagaku Zasshi*, **83**, 501 (1962).
65. A. Fry, *J. Am. Chem. Soc.*, **75**, 2686 (1953).
66. P. A. S. Smith and R. O. Kan, *J. Am. Chem. Soc.*, **82**, 4753 (1960).

67. P. F. Wiley, *J. Am. Chem. Soc.*, **71**, 3747 (1949).
68. P. F. Wiley, *J. Am. Chem. Soc.*, **71**, 1310 (1949).
69. A. Ebnöether, E. Jucker, E. Rissi, J. Rutschmann, E. Schreir, R. Steiner, R. Süess, and A. Vogel, *Helv. Chim. Acta*, **42**, 918 (1959).
70. A. C. Poshkus and J. E. Herweh, *J. Org. Chem.*, **30**, 2466 (1965).
71. D. J. Hadley and B. Wood, *U.S. Pat.* 2,838,558, June 10, 1958.
72. D. J. Hadley and E. J. Gasson, *Brit. Pat.* 803,901, Nov. 5, 1958.
73. I. S. Kolodina and B. V. Suvorov, *Izv. Akad. Nauk. Kaz. SSSR, Ser. Khim.*, 92 (1962).
74. S. D. Mekhtiev, G. N. Suleimanov, Z. Y. Magerromova, R. Y. Magerromova, and S. F. Mamedova, *Azerb. Khim. Zh.*, 77 (1964).
75. S. Saito, H. Iwasaki, and N. Ota, *Yuki Gosei Kagaku Kyokai Shi*, **22**, 828 (1964).
76. S. Saito and N. Ota, *Yuki Gosei Kagaku Kyokai Shi*, **22**, 730 (1964).
77. S. J. Green, *J. Soc. Chem. Ind.*, **51**, 123T (1964).
78. H. Kwart and I. Burchuk, *J. Am. Chem. Soc.*, **74**, 3094 (1952).
79. P. de Mayo and S. T. Reid, *Chem. Ind.* (*London*), 1576 (1962).
80. R. Scarpati and C. Santacroce, *Rend. Acad. Sci., Fis. Mat.*, **28**, 27 (1961).
81. J. Falbe and F. Korte, *Angew. Chem.*, **74**, 291 (1962).
82. Shell Int. Res. *Belg. Pat.* 623,333; April 8, 1963.
83. L. K. Freidlin, T. A. Sladkova, and F. E. Englina, *Izv. Akad. Nauk SSSR, Ser. Khim.*, 1248 (1965).
84. T. L. Gresham, J. E. Jansen, F. W. Shaver, and W. L. Beears, *J. Am. Chem. Soc.*, **76**, 486 (1954).
85. H. L. Needles and R. F. Whitfield, *J. Org. Chem.*, **31**, 341 (1966).
86. G. E. Ficken and R. P. Linstead, *J. Chem. Soc.*, 3525 (1955).
87. R. E. Linstead and M. Whalley, *J. Chem. Soc.*, 3530 (1955).
88. H. D. K. Drew and H. H. Hatt, *J. Chem. Soc.*, 586 (1937).
89. T. Matsuo, *Kogyo Kagaku Zasshi*, **68**, 1422 (1965).
90. H. Hopff and B. Muhlethaler, *Kunststoffe Plastics*, **4**, 257 (1957).
91. H. Hopff, *Bull. Soc. Chim. France*, 1283 (1958).
92. H. Hopff and P. Muhlethaler, *Chimia*, **11**, 336 (1957).
93. J. Furukawa, A. Onishi, and T. Tsurata, *Kogyo Kagaku Zasshi*, **60**, 350 (1957).
94. N. Tokura, R. Tada, and K. Yokoyama, *Bull. Chem. Soc. Japan*, **34**, 1812 (1961).
95. P. M. Brown, D. B. Spiers, and M. Whalley, *J. Chem. Soc.*, 2882 (1957).
96. B. E. Leach, J. H. Ford, and A. J. Whiffen, *J. Am. Chem. Soc.*, **69**, 474 (1947).
97. E. C. Kornfeld, R. G. Jones, and T. V. Parke, *J. Am. Chem. Soc.*, **71**, 150 (1949).
98. B. C. Lawes, *J. Am. Chem. Soc.*, **84**, 239 (1962).
99. N. A. Starkovsky and F. Johnson, *Tetrahedron Letters*, 919 (1964).
100. F. Johnson, N. A. Starkovsky, A. C. Paton, and A. A. Carlson, *J. Am. Chem. Soc.*, **88**, 149 (1966).
101. S. H. McElvain and D. H. Clemens, *J. Am. Chem. Soc.*, **80**, 3915 (1958).
102. A. A. Liebmann and F. E. De Gangi, *J. Pharm. Sci.*, **52**, 395 (1963).
103. C. M. Lee and W. D. Kumler, *J. Am. Chem. Soc.*, **83**, 4586 (1961).
104. C. M. Lee and W. D. Kumler, *J. Am. Chem. Soc.*, **84**, 565 (1962).

105. C. M. Lee and W. D. Kumler, *J. Org. Chem.*, **27**, 2055 (1962).
106. N. G. Bakhshiev, *Opt. i Spektroskopiya*, **13**, 192 (1962).
107. C. M. Lee and W. D. Kumler, *J. Am. Chem. Soc.*, **84**, 571 (1962).
108. C. Fayot and A. Foucaud, *Compt. Rend.*, **261**, 4018 (1965).
109. H. M. Randal, R. G. Fowler, N. Fuson, and R. Porgl, *Infrared Determination of Organic Structures*, Van Nostrand, New York, N.Y., 1949.
110. J. Uno and K. Machida, *Bull. Chem. Soc. Japan*, **34**, 545 (1961).
111. J. Uno and K. Machida, *Bull. Chem. Soc. Japan*, **34**, 551 (1961).
112. R. A. Abramowitch, *J. Chem. Soc.*, 1413 (1957).
113. N. A. Borisevitch and N. N. Khowratovitch, *Opt. Spectry. (USSR), (Engl. Transl.)*, **10**, 309 (1961).
114. J. Uno and K. Machida, *Bull. Chem. Soc. Japan*, **35**, 1226 (1962).
115. J. Uno, K. Machida, and I. Hamanuka, *Bull. Chem. Soc. Japan*, **34**, 1448 (1961).
116. H. K. Hall and R. Zbinden, *J. Am. Chem. Soc.*, **80**, 6428 (1958).
117. P. Bassignana, C. Cogrossi, S. Franco, and G. Polla-Mattiot, *Spectrochim. Acta*, **21**, 677 (1965).
118. T. Matsuo, *Bull. Chem. Soc. Japan*, **37**, 1844 (1964).
119. A. D. Walsh, *Discussions Faraday Soc.*, **2**, 18 (1947).
120. S. Nishizaki, *Nippon Kagaku Zasshi*, **86**, 696 (1965).
121. W. R. Roderick and P. L. Bhatia, *J. Org. Chem.*, **28**, 2018 (1963).
122. N. B. Calthup, L. H. Daley, and S. E. Wiberley, *Introduction to Infrared and Raman Spectroscopy*, Academic Press, New York, 1964, p. 266.
123. D. E. Ames and T. F. Grey, *J. Chem. Soc.*, 3518 (1955).
124. W. R. Roderick and W. G. Brown, *J. Am. Chem. Soc.*, **79**, 5196 (1957).
125. J. Schurz, A. Ullrich, and H. Bayzer, *Monatsh. Chem.*, **90**, 29 (1959).
126. W. D. Turner, *J. Chem. Soc.*, 4555 (1957).
127. D. E. Ames and T. F. Grey, *J. Chem. Soc.*, 631 (1955).
128. L. F. Gladchenko, L. G. Pikulik, and N. L. Belozerevich, *Opt. i Spectroskopiya*, **17**, 209 (1964).
129. A. Arcoria and G. Scarlata, *Ann. Chim. (Rome)*, **54**, 128 (1964).
130. A. Arcoria, H. Lumbroso, and R. Passerini, *Boll. Sedute Accad. Gioenia Sci. Nat. Catania*, **3**, 537 (1957).
131. A. Arcoria and R. Passerini, *Boll. Sci. Fac. Chim. Ind. Bologna*, **15**, 121 (1957).
132. A. Arcoria and R. Passerini, *Boll. Sci. Fac. Chim. Ind. Bologna*, **15**, 124 (1957).
133. A. Arcoria and F. Bottini, *Boll. Sci. Fac. Chim. Ind. Bologna*, **15**, 127 (1957).
134. L. G. Pikulik, *Fiz. Probl. Specktroskopii, Akad. Nauk SSSR, Materialy 13-go Soveshch.*, Leningrad, 1960, p. 297.
135. A. N. Sevchenko, L. G. Pikulik, and M. Y. Kostko, *Dokl. Akad. Nauk SSSR*, **162**, 57 (1965).
136. L. G. Pikulik, L. F. Gladchenko, and M. Y. Kostko, *Zh. Prikl. Spectroskopii Akad. Belorussk. USSR*, **2**, 160 (1965).
137. L. A. Kiyanskaya, *Izv. Akad. Nauk SSSR, Ser. Fiz.*, **29**, 1357 (1965).
138. D. Herbison-Evans and R. E. Richards, *Mol. Phys.*, **8**, 19 (1964).
139. A. M. Duffield, H. Budzikiewicz, and C. Djerassi, *J. Am. Chem. Soc.*, **87**, 2913 (1965).
140. J. H. Bowie, R. G. Cooks, S. O. Lawesson, P. Jakobsen, and G. Schrall, *Chem. Commun.*, 539 (1966).

141. J. S. Fritz and N. M. Leisichi, *Anal. Chem.* **23**, 589 (1951).
142. K. Auwers, *Ann. Chem.*, **309**, 316 (1899).
143. W. S. Owen, *Mikrochim. Acta*, 19 (1963)
144. M. A. Devereux and H. B. Donahoe, *J. Org. Chem.*, **25**, 457 (1960).
144a. J. T. Edwards and K. A. Terry, *J. Chem. Soc.*, 3527 (1957).
145. H. K. Hall, M. K. Brandt, and R. M. Mason, *J. Am. Chem. Soc.*, **80**, 6420 (1958).
146. I. Wadso, *Acta Chem. Scand.*, **19**, 1079 (1965).
147. A. K. Herd, L. Eberson, and T. Higuchi, *J. Pharm. Sci.*, **55**, 162 (1966).
148. A. Fouc, *Bull. Soc. Chim. Bretagne*, **35**, 88 (1960).
149. P. Crooy and A. Bruylants, *Bull. Soc. Chim. Belges*, **73**, 44 (1964).
150. A. Bruylants and F. de Kemmeter, *Bull. Soc. Chim. Belges*, **73**, 637 (1964).
151. B. Zerner and M. L. Bender, *J. Am. Chem. Soc.*, **83**, 2267 (1961).
152. U. Mazzucato, A. Foffani, and G. Cauzzo, *Ann. Chim. (Rome)*, **50**, 521 (1960).
153. S. Champy-Hatem, *Compt. Rend.*, **262**, 18 (1966).
154. S. Champy-Hatem, *Compt. Rend.*, **261**, 271 (1965).
155. E. R. Garrett and R. E. Notari, *J. Org. Chem.*, **31**, 425 (1966).
156. R. M. Peck, *J. Org. Chem.*, **27**, 2677 (1962).
157. J. H. M. Hall, *J. Org. Chem.*, **30**, 620 (1965).
158. L. Otvos, F. Dutka, and H. Tudos, *Atompraxis*, **10**, 536 (1964).
159. F. Dutka, H. Tudos, and L. Otvos, *Magy. Tud. Akad. Kozp. Kem. Kut. Int. Kozlemen.*, 11 (1963; publ. 1965).
160. L. Otvos, F. Dutka, and H. Tudos, *Acta Chim. Acad. Sci., Hung.*, **43**, 53 (1965).
161. K. C. Schreiber and V. P. Fernandez, *J. Org. Chem.*, **26**, 1744 (1961).
162. L. M. Rice, E. E. Reid, and C. H. Grogan, *J. Org. Chem.*, **19**, 884 (1954).
163. L. M. Rice, C. H. Grogan, and E. E. Reid, *J. Am. Chem. Soc.*, **75**, 4304 (1953).
164. L. M. Rice, C. H. Grogan, and E. E. Reid, *J. Am. Chem. Soc.*, **75**, 4911 (1953).
165. R. Griot, *Helv. Chim. Acta*, **42**, 67 (1959).
166. I. S. Rossi and C. Valvo, *Farmaco (Pavia), Ed. Sci.*, **12**, 1008 (1957).
167. M. Ferles, *Chem. Listy*, **52**, 2184 (1958).
168. W. Flitsh, *Chem. Ber.*, **97**, 1542 (1961).
169. E. Tagman, E. Sury, and K. Hoffman, *Helv. Chim. Acta*, **37**, 185 (1954).
170. Z. I. Horii, C. Iwata, and Y. Tamura, *J. Org. Chem.*, **26**, 2273 (1961).
171. A. A. Liebman and F. de Gangi, *J. Pharm. Sci.*, **52**, 276 (1963).
172. A. J. McAless and R. McCrindle, *Chem. Ind. (London)*, 1869 (1965).
173. J. N. Ashley, R. F. Collins, N. Davis, and N. E. Serett, *J. Chem. Soc.*, 3880 (1959).
174. J. H. Paden and H. Adkins, *J. Am. Chem. Soc.*, **58**, 2487 (1936).
175. A. Dunet, R. Ratouis, P. Cadiot, and A. Willemart, *Bull. Soc. Chim. France*, 906 (1956).
176. J. C. Brockmann, *Electro Organic Chemistry*, John Wiley and Sons, New York, 1926, p. 28.
177. B. Sakurai, *Bull. Chem. Soc. Japan*, **7**, 755 (1932).
178. E. W. Cook and W. G. France, *J. Phys. Chem.*, **36**, 2383 (1932).
179. B. Sakurai, *Shinshu Daigaku Bunviga Kubu Kiyo*, **5**, 11 (1955).

180. M. J. Allen and J. Ocampo, *J. Electrochem. Soc.*, **103**, 452 (1956).
181. R. Lukes and Z. Linhartova, *Collection Czech. Chem. Commun.*, **25**, 50 (1960).
182. R. Lukes, A. Fabryova, S. Dolezal, and L. Novotny, *Collection Czech. Chem. Commun.*, **25**, 1063 (1960).
183. R. Lukes and M. Cerny, *Collection Czech. Chem. Commun.*, **24**, 2722 (1959).
184. W. I. Awad, F. G. Baddar, M. A. Omara, and S. M. A. R. Omran, *J. Chem. Soc.*, 2040 (1965).
185. K. Heidenbluth, H. Toenjes, and R. Scheffler, *J. Prakt. Chem.*, **30**, 204 (1965).
186. F. Sachs and H. Ludwig, *Chem. Ber.*, **37**, 385 (1904).
187. L. Horner and E. H. Winkelmann, *Angew. Chem.*, **71**, 349 (1959).
188. C. Djerassi, *Chem. Rev.*, **43**, 271 (1948).
189. R. E. Pearson and J. C. Martin, *J. Am. Chem. Soc.*, **85**, 3142 (1963).
190. A. L. Henne and W. F. Zimmer, *J. Am. Chem. Soc.*, **73**, 1103 (1951).
191. K. Ziegler, A. Späth, E. Schaaf, W. Schumann, and E. Winkelmann, *Ann. Chem.*, **551**, 80 (1942).
192. G. F. Bloomfield, *J. Chem. Soc.*, 114 (1944).
193. M. F. Hibbelynch and R. H. Martin, *Bull. Soc. Chim. Belges*, **59**, 193 (1950).
194. M. F. Hibbelynch and R. H. Martin, *Experientia*, **51**, 69 (1949).
195. S. D. Ross, M. Finkelstein, and R. C. Petersen, *J. Am. Chem. Soc.*, **80**, 4327 (1958).
196. R. Filler, *Chem. Rev.*, **63**, 21 (1963).
197. J. Lecomte and C. Dufour, *Compt. Rend.*, **234**, 1887 (1952).
198. A. Schönberg, R. Moubasher, and M. Z. Barakat, *J. Chem. Soc.*, 2504 (1951).
199. C. Walling, A. L. Rieger, and D. D. Tanner, *J. Am. Chem. Soc.*, **85**, 3129 (1963).
200. R. E. Lovins, L. J. Andrews, and R. M. Keefer, *J. Org. Chem.*, **29**, 1616 (1964).
201. R. E. Lovins, L. J. Andrews, and R. M. Keefer, *J. Org. Chem.*, **30**, 1577 (1965).
202. G. A. Russell and Y. R. Vinson, *J. Org. Chem.*, **31**, 1994 (1966).
203. R. E. Lovins, L. J. Andrews, and R. M. Keefer, *J. Org. Chem.*, **30**, 4150 (1965).
204. F. L. Lambert, W. D. Ellis, and R. J. Parry, *J. Org. Chem.*, **30**, 304 (1965).
205. C. Walling and A. L. Rieger, *J. Am. Chem. Soc.*, **85**, 3134 (1963).
206. G. A. Russell and K. M. Desmond, *J. Am. Chem. Soc.*, **85**, 3139 (1963).
207. P. S. Skell, D. L. Tuleen, and P. D. Readio, *J. Am. Chem. Soc.*, **85**, 2850 (1963).
208. C. Rappe and R. Kumar, *Arkiv. Kemi*, **23**, 475 (1965).
209. A. Guillemonat, G. Peiffer, J. C. Traynard, and A. Leger, *Bull. Soc. Chim. France*, 1192 (1964).
210. G. Peiffer, J. C. Traynard, and A. Guillemonat, *Bull. Soc. Chim. France*, 1910 (1966).
211. H. T. Clarke and L. D. Behr, *Org. Syn.* Coll. Vol. II, 19 (1943).
212. E. Chapman and H. Stephen, *J. Chem. Soc.*, **127**, 1791 (1925).
213. J. C. Brockmann, *Electro Organic Chemistry*, John Wiley and Sons, New York, 1926, p. 78.
214. C. W. Huffmann and W. G. Skelly, *Chem. Rev.*, **63**, 631 (1963).

215. O. A. Moe and D. T. Warner, *J. Am. Chem. Soc.*, **71**, 1251 (1949).
216. W. W. Grouch and H. L. Lochte, *J. Am. Chem. Soc.*, **65**, 270 (1943).
217. R. W. Meikle and E. A. Williams, *Nature*, **210**, 523 (1966).
218. S. Gabriel and T. Posner, *Chem. Ber.*, **27**, 2492 (1894).
219. M. V. Kalnins, *Can. J. Chem.*, **44**, 2111 (1966).
220. K. Yanagi and S. Akiyoshi, *J. Org. Chem.*, **24**, 1122 (1959).
221. S. A. Abbas and W. J. Hickinbottom, *J. Chem. Soc.* (*C*), 1305 (1966).
222. L. W. Kissinger and H. E. Ungnade, *J. Org. Chem.*, **23**, 815 (1958).
223. M. B. Winstead, K. V. Anthony, and L. Leddich, *J. Chem. Eng. Data*, **7**, 414 (1962).
224. N. R. Ing and R. H. F. Manske, *J. Chem. Soc.*, 2348 (1926).
225. R. H. F. Manske, *Org. Syn.*, Coll. Vol. II, 83 (1943).
226. G. Choudhary, A. M. Cameron, and R. A. Black, *J. Phys. Chem.*, **72**, 2289 (1968).
227. C. Barkenbus and P. S. Landis, *J. Am. Chem. Soc.*, **70**, 684 (1948).

CHAPTER **8**

The chemistry of thioamides

W. Walter and J. Voss

University of Hamburg, Germany

383

I. GENERAL AND THEORETICAL ASPECTS OF THE THIOAMIDE GROUP

A. Historical Remarks

As early as 1815 Gay-Lussac obtained the first thioamide[1] by reacting hydrogen sulphide with cyanogen. It was called 'Flaveanwasserstoff'[2] (flaveanic acid) by Berzelius owing to its bright-yellow colour, and, as is well known, has the chemical structure of oxalic acid nitrile thioamide. Its counterpart, the red 'Rubeanwasserstoff'[2] (rubeanic acid), was prepared in 1825 by Wöhler and Liebig[3,4] in a similar manner, and was investigated more accurately by Völckel[5], who pointed out the chemical analogy of this compound with oxamide, thus performing the first structural evidence in the class of thioamides. In 1848 Cahours[6] and later on Hofmann[7] obtained a series of other thioamides from the corresponding carboxylic acid nitriles and hydrogen sulphide. This method, in addition to the thionation of amides, introduced in 1878 by Hofmann[8], has become most useful for the preparation of thioamides.

Since their discovery thioamides have turned out to be most versatile reagents especially in the field of heterocyclic chemistry. For some 20 years the investigation of their chemical properties has been developing rapidly, mainly on account of their enlarged technical application, and several reviews have been published in this period[9-13,357]. The methods of physical organic chemistry, on the other hand, have been extensively applied to thioamides only very recently, and no comprehensive publication about this matter has appeared as yet. We therefore will especially concentrate on this subject.

B. Nomenclature

The correct nomenclature of the thioamides conforming to the I.U.C. rules is discussed in detail by Hurd and DeLaMater[11]. Accordingly these compounds have to be named by substituting the ending 'thionamide' for '-ic acid' or '-oic acid' of the name of the corresponding acid, or by using the prefix 'thiol' or 'thion,' respectively, before the name of the corresponding amide, the latter being simpler and more customary for naming substituted species such as anilides, toluidides etc. In practice however the prefixes 'thiol' and 'thion' are rarely discriminated but generally replaced by 'thio', which, in fact, is sufficient for characterizing the functional group $CSNR_2$, because thiolamides $RC(NR)SH$, unknown as free molecules,

should be better called imidothiolic acids, and compounds like
$HSCH_2CONH_2$ or $CH_3CSCH_2CONH_2$ are named mercaptoacet-
amide and β-thionobutyramide, respectively. Throughout this
chapter this simplified nomenclature will be used. Corresponding to
the nomenclature of amides, $RCSNH_2$ is called a primary, RCSNHR
a secondary, and $RCSNR_2$ a tertiary thioamide.

Compounds bearing the $-C(=S)N\!\!<$ functional group at an atom
other than carbon (thioureas, thiocarbamic esters etc.) are not discussed here, except in certain cases for comparison purposes. Nor will
we deal in general with N-heterosubstituted compounds like thiohydroxamic acids, thiohydrazides (cf. Chapter 9 of this book), etc.,
or integrated heteroaromatic thioamide systems such as thiazole and
pyridine–thione.

C. Topology and Electronic Structure of the Thioamide Group

I. X-ray diffraction

The most instructive view of the arrangement of atoms within a
molecule, though in the crystalline state only, can be achieved by
x-ray diffraction, and several thioamides have been studied by this
means (Table 1).

From all available x-ray results it follows unambiguously that the
key atoms of the thioamide group (**3b**) are situated in a plane like the
ligands at normal olefinic double bonds, thus suggesting at least large
contributions from sp^2-hybrid atomic orbitals of the central carbon
and nitrogen atoms to the thioamide molecular orbital (**3a**), or, in
terms of the VB theory, strong preference of polar resonance structures
like **3c**, implying a partial carbon–nitrogen double bond.

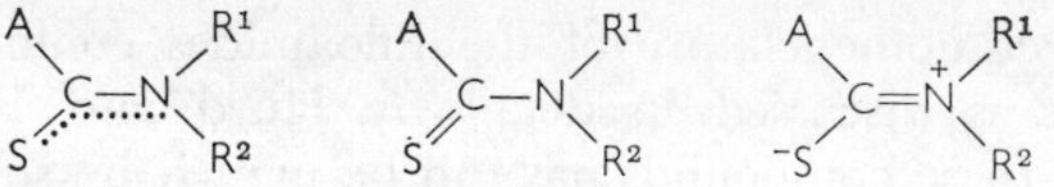

This agrees well with the data of Table 1. In particular all
valence angles at the central atoms are in the region of 120° giving
evidence for trigonal rather than tetrahedral (109. 5°) hybridization.
The torsional angles Θ, shown in Figure 1, between the plane of the
functional group $\mathbf{P}_F$ and planes $\mathbf{P}_1$ and $\mathbf{P}_2$ of aryl rings bound to the
C or N atom, however, are not 0°. For instance $\Theta_1 = 38°$ in $\mathbf{1}$[18]
and $\Theta_2 = 45°$ in $\mathbf{2}$[22]. The atomic distances fall between the known

TABLE 1. Bond lengths and angles in thioamides and similar compounds as determined by x-ray diffraction.

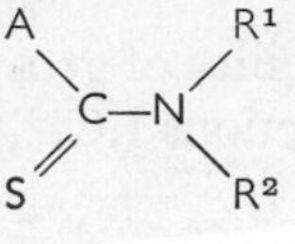

General structure:

$$\underset{S}{\overset{A}{\diagdown}}C=N\underset{R^2}{\overset{R^1}{\diagup}}$$

Compound	Bond lengths (Å)		Bond angles (degrees)				Refs.
	C—S	C—N	A ĈS	S ĈN	$R^1\hat{N}C$ $R^2\hat{N}C$	$R^1\hat{N}R^2$	
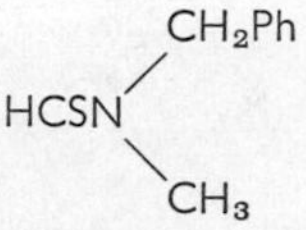 HCSN(CH₂Ph)(CH₃)	1·66	1·35		123	122 119	119	14
CH_3CSNH_2	1·731	1·324	120·7	117·7	117 116	123	15
$(CH_3CSNH_2)_4Cu$	1·683	1·302	123·7	122			16
$H_2NCSCSNH_2$	1·665 1·633	1·336 1·311	119·6 120·5	125·4 124·6			17
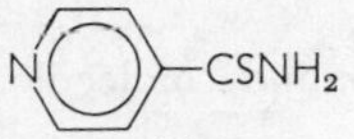 pyridyl–$CSNH_2$ (1)	1·65	1·32	116·7	124·4			18
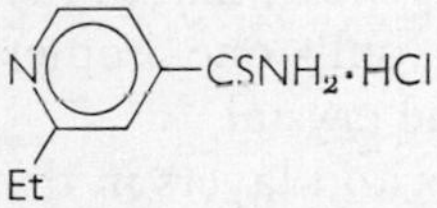 Et-pyridyl–$CSNH_2$·HCl	1·69	1·29	115·5	124·3			18a
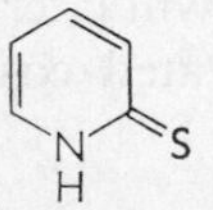 pyridine-2-thione (N–H, =S)	1·68	1·39	127	119	122		19
H_2NCSNH_2	1·71	1·33	122·2	122·2			20
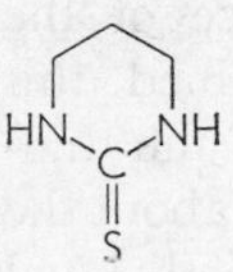 HN–C(=S)–NH (cyclic)	1·722	1·334	119·8	119·8	122·9		21
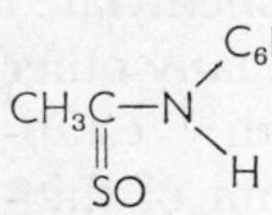 $CH_3C(=SO)$–N(C₆H₅)(H) (2)	1·646	1·341	117·6	118·1	126·7 117·9	115	22

values for single and double bonds: C—N 1·47 Å; C=N (from oximes) 1·27 Å[23]; and C—S 1·81 Å; C=S 1·60 Å[23,24], respectively, indicating marked electron delocalization (**3a**).

The planarity of the functional group implies the possibility of geometrical isomerism. This point is discussed in detail in sections I.C.2,

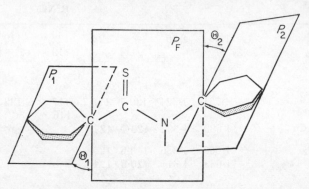

FIGURE 1. Angles Θ between the various planes of the thioamide molecule.

I.C.3, and I.C.5. The x-ray diffraction experiments show, that in the case of *N*-benzyl-*N*-methylthioformamide there is only one isomer (the so-called '*trans*', cf. section I.C.2) present in the crystal[14].

Thioamides form both dimers and polymer chains and layers in the lattice, held by intermolecular hydrogen bonds[15,18,19]. We will later mention some effects, that considerably disturb this delineated coplanar arrangement.

2. Nuclear magnetic resonance spectroscopic evidence

Sandström[25] discovered the magnetical non-equivalence of the CH_3 groups in *N,N*-dimethylthiobenzamide, and explained this phenomenon by suggesting a fixed planar configuration of the thioamide molecule accompanied by restricted internal rotation about the central C—N bond, as has been done in the case of *N,N*-dimethylated formamide and acetamide by Phillips[26] and later on for many other amides (Chapter 1). Independently Speziale and Smith[27] established a splitting of the CH_3 signal in the n.m.r. spectrum of thioacetanilide, which of course is due to the presence of two different species, but erroneously was referred to thion–thiol tautomerism rather than *cis–trans* isomerism.

Recently a variety of thioamides has been studied by n.m.r. with regard to *cis–trans* isomerism. These investigations have been particularly facilitated by the fact, that in some cases the *cis–trans* mixtures could be resolved into the two pure isomers[28-32]. In secondary thioamides other than thioformamides frequently one isomer only occurs. Table 2 shows equilibrium ratios of unsymmetrically substituted thioamides, usually determined by integration of suitable peak areas in the n.m.r. spectra, aided by the possibility of separation of isomers. The two forms are no longer called '*cis*' and '*trans*', because this nomenclature is ambiguous, but are designated corresponding to the suggestions of Blackwood and coworkers[39]*, i.e. any conformation with two substituents of higher priority (according to the sequence rules of Cahn and coworkers[40]) on opposite sites of a molecule is called (*E*) (from the German 'entgegen') and its counterpart (*Z*) (from 'zusammen') as is demonstrated in the examples below. This designation, moreover, brings about greater consistency in the (*Z*)/(*E*) ratios than the *cis–trans* terminology, because bulky groups on nitrogen generally are the ones with the higher priority (see Table 2) and on the thioacyl side the sulphur atom has priority.

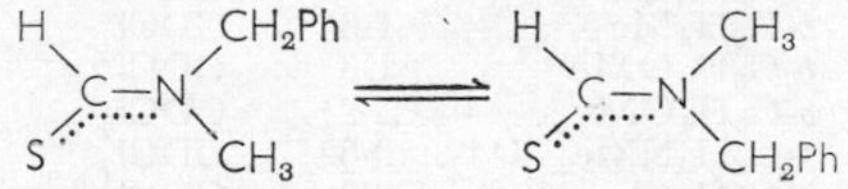

(*E*) (formerly '*trans*') (*Z*) (formerly '*cis*')

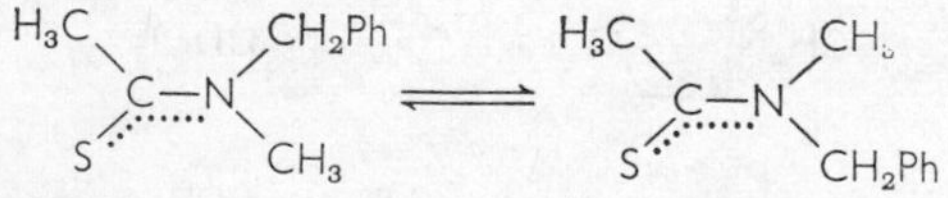

(*E*) (formerly dubious) (*Z*) (formerly dubious)

The assignment of n.m.r. lines to the conformers in question may be achieved by measuring the dependence of the chemical shifts on dilution with benzene, a viable method for thioamides[33-36,41,41a] as well as for amides[42], and in thioformamides and thioacetamides by means of the different coupling constants across the partial double bond[30,32,41]. Interestingly, the respective coupling constants in secondary thioformamides are always higher than in the corresponding formamides[32,33]. This may be explained by the increased double-bond character of the C—N bond in the former. In dimethylthioacetamide (DMTA) the methyl group near the S atom ('*cis*') is less shielded than the '*trans*'-methyl group, whereas the reverse is true for

* This terminology is used in the *Chemical Abstracts Index* too.

TABLE 2.　Isomer ratios of unsymmetrically substituted thioamides $R^1CSNR^2R^3$.

R^1	R^2	R^3	$[(Z)]/[(E)]$	Solvent	Separation of isomers	Refs.
H	H	Me	6·9	C_6H_6		33
H	H	Et	8·1	DMSO[a]		32
H	H	i-C_3H_7	2·3	C_6H_6	TLC[b]	31, 34
H	H	i-C_4H_9	2·5	C_6H_6	TLC	31, 34
H	H	t-C_4H_9	0·04	C_6H_6		34
H	H	CH_2Ph	5·2	C_6H_6	TLC	31, 34
H	H	CHMePh	2·8	C_6H_6		34
H	H	CH_2CH_2OMe	3·0	neat		32
H	H	CH_2CH_2OEt	3·8	neat		32
H	H	$CH_2CH_2NMe_2$	13·3	DMSO		32
H	Me	CH_2Ph	0·64	neat	spontaneous	28, 34, 35
H	Me	CHMePh	0·37	neat		34
H	Me	CH_2CH_2OH	0·33	C_6H_6	TLC	30
H	C_2H_5	CH_2CH_2OH	0·64	C_6H_6	TLC	30
H	i-C_3H_7	CH_2CH_2OH	0·32	C_6H_6	TLC	30
H	i-C_3H_7	CH_2Ph	0·25	C_6H_6		34
H	t-C_4H_9	CH_2CH_2OH	0·00	C_6H_6	pure	30
Me	H	Me	36	C_6H_6		33
Me	H	Ph	1·5[c]	$CDCl_3$		36
Me	H	Ph	1·7	$CDCl_3$		38
Me	H	p-C_6H_4Me	1·1[c]	$CDCl_3$		36
Me	H	p-C_6H_4Me	1·3	$CDCl_3$		38
Me	H	p-C_6H_4OMe	1·3	$CDCl_3$		38
Me	H	o-C_6H_4OMe	3·2	$CDCl_3$		38
Me	H	p-$C_6H_4NMe_2$	0·82	$CDCl_3$		38
Me	H	p-C_6H_4Cl	2·3	$CDCl_3$		38
Me	H	p-$C_6H_4NO_2$	5·1	$CDCl_3$		38
Me	H	o-C_6H_4COMe	d	$CDCl_3$		38
Me	H	CH_2N⟨morpholino⟩	d	$CDCl_3$	pure	32
N-Thioacetyl-indoline (4)[e]			3·8[c]	CCl_4		37
Et	H	Me	d	$CDCl_3$	pure	33
i-C_3H_7	H	Me	d	$CDCl_3$	pure	33
t-C_4H_9	H	Me	d	$CDCl_3$	pure	33
Ph	H	CH_2OH	d	$CDCl_3$	pure	32
p-$NO_2C_6H_4$	H	CH_2OH	d	$CDCl_3$	pure	32
2,4,6-$Me_3C_6H_2$	Me	CH_2Ph	2·1	?	fractional crystallization	29

[a] DMSO = hexadeuterodimethyl sulphoxide.

[b] TLC = thin-layer chromatography.

[c] The $[(Z)]/[(E)]$ ratio varies markedly with increasing dielectric constant of the solvent.　(4) for instance has $(Z)/(E) = 0·75$ in DMSO[37].

[d] No (E) isomer could be detected in these compounds.

[e]

(4)

dimethylthioformamide (DMTF); the latter behaving analogously to tertiary amides the magnetic anisotropy of which may be described by the model of Paulsen and Todt[43,44]. By the aid of sterically fixed N-thioacylpiperidines analogous models for the spheres of anisotropy of the thioformamide and thioacetamide systems have been developed[41a]. It appears that the inversion of magnetic shielding between DMTF and DMTA is due to steric effects[41a].

Inspection of Table 2 shows that in the case of tertiary thioamides the (E) conformation predominates unless R^1 is bigger than the S atom as is expected from sterical considerations. Surprisingly enough, most secondary thioformamides behave differently: in the dominant isomer the N-substituent shares the same side with the large S atom. This finding cannot be explained by assuming chain association of the (Z) isomer via intermolecular hydrogen bonds, because this conformation is quite stable in highly dilute solutions too as is shown by i.r. spectroscopy (section I.C.3). Intramolecular electrostatical interaction[31] as indicated in 5 can explain this

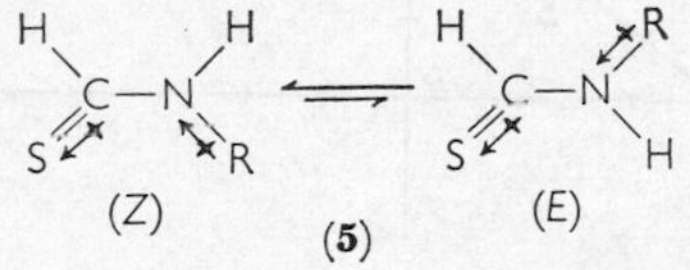

behaviour. The (Z) isomer is the one with the lower total dipole moment, and thus the lower free energy.

In some cases the isomer ratios are markedly dependent on solvent effects[36] (cf. footnote c of Table 2) but there are compounds which show only small changes[34].

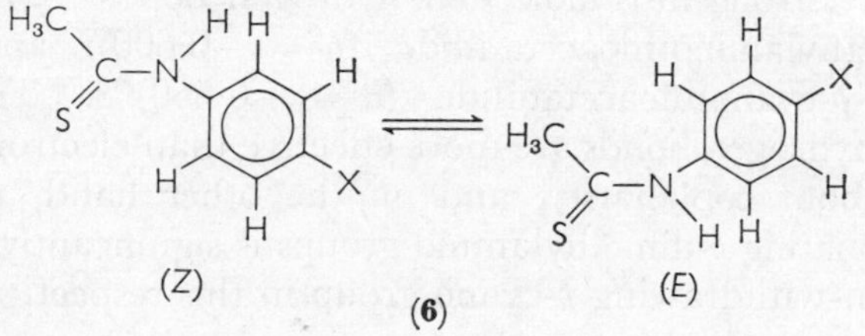

Linear correlation between the logarithmic isomer ratios of para-substituted thioacetanilides (6) and the Hammett constants of the respective substituents X is obtained ($\rho = 0.76$ in CCl_4) (Figure 2)[38]. This is explained by the assumption that an increase of π-electron density on the benzene ring (i.e. decrease of the σ value) will promote the orbital overlapping between the thioamide and the aryl group. The resulting more coplanar arrangement will however suffer from steric hindrance of the *ortho* protons and the large S atom in the (Z) conformer, and thus the less hindered (E) conformer will be favoured in the equilibrium.

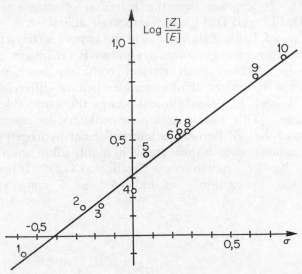

FIGURE 2. Relationship between $[(Z)]/[(E)]$ and σ in p-substituted thioacet-anilides[38] $MeCSNHC_6H_4X\text{-}p(\mathbf{6})$.

No.	1	2	3	4	5	6	7	8	9	10
X	NMe_2	OMe	Me	H	F	Cl	Br	I	CN	NO_2

The outlined interaction is also apparent from the difference of the chemical shifts $\Delta\tau = \tau_o - \tau_m$ between the *ortho* protons and *meta* protons, because the deshielding effect of the S atom increases with increasing coplanarity of the molecule. For instance $\Delta\tau = 140$ c.p.s. in N-thioacetylindoline (**4**), which is totally coplanar in the relevant parts; $\Delta\tau = 115$ c.p.s. for o-methoxythioacetanilide (enforced co-planarity by strong intramolecular hydrogen bonds), $\Delta\tau = 45$ c.p.s. for p-dimethylaminothioacetanilide $(\sigma = -0.600)$, and $\Delta\tau = 14$ c.p.s. for p-cyanothioacetanilide $(\sigma = +0.660)$[38]. These results show that hydrogen bonds are more effective than electronic effects in bringing about coplanarity, and on the other hand, the strongly electron-donating p-dimethylamino groups is significantly better than the electron-withdrawing p-cyano group in this respect.

The enthalpy differences ΔH and free-enthalpy differences ΔG between the (Z) and (E) conformers of unsymmetrical thioamides are quite small. For instance $\Delta G = 0.29$ kcal/mole for N-benzyl-N-methylthioformamide[45]. $\Delta G = 1.2$ kcal/mole for N-methyl-thioformamide[33]. Loewenstein and coworkers[46] reported $\Delta H = 2.4$ kcal/mole for the former compound, which unusually high value has to be corrected. On the other hand, the energies of activation E_a and especially the free enthalpies of activation $\Delta G^{\ddagger}$ for the (Z)–(E) isomerization in several cases are found in the range of 23 kcal/mole or even higher, which may be taken as the lower limit for the possible preparative separation of conformers at room temperature. Characteristic energy values of rotation around the C—N bond as determined by various n.m.r. techniques (approached line-shape equations and equilibration measurements) are compiled in Table 3. It is obvious from the table that the values of $\Delta G^{\ddagger}$ are throughout higher for thio-amides than for amides* (see section I.C.9 and Chapter 1), which might be connected with marked electron delocalization and increased double-bond character of the C—N bond. This explanation of course is a rather qualitative one and cannot be confirmed by quanti-tative computations[48] (see section I.C.8). Nor can the barrier height be directly correlated with the resonance energy. Increased contribution of dipolar resonance formulas is also indicated by lowered ^{14}N chemical shifts in thioamides[52] as compared with their oxygen analogues. Thioformamides show even larger values of $\Delta G^{\ddagger}$ than other thioamides, a fact not yet fully understood. In part this may be due to steric as well as mesomeric and inductive effects. Apart from this, thioamides with electron-withdrawing substituents ($R^1 = CN$, COOR) exhibit higher barrier heights than those with electron-releasing substituents $R^1 = Me$ and—not shown in Table 3—OMe[48], SMe[48,53], NR_2[48,54], Cl[48,55]). Crowding in the ground state, for instance in tertiary pivalic acid thioamides (Table 3), lowers the energy of internal rotation, whereas steric interactions in the tran-sition state give rise to higher barrier heights, as in mesitylene deriva-tives (Table 3).

Elam and coworkers observed[56] splitting of the CH signal in the n.m.r. spectrum of α,α-di-t-butylthioacetanilide (7). It is however not quite clear whether this phenomenon is due to restricted rotation around the $C_{(1)}$—$C_{(2)}$ bond[56] or around the C—N bond. Restricted

---

* $\Delta G^{\ddagger}$ values recently are preferred over E_a values for barrier heights to internal rotation, on account of experimental and theoretical reasons[45,47].

TABLE 3. Energies and free enthalpies of activation for the (Z)–(E) isomerization of thioamides, $R^1CSNR^2R^3$, as determined by n.m.r. spectroscopy.

R^1	R^2	R^3	E_a (kcal/mole)	$\Delta G^{\ddagger a}$ (kcal/mole)[a]	Solvent[b]	Refs.
H	Me	Me	27.9	26.7	neat	46
			36.2	27.6	ODC	46
				24.0	ODC	48
				26.6	neat	48
H	i-Pr	i-Pr	31.8	28.9	neat	46
			24.2	24.2	ODC	46
H	Me	CH$_2$Ph	25.1	25.8	neat	35
H	Et	CH$_2$CH$_2$OH	25.2	25.3	neat	30
Me	Me	Me	43.7	29.7	HCONH$_2$	49
			21.0	21.6	ODC	48
Cyclo-C$_3$H$_5$	Me	Me	17.1	18.4	ODC	50
EtOCO	Me	Me	20.8	23.4	ODC	48
CN	Me	Me	23.2	23.4	ODC	48
Ph	Me	Me	13.3	18.4	ODC	48
			15.4		PhCl	51
2,4,6-Me$_3$C$_6$H$_2$	Me	CH$_2$Ph		27.0		29
t-Bu	Me	Me		13.0	CDCl$_3$	41a
t-Bu	—CH$_2$CH$_2$CHMeCH$_2$CH$_2$—[c]			12.0	CDCl$_3$	41a

[a] Average values between $\Delta G^{\ddagger}_1$ and $\Delta G^{\ddagger}_2$ for the isomerization reaction from either starting product.
[b] ODC = o-dichlorobenzene.
[c] N-Thiopivaloyl-4-methylpiperidine.

internal rotation in primary thioamides $RCSNH_2$ has been observed for benzoic and pivalic acid thioamide in dimethylsulphoxide but not

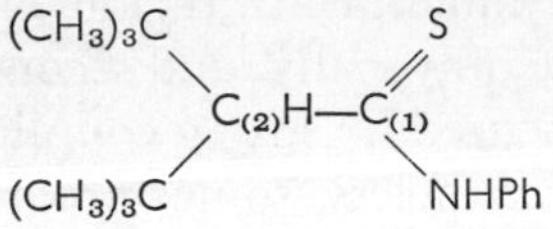

$$(CH_3)_3C \diagdown \qquad S$$
$$\qquad C_{(2)}H-C_{(1)} \diagup$$
$$(CH_3)_3C \diagup \qquad NHPh$$

(7)

in $CDCl_3$. This effect is explained by a specific astatic association of the solvent molecule with one proton, which is not shielded by the bulky group R of $8^{32,57}$. The magnetic non-equivalence of the two

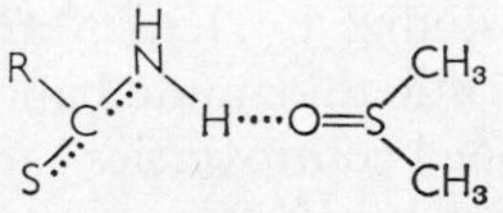

(8) 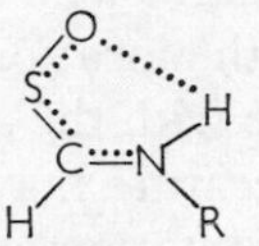 (9)

N—H-protons of ^{15}N-labelled thioacetamide is also evident from the occurrence of two different ^{15}NH coupling constants (91 and 94 c.p.s. respectively in $CDCl_3$) [57].

Finally some related compounds shall be mentioned. Thioformamide S-oxides (9) show the (E) conformation exclusively[58], which

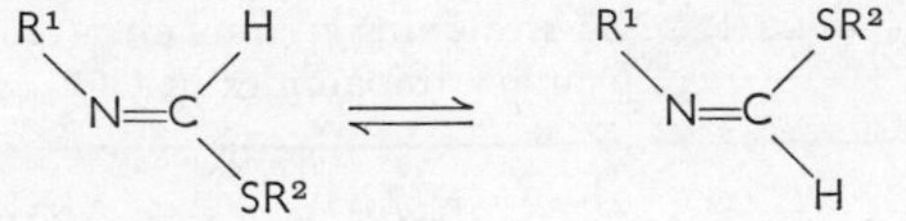

(E)   (10)   (Z)

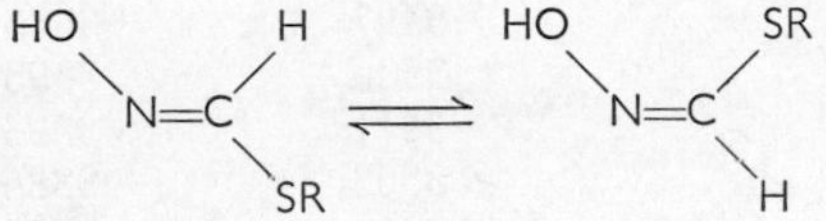

(E)   (11)   (Z)

is stabilized by intramolecular hydrogen bonds. This is supported by the x-ray diffraction results (cf. Table 1). Alkyl formimidothiolates (10) [45,358] and formhydroximidothiolic esters (11) [59] each consist of two isomers.

3. Infrared spectroscopic evidence

Since the pioneer work of Mecke and coworkers[60,61] the i.r. spectra of thioamides have given rise to much discussion. The remarkable differences in the interpretation and the assignment of the various frequencies arise from two facts. Firstly the typical i.r. bands of thioamides fall in the 'fingerprint' region of the spectra and often cannot be traced out unequivocally, and secondly the vibrations of the —CSN= group are obscured by strong coupling with other vibrations of the molecule. Nevertheless some progress has been recently achieved in this field, especially the theoretical calculations of the frequencies of thioformamides[62,63] and thioacetamide[64] by Suzuki, selective labelling of the thioacetamide molecule with ^{2}H, ^{13}C, and ^{15}N by Walter and Kubersky[65], and the extensive systematic work on numerous compounds, together with protonation, alkylation, complex formation, and seleno substitution studies by Jensen and Nielsen[66], now permit the location of characteristic thioamide bands. We shall not deal here with the above mentioned controversies. Summaries of the older literature may be found elsewhere[65,66].

The interpretation of bands in the NH region of thioamides is quite clear. In dilute solutions the symmetrical and the antisymmetrical NH$_2$ stretching vibration of primary thioamides are scarcely affected by electronic effects from other parts of the molecule and occur in a very small frequency interval (cf. Table 4). More concentrated solutions show additional bands, which are attributed to intermolecular

TABLE 4. NH stretching frequencies (cm^{-1})
of primary thioamides in CCl$_4$.

Compound	$\nu_{as}(\text{NH})$	$\nu_s(\text{NH})$	Refs.
HCSNH$_2$	3495	3374	67
MeCSNH$_2$	3497	3383	64, 65
EtCSNH$_2$	3506	3395	68
t-BuCSNH$_2$	3511	3393	68
CF$_3$CSNH$_2$	3503	3386	68
PhCH$_2$CSNH$_2$	3494	3375	68
PhCSNH$_2$	3508	3392	68
p-NO$_2$C$_6$H$_4$CSNH$_2$	3503	3386	68
p-FC$_6$H$_4$CSNH$_2$	3507	3391	68
p-MeC$_6$H$_4$CSNH$_2$	3509	3392	68
p-MeOC$_6$H$_4$CSNH$_2$	3510	3395	68
p-Me$_2$NC$_6$H$_4$CSNH$_2$	3513	3397	68
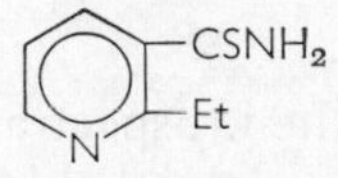	3504	3387	68

hydrogen bonds in chloroform and to specifical solvation via hydrogen bonds in acetonitrile. In the solid state bathochromic shifts of the two frequencies due to strong intermolecular association occur[65].

In secondary thioamides the sharp NH bands are split in two on account of (Z)–(E) isomerism, which was first observed by Russell and Thompson[69]. The results of later work on this subject are compiled in Table 5. Isomer ratios and some barrier heights of rotation as determined by i.r. spectroscopy, which generally agree well with n.m.r. data (cf. Tables 2 and 3), are shown in Table 6. The assignment of the various vibrations has been undertaken by means of dipole-moment measurements (cf. section I.C.5) and of the typical behaviour on dilution[70]. Both the (Z) and the (E) conformers show association frequencies at elevated concentration, but the cyclic dimers (12) built from (E) molecules are more resistant to dissociation than the chain polymers (13) formed by the (Z) isomer, and the correspond-

(12)   (13)

ing association bands depend differently on concentration. Recently the possible occurrence of a non-planar isomer of N-t-butylthio-isobutyramide in equilibrium with the predominating (Z) configuration has been concluded from asymmetries of the NH stretching band of this compound[71b].

The relation between isomer ratios and Hammett constants mentioned in section I.C.2 has been realized by i.r. measurements too $(\rho = 0.57)$[38]. In these experiments intermolecular interactions are practically absent due to the low concentrations $(10^{-3}$ M$)$ and non-polar solvent (CCl_4) used.

Frequency shifts of the NH, OH, and the thioamide 'B' bands corresponding to strong intramolecular hydrogen bonds are observed in salicylic acid thioanilides (14)[72]. These compounds may be resolved into two isomers, thus exhibiting another interesting example of restricted rotation about the $C_{(1)}$—$C_{(2)}$ bond (cf. section I.C.2). Conformations, fixed by intramolecular NH$\cdots$O=S bridges, may also be derived from the i.r. spectra of thioamide S-oxides[71] (Table 5).

The i.r. bands occurring between 700 and 1600 cm^{-1} have been classified by Jensen and Nielsen[66]. Their assignments (Table 7) are

TABLE 5. NH stretching frequencies (cm^{-1}) of secondary thioamides and S-oxides in CCl$_4$.

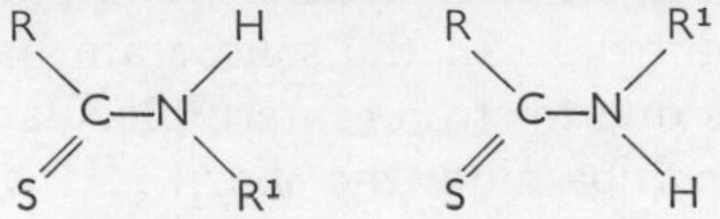

Structures: (Z) isomer — $\begin{smallmatrix}R\\C\\\|\\S\end{smallmatrix}$–$N\begin{smallmatrix}H\\\\R^1\end{smallmatrix}$; (E) isomer — $\begin{smallmatrix}R\\C\\\|\\S\end{smallmatrix}$–$N\begin{smallmatrix}R^1\\\\H\end{smallmatrix}$

Compound	(Z) isomer ν(NH)	ν(NH)$_{assoc}$	(E) isomer ν(NH)	ν(NH)$_{assoc}$	Refs.
HCSNHMe	3436		3396		63, 68
HCSNHCH$_2$Ph	3414		3383		68
HCSNHPh	3404		3374	3182	36, 70
MeCSNHMe[a]	3417				68, 71
MeCSNHCH$_2$Ph	3410				68
MeCSNHPh	3399	3250	3368	3200	70
MeCSNHC$_6$H$_4$NMe$_2$-p	3398		3367		68
MeCSNHC$_6$H$_4$OMe-p	3399		3368		36, 68
MeCSNHC$_6$H$_4$Me-p	3400		3359		36, 68
MeCSNHC$_6$H$_4$Cl-p	3401		3369		68
MeCSNHC$_6$H$_4$F-p	3401		3369		68
MeCSNHC$_6$H$_4$NO$_2$-p	3399		3364		68
Me$_3$CCSNHC$_6$H$_4$Me-p	3400				36
PhCH$_2$CSNHPh	3394		3352		68
PhCSNHMe[a]	3415				71
PhCSNHCMe$_3$	3395				68
PhCSNHPh	3394		3369		68
p-MeOC$_6$H$_4$CSNHMe	3428				68
p-MeC$_6$H$_4$CSNHPh[a]	3385	3245			71
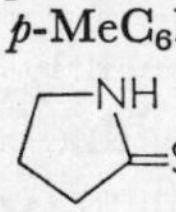 2-thioxopyrrolidine (5-membered ring, NH, =S)			3440	3225	71a
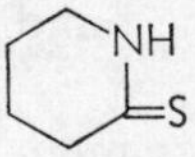 2-thioxopiperidine (6-membered ring, NH, =S)			3390	3177	71a
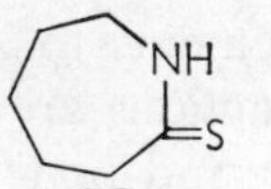 7-membered ring thione (NH, =S)			3410	3196	71a
MeC(SO)NH$_2$[a]			3487	3326	71
PhC(SO)NH$_2$			3480	3302	71
MeC(SO)NHPh[a]				3242	71
PhC(SO)NHPh[a]				3245	71

[a] In CHCl$_3$.

398

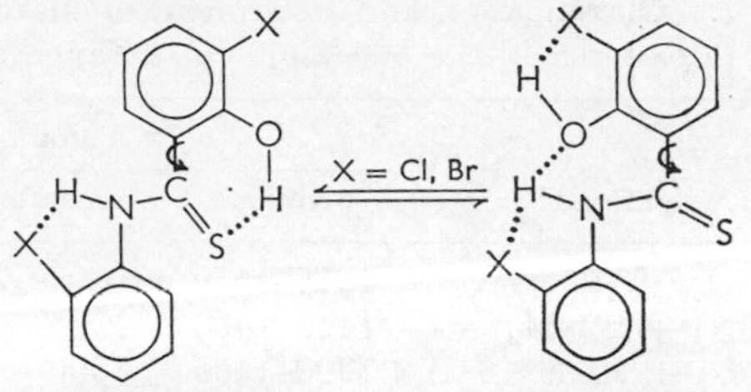

$$v(NH)\ 3325–3340 \qquad 3240–3300$$
$$v(OH)\ 2600–2800 \qquad 3320–3430$$
$$v('B')\ 1490–1536 \qquad 1518–1555$$

(14)

not always quite clear-cut but generally agree sufficiently with the suggestions of Suzuki[62–64], Walter and Kubersky[65], and Desseyn and Herman[73]. The data given in Table 7 may be taken as representative but there are significant deviations in special cases. It is notable that the position and shape of the class B band is strikingly different in the two isomers of secondary thioformamides. The B band of the (E) isomer is much more intense and hypsochromically shifted by ca. 35 cm^{-1} with respect to the less polar (Z) isomer, on account of increased double-bond character of the (E) isomer[31].

The most important general statement arising from these data presumably is the appearance of the CN stretching frequency in the region of C=N double bonds as well as that of the CS stretching vibration in the region of C—S single bonds. This fact further supports the concept of the resonance-stabilized planar S—C—N skeleton of

TABLE 6. Isomer ratios and isomerization free enthalpies of activation of secondary thioamides $R^1CSNR^2R^3$, in CCl_4.

R^1	R^2	R^3	$[(Z)]/[(E)]$	$\Delta G^{\ddagger}$ (kcal/mole)	Refs.
H	H	i-C_3H_7	≈ 10	22·4	32
H	H	i-C_4H_9	≈ 10	23·0	32
H	H	CH_2Ph	≈ 20	22·6	32
H	H	Ph	0·05		70
Me	H	Ph	1·5		69
Me	H	Ph	1·74		38
Me	H	$C_6H_4NMe_2$-p	0·80		38
Me	H	C_6H_4OMe-p	1·41		38
Me	H	C_6H_4Cl-p	3·26		38
Me	H	$C_6H_4NO_2$-p	8·5		38

TABLE 7. Classification and assignment of characteristic i.r. frequencies of thioamides in the 600–2000 cm^{-1} range (in KBr pellets)[66].

Class	Assignment[a]	Thioamide		
		primary	secondary	tertiary
A	ν(NH) + ν(CN)	1615–1650		
B	ν(CN) + b(NH$_2$)	1415–1480	1525–1565	1490–1530
C	ν(CC) + ν(CN) and others	1300–1400	1300–1400	1300–1400
D	ν(NCS) + r(NH$_2$)	1200–1300	950–1150	1000–1200
E	w(NH$_2$)	900–1000		
F	t + w(NH)	700–800	700–800	
G	ν(CS)	700–850	700–800	850–1000

[a] b = bending, r = rocking, w = wagging, t = twisting.

the thioamide molecule. Apart from this, i.r. spectra, and especially the easily traceable B, C, and D bands, represent a very valuable tool for the identification and characterization of this functional group.

4. Ultraviolet spectroscopic evidence

The remarkable chromophoric properties of the C=S group of thioketones extend to the thioamides. Compounds as simple as thiobenzamide or acetylthioacetamide show a bright-yellow colour whereas their oxygen analogues are colourless. This interesting fact led to the early investigation of the u.v. and visible spectra by Hantzsch and his school since 1930[74-76]. Hantzsch showed[75], by comparing the u.v. spectra of thioamides and their S- and N-substituted derivatives, that these compounds exist as thiones rather than imidothiols. Burawoy, on the other hand, was the first one to make systematic observations in this field[74]. He observed the characteristic longwavelength, low-intensity 'R' band (from 'Radikal') and the more intense 'K' band (from 'Konjugation') at shorter wavelengths.

Hosoya, Tanaka, and Nagakura[77] tracing back to Katagiri and coworkers[78] and independently Janssen[79,80], classified Burawoys 'R' and 'K' bands according to Kasha's terminology[81] as $n \rightarrow \pi^*$[77-80,82] and $\pi \rightarrow \pi^*$[77,78] bands respectively, on account of MO calculations, polarized u.v. absorption spectra[77], solvent shifts, intensities, and behaviour on protonation. Later on, Sandström especially, supported these assignments in a series of papers[25,50,81-86]. He studied the spectra of a large number of systematically substituted thioamides and their solvent dependence. Tables 8 and 9 show some data recently obtained by various authors.

TABLE 8. Ultraviolet spectra of thioamides.

Compound	Solvent[a]	$n \to \pi^*$		$\pi \to \pi^*$		b		Refs.
		λ_{max}(nm)	log ε	λ_{max}(nm)	log ε	λ_{max}(nm)	log ε	
MeCSNH₂	A	327	1.72	266	4.10	210	3.63	79
	B	361	1.38	266.9	4.08	231	3.77	
MeCSNHMe	A	321	1.69	261	4.13			79
	C	360	1.41	264	4.06			
MeCSNMe₂	A	330	1.75	269	4.18			
	C	365	1.61	272	4.17			79
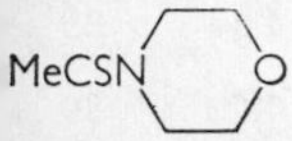 MeCSN⟨morpholine⟩	A			278	4.20			87
MeCSNHPh	C	392	1.74	330	3.49			88
 ▷—CSNMe₂	C	358	1.68	277	4.09			50
MeCSNHCOMe	A	429	1.50	282	4.30			83
	C	425	1.58	278.5	4.36	212	3.63	
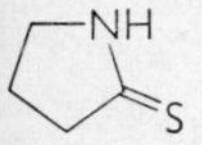 (pyrrolidine-2-thione)	A	319	1.82	266.5	4.16			85
	D	335	1.67	270	4.17			
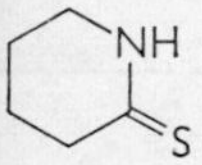 (piperidine-2-thione)	A			276	4.11			85
	D	340	1.73	281	4.08			
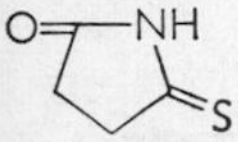 (5-oxo-pyrrolidine-2-thione)	A	392	1.32	269	4.32			85
	D	398.5	1.30	265	4.25			
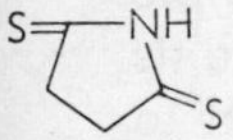 (5-thioxo-pyrrolidine-2-thione)	A	402	2.13	321	4.57	236.5	3.56	85
	D	406	2.20	315	4.55	234	3.69	
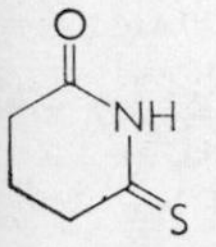 (6-oxo-piperidine-2-thione)	A	413.5	1.40	279	4.31			85
	D	417	1.43	276	4.25			
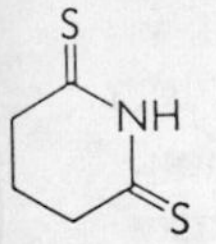 (6-thioxo-piperidine-2-thione)	A	422	2.23	336	4.51	237	3.84	85
	D	475.5	1.86	330.5	4.47	237	3.85	
PhCSNH₂	A	370	2.4	296	3.85	241	3.97	25
	C	418	2.33	298	3.81	239	3.94	
PhCSNHMe	A	390.5	2.41	286.5	3.86	238	4.03	25
	C	402	2.48	288	3.81	237	4.04	
PhCSNMe₂	A	366	2.47	281	3.97	239	3.99	25
	C	395	2.50	284	3.93	250	3.95	
 PhCSN⟨morpholine⟩	A	372	2.45	288	4.03	240	4.00	87

14 + c.o.a.

TABLE 8. (*Cont.*)

Compound	Solvent[a]	$n \to \pi^*$ λ_{max}(nm)	log ϵ	$\pi \to \pi^*$ λ_{max}(nm)	log ϵ	[b] λ_{max}(nm)	log ϵ	Refs.
$p\text{-}O_2NC_6H_4CSN$⟨⟩O	A	382	3·52	310	4·09	265	4·51	87
PhCSNHPh	A	402		317	3·91	239	4·11	89
	E	429	2·43					
PhCSNMePh	A			294	4·07	226	4·03	89
PhCSNHCOMe	A	480	2·18	273	3·91	226		89
	C	471	2·30					
PhCSNPhCOPh	A	467	2·21	279		237		89
	E	485	2·14					
PhCOCSN⟨⟩O	A	379	2·98	342	3·13	266	4·18	87
				330	3·25	256	4·19	
PhCOCH$_2$CSN⟨⟩O	A	325	3·76	282	4·17	245	4·11	87

[a] Solvents: A = ethanol, B = ether, C = saturated hydrocarbon, D = heptane–CH$_2$Cl$_2$ mixtures, E = benzene.
[b] Hosoya and coworkers[77] have attributed this band to a second $\pi \to \pi^*$ transition.

One can see from Table 8 that the long-wavelength bands of simple thioamides show, without exception, negative solvatochromic shifts and low intensities and thus unambiguously have to be assigned to $n \to \pi^*$ transitions. Remarkably the high-intensity absorption at ca. 270 nm shows slightly negative solvatochromism too, which has led Janssen[80] and Sandström[90] to the presumption that it might be due to an $n \to \sigma^*$ transition. Later on, however, Sandström demonstrated[25] by MO calculations that the 270 nm band of thioacetamide should be bathochromically displaced in thiobenzamide by conjugation if it were a $\pi \to \pi^*$ band as indeed is experimentally found (Table 8), whereas an $n \to \sigma^*$ band should remain unaffected by this substitution. Alkyl substituents cause reasonable hypsochromic shifts in many cases (Tables 8 and 9 and Figure 3). This holds especially for *N,N*-disubstituted molecules. Sandström showed that this can be ascribed to a steric inhibition of conjugation[25,84]. From his MO calculations angles Θ of rotation between the CSNR$_2$ group and the plane of the adjacent group (cf. Figure 1), shown in Table 9, may be derived[84]. His results agree qualitatively with the x-ray data for dithiooxamide[18]

TABLE 9. Substituent effects in the u.v. spectra of thiooxamide derivatives and twisting angles, Θ, for the C—C bond[84].

Compound	Solvent[a]	$\lambda_{n\to\pi^*}$(nm)			$\lambda_{\pi\to\pi^*}$(nm)			Θ		
		R = H	R = H, Me	R = Me	R = H	R = H, Me	R = Me	R = H	R = H, Me	R = Me
$N\equiv C$—$CSNR_2$	A	438		433	294		309			
	B	422		417	301		312			
$EtOOC$—$CSNR_2$	A	454	439	378	301	303	282	0°	0°	65°
	B	390	382	355	290	296	280			
H_2NCO—$CSNR_2$	A		399	370		300	290	0°	0°	50°
	B	411	394	360	301	298	278			
$MeHNCO$—$CSNR_2$	A	407	393	391	303	298	290	0°	10°	50°
	B	375	369	362	278	276	279			
Me_2NCO—$CSNR_2$	A	358	352	359	275	271	281	58°	55°	66°
	B			355			280			
H_2NCS—$CSNR_2$	A	483	486	370	312	310	301	0°	0°	50°
	B	486	464	350	310	307	270			
$MeHNCS$—$CSNR_2$	A	464	462		307	304		0°	0°	
	B		440			303				
Me_2NCS—$CSNR_2$	A	370	365		301	275	275	50°		89°
	B	350	345		270	272	272			

[a] A = heptane or other non-polar solvent, B = ethanol.

(section I.C.1). Sterically fixed cyclic thioamides therefore show only negligibly small hypsochromic or even bathochromic shifts on methylation[85,86]. On the other hand, N-aryl and especially N-acyl groups as well as corresponding α-substituents give rise to bathochromic shifts of both the $n \rightarrow \pi^*$ and the $\pi \rightarrow \pi^*$ bands by extending the conjugative system of the thioamide molecule (cf. Tables 8 and 9).

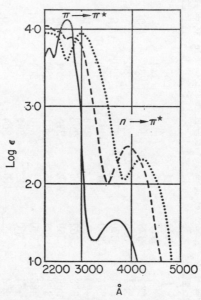

FIGURE 3. Ultraviolet spectra of thioacetamide (——), thiobenzamide (···), and N,N'-dimethylthiobenzamide (– – –) in heptane[25].

This holds for the thioamide S-oxides too[90a], whereas imidothiolic esters like S-protonated thioamides in strongly acidic solution[79,82] show no $n \rightarrow \pi^*$ bands, and their $\pi \rightarrow \pi^*$ bands are hypsochromically shifted with respect to the parent thioamides[89] (cf. Table 10).

Recently characteristic differences between the u.v. maxima of (Z) and (E) isomeric thioamides have been found[31]. The $\pi \rightarrow \pi^*$ bands of secondary thioformamides are bathochromically shifted in the more polar (E) isomers ($\lambda_{max} = 276$ nm in $CHCl_3$) relative to the (Z) isomers ($\lambda_{max} = 266$ nm).

The frequencies of the $n \rightarrow \pi^*$ bands of thioamides (as well as those of other thiones) may be obtained from empirically determined increments in a very simple manner as has been stated by Fabian, Viola, and

TABLE 10. $\pi \rightarrow \pi^*$ bands of thioamide S-oxides[90a] and imidothiolic esters[89].

Compound	λ_{max} (nm)
MeCSNH$_2$	267
MeCSONH$_2$	298
MeC(SMe)=NH	240
MeCSNHPh	299
MeCSONHPh	321
PhCSNHPh	317
PhCSONHPh	355
PhC(SMe)=NPh	296

Mayer[88]. The absorption maxima calculated by means of the formula

$$\lambda_{n \rightarrow \pi^*} = 10^4/(\tilde{\nu}_0 + a_X + a_Y)$$

where the wavelength is in nm units, agree well with experimental or quantum mechanical values. Thioacetone ($\lambda_0 = 10^4/\tilde{\nu}_0 = 499$ nm; $a_{Me} = 0$) is taken as reference compound; a_X and a_Y depend on the substituents X and Y in a molecule X—CS—Y. Some examples are given in Table 11. It would be worthwhile determining further data, especially of other N-alkyl and N-aryl groups.

The valuable chromophoric properties of the CSNR$_2$ group give rise to its use for configuration studies by means of circular dichroism (CD) and optical rotatory dispersion (ORD) measurements. Sjöberg and coworkers[91] have proposed the N-phenylthioacetyl- and N-thiobenzoyl-α-amino acids as suitable derivatives for establishing their

TABLE 11. Substituent-specific absorption increments, a, for the $n \rightarrow \pi^*$ transition of thiones according to Fabian and coworkers[88].

Substituent	a (cm^{-1})	Substituent	a (cm^{-1})
NH$_2$	6·8	CF$_3$	$-1·6$
NHMe	7·5	CF$_2$Cl	$-1·1$
NMe$_2$	7·3	CN	$-3·7$
NHPh	5·5	COOEt	$-5·3$
NHAc	3·1	CONH$_2$	$-2·5$
Me	0	CSNH$_2$	$-7·0$
$\tilde{\nu}_0$	20·05	Ph	$-2·4$

absolute configuration, since these compounds exhibit strong Cotton effects related to their $n \rightarrow \pi^*$ bands. Barrett[92,93] and Bach and co-workers[94] however stated that considerable caution has to be observed in assigning configuration to chiral molecules solely on basis of the sign of the Cotton effect, because it may be inverted on going from one solvent to another and even from one compound to another in a homologous series. This difficulty may be overcome by using the likewise optically active $\pi \rightarrow \pi^*$ transitions of the thiobenzoylamino acids instead of the $n \rightarrow \pi^*$ bands[95]. Empirical correlations between the CD and structure of peptides unfortunately are obscured by solvent-modified intramolecular interactions between the chromophore and the amino acid residues and thus no general spectroscopic method for N-terminal analysis of polypeptides can be based on it[96], except the location of N-terminal imino acid residues, e.g. proline[359]. On the other hand, the absolute configuration of an asymmetric centre directly linked to the functional group of a carboxylic acid can be determined from the CD, associated with the $n \rightarrow \pi^*$ transition of the corresponding N-methylthioamide[97]: (R) configurations produce a negative Cotton effect and (S) configurations a positive one. Anomalies arise if heteroatoms are present at the asymmetric centre.

5. Dipole moments

In Table 12 the dipole moments of thioamides determined as yet are compiled. All known values are considerably higher than those of the corresponding amides reflecting the marked electron delocalization in the thioamide molecule which has been repeatedly mentioned. This may be expressed by favouring the dipolar formula **3c** rather than **3b** for the description of a thioamide. Moreover, the dipole moments indicate that the C=S group is inherently more polarizable than the C=O group on account of the larger kernel of electrons in the S atom which inhibits the formation of double bonds.

It is seen from the data of the table that configuration plays an important role for the dipole moments of secondary thioamides. N-Alkyl derivatives occurring predominantly in the (Z) form generally exhibit lower values ($\mu \approx 4 \cdot 75$ D) than species present in the (E) form ($\mu > 5$ D). This fact has been adduced as striking evidence in support of the opinion that the preference of the (Z) configuration in thioamides might be due to electrostatic forces (section I.C.2). Thioformanilide though an almost pure (E) compound has a low dipole moment of $4 \cdot 36$ D. Whether this may be explained by assuming that the N-aryl and the N-alkyl dipoles have inverse direction or that the N-aryl

TABLE 12. Dipole moments of thioamides and corresponding amides.

Compounds	Solvent	$[(Z)]/[(E)]^a$	X = S μ (D)	X = O μ (D)	Refs.
HCXNHMe	CCl$_4$	6·9	4.53		97a
HCXNHBu-t	C$_6$H$_6$	0·04	5·17		97a
HCXNHPh	CCl$_4$	0·05	3·02	3·35	70, 100
	CCl$_4$	0·05	4·13		97a
HCXNMe$_2$	C$_6$H$_6$		4·74	3·86	97a
MeCXNH$_2$	dioxan		4·77	3·70	98^b, 99^c
MeCXNHMe	C$_6$H$_6$	36	4·76		97a
	CCl$_4$		4·64	3·55	100a
MeCXNHEt	C$_6$H$_6$		4·79		97a
	dioxan			3·90	100a
MeCXNHPr	C$_6$H$_6$		4·80		97a
MeCXNHBu-n	C$_6$H$_6$		4·77		97a
MeCXNHPh	CCl$_4$	1·5	4·54		100
	C$_6$H$_6$			3·65	100a
MeCXNMe$_2$	C$_6$H$_6$		4·74	3·74	97a^b, 103
MeCXN(morpholine)	C$_6$H$_6$		3·85		101
MeCXN(piperidine)	C$_6$H$_6$		5·03		101
	dioxan			4·07	104
EtCXNH$_2$	C$_6$H$_6$		3·86	3·47	100a^c, 101^b
EtCXNHMe	C$_6$H$_6$		4·44		101
	CCl$_4$		3·55		102
	dioxan		4·91		101
EtCXNMe$_2$	C$_6$H$_6$		4·65		101
EtCXN(morpholine)	C$_6$H$_6$		3·86		101
EtCXN(pyrrolidine)	C$_6$H$_6$		4·93		101
t-BuCXNMe$_2$	CCl$_4$		4·57		97a
$(CH_2)_n$ C=S, NH n = 3	dioxan	0	5·07	3·79	106
4	dioxan	0	5·15	3·83	106
5	dioxan	0	4·83	3·88	106
	C$_6$H$_6$		5·10	3·88	97a^b, 100a^c
11	C$_6$H$_6$	1^d	4·91	3·64	97a^b, 100a^c
PhCXNMe$_2$^{e}	C$_6$H$_6$		4·58	3·80	105

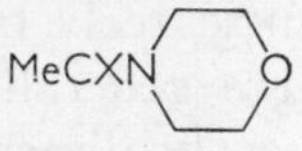

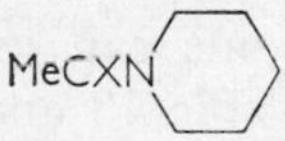

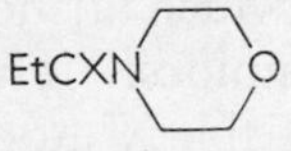

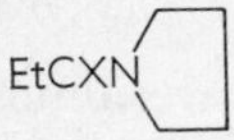

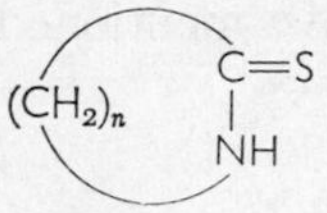

TABLE 12. (Cont.)

Compounds	Solvent	X = S [(Z)]/[(E)]^a	X = S μ (D)	X = O μ (D)	Refs.^{b,c}
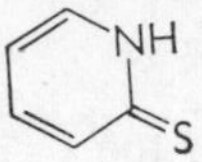 (structure: NH, S)	C$_6$H$_6$	0	2·78	1·95	107
	dioxan	0	5·29	2·94	107
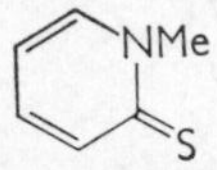 (structure: NMe, S)	C$_6$H$_6$	0	5·26	4·04	107
	dioxan	0	5·49	4·07	107

^{a} cf. Tables 2 and 6.
^{b} X = S only.
^{c} X = O only.
^{d} Determined by i.r. spectroscopy.
^{e} For X = Se: μ = 4·79 D [105].

dipole is only smaller but in the same direction, cannot be decided from the present data.

Exceptionally low values ($\mu < 4$ D; cf. the table) can scarcely be taken as real characteristics of the molecules. They are generally due to associated species, for instance hydrogen-bridged cyclic dimers (12) in the case of 1,2-dihydropyridine-2-thione[107], as one can see from the influence of solvents and concentrations (cf. Table 13 and reference 97a).

Lumbroso and coworkers[101,105] were able to determine mesomeric moments M of thioamide molecules from the formula $M = X - X'$, where X is the observed dipole moment and X' the σ moment. They found $M(R_2NCS—) = 2·45$ D, $M(RNHCS—) = 1·77$ D, $M(R_2NCO—) = 1·09$ D, and $M(RNHCO—) = 0·73$ D, which further supports the enlarged electron delocalization in thioamides. The calculations have been done under the assumption that the N atom is a planar, sp^2 hybrid.

The dipole moments of PhCSNMe$_2$ and MeCSNMe$_2$ are almost equal, the mesomeric moment of the phenyl ring being obviously negligible, due to its rotation out of the thioamide plane on account of steric hindrance[105], which is also evident from the above mentioned x-ray[18] and u.v. spectroscopic data.

6. Mass spectra

There is little knowledge about the mass spectra of thioamides, whereas thiourethanes[108,109] and thioureas[109,110] have been thoroughly studied. Walter and coworkers[111] examined the fragmentation processes occurring in *N*-arylthioamide molecules on electron impact. They established the following reactions:

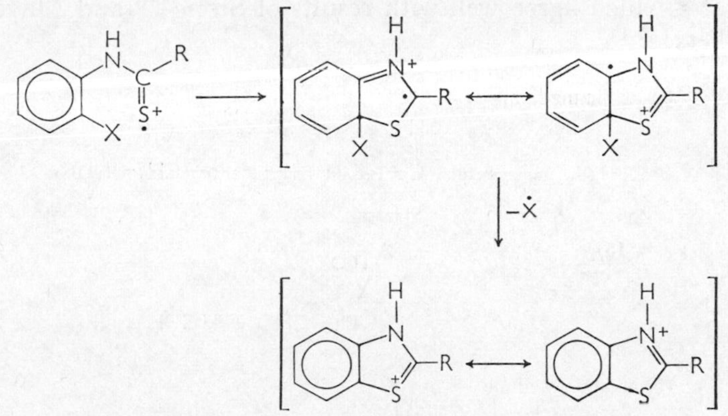

R = H, D, CH₃
X = H, D, CH₃, Cl

In *N*-phenyldithiophthalimide a similar cyclization has been observed (see below) by Anderson and coworkers[111a]. This cycliza-

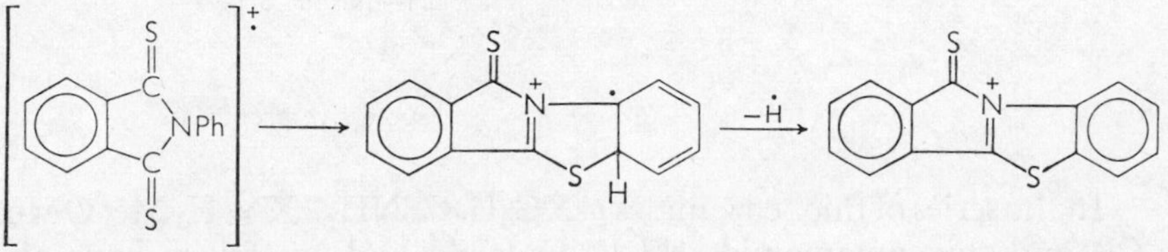

tion is not accompanied by statistical distribution of the H atoms. The positive charge therefore must be located almost exclusively on the S atom, rather than on the phenyl ring. No formation of heterocycles is indicated in the mass spectra of formanilide[111,111b] which is evidence for the peculiarity of the sulphur atom. Another characteristic reaction is the elimination of SH from the CHS group of thioformanilide[111], which must take place by rearrangement of the molecule to the thiolimide form, not known in the ground state of thioamides (cf. section IV).

7. Polarographic and electron paramagnetic resonance spectroscopic evidence

Thioamides generally yield two cathodic polarographic waves in aqueous systems each corresponding to transfer of two electrons. The

14*

half-wave potentials, $E_{\frac{1}{2}}$, as well as the products of the electrode processes depend on the pH value of the solution.   Lund has proposed the following mechanisms for the polarographic reduction of thio-amides[112] which agree well with results of Stone[113] and Mayer and coworkers[114,115].

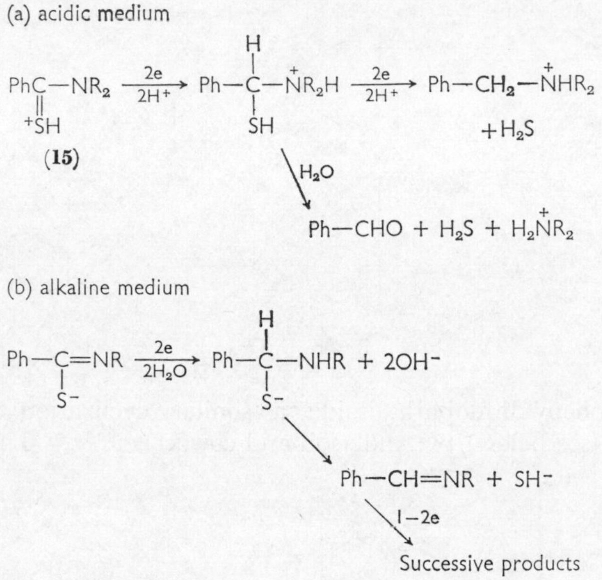

(a) acidic medium

$$\underset{\underset{\displaystyle (15)}{+SH}}{\overset{\displaystyle PhC-NR_2}{\|}} \xrightarrow[2H^+]{2e} \underset{SH}{\overset{\overset{\displaystyle H}{|}}{Ph-\underset{|}{C}-\overset{+}{N}R_2H}} \xrightarrow[2H^+]{2e} Ph-CH_2-\overset{+}{N}HR_2$$

$+H_2S$

$\downarrow H_2O$

$Ph-CHO + H_2S + H_2\overset{+}{N}R_2$

(b) alkaline medium

$$\underset{S^-}{\overset{\displaystyle Ph-C=NR}{|}} \xrightarrow[2H_2O]{2e} \underset{S^-}{\overset{\overset{\displaystyle H}{|}}{Ph-\underset{|}{C}-NHR}} + 2OH^-$$

$Ph-CH=NR + SH^-$

$\downarrow |-2e$

Successive products

   In the series of thiobenzamides $p\text{-}XC_6H_4CSNH_2$ (X = H, Me, OMe, Cl) and thiocinnamamides[360] Pappalardo and coworkers have observed linear relation between $E_{\frac{1}{2}}$ and Hammett constants.   They deduced from the appropriate ρ values of the two waves, $\rho_1 = 0.25$ and $\rho_2 = 0.36$, that the first wave in acidic solution should be due to reduction of the protonated molecule (15) corresponding to Lund's mechanism.   Thioamide S-oxides (16) exhibit a third wave at lower $E_{\frac{1}{2}}$ which is attributed to reduction of the S-oxide group[115,115a].

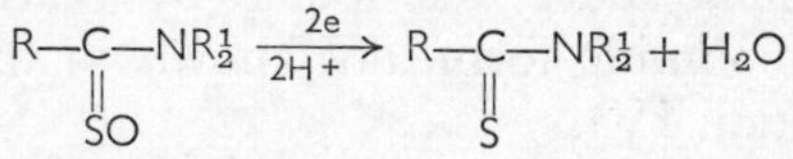

$$\underset{\underset{\displaystyle (16)}{SO}}{\overset{\displaystyle R-C-NR_2^1}{\|}} \xrightarrow[2H^+]{2e} \underset{S}{\overset{\displaystyle R-C-NR_2^1}{\|}} + H_2O$$

   An anodic wave is only obtained in the polarography of primary and secondary thioamides[112].   This is due to formation of a mercuric salt (17) and by-products.

Polarography in aprotic solvents has been studied by Voss and Walter[117]. Since their data are not influenced by prototropic processes, they give more information about the primary electron transfer.

$$PhCSNH_2 \xrightarrow[\substack{-e \\ -H^+}]{Hg} [PhCSNHHg^+] \longrightarrow PhCN + HgS + H^+$$

(17)

The first wave in this case is due to formation of thioamide radical anions (18) (transfer of one electron). It is seen from Table 13 that

TABLE 13. Half-wave potentials, $E_{\frac{1}{2}}$, of the polarographic reduction of thioamides in acetonitrile[117] [a,b].

X	MeCSX	p-MeOC$_6$H$_4$CSX		PhCSX		
NMe$_2$	1·62	1·53	1·9	1·44	1·68	
NHMe	1·64			1·40	1·7	
NH$_2$	1·63	1·35	1·75	1·23	1·66	
N(Me)C$_6$H$_4$OMe-p		1·25	1·72	1·21	1·58	
NHC$_6$H$_4$OMe-p	1·60			1·14		
NMePh		1·20	1·70	1·11	1·53	(18)
NHPh	1·62			1·10	1·7	(20)
N(C$_6$H$_4$OMe-p)$_2$		1·09	1·67	1·01	1·52	(17)
NPh$_2$	1·52	1·00	1·70	0·92	1·60	

[a] $E_{\frac{1}{2}}$ in volts.
[b] Related to internal Ag electrode rather than calomel electrode.

$E_{\frac{1}{2}}$ decreases with the electronegativity of the substituents on either the C or N atom of the thiocarboxamido group as one would expect. Thioamides of aromatic acids show one or two additional waves at more negative potentials (due to transfer of 1–2 electrons), which so far are unexplained since no product analyses have been undertaken.

The occurrence of 18 has been proved by e.p.r. spectroscopy[118]. From the high g values of the particular e.p.r. signals (PhCSNMe$_2$:

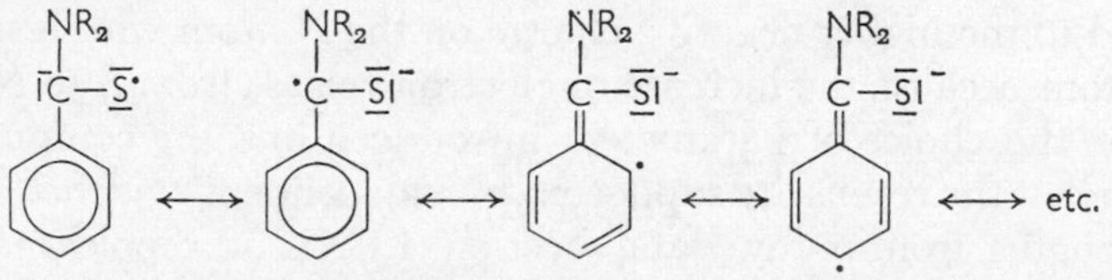

(18)

$g = 2·0059$; PhCSNPh$_2$ (19): $g = 2·0067$[117] one can deduce considerable spin density on the S atom. On the other hand, marked delocalization of the unpaired electron to the adjacent aryl group but scarcely to N-substituents is evident from the observed hyperfine structure. The equivalence of the two *ortho* protons in the anion **19**, as deduced from their equal coupling constants, indicates twisting of the phenyl ring to an extent of approximately 90° in the radical anion.

(19)

8. Quantum mechanical calculations

First quantum mechanical calculations in the field of thioamides dealt with u.v. bands[78]. The assignation and theoretical evaluation of absorption frequencies has remained the main purpose of application of the MO theory[25,77,83,86,88,90] (section I.C.4). Usually semi-empirical LCAO–MO calculations are carried out according to Naga-kura's iteration procedure[120] in order to account for the particular charges on the heteroatoms of the thioamide molecule, although Mehlhorn and Fabian[119] recently have pointed out that the simple HMO approximation often yields as good an agreement with experimental data as does the SCF method. Participation of $3d$ orbitals of the S atom has never been taken into consideration in this connexion.

Janssen[121], and Sandström and coworkers[25,50,83,84] have evaluated resonance energies and charge distributions of the thioamide systems as well as the $n \rightarrow \pi^*$ and $\pi \rightarrow \pi^*$ transition energies which generally agree well with observed data. Since thermodynamic values are not available, the resonance energies are not very substantial but yet they evidence stabilization of the thioamide group by electron delocalization and the possibility of conjugation with other π-electron systems (Table 14). Neither can the molecular diagrams of Table 14 be discussed on their own merit. For instance no significance may be attached to the higher negative charge on the N atom with respect to the S atom because the increase of electron release from S to N is the basis for the choice of parameters introduced into the computation. Nevertheless the results are supported by independent evidence such as bond lengths from x-ray data (section I.C.1) or dipole moments (section I.C.5), and may serve for comparison of properties of similar compounds.

Sandström could not find a correlation between the C—N π-bond

TABLE 14. π-Electron distributions, bond orders, and stabilization energies of thioamides.

Charge densities[a] and bond orders (underlined)	Resonance energies (β units)	Refs.

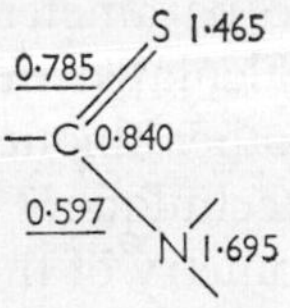

S 1·465
0·785
—C 0·840
0·597
N 1·695
$-1\cdot06^b$ 121

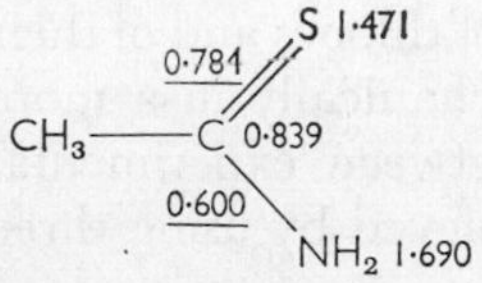

CH₃—C 0·839
0·784 S 1·471
0·600
NH₂ 1·690
$-1\cdot04$ 25, 121a

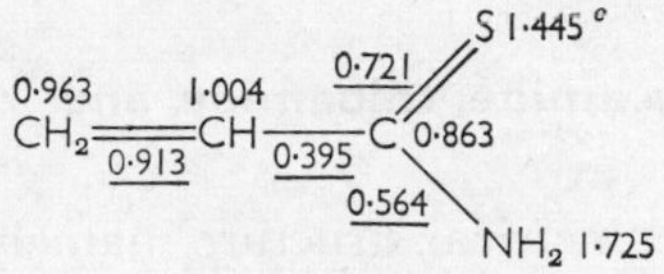

0·963   1·004   0·721   S 1·445 ^{c}
CH₂=CH—C 0·863
0·913   0·395
0·564
NH₂ 1·725
$-2\cdot436^c$ 25

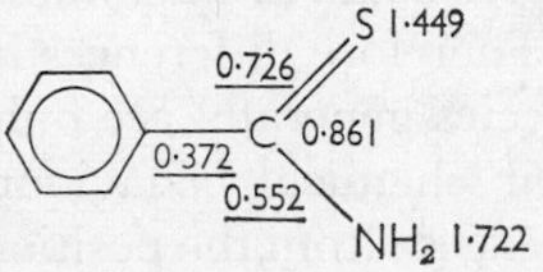

0·726   S 1·449
⟨ring⟩ —C 0·861
0·372
0·552
NH₂ 1·722
$-4\cdot108^d$ 25

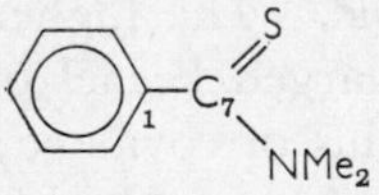

⟨ring⟩—C₇
S
NMe₂
1
$-3\cdot924^{d,\,e}$ 25

1·863 R₂N   0·387   0·766   S 1·436
0·874 C 0·234 C 0·854
0·877   0·576
1·272 O   NR₂ 1·701
$(-0\cdot206)^f$ 84

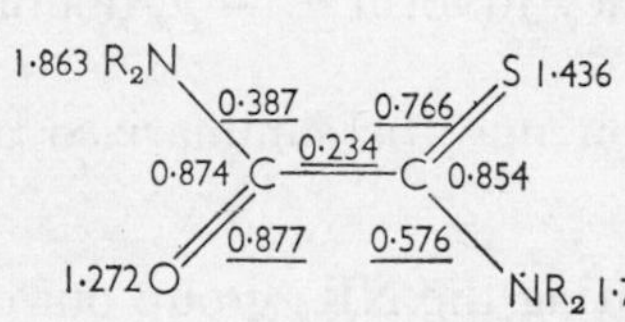

---

^{a} Charge densities calculated from the net charges given by Janssen[121].
^{b} The amide group exhibits $-0\cdot562\ \beta$[84].
^{c} Acrylic acid thioamide has not yet been prepared.
^{d} -2β included (resonance energy of the benzene ring).
^{e} $\beta_{1,7} = 0\cdot5\ \beta_{CC}$ on account of steric hindrance.
^{f} 'Excess stabilization energy', i.e. total π-electron energy of the molecule minus the π-electron energies of the two parts joined by the C—C bond.

orders of N,N-dimethylthiobenzamide and other dimethylthiocarba-moyl compounds with the free energies of activation $\Delta F^{\ddagger}$ for the internal rotation of the dimethylamino group[48], but the relation be-tween the loss of π-electron energy which occurs when the dimethyl-amino group is rotated out of conjugation, and $\Delta F^{\ddagger}$ is more defined[48].

Finally, attempts to estimate dipole moments from MO parameters may be mentioned. Janssen and Sandström[122] have calculated the dipole moments of N,N-dimethylthioacetamide among other thiones from π-electron densities obtained by means of the ω technique[123]. They employed constant σ moments and a simplified geometry of the molecule but could not generate a set of parameters of general validity. The differences $\Delta\mu$ between the dipole moments of thiones and of their oxygen analogues may be treated quantum mechanically in a more satisfactory way[122]. Fairly good agreement between experimental and theoretical dipole moments may also be achieved by using three empirical parameters in the calculations[123a].

9. Conclusion: comparison between the amide, thioamide, and selenoamide groups

The most important element of the geometrical structure, namely the planar arrangement of the functional group, occurs in thioamides as well as in amides (and in selenoamides). Therefore differences in the physical properties of these 'isologous' species generally are only gradual, although this does not hold for their chemical behaviour. The observed quantitative distinctions of course arise from the position of the sulphur in the third row of the periodic table. The higher atomic number of the S atom accounts for the enlarged kernel of electrons which for its part causes an increased radius of covalency (O: 0.74 Å; S: 1.04 Å[23]), decreased tendency to form $p_{\pi} - p_{\pi}$ double bonds, large polarizability, etc.

The main consequences of this situation may be summarized as follows:

(a) The energy of activation E_a for rotating the NR_2 group out of the plane of the R—CX—NR_2 molecule, i.e. the double-bond charac-ter of the central C—N bond, increases in going from $X = O$ to $X = S$ and $X = Se$ (although there are no significant changes in bond lengths). This is likely to be due to non-bonded repulsion forces[123b]. Example[51]:

$PhCXNMe_2$	$X = O$	$X = S$	$X = Se$
E_a (kcal/mole)	7.5	15.4	21.5

(b) The preference for the (Z) configuration is less marked in secondary and tertiary thioamides than in the corresponding amides on account of the enlarged steric requirement of the S atom. Example[34]:

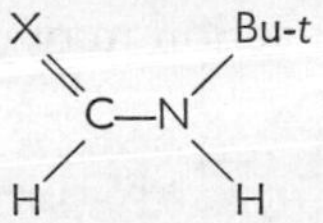

$$[(Z)]/[(E)]$$

X = O   2·34
X = S   0·04

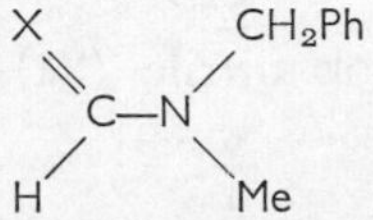

X = O   0·85
X = S   0·64

(c) The thioamide molecule is more polar and more polarizable. This is reflected in the particular spectroscopic properties but first of all in the enlarged dipole moments (cf. Table 12).

II. PREPARATION OF THIOAMIDES

Six reviews concerned with the formation of thioamides have appeared since 1949[9,11–13,124,357], and the reader is referred to these for more detailed accounts of the literature before 1965.

A. Thiolysis of Carboxylic Acid Derivatives

I. Thiolysis of nitriles

The long-known addition of hydrogen sulphide to a nitrile group[1,6,7] (reaction 1) is still widely used in many variations (cf.

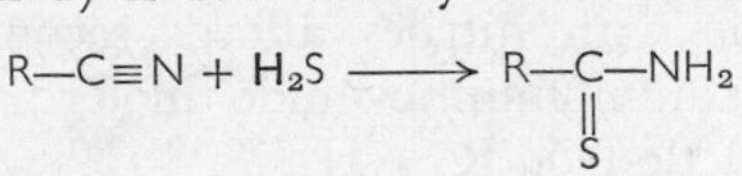

$$R—C≡N + H_2S \longrightarrow R—C(=S)—NH_2 \tag{1}$$

section I.A). The reaction is catalysed by bases and acids.

a. *Base-catalysed addition.* The reaction may proceed according to equation (2). M^+OH^- including ion exchangers[125], M^+OR^-,

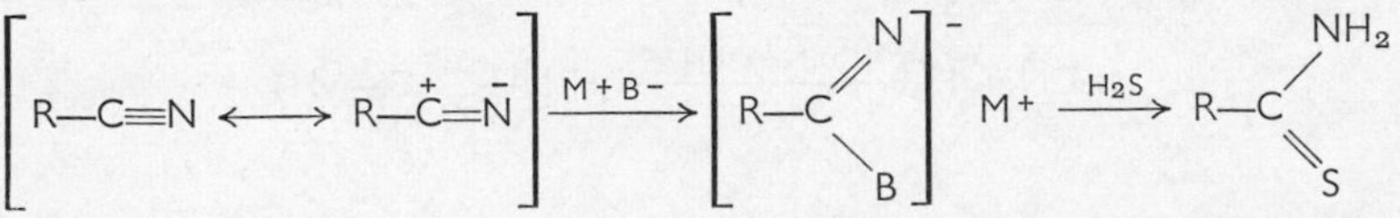

$$\left[R—C≡N \longleftrightarrow R—\overset{+}{C}=\overset{-}{N} \right] \xrightarrow{M^+B^-} \left[R—C\overset{N^-}{\underset{B}{\diagdown}} \right] M^+ \xrightarrow{H_2S} R—C\overset{NH_2}{\underset{S}{\diagdown}} \tag{2}$$

ammonia, and amines preferably pyridine are useful bases, and many aromatic thioamides are obtained in yields close to 90% from the corresponding nitrile in the presence of triethylamine and pyridine, an excess of which may serve as solvent[126].

If the nitrile has a hydrogen atom in the α-position this may be abstracted by the base (reaction 3). This reaction may explain the

$$RCH_2CN \xrightarrow{-H^+} \left[\begin{matrix} R-\overset{\shortmid}{\underset{\underset{H}{\shortmid}}{C}}-C\equiv N \end{matrix} \longleftrightarrow \begin{matrix} RC=C=N^- \\ \overset{\shortmid}{\underset{H}{}} \end{matrix} \right] \xrightarrow[+H^+]{H_2S} RCH_2-CSNH_2 \quad (3)$$

observation that optically active hydratropanitrile (**21**) with am-

$$\begin{matrix} PhCHCN \\ \overset{\shortmid}{\underset{Me}{}} \end{matrix}$$

(**21**)

monium bisulphide at 60° afforded a completely racemic thioamide[97]. Whether a course according to reaction (3) is the reason for the normally sluggish and unproductive reaction of aliphatic nitriles with hydrogen sulphide has not yet been studied in detail. Gilbert and Rumanowski[127] got aliphatic thioamides in yields of 19–50% using strong bases such as diethylamine, quarternary ammonium hydroxides or tetraalkylguanidines as catalysts in aprotic solvents such as dimethylformamide, dimethylsulphoxide or sulpholane.

Okumura and Moritani showed that a secondary amine may enter into the reaction product[128] (reaction 4). This is formally in line

$$RCN + Me_2NH_2^+Me_2NCS_2^- \xrightarrow{Benzene} RCSNMe_2 \quad (4)$$

with the experience of Kindler, that reaction of hydrogen sulphide and primary amine with nitriles affords secondary thioamides[129] (reaction 5). With ammonium cyanide thioformamides are obtained in good yields (reaction 5, R = H)[130,131,361] with sodium cyanide

$$RCN + R^1NH_2 + H_2S \longrightarrow RCSNHR^1 \quad (5)$$

dithiooxamides, when cupric tetrammine is used as a catalyst (reaction 5a)[362]. From DL-aminonitriles, e.g. $R^2CH(NH_2)CN$, the corres-

$$NaCN + H_2S \xrightarrow{[Cu(NH_3)_4]^{2+}} H_2N-\underset{\underset{S}{\parallel}}{C}-\underset{\underset{S}{\parallel}}{C}-NH_2 \quad (5a)$$

ponding thioamides are obtained according to reaction (5)[132]. An interesting deviation from this behaviour was recently observed for pentachlorobenzonitrile[132a], where the CN group remained un-affected, but substitution of Cl by SH occurred in the 4-position.

b. Acid-catalysed addition. If thiolysis is not accomplished by hydrogen sulphide but by another source of sulphur (e.g. reaction 4) acid catalysis may be successful. This is the case when sulphur is provided by thioacetamide[133] (reaction 6). The diimidoyl sulphide

$$RCN + MeCSNH_2 \underset{HCl}{\rightleftharpoons} RCSNH_2 + MeCN \tag{6}$$

22 is regarded as an intermediate, the equilibrium being shifted by

$$\underset{NH}{\overset{RC-S-CMe}{\underset{\|}{\overset{\|}{}}}} \quad \underset{NH}{\overset{\|}{}}$$

(22)

distillation of acetonitrile, the lowest boiling component[133].

The addition of O,O'-diethyl dithiophosphate (**23**) (readily obtained from ethanol and phosphorus pentasulphide) to nitriles is preparatively useful (reaction 7) and affords a route to thioamides which are not

$$RCN + HS\overset{S}{\underset{\|}{P}}(OEt)_2 \longrightarrow RC\overset{NH}{\underset{\|}{-}}S\overset{S}{\underset{\|}{-}}P(OEt)_2 \xrightarrow{HCl} RCSNH_2 + Cl\overset{S}{\underset{\|}{-}}P(OEt)_2$$

$$\text{(23)} \qquad\qquad \text{(24)} \qquad\qquad\qquad\qquad\qquad \text{(7)}$$

available by other methods[13]. The intermediate **24**, which is ana-logous to **22**, is cleaved by hydrogen chloride[134].

2. Thiolysis of imidoyl halogenides

An interesting example of reaction (8) ($R^1 = SO_2Ar$) has been published by Dubina and Burmistrov[134a].

There is some relationship between the reactions dealt with in the preceding section and the thiolysis of imidoyl chloride (**25**) (reaction 8),

$$\underset{X}{\overset{RC=NR^1}{\underset{|}{}}} + H_2S \longrightarrow RCSNHR^1 + HX \tag{8}$$

(25)

as is shown by the reaction of nitriles with thiocarboxylic acids and hydrogen chloride[135] (reaction 9). Therefore it is not surprising that

$$RCN + HCl \longrightarrow \underset{NH}{\overset{RC-Cl}{\underset{\|}{}}} \xrightarrow{R^1\overset{O}{\underset{\|}{C}}-SH} \left[\underset{}{\overset{NH}{\underset{\|}{}}} \underset{}{\overset{O}{\underset{\|}{}}} \\ RC-S-CR^1 \right] + HCl \longrightarrow$$

$$\text{(26)} \qquad\qquad\qquad\qquad\qquad\qquad\qquad\qquad RCSNH_2 + R^1COCl \quad (9)$$

23 is able to convert **25** (X = Cl, Br; R^1 = H) to thioamides in a preparatively useful manner[136].

The thiolysis may be extended to immonium chlorides (**27**) to afford N,N-disubstituted thioamides[137] (reaction 10). Compound **27** is

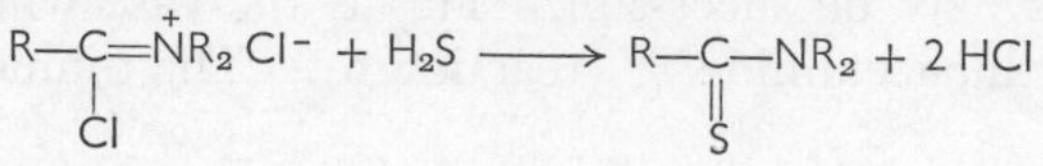

$$R—\overset{+}{C}=NR_2 \ Cl^- + H_2S \longrightarrow R—\underset{\underset{S}{\|}}{C}—NR_2 + 2\,HCl \tag{10}$$
$$\underset{Cl}{|}$$

(**27**)

easily obtained from N,N-disubstituted amides and phosgene[137] or from enamides with hydrogen chloride[27].

3. Thiolysis of amidines and imidic esters

Amidines and imidic esters are conveniently prepared from nitriles (reaction 11). As the latter are themselves able to react with hydrogen

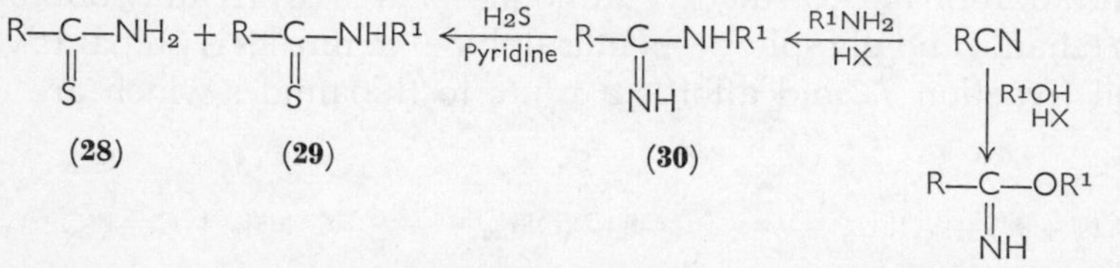

$$R—\underset{\underset{S}{\|}}{C}—NH_2 + R—\underset{\underset{S}{\|}}{C}—NHR^1 \xleftarrow[\text{Pyridine}]{H_2S} R—\underset{\underset{NH}{\|}}{C}—NHR^1 \xleftarrow[\text{HX}]{R^1NH_2} RCN$$

$$\quad\text{(28)}\qquad\qquad\text{(29)}\qquad\qquad\qquad\text{(30)}\qquad\qquad\Big\downarrow\begin{smallmatrix}R^1OH\\HX\end{smallmatrix}$$

$$R—\underset{\underset{NH}{\|}}{C}—OR^1 \tag{11}$$

sulphide, amidines and imidic esters have to show some advantage over nitriles as starting materials. Recently it has been shown that aliphatic thioamides are obtained in better yields when prepared according to equation (11)[138a] rather than (1), even if the latter is improved as indicated by Gilbert and Rumanowski[127].

When ammonia, primary, or secondary amines are used as bases, **20** can be taken to represent a deprotonated amidine. Reynaud and coworkers found that amidines prepared from alkylamines (**30**, R^1 = alkyl) preferably afford **28** on reaction with hydrogen sulphide[138a,138d], whereas the proportion of **28** and **29** depends on the basicity of R^1NH_2, when R^1 is aromatic as shown in Table 15[138d].

TABLE 15. Thiolysis of acetamidines (R = Me) according to reaction (11).

R^1	28 (%)	29 (%)	pK_a of R^1NH_2
Ph	52	46·5	4·60
p-MeC$_6$H$_4$	53·5	39·0	5·09
p-MeOC$_6$H$_4$	72	26	5·29
p-ClC$_6$H$_4$	20	74	3·99

N,N-Diethylamidines yield **28** exclusively when reacted with hydrogen sulphide, presumably due to steric reasons[138b] (reaction 12).

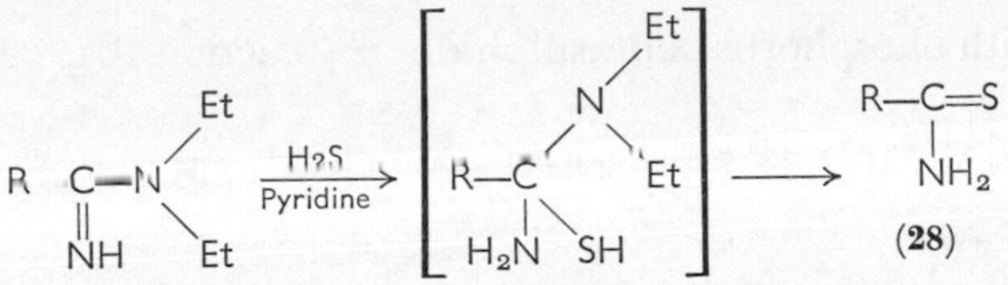

$$R-\underset{\underset{NH}{\|}}{\overset{\overset{Et}{\diagup}}{C}}-N\diagdown_{Et} \xrightarrow[\text{Pyridine}]{H_2S} \left[R-\underset{\underset{SH}{\diagup}}{\overset{\overset{Et}{\diagdown}N\diagup Et}{C}}\diagdown_{H_2N} \right] \longrightarrow \underset{\underset{\mathbf{(28)}}{NH_2}}{\overset{R-C=S}{}} \qquad (12)$$

This means that probably mechanisms of different type are operative in reactions (4) and (12).

4. Thionation* of amides

The most useful reagent for replacing the oxygen of amides by sulphur is phosphorus pentasulphide (reaction 13). The polarity of

$$RCONH_2 \xrightarrow{P_4S_{10}} RCSNH_2 \qquad (13)$$

solvents applied may be varied in the range between aromatic hydrocarbons and pyridine. Using the latter solvent it is possible to prepare thioamides directly from an amine and an acylating agent without isolating the amide[139] (reaction 14), in yields exceeding those of the

$$RCOX + R^1NH_2 \xrightarrow{Pyridine} RCONHR^1 \xrightarrow[P_4S_{10}]{Pyridine} RCSNHR^1 \qquad (14)$$

conventional two-step reaction. In xylene it is possible to obtain optically active thioamides[97] and in benzene α-amino-β-thiolactams (**31**)[140] from the appropriate amides.

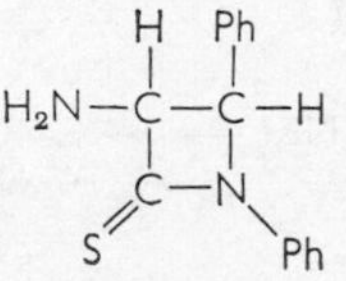

$$H_2N-\underset{\underset{\underset{S}{\diagdown}C-N}{\overset{\overset{H}{|}}{C}}}{\overset{\overset{H}{|}}{C}}-\overset{\overset{Ph}{|}}{\underset{}{C}}-H$$

(31)

Little is known about the mechanism by which amides are thionated. The observation that imidosulphonates which have to be located between the imidoyl–halogenides (section II.A.2) and the imidic esters (section II.A.3) are easily converted to thioamides by hydrogen sulphide[141] (reaction 15) suggests an analogous course for the

* Interconversion of a carbonyl group into a thiocarbonyl group by means of P_4S_{10} will be called 'thionation' rather than 'sulphuration' (introduction of sulphur into the molecule by means of the element).

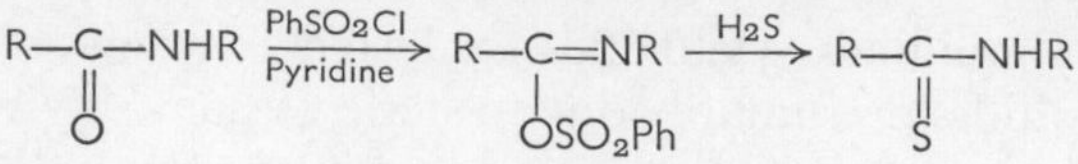

$$R-\underset{\underset{O}{\|}}{C}-NHR \xrightarrow[\text{Pyridine}]{PhSO_2Cl} R-\underset{\underset{OSO_2Ph}{|}}{C}=NR \xrightarrow{H_2S} R-\underset{\underset{S}{\|}}{C}-NHR \qquad (15)$$

reaction with phosphorus pentasulphide [142] (reaction 16). Reaction of

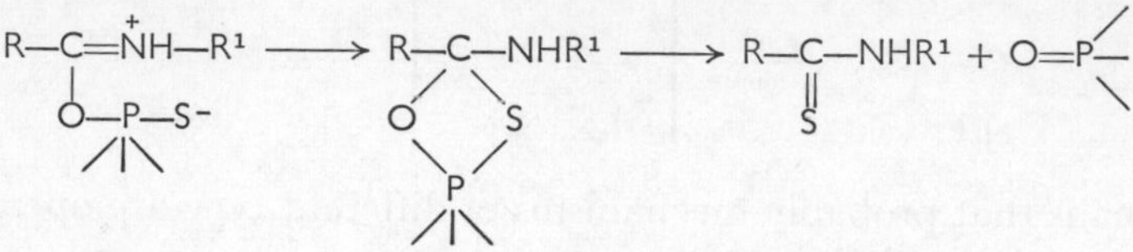

$$R-\underset{\underset{\overset{|}{O}-P-S^-}{\underset{/|\backslash}{}}}{\overset{+}{C}=NH}-R^1 \longrightarrow R-\underset{\underset{\overset{O}{}\;\;\overset{S}{}}{\underset{P}{\diamond}}}{C}-NHR^1 \longrightarrow R-\underset{\underset{S}{\|}}{C}-NHR^1 + O=P\overset{/}{\underset{\backslash}{}} \qquad (16)$$

salicylamide with phosphorus pentasulphide in pyridine (reaction 17) afforded **32** which is in favour of reaction (16) [13].

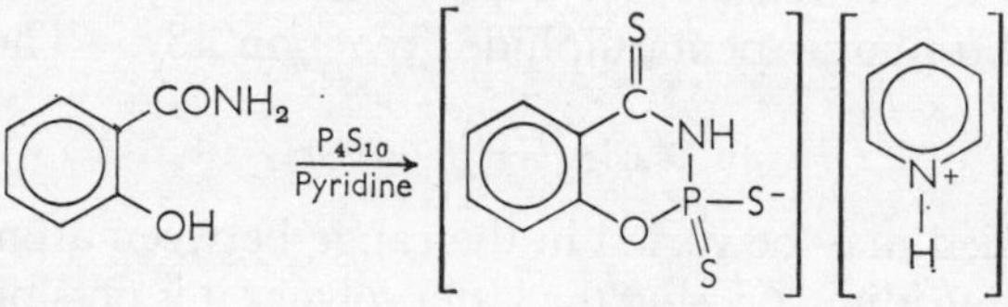

$$(32)$$

B. Thioacylation of Amines

Contrary to the acylation of amines, for which acyl halides and acid anhydrides are the predominant reagents, the analogous derivatives of thiocarboxylic acids (reactions 18, 19) are rarely used for thio-

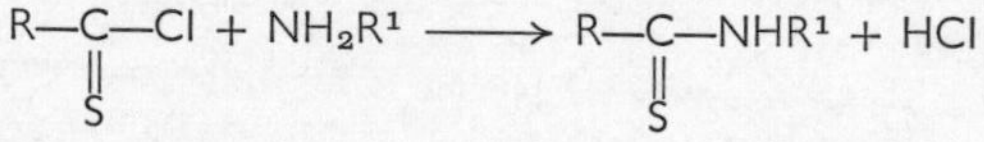

$$R-\underset{\underset{S}{\|}}{C}-Cl + NH_2R^1 \longrightarrow R-\underset{\underset{S}{\|}}{C}-NHR^1 + HCl \qquad (18)$$

$$R-\underset{\underset{S}{\|}}{C}-O-acyl + H_2NR^1 \longrightarrow R-\underset{\underset{S}{\|}}{C}-NHR^1 + HOacyl \qquad (19)$$

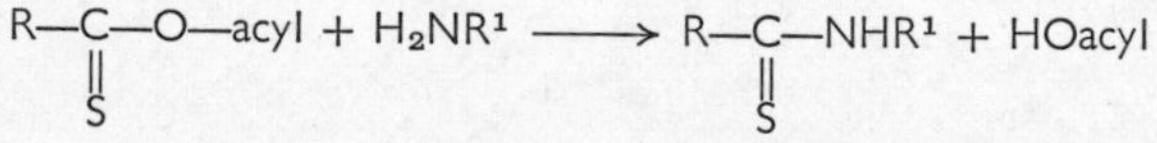

acylation, owing to the difficulties in their preparation. Nevertheless, the reagents treated in the following subsections are of sufficient thioacylating activity as to serve for synthesis.

I. Thionocarboxylates

The preparation of thioamides from thionocarboxylates and ammonia or primary amines (reaction 20) is somewhat restricted by a side-

$$R-\underset{\underset{S}{\|}}{C}-OEt + R^1NH_2 \xrightarrow{Ether} R-\underset{\underset{S}{\|}}{C}-NHR^1 + EtOH \qquad (20)$$

reaction (reaction 21) leading to amidines[138a]. The interference of

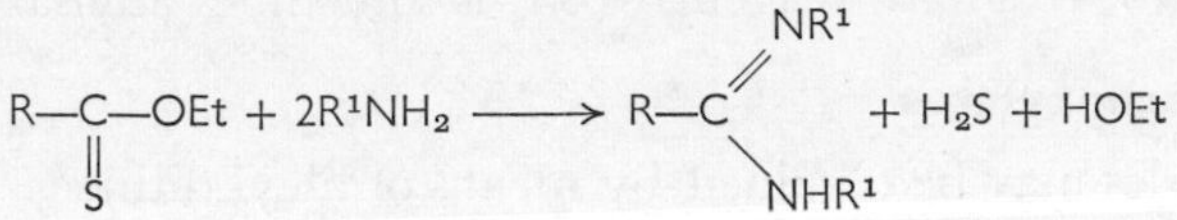

$$R-\underset{\underset{S}{\|}}{C}-OEt + 2R^1NH_2 \longrightarrow R-C\underset{\diagdown NHR^1}{\overset{\diagup NR^1}{}} + H_2S + HOEt \qquad (21)$$

reaction (21) is not possible with secondary amines, and excellent yields of tertiary thioamides result. The reduced reactivity of aromatic amines may be overcome by using their magnesium salts[138d] (reaction 22). With primary amines reaction (21) may be avoided

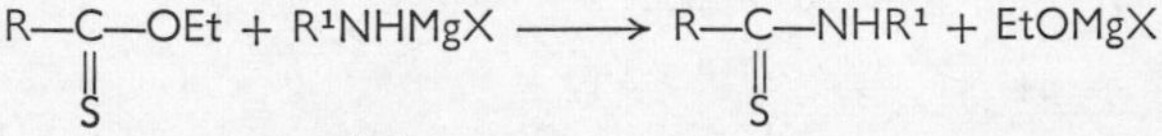

$$R-\underset{\underset{S}{\|}}{C}-OEt + R^1NHMgX \longrightarrow R-\underset{\underset{S}{\|}}{C}-NHR^1 + EtOMgX$$

$$\text{(33)} \qquad\qquad\qquad\qquad\qquad\qquad (22)$$

when tetrahydrofuran is used as solvent[138a,139]; and the best results are obtained when 2 moles of **33** are employed. With O-ethyl thioformate which is easily prepared from ethyl orthoformate and hydrogen sulphide (reaction 23)[143], no activation of the amine is

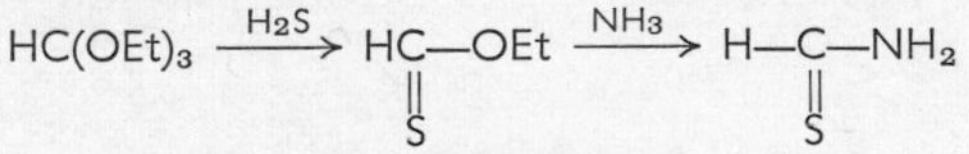

$$HC(OEt)_3 \xrightarrow{\text{H}_2\text{S}} HC\underset{\underset{S}{\|}}{-}OEt \xrightarrow{\text{NH}_3} H-\underset{\underset{S}{\|}}{C}-NH_2 \qquad (23)$$

necessary. This is the best method for preparing thioformamide (yield 90%)[144].

Up to now thioacylation with thionocarboxylates is the best available method of connecting two amino acids by a thiopeptide link[145] (reaction 24).

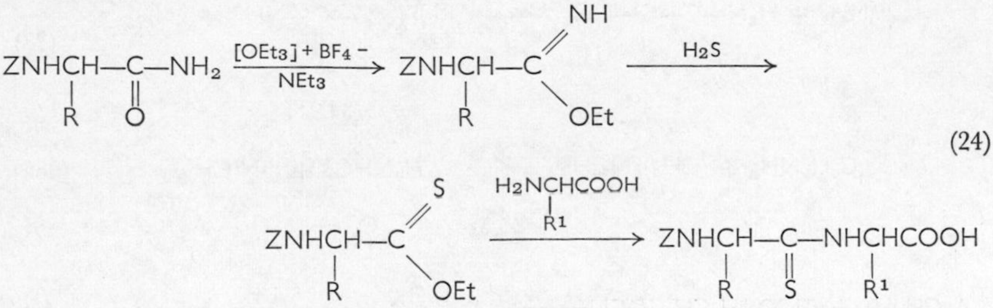

$$ZNHCH\underset{\underset{R}{|}\;\underset{O}{\|}}{-}C-NH_2 \xrightarrow[\text{NEt}_3]{\text{[OEt}_3\text{]}^+\,\text{BF}_4^-} ZNHCH\underset{\underset{R}{|}}{-}C\underset{\diagdown OEt}{\overset{\diagup NH}{}} \xrightarrow{\text{H}_2\text{S}}$$

$$(24)$$

$$ZNHCH\underset{\underset{R}{|}}{-}C\underset{\diagdown OEt}{\overset{\diagup S}{}} \xrightarrow[\underset{R^1}{|}]{\text{H}_2\text{NCHCOOH}} ZNHCH\underset{\underset{R}{|}\;\underset{S}{\|}\;\underset{R^1}{|}}{-}C-NHCHCOOH$$

2. Dithiocarboxylic acids

While thiocarboxylic acids produce amides when reacted with amines, thioamides are obtained from dithiocarboxylic acids (reaction 25) or their more stable alkali salts[11]. Aromatic dithioacids are less

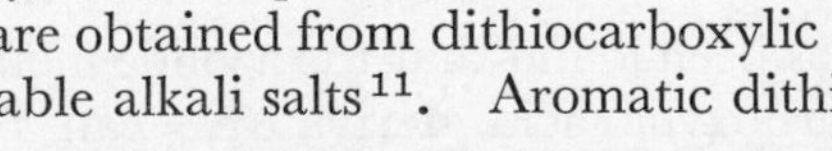

$$RCSSH + R^1NH_2 \longrightarrow RC\underset{\underset{S}{\|}}{}NHR^1 \qquad (25)$$

reactive than aliphatic ones. Therefore activation of the amines by conversion to 'Grignard reagents' (**33**) is sometimes advisable[146].

3. Dithiocarboxylates

Thioamides may be obtained by means of alkyl dithiocarboxylates from amines containing other labile functional groups, e.g. tryptamine, in reaction (26)[147].

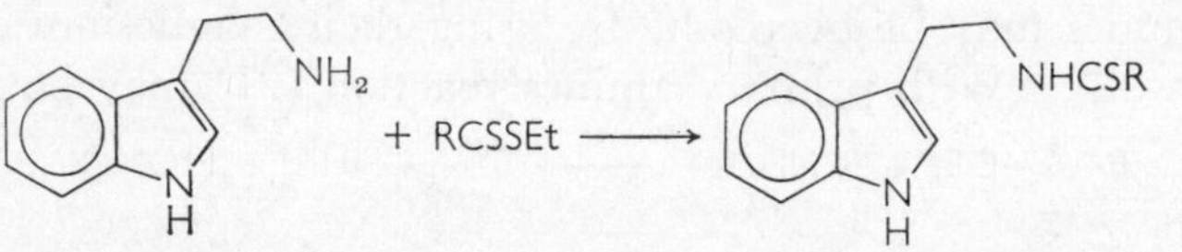

$$+ \text{RCSSEt} \longrightarrow \qquad\qquad (26)$$

For thioacylation of amino acids the carboxymethyl dithiocarboxylates introduced by Holmberg[148] have gained considerable importance[149,150] (reaction 27). The configuration of the assymetric carbon atom of the amino acid is retained in this reaction[92].

$$\text{HOOCCHNH}_2 + \text{RCSSCH}_2\text{COOH} \longrightarrow \text{HOOCCHNH}-\text{CR} + \text{HSCH}_2\text{COOH}$$
$$\underset{R}{\mid} \qquad\qquad\qquad\qquad\qquad \underset{R}{\mid}\ \underset{S}{\parallel} \qquad\qquad\qquad (27)$$

4. Thioamides

As in the case of alkyl dithiocarboxylates, thioacylation may be achieved by ordinary thioamides (reaction 28)[151], or by activated thioamides (reaction 29)[152]. Reaction 28 is catalysed by acids; even CO_2 is feasible (reaction 28a)[270,363].

$$2\text{RNH}_2 + \text{H}_2\text{N}-\underset{\underset{S}{\parallel}}{\text{C}}-\underset{\underset{S}{\parallel}}{\text{C}}-\text{NH}_2 \longrightarrow \text{RNH}-\underset{\underset{S}{\parallel}}{\text{C}}-\underset{\underset{S}{\parallel}}{\text{C}}-\text{NHR} + 2\text{NH}_3 \qquad (28)$$

$$(34)$$

$$\text{C}_4\text{H}_9\text{NH}_2 + (\text{CH}_3)_2\text{NCHS} \xrightarrow{\text{CO}_2} \text{C}_4\text{H}_9\text{NHCS} + \text{HN}(\text{CH}_3)_2 \qquad (28a)$$

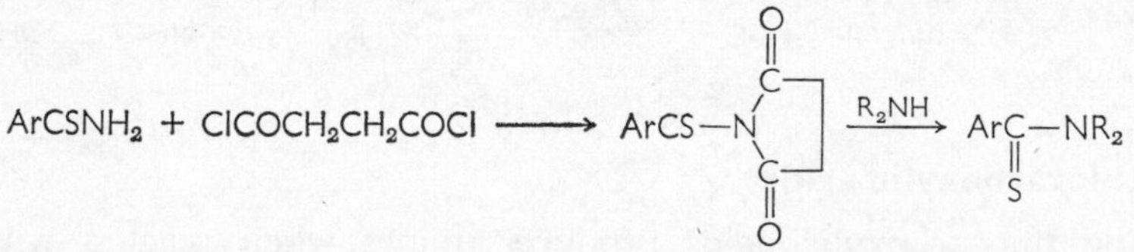

$$\text{ArCSNH}_2 + \text{ClCOCH}_2\text{CH}_2\text{COCl} \longrightarrow \text{ArCS}-\text{N} \xrightarrow{\text{R}_2\text{NH}} \text{Ar}\underset{\underset{S}{\parallel}}{\text{C}}-\text{NR}_2 \qquad (29)$$

Recently it has been found that the azolide method, familiar for the syntheses of carboxylic acid derivatives, can be extended to the preparation of thioamides[153] (reaction 30), and that N-thioaroyl-benzoxazolones are good thioacylating agents[153a] (reaction 30a).

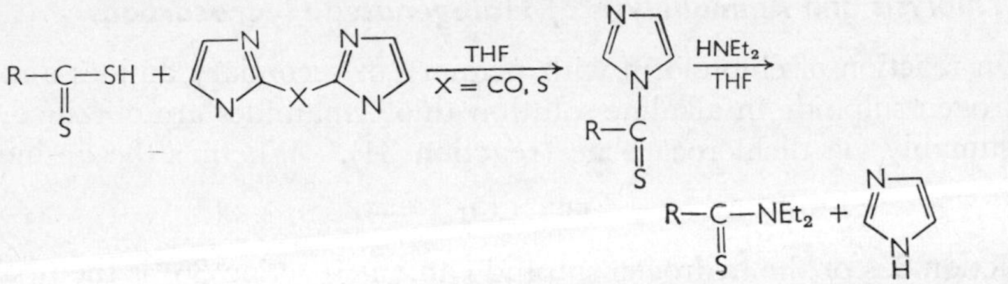

(30)

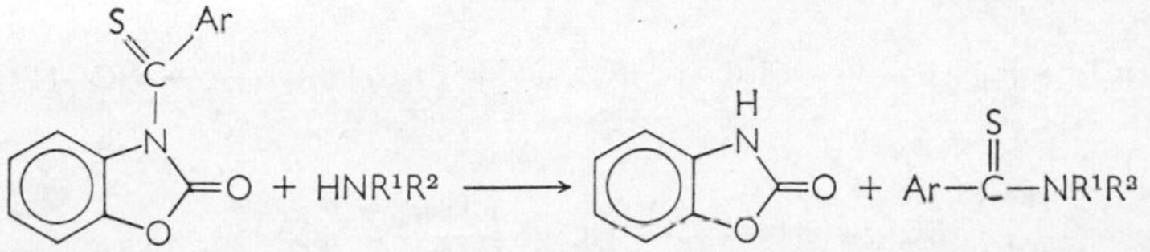

(30a)

5. Thioketenes

Thioketenes are regarded as intermediates in the reaction of acetylene thiolesters with amines (reaction 31)[154,155,364,365]. Recently thio-

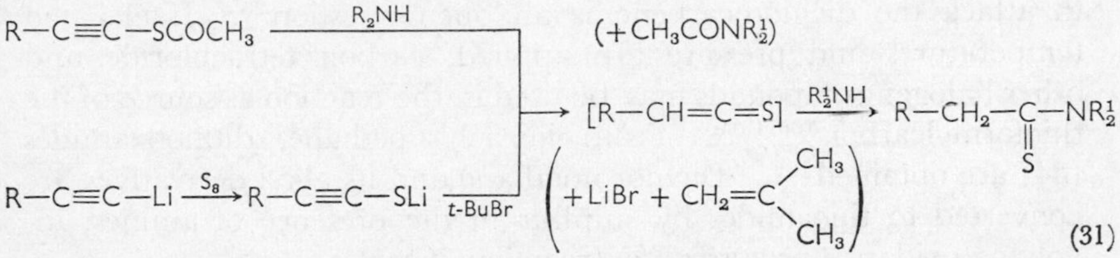

(31)

ketenes have been actually obtained and promise to become valuable reagents for the synthesis of thioamides[56] (reaction 32).

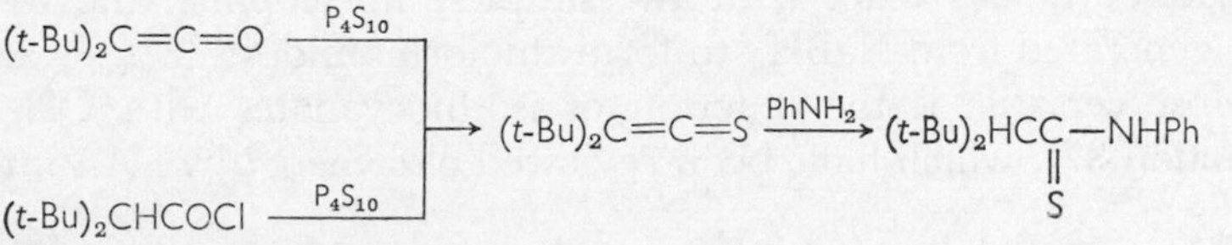

(32)

A derivative of thioketene is α-(dimercaptomethylene)camphor, which is obtained as disodium salt from the sodium compound of camphor with carbon disulphide. It forms a thioamide on reaction with amines[155a] (reaction 33).

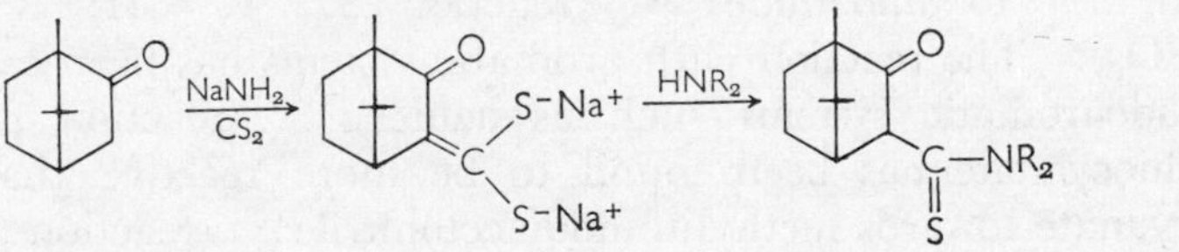

(33)

C. Thiolysis and Ammonolysis of Halogenated Hydrocarbons

On reaction of chloroform with primary or secondary amines and hydrogen sulphide in alkaline solution thioformamides are obtained, presumably via dichlorocarbene (reaction 34). Whether the amine

$$CHCl_3 + B^- \rightleftharpoons HB + CCl_3^- \longrightarrow CCl_2 + Cl^- \tag{34}$$

(reaction 35) or the hydrogen sulphide anion (reaction 36) is the first

$$CCl_2 + R_2NH \longrightarrow Cl_2\bar{C}-\overset{+}{N}HR_2 \longrightarrow Cl_2CHNR_2 \overset{SH^-}{\longrightarrow} \underset{S}{\overset{\|}{HC}}-NR_2 \tag{35}$$

$$CCl_2 + SH^- \longrightarrow \underset{S}{\overset{\|}{HC}}-S^- \overset{HNR_2}{\longrightarrow} \underset{S}{\overset{\|}{HC}}-NR_2 \tag{36}$$

$$R^1CHCl_2 + S + HNR^2R^3 \longrightarrow \underset{S}{\overset{\|}{R^1C}}-NR^2R^3 \tag{36a}$$

to attack the dichlorocarbene is an open question[28]. If elevated temperatures and pressure are applied, carbon tetrachloride and other halogen compounds may be used in the reaction as source of the thioformylcarbon[156,156a]. From hexachloroethane, dithiooxamides (**34**) are obtained[156]. Dichloromethane and its alkyl derivatives are converted to thioamides by sulphur in the presence of amines, involving oxidative processes[156a] (reaction 36a).

D. Addition of Nucleophiles to Isothiocyanates

Isothiocyanates react with the simplest nucleophile, the hydride ion, generated from $NaBH_4$ to form thioformamides[157].

Most versatile is the reaction of isothiocyanates with CH acids (reaction 37), which have been reviewed elsewhere[9,13]. An interest-

$$R^1\underset{R^2}{\overset{R}{\diagdown}}\!\!C^-Na^+ + R^3NCS \longrightarrow R^1\underset{R^2}{\overset{R}{\diagdown}}\!\!C-C\bar{N}R^3Na^+ \overset{HX}{\longrightarrow} R^1\underset{R^2}{\overset{R}{\diagdown}}\!\!C-C-NHR^3 \tag{37}$$

ing new example of the method is the conversion of disulphonylmethanes to thioamides[158] (reaction 37, R = H, $R^1 = R^2 = ArSO_2$). The reaction with aromatic systems has been extended to pseudoaromatic systems such as azulene[159] (reaction 38). Allyl isothiocyanate has been found to be more reactive than phenyl isocyanate towards methyliminodiacetonitrile[160] (reaction 39).

$$\text{RSO}_2\text{NCS} + \left[\begin{array}{c} \text{(azulene with CH}_3\text{ groups)} \\ \updownarrow \\ \text{(azulene cation/anion with CH}_3\text{ groups)} \end{array} \right] \xrightarrow[\text{(ether)}]{20°} \text{(azulene with CH}_3\text{ groups and RSO}_2\text{NH—C(=S)—)} \quad (38)$$

$$\text{NCCH}_2\underset{\underset{\text{Me}}{\mid}}{\text{N}}\text{CH}_2\text{CN} + \text{CH}_2{=}\text{CHCH}_2\text{NCS} \longrightarrow$$

$$\text{CH}_2{=}\text{CHCH}_2\text{NHC}\underset{\underset{\text{S}}{\parallel}}{\overset{\overset{\text{CN}}{\mid}}{—}\text{CH}}{—}\underset{\underset{\text{Me}}{\mid}}{\text{N}}{—}\overset{\overset{\text{CN}}{\mid}}{\text{CH}}{—}\text{C}\underset{\underset{\text{S}}{\parallel}}{}\text{NHCH}_2\text{CH}{=}\text{CH}_2 \quad (39)$$

Enamines (**35**) are useful partners in reactions with isothiocyanates (reaction 40) in which β-aminoalkene thioamides (**36**) are formed[13].

$$\underset{(35)}{\text{H}{—}\overset{\overset{\text{R}^1}{\mid}}{\text{C}}{=}\overset{\overset{\text{R}^2}{\mid}}{\text{C}}{—}\overset{\overset{\text{R}^3}{}}{\underset{\underset{\text{R}^4}{}}{\text{N}}}} \xrightarrow{\text{R}^5\text{NCS}} \underset{(36)}{\text{R}^5\text{NHC}\underset{\underset{\text{S}}{\parallel}}{\overset{\overset{\text{R}^1}{\mid}}{—}\text{C}}{=}\overset{\overset{\text{R}^2}{\mid}}{\text{C}}{—}\overset{\overset{\text{R}^3}{}}{\underset{\underset{\text{R}^4}{}}{\text{N}}}} \xrightarrow{\text{H}_3\text{O}+} \text{R}^5\text{NHC}\underset{\underset{\text{S}}{\parallel}}{\overset{\overset{\text{R}^1}{\mid}}{—}\text{CH}}{—}\overset{\overset{\text{R}^2}{\mid}}{\text{C}}{=}\text{O} \quad (40)$$

These vinylogous thioureas may be readily hydrolysed to β-oxo-thioamides.

With $\text{R}^3 = \text{H}$, **35** is an enolized imine, which may likewise react at the imino nitrogen. Such an ambivalence was found indeed with N-alkylcyclohexylideneamines (**37**), which form cyclohexenylthioureas (**38**) on reaction with isothiocyanate. A ready rearrangement to thioamides (**39**) takes place when **38** arises from aryl isothiocyanates[161]

$$\underset{(37)}{\overset{\text{NR}}{\bigcirc}} \rightleftharpoons \underset{(38)}{\overset{\text{NHR}}{\bigcirc}} \xrightarrow[\text{Benzene}]{\text{ArNCS}} \underset{(38)}{\overset{\text{RN}\overset{\overset{\text{S}}{\parallel}}{\text{C}}{-}\text{NHAr}}{\bigcirc}} \xrightarrow{\Delta} \underset{(39)}{\overset{\text{NHR}}{\bigcirc}\overset{\text{C}{-}\text{NHAr}}{\underset{\text{S}}{}}} \xrightarrow[\text{aq. HCl}]{\text{H}_3\text{O}+} \overset{\text{O}}{\bigcirc}\text{CSNHAr} \quad (41)$$

(reaction 41), so that thioamides form directly when reaction (41) is carried out at elevated temperatures.

E. Electrophilic Reaction of Thiocarbamoyl Chlorides

N,N-Disubstituted thiocarbamoyl chlorides react with homo- or hetcroaromatic compounds under Friedel–Crafts conditions to form N,N-disubstituted thioamides (reaction 42)[162]. For lack of sufficient

$$(CH_3)_2NCCl + HAr \xrightarrow{AlCl_3 \text{ or } SnCl_2} (CH_3)_2NCAr \qquad (42)$$
$$\overset{\parallel}{S} \qquad\qquad\qquad\qquad \overset{\parallel}{S}$$

activity benzene does not react according to (42), a yellow product is obtained instead, for which structure **40** seems to be most probable[162,163].

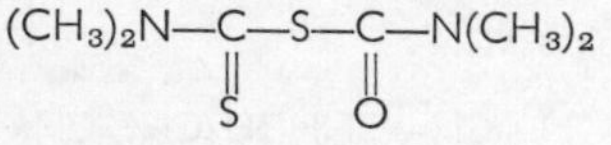

$$(CH_3)_2N-\overset{\parallel}{\underset{S}{C}}-S-\overset{\parallel}{\underset{O}{C}}-N(CH_3)_2$$

(40)

$$\underset{R^2}{\overset{R^1}{\diagdown}}N-\overset{\parallel}{\underset{S}{C}}-S-\overset{\parallel}{\underset{S}{C}}-N\underset{\diagdown R^2}{\overset{\diagup R^1}{}}$$

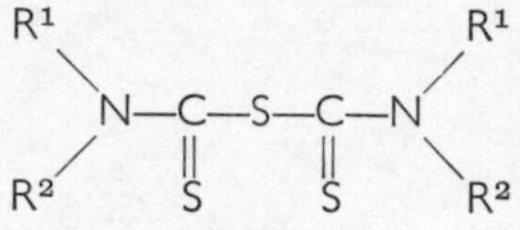

(40a)

Instead of the thiourethanes normally produced by solvolysis of thiocarbamoyl chlorides, sodium isopropoxide in propan-2-ol may form thioformamidcs, if R^1 and R^2 are bulky (reaction 42a)[366].

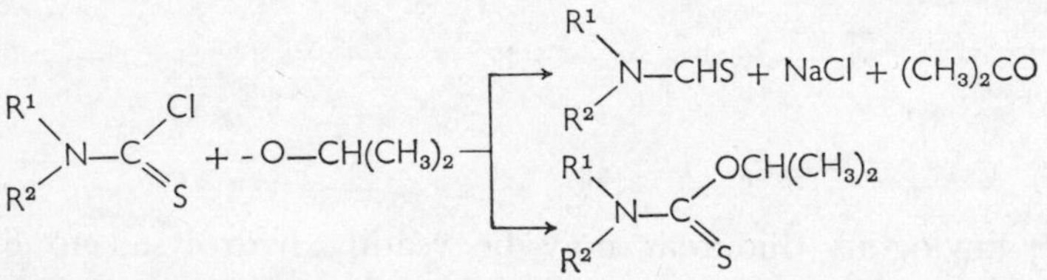

$$\underset{R^2}{\overset{R^1}{\diagdown}}N-\overset{\diagup Cl}{\underset{\diagdown S}{C}} + {}^-O-CH(CH_3)_2 \longrightarrow \begin{cases} \underset{R^2}{\overset{R^1}{\diagdown}}N-CHS + NaCl + (CH_3)_2CO \\[2mm] \underset{R^2}{\overset{R^1}{\diagdown}}N-\overset{\diagup OCH(CH_3)_2}{\underset{\diagdown S}{C}} \end{cases} \qquad (42a)$$

In some cases the yellow bis-thiocarbamoyl sulphides (**40a**) were isolated as by-products from the residues of distillation[366], cf. ref. 367).

F. The Willgerodt–Kindler Reaction

Alkyl aryl ketones are converted to ω-aryl thioamides by the Willgerodt–Kindler reaction, if the aliphatic chain of the product is unbranched, regardless of the position occupied by the carbonyl group in the ketones[164] (reaction 43). Normally the reaction is carried out

$$ArCOCH_2Me \xrightarrow[130^\circ]{HNR_2, S_8} ArCH_2CH_2\overset{\parallel}{\underset{S}{C}}NR_2 \qquad (43)$$

in boiling morpholine, frequently used as the amine in reaction (43); its wide range of variation may be marked by two recent observations. Dialkylamines and formaldehyde form thioformamides, when reacted with sulphur (reaction 43a)[368]. The dimethylamino group of formamide may be introduced into a thioamide by reacting unsaturated hydrocarbons with sulphur in dimethylformamide (reaction 43b)[369].

$$\begin{array}{c}R^1\\ \diagdown\\ \diagup\\ R^2\end{array}\!\!NH + CH_2O \longrightarrow \begin{array}{c}R^1\\ \diagdown\\ \diagup\\ R^2\end{array}\!\!N{-}CH_2OH \xrightarrow{\;S\;} \begin{array}{c}R^1\\ \diagdown\\ \diagup\\ R^2\end{array}\!\!N{-}\underset{\underset{S}{\|}}{C}{-}H \qquad (43a)$$

$$ArCH{=}CH_2 \xrightarrow[175°]{S,HCON(CH_3)_2} ArCH_2{-}\underset{\underset{S}{\|}}{C}{-}N(CH_3)_2 \qquad (43b)$$

The mechanisms by which the Willgerodt–Kindler reaction may proceed have been amply discussed[164]. The finding that thioamides result in the reaction of enamines with sulphur at 20°[165] (reaction 44),

$$\underset{\underset{NR_2^1}{|}}{R}C{=}CH_2 \xrightarrow[DMF]{S_8,20°} RCH{-}\underset{\underset{S}{\|}}{C}{-}NR_2^1 \qquad (44)$$

led Mayer to propose a simple and convincing mechanism for one of the main steps of this reaction (equation 45). This involves genera-

$$ArCOCH_2CH_3 + HNR_2 \underset{>100°}{\rightleftarrows} Ar\underset{\underset{NR_2}{|}}{C}{=}CHCH_3 \xrightleftharpoons{S_8} ArCH_2\underset{\underset{NR_2}{|}}{C}{=}CH_2 \xrightleftharpoons{S_8}$$

$$\qquad (45)$$

$$\left[ArCH_2\underset{\underset{NR_2}{|}}{C}{=}CH{-}SH \right] \longrightarrow \left[ArCH_2{-}\underset{\underset{NR_2}{|}}{C}\overset{\overset{H}{|}}{{-}C}{-}S^- \right] \longrightarrow ArCH_2{-}CH_2{-}\underset{\underset{NR_2}{|}}{C}{=}S$$

tion of an enamine at elevated temperature followed by reaction with sulphur leading to isomerization and thiolysis[166].

On the other hand it is difficult to formulate an enamine intermediate in reaction (46)[167]. In this case initial thiolation seems more

$$PhMe \xrightarrow{S,NH_3} Ph\underset{\underset{S}{\|}}{C}{-}NH_2 \qquad (46)$$

likely, as was suggested for the action of sulphur and cyclic amines[168] (reaction 47), although this reaction might well proceed via an

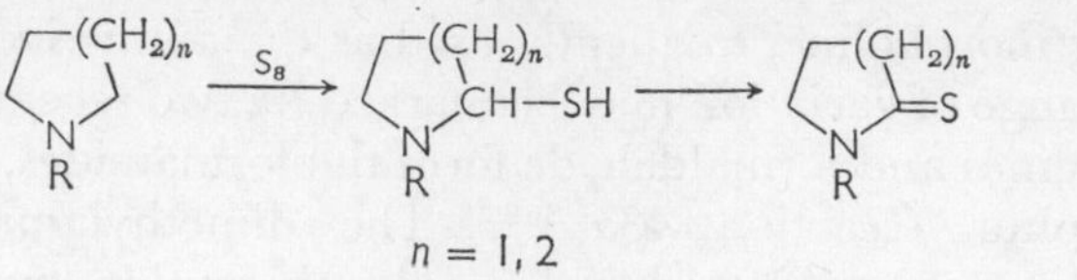

$$\underset{R}{\overset{\overbrace{(CH_2)_n}}{\big|}N} \xrightarrow{S_8} \underset{R}{\overset{\overbrace{(CH_2)_n}}{\big|}N}CH-SH \longrightarrow \underset{R}{\overset{\overbrace{(CH_2)_n}}{\big|}N}=S \tag{47}$$

$$n = 1, 2$$

enamine.   An investigation of this possibility seems worthwhile as well as for the conversion of a Mannich base to a thioamide[169] (reaction 48).

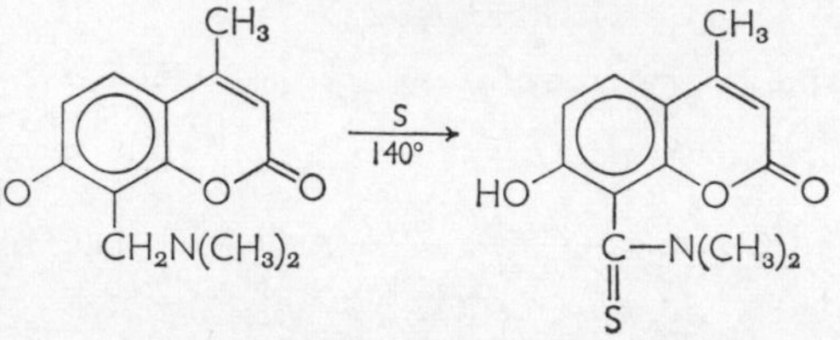

$$\tag{48}$$

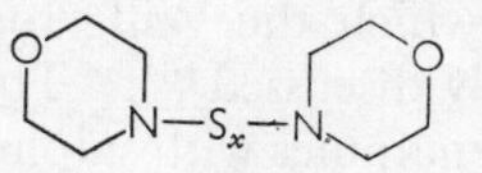

(41)

Temperature may be a critical parameter in deciding the reaction path, as formation of an enamine usually needs temperatures above 100°.   Thus reaction of styrylmorpholine (reaction 44, R = Ph, NR_2 = morpholyl) with **41** at 85° occurs according to (45) producing phenylthioacetomorpholide.   From acetophenone, phenylthioglyoxyl-morpholide is obtained under the same conditions (reaction 49) pointing to a different mechanism[170], cf. ref. 370.   Compound **42**

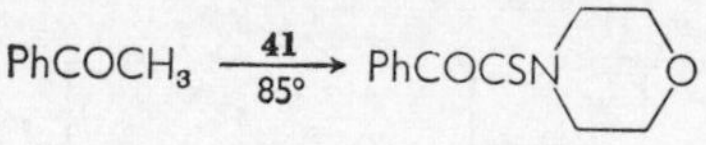

$$PhCOCH_3 \xrightarrow[85°]{41} PhCOCSN\!\!\big\langle\!\!\begin{array}{c}\\\\\end{array}\!\!\big\rangle\!\!O \tag{49}$$

(42)

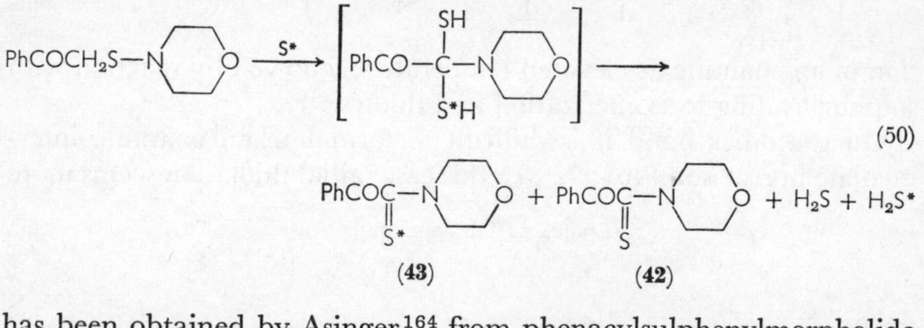

$$PhCOCH_2S\!-\!N\!\!\big\langle\!\!\begin{array}{c}\\\\\end{array}\!\!\big\rangle\!\!O \xrightarrow{S^*} \left[PhCO\!-\!\underset{S^*H}{\overset{SH}{\underset{|}{\overset{|}{C}}}}\!-\!N\!\!\big\langle\!\!\begin{array}{c}\\\\\end{array}\!\!\big\rangle\!\!O \right] \longrightarrow$$

$$\tag{50}$$

$$PhCO\underset{S^*}{\overset{\|}{C}}\!-\!N\!\!\big\langle\!\!\begin{array}{c}\\\\\end{array}\!\!\big\rangle\!\!O \;+\; PhCO\underset{S}{\overset{\|}{C}}\!-\!N\!\!\big\langle\!\!\begin{array}{c}\\\\\end{array}\!\!\big\rangle\!\!O \;+\; H_2S + H_2S^*$$

(43)   **(42)**

has been obtained by Asinger[164] from phenacylsulphenylmorpholide (reaction 50).   The reaction has been carried out with radioactive sulphur and 50% of the activity has been found in **43** and 50% in the hydrogen sulphide.

The foregoing discussion shows that further investigations are needed to elucidate the mechanisms operating in the Willgerodt–Kindler reaction.

G. Rearrangement of the Benzenesulphonates of Ketoximes

Benzenesulphonated ketoximes undergo spontaneous Beckmann rearrangement when hydrogen sulphide is present, and the resulting imidosulphonates (cf. section II.A.4) easily form thioamides *in situ*[171] (reaction 51). The steps subsequent to the esterification come close

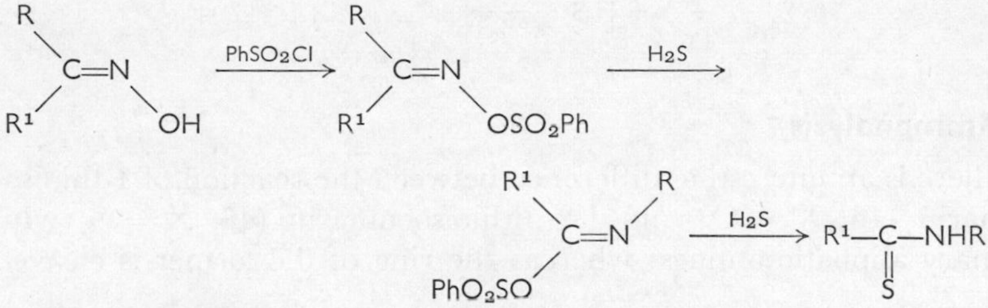

$$\begin{array}{ccc}
R\!\!\diagdown & R\!\!\diagdown \\
\diagup C\!=\!N\diagdown & \xrightarrow{\;PhSO_2Cl\;} & \diagup C\!=\!N\diagdown \\
R^1 \qquad OH & & R^1 \qquad OSO_2Ph
\end{array} \xrightarrow{\;H_2S\;}$$

$$\begin{array}{c}
R^1\!\!\diagdown \quad R \\
\diagup C\!=\!N\diagup \\
PhO_2SO
\end{array} \xrightarrow{\;H_2S\;} R^1\!\!-\!\!\underset{\underset{S}{\|}}{C}\!\!-\!\!NHR \qquad (51)$$

to the area of the Willgerodt–Kindler reaction. The conversion of benzaldoxime esters to thioamides[172] (reaction 52) affords an interesting analogue to reaction (51) in the aldehyde series.

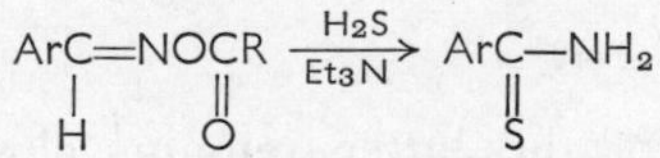

$$\underset{\underset{H}{|}}{ArC}\!=\!\underset{\underset{O}{\|}}{NOCR} \xrightarrow[Et_3N]{H_2S} Ar\underset{\underset{S}{\|}}{C}\!\!-\!\!NH_2 \qquad (52)$$

H. Cleavage of Heterocyclic Compounds

Notwithstanding the ability of thioamides to be starting materials for the synthesis of sulphur heterocycles there are heterocyclic compounds obtained by other routes, which afford thioamides on thiolytic, ammonolytic, hydrolytic or pyrolytic cleavage.

I. Thiolysis

The hetero ring of *N*-ethylbenzisoxazolium cation (**44**) is opened by bases to a keto ketenimine (**45**), which resembles the alkylation products of nitriles. Intermediate **45** adds nucleophiles such as hydrogen sulphide generating thioamides[173] (reaction 53).

Δ^2-Oxazolines are cleaved to thioamides under analogous conditions[174] (reaction 54).

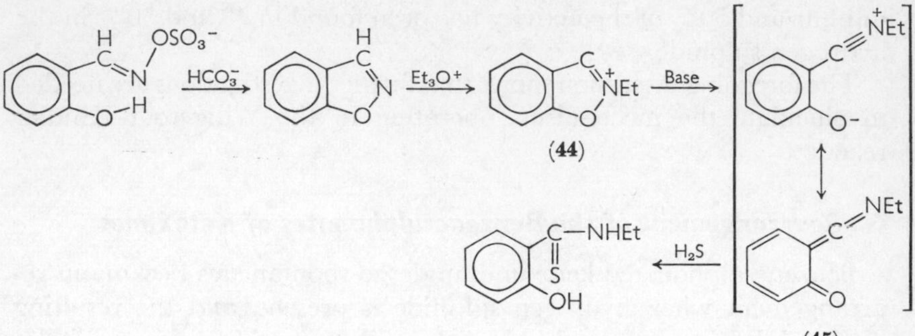

(44)

(45)

(53)

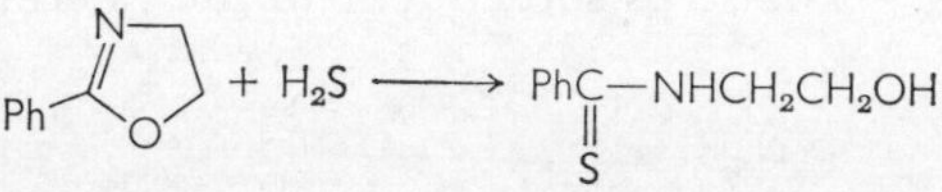

$$PhC-NHCH_2CH_2OH \qquad (54)$$

2. Ammonolysis

There is an interesting difference between the reaction of 1-thioiso-cumarin (**46**, X = O) and 1,2-dithioisocumarin (**46**, X = S) with primary aliphatic amines: whereas the ring of the former is cleaved

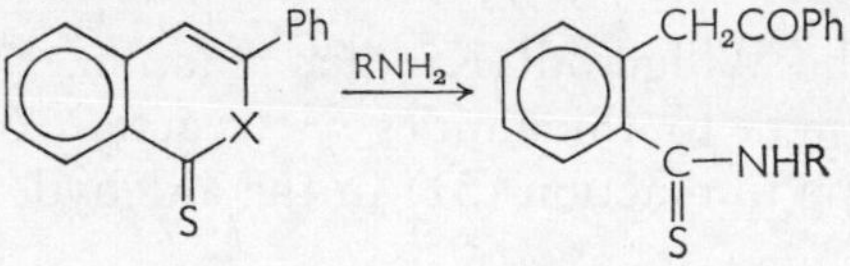

(55)

(**46**)

(reaction 55) that of the latter remains closed forming **47**. An

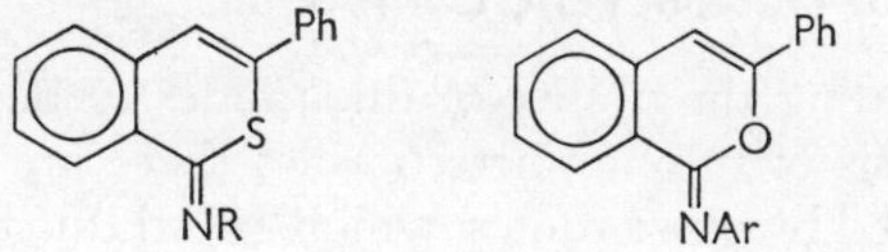

(**47**)   (**48**)

analogous compound (**48**) is obtained from 1-thioisocumarin (**46**,

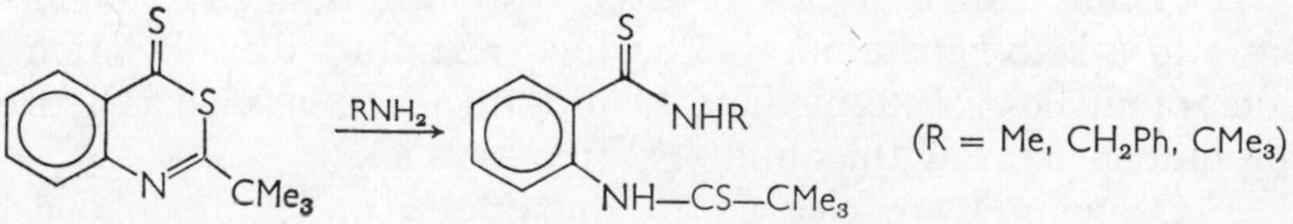

(48a)   (48b)   (R = Me, CH_2Ph, CMe_3)

X = O) with aromatic amines[175], whereas the benzothiazine-thione (**48a**) yields the bis-thioamide (**48b**) on reaction with amines[371].

The pyrimidine ring of 3-methyl-4-thiouracil is cleaved by dimethylamine and methanol to form *trans*-β-dimethylaminothioacrylic acid methylamide[176] (reactions 56).

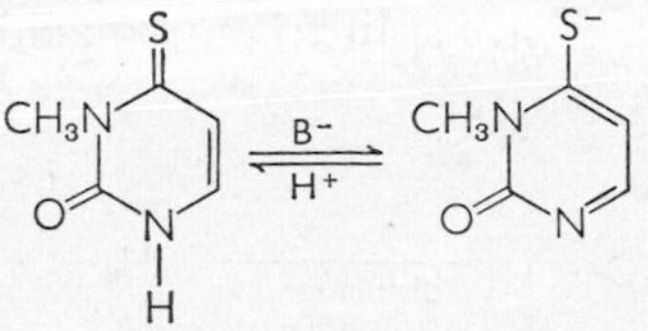

(56a)

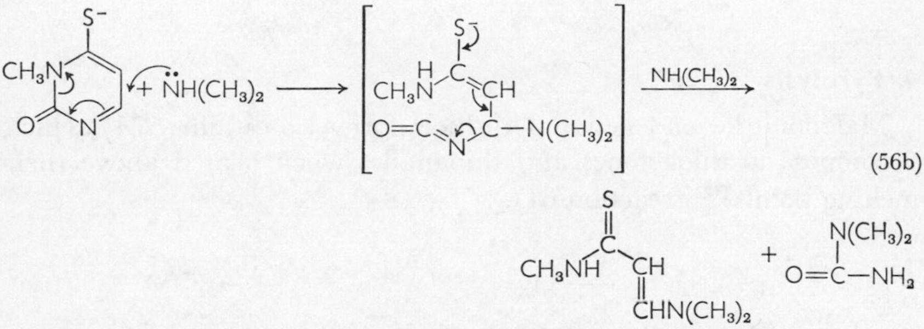

(56b)

Many heterocycles containing more than one sulphur atom are cleaved by amines to form thioamides. Reactions (57) and (58) may

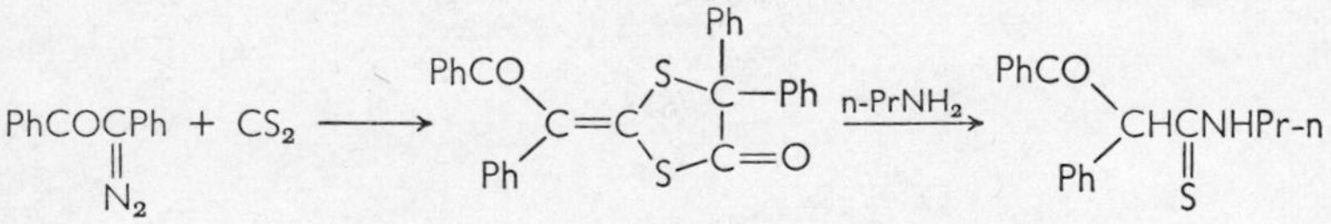

(57)

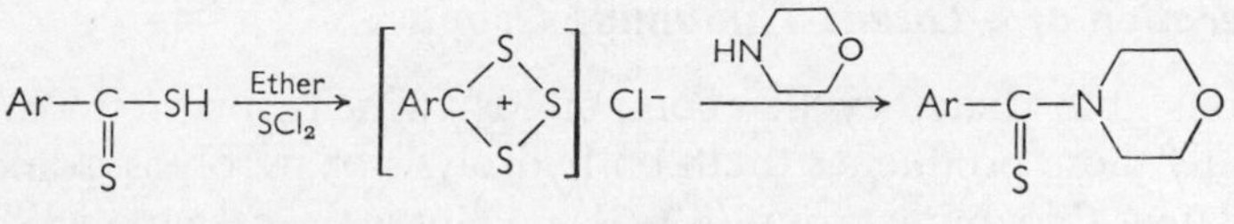

(58)

serve as examples, the first one involving a cyclic ketenedithioketal[177] (cf. section II.B.5), and the second one the capture of the trithietanylium ion shown in equation (58)[178].

3. Hydrolysis

Reaction of 4-benzylidene-2-phenyl-(4-*H*)-oxazole-5-one, with thioacetic acid results in the corresponding thiazolone (**49**) which is cleaved by dilute sodium hydroxide to form the thioamide (reaction 59)[179].

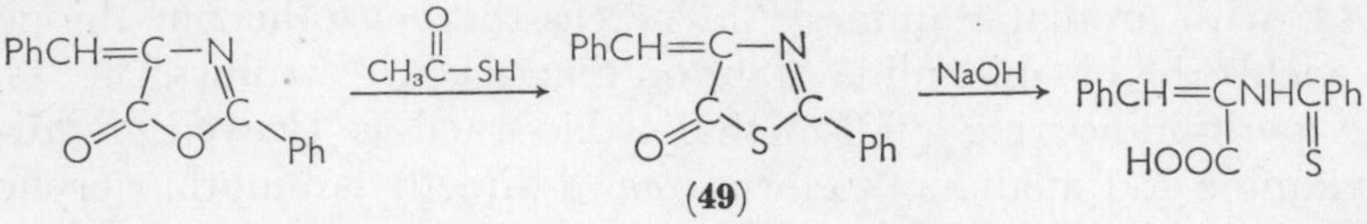

$$\text{(59)}$$

Imidazolidine-4-thiones (**50**) easily prepared from aliphatic ketones, sulphur, and ammonia[180] are split to amino acid thioamides by dilute acids[181] (reaction 60).

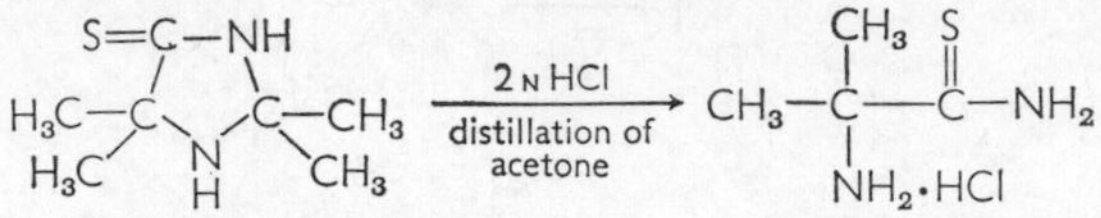

$$\text{(60)}$$

4. Pyrolysis

Malononitrile and geminal dithiols form 1,3-dithiins (**51**) which decompose to thioketones and thioamides when heated above their melting points[182] (reaction 61).

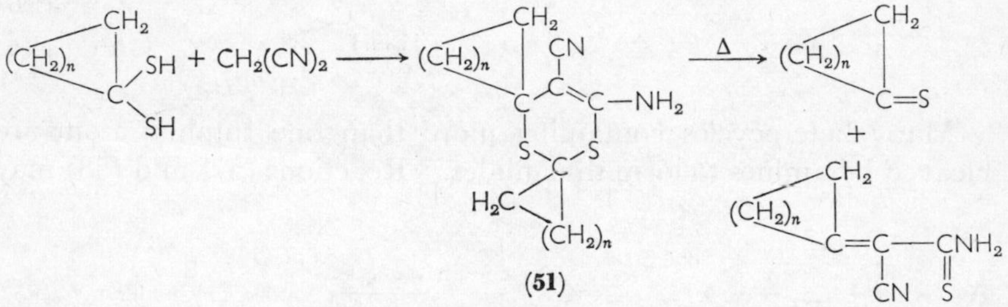

$$\text{(61)}$$

I. Liberation of a Latent Thioamide Group

Among the heterocyclic compounds which form thioamides, especially those producing them on hydrolysis or pyrolysis, some contain a latent thioamide group (**49–51**). This is the case too with some open-chain compounds such as thiohydroxamates (**52**), which slowly decompose forming thioamide and nitrile[183] (reaction 62).

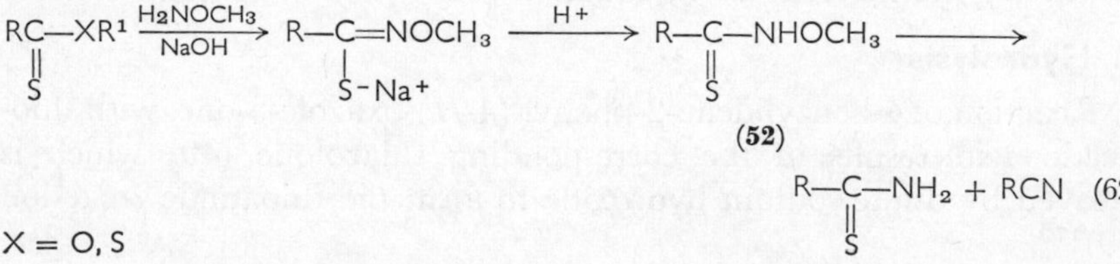

$$R-C(=S)-NH_2 + RCN \quad \text{(62)}$$

$$X = O, S$$

III. ANALYTICAL CHARACTERISTICS OF THIOAMIDES

A. Qualitative Determination

The i.r. spectra of thioamides are not as easily analysed as those of the amides but their B, C, and D bands[66] (section I.C.3) are specific enough for identification. I.r. spectroscopy in the 3000 cm^{-1} range (XH frequencies), or n.m.r. spectroscopy may be useful in special cases for the characterization of thioamides.

A number of chemical methods for identification are available. Gibbs' reagent (2,6-dichlorobenzoquinonechlorimine[184]), though not very specific, yields bluish colours and spots on chromatograms[185]. Grote's reagent (a solution of sodium nitroprussiate, hydroxylamine, and bromine in bicarbonate buffer) reacts with thioamides and other thiones to give a bright blue colour[186]. The products and mechanism of this reaction have not yet been fully explained, but probably iron(III) aquopentammine, formed from the nitroprussiate, plays an important role. Suitable for the location of thioamides on chromatograms is their oxidation by sodium periodate and subsequent treatment with benzidine, which causes colourless spots on a blue background (formation of benzidine blue is inhibited)[187]. A very sensitive proof is the decolouration of iodine–sodium azide in solution and on chromatograms, which is effected by the catalytic activity of thioamides[188,189], and of other bivalent sulphur compounds. The deficient specificity of this valuable method may be overcome by combining it with oxidation of the thioamides to their S-oxides, and formation of the characteristically coloured iron complexes[190,191], which are formed only by thiocarbamoyl derivatives but not by other thiones.

B. Quantitative Determination

Grote has proposed his colour reaction for photometric determination of thiones[186]. However, only chemical and electrochemical methods have been extensively applied for this purpose so far.

Kitamura has oxidized thioamides with hydrogen peroxide in the presence of sodium hydroxide and gravimetrically determined the amount of sulphate[192]. His procedure has been simplified by Wojahn[193], who titrated the sulphuric acid. This may be used for determination of thioamide S-oxides, too[194]. Thioacetamide may be titrated iodometrically, the end-point being detected visually, ampèrometrically, or by back titrating excess of iodine with thiosulphate[195]. Indirect iodometric methods have been applied by

15+c.o.a.

Jacob and Nair[196], and by Sarwar and Thibert[197], who used chloramine T and N-bromosuccinimide as reagents for the quantitative oxidation of thioacetamide. Vanadometric determination of thioacetamide (oxidation by excess vanadate(v)) has been managed by Nair and coworkers[198]. A chelatometric titration has been achieved by Washizuka[199]. He reacted thioacetamide with mercuric ethylenediaminetetraacetate and determined the free complexing agent with zinc.

Several electrochemical procedures are available. Ampèrometric titration of thioacetamide may be conducted by silver(I)[200] or copper(I) ions[201]. Coulometric titration of thioacetamide with electrogenerated silver, combined with potentiometric detection of the end-point, has been recommended by King and Eaton[202]. Finally, Kane has determined polarographically 2-ethylisonicotinic acid thioamide as well as its S-oxide in human and rabbit serum[115a].

IV. PROTOTROPIC AND CHELATING PROPERTIES

A. The Thionamide–Thiolimide Equilibrium

Two tautomeric forms of primary and secondary thioamides, **53a** and **53b**, have been suggested by Laar[203]. Chemical properties, mainly increased acidity as compared with amides (section IV.B), and attack of nucleophiles at the S atom (section V), have led to the opinion that the imidothiol form **53b** is markedly favoured. It is,

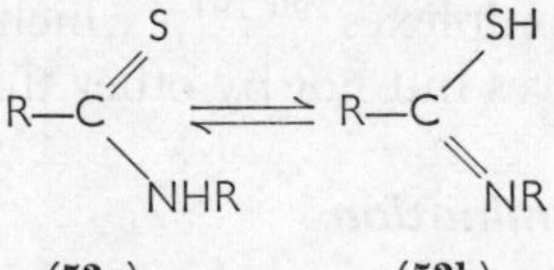

however, impossible to decide between these structures by chemical evidence of this type, since, (a) no information about a tautomeric equilibrium can be obtained from 'static' acidity[204], and (b) thioamide molecules, and especially their anions, are ambivalent systems, and reactions on either position may occur irrespective of the location of the proton, and depending only on the mechanism and conditions for the respective reaction[205]. Therefore only physical methods, such as those mentioned already in section I, can provide a reliable solution of the problem.

X-ray diffraction[15,18] as well as x-ray fluorescence[206] studies show

that the proton is linked to the N atom.   Occurrence of signals due to
S—H protons in n.m.r. spectra of thioamides has never been reported,
the observation of Speziale and Smith[27] obviously being misinter-
preted.   The i.r. spectra of thioamides, even N-acetylthioaceta-
amide[83], likewise exhibit only N—H rather than S—H bands (Tables
4 and 5).

The problem of thiono–thiol tautomerism has been investigated by
i.r. spectroscopy especially by Bacon and coworkers[207]. No exceptions
to the prevalence of the **53a** form in solution have been reported. These
results apparently have been supported by x-ray fluorescence lines of
two N-arylthiocarbamoylpyridines[206]. Walter and coworkers, however,
have recently proved that the mentioned compounds exist exclusively
in the thioamide form in solution as well as in the solid state. The
questionable shifts are due to especially strong hydrogen bonds between the NH_2 group and the nitrogen atom of the pyridine nucleus
and the so-called SH bonds represent overtones or combination
bands[372].

The u.v. spectra of methyl imidothiolates are quite different from
those of the isomeric N-methylthioamides.   The latter have spectra
which resemble closely those of unsubstituted thioamides.   This
observation was the earliest physical evidence for the existence of the
thione form in thioamides[75,209]. Infrared bands in the 2600 cm^{-1} and
3100 cm^{-1} regions and characteristic shifts of other bands have
been observed in solid thionicotine and thioisonicotine amides by
Jensen and Nielsen[66] and Sohár and Nemes[208], which have been
assigned to the S—H or N—H modes of the thiolimide form. Circular
dichroism studies likewise have shown that no portion of the thiolimide
tautomer is present in solutions of thiobenzoylamino acids[373].

At any rate, the proportion of **53b** in the equilibrium is below the
limit of detection even in 1,2-dihydro-2-pyridinethione[210] and similar
compounds, of which it was formerly believed that only by formation of
the thiol form would they provide an aromatic system.   Nevertheless,
well-known textbooks have accepted the dubious conception, and
numerous papers continue to appear that suggest the occurrence of
the thiol tautomer as self-evident in all cases in which any residue is
transferred to the sulphur in the course of a reaction.

The tautomerism phenomena may, however, be complicated by
hydrogen bonds of different types (intramolecular as well as intermolecular, cyclic and open-chain ones) as has been explained above.
Nuclear magnetic resonance[211] and infrared[211a] studies suggest that
the N—H group of the thioamides is a stronger proton donor and the

C=S group a weaker acceptor than the corresponding groups in amides.

Occurrence of S—H groups has been also excluded in β-ketothio-amides (54) in which only N—H and O—H groups are traceable[212].

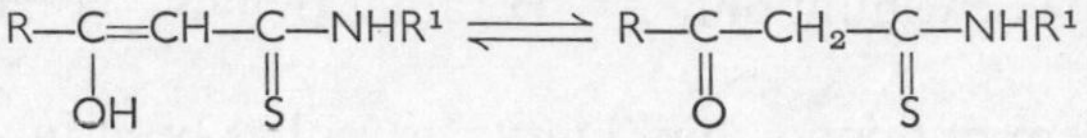

$$R-\underset{\underset{OH}{|}}{C}=CH-\underset{\underset{S}{\|}}{C}-NHR^1 \rightleftharpoons R-\underset{\underset{O}{\|}}{C}-CH_2-\underset{\underset{S}{\|}}{C}-NHR^1$$

$$(54)$$

α-Cyano-α-ethoxycarbonylthioacetanilides (54a), owing to an exceptional C—H acidity, exhibit pronounced S—H bands arising from the predominating en–thiole form 54b[374].

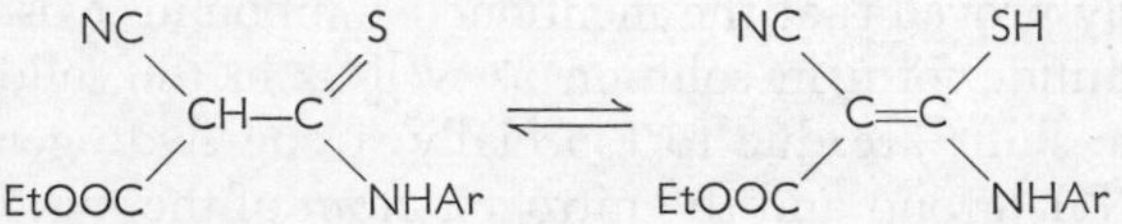

$$\underset{EtOOC}{\overset{NC}{>}}CH-C\underset{NHAr}{\overset{S}{\diagup}} \rightleftharpoons \underset{EtOOC}{\overset{NC}{>}}C=C\underset{NHAr}{\overset{SH}{\diagup}}$$

$$(54a) \qquad\qquad (54b)$$

B. Acidity and Basicity

Thioamides are amphoteric compounds and are converted to anions 55a and cations 55b by bases and acids. The salts of 55a[213]

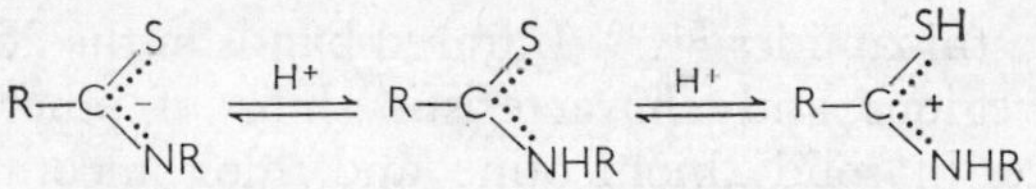

$$R-C\underset{NR}{\overset{S}{\diagdown}}^- \underset{}{\overset{H^+}{\rightleftharpoons}} R-C\underset{NHR}{\overset{S}{\diagdown}} \underset{}{\overset{H^+}{\rightleftharpoons}} R-C\underset{NHR}{\overset{SH}{\diagdown}}^+$$

$$(55a) \qquad\qquad (55) \qquad\qquad (55b)$$

and 55b[185,214] may even be prepared in the pure state. Although the solubility in alkaline solutions to form anions is a well-known property of thioamides pK_a values for the deprotonation have been determined only very recently[215,215a] (cf. Table 16). Thereafter thioamides are markedly stronger acids than amides which is in agreement with the qualitative observations and theoretical considerations about their electronic structure. The pK_a values of thioformanilides, thioacetanilides, and thiobenzanilides (56) conform well to the Hammett equation[215,215a]. From the pronounced influence of B groups ($\rho = 1\cdot74$) in 56 as compared to the influence of A groups ($\rho = 1\cdot16$)

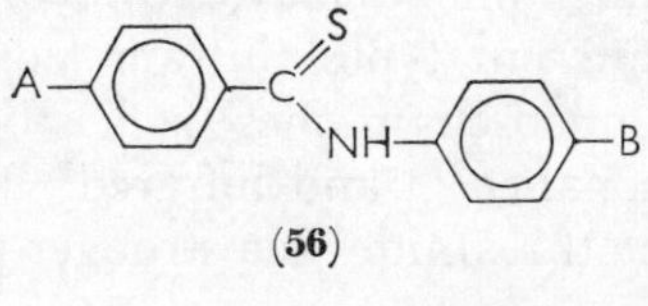

$$A-\langle\bigcirc\rangle-\underset{NH-\langle\bigcirc\rangle-B}{\overset{S}{\diagup}}$$

$$(56)$$

TABLE 16. pK values[a] of thioamides and the corresponding amides.

Compound	RCXNHR ⇌ RCXNR⁻ + H⁺ X = O	RCXNHR ⇌ RCXNR⁻ + H⁺ X = S	RCXNR₂ + H⁺ ⇌ RCXNHR₂[b,c] X = O	RCXNR₂ + H⁺ ⇌ RCXNHR₂[b,c] X = S
(E)-HCXNHCH₂CMe₃		16·2[l]		
(Z)-HCXNHCH₂CMe₃		15·6[l]		
HCXNHPh	13·6	11·44	(−634)	(−522)
DCXNHPh		11·50		(−520)
HCXNHC₆H₄NO₂-p	11·86	10·11		(−600)
HCXNHC₆H₄Me-p	13·7	11·64	(−628)	(−508)
HCXNMeC₄H₉	13·8	12·8	(−523)	(−453)
HCXNMe₂			−0·70[d]	−2·54[e]
MeCXNH₂	13·8	13·4	−0·6[f]	−1·76[g]
		13·4[h]	(−533)	(−480)
MeCXNHPh	13·8	11·56	+0·9[i]	(−545)
			(−588)	
MeCXNMe₂			+0·1[d]	−1·53[e]
EtCXNMc₂				−1·5[e]
Me₃CCXNMe₂				−1·51[e]
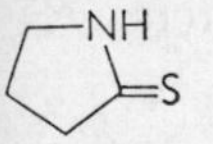 (5-membered ring) —NH / =S			−0·3[j]	−2·0[k]
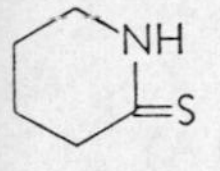 (6-membered ring) NH / =S			+0·6[j]	−1·4[k]
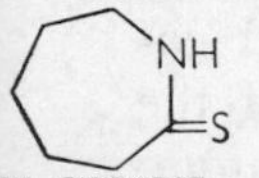 (7-membered ring) NH / =S			+0·3[j]	−1·6[k]
PhCXNH₂	13·8	12·85	−1·74[f]	(−516)
			(−592)	
PhCXNHPh	13·7	10·60	(−645)	(−610)
PhCXNHC₆H₄NO₂-p		9·23		
PhCXNHC₆H₄OMe-p		11·03		(−588)
p-MeOC₆H₄CXNH₂		13·05		(−495)
p-NO₂C₆H₄CXNH₂		11·84		(−591)

[a] Values are taken from references 215 and 215a unless otherwise indicated.

[b] Numbers in brackets denote potentials (mv) of half neutralization (in acetic anhydride–perchloric acid) which cannot be related quantitatively to pK values[215]. Increasing basicity is denoted by decrease of the absolute value of these (negative) potentials.

[c] For site of protonation see Chapter 3.

[d] Reference 216.

[e] Reference 217.

[f] Reference 218.

[g] Reference 219.

[h] Reference 220.

[i] Reference 221.

[j] Reference 222.

[k] Reference 223.

[l] Reference 375.

one can derive further evidence for the fact that proton abstraction takes place at the N atom, where the influence of B should be stronger than that of A[215].

The protonation of thioamides has been a matter of discussion but it has now become quite clear that the cations are thionium (**55b**)[214,217,224] rather than ammonium ions **57**[77,225], the latter, by

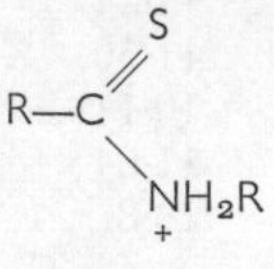

$$R—C\overset{\displaystyle S}{\underset{\displaystyle \overset{NH_2R}{+}}{\big\|}}$$

(**57**)

the way, lacking stabilization by conjugation. The pK_a values (Table 16) show that the thioamides are weaker Brönsted bases in aqueous solutions than amides, which corresponds to their increased acidity. They are, on the other hand, stronger Lewis bases towards the soft acid CH_3CO^+ in the acetic anhydride–perchloric acid system, which is seen from the data of Table 16. The basicity parameters exhibit linear relationship to Hammett constants, too[215].

C. Complex Formation

The formation of adducts of N,N-dimethylthioacetamide (DMTA) with iodine and phenol has been studied by Niedzielski and co-workers[226]. They deduced from the thermodynamic data of the interaction that coordination, like protonation, occurs at the S atom. It is noteworthy that the hydrogen-bonded phenol adduct of DMTA is less stable than the analogous dimethylacetamide (DMA) adduct whereas the charge-transfer complex with iodine is more stable. No difference in stability has been found in the adducts of DMTA and DMA with stannic chloride[227].

Thioamides form complexes of the composition [M·RCSNHR^1], [M·(RCSNHR^1)$_2$], and [M·(RCSNHR^1)$_4$] with various metal ions (M)[11,66]. The derivatives of dithiooxamide which are used in analytical chemistry (section VI.C.2), dithiomalonamides (**58**)[228],

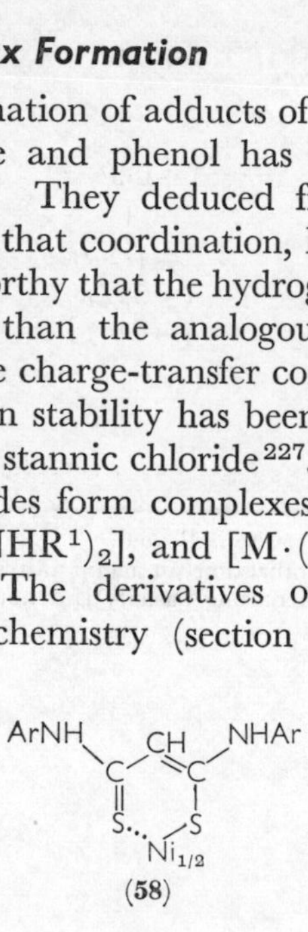

(**58**) 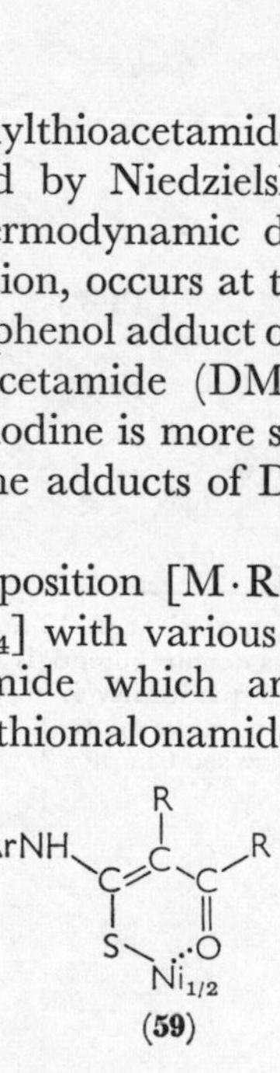   (**59**)

and β-ketothioamides (**59**)[229] are especially stabilized on account of cyclic resonance.

Two copper complexes of thioamides have been investigated by x-ray diffraction. Tetrahedral symmetry and coordination via the S atom have been found in $[Cu(MeCSNH_2)_4]^{16}$; a planar structure has been supported in the thiopicolinic acid anilide chelate (**60**) by e.p.r. spectroscopy[230]. Hexacoordination occurs in $[Ni(DMTA)_6]^{2+}$ $(ClO_4^-)_2$. Very marked differences in the chemical shifts of the pro-

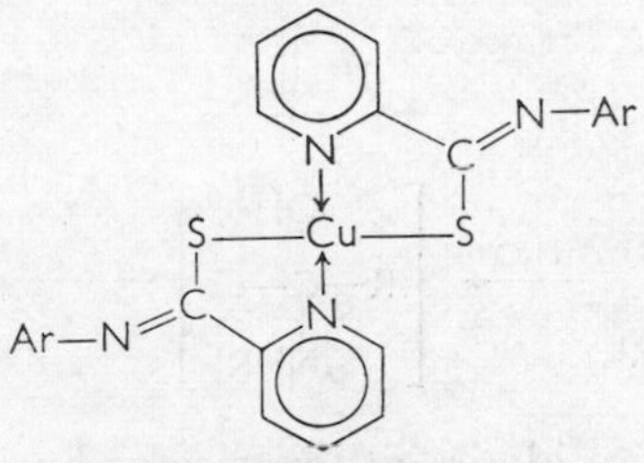

(**60**)

tons of the three methyl groups present in the DMTA ligand have been found in this complex[231]. This fact has been used for the assignation of the n.m.r. signals of the respective methyl groups. Mixed carbonyl or nitrosyl complexes of thioamides with manganese, rhenium and iron have been prepared by Hieber and coworkers[376–378] and by Alper and Edward[379].

V. CHEMICAL REACTIONS

A. Nucleophilic Attack at the Thioamide Group

I. Hydrolysis

Complete hydrolysis of thioamides yields carboxylic acids, hydrogen sulphide, and ammonia or an amine[9,11]. In many cases, however, nitriles, heterocycles, or products of oxidative degradation are formed[11] in a complicated reaction path.

Thioamides may, under certain conditions, especially in alkaline medium, be more difficult to hydrolyse than amides[11]. This does not hold, however, for thioacetamide, the mechanism of its hydrolysis having been thoroughly investigated. In acidic medium (equation 62a) the protonated species is attacked by water and the tetrahedral intermediate is cleaved forming hydrogen sulphide and amide[232] which is further hydrolysed at a much slower rate[232].

In alkaline solution, on the other hand (equation 62b), the prevailing intermediate is the relatively stable thiocarboxylate anion[233] rather than the amide[232]. The free thioacid may even be isolated in special cases[234]. These results do not imply the occurrence of the

$$
\begin{array}{c}
R-\underset{\overset{|}{NHR^1}}{\overset{SH}{C^{...}}} + \xrightarrow{H_2O} \left[R-\underset{\overset{|}{NHR^1}}{\overset{\overset{+}{O}H_2}{C}}-SH \right] \longrightarrow R-C\overset{O}{\underset{NHR^1}{\diagdown}} + H_2S + H^+ \quad (62a)
\end{array}
$$

$$
R-C\overset{S}{\underset{NHR^1}{\diagdown}}
$$

$$
R-\underset{\overset{|}{NR^1}}{\overset{S}{C^{...}-}} \xrightarrow{H_2O} \left[R-\underset{\overset{|}{NHR^1}}{\overset{OH}{C}}-S^- \right] \longrightarrow R-C\overset{O}{\underset{S}{\diagdown}} - + H_2NR^1 \quad (62b)
$$

$$
R-CO_2H
$$

thiol form of the thioamide in either case as has been assumed[219,234]. Ammonia as base probably yields an amidine as intermediate in the hydrolysis[235].

Monothiosuccinimides in water undergo hydrolysis to the corresponding imides. *N*-Alkyl derivatives react more slowly than *N*-aryl compounds, and the reaction rates of the latter show good correlation with Hammett constants. From the value $\rho = +0.348$ one can deduce the rate-determining step of the reaction to be the nucleophilic attack of a water molecule at the thiocarbonyl carbon atom[236].

Hydrolysis of thiolimidic esters, which generally yields thiolesters

$$
R-C\overset{SEt}{\underset{\underset{+}{NR^1_2}}{\diagdown}} \xrightarrow[k_1]{\underset{-H^+}{H_2O}} \left[R-\underset{\overset{|}{OH}}{\overset{SEt}{C}}-NR^1_2 \right] \overset{k_2}{\underset{k_3}{\diagup\diagdown}} \begin{array}{l} R-C\overset{O}{\underset{SEt}{\diagdown}} + NHR^1_2 \\[2ex] R-C\overset{O}{\underset{NR^1_2}{\diagdown}} + EtSH \end{array} \quad (63)
$$

$$
R = H^{237}, Me^{238}
$$

rather than amides (equation 63), takes place much more easily than hydrolysis of thioamides[11]. Recently kinetic evidence has been obtained which supports mechanism (63) where k_1 is rate determining and $k_2 > k_3$. In strong alkali a side-reaction (64) occurs[237].

$$
Me-C\overset{SR}{\underset{NH}{\diagdown}} + OH^- \longrightarrow Me-CN + RS^- + H_2O \quad (64)
$$

2. Substitution by other nucleophiles

Ammonia and amines may react with thioamides in a number of ways[9,11]. Both the substitution of one amino group by another one (cf. equation 28, section II.B.4) to form a modified thioamide, and the replacement of sulphur by an imino group yielding an amidine are well-known reactions, and are used for synthetic purposes.

Amidoximes are formed from thioamides and hydroxylamine according to equation (65)[9,11] which represents a suitable preparative method.

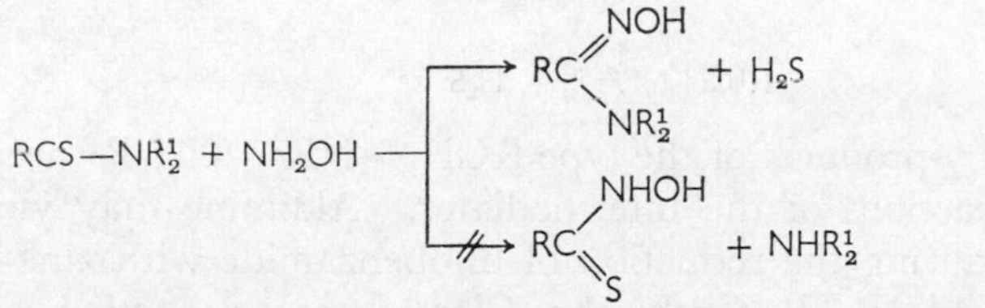

$$RCS-NR_2^1 + NH_2OH \longrightarrow \begin{cases} RC\begin{matrix} NOH \\ NR_2^1 \end{matrix} + H_2S \\ \\ \not\longrightarrow RC\begin{matrix} NHOH \\ S \end{matrix} + NHR_2^1 \end{cases} \tag{65}$$

Amidrazones (hydrazidines) result from thioamides and hydrazines (equation 66). Formation of thiohydroxamic acids or thiohydrazides

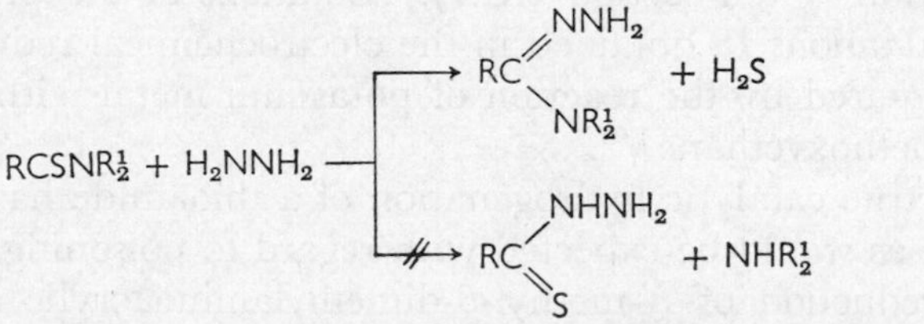

$$RCSNR_2^1 + H_2NNH_2 \longrightarrow \begin{cases} RC\begin{matrix} NNH_2 \\ NR_2^1 \end{matrix} + H_2S \\ \\ \not\longrightarrow RC\begin{matrix} NHNH_2 \\ S \end{matrix} + NHR_2^1 \end{cases} \tag{66}$$

(equations 65 and 66) have not yet been observed; for one exception see Chapter 9, section II.A.4.

N,N-Dimethylbenzamide dialkylmercaptoles (**61**) are obtained from sodium ethanethiolate and the imidothiolic ester salt **62**[238a,b].

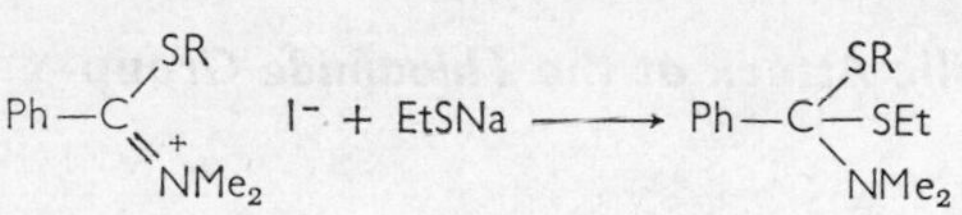

$$\text{Ph}-C\begin{matrix} SR \\ \overset{+}{N}Me_2 \end{matrix} \ I^- + \text{EtSNa} \longrightarrow \text{Ph}-C\begin{matrix} SR \\ SEt \\ NMe_2 \end{matrix}$$

(**62**, R = Me, Et)   (**61**)

[35]S-Labelled thioacetamide of 96% specific activity may be prepared by isotope exchange with elemental sulphur[380].

3. Reduction

Reduction of thioamides (equation 67) is generally easier than reduction of amides[9,11]. It may be achieved with several reagents:

(a) zinc or iron in acidic solution, (b) sodium, or better aluminium amalgam, (c) lithium hydride or lithium alanate[12], (d) Raney nickel, and (e) electrolytically. The reduction probably proceeds via α-aminothiols and aldimines yielding amines as products. In many

$$R^1-C\overset{S}{\underset{NR^2R^3}{}} \xrightarrow{[2H]} R^1-\underset{\underset{H}{|}}{\overset{\overset{SH}{|}}{C}}-NR^2R^3 \xrightarrow[-H_2S]{(R^3=H)} R^1-CH=NR^2 \xrightarrow{[2H]} R^1CH_2NHR^2$$

$$\downarrow [2H] \qquad\qquad\qquad\qquad\qquad (67)$$

$$R^1CH_2NR^2R^3 + H_2S$$

cases by-products of the type $RCH_2-NR^1-CH_2R^2$ are generated in side-reactions of the intermediates. Aldimine may yield benzaldehyde during the reduction of thiobenzamide with zinc in potassium hydroxide[11]. Recently the Clemmensen reduction of thioamides has been investigated. The ease with which the reaction takes place has been correlated to the half-wave potentials of the polarographic reduction[114] (cf. section I.C.7). Solutions of the purple thioamide radical anions 18 obtained in the electrochemical reduction may also be prepared by the reaction of potassium metal with the thioamide in dimethoxyethane[118].

No true catalytic hydrogenation of a thioamide has been reported so far, as would be expected with regard to poisoning of the catalyst. The reduction of N-methyl-β-dimethylaminoacrylic acid thioamide carried out by Watanabe and coworkers[176] needs large amounts of Raney nickel but no hydrogen, and thus cannot be called catalytic[239]. Remarkably, tin in hydrochloric acid does not reduce the thioamide group (cf. however, section V.D).

B. Electrophilic Attack at the Thioamide Group

I. Alkylation

Reaction of thioamides with alkyl halides or sulphates generally occurs at the S atom yielding imidothiolic esters[9,11,12,239,240]. This holds for tertiary thioamides, too, which form salts of N,N-disubstituted imidothiolic ester cations[11,150]. S-Alkylation of thioamides is the most suitable method for the preparation of imidothiolic esters, besides the reaction of thiols with imidic ester chlorides[241,242], and the synthesis from alkyl thiocyanates and polyphenols[243].

There are, on the other hand, a number of examples of N-alkylation

TABLE 17. Alkylation at the N atom of thioamides.

Substrate	Reagent[a]	Product	Refs.
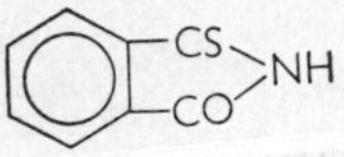 [structure: benzene ring with CS–NH–CO bridge]	R^1X^b	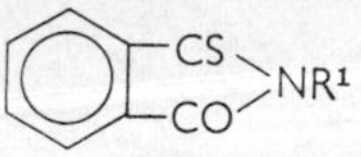 [structure: benzene ring with CS–NR^1–CO bridge]	244
$PhCSNH_2$	An_2CHCl^c	$PhCSNHCHAn_2$	245
$RCSNH_2{}^d$	Ph_3CCl	$RCSNHCPh_3$	246
$ArCSNH_2$	$XanOH^e$	$ArCSNHXan$	245, 247
$H_2NCSCSNH_2$	Ph_3CCl	$Ph_3CNHCSCSNHCPh_3$	248
$H_2NCSCSNH_2$	$XanOH^e$	$XanNHCSCSNHXan$	248
$HCSNHPh$	$XanOH^e$	$HCSN(Ph)Xan$	245
$MeCSNH_2$	$XanOH^e$	$MeCSNHXan$	245
$MeCSNH_2$	An_2CHOH^c	$MeCSNHCHAn_2$	245
$MeCSNH_2$	Ms_2CHOH^f	$MeCSNHCHMs_2$	245
$PhCSNH_2$	An_2CHOH^c	$PhCSNHCHAn_2$	245
$PhCSNH_2$	$AnCH(Ph)OH^c$	$PhCSNHCH(Ph)An$	245
$PhCSNH_2$	Ph_2CHOH	$PhCSNHCHPh_2{}^g$	245
$PhCSNH_2$	Ms_2CHOH^f	$PhCSNHCHMs_2$	245
$PhCSNHMe$	$XanOH^e$	$PhCSN(Me)Xan$	245
$PhCSNHPh$	$XanOH^e$	$PhCSN(Ph)Xan$	245

[a] The first two reactions are carried out in alkaline, all others in acidic medium.
[b] R^1 = alkyl or aralkyl.
[c] An = p-methoxyphenyl (anisyl).
[d] R = Me, CH_2Ph, aryl, but not H[249].

[e] Xan = xanthydryl: [structure of xanthydryl]

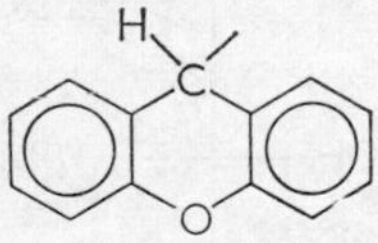

[f] Ms = 2,4,6-trimethylphenyl (mesityl).
[g] Only traces of this product were detected by thin-layer chromatography.

compiled in Table 17. In most cases the unusual N-alkylation obviously occurs under conditions that favour formation of carbonium ions, or as in the first item of Table 17, at thioamides with electronegative substituents. The reaction might well proceed via a kinetically controlled S-alkylation followed by rearrangement under suitable conditions. This mechanism (equation 68) has been supported recently by Walter and Krohn[245] who found that the N-aralkylthioamides (63b) are readily obtained by treating imidothiolic ester hydrochlorides 63a with dilute acid, or from the surprisingly stable free imidothiolic esters (64), simply by heating in inert solvents. Cross-reactions have proved this rearrangement to be intermolecular.

$$
\begin{array}{c}
\text{SCHAr}_2 \\
\\
\text{R—C} \quad\quad \text{Cl}^- \\
\overset{+}{\text{NH}_2} \\
\textbf{(63a)}
\end{array}
$$

Ar$_2$CHCl

K$_2$CO$_3$

$$
\text{R—CS—NH}_2 \quad\quad \text{H}_2\text{O/H}^+ \;\|\; \text{HCl/Ether}
$$

$$
\begin{array}{c}
\text{SCHAr}_2 \\
\\
\text{R—C} \\
\text{NH} \\
\textbf{(64)}
\end{array} \tag{68}
$$

Ar$_2$CHOH/H$^+$

$$
\begin{array}{c}
\text{S} \\
\\
\text{R—C} \\
\text{NHCHAr}_2 \\
\textbf{(63b)}
\end{array}
$$

Heat **(64)**

The hydrochloride **63a** on the other hand, may be reprecipitated by treating **63b** with hydrogen chloride in ether. Reaction (68) may provide a suitable tool for the preparation of *N*-alkylated thioamides of the type **63b**.

Reaction of thioamides, especially dithiooxamides with aldehydes, alone or together with amines, yields aminals[9,11]. *N*-(Hydroxymethyl)thioamides (**65**) are formed from thioamides and formaldehyde (equation 69)[250]. **65** is converted to *N*-aminomethyl (**66**)[250,251], *N*-alkoxymethyl (**67**)[251], or *N*-chloromethyl derivatives (**68**)[251] by amines, alcohols, or thionyl chloride, respectively (equation 69). Attack of aldehydes on the S atom has not yet been observed.

$$
\begin{array}{l}
\quad\quad\quad\quad\quad\quad \overset{\text{HNR}_2^1}{\longrightarrow} \text{RCSNH—CH}_2\text{NR}_2^1 \\
\quad\quad\quad\quad\quad\quad\quad\quad\quad\quad \textbf{(66)} \\
\text{RCSNH}_2 + \text{CH}_2\text{O} \longrightarrow \text{RCSNH—CH}_2\text{OH} \overset{\text{R}^1\text{OH}}{\longrightarrow} \text{RCSNH—CH}_2\text{OR}^1 \\
\quad\quad\quad\quad\quad\quad\quad\quad\quad \textbf{(65)} \quad\quad\quad\quad\quad\quad\quad\quad \textbf{(67)} \\
\quad\quad\quad\quad\quad\quad \overset{}{\underset{\text{SOCl}_2}{\longrightarrow}} \text{RCSNH—CH}_2\text{Cl} \\
\quad\quad\quad\quad\quad\quad\quad\quad\quad\quad \textbf{(68)}
\end{array} \tag{69}
$$

Finally, methylation by means of diazomethane shall be mentioned. *N*-Arylthioamides of aromatic carboxylic acids are methylated at the S atom (equation 70)[252]. Thioamides of minor 'dynamic' acidity

$$
\text{ArCSNHAr}^1 + \text{CH}_2\text{N}_2 \longrightarrow \begin{array}{c} \text{SMe} \\ \text{ArC} \\ \text{NAr}^1 \end{array} \tag{70}
$$

as thiobenzamide or its *N*-alkyl derivatives do not react with diazomethane at all[194] whereas their *S*-oxides which obviously exhibit increased dynamic acidity are attacked at the N atom (equation 71)[194].

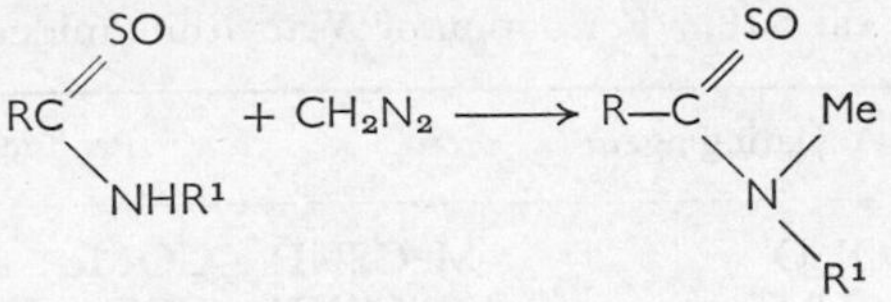

$$RC \overset{SO}{\underset{NHR^1}{\diagdown}} + CH_2N_2 \longrightarrow R-C \overset{SO}{\underset{N}{\diagdown}} Me \qquad (71)$$

$$\underset{R^1}{}$$

Arylation of cyanothioformanilide by means of aryl diazonium salts occurs at the S atom (equation 71a)[381].

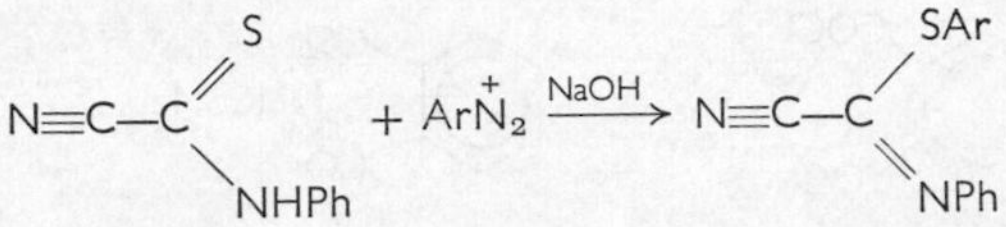

$$N{\equiv}C-C \overset{S}{\underset{NHPh}{\diagup}} + Ar\overset{+}{N_2} \xrightarrow{NaOH} N{\equiv}C-C \overset{SAr}{\underset{NPh}{\diagup}} \qquad (71a)$$

2. Acylation

a. Carboxylic and carbonic acid derivatives. Early reports about the acylation of thioamides have not provided a uniform understanding of this important reaction[9,11,12,213,253].

Bredereck and coworkers have observed the formation of *S*-benzoyl-formimidothiolic ester hydrochloride (**69**) from thioformamide

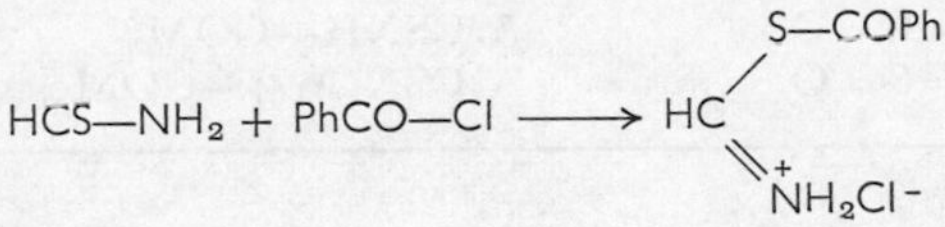

$$HCS-NH_2 + PhCO-Cl \longrightarrow HC \overset{S-COPh}{\underset{\overset{+}{N}H_2Cl^-}{\diagdown}} \qquad (72)$$

$$(69)$$

(equation 72)[249] which has remained the only unequivocal example of an *S*-acylation in the thioamide series. All other recent investigations, especially those of Goerdeler and coworkers, have established the exclusive formation of *N*-acylthioamides by the various acylating agents shown in Table 18. The mechanism of the reaction has not been studied in detail but it seems likely that primarily attack upon the S atom occurs which is followed by rearrangement.

Acylation of thioamides may be achieved by *N*-functional derivatives of carboxylic acids, too. Reaction of imidoyl chlorides with primary thioamides yields secondary thioamides by transfer of sulphur (equation 73, see also equation 8 in section II.A.2). The alkali salts of secondary thioamides form *N*-thioacylamidines (**70**, equation 74)[253]. **70** (Ar^1 = Ar^3 = Ph; Ar^2 = Ar^4 = α-naphthyl) is rearranged to the isomeric diimidoyl sulphide (**71**) on heating[253]. Diimidoyl sulphides may be also obtained by reaction of amidines on thioamides[12] (equation 76). *N*-Thioacylamidines of type **72** are the products of

TABLE 18. Formation of *N*-acylthioamides.

Substrate	Acylating agent	Product	Refs.
MeCSNH$_2$	(MeCO)$_2$O	MeCSNH—COMe	83, 254
MeCSNH$_2$	RCO—Cl	MeCSNH—COR	255
MeCSNH$_2$	Cl—CO(CH$_2$)$_n$CO—Cl^a	MeCSNHCO(CH$_2$)$_n$CONH—COMe	255
MeCSNH$_2$	ArCO—Cl	MeCSNH—COAr	255
MeCSNH$_2$	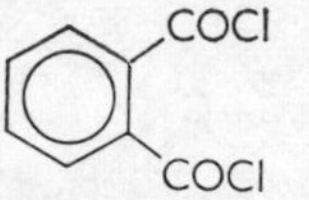	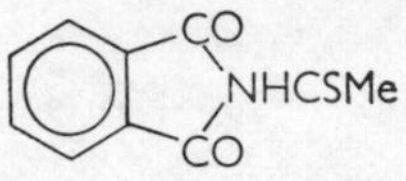	255
MeCSNH$_2$	MeC(SO)NH$_2$	MeCSNH—COMe	254
ArCSNH$_2$	RCO—Cl	ArCSNH—COR	256, 257
ArCSNH$_2$	Cl—CO(CH$_2$)$_n$CO—Cl^a	ArCSNHCO(CH$_2$)$_n$CONH—CSAr	152
ArCSNH$_2$	Cl—CO(CH$_2$)$_n$CO—Cl^b		152, 257
PhCSNHCH$_2$Ph	PhCO—Cl	PhCSN(CH$_2$Ph)—COPh	258
ArCSNHAr^1	MeCO—Cl	ArCSN(Ar^1)—COMe	259
ArCSNH$_2$	CH$_2$=C=O	ArCSNH—COMe	259
ArCSNHAr^1	CH$_2$=C=O	ArCSN(Ar^1)—COMe	259

^{a} n = 3,4.
^{b} n = 2,3.

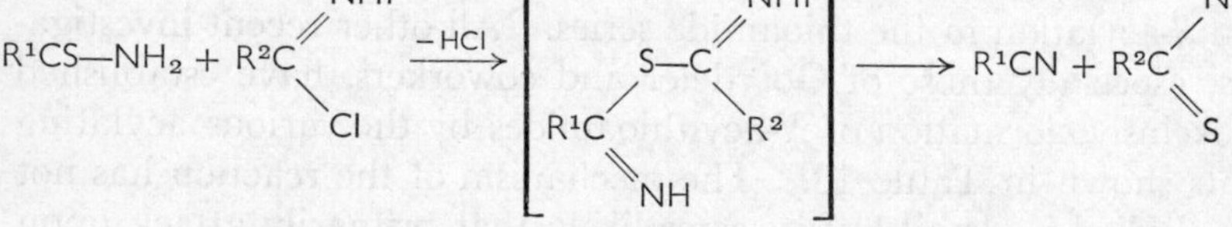

$$ \text{(73)} $$

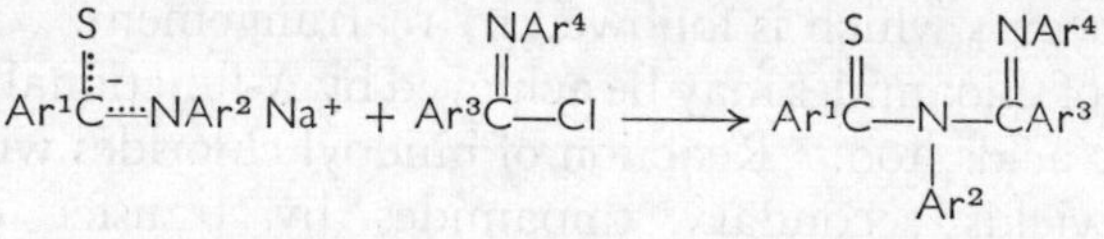

$$ \text{(74)} $$

$$ \text{(70)} $$

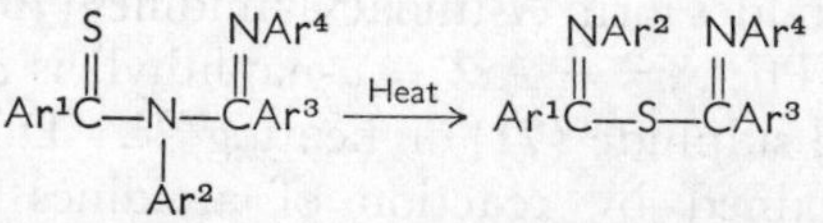

$$ \text{(75)} $$

$$ \text{(70)} \qquad\qquad \text{(71)} $$

$$RCS—NH_2 + R^1C \overset{NH}{\underset{NH_2}{\Big\backslash}} \longrightarrow R—\overset{NH}{\underset{\|}{C}}—S—\overset{NH}{\underset{\|}{C}}—R^1 \qquad (76)$$

the acid-catalysed reaction between thioamides and nitriles (equation 77) which has been extensively investigated by Ishikawa (see reference 11), and recently by Goerdeler and Porrmann[260]. Reaction (77)

$$RCS—NH_2 + R^1CN \xrightarrow{HCl} \overset{S}{\underset{\|}{R}C}—N=\overset{NH_2}{\underset{|}{C}}R^1 \xrightarrow{H_2O} \overset{S}{\underset{\|}{R}C}—NH—\overset{O}{\underset{\|}{C}}R^1 \qquad (77)$$

$$(72)$$

probably proceeds via intermediate imidoyl chlorides (cf. equation 73) and diimidoyl sulphides of type 71[260]. The yields depend on the nature of R and R^1. Compounds 72 may be partially hydrolysed (equation 77) yielding N-acylthioamides that are not always obtainable by other means[261].

Alkyl cyanates $ROCN$[262] and cyanamide H_2NCN[11] abstract hydrogen sulphide from thioamides forming thione carbamates or thiourea.

Chlorothioformates (73) yield N-thioacylthionecarbamates (74) when they react with the anions of thioamides[263], whereas chloroformamidines (75) form S-imidoylthioureas (76) which undergo rearrangement to the N-imidoylthioureas (77a), or thioacylguanidines (77b) (equation 79)[264].

Phenyl isocyanate yields degradation products only in reactions with thioamides but N-thiobenzoyl-N'-benzoylurea (78) is readily formed from thiobenzamide and benzoyl isocyanate[265].

b. Sulphenyl chlorides. Benzenesulphinyl chloride[266] and benzenesulphonyl chloride, as well as S_2Cl_2, $SOCl_2$, and SO_2Cl_2 yield imidoylthioamides of the types 70–72, heterocycles, and oxidation products rather than compounds that contain the residue of the attacking reagent[9,11]. N-(Arylsulphonyl)thiobenzamides may, however, be obtained indirectly by thiolysis of N-(arylsulphonyl)imido chlorides according to equation (78)[266a]. Sulphenyl chlorides, however,

$$Ph—C \overset{N—SO_2Ar}{\underset{Cl}{\Big\backslash}} \xrightarrow{Na_2S} Ph—C \overset{NH—SO_2Ar}{\underset{S}{\Big\backslash}} \qquad (78)$$

with secondary thioamides readily form iminomethane disulphides (79)[267,268], rather than N-thioacylsulphenamides (80). This has

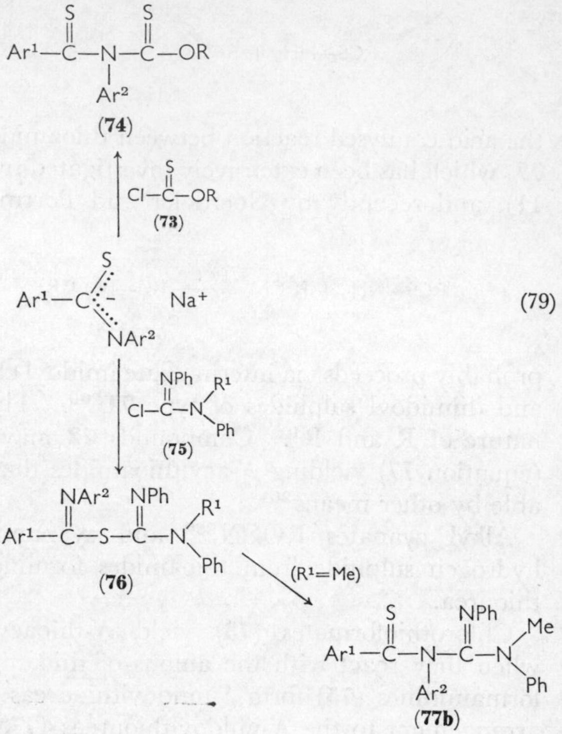

$$\underset{\underset{Ar^2}{\mid}}{Ar^1}-\overset{\overset{S}{\parallel}}{C}-N-\overset{\overset{S}{\parallel}}{C}-OR$$

$$(74)$$

$$\uparrow$$

$$Cl-\overset{\overset{S}{\parallel}}{C}-OR$$

$$(73)$$

$$Ar^1-\overset{\overset{S}{\parallel}}{\underset{NAr^2}{C}}\cdots\quad Na^+ \qquad\qquad\qquad (79)$$

$$\downarrow \quad Cl-\overset{\overset{NPh}{\parallel}}{C}-N\overset{R^1}{\underset{Ph}{}}$$

$$(75)$$

$$Ar^1-\overset{\overset{NAr^2}{\parallel}}{C}-S-\overset{\overset{NPh}{\parallel}}{C}-N\overset{R^1}{\underset{Ph}{}}$$

$(R^1=Et)$ ↙ (76) ↘ $(R^1=Me)$

$$\underset{\underset{Ph}{\mid}}{Ar^1}-\overset{\overset{NAr^2}{\parallel}}{C}-N-\overset{\overset{S}{\parallel}}{C}-N\overset{Et}{\underset{Ph}{}}$$

$$(77a)$$

$$Ar^1-\overset{\overset{S}{\parallel}}{C}-\underset{\underset{Ar^2}{\mid}}{N}-\overset{\overset{NPh}{\parallel}}{C}-N\overset{Me}{\underset{Ph}{}}$$

$$(77b)$$

$$Ph-\overset{\overset{S}{\parallel}}{C}-NH_2 + Ph-\overset{\overset{O}{\parallel}}{C}-N=C=O \longrightarrow Ph-\overset{\overset{S}{\parallel}}{C}-NH-\overset{\overset{O}{\parallel}}{C}-NH-\overset{\overset{O}{\parallel}}{C}-Ph$$

$$(78)$$

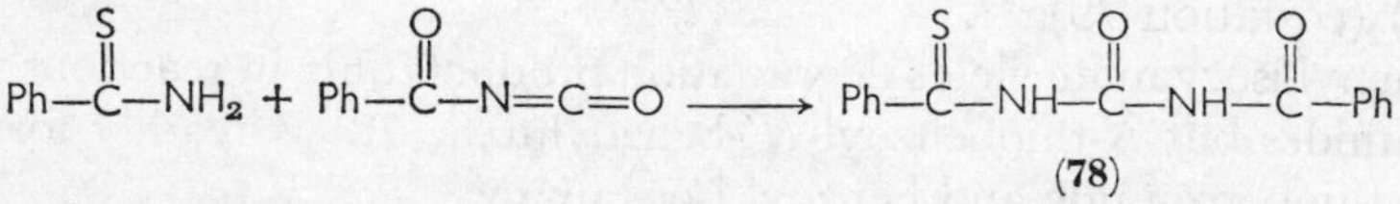

been proved by chemical and physical methods (i.r. spectroscopy of ^{15}N-labelled PhC(NPh)SSPh[267]). From unsubstituted thiobenz-amide only the hydrochloride of **79** ($R^1 = R^3 = Ph$; $R^2 = H$) may be obtained, the free base being unstable[268].

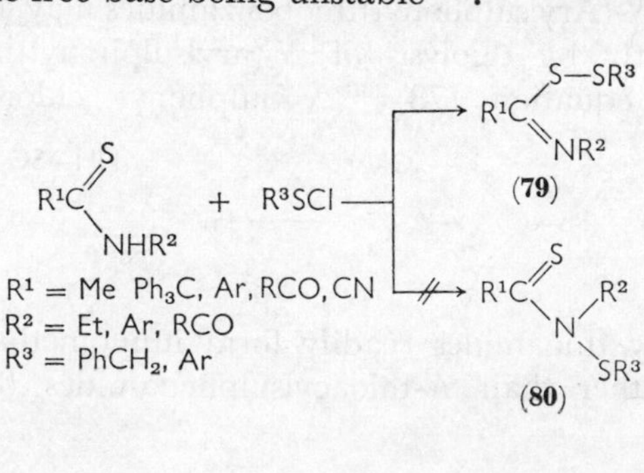

$$\overset{\overset{S}{\parallel}}{R^1C}\underset{NHR^2}{} + R^3SCl \longrightarrow$$

→ $R^1C\overset{S-SR^3}{\underset{NR^2}{\parallel}}$ (79)

↛ $R^1C\overset{S}{\underset{N}{\parallel}}\overset{R^2}{\underset{SR^3}{}}$ (80)

$R^1 = $ Me, Ph_3C, Ar, RCO, CN
$R^2 = $ Et, Ar, RCO
$R^3 = PhCH_2$, Ar

3. Oxidation

One of the most striking differences between the chemical properties of thioamides and amides is their behaviour on oxidation. Amides are hardly oxidized by mild oxidants. If forced, the oxidation reaction occurs at the carbon atoms of the side-chains. Thioamides, on the other hand, are readily attacked at the sulphur atom yielding a large variety of products quite characteristic in some cases.

Oxidation by means of ozone, iodine, hydrogen peroxide, sulphuric acid, nitrous and nitric acid derivatives, selenium dioxide, potassium permanganate, potassium hexacyanoferrate(III), mercuric oxide, N,N-dichlorocarbamates, and epoxides to form sulphur-free products and heterocycles has been reviewed formerly[9,11,12]. Recently the oxidative desulphuration of thioacetamide by alkaline hexacyanoferrate(III) has been studied kinetically[269]. The deprotonation of the substrate has been found to be rate determining, the electron transfer to the oxidant and the degradation to form acetamide being fast steps.

The preparation of diimidoyl disulphides (**81**) may be achieved by selective oxidation of thioamides. Hydrogen peroxide[270], hexacyanoferrate(III)[271], lead tetraacetate[272], phosgene oxime[270], N-chlorosuccinimide[270], dibenzoyl peroxide[273], and 2-bromo-2-nitropropane-1,3-diol[274] have been used as oxidants. The most favourable method seems to be oxidation by iodine[275]. Upon heating, the disulphides **81** are easily converted to the sulphides **82** with elimination of sulphur[275].

$$2R^1CS\!-\!NHR^2 \xrightarrow{\ [O]\ } \underset{\substack{\| \\ NR^2}}{R^1C}\!-\!S\!-\!S\!-\!\underset{\substack{\| \\ NR^2}}{CR^1} \xrightarrow{\ Heat\ } \underset{\substack{\| \\ NR^2}}{R^1C}\!-\!S\!-\!\underset{\substack{\| \\ NR^2}}{CR^1} + S$$

$$(81) \qquad\qquad\qquad (82)$$

The formation of thioamide S-oxides (the name 'sulphoxide' which has been occasionally used should be reserved for non-cumulative structures of the type R—SO—R) has been first observed by Kitamura[276] who oxidized thioamides by means of hydrogen peroxide (equation 80) and obtained substances that contained one oxygen atom more than the starting material. He assigned the structure **83b** to these compounds, and, consequently, called them thioperimidic acids. The systematic investigations of Walter and co-workers[58,71,189-191,194,277], however, have proved the oxidation products to be real S-oxides (**83a**). The isomeric structure **84** (thiohydroxamic acid) has been excluded too. Besides x-ray[22] and i.r.

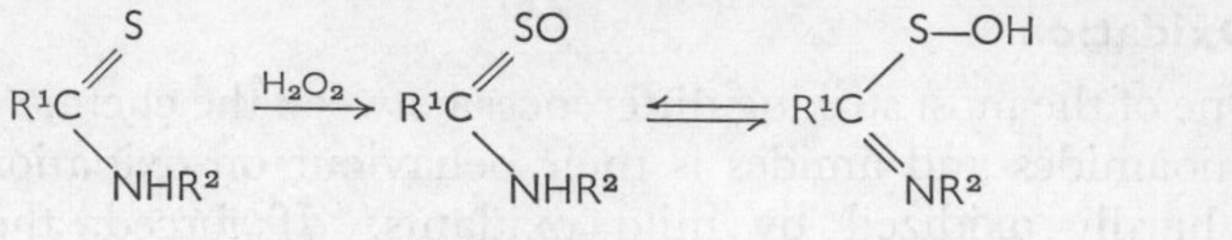

$$\underset{NHR^2}{\overset{S}{R^1C}} \xrightarrow{H_2O_2} \underset{\underset{\textbf{(83a)}}{NHR^2}}{\overset{SO}{R^1C}} \rightleftharpoons \underset{\underset{\textbf{(83b)}}{NR^2}}{\overset{S-OH}{R^1C}} \tag{80}$$

spectroscopic[277] results the structure **83** can be deduced from chemical evidence shown in equations $(81)^{191,278}$, $(82)^{191}$, and $(83)^{191,194}$.

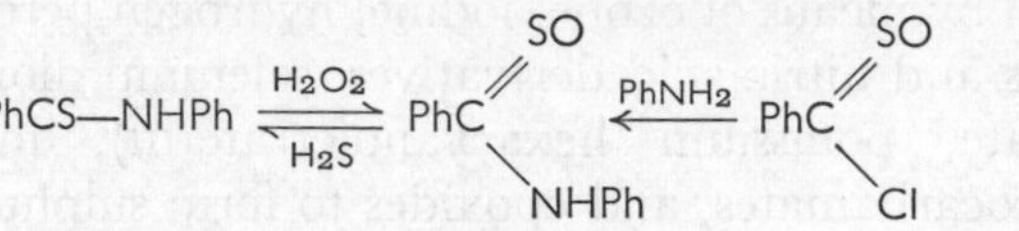

$$PhCS-NHPh \underset{H_2S}{\overset{H_2O_2}{\rightleftharpoons}} \underset{\underset{m.p.\ 149°}{NHPh}}{\overset{SO}{PhC}} \xleftarrow{PhNH_2} \underset{Cl}{\overset{SO}{PhC}} \tag{81}$$

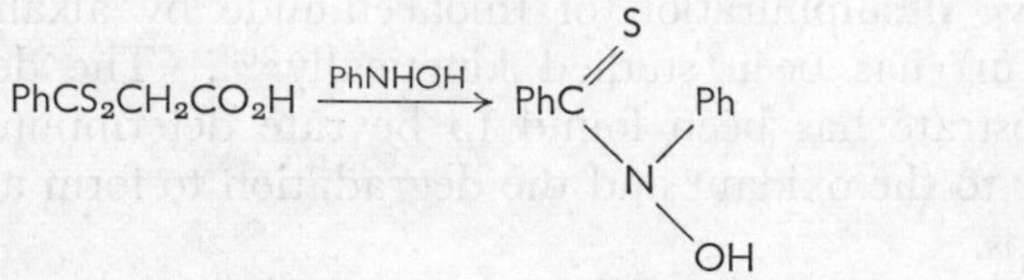

$$PhCS_2CH_2CO_2H \xrightarrow{PhNHOH} \underset{\underset{\underset{\underset{m.p.\ 101°}{OH}}{N}}{Ph}}{\overset{S}{PhC}} \tag{82}$$

(84)

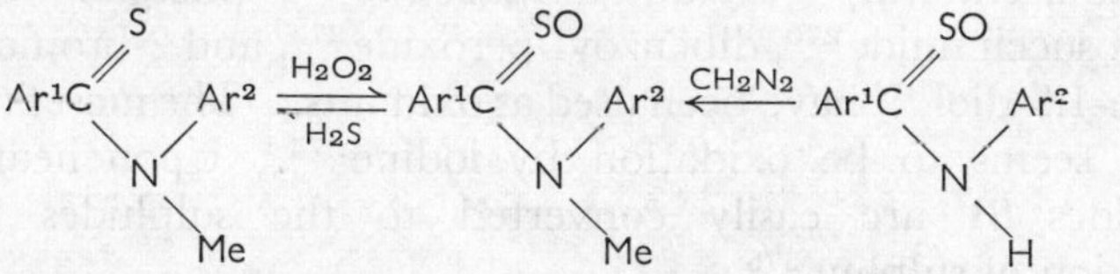

$$\underset{\underset{Me}{N}}{\overset{S}{Ar^1C}} Ar^2 \underset{H_2S}{\overset{H_2O_2}{\rightleftharpoons}} \underset{\underset{Me}{N}}{\overset{SO}{Ar^1C}} Ar^2 \xleftarrow{CH_2N_2} \underset{\underset{H}{N}}{\overset{SO}{Ar^1C}} Ar^2 \tag{83}$$

Although numerous thioamide *S*-oxides—primary[189,190,279], secondary[58,191,194,279,382], tertiary[191,194], aliphatic, aromatic, and heterocyclic[190,280-282,371] as well as bifunctional ones[283,284]—are known, only a few specific chemical reactions of this class of compounds have been detected so far. Acylating properties and the *N*-methylation with diazomethane[285] have been already mentioned (Tables 17, 18). It is noteworthy that diazomethane attacks 2-*t*-butylquinazolinethione 4-*S*-oxide (**85**) on its O atom to form the sulphenic ester **86**[285]. The interesting compound **87**, an amide of the iminosulphenic acid **83b** has been recently obtained according to equation $(84)^{270}$.

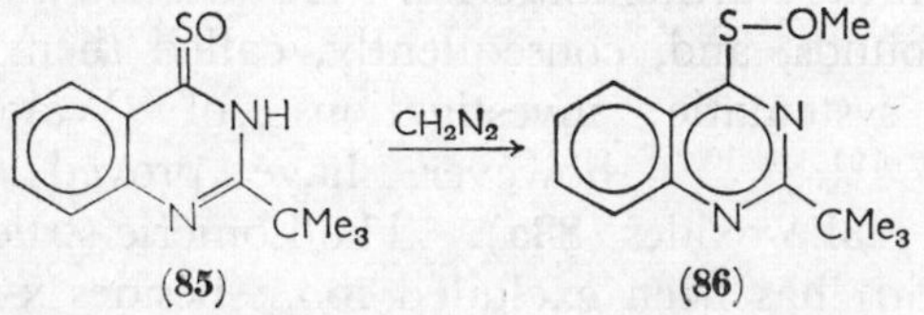

(85)   (86)

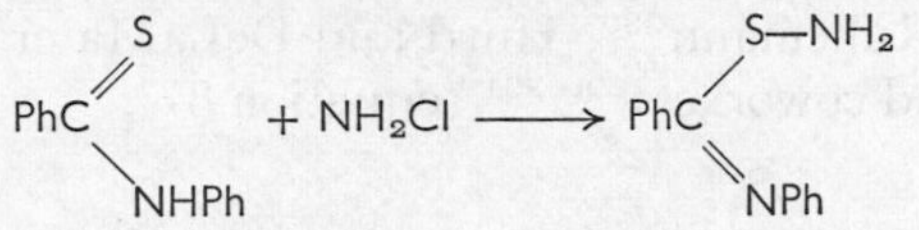

$$PhC\overset{\displaystyle S}{\underset{\displaystyle NHPh}{\diagdown}} + NH_2Cl \longrightarrow PhC\overset{\displaystyle S-NH_2}{\underset{\displaystyle NPh}{\diagup}} \qquad (84)$$

$$(87)$$

No formation of *S*-dioxides as in the case of thioureas has been observed as yet in the thioamide series[286].

Demethylation of *N*,*N*-dimethylthiopicolinamide (**88**) and *N*,*N*-dimethylpyrazinethiocarboxamide (**89**) occurs by oxidation with sulphur, the corresponding *N*-monomethyl derivative being formed

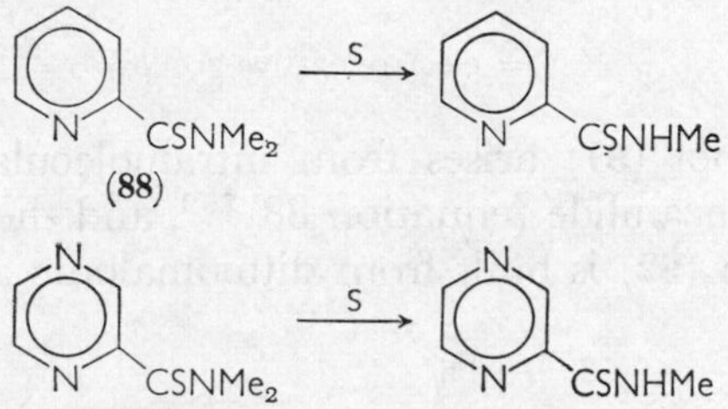

$$ \overset{}{\underset{N\,\,\,CSNMe_2}{}} \xrightarrow{\,\,S\,\,} \overset{}{\underset{N\,\,\,CSNHMe}{}} \qquad (85a)$$

$$(88)$$

$$ \overset{N}{\underset{N\,\,\,CSNMe_2}{}} \xrightarrow{\,\,S\,\,} \overset{N}{\underset{N\,\,\,CSNHMe}{}} \qquad (85b)$$

$$(89)$$

(equations 85a and 85b)[287]. High-temperature chlorination of *N*-methylthiobenzamide yields the dichloride of the corresponding iso-cyanate according to equation (85c)[287a].

$$PhCSNHMe \xrightarrow{Cl_2} PhCCl_2N{=}CCl_2 \qquad (85c)$$

C. Formation of Heterocycles

Intra- and intermolecular cyclizations can be actually considered to be the most versatile and important reactions of thioamides. The majority of these reactions might have been classified in the preceding sections as they exhibit at least one nucleophilic or electrophilic step, such as solvolysis, alkylation, acylation, or oxidation. Since the mechanisms of these complicated reactions are, however, scarcely known and possibly may include concerted attacks it seems expedient to discuss cyclizations in a separate section, subdivided according to the type of ring produced rather than the type of cyclization reaction.

I. Five-membered rings

a. Heterocycles containing only sulphur in the ring. 2-Aminothiophenes (**90**) are obtained from phenacyl bromides and tertiary thioamides[288] (equation 86). Formation of tetrahydrothiophenes has been mentioned by Ruhemann[289], Hurd and DeLaMater[11], and recently by Barnikow and coworkers[290,291] (equation 87).

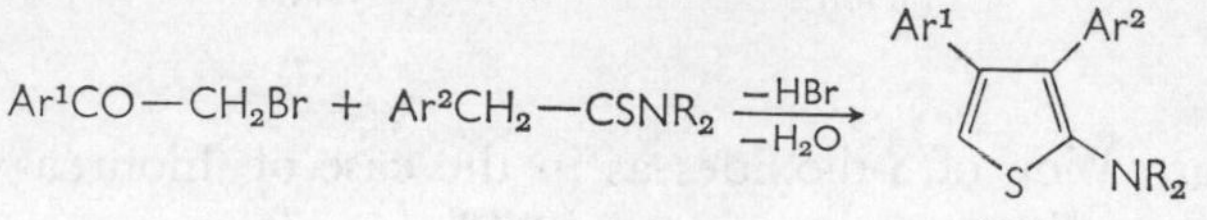

$$Ar^1CO-CH_2Br + Ar^2CH_2-CSNR_2 \xrightarrow[-H_2O]{-HBr}$$

(86)

(90)

$$\underset{\underset{Y}{|}}{\overset{\underset{X}{|}}{HCCSNHAr}} \xrightarrow[ClCH_2COOEt]{EtO^-}$$

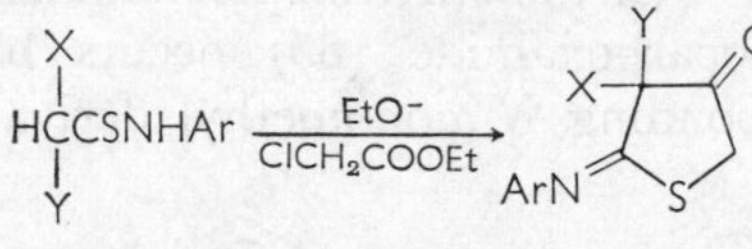

(87)

X Y = electronegative groups

The benzodithiol (**91**) arises from intramolecular cyclization of o-mercaptothiobenzanilide (equation 88)[292], and the pseudoaromatic dithiolium system (**92**) is built from dithiomalonic amides (equation 89)[283,293,294].

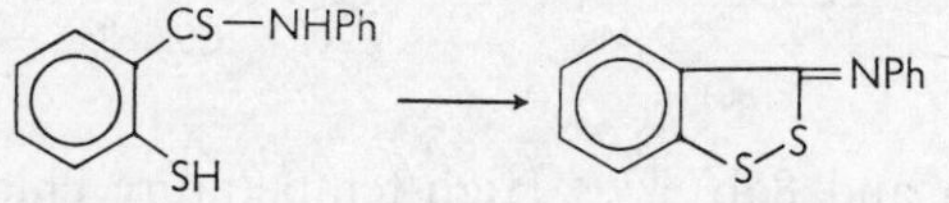

(88)

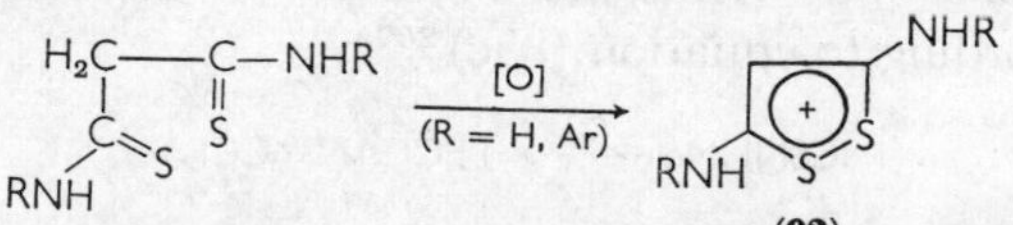

(91)

$$\underset{\underset{RNH}{|}}{\overset{H_2C-\overset{\overset{S}{\|}}{C}-NHR}{\underset{C}{\|}}} \xrightarrow[(R = H, Ar)]{[O]}$$

(89)

(92)

b. Heterocycles containing only nitrogen in the ring. Formation of isatin, imidazolines, oxazolines, and tetrazoles has been reviewed formerly[11]. Mercaptoimidazoles (**93**) can be prepared from α-acylaminothioamides (equation 90a)[11], α-ketothioamides (equation 90b)[295], or by

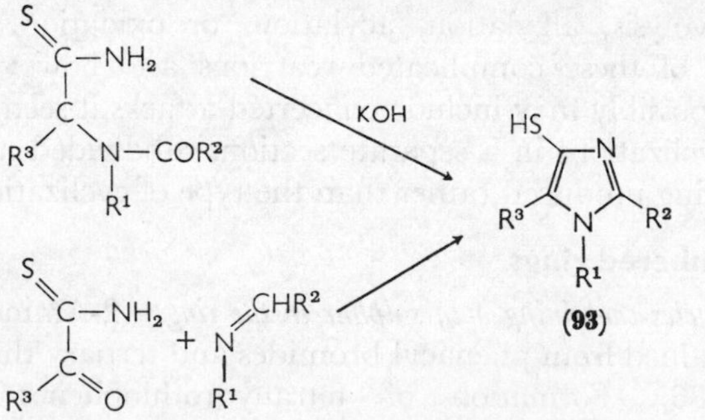

(90a)

(90b)

(93)

simultaneous reaction of carbonyl compounds and ammonia with α-ketothioamides[295a,b]. Similarly the known syntheses of isoxazoles (**94**)[11], and pyrazoles (**95**)[11] from phenylpropiolic acid thioamides (equations 91c and 92a) may now be achieved by different routes (equations 91b[296] and 92b[290,297]). **94** and **95** may be obtained from

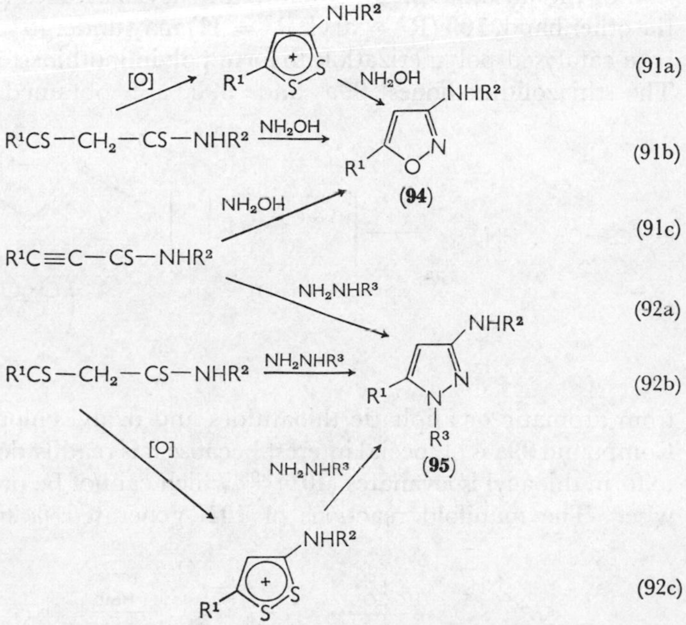

$$[O] \longrightarrow R^1 \underset{S}{\overset{NHR^2}{\diagdown}} \overset{+}{S} \qquad (91a)$$

$$R^1CS-CH_2-CS-NHR^2 \xrightarrow{NH_2OH} \qquad (91b)$$

$$\underset{R^1}{\overset{NHR^2}{\diagdown}} \underset{O}{\overset{N}{\diagup}} \quad (\textbf{94})$$

$$\xrightarrow{NH_2OH}$$

$$R^1C{\equiv}C-CS-NHR^2 \qquad (91c)$$

$$\xrightarrow{NH_2NHR^3} \qquad (92a)$$

$$R^1CS-CH_2-CS-NHR^2 \xrightarrow{NH_2NHR^3} \underset{R^1}{\overset{NHR^2}{\diagup}}\underset{R^3}{\overset{N}{\diagdown}} \quad (92b)$$

$$[O] \qquad \xrightarrow{NH_2NHR^3} \quad (\textbf{95})$$

$$R^1 \underset{S}{\overset{NHR^2}{\diagdown}} \overset{+}{S} \qquad (92c)$$

dithiomalonic amides via a dithiolium system (equations 91a and 92c) with yields even better than by the direct route[294]. 2,5,5-Trimethyl-4-oxoisoxazole-3-thione (**96**) results from **97** and acetone[279]. Regitz

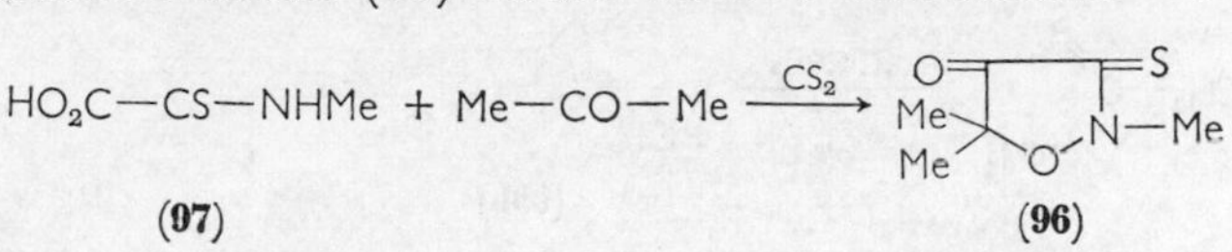

$$HO_2C-CS-NHMe + Me-CO-Me \xrightarrow{CS_2} \underset{Me}{\overset{O{=}\diagup\diagdown{=}S}{\underset{Me}{\diagdown}}}\underset{O}{\overset{}{\diagdown}}N-Me$$

$$(\textbf{97}) \qquad\qquad\qquad\qquad (\textbf{96})$$

and Liedhegener have prepared the 1,2,3-triazole (**98**) by reacting tosyl azide with α-acylthioacetamides[212].

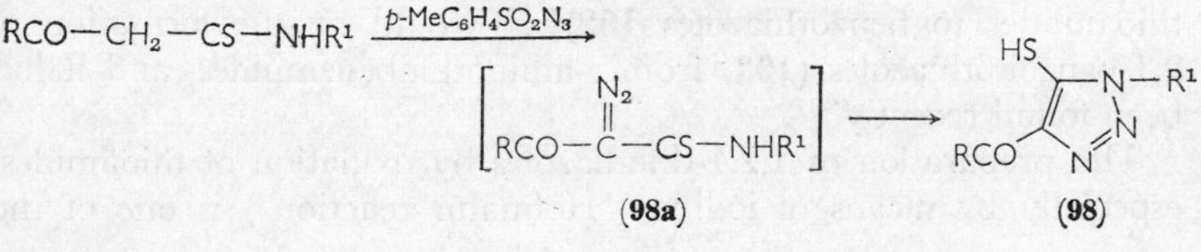

$$RCO-CH_2-CS-NHR^1 \xrightarrow{p\text{-}MeC_6H_4SO_2N_3} \left[\underset{RCO-\overset{N_2}{\overset{\|}{C}}-CS-NHR^1}{}\right] \longrightarrow \underset{RCO}{\overset{HS}{\diagdown}}\underset{N}{\overset{N-R^1}{\diagup\diagdown N}}$$

$$(\textbf{98a}) \qquad\qquad\qquad\qquad (\textbf{98})$$

c. Heterocycles containing nitrogen and sulphur. For the synthesis of thiazolines (**99**) from various α,β-bifunctional molecules of the type XCR_2CR_2Y (X,Y = halogen, OH, SH, NH_2) and thioamides, as well as for the preparation of thiazoles from compounds of the type $RCOCR_2X$, see ref. 11. Compound **99** is also formed by ring expansion of the unstable intermediate *N*-thioacylaziridines (**100**)[298]. On the other hand, **100** (R^1 = aryl, R^2 = H) may undergo spontaneous or base-catalysed polymerization to form polyiminothioesters (**100a**)[298a]. The thiazolidinediones **99a** and **99b** are obtained respectively

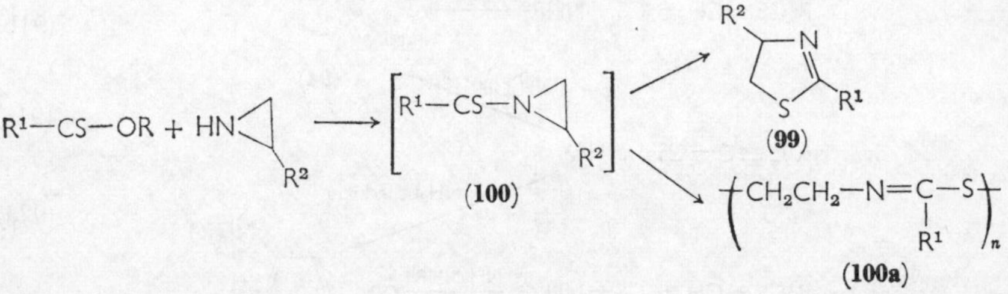

from aromatic or aliphatic thioamides and oxalyl chloride[152,299,383]. Compound **99a** is of special interest because it is readily decarbonylated to form thioacyl isocyanates (**101**)[299] which cannot be prepared otherwise. The manifold reactions of **101**, generated *in situ* from **99a**,

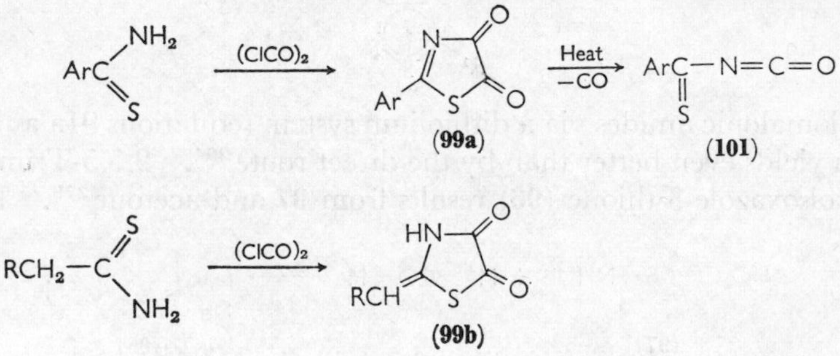

leading to heterocycles are discussed in the publications of Goerdeler and coworkers[300].

A pendant of the well-known 'Jacobson reaction' (oxidation of thioanilides to benzothiazoles **102**)[9,11] would be the formation of 2,1-benzoisothiazoles (**103**) from *o*-aminothiobenzamides, and it has been found recently[301].

The preparation of 1,2,4-thiadiazoles by oxidation of thioamides, especially by means of iodine ('Hofmann reaction') is one of the

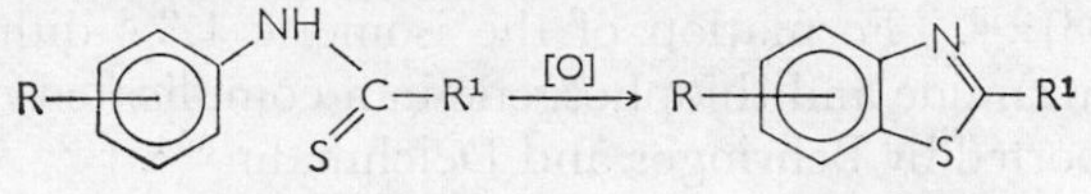

(102)

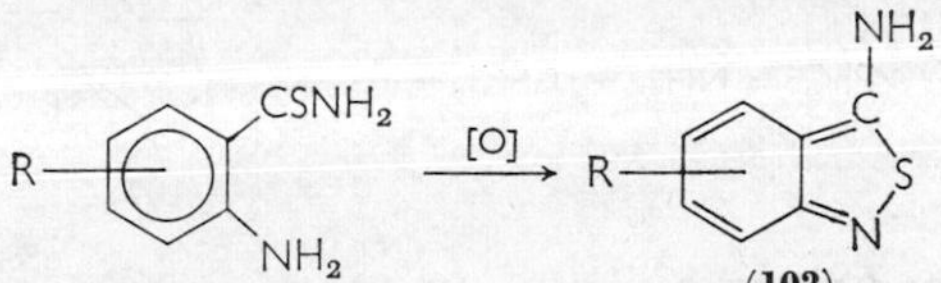

(103)

longest known reactions of thioamides[9,11,302]. 1,3,4-Thiadiazoles (104) are produced by 1,3-dipolar addition of thioamides to N-phenyl-nitrileimines[303].

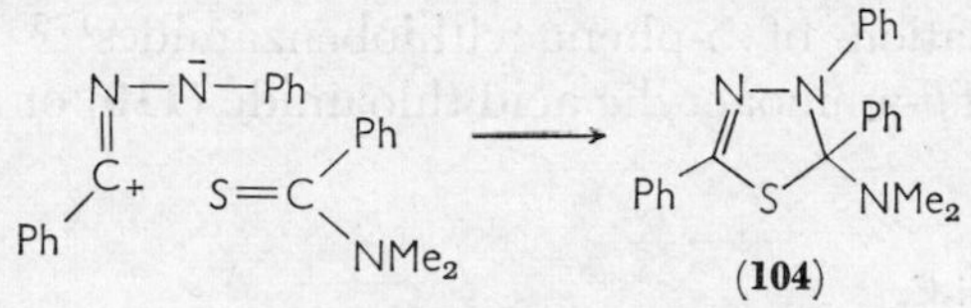

(104)

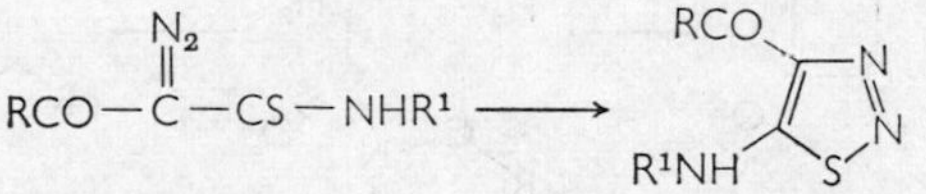

(98a)   (105)

The diazo compound 98a may be cyclized to the 1,2,3-thiadiazole (105)[212].

1,2,4-Dithiazoles (106a, b) on the other hand, have been obtained from thiobenzamides and suitable bifunctional sulphenyl chlorides[268]. 106b is formed by oxidation of N-phenyl-N'-thiobenzoylthiourea too

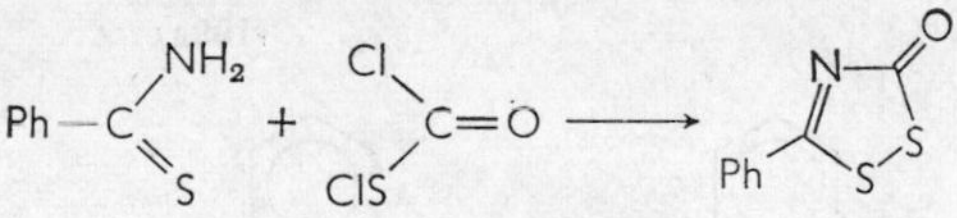

(106a)

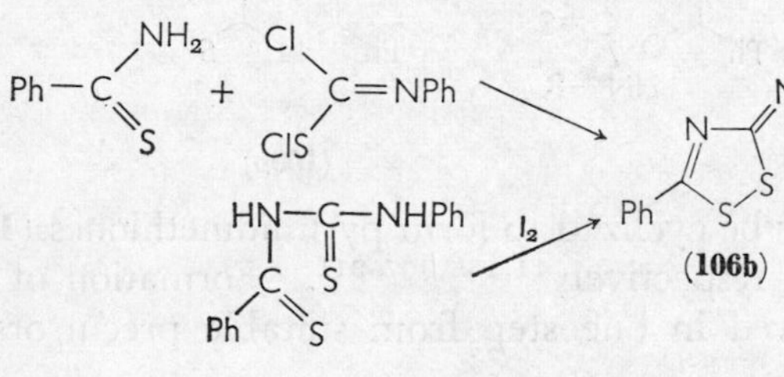

(106b)

(93)

(equation 93)[268]. Formation of the isomeric 1,3,4-dithiazole (**107**) from thiobenzamide and thiophosgene in a complicated reaction (94) has been reported by Behringer and Deichmann[304].

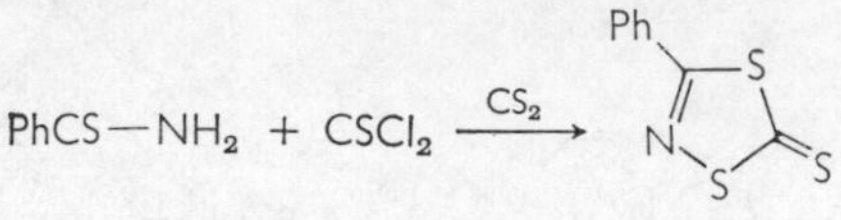

$$PhCS-NH_2 + CSCl_2 \xrightarrow{CS_2} \quad (94)$$

(**107**)

2. Six-membered rings

a. Heterocycles containing only nitrogen in the ring. Pyridones (**108**) result from the reaction of thioamides and diketene (equation 95)[305], and pyridinethiones (**109a**), or isoquinolinethiones (**109b**) may be obtained by condensation of cyanothioacetamide and 1,3-diketones[306], or by cyclization of *o*-phenacylthiobenzamides[175] (equation 96). Derivatives of β-aminoacrylic acid thioamide (**110**) or anthranilic acid

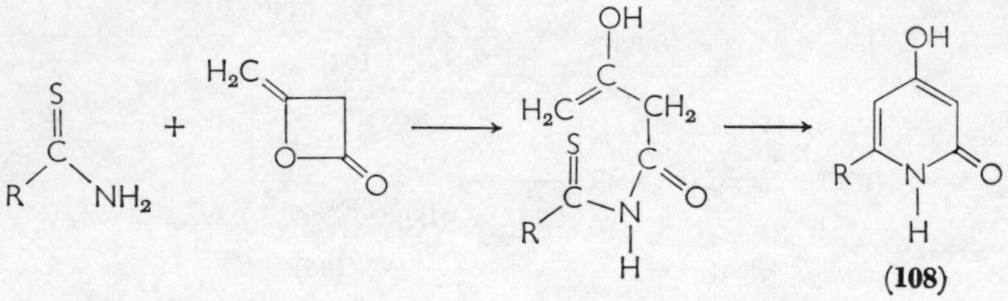

(95)

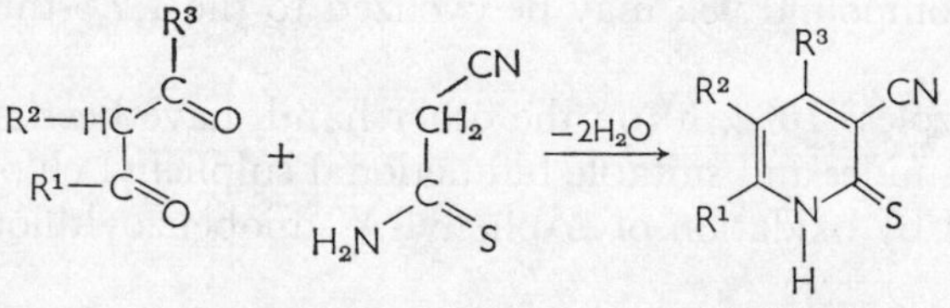

(**108**)

(**109a**)

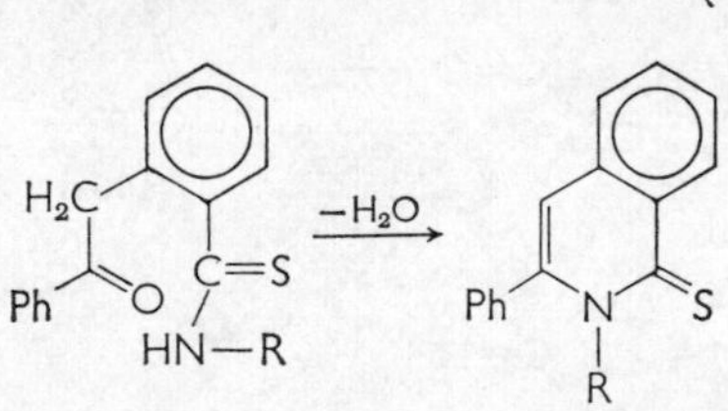

(96)

(**109b**)

thioamide may be cyclized to form pyrimidinethiones (**111**), or quinazolinethiones, respectively[11,155,307-311]. Formation of **110** and **111** may be achieved in one step from suitable precursors[11,307,309,311].

The formation of 1,2,4-triazines and dihydro-1,2,4,5-tetrazines has been treated formerly[11].

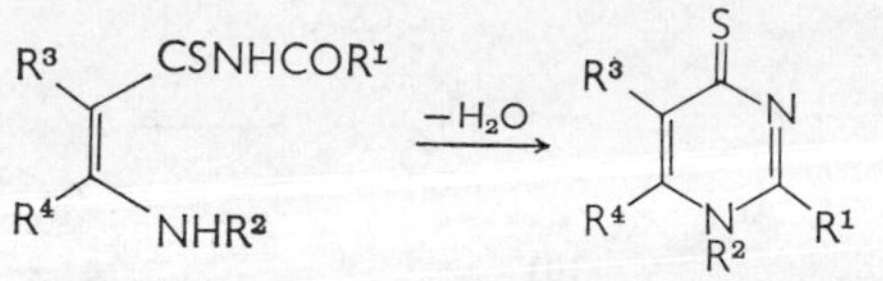

(110)   (111)

b. Heterocycles containing nitrogen and sulphur. The 5,6-dihydro-1,3,4-thiazines (**112**) may be synthesized from thioamides and γ-halo-amines[11], by ring expansion of *N*-thioacylacetidines (**113**; equation 97)[312] analogously to the formation of **99** from **100**, or 4-hydroxy

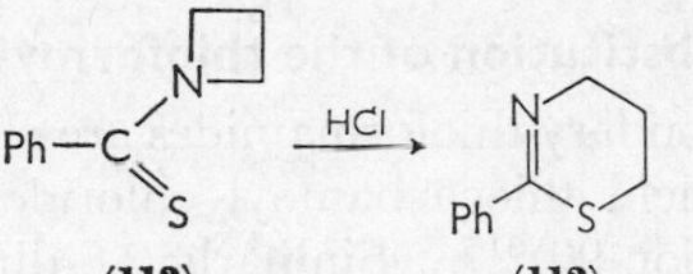

(113)   (112)   (97)

derivatives (**114**) from vinyl ketones (equation 98)[313]. Acylation of thioamides by means of malonyl chloride[152] or carbon suboxide[314]

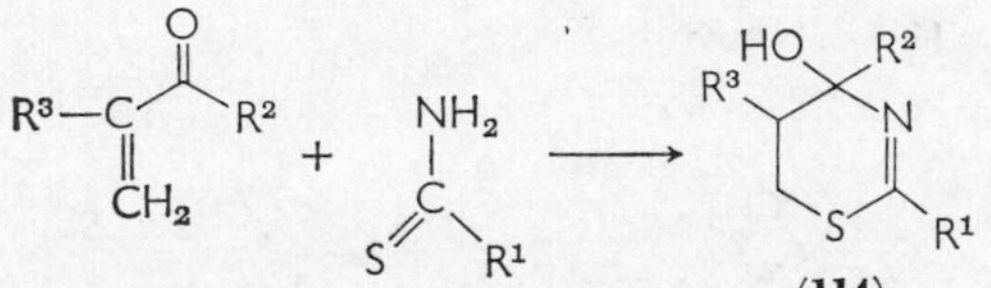

(114)   (98)

yields 1,3-thiazinones (**115**) (or tautomeric forms). 2,6-Diphenyl-1,3,5-thiadiazine (**116**) is formed on oxidation of methylenebis-thiobenzamide (**117**)[283]. 1,3,5-Thiadiazines of type **118** are the

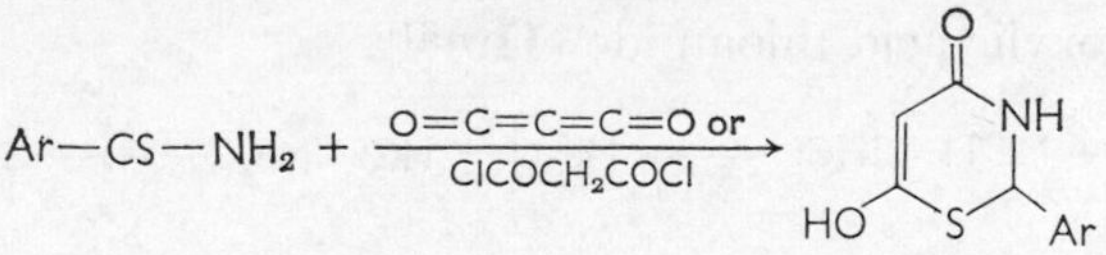

(115)

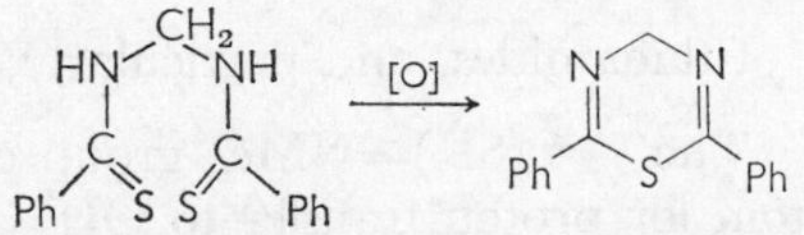

(117)   (116)

products of spontaneous dimerization of the thioacyl isocyanates (**101**) readily obtainable from thioamides via **99a** [299].

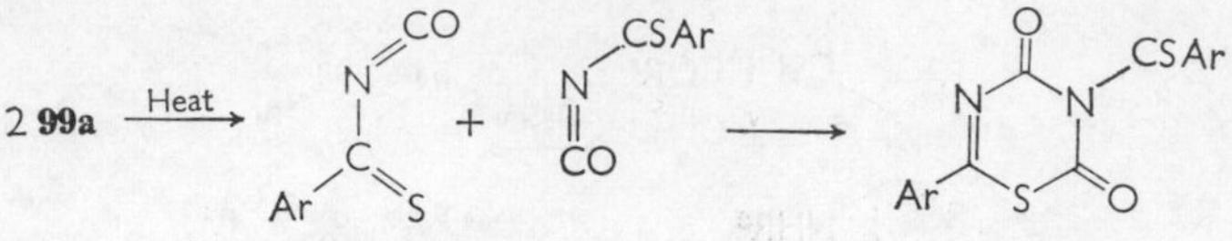

(**101**)   (**118**)

D. Reactions of Thioamides not Involving the Functional Group

In view of the reactivity of thioamides (equation 101)[318] it is worthwhile dealing with reactions on other parts of the molecule, leaving the thioamide group more or less unaffected.

1. Electrophilic substitution of the thioformyl proton

Tertiary and secondary thioformamides are chlorinated by reagents such as SCl_2 to yield thiocarbamoyl chlorides, or isothiocyanates, respectively (equation 99)[315]. Similarly 2,6-dimethylthioformanilide has been recently brominated yielding the first example of the hitherto unknown class of thiocarbamoyl bromides (equation 99a)[375]. Dime-

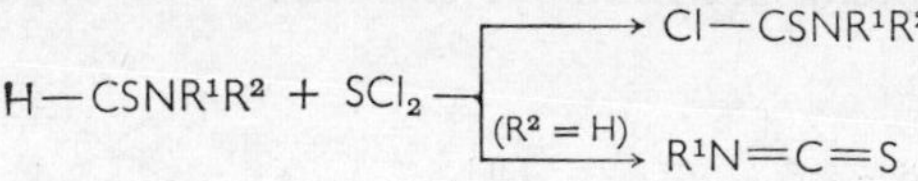

$$H—CSNR^1R^2 + SCl_2 \longrightarrow \begin{cases} Cl—CSNR^1R^2 \\ \xrightarrow{(R^2 = H)} R^1N{=}C{=}S \end{cases} \tag{99}$$

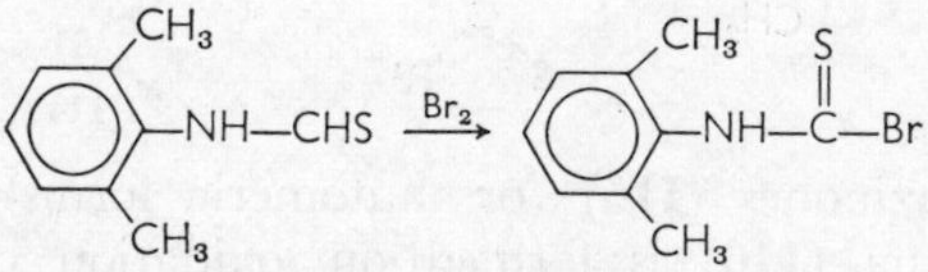

(99a)

thylthioformamide may be converted after Vilsmeyer–Haack into N,N-dimethylglyoxylic acid thioamide (**119**)[316].

$$HCSNMe_2 + Me_2\overset{+}{N}{=}CHCl \longrightarrow Me_2\overset{+}{N}{=}CHCSNMe_2 \xrightarrow{H_2O} HCOCSNMe_2$$

(**119**)

2. Reactions involving the activated α-position

The α-position of thioamides, and particularly, imidothiolic esters is quite reactive. The $—C(SR){=}\overset{+}{N}Me_2$ group especially, activates an α-hydrogen atom for proton transfer to OH^- ion about 2×10^4 times better than the $COSR$ group[238].

Thioacetamides readily undergo aldol condensations forming cinnamic acid thioamides (equation 100) [317]. Acetoacetic acid thio-

$$ArCHO + MeCSNR_2 \longrightarrow ArCH{=}CHCSNR_2 \qquad (100)$$

anilides react with aromatic or aliphatic amines to form β-amino-acrylic acid thioanilides (equation 101) [318].

$$\underset{\underset{NHR}{|}}{MeCOCH_2CSNHAr + RNH_2 \longrightarrow MeC{=}CHCSNHAr} \qquad (101)$$

Ketene S,N-acetals (**120**) are formed from thioamides by alkylation and reaction with bases (B$^-$) (equation 102) [319–321,384]. Similarly the

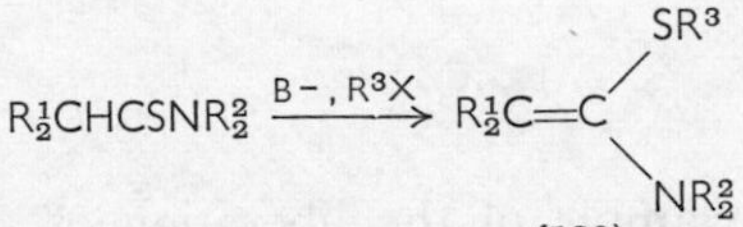

$$R_2^1CHCSNR_2^2 \xrightarrow{B^-, R^3X} R_2^1C{=}C\overset{\displaystyle SR^3}{\underset{\displaystyle NR_2^2}{\big\backslash}} \qquad (102)$$

(120)

S,N-acetals of o-quinone (**121**, X = O), p-quinone (**122**, X = O), or quinoneimine are built from hydroxy-[322,323] or aminothiobenzamides[324]

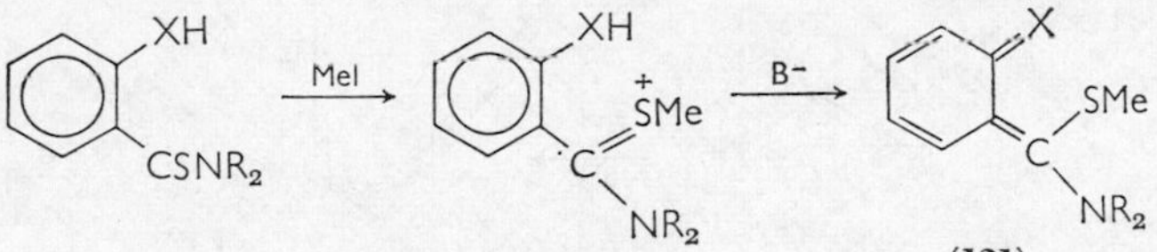

(103)

(121)

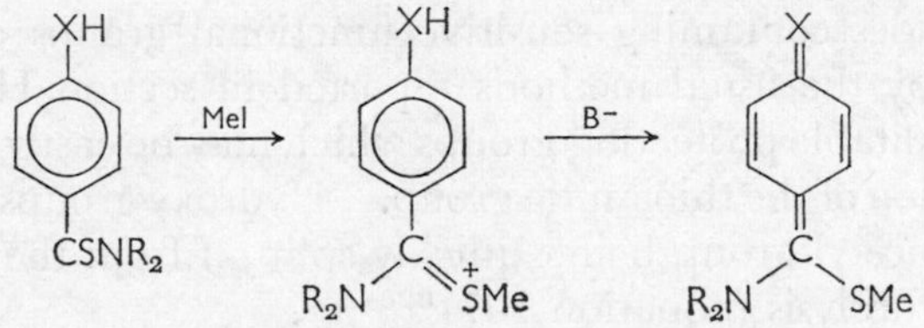

(104)

(122)

(equations 103 and 104; X = O, NH), and o-quinodimethane S,N-acetals (**123**) from the C—H active thioamides **124**[325]. The S-

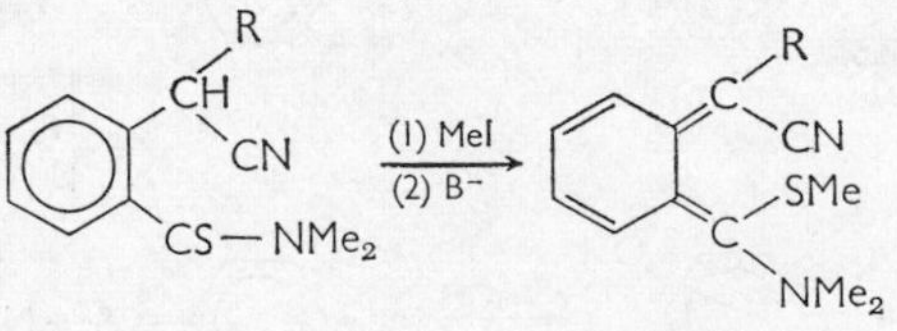

(124) **(123)**

allyl ketene S,N-acetal **125** undergoes Claisen rearrangement to **126** thus providing a method for lengthening a thioamide molecule by a C_3 residue (equation 105) [326]. The generation of **126** may be re-

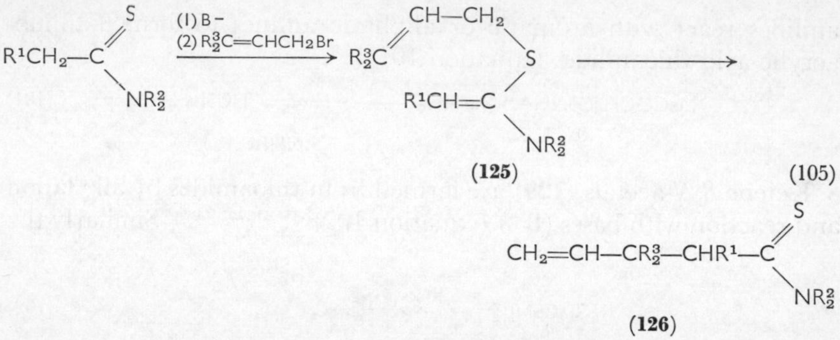

$$(105)$$

garded as another example of the liberation of a latent thioamide group (cf. section II.I).

Finally, the aromatization of polyhalogenated cycloaliphatic thioamides to form thiobenzamides (equation 106) [327] may be mentioned.

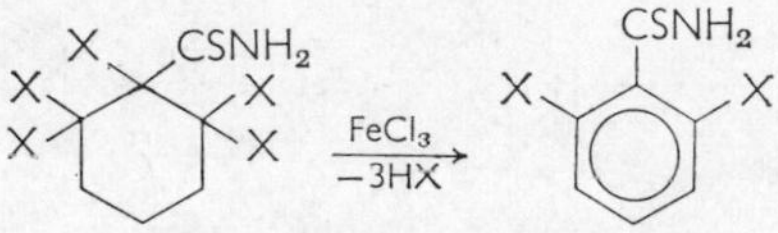

$$(106)$$

3. Preparation of hydroxy and amino thioamides

Thioamides containing sensitive functional groups can hardly be prepared by the usual methods reported in section II. There are, however, suitable protecting groups which may be easily removed after the formation of the thioamide group. Hydroxy groups may be acetylated, the acetyl group being quickly split off from the thioamide by alkaline hydrolysis (equation 107) [328-330].

Salicylic acid thioamides (**127**) may be prepared via the benzoxazinediones (**128**) [331,332]. Amino groups may be blocked by carbo-

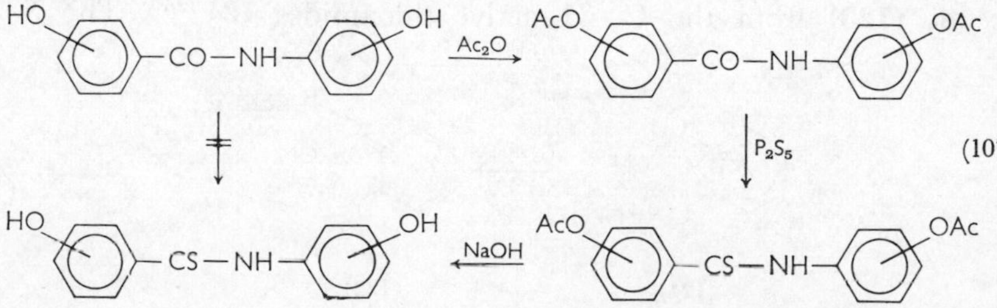

$$(107)$$

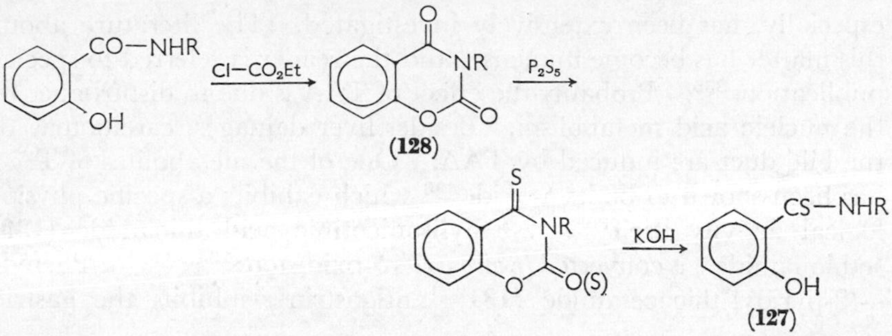

(128)

(127)

benzoxylation, the protecting group being removed by hydrogen bromide in acetic acid[333]. Interestingly p-aminothiobenzamides may be obtained by reduction of the corresponding nitro compound with stannous chloride[11] (cf. section V.A.3) or hydrogen sulphide[190] which leave the thioamide group unchanged.

VI. SPECIAL TOPICS

A. Natural Occurrence of Thioamides

Real thioamides as defined in section I.B have not yet been found in biological material, the thiazole residues of thiamine or the firefly luciferine not being taken into consideration here. 4-Thiouridylic acid (**129**) has been isolated as a minor constituent from the ribonucleic acids of *E. coli* and *S. typhimurium*[334]. The non-enzymatic degradation of some sulphur-containing glucosides yields, however, real open-chain thioamides[334a,334b]. For instance β-hydroxy-β-phenylpropionic acid thioamide (**128b**) is obtained from glucobarbarine (**128a**)[334a].

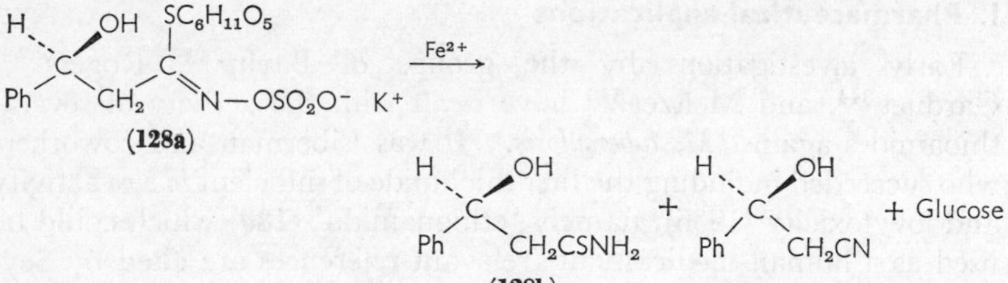

(128a)

(128b)

B. Physiological Activity

It is well known that thioacetamide (TAA) possesses marked physiological activity. The generation of liver cirrhosis by this reagent, especially, has been extensively investigated. The literature about this matter has become immense, and the reader is referred to special publications[335]. Probably the effect of TAA is due to disturbance of the nucleic acid metabolism. Besides liver damages, carcinomas of the bile duct are induced by TAA. One of the metabolites of TAA has been shown to be its S-oxide[336] which exhibits a specific physiological activity itself[337]. 2-Ethylisonicotinic acid thioamide (**130**, 'ethionamide') is converted *in vivo* to its S-oxide too[115a,338]. α-Phenyl-α-(2-pyridyl)thioacetamide (**131**, 'antigastrin') inhibits the gastric

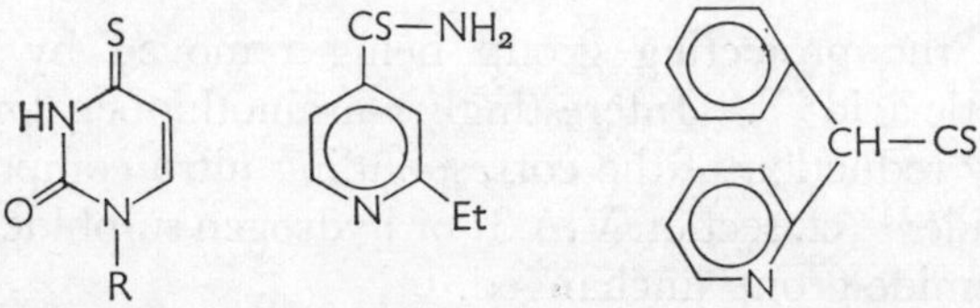

(**129**)   (**130**)   (**131**)

response to gastrin, and thus has anti-ulcer properties[339-341]. Thiobenzamides of type **131a** are strong antidiabetics[385]. The physio-

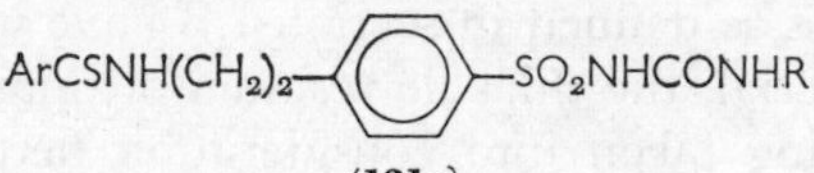

$$ArCSNH(CH_2)_2\text{—}\bigcirc\text{—}SO_2NHCONHR$$

(**131a**)

logical activity of thioamides has caused many attempts to use them for pharmaceutical applications as discussed in the next section.

C. Applications

I. Pharmaceutical applications

Early investigations by the groups of Bavin[342], Rogers[343], Gardner[344], and Meltzer[345] have dealt with the activity of several thioamides against *M. tuberculosis*. It was Liberman and coworkers who succeeded in finding the first thioamide of sufficient *in vivo* activity and low toxicity (1956), namely 'ethionamide' (**130**) which could be used as a human medicament (relevant references are cited by Seydel[346]). More recently, numerous thioamides derived from various types of acids as well as thioamide S-oxides have been tested, and the relations between the antitubercular activity and structural parameters of the thioamides such as u.v. and i.r. spectra, hydrolysis rates, and substituent effects have been studied[317,346-349]. It seems, however, that no thioamide of increased suitability with respect to **130** has been obtained so far.

Activity of thioamides against bacteria other than *M. tuberculosis* has been found by Weuffen and coworkers[350], the substituted thiobenzamide **132** being especially effective.   Salicylic acid thioanilide (**133**)

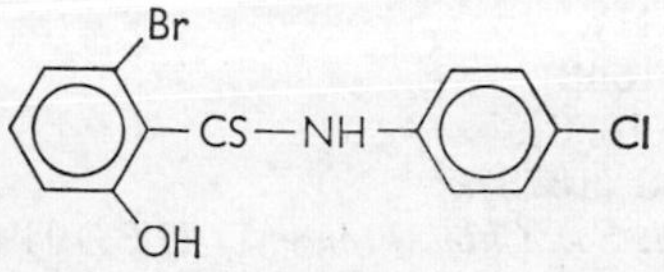

(**132**) 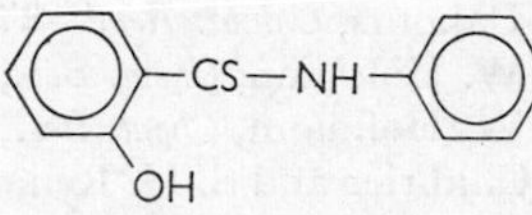   (**133**)

has fungistatic activity[351].   Secondary dithiooxamides are amoebicides[151].

2. Miscellaneous applications

Thioamides may be used as herbicides, and numerous patents which cannot be cited here deal with this application.   Especially polychlorothiobenzamides and cyclopropane and cyclobutane derivatives have been proposed for this purpose.

Thioamides have been technically applied as vulcanization promotors, antioxidants, and corrosion inhibitors.

Thioacetamide may be used as analytical reagent (generator of hydrogen sulphide)[352,353].   The use of dithiooxamide for the detection and determination of metal cations has been reviewed by Hurd and DeLaMater[11].   Pyridinethiocarboxamides[354] and thiocaprolactam[355,356] are suitable for the photometric determination of Fe^{II}, or Bi^{III}, respectively.

Sequence analysis of peptide chains may be achieved by thioacylating the peptide, followed by degradation and identification of the thioamide (**134**) obtained, which contains the *N*-terminal amino acid residue of the original peptide (equation 108)[386].

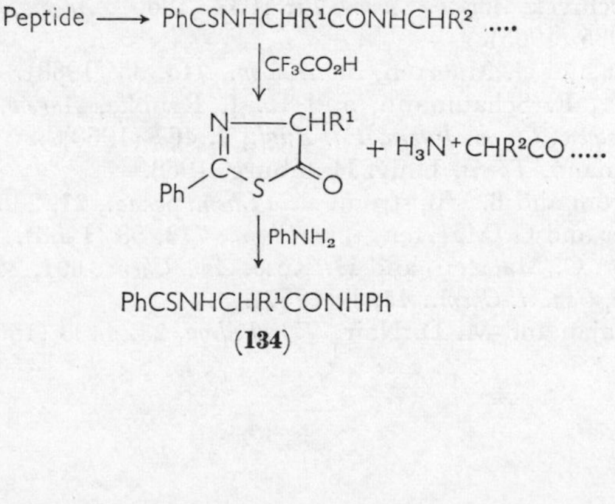

Peptide $\longrightarrow$ PhCSNHCHR^{1}CONHCHR^2 .....

$$\downarrow CF_3CO_2H$$

$+ H_3N^+CHR^2CO.....$ (108)

$$\downarrow PhNH_2$$

PhCSNHCHR^{1}CONHPh

(**134**)

VII. REFERENCES

1. M. Gay-Lussac, *Ann. Chim.*, **95**, 136, esp. 196 (1815).
2. J. J. Berzelius, *Lehrbuch der Chemie*, 5. Aufl. Dresden and Leipzig 1843, Bd. 1, pp. 840.
3. F. Wöhler, *Pogg. Ann.*, **3**, 177 (1825).
4. J. Liebig and F. Wöhler, *Pogg. Ann.*, **24**, 167 (1832).
5. C. Völckel, *Ann. Chem.*, **38**, 314 (1841).
6. A. Cahours, *Compt. Rend.*, **27**, 239 (1848).
7. A. W. Hofmann, *Chem. Ber.*, **1**, 38 (1868).
8. A. W. Hofmann, *Chem. Ber.*, **11**, 338 (1878).
9. P. Chabrier and S. H. Renard, *Bull. Soc. Chim. France*, D 272 (1949).
10. A. Schöberl and A. Wagner, in *Methoden der Organischen Chemie*, Vol. 9, 4th ed., (Eds. J. Houben and T. Weyl), Georg Thieme Verlag, Stuttgart, 1955, p. 741.
11. R. N. Hurd and G. DeLaMater, *Chem. Rev.*, **61**, 45 (1961).
12. E. E. Reid, *Organic Chemistry of Bivalent Sulfur*, Vol. IV, Chemical Publishing Co., New York, 1962, pp. 45–58.
13. W. Walter and K.-D. Bode, *Angew. Chem.*, **78**, 517 (1966); *Angew. Chem. Intern. Ed. Engl.*, **6**, 281 (1967).
14. A. M. Piazzesi, R. Bardi, M. Mammi, and W. Walter, *Ric. Sci. Rend.*, **34**, (II-A), 173 (1964); *Chem. Abstr.*, **63**, 2878d (1965).
15. M. R. Truter, *J. Chem. Soc.*, 997 (1960).
16. M. R. Truter, *Acta Cryst.*, **10**, 785 (1957).
17. P. J. Wheatley, *J. Chem. Soc.*, 396 (1965).
18. J.-C. Colleter and M. Gadret, *Bull. Soc. Chim. France*, 3463 (1967).
18a. J.-C. Colleter and M. Gadret, *Acta Cryst.*, **B24**, 513 (1968).
19. B. R. Penfold, *Acta Cryst.*, **6**, 707 (1953).
20. N. R. Kunchur and M. R. Truter, *J. Chem. Soc.*, 2551 (1958).
21. H. W. Dias and M. R. Truter, *Acta Cryst.*, **17**, 937 (1964).
22. O. H. Jarchow, *Acta Cryst.*, **B25**, 267 (1969).
23. G. W. Wheland, *Resonance in Organic Chemistry*, John Wiley and Sons, New York, Chapman and Hall, Ltd., London, 1955, p. 165.
24. S. C. Abrahams, *Quart. Rev. (London)*, **10**, 407 (1956).
25. J. Sandström, *Acta Chem. Scand.*, **16**, 1616 (1962).
26. W. D. Phillips, *J. Chem. Phys.*, **23**, 1363 (1955).
27. A. J. Speziale and L. R. Smith, *J. Org. Chem.*, **28**, 3492 (1963).
28. W. Walter and G. Maerten, *Ann. Chem.*, **669**, 66 (1963).
29. A. Mannschreck, *Angew. Chem.*, **77**, 1032 (1965); *Angew. Chem. Intern. Ed. Engl.*, **4**, 985 (1965).
30. W. Walter and G. Maerten, *Ann. Chem.*, **715**, 35 (1968).
31. W. Walter, E. Schaumann, and K. J. Reubke, *Angew. Chem.*, **80**, 448 (1968); *Angew. Chem., Intern. Ed. Engl.*, **7**, 467 (1968).
32. E. Schaumann, *Thesis*, Univ. Hamburg, 1968.
33. J. Sandström and B. Uppström, *Acta Chem. Scand.*, **21**, 2254 (1967).
34. W. Walter and G. Maerten, *Ann. Chem.*, **712**, 58 (1968).
35. W. Walter, G. Maerten, and H. Rose, *Ann. Chem.*, **691**, 25 (1966).
36. I. D. Rae, *Can. J. Chem.*, **45**, 1 (1967).
37. K. Nagarajan and M. D. Nair, *Tetrahedron*, **23**, 4493 (1967).

38. W. Walter and H.-P. Kubersky, *Spectrochim. Acta,* in press.

39. J. E. Blackwood, C. L. Gladys, K. L. Loening, A. E. Petrarca, and J. E. Rush, *J. Am. Chem. Soc.,* **90,** 509 (1968).

40. R. S. Cahn, C. Ingold, and V. Prelog, *Angew. Chem.,* **78,** 413 (1966); *Angew. Chem. Intern. Ed. Engl.,* **5,** 385 (1966).

41. R. C. Neuman, Jr. and L. B. Young, *J. Phys. Chem.,* **69,** 1777 (1965).

41a. W. Walter, E. Schaumann, and H. Paulsen, *Ann. Chem.,* **727,** 61 (1969).

42. J. V. Hatton and R. E. Richards, *Mol. Phys.,* **3,** 253 (1960).

43. H. Paulsen and K. Todt, *Angew. Chem.,* **78,** 943 (1966); *Angew. Chem. Intern. Ed. Engl.,* **5,** 899 (1966).

44. H. Paulsen and K. Todt, *Chem. Ber.,* **100,** 3385 (1967).

45. G. Maerten, *Dissertation,* University of Hamburg, 1967.

46. A. Loewenstein, A. Melera, P. Rigny, and W. Walter, *J. Phys. Chem.,* **68,** 1597 (1964).

47. A. Streitwieser, *Molecular Orbital Theory for Organic Chemists,* John Wiley and Sons, New York, London, 1961, p. 311.

48. J. Sandström, *J. Phys. Chem.,* **71,** 2318 (1967).

49. R. C. Neuman, Jr. and L. B. Young, *J. Phys. Chem.,* **69,** 2570 (1965).

50. G. Isaksson and J. Sandström, *Acta Chem. Scand.,* **21,** 1605 (1967).

51. G. Schwenker and H. Roßwag, *Tetrahedron Letters,* 4237 (1967).

52. P. Hampson and A. Mathias, *Mol. Phys.,* **13,** 361 (1967).

53. C. E. Holloway and M. H. Gitlitz, *Can. J. Chem.,* **45,** 2659 (1967).

54. T. H. Siddall, III and W. E. Stewart, *J. Org. Chem.,* **32,** 3261 (1967).

55. R. C. Neuman, Jr., D. N. Roark, and V. Jonas, *J. Am. Chem. Soc.,* **89,** 3412 (1967).

56. E. U. Elam, F. H. Rash, J. T. Dougherty, V. W. Goodlett, and K. C. Brannock, *J. Org. Chem.,* **33,** 2738 (1968).

57. W. Walter, H.-P. Kubersky, E. Schaumann, and K. J. Reubke, *Ann. Chem.,* **719,** 210 (1968).

58. W. Walter and G. Maerten, *Ann. Chem.,* **712,** 46 (1968).

59. J. H. Davies, R. H. Davis, and P. Kirby, *J. Chem. Soc. (C),* 431, (1968).

60. R. Mecke, Jr., R. Mecke, and A. Lüttringhaus, *Z. Naturforsch.,* **10b,** 367 (1955).

61. R. Mecke, Jr. and R. Mecke, *Chem. Ber.,* **89,** 343 (1956).

62. I. Suzuki, *Bull. Chem. Soc. Japan,* **35,** 1286 (1962); *Chem. Abstr.,* **57,** 11980b (1962).

63. I. Suzuki, *Bull. Chem. Soc. Japan,* **35,** 1456 (1962); *Chem. Abstr.,* **57,** 16003h (1962).

64. I. Suzuki, *Bull. Chem. Soc. Japan,* **35,** 1449 (1962); *Chem. Abstr.,* **57,** 16003c (1962).

65. W. Walter and H.-P. Kubersky, *Ann. Chem.,* **694,** 56 (1956).

66. K. A. Jensen and P. H. Nielsen, *Acta Chem. Scand.,* **20,** 597 (1966).

67. M. Davies and W. J. Jones, *J. Chem. Soc.,* 955 (1958).

68. W. Walter and H.-P. Kubersky, unpublished results.

69. R. A. Russell and H. W. Thompson, *Spectrochim. Acta,* 8, 138 (1956).

70. I. Suzuki, M. Tsuboi, T. Shimanouchi, and S. Mizushima, *Spectrochim. Acta,* **16,** 471 (1960).

71. W. Walter and M. Steffen, *Ann. Chem.,* **712,** 53 (1968).

71a. N. Kulevsky and P. M. Froehlich, *J. Am. Chem. Soc.,* **89,** 4839 (1967).

71b. R. L. Jones and R. E. Smith; *J. Mol. Structure*, **2**, 475 (1968).

72. W. Geiger and J. Kurz, *Naturwissenschaften*, **54**, 564 (1967).

73. H. O. Desseyn and M. A. Herman, *Spectrochim. Acta*, **23A**, 2457 (1967).

74. A. Burawoy, *Chem. Ber.*, **63**, 3155 (1930).

75. A. Hantzsch, *Chem. Ber.*, **64**, 661 (1931).

76. A. Burawoy, *J. Chem. Soc.*, 1177 (1939).

77. H. Hosoya, J. Tanaka, and S. Nagakura, *Bull. Chem. Soc. Japan*, **33**, 850 (1960).

78. S. Katagiri, Y. Amako, and H. Azumi, *Symp. Struct. Chem.*, Kyoto, 1958.

79. M. J. Janssen, *Rec. Trav. Chim.*, **79**, 454 (1960).

80. M. J. Janssen, *Rec. Trav. Chim.*, **79**, 464 (1960).

81. M. Kasha, *Discussions Faraday Soc.*, **9**, 14 (1950).

82. M. J. Janssen, *Spectrochim. Acta*, **17**, 475 (1961).

83. J. Sandström, *Acta Chem. Scand.*, **17**, 678 (1963).

84. B. Persson and J. Sandström, *Acta Chem. Scand.*, **18**, 1059 (1964).

85. U. Berg and J. Sandström, *Acta Chem. Scand.*, **20**, 689 (1966).

86. J. Sandström and B. Uppström, *Acta Chem. Scand.*, **19**, 2432 (1965).

87. G. Montaudo and G. Purello, *Ann. Chim. (Rome)*, **51**, 1369 (1961).

88. J. Fabian, H. Viola, and R. Mayer, *Tetrahedron*, **23**, 4323 (1967).

89. N. Stojanac and N. Trinajstić, *Monatsh. Chem.*, **98**, 2263 (1967).

90. J. Sandström, *Svensk Kem. Tidskr.*, **72**, 612 (1960).

90a. W. Walter, H. Christ, and J. Voss, unpublished results.

91. B. Sjöberg, B. Karlén, and R. Dahlbom, *Acta Chem. Scand.*, **16**, 1071 (1962).

92. G. C. Barrett, *J. Chem. Soc.*, 2825 (1965).

93. G. C. Barrett, *J. Chem. Soc.* (C), 1771 (1966).

94. E. Bach, A. Kjaer, R. Dahlbom, T. Walle, B. Sjöberg, E. Bunnenberg, C. Djerassi, and R. Records, *Acta Chem. Scand.*, **20**, 2781 (1966).

95. G. C. Barrett, *J. Chem. Soc.* (C), 1 (1967).

96. G. C. Barrett, *Chem. Commun.*, 40 (1968).

97. J. V. Burakevich and C. Djerassi, *J. Am. Chem. Soc.*, **87**, 51 (1965).

97a. W. Walter and H. Hühnerfuss, *J. Mol. Struct.*, in press.

98. S. Soundararajan, *Trans. Faraday Soc.*, **53**, 159 (1957).

99. C. M. Lee and W. D. Kumler, *J. Am. Chem. Soc.*, **84**, 571 (1962).

100. I. Suzuki, *Nippon Kagaku Zasshi*, **80**, 697 (1959); *Chem. Abstr.*, **53**, 21162 (1959).

100a. A. L. McClellan, *Tables of Experimental Dipole Moments*, W. H. Freeman and Co., San Francisco and London, 1963.

101. H. Lumbroso, C. Pigenet, and P. Reynaud, *Compt. Rend.*, **264**, 732 (1967).

102. H. Lumbroso and C. Pigenet, *Compt. Rend.*, **266**, 735 (1968).

103. A. Lüttringhaus and J. Grohmann, *Z. Naturforsch.*, **10b**, 365 (1955).

104. C. M. Lee and W. D. Kumler, *J. Am. Chem. Soc.*, **84**, 565 (1962).

105. H. Lumbroso, C. Pigenet, H. Rosswag, and G. Schwenker, *Compt. Rend. Ser. C*, **266**, 1479 (1968).

106. C. M. Lee and W. D. Kumler, *J. Org. Chem.*, **27**, 2052 (1962).

107. M. H. Krackov, C. M. Lee, and H. G. Mautner, *J. Am. Chem. Soc.*, **87**, 892 (1965).

108. J. B. Thomson, P. Brown, and C. Djerassi, *J. Am. Chem. Soc.*, **88**, 4049 (1966).

109. A. M. Duffield, C. Djerassi, and J. Sandström, *Acta Chem. Scand.*, **21**, 2167 (1967).

110. R. H. Shapiro, J. W. Serum, and A. M. Duffield, *J. Org. Chem.*, **33**, 243 (1968).

111. W. Walter, R. F. Becker, and H. F. Grützmacher, *Tetrahedron Letters*, 3515 (1968).

111a. C. M. Anderson, R. N. Warrener, and C. S. Barnes, *Chem. Commun.*, 166 (1968).

111b. M. A. Baldwin, A. G. Loudon, A. Maccoll, D. Smith, and A. Ribera, *Chem. Commun.*, 350 (1967).

112. H. Lund, *Collection Czech. Chem. Commun.*, **25**, 3313 (1960).

113. K. G. Stone, *J. Am. Chem. Soc.*, **69**, 1832 (1947).

114. R. Mayer, S. Scheithauer, and D. Kunz, *Chem. Ber.*, **99**, 1393 (1966).

115. P. Rosmus, D. Kunz, and R. Mayer, *Z. Anal. Chem.*, **231**, 360 (1967).

115a. P. O. Kane, *Nature*, **195**, 495 (1962).

116. S. Gurrieri and G. C. Pappalardo, *Ann. Chim. (Rome)*, **57**, 1136 (1967).

117. J. Voss and W. Walter, unpublished results.

118. J. Voss and W. Walter, *Tetrahedron Letters*, 1751 (1968).

119. A. Mehlhorn and J. Fabian, *Z. Chem.*, **5**, 420 (1965).

120. S. Nagakura, *Bull. Chem. Soc. Japan*, **25**, 164 (1952).

121. M. J. Janssen, *Rec. Trav. Chim.*, **79**, 1066 (1960).

121a. A. Ažman, M. Drofenik, D. Hadži, and B. Lukman, *J. Mol. Structure*, **1**, 181 (1967–1968).

122. M. J. Janssen and J. Sandström, *Tetrahedron*, **20**, 2339 (1964).

123. Ref. 17, p. 115.

123a. R. Mayer, J. Fabian, and A. Mehlhorn, private communication.

123b. D. O. Hughes, *Tetrahedron*, **24**, 6423 (1968).

124. Ref. 10, pp. 762–768.

125. H. D. Eilhauer and G. Reckling, *Ger. (East) Pat.*, 50,830 (1966); *Chem. Abstr.*, **66**, 55394 (1967).

126. A. E. S. Fairfull, J. L. Lowe, and D. A. Peak, *J. Chem. Soc.*, 742 (1952).

127. E. E. Gilbert and E. J. Rumanowski, *U.S. Pat.*, 3,336,381 (1967); *Chem. Abstr.*, **68**, 68747 (1968); E. E. Gilbert, E. J. Rumanowski, and P. E. Newallis, *J. Chem. Eng. Data*, **13**, 130 (1968).

128. F. S. Okumura and T. Moritani, *Bull. Chem. Soc. Japan*, **40**, 2209 (1967).

129. K. Kindler, *Ann. Chem.*, **431**, 187 (1923).

130. P. L. de Benneville, J. S. Strong, and V. T. Elkind, *J. Org. Chem.*, **21**, 772 (1956).

131. M. Seefelder, *Chem. Ber.*, **99**, 2678 (1966).

132. J. Jenni, H. Kühne, and B. Prijs, *Helv. Chim. Acta*, **45**, 1163 (1962).

132a. G. Beck, E. Degener, and H. Heitzer, *Ann. Chem.*, **716**, 46 (1968).

133. E. C. Taylor and J. A. Zoltewicz, *J. Am. Chem. Soc.*, **82**, 2656 (1960).

134. H. G. Schicke and G. Schrader, *Ger. Pat.* 1,111,172 (1960); *Chem. Abstr.*, **56**, 2474 (1962).

134a. V. L. Dubina and S. I. Burmistrov, *Zh. Organ. Khim.*, **2**, 1845 (1966); *Chem. Abstr.*, **66**, 55172 (1967).

135. S. Ishikawa, *Sci. Papers. Inst. Chem. Res. (Tokyo)*, **7**, 293 (1927); *Chem. Abstr.*, **22**, 1343 (1928).

136. H. G. Schicke, *Belg. Pat.*, 664,091 (1965); *Chem. Abstr.*, **65**, 659e (1966).

137. H. Eilingsfeld, M. Seefelder, and H. Weidinger, *Chem. Ber.*, **96**, 2671 (1963).

138a. P. Reynaud, R. C. Moreau, and J.-P. Samama, *Bull. Soc. Chim. France*, 3623 (1965).

138b. P. Reynaud, R. C. Moreau, and P. Fodor, *Compt. Rend. Ser. C*, **264**, 1414 (1967).

138c. P. Reynaud, R. C. Moreau, and P. Fodor, *Compt. Rend.*, **263**, 788 (1966).

138d. P. Reynaud, R. C. Moreau, and J.-P. Samama, *Bull. Soc. Chim. France*, 3628 (1965).

139. J. Voss and W. Walter, *Ann. Chem.*, **716**, 209 (1968).

140. K. R. Henery-Logan, H. P. Knoepfel, and J. V. Rodricks, *J. Heterocyclic Chem.*, **5**, 433 (1968).

141. J. Witte and R. Huisgen, *Chem. Ber.*, **91**, 1129 (1958).

142. A. G. Long and A. Tulley, *J. Chem. Soc.*, 1190 (1964).

143. R. Mayer and H. Berthold, *Z. Chem.*, **3**, 310 (1963).

144. R. Mayer and J. Orgis, *Z. Chem.*, **4**, 457 (1964).

145. W. Ried and E. Schmidt, *Ann. Chem.*, **695**, 217 (1966).

146. G. Alliger, G. E. P. Smith, Jr., E. L. Carr, and H. P. Stevens, *J. Org. Chem.*, **14**, 962 (1949).

147. A. Mohsen, M. E. Omar, and S. Yamada, *Chem. Pharm. Bull. (Tokyo)*, **14**, 856 (1966); *Chem. Abstr.*, **65**, 18567h (1966).

148. B. Holmberg, *Arkiv. Kemi, Mineral. Geol. Ser. A*, **17**, 23 (1944); *Chem. Abstr.*, **39**, 4065^2 (1945).

149. F. Kurzer, *Chem. Ind. (London)*, 1333 (1961).

150. K. A. Jensen and C. Pedersen, *Acta Chem. Scand.*, **15**, 1087 (1961).

151. G. R. Wendt and E. Hertz, *U.S. Pat.*, 3,354,156 (1967); *Chem. Abstr.*, **68**, 114638 (1968).

152. J. Goerdeler and H. Horstmann, *Chem. Ber.*, **93**, 671 (1960).

153. W. Walter and M. Radke, *Angew. Chem.*, **80**, 315 (1968); *Angew. Chem. Intern. Ed. Engl.*, **7**, 302 (1968).

153a. G. Wagner and S. Leistner, *Z. Chem.*, **8**, 376 (1968).

154. H. E. Wijers, C. H. C. van Ginkel, L. Brandsma, and J. F. Arens, *Rec. Trav. Chim.*, **86**, 907 (1967).

155. P. W. J. Schuijl and L. Brandsma, *Rec. Trav. Chim.*, **87**, 38 (1968).

155a. J. Sotiropoulos, A.-M. Lamazouère, and P. Bédos, *Compt. Rend. Ser. C*, **265**, 99 (1967).

156. R. T. Wragg, *Ger. Pat.*, 1,227,452 (1966); *Chem. Abstr.*, **66**, 10599 (1967); *Ger. Pat.*, 1,227,451 (1966); *Chem. Abstr.*, **66**, 10601 (1967).

156a. F. Becke and H. Hagen, *Chemiker Z.*, **93**, 474 (1969).

157. S. E. Ellzey, Jr., and C. H. Mack, *J. Org. Chem.*, **28**, 1600 (1963).

158. G. Barnikow, K. Krüger, and G. Hilgetag, *J. Prakt. Chem.*, [4], **35**, 302 (1967).

159. F. Effenberger, R. Gleiter, L. Heider, and R. Niess, *Chem. Ber.*, **101**, 502 (1968).

160. R. S. Lévy, *Bull. Soc. Chim. France*, 693 (1967).

161. J. P. Chupp and E. R. Weiss, *J. Org. Chem.*, **33**, 2357 (1968).

162. H. Viola, S. Scheithauer, and R. Mayer, *Chem. Ber.*, **101**, 3517 (1968).

163. R. W. White, *Can. J. Chem.*, **32**, 867 (1954).

164. F. Asinger, W. Schäfer, K. Halcour, A. Saus, and H. Triem, *Angew. Chem.*, **75**, 1050 (1963); *Angew. Chem. Intern. Ed. Engl.*, **3**, 19 (1964); R. Wegler, E. Kühle, and W. Schäfer, *Angew. Chem.*, **70**, 351 (1958).

165. R. Mayer and J. Wehl, *Angew. Chem.*, **76**, 861 (1964); *Angew. Chem. Intern. Ed. Engl.*, **3**, 705 (1964).

166. R. Mayer in *Organo-Sulfur Chemistry*, (Ed. M. J. Janssen), Interscience Publishers, New York, London, Sydney, 1967, p. 231.

167. W. G. Toland, *J. Org. Chem.*, **27**, 869 (1962).

168. S. Wawzonek and G. R. Hansen, *J. Org. Chem.*, **31**, 3580 (1966).

169. K.-II. Boltze and H.-D. Dell, *Angew. Chem.*, **78**, 114 (1966); *Angew. Chem. Intern. Ed. Engl.*, **5**, 125 (1966).

170. G. Purello, *Gazz. Chim. Ital.*, **97**, 539 (1967).

171. R. Huisgen and J. Witte, *Chem. Ber.*, **91**, 972 (1958).

172. J. Yates and E. Haddock, *Brit. Pat.*, 1,028,912 (1966); *Chem. Abstr.*, **65**, 3804c (1966).

173. D. S. Kemp and R. B. Woodward, *Tetrahedron*, **21**, 3019 (1965).

174. A. A. Goldberg and W. Kelly, *J. Chem. Soc.*, 1919 (1948).

175. L. Legrand and N. Lozac'h, *Bull. Soc. Chim. France*, 3828 (1966).

176. K. A. Watanabe, H. A. Friedman, R. J. Cushley, and J. J. Fox, *J. Org. Chem.*, **31**, 2942 (1966).

177. P. Yates and L. L. Williams, *Tetrahedron Letters*, 1205 (1968).

178. E. Campaigne, *J. Heterocyclic Chem.*, **5**, 141 (1968).

179. S. I. Lur'e and L. G. Gatsenko, *Zh. Obshch. Khim.*, **22**, 262 (1952); *Chem. Abstr.*, **47**, 2168c (1953).

180. F. Asinger, W. Schäfer, H. Meisel, H. Kersten, and A. Saus, *Monatsh. Chem.*, **98**, 338 (1967).

181. F. Asinger, W. Schäfer, H. Kersten, and A. Saus, *Monatsh. Chem.*, **98**, 1843 (1967).

182. R. Mayer and J. Jentzsch, *Angew. Chem.*, **74**, 292 (1962); *Angew. Chem. Intern. Ed. Engl.*, **1**, 217 (1962).

183. S. Mizukami and K. Nagata, *Chem. Pharm. Bull.* (*Tokyo*), **14**, 1249 (1966); *Chem. Abstr.*, **66**, 65255u (1967).

184. D. Gibbs, *J. Biol. Chem.*, **72**, 649 (1927).

185. W. Walter, unpublished results.

186. I. W. Grote, *J. Biol. Chem.*, **93**, 25 (1931).

187. R. Stephan and J. G. Erdman, *Nature*, **203**, 749 (1964).

188. E. Chargaff, C. Levine, and C. Green, *J. Biol. Chem.*, **175**, 67 (1948).

189. W. Walter, *Ann. Chem.*, **633**, 35 (1960).

190. W. Walter and J. Curts, *Chem. Ber.*, **93**, 1511 (1960).

191. W. Walter, J. Curts, and H. Pawelzik, *Ann. Chem.*, **643**, 29 (1961).

192. R. Kitamura and F. Masuda, *Yakugaku Zasshi*, **58**, 251 (1938); *Chem. Zentr.*, **I**, 4607 (1939).

193. H. Wojahn, *Arch. Pharm.*, **284**, 243 (1951).

194. W. Walter, J. Voss, J. Curts, and H. Pawelzik, *Ann. Chem.*, **660**, 60 (1962).

195. E. H. Swift and F. C. Anson in *Advances in Analytical Chemistry and Instrumentation*, Vol. 1 (Ed. Ch. N. Reilley), Interscience Publishers, New York, 1960, p. 340.

196. T. J. Jacob and C. G. R. Nair, *Talanta*, **13**, 154 (1966).

197. M. Sarwar and R. J. Thibert, *Anal. Letters*, **1**, 381 (1968).

198. C. G. R. Nair, S. Geetha, and P. T. Joseph, *Indian J. Appl. Chem.*, **30**, 60 (1967); *Chem. Abstr.*, **67**, 122063 (1967).
199. S. Washizuka, *Bunseki Kagaku*, **16**, 963 (1967); *Chem. Abstr.*, **68**, 9198 (1968).
200. M. Pryszczewska, *Talanta*, **12**, 569 (1965).
201. M. Pryszczewska, *Talanta*, **13**, 1700 (1966).
202. D. M. King and W. S. Eaton, *Talanta*, **15**, 347 (1968).
203. C. Laar, *Chem. Ber.*, **19**, 730 (1886).
204. F. Arndt, *Abhandl. Braunschweig. Wiss. Ges.*, **8**, 1 (1956).
205. R. Gompper, *Angew. Chem.*, **76**, 412 (1964); *Angew. Chem. Intern. Ed. Engl.*, **3**, 560 (1964).
206. A. T. Shuvaev, A. V. Landyshev, A. K. Belousov, and G. S. Baida, *Izv. Akad. Nauk. USSR, Ser. Fiz.*, **31**, 898 (1967); *Chem. Abstr.*, **68**, 82619 (1968).
207. N. Bacon, A. J. Boulton, R. T. C. Brownlee, A. R. Katritzky, and R. D. Topsom, *J. Chem. Soc.*, 5230 (1965).
208. P.Sohár and A. Nemes, *Acta Chim. Acad. Sci. Hung.*, **56**, 25 (1968); *Chem. Abstr.*, **69**, 31706 (1968).
209. P. May, *J. Chem. Soc.*, 2272 (1913).
210. A. R. Katritzky and R. A. Jones, *J. Chem. Soc.*, 2947 (1960).
211. E. P. Dudek and G. Dudek, *J. Org. Chem.*, **32**, 823 (1967).
211a. T. Gramstad and J. Sandström, *Spectrochim. Acta*, **A25**, 31 (1969).
212. M. Regitz and A. Liedhegener, *Ann. Chem.*, **710**, 118 (1967).
213. R. Boudet, *Bull. Soc. Chim. France*, 377 (1951).
214. W. Kutzelnigg and R. Mecke, *Spectrochim. Acta*, **17**, 530 (1961).
215. R. F. Becker, *Thesis*, University of Hamburg, 1968.
215a. W. Walter and R. F. Becker, *Ann. Chem.*, **727**, 71 (1969).
216. R. L. Adelman, *J. Org. Chem.*, **29**, 1837 (1964).
217. M. J. Janssen, *Rec. Trav. Chim.*, **81**, 650 (1962).
218. K. Yates and J. B. Stevens, *Can. J. Chem.*, **43**, 529 (1965).
219. D. Rosenthal and T. I. Taylor, *J. Am. Chem. Soc.*, **79**, 2684 (1957).
220. J. T. Edward and I. C. Wang, *Can. J. Chem.*, **40**, 399 (1962).
221. H. Lemaire and H. J. Lucas, *J. Am. Chem. Soc.*, **73**, 5198 (1951).
222. R. Huisgen, H. Brade, H. Walz, and I. Glogger, *Chem. Ber.*, **90**, 1437 (1957).
223. J. T. Edward and H. Stollar, *Can. J. Chem.*, **41**, 721 (1963).
224. T. Birchall and R. J. Gillespie, *Can. J. Chem.*, **41**, 2642 (1963).
225. E. Spinner, *Spectrochim. Acta*, **15**, 95 (1959).
226. R. J. Niedzielski, R. S. Drago, and R. L. Middaugh, *J. Am. Chem. Soc.*, **86**, 1694 (1964).
227. M. Zackrisson, *Acta Chem. Scand.*, **15**, 1785 (1961).
228. G. Barnikow and H. Kunzek, *Z. Chem.*, **6**, 343 (1966).
229. G. Barnikow and H. Kunzek, *Ann. Chem.*, **700**, 36 (1966).
230. V. F. Anufrienko, E. K. Mamaeva, E. G. Rukhadze, and I. G. Il'ina, *Teor. Eksp. Khim.*, **3**, 363 (1967); *Chem. Abstr.*, **68**, 73939 (1968).
231. B. B. Wayland, R. S. Drago, and H. F. Henneike, *J. Am. Chem. Soc.*, **88**, 2455 (1966).
232. E. A. Butler, D. G. Peters, and E. H. Swift, *Anal. Chem.*, **30**, 1379 (1958).
233. M. Ccfola, S. Peter, P. S. Gentile, and A. V. Celiano, *Talanta*, **9**, 537 (1962).
234. J. Seydel, *Tetrahedron Letters*, 1145 (1966).

235. D. G. Peters and E. H. Swift, *Talanta*, **1**, 30 (1958).
236. D. T. Witiak, T.-F. Chin, and J. L. Lack, *J. Org. Chem.*, **30**, 3721 (1965).
237. R. K. Chaturvedi, A. E. MacMahon, and G. L. Schmir, *J. Am. Chem. Soc.*, **89**, 6984 (1967).
238. G. E. Lienhard and T.-Ch. Wang, *J. Am. Chem. Soc.*, **90**, 3781 (1968).
238a. T. Mukayama, T. Yamaguchi, and H. Nohira, *Bull. Chem. Soc. Japan*, **38**, 2107 (1965).
238b. T. Mukayama and T. Yamaguchi, *Bull. Chem. Soc. Japan*, **39**, 2005 (1966).
239. P. Reynaud, R. C. Moreau, and N. H. Thu, *Compt. Rend.*, **253**, 1968 (1961).
240. P. Reynaud, R. C. Moreau, and T. Gousson, *Compt. Rend.*, **259**, 4067 (1964).
241. A. W. Chapman, *J. Chem. Soc.*, 2296 (1926).
242. W. Walter, J. Voss, and J. Curts, *Ann. Chem.*, **695**, 77 (1966).
243. R. J. Kaufmann and R. Adams, *J. Am. Chem. Soc.*, **45**, 1744 (1923).
244. W. Köhler, M. Bubner, and G. Ulbricht, *Chem. Ber.*, **100**, 1073 (1967).
245. W. Walter and J. Krohn, *Chem. Ber.*, **102**, 3786 (1969).
246. H. Bredereck, R. Gompper, and D. Bitzer, *Chem. Ber.*, **92**, 1139 (1959).
247. G. Bergson, *Arkiv. Kemi*, **16**, 315 (1961).
248. R. N. Hurd, G. DeLaMater, and J. P. McDermott, *J. Org. Chem.*, **27**, 269 (1962).
249. H. Bredereck, R. Gompper, and H. Seiz, *Chem. Ber.*, **90**, 1837 (1957).
250. H. Böhme and H.-H. Hotzel, *Arch. Pharm.*, **300**, 241 (1967).
251. J. Wijma, *U.S. Pat.*, 3,374,084 (1968); *Chem. Abstr.*, **69**, 35783 (1968).
252. N. Stojanac and V. Hahn, *Bull. Sci. Conseil Acad. RSF Yugoslavie*, **11**, 98 (1966); *Chem. Abstr.*, **65**, 20084g (1966).
253. H. Rivier and C. Schneider, *Helv. Chim. Acta*, **3**, 115 (1920).
254. W. Walter, *Ann. Chem.*, **633**, 49 (1960).
255. J. Goerdeler and K. Stadelbauer, *Chem. Ber.*, **98**, 1556 (1965).
256. J. Goerdeler and H. Horstmann, *Chem. Ber.*, **93**, 663 (1960).
257. Shell, *Fr. Pat.*, 1,422,405 (1965).
258. H. Busse, *Dissertation*, University of Hamburg, 1966.
259. N. Stojanac and V. Hahn, *Chimia (Aarau)*, **20**, 175 (1966).
260. J. Goerdeler and H. Porrmann, *Chem. Ber.*, **94**, 2856 (1961).
261. J. Goerdeler and H. Porrmann, *Chem. Ber.*, **95**, 627 (1962).
262. D. Martin, A. Weise, H.-J. Niclas, and S. Rackow, *Chem. Ber.*, **100**, 3756 (1967).
263. H. Rivier and J. Schalch, *Helv. Chim. Acta*, **6**, 605 (1923).
264. H. Rivier and M. Langer, *Helv. Chim. Acta*, **26**, 1722 (1943).
265. J. Goerdeler and H. Schenk, *Chem. Ber.*, **99**, 782 (1966).
266. W. Walter and P.-M. Hell, *Ann. Chem.*, **727**, 50 (1969).
266a. V. L. Dubina and S. I. Burmistrov; *U.S.S.R. Pat.*, 181,085 (1966); *Chem. Abstr.*, **65**, 8831ᵃ (1966).
267. W. Walter and P.-M. Hell, *Angew. Chem.*, **77**, 720 (1965); *Angew. Chem. Intern. Ed. Engl.*, **4**, 696 (1965).
268. W. Walter and P.-M. Hell, *Ann. Chem.*, **727**, 22 (1969).
269. M. C. Agrawal and S. P. Mushran, *J. Phys. Chem.*, **72**, 1497 (1968).
270. W. Walter, J. Holst, W.-R. Knabjohann, J. Voss, D. Lentfer, and U. Sewekow, unpublished results.

271. K. Fries and W. Buchler, *Ann. Chem.*, **454**, 233 (1927).
272. V. Hahn, Z. Stojanac, and D. Emer, *XIV. Intern. Congr. Pure Appl. Chem.*, *Zurich*, 1955, ref. No. 474.
273. F. Hodosan, *Bull. Soc. Chim. France*, 633 (1957).
274. F. Hodosan, *Rev. Chim. (Bucarest)*, **4**, 105 (1959); *Chem. Abstr.*, **53**, 18855b (1959).
275. J. R. Schaeffer, C. T. Goodhue, H. A. Risley, and R. E. Stevens, *J. Org. Chem.*, **32**, 392 (1967).
276. R. Kitamura, *J. Pharm. Soc. Japan*, **58**, 246, 809 (1938).
277. W. Walter and H.-P. Kubersky, *Ann. Chem.*, **694**, 70 (1966).
278. J. F. King and T. Durst, *Tetrahedron Letters*, 585 (1963).
279. W. Walter and K.-D. Bode, *Ann. Chem.*, **698**, 131 (1966).
280. Rhone-Poulenc S. A., *Belg. Pat.*, 616,752 (1962); *Chem. Abstr.*, **58**, 13923b (1963).
281. Rhone-Poulenc S. A., *Belg. Pat.*, 618,276 (1962); *Chem. Abstr.*, **59**, 5142a (1963).
282. D. Liberman, N. Rist, and F. Grumbach, *Compt. Rend.*, **257**, 307 (1963).
283. W. Walter and J. Curts, *Ann. Chem.*, **649**, 88 (1961).
284. W. Walter and K.-D. Bode, *Ann. Chem.*, **660**, 74 (1962).
285. W. Walter and J. Voss, *Ann. Chem.*, **698**, 113 (1966).
286. W. Walter and G. Randau, *Ann. Chem.*, **722**, 80 (1969).
287. T. Taguchi and K. Yoshihira, *Chem. Pharm. Bull. (Tokyo)*, **11**, 430 (1963).
287a. H. Holtschmidt, E. Degener, H.-G. Schmelzer, H. Tarnow, and W. Zecher; *Angew. Chem.*, **80**, 942 (1968); *Angew. Chem. Intern. Ed. Engl.*, **7**, 856 (1968).
288. H. Hartmann and R. Mayer, *Z. Chem.*, **6**, 28 (1966).
289. S. Ruhemann, *J. Chem. Soc.*, 621 (1908).
290. G. Barnikow, *Ann. Chem.*, **700**, 46 (1966).
291. G. Barnikow and H. Niclas, *Z. Chem.*, **6**, 417 (1966).
292. G. Wagner and P. Richter, *Z. Chem.*, **6**, 220 (1966).
293. U. Schmidt, *Chem. Ber.*, **92**, 1171 (1959).
294. G. Barnikow, *Chem. Ber.*, **100**, 1389 (1967).
295. H. Offermanns, P. Krings, and F. Asinger, *Tetrahedron Letters*, 1809 (1968).
295a. F. Asinger, W. Schäfer, and A. Saus, *Monatsh. Chem.*, **96**, 1278 (1965).
295b. F. Asinger, A. Saus, H. Offermanns, and H.-D. Hahn, *Ann. Chem.*, **691**, 92 (1966).
296. L. Birkofer and A. Widdig, *Tetrahedron Letters*, 4299 (1965).
297. G. Barnikow, *Z. Chem.*, **6**, 109 (1966).
298. P. Reynaud, R. C. Moreau, and P. Fodor, *Compt. Rend., Ser. C*, **266**, 632 (1968).
298a. Y. Iwakura, A. Nabeya, and T. Nishiguchi, *J. Polymer Sci.*, *A-1*, **6**, 2591 (1968).
299. J. Goerdeler and H. Schenk, *Chem. Ber.*, **98**, 2954 (1965).
300. J. Goerdeler and R. Weiss, *Chem. Ber.*, **100**, 1627 (1967), and the preceding papers of this series cited herein.
301. M. Seefelder and H. Armbrust, *Belg. Pat.*, 670,652 (1966); *Chem. Abstr.*, **66**, 65467 (1967).
302. A. W. Hofmann, *Chem. Ber.*, **2**, 645 (1869).

303. R. Huisgen, R. Grashey, M. Seidel, H. Knupfer, and R. Schmidt, *Ann. Chem.*, **658**, 169 (1962).
304. H. Behringer and D. Deichmann, *Tetrahedron Letters*, 1013 (1967).
305. T. Kappe, I. Maninger, and E. Ziegler, *Monatsh. Chem.*, **99**, 85 (1968).
306. U. Schmidt and H. Kubitzek, *Chem. Ber.*, **93**, 1559 (1960).
307. G. DeStevens, B. Smolinsky, and L. Dorfman, *J. Org. Chem.*, **29**, 1115 (1964).
308. R. W. J. Carney, J. Wojtkunski, and G. DeStevens, *J. Org. Chem.*, **29**, 2887 (1964).
309. J. Schoen and K. Bogdanowicz-Szwed, *Roczniki Chem.*, **40**, 307 (1966); *Chem. Abstr.*, **65**, 709d (1966).
310. J. Goerdeler and D. Wieland, *Chem. Ber.*, **100**, 47 (1967).
311. J. A. Zoltewicz and T. W. Sharpless, *J. Org. Chem.*, **32**, 2681 (1967).
312. Y. Iwakura, A. Nabeya, T. Nishiguchi, and K.-H. Ohkawa, *J. Org. Chem.*, **31**, 3352 (1966).
313. G. C. Barrett, S. H. Eggers, T. R. Emerson, and G. Lowe, *J. Chem. Soc.*, 788 (1964).
314. E. Ziegler and R. Wolf, *Monatsh. Chem.*, **95**, 1061 (1964).
315. U. Hasserodt, *Chem. Ber.*, **101**, 113 (1968).
316. E. Günther, F. Wolf, and G. Wolter, *Z. Chem.*, **8**, 63 (1968).
317. G. Pappalardo, B. Tornetta, and G. Scapini, *Farmaco Ed. Sci.*, **21**, 740 (1966); *Chem. Abstr.*, **66**, 46363 (1967).
318. A. N. Borisevich and P. S. Pel'kis, *Zh. Organ. Khim.*, **3**, 1339 (1967); *Chem. Abstr.*, **67**, 99814 (1967).
319. R. Gompper and W. Elser, *Tetrahedron Letters*, 1971 (1964).
320. P. J. W. Schuijl, H. J. T. Bos, and L. Brandsma, *Rec. Trav. Chim.*, **87**, 123 (1968).
321. R. Raap, *Can. J. Chem.*, **46**, 2255 (1968).
322. R. Gompper and R. R. Schmidt, *Z. Naturforsch.*, **17b**, 851 (1962).
323. R. Gompper and R. R. Schmidt, *Chem. Ber.*, **98**, 1385 (1965).
324. R. Gompper and H.-D. Lehmann, *Angew. Chem.*, **80**, 38 (1968); *Angew. Chem. Intern. Ed. Engl.*, **7**, 74 (1968).
325. R. Gompper, E. Kutter, and H. Kast, *Angew. Chem.*, **79**, 147 (1967); *Angew. Chem. Intern. Ed. Engl.*, **6**, 171 (1967).
326. P. J. W. Schuijl and L. Brandsma, *Rec. Trav. Chim.*, **87**, 929 (1968).
327. N. N. Philips Gloeilampenfabrieken, *Neth. Pat.*, 6,511,982 (1965).
328. M. Jancevska, K. Jakopčić, and V. Hahn, *Croatica Chim. Acta*, **37**, 67 (1965).
329. M. M. Jancevska, *Glasnik Hem. Drustva, Beograd*, **31**, 149 (1966); *Chem. Abstr.*, **69**, 2773 (1968).
330. M. M. Jancevska, *Glasnik Hem. Drustva, Beograd*, **31**, 255 (1966); *Chem. Abstr.*, **69**, 26958 (1968).
331. G. Wagner and D. Singer, *Z. Chem.*, **2**, 306 (1962).
332. G. Wagner and D. Singer, *Z. Chem.*, **3**, 148 (1963).
333. K. Jakopčić and V. Hahn, *Naturwissenschaften*, **51**, 482 (1964).
334. M. N. Lipsett, *J. Biol. Chem.*, **240**, 3975 (1965).
334a. F. L. Austin, C. A. Gent, and I. A. Wolff, *J. Agr. Food Chem.*, **16**, 752 (1968).
334b. F. L. Austin and C. A. Gent, *Chem. Commun.*, 71 (1967).
335. J. Brodehl, *Klin. Wochschr.*, **39**, 956 (1961).

*

474   W. Walter and J. Voss

336. R. Ammon, H. Berninger, H. J. Haas, and I. Landsberg, *Arzneimittel-Forsch.*, **17**, 521 (1967).
337. V. Becker and W. Walter, *Acta Hepato-Splenol.*, **12**, 129 (1965); *Chem. Abstr.*, **63**, 12211e (1965).
338. J. P. Johnston, P. O. Kane, and M. R. Kibby, *J. Pharm. Pharmacol.*, **19**, 1 (1967).
339. H. W. Sause, *Belg. Pat.*, 669,165 (1965); *Chem. Abstr.*, **65**, 13666c (1966).
340. B. S. Bedi, G. Gillespie, and I. E. Gillespie, *Lancet*, 1240 (1967).
341. G. Gillespie, V. I. McCusker, B. S. Bedi, H. T. Debas, and I. E. Gillespie, *Gastroenterology*, **55**, 81 (1968); *Chem. Abstr.*, **69**, 42633 (1968).
342. E. M. Bavin, D. J. Drain, M. Seiler, and D. E. Seymour, *J. Pharm. Pharmacol.*, **4**, 844 (1952).
343. E. F. Rogers, W. J. Leanza, H. J. Becker, A. R. Matzuk, R. C. O'Neill, A. J. Basso, G. A. Stein, M. Solotorovsky, F. J. Gregory, and K. Pfister, 3rd., *Science*, **116**, 253 (1952).
344. T. S. Gardner, E. Wenis, and J. Lee, *J. Org. Chem.*, **19**, 753 (1954).
345. R. I. Meltzer, A. D. Lewis, and J. A. King, *J. Am. Chem. Soc.*, **77**, 4062 (1955).
346. J. Seydel, *Chemotherapia*, **5**, 46 (1962).
347. J. Seydel, *Z. Naturforsch.*, **16b**, 419 (1961).
348. G. Pappalardo, B. Tornetta, P. Condorelli, and A. Bernardini, *Farmaco Ed. Sci.*, **22**, 808 (1967); *Chem. Abstr.*, **68**, 21141 (1968).
349. J. Seydel, E. Wempe, and H. J. Nestler, *Arzneimittel-Forsch.*, **18**, 362 (1968).
350. W. Weuffen, G. Wagner, D. Singer, and L. Hellmuth, *Pharmazie*, **21**, 477 (1966).
351. W. Weuffen, G. Wagner, D. Singer, and M. Petermann, *Pharmazie*, **21**, 613 (1966).
352. Reference 195, p. 293.
353. M. Pryszczwska, *Proc. Conf. Appl. Phys.-Chem., Methods Chem. Anal.*, Budapest, **1**, 256 (1966); *Chem. Abstr.*, **69**, 8120 (1968).
354. R. T. Pflaum, G. F. Brunzie, and L. E. Cook, *Proc. Iowa Acad. Sci.*, **72**, 123 (1965); *Chem. Abstr.*, **69**, 15800 (1968).
355. H. Sikorska-Tomička, *Chem. Anal. Warschau*, **12**, 1291 (1967); *Chem. Abstr.*, **68**, 119106 (1968).
356. H. Sikorska-Tomička; *Chem. Anal. Warschau*, **13**, 341 (1968); *Chem. Abstr.*, **69**, 73646 (1968).
357. K. A. Petrov and L. N. Andreev, *Usp. Khim.*, **38**, 41 (1969); *Russ. Chem. Rev. (English Transl.)* **31**, 21 (1969).
358. T. Saegusa, S. Kobayashi, K. Hirota, Y. Okumura, and Y. Ito, *Bull. Chem. Soc. Japan*, **41**, 1638 (1968).
359. G. C. Barrett, *J. Chem. Soc. (C)*, 1123 (1969).
360. G. C. Pappalardo, *Ann. Chim. (Rome)*, **58**, 756 (1968).
361. R. Tull and L. M. Weinstock, *Angew. Chem.*, **81**, 291 (1969); *Angew. Chem. Intern. Ed. Engl.*, **8**, 278 (1969).
362. P. D. Schickedantz, *U.S. Pat.* 3,408,394 (1968); *Chem. Abstr.*, **70**, 46880 (1969).
363. Y. Otsuji, N. Matsumura, and E. Imoto; *Bull. Chem. Soc. Japan*, **41**, 1485 (1968).

364. H. E. Wijers, C. H. D. Van Ginkel, P. J. W. Schuijl, and L. Brandsma, *Rec. Trav. Chim.*, **87**, 1236 (1968).

365. M. L. Petrov, B. S. Kupin, and A. A. Petrov, *Zh. Organ. Khim.*, **4**, 2053 (1968); *Chem. Abstr.*, **70**, 28588 (1969).

366. W. Walter and R. F. Becker, *Ann. Chem.*, **725**, 234 (1969).

367. E. Lieber and J. P. Trivedi, *J. Org. Chem.*, **25**, 650 (1960).

368. L. Maier, *Angew. Chem.*, **81**, 154 (1969); *Angew. Chem. Intern. Ed. Engl.*, **8**, 141 (1969).

369. J. Chauvin and Y. Mollier, *Compt. Rend. Ser. C*, **268**, 294 (1969).

370. F. Asinger, A. Saus, H. Offermann, and F. A. Dagga, *Ann. Chem.*, **723**, 119 (1969).

371. W. Walter and J. Voss, *Ann. Chem.*, **695**, 87 (1966).

372. W. Walter, H.-P. Kubersky, and D. Ahlquist, *Ann. Chem.*, in press.

373. G. C. Barrett and A. R. Khokhar, *J. Chem. Soc. (C)*, 1120 (1969).

374. H. Kunzek and G. Barnikow, *Chem. Ber.*, **102**, 351 (1969).

375. W. Walter and R. F. Becker, unpublished results.

376. W. Hieber and M. Gscheidmeier, *Chem. Ber.*, **99**, 2312 (1966).

377. W. Hieber and K. Kaiser, *Z. Anorg. Allgem. Chem.*, **358**, 271 (1968).

378. W. Hieber and W. Rohm, *Chem. Ber.*, **102**, 2787 (1969).

379. H. Alper and J. T. Edward, *Can. J. Chem.*, **46**, 3112 (1968).

380. C. Suarez Contreras, *An. Quim.*, **64**, 819 (1968); *Chem. Abstr.*, **70**, 56896 (1969).

381. E. P. Nesynov, M. M. Besprozvannaya, and P. S. Pel'kis, *Zh. Organ. Khim.*, **5**, 58 (1969); *Chem. Abstr.*, **70**, 87236 (1969).

382. W. E. Kingsbury and C. R. Johnson, *Chem. Commun.*, 365 (1969).

383. G. Barnikow and G. Saeling, *Z. Chem.*, **9**, 145 (1969).

384. R. Gompper and W. Elser, *Ann. Chem.*, **725**, 24, 73 (1969).

385. H. Weber, W. Aumüller, R. Weyer, and K. Muth, *S. African Pat.*, 6,706,091 (1968); *Chem. Abstr.*, **70**, 57462 (1969).

386. G. C. Barrett and A. R. Khokhar, *J. Chromatog.*, **39**, 47 (1969).

CHAPTER **9**

The chemistry of the thiohydrazide group

W. WALTER and K. J. REUBKE

University of Hamburg, Germany

I. GENERAL REMARKS

Whereas the thioamides have attracted great attention for a long time the chemistry of the closely related thiohydrazides is relatively young. Difficulties in the preparation of these compounds together with their often little stability may be the reason for the comparatively small number of publications on this topic. A review on the subject has not as yet been published. Only compounds of the general formula 1

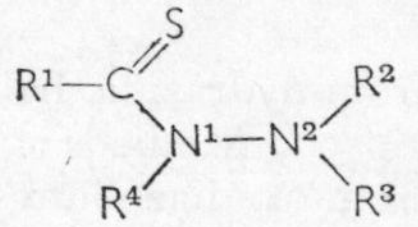

$$
R^1-C\underset{\underset{R^4}{\diagup}}{\overset{\overset{\displaystyle S}{\parallel}}{\diagdown}}\underset{N^1-N^2}{\overset{}{}}\underset{R^3}{\overset{R^2}{\diagdown}}
$$

(1)

are regarded here as thiohydrazides where R^1 is hydrogen or a residue linked by carbon. The nitrogen adjoining the —C=S group is designated as N^1 both in the thionohydrazide—as well as in the zwitterionic form (section III.A).

Great differences as to ease of synthesis, physical properties and chemical reactions are to be expected for the various types of substituted thiohydrazides. This paper will deal mainly with open-chain derivatives though the cyclic products will not be completely disregarded.

Thiohydrazides have often been treated along with thioamides and the close relation between both classes of compounds concerning a number of physical as well as chemical properties is well established. An object of special interest will therefore be the question as to whether there are generally only minor differences or if profound ones are also to be found between the two classes.

The first to our knowledge to report the synthesis of thiohydrazides was Sakurada[1]. As he has given neither yields nor physical properties of any kind except an analytical value for sulphur content these first results seem of little value. A systematic investigation of methods of preparation, physical properties and chemical reactions of thiohydrazides began in 1929 with the work of Wuyts and his school[2].

II. THE FORMATION OF THIOHYDRAZIDES

A. Thioacylation of Hydrazines

1. Thioacylation with dithioacids

The reaction of hydrazine and some substituted hydrazines with dithioacids was first reported by Wuyts and coworkers[2-8] and was later reexamined by Jensen and coworkers[9-11]. If a dithioacid 2 is added to an ethereal solution of a hydrazine at low temperature a salt is formed which in some cases can be isolated[2,5]. The decomposition of the salts or of the mixture of dithioacid and hydrazine, where a salt can not be isolated, yields various compounds depending on the substitution in the acid and the hydrazine as well as on the reaction conditions.

The thiohydrazide (reaction 1) is predominantly formed by the reaction of aliphatic dithioacids (R^1 = alkyl) on 1-methyl-1-phenyl-hydrazine (R^2 = Me; R^3 = Ph) if there is no excess of base and if a polar non-basic solvent is used[4,5]. With aromatic dithioacids (R^1 = aryl) 1-methyl-1-phenylhydrazine yields hydrazones (reaction 2) as main products[4] and with phenylhydrazine (R^2 = H; R^3 = Ph) mixtures of thiohydrazide and hydrazone are formed[2,3]. Whether the formation of hydrazone involves the reduction of primarily formed thiohydrazide or whether another mechanism is operative cannot be decided by the given facts. Reactions (1) and (2) may be regarded as the principal reactions. Another complication arises if unsubstituted hydrazine is reacted with an aromatic dithioacid under these conditions, for then a second thioacylation on N^2 and elimination of hydrogen sulphide might result in thiadiazoles (reaction 5)[7] though in some cases good yields of thiohydrazides were reported[9]. According to reactions (3), (8), and (9) azines, dihydrotetrazines or diamidrazones might be obtained from aromatic dithioacids and hydrazine hydrate [8]. If reactions (1) and (2) proceed on the same molecule of hydrazine (reaction 4) as is the case with methylhydrazine and aromatic dithioacids, thioacylated hydrazones are formed [7,13].

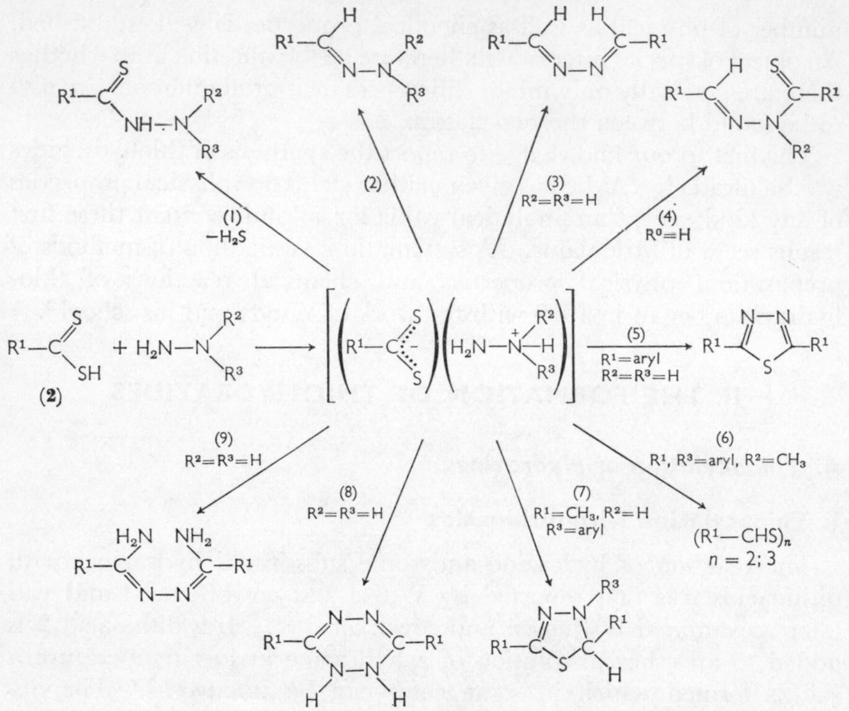

Reactions (2), (3), and (4) involve reduction of the dithioacid. In one case viz. dithio-*o*-toluic acid, the reaction with 1-methyl-1-phenylhydrazine led to thioaldehyde (isolated as its polymer) as a minor by-product (reaction 6). This may be due to the fact, that with this hydrazine only reactions (1) and (2) are possible. As the reactions of N^2-substituted thiohydrazides with carbonyl compounds lead to dihydrothiadiazoles (section IV.F.1.a) the formation of 2,5-dimethyl-3-phenyl-2,3-dihydro-1,3,4-thiadiazole from phenylhydrazine and dithioacetic acid is not surprising (reaction 7) [5,6]. No thiohydrazides could be obtained by the reaction of aliphatic dithioacids with unsubstituted hydrazine [10,11], 3-6-dialkyldihydrotetrazines are obtained instead (reaction 8) [11,12].

The more stable metal salts of dithioacids may be reacted with hydrazines[13-19], too. This method is most useful when the potassium or sodium salts of dithioacids may easily be obtained from trichloromethyl derivatives, e.g. potassium dithioformate from chloroform[14-16] and potassium salt of dithioisonicotinic acid from p-trichloromethylpyridine[17,18]. Whereas the reaction of the potassium salt of dithioisonicotinic acid leads to the corresponding thiohydrazide with unsubstituted hydrazine and to the corresponding hydrazone according to reaction (2) with phenylhydrazine, potassium dithioformate is reported to give the products of reaction (1) with phenylhydrazine or other monoarylhydrazines[14-16], and to produce a 1,3,4-thiadiazole with unsubstituted hydrazine according to reaction (5)[12].

2. Thioacylation with dithioacid esters

Less complex is the reaction of dithioacid esters with hydrazines. Whereas the reaction with methyl and ethyl esters of dithioacids with hydrazine often leads to thiohydrazides in good yields[9,19a], the higher alkyl esters react slowly or not at all[10]. Thiadiazoles are obtained in some cases, presumably due to a second thioacylation and elimination of hydrogen sulphide (cf. reaction 49)[19a].

With monosubstituted hydrazines mixtures of N^1- and N^2-thioacylated products are obtained, e.g. reaction of methyl dithiobenzonate with benzylhydrazine yielded 23% of N^1- and 72% of N^2-thiobenzoylbenzylhydrazine[20] (reactions 10 and 11).

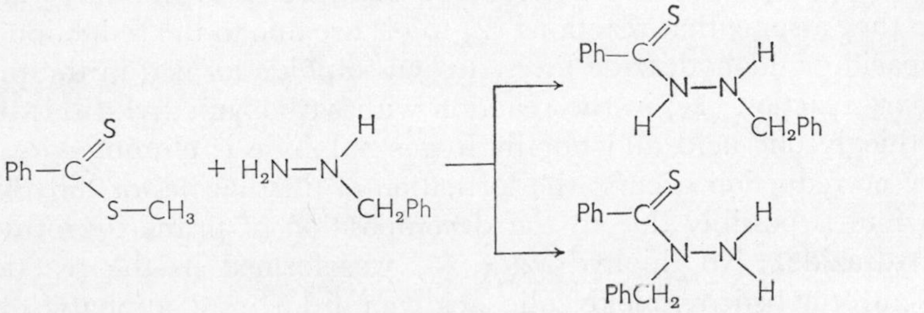

From β-carbonyldithioacid esters on reaction with hydrazine or phenylhydrazine no thiohydrazides were isolated and diazoles were formed instead (equation 12)[21].

Best results are obtained with the activated carboxymethyl esters of dithioacids which were first introduced as thioacylating reagents by Holmberg[22] and later were further investigated by Jensen and coworkers[10,23]. The reaction of unsubstituted hydrazine with carboxymethyl esters (4) of aromatic dithioacids produces satisfactory yields of unsubstituted thiohydrazides (5)[10,24,25] (equation 13).

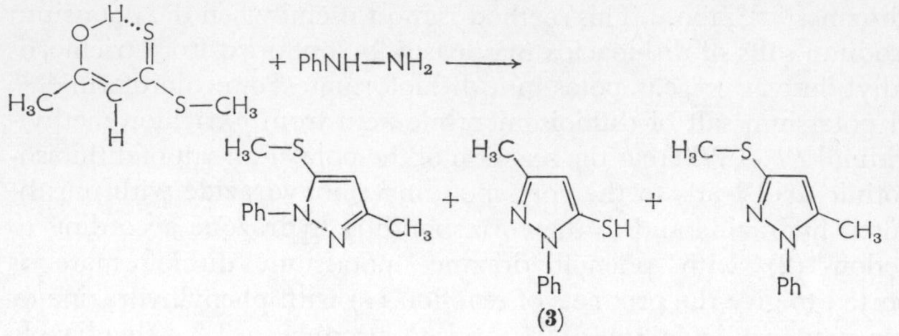

$$R = t\text{-Bu}, \; o\text{-}C_6H_4OBu\text{-i} \;\; o\text{-}C_6H_4OC_5H_{11}\text{-n}$$

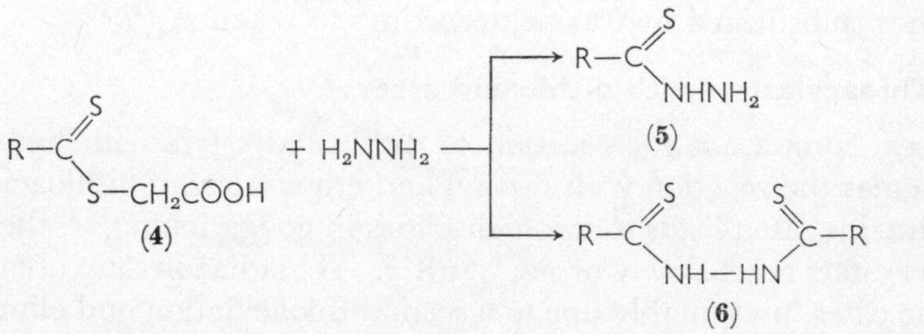

The formation of thioaldehyde, hydrazone or thioacylhydrazone has not been observed, whereas thiadiazoles or dihydrotetrazines have sometimes been isolated as main products (analogous to reactions 5 or 8)[10,11]. This proves that Wuyts and Lacourt[6] are probably right when they assume that reactions (2) to (4) are due to the reduction of dithioacid or thiohydrazide by hydrogen sulphide formed in the preliminary reaction (1); in the reaction with carboxymethyl dithioates (4), thioglycolic acid and not hydrogen sulphide is eliminated and hence no reduction occurs; the formation of thiadiazoles or dihydrotetrazines is possibly due to the decomposition of primarily formed thiohydrazide. No thiohydrazide (5) was formed in the reaction of *m*-nitrothiobenzoylthioglycolic acid nor from most aliphatic carboxymethyl dithioates. The only aliphatic *N*-unsubstituted thiohydrazide obtained by this method is thiopivalic acid hydrazide (5, R = *t*-Bu), N^1,N^2-dithioacylated hydrazines (6) were obtained only with (R = *o*-C$_6$H$_4$OBu-i and *o*-C$_6$H$_4$OC$_5$H$_{11}$-n). In all other cases excess of 4 led to the thiadiazole[10] (reaction 49, section IV.F.1.a).

N-Monosubstituted hydrazines react with carboxymethyl dithioates (4) to yield mixtures of N^1- and N^2-thioacylated products, by reactions similar to (10) and (11). The ratio of the two isomers depends largely on the steric requirements. With n-alkylhydrazines the N^1-thioacylation prevails, with t-alkyl as well as with phenyl- or (substituted phenyl)-hydrazines only the N^2-thioacylated product is formed.

The influence of the substituent of the carboxymethyl dithioate seems to be much less important. With N,N-disubstituted hydrazines the reaction proceeds well irrespective of the nature of the alkyl substituent on the hydrazine though in some cases with rather low yields (2–3% for N,N-dimethylthiopivalic acid hydrazide)[23].

N^1,N^2-Disubstituted hydrazines show a different behaviour. Only N^1,N^2-primary alkyl or benzyl-substituted thiohydrazides are obtainable. No reaction could be observed with N^1,N^2-diphenyl-hydrazine (hydrazobenzene)[23].

3. Thioacylation with thionoacid esters

There is some controversy about the thioacylation of hydrazines with thionoacid esters. Sakurada[1] reported the synthesis of some N^2-phenylthiohydrazides from phenylhydrazine and thionoacid esters together with the analogous reaction of amines to yield thioamides. Whereas the latter reaction has been critically reexamined and checked (Chapter 8, section II.B) the results concerning the reaction with phenylhydrazine seem not to have been checked. The reaction of ethyl thionoisonicotinate with hydrazine to yield thioisonicotinic acid hydrazide has been reported in a patent[26] but Jensen and coworkers[10] tried to repeat the reaction without success. They also reported[10] that the reaction of ethyl thionobenzoate with hydrazine failed to give the corresponding thiohydrazide but that diphenyldihydrotetrazine and diphenyltriazole were isolated (reaction 15).

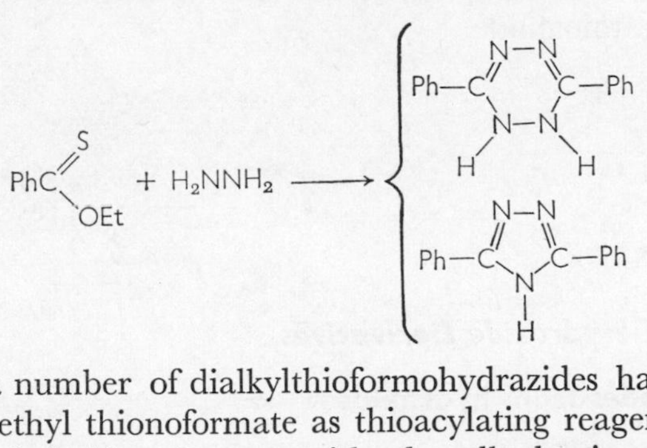

$$\text{PhC}\overset{S}{\underset{OEt}{\big\backslash}} + H_2NNH_2 \longrightarrow \qquad (15)$$

Recently a number of dialkylthioformohydrazides have been prepared using ethyl thionoformate as thioacylating reagent[27,28]. The reaction of ethyl thionoformate with phenylhydrazine resulted in a complex mixture of products, and with 1-methyl-1-phenylhydrazine N^1,N^3-dimethyl-N^1,N^3-diphenylformamidrazone **6c**, and **6a** together with thiohydrazide and the hydrazonoester **6b** were obtained (reaction 16). The yield of the different compounds depends largely on reaction conditions. In basic medium and at moderate temperature the thiohydrazide is the main product, whereas **6b** predominates when the reaction is carried out in ethanol at low temperature.

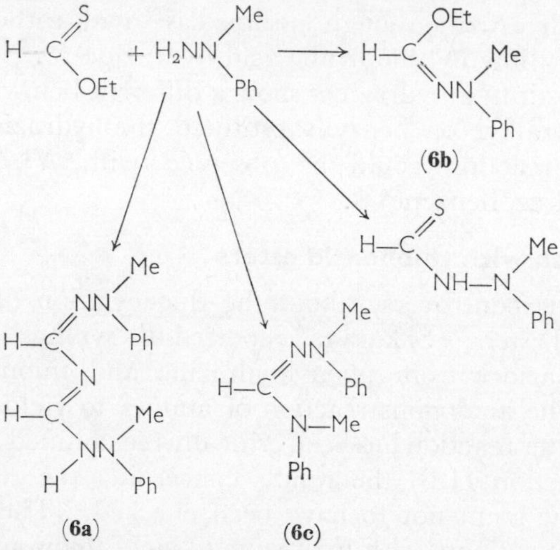

$$(16)$$

4. Thioacylation with thioamides

In a Japanese patent[98] the synthesis of 2,6-dichlorothiobenzoylhydrazones from 2,6-dichlorothiobenzamide, hydrazine hydrate and ketones according to equation (16a) is described. This reaction is of special interest since it represents the thioacylation of hydrazine or hydrazone by a thioamide.

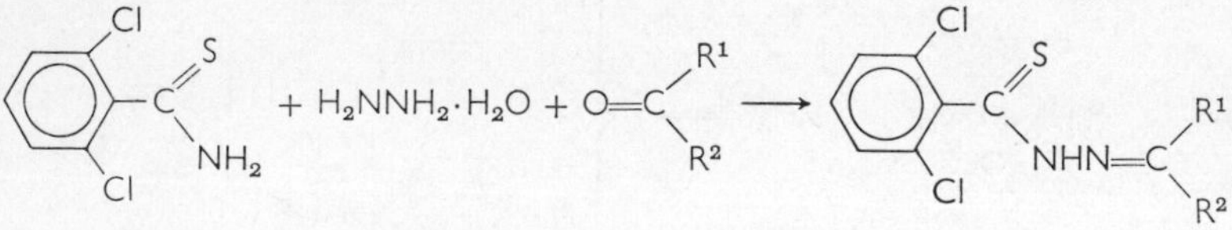

B. Thiolysis of Hydrazide Derivatives

I. Thiohydrazides from hydrazides

a. Phosphorus pentasulphide method. Amides are easily converted to thioamides with phosphorus pentasulphide in many cases (Chapter 8, section II.A). The analogous reaction in the hydrazide series is more complex. Whereas Profft and coworkers[30] obtained a small yield (14%) of thiopicolinic hydrazide, Jensen and coworkers[10] could not prepare thionicotinic or thioisonicotinic hydrazide by this procedure and only a small amount (6%) of thiobenzohydrazide was isolated from benzohydrazide.

Bredereck and coworkers[31] synthesized N,N-diethyl- and N,N-diisopropylthioformohydrazide from the corresponding formohydrazides with phosphorus pentasulphide in benzene solution, in satisfactory yields (61 and 37% respectively); N^2-alkylthiohydrazides were prepared from the oxygen analogues[67] by using methylene chloride as solvent. Recently the conversion of hydrazides into thiohydrazides, in moderate yields by phosphorus pentasulphide in toluene has been reported in a patent[33] for some N^1,N^2-dialkyl derivatives of 2,6-dichlorobenzohydrazide.

Conversion of the oxygen analogues into thiohydrazides by the phosphorus pentasulphide method is a well-known reaction for cyclic compounds such as $2H$-pyridaz-3-thione[34] (7) (reaction 17); or

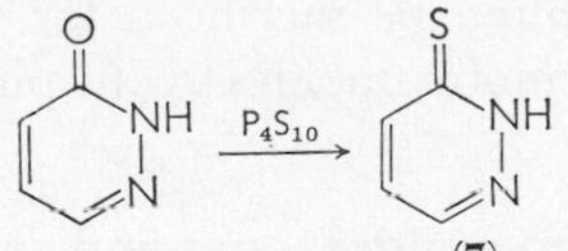

$$\text{(17)}$$

(7)

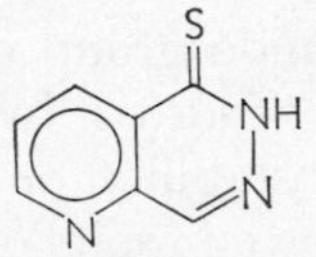

(8)

pyrido-[2,3-d]-$2H$-pyridaz-3-thione (8)[35].

Diacylhydrazines are converted into thiadiazoles by phosphorus pentasulphide. The first step in this reaction presumably is the formation of acyl thiohydrazides[36] (reaction 18) (cf. section IV.F.1.a).

$$\underset{\text{RCNHNHCR}}{\overset{O \quad\quad O}{||\quad\quad ||}} \xrightarrow{\text{P}_4\text{S}_{10}} \left[\underset{\text{RCNHNHCR}}{\overset{O \quad\quad S}{||\quad\quad ||}} \right] \longrightarrow \text{reaction (49)} \qquad \text{(18)}$$

b. Thioacylhydrazones from hydrazides. Thiohydrazides were often prepared in order to be condensed with aldehydes to thioacylhydrazones[9]. The latter were supposed to exhibit interesting physiological properties (section V). A method for preparing certain of these compounds in one step from heterocyclic methyl compounds, hydrazides and sulphur according to equation (19) has been presented in a patent[37].

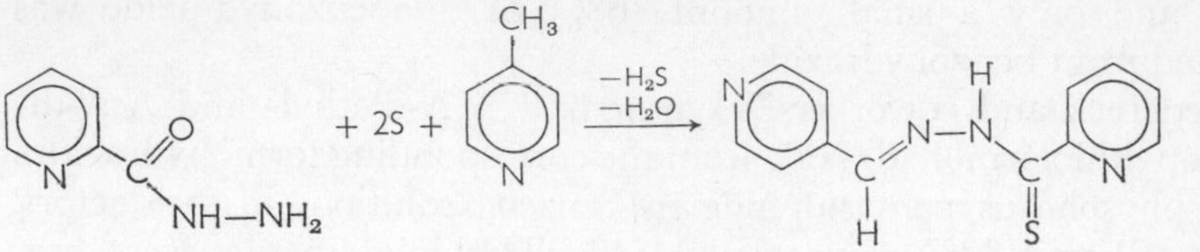
(19)

c. Thioformohydrazides from formohydrazides via N-*isocyanodialkylamines.* An elegant method for preparing dialkylthioformohydrazides is the acid-catalysed addition of hydrogen sulphide to *N*-isocyanodialkylamines, with excellent yields (reaction 20). Since *N*-

$$\mathrm{H\overset{O}{\overset{\|}{C}}NHNR_2 \xrightarrow{-H_2O} \bar{C}{\equiv}\overset{+}{N}NR_2 \xrightarrow{H_2S} H\overset{S}{\overset{\|}{C}}NHNR_2} \qquad (20)$$

isocyanodialkylamines are prepared from formohydrazides, the gain in overall yield may often be sacrificed for the simplicity of direct conversion of dialkylformohydrazides by the phosphorus pentasulphide method.

2. Thiohydrazides from amidrazones and related compounds

The substitution of the imido group of acetamidrazone failed to give thioacetohydrazide as reported by Jensen and coworkers[10]. This may be due to the instability of *N*-unsubstituted thioacetohydrazide which is still unknown in spite of a patent[19] claiming to give a method for preparing this compound.

The N^2-benzoylhydrazone of *N,N*-dimethylformamide (**9**) was converted by hydrogen sulphide into N^2-thioformylbenzohydrazide[36] (reaction 21).

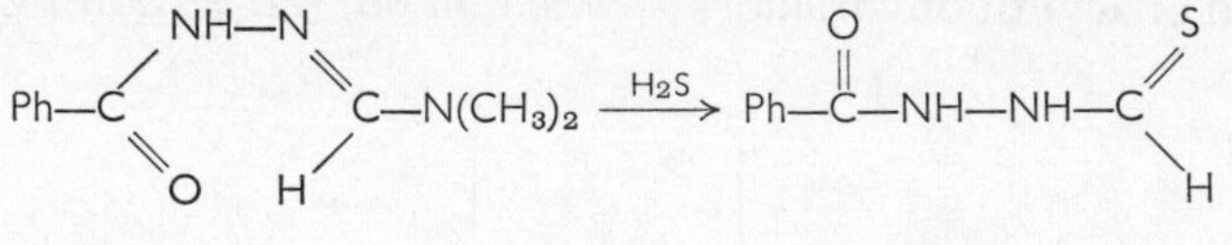
(21)

(**9**)

The conversion of amidrazones to thiohydrazides by hydrogen sulphide at pressures of several atmospheres was possible in the perfluoroalkyl series[38] (reaction 22). Reaction conditions had to be

$$\mathrm{R_F\overset{NH_2}{\overset{|}{C}}{=}NNR_2 \xrightarrow{H_2S} R_F\overset{S}{\overset{\|}{C}}NHNR_2} \qquad (22)$$

carefully tested in every case in order to prevent side-reactions due to solvolytic or reductive properties of hydrogen sulphide.

Formazanes of aldonic acids have been converted into thiohydrazides according to equation (23) [39,40].

$$\begin{array}{c} \text{C} \overset{N-NH-Ph}{\underset{N=N-Ph}{\diagup}} \\ | \\ HO-CH \\ | \end{array} \xrightarrow[\text{EtOH}]{\text{H}_2\text{S}} \begin{array}{c} \text{C} \overset{S}{\diagup} \\ | \qquad \quad \text{H} \\ NH-N \diagdown \\ HO-CH \qquad Ph \\ | \end{array} \tag{23}$$

3. Thiohydrazides from hydrazidic halides

Thiohydrazides can in some cases be prepared from hydrazidic bromides [41]. The latter are easily obtained by bromination of aldehyde hydrazones [42,43] (reaction 24). The hydrazidic bromides need

$$\underset{\text{PhC}=NNR_2}{\overset{H}{|}} \xrightarrow{\text{Br}_2} \underset{\text{PhC}=NNR_2}{\overset{Br}{|}} \xrightarrow{\text{H}_2\text{S}} \underset{\text{PhCNHNR}_2}{\overset{S}{||}} \tag{24}$$

not be isolated but are reacted with hydrogen sulphide solution immediately after adding in succession bromine and triethylamine to the hydrazone. The method has as yet been tried only for dialkylhydrazones of benzaldehyde and nitro-substituted benzaldehydes.

The reaction of o-nitrophenylazochloroacetic acid with potassium sulphide yields thiooxalic acid N^2-o-nitrophenylhydrazide as described in reaction (25) [44]. **10** is formulated merely to clarify the

$$o\text{-NO}_2C_6H_4\text{—N=N—CH} \overset{COOH}{\underset{Cl}{\diagdown}} \quad \rightleftharpoons \quad \left[o\text{-NO}_2C_6H_4\text{—NH—N=C} \overset{COOH}{\underset{Cl}{\diagdown}} \right] \xrightarrow{K_2S}$$

$$\textbf{(10)}$$

$$o\text{-NO}_2C_6H_4\text{—NH—NH—C} \overset{COOH}{\underset{S}{\diagdown\diagdown}} \tag{25}$$

analogy of the overall process to reaction (24) and is not believed to be a reaction intermediate.

In cyclic compounds the thiohydrazide group may be formed by substitution of chlorine as in 3,6-dichloropyridazine converted to 6-mercapto-2H-pyridazine-3-thione (**11**) by potassium hydrogen sulphide (reaction 26) [45].

Instead of hydrogen sulphide, thiourea can be used in some cases

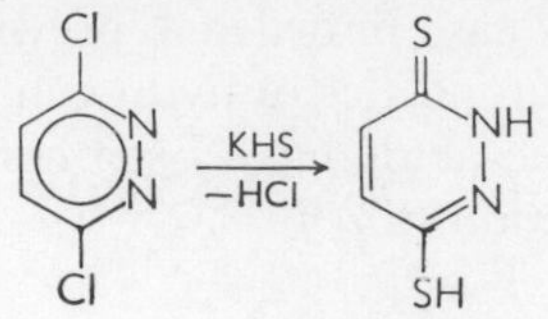

$$\text{(11)} \qquad\qquad \xrightarrow[-\text{HCl}]{\text{KHS}} \qquad\qquad \text{(26)}$$

(11)

to substitute sulphur for chlorine in pyridazine derivatives as in thieno-[2,3-*d*]-pyridazine (**12**)[46] and furo-[2,3-*d*]-pyridazine (**13**)[47] derivatives (reactions 27a,b).

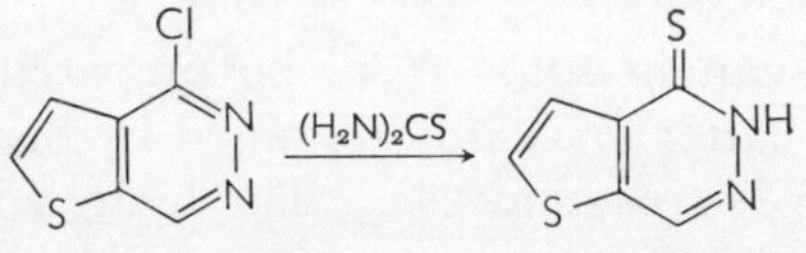

$$\xrightarrow{\text{(H}_2\text{N)}_2\text{CS}} \qquad\qquad \text{(27a)}$$

(12)

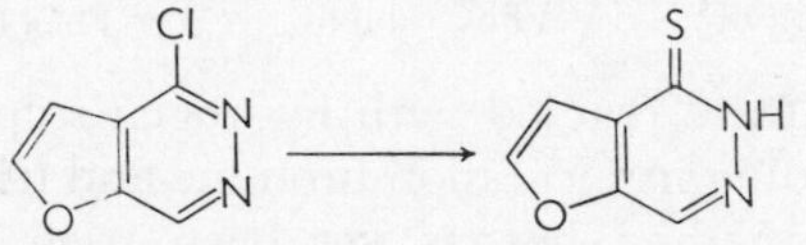

$$\longrightarrow \qquad\qquad \text{(27b)}$$

(13)

C. Thiohydrazides by Special Methods

Recently a number of reactions has been reported to produce thiohydrazides. The reactions outlined in this section were neither designed as preparative methods nor were they tested for wider applicability.

I. Thiohydrazides from oxadiazolium salts

Recently it has been found[48] that upon nucleophilic attack of hydrogen sulphide on 1,3,4-oxadiazolium salts (**14**) ring scission occurs with formation of acylated thiohydrazides according to reaction (28).

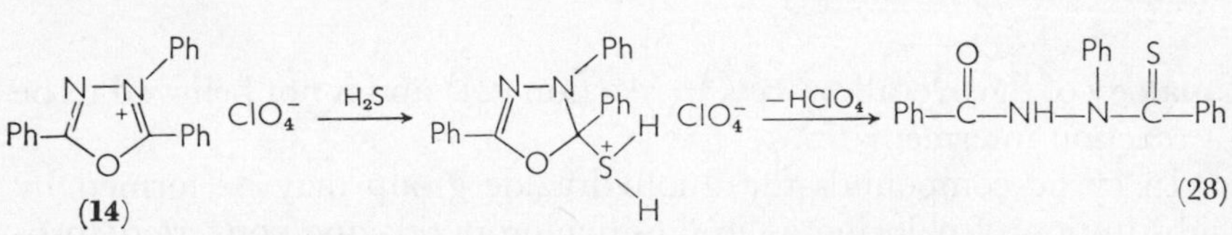

$$\xrightarrow{\text{H}_2\text{S}} \qquad\qquad \xrightarrow{-\text{HClO}_4} \qquad \text{Ph}-\overset{O}{\underset{}{C}}-\text{NH}-\overset{Ph}{\underset{}{N}}-\overset{S}{\underset{}{C}}-\text{Ph} \qquad \text{(28)}$$

(14)

This reaction is related to the formation of thiohydrazides from hydrazidic derivatives, for the oxadiazolium ring may be looked at as the anhydro form of a doubly enolized N^1,N^2-diacylhydrazine.

2. Thiohydrazides from isosydnones and sydnones

Closely related to reaction (28) is the ring opening of isosydnones (**15**) by hydrogen sulphide to yield N^2-substituted thiohydrazides[49].

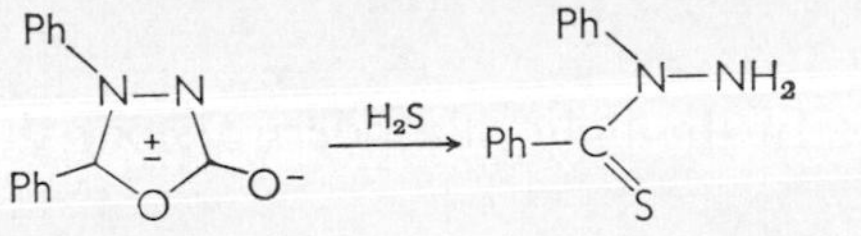

$$\text{(15)} \xrightarrow{H_2S} Ph-C \begin{array}{c} N-NH_2 \\ S \end{array}$$

(29)

Reaction (29) is of some interest as it allows the preparation of N^1-substituted isomers in those cases where thioacylation yields only the N^2-isomer (section II.A.2). N^1-Phenylthiobenzohydrazide was previously unknown(cf. ref. 32).

N-Phenylsydnone (**16**) reacts with 4,4'-dimethoxythiobenzophenone in a 1,3-dipolar addition. The bicyclic intermediate is transformed into the thioformylhydrazone **17**[50] (reaction 30).

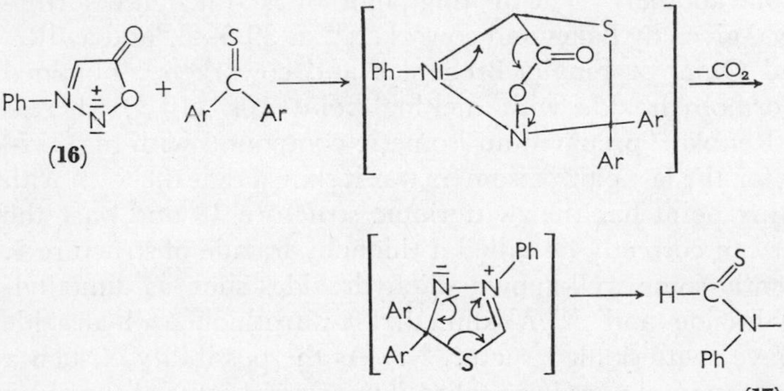

(30)

(17)

III. STRUCTURE AND PHYSICAL PROPERTIES OF THIOHYDRAZIDES

A. Thiohydrazide Tautomerism

A fundamental difference between thiohydrazides and thioamides has been pointed out when it was shown that thiohydrazides can in some cases exist in a tautomeric zwitterionic form[27]. Structure **18** differs from the earlier discussed **19**[2,51] by the position of the hydrogen atom. Wuyts and Lacourt[51] cited the solubility in alkali, the easy formation of disulphides and the reaction with carbonyl compounds as evidence for the existence of a tautomeric 'mercapto' form of

(1)   (18)   (19)

thiohydrazides. In the light of modern theory the argument is less convincing.

Brown and coworkers[38] concluded from i.r. spectroscopic evidence (cf. section III.B.2) that in some perfluoroalkylthiohydrazides the proton might best be represented as situated between S and N^2. Nuclear magnetic resonance spectroscopic evidence (cf. section III.B.4) allowed the distinction between zwitterionic (18) and neutral thiolimidic form (19). Three thiohydrazides which seem to exist in two different forms in the solid state have been reported in the literature. Holmberg found that thiobenzohydrazide is 'dimorphous' one form melting at 70·5–71·5°, the other at 81–82°. Both forms were converted into one another. The melting point for N^2-phenylthioformohydrazide was given by Baker and coworkers[15] as 39·5–41° and as 102° by Sato and Ohta[14]. Finally Bredereck and coworkers[31] obtained diethylthioformohydrazide with melting point 108–110°*, whereas Walter and Reubke[27] prepared an isomeric compound with m.p. 77–78°. Only for the last pair of isomers was it shown that the form with higher melting point has the zwitterionic structure 18 and only the other isomer can correctly be called a thionohydrazide of structure 1.

Very recently some well-known thiohydrazides such as dimethyl-thiobenzohydrazide and N^2,N^2-dimethyl-m-nitrothiobenzohydrazide proved to have zwitterionic structure[41]. As the possibility of such a structure has previously not been taken into account some of the older assignment of physical properties will have to be critically revised.

In many cases it is not obvious whether the examined thiohydrazide has zwitterionic or thionohydrazide structure or whether mixtures of both forms are involved.

B. Molecular and Electronic Structure of Thiohydrazides

I. X-ray diffraction

There are as yet no x-ray diffraction data available for any open-chain thiohydrazides. Only the structures of 2H-pyridaz-3-thione

* Bredereck and coworkers[31] gave the m.p. 102–105°, due probably to a small amount of the other isomer, not detectable by analysis and not easily removed by recrystallization.

(7)[52] and of the copper complex of dimethylthioformohydrazide[53] (cf. section IV.C) have been elucidated.   Even though in 2H-pyridaz-3-thione proton migration to the sulphur atom would be favoured by the formation of a heteroaromatic system, as was formerly believed, this molecule exists in the solid state in the thionohydrazide form.   As for thioamides the formation of the thiolimidic form is not necessary for providing the aromatization of the heterocycle[57].   Bond lengths and angles of the true thionohydrazides are comparable to those of thioamides (cf. Chapter 8).

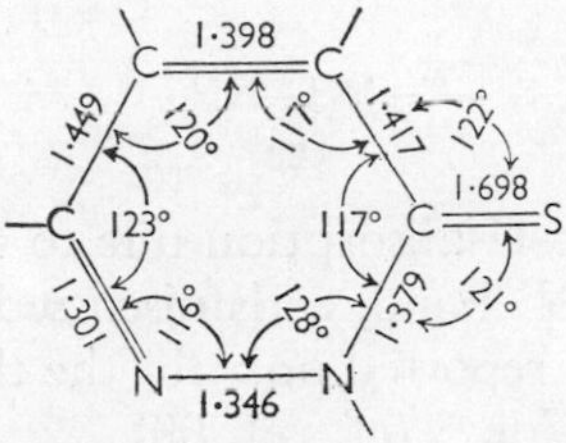

(7)

2. Infrared spectroscopic evidence

No publication completely devoted to the i.r. spectra of thiohydrazides is known to us, data being scattered in papers mostly concerned with other topics.   In their comprehensive article on the i.r. spectra of thioamides and selenoamides Jensen and Nielsen[54] (cf. Chapter 8) report that the A band due to vibrations of the NH_2 group is found at lower frequencies in thiohydrazides than in thioamides.   This is explained by the fact that the NH_2 group is not directly attached to the C—S group.   The B band (1400–1600 cm^{-1}) is reported to be present in all, the C band (1200–1400 cm^{-1}) in most thiohydrazides, nothing being said about the D (1000–1200 cm^{-1}), E, F, and G bands in these compounds.

Infrared spectra of N-unsubstituted thiohydrazides of eight aromatic acids (thiobenzohydrazide of m.p. 70–71°, cf. section III.A) have been reported by Rao and coworkers[55].   Three bands were given as characteristic for these compounds: I (1545–1495 cm^{-1}), II (1325–1300 cm^{-1}), and III (1050–1000 cm^{-1}) which correspond to the B, C, and D bands of Jensen and Nielsen[54].   It was inferred from the spectra that no tautomerism with the thiolimidic form as discussed for thioamides (cf. Chapter 8) occurred, all of the compounds exhibiting characteristic N-H absorption.   For the same reason Lieber[56], too, assigned the thionohydrazide structure to these compounds.

In N^2-acylthiohydrazides and N^1,N^2-dithioacylhydrazines the simultaneous resonance in amidic and thioamidic, or in both thion-amidic groupings respectively is not favoured because of the adjacent-charge rule (dication effect[57]).   The thiolimidic form is therefore more likely to be encountered in these compounds.   No data are available on the few dithioacylhydrazines known.   Some acylthiohydrazides have been studied by Sandström[58] and from the presence of two bands in the stretching region of associated N—H, a strong carbonyl absorption, and the absence of absorption assignable to S—H stretching vibration, an amidic–thionamidic structure (**20**) was assigned.

$$\underset{\textbf{(20)}}{R^1\!-\!\overset{\displaystyle S}{\overset{\|}{C}}\!-\!NH\!-\!NH\!-\!\overset{\displaystyle O}{\overset{\|}{C}}\!-\!R^2}$$

A splitting of the N—H absorption due to the presence of *cis–trans* isomers about the C—N^1 bond, as is discussed in Chapter 8 for thioamides, has as yet been reported only for the thionohydrazide form of diethylthioformohydrazide[59] (cf. ref. 99).

Only in 6-mercapto-2H-pyridazin-3-thione (**11**) does the extra stabilizing conjugation with the ring double bond and the dication effect hindering simultaneous resonance in the two thioamide groups lead to enolization of one thionamide half[57].   Structure **11** is analogous to the one found for the oxygen analogue, cyclic maleic hydrazide. The S—H stretching vibration for **11** is reported[60] at 2360 cm^{-1}.

Whereas the i.r. spectrum of heptafluorothiobutyrohydrazide shows strong N—H absorption in the spectrum of its N^2,N^2-dimethyl derivative no such absorption is observed, but instead strong bands appear at 2530–2570 cm^{-1} which do not shift in diluted carbon tetrachloride solution[38].   It was inferred that the proton in the latter compound was involved in a strong hydrogen bond between S and N^2 according to structure **19**.   The absence of N—H absorption and a band at 2776 cm^{-1} (in sodium bromide) in the spectrum of the one isomer of N^2,N^2-diethylthioformohydrazide led Bredereck and co-workers[31] to the assumption that a thiolimide tautomer was present. As the band at 2776 cm^{-1} is observed at an uncommonly high frequency for S—H stretching vibration and participation in a hydrogen bond could explain only a shift in the opposite direction, Walter and Reubke[27,59], therefore, assigned structure **18** ($R^1 = H$; $R^2 = R^3 = C_2H_5$) to this compound.   The band under discussion is then to be ascribed to the ammonium hydrogen in agreement with data on ammonium[61] and hydrazonium salts[62].   Recently Anthoni and coworkers[63] deduced a similar structure for the closely related thiosemicarbazide derivative **21** which shows relatively weak and broad bands in the 2500–2900 cm^{-1} region ascribed to the ammonium group.

(21)

3. Ultraviolet spectroscopic evidence

Ultraviolet spectra of some thiohydrazides have been recorded and discussed by Sandström and coworkers[64-66]. The $n \rightarrow \pi^*$ and $\pi \rightarrow \pi^*$ transitions in N-unsubstituted and N-alkyl-substituted thiohydrazides are shifted hypsochromically compared to the corresponding thioamides and are found at 381 to 325 nm and 266 to 289 nm respectively. The shifts are much larger than expected from simple LCAO–MO calculations. It is concluded that the effect is not purely inductive but that a strong electron repulsion is operative. The absorption maximum of '3-mercapto-pyridazine' (**7**) is found in the same region (bands at 282 and 355 nm) showing, that it is no mercapto compound but has the thionamidic structure also in solution and therefore is better called 2H-pyridaz-3-thione[67]. Somewhat more complex are the results for 6-mercapto-2H-pyridazin-3-thione (**11**)[67] because of the conjugation with the ring double bonds. The observed absorptions are in good agreement with the assigned structure **11**.

In accord with structure **18** proposed for one isomer of diethyl-thioformohydrazide the $\pi \rightarrow \pi^*$ absorption is observed at still shorter wavelength—262 nm—whereas that for the thionohydrazide form is found at 274 nm (in chloroform solution)[59].

For heptafluorothiobutyrohydrazide, apparently a true thiohydrazide (section III.B.2), the absorption maximum is found at 284 nm[38]. It is shifted to 276 nm in the N^2,N^2-dimethylated derivative, which most certainly has zwitterionic structure. It is assumed that the $\pi \rightarrow \pi^*$ transition is aided by the electron-withdrawing perfluoroalkyl groups.

For N^2-phenylthiohydrazides the strong absorption due to the $\pi \rightarrow \pi^*$ transition is shifted to the red compared with the corresponding thioamides, 287–333 nm (PhCH$_2$CSNHNHPh and PhCSNHNHPh), the same holding for the N^2-acylthiohydrazides[58] in good agreement with theoretical predictions, indicating that some interaction is involved between the amide and the thioamide half of the molecule. The red shift of the $\pi \rightarrow \pi^*$ band actually increases with increasing conjugation. In thioacylhydrazones the corresponding band is found at 312.5 nm ($PhCSNHN=C(CH_3)_2$) to 365 nm ($PhCSNHN=CHCH=CHPh$)[65]. $n \rightarrow \pi^*$ transitions are not easily detected in the spectra of these compounds, the $\pi \rightarrow \pi^*$ transitions causing broad and strong bands.

4. Nuclear magnetic resonance spectroscopic evidence

Nuclear magnetic resonance spectra first allowed a conclusive decision about the position of the hydrogen atom in the zwitterionic form of thiohydrazides to be made[59]. There is a broad signal observed in the spectrum of the diethylthioformohydrazide zwitterion at $\tau = -1.5$ to -1.0 p.p.m., the exact position depending on the solvent. The broadening indicates the position on an N atom, the position at very low field is characteristic for hydrogen-bonded protons. For an S—H proton a sharp signal would be expected as is observed in compound **3**, for which the thioamidic form would be very unfavourable. The S—H proton in this compound is reported at $\tau = 6.72$ p.p.m.[21]. Furthermore a coupling (4 to 6 Hz) of the N—H proton with the α-hydrogens of the N^2-alkyl groups confirms the assignment. The spectrum of N,N-dimethyl-m-nitrothiobenzohydrazide is given as an example (Figure 1). The splitting of the methyl signal (5 Hz) indicates that the coupled proton is geminal to the methyl groups at N^2. The N—H – α-C—H coupling was observed also for the diethylthioformohydrazide zwitterion (6 Hz), N,N-dimethylthiobenzohydrazide (4 Hz), and N,N-dimethyl-p-nitrothiobenzohydrazide (5 Hz). The splitting of the methyl signal disappears on addition of a polar solvent as well as on deuteration[41].

From the presence of two C—H and N—H signals, apart from those for the zwitterionic form in the n.m.r. spectrum of an equilibrium mixture of diethylthioformohydrazide at low temperature, the occurrence of two rotational isomers was inferred[67]. The pair of protons belonging to the more abundant isomer is observed as an AB system with a coupling constant of 12.8 Hz which proves an (E) configuration with the protons at C and N '*trans*' to each other, i.e. **21a**[59].

(21a)

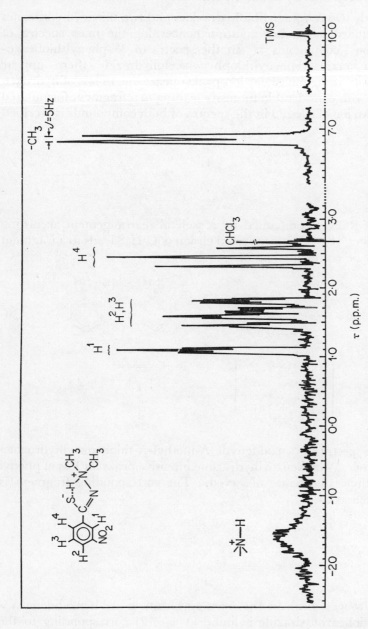

FIGURE 1. Nuclear magnetic resonance spectrum of N,N-dimethyl-m-nitrothiobenzohydrazide in $CDCl_3$. Left part of spectrum with amplitude four times greater than right part.

5. Mass spectrometry of thiohydrazides

Recently two N^2-phenylthiohydrazides and two thioacylhydrazones were included in an investigation concerning the mass spectra of thiocarbonyl compounds[68]. In the spectra of N^2-phenylthiobenzo-hydrazide and N^2-phenylthiophenylacetohydrazide there appear (apart from strong molecular ion peaks) peaks due to loss of hydrogen sulphide with considerable intensity leading to a fragment formulated as **22**. An ion of m/e 125 in the spectra of both compounds is ascribed

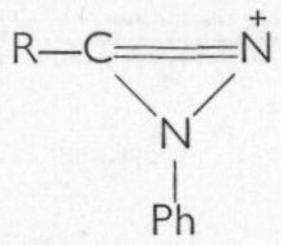

(**22**)

to loss of RCN accompanied by a skeletal rearrangement according to equation (31). An unexpected elision of C_6H_5S leads to a fragment that may be represented as **23**.

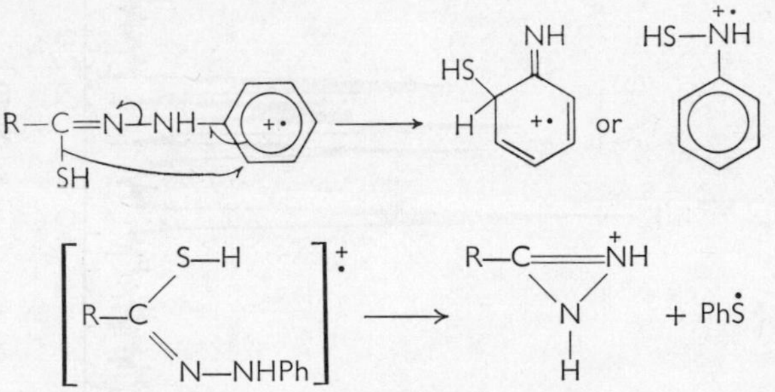

(31)

(**23**)

In the spectra of benzaldehyde N-methyl-N-thiobenzoylhydrazone and acetone N-thiobenzoylhydrazone intensive peaks for loss of phenyl and methyl radicals are observed. The corresponding fragment is formulated as **24**.

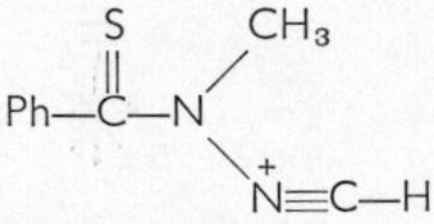

(**24**)

The strongest peak in the mass spectrum of N^2-benzylidene-N^1-methylthiobenzohydrazide is found at m/e 121 corresponding to the fragmentation reaction (32).

$$Ph-C \underset{\underset{N \equiv C-H}{+}}{\overset{\overset{|S|}{\parallel}}{N}} CH_3 \longrightarrow Ph-C \equiv \overset{+}{S} + \underset{\overset{|}{CH_3}}{\overset{N=CH}{N}} \qquad (32)$$

$$m/e \ 121$$

As only a few thiohydrazides, which are not representative of the whole group, have been investigated so far, no conclusions can easily be drawn from these results concerning the thiohydrazido grouping.

6. Theoretical aspects of molecular and electronic structure of thiohydrazides

It seems reasonably safe to draw a number of conclusions from the physical evidence outlined in the preceding sections as to the molecular structure of thiohydrazides.

Thiohydrazides can exist in either the thionohydrazide or the zwitterionic form. In some cases one compound can be isolated in both forms, or the less stable tautomer can be detected in solution by physical methods. The thiolimidic form has never been detected except for cyclic compounds (**3** and **11**). The preference of the thiono-hydrazide over the thiolimidic form is probably due to resonance stabilization in the N—C=S group (cf. Chapter 8). Resonance is possible only if the N—N—C=S group is planar or nearly planar. If therefore N^1 is trigonally hybridized hindered rotation about the C—N^1 bond is to be expected as is the case for thioamides. Indications for *cis–trans* isomerism have been found by i.r. and n.m.r. spectroscopy (sections III.B.2 and 4). For the one thionohydrazide investigated to some extent, diethylthioformohydrazide, predominance of the *(E)* form has been established beyond doubt, in agreement with a hypothesis put forward by Walter and coworkers[69] according to which preference of one of the *cis–trans* isomers is at least partly due to a compensation of the σ bond moments:

$$\underset{S}{\overset{H}{\diagdown}} C-N \underset{H}{\overset{NEt_2}{\diagup}}$$

Proton migration to N^2 is possible in these compounds whereby the zwitterion is formed if the basicity of N^2 is great enough. It is enhanced by resonance and by hydrogen bonding:

$$\left[\underset{N}{\overset{S \cdots H}{\parallel}} -C \diagdown \overset{+}{N} \diagup \longleftrightarrow -C \underset{N}{\overset{S^- \cdots H}{\diagdown}} \overset{+}{N} \diagup \right]$$

Little is known about the electronic structure of thiohydrazides. The close resemblance of the ultraviolet spectra of thiohydrazides and thioamides seemed to justify the application of the simple LCAO–MO method for the calculation of bond orders and electronic transitions of thiohydrazides (cf. Chapter 8). Such a calculation has been performed for N^2-phenylthioacetohydrazide and for some alkylidenethiohydrazides[58,64,65]. The results did not agree well with experimental values and are therefore not reported here in detail. Apparently the influence of the configuration at the C—N^1 bond has to be taken into account.

IV. CHEMICAL REACTIONS OF THIOHYDRAZIDES

A. General Features, Stability, and Analytical Characteristics of Thiohydrazides

Thiohydrazides are mostly colourless solids with sharp melting points but it is often quite difficult to obtain pure products. All thiohydrazides with hydrogen at N^1 are readily soluble in alkali (section IV.B).

Most thiohydrazides are sensitive to light and heat[23]. Thiohydrazides of aliphatic acids are quite unstable. Many attempts to prepare N-unsubstituted thiohydrazides failed and only one compound of this class is known (section II.A). Even at $-30°$ the reaction of 1-hexyn-1-yl thiolacetate (25) with hydrazine hydrate resulted in the formation of 3,6-di-n-pentyldihydrotetrazine instead of the thiohydrazide, which most probably is an intermediate in the reaction, whereas reaction of 25 with amines leads to thioamides (cf. Chapter 8)[12].

$$
\underset{(25)}{\text{n-BuC}\equiv\text{CS}\overset{\overset{\text{O}}{\|}}{\text{C}}\text{CH}_3} \xrightarrow{\text{N}_2\text{H}_4} \text{CH}_3\text{CONHNH}_2 + \left[\text{n-BuCH}_2\overset{\overset{\text{S}}{\|}}{\text{C}}\text{NHNH}_2\right] \xrightarrow{-20°}
$$

$$
\longrightarrow \text{n-BuCH}_2-\text{C} \underset{\underset{\text{N—NH}}{\diagdown\diagup}}{\overset{\overset{\text{NH—N}}{\diagup\diagdown}}{}} \text{C—CH}_2\text{Bu-n} \qquad (33)
$$

In the perfluoroalkyl series the N-unsubstituted thiohydrazides are more stable, but upon storage at room temperature they decompose with loss of hydrogen sulphide and sulphur, the latter being due to the Wuyts reduction (reaction 2).

Thiohydrazides can be detected on chromatoplates by the iodine–azide reaction (cf. Chapter 8). A special test for N^2-monosubstituted thiohydrazides is the Wuyts reaction. From these compounds and benzaldehyde, dihydrothiadiazoles are formed (section IV.D, reaction 54). Upon addition of hydrogen peroxide[41,70] or sodium nitrite[71] in concentrated sulphuric acid solution the presence of thiadiazolines is indicated by a blue or green colour. As in thioamides compared to amides, the sulphur atom in thiohydrazides has far greater nucleophilicity than the oxygen atom in hydrazides. Therefore a number of analogies between thioamides and thiohydrazides are observed in their reactions. On the other hand, the presence of another centre for electrophilic attack in the thiohydrazide molecule brings about a number of differences between both classes. In many cases electrophilic attack on thiohydrazide seems to be equally or almost equally easy at the sulphur and the N^2 atom.

B. Acidity and Basicity of Thiohydrazides

Acidity constants are of great importance for the decision as to whether formation of a zwitterionic form is possible at all. On the other hand it is reasonably safe to suppose that the two isomeric forms

TABLE 1. Equilibrium constants for protonation and deprotonation of thiohydrazides.

Compound		$pK_{a1}(+H^+)$	$pK_{a2}(-H^+)$	Refs.
HCSNHNEt$_2$	thiono form	2·8	10·4	27
	betaine form	3·4	11·7	27
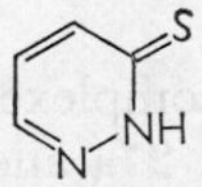 (pyridine-2-thione structure)	thiono form	−2·68	8·30	66
C$_6$H$_5$CSNHNHCH$_2$C$_6$H$_5$			7·63[a]	20
C$_6$H$_5$CH$_2$CSNHNHCH$_2$C$_6$H$_5$			8·7[b]	64
RCSNHNH$_2$		5·0–6·0	10·2–11·4	23
Alkyl—CSNHNH—aryl		4·4–4·7	9·3–10·1	23
Alkyl—CSNHNH—alkyl		5·6–6·2	10·7–11·1	23
Aryl—CSNHNH—alkyl		4·9–5·7	9·9–10·4	23
Aryl—CSNHNH—aryl		4·2–4·3	9·2–9·6	23
RCSNHNR$_2^1$		6·5–6·7		23
RCSNR^{1}NH$_2$		7·2–7·5		23

[a] 3% EtOH, 97% H$_2$O (w/w).
[b] 20% EtOH, 80% H$_2$O (w/w).

of a given thiohydrazide will show different pK_a values, as was found
for diethylthioformohydrazide[27]. In Table 1 pK_a values of thiohydrazides are listed. The ranges given in the lower part are those
reported by Jensen and coworkers[23] for classes of compounds. The
data show that the basicity ranges from $pK_{a1} = 2\cdot68$ to $7\cdot5$, the
acidity from $pK_{a2} = 7\cdot63$ to $11\cdot7$. There are as yet insufficient data
available for individual thiohydrazides of known structure to explain
these rather large differences.

C. Complex Formation of Thiohydrazides

Holmberg[72] prepared a number of complex salts of thiobenzohydrazide with transition metal ions. Later Jensen and Miquel[73]
assigned the N^2-methylthiohydrazide structure to the ligand of the
nickel complex of a methylated thiobenzobydrazide, for it was then
thought that only compounds able to enolize to an imidothiol form
could form complexes. Later Holmberg[13] demonstrated that the
ligand in this complex was actually the N^1 isomer. Holmberg[13] also
synthesized the N^2 isomer. The latter, too, forms a nickel complex as
was reported by Bähr and Schleitzer[74] who assigned the structures
26a and **26b** to the isomeric complexes.

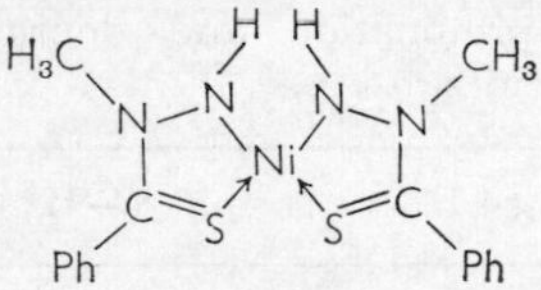

ochre   olive green
(26a)   (26b)

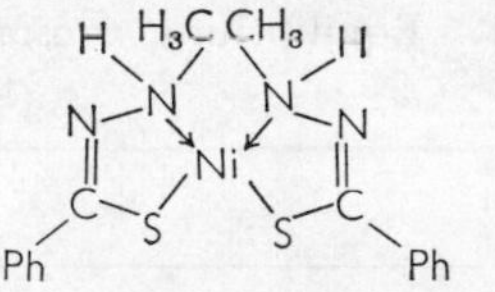

Diethylthioformohydrazide is reported to yield copper complexes
of the thionohydrazide and the zwitterionic tautomer forms, **27a** and
27b respectively[59]. The formation of the complex of the thionohydrazide tautomer was only deduced from the ultraviolet spectrum,
the other complex was isolated. Whereas structure **27a** was derived

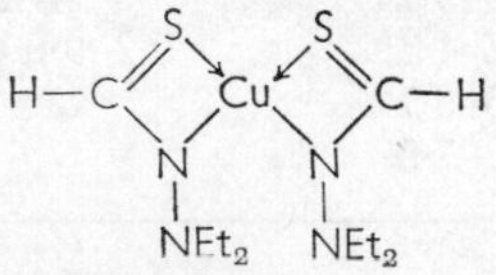

(27a) 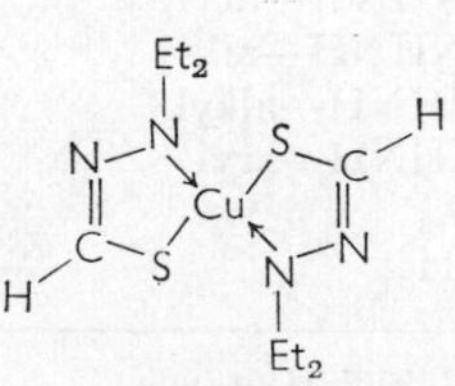   (27b)

from ultraviolet spectral analogy with complexes of thioamides, assignment of structure **27b** was based on the close resemblance of all physical properties of the isolated complex with the copper complex of dimethylthioformohydrazide, for which the structure was determined by x-ray diffraction[53]. As in space group $P_{2_1/c}$ there are four general positions in the unit cell but only two molecules of the N^2,N^2-dimethylthioformohydrazide copper complex are found in it, the molecule must possess central symmetry. The bond lengths Cu—S (2·24 Å) and C—N^1 (1·29 Å) as well as the angle at N^2 (114°) indicate trigonal hybridization of N^1 and a practically localized C—N double bond.

D. Alkylation of Thiohydrazides and Addition to Double Bonds

I. Alkylation of Thiohydrazides

Whereas the question whether *O*-alkylation of hydrazides is possible seems not to be settled (cf. Chapter 10) *S*-alkylation of thiohydrazides is the general reaction, *N*-alkylation being the exception. If *N*-unsubstituted thiohydrazides are treated with alkyl halides in alkaline solution *S*-alkylation produces unstable α-hydrazono sulphides which are converted into dihydrotetrazines with elision of mercaptan (cf. reaction 64). Only in the perfluoroalkylthiohydrazide series *N*-unsubstituted α-hydrazono sulphide (**28**) was obtained[38] (reaction 34). Upon heating in hydrochloric acid solution **28** was decomposed

$$\overset{\text{S}}{\overset{\|}{\text{R}_\text{F}\text{CNHNH}_2}} + \text{CH}_3\text{I} \xrightarrow{-\text{HI}} \overset{\text{SCH}_3}{\overset{|}{\text{R}_\text{F}\text{C}}}=\text{NNH}_2 \xrightarrow[-\text{N}_2]{\text{HCl},\Delta} \text{R}_\text{F}\text{CH}_2\text{SCH}_3 \qquad (34)$$

$$(28) \qquad\qquad\qquad (29)$$

with loss of nitrogen to form the sulphide **29**[38], the reaction being analogous to the Wolff–Kischner reduction.

The *S*-alkylation products of N^2-substituted thiohydrazides seem to be more stable. Sato and Ohta[75] reported the formation of ethyl phenylhydrazonomethyl sulphide from N^2-phenylthioformohydrazide and ethyl iodide in sodium ethoxide solution (reaction 35).

$$\overset{\text{S}}{\overset{\|}{\text{HCNHNHPh}}} + \text{EtI} \xrightarrow{\text{NaOEt}} \overset{\text{SEt}}{\overset{|}{\text{HC}}}=\text{NNHPh} \qquad (35)$$

S-Alkylation is probably the first step in the formation of thiadiazines from *N*-monosubstituted or *N*-unsubstituted thiohydrazides and α-chlorocarbonyl compounds (cf. reactions 61 and 62). As in the case of N^2-phenylthiohydrazides open-chain S-alkylated products are formed[14,76] (reaction 36). In stronger alkaline solution ben-

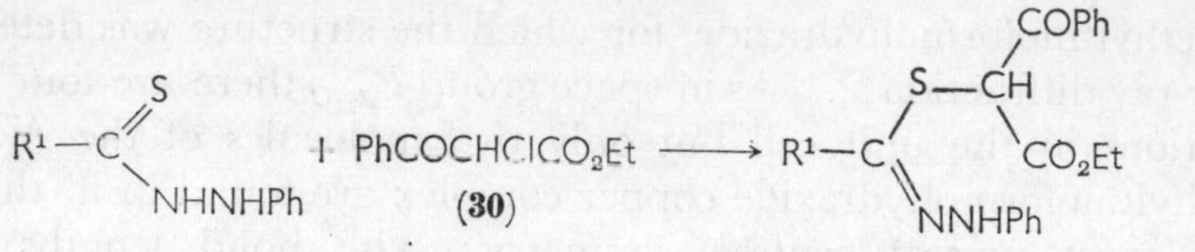

$$R^1-C\underset{NHNHPh}{\overset{S}{\diagdown}} + PhCOCHClCO_2Et \longrightarrow R^1-C\underset{NNHPh}{\overset{S-CH\overset{COPh}{\diagup}}{\diagdown}}\overset{COPh}{\underset{CO_2Et}{}} \tag{36}$$

(30)

zoylhydrazones of ethoxycarbonylmethyl thiolcarboxylates (**31**) are obtained from **30** with thiohydrazides[76] (reaction 37). In an acid-

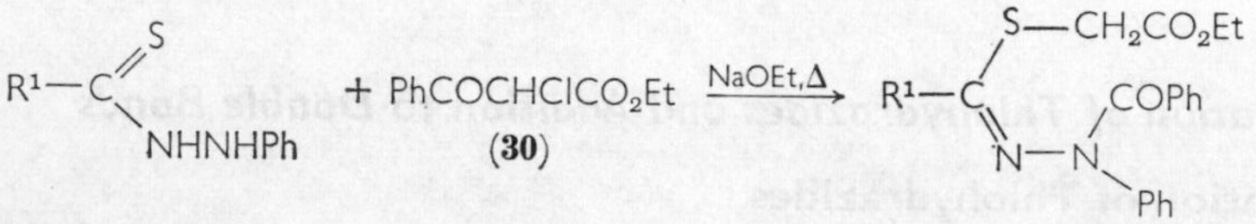

$$R^1-C\underset{NHNHPh}{\overset{S}{\diagdown}} + PhCOCHClCO_2Et \xrightarrow{NaOEt,\Delta} R^1-C\underset{N-N}{\overset{S-CH_2CO_2Et}{\diagdown}}\underset{\underset{Ph}{}}{\overset{COPh}{}} \tag{37}$$

(30) (**31**)

catalysed reaction, compounds **31** are converted to 1,3,4-oxadiazoles[77] (reaction 38).

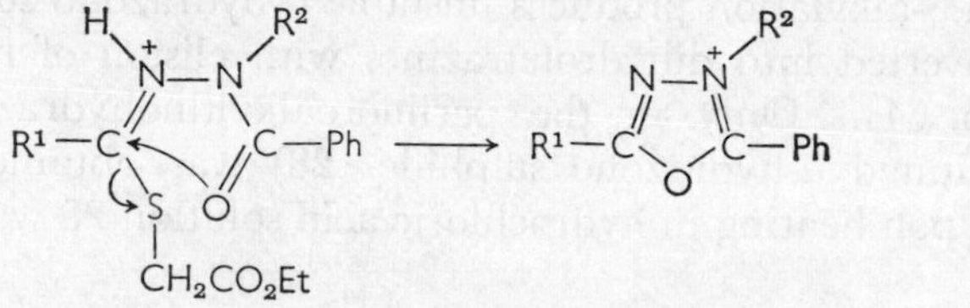

$$R^1-C\overset{\overset{H}{\underset{+}{N}}-\overset{R^2}{N}}{\underset{\underset{CH_2CO_2Et}{S}}{}}C-Ph \longrightarrow R^1-C\overset{\overset{N}{}-\overset{R^2}{\overset{+}{N}}}{\underset{O}{}}C-Ph \tag{38}$$

Two isomeric S-methyl derivatives (**32** and **33**) were obtained by methylation of the two isomers of diethylthioformohydrazide in sodium ethoxide solution[59] (reactions 39 and 40). The isomers **32** and **33**

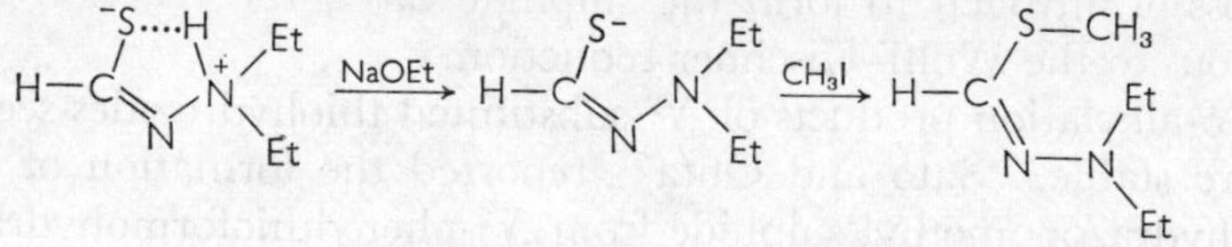

$$H-C\overset{\overset{-S\cdots H}{}}{\underset{N}{}}\overset{Et}{\underset{\overset{+}{N}}{\diagup}}\overset{Et}{\underset{Et}{}} \xrightarrow{NaOEt} H-C\overset{S^-}{\underset{N}{\diagdown}}\overset{Et}{\underset{Et}{\diagup}} \xrightarrow{CH_3I} H-C\overset{S-CH_3}{\underset{N-N}{\diagdown}}\overset{Et}{\underset{\underset{Et}{}}{}} \tag{39}$$

(**32**)

$$S=C\overset{H}{\underset{NH}{\diagdown}}\overset{Et}{\underset{Et}{\diagup}} \xrightarrow{NaOEt} S^-—C\overset{H}{\underset{N}{\diagdown}}\overset{Et}{\underset{Et}{\diagup}} \xrightarrow{CH_3I} H_3C-S-C\overset{H}{\underset{N-N}{\diagdown}}\overset{Et}{\underset{\underset{Et}{}}{}} \tag{40}$$

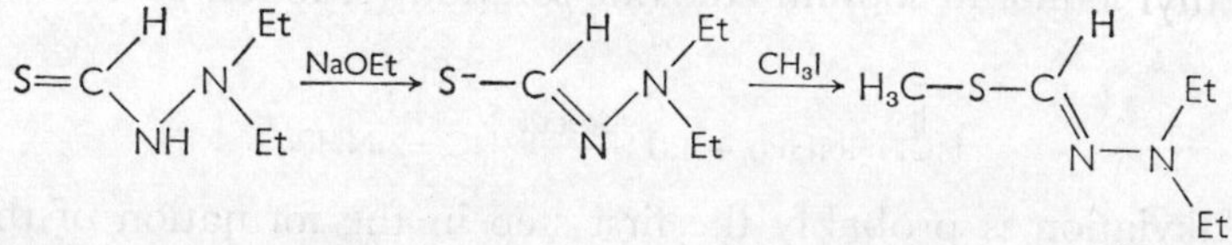

(**33**)

have different boiling points and different i.r. and n.m.r. spectra.

In sodium ethoxide solution slow interconversion of both into an equilibrium mixture is observed which becomes fast upon heating.

N^1-Methylthiobenzohydrazide is also S-methylated to give compound **34** which may be formulated as a salt. This decomposes in hot aqueous solution with loss of methylhydrazinium iodide to form methyl thiolbenzoate[13] (reaction 41). In no case has N-alkylation of

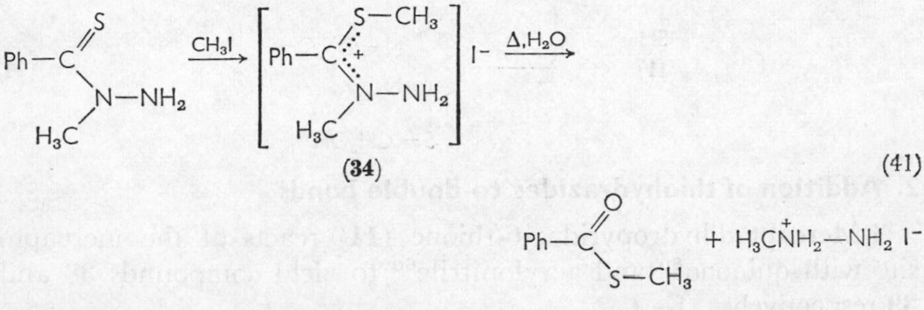

(41)

an open-chain thiohydrazide been reported. In the cyclic series S-alkylation dominates as well. The reaction of thieno-[2,3-d]-dihydropyridazthiones[46] (**35**, X = S) and of furo-[2,3-d]-dihydropyridazthiones[47] (**35**, X = O) with chloroacetic acid produces carboxymethylthio derivatives (reaction 42).

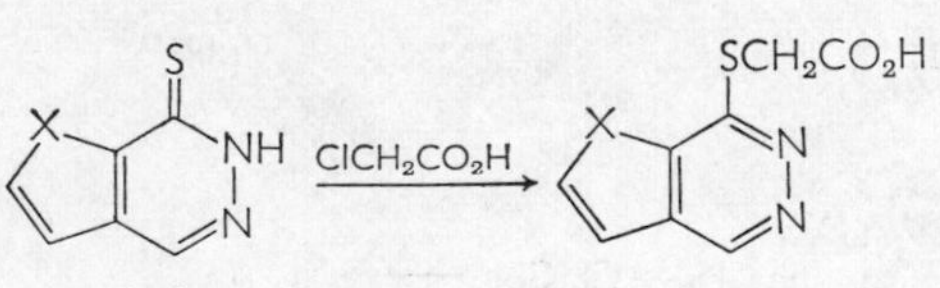

(42)

Only the reaction of acetobromoglucose (**36**) with the cyclic thiohydrazides 3-mercaptophthalazine (**37a**), dihydropyridaz-3-thione (**7**)[78], and cinnol-3-thione (**37b**)[79] yields mixtures of S- and N-alkylated products.

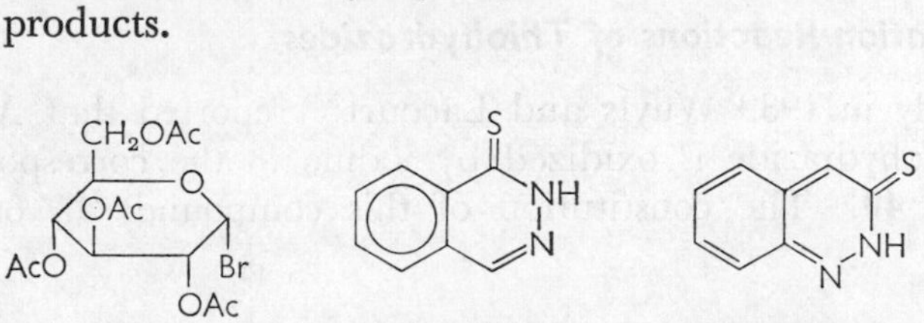

Aminomethylation or hydroxymethylation of **11** is possible at both one sulphur atom and at nitrogen[81] (reactions 43 and 44).

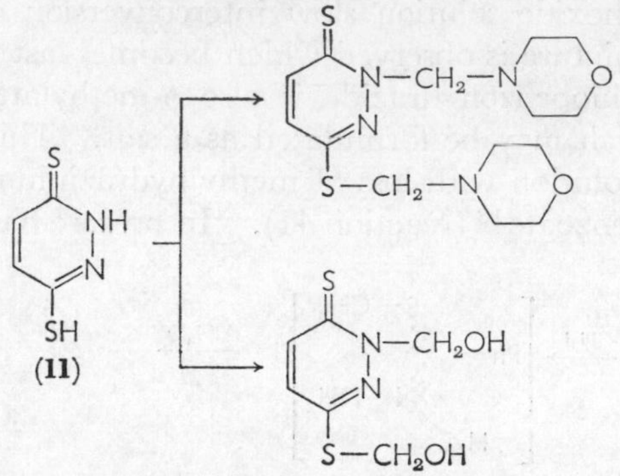

(43)

(44)

(11)

2. Addition of thiohydrazides to double bonds

3-Mercaptodihydropyridaz-6-thione (**11**) reacts at the mercapto site with quinone[80] and acrylonitrile[60] to yield compounds **38** and **39** respectively.

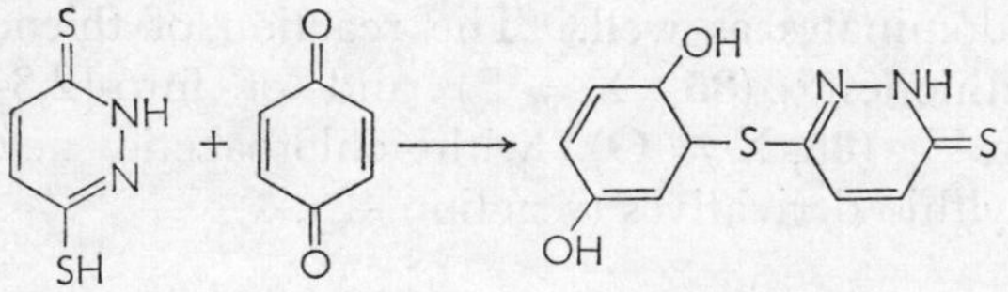

(11)   (38)

(45a)

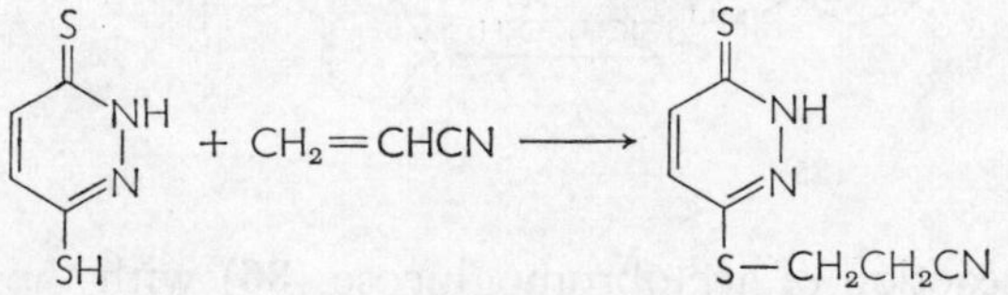

(11)   (39)

(45b)

E. Oxidation Reactions of Thiohydrazides

Already in 1939 Wuyts and Lacourt[51] reported that N^2-phenyl-thiobenzohydrazide is oxidized by iodine to the corresponding di-sulphide **40**. The constitution of this compound, an orange–red

$$2Ph\overset{S}{\overset{\|}{C}}NHNHPh \xrightarrow{\ I_2\ } PhNHN=C\overset{Ph}{\overset{|}{S}}S\overset{Ph}{\overset{|}{C}}=NNHPh$$

(46)

(40)

solid of m.p. 149°, was proved by reduction to thiohydrazide with stannous chloride (reaction 47) and by reaction with methyl iodide to yield the S-methylated compound **41**. The m.p. of **40** was later given

$$\textbf{40} \xrightarrow{\text{SnCl}_2} \overset{\overset{\text{S}}{\|}}{\text{Ph—C—NHNHPh}} \qquad (47)$$

$$\textbf{40} \xrightarrow{\text{MeI}} \overset{\overset{\text{SCH}_3}{|}}{\text{PhC}=\text{NNHPh}} \qquad (48)$$

$$\textbf{(41)}$$

by Holmberg[82] as 135–136° on slow heating and 141–142° on rapid heating with decomposition.

Formation of disulphide upon oxidation of the perfluoroalkylthiohydrazides was also observed by Brown and Pater[38]. The reaction proceeds with great ease with hydrogen peroxide or iodine. Even N-unsubstituted heptafluorothiobutyrohydrazide yielded a disulphide stable at room temperature. Only upon heating in acid medium was the latter compound converted into the corresponding 1,3,4-thiadiazole (section IV.F). The u.v. absorption maxima of these disulphides are reported at 300–389, 247–291, and 232–238 nm.

Holmberg[82–84] also isolated disulphides upon oxidation of thiohydrazides but only in a few cases, the most common oxidation products being heterocyclic compounds (section IV.F).

In connexion with a study concerning MO calculations of phenyl-azocarboxylic acid derivatives, Bock and coworkers[85] oxidized phenylthiobenzohydrazide with hypobromite. They obtained an orange–red solid of m.p. 136° for which they give the formula $(\text{C}_6\text{H}_5\text{N}=\text{NCSC}_6\text{H}_5)_2$. However, the analytical values given agree much better with the calculated values for the disulphide **40**. Considering also the u.v. absorption maxima given for this compound at 394, 324, and 268 nm, it seems quite certain, that in this case, again, **40** was obtained.

S-Oxides as obtained upon oxidation of thioamides (Chapter 8) and thioureas, have not yet been mentioned in the thiohydrazide series.

F. Formation of Heterocyclic Compounds from Thiohydrazides

Thiohydrazides have been used as starting materials for a variety of syntheses of heterocycles. These reactions will only be outlined in general here. As in many cases little is known about the reaction mechanisms, the classification takes account only of the structure of the reaction product.

17*

I. Sulphur-containing heterocycles

a. Thiadiazoles and thiadiazolines. The formation of thiadiazoles from compounds containing the thiohydrazide group has been used preparatively long before a thiohydrazide was isolated, for it is quite certain that Stollé's[86,87] thiadiazole synthesis proceeds via an acylthiohydrazide intermediate that can not be isolated under the somewhat rough reaction conditions. If acylthiohydrazides prepared by acylation of thiohydrazides or thioacylation of hydrazides are heated in acidic solution, thiadiazoles are formed in excellent yields[36] (reaction 49).

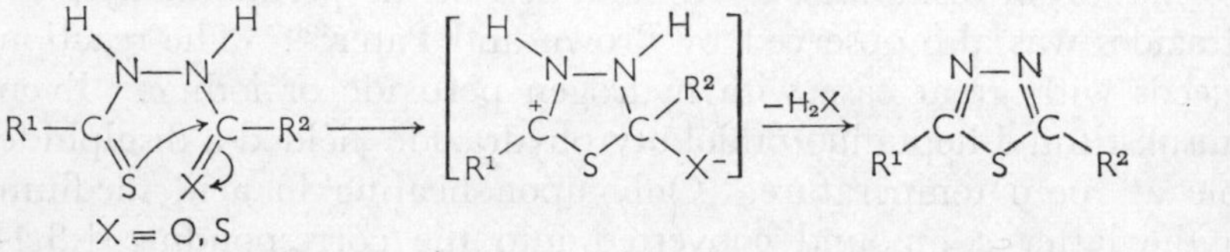

$$\text{(49)}$$

$$X = O, S$$

The reaction of dithioacids with unsubstituted hydrazine (reaction 5)[7,11], as well as the reaction of *N*-unsubstituted thiohydrazides with carboxymethyldithioates[9,11] to yield thiadiazoles, is completely analogous to reaction (49).

Nucleophilic attack of sulphur on the hydrazone carbon atom of an aldehyde *N*-thioacylhydrazone under oxidative conditions also leads to thiadiazoles[83] (reaction 50).

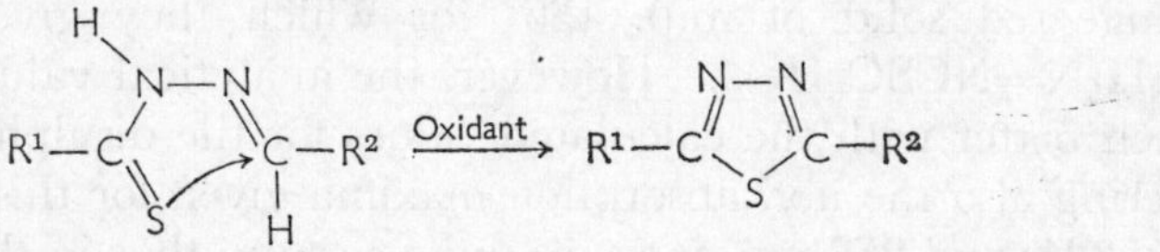

$$\text{(50)}$$

Another mechanism seems to be operative in the formation of 2,5-disubstituted thiadiazoles from *N*-unsubstituted thiohydrazides upon heating, or upon oxidation[22,72] (reaction 51).

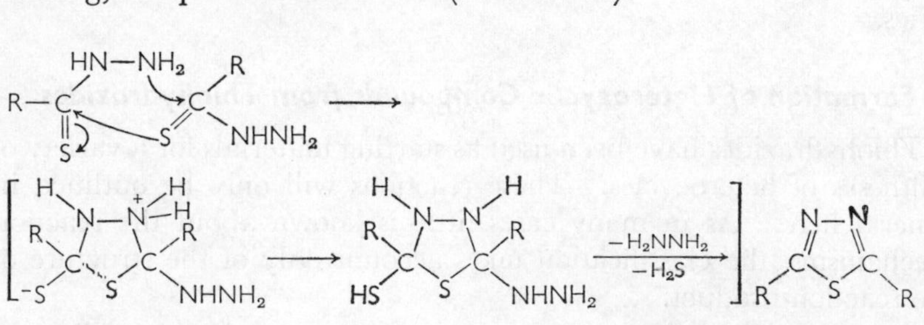

$$\text{(51)}$$

Similarly Brown and Pater[38] found that bis(α-hydrazonohepta-fluorobutyl) disulphide (**42**) is converted to the thiadiazole in acid solution (reaction 52).

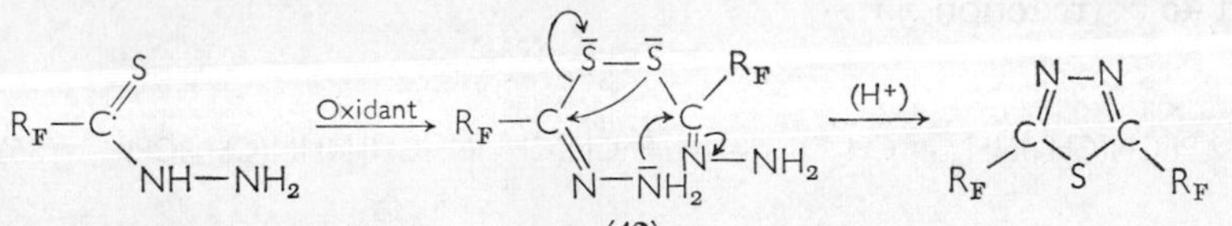

$$
R_F-C\overset{S}{\underset{NH-NH_2}{}} \xrightarrow{Oxidant} R_F-C\underset{N-NH_2}{\overset{S-S}{}}C\underset{}{\overset{R_F}{N^2-NH_2}} \xrightarrow{(H^+)} \underset{R_F}{\overset{N-N}{C-S-C}}R_F \tag{52}
$$

(**42**)

If a thiohydrazide is treated with an acylated carbodiimide, an intermediate presumably of structure **43** is obtained, which yields a 2-acylaminothiadiazole (**44**) upon standing in acidic solution[88] (reaction 53).

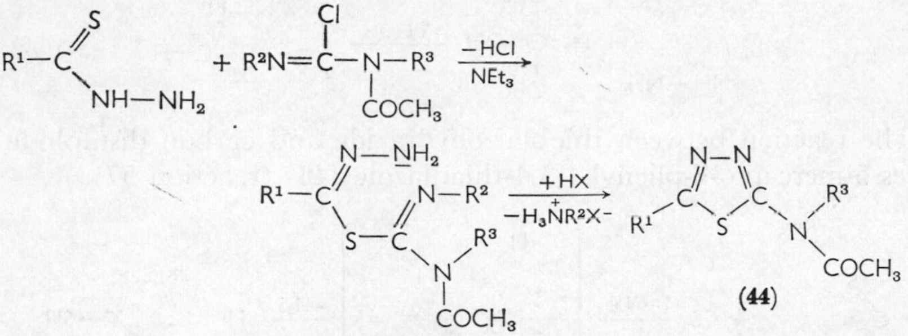

$$
R^1-C\overset{S}{\underset{NH-NH_2}{}} + R^2N=\underset{COCH_3}{\overset{Cl}{C}}-N-R^3 \xrightarrow[NEt_3]{-HCl}
$$

$$
R^1-C\underset{S-C}{\overset{N-NH_2}{}}\underset{\underset{COCH_3}{N}}{\overset{N-R^2}{R^3}} \xrightarrow[-H_3\overset{+}{N}R^2X^-]{+HX} R^1-C\underset{S}{\overset{N-N}{}}C\underset{\underset{COCH_3}{N}}{R^3} \tag{53}
$$

(**43**) (**44**)

Nucleophilic attack of both the N^2 and the S at a carbon atom is involved in the formation of dihydrothiadiazoles[13,71,89] from N^2-substituted thiohydrazides and aldehydes or ketones (reaction 54).

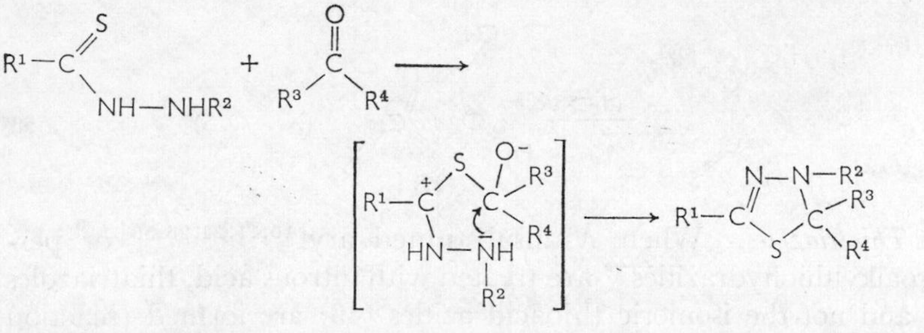

$$
R^1-C\overset{S}{\underset{NH-NHR^2}{}} + \underset{R^3}{\overset{O}{C}}R^4 \longrightarrow
$$

$$
\left[R^1-\overset{+}{C}\underset{\underset{R^2}{HN-NH}}{\overset{S}{}}\underset{R^4}{\overset{O^-}{C}}R^3 \right] \longrightarrow R^1-C\underset{S}{\overset{N-N-R^2}{}}C\underset{R^4}{R^3} \tag{54}
$$

The formation of dihydrothiadiazoles in the reaction of dithioacid and substituted hydrazines[6] (reaction 7, section II.A.1) may be formulated accordingly with thioaldehyde instead of the O-analogue of reaction (54), as thioaldehydes are formed by reduction of dithioacids by hydrogen sulphide (cf. reaction 6).

If on the other hand an N^1-alkylated thiohydrazide is reacted with aldehyde, no thiadiazoline is formed. N^1-Methylthiobenzohydrazide, for example, reacts with formaldehyde to yield the open-chain compound **45**[13] (reaction 55).

$$2 \text{ Ph}\overset{\text{S}}{\overset{\|}{\text{C}}}\text{N(CH}_3\text{)NH}_2 + \text{CH}_2\text{O} \longrightarrow \text{Ph}\overset{\text{S}}{\overset{\|}{\text{C}}}\text{N(CH}_3\text{)NHCH}_2\text{NHN(CH}_3\text{)}\overset{\text{S}}{\overset{\|}{\text{C}}}\text{Ph} \quad (55)$$

$$(45)$$

The action of thiohydrazides on orthoesters[90] (reaction 56) to yield thiadiazole appears to take place in the same way as that of carboxylic acid hydrazides.

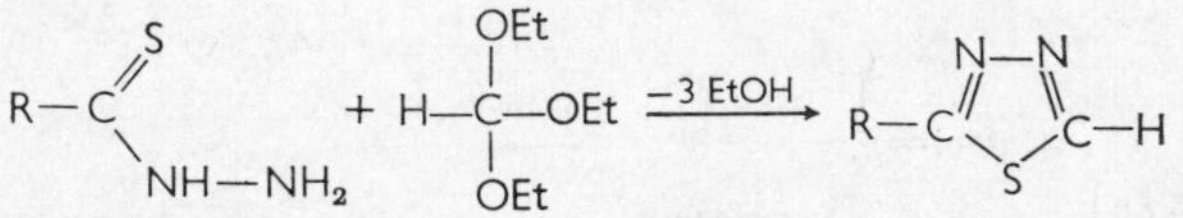

$$(56)$$

The reaction between thiobenzohydrazide and carbon disulphide gives 5-mercapto-2-phenyl-1,3,4-thiadiazole (**46**) (reaction 57).

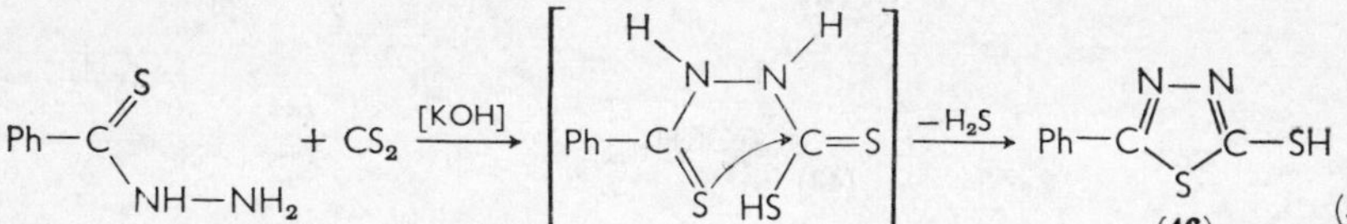

$$(57)$$

With the more reactive phosgene[32,49,91,92] or thiophosgene[91] mesoionic 1,3,4-thiadiazoles **47** can be obtained from N^1-monosubstituted thiohydrazides (reaction 58).

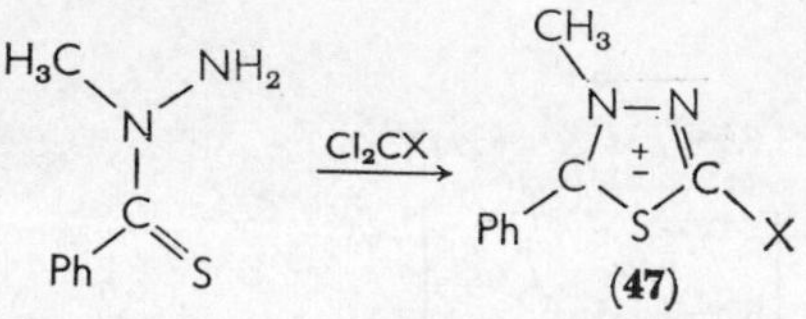

$$(58)$$

b. Thiatriazoles. When N-unsubstituted aryl[10,11,24,25,56] or perfluoroalkylthiohydrazides[38] are treated with nitrous acid, thiatriazoles (**48**) and not the isomeric thioacid azides (**49**) are formed (reaction 59). The same products are obtained from carboxymethyl dithioates and sodium azide[24,25] (reaction 60). Reaction (59) is analogous to the well-known formation of 4-aminothiatriazole from thiosemicarbazide and nitrous acid[93]. The cyclic character of these compounds has been established unequivocally by i.r. spectroscopy.

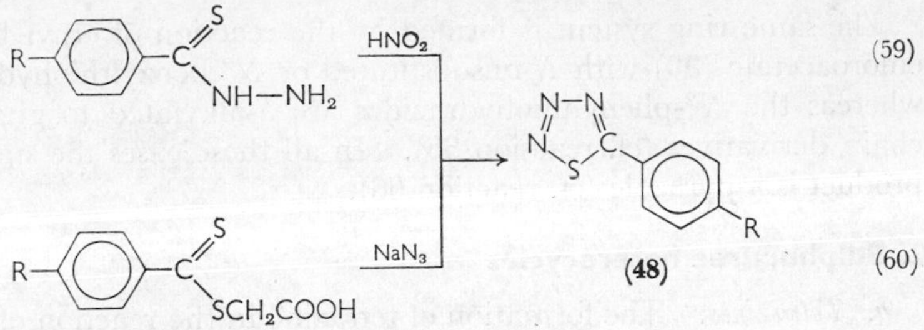

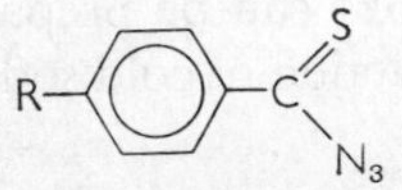

R—⟨benzene⟩—C(=S)—NH—NH₂ →(HNO₂) (59)

(structure 48) (60)

R—⟨benzene⟩—C(=S)—SCH₂COOH →(NaN₃) **(48)**

R—⟨benzene⟩—C(=S)—N₃

(49)

c. Thiadiazines. Reactions of thiohydrazides with α-halogeno-carbonyl compounds leading to the thiadiazine ring system are closely related to the analogous reactions of thiosemicarbazides[94]. Thio-benzohydrazide reacts with α-halogenoacyl chlorides to yield 2-phenyl-1,3,4-thiadiazin-5-one[72] **(50)** (reaction 61).

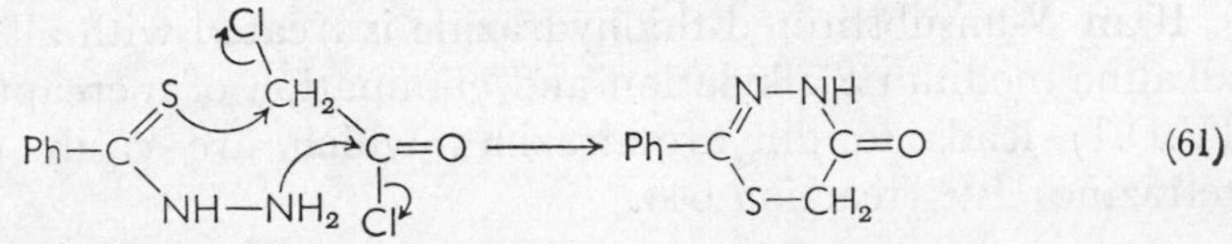

Ph—C(=S)(NH—NH₂) + ClCH₂—C(=O)Cl → Ph—C(N—NH)(S—CH₂)(C=O)   (61)

(50)

In the reaction of α-halogenoketones with thiohydrazides, e.g., reaction (62), formation of the thiadiazine ring system **51** has been observed[77]. The thiadiazines undergo ring contraction (cf. reaction 66) and were not isolated in all cases.

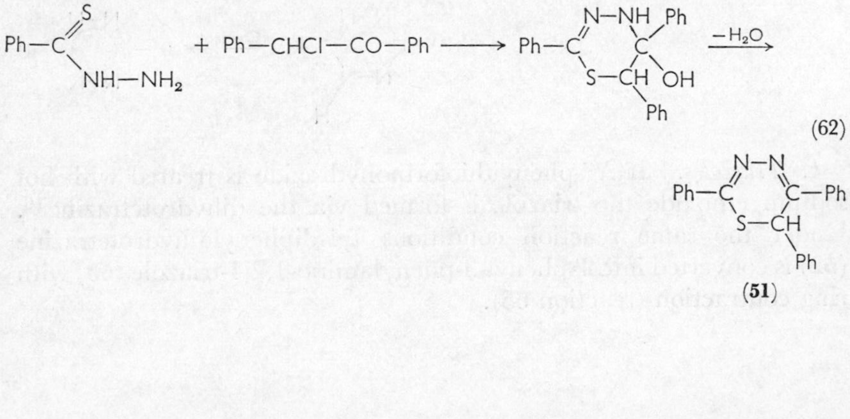

Ph—C(=S)(NH—NH₂) + Ph—CHCl—CO—Ph ⟶ Ph—C(N—NH)(S—CH)(C(Ph)(OH))Ph —(−H₂O)→

(62)

Ph—C(N—N)(S—CH)(C—Ph)Ph

(51)

The same ring system is formed by the reaction of ethyl benzoyl-chloroacetate (**30**) with N-unsubstituted or N^2-benzylthiohydrazides, whereas the N^2-phenylthiohydrazides are S-alkylated to give open-chain derivatives (cf. reaction 36). In all these cases the stable end product is a pyrazole (cf. reaction 66).

2. Sulphur-free heterocycles

a. Tetrazines. The formation of tetrazine in the reaction of hydrazine with dithioacids was mentioned above in section II.A.1. 1,4-Diphenyldihydrotetrazine (**52**) can be prepared from the N^2-phenyl-thioformohydrazide by the action of cold sodium ethoxide[16] (reaction 63).

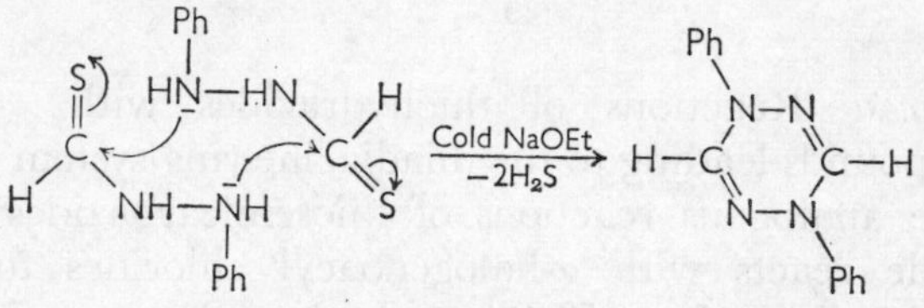

$$(63)$$

If an N-unsubstituted thiohydrazide is treated with alkyl halide in alkaline medium S-alkylation and elimination of mercaptide (section IV.D.1) leads to dihydrotetrazines which are easily oxidized to tetrazines[11,72] (reaction 64).

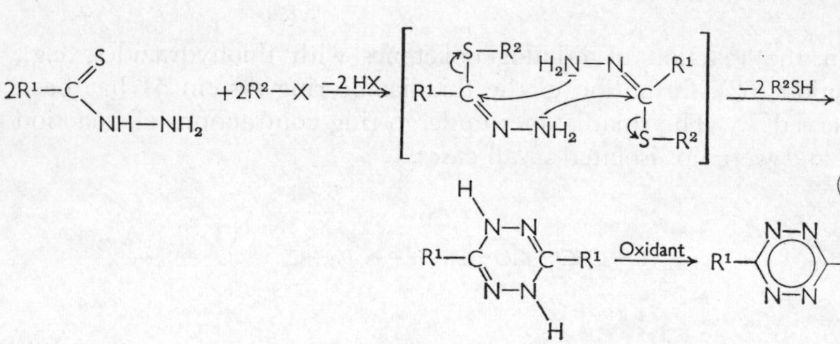

$$(64)$$

b. Triazoles. If N^2-phenylthioformohydrazide is treated with hot sodium ethoxide the triazole is formed via the dihydrotetrazine[16]. Under the same reaction conditions 1,4-diphenyldihydrotetrazine (**52**) is converted into 1-phenyl-3-phenylamino-1,2,4-triazole (**53**) with ring contraction (reaction 65).

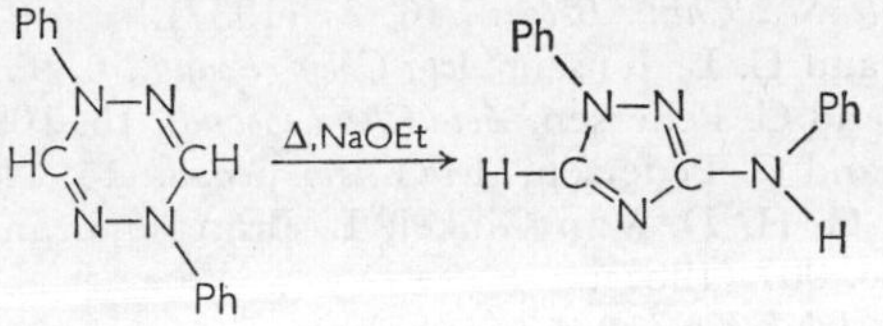

$$\text{(52)} \qquad \xrightarrow{\Delta,\text{NaOEt}} \qquad \text{(53)} \qquad\qquad (65)$$

c. Pyrazoles. Thiohydrazides often react with α-halogenocarbonyl compounds to yield pyrazoles. The reaction proceeds via thiadiazines (cf. reactions 61 and 62) which rearrange with loss of sulphur[89] (reaction 66). The ring contraction with loss of sulphur is familiar,

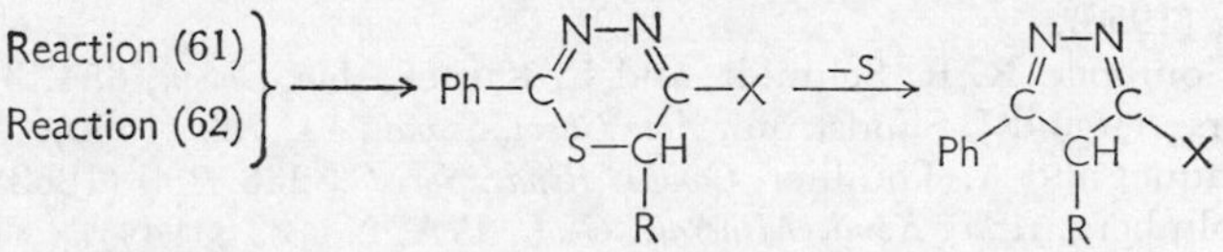

$$\left.\begin{array}{l}\text{Reaction (61)}\\\text{Reaction (62)}\end{array}\right\} \longrightarrow \quad \xrightarrow{-S} \qquad\qquad (66)$$

too, for thiadiazines prepared from thiosemicarbazides[94], thiocarbazides[95], and dithiocarbazic acid esters[96].

V. APPLICATIONS OF THIOHYDRAZIDES

Although a number of investigations on thiohydrazides were undertaken with the aim of obtaining substances of pharmacological use[9,39,44,97], these compounds seem to have no advantage over known compounds in their physiological activity, so that they have not been introduced as pharmaceutics, in contrast to thiosemicarbazides.

A number of patents is concerned with thiohydrazides as pesticides[33,44,97]. Their application in this field, however, seems to be not very wide.

Thiohydrazides are of increasing importance in the synthesis of heterocycles as was shown above.

VI. REFERENCES

1. Y. Sakurada, *Bull. Chem. Soc. Japan*, **2**, 307 (1927).
2. H. Wuyts, *Bull. Soc. Chim. Belges*, **38**, 195 (1929).
3. H. Wuyts, *Bull. Soc. Chim. Belges.*, **39**, 58 (1930).
4. H. Wuyts and M. Goldstein, *Bull. Soc. Chim. Belges.*, **40**, 497 (1931).
5. H. Wuyts and A. Lacourt, *Bull. Soc. Chim. Belges.*, **42**, 1 (1933).
6. H. Wuyts and A. Lacourt, *Bull. Soc. Chim. Belges.*, **42**, 376 (1933).
7. H. Wuyts and Li Chia Kuang, *Bull. Soc. Chim. Belges.*, **42**, 153 (1933).

8. H. Wuyts, *Bull. Soc. Chim. Belges.*, **46**, 27 (1937).
9. K. A. Jensen and C. L. Jensen, *Acta Chem. Scand.*, **6**, 957 (1952).
10. K. A. Jensen and C. Pedersen, *Acta Chem. Scand.*, **15**, 1097 (1961).
11. K. A. Jensen and C. Pedersen, *Acta Chem. Scand.*, **15**, 1124 (1961).
12. H. E. Wijers, C. H. D. Van Ginkel, L. Brandsma, and J. F. Arens, *Rec. Trav. Chim.*, **86**, 907 (1967).
13. B. Holmberg, *Arkiv Kemi*, **9**, 47 (1955).
14. T. Sato and M. Ohta, *J. Pharm. Soc. Japan*, **74**, 821 (1954); *Chem. Abstr.*, **49**, 9537 (1955).
15. W. Baker, W. D. Ollis, and V. D. Poole, *J. Chem. Soc.*, 1542 (1950).
16. W. Baker, W. D. Ollis, and V. D. Poole, *J. Chem. Soc.*, 3389 (1950).
17. H-B. König, W. Siefken, and H. A. Offe, *Chem. Ber.*, **87**, 825 (1954).
18. Farbenf. Bayer, *Brit. Pat.*, 758,149 (1956); *Chem. Abstr.*, **51**, 9712f (1957).
19. Aktieselskabet 'Ferrosan', *Danish Pat.*, 79,776 (1955); *Chem. Abstr.*, **50**, 8730g (1956).
19a. R. Gompper, R. R. Schmidt, and E. Kutter, *Ann. Chem.*, **684**, 37 (1965).
20. B. Forsgren and J. Sandström, *Acta Chem. Scand.*, **14**, 789 (1960).
21. M. Saquet and A. Thuillier, *Compt. Rend.*, *Ser. C.*, **266**, 290 (1968).
22. B. Holmberg, *Arkiv Kemi, Mineral.*, *Geol.*, **17A**, No. 23 (1944).
23. K. A. Jensen, H. R. Baccaro, O. Buchardt, G. E. Olsen, C. Pedersen, and J. Toft, *Acta Chem. Scand.*, **15**, 1109 (1961).
24. P. A. S. Smith and D. H. Kenny, *J. Org. Chem.*, **26**, 5221 (1961).
25. E. Lieber, C. N. R. Rao, and R. C. Orlowski, *Can. J. Chem.*, **41**, 926 (1963).
26. N. Sugimoto and S. Shigematsu, *Japan Pat.*, 7,481 (1954); *Chem. Abstr.*, **50**, 4236i (1956).
27. W. Walter and K. J. Reubke, *Angew. Chem.*, **79**, 381 (1967); *Angew. Chem. Intern. Ed. Engl.*, **6**, 368 (1967).
28. K. J. Reubke, *Thesis*, University of Hamburg, 1967.
29. W. Walter, K. J. Reubke, and H. Weiss, unpublished results.
30. E. Profft, F. Schneider, and H. Beyer, *J. Prakt. Chem.*, [4], **2**, 147 (1955).
31. H. Bredereck, B. Föhlisch, and K. Walz, *Ann. Chem.*, **688**, 93 (1965).
32. R. Grashey, M. Baumann, and W. D. Lubos, *Tetrahedron Letters*, 5877, 5881 (1968); R. Grashey and M. Baumann, *Angew. Chem.*, **81**, 115 (1969); *Angew. Chem. Intern. Ed. Engl.*, **8**, 133 (1969).
33. Shell International Research Maatschappij N.V., *Neth. Pat.*, 6,608,261, *Chem. Abstr.*, **69**, 35785 (1968).
34. G. F. Duffin and J. D. Kendall, *J. Chem. Soc.*, 3789 (1959).
35. S. Kakimoto and S. Tonooka, *Bull. Chem. Soc. Japan*, **40**, 153 (1967).
36. H. Eilingsfeld, *Chem. Ber.*, **98**, 1308 (1965).
37. E. Kuhle and R. Wegler, *U.S. Pat.*, 2,774,757 (1956); *Chem. Abstr.*, **51**, 6705 (1957).
38. H. C. Brown and R. Pater, *J. Org. Chem.*, **30**, 3739 (1965).
39. I. Ya. Postowskii and M. I. Ermakova, *Zh. Obshch. Khim.*, **29**, 1333 (1959); *Chem. Abstr.*, **54**, 8796c (1960).
40. G. Zemplén, L. Mester, and A. Messmer, *Chem. Ber.*, **86**, 697 (1953).
41. W. Walter and K. J. Reubke, *Tetrahedron Letters*, 5973 (1968).
42. F. L. Scott, F. A. Groeger, and A. F. Hegarty, *Tetrahedron Letters*, 2463 (1968).
43. A. F. Hegarty and F. L. Scott, *J. Org. Chem.*, **33**, 753 (1967).

44. M. O. Lozinskii and P. S. Pel'kis, *U.S.S.R. Pat.*, 169,513 (1965); *Chem. Abstr.*, **63**, 8259f (1965).
45. J. Druey, Kd. Meier, and K. Eichenberger, *Helv. Chim. Acta*, **37**, 121 (1954).
46. M. Robba, B. Roques, and Y. Le Guen, *Bull. Soc. Chim. France*, 4220 (1967).
47. M. Robba and M. C. Zaluski, *Compt. Rend., Ser. C.*, **266**, 31 (1968).
48. G. V. Boyd and A. J. H. Summers, *Chem. Commun.*, 549 (1968).
49. A. R. McCarthy, W. D. Ollis, and C. A. Ramsden, *Chem. Commun.*, 499 (1968).
50. H. Gotthardt, R. Huisgen, and R. Knorr, *Chem. Ber.*, **101**, 1056 (1968).
51. H. Wuyts and A. Lacourt, *Bull. Soc. Chim. Belges.*, **48**, 193 (1939).
52. C. H. Carlisle and M. B. Hossain, *Acta Cryst.*, **21**, 249 (1966).
53. W. Walter, J. Holst, O. Jarchow, and H. Junge, *Naturwissenschaften*, **55**, 227 (1968); W. Walter and J. Holst, *Naturwissenschaften*, **56**, 327 (1969).
54. K. A. Jensen and P. H. Nielsen, *Acta Chem. Scand.*, **20**, 597 (1966).
55. C. N. R. Rao, R. Venkataraghavan, and T. R. Kasturi, *Can. J. Chem.*, **42**, 36 (1964).
56. E. Lieber, *U.S. Dept. Comm., Office, Tech. Serv. P.B. Rept.* 148,532 (1960); *Chem. Abstr.*, **58**, 2446g (1963).
57. F. Arndt, *Angew. Chem.*, **61**, 397 (1949).
58. J. Sandström, *Acta Chem. Scand.*, **17**, 1380 (1963).
59. W. Walter and K. J. Reubke, *Chem. Ber.*, **102**, 2117 (1969).
60. B. Stanovnik and M. Tisler, *Croat. Chem. Acta*, **36**, 81 (1964).
61. C. N. R. Rao, *Chemical Applications of Infrared Spectroscopy*, Academic Press, New York and London, 1963, p. 246.
62. *Sadtler Standard Spectra*, No. 7649.
63. U. Anthoni, Ch. Larsen, and P. H. Nielsen, *Acta Chem. Scand.*, **22**, 1050 (1968); **23**, 1231 (1969).
64. J. Sandström and S. Sunner, *Acta Chem. Scand.*, **17**, 731 (1963).
65. J. Sandström, *Acta Chem. Scand.*, **17**, 937 (1963).
66. A. Albert and G. B. Barlin, *J. Chem. Soc.*, 3129 (1962).
67. W. Walter and K. J. Reubke, unpublished results.
68. A. M. Duffield, C. Djerassi, and J. Sandström, *Acta Chem. Scand.*, **21**, 2167 (1967).
69. W. Walter, E. Schaumann, and K. J. Reubke, *Angew. Chem.*, **80**, 448 (1968); *Angew. Chem. Intern. Ed. Engl.*, **7**, 467 (1968).
70. H. Wuyts, *Compt. Rend.*, **196**, 1678 (1933).
71. H. Wuyts and A. Lacourt, *Bull. Acad. Roy. Belg. Kl. Sci.*, 5 **20**, 156 (1934), *C* 1934, II, 945.
72. B. Holmberg, *Arkiv Kemi Mineral. Geol.*, **25A**, 18 (1947).
73. K. A. Jensen and J. F. Miquel, *Acta Chem. Scand.*, **6**, 189 (1952).
74. G. Bähr and G. Schleitzer, *Angew. Chem.*, **68**, 375 (1956).
75. T. Sato and M. Ohta, *Bull. Chem. Soc. Japan*, **27**, 624 (1954); *Chem. Abstr.*, **50**, 213 (1956).
76. J. Sandström, *Acta Chem. Scand.*, **16**, 2395 (1962).
77. J. Sandström, *Acta Chem. Scand.*, **17**, 95 (1963).
78. G. Wagner and D. Heller, *Z. Chem.*, **4**, 28 (1964).
79. G. Wagner and D. Heller, *Arch. Pharm.*, **300**, 783 (1967).
80. A. Pollak, B. Stanovnik, and M. Tisler, *Monatsh. Chem.*, **97**, 1523 (1966).
81. A. Pollak and M. Tisler, *Monatsh. Chem.*, **96**, 642 (1965).

82. B. Holmberg, *Arkiv Kemi*, **7**, 517 (1954).
83. B. Holmberg, *Arkiv Kemi*, **9**, 65 (1955).
84. B. Holmberg, *Arkiv Kemi*, **4**, 33 (1952).
85. H. Bock, E. Baltin, and J. Kroner, *Chem. Ber.*, **99**, 3337 (1966).
86. R. Stollé, *Chem. Ber.*, **32**, 797 (1899).
87. R. Stollé, *J. Prakt. Chem. Ser.*, 2, **69**, 145 (1904).
88. K. Hartke and A. Birke, *Arch. Pharm.*, **299**, 921 (1966).
89. A. Lacourt, *Bull. Soc. Chim. Belges.*, **43**, 206 (1934).
90. C. Ainsworth, *J. Am. Chem. Soc.*, **77**, 1148 (1955).
91. K. T. Potts and C. Sapino, Jr., *Chem. Commun.*, 672 (1968).
92. A. Lazaris, *Zh. Organ. Chim.*, **3**, 1902 (1967); *Chem. Abstr.*, **68**, 12910 (1968); *Zh. Organ. Khim.*, **4**, 1849 (1968); *Chem. Abstr.*, **70**, 19992 (1969).
93. M. Freund and A. Schander, *Chem. Ber.*, **29**, 2500 (1896).
94. H. Beyer and G. Wolter, *Chem. Ber.*, **89**, 1652 (1956).
95. H. Beyer, G. Wolter, and H. Lemke, *Chem. Ber.*, **89**, 2550 (1956).
96. H. Beyer, E. Bulka, and F.-W. Beckhaus, *Chem. Ber.*, **92**, 2593 (1959).
97. S. A. Heininger, *U.S. Pat.*, 2,909,556 (1959); *Chem. Abstr.*, **54**, 2169a (1960).
98. S. Wakayama and G. Takahashi, *Japan Pat.*, 6819295 (1968); *Chem. Abstr.*, **70**, 57474 (1969).
99. W. Walter and H. Weiss, *Angew. Chem.*, in press.

CHAPTER **10**

The chemistry of hydrazides

Hans Paulsen and Dieter Stoye

University of Hamburg, Germany

515

I. INTRODUCTION

In view of their high reactivity, hydrazides are important starting materials and intermediates in the synthesis of certain amines, aldehydes, and heterocyclic compounds that are otherwise difficult to prepare. Many hydrazides, and in particular variously modified aromatic carboxylic acid hydrazides, have been tested for physiological effects since isonicotinic acid hydrazide (isoniazid) was found to be tuberculostatic. In analytical organic chemistry, hydrazides are used to identify carboxylic acids and to detect carbonyl compounds that form acylhydrazones.

The chemistry of hydrazides which is summarized in several books of a general nature[1-4], is described in the present work from the viewpoint of reaction mechanisms and related theoretical considerations. Carbonic acid derivatives, sulphonylhydrazines, and sulphinylhydrazines will be mentioned only when they are directly connected with the reactions of carboxylic acid hydrazides.

II. NOMENCLATURE

Hydrazides can be regarded both as derivatives of carboxylic acids and as derivatives of hydrazine. Simple members are described as carboxylic acid hydrazides or acylhydrazines. The former is the preferred name, and is used by *Chemical Abstracts*, as for example in 'acetic acid hydrazide'. However, it is usual to employ a shorter form where the '-ic' ending of the acid is replaced by '-hydrazide' or '-ohydrazide' as in acethydrazide or butyrohydrazide.

The nitrogens are designated as 1 and 2, or α and β, or N and N', the first member of each pair denoting the nitrogen where the acyl group is inserted. The use of 1 and 2 is generally preferred, provided there is no possibility of confusion with the numbers of other residues of the molecule. Substituted hydrazides are named as carboxylic acid hydrazides, e.g. acetic acid 2-phenylhydrazide (in the *Chemical Abstracts Index*), or else as acylhydrazines, e.g. 1-acetyl-2-phenyl-hydrazine. According to *Chemical Abstracts*, the naming of multiply acylated hydrazines is based on hydrazine, e.g. 1,2-diacetylhydrazine, the term 'diacethydrazide' being less desirable. Diacylhydrazines are either symmetrical (1,2) or asymmetrical (1,1). The prefixes *sym* and *asym* are also used with alkyl and aryl substituents in simple carboxylic acid hydrazides. Thus, *sym*-acylmethylhydrazine is 1-acyl-2-methylhydrazine; however, the latter form is preferred.

III. GENERAL AND PHYSICOCHEMICAL CHARACTERISTICS

A. General Characteristics

Unsubstituted hydrazides are generally easily crystallizable solids, their melting points increasing steadily in a given homologous series[5]. Diacylhydrazines are also crystalline, but triacetylhydrazine and tetrakis(trifluoroacetyl)hydrazine can be obtained only as oily substances[6]. Tetraacetylhydrazine is crystalline, but its melting point is lower than that of 1,1- and 1,2-diacetylhydrazine. The melting points of 1,2-diacylhydrazines are consistently higher than those of 1,1-diacylhydrazines. Similarly, 2-alkyl-substituted hydrazides melt at a higher temperature than 1-alkyl-substituted ones[7]. The substitution of an alkyl group for the amide hydrogen generally lowers the melting point of the hydrazide[8,9].

The hydrazides of lower carboxylic acids readily dissolve in water. As the molecular weight increases the solubility in water decreases, because the hydrophobic nature of the substituents eventually outweighs the hydrophilic nature of the hydrazide group.

B. X-ray Structure Analysis

The structure of isonicotinic acid hydrazide[10], n-heptanoic acid hydrazide[11], and n-dodecanoic acid hydrazide[5] has been determined by x-ray crystallography. The N—N bond length is always between 1·39 and 1·42 Å, which is shorter than in hydrazine itself (1·46–1·47 Å). This contraction is ascribable to the formal charge effect and to the fact that the electron-attracting acyl group reduces the repulsion between the lone pairs of the nitrogens. The C—N bond length is 1·33 Å, which is the same as in the pyridine ring. This bond must therefore have roughly a 50% double-bond character. The substituents on the terminal or β-nitrogen atom have a pyramidal arrangement, with bond angles as depicted. The two hydrogens point in the direction of the carbonyl oxygen. All six atoms in the group lie almost exactly in the same plane.

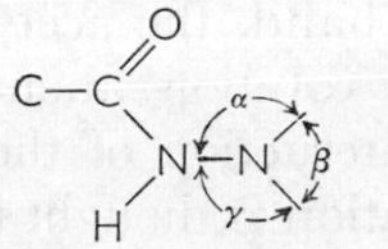

$$\alpha = 101°$$
$$\beta = 98°$$
$$\gamma = 109°$$

In the crystalline state, the hydrazide molecules are linked together by intermolecular N—H···O and N—H···N linkages. In n-dodecanoic acid hydrazide, these hydrogen bonds give rise to ribbon-shaped macromolecules, which pair up to form molecular double layers[5]. The latter have the same structure in a given homologous series of hydrazides (though of course they increase in thickness with increasing size of the aliphatic acid residue), and this is why the melting points of homologues form a continuously increasing series.

Like the monoacylhydrazines, diacylhydrazines such as 1,2-diformylhydrazine and 1,2-diacetylhydrazine are centrosymmetric planar molecules with an N—N bond length of 1·39 Å[12,13]. The structure of N,N'-disuccinimide (1) has been determined by three-dimensional x-ray structure analysis[14]. The molecule possesses a two-fold symmetry axis parallel to the N—N bond. The angle between the planes of the two rings is 65°. The N—N bond length is 1·37 Å, which is in agreement with expectations. On the other hand, the C—N distance is 1·39 Å, which is somewhat longer than in acyclic hydrazides, but shorter than in the case of an ordinary C—N bond, as in aliphatic amines. The twist between the two imide rings is explained by non-bonding electron repulsion between acyl carbonyl groups.

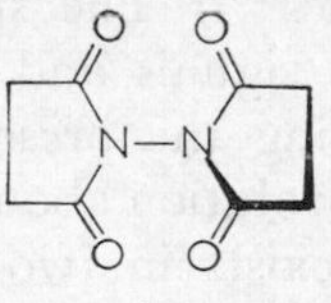

(1)

C. Nuclear Magnetic Resonance Spectroscopy

The chemical shifts of the protons of simple hydrazides are similar to those of protons with a comparable chemical environment in other compounds, e.g. in amides.

The n.m.r. spectroscopy of trisubstituted hydrazides has given some interesting information about their conformation[15a,15b]. Thus, the observation that at room temperature the two acetyl groups in N-(diacetylamino)tetramethylsuccinimide give a singlet at $\tau = 7 \cdot 91$ indicates free rotation on the part of these groups. In the case of compounds 2 and 3, on the other hand, the acetyl signals show a slight splitting, indicating hindered rotation about the N—N bond[15a]. Owing to the non-bonding interaction of the four amide carbonyl groups, the preferred conformation is thought to be that in which the plane of the diacetylamino group is normal to the plane of the succinimide ring, as a result of which, the methyl groups become magnetically non-equivalent. This is particularly noticeable in compound 4, where, if the planes of the two diacylamino groups are normal to each other, the lower acetyl methyl group is situated in a region where it is shielded by a benzene ring. Therefore, there is a large difference in the chemical shift ($1 \cdot 46$ p.p.m.) between the signals of the methyl groups[15a]. Similar results have been obtained with N-(diacetyl-amino)-3-methyl-3-phenylsuccinimide[15b]. The activation free enthalpy for the free rotation about the N—N bond is estimated as $\Delta G^{\ddagger} = 20$–23 kcal/mole[15b].

(2) X = CH$_2$CH=CHCH$_2$
(3) X = CH$_2$CH$_2$CH$_2$CH$_2$

(4)

Restricted rotation about the N—N bond is also found in 1,2-diacyl-1,2-dibenzylhydrazines[16]. The spectrum of 5 (R = CH$_3$) shows four different acetyl signals and four AB systems of benzyl methylene protons, indicating the presence of three conformations (5–7). Owing to hindered rotation about the N—N and the N—CO bond, each conformation exists in two chiral forms. The benzyl methylene protons are therefore non-equivalent, and AB systems are thus formed. As the temperature is raised, the four acetyl signals coalesce into a singlet, and the four methylene signals into one AB system. The temperature at which this happens is where the hindered rotation of the N-acetyl groups changes into free rotation in the sense of n.m.r. time scale. The fact that an AB system is retained shows that the chirality and thus the hindered rotation about the N—N bond still persist.   The AB system of the methylene groups gives way to a singlet only at higher temperatures (190°c), where the rotation about the N—N bond becomes unrestricted.   On the basis of the coalescence temperature, the activation free enthalpy for the rotation about the N—N bond is $\Delta G^{\ddagger} = 23.4$ kcal/mole at $461°$ к in this case.

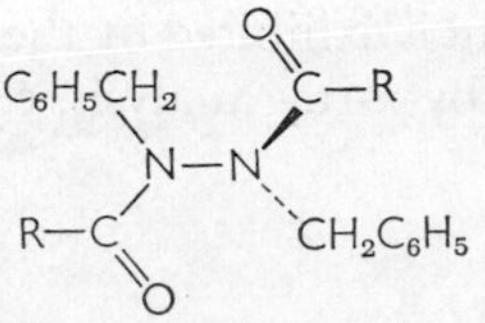

(5) 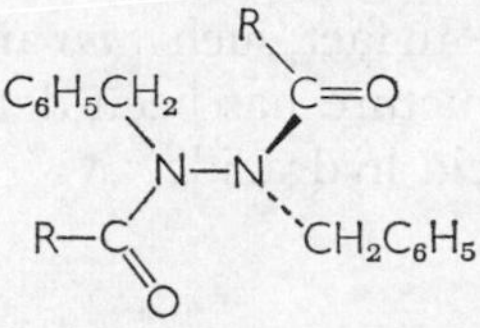   (6)

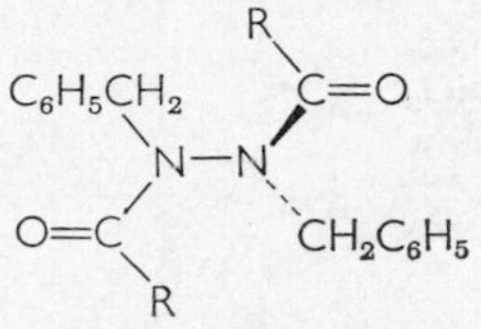

(7) 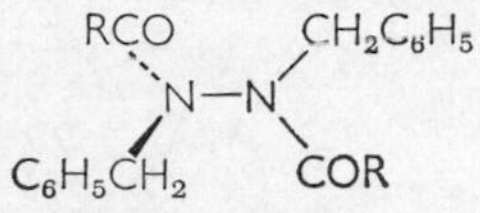   (8)

In 1,2-diacetyl-1-benzylhydrazine, the splitting of the benzyl methylene groups into an AB system occurs at a considerably lower temperature, and the activation free energy for the rotation in this compound is estimated at only $\Delta G^{\ddagger} = 13$ kcal/mole at $277°$к[16].   It is concluded from these results that the twist form **8** is the preferred conformation of 1,2-diacyl-1,2-dialkylhydrazines in the ground state: it is in this form that the repulsion between the substituents on the nitrogen atoms is at its minimum.   A similar effect has been found in cyclic diacylhydrazines of the tetrahydropyridazine type[15a,16,17], exemplified by compounds **9** and **10**.   The outcome of the effect in this case is that the ring inversion is greatly slowed down and has a relatively high-energy barrier.

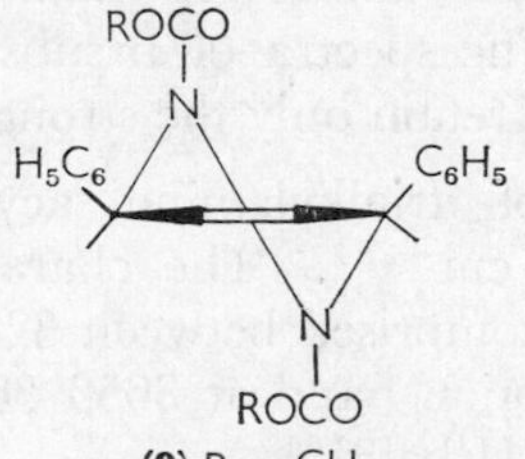

(**9**) R = CH₃
(**10**) R = C₂H₅

D. Dipole Measurements

The conformation of the hydrazide group has been determined from the measured dipole moments of aliphatic and aromatic hydrazides[18]. According to this, the R group and the $N_{(\beta)}H_2$ group are *trans* with respect to the $C-N_{(\alpha)}$ axis, as shown in **11**. The hydrogens of the amino group form one or two hydrogen bonds with the carbonyl oxygen. In fact, such *trans* arrangement is exhibited by the hydrazides whose structure has been determined by x-ray analysis, e.g. by isonicotinic acid hydrazide.

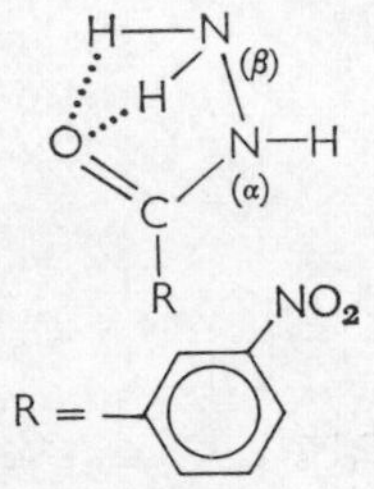

(11)

It appears that 1,2-dibenzoylhydrazine is not a *trans* planar compound, but exists in a staggered conformation on account of the electrostatic repulsion between the lone pairs on the nitrogens[18]. This result agrees with x-ray and n.m.r. findings.

E. Infrared and Ultraviolet Spectroscopy

The i.r. spectra of crystalline hydrazides show an amide I band at 1625–1670 cm^{-1}, due to the carbonyl group whose double-bond character is reduced by the mesomeric effect of the amide system. A weak band at 1610–1620 cm^{-1} is attributed to NH_2 deformation[19,20]. The region 1530–1570 cm^{-1} contains a strong amide II band, which is ascribed to a $C-N-H$ vibration comprising $N-H$ deformation and $C-N$ stretch[21]. A weak amide III band occurs in the range of 1200–1305 cm^{-1}[22]. The spectra of trisubstituted hydrazides lack the amide II band and retain only the strong amide I band[23]. The band characteristic of trialkylamine acylimides ($R_3^1\overset{+}{N}\overset{-}{N}COR^2$) appears at 1555–1590 cm^{-1}[24]. The characteristic frequencies for the $N-H$ stretch are comprised between 3200 and 3250 cm^{-1}[21,22]. A much weaker absorption band at 3050–3070 cm^{-1} is probably a harmonic of the amide II band[21].

The spectra recorded for hydrazides in solution are different as regards the position and the number of the absorption bands[20]. Thus, in a chloroform solution the amide I band is displaced by 20 cm^{-1} towards higher frequencies, while the amide II band occurs around 1500 cm^{-1}. When the solution is dilute, the N—H stretch band, situated at about 3250 cm^{-1} for crystalline hydrazides, appears at 3450 cm^{-1}. As the concentration of the solution is increased, a band appears gradually at 3340 cm^{-1}[20], probably owing to the formation of intermolecular hydrogen bonds in concentrated solutions. Hydrazides are fully associated in the solid state, owing to the establishments of NH$\cdots$O and NH$\cdots$N bonds. This also explains the shift of the amide I and the amide II band that occurs when solid hydrazides are dissolved[20]. Since the spectra of solid 1,2-diacylhydrazines are characterized by the presence of associated NH bands only and the absence of non-associated bands, it is assumed that not only intermolecular but also intramolecular hydrogen bonds are formed as in **12**[22]. Diacylhydrazines do not exhibit an absorption band at the stretching frequency of a free OH group in the solid state and neutral solutions[22], in spite of other evidence pointing to the occurrence of enolization (cf. section V.A).

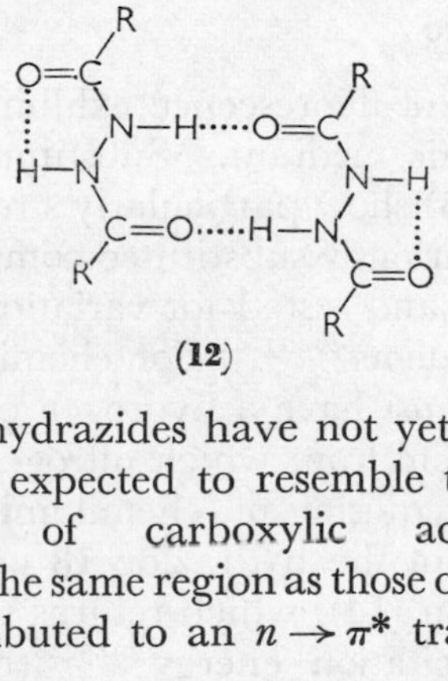

(12)

The u.v. spectra of hydrazides have not yet been investigated extensively, but they are expected to resemble those of amides. The absorption maxima of carboxylic acid diarylhydrazides (RCONHNAr$_2$) are in the same region as those of carbonyl compounds. This absorption is attributed to an $n \rightarrow \pi^*$ transition[25].

F. Polarography and Electrochemistry

The phenylhydrazides of dibutylglycolic and diphenylglycolic acids have been investigated polarographically in the pH range of 2–12, and so have their *N*-methyl and *N*-acetyl derivatives. As the pH is raised, the half-wave potential is displaced towards negative values. A number of these hydrazides exhibit a second anode wave at pH > 12. At concentrations between 5×10^{-6} and 5×10^{-4} mole/l, there is a linear relationship between the anode wave height and the concentration, so that polarography can be used for quantitative analysis in this domain[26].

The electrical conductivity of 1,2-diacylhydrazines increases considerably as the temperature is raised, and reaches a particularly high value on melting[22]. This is explained by assuming that, as the temperature is raised, 1,2-diacylhydrazines change into the enolic form, which has a higher conductivity because of dissociation. In an oxygen-free alkaline solution, the hydrazide group of phthalic acid hydrazides suffers irreversible oxidation at a platinum electrode[27].

G. Hudson's Phenylhydrazide Rule

To determine the configuration of aldonic acids at $C_{(2)}$, Hudson has formulated a phenylhydrazide rule and an analogous amide rule, which are of general applicability: the phenylhydrazide of an aldonic acid exhibits a more positive or a more negative optical rotation than the corresponding free acid, according to whether the OH group at $C_{(2)}$ conforms to a D- or an L-configuration[28].

H. Chemiluminescence

Hydrazides capable of fluorescence exhibit chemiluminescence on oxidation in an alkaline medium. Substituted cyclic hydrazides of the type of luminol (13) show particularly strong chemiluminescence. Many modified and variously substituted compounds of the type of 13 have been synthesized and tested for variation of the chemiluminescence with the substituents[29-31]. For chemiluminescence to occur, the hydrazide group must have a hydrogen on both nitrogen atoms. The presence of a system from which nitrogen can be easily cleaved out is clearly a prerequisite of chemiluminescence. Nitrogen is probably cleaved out of the hydrazide 13 oxidatively, to leave behind an excited dianion (14), which returns to the ground state (15) after radiating its excitation energy[29-31] (reaction 1). It is not yet known with certainty what intermediates are formed in the oxida-

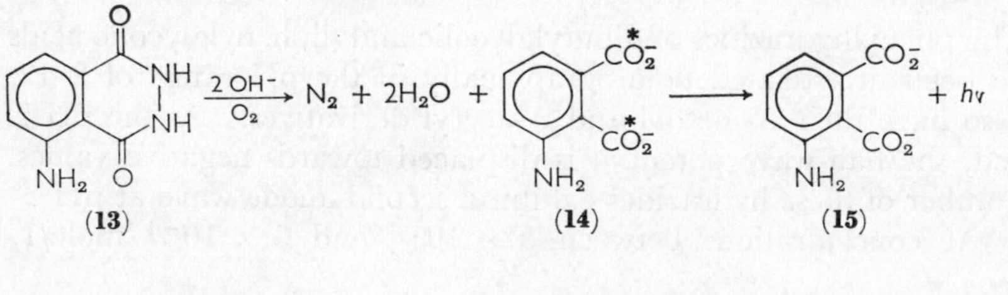

$$\text{(13)} \quad \xrightarrow[O_2]{2\ OH^-} \quad N_2 + 2H_2O + \text{(14)} \quad \longrightarrow \quad \text{(15)} \quad + \ h\nu \qquad (1)$$

tion of the hydrazide into the dianion[29]. A general survey of the chemiluminescence of organic compounds has been given by Gundermann[29].

IV. PREPARATION OF HYDRAZIDES

A. Hydrolytic Methods

I. Hydrolysis of nitriles

Hydrolytic methods are not of great practical importance for the preparation of hydrazides, because they are often accompanied by side-reactions. Hydrazides can be prepared by partial hydrolysis of nitriles into amides, followed by the reaction of the latter with hydrazine. 4-Cyanopyridine (16) can be converted into isonicotinic acid hydrazide (18) in a single operation by heating it with hydrazine hydrate in an aqueous alkaline solution[32,33]. However, the yield is not very good, because 3,5-di(4-pyridyl)-1,2,4-triazole (19) is formed as a by-product[34] in a ring-forming condensation of the hydrazide with the unreacted nitrile. The yield of the hydrazide 18 can be raised to 65% if the nitrile (16) is first converted into the amide (17) with dilute NaOH, and then the resulting reaction mixture is heated with hydrazine hydrate[34] (reaction 2).

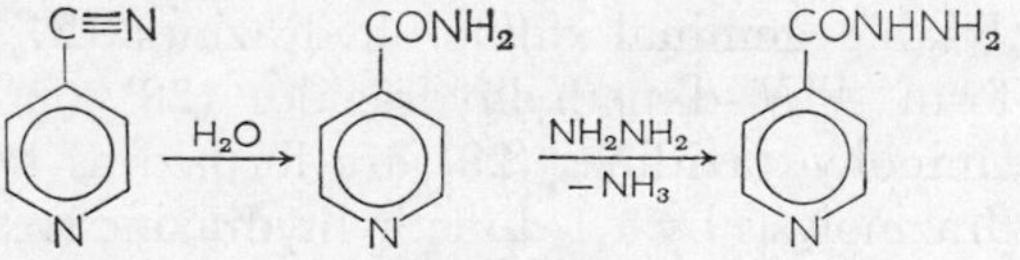

(2)

$$
\begin{array}{ccc}
\text{C}\!\equiv\!\text{N} & \text{CONH}_2 & \text{CONHNH}_2 \\
(16) & (17) & (18)
\end{array}
$$

$$\xrightarrow{\text{H}_2\text{O}} \qquad \xrightarrow[-\text{NH}_3]{\text{NH}_2\text{NH}_2}$$

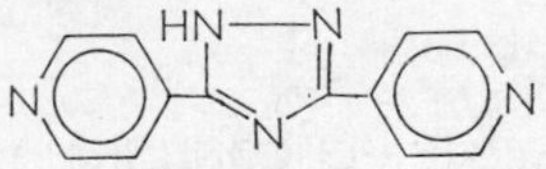

(19)

2. Hydrolysis of hydrazidic halides

Carboxylic acid hydrazides 24 are obtained by hydrolysis of hydrazidic bromides 20 (reaction 3), which are readily accessible by the bromination of aldehyde hydrazones in a mixture of glacial acetic acid and acetic anhydride[35-38]. N'-Monosubstituted and N',N'-disubstituted aromatic and aliphatic carboxylic acid hydrazides can thus be prepared in good yields from aldehyde hydrazones[35-38].

Alkylhydrazidic bromides often hydrolyse on dissolving in aqueous acetone or on being heated to 70°C. The rate of hydrolysis is generally higher for N',N'-disubstituted compounds (**20**) than for N'-monosubstituted ones (**21**)[35,37]. Arylhydrazidic bromides hydrolyse only at 100°C in 50% dioxan[36] or at 150°C in dimethylformamide in the presence of $KHCO_3$. It is assumed that the bromine is displaced by the bicarbonate anion to form an α-bicarbonate (**22**), which decomposes into the hydrazide on decarboxylation[36].

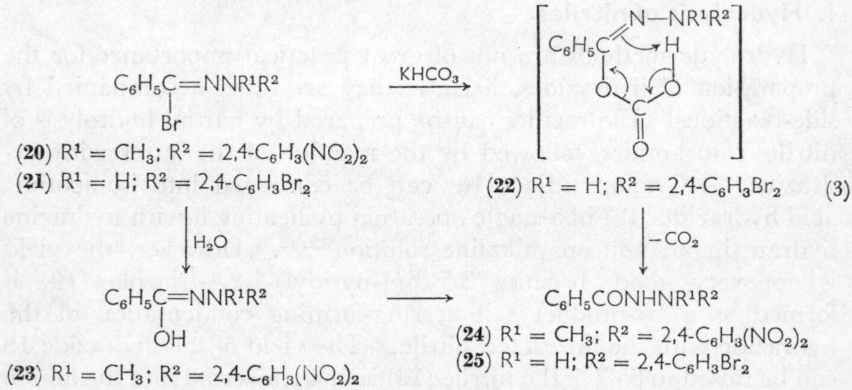

$$
\begin{array}{ccc}
\underset{\underset{Br}{|}}{C_6H_5C}{=}NNR^1R^2 & \xrightarrow{KHCO_3} & \left[\begin{array}{c} C_6H_5C{\overset{N-NR^1R^2}{\diagdown}}H \\ \underset{O}{\overset{O}{\diagup}}{\diagdown}C{\diagup} \end{array} \right]
\end{array}
$$

(**20**) $R^1 = CH_3$; $R^2 = 2,4$-$C_6H_3(NO_2)_2$

(**21**) $R^1 = H$; $R^2 = 2,4$-$C_6H_3Br_2$ (**22**) $R^1 = H$; $R^2 = 2,4$-$C_6H_3Br_2$ (3)

$\downarrow H_2O$ $\downarrow -CO_2$

$\underset{\underset{OH}{|}}{C_6H_5C}{=}NNR^1R^2 \qquad\qquad\longrightarrow\qquad\qquad C_6H_5CONHNR^1R^2$

(**24**) $R^1 = CH_3$; $R^2 = 2,4$-$C_6H_3(NO_2)_2$

(**23**) $R^1 = CH_3$; $R^2 = 2,4$-$C_6H_3(NO_2)_2$ (**25**) $R^1 = H$; $R^2 = 2,4$-$C_6H_3Br_2$

3. Hydrolysis of *gem*-difluorohydrazines

1,1-Dimethylhydrazine adds on to 1,1-difluoroolefins (**26**) to give readily hydrolysable geminal difluorohydrazines (**27**). The latter immediately form $N'N'$-dimethylhydrazides (**28**) on contact with water[39]. N-Aminohydrazidines (**29**) are formed as by-products on account of hydrazinolysis by 1,1-dimethylhydrazine (reaction 4).

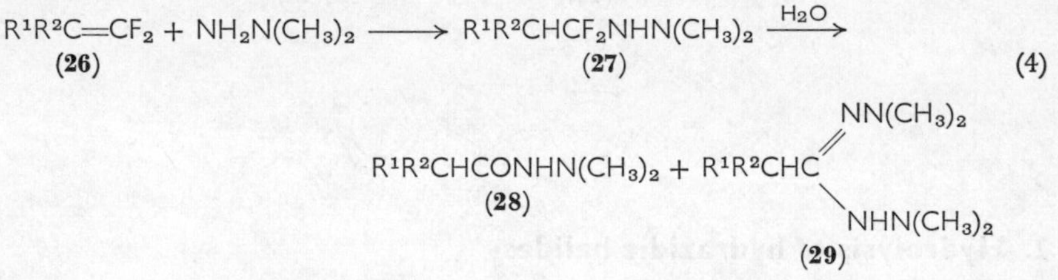

$$
\underset{(\mathbf{26})}{R^1R^2C{=}CF_2} + NH_2N(CH_3)_2 \longrightarrow \underset{(\mathbf{27})}{R^1R^2CHCF_2NHN(CH_3)_2} \xrightarrow{H_2O}
$$

(4)

$$
\underset{(\mathbf{28})}{R^1R^2CHCONHN(CH_3)_2} + R^1R^2CHC{\overset{\displaystyle NN(CH_3)_2}{\underset{\displaystyle NHN(CH_3)_2}{\diagup}}}
$$

(**29**)

B. Acylation of Hydrazines

The reaction of hydrazine and its aryl and alkyl derivatives with acylating agents is the most important method of preparing hydrazides[40].

I. Hydrazinolysis of amides

Amides can be converted into hydrazides by heating them with hydrazine hydrate[40] or anhydrous hydrazine[41]. This reaction requires fairly high temperatures and often a long time. The hydrazinolysis of amides is used only in exceptional cases, because amides are generally obtained by acylating amines with esters, anhydrides, or acid chlorides, and hydrazides can be prepared by hydrazinolysis of these reagents. Akabori[42] has made the important observation that the peptide linkages of proteins can also be cleaved by hydrazinolysis, which can therefore be used for the determination of terminal carboxyl groups. In fact, this method has found extensive application in protein chemistry[43].

Activated amides can be converted into hydrazides with the aid of hydrazine hydrate under very mild conditions. In the imidazolide method, aromatic carboxylic acids, such as benzoic acid, are converted in a single operation into hydrazides with the aid of N,N'-carbonyldiimidazole in tetrahydrofuran containing hydrazine hydrate[44] (reaction 5).

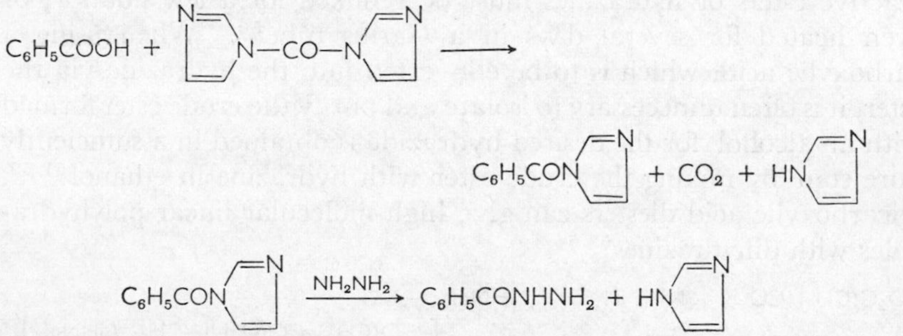

$$\tag{5}$$

N,N-Dialkylcarboxylic acid amides, such as **30**, react with sodium hydrazide in ether at $0°c$ to give hydrazides[45,46]. It is assumed that the hydrazine anion becomes attached to the carbonyl group in a nucleophilic reaction, after which the intermediate **31** decomposes into dialkylamine and the hydrazide **33** via the cyclic intermediate **32** (reaction 6). N-Monosubstituted amides (**34**) do not react with sodium hydrazide in the same way, but form instead a resonance-stabilized amide anion (**35**) through deprotonation, and this anion regenerates the amide during processing[46] (reaction 7).

2. Acylation with esters and lactones

The best method to prepare hydrazides is to react carboxylic acid esters with hydrazine hydrate[1,3,40]. This reaction proceeds both

$$C_6H_5CON\langle\rangle \longrightarrow C_6H_5\overset{NHNH_2}{\underset{O^-}{C}}-N\langle\rangle \longrightarrow \left[\begin{matrix} & \overset{H}{\underset{N\cdots H}{N}} \\ C_6H_5C & \underset{O^-}{\overset{\|}{\cdots\cdots}}N\langle\rangle \end{matrix}\right]H^+ \longrightarrow$$

(30)   (31)   (32)

+ NaNHNH₂   (6)

$$HN\langle\rangle + C_6H_5\overset{}{\underset{O^-}{C}}=NNH_2 \longrightarrow C_6H_5CONHNH_2$$

(33)

$$C_6H_5CONHR \rightleftharpoons C_6H_5\overset{}{\underset{O}{C}}\cdots NR + H^+ \qquad (7)$$

(34)   (35)

without a solvent and in the presence of alcohol, dimethylformamide[47,48], and other organic solvents. It often takes place spontaneously, with evolution of heat. Reaction mixtures involving less
reactive esters or hydrazines must be refluxed for a few hours[49], or
even heated for several days in a Carius tube[50]. When using a
carboxylic acid, which is to be converted into the hydrazide via the
ester, it is often unnecessary to isolate and purify the crude ester formed
with an alcohol, for the desired hydrazide is obtained in a sufficiently
pure state by mixing the crude ester with hydrazine in ethanol[51,52].
Dicarboxylic acid diesters can give high-molecular linear polyhydrazides with dihydrazines[53,54].

$$RO_2C(CH_2)_xCO_2R + NH_2NH(CH_2)_yNHNH_2 \longrightarrow$$
$$-\!\!\left[CO(CH_2)_xCONHNH(CH_2)_yNHNH\right]\!\!-_n$$

Few systematic investigations have so far been done to find out
which nitrogen is acylated by carboxylic acid esters in the case of
unsymmetrically alkylated or arylated hydrazines[7]. With methylhydrazine, esters react to give preferentially the 1-acyl-2-methylhydrazine
(36), besides a small amount of the 1-acyl-1-methylhydrazine
(37) (reaction 8). The larger the R^2 group, the slower the
reaction and the smaller is the amount of 37 compared with 36. In
fact, with large R^2 groups only 36 is found[7,55-57]. The reaction of
esters with other monoalkylhydrazines similarly leads to more 36 than
37. The larger the R^1 group of the hydrazine[58], the smaller the
amount of 37. Steric effects clearly play a decisive part in the hydrazinolysis of esters. Methyl formate reacts with monoalkyl-substituted

$$R^1NHNH_2 + R^2CO_2CH_3 \longrightarrow R^1NHNHCOR^2 + \left(\begin{array}{c} R^1NNH_2 \\ | \\ COR^2 \end{array} \right) \quad (8)$$

$$(36)$$

$$R^1 = alkyl \qquad\qquad (37)$$

hydrazines (**38a–38c**) in an anomalous manner to give 1-alkyl-1-formylhydrazines[56,57] (**39a–39c**). It is only in the presence of a bulky substituent, such as the cyclohexyl group in **38d**, that the reaction takes place at unsubstituted nitrogen and gives **40**. The small formyl group is therefore less hindered by alkyl groups in its attack on the substituted nitrogen atom.

$$RNHNH_2 + HCO_2CH_3 \longrightarrow \begin{array}{l} \overset{\displaystyle RNNH_2}{\underset{\displaystyle CHO}{|}} \\ (\mathbf{39a})\ R = n\text{-Bu;} \\ (\mathbf{39b})\ R = CH_2C_6H_5; \\ (\mathbf{39c})\ R = CH_2CH_2CN \\[6pt] \overset{\displaystyle RNHN}{\underset{\displaystyle CHO}{|}} \\ (\mathbf{40})\ R = cyclohexyl \end{array} \quad (9)$$

(**38a**) R = n-Bu;
(**38b**) R = CH₂C₆H₅;
(**38c**) R = CH₂CH₂CN;
(**38d**) R = cyclohexyl

1,1-Dimethylhydrazine does not form hydrazides with acetates and benzoates[7], and can be converted into 2,2-dimethylhydrazides only by the use of esters containing, near the CO group, a strongly electron-attracting group such as CN, NO₂ or halogen, which promotes the nucleophilic attack on the CO group[7,24]. The nucleophilic character of unsymmetrical dimethylhydrazine is clearly not strong enough to ensure a reaction with non-activated esters[59]. The formate is again an exception, because it does form the hydrazide with 1,1-dimethylhydrazine[7] (reaction 10). Symmetric dimethylhydrazine reacts with esters only with great difficulty, and the starting materials are generally recovered unchanged[23].

$$(CH_3)_2NNH_2 \begin{array}{l} \xrightarrow[]{+\ R^1CO_2R^2} R^1CONHN(CH_3)_2 \\ \qquad R^1=CH_3 \\ \qquad R^2=C_6H_5 \\[10pt] \xrightarrow[+\ HCO_2CH_3]{} HCONHN(CH_3)_2 \end{array} \quad (10)$$

Aryl-substituted hydrazines yield only 1-acyl-2-arylhydrazines with esters[4,60] (reaction 11). The specific acylation of aryl-substituted hydrazines on the free amino group may be attributed to the fact that the mesomeric effect of the aryl group reduces the nucleophilic

18—c.o.a.

character of the substituted nitrogen atom and thus favours the acylation of the adjacent nitrogen. However, the possibility that steric effects play a part cannot be excluded.

$$C_6H_5NHNH_2 + R^1CO_2R^2 \longrightarrow R^1CONHNHC_6H_5 \qquad (11)$$

$$(R^1 = H, C_6H_5CH{=}CH; R^2 = CH_3)$$

Alkyl esters that do not undergo alkaline hydrolysis readily, often react with hydrazines only with difficulty or not at all. In such cases good results are frequently obtained by preparing and reacting with hydrazine the activated esters such as p-nitrobenzyl[61,62] or cyano-methyl esters[50,63,64], as in the case of p-nitrobenzyl pyrroleacetate[62] and the cyanomethyl ester of benzoylglycine[63]. Under the normal conditions of the reaction of hydrazines with esters, diacylhydrazines are not formed in an appreciable quantity; their formation generally requires longer reaction times and higher temperatures.

The reaction kinetics have been thoroughly investigated in the case of the hydrazinolysis of substituted ethyl phenylacetates[65]. The rate depends on the concentration of the conjugate acid $NH_2NH_3^+$ and the concentration of the base NH_2NH_2 according to the equation

$$- \frac{d[\text{Ester}]}{dt} = (k_n + k_b[NH_2NH_2] + k_a[NH_2NH_3^+])[NH_2NH_2][\text{Ester}]$$

where k_n is a second-order rate constant for the nucleophilic substitution of the alkoxy group of the ester, and k_a and k_b are third-order rate constants for the general acid and base catalysis of the reaction.

Hydrazine is more strongly nucleophilic than could be expected from its basicity[66]. This is called an α-effect, because it is attributed to stabilization of the hydrazine group (acquiring a partial positive charge in the transition state **41** during the nucleophilic reaction)

$$NH_2NH_2 + RCH_2C\underset{OC_2H_5}{\overset{O}{\big\langle}} \longrightarrow$$

$$(R = XC_6H_4; X = H,$$
$$m\text{-}, p\text{-}NO_2, p\text{-}CH_3, p\text{-}CH_3O)$$

$$\left[H_2N\overset{H}{\underset{H}{\overset{|}{N}}}\cdots\overset{O}{\underset{CH_2R}{\overset{\|}{C}}}\cdots OC_2H_5 \right] \longrightarrow NH_2NHC\underset{CH_2R}{\overset{O}{\big\langle}} + HOC_2H_5 \qquad (12)$$

$$(41)$$

by the lone pair of electrons on the α-nitrogen atom[66]. This effect influences k_a and k_b more than k_n, which is modified by it only to a small extent[65].

The significance of the general base catalysis is also manifested in the rate equation for the reaction of ethyl phenylacetate with mono-methylhydrazine[67]. The kinetics of the reaction between ethyl phenylacetate and (dimethylaminoalkyl)hydrazines suggest that the hydrazinolysis is catalytically assisted by intramolecular attack of the dimethylamino group. The transition state **42** has been proposed to explain this intramolecular base catalysis[67].

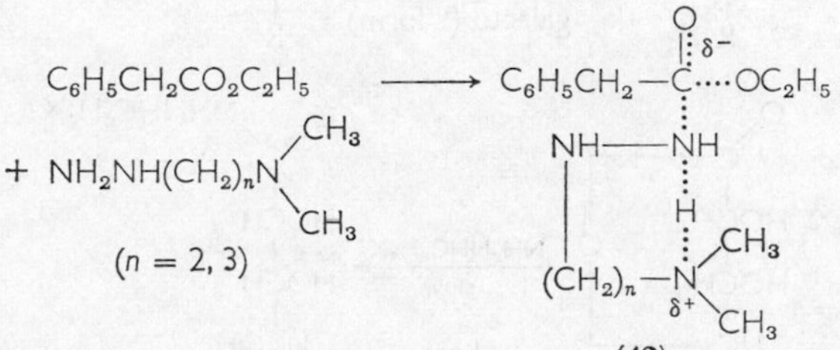

$$C_6H_5CH_2CO_2C_2H_5$$
$$+ \ NH_2NH(CH_2)_n N \overset{CH_3}{\underset{CH_3}{\diagdown}}$$
$$(n = 2, 3)$$

$$\longrightarrow$$

$$\underset{(42)}{}$$

(42)

The reaction of lactones with hydrazines is generally accompanied by ring opening and leads to hydroxycarboxylic acid hydrazides. Thus, β-trichloromethyl-β-propiolactone (**43**) reacts with hydrazine or phenylhydrazine to give 4,4,4-trichloro-3-hydroxybutyrohydrazide (**44**)[68,69] (reaction 13). Readily crystallizable phenylhydrazides can be used to identify naturally occurring lactones such as D-digitoxonic acid lactone[70]. The preparation of crystalline phenylhydrazides from aldonic acid lactones and aldaric acid lactones is similarly used in the characterization of these groups of compounds[28,71,72].

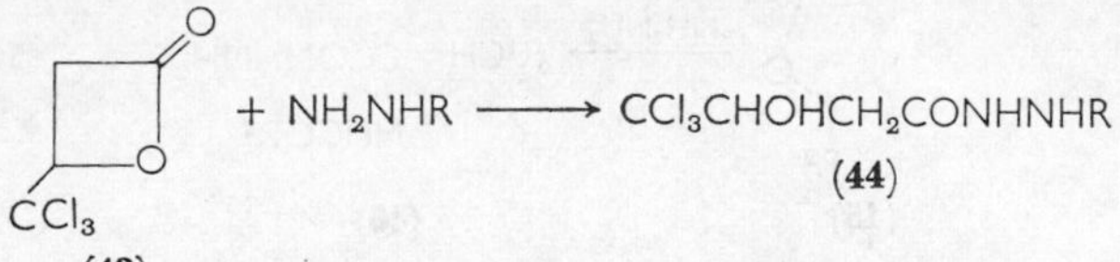

$$\underset{(43)}{} \quad + \ NH_2NHR \longrightarrow CCl_3CHOHCH_2CONHNHR \qquad (13)$$
$$(44)$$

Stroh and Henning[71] investigated polarimetrically the way in which the rate of arylhydrazide formation from arylhydrazines and aldonic acid γ-lactones varies with the configuration of the lactone and the substituents of the arylhydrazine. Aldonic acid γ-lactones in which the OH groups in the lactone ring have the same configuration react to give hydrazides at the same rate. The rate of reaction is higher for lactones with the arabino configuration (D-galactono-lactone and D-arabonolactone, reaction 14a) than for lactones with a lyxo configuration (D-mannonolactone and D-lyxonolactone, reaction 14b)[72]. This difference is ascribed to a difference in the formation of

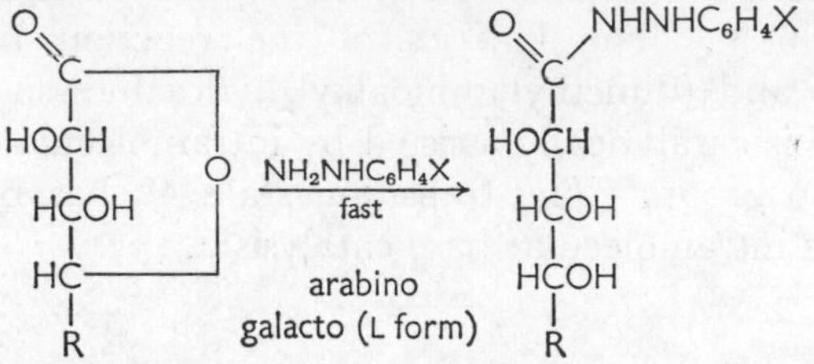

$$(14a)$$

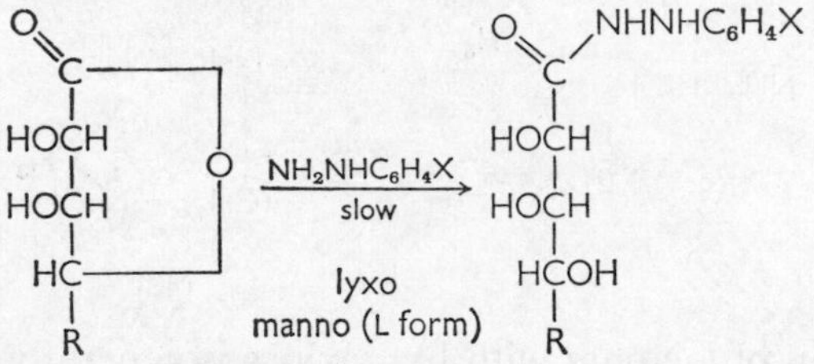

$$(14b)$$

hydrogen bonds going from the ring OH groups to the carbonyl oxygen, which is influenced by steric factors. However, no connexion has been found between the reaction rate and the basicity of the variously substituted phenylhydrazines. Unsubstituted phenylhydrazine (X = H in $XC_6H_4NHNH_2$) reacts with all the aldonic acid lactones faster than the substituted ones (X = p-$CO_2C_2H_5$, p-Br, m-OCH_3, m-CH_3, p-CH_3, and p-OCH_3)[71].

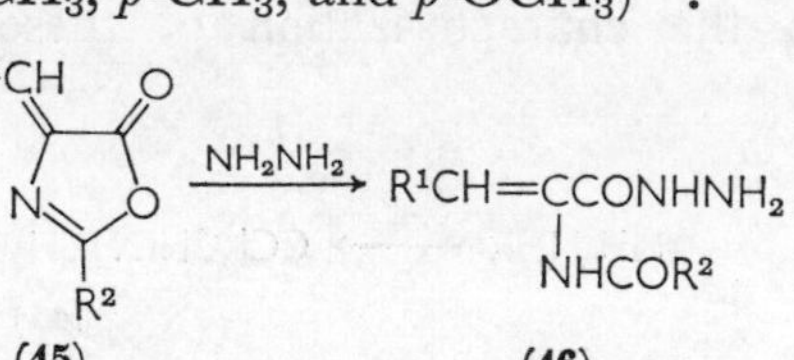

(R^1 = XC_6H_4; X = H, o-, p-CH_3O, m-NO_2, m-, p-CH_3, o-Cl; R^2 = CH_3, C_6H_5, p-$(CH_3)_2NC_6H_4$)

Unsaturated azlactones (45) having an oxazolinone structure react with hydrazine to form α-acylaminoacrylic acid hydrazides (46). Variously substituted acrylohydrazides having structure 46 can thus be prepared[73-75] by varying the substituents R^1 and R^2.

3. Acylation with acyl chlorides and anhydrides

Carboxylic acid chlorides and anhydrides generally react very vigorously with hydrazine to form acylhydrazines, which often immediately react further to give 1,2-diacylhydrazines[40]. This secondary reaction can be suppressed by diluting the acylating agent with ether[40], benzene[23], or hexane[76,77], and by adding it dropwise to the hydrazine solution at low temperatures[76]. In the case of unreactive carboxylic acid esters[50] or low-basicity hydrazines, acylation with acid anhydrides or acyl chlorides might be the only possible way of preparing the hydrazides. This acylation is often carried out in pyridine[78,79], in an aqueous alkaline solution[23], or in an organic solvent containing sodium carbonate[80]. With unreactive substances, the reaction mixture must be boiled for some time[37,81,82].

Acylation of alkyl-substituted hydrazines with acyl chlorides and anhydrides occurs preferentially on the substituted nitrogen atom[7,55]. Thus, methylhydrazine reacts with benzoic anhydride to give 1-benzoyl-1-methylhydrazine[7,83] (reaction 15), since the CH_3 group should enhance the nucleophilicity of $N_{(1)}$. On the other hand, phenylhydrazine, in which the mesomeric effect of the phenyl group reduces the nucleophilicity of $N_{(1)}$, reacts with acid anhydrides[82] and acid chlorides[80,78] to form N'-phenylhydrazides (reaction 16). The nucleophilic character of the nitrogens in 1,2-dimethylhydrazine is so strong that acylation leads to the diacylhydrazine as the main product[23] (reaction 17). With benzoyl chloride under normal conditions, the much less nucleophilic hydrazobenzene forms only 1-benzoyl-1,2-diphenylhydrazine[78] (reaction 18). Asymmetrically disubstituted hydrazines, such as 1,1-dimethylhydrazine[23,77], 1,4-diaminopiperazine[81], and hydrazones[79] can be acylated on the unsubstituted amino group to give β,β-disubstituted carboxylic acid hydrazides (reaction 19). In contrast to the acylation of hydrazines with esters, where

$$CH_3NHNH_2 + (C_6H_5CO)_2O \longrightarrow CH_3NNH_2 \qquad (15)$$
$$\underset{\displaystyle COC_6H_5}{|}$$

$$C_6H_5NHNH_2 + (RCO)_2O \text{ (or RCOCl)} \longrightarrow C_6H_5NHNHCOR \qquad (16)$$

$$CH_3NHNHCH_3 + (RCO)_2O \longrightarrow CH_3N\underset{\displaystyle COR}{\underset{|}{\quad}}N\underset{\displaystyle COR}{\underset{|}{\quad}}CH_3 \qquad (17)$$

$$C_6H_5NHNHC_6H_5 + (RCO)_2O \text{ (or RCOCl)} \longrightarrow C_6H_5NHNC_6H_5 \qquad (18)$$
$$\underset{\displaystyle COR}{|}$$

$$(CH_3)_2NNH_2 + (RCO)_2O \text{ (or RCOCl)} \longrightarrow (CH_3)_2NNHCOR \qquad (19)$$

steric effects were decisive, on acylation with acyl chlorides and anhydrides the electronic effects of the substituents seem to be the important ones.   Steric effects operate here only with very large groups,
as in the case of hydrazobenzene.

Acylations with acyl chlorides and acid anhydrides offer the best
means of preparing diacyl-, triacyl-, and tetraacylhydrazines (section
VI.B.2).

Chlorides and anhydrides of certain dicarboxylic acids can form
cyclic hydrazides[1,3].   Thus, oxalyl chloride reacts with N,N'-diiso-
propylhydrazine or N,N'-di-t-butylhydrazine to give the corresponding
very   unstable   1,2-diacetidinediones[84]   (**47**)   (reaction 20).   Otherwise cyclic hydrazides are obtained only when the resulting ring contains a double bond.   Thus, succinic anhydride does not form a cyclic
hydrazide, while maleic anhydride and phthalic anhydride do[85-89].

$$\text{RNHNHR + CICOCOCI} \longrightarrow \underset{\substack{\text{RN}\text{—NR} \\ (\textbf{47})}}{\overset{\text{O}\qquad\text{O}}{\big|\qquad\big|}} \tag{20}$$
$$(R = \text{i-Pr},\ t\text{-Bu})$$

It is possible to acylate the less nucleophilic hydrazine nitrogen with
the aid of an acyl chloride by utilizing the stronger acidity of the N—H
bond involving the less nucleophilic nitrogen.   Thus, the reaction of
phenylhydrazine with sodium leads preferentially to the sodium compound **48**, and on treatment with acyl chloride, 1-acyl-1-phenyl-
hydrazine (**49**) is obtained as the main product[90,91] (reaction 21).
However, the yield is often low, since the reaction is difficult to control.
By-products are thus obtained, particularly after a long reaction
period, in the form of 1-acyl-2-phenylhydrazine, 1,2-diacyl-1-
phenylhydrazine, and—owing to reduction and N—N cleavage—
aniline and ammonia.

$$C_6H_5NHNH_2 + Na \longrightarrow \underset{(\textbf{48})}{C_6H_5NNaNH_2} \xrightarrow{\text{RCOCl}} \underset{\substack{| \\ \text{COR} \\ (\textbf{49})}}{C_6H_5NNH_2} \tag{21}$$

Another useful method for the preparation of hydrazides having an
acyl group on the less nucleophilic nitrogen is the diacylation of substituted hydrazines to symmetric diacylhydrazines, followed by partial
acid hydrolysis.   Thus, phenylhydrazine can be converted into 1,2-
dibenzoyl-1-phenylhydrazine (**50**) with 2 moles of benzoyl chloride.

In the partial acid hydrolysis, the benzoyl group on the monosubstituted nitrogen cleaves off faster, so that 1-benzoyl-1-phenylhydrazine (**51**) is obtained[91a] (reaction 22).

$$C_6H_5NHNH_2 + 2\,C_6H_5COCl \longrightarrow C_6H_5NNHCOC_6H_5 \xrightarrow{H_2O}$$
$$\underset{COC_6H_5}{|}$$
$$(\mathbf{50}) \qquad C_6H_5NNH_2 + C_6H_5COOH \quad (22)$$
$$\underset{COC_6H_5}{|}$$
$$(\mathbf{51})$$

The primary amino group in substituted hydrazines can also be blocked by the formation of hydrazones, whereupon acylation can occur only on the other nitrogen[79,92,93]. This process first gives the hydrazone (**52**), which then liberates 1-acyl-1-phenylhydrazine (**53**) on acid hydrolysis (reaction 23).

$$C_6H_5NHNH_2 + CO(CH_3)_2 \longrightarrow C_6H_5NHN{=}C(CH_3)_2 \xrightarrow{RCOCl}$$
$$(\mathbf{52})$$
$$C_6H_5NN{=}C(CH_3)_2 \xrightarrow{H^+} C_6H_5NNH_2 \quad (23)$$
$$\underset{COR}{|} \qquad\qquad\qquad \underset{COR}{|}$$
$$(\mathbf{53})$$

4. Acylation with ketenes

In ether solution, phenylhydrazines give quantitative yields of 1-acyl-2-phenylhydrazines when reacted with ketene, dimethylketene, or diphenylketene[94] (reaction 24). In the presence of hydrazine, the ketene formed *in situ* by the fragmentation of chloroacetic acid hydrazide (reaction 25) gives acethydrazide[41,95]. With diphenylketene, methylhydrazine immediately gives the *N,N'*-diacyl derivative (**54**), and the monoacyl derivative cannot be intercepted (reaction 26).

$$R_2C{=}C{=}O + ArNHNH_2 \longrightarrow \left[R_2C{=}C\underset{OH}{\overset{NHNHAr}{\diagup}} \right] \longrightarrow R_2CHC\underset{O}{\overset{NHNHAr}{\diagup}} \quad (24)$$

$$(R = H, CH_3, C_6H_5)$$

$$CH_2ClCONHNH_2 \longrightarrow [CH_2{=}C{=}O] \xrightarrow{NH_2NH_2} CH_3C\underset{NHNH_2}{\overset{O}{\diagup}} \quad (25)$$

$$CH_3NHNH_2 + 2\,O{=}C{=}C(C_6H_5)_2 \longrightarrow CH_3NCOCH(C_6H_5)_2 \quad (26)$$
$$\underset{HNCOCH(C_6H_5)_2}{|}$$
$$(\mathbf{54})$$

C. Reaction between Carboxylic Acids and Hydrazines

I. Thermal dehydration of hydrazinium salts

The thermal dehydration of hydrazinium salts is rarely used for the synthesis of hydrazides, because it requires drastic conditions. The monoacylhydrazines formed primarily may disproportionate into symmetric diacylhydrazines, which often constitute the main product[40]. The hydrazide can frequently be obtained in a good yield, as for example in the reaction of 2-quinolylhydrazine with isobutyric acid[96,97]. Refluxing of acetic acid with 1-methyl-2-phenylhydrazine gives 1-acyl-1-methyl-2-phenylhydrazine[98] (reaction 27). Isonicotinic acid hydrazide can be prepared in good yield from hydrazinium isonicotinate, by removing the water azeotropically with pentanol[99] (reaction 28). On the other hand, hydrazobenzene reacts with crotonic acid to form crotonic acid 1,2-diphenylhydrazide only in low yield[60].

$$CH_3CO_2^-\overset{+}{N}H_2NHC_6H_5 \underset{CH_3}{|} \xrightarrow{\Delta} CH_3CONNHC_6H_5 \underset{CH_3}{|} \quad (27)$$

$$N\langle\bigcirc\rangle-CO_2^-\overset{+}{N}H_3NH_2 \xrightarrow{\Delta} N\langle\bigcirc\rangle-CONHNH_2 \quad (28)$$

2. The carbodiimide method

N,N'-Dicyclohexylcarbodiimide can be used as a dehydrating agent in the reaction of carboxylic acids with hydrazines. The reaction is carried out in methylene chloride at room temperature; it takes a few hours and gives a good yield of the hydrazide. In fact, the yield is often better than in the ester hydrazinolysis[100].

The reaction is thought to proceed as shown in reaction (29). The carboxylic acid first adds to the dicyclohexyldiimide to form O-acylisourea (**55**), which acylates hydrazine into hydrazide (**58**) giving also dicyclohexylurea (**59**). However, **55** can isomerize into N-acylurea (**56**). Thus, **56** has been isolated as a by-product in the reaction with p-nitrobenzoic acid and N,N'-dimethylhydrazine. A side-reaction in which **55** interacts with carboxylic acid to give the acid anhydride (**57**) and dicyclohexylurea (**59**) is also possible.

The acylation of monosubstituted hydrazines by the diimide method follows the same rule as acylation with acyl chlorides and anhydrides[100]. The electronic effects of the substituents are again

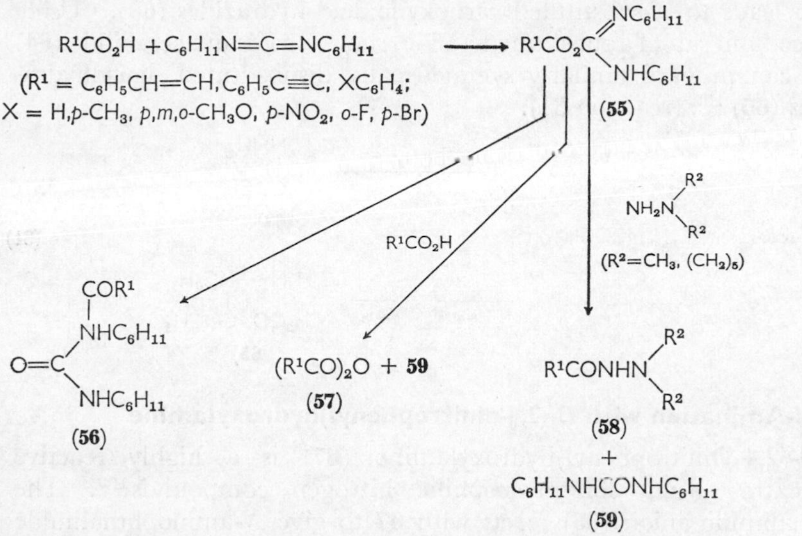

$$R^1CO_2H + C_6H_{11}N=C=NC_6H_{11} \longrightarrow R^1CO_2C\overset{NC_6H_{11}}{\underset{NHC_6H_{11}}{}}$$

$$(R^1 = C_6H_5CH=CH, C_6H_5C\equiv C, XC_6H_4;$$
$$X = H, p\text{-}CH_3, p,m,o\text{-}CH_3O, p\text{-}NO_2, o\text{-}F, p\text{-}Br)$$

$$(55)$$

$$NH_2N\overset{R^2}{\underset{R^2}{}} \qquad (29)$$

$$(R^2=CH_3, (CH_2)_5)$$

$$R^1CO_2H$$

$$O=C\overset{\overset{\displaystyle COR^1}{|}}{\underset{NHC_6H_{11}}{\overset{NHC_6H_{11}}{}}}$$

$$(56)$$

$$(R^1CO)_2O + 59$$

$$(57)$$

$$R^1CONHN\overset{R^2}{\underset{R^2}{}}$$

$$(58)$$
$$+$$
$$C_6H_{11}NHCONHC_6H_{11}$$

$$(59)$$

more important than steric factors. Thus, methylhydrazine is acylated by various acids always into 1-acyl-1-methylhydrazine. Analogously, aromatic hydrazines, such as phenylhydrazine, are acylated by benzoic acid on the unsubstituted amino group (cf. reactions 15 and 16).

D. N-*Amination of Amides*

1. Schestakov's reaction

Similarly to Hofmann's amide degradation, monoacylated ureas can be converted into hydrazides with the aid of sodium hypochlorite[101]. Accordingly, benzoylurea (60) reacts with sodium hypochlorite to give benzoic acid hydrazide (61) (reaction 30).

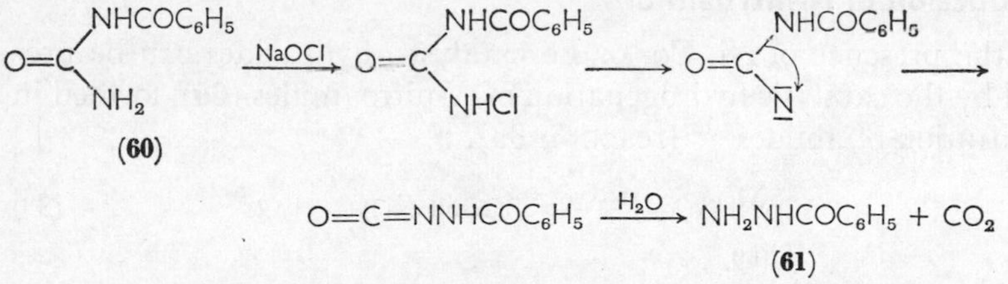

$$O=C\overset{NHCOC_6H_5}{\underset{NH_2}{}} \xrightarrow{NaOCl} O=C\overset{NHCOC_6H_5}{\underset{NHCl}{}} \longrightarrow O=C\overset{NHCOC_6H_5}{\underset{\underline{N}}{}} \longrightarrow$$

$$(60)$$

$$O=C=NNHCOC_6H_5 \xrightarrow{H_2O} NH_2NHCOC_6H_5 + CO_2 \qquad (30)$$

$$(61)$$

2. N-Amination with sodamide

N-Chloroamides (62) react with sodamide to give carboxylic acid hydrazides[102]. Only secondary amides can be used here as the starting materials, since they readily give 62. The reaction with sodamide

18*

then leads to N-substituted carboxylic acid hydrazides (**63**).   Using
the sodium salt of a substituted amide, such as sodium acetanilide (**64**)
one can prepare similarly symmetrically disubstituted diacylhydrazines (**65**) [102] (reaction 31).

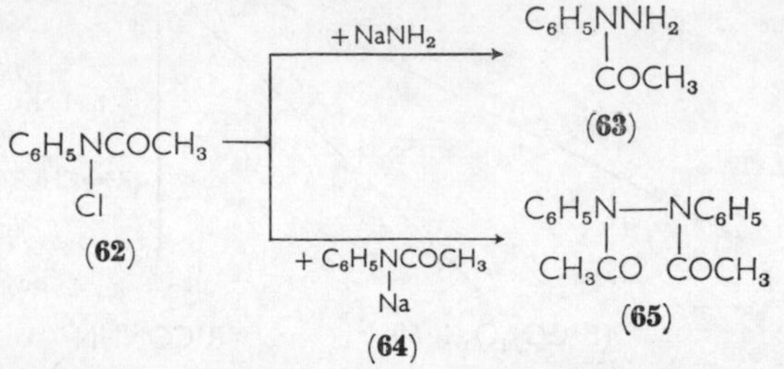

(31)

3. *N*-Amination with *O*-(2,4-dinitrophenyl)hydroxylamine

O-(2,4-Dinitrophenyl)hydroxylamine (**67**) is a highly reactive
aminating agent for nucleophilic nitrogen compounds[103]. The
phthalimide anion (**66**) reacts with **67** to give N-aminophthalimide
(**68**) in 88% yield (reaction 32).

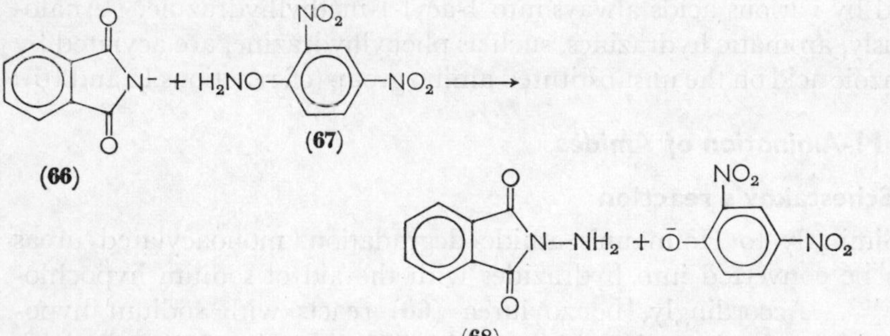

(32)

4. Reduction of *N*-nitroamides

In the presence of Ni, Co, or Fe catalysts, hydrazides can be prepared by the catalytic hydrogenation of N-nitroamides (**69**), formed in
the nitration of amides[104] (reaction 33).

$$RCONHNO_2 \xrightarrow{H_2/Cat.} RCONHNH_2 + H_2O$$

(33)

(**69**)

E. Conversion of Azo Compounds

Carbonylazo compounds, such as **70**, can be converted into hydrazides by hydrogenation of the N=N bond.   However, since the car-

bonylazo compounds are prepared from hydrazides by oxidation, this method is hardly practical. Nevertheless, carbonylazo compounds may be useful for the synthesis of substituted hydrazides.

Benzoylazobenzene (**70**) reacts with Grignard reagents to form by 1,4-addition the Grignard compound (**71**), which can then be hydrolysed into a 2,2-disubstituted hydrazide (**72**)[105,106] (reaction 34).

$$C_6H_5N{=}NCOC_6H_5 + RMgX \longrightarrow C_6H_5NRN{=}CC_6H_5 \xrightarrow{H_2O}$$

$$\text{(70)} \qquad\qquad \underset{\text{OMgX}}{|}$$

$$\text{(71)} \qquad C_6H_5NRNHCOC_6H_5 \quad (34)$$

$$\text{(72)}$$

The azo compound **74**, which is obtained by reacting benzaldehyde phenylhydrazone with dipotassium nitrosobisulphate (**73**), decomposes in aqueous solution into 1-benzoyl-2-phenylhydrazine and hydroxylimidobissulphuric acid[107] (reaction 35).

$$C_6H_5CH{=}NNHC_6H_5 + 2\,ON(SO_3K)_2 \longrightarrow C_6H_5CHN{=}NC_6H_5 \longrightarrow$$

$$\text{(73)} \qquad\qquad\qquad \underset{\text{ON(SO}_3\text{K)}_2}{|}$$

$$\text{(74)} \qquad\qquad\qquad \text{(35)}$$

$$C_6H_5\underset{\underset{O}{\|}}{C}NHNHC_6H_5 + HON(SO_3K)_2$$

The treatment of hydrazones of the type of **75** with an ethereal 40% peracetic acid solution leads to carboxylic acid hydrazides (**80**) in good yields[108]. The *cis*-azoxy compound (**76**) formed in the first step rearranges into an *N*-hydroxyhydrazone (**77**), then by addition and elimination gives the α-hydroxyazo compound (**78**), which finally tautomerizes into **79** and the hydrazide (**80**) (reaction 36). This

$$\qquad\qquad\qquad O \qquad\qquad\qquad OH$$
$$\qquad\qquad\qquad \uparrow \qquad\qquad\qquad |$$
$$ArCH{=}NNHCH_3 \longrightarrow ArCH_2N{=}NCH_3 \longrightarrow ArCH{=}NNCH_3 \xrightarrow[H_2O]{H+}$$

$$\text{(75)} \qquad\qquad \text{(76)} \qquad\qquad\qquad \text{(77)}$$

$$\underset{\text{OH}}{|} \qquad\qquad\qquad \underset{\text{OH}}{|}$$
$$ArCHN{=}NCH_3 \longrightarrow Ar\overset{|}{C}{=}NNHCH_3 \longrightarrow ArCONHNHCH_3 \quad (36)$$

$$\text{(78)} \qquad\qquad\qquad \text{(79)} \qquad\qquad\qquad \text{(80)}$$

mechanism, and particularly the step **76** → **77** proceeds only with aromatic hydrazones, due to resonance stabilization in **77**. However, in some cases, hydrazides can be prepared from aliphatic hydrazones as well[108].

F. Cleavage of Cyclic Compounds

I. C—C cleavage

The C—C cleavage of cyclic systems to form hydrazides is known only in the case of the diazo cleavage of enediols and the hydrazinolytic cleavage of the cyclobutanone ring. A benzenediazonium cation adds in the enediol **81** on to the carbon atom that is next to the electron-attracting carbonyl group[109]. With the assistance of the primary OH group of the glycol side-chain, the resulting azo compound (**82**) cleaves into an α-hydroxyazo compound (**83**), which then rearranges into the hydrazide (**84**) (reaction 37).

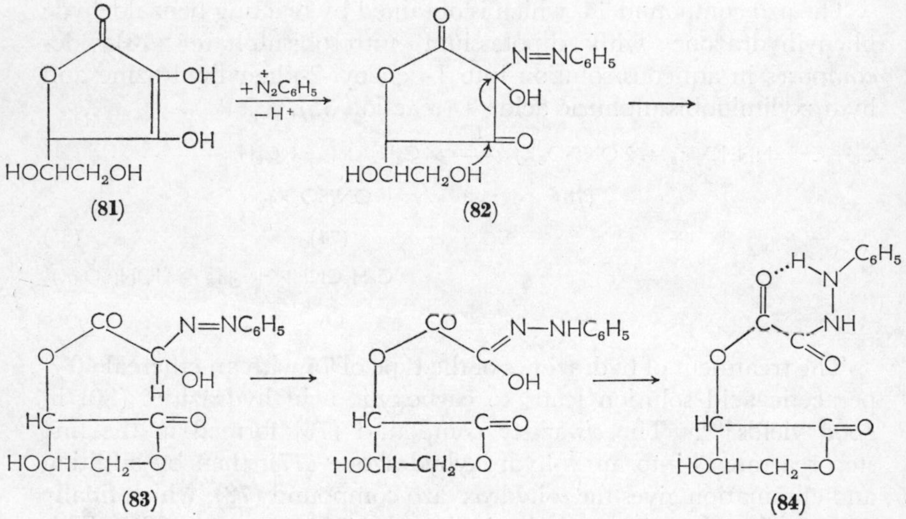

(37)

Acyclic enediols can be converted into hydrazides with diazonium ions, analogously to the cleavage of the cyclic enediol described above. Thus, 1-benzoyl-2-α-pyridylethenediol (**85**) reacts with *p*-chlorobenzenediazonium sulphate in sulphuric acid solution, giving rise to α-picolinic acid 2-*p*-chlorophenylhydrazide (**87**) via an α-hydroxyazo compound (**86**) (reaction 38)[110].

When 7,7-diphenylbicyclo[3,2,0]hept-2-en-6-one (**88**) is heated with hydrazine hydrate for a fairly long time, the strained cyclobutanone ring opens to give *cis*-3-benzhydrylcyclopentene-4-carboxylic acid hydrazide (**89**)[111], which rearranges under the reaction conditions into the thermodynamically more stable *trans* form (**90**) (reaction 39).

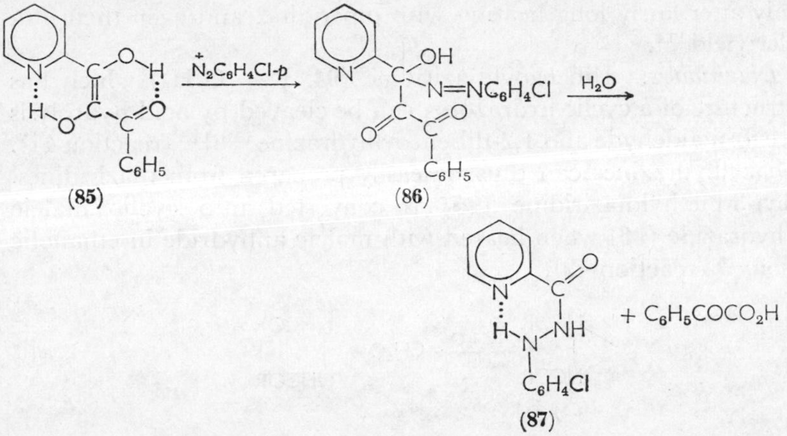

(85) → (86)

$$\xrightarrow{\overset{+}{N}_2C_6H_4Cl\text{-}p}$$

$$\xrightarrow{H_2O}$$

(87)

$$+ \; C_6H_5COCO_2H \quad (38)$$

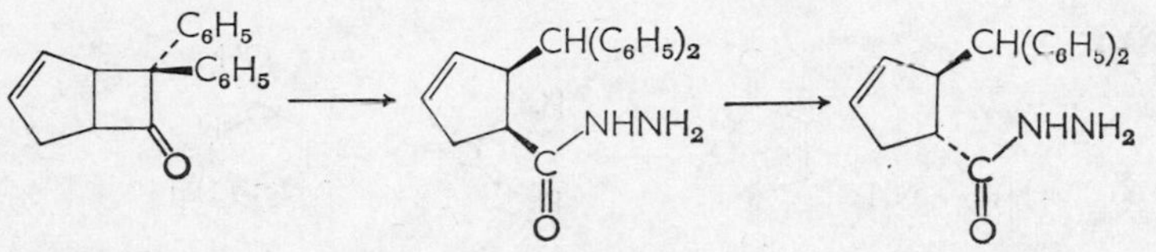

(88) ⟶ (89) ⟶ (90) (39)

2. Cleavage of heterocyclic compounds

a. Acyloxaziridines. Oxaziridines, which are readily accessible compounds, can be acylated on the nitrogen atom[112]. Acyloxaziridines (**91**) are very reactive towards amines, e.g. with piperidine the hydrazide **93** is formed in an excellent yield[113,114] (reaction 40). However, mild conditions are sufficient only when $R^1 = H$ and $R^2 = C_6H_5$. Thus, 2-benzoyl-3,3-pentamethyleneoxaziridine (**91b**) yields

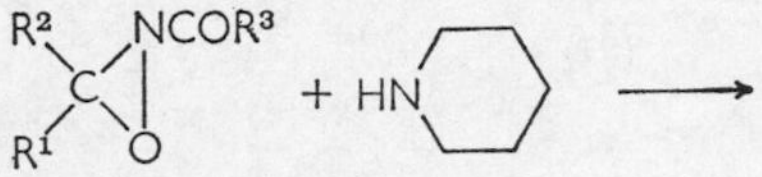

(**91a**) $R^1 = H$; $R^2 = C_6H_5$;
  $R^3 = p\text{-}C_6H_4NO_2$;

(**91b**) $R^1, R^2 = $ pentamethylene;
  $R^3 = C_6H_5$

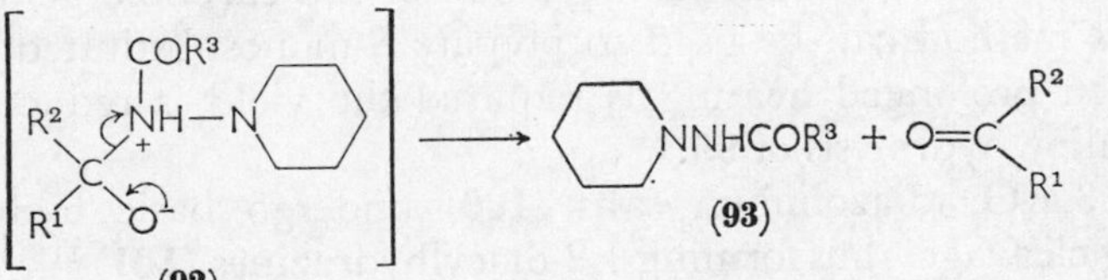

$$\longrightarrow \quad \text{NNHCOR}^3 + O=C\begin{smallmatrix}R^2\\R^1\end{smallmatrix} \quad (40)$$

(**92**)   (**93**)

93 only after fairly long heating with piperidine, and even then in a smaller yield[114].

b. Diaziridines. Dibenzoyldiaziridine (**94**, $R = C_6H_5$) which has the structure of a cyclic hydrazone, can be cleaved by acid hydrolysis to give formaldehyde and 1,2-dibenzoylhydrazine[113,115] (reaction 41). 1,2-Diacylhydrazines can thus be easily prepared from diaziridines. 3-Ethyl-3-methyldiaziridine (**95**) is converted into cyclic maleic acid hydrazide (**96**) when heated with maleic anhydride in ethanolic solution[116] (reaction 42).

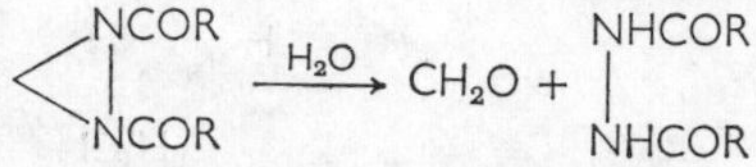

$$
\begin{array}{c}
\overset{NCOR}{\underset{NCOR}{\triangleleft}} \quad \xrightarrow{\ H_2O\ } \quad CH_2O \ + \ \overset{NHCOR}{\underset{NHCOR}{|}}
\end{array}
\qquad (41)
$$

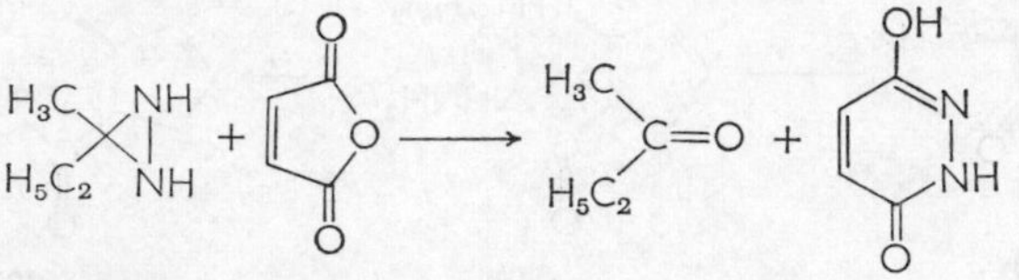

$$(\mathbf{94})$$

$$
\begin{array}{c}
\overset{H_3C}{\underset{H_5C_2}{>}}\!\!\underset{NH}{\overset{NH}{<}} \ + \ \text{(maleic anhydride)} \ \longrightarrow \ \overset{H_3C}{\underset{H_5C_2}{>}}\!C{=}O \ + \ \text{(maleic hydrazide)}
\end{array}
\qquad (42)
$$

$$(\mathbf{95}) \qquad\qquad\qquad (\mathbf{96})$$

c. 1,2-Diacetidinedione. 1,2-Diacetidinediones (**47**), formed according to reaction 20, spontaneously react in alcohol in the presence of catalytic amounts of an inorganic acid, giving oxalic acid monoester hydrazides (**97**)[84] (reaction 43).

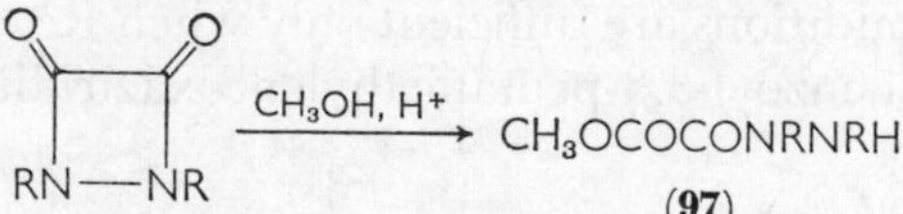

$$
\underset{\underset{(\mathbf{47})}{RN\!-\!NR}}{\overset{O\quad\ O}{\|\quad\ \|}} \quad \xrightarrow{\ CH_3OH,\ H^+\ } \quad CH_3OCOCONRNRH
\qquad (43)
$$

$$\qquad\qquad\qquad\qquad\qquad (\mathbf{97})$$

$$(R = alkyl)$$

d. Oxadiazoles. On heating a 2,4-substituted 1,3,4-oxadiazolin-5-one (**98**) in aqueous NaOH, the hydrazide **99** is obtained (reaction 44)[117]. The rupture of the ring is due to hydrolysis of the lactone group, followed by decarboxylation of the carbonic acid hydrazide. This method can be used to prepare β-monosubstituted hydrazides. When prolonged heating is required the yield is reduced owing to alkaline hydrolysis of **99**.

1,3,4-Oxadiazolinium salts (**100**) undergo basic hydrolysis with ring cleavage, thus forming 1,2-diacylhydrazines (**101**)[118,119] (reaction

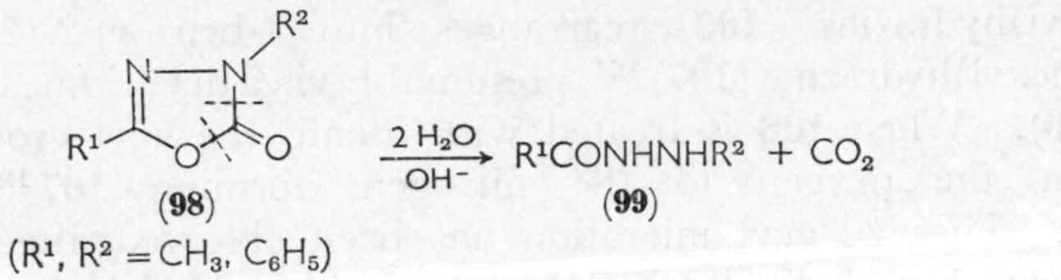

$$R^1CONHNHR^2 + CO_2 \qquad (44)$$

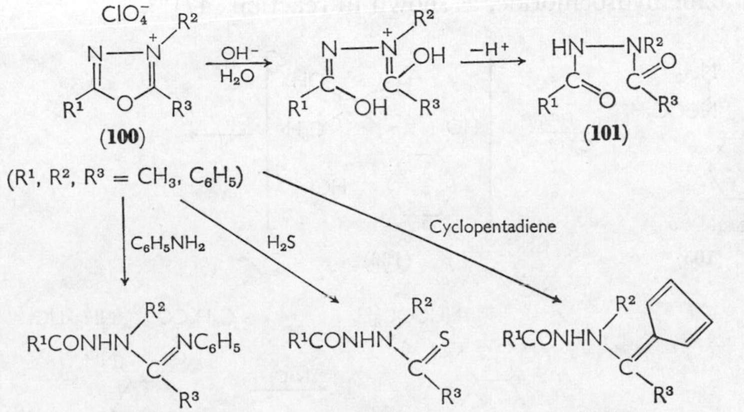

$$(98) \xrightarrow[OH^-]{2 H_2O} (99)$$

$$(R^1, R^2 = CH_3, C_6H_5)$$

$$(100) \qquad \xrightarrow[H_2O]{OH^-} \qquad \xrightarrow{-H^+} \qquad (101)$$

$$(R^1, R^2, R^3 = CH_3, C_6H_5)$$

$$\qquad (45)$$

Cyclopentadiene

$C_6H_5NH_2 \qquad H_2S$

R^1CONHN ... NC_6H_5 R^1CONHN ... S R^1CONHN

$(102) \qquad (103) \qquad (104)$

45). The hydroxyl ion effects a nucleophilic displacement of the OC bond, followed by deprotonation and rearrangement into diacylhydrazine. By varying the R^2 group, one can thus prepare various monoalkyl- or monoaryl-substituted diacylhydrazines. The reaction of **100** with other nucleophilic reagents capable of removing a proton (e.g. aniline, H_2S, and sodiocyclopentadiene) leads to ring opening and the formation of hydrazides **102–104** respectively (reaction 45)[119].

The recently described hydrogenolytic cleavage of 2,5-bis(trichloromethyl)-1,3,4-oxadiazole[120] in perchloric acid to give 1,2-bis(dichloroacetyl)hydrazine (**101**, $R^1 = R^3 = CHCl_2$; $R^2 = H$) may well proceed by a similar mechanism, via the formation of an oxadiazolinium perchlorate (**100**, $R^1 = R^3 = CCl_3$; $R^2 = H$), which is transformed into the 1,2-bis(dichloroacetyl)hydrazine after hydrogenolytic removal of a chlorine atom from each CCl_3 group.

G. Interconversion of Hydrazides

I. Acyl migration

Acyl migration has only rarely been encountered in hydrazides. Under conditions of acid catalysis, 1-benzoyl-1-(2-hydroxycyclohexyl)hydrazine (**105**) rearranges into 1-benzoyl-2-(2-hydroxycy-clohexyl)hydrazine (**107**) [121], presumably via a diaziridine intermediate (**106**). When **108** is treated with alkali, the acyl group migrates from the oxygen to the nitrogen, forming **107**[121] (reaction 46). Thermal acyl migration has been observed after heating 1-benzoyl-1-methylhydrazine hydrochloride or 1-benzoyl-1-phenyl-hydrazine hydrochloride, as shown in reaction (47)[83].

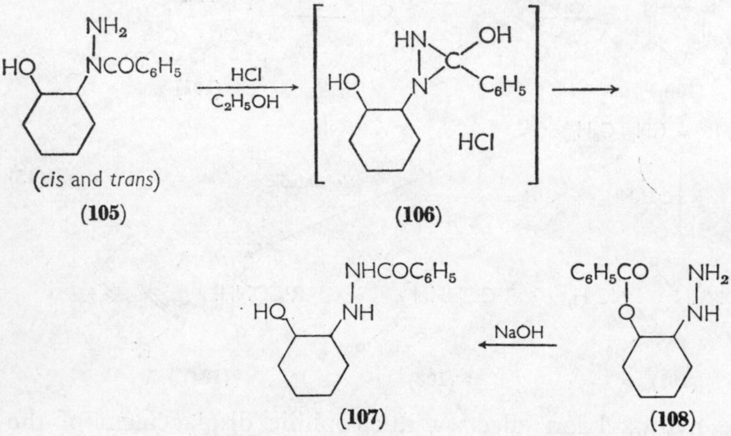

(46)

$$C_6H_5CONRNH_2 \cdot HCl \xrightarrow{\Delta} NHRNHCOC_6H_5 + HCl \qquad (47)$$

(R = CH$_3$, C$_6$H$_5$)

Cyclic hydrazides can rearrange reversibly from a six-membered ring **109** into a five-membered ring **110**, as shown in reaction (48). Acid media favour the six-membered ring, and alkaline media the five-membered one[122,123].

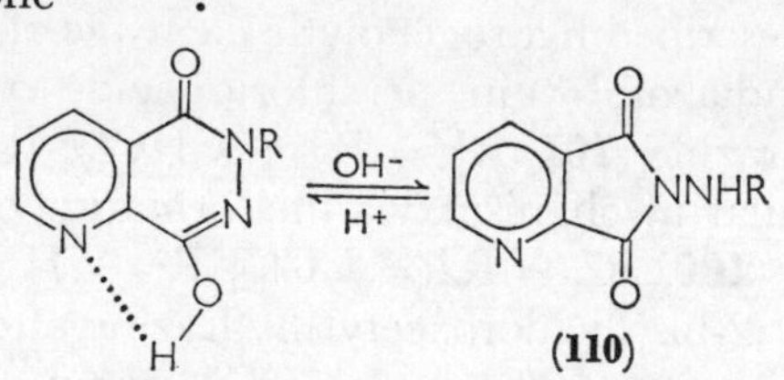

(48)

(R = p-C$_6$H$_4$NO$_2$ 2,4-C$_6$H$_3$(NO$_2$)$_2$)

2. Reactions not involving the hydrazide function

Oxalic monoester hydrazides (**111**) with various aryl and alkyl groups on the terminal nitrogen can be converted into diaryl- and dialkylglycolic acid hydrazides (**112**) with the aid of Grignard reagents[124-126]. Other reactions involving the acid group are often difficult to carry out, since the very reactive hydrazide group easily interferes in the process. It is therefore generally advisable to effect these reactions before the hydrazide group is introduced.

$$R^1R^2NNHCOCOC_2H_5 + 2 R^3MgBr \longrightarrow R^1R^2NNHCOCR^3_2OH$$

(**111**) (**112**)

V. PROTOTROPIC AND COMPLEXING CHARACTERISTICS

A. Enolization of the Hydrazide Group

A hydrazide can in principle change from its resonance-stabilized amide form (**113**) to the tautomeric enol form (**114**) by the shift of a hydrogen from nitrogen to oxygen. However, monoacylhydrazines behave as amides, and no enolization can be detected.

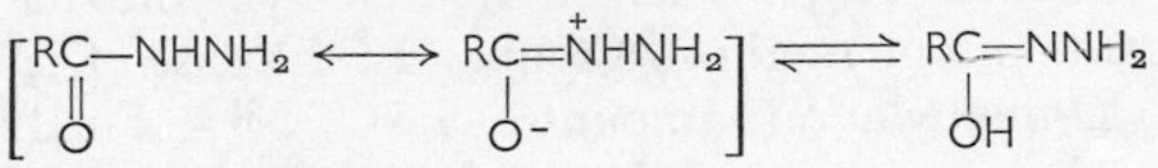

$$\left[\underset{O}{RC}-NHNH_2 \longleftrightarrow RC\overset{+}{=}NHNH_2 \right] \rightleftharpoons \underset{OH}{RC}=NNH_2$$

(**113**) (**114**)

For *p*-nitrobenzhydrazide (**116**), which is used as an acid–base indicator, two pK_a values have been found photometrically[127],

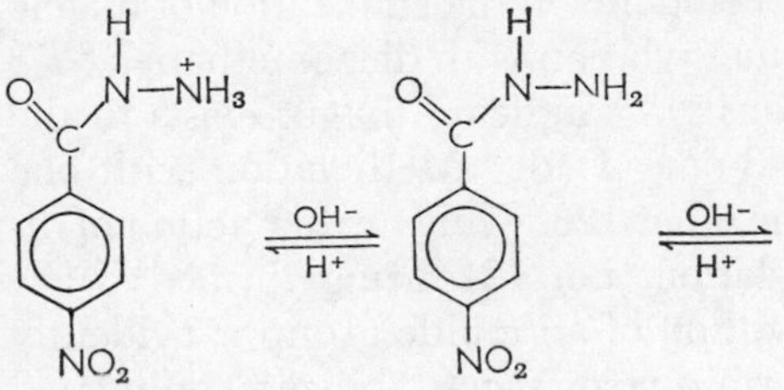

(**115**) (**116**)

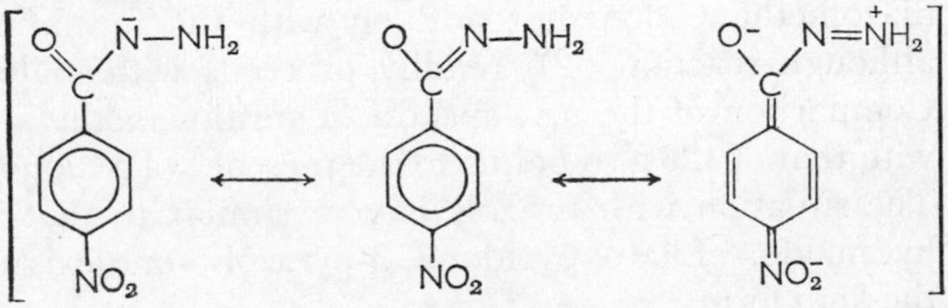

(**117**)

namely 2·77 and 11·17. It is believed that a protonated form **115** exists in acid solutions, while in alkaline solutions the abstraction of a proton leads to a mesomeric anion **117**.

One of the hydrazide groups in vicinal bis-benzoylhydrazones probably exists in the enolic form (its proton gives an n.m.r. signal at $\tau \simeq 4·25$), while the other hydrazide group is present in the normal amide form. The formation of a hydrogen-bonded species (**118**) is assumed to fix the bis-benzoylhydrazone groups in these forms[128].

(**118**)

As regards 1,2-diacylhydrazines, however, certain observations indicate that one of the acid hydrazide groups may be enolized. Thus, the product formed between 1,2-diacetylhydrazine and diazomethane contains one methoxy group. In the case of 1,2-dibenzoyl-hydrazine, *O*-methylation amounts to 50%[129]. Furthermore, the increase in the conductivity observed when a melt of 2-benzoyl-1-methacryloylhydrazine is heated, can be explained by enolization[22]. On the other hand, the i.r. spectra show no enol form in diacylhydra-zines in the solid state or in solution[22].

Considerably more work has been done on the enolization of maleic acid hydrazide. The results of methylation with diazomethane[129,130] and of spectroscopic investigations[86,131] indicate that it exists in the form of 2*H*-6-hydroxypyridazin-3-one (**119**). Methylation with one mole of diazomethane leads to the ether **120**, while further action of di-azomethane results in *N*-methylation, i.e. **121** being formed[129,130]. In maleic acid hydrazide, enolization of an amide group is evidently promoted by the tendency to form a resonance-stabilized conjugated system. In fact, the C=C bond in **119** has no olefinic character: diazomethane does not add on with the formation of pyrazoline, although reaction (49) readily proceeds with maleic acid imide[129]. Comparison of the u.v. spectra of various methylated pyridazinones with that of **120** also points to the presence of a monoenolic form[86,131]. The situation with urazole is very similar to that with maleic acid hydrazide. Like 2-pyridone, 3-pyrazol-5-one exists preferentially in the keto form[129a].

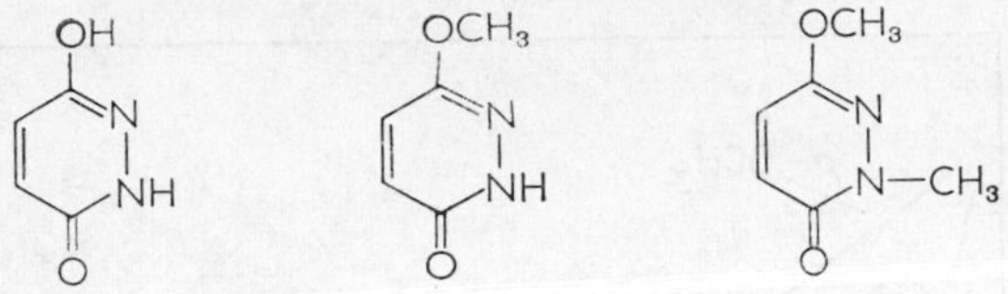

(119)   (120)   (121)

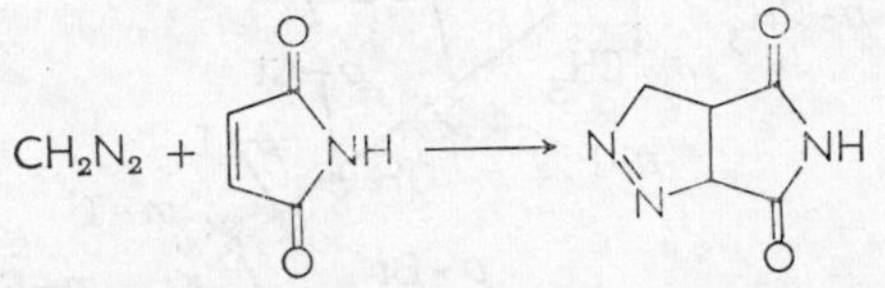

(49)

B. Acidity and Basicity

Monoacylhydrazines are weakly basic, and therefore form soluble salts with inorganic acids. Acetic acid hydrazide has a pK_a of 3·24, about five units lower than that of hydrazine[132].

The electron-attracting phenyl group lowers the basicity of hydrazides so much that the compounds assume an acidic character. Thus, 1-benzoyl-2-phenylhydrazine no longer dissolves in acids, but it is soluble in bases.

The pK_b values of a series of aroylhydrazines of the type $XC_6H_4CONHNH_2$ obey Hammett's equation[133], as can be seen from Figure 1. The influence of the substituents in aroylhydrazines ($\rho = -0·69$) is only about half as much as in arylhydrazines[133] ($\rho = -1·21$).

The hydrogen on the nitrogen linked to the acyl group is weakly acidic and diacylhydrazines dissolve in bases and are capable of forming salts. Thus, diformylhydrazine forms both a monosodium and a disodium salt[3], and various salts of dibenzoylhydrazine have been prepared[134]. The mercury salt of 1,2-bis(trifluoroacetyl)-hydrazine is used as a starting material for the synthesis of tetrakis-(trifluoroacetyl)hydrazine[6].

Maleic acid hydrazide dissociates in two steps. The pK value for the first dissociation is 5·67, while a value of 5×10^{-14} has been found for the dissociation constant for the removal of the second proton[135].

C. Complex Formation

Aromatic carboxylic acid hydrazides such as the hydrazides of isonicotinic, benzoic, and salicylic acid react with metal salts to form

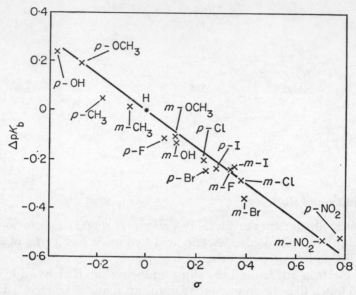

FIGURE 1. Hammett's relationship for ring-substituted benzhydrazides of the type $XC_6H_4CONHNH_2$. [Reproduced by permission of Verlag Chemie, GmbH, from *Chem. Ber.*, **101**, 751 (1968).]

complexes of general structure **122**[136,137]. Isopropylidenehydrazones of the hydrazides are also capable of complexing[138]. In all these complexes the hydrazides are normally present in the keto and not the enol form[50], so that ketonic chelation involving the C=O group and

$$[M(ArCONHNH_2)_n]A_m$$
(**122**)

(Ar = C_5H_4N, C_6H_5, C_6H_4OH;
M = Cu^{2+}, Cd^{2+}, Zn^{2+}, Fe^{2+}, Co^{2+}, Ni^{2+}, Mn^{2+};
A = SCN^-, $C_6H_5CO_2^-$, $C_4H_2O_4^{2-}$, SO_4^{2-}; m, n = 1, 2)

$$\left[\begin{array}{c} R-C=O \quad\quad \underset{H}{N}-NH \\ M \\ HN-N \quad\quad O=C-R \\ H \quad\quad H \end{array} \right]^{2+} SO_4^{2-}$$

(**122a**)

(M = Cu^{II}, Cd^{II}; R = —C₅H₄N pyridyl)

the primary NH_2 group occurs (e.g. **122a**) [137,138]. The pH-dependence of the complexing in the presence of formaldehyde has been investigated by u.v. spectroscopy on the copper chelates of semioxamido-hydrazide, oxalic acid dihydrazide, and N'-monophenyloxalodi-hydrazide [139]. The keto form has been found for the hydrazides at acidic pH values, but it is suspected that as the pH is raised from 2 to 8 a semi-enol form gains ground, and a hydrazide–enol form exists at pH > 8.

Polarographic work on the zinc complexes of acetylhydrazine, 1,2-diacetylhydrazine, and succinic acid hydrazide shows that dihydrazide complexes are more stable than monohydrazide ones. Dihydrazides also form higher complexes more easily [140]. The complexing of hydrazides with metal ions may be of importance for the biological action of these compounds [50]. Lewis acids such as $SbCl_5$, $TiCl_4$, $SnCl_4$, and $SnBr_4$ react with mono- and diacylhydrazines to form relatively stable complexes, in which the carbonyl oxygen functions as an electron donor to the metal atom [141].

The action of triethylaluminium on hydrazides leads to diethyl-aluminium carboxylic acid hydrazides (**123**) with the elimination of ethane. When 2 moles of triethylaluminium are used, bis(diethyl-aluminium) carboxylic acid hydrazides (**124**) can be isolated [142] (reaction 50). The organophosphorus compound **125** reacts with

$$RCONHNH_2 \longrightarrow [(RCONNH_2)Al(C_2H_5)_2] + [(RCONNHAl(C_2H_5)_2)Al(C_2H_5)_2]$$
$$+ Al(C_2H_5)_3 \qquad\qquad (123) \qquad\qquad\qquad (124) \qquad\qquad (50)$$

2 moles of benzoic acid hydrazide to form the cyclic compound **126** [143] (reaction 51).

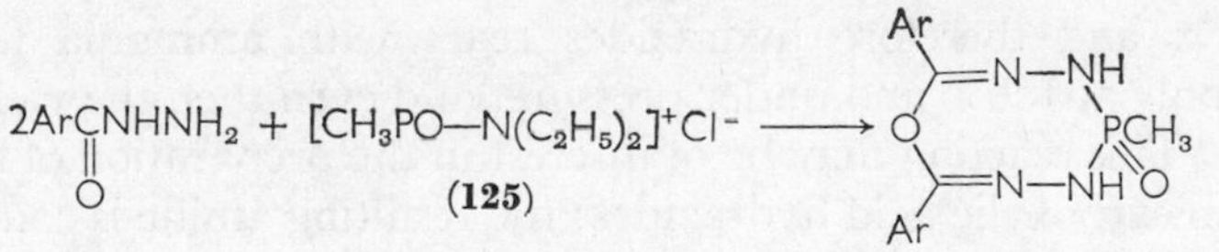

$$2ArCNHNH_2 + [CH_3PO{-}N(C_2H_5)_2]^+Cl^- \longrightarrow$$

(125)

(126)

VI. REACTIONS OF HYDRAZIDES

Hydrazides can react both at the carbonyl group and at their hydrazino group. Owing to the polarization of the carbonyl group, hydrazides are expected to be subject to both electrophilic attacks on the oxygen and nucleophilic attacks on the carbon of the CO groups. However, the reactions of the hydrazino group are much more important. These rely on the pronounced nucleophilic character of the nitrogen. The formation of heterocyclic compounds from hydrazides, involving all types of nucleophilic and electrophilic reactions, is sufficiently important to warrant discussion in a separate section.

A. Reactions Involving the Carbonyl Group

I. Hydrolysis

Hydrazides are generally stable to acids and bases in the cold[3], and hydrolytic cleavage to form the free carboxylic acid in an acidic medium occurs only when the latter is strong and the hydrazide is heated for a fairly long time, e.g. for 8 hr in the case of Δ^3-1,2-diacetyl-pyrazoline[144] (with concentrated HCl) and N'-(1-methyl-4-piperidinyl)acethydrazide[145] (with 23% HCl). Alkaline hydrolysis, e.g. with a concentrated solution of $Ba(OH)_2$, also requires heating for many hours. Hydrolysis with mineral acids[144–147] is preferred to alkaline hydrolysis[148,149], since redox reactions may accompany the latter[150].

The acid-catalysed partial hydrolysis of diacylhydrazines gives monoacylhydrazines in which the acyl group is on the nitrogen that is the less nucleophilic on account of the electronic effects of the substituents[91a]. The hydrolysis of hydrazides is likely to proceed by a mechanism similar to that of the hydrolysis of amides, but no detailed studies have yet been done on the kinetics and on the mechanism of the cleavage.

2. Ammonolysis

In comparison with hydrazine, ammonia has a weakly nucleophilic character, and therefore hydrazides react with ammonia to give amides only at 150°c and under pressure, and even then the yields are low[104]. This reaction may be of interest in the preparation of hydrazine from carboxylic acid hydrazides: the resulting amide is converted into N-nitroamide, which is recycled after reduction to the hydrazide[104]. On the other hand, the reaction between amides and hydrazine leads to good yields (section IV.B.1).

3. Reduction with complex metal hydrides

When heated with $LiAlH_4$ in ether or tetrahydrofuran, 1,2-diacyl-1,2-dimethylhydrazines (**127a**) are converted into tetraalkylhydrazines (**128a**) in yields of up to 57%. Diacylhydrazines in which the nitrogens carry hydrogen atoms (**127b**) are reduced only with difficulty with the aid of a fairly large excess of $LiAlH_4$. The yield of the resulting dialkylhydrazines (**128b**) is up to 35%, and much of the diacyl compound (**127b**) is recovered unchanged[23] (reaction 52). Lithium aluminium hydride reduces only the monoacylhydrazines with no hydrogen at $N_{(1)}$, e.g. $C_6H_5CON(CH_3)NH_2$ and $C_6H_5CON(C_6H_5)NH_2$[49].

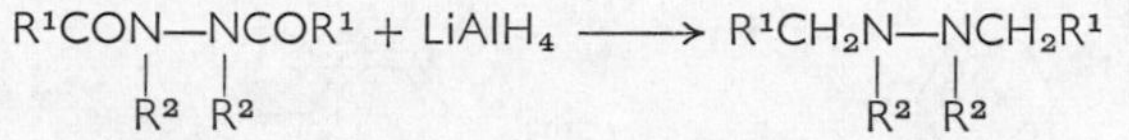

$$R^1CON-NCOR^1 + LiAlH_4 \longrightarrow R^1CH_2N-NCH_2R^1 \qquad (52)$$
$$\qquad\quad |\quad\ |\qquad\qquad\qquad\qquad\qquad\ |\quad\ |$$
$$\qquad\quad R^2\ \ R^2\qquad\qquad\qquad\qquad\qquad R^2\ R^2$$

(**127a**) $R^1 = C_6H_5$, CH_3; $R^2 = CH_3$;   (**128a**)

(**127b**) $R^1 = C_6H_5$, CH_3, C_2H_5O; $R^2 = H$ (**128b**)

As regards the reduction by complex metal hydrides, it is assumed[151-153] that, in the rate-determining step, the carbonyl oxygen and the metal atom form the complex **130** and a hydride ion is transferred to the carbon with a partial positive charge. This adduct is then reduced by $LiAlH_4$ to give the product (**131**) (reaction 53). The difference between the reactivity of R^1CONR^2N (**129**) and that of R^1CONHN (**132**) is due to the mobility of the hydrogen on $N_{(1)}$ which gives rise to the complex **133** and resists further reduction by $LiAlH_4$. **133** reforms **132** on hydrolysis[49].

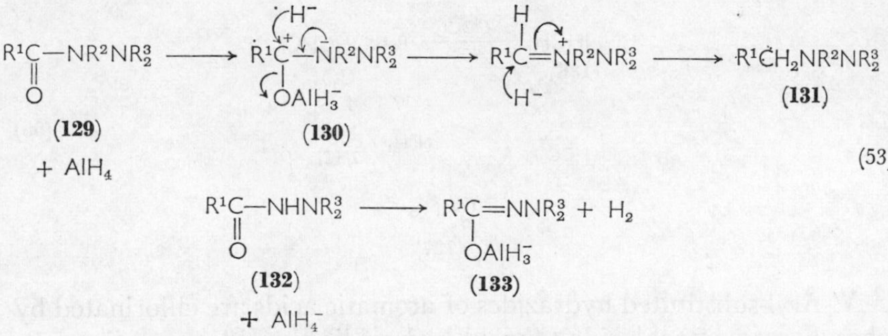

$$R^1C-NR^2NR_2^3 \longrightarrow R^1C\overset{+}{-}NR^2NR_2^3 \longrightarrow R^1C=\overset{+}{N}R^2NR_2^3 \longrightarrow R^1CH_2NR^2NR_2^3$$

(**129**)   (**130**)   (**131**)

\+ AlH_4 (53)

$$R^1C-NHNR_2^3 \longrightarrow R^1C=NNR_2^3 + H_2$$

(**132**)   (**133**)

\+ AlH_4^-

This method is thus particularly suitable for the reduction of carbonyl groups in diacyldialkylhydrazines[23,153]. Cyclic N-alkyl-hydrazides of the type **134**[144] and N-aminoimides of dicarboxylic acids of the type **135**[154] are also easy to convert into the corresponding hydrazine derivatives by reduction with $LiAlH_4$[144] or $NaBH_4$[154] (reactions 54 and 55).

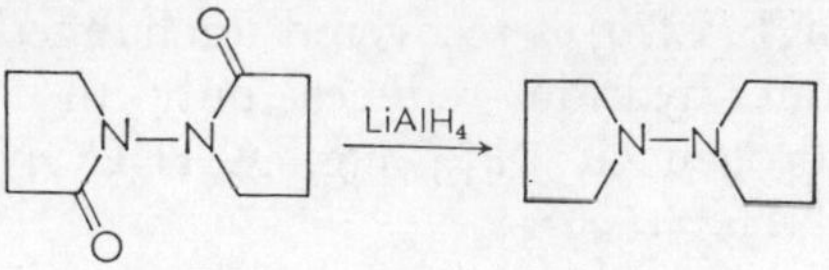

$$(54)$$

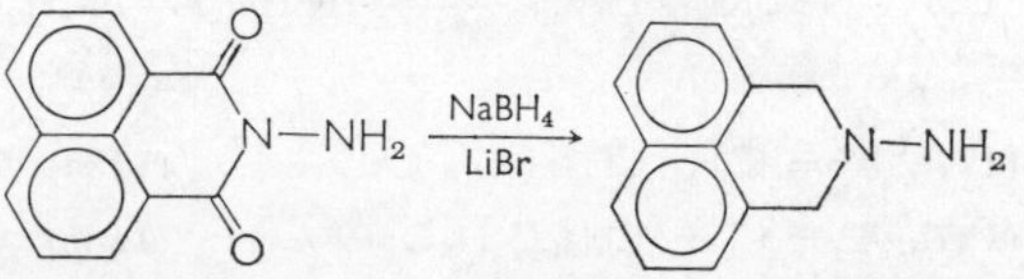

$$(55)$$

(**134**)

(**135**)

The solvent in which the reduction is carried out plays a significant part. Thus, as expected, 1-acetyl-2-phenylhydrazine is not reduced by $LiAlH_4$ in diethyl ether, whereas 1-ethyl-2-phenylhydrazine is formed in a 85–95% yield in dimethoxymethane[98].

4. Reaction with chlorinating agents

The formyl group of N',N'-disubstituted formic acid hydrazides (**136**) is chlorinated by phosgene in organic solvents at low temperatures. Dehydrochlorination in the presence of trimethylamine leads to N-isocyanodialkylamine (**137**). The latter reacts with formic acid, in the course of which CO is evolved and compound **136** is reformed[155] (reaction 56).

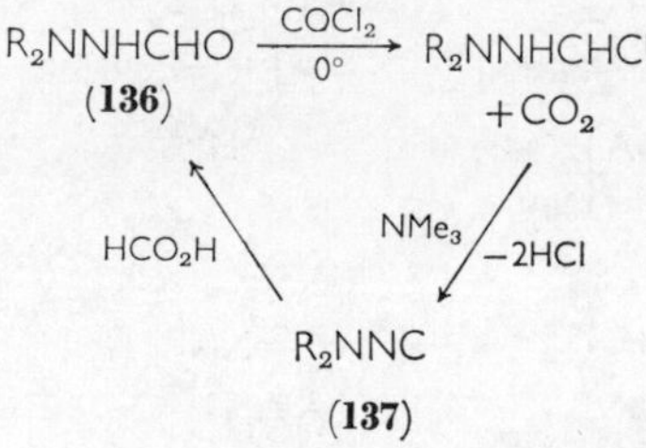

$$(56)$$

N'-Aryl-substituted hydrazides of aromatic acids are chlorinated by phosphorus pentachloride to yield hydrazidic chlorides (reaction 57), while 1,2-diacylhydrazines give rise to 1,1'-bis-chloroazines (reaction 58)[156,157]. The chlorination presumably occurs in the enolic form of the hydrazide[129]. Cyclic hydrazides are converted into chloroazines by treatment with phosphorus oxychloride (reaction 59)[158].

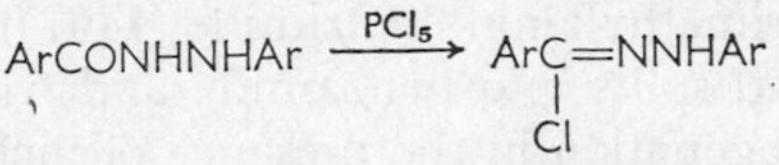

$$\text{ArCONHNHAr} \xrightarrow{\text{PCl}_5} \underset{\underset{\text{Cl}}{|}}{\text{ArC}}{=}\text{NNHAr} \tag{57}$$

$$\text{RCONHNHCOR} \xrightarrow{\text{PCl}_5} \underset{\underset{\text{Cl}}{|}}{\text{RC}}{=}\text{NN}{=}\underset{\underset{\text{Cl}}{|}}{\text{CR}} \tag{58}$$

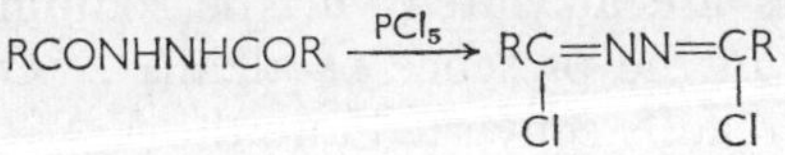

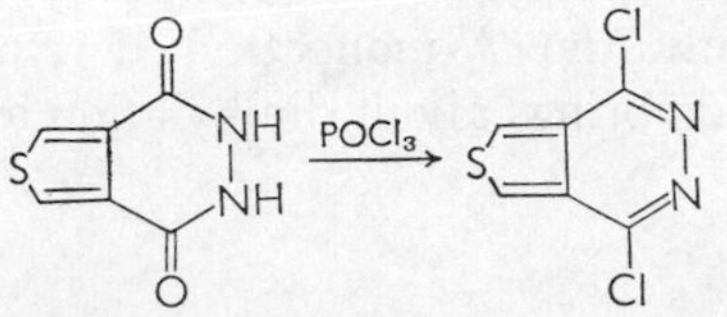

$$\tag{59}$$

B. Reactions Involving the Hydrazine Group

I. Alkylation

The sodium salts of hydrazides, formed by the action of metallic sodium on hydrazides, are alkylated by alkyl halides on the acylated nitrogen to give **138** in non-polar solvents such as ether and benzene (reaction 60). The reaction proceeds preferentially on the acylated nitrogen atom[8,159]. On the other hand, in neutral solution or in ethanolic solution in the presence of sodium alkoxide, alkyl halides alkylate the non-acylated nitrogen of the hydrazide and thus lead to **139** (reaction 61)[160-162]. In the hydrazide, the electron-attracting effect of the acyl group confers a partial positive charge on the acylated nitrogen, and this favours the attack on the terminal nitrogen[145].

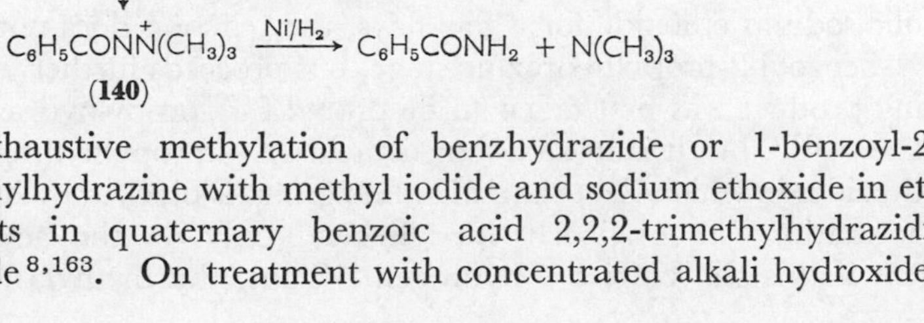

Exhaustive methylation of benzhydrazide or 1-benzoyl-2,2-di-methylhydrazine with methyl iodide and sodium ethoxide in ethanol results in quaternary benzoic acid 2,2,2-trimethylhydrazidinium iodide[8,163]. On treatment with concentrated alkali hydroxides, the latter gives basic trimethylaminobenzimide (**140**) (cf. section VI.B.13), and this compound splits into benzamide and trimethylamine when subjected to hydrogenation in the presence of nickel (reaction 62).

Diacylhydrazines are alkylated via the sodium salts[164] or in an ethanolic solution in the presence of alkalis[9]. Cyclic succinic acid hydrazide (**141**) is readily converted into the N,N'-dimethyl derivative (**142**) with the aid of methyl iodide; with ethyl iodide, on the other hand, it forms only the N-monoethyl derivative (**143**). Tetramethylene dibromide brings about dialkylation and gives the bicyclic product **144**[9].

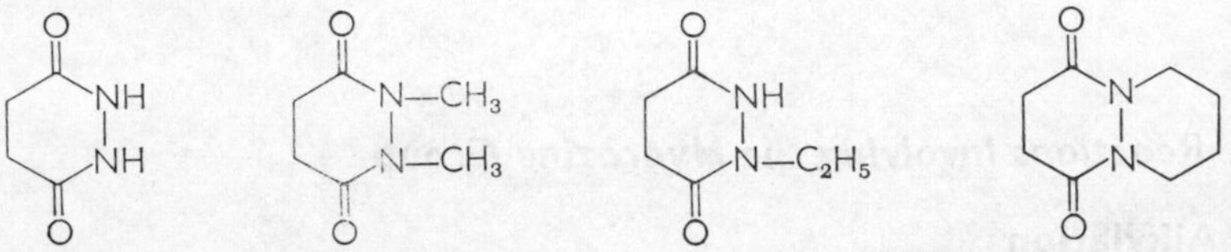

(**141**)   (**142**)   (**143**)   (**144**)

Alkylation competes with acylation when an acylhydrazine reacts with a halogenated acid chloride. Thus, the reaction of 3-bromo-2,2-dipropylpropionyl chloride (**145**) with 1-acetyl-2-benzylhydrazine (**146**) in benzene in the presence of triethylamine leads to 1-acetyl-2-benzyl-4,4-di-n-propylpyrazolidin-3-one (**147**). The more basic, non-acylated nitrogen is the target both for acylation by the acid chloride group and for alkylation by the alkyl bromide group. The product shows that acylation predominates[165].

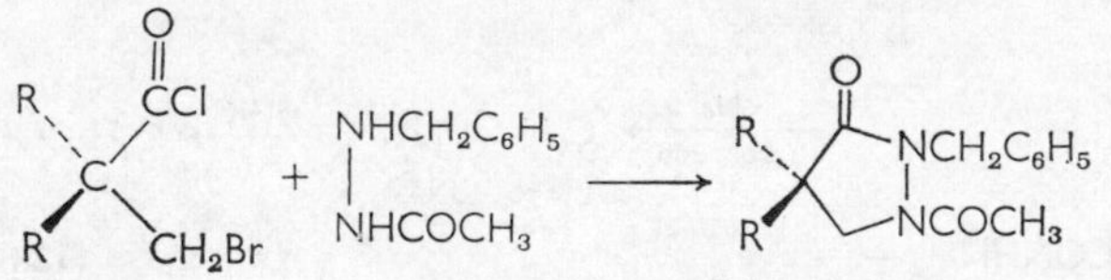

(**145**)   (**146**)   (**147**)
(R = n-Pr)

When benzoic acid hydrazide is treated with propyl bromide in ethanolic sodium ethoxide for a few days, the reaction does not stop at the 1-benzoyl-2-propylhydrazine stage, but proceeds further. The resulting product was first taken to be propyl 2-N-propylhydrazonobenzoate[166,167], but it is in fact benzoic acid 2,2-dipropylhydrazide[8].

Hydrazides in which the terminal amino group is blocked by hydrazone formation can obviously be alkylated only on the acylated nitrogen, if this still carries a hydrogen. Thus, the methylation of

2,3-butanedione monoacethydrazone leads after hydrazone cleavage to 1-acetyl-1-methylhydrazine[168] (reaction 63).

$$CH_3CONHN{=}CCH_3 \xrightarrow{CH_3I} \begin{array}{c} CH_3CO \\ {\diagdown} \\ NN{=}CCH_3 \\ {\diagup} \\ CH_3 \end{array} \xrightarrow{H_2O} \begin{array}{c} CH_3CO \\ {\diagdown} \\ NNH_2 + CH_3CO \\ {\diagup} \\ CH_3 \end{array}$$

with $O{=}CCH_3$ below on left portions and CH_3CO below on right.

(63)

Cyanomethyl benzenesulphonate transfers the cyanomethyl group onto the basic amino group of aromatic carboxylic acid hydrazides[169] to give rise to compounds of the type **148** (reaction 64). Cyanoethylation and sulphoethylation of the hydrazide group are discussed in section VI.B.7.

$$C_6H_5SO_3CH_2CN + NH_2NHCOAr \longrightarrow C_6H_5SO_3H + ArCONHNHCH_2CN \quad (64)$$

(148)

Berdinskii and coworkers[162] investigated the kinetics of the alkylation of dibutylglycolic acid 2-arylhydrazide with ethyl iodide in absolute alcohol in the presence of sodium ethoxide, by determining the concentration of the products with the aid of a polarograph. The alkylation obeys second-order kinetics, as expected for an S_N2 mechanism. The rate depends on the aryl substituents of the reacting nitrogen. In comparison with the phenyl group, the rate is increased by electron-repelling substituents in the *para* position (e.g. CH_3 and CH_3O), and is strongly decreased by the electron-attracting bromine atom in the *para* position. A *m*-CH_3 group seems to have a small influence on the rate, while the latter is lowered by a *m*-CH_3O group. According to these investigations, the alkylation of hydrazides obeys Hammett's equation.

2. Acylation

a. Acylation with carboxylic acid derivatives. Monoacylhydrazines are readily acylated by acyl chlorides or anhydrides to give 1,2-diacylhydrazines[40] (reaction 65). Acyl chlorides react rapidly even at low temperatures in ether[76], in aqueous solution containing NaOH[121], in triethylamine[165], chloroform[170], pyridine[171,172] and xylene[173], the products being obtained in good yields. The yields are also good in the reactions with acid anhydrides[170,172,173]. Benzoylhydrazine is converted into 1,2-dibenzoylhydrazine in a moderate yield by the diimide method[100] (section IV.C.2). 1,1-Dibenzoyl-2,2-dimethylhydrazine is formed by two consecutive acylations of 1,1-dimethylhydrazine with benzoyl chloride in a cooled aqueous solution in the presence of sodium carbonate (reaction 66)[23].

Further acylation of 1,2-diacylhydrazines is more difficult, since the remaining amide NH groups are only weakly nucleophilic. Tetraacylhydrazines can be prepared only by the use of a large excess of anhydride[2,6] (reaction 67). Tribenzoylhydrazine has been obtained from the monosodium salt of dibenzoylhydrazine[2] (reaction 68).

$$R^1CONHNH_2 + R^2COCl \text{ (or } (R^2CO)_2O) \longrightarrow R^1CONHNHCOR^2 \tag{65}$$

$$NH_2N(CH_3)_2 \xrightarrow{C_6H_5COCl}$$

$$C_6H_5CONHN(CH_3)_2 \xrightarrow{C_6H_5COCl} (C_6H_5CO)_2NN(CH_3)_2 \tag{66}$$

$$RCONHNHCOR + (RCO)_2O \xrightarrow{Excess} (RCO)_2NN(COR)_2 \tag{67}$$

$$C_6H_5CONNaNHCOC_6H_5 + C_6H_5COCl \longrightarrow (C_6H_5CO)_2NNHCOC_6H_5 \tag{68}$$

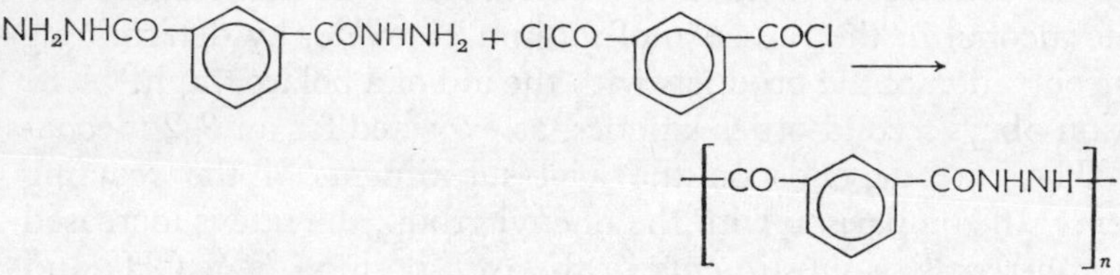

$$(69)$$

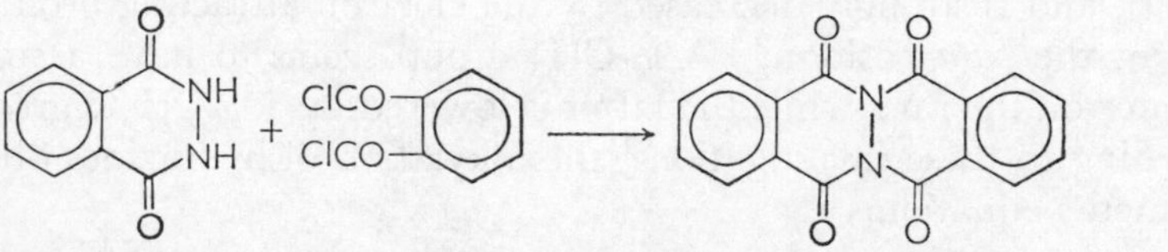

$$(70)$$

If the acyl residue of the hydrazide contains an acylatable hydroxyl group, as in the much investigated glycolic acid hydrazides[174,175], then the more basic $N_{(2)}$ is acylated preferentially before the OH group.

Dihydrazides of aromatic dicarboxylic acids (e.g. isophthalic acid) react with aromatic dicarboxylic acid dichlorides (e.g. isophthalic acid dichloride) in hexamethylenephosphonamide or N-methylpyrrolidone at 0°c to form high-molecular polyhydrazides[176–177a] (reaction 69). Cyclic hydrazides of dicarboxylic acids react with dicarboxylic acid chlorides, giving bicyclic compounds. Thus, phthalic acid hydrazide and phthalic acid dichloride lead to the corresponding bisphthalic acid hydrazide[178] (reaction 70). Further combinations are possible when succinic and malonic acid derivatives are used[178].

b. Kinetic investigations. Berdinskii and coworkers[175] have studied the kinetics of the acylation of glycolic acid hydrazides $R^1C_6H_4NHNHCOC(C_6H_5)_2OH$ with various substituted aromatic acid chlorides $R^2C_6H_4COCl$. The reaction is second order. The rate depends on the nature of R^2 in the attacking acylium ion. In comparison with the phenyl group, the rate is retarded when R^2 is electron repelling (p-CH$_3$) and accelerated when R^2 is electron attracting (p-Cl). On the other hand, experiments in which the R^1 group is varied show that benzhydrazides that contain the electron-releasing groups, $R^1 = p$-N(CH$_3$)$_2$ and p-OCH$_3$, are easier to acylate than hydrazides containing electron-attracting groups such as $R^1 = o$-NO$_2$, m-NO$_2$, and p-NO$_2$. A similar influence on the rate has also been found in the acylation of similarly substituted benzhydrazides $R^1C_6H_4CONHNH_2$ with benzoyl chloride[179]. It can thus be seen that the aryl substituents affect the nucleophilic character of the reacting hydrazide nitrogen, this influence being describable by Taft's equation[175].

Taft's equation can also be used in the case of aliphatic hydrazides. When $R^3 = CH_3$, C_2H_5, or C_6H_5 is substituted instead of $R^3 = H$ in the hydrazide $R^3CH_2CONHNH_2$, the effect on the acylation rate is small. When, however, R^3 is OCH$_3$ or OC$_6$H$_5$, the rate of the acylation is considerably lowered, in accordance with the higher negative inductive effect of these groups[179].

The acylation kinetics of benzhydrazides and butyrohydrazide with anhydrides and particularly succinic anhydride in benzene have also been investigated[180–184]. It has thus been found that the reaction undergoes autocatalysis by the acid formed from the anhydride during the process. The reaction is also catalysed when other organic acids are added to the mixture. There is a linear relationship between the pK_a of the catalysing acid and the logarithm of the catalytic rate constant (log k_a = const. $\times$ pK_a). The experimental rate constants of the reaction obey the relationship: $k = k_0 + k_a m\alpha$, where m is the molarity of the acid, α is the degree of dissociation of the acid[183], and k_0 and k_a are the rates of the uncatalysed and the catalysed reactions, respectively.

c. Acylation with chlorides of sulphur-containing acids. Sulphonyl chlorides acylate hydrazides giving 1-acyl-2-sulphonylhydrazines (**149**) (reaction 71a), which constitute the starting compounds for the preparation of aldehydes by the McFadyen–Stevens reaction[185].

The sulphonation generally proceeds in pyridine as an exothermic reaction[185-187]. In organic solvents, aromatic sulphinyl chlorides react with hydrazides to form 1-acyl-2-arenesulphinylhydrazines (**150**)[188] (reaction 71b). The action of thionyl chloride leads to the replacement of two amino hydrogens and to the formation of 1-acyl-2-sulphinylhydrazines[189,190] (reaction 72) whose stability depends to a large extent on the structure of the hydrazide acyl group (cf. section VI.B.9). 1-Phenylacetyl-2-sulphinylhydrazine (**151**) is a very stable compound[189], but the stability is considerably lower with other alkyl and aryl groups in the acid residue; hydrolysis with water then leads back to the starting hydrazides, while heating results in the elimination of carboxylic acids[189].

$$R^1CONHNH_2 \begin{cases} \xrightarrow{ClSO_2R^2} & R^1CONHNHSO_2R^2 \quad (149) \\ \\ \xrightarrow{ClSOR^2} & R^1CONHNHSOR^2 \quad (150) \end{cases}$$

(71a)

(71b)

$$C_6H_5CH_2CONHNH_2 + SOCl_2 \longrightarrow C_6H_5CH_2CONHNSO \quad (151)$$

(72)

Sulphamyl chlorides react with aromatic hydrazides to give 1-acyl-2-sulphamylhydrazines (**152**)[191] (reaction 73a). When benzoic acid hydrazide is heated in chlorobenzene with 1 mole of PCl_5, N-(trichlorophosphaza)benzamide (**153**) is obtained[192] (reaction 73b). The corresponding compounds with Ar = o-$NO_2C_6H_4$ and p-$CH_3C_6H_4$ are of interest as intermediates in the reductive chlorination of arylcarboxylic acid hydrazides (cf. section VI.B.9).

$$ArCONHNH_2 \begin{cases} \xrightarrow{ClSO_2NR^2} & ArCONHNHSO_2NR_2 \quad (152) \\ \\ \xrightarrow{PCl_5} & ArCONHN=PCl_3 \quad (153) \end{cases}$$

(73a)

(73b)

$$(Ar = C_6H_5, o\text{-}NO_2C_6H_4, p\text{-}CH_3C_6H_4)$$

3. Nitrosation

Nitrosation of unsubstituted hydrazides is the method used most frequently for the preparation of acyl azides, whose Curtius degradation leads to primary amines[40,193]. Acyl azides are important intermediates in peptide syntheses [194-201]. Hydrazides are usually converted into azides by sodium nitrite in strongly acidic aqueous solutions at low temperatures [201-203]. When the temperature is insufficiently low and the acid concentration insufficiently high, not only azides but also amides and nitrous oxide are formed on account of N—N cleavage [197]. This is understandable if one assumes that the hydrazide first gives rise to an N-nitroso derivative, which can then form the azide 155 on dehydration, or the amide 154 on the elimination of nitrous oxide [197] (reaction 74).

$$RCONHNH_2 \xrightarrow{HNO_2} RCONHNHN{=}O \longrightarrow RCONH_2 + N_2O$$
$$(154) \qquad\qquad (74)$$

$$RCONHN{=}NOH \longrightarrow RCON_3 + H_2O$$
$$(155)$$

2,2-Dialkyl-substituted hydrazides are attacked by nitrous acid at the tertiary nitrogen. The reaction products that can be isolated in the nitrosation of N-benzamidopiperidine (156) are the carboxylic acid 161 and the aldehyde 160 [204]. It is assumed that the nitroso-ammonium ion 157 is formed first, which changes into the immonium ion 158 after the elimination of NOH. Hydrolysis of 158 and re-nitrosation on the alkylated hydrazide nitrogen of the resulting 159 lead to the nitroso compound 160. The latter is partly oxidized into the acid 161. Compounds 160 and 161 can be converted into the

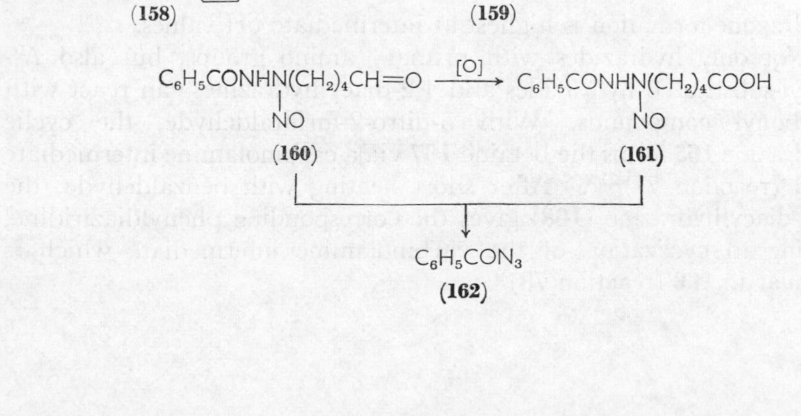

$$C_6H_5CONHN\!\!\left\langle\;\right\rangle \xrightarrow{HNO_2} C_6H_5CONH\overset{+}{N}\!\!\left\langle\;\right\rangle \xrightarrow{-NOH}$$
$$\qquad\qquad\qquad\qquad\qquad\quad |$$
$$\qquad\qquad\qquad\qquad\qquad\;\; NO$$
$$(156) \qquad\qquad\qquad\qquad (157)$$

$$C_6H_5CONH\overset{+}{N}\!\!\left\langle\;\right\rangle \xrightarrow[-H^+]{H_2O} C_6H_5CONHNH(CH_2)_4CH{=}O \xrightarrow{HNO_2}$$
$$(158) \qquad\qquad\qquad\qquad\qquad (159)$$

$$C_6H_5CONHN(CH_2)_4CH{=}O \xrightarrow{[O]} C_6H_5CONHN(CH_2)_4COOH \qquad (75)$$
$$\qquad\quad |\qquad\qquad\qquad\qquad\qquad\qquad\qquad |$$
$$\qquad\quad NO\qquad\qquad\qquad\qquad\qquad\qquad\;\; NO$$
$$(160) \qquad\qquad\qquad\qquad\qquad\qquad (161)$$

$$C_6H_5CON_3$$
$$(162)$$

azide **162** by repeated nitrosating dealkylation, as is known in the case of tertiary amines[204] (reaction 75).

4. Reactions with carbonyl and thiocarbonyl compounds

a. Hydrazone formation. The free NH_2 group in hydrazides as a rule reacts readily with carbonyl compounds such as ketones and aldehydes in the presence of catalytic amounts of acids, with the formation of hydrazones[8,83,92,93,165,205-209]. The hydrazones often crystallize very easily, and can therefore be used for the identification and purification of carbonyl compounds[21,210]. Sterically hindered ketones and aldehydes form hydrazones only with difficulty or not at all. Hydrazides carrying a quaternary ammonium group on the α-carbon of the acid residue, such as Girard's reagent T $(Cl^- (CH_3)_3\overset{+}{N}CH_2CONHNH_2)$[211] give with aldehydes and ketones water-soluble hydrazones, which can be easily separated from complex reaction mixtures by shaking with water[212]. Optically active hydrazides, such as L-menthylglycine hydrazide, are used to convert DL-carbonyl compounds into their resolvable enantiomers via hydrazone formation, the carbonyl compounds being then recovered by hydrazone cleavage after separation of the optical isomers[210,213].

It is believed that the first step in the hydrazone formation is a nucleophilic attack of the hydrazide nitrogen on the carbonyl group of the ketone, which leads to carbinolamine (**163**); the latter then loses water under the reaction conditions and thus forms the hydrazone **164** (reaction 76). As in the formation of semicarbazones, the rate of

$$R^1R^2C{=}O + H_2NNHCOR^3 \longrightarrow R^1R^2C\overset{\displaystyle OH}{\underset{\displaystyle NHNHCOR^3}{<}} \xrightarrow{-H_2O} R^1R^2C{=}NNHCOR^3 \quad (76)$$

$$(163) \qquad\qquad (164)$$

hydrazone formation is highest at intermediate pH values.

Not only hydrazides with primary amino groups, but also N'-alkyl-substituted hydrazides and 1,2-diacylhydrazines can react with carbonyl compounds. With 5-nitro-2-furfuraldehyde, the cyclic hydrazide **165** forms the betaine **167** via a carbinolamine intermediate **166** (reaction 77)[214]. After short heating with benzaldehyde, the 1,2-diacylhydrazine (**168**) gives the corresponding phenyldiaziridine, owing to cyclization of the carbinolamine intermediate which is similar to **166** (reaction 78)[215].

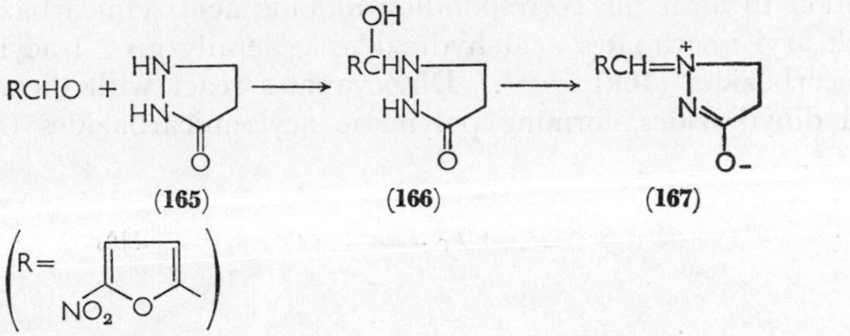

$$\text{RCHO} + \overset{HN-}{\underset{HN-}{\Big|}}\underset{O}{\diagdown} \longrightarrow \overset{OH}{\underset{HN-}{\underset{O}{\overset{RCHN-}{\Big|}}}}\diagdown \longrightarrow RCH=\overset{+}{N}\diagdown\underset{O^-}{\diagup} \tag{77}$$

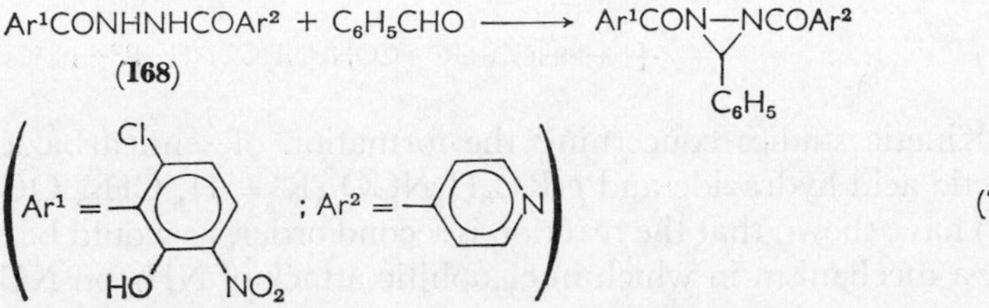

$$\text{(165)} \qquad\qquad \text{(166)} \qquad\qquad \text{(167)}$$

$$\left(R = \underset{NO_2}{\diagdown}\diagup O \diagup \right)$$

$$\text{Ar}^1\text{CONHNHCOAr}^2 + C_6H_5\text{CHO} \longrightarrow \text{Ar}^1\text{CON}\underset{C_6H_5}{\overset{|}{-}}\text{NCOAr}^2$$

$$\text{(168)}$$

$$\left(Ar^1 = \underset{HO}{\overset{Cl}{\diagdown}}\diagdown\underset{NO_2}{\diagup} \quad ; \quad Ar^2 = \diagup\diagdown N \right) \tag{78}$$

The condensation of monosaccharides with various benzhydrazides substituted in the ring leads to hydrazones. Paper chromatographic investigation of the way in which the rate of this condensation varies with the reaction conditions and with the basicity of the benzhydrazide has shown that more weakly basic acylhydrazines react faster with monosaccharides in a neutral medium than do the more strongly basic ones. In acetic acid solution, on the other hand, the rate increases with the basicity of the acylhydrazine[133].

Hydrogenation of acylhydrazones is a good general method for preparing N'-alkylhydrazides that are otherwise difficult to obtain (reaction 79). This reaction is usually carried out with hydrogen and a noble metal catalyst[216-218], i.e. under conditions where the N—N and the C=O bonds of the hydrazide are not attacked[146-149,165,205-208].

$$R^1\text{CONHN}=CR^2R^3 \xrightarrow{\text{H}_2/\text{Cat.}} R^1\text{CONHNHCHR}^2R^3 \tag{79}$$

Hydrazones are readily cleaved by acid hydrolysis, and yield the original hydrazide and the carbonyl compound[93,168,219]. In difficult cases, and where the substances are sensitive to acids, the carbonyl compound can be recovered by reaction of the acylhydrazone with benzaldehyde in dioxan containing a small amount of acetic acid[220,221].

b. *Semicarbazide formation.* Aldonic acid hydrazides, such as D-gluconohydrazides, react with potassium cyanate in a strong HCl solution to form the corresponding aldonic acid semicarbazides[222]. With aryl isocyanates acid hydrazides generally give 1-acyl-4-aryl-semicarbazides (**169**)[223,224]. Diisocyanates react with dicarboxylic acid dihydrazides, forming polymeric acylsemicarbazides (reaction 80)[225].

$$RCONHNH_2 + O{=}C{=}NAr \longrightarrow RCONHNHCONHAr$$

$$(\textbf{169})$$

$$NH_2NHCO{-}R^1{-}CONHNH_2 + OCN{-}R^2{-}NCO \longrightarrow$$

$$\text{-}\!\!\{NHNHCO{-}R^1{-}CONHNHCONH{-}R^2{-}NHCO\}_n \qquad (80)$$

Kinetic studies concerning the formation of semicarbazides from acetic acid hydrazide and p-RC_6H_4NCO (R = H, CH_3, OCH_3, and Cl) have shown that the reaction is second order, as would be expected for a mechanism in which nucleophilic attack of NH_2 on NCO is the rate-determining step[226].

1-Acyl-4-ureidosemicarbazides (**172**) arise when cyanogen bromide acts on monoacylhydrazines in an aqueous solution of hydrazine[227]. The reaction presumably proceeds via an intermediate formed between hydrazine and cyanogen bromide, namely 1,2-dicyanohydrazine (**170**). This in turn reacts with the hydrazide by using a nitrile group to give the intermediate **171** and the latter is finally hydrolysed into **172** (reaction 81).

$$NCNHNHCN + RCONHNH_2 \longrightarrow RCONHNHCNHNHCN \xrightarrow[-NH_3]{+H_2O}$$

$$(\textbf{170}) \qquad\qquad\qquad\qquad\qquad \underset{NH}{\overset{\|}{}}$$

$$(\textbf{171})$$

$$RCONHNHCNHNHCONH_2 \quad (81)$$
$$\underset{O}{\overset{\|}{}}$$

$$(\textbf{172})$$

c. Thiosemicarbazide formation. Thiosemicarbazides are formed from thioisocyanates and acid hydrazides analogously to the semicarbazides[223,228–230] (reaction 82). For example, acetylated monosaccharides carrying a thioisocyanate residue on $C_{(1)}$ react with isonicotinic acid hydrazide to give derivatives such as 1-isonicotinoyl-4-[2,3,4,6-tetra-O-acetyl-β-D-glucosyl]thiosemicarbazide[231].

$$R^1CONHNH_2 + S{=}C{=}NR^2 \longrightarrow R^1CONHNHCSNHR^2 \qquad (82)$$

5. Tetrazanes and tetrazenes

a. Coupling with diazonium salts. Diazonium ions attack the more strongly nucleophilic, terminal nitrogen of monoacylhydrazines, leading to acyltetrazenes (**173**). However, these have not been isolated, but have instead been cyclized into tetrazoles (**174**) with the aid of NaOH (reaction 83) [232] (cf. section VI.C.5).

1,2-Diacylhydrazines react with equivalent amounts of aryldiazonium salts, also to give tetrazenes. In the presence of two equivalents of the aryldiazonium salt both hydrazide nitrogens suffer electrophilic attack, and hexaazadienes (**175**) are formed [232,233]. On being warmed in an alcoholic solution of NaOH, these lose an acylium ion and decompose into the azide, aniline, and nitrogen. The acyltriazene RCONHN=NAr can be intercepted by treating **175** with alkali at a low temperature (reaction 84) [233].

$$
\begin{array}{c}
\text{RCONHNH}_2 \longrightarrow \text{RCON}\!-\!\text{N} \xrightarrow{\text{NaOH}} \text{R}\!-\!\text{C}\underset{\text{N}-\text{N}}{\overset{\text{N}-\text{N}}{\big|\big|}}\text{Ar} \qquad (83) \\
+ \ \text{Ar}\overset{+}{\text{N}}_2 \qquad\quad \text{ArN}\!=\!\text{N} \\
\qquad (173) \qquad\qquad\qquad (174)
\end{array}
$$

$$
\begin{array}{c}
\text{RCONHNHCOR} \longrightarrow \text{RCON}\!-\!\text{NCOR} \xrightarrow{\text{Ar}\overset{+}{\text{N}}_2} \text{RCON}\!-\!\text{NCOR} \xrightarrow{\text{NaOH}} \\
+ \ \text{Ar}\overset{+}{\text{N}}_2 \qquad\qquad \text{N}\!=\!\text{NAr} \qquad\qquad \text{ArN}\!=\!\text{N} \ \ \text{N}\!=\!\text{NAr} \\
\qquad\qquad\qquad\qquad\qquad\qquad\qquad (175)
\end{array}
$$

(R = H, CH₃; Ar = p-ClC₆H₄,
p-BrC₆H₄, p-NO₂C₆H₄)

$$(84)$$

$$
\begin{bmatrix}
\text{RCON}\!-\!\bar{\text{N}} \\
\text{ArN}\!=\!\text{N} \quad \text{N}\!=\!\text{NAr}
\end{bmatrix}
\xrightarrow{-\text{ArN}_3}
[\text{RCON}\!-\!\overset{-}{\text{N}}\!=\!\text{NAr}]
\xrightarrow{\text{H}_2\text{O}}
\text{ArNH}_2 + \text{N}_2 + \text{RCOO}^-
$$

$$\downarrow$$

$$\text{RCONH}\!-\!\text{N}\!=\!\text{NAr}$$

b. Addition to azo compounds. Monoacylhydrazines react with azodicarboxylic esters with vigorous evolution of nitrogen and the formation of 1,2-diacylhydrazines (**179**) [156]. According to the mechanism proposed for this reaction, the hydrazide adds to the azo compound to form a tetrazane **176**, hydrazodicarboxylic acid ester is cleaved out, and the liberated fragment **177** dimerizes into diacyltetrazene (**178**). The latter loses nitrogen and changes into 1,2-diacylhydrazine (**179**) (reaction 85) [156].

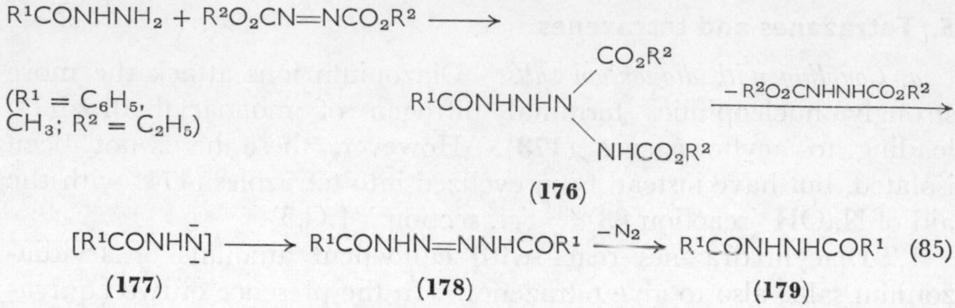

$R^1CONHNH_2 + R^2O_2CN{=}NCO_2R^2 \longrightarrow$

$$
(R^1 = C_6H_5, \qquad\qquad R^1CONHNHN\!\!\begin{array}{c} CO_2R^2 \\[2pt] \diagup \\[-4pt] \diagdown \\[2pt] NHCO_2R^2 \end{array}\!\! \xrightarrow{\;-R^2O_2CNHNHCO_2R^2\;}
$$
$CH_3; R^2 = C_2H_5)$

$$(176)$$

$$[R^1CONH\bar{N}] \longrightarrow R^1CONHN{=}NNHCOR^1 \xrightarrow{-N_2} R^1CONHNHCOR^1 \quad (85)$$

$$\quad(177)\qquad\qquad\qquad(178)\qquad\qquad\qquad\qquad(179)$$

Neither the tetrazane (**176**) nor the tetrazene (**178**) have been isolated.

c. Reaction with sulphur monochloride. The reaction of sulphur mono-chloride with 1-acyl-2,2-dimethylhydrazine, carried out by heating in benzene, gives a good yield of 2,3-diacyltetrazane (**182**) and small amounts of 1,3,4-oxadiazoline-2-thione[80]. As regards the mechanism, it is assumed that the chlorodithio cation derived from S_2Cl_2 first delivers an electrophilic attack on the oxygen of the carbonyl group, and the *O*-chlorothio compound (**180**) rearranges into the *N*-chloro-thio isomer (**181**); two molecules of the latter eliminate the S_2Cl group, form an N—N bond, and give rise to **182** (reaction 86)[80]. Unsub-stituted hydrazides react with S_2Cl_2 to give heterocyclic compounds (cf. section VI.C).

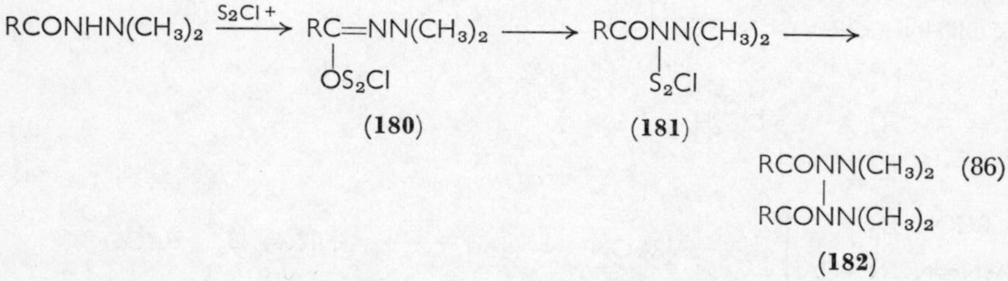

$$RCONHN(CH_3)_2 \xrightarrow{S_2Cl^+} \underset{\underset{OS_2Cl}{\big|}}{RC{=}NN(CH_3)_2} \longrightarrow \underset{\underset{S_2Cl}{\big|}}{RCONN(CH_3)_2} \longrightarrow$$

$$(180)\qquad\qquad\qquad(181)$$

$$\underset{\underset{RCONN(CH_3)_2}{\big|}}{RCONN(CH_3)_2} \quad (86)$$

$$(182)$$

d. Oxidative tetrazene formation. The oxidation of 1-acetyl-1-methylhydrazine with potassium permanganate or hypobromite at 0°c leads to 1,4-diacetyl-1,4-dimethyl-2-tetrazene (**183**) (reaction 87)[168].

$$2\,CH_3CONCH_3NH_2 \xrightarrow{[O]} CH_3CONCH_3N{=}NNCH_3COCH_3 \qquad (87)$$

$$(183)$$

6. Amidrazone formation

Amidrazones (**185**) are reactive intermediates formed in the syn-thesis of triazoles (cf. section VI.C.3) from imidic esters (**184**) and acylhydrazines[51,234] (reaction 88). However, it has been possible to isolate intermediate acylamidrazones (**185**), such as that formed from benzhydrazide and benzimidic ethyl ester (**185**, R^1, R^2 = C_6H_5)[235], and the one formed from caprolactim methyl ether and isonicotinic acid hydrazide[236].

$$R^1COCH_3 + NH_2NHCOR^2 \xrightarrow{-CH_3OH} R^1CNHNHCOR^2 \tag{88}$$
$$\underset{\text{NH}}{\|} \qquad\qquad\qquad\qquad \underset{\text{NH}}{\|}$$
$$\text{(184)} \qquad\qquad\qquad\qquad\qquad \text{(185)}$$

7. Reactions involving C=C bonds

In view of the predominating nucleophilic character of the hydrazide group, additions to ordinary carbon–carbon double bonds are not likely to occur. However, the C=C bonds activated by the presence of strongly electron-attracting neighbouring groups can react with hydrazides. Thus, alkenylnitriles (**186**) react with hydrazides to form carboxylic acid 2-cyanoethylhydrazides (**187**) (reaction 89)[145]. Ethylenesulphonyl compounds (**188**), which carry the strongly electron-attracting sulphonyl group next to the olefinic bond, can sulphoethylate cyclic hydrazides such as phthalic and maleic acid hydrazide to give products like **189** (reaction 90)[237].

$$NCCH{=}CH_2 + NH_2NHCOR \longrightarrow NCCH_2CH_2NHNHCOR \tag{89}$$
$$\text{(186)} \qquad\qquad\qquad\qquad\qquad \text{(187)}$$

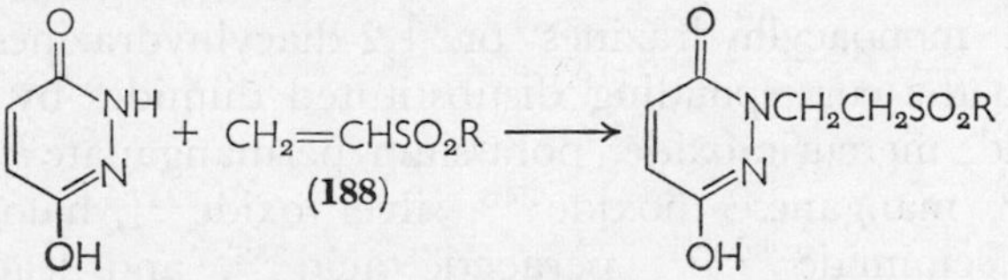

$$\text{(188)} \qquad\qquad\qquad\qquad \text{(189)} \tag{90}$$

8. Hydrogenolytic rupture of N—N bonds

Under normal conditions of hydrogenation with platinum or palladium, when the C=C and the C=N bonds become hydrogenated, the N—N bond does not break[238]. Hydrogenation in the presence of catalytic amounts of Raney nickel generally does not lead to rupture of the N—N bond of the hydrazide. However, if the latter is heated in ethanolic solution with an excess of Raney nickel, the N—N bond breaks and the amide is formed in a yield of 60–80%[238,239].

1,2-Diacylhydrazines are also reduced by Raney nickel, but require a longer time (reaction 91). The reaction is incomplete if one of the nitrogens is alkylated. Treatment of N,N'-dialkyl-N,N'-diacyl-hydrazines with Raney nickel does not cause bond fission, nor can the asymmetric N',N'-dialkyl-N,N-diacylhydrazines be cleaved under these conditions. Evidently the substituents on the nitrogens protect the N—N bond from contact with the surface of the nickel catalyst on which the reaction should take place[240].

$$R^1CONHNR^2COR^3 \xrightarrow{\text{H}_2/\text{Raney Ni}} R^1CONH_2 + R^3CONHR^2 \qquad (91)$$

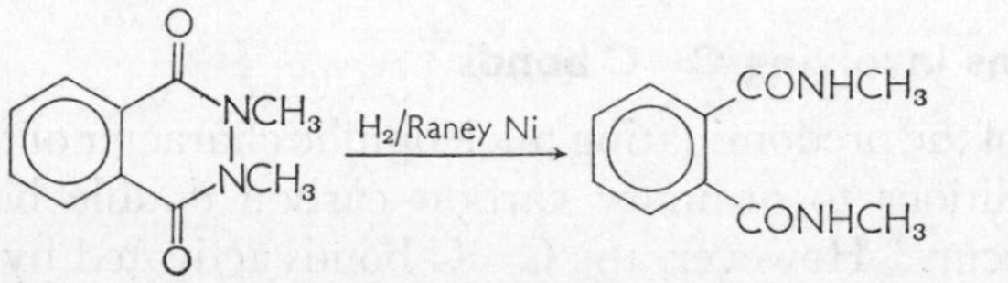

$$(92)$$

Cyclic hydrazides, such as phthalic and maleic acid hydrazide, do not suffer bond fission when treated with Raney nickel. However, phthalic acid 1,2-dimethylhydrazide is cleaved in this manner as shown in reaction 92[240]. Raney nickel and hydrazine form another very active mixture for the reduction of hydrazides into amides[241]. This mixture can split maleic acid hydrazide, but not phthalic acid hydrazide[241].

9. Oxidation

a. General methods. Hydrazides that carry hydrogens on their nitrogen atoms are very sensitive to oxidizing agents. Thus, 2-substituted monoacylhydrazines or 1,2-diacylhydrazines are converted into the corresponding disubstituted diimides by oxygen[242], nitrous acid, mercuric oxide, potassium permanganate[243,244], ferric chloride[245], manganese dioxide[246], silver oxide[247], halogens[248,249], N-bromosuccinimide[250,251], peracetic acid[252], and lead tetraacetate[78,253,253a] (reaction 93). The diimides can generally be isolated[78,242,247,249].

$$R^1CONHNHCOR^2 \xrightarrow{[O]} R^1CON{=}NCOR^2 \qquad (93)$$

The monoacyldiimides formed by the oxidation of monoacyl-hydrazines have a low stability; they function as acylating agents and form 1,2-diacylhydrazines with unoxidized hydrazide, with the elimination of nitrogen[150,251,252]. Thus, the oxidation of acetic acid hydrazide with peracetic acid leads to 1,2-diacetylhydrazine in 95% yield[252].

Other strongly nucleophilic compounds, such as amino compounds, are acylated by the acyldiimide formed in the oxidation of monoacyl-hydrazines. This reaction can be used for peptide synthesis[250,251]. A benzyloxycarbonylamino acid hydrazide (e.g. Cbz$_2$-L-lysine hydrazide) is oxidized by N-bromosuccinimide or iodine in tetrahydrofuran in the presence of a second amino acid ester (e.g. Gly-OC$_6$H$_4$NO$_2$), the resulting diimide stage reacting to give the peptide ester (e.g. Cbz$_2$-L-Lys-Gly-OC$_6$H$_4$NO$_2$) in 90–99% yield[251]. 2-Phenylhydrazides can also be converted in this manner. Thus **190** is oxidized by N-bromosuccinimide to **191**, bringing about terminal N-acylation of a second peptide hydrazide (**192**)[254,255] to yield **193**. The hydrazide group in **193** can be easily removed by oxidation to the diimide **194**. This reacts with water to yield the acid, or with alcohol to yield the corresponding ester[246,254,255] (reaction 94).

Cbz—Gly—L-Phe—NH—NH—C$_6$H$_5$ $\xrightarrow{\text{NBS}}$ Cbz—Gly—L-Phe—N=N—C$_6$H$_5$
        (**190**)   (**191**)

191 + Gly—L-Phe—NH—NH—C$_6$H$_5$ $\longrightarrow$
       (**192**)

                                   Cbz—Gly—L-Phe—Gly—L-Phe—NH—NH—C$_6$H$_5$
                                           (**193**)

193 $\xrightarrow{\text{[O]}}$ Cbz—Gly—L-Phe—Gly—L-Phe—N=N—C$_6$H$_5$
                                      (**194**)

                   H$_2$O   ROH

Cbz—Gly—L-Phe—Gly—L-Phe—OH   Cbz—Gly—L-Phe—Gly—L-Phe—OR
                                                        (94)

b. Kalb–Gross oxidation. In the oxidation of monoacylhydrazines with potassium ferricyanide in ammoniacal solution, the acyldiimide formed as an intermediate decomposes with the evolution of nitrogen to give aldehydes[248,256] (reaction 95) (see also section VI.B.10 for another reaction leading to aldehydes).

RCONHNH$_2$ $\xrightarrow[\text{OH}^-]{\text{K}_3[\text{Fe(CN)}_6]}$ RCON=NH $\xrightarrow{-\text{N}_2}$ RCHO   (95)

c. Carpino reaction. The oxidation of hydrazides with gaseous chlorine in organic solvents in the presence of excess hydrogen chloride does not lead to 1,2-diacylhydrazines, but to the formation of the corresponding carboxylic acid chlorides accompanied by liberation of nitrogen[257,258]. When the reaction is carried out with benzhydrazide and hydrogen bromide and bromine, one can isolate an intermediate hydrazide hydrobromide perbromide (**195**), which forms benzoyl bromide on treatment with cold water (reaction 96). The formation of acyl chlorides from hydrazides probably takes place through an analogous intermediate.

$$C_6H_5CONH\overset{+}{N}H_3\overset{-}{Br} \xrightarrow{Br_2} C_6H_5CONH\overset{+}{N}H_3\overset{-}{Br_3} \longrightarrow C_6H_5COBr + N_2 \quad (96)$$

$$(195)$$

d. Related reactions. Various reactions can be considered as variants of the Carpino reaction. The sulphinylhydrazides formed in the interaction of hydrazides with thionyl chloride (cf. section VI.B.2) give carboxylic acids spontaneously, or on warming, and the acids thus formed are chlorinated into acyl chlorides by the excess of thionyl chloride (reaction 97). The acyl chloride can react further with the unchanged hydrazide to give a 1,2-diacylhydrazine. The final product is the carboxylic acid, acyl chloride, diacylhydrazine, or a mixture of all three, according to the conditions of the reaction and the nature of the hydrazide[189].

$$\underset{\underset{O}{\|}}{RCNHNSO} \rightleftharpoons \underset{\underset{OH}{|}}{RC=NNSO} \xrightarrow{\Delta} RCOH + N_2 + S \xrightarrow{SOCl_2} RCOCl \quad (97)$$

With two equivalents of sulphur monochloride, hydrazides form carboxylic acid chlorides. According to the proposed mechanism, the electrophilic chlorodithio cation first attacks the amino group of the hydrazide to yield the N'-chlorodithio compound **196**, which then forms the intermediate **197** on combining with a second chlorodithio cation. **197** decomposes into the azo compound **198** and finally into the acyl chloride on elimination of nitrogen and sulphur[259] (reaction 98). The formation of tetrazane from S_2Cl_2 and 2,2-dialkyl-substituted hydrazides in which the terminal nitrogen cannot be attacked was discussed in section VI.B.5.

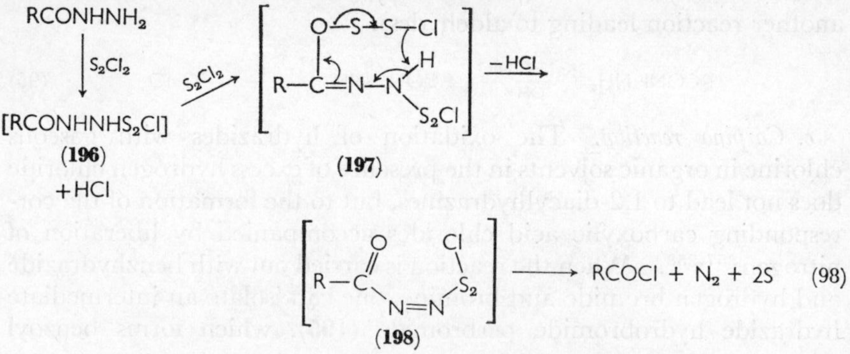

$$\longrightarrow RCOCl + N_2 + 2S \quad (98)$$

On being heated in chlorobenzene in the presence of PCl_5, N-(trichlorophosphaza)arylamides (e.g. $C_6H_5CONHN=PCl_3$, prepared from benzoic acid hydrazide and PCl_5) decompose into the corresponding benzylidene chlorides (e.g. $C_6H_5CHCl_2$)[192]. However, nothing is known about the mechanism of this reaction.

10. McFadyen–Stevens reaction

This is a base-induced elimination of sulphonic acid from carboxylic acid N'-sulphonylhydrazides (199) to give an aldehyde via the acyldiimide 200. The transformation occurs in alkaline medium analogously to the Kalb–Gross reaction (cf. section VI.B.9.b) with liberation of nitrogen[185–187,260]. By a bimolecular mechanism the reaction (99a) may lead directly to the diimide 200. Alternatively deprotonation into the anion 201 (reaction 99b) has also been considered as the first step[261]. The anion 201 then eliminates a sulphonic acid anion and gives the fragment 202, which might rearrange into 200.

$$Ar^1CON{-}NH{-}SO_2Ar^2 \longrightarrow Ar^1CON{=}NH + (Ar^2SO_2^- + BH) \longrightarrow Ar^1CHO + N_2$$

$$\underset{(199)}{B{-}H} \qquad \qquad (200) \qquad \qquad \qquad (99a)$$

$$Ar^1CONH{-}N{-}SO_2Ar^2 \longrightarrow Ar^1CONH{-}\bar{N}SO_2Ar^2 \longrightarrow Ar^1CONH{-}\bar{N} \qquad (99b)$$

$$\underset{(199)}{\overset{H\ B^-}{}} \qquad \qquad (201) \qquad \qquad \qquad (202)$$

$$+ HB$$

In homogeneous solution, aldehydes are formed with great difficulty or not at all. By contrast, in a heterogeneous medium involving sodium carbonate or glass powder, the reaction is generally very fast. Evidently, it is catalysed by the solid surface[185]. While the reaction proceeds very well in the case of aromatic hydrazides, aliphatic hydrazides with hydrogen atoms on the α-carbon of the acyl group present some difficulties, due to secondary reactions of the derived aldehydes in alkaline solution. Aliphatic hydrazides in which these hydrogens were replaced by alkyl groups undergo the McFadyen–Stevens reaction to give aldehydes in the same way as aromatic hydrazides[187].

A variant of the McFadyen–Stevens reaction is represented by the reductive cleavage of 2-acyl-1-sulphonylhydrazines[262]. The reduction of sulphonylhydrazides with $LiAlH_4$ leads to intermediate sulphonyl-alkylhydrazino compounds, which form a hydrocarbon on elimination of nitrogen and sulphinic acid (reaction 100). Good

19*

results are obtained with aliphatic sulphonylhydrazides, while the yields with aromatic hydrazides are very low. Alcohols and aldehydes have been detected as by-products.

$$RCONHNHSO_2R \xrightarrow{\text{LiAlH}_4} [RCH_2NHNHSO_2R] \longrightarrow RCH_3 + N_2 + RSO_2H \quad (100)$$

II. Kametani reaction

Hydrazides react with chloral to give chloral hydrazones[263] (**203**) which are good starting materials for the preparation of various carboxylic acid derivatives. Amines add on to the carbonyl group of **203** to form the amide **204**. Treatment of **203** with ethanol leads to the ester **205**[264].

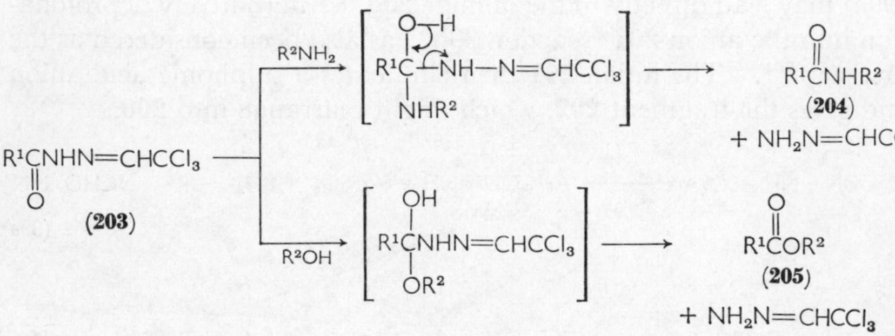

12. Fragmentation

Hydrazides which carry on the α-carbon of the acyl residue a nucleofugic group that readily cleaves off as an anion undergo base-induced cleavage of the type of Grob's fragmentation[41,95,265a]. Thus, chloroacetic acid hydrazide is fragmented on treatment with alkali; the chlorine anion splits off as a nucleofugic fragment, ketene as an olefinic fragment, and diimine as an electrofugic fragment (reaction 101)[95]. The diimine disproportionates into nitrogen and hydrazine, while the ketene gives acetic acid with water[95].

$$Cl-CH_2CO-NHNH_2 \xrightarrow{OH^-} Cl^- + CH_2{=}CO + NH{=}NH(\rightarrow\tfrac{1}{2}N_2 + \tfrac{1}{2}NH_2NH_2)$$

$$(101)$$

When heated with hydrazine, derivatives of 5-*O*-methylsulphonyl-α-D-glucuronic acid hydrazide (**206**) exhibit a similar fragmentation: diimine splits off and the sugar ketene **207** is formed. The latter forms either the lactone **208** (by intramolecular reaction with an OH group of the acid residue) or, with excess hydrazine, the 5-deoxyglucuronic acid hydrazide[41] (**209**).

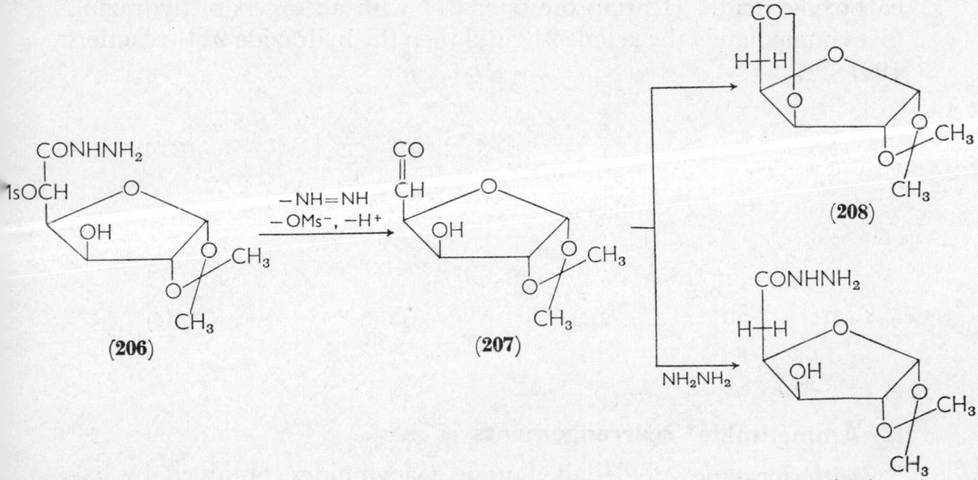

(208)

(206)   (207)   (209)

Investigations on the influence of N'-substituents have shown that if both the hydrogens on this atom are replaced by alkyl groups the fragmentation does not take place. An N'-methyl group retards the reaction, while a phenyl group causes it to proceed very quickly. On the basis of these results, it is assumed that the first step is abstraction of a proton from the N'-position. This is slowed down by the inductive effect of a methyl group, and is accelerated by the mesomeric effect of the phenyl group. The resulting hydrazide anion **210** splits out of a preferred conformation that satisfies the stereoelectronic conditions of the antiparallel orientation of the electron pairs participating in the reaction, these being the conditions for a synchronous process[41] (reaction 102).

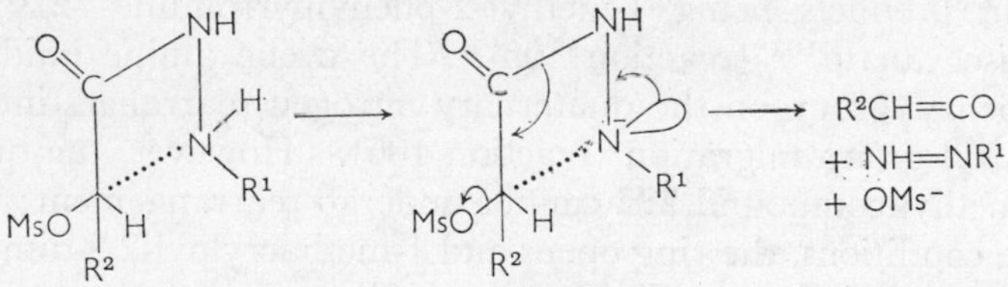

(210)

The cleavage of aziridine-2-carboxylic acid hydrazide (**212**) bears a certain resemblance to the fragmentation reactions above. When **212** is heated in water the three-membered ring opens and ketene (**213**) and diimine are formed. The ketene reacts further to give a carboxylic acid.   Heating the ester **211** with an excess of hydrazine gives immediately the ketene **213** and then the hydrazide **214** (reaction 103)[265b].

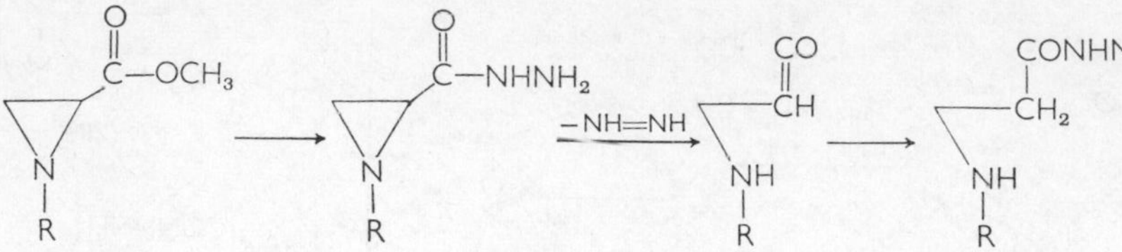

(103)

(211)   **(212)**   **(213)**   **(214)**

(R = t-Bu,
CH$_2$C$_6$H$_5$)

13. Amine imide* rearrangements

Zwitterion-type 2,2,2-trialkylamine 1-acylimides, obtained by exhaustive alkylation of hydrazides[8,266–267a] or from carboxylic acid esters[24] or chlorides[268,269] with 1,1,1-trialkylhydrazinium salts in the presence of alkalis, can undergo various rearrangements.   Thus, **215** rearranges into **216** by migration of the benzyl residue, in a thermolytic reaction of the type of a Stevens rearrangement[268] (reaction 104).

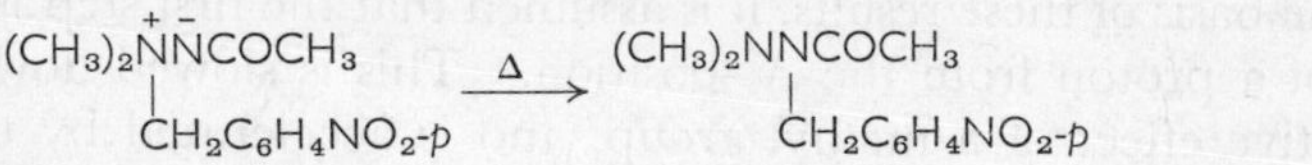

$$(CH_3)_2\overset{+}{N}\overset{-}{N}COCH_3 \quad \xrightarrow{\Delta} \quad (CH_3)_2NNCOCH_3$$
$$\overset{|}{C}H_2C_6H_4NO_2\text{-}p \qquad\qquad \overset{|}{C}H_2C_6H_4NO_2\text{-}p$$

(104)

(215)   **(216)**

In the amine imide–hydrazide rearrangement only benzyl groups migrate, other groups being incapable of rearranging[163].   Thus, 1-methyl-2-phenylpyrrolidine-1-acetimide (**218**) gives only a small amount of the rearrangement product **217** under pyrolytic conditions, the main products being 1-methyl-2-phenylpyrrolidine (**219**) and methyl isocyanate[163] (reaction 105).   The cyclic amine imide **220** with a benzyl group on the quaternary nitrogen rearranges into **221** with benzyl group migration (reaction 106).   However, the quaternary dimethyl compound **222** cannot undergo rearrangement; under the same conditions, the ring opens and 1-methacryloyl-2,2-dimethyl-hydrazine (**223**) is formed[270] (reaction 107).

In acyclic amine imides in which the quaternary nitrogen does not

---

* The nomenclature of these compounds can be derived by analogy to (CH$_3$)$_3\overset{+}{N}\overset{-}{N}$H, trimethylamine imine, or (CH$_3$)$_3\overset{+}{N}\overset{-}{N}$COCH$_3$, trimethylamine acetimide.

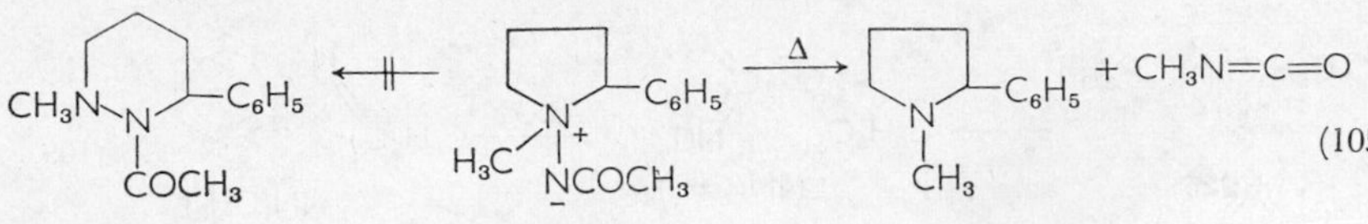

$$CH_3N-N-C_6H_5 \overset{+}{\longleftarrow}|\!\!\!|\!\!\!- \underset{H_3C}{\overset{+}{N}}\underset{NCOCH_3}{\overset{|}{N}}-C_6H_5 \xrightarrow{\Delta} \underset{CH_3}{N}-C_6H_5 + CH_3N{=}C{=}O$$

$$\underset{COCH_3}{|}$$

(105)

(217)   **(218)**   **(219)**

$$\underset{C_6H_5CH_2}{H_3C}\overset{CH_3}{\underset{N}{\overset{+}{N}}}{=}O \xrightarrow{\Delta} CH_3N\overset{CH_3}{\underset{CH_2C_6H_5}{N}}{=}O$$

(106)

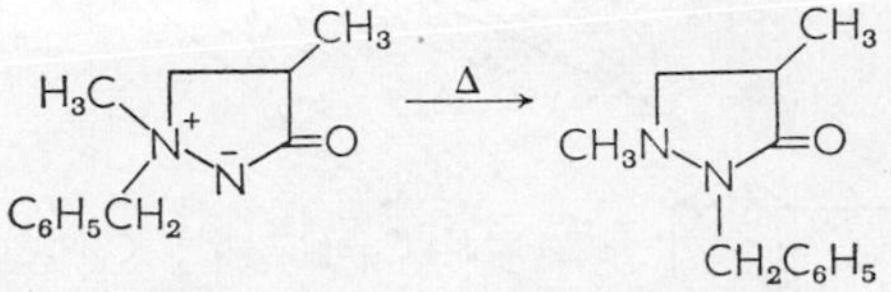

(220)   **(221)**

$$\underset{H_3C}{H_3C}\overset{CH_3}{\overset{+}{N}}{=}O \xrightarrow{\Delta} CH_2{=}\underset{CH_3}{\overset{|}{C}}CONHN(CH_3)_2$$

(107)

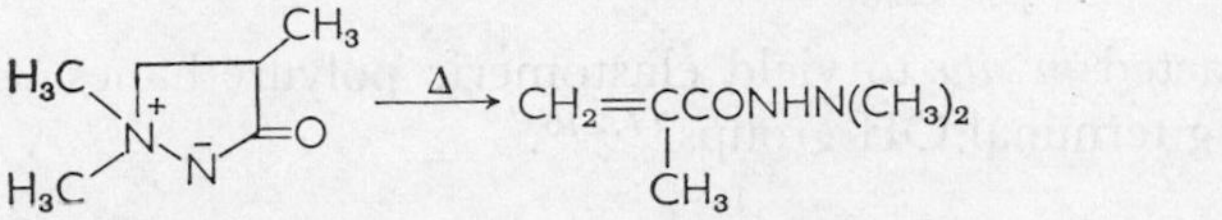

(222)   **(223)**

carry a benzyl group, (**224**), pyrolytic conditions produce cleavage of the N—N bond, as in the Curtius–Hofmann degradation. The products are trimethylamine and an isocyanate (reaction 108), which undergoes the usual further reactions[271–273]. In the thermolysis of 2,2,2-trimethyl-

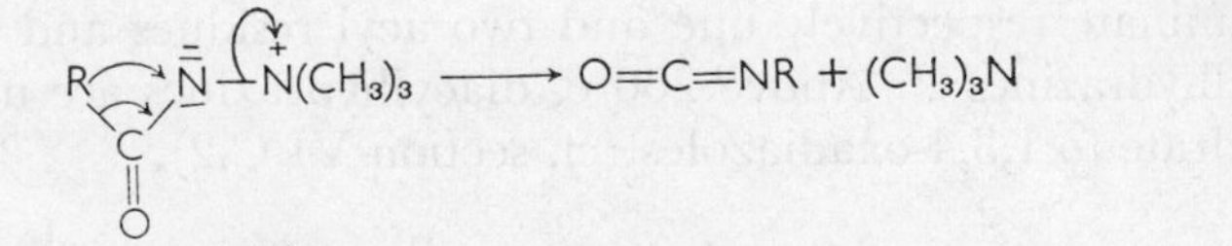

$$R-\overset{-}{\overset{..}{N}}-\overset{+}{N}(CH_3)_3 \longrightarrow O{=}C{=}NR + (CH_3)_3N$$

(108)

(224)

amine benzimide (**225**), the formation of the main products—trimethylamine and phenyl isocyanate—is accompanied by a number of by-products, whose formation is explained by the reaction of **225** with phenyl isocyanate to give the intermediate **227** via **226**[274]. Thermolysis of **227** leads to elimination of trimethylamine and ring closure, forming 2-phenylbenzimidazole (**228**), or—after migration of the phenyl group from the carbon to the nitrogen—the N,N'-diphenylcarbodiimide, which can be isolated as N,N'-diphenylurea after reaction with water[274]. Benzanilide is formed in a small amount as a hydrolysis product of **227** (reaction 109).

Trimethylamine imides of acyclic, alicyclic, and aromatic dicarboxylic acids pyrolyse to diisocyanates, which can either be isolated

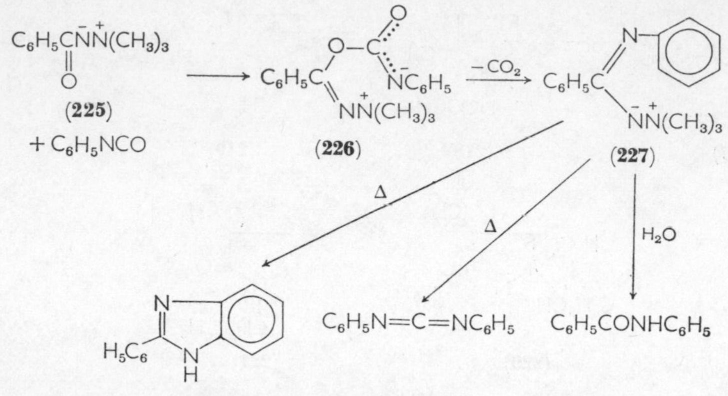

$$\text{(109)}$$

The reaction scheme showing structures (225), (226), (227), (228), with:

$C_6H_5C\overset{-}{N}\overset{+}{N}(CH_3)_3$ with $\|O$ below
(225)
$+ C_6H_5NCO$

$\longrightarrow$

C_6H_5C — O—C(=O) ... $\overset{-}{N}C_6H_5$, $\overset{+}{N}N(CH_3)_3$
(226)

$\xrightarrow{-CO_2}$

C_6H_5C ... with $\overset{-}{N}\overset{+}{N}(CH_3)_3$
(227)

Δ Δ H_2O

H_5C_6 ... benzimidazole ring with N, N—H
(228)

$C_6H_5N{=}C{=}NC_6H_5$ $C_6H_5CONHC_6H_5$

or reacted *in situ* to yield elastomeric polyurethanes with polyesters having terminal OH groups[77,275].

14. Thermal reactions

As can be seen from the formation of hydrazides on heating of hydrazinium salts of carboxylic acids (section IV.C.1), monoacyl-hydrazines have a high thermal stability. On strong heating, however, hydrazine is eliminated and 1,2-diacylhydrazines are formed[150], which have the highest thermal stability among hydrazides. Triacyl- and tetraacylhydrazines readily cleave in aqueous or acidic solutions to eliminate respectively one and two acyl residues and to give 1,2-diacylhydrazines[2]. Above 200°c, diacylhydrazines are unstable and dehydrate to 1,3,4-oxadiazoles (cf. section VI.C.2).

C. Formation of Heterocyclic Compounds

I. General aspects of the cyclization of hydrazides

The cyclization of hydrazides to yield heterocyclic compounds, is particularly important for the synthesis of five-membered hetero-cyclic rings. 1,3,4-Oxadiazoles, 1,3,4-thiadiazoles, *sym*-triazoles, pyra-zoles, and their substituted, hydrogenated, and oxidized derivatives can be readily obtained from hydrazides.

Five-membered heterocyclic rings can generally be prepared from compounds of type I as shown in reaction (110), when an activated nucleophilic group X attacks the electrophilic carbon of the carbonyl group (route 1). Nucleophilic ring closure of the carbonyl oxygen to form oxadiazoles (route 2) is expected as a competing reaction. The actual route taken depends on the substituents and on the reaction conditions. Structures of type II contain on the β-carbon atom of

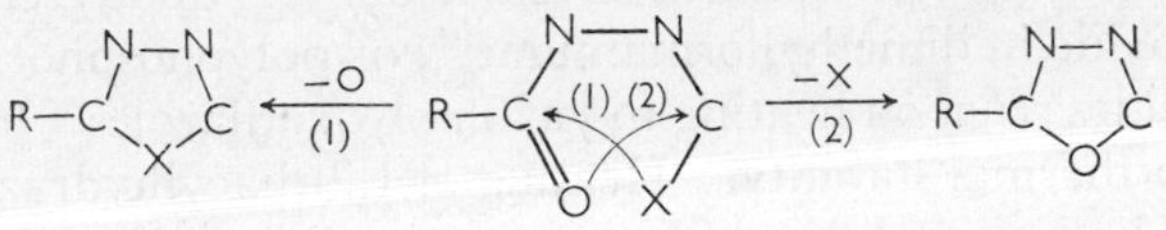

$$\text{Type I} \qquad (110)$$

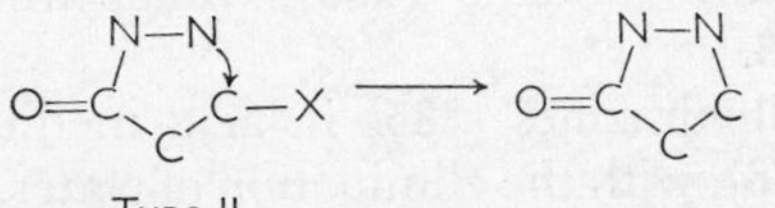

$$\text{Type II} \qquad (111)$$

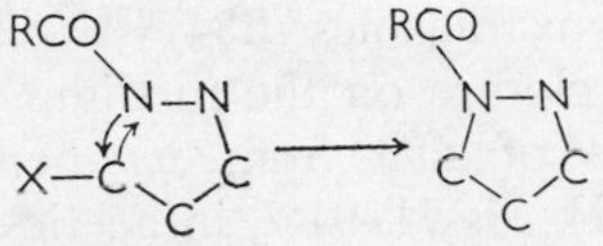

$$\text{Type III} \qquad (112)$$

the acyl residue a reaction centre susceptible to nucleophilic attack by the N' atom of the hydrazide, leading to the formation of a pyrazole system (reaction 111). The most important syntheses of heterocyclic compounds with a five-membered ring start from these two types. Less frequently, the reaction involves compounds of type III, in which a heterocyclic compound is formed by interaction of $N_{(1)}$ with a side-chain attached to $N_{(2)}$ (reaction 112).

The four systems **229–232** can be used for the preparation of six-membered heterocyclic rings in corresponding ways (section VI.C.6). To the authors' knowledge no piperidazine synthesis has been accomplished by cyclization of hydrazides, in a manner similar to type III.

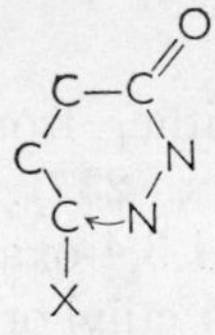

(229) 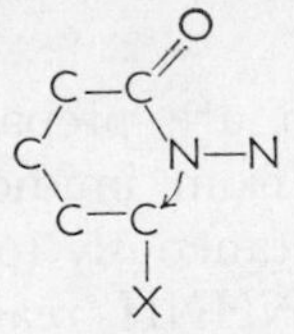 **(230)** 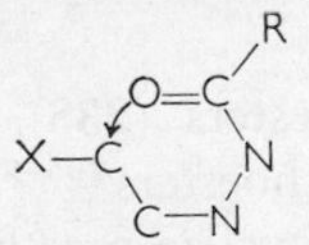 **(231)** 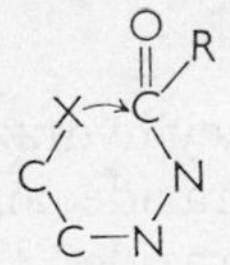 **(232)**

2. 1,3,4-Oxadiazole and 1,3,4-thiadiazole

2,5-Disubstituted 1,3,4-oxadiazoles (**234**) are formed by the dry heating of 1,2-diacylhydrazines (**233**), accompanied by the elimination of water[6,150,276]. The reaction is facilitated by the addition of water-absorbing materials such as phosphorus pentoxide[170], phosphorus pentachloride[120,277], phosphorus oxychloride[7,278], thionyl chloride[279], sulphur trioxide in dimethylformamide[280], or polyphosphoric acid[281]. Polyhydrazides react on heating to yield polyoxadiazoles[282,283], which have a high thermal stability. Heating of 1,2-diacylhydrazines (**233**) with phosphorus pentasulphide results in 2,5-disubstituted 1,3,4-thiadiazoles (**236**)[235,284]. It may be assumed that the intermediate is a monothiodiacylhydrazine (**235**), which then undergoes ring closure to give **236**.

Monothiodiacylhydrazines (**235**) in acid medium lead exclusively to thiadiazoles (**236**), with the elimination of water, while in the case of base catalysis, the preferred course is elimination of hydrogen sulphide and the formation of oxadiazoles (**234**)[235]. The dependence of the direction of the ring closure on the medium has been examined in detail[235]. The yield of **236** from *sym*-benzoylthiobenzhydrazide (**235**, $R^1 = R^2 = C_6H_5$) is 71–91% in acetic acid, *N*-methylpyrrolidone, 'butyl glycol', and a mixture of 'butyl glycol' and dimethylaniline. In a mixture of 'butyl glycol' and tripropylamine, on the other hand, compound **234** is obtained in a yield of 87%.

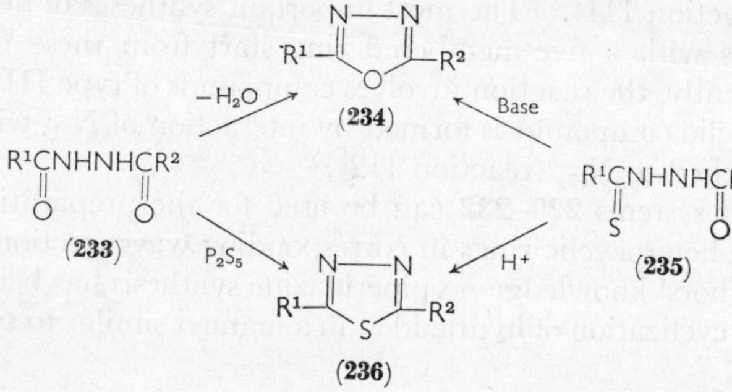

Acylhydrazidic esters (**238**), which are prepared either from a hydrazide and orthoesters[285–287] or from imidic esters (**237**) and hydrazides[288–290], cyclize mostly spontaneously to give 1,3,4-oxadiazoles (**234**). When hydrazides $R^2CONHNH_2$ react with ethyl orthoformate, the corresponding monosubstituted 1,3,4-oxadiazoles (**234**, $R^1 = H$) are obtained[285,286]. However, the intermediate **238** can be isolated when $R^1 = R^2 = H$[286]. Also when $R^1 = R^2 = CH_3$, **238** can be isolated from the reaction mixture of the corresponding imidic ester (**237**, $R^1 = CH_3$) and acetic acid hydrazide[288]. The action of the hydrochlorides of bifunctional imidic esters on dicarboxylic acid hydrazides leads to polyoxadiazoles[288].

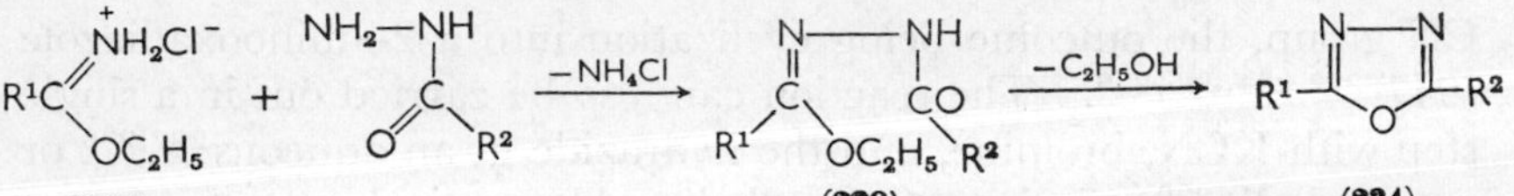

$$R^1C\overset{\overset{+}{N}H_2Cl^-}{\underset{OC_2H_5}{}} + \underset{O}{\overset{NH_2-NH}{\underset{R^2}{C}}} \xrightarrow{-NH_4Cl} R^1\overset{N-NH}{\underset{OC_2H_5}{\underset{R^2}{C}}}CO \xrightarrow{-C_2H_5OH} R^1\overset{N-N}{\underset{O}{}}R^2$$

$$\quad\text{(237)}\qquad\qquad\qquad\qquad\text{(238)}\qquad\qquad\qquad\text{(234)}$$

Hydrazides react with phenyl isocyanide dichloride (**239**, R^2 = C_6H_5) to form 2-phenylamino-1,3,4-oxadiazoles (**240**) with the elimination of hydrogen chloride[291]. The reaction of hydrazides with N-sulphonylisocyanide dichlorides (**239**, R^2 = $SO_2C_6H_5$) giving 2-N-sulphonylamino-1,3,4-oxadiazoles, is entirely analogous (reaction 113)[292,293].

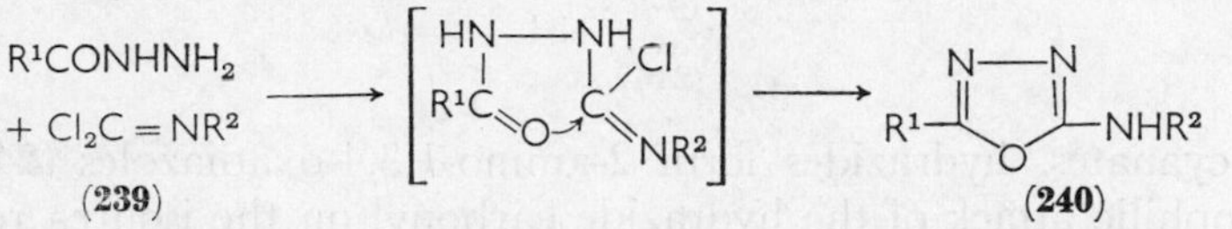

$$\begin{array}{c} R^1CONHNH_2 \\ + Cl_2C=NR^2 \end{array} \longrightarrow \left[R^1C\overset{HN-NH}{\underset{O}{}}\overset{}{\underset{NR^2}{C}}{}^{Cl} \right] \longrightarrow R^1\overset{N-N}{\underset{O}{}}NHR^2 \quad (113)$$

$$\quad\text{(239)}\qquad\qquad\qquad\qquad\qquad\text{(240)}$$

The type of reaction (113) comprises also the oxadiazole formation from p-nitrobenzoic acid hydrazides with methylmercaptochloromethylene-N,N-pentamethyleneimmonium chloride (**241**)[294]. The reaction conditions determine whether the product is a 2-(N-piperidyl)-1,3,4-oxadiazole (**243**) or a 2-methylmercapto-1,3,4-oxadiazole (**244**), resulting from the competing elimination of methyl mercaptan or piperidine. The analogous reaction of benzoic acid hydrazide with methylmercaptochloromethylene-N,N-diphenylimmonium chloride can be stopped at the corresponding intermediate analogous to **242** by carrying it out at a relatively low temperature[294].

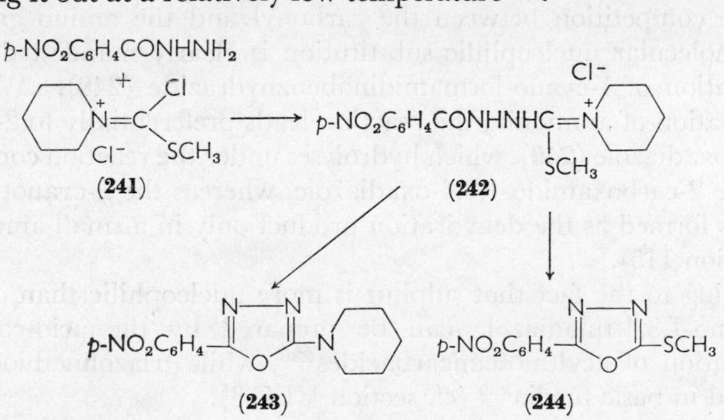

$$p\text{-}NO_2C_6H_4CONHNH_2$$

$$\text{(241)} \qquad \text{(242)}$$

$$\text{(243)} \qquad \text{(244)}$$

The carboxylic acid N'-cyanohydrazide (**245**), formed by re-action of hydrazides with cyanogen bromide, immediately rearranges by nucleophilic attack of the hydrazide carbonyl on the carbon of the CN group, the outcome being cyclization into a 2-aminooxadiazole (**247**)[222,224,295,296]. The reaction can also be carried out in a single step with KCN, bromine, and the hydrazide in an aqueous[297,298] or an alcoholic[230] solution of an alkali. Alternatively, on treatment

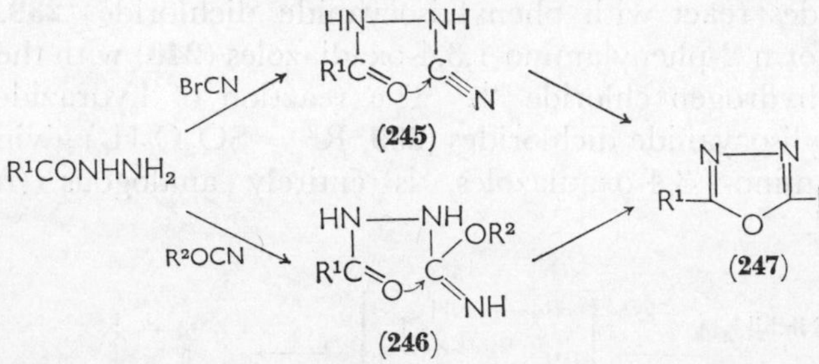

(**245**)

(**246**)

(**247**)

with cyanates, hydrazides form 2-amino-1,3,4-oxadiazoles (**247**) by nucleophilic attack of the hydrazide carbonyl on the isourea residue of intermediate **246**[299]. With dihydrazides and dicyanates the pro-ducts are bis(2-amino-1,3,4-oxadiazoles) (reaction 114)[299,299a].

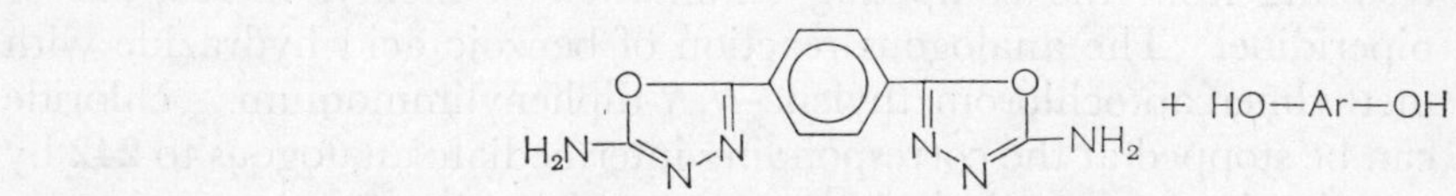

(114)

The competition between the carbonyl and the amino group in intramolecular nucleophilic substitution is clearly manifested in the cyclization of 1-cyano-formamidinobenzhydrazide (**248**). With the elimination of ammonia, the reaction leads preferentially to 2-cyano-1,3,4-oxadiazole (**249**), which hydrolyses under the reaction conditions to give 2-carboxamido-1,3,4-oxadiazole, whereas the 2-cyanotriazole, **250**, is formed as the dehydration product only in a small amount[300] (reaction 115).

Owing to the fact that sulphur is more nucleophilic than oxygen, 2-amino-1,3,4-thiadiazole can be prepared by the acid-catalysed cyclization of acylthiosemicarbazides[230], while triazolinethiones are formed in basic media[229] (cf. section VI.C.3).

$$C_6H_5\text{-(1,3,4-oxadiazole)-CN} \quad \xleftarrow{-NH_3} \quad C_6H_5\underset{\underset{O}{\|}}{C}NHN=\underset{\underset{CN}{}}{C}NH_2 \quad \xrightarrow{-H_2O} \quad C_6H_5\text{-(1,2,4-triazole,NH)-CN} \quad (115)$$

$$(249) \qquad\qquad (248) \qquad\qquad (250)$$

Phosgene or thiophosgene form with hydrazides the intermediates **251** and **252**, which cyclize into 1,3,4-oxadiazolin-5-ones (**253**) [230,301,302] and 1,3,4-oxadiazolin-5-thiones (**254**) [230,303] on being heated in toluene, dioxan, chloroform, or acetic ester (reaction 116) [80,230]. These reactions also proceed when 2,2-dialkyl-substituted hydrazides are brought into contact with phosgene or thiophosgene [117,303]. The intermediate is thought to be a quaternary ammonium salt (**255, 256**) which then loses alkyl chloride and forms **257** or **258** (reaction 117).

$$\text{RCONHNH}_2 + \underset{Cl}{\overset{Cl}{\diagdown}}C{=}X \quad\longrightarrow\quad \left[\begin{array}{c} HN{-}NH \\ RC\underset{O}{\diagup}\quad\underset{Cl}{\diagdown}C{=}X \end{array}\right] \quad\longrightarrow\quad \begin{array}{c} N{-}NH \\ R{-}\underset{O}{\diagup}{=}X \end{array} \qquad (116)$$

$$(251)\ X{=}O \qquad\qquad (253)\ X{=}O$$
$$(252)\ X{=}S \qquad\qquad (254)\ X{=}S$$

$$\text{R}^1\text{CONHNR}^2_2 + \underset{Cl}{\overset{Cl}{\diagdown}}C{=}X \quad\longrightarrow\quad \left[\begin{array}{c} N{-}\overset{+}{N}R^2_2\ Cl^- \\ R^1{-}\underset{O}{\diagup}{=}X \end{array}\right] \quad\xrightarrow{-R^2Cl}\quad \begin{array}{c} N{-}NR^2 \\ R^1{-}\underset{O}{\diagup}{=}X \end{array} \qquad (117)$$

$$(255)\ X{=}O \qquad\qquad (257)\ X{=}O$$
$$(256)\ X{=}S \qquad\qquad (258)\ X{=}S$$

Hydrazides having a substituent on the α-nitrogen cannot form 1,3,4-oxadiazolin-5-one with phosgene, thus, when 1-benzoyl-1-methylhydrazine is reacted with phosgene, one obtains 4-methyl-5-phenylisosydnone (**260**) by cyclization via the intermediate **259** [83].

$$\text{C}_6\text{H}_5\text{CONCH}_3\text{NH}_2 + \underset{Cl}{\overset{Cl}{\diagdown}}C{=}O \quad\longrightarrow\quad \begin{array}{c} CH_3N{-}NH \\ C_6H_5C\underset{O}{\diagup}\quad\underset{Cl}{\diagdown}C{=}O \end{array}$$

$$(259)$$

$$\begin{array}{c} CH_3N{-}N \\ C_6H_5{-}\underset{O}{\diagup}{\overset{+}{}}{-}O^- \end{array} \quad\xrightarrow{+COCl_2}\quad \begin{array}{c} N{-}N \\ C_6H_5{-}\underset{O}{\diagup}{-}Cl \end{array} \qquad (118)$$

$$(260) \qquad\qquad\qquad (261)$$

The chlorination of **260** with phosgene gives 2-chloro-5-phenyl-1,3,4-oxadiazole (**261**), which is presumably formed by the elimination of methyl chloride from an intermediate quaternary salt[83] (reaction 118).

This course corresponds to the synthesis of a meso-ionic 1,3,4-oxadiazole-2-thione (**264**) by the treatment of ammonium dithiocarbazinate (**262**) with phosphorus oxychloride and triethylamine in

$$
\underset{(262)}{C_6H_5\overset{\underset{\displaystyle C_6H_5}{|}}{\underset{\underset{\displaystyle S_-\ \ NH_4^+}{\|}}{\underset{\displaystyle O}{C}}}NNHC=S} \xrightarrow{POCl_3} \underset{(263)}{[C_6H_5\overset{\underset{\displaystyle C_6H_5}{|}}{\underset{\underset{\displaystyle O}{\|}}{C}}NN=C=S]} \longrightarrow \underset{(264)}{} \tag{119}
$$

ether, in which the isothiocyanate **263** has been postulated as an intermediate[304] (reaction 119).

3. 4H-1,2,4-Triazole

Symmetric triazoles can be prepared from hydrazides having an $N_{(2)}$—C—N partial structure. The latter has to be sufficiently nucleophilic, so that the ring can be closed by attack on the hydrazide carbonyl carbon. Oxadiazole formation may proceed as a competing reaction. Acylamidrazones (**265**) thus give 4H-1,2,4-triazoles (**266**) in a good yield when they are heated in an alkaline or a neutral medium, the ring closure being accompanied by loss of water[51,234,236,289,290,305]. In a strongly acidic medium, on the other hand, the preferred reaction is elimination of ammonia, leading to 1,3,4-oxadiazoles (**234**) (reaction 120)[235]. In the latter case the strongly basic nitrogen is protonated, giving an ammonium ion, which

$$
\underset{(234)}{R^1\text{---}O\text{---}R^2} \xleftarrow[-NH_3]{H^+} \underset{(265)}{R^1C\overset{\displaystyle N\text{---}NH}{\underset{\displaystyle H_2N\ \ O}{\diagdown}}CR^2} \xrightarrow[-H_2O]{OH^-} \underset{(266)}{R^1\text{---}\overset{\displaystyle N\text{---}N}{\underset{\underset{\displaystyle H}{N}}{}}\text{---}R^2} \tag{120}
$$

is cleaved off by the nucleophilic carbonyl oxygen on account of its good nucleofugic character.

Hydrazides of the structure **267**, in which the C—N attached to

$$
\underset{(267)}{} \longrightarrow \underset{(268)}{} \tag{121}
$$

$N_{(2)}$ forms part of a heterocyclic system, form bicyclic triazole systems (268) in weakly acidic solutions (reaction 121) [96,97,306].

Aromatic 1,2-diacylhydrazines are converted with phosphaza derivatives (269) of various aromatic amines into 3,4,5-triaryl-1,2,4-triazoles (270) [307] (reaction 122). This reaction does not take place with aliphatic compounds.

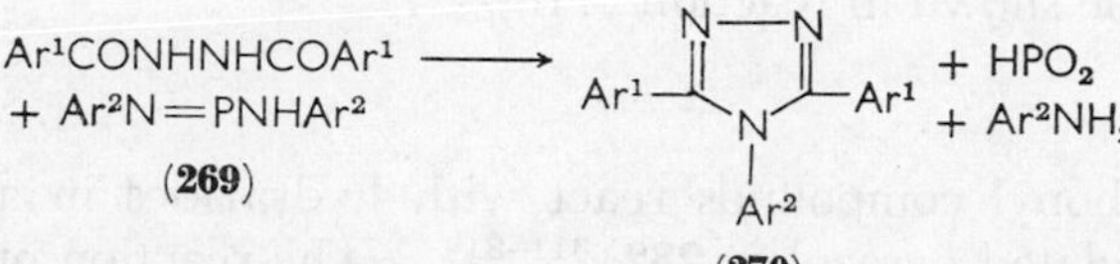

$$Ar^1CONHNHCOAr^1 \longrightarrow \quad + HPO_2 \quad (122)$$
$$+ Ar^2N=PNHAr^2 \qquad Ar^1 \underset{Ar^2}{\overset{N-N}{\bigtriangleup}} Ar^1 \quad + Ar^2NH_2$$
$$(269) \qquad\qquad\qquad (270)$$

The condensation of two hydrazide molecules gives N-aminotriazoles (271) (reaction 123) [308]. C-Aminotriazoles (273) are formed in the cyclization of acylaminoguanidines (272) (reaction 124) [309].

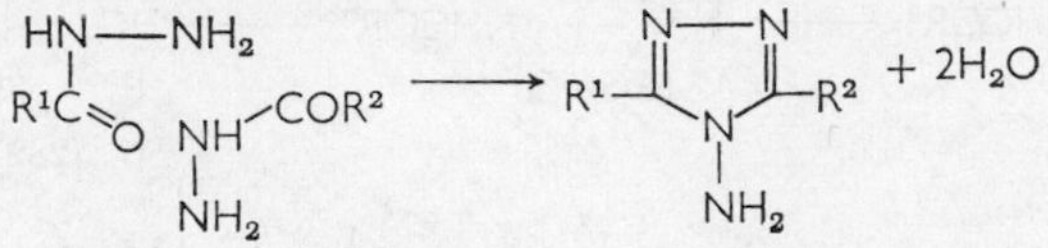

$$\begin{array}{c} HN{-}NH_2 \\ R^1C{\overset{}{\underset{O}{\,}}}{\underset{NH}{\,}}{\overset{COR^2}{\,}} \\ \quad NH_2 \end{array} \longrightarrow R^1 \underset{NH_2}{\overset{N-N}{\bigtriangleup}} R^2 \; + 2H_2O \quad (123)$$
$$(271)$$

$$\begin{array}{c} HN{-}NH \\ R^1C{\underset{O}{\,}} \; {\underset{NH}{C}}NHR^2 \end{array} \longrightarrow R^1 \underset{H}{\overset{N-N}{\bigtriangleup}} NHR^2 \quad (124)$$
$$(272) \qquad\qquad\qquad (273)$$

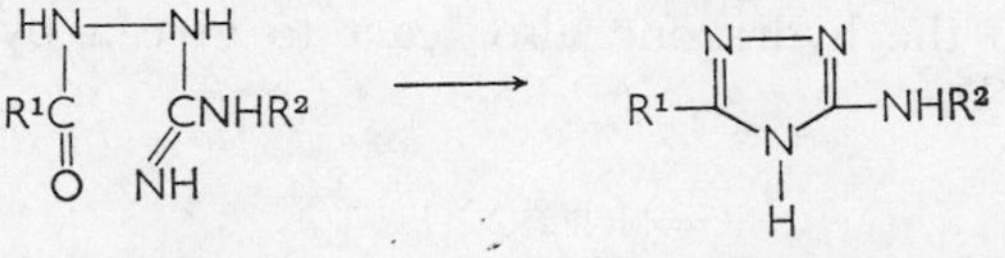

Triazolin-5-ones (276) and triazolin-5-thiones (277) can be prepared by alkaline internal condensation of acylsemicarbazides (274) and acylthiosemicarbazides (275) (reaction 125), since in alkaline media

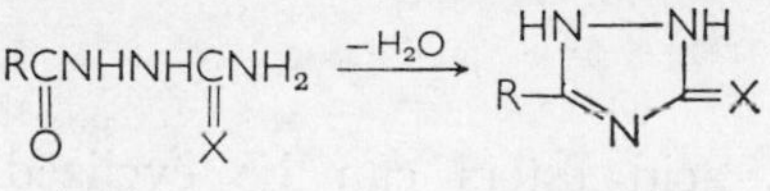

$$RCNHNHCNH_2 \xrightarrow{-H_2O} R \underset{N}{\overset{HN-NH}{\bigtriangleup}} X \quad (125)$$
$$\overset{\|}{O} \quad \overset{\|}{X}$$
$$(274) \; X=O \qquad\qquad (276) \; X=O$$
$$(275) \; X=S \qquad\qquad (277) \; X=S$$

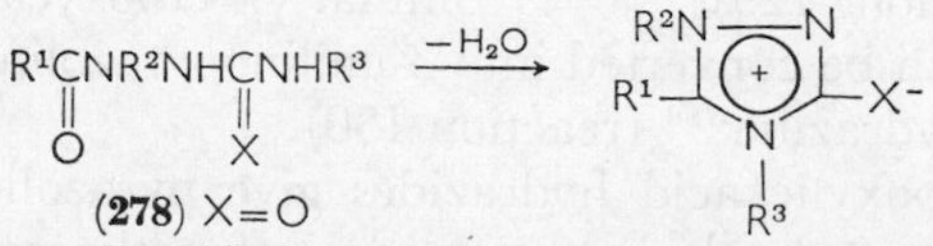

$$R^1CNR^2NHCNHR^3 \xrightarrow{-H_2O} R^1 \underset{R^3}{\overset{R^2N-N}{\underset{N}{\bigoplus}}} X^- \quad (126)$$
$$\overset{\|}{O} \quad\; \overset{\|}{X}$$
$$(278) \; X=O$$
$$(279) \; X=S$$
$$(280) \; X=O; \; R^1=Me; \; R^2=Me; \; R^3=Ph$$
$$(281) \; X=S; \; R^1=Me, \, Ar;$$
$$\qquad R^2=Me, \, Ar; \; R^3=Ar, \, Me$$

the amino group is more nucleophilic than either the oxygen of the carbonyl group or the sulphur of the thiocarbonyl group[228,229,295]. Acylsemicarbazides (278) and acylthiosemicarbazides (279) the α-nitrogen of which carries an alkyl or an aryl substituent (R^2) often cyclize rapidly to a meso-ionic triaza system (280, 281) (reaction 126), analogously to the isosydnone formation from α-substituted hydrazides and phosgene shown in reaction (118)[310].

4. Pyrazole

1,3-Dicarbonyl compounds react with hydrazides in acid media to give 1-acylated pyrazoles (282)[311-314]. The reaction proceeds via the monohydrazone, which is then cyclized with loss of water (reaction 127). It has been found with a steroid (283) that under certain

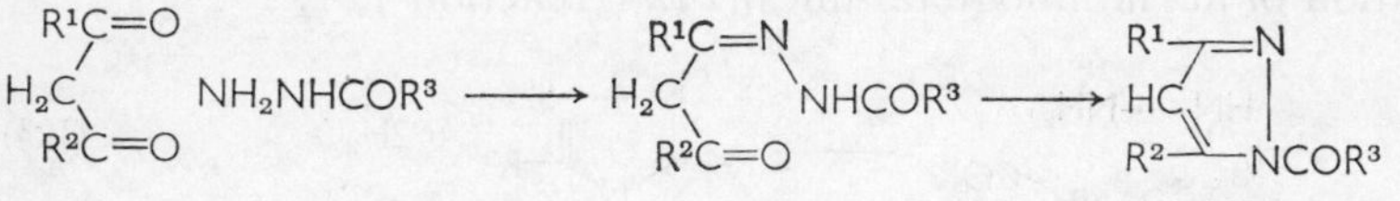

$$
\begin{array}{c}
R^1C{=}O \\
H_2C \\
R^2C{=}O
\end{array}
\quad NH_2NHCOR^3 \longrightarrow
\begin{array}{c}
R^1C{=}N \\
H_2C \quad\quad NHCOR^3 \\
R^2C{=}O
\end{array}
\longrightarrow
\begin{array}{c}
R^1{-}\!=\!N \\
HC \quad\quad | \\
R^2{-}\!-NCOR^3
\end{array}
\quad (127)
$$

$$(282)$$

conditions acylhydrazones with an epoxide ring in the α,β-position with respect to the hydrazone also react to give a pyrazole (284) (reaction 128)[315].

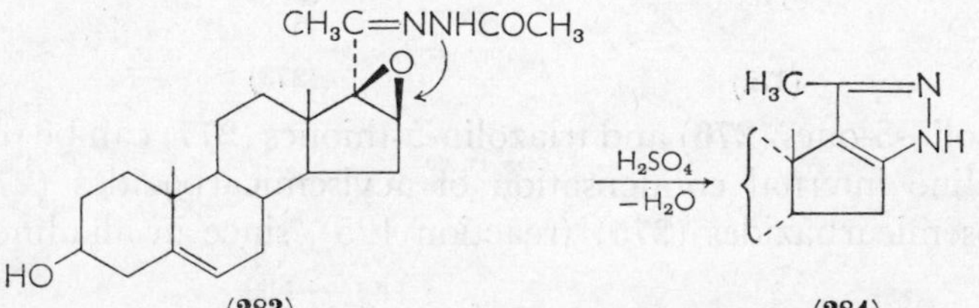

$$(283) \qquad\qquad (284) \qquad (128)$$

β-Ketocarboxylic acid esters can be cyclized with 1-acetyl-1-phenylhydrazines in the presence of phosphorus trichloride. The reaction is accompanied by the loss of acetic acid, and leads to pyrazolinones (285) (reaction 129)[316,317]. Similarly, ethoxycarbonyl-thioacetanilide (286) can be converted into 3-anilinopyrazoline-5-one (287) with the aid of hydrazine[318] (reaction 130).

α,β-Unsaturated carboxylic acid hydrazides give pyrazolidinones on heating[72,73,319]. Thus, the thermal treatment of methacrylic acid 2-methylhydrazide leads to 1,4-dimethyl-3-pyrazolidinone (288)[270] (reaction 131). The acylation of asym-dimethylhydrazine with methacrylic acid chloride gives the cyclic ammonium salt **289**, which can be converted into the cyclic amine imide **290** in the presence of bases[270] (reaction 132).   When **290** is heated to 220°c under vacuum,

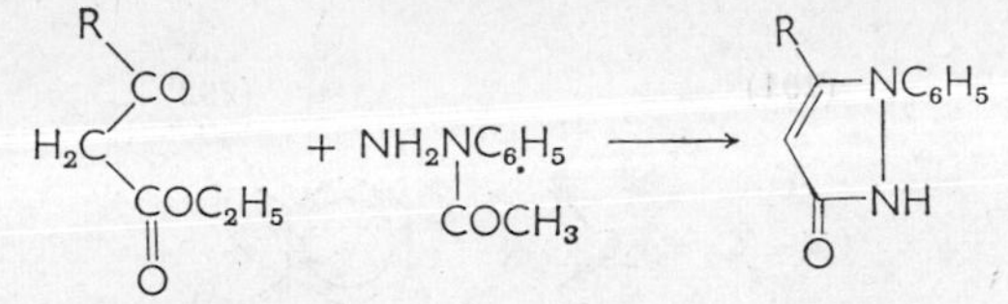

$$R-CH(CO)(COC_2H_5)=O \cdot H_2C + NH_2NC_6H_5 \cdot COCH_3 \longrightarrow R-C(=CH)(NC_6H_5) \cdot C(=O)-NH \tag{129}$$

(R = CH$_3$,C$_6$H$_5$)   (**285**)

$$C_6H_5NH-C(CS)-H_2C-COC_2H_5=O + NH_2NH_2 \longrightarrow C_6H_5NH-C(=N)-CH_2-C(=O)-NH \tag{130}$$

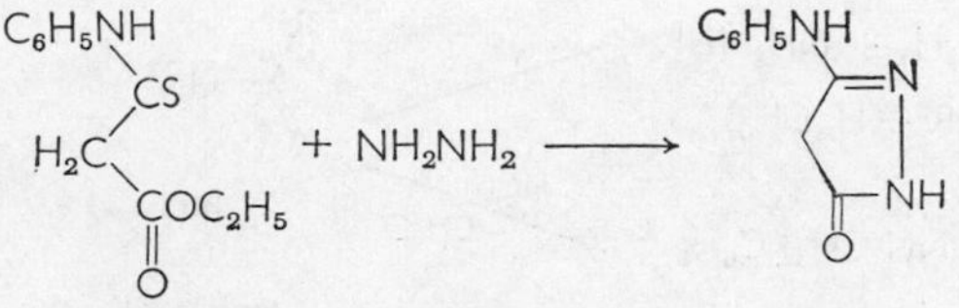

(**286**)   (**287**)

the ring opens and acrylic acid 2,2-dimethylhydrazide is formed (reaction 132).   This compound can be reconverted into **290** by heating to 120°c in a sealed tube[267,320].

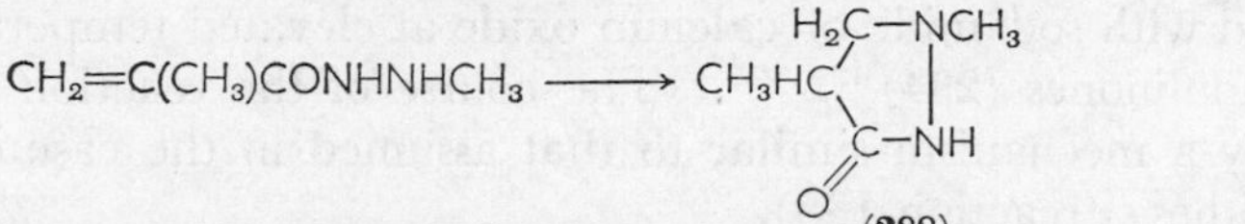

$$CH_2{=}C(CH_3)CONHNHCH_3 \longrightarrow CH_3HC(-NCH_3)(C{-}NH){=}O \tag{131}$$

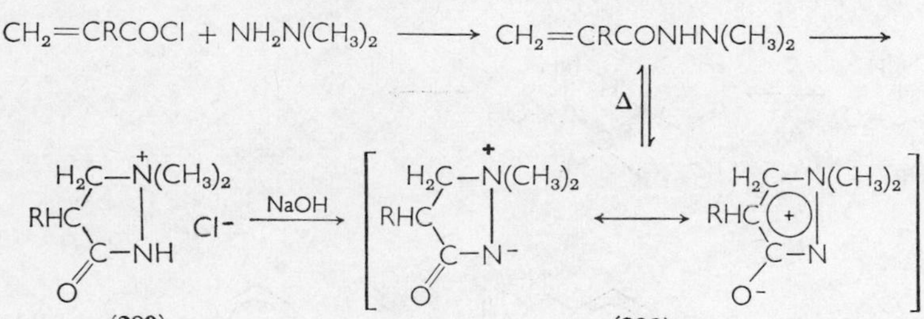

(**288**)

$$CH_2{=}CRCOCl + NH_2N(CH_3)_2 \longrightarrow CH_2{=}CRCONHN(CH_3)_2 \longrightarrow$$

$$\Delta \updownarrow$$

$$H_2C{-}\overset{+}{N}(CH_3)_2 \; RHC(C{-}NH){=}O \; Cl^- \xrightarrow{NaOH} \left[H_2C{-}\overset{+}{N}(CH_3)_2 \; RHC(C{-}N^-){=}O \longleftrightarrow H_2C{-}N(CH_3)_2 \; RHC(\overset{+}{C}{-}N)(O^-) \right] \tag{132}$$

(**289**)   (**290**)

Thermolysis of *N*-acrylyl derivatives of cyclic hydrazides (**291**) leads to condensed pyrazolidinones (**292**) (reaction 133)[321]. With alkylated hydrazines and unsubstituted hydrazine, propargylates form pyrazolinones (**293a,b**)[322].   In the case of phenylpropargylic acid the pyrazolinone (**293c**) can also be obtained in a good yield from the corresponding amine imide[323] (reaction 134).

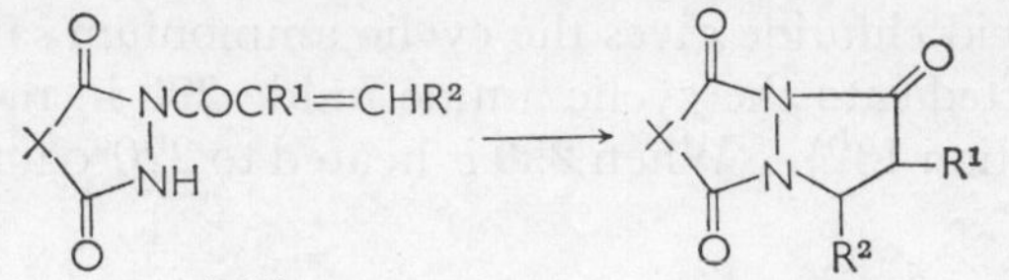

$$(291) \qquad (292) \qquad (133)$$

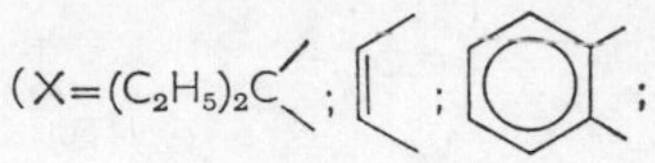

$$(X=(C_2H_5)_2C\diagup \; ; \; \diagup \; ; \; \diagup\text{(xylyl)} \; ;$$

$$R^1=H \text{ or } CH_3;$$
$$R^2=H, CH_3 \text{ or } C_6H_5)$$

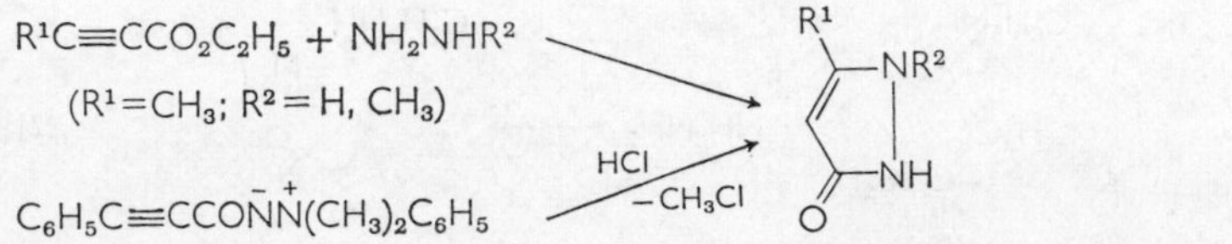

$$R^1C{\equiv}CCO_2C_2H_5 + NH_2NHR^2$$
$$(R^1=CH_3; R^2=H, CH_3)$$

$$C_6H_5C{\equiv}CCON\bar{}N^+(CH_3)_2C_6H_5 \xrightarrow[-CH_3Cl]{HCl}$$

$$(134)$$

(293a) $R^1=CH_3; R^2=H$
(293b) $R^1=CH_3; R^2=CH_3$
(293c) $R^1=C_6H_5; R^2=CH_3$

5. Other five-membered heterocyclic compounds

Analogously to Fischer's indole synthesis, phenylhydrazides can be cyclized with sodamide or calcium oxide at elevated temperatures to give indolinones (294)[324,325]. The course of the reaction is explained by a mechanism similar to that assumed in the case of the indole synthesis (reaction 135).

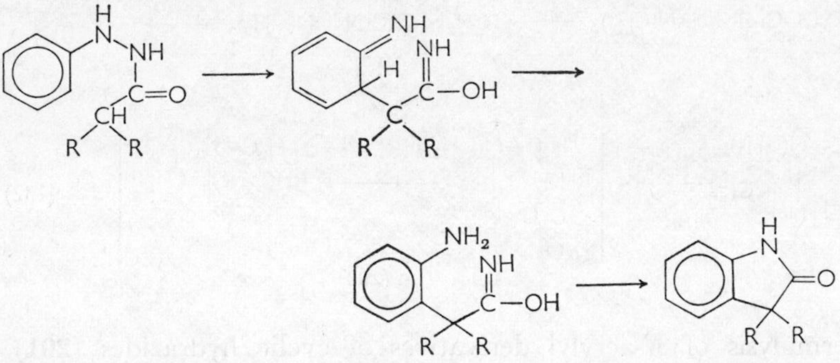

$$(135)$$

(294)

The heating of o-aminothiophenol, o-aminophenol, or o-phenylene-diamine with benzoic acid hydrazide leads to loss of water and hydrazine, and to the formation of the corresponding benzazole derivatives (295)[326] (reaction 136).

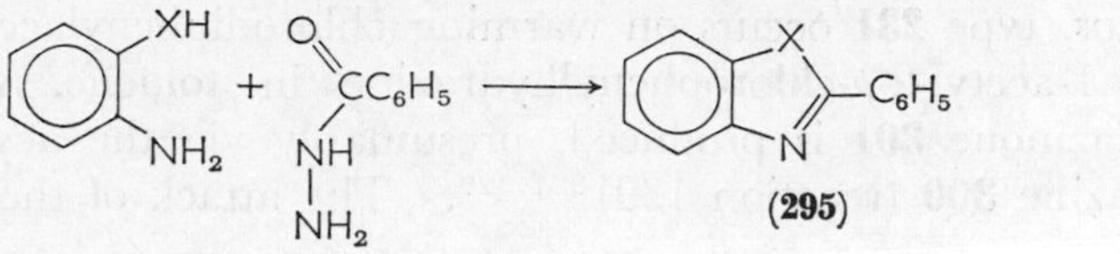

$$\text{(136)}$$

$$(X = O, S, NH)$$

Tetrazoles (**296**) can be obtained from tetrazenes (cf. section VI.B.5) by treatment with alkalis[233] (reaction 137).

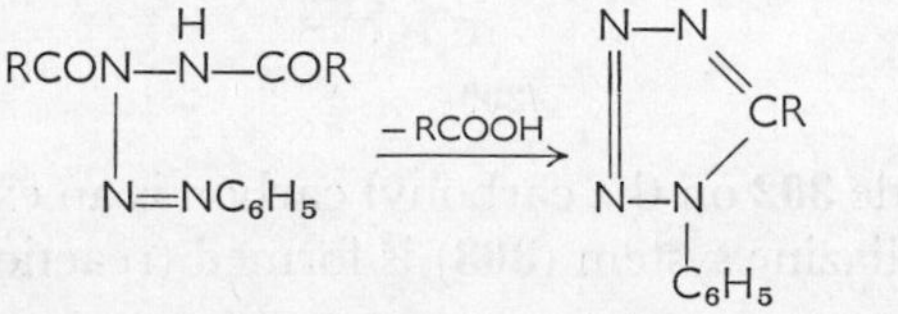

$$\text{(137)}$$

$$\text{(296)}$$

Vicinal bis(benzoylhydrazones) (**297**) rearrange into 1-(benzoyloxybenzylideneamino)-1,2,3-triazoles (**299**) on oxidation with mercuric oxide or iodine[128] (reaction 138). It is possible that the bis(acyldiazo) compound **298** which is formed in the oxidation and which behaves as a strong acylating agent similar to acyldiimides, acylates the carbonyl group of an acyl residue intramolecularly into a benzoyloxy group. A diimine residue with a partial negative charge attacks the partially positive nitrogen of the other azo group, and the ring closes to form a 1,2,3-triazole (**299**)[128].

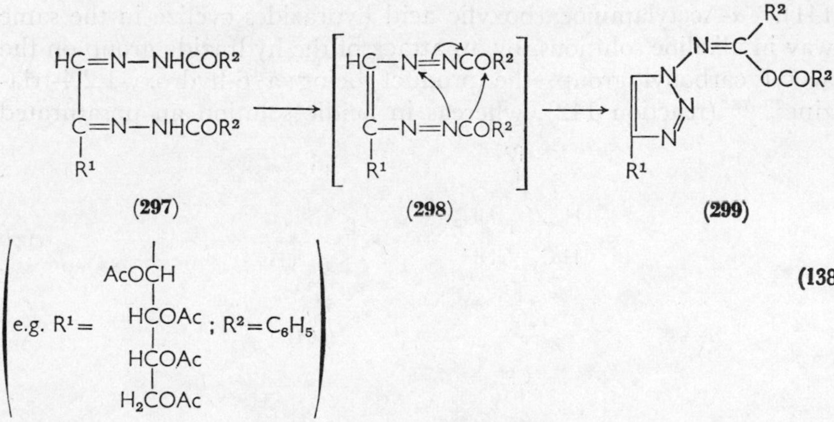

$$\text{(297)} \qquad \text{(298)} \qquad \text{(299)}$$

$$\left(\text{e.g. } R^1 = \begin{array}{c} \text{AcOCH} \\ | \\ \text{HCOAc} \\ | \\ \text{HCOAc} \\ | \\ \text{H}_2\text{COAc} \end{array} ; R^2 = C_6H_5 \right) \tag{138}$$

6. Six-membered heterocyclic compounds

All four reaction types (**229–232**) can be used to prepare six-membered heterocyclic compounds from hydrazides.

Thus, type **231** occurs on warming chlorodiphenylacetyl chloride with 1-acetyl-2-*p*-chlorophenylhydrazine in toluene, whence the oxadiazinone **301** is produced, presumably via the acyclic diacyl-hydrazine **300** (reaction 139)[327,328]. The attack of the OH group

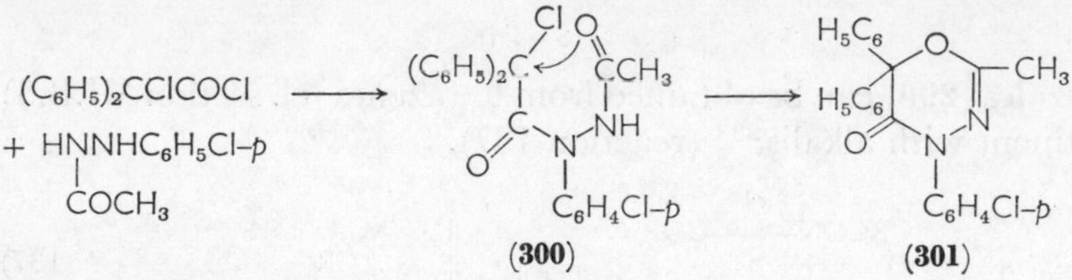

$$(C_6H_5)_2CClCOCl \longrightarrow \quad \longrightarrow \quad (139)$$
$$+ \; HNNHC_6H_5Cl\text{-}p$$
$$\qquad COCH_3$$

(**300**)   (**301**)

in the hydrazide **302** on the carbonyl carbon is an example of type **232** where an oxadiazine system (**303**) is formed (reaction 140)[121].

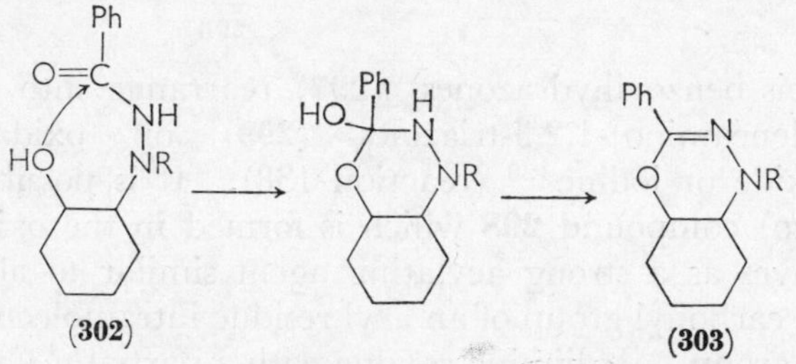

$$\longrightarrow \qquad \longrightarrow \qquad (140)$$

(**302**)   (**303**)

Following a reaction of type **229**, γ-ketocarboxylic acid hydrazides form pyridazinones (**304**) with the elimination of water[329,330] (reaction 141). α-Acetylaminocarboxylic acid hydrazides cyclize in the same way in alkaline solutions, by an attack of the hydrazide group on the acetyl carbonyl group, the product being a 6-hydroxy-1,2,4-tria-zine[73-75] (reaction 142), whereas in acidic solution an unsaturated

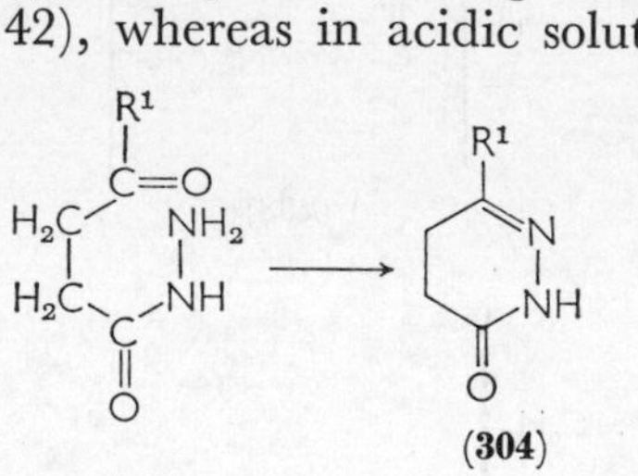

$$\longrightarrow \qquad (141)$$

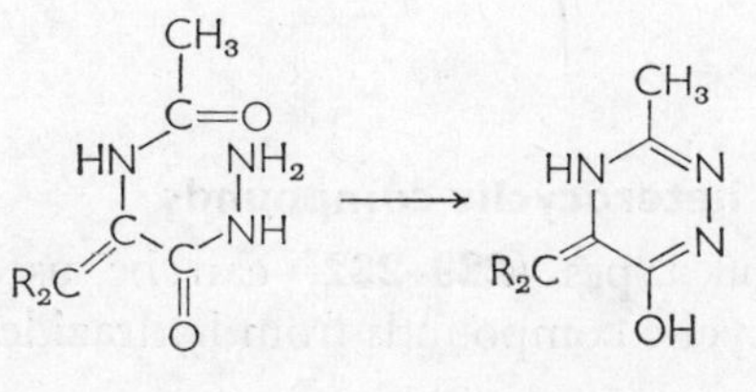

(**304**)

$$\longrightarrow \qquad (142)$$

azlactone like **45** is formed[330a]. The preparation of a pyridazine (**306**) by the action of 10% HCl on a dicarboxylic acid dihydrazide (**305**) also belongs to reactions of this type[331] (reaction 143). However, it is not yet certain whether **306** exists in a semi-enolic form (like **119**).

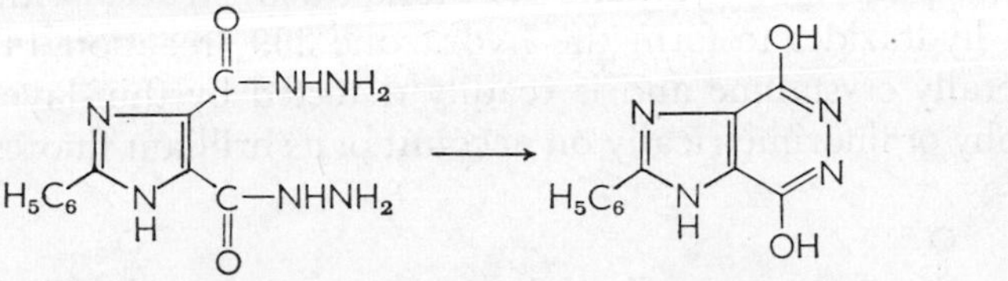

$$(143)$$

(**305**)   (**306**)

Cyanoacetic acid hydrazide, in which the methylene group is rendered acidic by the neighbouring cyano group, condenses with 1,3-diketones to form 2,4-dialkyl-1-amino-5-cyanopyrid-6-one (**307**)[332]. This is an example of the less frequent cyclizations of type **230** (reaction 144).

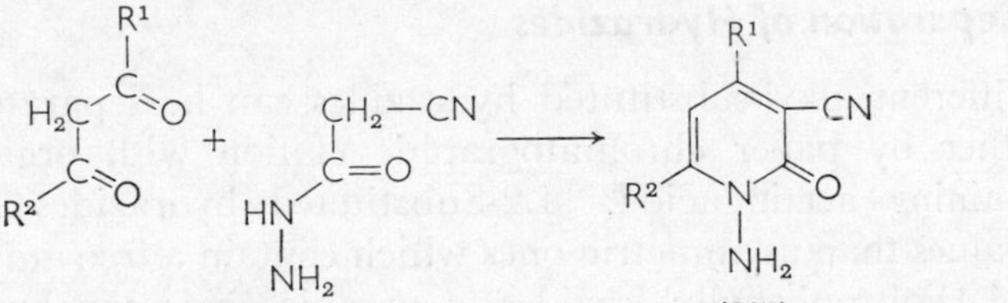

$$(144)$$

(**307**)

$(R^1 = CH_3, C_6H_5; R^2 = CH_3, C_6H_5)$

VII. ANALYSIS OF HYDRAZIDES

A. Qualitative Determination

Hydrazides can be detected by an examination of i.r. spectra for characteristic bands (cf. section III.E), by carrying out the specific Bülow reaction[72,333] (in which a characteristic colour is formed in the presence of ferric chloride and concentrated sulphuric acid), or by utilizing the less specific colour reaction involving picryl chloride and ammonia[334].

Easily hydrolysable hydrazides are detected on paper by spraying with an ethanolic solution of p-dimethylaminobenzaldehyde hydrochloride[7,335], which gives an intense red colour owing to the formation of the azine from free hydrazine and p-dimethylaminobenzaldehyde. Aqueous solutions of the salt of the azine are yellow. A stable colour of maximum intensity which obeys Beer's law is reached in 15 min which can be used for quantitative determinations. Stable hydrazides must first be hydrolysed into the acid and hydrazine before the colour reaction is carried out[336].

2-(Diphenylacetyl)indane-1,3-dione (**308**) reacts with carboxylic acid hydrazides to form the hydrazone **309** (reaction 145), which is generally crystalline and is readily detected by thin-layer chromatography or fluorimetrically on account of its brilliant fluorescence[337].

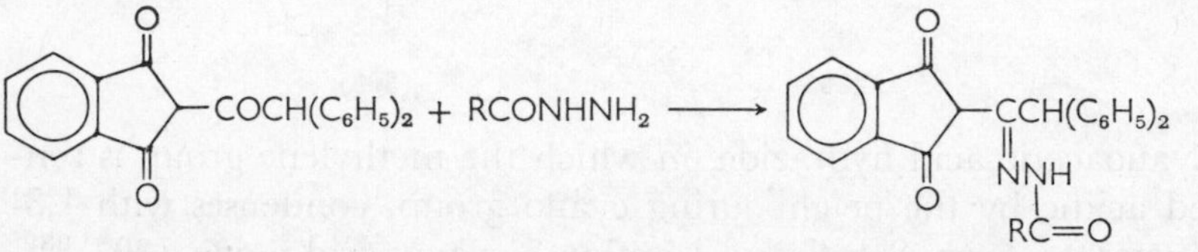

$$\text{(308)} \qquad \text{COCH(C}_6\text{H}_5)_2 + \text{RCONHNH}_2 \longrightarrow \qquad \text{CCH(C}_6\text{H}_5)_2 \quad (145)$$

(**308**)

(**309**)

B. Separation of Hydrazides

Different alkyl-substituted hydrazides can be separated from one another by paper chromatographic elution with organic solvents containing acetic acid[7]. 1,2-Substituted hydrazides have higher R_f values than asymmetric ones which contain a free amino group.

2,4-Dinitrophenylhydrazides of organic acids can be resolved by thin-layer chromatography on silica gel, alumina, and polyamide plates. The best results are obtained on polyamide layers, particularly in the case of carboxylic acid hydrazides with 6–18 carbon atoms[338]. Hydrazides have also been separated electrophoretically on paper after buffering the latter with an acid, so that the hydrazide cation could be formed[339].

C. Quantitative Determination

Hydrazides have been determined semi-quantitatively by measuring the size of the spots obtained after chromatography on thin layers[7,340]. On the basis of the colour reactions with heavy metal salts (complexing) or with *p*-dimethylaminobenzaldehyde, hydrazides can be easily determined colorimetrically and spectrophotometrically[336,341,342]. The polarographic determination of hydrazides has been employed in kinetic investigations[26,343]. Adipic acid dihydrazide and oxalic acid dihydrazide have been titrated potentiometrically with sodium nitrite[344] and with potassium iodate[345].

VIII. PHYSIOLOGICAL ACTIVITY

Since the discovery that isonicotinic acid hydrazide has a strong antituberculotic action[206], many derivatives of this compound have been synthesized and tested for antibacterial properties[50,171,206,209,346–349]. It is not yet known with certainty how isonicotinic acid hydrazide derivatives exert this effect. The particularly high activity of certain derivatives is assumed to be due to the diacylhydrazine group as the biologically active centre[350].

Carboxylic acid 1,2-diarylhydrazides have been reported to possess anti-inflammatory properties[351,352], and a diuretic action has been ascribed to benzoic acid hydrazide derivatives[302]. Isoxazolecarboxylic acid hydrazides[301] are active against leprosy, and an anticonvulsive action has been reported for phenothiazinecarboxylic acid hydrazides[353]. Hydrazones of hydrazides frequently act as monoamine oxidase inhibitors[207,208,216], which find application as psychopharmaceutical preparations. Dihydrazides have recently been introduced as anthelmintics[354]. A fungicidal action is ascribed to pentachlorobenzoic acid hydrazide[355]. Maleic acid hydrazide is used to regulate and inhibit the growth of plants[336].

IX. MISCELLANEOUS APPLICATIONS

Hydrazides are used for the heat and corrosion stabilization of cellulose and cellulose derivatives[356], and as antioxidants for polyolefins and polyurethanes, which are otherwise oxidized in the presence of copper. The incorporation of hydrazides has improved the applicability of these plastics as cable insulation[357–359]. Small amounts of hydrazides sensitize electrophotographic layers made of poly(vinylcarbazole)[360]. Dihydrazides can be used in cigarette filters for the selective removal of aldehydes from tobacco smoke[361]. Dihydrazides can be reacted with other bifunctional compounds to give polyhydrazides, which can be spun from dimethyl sulphoxide solutions[53,54,177]. Ion-exchange resins for the separation of Cu, Ni, Co, Mg, and transition metal ions have been prepared from copolymers of 2-methyl-5-vinylpyridine and hydrazides of 1,2-ethylenedicarboxylic acids[362].

X. REFERENCES

1. E. Müller, Ed., *Houben-Weyl's Methoden der Organischen Chemie*, Vol. VIII, Georg Thieme Verlag, Stuttgart, 1952, pp. 676–80.
2. E. Müller, Ed., *Houben-Weyl's Methoden der Organischen Chemie*, Vol. X/2, Georg Thieme Verlag, Stuttgart, 1967, pp. 123, 169.

3. P. A. S. Smith, *The Chemistry of Open-chain Organic Nitrogen Compounds*, Vol. II, W. A. Benjamin, Inc., New York, 1966, p. 173.

4a. E. H. Rodd, *Chemistry of Carbon Compounds*, Vol. I/A, Elsevier Publishing Co., Amsterdam, 1951, p. 600

4b. E. H. Rodd, *Chemistry of Carbon Compounds*, Vol. III/A, Elsevier Publishing Co., Amsterdam, 1954, p. 365.

5. L. H. Jensen, *J. Am. Chem. Soc.*, **78**, 3993 (1956).

6. J. A. Young, W. S. Durrell, and R. D. Dresdner, *J. Am. Chem. Soc.*, **84**, 2105 (1962).

7. R. L. Hinman and D. Fulton, *J. Am. Chem. Soc.*, **80**, 1895 (1958).

8. R. L. Hinman and M. C. Flores, *J. Org. Chem.*, **24**, 660 (1959).

9. R. L. Hinman and R. J. Landborg, *J. Org. Chem.*, **24**, 724 (1959).

10. L. H. Jensen, *J. Am. Chem. Soc.*, **76**, 4663 (1954).

11. L. H. Jensen and E. C. Lingafelter, *Acta Cryst.*, **6**, 300 (1953).

12. Y. Tomiie, C. H. Koo and I. Nitta, *Acta Cryst.*, **11**, 774 (1958).

13. R. Shintani, *Acta Cryst.*, **13**, 609 (1960).

14. G. S. D. King, *J. Chem. Soc.*, (*B*), 1224 (1966).

15a. B. H. Korsch and N. V. Riggs, *Tetrahedron Letters*, 5897 (1966).

15b. A. Foucaud and R. Roudaut, *Compt. Rend.*, **266**, 726 (1968).

16. G. J. Bishop, B. J. Price and I. O. Sutherland, *Chem. Commun.*, 672 (1967).

17a. J. E. Anderson and J. M. Lehn, *Tetrahedron*, **24**, 123 (1968).

17b. J. E. Anderson and J. M. Lehn, *Tetrahedron*, **24**, 137 (1968).

18. H. Lumbroso and J. Barassin, *Bull. Soc. Chim. France*, 3190 (1964).

19. H. E. Baumgarten, P. L. Creger, and R. L. Zey, *J. Am. Chem. Soc.*, **82**, 3977 (1960).

20. J. B. Jensen, *Acta Chem. Scand.*, **10**, 667 (1956).

21. D. M. Wiles and T. Suprunchuk, *Can. J. Chem.*, **46**, 701 (1968).

22. B. A. Zadorozhnyi, I. K. Ishchenko, T. R. Mnatsakanova, and O. P. Shvaika, *Zh. Organ. Khim.*, **2**, 432 (1966).

23. R. L. Hinman, *J. Am. Chem. Soc.*, **78**, 1645 (1956).

24. W. J. McKillip and R. C. Slagel, *Can. J. Chem.*, **45**, 2619 (1967).

25. E. P. Nesynov and P. S. Pel'kis, *Zh. Obshch. Khim.*, **37**, 1051 (1967); *Chem. Abstr.*, **68**, 25257f (1968).

26. I. S. Berdinskii, G. S. Posyagin, and V. F. Ust-Kachkintsev, *Uch. Zap. Permsk. Gos. Univ.*, **141**, 320 (1966); *Chem. Abstr.*, **68**, 110828n (1968).

27. B. Epstein and T. Kuwana, *J. Electroanal. Chem.*, **15**, 389 (1967).

28. C. S. Hudson, *J. Am. Chem. Soc.*, **39**, 462 (1917).

29. K.-D. Gundermann, *Angew. Chem.*, **77**, 572 (1965); *Angew. Chem. Intern. Ed. Engl.*, **4**, 566 (1965).

30. E. H. White, M. M. Bursey, D. F. Roswell, and J. H. M. Hill, *J. Org. Chem.*, **32**, 1198 (1967).

31. E. H. White and K. Matsuo, *J. Org. Chem.*, **32**, 1921 (1967).

32. E. J. Gasson (Distillers Co. Ltd.), *Brit. Pat.*, 787,282 (1957); *Chem. Abstr.*, **52**, 6729h (1958).

33. S. P. Dutta and A. K. Acharyya, *Indian J. Appl. Chem.*, **29**, 99 (1966); *Chem. Abstr.*, **68**, 114389z (1968).

34. K. K. Moll, H. Seefluth, L. Brüsehaber, and G. Schrattenholz, *Pharmazie*, **23**, 36 (1968).

35. F. L. Scott and J. B. Aylward, *Tetrahedron Letters*, 841 (1965).

36. I. T. Barnish and M. S. Gibson, *J. Chem. Soc.*, 2999 (1965).
37. A. F. Hegarty and F. L. Scott, *J. Org. Chem.*, **33**, 753 (1968).
38. F. L. Scott, F. A. Groeger, and A. F. Hegarty, *Tetrahedron Letters*, 2463 (1968).
39. A. V. Fokin, Y. N. Studnev, and N. A. Prošhin, *Zh. Obshch. Khim.*, **37**, 1725 (1967).
40. P. A. S. Smith, *Org. Reactions*, **3**, 366 (1946).
41. H. Paulsen and D. Stoye, *Chem. Ber.*, **99**, 908 (1966).
42. S. Akabori, K. Ohno, and K. Narita, *Bull. Chem. Soc. Japan*, **25**, 214 (1952).
43. C. I. Niu and H. Fraenkel-Conrat, *J. Am. Chem. Soc.*, **77**, 5882 (1955).
44. H. A. Staab, M. Lücking, and F. H. Dürr, *Chem. Ber.*, **95**, 1275 (1962).
45. Th. Kauffmann and J. Sobel, *Angew. Chem.*, **75**, 1177 (1963).
46. Th. Kauffmann, *Angew. Chem.*, **76**, 206 (1964); *Angew. Chem. Intern. Ed. Engl.*, **3**, 342 (1964).
47. O. L. Salerni, B. E. Smart, A. Post, and C. C. Cheng, *J. Chem. Soc. (C)*, 1399 (1968).
48. E. C. Kornfeld, E. J. Fornefeld, G. B. Kline, M. J. Mann, D. E. Morrison, R. G. Jones, and R. B. Woodward, *J. Am. Chem. Soc.*, **78**, 3087 (1956).
49. R. L. Hinman, *J. Am. Chem. Soc.*, **78**, 2463 (1956).
50. Th. Rinderspacher and B. Prijs, *Helv. Chim. Acta*, **41**, 22 (1958).
51. E. J. Browne and J. B. Polya, *J. Chem. Soc.*, 5149 (1962).
52. E. D. Nicolaides, *J. Org. Chem.*, **32**, 1251 (1967).
53. D. F. Loncrini, (Compagnie Francaise Thomson-Houston), *Fr. Pat.*, 1,412,896 (1965); *Chem. Abstr.*, **65**, 5556e (1966).
54. M. Hasegawa and H. Takahashi, *J. Polymer Sci. B*, **4**, 369 (1966).
55. W. J. Theuer and J. A. Moore, *J. Org. Chem.*, **29**, 3734 (1964).
56. H. Dorn, A. Zubek, and K. Walter, *Ann. Chem.*, **707**, 100 (1967).
57. H. Dorn, A. Zubek, and K. Walter, *Z. Chem.*, **7**, 150 (1967).
58. K. A. Jensen, H. R. Baccaro, O. Buchardt, G. E. Olsen, C. Pedersen, and J. Toft, *Acta Chem. Scand.*, **15**, 1109 (1961).
59. W. P. Jencks and J. Carriuolo, *J. Am. Chem. Soc.*, **82**, 1778 (1960).
60. P. Bouchet, J. Elguero, and R. Jacquier, *Bull. Soc. Chim. France*, 3502 (1967).
61. H. Gross, *Chem. Ber.*, **95**, 2270 (1962).
62. J. Gloede, K. Poduška, H. Gross, and J. Rudinger, *Collection Czech. Chem. Commun.*, **33**, 1307 (1968).
63. P. Grudzinska, *Roczniki Chem.*, **34**, 1687 (1960); *Chem. Abstr.*, **56**, 5863c (1962).
64. J. Strumillo and A. Kotelko, *Acta Polon. Pharm.*, **24**, 657 (1967); *Chem. Abstr.*, **68**, 68621n (1968).
65. T. C. Bruice and S. J. Benkovic, *J. Am. Chem. Soc.*, **86**, 418 (1964).
66. J. O. Edwards and R. G. Pearson, *J. Am. Chem. Soc.*, **84**, 16 (1962).
67. T. C. Bruice and R. G. Willis, *J. Am. Chem. Soc.*, **87**, 531 (1965).
68. F. L. Lukintskii and B. A. Vovsi, *Zh. Organ. Khim.*, **3**, 794 (1967).
69. F. L. Lukintskii and B. A. Vovsi (Leningrad Chemical-Pharmaceutical Institute), *U.S.S.R. Pat.*, 190,883 (1967); *Chem. Abstr.*, **68**, 49124m (1968).
70. H. Allgeier, *Helv. Chim. Acta*, **51**, 668 (1968).
71. H.-H. Stroh and D. Henning, *Chem. Ber.*, **100**, 388 (1967).
72. H.-H. Stroh and D. Henning, *Z. Chem.*, **6**, 378 (1966).

73. K. Nálepa and J. Slouka, *Monatsh. Chem.*, **98**, 412 (1967).

74. K. Nálepa, *Monatsh. Chem.*, **98**, 1230 (1967).

75. A. Mustafa, W. Asker, A. H. Harhash, M. A. E. Khalifa, and E. M. Zayed, *Ann. Chem.*, **713**, 151 (1968).

76. A. T. Prudchenko, S. Λ. Vereshchagina, V. A. Barkhash, and N. N. Vorozhtsov, Jr., *Zh. Obshch. Khim.*, **37**, 2195 (1967); *Chem. Abstr.*, **68**, 78211t (1968).

77. W. J. McKillip, L. M. Clemens, and R. Haugland, *Can. J. Chem.*, **45**, 2613 (1967).

78. S. G. Cohen and J. Nicholson, *J. Org. Chem.*, **30**, 1162 (1965).

79. L. M. Grivnak and B. G. Boldyrev, *Zh. Organ. Khim.*, **3**, 2160 (1967).

80. P. Hope and L. A. Wiles, *J. Chem. Soc.*, (*C*), 2636 (1967).

81. J. Dick and A. Lupea, *Pharmazie*, **22**, 555 (1967).

82. A. S. Endler and E. J. Becker, *Org. Syn.*, **37**, 60 (1957).

83. C. Ainsworth, *Can. J. Chem.*, **43**, 1607 (1965).

84. J. C. Stowell, *J. Org. Chem.*, **32**, 2360 (1967).

85. T. Curtius and H. A. Foersterling, *J. Prakt. Chem.*, **51**, [2], 371 (1895).

86. D. M. Miller and R. W. White, *Can. J. Chem.*, **34**, 1510 (1956).

87. R. M. Gel'shtein and F. V. Aleksandrova, *Metody Poluch. Khim. Reaktivov Prep.*, **15**, 40 (1967); *Chem. Abstr.*, **68**, 105137g (1968).

88. K. Belniak, E. Domagalina and H. Hopkala, *Roczniki Chem.*, **41**, 831 (1967); *Chem. Abstr.*, **68**, 2870m (1968).

89. E. Domagalina and I. Kurpiel, *Roczniki Chem.*, **40**, 1869 (1966); *Chem. Abstr.*, **67**, 32504s (1967).

90. A. Michaelis and F. Schmidt, *Chem. Ber.*, **20**, 43, 1713 (1887).

91a. A. Michaelis and F. Schmidt, *Ann. Chem.*, **252**, 300 (1889).

91b. O. Widman, *Chem. Ber.*, **27**, 2965 (1894).

92. A. Michaelis and E. Hadanck, *Chem. Ber.*, **41**, 3285 (1908).

93. G. Lockemann, *Chem. Ber.*, **43**, 2223 (1910).

94. J. van Alphen, *Rec. Trav. Chim.*, **43**, 825 (1924).

95. R. Buyle, *Helv. Chim. Acta*, **47**, 2449 (1964).

96. G. A. Reynolds and J. A. Van Allan, *J. Org. Chem.*, **24**, 1478 (1959).

97. R. G. Glushkov and O. Y. Magidson, *Zh. Obshch. Khim.*, **30**, 649 (1960); *Chem. Abstr.*, **55**, 24712a (1961).

98. K. Kratzl and K. P. Berger, *Monatsh. Chem.*, **89**, 83 (1958).

99. O. Frehden and G. Lazarescu, *Rev. Chim. (Bucharest)*, **13**, 91 (1962); *Chem. Abstr.*, **57**, 11155i (1962).

100. R. F. Smith, A. C. Bates, A. J. Battisti, P. G. Byrnes, C. T. Mroz, T. J. Smearing, and F. X. Albrecht, *J. Org. Chem.*, **33**, 851 (1968).

101. P. Schestakov, *Chem. Ber.*, **45**, 3273 (1912).

102. W. F. Short, *J. Chem. Soc.*, **120**, 1445 (1921).

103. T. Sheradsky, *Tetrahedron Letters*, 1909 (1968).

104. W. E. Hanford, *U.S. Pat.*, 2,717,200 (1955); *Chem. Zentr.*, 7336 (1956).

105. R. Stollé and W. Reichert, *J. Prakt. Chem.*, **122**, [2], 344 (1929).

106. L. A. Carpino, P. H. Terry and P. J. Crowley, *J. Org. Chem.*, **26**, 4336 (1961).

107. H.-J. Teuber and K.-H. Dietz, *Angew. Chem.*, **78**, 1101 (1966); *Angew. Chem. Intern. Ed. Engl.*, **5**, 1049 (1966).

108. B. T. Gillis and K. F. Schimmel, *J. Org. Chem.*, **27**, 413 (1962).

109. B. Eistert and H. Munder, *Chem. Ber.*, **88**, 226 (1955).
110. B. Eistert and H. Munder, *Chem. Ber.*, **91**, 1415 (1958).
111. R. N. McDonald and A. C. Kovelesky, *J. Org. Chem.*, **28**, 1433 (1963).
112. E. Schmitz, R. Ohme and S. Schramm, *Z. Chem.*, **3**, 190 (1963).
113. E. Schmitz, R. Ohme and S. Schramm, *Tetrahedron Letters*, 1857 (1965).
114. E. Schmitz, S. Schramm and R. Ohme, *J. Prakt. Chem.*, **36**, [4], 86 (1967).
115. E. Schmitz, *Organische Chemie in Einzeldarstellungen*, Vol. 9, Springer Verlag, Berlin, Heidelberg, New York, 1967, p. 67.
116. M. D. Hinchliffe, J. Miller, B. J. Needham, and M. A. Smith (Whiffen & Sons, Ltd.), *Brit. Pat.*, 1,066,054 (1967).
117. R. F. Meyer, *J. Heterocyclic Chem.*, **2**, 305 (1965).
118. G. V. Boyd, *Chem. Commun.*, 954 (1967).
119. G. V. Boyd and A. J. H. Summers, *Chem. Commun.*, 549 (1968).
120. C. N. Yiannios, A. C. Hazy, and J. V. Karabinos, *J. Org. Chem.*, **33**, 2076 (1968).
121. T. Taguchi, J. Ishibashi, T. Matsuo, and M. Kojima, *J. Org. Chem.*, **29**, 1097 (1964).
122. S. Baloniak, *Roczniki Chem.*, **41**, 1143 (1967).
123. E. Domagalina and I. Kurpiel, *Roczniki Chem.*, **41**, 1241 (1967); *Chem. Abstr.*, **68**, 39574u (1968).
124. I. S. Berdinskii and S. V. Kalugina, *Zh. Organ. Khim.*, **3**, 118 (1967).
125. I. S. Berdinskii and G. N. Kazanceva, *Zh. Organ. Khim.*, **3**, 121 (1967).
126. I. S. Berdinskii and V. A. Mal'ceva, *Zh. Organ. Khim.*, **3**, 539 (1967).
127. G. J. Papariello and S. Commanday, *Anal. Chem.*, **36**, 1028 (1964).
128. H. El Khadem, M. A. M. Nassr, and M. A. E. Shaban, *J. Chem. Soc.*, (*C*), 1465 (1968).
129a. F. Arndt, *Angew. Chem.*, **61**, 397 (1949).
129b. F. Arndt, L. Loewe, and L. Ergener, *Istanbul Univ. Fen. Fac. Mecmuasi*, **13A**, 103 (1948); *Chem. Abstr.*, **43**, 579 (1949).
130. K. Eichenberger, A. Staehelin, and J. Druey, *Helv. Chim. Acta*, **37**, 837 (1954).
131a. K. Eichenberger, R. Rometsch and J. Druey, *Helv. Chim. Acta.*, **37**, 1298 (1954).
131b. G. Adembri, F. DeSio, R. Nesi, and M. Scotton, *J. Chem. Soc.* (*C*), 2857 (1968).
132. C. R. Lindegren and C. Niemann, *J. Am. Chem. Soc.*, **71**, 1504 (1949).
133. H.-H. Stroh and H. Tengler, *Chem. Ber.*, **101**, 751 (1968).
134. A. Benrath, *J. Prakt. Chem.*, **107**, [2], 211 (1924).
135. O. Makitie, *Suomen Kemistilehti*, (*B*), **39**, 282 (1966); *Chem. Abstr.*, **66**, 99067z (1967).
136. I. Gercu, *Omagiu Raluca Ripan*, 263 (1966); *Chem. Abstr.*, **67**, 49969v (1967).
137. R. M. Issa, M. F. El-Shazly, and M. F. Iskander, *Z. Anorg. Allgem. Chem.*, **354**, 90 (1967).
138. R. M. Issa, M. F. Iskander, and M. F. El-Shazly, *Z. Anorg. Allgem. Chem.*, **354**, 98 (1967).
139. A. Badinand and J. J. Vallon, *Chim. Anal. (Paris)*, **48**, 313, 396 (1966).
140. G. R. Supp, *Anal. Chem.*, **40**, 981 (1968).
141. R. C. Paul and S. L. Chadha, *Spectrochim. Acta*, **23A**, 1249 (1967).

142. T. Kauffmann, L. Bán, and D. Kuhlmann, *Angew. Chem.*, **79**, 243 (1967); *Angew. Chem. Intern. Ed. Engl.*, **6**, 256 (1967).

143. A. F. Grapov, N. N. Mel'nikov, S. L. Portnova, and L. V. Razvodovskaya, *Zh. Obshch. Khim.*, **36**, 2222 (1966); *Chem. Abstr.*, **66**, 76085q (1967).

144. H. Stetter and H. Spangenberger, *Chem. Ber.*, **91**, 1982 (1958).

145. A. Ebnöther, E. Jucker, A. Lindenmann, E. Rissi, R. Steiner, R. Süess, and A. Vogel, *Helv. Chim. Acta*, **42**, 533 (1959).

146. H. Kloes and H. A. Offe (Farbenfabriken Bayer AG), *Ger. Pat.*, 1,095,841 (1959); *Chem. Abstr.*, **57**, 11019h (1962).

147. H. Röhnert, *Z. Chem.*, **5**, 302 (1965).

148. Y. Nitta and K. Okui, *Japan Pat.*, 13972 (1962); *Chem. Abstr.*, **59**, 9802f (1963).

149. C. Simon, (J. R. Geigy A.G.), *Swiss Pat.*, 307,629 (1955); *Chem. Abstr.*, **51**, 5113f (1957).

150. T. Curtius and G. Struve, *J. Prakt. Chem.*, **50**, [2], 295 (1894).

151. E. Schenker, *Angew. Chem.*, **73**, 81 (1961).

152. N. G. Gaylord, *Reduction with Complex Metal Hydrides*, Interscience Publishers, New York, London, 1956, p. 544.

153. A. Hajós, *Komplexe Hydride*, VEB Deutscher Verlag der Wissenschaften, Berlin, 1966.

154. H. C. Brown and B. C. Subba Rao, *J. Am. Chem. Soc.*, **78**, 2582 (1956).

155. H. Bredereck, B. Föhlisch, and K. Walz, *Angew. Chem.*, **74**, 388 (1962).

156. E. Fahr and H. Lind, *Angew. Chem.*, **78**, 376 (1966); *Angew. Chem. Intern. Ed. Engl.*, **5**, 372 (1966).

157. R. Stollé, *J. Prakt. Chem.*, **75**, [2], 416 (1907).

158. M. Robba, B. Roques, and J. Le Guen, *Bull. Soc. Chim. France*, 4220 (1967).

159. J. Zschiedrich and H. Füller, *Ger. (East) Pat.*, 53,369 (1967).

160. J. Tafel, *Chem. Ber.*, **18**, 1739 (1885).

161. E. Fischer, *Ann. Chem.*, **190**, 125 (1878).

162. I. S. Berdinskii, E. Y. Posyagin, G. S. Posyagin, and V. F. Ust-Kachkintsev, *Zh. Organ. Khim.*, **4**, 91 (1968); *Chem. Abstr.*, **68**, 87000a (1968).

163. S. Wawzonek and R. C. Gueldner, *J. Org. Chem.*, **30**, 3031 (1965).

164. C. Harries and T. Haga, *Chem. Ber.*, **31**, 56 (1898).

165. E. Bellasio, A. Ripamonti, and E. Testa, *Gazz. Chim. Ital.*, **98**, 3 (1968).

166. D. Libermann, F. Grumbach, and N. Rist, *Compt. Rend.*, **237**, 338 (1953).

167. D. Libermann (Chimie et Atomistique), *Fr. Pat.*, 1,159,157 (1958); *Chem. Abstr.*, **54**, 18326g (1960).

168. K. Ronco, B. Prijs, and H. Erlenmeyer, *Helv. Chim. Acta*, **39**, 1253 (1956).

169. S. Grudzinski, *Acta Polon. Pharm.*, **24**, 279 (1967); *Chem. Abstr.*, **68**, 49272h (1968).

170. H. C. Brown, M. T. Cheng, L. J. Parcell, and D. Pilipovich, *J. Org. Chem.*, **26**, 4407 (1961).

171. A. Winterstein, B. Hegedüs, B. Fust, E. Böhni, and A. Studer, *Helv. Chim. Acta*, **39**, 229 (1956).

172. Chas. Pfizer & Co., Inc., *Brit. Pat.*, 1,065,938 (1967); *Chem. Abstr.*, **68**, 39478r (1968).

173. E. Klingsberg, *J. Am. Chem. Soc.*, **80**, 5786 (1958).

174. I. S. Berdinskii, *Zh. Organ. Khim.*, **3**, 1300 (1967).

175. I. S. Berdinskii, G. F. Piskunova, G. S. Posyagin, E. S. Ponosova, and I. M. Shevaldina, *Zh. Organ. Khim.*, **3**, 1645 (1967); *Chem. Abstr.*, **68**, 39312g (1968).

176. A. L. Rusanov, V. V. Korshak, E. S. Krongauz, and I. B. Nemirovskaya, *Vysokomolekul. Soedin*, **8**, 804 (1966); *Chem. Abstr.*, **65**, 9035e (1966).

177. A. H. Frazer (E. I. du Pont de Nemours & Co.), *U.S. Pat.*, 3,238,183 (1966); *Chem. Abstr.*, **65**, 12355f (1966).

177a. A. J. Yakubovič, *Vysokomolekul. Soedin.*, **10A**, 2172 (1968).

178. A. Le Berre and J. Godin, *Compt. Rend.*, **260**, 5296 (1965).

179. A. P. Grekov and M. S. Marakhova, *Stsintillyatory i Stsintillyats. Materialy (Kharkov)*, 24 (1963); *Chem. Abstr.*, **63**, 8144g (1965).

180. G. S. Posyagin, V. F. Ust-Kachkintsev, and I. S. Berdinskii, *Uch. Zap. Perm. Gos. Univ.*, **141**, 327 (1966); *Chem. Abstr.*, **68**, 110829p (1968).

181. A. P. Grekov and V. K. Skripchenko, *Zh. Organ. Khim.*, **3**, 1251 (1967).

182. A. P. Grekov and V. K. Skripchenko, *Zh. Organ. Khim.*, **3**, 1287 (1967).

183. A. P. Grekov and V. K. Skripchenko, *Zh. Organ. Khim.*, **3**, 1844 (1967); *Chem. Abstr.*, **68**, 38687w (1968).

184. A. P. Grekov and V. K. Skripchenko, *Zh. Organ. Khim.*, **4**, 243 (1968); *Chem. Abstr.*, **68**, 95084w (1968).

185. M. S. Newman and E. G. Caflisch, Jr., *J. Am. Chem. Soc.*, **80**, 862 (1958).

186. M. S. Newman and I. Ungar, *J. Org. Chem.*, **27**, 1238 (1962).

187. M. Sprecher, M. Feldkimel, and M. Wilchek, *J. Org. Chem.*, **26**, 3664 (1961).

188. H.-H. Stroh and G. Jähnchen, *Z. Chem.*, **8**, 24 (1968).

189. P. Hope and L. A. Wiles, *J. Chem. Soc.*, 5386 (1965).

190. D. Klamann, U. Krämer, and P. Weyerstahl, *Chem. Ber.*, **95**, 2694 (1962).

191. F. L. Scott and J. A. Barry, *Tetrahedron Letters*, 513 (1968).

192. G. I. Matyushecheva, A. V. Narbut, G. I. Derkach, and L. M. Yagupol'-skii, *Zh. Organ. Khim.*, **3**, 2254 (1967); *Chem. Abstr.*, **68**, 59236p (1968).

193. M. Semonský and N. Kucharczyk, *Collection Czech. Chem. Commun.*, **33**, 577 (1968).

194. H. Zahn and H. Determann, *Chem. Ber.*, **90**, 2176 (1957).

195. F. Weygand and W. Steglich, *Chem. Ber.*, **92**, 313 (1959).

196. B. F. Erlanger, W. V. Curran, and N. Kokowsky, *J. Am. Chem. Soc.*, **81**, 3055 (1959).

197. J. Honzl and J. Rudinger, *Collection Czech. Chem. Commun.*, **26**, 2333 (1961).

198. E. A. Morozova, E. S. Oksenojt, and I. A. Grava, *Zh. Obshch. Khim.*, **37**, 1764 (1967).

199. K. L. Agarwal, G. W. Kenner, and R. C. Sheppard, *J. Chem. Soc.*, (*C*), 1384 (1968).

200. T. Wieland and B. Heinke, *Ann. Chem.*, **615**, 184 (1958).

201. G. Ehrhardt, W. Siedel, and H. Nahm, *Chem. Ber.*, **90**, 2088 (1957).

202. W. R. Vaughan and J. L. Spencer, *J. Org. Chem.*, **25**, 1160 (1960).

203. L. Bernardi and O. Goffredo, *Gazz. Chim. Ital.*, **94**, 947 (1964).

204. P. A. S. Smith and H. G. Pars, *J. Org. Chem.*, **24**, 1325 (1959).

205. D. Evans and T. F. Grey, *J. Chem. Soc.*, 3006 (1965).

206. W. Wenner, *J. Org. Chem.*, **18**, 1333 (1953).

207. A. Alemany, M. Bernabé, E. F. Alvarez, M. Lora-Tamayo, and O. Nieto, *Bull. Soc. Chim. France*, 780 (1967).

596   Hans Paulsen and Dieter Stoye

208. D. Libermann and J.-C. Denis, *Bull. Soc. Chim. France*, 1952 (1961).
209. A. L. Mndzhoyan, *Arm. Khim. Zh.*, **19**, 793 (1966); *Chem. Abstr.*, **67**, 53943z (1967).
210. N. J. Leonard and J. H. Boyer, *J. Org. Chem.*, **15**, 42 (1950).
211. A. Girard and G. Sandulesco, *Helv. Chim. Acta*, **19**, 1095 (1936).
212. H. H. Inhoffen, F. Bohlmann, K. Bartram, G. Rummert, and H. Pommer, *Ann. Chem.*, **570**, 58 (1950).
213. R. B. Woodward, T. P. Kohman, and G. C. Harris, *J. Am. Chem. Soc.*, **63**, 120 (1941).
214. J. C. Howard, G. Gever, and P. H.-L. Wei, *J. Org. Chem.*, **28**, 868 (1963).
215. C. v. Plessing, *Arch. Pharm.*, **297**, 240 (1964).
216. M. H. Weinswig and E. B. Roche, *J. Pharm. Sci.*, **54**, 1216 (1965); *Chem. Abstr.*, **63**, 13171d (1965).
217. M. Freifelder, W. B. Martin, G. R. Stone, and E. L. Coffin, *J. Org. Chem.*, **26**, 383 (1961).
218. G. F. Bettinetti, *Farmaco, (Pavia) Ed. Sci.*, **16**, 823 (1961); *Chem. Abstr.*, **57**, 7165a (1962).
219. P. A. S. Smith, J. M. Clegg, and J. Lakritz, *J. Org. Chem.*, **23**, 1595 (1958).
220. H. Paulsen and D. Stoye, *Angew. Chem.*, **80**, 120 (1968); *Angew. Chem. Intern. Ed. Engl.*, **7**, 134 (1968).
221. H. Paulsen and D. Stoye, *Chem. Ber.*, **102**, 834 (1969).
222. H. Gehlen and R. Zeiger, *J. Prakt. Chem.*, **37**, [4], 269 (1968).
223. N. P. Buu-Hoi, N. D. Xuong, and E. Lescot, *Bull. Soc. Chim. France*, 441 (1957).
224. H.-H. Stroh and H. Beitz, *Ann. Chem.*, **700**, 78 (1966).
225. S. Ishida and R. Moriya (Asahi Chemical Industry Co., Ltd.), *Japan. Pat.*, 17,518 (1967); *Chem. Abstr.*, **68**, 3486c (1968).
226. A. P. Grekov and V. K. Skripchenko, *Zh. Organ. Khim.*, **3**, 1294 (1967).
227. H. Gehlen and G. Dase, *Ann. Chem.*, **646**, 78 (1961).
228. E. F. Godefroi and E. L. Wittle, *J. Org. Chem.*, **21**, 1163 (1956).
229. G. J. Durant, *J. Chem. Soc. (C)*, 92 (1967).
230. W. R. Sherman, *J. Org. Chem.*, **26**, 88 (1961).
231. F. Micheel and W. Brunkhorst, *Chem. Ber.*, **88**, 481 (1955).
232. O. Dimroth and G. De Montmollin, *Chem. Ber.*, **43**, 2904 (1910).
233. J. P. Horwitz and V. A. Grakauskas, *J. Am. Chem. Soc.*, **79**, 1249 (1957).
234. E. J. Browne and J. B. Polya, *J. Chem. Soc. (C)*, 824 (1968).
235. H. Eilingsfeld, *Chem. Ber.*, **98**, 1308 (1965).
236. S. Petersen and E. Tietze, *Chem. Ber.*, **90**, 909 (1957).
237. A. Le Berre and B. Dumaitre, *Compt. Rend.*, *Ser. C*, **265**, 642 (1967).
238. C. Ainsworth, *J. Am. Chem. Soc.*, **76**, 5774 (1954).
239. C. Ainsworth, *J. Am. Chem. Soc.*, **78**, 1636 (1956).
240. R. L. Hinman, *J. Org. Chem.*, **22**, 148 (1957).
241. F. P. Robinson and R. K. Brown, *Can. J. Chem.*, **39**, 1171 (1961).
242. S. B. Needleman and M. C. Chang Kuo, *Chem. Rev.*, **62**, 405 (1962).
243. G. Ponzio, *Atti Acad. Sci. Torino*, **44**, 295 (1909).
244. M. Giua, *Atti Acad. Sci. Torino*, **63**, 259 (1928).
245. R. B. Kelly, E. G. Daniels, and J. W. Hinman, *J. Org. Chem.*, **27**, 3229 (1962).
246. R. B. Kelly, *J. Org. Chem.*, **28**, 453 (1963).

247. D. Mackay, U. F. Marx, and W. A. Waters, *J. Chem. Soc.*, 4793 (1964).
248. L. Kalb and O. Gross, *Chem. Ber.*, **59**, 727 (1926).
249. J. E. Leffler and W. B. Bond, *J. Am. Chem. Soc.*, **78**, 335 (1956).
250. Y. Wolman, P. M. Gallop, and A. Patchornik, *J. Am. Chem. Soc.*, **83**, 1263 (1961).
251. Y. Wolman, P. M. Gallop, A. Patchornik, and A. Berger, *J. Am. Chem. Soc.*, **84**, 1889 (1962).
252. L. Horner and H. Fernekess, *Chem. Ber.*, **94**, 712 (1961).
253. S. Hünig and G. Kaupp, *Ann. Chem.*, **700**, 65 (1966).
253a. J. B. Aylward and R. O. C. Norman, *J. Chem. Soc.* (*C*), 2399 (1968).
254. H. B. Milne and W. Kilday, *J. Org. Chem.*, **30**, 64 (1965).
255. H. B. Milne and C. F. Most, Jr., *J. Org. Chem.*, **33**, 169 (1968).
256. E. Mosettig, *Org. Reactions*, **8**, 233 (1954).
257. L. A. Carpino, *Chem. Ind.* (*London*), 123 (1956).
258. L. A. Carpino, *J. Am. Chem. Soc.*, **79**, 96 (1957).
259. P. Hope and L. A. Wiles, *J. Chem. Soc.*, 5837 (1964).
260. J. S. McFadyen and T. S. Stevens, *J. Chem. Soc.*, 584 (1936).
261. U. M. Brown, P. H. Carter, and M. Tomlinson, *J. Chem. Soc.*, 1843 (1958).
262. L. Caglioti, *Tetrahedron*, **22**, 487 (1966).
263. T. Kametani and O. Umezawa, *Chem. Pharm. Bull.* (*Tokyo*), **14**, 369 (1966); *Chem. Abstr.*, **65**, 5361d (1966).
264. T. Kametani and O. Umezawa, *Yakugaku Zasshi*, **85**, 514, 518 (1965); *Chem. Abstr.*, **63**, 6851d, 6959h (1965).
265a. C. A. Grob and P. W. Schiess, *Angew. Chem.*, **79**, 1 (1967); *Angew. Chem. Intern. Ed. Engl.*, **6**, 1 (1967).
265b. J. A. Deyrup and S. C. Clough, *J. Am. Chem. Soc.*, **90**, 3592 (1968).
266. L. A. Ovsyannikova, T. A. Sokolova, and N. P. Zapevalova, *Zh. Organ. Khim.*, **4**, 459 (1968).
267. T. A. Sokolova, L. A. Ovsyannikova, and N. P. Zapevalova, *Zh. Organ. Khim.*, **2**, 818 (1966).
267a. B. M. Culbertson, *J. Polymer Sci. A.*, **16**, 2197 (1968).
268. S. Wawzonek and E. Yeakey, *J. Am. Chem. Soc.*, **82**, 5718 (1960).
269. R. Appel, H. Heinen, and R. Schöllhorn, *Chem. Ber.*, **99**, 3118 (1966).
270. W. S. Wadsworth, *J. Org. Chem.*, **31**, 1704 (1966).
271. M. S. Gibson and A. W. Murray, *J. Chem. Soc.*, 880 (1965).
272. R. C. Slagel, *J. Org. Chem.*, **33**, 1374 (1968).
273. R. C. Slagel and A. E. Bloomquist, *Can. J. Chem.*, **45**, 2625 (1967).
274. M. S. Gibson, P. D. Callaghan, R. F. Smith, A. C. Bates, J. R. Davidson, and A. J. Battisti, *J. Chem. Soc.*, (*C*), 2577 (1967).
275. Archer Daniels Midland, Co., *Neth. Appl.*, 6,612,603 (1967); *Chem. Abstr.*, **67**, 53721a (1967).
276. G. Pellizzari, *Atti Reale Acad. Lincei*, **8**, [5], 327 (1899).
277. R. Stollé and F. Helwerth, *J. Prakt. Chem.*, **88**, [2], 315 (1913).
278. A. P. Grekov, L. N. Kulakova, and P. O. Shvǎika, *Zh. Obshch. Khim.*, **29**, 3054 (1959); *Chem. Abstr.*, **54**, 13108g (1959).
279. J. C. Thurman, *Chem. Ind.* (*London*), 752 (1964).
280. E. Baltazzi and A. J. Wysocki, *Chem. Ind.* (*London*), 1080 (1963).
281. F. D. Popp, *J. Chem. Soc.*, 3503 (1964).
282. M. Hasegawa and T. Unishi, *Polymer Letters*, **2**, 237 (1964).

283. F. T. Wallenberger, *Angew. Chem.*, **76**, 484 (1964); *Angew. Chem. Intern. Ed. Engl.*, **3**, 460 (1964).
284. W. J. Chambers and D. D. Coffman, *J. Org. Chem.*, **26**, 4410 (1961).
285. C. Ainsworth, *J. Am. Chem. Soc.*, **77**, 1148 (1955).
286. C. Ainsworth, *J. Am. Chem. Soc.*, **87**, 5800 (1965).
287. M. Ito, *Yakugaku Kenkyu*, **34**, 410 (1962); *Chem. Abstr.*, **58**, 11346h (1963).
288. H. Weidinger and J. Kranz, *Chem. Ber.*, **96**, 1049 (1963).
289. M. Pesson, S. Dupin, and M. Antoine, *Bull. Soc. Chim. France*, 1364 (1962).
290. R. Kraft, H. Paul, and G. Hilgetag, *Chem. Ber.*, **101**, 2028 (1968).
291. K. Möckel and H. Gehlen, *Z. Chem.*, **4**, 388 (1964).
292. R. Neidlein and W. Haussmann, *Z. Naturforsch.*, **21**, 898 (1966).
293. R. Neidlein and W. Haussmann, *Arch. Pharm.*, **300**, 180 (1967).
294. H. Eilingsfeld and L. Möbius, *Chem. Ber.*, **98**, 1293 (1965).
295. H. Gehlen and G. Blankenstein, *Ann. Chem.*, **638**, 136 (1960).
296. K. Futaki and S. Tosa, *Chem. Pharm. Bull.* (*Tokyo*), **8**, 908 (1960).
297. A. P. Swain, *U.S. Pat.*, 2,883,391 (1959); *Chem. Abstr.*, **53**, 16157 (1959).
298. G. Cipens and V. J. Grinstein, *Izv. Akad. Nauk. Latv. SSR, Ser. Khim.*, **2**, 255 (1962); *Chem. Abstr.*, **59**, 12789g (1963).
299. M. Hedayatullah, *Bull. Soc. Chim. France*, 1572 (1968).
299a. E. Grigat and R. Pütter, *Chem. Ber.*, **97**, 3560 (1964).
300. K. Matsuda and L. T. Morin, *J. Org. Chem.*, **26**, 3783 (1961).
301. T. S. Gardner, E. Wenis, and J. Lee, *J. Org. Chem.*, **26**, 1514 (1961).
302. E. Jucker and A. Lindenmann, *Helv. Chim. Acta*, **45**, 2316 (1962).
303. R. F. Meyer and B. L. Cummings, *J. Heterocyclic Chem.*, **1**, 186 (1964).
304. A. J. Lazaris, *Zh. Organ. Khim.*, **3**, 1902 (1967).
305. K. T. Potts, *J. Chem. Soc.*, 3461 (1954).
306. A. Pollak and M. Tišler, *Tetrahedron*, **22**, 2073 (1966).
307. E. Klingsberg, *J. Org. Chem.*, **23**, 1086 (1958).
308. R. Stollé, *J. Prakt. Chem.*, **69**, [2], 155, 483, 488, 498, 505 (1904).
309. H. Gehlen and E. Benatzky, *Ann. Chem.*, **615**, 60 (1958).
310. K. T. Potts, S. K. Roy, and D. P. Jones, *J. Org. Chem.*, **32**, 2245 (1967).
311. K. v. Auwers and K. Dietrich, *J. Prakt. Chem.*, **139**, [2], 65 (1934).
312. W. Ried and B. Schleimer, *Ann. Chem.*, **619**, 43 (1958).
313. W. Ried and F.-J. Königstein, *Ann. Chem.*, **622**, 37 (1959).
314. W. Ried and F.-J. Königstein, *Chem. Ber.*, **92**, 2532 (1959).
315. A. A. Achrem, A. V. Kamenikii, and A. V. Skorova, *Izv. Akad. Nauk. SSR*, 1807 (1967).
316. A. Michaelis, *Ann. Chem.*, **358**, 127 (1908).
317. W. Willert, *Ann. Chem.*, **358**, 159 (1908).
318. G. Barnikow, H. Kunzek, and D. Richter, *Ann. Chem.*, **695**, 49 (1966).
319. L. Knorr, *Chem. Ber.*, **20**, 1107 (1887).
320. T. A. Sokolova, I. N. Osipova, and L. A. Ovsyannikova, *Zh. Organ. Khim.*, **3**, 2252 (1967); *Chem. Abstr.*, **68**, 39533e (1968).
321. A. Le Berre, M. Dormoy, and J. Godin, *Compt. Rend.*, **261**, 1872 (1965).
322. F. Lingens and H. Schneider-Bernlöhr, *Ann. Chem.*, **686**, 134 (1965).
323. H. W. Schiessl and R. Appel, *J. Org. Chem.*, **31**, 3851 (1966).
324. K. Brunner, *Monatsh. Chem.*, **18**, 531 (1897).
325. J. Staněk and D. Rybář, *Chem. Listy*, **40**, 173 (1946); *Chem. Abstr.*, **45**, 5147e (1951).

326. K. Hideg and H. O. Hankovszky, *Tetrahedron Letters*, 2365 (1965).
327. C. W. Bird, *J. Chem. Soc.*, 674 (1963).
328. C. W. Bird, *J. Chem. Soc.*, 5284 (1964).
329. O. Poppenberg, *Chem. Ber.*, **34**, 3263 (1901).
330. L. Wolff, *Ann. Chem.*, **394**, 98 (1912).
330a. C. Bodea and J. Oprean, *Rev. Roumaine Khim.*, **13**, 1647 (1968).
331. M. Malm and R. N. Castle, *J. Heterocyclic Chem.*, **1**, 182 (1964).
332. W. Ried and A. Meyer, *Chem. Ber.*, **90**, 2841 (1957).
333. E. Fischer and F. Passmore, *Chem. Ber.*, **22**, 2730 (1889).
334. F. H. Pollard and A. J. Banister, *Anal. Chim. Acta*, **14**, 70 (1956).
335. R. L. Hinman, *Anal. Chim. Acta*, **15**, 125 (1956).
336. P. R. Wood, *Anal. Chem.*, **25**, 1879 (1953).
337. W. A. Mosher, I. S. Bechara, and E. J. Pozomek, *Talanta*, **15**, 482 (1968).
338. A. C. Thompson and P. A. Hedin, *J. Chromatog.*, **21**, 13 (1966).
339. D. W. Russell, *J. Chromatog.*, **19**, 199 (1965).
340. C. F. Most and H. B. Milne, *J. Chromatog.*, **34**, 551 (1968).
341. K. Maroszynska and T. Lipiec, *Acta Polon. Pharm.*, **25**, 21 (1968).
342. A. P. Grekov and S. A. Malyutenko, *Zh. Anal. Khim.*, **23**, 639 (1968).
343. L. I. Kovalenko, *Farm. Zh. (Kiev)*, **23**, 55 (1968).
344. A. P. Grekov and V. A. Yanchevskii, *Metody Anal. Khim. Reaktivov Prep.*, **12**, 103 (1966); *Chem. Abstr.*, **67**, 17638v (1967).
345. A. P. Grekov and S. A. Sukhorukova, *Metody Anal. Khim. Reaktivov Prep.*, **12**, 109 (1966); *Chem. Abstr.*, **67**, 17660w (1967).
346. O. Isler, H. Gutmann, O. Straub, B. Fust, E. Böhni, and A. Studer, *Helv. Chim. Acta*, **38**, 1033, 1046 (1955).
347. K. U. M. Prasad, B. H. Iyer, and M. Sirsi, *J. Indian Chem. Soc.*, **43**, 784 (1966).
348. S. S. Parmar and R. Kumar, *J. Med. Chem.*, **11**, 635 (1968).
349. P. Gheorghiu, V. Stroescu, C. Demetrescu, H. Benes, R. Ciorbaru, and A. Granici, *Pharmazie*, **23**, 23 (1968).
350. C. Demetrescu, C. Chirită, and V. Ion, *Pharm. Zentr.*, **107**, 127 (1968).
351. R. Pfister, A. Sallmann, and W. Hammerschmidt (J. R. Geigy AG), *Swiss Pat.*, 421,132 (1967); *Chem. Abstr.*, **68**, 49283n (1968).
352. W. H. Hunter, J. King, and B. J. Millard (Benger Laboratories, Ltd.), *Brit. Pat.*, 1,086,636 (1967); *Chem. Abstr.*, **68**, 39491q (1968).
353. J. Renz, J. P. Bourquin, H. Winkler, C. Bruieschweiler, L. Ruesch, and G. Schwarb (Sandoz Ltd.), *Swiss Pat.*, 419,136 (1967); *Chem. Abstr.*, **68**, 29709c (1968).
354. R. Cavier and R. Rips, *J. Med. Chem.*, **8**, 706 (1965).
355. E. Degener, H. Scheinpflug, and H. G. Schmelzer (Farbenfabriken Bayer AG), *Brit. Pat.*, 1,085,474 (1967); *Chem. Abstr.*, **68**, 95567f (1968).
356. L. R. B. Hervey and R. P. Tschirch (Olin Mathieson Chemical Corp.), *U.S. Pat.*, 3,284,234 (1966); *Chem. Abstr.*, **66**, 12077f (1967).
357. Farbenfabriken Bayer AG, *Fr. Pat.*, 1,400,720 (1965); *Chem. Abstr.*, **64**, 3783g (1966).
358. C. E. Tholstrup and J. C. Ownby (Eastman Kodak Co.), *Brit. Pat.*, 1,093,383 (1967); *Chem. Abstr.*, **68**, 22546p (1968).
359. S. Kawawata, S. Nakagawa, K. Masuko, and R. Ito (Hitachi Wire & Cable, Ltd.), *Japan. Pat.*, 17,649 (1967); *Chem. Abstr.*, **68**, 40568b (1968).

360. Gevaert Photo-Producten N.V., *Brit. Pat.*, 988,363 (1965); *Chem. Abstr.*, **65**, 1662f (1966).
361. R. E. Leonard and G. P. Touey (Eastman Kodak Co.), *U.S. Pat.*, 3,359,990 (1967); *Chem. Abstr.*, **68**, 85070z (1968).
362. E. M. Vasil'eva, E. K. Podval'naya, O. P. Kolomeitsev, and R. K. Gavurina, *Vysokomolekul. Soedin, Ser. A*, **9**, 1499 (1967); *Chem. Abstr.*, **68**, 30524v (1968).

CHAPTER 11

Biological formation and reactions of the amido group

J. E. Reimann and R. U. Byerrum

Michigan State University, East Lansing, Michigan, U.S.A.

20*

I. INTRODUCTION

Ammonia is the common precursor for cellular nitrogen compounds and central to the utilization of nitrogen for the synthesis of proteins and nitrogen-containing biological compounds essential to the structural and functional integrity of all organisms. Many microorganisms are able to derive all of their cellular nitrogen from ammonium salts of the culture medium. Inorganic nitrogen from atmospheric nitrogen, nitrate and nitrite salts can serve as a source of ammonia for certain organisms. A number of bacteria which include *Azotobacter*, *Clostridium*, *Rhodosphrillum* and blue–green algae are able to meet their nitrogen requirements by the conversion of nitrogen gas to ammonia. Bacteria of the genus *Rhizobium* live in the root nodules of legumes in a symbiotic relationship with the plant. Neither the bacteria nor plant alone is capable of fixing nitrogen but combined they are responsible for the conversion of large quantities of atmospheric nitrogen to ammonia. Plants, fungi and some bacteria obtain ammonia for incorporation into organic compounds through the reduction of nitrate or nitrite. Animals can derive a major portion of their cellular nitrogen through the assimilation of ammonia but some preformed amino acids must be supplied in the diet. The requirement for a dietary source of the essential amino acids arises from an inability to synthesize the appropriate carbon skeleton rather than an inability to utilize ammonia in amino acid synthesis.

There are three quantitatively important pathways for the incorporation of the inorganic nitrogen of ammonia into organic compounds. Glutamic acid derives its α-amino nitrogen directly from ammonia and in subsequent transamination reactions with α-keto acids indirectly provides amino nitrogen for all other amino acids. The formation and transamination reactions of glutamic acid were considered in the volume of this series on the chemistry of the amino group.

The incorporation of ammonia into the amide function of glutamine, catalysed by the enzyme glutamine synthetase, is a second important pathway. The third route for the incorporation of inorganic nitrogen into organic compounds results from the reaction of ammonia, carbon dioxide and adenosine triphosphate to form carbamoyl phosphate. Because of its participation in the assimilation of ammonia and its precursor role in the synthesis of the carbamic acid amide (urea) and amidine derivatives, the formation and reactions of carbamoyl phosphate will be considered in this chapter.

Although the quantitative importance of glutamine and carbamoyl phosphate formation in ammonia utilization has not been investigated as thoroughly as the incorporation of ammonia into the α-amino nitrogen of amino acids, there is a good deal of evidence that a significant quantity of ammonia is assimilated through the formation of these compounds. Metabolic studies[1] employing ^{15}N-labelled inorganic compounds show that the α-amino nitrogen of glutamic acid and amide nitrogens are the first groups to be labelled. The percent of total ^{15}N incorporated into amide groups of the protein decreased with time as other nitrogen-containing compounds increased in relative ^{15}N content, indicating that ammonia is 'fixed' in an amide linkage then utilized in subsequent reactions for the synthesis of other nitrogen compounds. The rate of glutamine synthesis in exponentially growing yeast cells is far more than would be required for protein synthesis[2], suggesting an additional function in the assimilation of ammonia. The combined administration of glutamine and asparagine, and a carbohydrate, as a carbon source, adequately supports the growth of barley embryos[3]. The amide and amino nitrogens of glutamine and asparagine apparently can supply all the nitrogen requirements for the growing embryo. The rapid incorporation of $^{15}NH_3$ into the uracil of growing bacteria[4] and the guanidine nitrogens of arginine in yeast cells[2] indicates that the synthesis of carbamoyl phosphate, a precursor to these compounds, accounts for a part of ammonia assimilation in these organisms. A quantitative analysis of ammonia assimilation shows that 61% of the total ammonia utilized in growing yeast cells is associated with the α-amino groups of free and protein amino acids. Utilization of the remaining 39% probably can be accounted for in part by the synthesis of the amino acid amides, glutamine and asparagine, and carbamoyl phosphate. The rapid assimilation of ammonia into these compounds might be expected since they participate in the biosynthesis of many compounds requisite for growth.

These compounds play an equally important role in the transport and elimination of nitrogen compounds. Degradative metabolism of nitrogenous compounds in organisms leads to the formation of the major excretion products ammonia, uric acid and urea. Higher plants incorporate into and store ammonia in the amide groups of asparagine and glutamine. Uricotelic animals (birds, reptiles) eliminate nitrogen by the formation of uric acid. Two of the four nitrogen groups of uric acid are derived from the amide nitrogen of glutamine. Free ammonia in mammals is highly toxic and the very low concentration found in tissues and fluids testifies to an efficient mechanism for its removal. The formation of glutamine from glutamic acid and ammonia appears to serve an important function in the trapping and transport of free ammonia. Glutamine concentrations in the blood are relatively high compared to tissue levels and glutamine amide nitrogen accounts for most of the ammonia excreted by mammals. In man and other terrestrial vertebrates most of the ammonia is converted to and excreted as urea (ureotelism). The ammonia utilized for urea formation is released from the amides of glutamine and asparagine, and from amino acids by deamination reactions, to form carbamoyl phosphate. The carbamoyl phosphate enters into a cyclic reaction with ornithine to yield, ultimately, urea.

The biochemical utilization of ammonia in the formation of amide groups and carbamoyl phosphate and the reactions of these compounds in the biosynthesis of amines, other amides, and amidine and guanidine derivatives will be considered in detail. Consideration has also been given to the mechanism of these reactions, albeit incompletely understood, in hopes of acquainting the organic chemist with a few of the current problems and complexities of enzyme-catalysed reactions.

II. FORMATION OF AMIDES AND AMIDINE DERIVATIVES: ATP → ADP + Pi REACTIONS

In sections II and III the biosynthesis of biological acyl derivatives of ammonia, which require adenosine triphosphate (ATP), are considered. Some attention has been devoted to the yet unresolved mechanisms of these reactions in the hope that common features will permit a meaningful organization of the biosynthetic formation of these compounds. Grouping of the ATP-dependent synthesis of amide, amidine and guanidine groups is based on the product formed from the nucleotide on the assumption that its cleavage to ADP and Pi or to AMP and PPi may result from a common mechanism. Generally, cleavage of ATP to ADP–Pi is associated with the biosynthesis of amide groups and amidine derivatives, whereas cleavage to AMP–PPi occurs in the biosynthesis of guanidine groups. The two known exceptions are the synthesis of the amide bonds catalysed by the bacterial asparagine synthetase and nicotinamide–adenine dinucleotide (NAD) synthetase. The formation of AMP–PPi in asparagine biosynthesis, catalysed by the bacterial enzyme, indicates a mechanism of catalysis which differs from the synthesis of other biological amide compounds. The deviation of NAD synthetase from this organization is discussed in section III.

The evidence from studies of the nucleoside triphosphate dependent synthesis of amides and amidine derivatives, which are accompanied by the formation of the corresponding nucleoside diphosphate and Pi, generally suggest a mechanism of carboxyl-group activation by the formation of an acyl phosphate intermediate or by electrophilic catalysis of a quaternary complex which results in the cleavage of the terminal phosphate of ATP.

A. Glutamine Synthesis

Glutamine synthetase in the presence of adenosine triphosphate (ATP) and metal ions (Mg^{2+} or Mn^{2+}) catalyses the enzymatic synthesis of glutamine from glutamate and ammonia (equation 1).

$$
\begin{array}{cc}
\text{COO}^- & \text{CONH}_2 \\
| & | \\
\text{CH}_2 & \text{CH}_2 \\
| & | \\
\text{CH}_2 + \text{NH}_3 + \text{ATP} \underset{\phantom{Mg^{2+}}}{\overset{Mg^{2+}}{\rightleftharpoons}} \text{CH}_2 + \text{ADP} + \text{Pi} \\
| & | \\
\text{CHNH}_3^+ & \text{CHNH}_3^+ \\
| & | \\
\text{COO}^- & \text{COO}^- \\
\text{L-glutamate} & \text{L-glutamine}
\end{array}
\qquad (1)
$$

The energy requirement for glutamine synthesis can be derived from the hydrolysis of ATP; that is, the free energy resulting from the ultimate hydrolysis of the pyrophosphate linkage can be utilized in a coupled reaction to shift the equilibrium of an otherwise unfavourable reaction more nearly to completion. Reaction (1) can be formulated as the sum of two reactions: (a) the endergonic reaction [5] of glutamate and ammonia to form glutamine (equation 2a) and (b) the exergonic hydrolysis of ATP to adenosine diphosphate (ADP) and inorganic phosphate (Pi) (equation 2b). A standard free-energy change of

$$\text{Glutamate} + \text{NH}_3 \rightleftharpoons \text{Glutamine} + \text{H}_2\text{O} \quad \Delta F + 3400 \text{ cal} \quad (2a)$$

$$\text{ATP} + \text{H}_2\text{O} \rightleftharpoons \text{ADP} + \text{Pi} \quad \Delta F - 7770 \text{ cal} \quad (2b)$$

−4300 cal associated with the overall reaction has been calculated from the experimentally determined equilibrium constant [6]. The mechanism by which the energy is made available to couple these reactions is not completely understood. However, there is considerable information regarding glutamine synthetase and the enzyme has been the subject of several reviews [7-9].

I. Reaction mechanism

Carboxylic acids at physiological pH values are primarily in the ionized form as carboxylate ions and are chemically unreactive toward nucleophilic substitution at the carboxylic site. Biochemical activation is accomplished by conversion of the carboxylate ion to an ester, a thioester or an anhydride.

There are two general mechanisms responsible for carboxyl activation utilizing the phosphate bond energy of ATP. Both mechanisms involve the formation of an intermediate by the transfer of some portion of the ATP molecule to the substrate. Direct carboxyl-group activation is achieved by one group of activating enzymes by the transfer of a phosphate group from ATP to form an acyl phosphate intermediate in the biosynthetic pathway (equation 3). The free intermediate is catalysed by a second enzyme to liberate the inorganic phosphate and form the final product. β-Aspartyl kinase, an enzyme of this type, catalyses the formation of β-aspartyl phosphate an intermediate in the biosynthesis of threonine [10].

$$^-\text{OOC}-\underset{\underset{\text{NH}_3^+}{|}}{\text{CH}}-\text{CH}_2-\text{COO}^- + \text{ATP} \rightleftharpoons {}^-\text{OOC}-\underset{\underset{\text{NH}_3^+}{|}}{\text{CH}}-\text{CH}_2-\overset{\overset{\text{O}}{\|}}{\text{C}}-\text{OPO}_3\text{H}^- + \text{ADP} \quad (3)$$

A second group of activating enzymes catalyse a two-step reaction sequence which includes the synthesis of an enzyme-bound acyl adenylate intermediate followed by acyl transfer from the ATP moiety to an acceptor molecule (equation 4). Enzymes that catalyse the formation of acyl adenylates are associated with AMP–PPi cleavage of ATP and are responsible for the activation of amino acids in protein synthesis, acetate in acetyl coenzyme A formation and alkyl carboxylates in the synthesis of fatty acids.   In most cases evidence

$$RCOO^- + ATP \underset{}{\overset{Enzyme}{\rightleftharpoons}} E \ldots R-\overset{\overset{O}{\|}}{C}-O-\overset{\overset{O}{\|}}{\underset{\underset{O^-}{|}}{P}}-O-Ad + PPi \overset{R^1X}{\longrightarrow}$$

$$R-\overset{\overset{O}{\|}}{C}-X-R^1 + AMP \quad (4)$$

for an acyl adenylate intermediate is indirect and consists of trapping the intermediate by formation of acid hydroxamates on treatment of enzyme–substrate reaction mixtures with hydroxylamine, and showing that the synthetic acyl adenylates participate in the forward and reverse enzymatic reaction.   Other evidence includes observation of the reversible step of reaction (4) as evidenced by an ATP–PPi exchange. In a few cases, this intermediate has been demonstrated unequivocally by chemical isolation and identification[11–14].

The mechanism of activating enzymes associated with the cleavage of ATP to yield ADP and inorganic phosphate is less well understood but at least for some of the enzymes in this group, probably involves the formation of an enzyme-bound activated complex intermediate. A detailed consideration will be given to the activation of the carboxyl group by this mechanism since this is important in reactions in which an amide bond is formed.

Glutamine synthetase is the most extensively studied enzyme of the group.   Early studies established the general reaction catalysed by glutamine synthetase as shown in equation (1).   Krebs[15] investigated the synthetic activity in a number of tissues from mammals and birds and concluded that the formation of glutamine was dependent on energy-giving reactions, and the tissue slices must contain a factor concerned with this transmission of energy.   Speck[16] demonstrated that the reactions of glutamate and ammonia in pigeon liver extracts required ATP, and the disappearance of ammonia was stoicheiometric with the formation of 'acid-labile ammonia' and inorganic phosphate.   Similar findings were reported for preparations from *Staphylococcus aureus*[17], sheep brain and lupine seedlings[18].   Participation of ATP in glutamine synthesis was further substantiated by the demonstration that ^{32}P-labelled phosphate was incorporated into ATP when the reaction was studied in the reverse direction[19].

Radioactive tracer studies[20,21] employing ^{18}O-labelled glutamate indicated that ATP coupled the reaction in some manner which resulted in the transfer of one oxygen atom from glutamate to inorganic phosphate.   In the reverse reaction Varner and associates[22], observed the incorporation of ^{18}O from inorganic phosphate into glutamate.   The transfer of oxygen from glutamate to inorganic phosphate suggested a direct reaction between ATP and the γ-carboxyl group.   These facts led to the hypothesis that ATP reacted with glutamate to yield a γ-glutamyl phosphate intermediate.

Although no evidence could be obtained for the formation of a free intermediate or for the utilization of added γ-glutamyl phosphate[23] further investigation indicated an activated carboxyl was formed but remained bound to the enzyme.   Glutamine synthetase also catalyses a transfer reaction first reported by Stumpf and Loomis[24], where the γ-glutamyl moiety is transferred from glutamine to hydroxylamine to form the hydroxamate according to equation (5).

$$\text{Glutamine} + \text{Hydroxylamine} \longrightarrow \text{Glutamyl hydroxamate} + \text{Ammonia} \qquad (5)$$

When this reaction was carried out in the presence of ^{14}C-labelled glutamate little isotope was incorporated into the synthesized γ-glutamyl hydroxamate[25].   The results of this experiment suggested that the activated form of glutamate was firmly bound to the enzyme and not in equilibrium with glutamate in solution.

More direct evidence was obtained for the formation of an activated carboxyl intermediate in experiments based on the tendency of the γ-glutamyl derivative to undergo a rapid cyclization reaction to form pyrrolidonecarboxylate[26,27].   A reaction mixture containing enzyme, glutamate, ATP and magnesium ions, but no ammonia, was allowed to react, then heated at 60°c.   Significant formation of pyrrolidonecarboxylate was observed.   Little or no cyclization product was formed in control experiments in which the enzyme was heat-inactivated prior to the addition of substrates.   On addition of ammonia to the system, glutamine was formed, accompanied by a marked decrease in pyrrolidonecarboxylate formation.

The nature of this activated glutamyl intermediate was not evident from the above findings and a number of mechanisms were considered in an attempt to explain the participation of ATP in glutamine synthesis.   Proponents of a consecutive mechanism postulated the formation of an enzyme-bound γ-glutamyl phosphate from glutamate and ATP (equation 6a) followed by nucleophilic attack on the activated intermediate to form glutamine (equation 6b).   A number of experimental facts were not explicable in terms of phosphoanhydride inter-mediate formation or consistent with the above formulation. The arguments were based primarily on the role of ADP and the inability

Enzyme + Glutamate + ATP $\rightleftharpoons$ Enzyme ... γ-Glutamyl phosphate + ADP   (6a)

Enzyme ... γ-Glutamyl phosphate + NH_3 $\rightleftharpoons$ Enzyme + Glutamine + Pi   (6b)

of the enzyme to catalyse partial reactions. When studied in the forward direction, free ADP is not detected unless ammonia is present in the reaction mixture. This suggested that the formation of the amide bond occurred concomitantly with the cleavage of ATP in contrast with equation (6a) in which ADP is liberated in the first step of intermediate formation.

Studies with ^{32}P label showed that all three substrates had to be present for ADP–ATP or Pi–ATP exchange reactions to take place[27]. From equations (6a and 6b) one might expect that catalysis of partial reactions could occur, however, such interpretations can be misleading. The failure to observe exchange reactions in the absence of one com-ponent is neither evidence for nor against the formation of an acyl phosphate intermediate. A number of enzymes[28] exhibit activity only in the presence of a second substrate which is not involved in covalent-bond formation. The binding of substrates may result in conformation or electronic changes necessary for the catalytic step to occur.

In the reverse reaction catalysed by glutamine synthetase, arsenate can replace phosphate. According to a consecutive mechanism a glutamyl arsenate intermediate is formed which would spontaneously react with water since arsenate anhydrides are inherently unstable. The enzyme is then regenerated for further reaction. The arsenolysis reaction, however, requires catalytic amounts of ADP indicating that the nucleotide is a necessary component of the catalytic reaction[6].

To account for the apparent role of ADP as part of the 'activated intermediate' Buchanan and Hartman[29] postulated a concerted mechanism in which all three substrates, the nucleotide, glutamate and ammonia, form a complex on the enzyme. The reaction occurs with the simultaneous cleavage of the carbon-to-oxygen and oxygen-to-phosphorus bonds and formation of the nitrogen-to-carbon bond of glutamine as depicted in equation (7). Although the free energy of activation is larger than in the one involving an acyl phosphate inter-mediate, this concerted mechanism has not been unequivocally ruled out.

The nature of the activated glutamyl intermediate has been the

$$\text{Enzyme} \left[\begin{array}{c} \overset{H}{\underset{H}{R-N}} \cdots \overset{O^-}{\underset{R^1}{C}} \cdots O \cdots \overset{O^-}{\underset{O^-}{P}} \cdots O - \overset{O}{\underset{O^-}{P}} - O - \overset{O}{\underset{O^-}{P}} - O - \text{Adenine} \end{array} \right]$$

$$\downarrow \tag{7}$$

$$\text{Enzyme} + \text{RNHCR}^1 + \text{ADP} + \text{Pi}$$

subject of a continued investigation, principally by Meister and associates. Their efforts were directed towards the isolation and utilization of phosphate intermediates in support of a consecutive mechanism.

Glutamine synthetase can catalyse amide formation from a number of isomeric analogues of glutamine. The position of the amino group is not critical for enzymatic activity so that β-aminoglutarate is enzymatically converted to β-aminoglutaramate by glutamine synthetase. Since the proposed intermediate for this substrate, β-aminoglutaryl phosphate, is less prone to cyclization than γ-glutamyl phosphate, it has been possible to synthesize and test it as a substrate for the enzyme[30]. Glutamine synthetase catalysed the conversion of β-aminoglutaryl phosphate to β-aminoglutaramate. The synthesized phosphate intermediate was also utilized by the enzyme in the back reaction as indicated by the formation of ATP. Although these results are in accordance with a consecutive mechanism, they do not constitute direct evidence that an acyl phosphate is in fact an intermediate in the enzymatically catalysed process. β-Aminoglutaryl phosphate has not been isolated as an enzyme-catalysed intermediate.

Methionine sulphoximine, methionine sulphone and methionine sulphoxide are effective inhibitors of glutamine synthetase. In recent studies, Ronzio and Meister[31] have isolated a phosphorylated derivative of methionine sulphoximine from incubation mixtures containing glutamine synthetase, ATP and $MnCl_2$. The structure of the derivative has not been established, but from a consideration of the chemical properties these investigators suggest the phosphate is linked to the sulphoximine sulphur atom as in **1**. Methionine sulphoximine is a

$$^-O - \overset{O}{\underset{}{C}} - \underset{\underset{NH_3^+}{|}}{CH} CH_2 CH_2 \overset{NH}{\underset{\underset{CH_3}{|}}{S}} - OPO_3H^-$$

(1)

convulsion-producing agent and its antagonism towards glutamine synthetase suggests an important function of this enzyme in brain metabolism. Since the enzyme is irreversibly inhibited with this non-physiological substrate, secondary reactions may be responsible for the product formed. More convincing evidence for the formation of an acyl phosphate intermediate was obtained in Meister's laboratory with the enzyme glutathione synthetase (section II.D).

According to a consecutive-mechanism hypothesis glutamate and ATP react to form an acyl intermediate in the absence of a nucleophile, in contrast to a concerted mechanism which requires the presence of a nucleophile for a reaction to occur. Isotopic dilution studies[27] have provided evidence that the formation of a carboxyl-activated intermediate does in fact occur in the absence of a nucleophile. Reaction mixtures containing ^{14}C-labelled glutamate and ATP were allowed to react with glutamine synthetase in the absence of hydroxylamine. At the end of a two-minute incubation period, a mixture of excess unlabelled D-glutamate and hydroxylamine was added. The protein was precipitated after 15 seconds and the radioactivity of the γ-glutamyl hydroxamate was determined. The γ-glutamyl hydroxamate formed contained a greater amount of ^{14}C label than would be expected if the labelled and unlabelled glutamate had equilibrated prior to reaction. The preferential conversion of labelled glutamate was interpreted as evidence for a γ-glutamyl phosphate intermediate since the glutamate bound in the 2 minute incubation period was immediately available for reaction with hydroxylamine, presumably as the ATP-activated intermediate.

Evidence was also obtained for the cleavage of ATP in the absence of a nucleophile which was associated with glutamate binding[27]. The formation of equimolar quantities of ADP and Pi was demonstrated in reaction mixtures of glutamine synthetase, ATP and glutamate, however, ADP appears to remain bound to the enzyme. In order to detect these products the protein first had to be denatured. If ADP, resulting from the formation of an acyl phosphate intermediate does in fact remain bound to the enzyme, some of the observations that appeared inconsistent for a consecutive-mechanism hypothesis would be clarified. The failure to detect ADP prior to the reaction with a nucleophile would be a consequence of its remaining associated with the enzyme after the reaction of ATP and glutamate. Similarly, exchange reactions of Pi–ATP and ADP–ATP would not be expected to occur if the formed ADP was not in equilibrium with components in the solution.

Since the binding of nucleotides appeared to be of some significance, the binding of glutamine to the enzyme in the reverse reaction was investigated[27,32]. In protein separation experiments, it was found that ^{14}C-labelled L-glutamine sedimented with the protein only in the presence of ADP. The need for ADP in order to bind the substrate

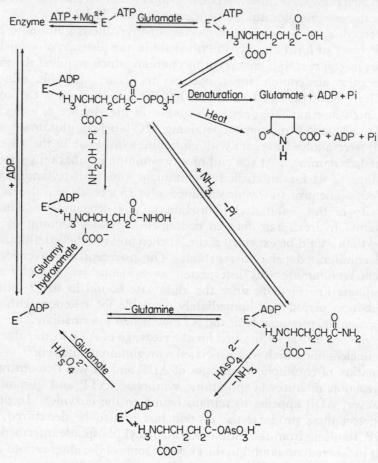

FIGURE 1. Reactions catalysed by glutamine synthetase.

in the back reaction explains the requirement for catalytic amounts of ADP in the transfer and arsenolysis reactions.

With the evidence that ADP is a part of the activated complex, the formulation of a consecutive mechanism is more accurately described by equation (8).

Enzyme + Glutamate + ATP $\rightleftharpoons$

$$\text{Enzyme} \underset{\diagdown \text{ADP}}{\overset{\diagup \gamma\text{-Glutamyl phosphate}}{}} \quad \overset{\text{NH}_3}{\rightleftharpoons} \quad \text{Enzyme} + \text{Glutamine} + \text{ADP} + \text{Pi} \quad (8)$$

On the basis of accumulated evidence Meister and coworkers[8,9] have proposed the scheme in Figure 1 to account for the known reactions of glutamine synthesis.

These investigators have proposed that the synthesis of glutamine occurs in a step-wise reaction sequence consisting of an initial binding of ATP to the enzyme in the presence of metal ions (Mg^{2+} or Mn^{2+}) which then reacts with glutamate to form an enzyme-bound activated γ-carboxyl intermediate, presumably glutamyl phosphate. ADP remains bound to the protein. The activated intermediate then reacts with ammonia to yield glutamine, inorganic phosphate and ADP.

Conclusive evidence for the formation of an intermediate should also include evidence that the rates of formation and utilization of the proposed intermediate are sufficient to account for the observed rates of product synthesis in order to eliminate the possibility that the proposed intermediate is formed in a side-reaction.

Distinguishing between a concerted and a consecutive mechanism must wait until measurements of the timing of the transformations involved can be made. The reservations concerning the interpretations of exchange studies, catalytic activity with non-physiological substrates and the isolation of intermediate products should be kept in mind for the mechanism discussions in the following sections.

2. Stereospecificity

The stereochemical requirements of glutamine synthetase are relatively non-specific in that a number of glutamic acid derivatives, including optical and structural isomers, are enzymatically active. The enzyme acts on both the L- and D-isomers of glutamate[33] and the derivatives, α-methylglutamate[34,35], β-methyl- and β-hydroxyglutamate, γ-methyl-[25] and γ-hydroxyglutamate[36]. β-Glutamic acid is converted to D-β-glutamine[30].

Kagan and Meister[37,38] carried out comparative studies on the enzymatic activity of the resolved optical isomers of these compounds. The Michaelis constants (K_m) and the relative maximum velocities were measured for amide and hydroxamate synthesis. The optical isomers of these glutamate derivatives that were subject to enzymatic catalysis are listed in Table 1. These authors[38,39] have presented a

TABLE 1. Optical isomers of glutamate derivatives subject to enzymatic catalysis.

L-Glutamic acid
D-Glutamic acid
α-Methyl-L-glutamic acid
threo-β-Methyl-D-glutamic acid
threo-β-Hydroxy-D-glutamic acid
threo-γ-Methyl-L-glutamic acid
threo-γ-Hydroxy-L-glutamic acid
erythro-γ-Hydroxy-L-glutamic acid
threo-γ-Hydroxy-D-glutamic acid
erythro-γ-Hydroxy-D-glutamic acid

hypothesis concerning the conformation of the enzyme-bound substrates to account for the observed activity of certain substituted glutamic acid optical isomers. It is assumed that substrates are oriented on the enzyme in an extended conformation and that the α-carboxyl and amino groups are bound to the same respective sites. Examination of a constructed isomer model shows that the α-hydrogen of L-glutamic acid (**2a**), so oriented on a hypothetical enzyme surface, is directed away from the enzyme surface. The α-methyl derivative of L-glutamic acid is an active substrate. In contrast, when D-glutamic acid (**2b**) is rotated 69° to the right about an axis formed by a line

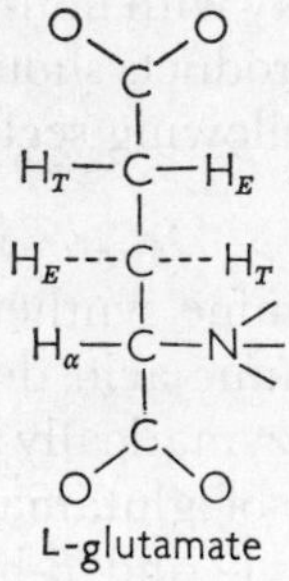

L-glutamate
(**2a**)

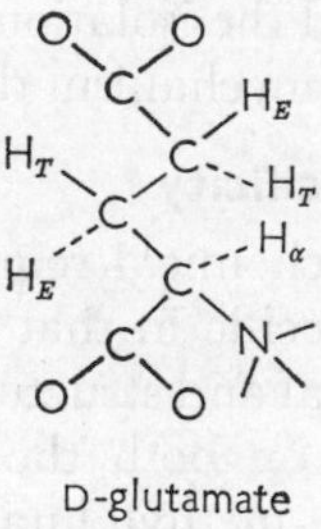

D-glutamate
(**2b**)

intersecting carbon atoms 1, 3, and 5 to permit binding of the α-carboxyl and amino groups, the α-hydrogen is directed towards the enzyme and substitution of a methyl group in this position results in loss of enzymatic susceptibility. The hypothesis states that the introduction of a bulky group in the substrate in a position directed towards the enzyme results in loss or marked reduction of the enzymatic activity.

Substrate specificities observed for the iomers of β- and γ-substituted derivatives of glutamic acid are consistent with this proposal. For example, models show that both the *erythro*- and *threo*-β-hydrogens of L-glutamic acid and the *erythro*-β-hydrogen of D-glutamic acid are oriented approximately in the same direction as the α-hydrogen of D-glutamic acid, towards the enzyme. *Erythro*- and *threo*-hydrogen assignments were made with respect to a staggered conformation shown in **2a** and **2b** so that substitution of *threo* hydrogen leads to the formation of a *threo* isomer. Substitution of a methyl or hydroxyl group into these positions leads to a loss of activity. On the other hand, replacement of the *threo*-β-hydrogen of D-glutamic acid, which is directed away from the enzyme, does not destroy enzymatic activity. Steric hindrance with the enzyme surface is consistent with the fact that of the four γ-methylglutamic isomers only *threo*-γ-methyl-L-glutamic acid serves as a substrate. Similarly it was found that *threo*-γ-hydroxyl-L-glutamic acid was the most active of the four possible isomers of γ-hydroxyglutamic acid, although substantial activity was exhibited by the *erythro*-γ-hydroxy isomers of both L- and D-glutamic acid.

Meaningful conclusions regarding the specific effects of structural and stereo alterations of the substrate on the various parameters that constitute the overall enzymatic activity are difficult to evaluate unless a detailed study of the kinetic constants is undertaken. Studies relating structural requirements for catalysis to the kinetic constants of the reaction have been carried out with acetylcholinesterase[40] and chymotrypsin[41,42]. Introduction of a bulky group may influence the conformation of the enzyme–substrate complex in two general ways. (1) Steric hindrance may prevent binding of substrates to the enzyme or promote different modes of binding which lead to unproductive complexes, or (2) may result in an unfavourable orientation of the substrate on the enzyme for catalysis or for nucleophilic attack in the subsequent reaction steps. A comparison of the 'enzymatic activity' with various substrates may reflect differences in the affinity of the enzyme for a substrate, changes in the rate constants of the catalysed reaction or a combination of both factors. Moreover an interpretation of the effects of structural parameters on the individual reaction steps requires that the experimentally measured kinetic constants are properly related to the constants of the rate equation. For example, the meaning of the measured Michaelis constant, $K_m(\text{app})$, may be a true dissociation constant (K_s) or a complex term containing K_s and rate constants depending on the rate-determining step of the reaction[42,43].

A consideration of two aspects of structural alterations on glutamine synthetase activity indicates some of the problems encountered in making semi-quantitative interpretations. The experimentally measured K_m(app) values for a series of substituted glutamic acid derivatives in hydroxamate synthesis are quite constant, whereas the K_m(app) values for the same derivatives in amide synthesis show considerable variation. The experimentally measured K_m(app) for amide formation may not be a true dissociation constant but rather a complex constant containing equilibrium and rate terms. The larger variations observed in this constant then may be attributable in part to changes in rate constants. The relatively constant K_m(app) values for various substrates in hydroxamate synthesis suggest that the corresponding K_s is perturbed to a lesser extent by the rate constants. This may be due to the non-enzymatic reaction of hydroxylamine known to occur with activated carboxyl groups [234].

The ratio of rates for nucleophilic attack (V-hydroxylamine/V-ammonia) on a common intermediate should be reasonably constant for a series of substrates. The measured ratios, however, show considerable variation. For example, maximum-velocity measurements of L-glutamate with ammonia and hydroxylamine are nearly the same, whereas the isomers of α-aminoadipate exhibit substantial activities in hydroxamate synthesis but are less than 3% as active in the reaction with ammonia [44].

Meister [7] has interpreted these findings in terms of a relatively non-specific activation step in the formation of a glutamyl intermediate followed by an optically specific reaction of ammonia (which is presumably enzyme bound). Hydroxamate formation exhibits less specificity presumably due to the known non-enzymatic reaction of hydroxylamine with an activated intermediate.

Determination of inhibition constants (K_i) for the optical isomers, which are considered true binding constants, would provide useful information in the separation of the effects of rate constants on the true dissociation constant.

Changes in the substrate adjacent to the reacting centre of the glutamate derivatives might be expected to produce two different effects. Substitution at the γ-carbon of glutamic acid may provide steric hindrance for the formation of an optimal conformation of the enzyme–substrate complex or it may influence the rate constants of the bond-breaking and bond-making processes through inductive or electron-withdrawing effects. Contributions of such combined effects are impossible to separate simply in terms of overall activity. For example, of the four γ-methylglutamic acids, only the *threo*-γ-methyl isomer had measurable activity in amide synthesis. In contrast, substantial rates were observed for three of the γ-hydroxyglutamic acid isomers. These relative activities were interpreted in terms of a critical 'available' space for substituents near the enzyme surface. The available space could accommodate a hydroxyl group but not a methyl group. Alternatively, the inhibition effects on enzyme activity by steric hindrance may be offset by the substitution of an electron-withdrawing group adjacent to the reacting carbonyl, thereby increasing its susceptibility to nucleophilic attack.

3. Control mechanisms

Highly complex cellular activities require regulatory mechanisms which enable the organism to maintain appropriate biological balance of metabolic systems. Cellular balance is achieved by the regulation of certain critical enzymes through a number of regulatory mechanisms. The amide group of glutamine is the preferred source of nitrogen for the end-product of several important biosynthetic pathways including pyridine nucleotides (3), purines (4,5), pyrimidines (6), certain amino acids (7,8), amino sugars (9) and carbamoyl phosphate (10) as illustrated in Figure 2. Multiple pathways with a common intermediate must maintain a delicate balance and the regulatory processes include repression or genetic control of *de novo* synthesis of enzymes, cumulative feedback inhibition by the end-products of glutamine metabolism, alterations of glutamine synthetase by enzyme-catalysed reactions, and modulation of enzyme activity by conformational changes mediated by divalent cations.

Glutamine synthetase in *Escherichia coli* is subject to feedback inhibition by the end-products of six compounds that derive nitrogen from the amide group of glutamine in addition to glycine and alanine. Inhibition at saturating concentrations of the compounds tested separately is only partial, but collectively they cause almost complete loss of activity. A mechanism of limited maximum inhibition of glutamine synthetase is biologically advantageous, otherwise total inhibition of an intermediate common to several pathways might result from an excess of only one of the end-products. The nature of the inhibition is attributed[45] to specific, separate, non-interacting binding sites on the enzyme for each of the eight inhibitors. Alternative models such as a single enzyme with a single non-specific allosteric site, isoenzymes with sites differing in inhibitor susceptibility or a single enzyme with multiple sites of different affinities for the

618   J. E. Reimann and R. U. Byerrum

inhibitor were ruled out on the basis of kinetic analysis and establishment of the cumulative nature of inhibition by combinations of the
end-products.

Evidence for separate binding sites was obtained by kinetic measurements which showed that the eight inhibitors could be grouped into

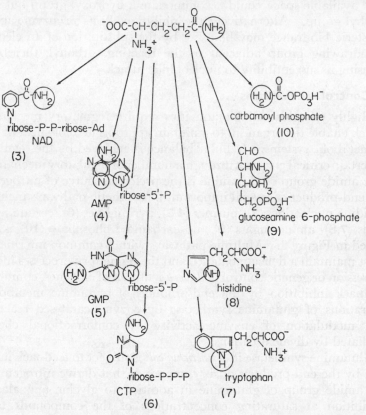

FIGURE 2.   Fate of amidic nitrogen of glutamine.

three inhibitions types[45].   Alanine, AMP and carbamoyl phosphate
were non-competitive inhibitors with respect to substrate; glycine,
tryptophan and cytidylic acid were partially competitive with respect
to glutamate; histidine and glucosamine 6-phosphate were partially
competitive with respect to ammonia. Effects of modifying the
enzyme by aging, acetone or urea treatment on the behaviour of the
inhibitor indicated different sites were involved for each compound
within a kinetic category.

The independent nature of the binding sites was established by demonstrating that the fraction of enzymatic activity measured in the presence of two or more inhibitors was equal to the product of residual activities for each inhibitor measured separately. The mechanism by which cumulative inhibition is effected by a number of chemically diverse products is not understood.

Glutamine synthetase from bacterial sources exists in two distinct forms which differ in specific activity, end-product inhibition patterns and specificity for metal ions. Mecke and coworkers[46] found that the active enzyme was converted enzymatically to a second form which had markedly reduced glutamine synthesizing ability although the transfer activity was unaffected. The two forms of the enzyme have the same amino acid composition and sedimentation behaviour, but differ, in that the altered form contains a covalently bound AMP molecule[47]. A comparison[48] of the activities showed that the adenylation reaction converts a more active form of the enzyme, specific for Mg^{2+} ions, to a less active species that is specific for Mn^{2+} ions. The adenylated enzyme is more sensitive to inhibition by histidine, tryptophan and AMP. Regulation of synthesizing activity in this case is achieved by enzymatic modification of the enzyme itself.

The enzyme catalysing the adenylation reaction has been partially purified[48] and the reaction probably proceeds according to equation (9). The modifying enzyme also appears to be subject to a control

$$\text{Glutamine synthetase + ATP} \xrightarrow[\text{Mg}^{2+}]{\text{Adenyl transferase}} \text{Adenyl glutamine synthetase + PPi}$$

$$(9)$$

mechanism in that the substrate and product can regulate the synthesizing activity. The adenyl transfer reaction is stimulated by glutamine and retarded by glutamate. In addition the predominant form of the enzyme found *in vivo* is dependent on the constituents of the culture medium. The adenylated enzyme is the principal form isolated from cultures containing an excess of ammonia, whereas growth conditions limiting in ammonia, lead to the formation of the more reactive form.

A third control mechanism for the bacterial glutamine synthesis, which involves conformational changes of the enzyme is suggested by the recent studies of Stadtman and associates[49-51]. In an interesting series of studies, these investigators have demonstrated the interconversions of a catalytically active form to an inactive glutamine synthetase. It was noted that some preparations of the enzyme from *E. coli* exhibited a lag period in the rate measurements of the enzymatic reaction before maximal velocity was obtained. The lag phase was attributed[49] to a time-dependent conversion of an inactive form, produced by the removal of divalent cations, to a catalytically active form of the enzyme on exposure to cations in the assay mixture. Treatment of the enzyme with 0·5 mM concentrations of Mg^{2+} or Mn^{2+} ions or higher concentrations of Ca^{2+} ions for an interval prior to rate measurements eliminated the lag phase. The enzyme was also activated by prior incubation of the enzyme at alkaline pH and high ionic strength, or at pH 7·1 with glutamate. These effects are thought to activate the enzyme by the concerted action of glutamate or alkaline buffers and trace amounts of divalent cations since EDTA prevented this activation.

Evidence was obtained that activation was associated with conformational alterations of the protein. Removal of divalent cation from the enzyme produces a catalytically inactive form which has a lower sedimentation coefficient, high viscosity and a higher apparent specific volume than the active form[50] indicating a less compact and somewhat more asymmetric structure. Conformational changes on removal of cations from the protein are also indicated by the increased susceptibility to inactivation by sulphydryl reacting reagents and the change in environment of the aromatic amino acid groups. Difference spectra between the active and inactive forms show that tyrosine and tryptophan are transferred from a non-polar to a more polar environment; that is, some of these groups are exposed to the solvent. The inactive form is much less stable to mild protein denaturing agents. Treatment with 1 M urea at alkaline pH induces complete dissociation to twelve subunits. Electron microscopic examination[51] of the active and inactive forms of the enzyme show the subunits are arranged in two hexagonally stacked layers.

The physiological significance of conformational alterations mediated by metal ions and substrate as a control mechanism is difficult to evaluate, since little is known about the intracellular concentrations or fluxes of divalent cations or glutamate.

B. Asparagine

Enzyme activity capable of catalysing the synthesis of asparagine has been demonstrated in animal tissues, yeast and plants. Investigation of the synthetic activity in these diverse systems has provided evidence for several mechanisms for the biosynthesis of asparagine.

An enzyme, isolated and purified 100-fold from yeast cells, was reported to catalyse the formation of asparagine from ammonia and aspartate with the concomitant cleavage of ATP to ADP and inorganic phosphate[52]. The synthesis of asparagine in yeast appears to be completely analogous to the pathway for glutamine synthesis.

Preparations from two bacterial sources, *Lactobacillus arabinosus*[53] and *Streptococcus bovis*[54] catalyse asparagine synthesis from aspartate and ammonia but unlike the yeast enzyme and glutamine synthetase, ATP is converted to AMP and inorganic pyrophosphate. AMP was identified as the product by chromatography. Furthermore, ^{32}P-labelled pyrophosphate is incorporated into ATP during the course of the reaction. The activation of the substrate by the bacterial enzyme may occur by the formation of an enzyme-bound aspartyl adenylate.

Evidence for a novel pathway for asparagine biosynthesis has been observed in a number of plants. It was noted during *in vivo* experiments using sorghum, barley, pea, flax and clover seedlings administered with ^{14}C-labelled cyanide that considerable radioactivity was incorporated into asparagine[55]. In decarboxylation studies of asparagine, and of aspartic acid obtained by enzymatic hydrolysis of the isolated asparagine, essentially all the ^{14}C label was found in the amide carbon atom.

The more immediate precursor for the biogenesis of asparagine by this pathway is β-cyano-L-alanine (equation 10a) and γ-glutamyl-β-cyanoalanine (equation 10b)[56]. Radioactive *N*-γ-glutamyl-β-cyano-

$$
\begin{array}{c}
\text{CN} \\
|\\
\text{CH}_2 \\
|\\
\text{CHNH}_3^+ \;+\; \text{H}_2\text{O} \;\longrightarrow\; \\
|\\
\text{COO}^- \\
\beta\text{-cyano-L-alanine}
\end{array}
\qquad
\begin{array}{c}
\text{CONH}_2 \\
|\\
\text{CH}_2 \\
|\\
\text{CHNH}_3^+ \\
|\\
\text{COO}^- \\
\text{asparagine}
\end{array}
\qquad \text{(10a)}
$$

$$
\begin{array}{c}
\text{CN} \\
|\\
\text{CH}_2 \\
|\\
\text{CH}_2\text{CONH—CH} \\
|\qquad\qquad|\\
\text{CH}_2 \qquad \text{COO}^- \;+\; \text{H}_2\text{O} \;\longrightarrow\; \\
|\\
\text{CHNH}_3^+ \\
|\\
\text{COO}^- \\
\gamma\text{-glutamyl-}\beta\text{-cyanoalanine}
\end{array}
\qquad
\begin{array}{c}
\text{CONH}_2 \\
|\\
\text{CH}_2 \\
|\\
\text{CHNH}_3^+ \\
|\\
\text{COO}^- \\
\text{asparagine}
\end{array}
\qquad \text{(10b)}
$$

alanine was isolated from common vetch seedlings to which ^{14}C-labelled cyanide had been administered. On hydrolysis of the peptide, 99·8% of the recovered activity was found in the formed aspartic acid. The biosynthesized N-γ-glutamyl-β-[^{14}C]cyano-alanine and chemically synthesized β-[^{14}C]cyano-L-alanine served as excellent precursors of asparagine when administered to species of *Lathyrus*. That the cyano carbon of β-cyanoalanine provided the amide carbon of asparagine was established by degradation of the isolated radio-active asparagine to alanine which contained less than 1% of the specific activity of aspartic acid.

Studies on the HeLa strain human carcinoma cells indicated that the amide nitrogen of glutamine is utilized directly for asparagine synthesis. Levintow[57] found that significant quantities of isotope were incorporated in protein asparagine when cells were grown on a medium containing glutamine ^{15}N-labelled in the amide nitrogen. In similar experiments using ^{15}N-labelled ammonia no significant incorporation occurred.

The inability of certain malignant cells to biosynthesize asparagine has provided an important chemotherapeutic approach for the treatment of cancer. Certain types of cancer cells require an external source of L-asparagine whereas normal cells do not. Treatment of animals with transplanted or induced tumours with L-asparaginase an enzyme which hydrolyses the amide group of asparagine, deprives the malignant cells of an extracellular source of asparagine and they die.

Observations regarding this metabolic defect of some cancer cells were first made by Kidd[58]. In experiments designed to measure immunological responses he noted that transplanted lymphomas were suppressed in control animals injected with guinea pig serum. Neuman and McCoy[59], in testing amino acid requirements for the cultivation of Walker carcinosarcoma 256 in tissue culture, found an absolute requirement for L-asparagine and L-glutamine. The addition of the free dicarboxylic amino acids, ammonia and ATP, did not replace the amide requirement. Broome[60] correlated these observations by demonstrating that the antileukemia factor in guinea pig serum was L-asparaginase. The ability to suppress the growth of implanted mouse lymphomas paralleled L-asparaginase activity in protein purification studies, and pH or temperature inactivation of the enzyme resulted in the loss of this ability[61]. Tumour cells which lacked an asparagine requirement for optimal growth in tissue culture were also resistant to the effects of guinea pig serum.

Since that time, asparaginase isolated from *E. coli* has been found effective in suppressing growth of a variety of cancer cells[62]. Preliminary experiments indicate that human leukemias are sensitive to this treatment, but more extensive clinical studies have been limited by the availability of the purified enzyme.

C. Glutathione

The tripeptide, γ-glutamylcysteinylglycine, is biosynthesized in two reaction steps (11a,b). Equation (11a) is discussed in section IV.E. The activation of the dipeptide carboxyl group (equation 11b)

$$\text{L-Glutamate} + \text{L-Cysteine} + \text{ATP} \overset{Mg^{2+}}{\rightleftharpoons} \text{L-}\gamma\text{-Glutamylcysteine} + \text{ADP} + \text{Pi} \qquad (11a)$$

$$\text{L-}\gamma\text{-Glutamylcysteine} + \text{Glycine} + \text{ATP} \overset{Mg^{2+}}{\rightleftharpoons} \text{L-Glutathione} + \text{ADP} + \text{Pi} \qquad (11b)$$

involves the formation of an α-carboxyl amide rather than the γ-carboxyl amide.

Analogous to glutamine synthesis, the formation of glutathione is associated with breakdown of ATP to ADP and Pi; isotopic oxygen is transferred from the free carboxyl group of the dipeptide to inorganic phosphate; and when glycine is replaced with hydroxylamine, a hydroxamate is formed. In the reverse reaction the enzyme catalyses the arsenolysis of the tripeptide, hydroxamate formation and an exchange reaction of glycine. These reactions require a catalytic amount of nucleotide[64,65]. Efforts to obtain evidence for a carboxyl-activated intermediate were successful in the glutathione synthetase catalysed formation of opthalmic acid from the dipeptide γ-glutamyl-α-aminobutyrate and glycine. Pulse labelling experiments[66], analogous to those carried out with glutamine synthetase, indicated hydroxylamine reacted preferentially with a preformed intermediate. The presumed intermediate was isolated from reaction mixtures which contained ^{14}C-labelled γ-glutamyl-α-aminobutyrate and ATP-γ-^{32}P, and substrate quantities of enzyme, but no glycine or hydroxylamine. Following treatment of the reaction mixture with perchloric acid the protein-free solution was subjected to paper electrophoresis and a 'dipeptide derivative', containing both ^{14}C and ^{32}P, was separated from the other components. The 'dipeptide derivative' on treatment with hydroxylamine was converted to γ-glutamyl-α-aminobutyryl hydroxamate as judged by its electrophoretic mobility. Properties of the chemically synthesized intermediate were the same as those of the enzymatically formed 'dipeptide derivative' with respect to perchloric acid stability, reactivity with hydroxylamine and electrophoretic behaviour. Moreover, the chemically synthesized acyl phosphate intermediate was utilized by the enzyme for opthalmic acid synthesis in the forward reaction and ATP formation in the back reaction[67].

Studies [68] on glutathione synthetase from yeast, which has been purified to homogeneity by electrophoretic and ultracentrifugal criteria, have revealed interesting mechanistic aspects of the enzyme. The ability of enzyme preparations to catalyse an ADP–ATP exchange reaction is markedly decreased on purification, indicating this activity is not a function of the enzyme. However, a ^{14}C-labelled intermediate, presumably γ-glutamyl-α-aminobutyryl phosphate, exchanges rapidly with unlabelled dipeptide in the absence of a nucleophile. The exchange rate is rapid relative to the breakdown of the activated intermediate. If the activated intermediate is γ-glutamyl-α-aminobutyryl phosphate, as the previous studies have suggested, one would not expect to observe this exchange reaction, but a transfer of the activated acyl moiety to water with a loss in the group potential gained in its reaction with ATP.

The exchange of dipeptide into the activated intermediate in the absence of a concomitant ADP–ATP exchange is consistent with a concerted-reaction mechanism hypothesis. Exchange was demonstrated by preincubation of the enzyme, ^{14}C-labelled dipeptide and ATP so that the activated intermediate was isotopically labelled. Unlabelled dipeptide was added to the reaction mixture and allowed to equilibrate for varying periods of time prior to the addition of hydroxylamine and formation of the hydroxamate. The isotope content of the γ-glutamyl-α-aminobutyryl hydroxamate decreased with the time allowed for equilibration. The complete complex envisaged by a concerted mechanism is shown in equation (7) of section II.A. These results indicate that the substrates, ATP and dipeptide, are bound to the enzyme and that the bound dipeptide is in equilibrium with dipeptide in the medium.

An alternative mechanism involving a phosphorylated enzyme intermediate might also be consistent with these results. The pathway for this mechanism, depicted in equations (12a,b) would proceed by an ATP reaction with some group on the enzyme, with release of ADP which remains firmly bound to the enzyme and consequently is not in equilibrium with ADP in solution. The dipeptide reacts with the enzyme phosphate to form an activated enzyme-bound acyl phosphate

$$ATP + Enzyme{-}X^- \longrightarrow E{-}X{-}PO_3H^- + ADP \tag{12a}$$

$$E{-}X{-}PO_3H^- + R{-}COO^- \rightleftharpoons E\underset{X{-}PO_3H}{\overset{R{-}CO}{\diagdown\diagup}}O^- \longrightarrow E\underset{X^-}{\overset{R{-}\overset{\overset{O}{\|}}{C}{-}OPO_3H^-}{\diagup}} \tag{12b}$$

intermediate. The freely reversible reaction (12b) would permit the exchange of the preformed ^{14}C-labelled dipeptide intermediate with unlabelled dipeptide of the solution. Attempts to isolate a phosphorylated enzyme intermediate were not successful[68].

D. 5'-Phosphoribosyl-glycineamide

The ability to synthesize purines (e.g. **11**) is nearly universal in that only a few organisms must be supplied these bases from an exogenous source. The synthesis of carbon-to-nitrogen bonds through formation of an amide linkage in the purine biosynthetic pathway will be considered in the present and following sections.

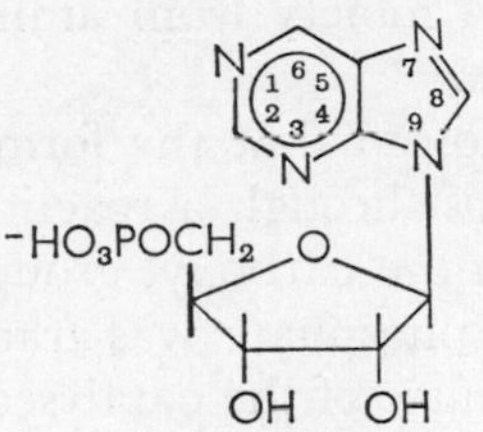

(11)

The enzyme phosphoribosyl-glycineamide synthetase catalyses the reaction of ribosylamine 5-phosphate (**12**) and glycine[70] to yield **13** according to equation (13). The characteristics of this reaction are

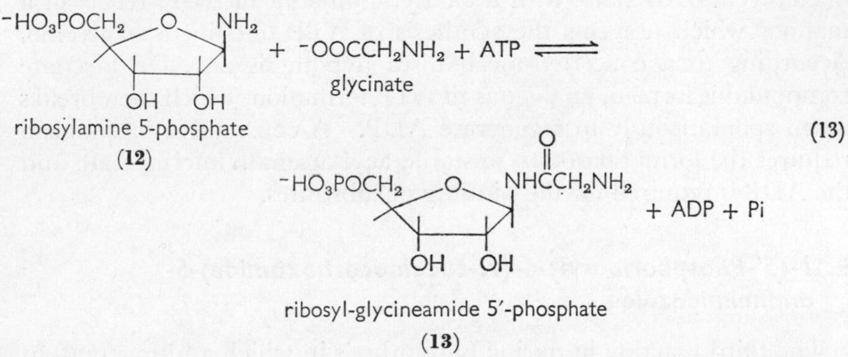

ribosylamine 5-phosphate
(12)

(13)

ribosyl-glycineamide 5'-phosphate
(13)

so similar to the glutamine and glutathione synthesizing enzymes that in all probability the same general mechanism is operative in all three reactions. Buchanan and Hartman[69,72] postulated a concerted reaction of three substrates in contrast to a consecutive-reaction

21+c.o.a.

mechanism inherent in the formation of an enzyme-bound acyl phosphate intermediate (as discussed in section II.A.1).

Studies[72] using an enzyme preparation partially purified from avian liver showed that catalysis was ATP dependent, ADP and inorganic phosphate were the products of the ATP utilized, and there was no evidence for the participation of more than one enzyme or for the formation of a free intermediate. The formation of an activated bound carboxyl intermediate was evidenced by synthesis of glycyl hydroxamate from a reaction mixture containing glycine, ATP and hydroxylamine. Reversibility of the reaction was demonstrated, and in the back reaction, the same hydroxamate was formed when **13**, inorganic phosphate, ADP and hydroxylamine were allowed to react with the enzyme. The latter reaction is analogous to the catalysed transfer of the γ-glutamyl moiety from ammonia to hydroxylamine with glutamine synthetase.

Isotope experiments showed that the formation of the amide was coupled to ATP hydrolysis through a reaction of the terminal phosphate group of ATP with the carboxyl group of glycine. One ^{18}O-labelled oxygen atom in phosphate was transferred to the carboxyl group of glycine in the course of the catalysed reaction.

Another common property of the two enzymes is that they are both subject to arsenolysis reactions. Arsenate was found to cause cleavage of the glycineamide to glycine, and the presence of ADP was an absolute requirement. The incorporation of ^{32}P into ATP in the reverse reaction is inhibited by arsenate although the production of glycine is unchanged. Arsenate, with a catalytic amount of ADP, reacts in a manner which prevents the synthesis of ATP but liberates glycine. According to a concerted-mechanism hypothesis an ADP–arsenate compound is formed, analogous to ATP formation, which then breaks down spontaneously to regenerate ADP. A consecutive mechanism requires the formation of an unstable acyl arsenate intermediate and the ADP is required for the binding of substrates.

E. 1-(5′-Phosphoribosyl)-4-(N-succinocarboxamide)-5-aminoimidazole

The third reaction in purine biosynthesis in which a nitrogen atom is incorporated by formation of an amide group is the synthesis of **15** (equation 14).

The product was isolated and characterized from a system in which 1-(5′-phosphoribosyl)-5-aminoimidazole was the precursor, and its conversion to **15** required two fractions obtained from partially purified chicken liver extracts[73]. One fraction was responsible for the formation of the 4-carboxy intermediate **14**[74]. Conversion to the corresponding succinocarboxamide derivative was dependent

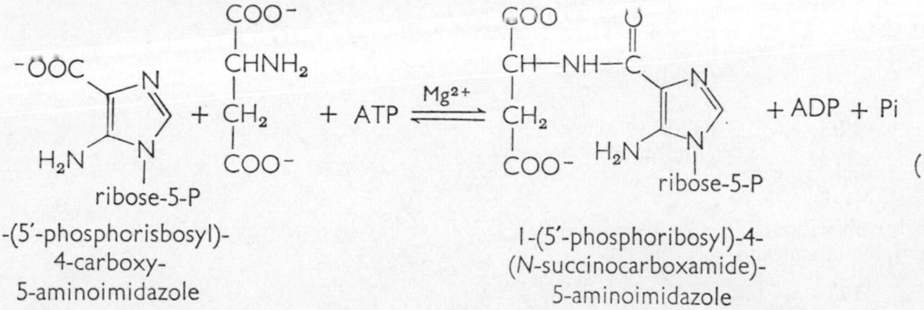

$$
\begin{array}{c}
\text{1-(5'-phosphorisbosyl)-}\\
\text{4-carboxy-}\\
\text{5-aminoimidazole}\\
\textbf{(14)}
\end{array}
\quad + \text{ATP} \xrightleftharpoons{\text{Mg}^{2+}} \quad
\begin{array}{c}
\text{1-(5'-phosphoribosyl)-4-}\\
\text{(N-succinocarboxamide)-}\\
\text{5-aminoimidazole}\\
\textbf{(15)}
\end{array}
\quad + \text{ADP} + \text{Pi}
\tag{14}
$$

on aspartic acid[71], ATP and Mg^{2+} ions; the breakdown of ATP to ADP and inorganic phosphate was equivalent to the amount of succinocarboxamide synthesized, thus establishing the stoicheiometry of the reaction. Although no mechanistic studies were done, the role of nucleoside phosphates appears to be similar to that found in glycineamide synthesis. In the cleavage of the amide bond of the succinocarboxamide in the reverse reaction, phosphate can be replaced by arsenate and only catalytic amounts of ADP are needed.

In the next step of the biosynthetic pathway the enzyme adenylosuccinase catalyses the conversion of **15** to **16**, another amide-containing intermediate in purine biosynthesis[75].

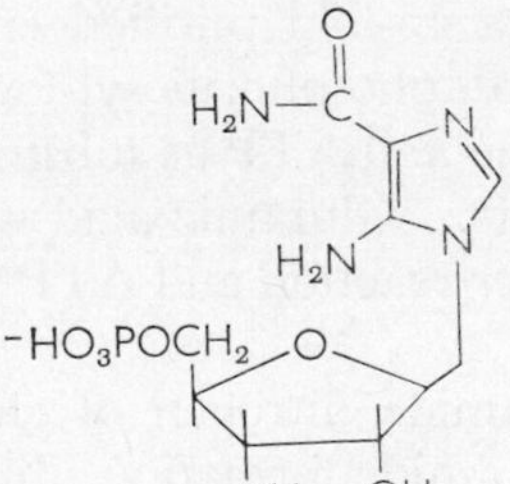

(16)

F. 5'-Phosphoribosyl-formylglycineamidine(18)

Formation of the amidine nitrogen atom corresponding to the $\text{N}_{(3)}$ position of the purine ring is biosynthetically accomplished by the formation of an amidine derivative (**18**) from 5'-phosphoribosyl-formylglycineamide (**17**) as shown in equation (15).

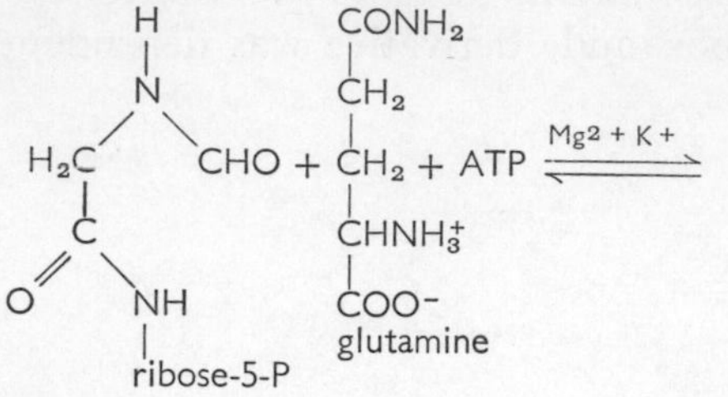

$$
\begin{array}{c}
H \\
| \\
N \\
H_2C \quad\diagdown\quad CHO \\
\diagdown C \diagup \\
\| \\
O \quad NH \\
| \\
\text{ribose-5-P}
\end{array}
+
\begin{array}{c}
CONH_2 \\
| \\
CH_2 \\
| \\
CH_2 \\
| \\
CHNH_3^+ \\
| \\
COO^- \\
\text{glutamine}
\end{array}
+ ATP \xrightarrow{\;Mg^{2+} + K^+\;}
$$

5'-phosphoribosyl-
formylglycineamide

(**17**)

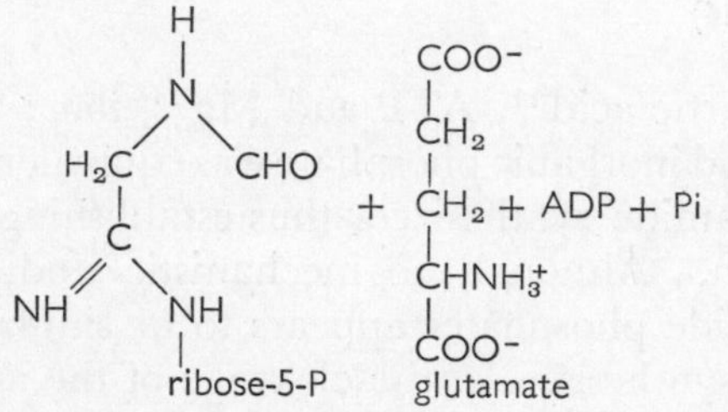

$$
\begin{array}{c}
H \\
| \\
N \\
H_2C \quad\diagdown\quad CHO \\
\diagdown C \diagup \\
\| \\
NH \quad NH \\
| \\
\text{ribose-5-P}
\end{array}
+
\begin{array}{c}
COO^- \\
| \\
CH_2 \\
| \\
CH_2 \\
| \\
CHNH_3^+ \\
| \\
COO^- \\
\text{glutamate}
\end{array}
+ ADP + Pi \qquad (15)
$$

5'-phosphoribosyl-
formylglycineamidine

(**18**)

The substituted amidine **18** was separated[76] from a reaction mixture of substrates and pigeon liver extract by an anion resin exchange column. Studies[77] on a 45-fold purified enzyme preparation from pigeon liver showed that phosphoribosyl-formylglycineamidine synthetase utilized glutamine and ATP as substrates, and Mg^{2+} and K^+ were needed as activators. Glutamic acid was identified as the product in the amide transfer reaction and ATP was cleaved to ADP and Pi.

The transfer of the amide nitrogen of glutamine to an acceptor molecule, in contrast to amide formation from ammonia, involves an additional carbon-to-nitrogen bond-breaking process, thus increasing the complexity of the catalytic function of the protein. For these transfer reactions which require a nucleotide, the activation of the acceptor molecule for reaction with the amide group presumably occurs by interaction with ATP by a mechanism similar to those discussed under glutamine synthetase. The activated carbonyl must then react with the amide nitrogen of glutamine. This reaction may take place before, after or concomitantly with the C—N bond-breaking step resulting in the expulsion of glutamate. Some insight into the mechanism for the enzymatic catalysis of this step, has been gained through the use of the inhibitor azaserine.

In the course of experiments designed to identify intermediates in biosynthesis of purines, it was found[78] that the addition of the antibiotic azaserine (O-diazoacetyl-L-serine, **19**) to reaction mixtures resulted in the accumulation of ribosyl-formylglycineamide. Kinetic analysis[79] showed competitive inhibition when glutamine and azaserine were incubated with the enzyme simultaneously. However, when azaserine was preincubated with the enzyme, inactivation occurred which could not be reversed by glutamine. These findings suggested azaserine reacted with the enzyme in an irreversible manner, probably at the glutamine binding site. Other structural analogues of glutamine, such as 6-diazo-5-oxo-6-norleucine and γ-glutamylhydrazine also inhibit activity. Essentially all the enzymes which catalyse the transfer of the amide nitrogen from glutamine are to some extent inhibited by azaserine.

Irreversible binding of an inhibitor at the active site of an enzyme provides a 'marker' permitting identification of the functional group participating in catalysis and thus provides a technique that has been used to study a number of enzymes. Dawid and coworkers[80] demonstrated (equation 16) that ^{14}C-labelled azaserine was bound covalently to phosphoribosyl-formylglycineamidine synthetase (**20**) purified from *Samonella typhimurium*[81]. Degradation of the inactivated enzyme (**21**) employing enzyme peptidase yielded a radioactive compound, postulated to be N-[2-(L-2-amino-2-carboxyethylthio)acetyl]-DL-serine (**22**), which was subsequently confirmed[82] by comparison with the synthesized compound. This was converted to serine and labelled S-carboxymethylcysteine (**23**) on acid hydrolysis. The investigators concluded this compound resulted from an alkylation reaction of azaserine with a sulphydryl group of a cysteine residue of the enzyme. The formation of **22** on enzymatic hydrolysis involves an acyl shift from oxygen to nitrogen taking place in the serine residue.

Azaserine presumably is bound in a conformation favourable for an alkylation reaction with the active site of the enzyme. The alkylation reaction is initiated by a proton transfer from the enzyme sulphydryl group to $C_{(5)}$ of azaserine. The formation of the diazonium salt generates a highly electrophilic diazomethine carbon atom which then may react with the nucleophilic sulphydryl anion to form the product (equation 17).

Conclusive evidence that the inhibitor reacts with only the sulphydryl group at the active site to the exclusion of all other sulphydryl groups of the enzyme is difficult to obtain in studies of this nature.

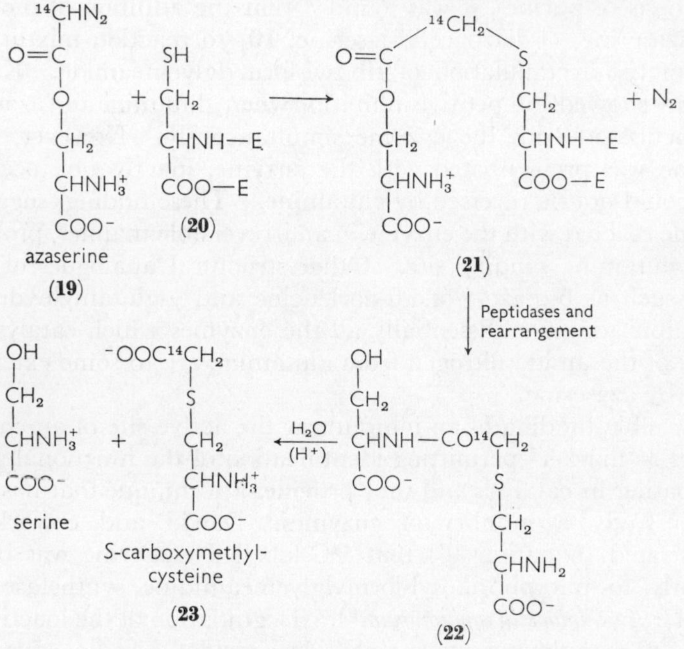

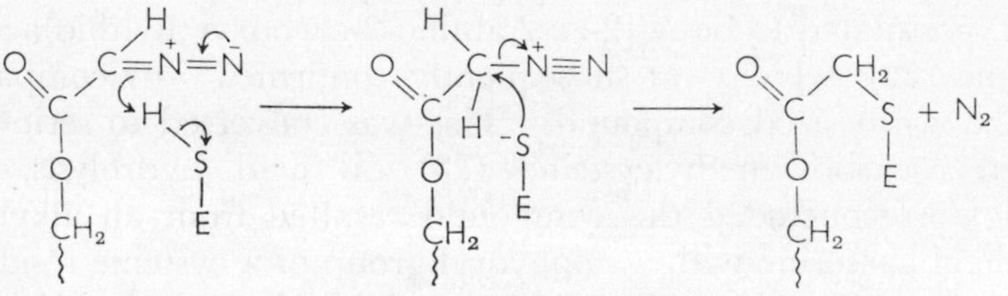

Failure to account for all the radioactivity incorporated into the protein by ^{14}C-labelled inhibitor or the observation of other radioactive products is due, in part, to secondary reactions that may occur during degradation of the protein and isolation of products. Azaserine, however, does not react with free cysteine suggesting that the enzyme sulphydryl has special catalytic properties. It was further proposed that this same sulphydryl group, in the normal catalytic function of the enzyme, displaces the amide nitrogen from the carbonyl carbon of glutamine with the formation of a thioester intermediate.

G. Adenylosuccinate

Adenylosuccinate (**25**), an intermediate[83,84] in the biosynthesis of adenosine monophosphate (AMP) is formed from inosine monophosphate (IMP) and aspartate (equation 18). The condensation

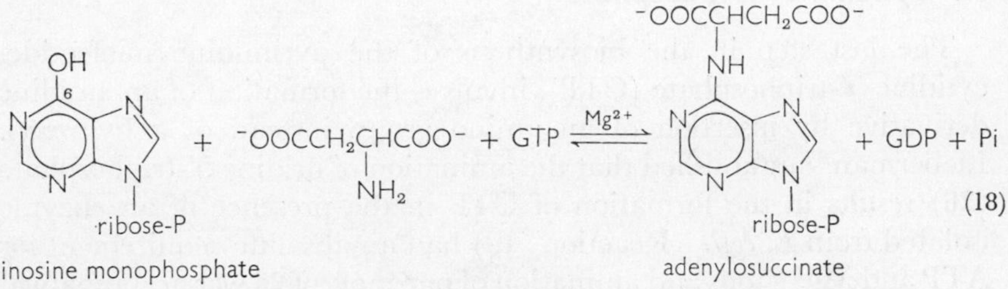

$$\text{inosine monophosphate} + {}^-\text{OOCCH}_2\text{CHCOO}^- + \text{GTP} \underset{\longleftarrow}{\overset{\text{Mg}^{2+}}{\rightleftharpoons}} \text{adenylosuccinate} + \text{GDP} + \text{Pi} \tag{18}$$

inosine monophosphate

(**24**)

adenylosuccinate

(**25**)

product **25** was isolated from a reaction mixture catalysed by a yeast extract[83], and its structure was deduced from u.v. spectra and enzymatic degradation methods: 5′-nucleotidase hydrolysed the phosphate moiety, and AMP synthetase degraded the adenylosuccinate to fumarate and AMP. Confirmation of the structure was obtained[85] by a comparison of the chemical properties of chemically synthesized 6-succinoaminopurine with the natural aglycone produced from mild hydrolysis of **25**.

In a detailed study on the purified *E. coli* enzyme, Lieberman[86] established the stoicheiometry of the reaction with respect to all reactants and products. The nitrogen donor, aspartate, could not be replaced in the condensation reaction by ammonia, the amides of glutamate and aspartate or by other free amino acids. Unlike other amide- and amidine-synthesizing reactions, guanosine triphosphate (GTP) rather than ATP, couples the formation of adenylosuccinate. The triphosphates of cytidine, uridine and inosine were essentially inactive.

In experiments[86] employing an ^{18}O label on $C_{(6)}$—OH of **24**, it was found that the oxygen atom was transferred to Pi. These results suggested the formation of a phosphorylated intermediate, and the author postulated an enzyme-catalysed formation of 6-phosphoryl-IMP which yielded **25** by displacement of the phosphate group by aspartate. Further evidence consistent with the formation of an activated intermediate was obtained from experiments in which the aspartate was replaced by hydroxylamine. Adenylosuccinate synthetase catalysed the formation of a new compound which was a derivative of IMP. The ^{14}C content was the same as the ^{14}C-labelled IMP substrate, and analysis revealed appropriate molar ratios of purine base, pentose and phosphate. The compound, presumably the hydroxamate, formed a chromogen with acid $FeCl_3$.

H. Cytidine 5'-Triphosphate

The last step in the biosynthesis of the pyrimidine nucleotide, cytidine 5'-triphosphate (CTP), involves the formation of an amidine derivative by insertion of an amino group instead of a hydroxyl. Lieberman[87] established that the amination of uridine 5'-triphosphate (**26**) results in the formation of CTP in the presence of an enzyme isolated from *E. coli*. Reaction (19) had an absolute requirement for ATP and Mg^{2+} ions and amination of one mole of **26** was accompanied by the release of one mole of ^{32}P-labelled Pi derived from labelled ATP. CTP synthetase from HeLa carcinoma cells[88] and rat liver[89]

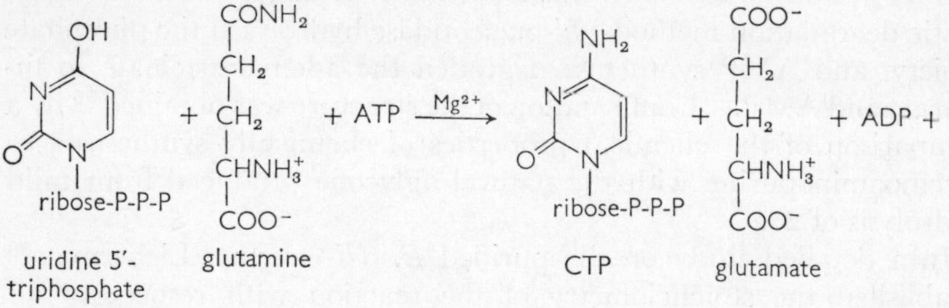

utilized glutamine as the nitrogen donor. ^{15}N-Label studies showed that the cytosine amino group was derived from the amide nitrogen of glutamine. Formation of cytidine nucleotide was strongly inhibited by the glutamine antagonist 6-diazo-5-oxo-L-norleucine and to a lesser extent by azaserine. Reinvestigation of the *E. coli* preparations, which utilized only ammonia, established that glutamine was the primary amino donor for both the bacterial and mammalian systems[90]. The glutamine site of the bacterial enzyme, important for binding or catalytic activity of the carbon–nitrogen bond breaking process, presumably was labile under the conditions used for purification in the original studies. Ammonia can partially substitute for glutamine at higher concentrations and pH values. The inhibition by a glutamine antagonist and stimulation by GTP of enzymatic activity, which was characteristic of the glutamine reaction, was not observed when ammonia was used as the nitrogen donor. The existence of two separate enzymes, utilizing different nitrogen donors, would also explain these observations.

Hydroxylamine can either replace the substrate in the synthetase-catalysed reaction or it reacts with an activated intermediate in a non-enzymatic reaction to form a hydroxamate. When hydroxylamine was substituted for a nitrogen donor two products were formed that were tentatively identified as the 6-*N*-hydroxy analogues of CTP and cytidine 5′-diphosphate.

III. FORMATION OF GUANIDINE DERIVATIVES: ATP → AMP + PPi REACTIONS

The biosynthesis of guanidine derivatives is associated with the cleavage of ATP to AMP and PPi. This group of reactions includes the formation of guanosine monophosphate (GMP) and argininosuccinate. Although the manner in which ATP is coupled to these reactions is not known, two possible mechanisms appear reasonable. Activation may occur by transfer of the AMP moiety to the substrate to form an adenyl intermediate $COPO_2$—O—Adenine. Alternatively, a group on the enzyme may be activated by a reaction with ATP which then reacts with substrate to form an intermediate covalently bound to the enzyme, Enzyme—OC. The main evidence in support of this mechanism consists of oxygen transfer from the substrate to AMP and inactivation of the enzyme by treatment with hydroxylamine. This type of evidence however, is not unambiguous.

The biosynthesis of the alkylamine, 5-phosphoribosylamine, also occurs with the formation of AMP and PPi. In this case, the ATP activation step is known to result in the formation of a free phosphoribosyl pyrophosphate intermediate. Furthermore the mechanism is unique, in terms of our present state of knowledge, since the activation and transfer steps require catalysis by two distinct enzymes.

The AMP–PPi cleavage in NAD synthesis may reflect the energy requirements for the amidation of a carboxyl group attached to a pyridinium ring. If the appropriate information were available, organization of the amidation reactions might be more meaningful based on the energy needed for their formation rather than indirectly by the ATP split or by the group synthesized. The relative energy

21*

requirements for the synthesis of an alkylamide and an amide attached to a strong electron-withdrawing substituent, such as the positively charged pyridinium ring, can be evaluated in terms of the relative stabilities of reactants and products. Electron withdrawal by the pyridinium group will stabilize the carboxylate anion of the reactant, whereas the same forces will tend to destabilize the amide product by further increasing the positive charges on the carbonyl carbon atom. The greater stabilization of reactant relative to the product may be reflected in a larger energy requirement for synthesis of the amide group of NAD. For example the free-energy change for the synthesis of glutamine[5] according to equation (2a) is 3400 cal/mole whereas 8200 cal/mole has been calculated for the analogous synthesis of the guanidine group of argininosuccinate[63].

These biosynthetic reactions are thermodynamically coupled to the hydrolysis of ATP. Since the free energy of hydrolysis of ATP to ADP and Pi is $-7,700$ cal/mole the synthesis of argininosuccinate, on the bais of the above figures, is energetically not feasible when coupled to ADP–Pi cleavage. The value of $-10,300$ cal/mole for ATP hydrolysis to AMP and PPi has been calculated in conjunction with argininosuccinate biosynthesis[63] which is considerably higher than other published values. However, regardless of the absolute values, the AMP–PPi split provides an additional driving force for synthetic reactions in that pyrophosphatase in the cell pulls the reaction towards completion by removing one of the products. If one assumes a mechanism of acyl activation through the formation of intermediates, a similar comparison can be made; the group transfer potential of acyl–AMP is higher than that for acyl phosphate. The free energies of hydrolysis, i.e. the tendency to transfer a donor molecule to a common acceptor, water, are $-10,500$ cal/mole for acetyl phosphate and $-13,300$ cal/mole for acetyl–AMP[28].

A. Nicotinamide–Adenine Dinucleotide

The final step in the biosynthesis of nicotinamide–adenine dinucleotide (NAD) is the formation of the amide group derived from the amide nitrogen of glutamine (equation 20).

Preiss and Handler[91] partially purified the enzyme catalysing this reaction, NAD synthetase, from yeast and liver, and established the identity of the products. Formation of NAD requires Mg^{2+} ions and is specific for ATP. In contrast to the reactions considered so far, in which activation of the carboxyl group for amide formation results

$$
\begin{array}{l}
\text{deamido-NAD} \quad + \quad \text{glutamine} \quad + \quad \text{ATP} \xrightarrow{\text{Mg}^{2+}} \\
\end{array}
$$

(structures: deamido-NAD with CONH$_2$ omitted — COO$^-$ on ribose-P-P-ribose-AD pyridinium; glutamine with CONH$_2$, CH$_2$, CH$_2$, CHNH$_3^+$, COO)

$$
\text{NAD} \quad + \quad \text{glutamate} \quad + \quad \text{AMP} + \text{PPi}
$$

(NAD structure: CONH$_2$ on ribose-P-P-ribose-AD pyridinium; glutamate: COO$^-$, CH$_2$, CH$_2$, CHNH$_3^+$, COO$^-$)

(20)

in the cleavage of ATP to ADP–Pi, ATP in this reaction is converted to AMP and pyrophosphate (PPi). Substitution of hydroxylamine for glutamine led to the formation of a small amount of a hydroxamate suggesting that activation in this case is accomplished by the formation of an enzyme-bound acyl adenylate intermediate.

Ammonia can also serve as a nitrogen donor for NAD synthetase. The pH–rate profile for glutamine as a nitrogen donor exhibits a broad maximum from pH 6·2 to 7·6 whereas the activity with ammonia increases with increasing pH to an optimum pH of 8·3, suggesting the unionized form of ammonia is the active nucleophile. A comparison of the concentrations of glutamine and ammonia required in order to obtain one-half maximal activity (K_m) is consistent with this supposition. The K_m value for ammonia is considerably larger than that for glutamine but when compared in terms of the unionized concentration of NH$_3$ the K_m's are of the same order. Common to other enzymes utilizing glutamine, azaserine inhibits NAD synthetase.

An enzyme preparation from a bacterial source, E. coli, differs from the yeast and mammalian enzymes in that ammonia is found to be a much more efficient donor than glutamine[92]. Greater reactivity is reflected in both the rate of catalysis and the affinity for the enzyme. At saturating concentrations of the nucleophile, NAD synthesis with glutamine is about 30% of the rate observed with ammonia. The concentration needed for saturation of the enzyme is 200 times higher for glutamine than for ammonia. Moreover, the bacterial enzyme is not inhibited by the glutamine antagonist azaserine. Inhibition is observed, however, with psicofuranine, deocyinine and adenosine.

B. Guanosine 5'-Phosphate

Two enzymes have been reported which catalyse the conversion of xanthosine 5'-phosphate (XMP) to guanosine 5'-phosphate (GMP). GMP synthetase isolated from mammalian tissues utilized the amide nitrogen of glutamine as the nitrogen donor[93,94] whereas ammonia is the more active substrate for the bacterial enzyme[95] (equation 21).

The mammalian enzymes purified from pigeon liver[93] and calf thymus[94] are essentially the same with regards to stoicheiometry, nature of the nitrogen source and fate of the displaced xanthosine oxygen atom. Amination of XMP requires glutamine, ATP and Mg^{2+} ions. A sulphydryl group is essential for both enzymes as evidenced by either inhibition in the presence of p-chloromercuri-benzoate or by the protection against inactivation provided by sulphydryl-reacting compounds such as 2-mercaptoethanol or cysteine.

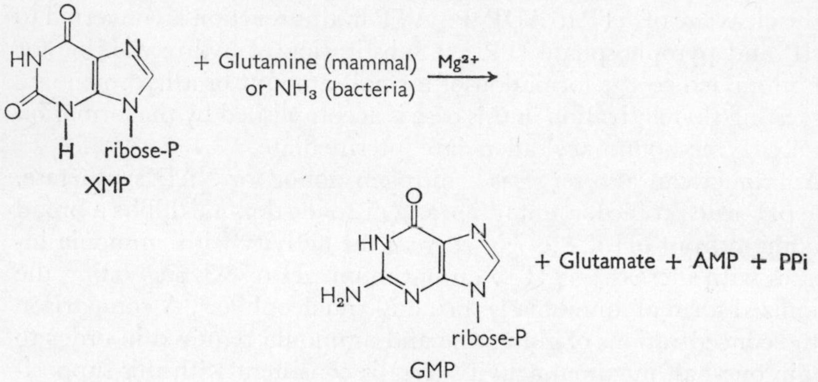

The stoicheiometry of the reaction was established by demonstrating that the formation of GMP, PPi and glutamate were equimolar with the amount of glutamine consumed. Isotopic labelling experiments showed that the amide nitrogen of glutamine was incorporated into GMP essentially without dilution. Relatively high concentrations of ammonium chloride also served as a nitrogen source. Glutamine was probably the preferred substrate for the mammalian enzyme since at saturating concentrations the rate of GMP synthesis with ammonia was only 15% of that observed with glutamine. Moreover, the enzyme was inhibited by azaserine and 6-diazo-5-oxo-L-norleucine.

The mechanism by which ATP couples the transfer of an amide group to form a guanidine is not known. The ATP cleavage to AMP and PPi is the same as that observed in the ATP-dependent activation of amino acids.   The carboxyl group of amino acids is activated by the formation of an acyl adenylate intermediate, and the results of ^{18}O studies suggest a similar mechanism may be operative in the biosynthesis of GMP.   The ^{18}O-labelled oxygen atom of XMP, which is replaced by a nitrogen atom in the formation of GMP is transferred to AMP essentially without dilution in radioactivity[93].   A direct reaction between ATP and XMP to form an adenyl intermediate could account for this ^{18}O transfer.   A mechanism involving the formation of an adenyl intermediate is depicted in equation (22).

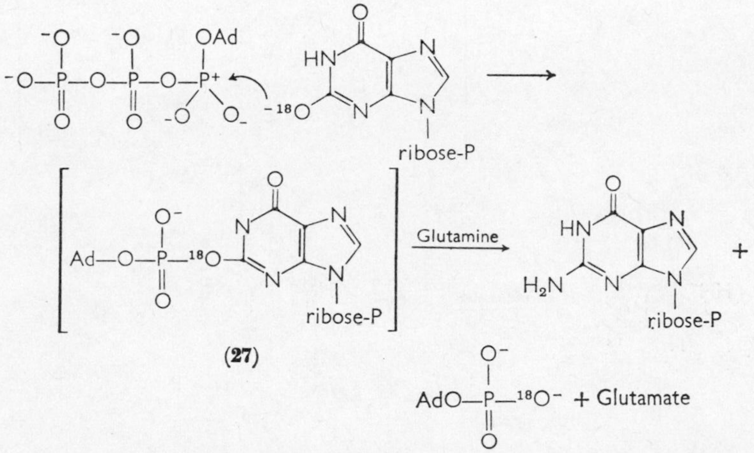

$$(22)$$

This reaction may occur by a consecutive mechanism with the formation of an adenyl intermediate (27) or by a concerted mechanism in which the extraction of the oxygen atom by ATP occurs without covalent-bond formation.   Enzymes that catalyse the formation of acyl adenylate intermediates also catalyse partial-exchange reactions. No exchange reaction was observed on incubation of the purified calf thymus enzyme with ATP, Mg^{2+} and ^{32}P-labelled PPi.   Reversibility of the overall reaction, however, could not be demonstrated.

An alternative mechanism could also account for transfer to AMP of ^{18}O from XMP, labelled in the 2-hydroxyl group.   In equation (23) ATP reacts with some group on the enzyme, for example a carboxylate, to form an adenylated enzyme (28).   XMP then reacts with the activated enzyme complex with the release of AMP followed by nucleophilic attack of ammonia or glutamine on the XMP–enzyme adduct (29) to yield GMP and enzyme which contains the ^{18}O atom derived from XMP.   During the next cycle of the enzyme-catalysed reaction the ^{18}O atom would be transferred from the enzyme to AMP. This reaction differs from the mechanisms previously discussed in that the first activation step involves a group on the enzyme.

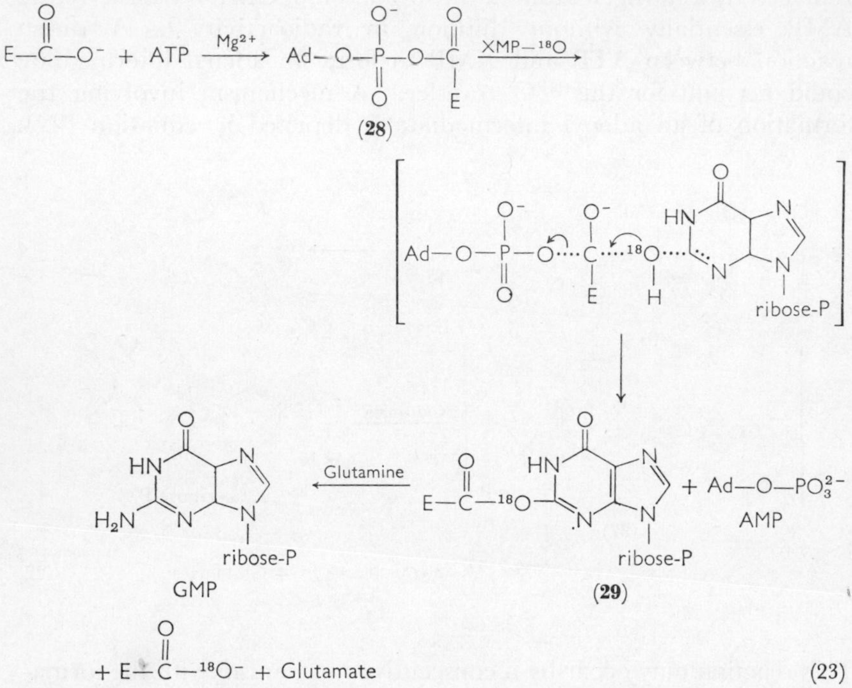

(28)

(29)

$$+ E-\overset{O}{\underset{\|}{C}}-^{18}O^- + \text{Glutamate} \tag{23}$$

A possible mechanism involving an AMP–nitrogen interaction was ruled out on the basis that phosphoramidate was not utilized for GMP synthesis[94].

Moyed and Magasanik[95] found that with the purification of the enzyme from *Aerobacter aerogenes*, glutamine became a progressively poorer substrate. When the enzyme was purified 300-fold, only ammonia served as a nitrogen donor. The partially purified preparations probably generated ammonia from glutamine by the action of glutaminase, an enzyme that hydrolyses the amide nitrogen of glutamine.

In addition to the AMP–PPi cleavage of ATP the enzymes which catalyse the synthesis of NAD and GMP bear other similarities. The synthetases isolated from a mammalian source utilize glutamine as substrate and are inhibited by azaserine, whereas the bacterial enzymes of both synthetases utilize ammonia and are inhibited by psicofuranine. It is conceivable that in the course of adaption from ammonia-utilizing to glutamine-utilizing enzymes an active site develops which catalyses the cleavage of the carbon–nitrogen amide bond by a mechanism common to both enzymes.

The synthesis of GMP, catalysed by the bacterial enzyme, is dependent on XMP, ATP, NH_3, and Mg^{2+} ions. When the reaction was carried out in the presence of excess ammonia and XMP was quantitatively converted to GMP, ATP and equimolar amounts of AMP and PPi, were formed.

The results of inhibition studies[95] suggest that ATP couples with the amide transfer reaction by the formation of an activated intermediate covalently bound to some group on the enzyme. Hydroxylamine inhibits GMP synthesis by inactivating the enzyme. Maximum inhibition was observed when ATP, XMP, and Mg^{2+} ions were preincubated with the enzyme. Inhibition was irreversible since the removal of hydroxylamine by reacting it with biacetyl did not restore the activity of the enzyme. The rate of increase of the observed inactivation follows first-order kinetics and is not affected by ammonia. These findings suggest that ATP and XMP react with the enzyme to form an activated complex. This complex can then react with ammonia to yield GMP, PPi, AMP, and free enzyme, or it can react with hydroxylamine which leads to inactivation of the enzyme. The hydroxylamine inactivation strongly suggests that an intermediate is bound to the enzyme by a covalent linkage, possibly a phosphorylated or an adenylated enzyme (equation 23).

The role of XMP as part of the activated complex 29 is not explicable directly in terms of the postulated mechanism. As can be seen from equation (23) a mechanism which involves the activation of an enzyme group and subsequently an acceptor molecule, requires the formation of two intermediates vulnerable to inactivation on nucleophilic attack by hydroxylamine. Attack of the activated purine molecule 29 by hydroxylamine would presumably take place at the carbon atom that ammonia attacks which would yield a purine hydroxamate and free enzyme. Attack of the enzyme–ATP derivative (28), however, would result in the formation of an inactive enzyme hydroxamate. In order to conform to the mechanism described, the requirement for XMP may be related to binding of substrates, analogous to the AMP requirement for the binding of glutamine to glutamine synthetase, or XMP may effect a conformational change of the protein which influences the reactivity of the active site.

The results of studies of GMP synthetase inhibition by the antibiotic psicofuranine can also be interpreted in terms consistent with a
hypothesis of a conformational change induced by XMP binding.
The inactivation by psicofuranine has been studied in enzyme preparations obtained from *E. coli* and from a mutant strain isolated after
ultraviolet irradiation of the parental organism[96]. The enzyme from
the parental strain is irreversibly inactivated by psicofuranine in the
presence of XMP and PPi. Inhibition of the enzyme obtained from
the mutant or resistant strain is only partial, requires the presence of
PPi, but not XMP and is reversible. These results indicated that
more than one type of inhibition was being observed and the investigators proposed a two-step process formulated by equations (24 and 25).

$$\text{Enzyme} + \text{Psicofuranine} + \text{PPi} \overset{\text{Slow}}{\rightleftharpoons} \text{Enzyme}_1 \text{ (inhibited)} \qquad (24)$$

$$\text{Enzyme}_1 \text{ (inhibited)} + \text{XMP} \overset{\text{Irreversible}}{\longrightarrow} \text{Enzyme}_2 \text{ (inhibited)} \qquad (25)$$

Step (24) is dependent on PPi and can be reversed by dilution or
through the removal of PPi by treatment of the reaction mixture with
pyrophosphatase. Step (25) is observed in the enzyme from parental
E. coli but not in the resistant strain. The reaction is dependent on
XMP as well as PPi, is irreversible and is a relatively slow reaction.
When the enzyme is incubated with $1 \times 10^{-7}\text{M}$ psicofuranine, XMP
and PPi, a time-dependent inactivation occurs in which one-half the
maximum inhibition is achieved in 20 minutes. Kinetic results indicate that psicofuranine is a non-competitive inhibitor with respect to
XMP, ATP, and ammonia. The differences between the parental
and mutant strains cannot be accounted for by a change in the affinity
of enzyme for substrates since the measured binding constants are
essentially the same.

The XMP requirement, in order for the irreversible inhibition step
to occur, may be due to either an interaction of psicofuranine with
XMP or to a change in the conformation of the protein effected by
XMP which permits an interaction of psicofuranine with the catalytic
site of the enzyme.

Although it is premature to postulate a mechanism based on the
available information, two additional observations are consistent with
a change in conformation-active site hypothesis. If it is assumed irreversible inactivation results from a conformational change of the
enzyme effected by XMP, the resistant strain either does not undergo a
conformational change or this change differs from that which occurs
in the parent strain where no inhibition by psicofuranine occurs.

Tertiary structural differences between the enzymes may account for these differences in reactivity, and heat-inactivation studies indicate that the conformation stabilities are in fact different. The resistant strain is more easily denatured, and protection against heat inactivation provided by ATP and XMP is more effective for the parental strain.

The substrates ATP and ammonia protect the enzyme against the irreversible inactivation step. Psicofuranine (**30**) although structurally similar to ATP, is probably binding at a site distinct though possibly not distant from that which binds ATP since inhibition kinetics were non-competitive.

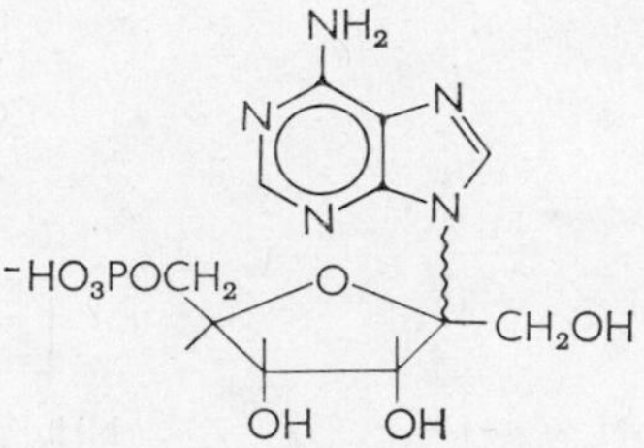

(**30**)

AMP and adenosine inhibit enzymatic activity in a qualitatively similar manner to that of psicofuranine. The decrease in activity observed in the presence of GMP or its analogues suggests this inhibition may have a regulatory function.

C. Argininosuccinate

The major end-product of nitrogen metabolism in terrestrial vertebrates is urea. The principle sources of ammonia for urea production arise from the deamination of α-amino acids and the hydrolysis of the amides of glutamic and aspartic acids. The terminal step in urea production is the hydrolysis of the guanido group of arginine. Krebs and Henseleit[97] discovered that ornithine functioned catalytically in the conversion of ammonia to urea in rat liver slices and they postulated a series of reactions known as the ornithine or urea cycle shown in Figure 3. In this cyclic process ammonia and CO_2 are incorporated into organic compounds at one site and urea leaves at another. The conversion of citrulline to arginine, urea-cycle components, has been elucidated by Ratner and her associates[98-100]. The formation of the guanidine group of arginine, which requires two enzymes, results from a condensation reaction involving citrulline and aspartate to form argininosuccinate. The latter compound is cleaved in a subsequent reaction step to yield arginine and fumarate. The requirement for two enzymes to transfer the α-amino group of aspartate is a common characteristic of the enzymes which utilize the α-amino group as a nitrogen donor.

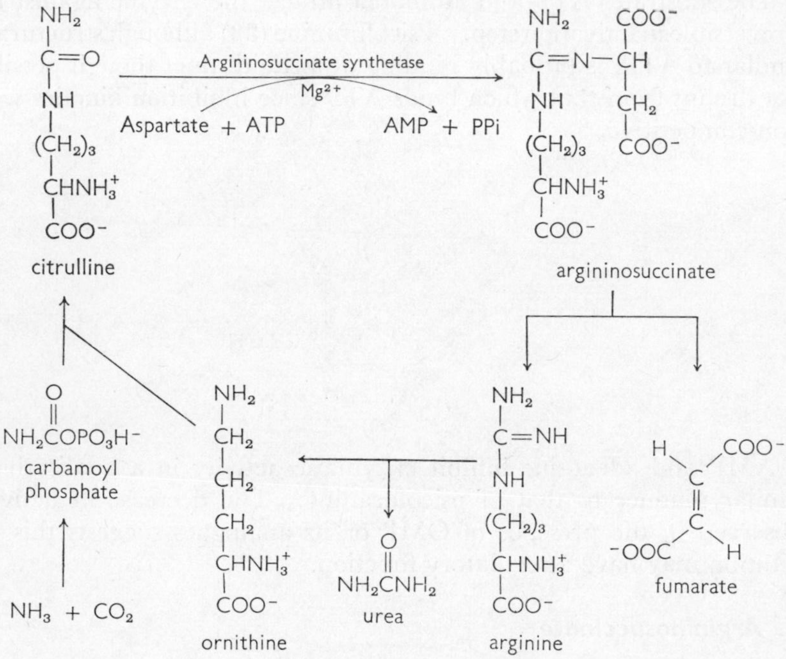

FIGURE 3. Urea cycle.

The condensation reaction, catalysed by argininosuccinate synthetase, required ATP which is converted to AMP and PPi during the course of the reaction[101]. In the absence of pyrophosphatases, which were present in earlier preparations, PPi accumulated in amounts equivalent to the utilization of citrulline. The position of the equilibrium does not favour synthesis so the stoicheiometry was determined for the back reaction and the specificity of AMP was established.

Argininosuccinate was also reported to occur in plant extracts[102,103] and algal cells[104], and fumarate was established as the product rather than malate as originally reported. The intermediate was biosynthetically prepared from arginine, fumarate and the mammalian argininosuccinase and was characterized by titration data, chemical reactions and physical properties[105].

In a mechanism study, Rochovansky and Ratner[106] demonstrated that ^{18}O is transferred from the ureido group of citrulline to AMP. This transfer may result from a direct interaction between the citrulline oxygen atom and the AMP moiety of ATP, or transfer may be an indirect one involving a group on the enzyme. Attempts to demonstrate partial reactions by ^{32}PPi–ATP isotope-exchange studies were unsuccessful. All three substrates must be present to detect any reaction. These findings indicate either a concerted mechanism in which the PPi bond is broken simultaneously with C—N bond formation, or the products of the activation reaction remain bound to the enzyme until nucleophilic attack of the intermediate occurs. A free-energy change for the ATP-dependent, enzyme-catalysed reaction at pH 7·5 of -2100 cal/mole was calculated from the equilibrium constant[63].

D. 5-Phosphoribosylamine

Formation of 5-phosphoribosylamine (**12**) is the first nitrogen-transfer step in a series of biosynthetic reactions which lead to the formation of the purine nucleotides. The nitrogen atom of **12** corresponds to $N_{(9)}$ of the purine ring (see **11**). The biosynthesis of **12** from ribose 5-phosphate (**31**) and glutamine occurs in a two-step sequence shown in equation (26).

In the amide-transfer reactions considered in the previous sections, the activation step and the carbon–nitrogen bond-breaking and bond-making steps are catalysed by a single enzyme. In contrast, the biosynthesis of phosphoribosylamine, which occurs by an analogous ATP-dependent activation of an acceptor molecule and nucleophilic attack by the amide nitrogen atom on the activated pyrophosphate derivative (**32**), is carried out in two discrete steps by two distinct enzymes and the free intermediate is isolable.

The activated intermediate is synthesized by the transfer of the pyrophosphate moiety of ATP to ribose 5-phosphate. 5-Phosphoribosyl 1-pyrophosphate has been isolated and characterized as the product of the enzyme-catalysed reaction, and an equimolar stoicheiometry of reactants and products was established[107]. Mg^{2+} ions were essential for activity. The enzyme was purified from pigeon liver and was also found in mammalian liver and *E. coli*.

Khorana and coworkers[108] have demonstrated that the reaction

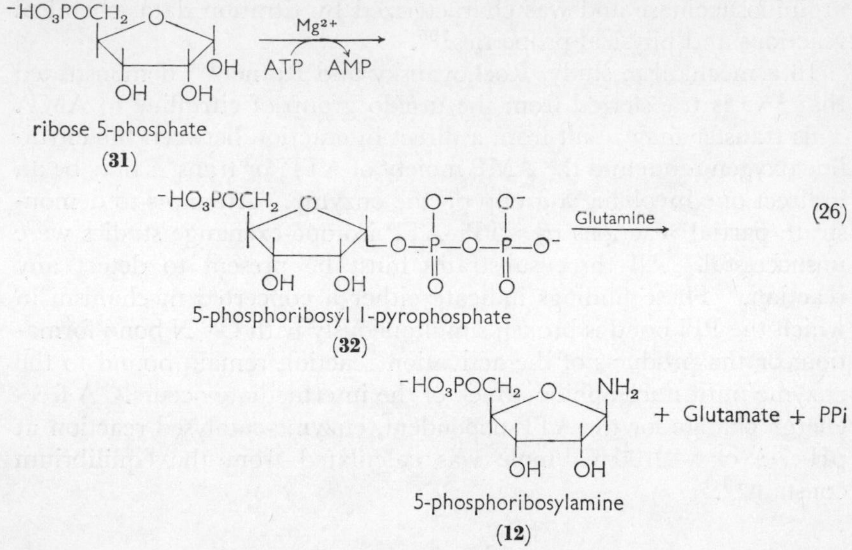

$^-HO_3POCH_2$

ribose 5-phosphate

(31)

$\xrightarrow[\text{ATP} \quad \text{AMP}]{\text{Mg}^{2+}}$

$^-HO_3POCH_2$

5-phosphoribosyl 1-pyrophosphate

(32)

$\xrightarrow{\text{Glutamine}}$ (26)

$^-HO_3POCH_2$ NH_2

+ Glutamate + PPi

5-phosphoribosylamine

(12)

proceeds by a direct transfer of the pyrophosphate moiety. A pyrophosphate transfer rather than the transfer of single phosphate groups was indicated from the finding that ribose 1,5-diphosphate would not substitute for the substrate. Furthermore, it was shown that the pyrophosphate group of **32** originated from the terminal and middle phosphate groups of ATP. Identification of the individual phosphate groups was possible by employing ^{32}P-labelled ATP in specific phosphate groups and a degradation procedure of **32** which liberated the terminal phosphate group and formed 5-phosphoribose 1,2-cyclic phosphate (**33**) (equation 27).

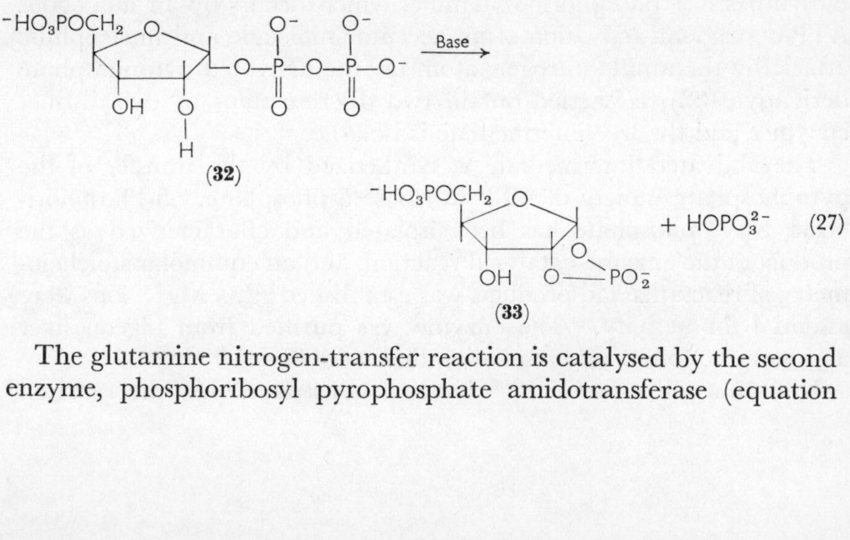

$^-HO_3POCH_2$

(32)

$\xrightarrow{\text{Base}}$

$^-HO_3POCH_2$

+ HOPO$_3^{2-}$ (27)

(33)

The glutamine nitrogen-transfer reaction is catalysed by the second enzyme, phosphoribosyl pyrophosphate amidotransferase (equation

26). Goldthwait[109] partially purified the enzyme from pigeon liver and showed that for each mole of **32** utilized, an equimolar amount of glutamine was converted to glutamate. Azaserine inhibited the reaction and was competitive with respect to glutamine.

The product (**12**), due to its instability, could not be isolated. However, indirect evidence strongly supports the contention that 5-phosphoribosylamine is enzymatically formed. Chemically synthesized **12** was found to be unstable even under mild conditions. When the enzymatic reaction was carried out with **32** labelled with ^{32}P in the 5-position, the main radioactive product was indistinguishable from ribose 5-phosphate, which presumably resulted from the hydrolysis of phosphoribosylamine. Chemically prepared **12** was active in the enzymatic synthesis of 5'-phosphoribosyl-glycineamide (**13**), the subsequent step in purine biosynthesis (section II.E), thus suggesting that phosphoribosylamine is in fact a true product.

The transfer of the amide nitrogen probably consists of a single displacement reaction with an inversion at $C_{(1)}$ of ribose. **32** has the α-configuration whereas the naturally occurring purine nucleotides possess the β-configuration.

IV. ADDITIONAL BIOSYNTHETIC REACTIONS OF GLUTAMINE

A. Histidine

All mammals studied, with the exception of man, require a dietary source of histidine for growth or maintenance of nitrogen balance. One of the nitrogen atoms of the imidazole ring of histidine is derived from the amide nitrogen of glutamine. In studies on cultures of *E. coli*, Neidle and Waelsch[110] demonstrated the incorporation of ^{15}N-labelled amide nitrogen of glutamine into histidine. Ammonia, the α-amino nitrogens of glutamate and aspartate, and asparagine amide nitrogen did not compete with glutamine as a nitrogen source. By enzymatic degradation of the isolated histidine, it was shown that 80% of the incorporated ^{15}N was in $N_{(1)}$ of the imidazole ring. The biosynthesis of hisitidine is a multistep process involving the formation of at least nine intermediates. Elucidation of part of the pathway has been achieved through studies on the synthesis of imidazoleglycerol phosphate (**35**), one of these intermediates. The carbon–nitrogen structural skeleton of **35** is synthesized from ATP, ribose 5-phosphate and glutamine. In a condensation reaction ribose 5-phosphate forms a covalent bond with $N_{(1)}$ of ATP. The phosphoribosyl–ATP intermediate is converted in a series of reactions to **34**, which then reacts with glutamine to form imidazoleglycerol phosphate (**35**).

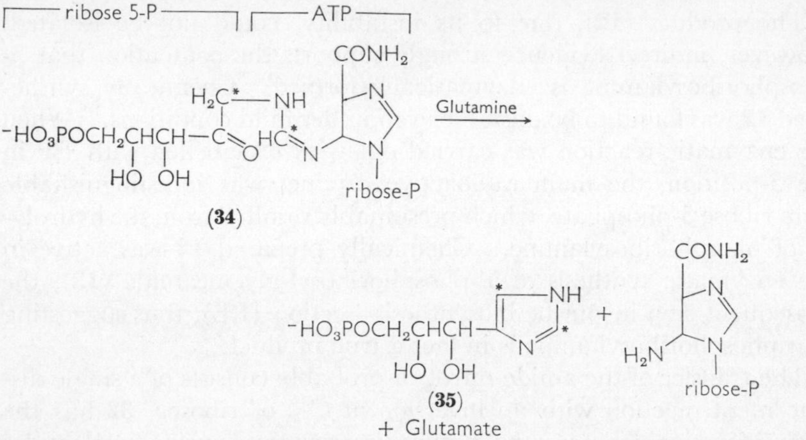

Isotopic tracer studies demonstrated that $C_{(2)}$ of adenine and $C_{(1)}$ of ribose are incorporated into **35** without dilution[111].

Bacterial extracts containing ATP, ribose 5-phosphate and Mg^{2+}, allowed to react in the absence of glutamine, accumulate **34**, which was isolated and identified by Smith and Ames[112]. The intermediate **34** contained a reducing group, lacking in earlier intermediates, and the enzyme responsible for its formation catalysed an Amadori-type rearrangement on the aminoaldolose of the formimino side-chain to form the aminoketose.

Little is known concerning the amide transfer mechanism in the conversion of **34** to **35** or the requirements for additional cofactors or metal ions, since relatively impure enzyme preparations have been used to carry out the reaction. Some information has been obtained in studies[112] with *Salmonella typhinurium* mutants, which lack certain enzymes necessary for the biosynthesis of **35**. The combined extracts of two of these mutants are capable of converting the intermediate and glutamine to **35**, however, when tested separately the extracts were inactive. The findings suggest two factors, and a two-step reaction sequence occurs in the amide transfer and ring-closure reactions leading to synthesis of the imidazole ring. Two general mechanisms could account for a two-step reaction. An intermediate Schiff base may be formed by transfer of the amide group to the ribulose moiety, or an intermediate might have the structure of activated substrate bound to the enzyme. The deficiency resulting from a blocked gene of one mutant can be overcome by using high concentrations of ammonia instead of glutamine as the nitrogen donor. Since glutamine is probably the normal substrate, it is conceivable that the binding or catalytic functions concerned with the amide transfer reaction have been modified in the mutant. Clarification of the mechanism for amide transfer in histidine biosynthesis must await further studies.

Participation of histidine in the catalytic site has been reported for chymotrypsin, succinic thiokinase, ribonuclease, and hexokinase-catalysed reactions, and has been implicated in many others. The chemical properties of the cyclic amidine imidazole portion of the histidine molecule are uniquely appropriate for general acid–base[113] and nucleophilic[114] catalysis. Detailed considerations of enzyme reactions involving imidazole catalysis can be found in publications by Bender and Kézdy[115], Bruice, and Benkovic[116], and Anderson and coworkers[117].

B. Anthranilate

Anthranilic acid (**37**) arises from the biosynthetic pathway responsible for the formation of the aromatic amino acids phenylalanine, tyrosine, tryptophan and *p*-aminobenzoic acid. It is an intermediate in the formation of tryptophan in many organisms and from evidence obtained from nutritional, genetic and end-product inhibition studies, anthranilate synthetase is probably the first enzyme in the pathway specifically involved in tryptophan biosynthesis. The carbon skeleton is derived from a product of carbohydrate metabolism, 2-keto-3-deoxy-7-phospho-D-glucoheptonic acid. A series of enzymatic conversions culminate in the formation of shikimic acid (**38**), one of the first recognized intermediates in the biosynthesis of aromatic amino acids. Only recently has attention been focused on the sequence of reactions for the conversion of this intermediate to the aromatic amino acids. Shikimate in three reaction steps is converted to chorismic acid (see **36**), the immediate precursor of anthranilate. Anthranilate synthetase catalyses the reaction of chorismate and glutamine in the presence of Mg^{2+} ions as shown in equation (29).

Glutamine was found to be the most efficient nitrogen donor in the conversion of shikimate 5-phosphate[118] and chorismate[119] to anthranilate. The transfer of the amide nitrogen rather than the α-amino nitrogen of glutamine was established in experiments employing glutamine labelled with ^{15}N in the amide group[118]. Isolation and

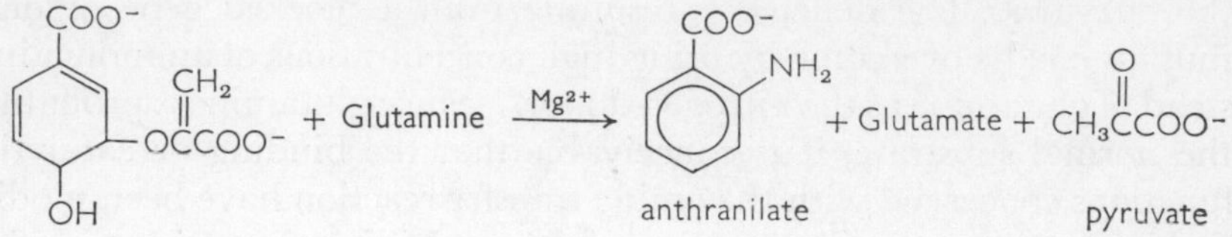

$$\text{chorismate (36)} + \text{Glutamine} \xrightarrow{\text{Mg}^{2+}} \text{anthranilate} + \text{Glutamate} + \text{CH}_3\text{CCOO}^- \text{ (pyruvate)} \tag{29}$$

analysis of anthranilate showed that the ^{15}N nitrogen atom was incorporated essentially without dilution. Ammonia, as in other glutamine enzymes, can also serve as a nitrogen source provided relatively high concentrations are used at alkaline pH values. Anthranilate accumulated more rapidly in whole cells when glutamine was the substrate, hence the amide nitrogen was considered to be the physiological substrate. The glutamine antagonists azaserine and 6-diazo-5-oxo-L-norleucine inhibited the synthesis of anthranilate in bacterial extracts[119] and whole cells[120] when glutamine was used as the nitrogen source. No inhibition by these compounds was observed, however, in whole-cell studies using ammonia as the nitrogen donor.

Srinivasan found that the radioactivity from [3,4-^{14}C]glucose was incorporated into the carboxyl carbon and the 3 and 4 ring positions of shikimate. Comparison[121] of the labelling pattern of anthranilate, biosynthesized from the same labelled glucose, with shikimate showed that aromatization of the ring takes place without rearrangement of the carbon skeleton as shown in 37 and 38. With this information

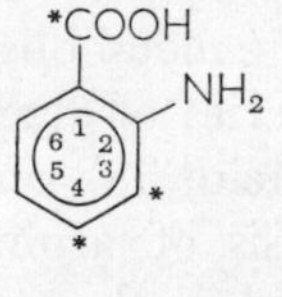

anthranilic acid
(37)

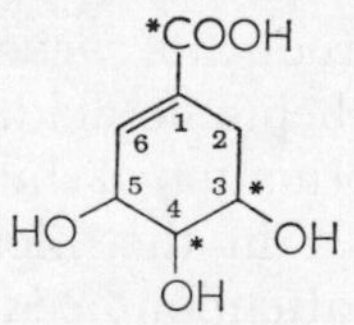

shikimic acid
(38)

and the known reactions used in the degradation of the aromatic ring, it was possible to establish that the amination reaction occurred at $C_{(2)}$ of the intermediate rather than at $C_{(6)}$.

Chorismic acid, was isolated[122] as an intermediate in the biosynthesis of aromatic amino acids by bacterial mutants. These mutants lacked enzymes capable of utilizing the intermediate in further reactions and as a consequence accumulated an isolable quantity of the intermediate. Chorismate (**36**) appears to be a 'branch compound' or the last common intermediate of the various specific pathways of aromatic amino acid biosynthesis. The ability of this compound to serve as a precursor for anthranilate synthesis has been demonstrated with the enzymes isolated from *E. coli*[123] and *Neurospora crassa*[124] and with extracts of *Aerobacter aerogenes*[122]. The relative stereochemistry of chorismate was deduced from the known structures of shikimic and prephenic acids and the n.m.r. spectrum[125].

Little is known concerning the mechanism of the reaction. Purified enzyme preparations[123,124] require only Mg^{2+} ions to catalyse reaction (29). Inhibition of the enzyme was observed[123] in the presence of sulphydryl-reacting reagents. In view of the apparent complexity of the reaction, recent investigations have sought to ascertain whether the reaction is catalysed by a single enzyme or an enzyme complex. DeMoss[124] achieved an 83-fold purification of the *Neurospora* enzyme and suggested that this reaction, or series of reactions, is catalysed by a single enzyme. Ultracentrifugation and gel electrophoresis studies[123], carried out on a purified enzyme obtained from *E. coli*, indicated the synthesizing activity was associated with a predominant protein in the preparation, but did not distinguish between a single enzyme or an enzyme complex. Egan and Gibson[126] have reported that a protein purified about 1400-fold with respect to anthranilate synthetase activity also has phosphoribosyl transferase activity, the next enzyme in the biosynthetic sequence of tryptophan formation from anthranilate. The presence of two enzymatic activities on the purified protein suggests that a molecular aggregate is involved.

A number of compounds (Figure 4) have been proposed as possible intermediates in the conversion of chorismate to anthranilate. If it is assumed that the first step in the synthesis of anthranilate proceeds by a nucleophilic attack of the amide nitrogen of glutamine at $C_{(2)}$ of chorismate with the simultaneous cleavage of the *para*-hydroxyl group, an intermediate having the structure **39a** would be formed. Subsequent reactions of this intermediate may involve the expulsion of pyruvate to form **39b**, the expulsion of glutamate followed by migration of the pyruvyl moiety to form **40**[127], or the expulsion of glutamate to form **41**[128]. Lingen and coworkers[129], synthesized compounds **39b** and **40** and found that they were inactive in extracts capable of synthesizing anthranilate from shikimate, pyruvate and glutamine.

Levin and Sprinson[128] have proposed a scheme which involves the formation of intermediate **41**. A *Streptomyces aureofaciens* mutant was reported[130] to accumulate large quantities of *trans*-2,3-dihydro-3-hydroxyanthranilic acid. It was not known if the anthranilate biosynthetic pathway was genetically blocked in this organism. Compound **41** presumably could undergo elimination of pyruvate to yield anthranilate or hydrolysis to accumulate the reported hydroxyamino

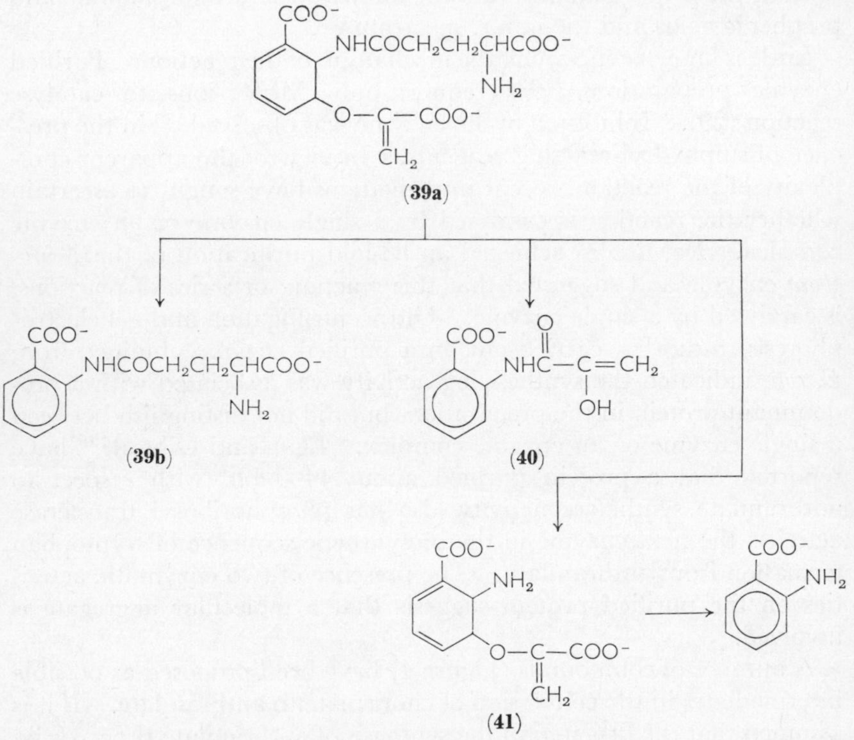

FIGURE 4.   Possible derivatives of the chorismate–glutamine reaction.

acid.   Srinivasan[121] found 2,3-dihydro-3-hydroxyanthranilate would not serve as a precursor for the biosynthesis of anthranilate and suggested the true intermediate still bears the enolpyruvyl side-chain. Consistent with these observations, an intermediate of anthranilate biosynthesis has been isolated which corresponds to structure **39a**[131]. The results of kinetic studies[123] indicate a sequential mechanism involving two substrates and three products for the catalysed formation of anthranilate and are consistent with the mechanism postulated by Levin and Sprinson.

C. p-Aminobenzoate

The biosynthetic pathway for aromatic amino acids also gives rise to p-aminobenzoate. Shikimate 5-phosphate and glutamine were found to be precursors[132]. Isotopic labelling studies demonstrated that the amino nitrogen was derived from the amide nitrogen atom of glutamine[133]. First recognized as a growth factor, p-aminobenzoate is a constituent of the folic acid vitamins.

D. Glucosamine 6-Phosphate

The transfer of an amino group from glutamine to hexose phosphate occurs according to equation (30). Studies on the capsular poly-

$$\text{D-Fructose-6-P} + \text{L-Glutamine} \longrightarrow \text{2-Amino-2-deoxy-D-glucose-6-P} + \text{L-Glutamate}$$
(30)

saccharide material of *Streptococcus* species which contains glucosamine as part of the hyaluronic acid structure, demonstrated that the carbon skeleton is derived from glucose[134,135]. Analysis of glucosamine isolated from cell cultures administered with 1-^{14}C-glucose showed the radioactivity was present almost exclusively at $C_{(1)}$. Similar results were reported for blood glucosamine isolated from rats following administration of labelled glucose[136]. A partially purified enzyme from *Neurospora* utilized either glucose-6-P or fructose-6-P as a nitrogen acceptor[137], but the preparation probably was contaminated with phosphohexoisomerase since subsequent studies[138] showed fructose-6-P was the only substrate.

Glutamine served as the only nitrogen donor for the purified enzymes from *E. coli*, rat liver and *Neurospora*[138]. A decrease in amide nitrogen paralleled an increase in glucosamine and glutamate concentrations during the course of the enzyme-catalysed reaction[137]. Enzymatic activity was strongly inhibited by 6-diazo-5-oxo-L-norleucine[138].

The stoicheiometry of the equation was established for the *E. coli* and rat liver enzymes. However, measurements with the *Neurospora* preparation gave high values for glutamine utilization due to the presence of glutaminase.

No evidence could be obtained for a cofactor requirement by dialysis or ion exchange treatment of the enzyme. Elucidation of the mechanism for this reaction awaits further studies.

E. γ-Glutamylamides

A large number of naturally occurring dipeptides, formed by the reaction of the γ-carboxyl group of glutamic acid with the α-amino nitrogen of various amino acids, have been isolated mainly from higher plants. Virtanen and his associates[139] have reported the isolation of γ-glutamyl derivatives of valine, isoleucine, leucine, phenylalanine, methionine and others from plant sources. Dipeptides, containing amino acid derivatives, such as γ-glutamyl-S-(1-propenyl)cysteine sulphoxide and γ-glutamyl-S-methylcysteine have also been found. For a comprehensive list of γ-glutamyl peptides and sources for their isolation see Thompson and coworkers[140] and Fowden[141]. Several γ-glutamyl derivatives in which the amide moiety stems from a nitrogen compound other than an amino acid have been isolated and characterized. Theanine (**42**), a component of tea leaves, is the amidic N-ethyl derivative of glutamine. More recently γ-glutamyl-p-hydroxyaniline (**43**)[142] and β-N-(γ-glutamyl)-p-hydroxymethyl-phenylhydrazine (**44**)[143] have been isolated from mushrooms.

COO^-	COO^-	COO^-
$CHNH_3^+$	$CHNH_3^+$	$CHNH_3^+$
CH_2	CH_2	CH_2
CH_2	CH_2	CH_2
$CONHEt$	$CONH—C_6H_4OH$-p	$CONH—NH—C_6H_4CH_2OH$-p
(**42**)	(**43**)	(**44**)

Glutathione, γ-glutamyl-cysteinylglycine (section II.D), is widely distributed in nature and functions as a cofactor in a number of enzymatic reactions. Two related γ-glutamyl dipeptides, in which the cysteinyl residue is replaced, have been isolated from calf lens; opthalmic acid containing the α-amino-n-butyryl group[144] and noropthalmic acid with a substituted alanyl residue[145].

The physiological significance of the γ-glutamyl dipeptides is not clear. The occurrence of these compounds in the storage organs of plants, bulbs and seeds, suggests a role in transport and storage for non-protein amino acids. In mammals, dipeptides are excreted in the urine. Three γ-glutamyl and fifteen β-aspartyl dipeptides have been isolated from human urine[146]. The localization of γ-glutamyl transpeptidase, an enzyme catalysing the formation of dipeptides, in mammalian kidney tissue, indicates a possible function in the excretion of amino acids. A role in the synthesis of structural components may be of biological importance as, for example, the capsular material of some microorganisms contains a γ-glutamyl polypeptide.

The biosynthesis of γ-glutamyl dipeptides may occur by a mechanism involving the direct formation of the amide bond between glutamate and an amino acid or by the transfer of a γ-glutamyl residue from a dipeptide already formed to a free amino acid or polypeptide in a transpeptidation reaction.

The formation of γ-glutamylcysteine from glutamate and cysteine, the first step in the biosynthesis of the tripeptide glutathione, occurs by direct synthesis (equation 11a).

The mechanism of γ-glutamylcysteine synthetase catalysed reaction appears to be quite similar to the mode of carboxyl activation previously described for glutamine synthesis (section II.A). Catalysis by the hog kidney enzyme, purified 2500-fold, is Mg^{2+} ion and ATP-dependent, the nucleotide is cleaved to ADP and inorganic phosphate and an oxygen atom of glutamate is transferred to inorganic phosphate during the course of the reaction[65]. This enzyme, in contrast to the glutamine-synthesizing enzyme, catalyses the rapid transfer of ADP into ATP, which is dependent on Mg^{2+} ions but not on glutamate or cysteine. However, this reaction may be a consequence of an impure enzyme since an analogous exchange reaction was also observed[19] in partially purified preparations of glutamine synthetase. On careful exclusion of ammonia from the enzyme, the exchange, in absence of the other substrates, was no longer observed.

Much less is known regarding the mechanism of γ-glutamyl transpeptidation reactions. Hanes and coworkers[147] demonstrated that a mammalian kidney enzyme catalysed the transpeptidation reaction of glutathione as well as its hydrolysis. The two activities are frequently associated with hydrolytic enzymes (phosphatases, glycosidases, proteases, amidases) in which the formation of an activated enzyme–substrate complex may react with water in a hydrolytic reaction or with an amine in a transfer reaction. The kidney enzyme catalyses the transfer of the γ-glutamyl residue from glutathione to an amino acid acceptor molecule (equation 31).

γ-L-Glutamyl-L-cysteinylglycine + L-Amino acid $\longrightarrow$

$\qquad\qquad$ γ-L-Glutamyl-L-amino acid + L-Cysteinylglycine   (31)

The specificity for the acceptor molecule is low and the γ-glutamyl group of glutathione may be transferred to valine, phenylalanine, leucine[148], methionine, arginine and glutamate as well as certain dipeptides[149]. Similarly a number of γ-glutamyl dipeptides may serve as donors and this lack of specificity permits the use of N-(γ-glutamyl)-α-naphthylamine[150] and N-(γ-glutamyl)-p-nitroaniline[151] as the γ-glutamyl donor and provides for the convenient determination of the liberated amine by spectrophotometric means. Because the kidney enzyme is bound to tissue particles, difficulties have been encountered in obtaining soluble preparations, and therefore its purification requires extraction with solvents such as deoxycholate. Preparations purified about 1000-fold are glycoproteins in that they contain neutral sugars, amino sugars and sialic acid[150]. Treatment of the enzyme with neuraminidase did not significantly alter the activity, thus sialic acid is not a component of the catalytic centre. Activity is lost, however, when the enzyme is treated with periodate and with sulphydryl-reacting reagents such as N-ethylmaleimide, mercury and iodoacetate. In time studies[150] employing the substrate N-(γ-L-glutamyl)-α-naphthylamine, 40% of the theoretical α-naphthylamine appeared in the first few minutes of the reaction and N-(γ-glutamyl-γ-glutamyl)-α-naphthylamine was the only product (equation 32). On prolonged incubation poly-γ-glutamyl peptides were formed which subsequently could be hydrolysed to glutamic acid.

N-(γ-L-Glutamyl)-α-naphthylamine $\longrightarrow$

$$N\text{-}(\gamma\text{-L-Glutamyl-}\gamma\text{-L-glutamyl)-}\alpha\text{-naphthylamine} + \alpha\text{-Naphthylamine} \quad (32)$$

An enzyme which catalyses the hydrolysis or transfer of the γ-glutamyl residue from **44** to p-hydroxyaniline has been partially purified from commercial mushrooms[152]. This enzyme is distinct from the mammalian enzyme in that it is not active with amino acid acceptors.

High molecular weight (up to 53,000) polyglutamic acid peptides have been isolated from the capsular material of *Bacillus* microorganisms[153] and from exocellular material in the cultural medium. Polyglutamate synthetase activity in cell-free extracts is not inhibited by chloramphenicol, streptomycin, penicillin, deoxyribonuclease or ribonuclease[154], reagents which inhibit protein synthesis. In contrast to the α-linkage of the peptide bond of proteins, the glutamate residues of the polymer secreted into the culture medium are joined through an amide bond of the γ-carboxyl group[155,156]. Growth of the polymers occurred by the transfer of a γ-glutamyl moiety to the free amino end-group of the polypeptide. By labelling a dipeptide substrate and determining the location of the ^{14}C label in the product, Williams and Thorne[157] postulated the sequence of reactions (33). Partially puri-

$$\begin{aligned}
\text{E} + \gamma\text{-D-Glutamyl-D-glu*} &\longrightarrow \text{E—glu} + \text{D-glu*} \\
\text{E—glu} + \gamma\text{-D-Glutamyl-D-glu*} &\longrightarrow \text{E} + \gamma\text{-D-glu-D-glu-D-glu*}
\end{aligned} \quad (33)$$

fied enzyme preparations from culture filtrates were capable of synthesizing short-chain polypeptides[158]. Accumulation of high molecular weight polymers was probably not observed due to competing hydrolytic reactions or an unfavourable equilibrium resulting from the uncoupling of subsequent reaction of the polypeptide. Transpeptidation reactions appear to constitute an alternative pathway for the biosynthesis of polypeptides.

Synthesis requires an energy source (ATP or other nucleotide triphosphate), Mn^{2+} and β-mercaptoethanol[154]; hydroxamate formation and appropriate exchange reactions have been observed suggesting the formation of an activated enzyme–substrate complex[157].

Two observations have been reported that indicate the α-hydrogen atom is removed during the process of catalysis. Cell-free extracts of *Bacillus licheniformis* catalyse the synthesis of polyglutamic acid which contains about 40% of D-isomer residues. Isotope dilution studies, however, indicate that only the L-isomer was used and without being converted to the D-isomer prior to incorporation[154]. The utilization of ^{14}C-labelled glutamate was markedly decreased by the addition of the L-isomer but not by addition of the D-isomer. Earlier *in vivo* studies, employing glutamic acid labelled in the α-amino nitrogen and in the α-hydrogen atom demonstrated that while 30% of the α-amino group was incorporated into the polymer no α-deuterium was detected[155]. Although the mechanistic significance of bond cleavage of the α-hydrogen atom is not understood, these findings suggest an alternative mechanism for catalysing the reaction between a relatively unreactive carboxyl group and an amine, analogous to nucleophilic catalysis of ester hydrolysis by imidazole. The tetrahedral addition intermediate **45** becomes stabilized on losing the α-proton, and the

$$\begin{array}{ccc} \overset{O^-}{\underset{\underset{OH}{\overset{|}{C}}\,H}{\overset{|}{-C}}}\overset{H}{\underset{|}{N}}\overset{H}{\underset{|}{C}}- & \longrightarrow & -\overset{O}{\overset{\|}{C}}-NH-\overset{H}{\underset{|}{C}}- \\ \textbf{(45)} & & \end{array} \qquad (34)$$

hydroxide ion is then preferentially expelled leading to amide formation (equation 34).

V. AMIDE HYDROLYSIS

Several enzymes are known which hydrolyse the amide group of glutamine and asparagine. The hydrolases vary considerably with respect to substrate specificity, transfer activity, pH optimum, inhibition patterns, anion activation and antitumour activity in the case of asparaginase.

Greenstein and collaborators[159,160] orignally described two activities in rat liver digests which catalysed the deamidation of glutamine. One of the activities required the presence of an α-keto acid for deamidation to occur. This enzyme was found only in heptatic tissues. The second activity was stimulated by phosphate, sulphate, and arsenate. Anion-activated glutaminase was found in brain and spleen of rats, mice, rabbits, and guinea pigs and in the kidney of rats, dogs, and cats. Errera[161] established that the two activities observed in liver digests were separate enzymes. Activation of hydrolytic activity by anions, however, was not observed in rabbit and guinea pig kidney extracts[162] indicating that three types of deamidating enzymes occur in animal tissue. Two asparagine deamidating activities were also found in rat liver extracts; an α-keto acid dependent one and an asparaginase which was not activated by either phosphate or by keto acids.

A. ω-Amidase

The enzyme ω-amidase catalyses the deamidation of α-keto-glutaramate and α-ketosuccinamate. In liver tissues the ω-amidase activity was coupled with transaminases for glutamine and asparagine, which in consort effected the conversion of these compounds to α-ketoglutarate and oxaloacetate respectively. The observed α-keto acid dependency was due to its role in the transamination reaction. The combined reactions are shown in equation (35).

$$
\begin{array}{c}
\begin{array}{l}
\text{CONH}_2 \\
| \\
\text{CH}_2 \\
| \\
\text{CH}_2 \quad\quad \text{COO}^- \\
| \quad\quad\quad\quad | \\
\text{CHNH}_3^+ \quad \text{CH}_3 \\
| \\
\text{COO}^- \\
\text{glutamine} \quad \text{pyruvate}
\end{array}
\xrightarrow{\text{Transaminase}}
\begin{array}{l}
\text{COO}^- \quad \text{CONH}_2 \\
| \quad\quad\quad\quad | \\
\text{CHNH}_3^+ \quad \text{CH}_2 \\
| \quad\quad\quad\quad | \\
\text{CH}_3 \quad\quad \text{CH}_2 \\
\quad\quad\quad\quad | \\
\quad\quad\quad\quad \text{C}=\text{O} \\
\quad\quad\quad\quad | \\
\quad\quad\quad\quad \text{COO}^- \\
\text{alanine} \quad \alpha\text{-ketoglutaramate}
\end{array}
\xrightarrow{\omega\text{-Amidase}}
\begin{array}{l}
\text{COO}^- \\
| \\
\text{CH}_2 \\
| \\
\text{CH}_2 + \text{NH}_3 \\
| \\
\text{C}=\text{O} \\
| \\
\text{COO}^- \\
\alpha\text{-ketoglutarate}
\end{array}
\end{array}
$$

$$(35)$$

In reaction mixtures which contained an α-keto acid (pyruvate), ^{15}N-labelled glutamine and a partially purified enzyme preparation, Meister and Tice[163,164] demonstrated that the ammonia produced arose from the amide nitrogen of glutamine. α-Ketoglutarate was isolated as a product which showed that during the course of the reaction glutamine had lost both nitrogen atoms. Subsequently the enzyme responsible for amide hydrolysis was separated from the transamination enzyme and it was demonstrated that α-ketoglutaramate and α-ketosuccinamate were the substrates for the deamidation reaction.

The ω-amidase is found in a large number of animal tissues, in microorganisms and plants.

B. Glutamine

The glutaminase enzyme, activated by anions, has been purified from pig[165] and dog kidney cortex[166]. Sulphydryl reagents including mercuric chloride, p-mercuribenzoate, iodoacetamide and N-ethylmaleimide were found to be potent inhibitors of enzymatic activity. Inhibition induced by mercurials was prevented or reduced by the addition of sulphydryl compounds such as glutathione and cysteine. Sayre and Roberts[166] postulated that an enzyme sulphydryl group at the active site participates in general acid–base or nucleophilic catalysis. The pH optimum for enzyme activity at 8·1 further suggests the involvement of a sulphydryl group since it would be sufficiently ionized at this pH.

Polyvalent anions enhance kidney glutaminase activity markedly. Phosphate, arsenate, sulphate, and nitrate[160], in decreasing order of effectiveness activated the enzyme, whereas chloride, bromide, and cyanide[166] were inhibitory in the presence of phosphate. Anion activation appears to be related to binding of the substrate. The Michaelis (dissociation) constant decreased about 10-fold with an increase in phosphate concentration to 0·2 M. Bromosulphalein and 2,4-dinitro-1-naphthol-7-sulphonic acid were found to inhibit the enzyme. These results suggested to the authors that the catalytic portion of the enzyme consisted of phosphate- and substrate-binding cationic sites and an enzyme sulphydryl group as depicted in **46**.

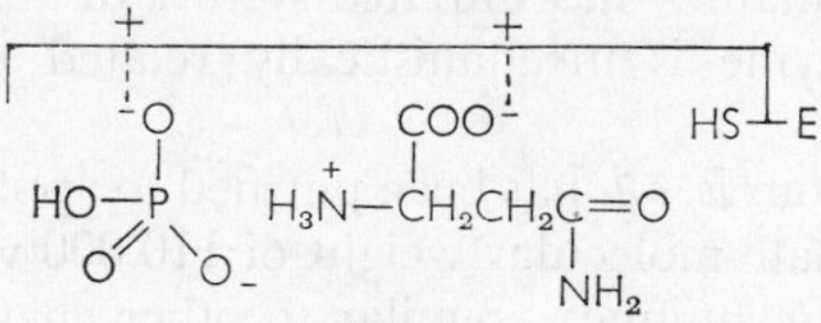

$$
\begin{array}{ccc}
\overset{+}{} & \overset{+}{} & \\
\,^-O & COO^- & HS-E \\
\mid & \mid & \\
HO-P & H_3\overset{+}{N}-CH_2CH_2C=O & \\
O^{\diagdown}O^- & \mid & \\
& NH_2 &
\end{array}
$$

(46)

According to this model the negatively charged groups of the dye molecules, by binding with the enzyme, inhibit cationic sites.

The relatively high concentrations of phosphate required for

22 + c.o.a.

activation, which certainly exceeds cellular concentrations makes questionable a phosphate–substrate complex requirement for catalytic activity. The order of anion activation suggests that the increase in binding at the 'active site' could result from conformational changes induced by salt interactions with the protein. The relative effectiveness of anions in activating the enzyme follows the same order of activity of ions reported by Robinson and Jencks[167] in studies on the effects of various salts on the solubility of a model peptide acetyltetraglycine ethyl ester. These authors observed that those ions which are most effective in causing protein precipitation are the most effective in preventing the denaturation of proteins. Anion activation or inhibition of a number of enzymes—glutaminase inclusive—can be correlated with the tendency of salts to inhibit or to increase denaturation of the protein. Klingman and Handler[165] concluded from their studies of glutaminase, that phosphate 'protected' the enzyme against denaturation at the assay temperature of 37°c.

The results of exchange studies catalysed by the kidney enzyme were interpreted in terms of an enzyme–glutamate complex. In the presence of glutamine and Pi, the enzyme catalysed the exchange of $^{15}NH_3$ into glutamine but not ^{14}C-labelled glutamate[165]. Similar results were obtained with an amidase isolated from *Pseudomonas fluorescens*, which catalysed deamidation of 2- and 3-carbon primary amides[168]. Presumably the complex, which is in equilibrium with reactants, is hydrolysed in an irreversible step according to equation (36).

$$\text{Enzyme + Glutamine} \rightleftharpoons \text{Enzyme–glutamate + NH}_3 \xrightarrow{\text{H}_2\text{O}}$$
$$\text{Enzyme + Glutamate} \quad (36)$$

The glutaminase from *E. coli* bears a resemblance to the esterases and peptidases, which presumably involve acyl–enzyme intermediates, in that it catalyses the transfer reaction of an acyl group from a number of donors including amides, esters and thioesters to suitable acceptors. Hartman[169] has initiated studies in an attempt to determine if this enzyme is mechanistically related to the proteolytic enzymes.

Glutaminase from *E. coli* has been purified to apparent homogeneity and an approximate molecular weight of 110,000 was determined by the gel filtration technique. Similar to other enzymes metabolizing glutamine, 6-diazo-5-oxo-L-norleucine irreversibly inhibited glutaminase. Glutamine provided protection against this inhibition, indicating that the substrate and inhibitor react with the same site. ^{14}C-Labelled inhibitor was covalently bound to the enzyme. Assuming a one-to-one ratio of inhibitor to active site, titration of the enzyme with inhibitor showed two active sites per molecule.

In addition to amides, the enzyme catalysed the hydrolysis of γ-glutamyl derivatives including N-substituted amides, hydroxamic acid, esters and thioesters. Binding constants were determined for the catalytically active substrates from the Michaelis constant, which was shown to be equal to the dissociation constant[171], and from the inhibition constants for substrates that were bound but did not undergo reaction. Little specificity was found for the substituent on the acyl group since amides, oxygen and thiolesters served as substrates, and binding was limited only by size. The L-glutamyl isomer and unsubstituted α-amino and carboxyl groups were required for binding. Compounds that contained an oxygen atom in the principal chain were not bound and this observation accounts for the failure of azaserine to inhibit the enzyme.

Glutaminase is primarily a hydrolase. The acyl group can be transferred to hydroxylamine and methanol but at relatively slow rates. In view of the relative nucleophilicities of hydroxylamine, methanol and water, these results were interpreted to indicate a water-binding site on the enzyme.

The proteolytic enzymes may be grouped according to a common catalytic entity: the serine esterases chymotrypsin, trypsin, subtilisin; the thio acylases papain, ficin and pepsin; and the metalloenzyme carboxypeptidase. A comparison of glutaminase with these groups of enzymes did not reveal characteristics that would suggest a common mechanism was operative. Diisopropylphosphorofluoridate which inhibits the serine esterases by reacting irreversibly with the alcohol oxygen of the serine active site, has no effect on glutaminase. There is no convincing evidence that glutaminase is a thiol enzyme; although inhibited by organic mercurials, it is insensitive to more specific sulphydryl reagents such as iodoacetate and N-ethylmaleimide. Glutaminase behaviour also differs from papain with respect to inactivation by hydrogen peroxide and photooxidation. Papain is immediately inactivated by treatment with hydrogen peroxide, whereas glutaminase has a half-life of about 3 hours. Glutaminase undergoes irreversible photoinactivation in contrast to papain which after similar treatment is totally reactivated by the presence of a dithiol reducing agent. Metal-binding agents, which inactivate metallohydrolases, do not inhibit glutaminase. The mechanism of action for glutaminase appears to differ from that of the esterases and peptidases.

The studies on glutaminase activity were extended to experiments designed to distinguish between a one-step mechanism in which the substrate is hydrolysed in a single displacement reaction, or the formation of an acyl–enzyme intermediate in a two-step displacement sequence analogous to the mechanism of serine and thiol acylases [170,171]. The results of these investigations suggested that a common intermediate results from the amide, ester and thioester substrates, and the formation of the intermediate is the slowest step in the catalytic cycle. Attempts to detect a covalently bound intermediate were unsuccessful. However, if hydrolysis of the intermediate is faster than its formation, detection may not be possible by the usual techniques. Studies of the kinetic parameters as a function of pH, temperature, and deuterium isotope effects did not indicate whether covalent or non-covalent (conformational) processes were involved in the rate-determining step of the glutaminase reaction.

C. Asparagine

The finding that asparaginase of guinea pig serum is an effective chemotherapeutic agent in the treatment of certain types of cancer has led to a search for a convenient source of the enzyme.

Not all asparaginases are capable of antitumour activity. The enzymes isolated from yeast and *Bacillus coagulans* were found to be inactive. The report that asparaginase of *E. coli* was effective against mouse leukemia initiated purification studies of the enzyme since a bacterial source could provide large quantities of the enzyme for antitumour studies and clinical treatment. Campbell and co-workers [172], on purification of the *E. coli* enzyme reported two asparaginases, EC-1 and EC-2, were present.

Activities of the two enzymes as a function of hydrogen ion concentration is distinctly different. The rate of asparagine hydrolysis for EC-1 falls off rapidly below a pH value of 8·4 whereas EC-2 exhibits a plateau of maximum activity between pH 6·0–8·4. Due to this difference in pH optima it is possible to determine the activity attributable to EC-1 and EC-2 in a mixture of the two enzymes.

EC-1 preparations showed no antilymphoma activity. In contrast, EC-2 preparations at high dosage levels frequently resulted in cures of mouse lymphomas and at intermediate dosage levels caused the disappearance of subcutaneous tumour mass and extended survival times. There appears to be no correlation between optimal activity at certain pH values and antitumour activity since the pH–rate profile of guinea pig serum asparaginase[173] is intermediate between the curves for EC-1 and EC-2. The enzyme's effectiveness as an anti-tumour agent is probably related to its stability or activity when injected into the tumour mass.

VI. FORMATION OF CARBAMOYL PHOSPHATE

The second major pathway for the incorporation of ammonia into organic compounds considered in this chapter involves the biosynthesis of carbamoyl phosphate. This compound is formed by the combination of ammonia, carbon dioxide and phosphate and is the biochemical reagent for the carbamoylation of amino compounds. The analogous reaction in organic chemistry is the introduction of the NH_2CO group by the use of cyanic acid. Carbamoyl phosphate is primarily utilized for the biosynthesis of the pyrimidines and of arginine in the urea cycle (Figure 3, section III.C).

Jones and coworkers[174] first demonstrated that synthetic carbamoyl phosphate, prepared by the reaction of dihydrogen phosphate with cyanate, reacts with ornithine in the presence of liver extracts to form citrulline. Investigations on the biosynthesis of carbamoyl phosphate have established that three general enzyme systems in a variety of organisms are responsible for its synthesis. Carbamate kinase is found in microorganisms and plants, and carbamoyl phosphate synthetase occurs in animal tissues, microorganisms and mushrooms[175,176]. Two types of synthetases have been described which differ primarily in that one utilizes ammonia as substrate and requires N-acetylglutamate, or an analogue, for activity and the other utilizes glutamine and is not activated by N-acetylglutamate.

A. Carbamate Kinase

Carbamate kinase catalyses the biosynthesis of carbamoyl phosphate according to equation (37). Carbamate is considered the true

$$NH_4^+ + HCO_3^- \rightleftharpoons \underset{\text{carbamate}}{NH_2COO^-} + ATP \rightleftharpoons NH_2\overset{\overset{O}{\|}}{\underset{\underset{O_-}{|}}{C}}\overset{O}{\underset{}{\overset{\|}{P}}}—O^- + ADP \quad (37)$$

substrate and is formed non-enzymatically. The equilibrium is in favour of ATP formation and this may be of the greater importance in microorganisms which are capable of supplying part of their energy requirements by the generation of ATP from carbamoyl phosphate.

A kinetic study[177] on the crystallized enzyme supported a mechanism described by the following series of reactions (38a–38c). The

$$\text{Enzyme} + \text{MgATP} \underset{\text{Slow}}{\rightleftharpoons} \text{E—MgATP} \tag{38a}$$

$$\text{E—MgATP} + \text{Carbamate} \rightleftharpoons \text{E} \overset{\text{MgATP}}{\underset{\text{carbamate}}{\diagup\diagdown}} \rightleftharpoons \text{E—MgADP} + \text{Carbamoyl phosphate} \tag{38b}$$

$$\text{E—MgADP} \overset{\text{Slow}}{\rightleftharpoons} \text{Enzyme} + \text{MgADP} \tag{38c}$$

kinetic behaviour indicated that the nucleotides were the first substrates to be bound and their dissociation from the enzyme was the rate-limiting step in both the forward and reverse directions.

B. Carbamoyl Phosphate Synthetase I

The catalysed formation of carbamoyl phosphate by the liver enzyme is represented by the overall reaction of equation (39). This reaction

$$\text{NH}_4^+ + \text{HCO}_3^- + 2\text{ATP} \xrightarrow[N\text{-acetylglutamate}]{\text{Mg}^{2+}} \text{NH}_2\overset{\overset{\text{O}}{\|}}{-}\text{C}-\text{O}-\text{PO}_3^{2-} + 2\text{ADP} + \text{Pi} \tag{39}$$

differs from the carbamate kinase-catalysed reaction in that two moles of ATP and one of N-acetylglutamate are required, ammonium bicarbonate is the substrate and the overall reaction is not readily reversible. The enzyme has been demonstrated in intestinal mucosa of some animals, but the level of activity was low compared to that of liver.

Metzenberg and coworkers[178] obtained evidence that the reaction takes place in two steps and postulated the intermediate formation of an 'active CO_2'. When relatively large amounts of enzyme were incubated with ATP, N-acetylglutamate and bicarbonate, orthophosphate was released in the absence of ammonium ions. Reversal of one reaction step was demonstrated by incubating ^{14}C-labelled carbamoyl phosphate and ADP in the presence of the synthetase. The ATP synthesized was equivalent to the carbamoyl phosphate that disappeared during the course of the reaction.

The synthetase catalyses the cleavage of Pi from ATP in the presence of amines other than ammonia, namely: hydroxylamine, hydrazine and O-methylhydroxylamine. In the presence of ornithine and ornithine carbamoyltransferase some evidence was obtained for the formation of a citrulline analogue with O-methylhydroxylamine as the amine donor[179]. The enzyme-bound carbamate analogues formed from hydrazine and hydroxylamine presumably were unstable.

Jones and Spector[180] showed that ^{18}O-labelled atoms of bicarbonate were transferred to the orthophosphate released and the oxygen bridge atom (C—O—P) of carbamoyl phosphate. The reaction sequence (40a–40c) is consistent with the above findings[181].

$$\text{Enzyme} + \text{ATP} + \text{CO}_2 \rightleftharpoons \text{E—[}^-\text{OOCOPO}_3^{2-}\text{]} + \text{ADP} \qquad (40a)$$

$$\text{E—[}^-\text{OCOOPO}_3^{2-}\text{]} + \text{NH}_3 \longrightarrow \text{E—[NH}_2\text{COO}^-\text{]} + \text{Pi} \qquad (40b)$$

$$\text{E—[NH}_2\text{COO}^-\text{]} + \text{ATP} \rightleftharpoons \text{E} + \text{NH}_2\text{COOPO}_3^{2-} + \text{ADP} \qquad (40c)$$

The role of N-acetylglutamate, is probably that of an activator. Phosphorylated and CO_2 derivatives of acetyl glutamate have been postulated to account for its requirement in the carbamoyl phosphate synthetase reaction[175]. The results of more recent studies suggest that N-acetylglutamate activates the enzyme by inducing a conformational change. Binding of the activator to the enzyme resulted in spectral changes, the appearance of a second ATP binding site and an increased susceptibility to heat inactivation[182,183].

C. Carbamoyl Phosphate Synthetase II

A second type of synthetase has been found in yeast[184], $E.\ coli$[185], Ehrlich ascites cells[186], foetal rat liver and pigeon liver[187], which differs from synthetase I in that acetyl glutamate is not required, and either glutamine or NH_3 can serve as the nitrogen donor. Studies on purified preparations obtained from $E.\ coli$ indicated that glutamine is probably the true physiological donor and established that one molecule of carbamoyl phosphate is associated with the cleavage of two molecules of ATP to ADP[188,189]. The partial reverse reaction, the formation of ATP from ADP and carbamoyl phosphate was also demonstrated.

Wellner and coworkers[190] have reported that the $E.\ coli$ synthetase is a biotin-containing enzyme. Biotin (50) is a cofactor associated with carboxylation and transcarboxylation reactions such as the carboxylation of acetyl–coenzyme A and the carboxyl transfer reaction in the biosynthesis of oxaloacetate. The mechanism of biotin activation of CO_2 involves the intermediate formation of an enzyme-bound carboxybiotin compound. Avidin, a biotin inhibitor, inhibited the glutamine-dependent synthetase activity, addition of biotin to crude enzyme preparations enhances activity and, finally, analysis of the purified enzymes showed that it contained bound biotin. The authors postulated the formation of an enzyme-bound carbonic acid–phosphoric acid anhydride intermediate (47) which reacted with the enzyme-bound biotin to yield a carboxybiotin enzyme (48). The reaction sequence (41a–41c) was proposed for the biosynthesis of carbamoyl phosphate catalysed by synthetase II. Huston and Cohen[190a] recently reported that biotin was not significantly present nor did it function as a coenzyme in purified preparations of carbamoyl phosphate synthetase from *E. coli* or liver mitochondria.

$$
\underset{\text{Enzyme}}{\overset{\text{biotin}}{\diagup}} + \text{ATP} + \text{HCO}_3^- \rightleftharpoons \underset{[^-\text{OCOOPO}_3^{2-}]}{\overset{\text{biotin}}{\text{E}\diagup}} + \text{ADP} \rightleftharpoons
$$

$$(47)$$

$$
\text{E—biotin—CO}_2 + \text{ADP} + \text{Pi} \quad (41a)
$$

$$(48)$$

$$
\text{E—biotin—CO}_2 + \text{Glutamine} + \text{H}_2\text{O} \longrightarrow \underset{[\text{NH}_2\text{CO}_2^-]}{\overset{\text{biotin}}{\text{E}\diagup}} + \text{Glutamate} \quad (41b)
$$

$$
\underset{[\text{NH}_2\text{CO}_2^-]}{\overset{\text{biotin}}{\text{E}\diagup}} + \text{ATP} \rightleftharpoons \text{E—biotin} + \text{NH}_2\text{COOPO}_3^{2-} + \text{ADP} \quad (41c)
$$

Hager and Jones[187] have noted that in mammalian tissues, synthetase II and the enzymes responsible for pyrimidine synthesis, are located in the soluble fraction, whereas, synthetase I and the enzymes specific for urea biosynthesis are present in liver mitochondria. They suggest that the two pathways derive their carbamoyl phosphate from separate synthetases and that pyrimidine biosynthesis is catalysed by the glutamine-dependent enzyme.

D. Miscellaneous Pathways for Carbamoyl Phosphate Formation

Carbamoyl phosphate can also be formed in the course of degradation reactions of certain nitrogen-containing compounds. Some organisms are capable of deriving part of their energy requirement from carbamoyl phosphate formed from essentially catabolic processes. The phosphate group of carbamoyl phosphate can be transferred to ADP in an endergonic reaction to form ATP. The ureido compounds that probably contribute to ATP biosynthesis include citrulline from the urea cycle (Figure 3) and degradation products which arise from creatinine, purine and pyrimidine catabolism.

For example, in *Streptococcus faecalis* arginine is hydrolysed to citrulline and ammonia[191,192]. The citrulline in turn is phosphorolysed in an enzyme-catalysed reaction to yield carbamoyl phosphate. In the presence of ADP the phosphate group is transferred to form ATP in a reverse of the reaction described for carbamoyl phosphate synthesis catalysed by the kinase[179] (equation 37).

ATP biosynthesis has been reported to occur by the reaction of ADP with carbamoyl phosphate in extracts from *S. allantoicus*[193]. In this case carbamoyl phosphate is formed by the phosphorolysis of carbamoyl oxamate, a product of purine degradation (equation 42).

$$\overset{O\,O}{\underset{}{-O\overset{\parallel}{C}C NH-}}\overset{O}{\overset{\parallel}{C}NH_2} + Pi \longrightarrow \overset{O\,O}{\underset{}{-O\overset{\parallel}{C}C NH_2}} + NH_2\overset{O}{\overset{\parallel}{C}}OPO_3H^- \xrightarrow{ADP}$$

carbamoyl oxamate   oxamate

$$ATP + CO_2 + NH_3 \quad (42)$$

Degradation reactions catalysed by a liver enzyme give rise to carbamoyl-β-alanine and carbamoyl-β-aminoisobutyric acid from the pyrimidines uracil and thymine respectively[194]. The subsequent degradation of these compounds to β-alanine or β-aminoisobutyric acid and CO_2 plus ammonia probably proceeds through the intermediate

$$H_2N\overset{O}{\overset{\parallel}{C}}\overset{H}{\underset{}{N}}CH_2CH_2COO^- + Pi \longrightarrow H_2N\overset{O}{\overset{\parallel}{C}}OPO_3H^- + H_2NCH_2CH_2COO^- \quad (43)$$

carbamoyl-β-alanine β-alanine

formation of carbamoyl phosphate[195] (e.g. equation 43). *Eucobacterium sarcosinogenum* extracts have been reported to degrade creatinine in a multistep reaction sequence (44) to carbamoyl phosphate[196].

$$\underset{\substack{| \\ CH_3NCH_2COO^-}}{\overset{\substack{NH_2 \\ | \\ C=NH \\ |}}{}} \longrightarrow \longrightarrow \longrightarrow NH_2\overset{O}{\overset{\parallel}{C}}OPO_3H^- + NH_3 + CH_3NHCH_2COO^-$$

creatinine   sarcosine

$$(44)$$

VII. REACTIONS OF CARBAMOYL PHOSPHATE

A. Citrulline Biosynthesis

Ornithine carbamoyl transferase catalyses the formation of citrulline according to equation (45). The enzyme has been found in the livers

22*

$$
\begin{array}{c}
\text{NH}_2 \\
| \\
\text{CH}_2 \\
| \\
\text{CH}_2 \\
| \\
\text{CH}_2 \\
| \\
\text{CHNH}_3^+ \\
| \\
\text{COO}^-
\end{array}
\quad + \text{H}_2\text{NCOOPO}_3\text{H}^- \;\rightleftharpoons\;
\begin{array}{c}
\text{NH}_2 \\
| \\
\text{C}{=}\text{O} \\
| \\
\text{NH} \\
| \\
\text{CH}_2 \\
| \\
\text{CH}_2 \\
| \\
\text{CH}_2 \\
| \\
\text{CHNH}_3^+ \\
| \\
\text{COO}^-
\end{array}
\quad + \text{Pi} \qquad (45)
$$

ornithine　　　　　　　　　　　　　citrulline

of all vertebrates that eliminate ammonia as urea, i.e. the ureotelic animals[197]. Citrulline participates in the urea cycle (Figure 3, section III.C), a series of reactions which convert ammonia, resulting from the catabolism of nitrogen-containing compounds, into urea for elimination.

Studies on the rat liver[198] and bovine liver[199] enzymes established the specificity for the substrate since ornithine could not be replaced by naturally occurring amino acids or by alkylamines such as spermine, spermidine, diaminopentane and others. The pH optimum for enzymatic activity and the more favourable Michaelis constant at alkaline pH values suggested that the unionized δ-amino group of ornithine was probably the active form. Participation of a sulphydryl group was suggested by the inhibition of the enzyme at low concentrations of p-mercuribenzoate. Either ornithine or carbamoyl phosphate provided partial protection against this inhibition. ^{32}P-Labelled phosphate did not exchange with carbamoyl phosphate in the absence of ornithine. Since there was no indication for partial reactions which would be expected for a mechanism involving the intermediate formation of carbamoyl–enzyme, Reichard[198] concluded that the reaction occurred by a single-displacement mechanism.

B. Carbamoyl Aspartate Biosynthesis

The catalysed transfer of the carbamoyl group to L-aspartate yields carbamoyl-L-aspartate (equation 46), a step in the biosynthetic pathway which leads to the formation of orotic acid and the pyridine nucleotides uracil, cytosine and thymine. The transfer of the carbamoyl group to aspartate is analogous to the transcarbamoylation reaction to form citrulline. Experiments with 100-fold purified enzyme from E. coli demonstrated a strict substrate specificity as only the physiological reactants of equation (46) were catalytically reactive[200]. Again, the enzyme was inhibited by p-chloromercuribenzoate but the presence of both substrates was required for protection against this inhibition. Isotopic exchange studies using ^{32}P and $H_2{}^{18}O$ gave no evidence for a carbamoylated enzyme intermediate or for the

$$
\underset{\text{aspartate}}{\overset{\displaystyle O}{\underset{\displaystyle NH_2\overset{\|}{C}OPO_3H^-}{}} + \overset{\displaystyle COO^-}{\underset{\displaystyle COO^-}{\overset{|}{\underset{|}{\overset{\displaystyle CH_2}{\underset{\displaystyle CHNH_3^+}{}}}}}} \xrightarrow[\text{carbamoyl transferase}]{\text{Aspartate}} \underset{\text{carbamoyl aspartate}}{\overset{\displaystyle COO^-}{\underset{\displaystyle COO^-}{\overset{|}{\underset{|}{\overset{\displaystyle CH_2 \quad O}{\underset{\displaystyle CHNH\overset{\|}{C}NH_2}{}}}}}} + Pi \quad (46)
$$

participation of water in the reaction. The transfer of the carbamoyl moiety probably occurs by a single-displacement mechanism.

Aspartate carbamoyl transferase catalyses the first specific step in the biosynthetic pathway leading to the formation of pyrimidines, and is functionally classified as a regulatory enzyme. The production of pyrimidine nucleotides is controlled by feedback inhibition of this enzyme by cytidine triphosphate (CTP), an end-product of the biosynthetic pathway. CTP is structurally dissimilar to the substrates aspartate and carbamoyl phosphate, and it would not be expected to compete with these compounds for an enzyme site. Gerhart and Schachman[201] concluded from studies with the purified *E. coli* enzyme that the inhibitor was bound at a specific regulatory site. The regulatory enzymes are referred to as *allosteric* (other site) enzymes. Inhibition by the allosteric effector presumably is caused by a change in the conformational state of the enzyme, accompanied by a decreased affinity for the substrate.

A second characteristic of aspartate carbamoyl transferase, and a number of other regulatory enzymes, is manifested in its activity as a function of substrate concentration. A plot of reaction velocity against substrate concentration, for most enzymes, gives a hyperbolic saturation curve. Similar plots of aspartate carbamoyl transferase activity as a function of aspartate concentration yield sigmoid curves, indicating that more than one substrate molecule interacts with the enzyme, and the binding of one molecule exerts cooperative effects, i.e. facilitates the binding of the second molecule. Aspartate carbamoyl transferase is the most extensively studied allosteric enzyme, and observations on the enzymes behaviour thus far are consistent with the Monod–Wyman–Changeux model[202] for the mechanism of allosteric enzymes. According to this hypothesis allosteric enzymes are composed of two or more identical subunits which exist in two conformational states. At low concentrations the binding of a substrate molecule to one conformational state, which has a greater affinity for substrate, displaces the equilibrium to that state which favours binding. The transition from one state to another results in the simultaneous change of all the identical subunits and hence facilitates the subsequent binding of additional substrate molecules.

Briefly, aspartate carbamoyl transferase has been dissociated into subunits and the existence of distinct substrate and inhibitor sites located on separate subunits has been demonstrated[201,203]. The catalytic subunit of the dissociated enzyme is enzymatically active and not susceptible to inhibition by CTP, which is bound to a regulator subunit. The cooperative effects observed kinetically with the native enzyme are not seen with the dissociated enzyme. The proposed alterations of protein conformation which result with ligand binding have been substantiated by chemical and physical measurements[204,205].

As we have seen, the formation of amide, amidine and guanidine groups play an important role in the biosynthesis of purine and pyrimidine compounds. These compounds, as constituents of the helical deoxyribonucleic acids[206] (DNA), proved a vital function in the storage and transmission of genetic information.

C. Biotin

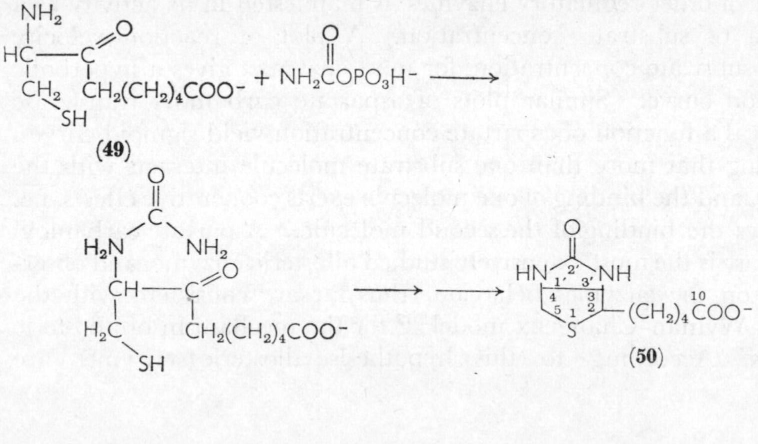

A biosynthetic pathway has been proposed[207] in which carbamoyl phosphate contributes a carbon and nitrogen atom to the formation of the amidine group of biotin (**50**). From the results of studies in *Achromobacter*, employing isotopically labelled precursors, the investigators postulated that pimelic acid, cysteine and carbamoyl phosphate condensed in a series of reactions to form biotin. Carbamoyl phosphate presumably reacted with the hypothetical intermediate **49** as shown in equation (47). It was known that pimelic acid stimulated biotin synthesis in microorganisms and isotope studies showed that the label from $[3\text{-}^{14}C]$-cysteine was incorporated into $C_{(5)}$ of biotin. Radioactivity originating from $^{14}CO_2$ was found in the $C_{(2')}$ and $C_{(10)}$ positions. According to this scheme CO_2 was utilized for the synthesis of carbamoyl phosphate and subsequently incorporated into the $C_{(2')}$ position of biotin.

Desthiobiotin (**51**) is also utilized by a number of microorganisms in the biosynthesis of biotin, but it is not clear how the proposed scheme accounts for its utilization.

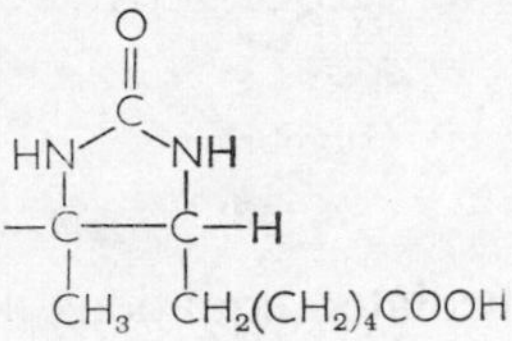

(**51**)

VIII. FORMATION OF GUANIDINES

For the most part, guanidino compounds are biosynthesized by the transfer of the arginine group to an amino acceptor. Biological compounds that derive an amidine group in transamidination reactions include guanidinoacetate the immediate precursor to creatine (equation 48), hypotaurocyamine (**52**), lombricine (**53**), streptomycin and probably canavanine (**54**) found in jack bean and γ-guanidino-butyric acid which is biosynthesized in brain tissue[208,209].

Although arginine is probably the donor for synthesis of most guanidines in nature the enzyme amidino transferase, purified from mammalian kidney[210] and streptomyces[211], does not show strict substrate specificity. Arginine, guanidinoacetate, canavanine, 4-guanidinobutyrate and 3-guanidinopropionate can serve as amidine donors and ornithine, glycine, canaline, 4-aminobutyrate, 3-amino-propionate and hydroxylamine can act as amidine acceptors with the kidney enzyme.

Hypotheses for a single-displacement[210] and a double-displacement mechanism[211] have been postulated based on competitive inhibition kinetics and the hydroxylamine reaction. More recently direct evidence was obtained by Grazi and coworkers[212] for the formation of an

$$
\begin{array}{c}
\text{NH}_2 \\
| \\
\text{C}=\text{NH} \\
| \\
\text{NH} \\
| \\
(\text{CH}_2)_3 \\
| \\
\text{CHNH}_3^+ \\
| \\
\text{COO}^-
\end{array}
\;+\;
\begin{array}{c}
\text{NH}_2 \\
| \\
\text{CH}_2 \\
| \\
\text{COO}^-
\end{array}
\;\rightleftharpoons\;
\begin{array}{c}
\text{NH}_2 \\
| \\
\text{CH}_2 \\
| \\
\text{CH}_2 \\
| \\
\text{CH}_2 \\
| \\
\text{CHNH}_3^+ \\
| \\
\text{COO}^-
\end{array}
\;+\;
\begin{array}{c}
\text{NH}_2 \\
| \\
\text{C}=\text{NH} \\
| \\
\text{NH} \\
| \\
\text{CH}_2 \\
| \\
\text{COO}^-
\end{array}
\;\xrightarrow{\text{CH}_3}\;
\begin{array}{c}
\text{NH}_2 \\
| \\
\text{C}=\text{NH} \\
| \\
\text{N—CH}_3 \\
| \\
\text{CH}_2 \\
| \\
\text{COO}^-
\end{array}
\qquad (48)
$$

arginine   glycine   ornithine   guanidinoacetate   creatine

$$
\begin{array}{c}
\text{NH}_2 \\
| \\
\text{C}=\text{NH} \\
| \\
\text{HNCH}_2\text{SO}_2\text{H}
\end{array}
\qquad\qquad
\begin{array}{c}
\text{NH}_2 \\
| \\
\text{C}=\text{NH} \qquad \text{O} \quad \overset{\text{COO}^-}{\underset{}{\text{CHNH}_2}} \\
| \qquad\qquad || \;\; | \\
\text{HNCH}_2\text{CH}_2\text{OPOCH}_2 \\
\qquad\qquad\quad | \\
\qquad\qquad\quad \text{O}^-
\end{array}
\qquad\qquad
\begin{array}{c}
\text{NH} \\
\qquad || \\
\text{O—NHCNH}_2 \\
| \\
(\text{CH}_2)_3 \\
| \\
\text{H}_2\text{NCHCOO}^-
\end{array}
$$

hypotaurocyamine   lombricine   canavanine

(52)   **(53)**   **(54)**

enzyme–amidine intermediate (equations 49a,b). The hog kidney enzyme was purified to homogeneity as judged by chromatography and centrifugation in a sucrose density gradient[213]. The purified

$$\text{Enzyme} + \text{Arginine} \rightleftharpoons \text{Ornithine} + \text{Enzyme–amidine} \qquad (49\text{a})$$

$$\text{Enzyme–amidine} + \text{Glycine} \rightleftharpoons \text{Enzyme} + \text{Guanidinoacetate} \qquad (49\text{b})$$

transamidinase retained hydrolytic activity corresponding to 1% of the transfer activity. The ratio of hydrolytic to transfer activity was constant through a 10-fold increase in purification indicating that this activity was an intrinsic property of the enzyme and in retrospect resulted from hydrolysis of an enzyme intermediate. In the absence of an acceptor compound, an enzyme–amidine complex was formed from a reaction mixture of enzyme and arginine labelled in the carbon atom of the guanidino group. The complex, separated on a sephadex G25 column, decomposed on heating, liberating urea. When incubated with glycine or ornithine the enzyme complex transferred the amidine group to the acceptor and ^{14}C-labelled products were identified by cochromatography with authentic samples.

Walker[211] postulated that a thioamidine derivative was formed by reaction of the amidine moiety with an essential sulphydryl group of the enzyme. Amidino transferase was inhibited by *p*-mercuribenzoate and thio compounds, such as 2-thiouracil and 2-thiohistidine, provided an oxidant was present. The inhibition was attributed to the formation of a mixed disulphide with the enzyme sulphydryl. Consistent with this hypothesis, formamidine disulphide (**55**) containing both amidine and disulphide groups, was reported to be a potent inhibitor (equation 50).

$$\text{Enzyme—SH} + \underset{\substack{\| \\ \text{NH}}}{H_2N—C}—S—S—\underset{\substack{\| \\ \text{NH}}}{C—NH_2} \longrightarrow$$

$$(\textbf{55})$$

$$\text{Enzyme—S—S—}\underset{\substack{\| \\ \text{NH}}}{C}—NH_2 + \underset{\substack{\| \\ \text{S}}}{H_2N—C}—NH_2 + H^+ \quad (50)$$

Grazi and associates[214] observed that the amidino transferase, in the presence of labelled amidine donor substrates, was irreversibly inactivated by the addition of bicarbonate. Precipitation of the inactivated enzyme by ammonium sulphate and dialysis of the enzyme solution did not restore activity. This finding suggested that the enzyme–amidine intermediate had interacted to form a stable covalent bond. Enzymatic hydrolysis of the protein yielded fragments which retained the radioactive amidine. The amount of amidine moiety 'fixed' to the enzyme paralleled the extent of inactivation. These investigators proposed that the bicarbonate-catalysed inactivation resulted from an intramolecular cyclization reaction with a vicinal amino group on the enzyme to form a stable product. An analogous non-enzymatic conversion of S-2-aminoethylisothiourea to 2-amino-thiazoline is shown in equation (51)[215]. The amino group of the

$$\begin{array}{c}\underset{H_2N}{H_2C}—\underset{S}{CH_2} \\ \underset{HN}{\overset{|}{C}}\underset{NH_2}{} \end{array} \underset{H^+}{\overset{OH^-}{\rightleftharpoons}} \left[\begin{array}{c}H_2C—CH_2 \\ HN \quad S \\ \underset{H_2N}{\overset{C}{}}\ NH_2\end{array}\right] \longrightarrow \begin{array}{c}H_2C—CH_2 \\ N \quad S \\ \underset{NH_2}{\overset{C}{}}\end{array} + NH_4^+ \quad (51)$$

enzyme is probably involved in the binding of the acceptor molecule since ornithine prevents the carbonate inactivation. In the absence of acceptor molecules, a maximum inhibition of 60% was obtained when the enzyme was incubated with arginine and bicarbonate. Complete inactivation of the enzyme presumably did not occur because as the reaction proceeded, ornithine was formed from the substrate and at a sufficient concentration prevented further inactivation of the enzyme.

IX. FORMATION OF N-ACYL COMPOUNDS

Many of the biological amines are known to occur as N-acyl derivatives. Classes of compounds that contain N-acylated groups include the amino acids, alkaloids, hormones, proteins, carbohydrate amines and lipids. The biological function of acylated amino groups is not understood, however, two consequences of chemically modifying a molecule by amine substitution are obvious. The blocked amine is no longer free to react and the protonation of the nitrogen atom is avoided. A decrease in the potential electrostatic interactions may be of importance in reactions with other proteins (enzymes) and with interactions with biological structures such as cell walls.

Several of the many acylamino acids, which occur in nature, are eliminated as waste products hence a mechanism for 'detoxification' is considered one function of N-acylation reactions. Oral administration of benzoate and phenyl acetate to higher primates results in the excretion of benzoylglycine and phenylacetylglycine. Phenylacetylglutamine is a normal constituent of human urine and the phenylacetyl moiety probably arises from the amino acid phenylalanine.

A unique role for N-acetylaspartate in the metabolism of nervous tissue is suggested by the fact that it is found only in the brain. The distribution pattern of N-acetylaspartate correlates with enzymatic respiratory activity and increased vascularity or blood supply. Tallan[216] has proposed that the significant concentration of this compound makes up in part for the known anion deficit in brain tissue.

A number of proteins have been found to contain an N-terminal amino acid which is acetylated, and more recent reports indicate that an amino acid with a blocked amino group is the starting point for protein synthesis. The formation of polyphenylalanine, catalysed by bacterial extracts, requires N-acetylphenylalanine–transfer-RNA for the initiation of synthesis of the peptide chain[217].

In addition to the N-acetylamino acids, derivatives of malonyl, cinnamoyl and indolacetyl groups are found conjugated to either amino acids or amino acid degradation products in plants[218]. Mucopeptides, constituents of cell walls of bacteria and other organisms, are polymers which contain N-acetylmuramic acid and N-acetylglucosamine. Amide bonds of lipids occur in sphingomyelins and cerebrosides which are synthesized from fatty acids and sphingosine. Fatty acids also form acyl derivatives with ethanolamine, phenylalanine and other amino acids.

The most common mechanism for the formation of N-acylamines is represented by the reaction sequence (52a–52c). Initial activation of the carboxyl group is achieved by formation of an acyl-AMP derivative from the reaction of an acid and ATP in a mechanism analogous to the acetate activating system first described by Berg[219]. In a second

step acyl-AMP reacts with coenzyme A to form the acyl-CoA derivative which is the actual acylating reagent.   The acyl moiety is transferred to an amine acceptor in a third reaction.   The enzyme systems

$$\text{Enzyme} + \text{RCOO}^- + \text{ATP} \longrightarrow \text{E}{-}\text{R}{-}\overset{\displaystyle O}{\overset{\|}{\text{C}}}{-}\text{AMP} + \text{PPi} \qquad (52a)$$

$$\text{E}{-}\text{R}{-}\overset{\displaystyle O}{\overset{\|}{\text{C}}}{-}\text{AMP} + \text{CoA}{-}\text{SH} \longrightarrow \text{E} + \text{R}{-}\overset{\displaystyle O}{\overset{\|}{\text{C}}}{-}\text{S}{-}\text{CoA} + \text{AMP} \qquad (52b)$$

$$\text{R}{-}\overset{\displaystyle O}{\overset{\|}{\text{C}}}{-}\text{S}{-}\text{CoA} + \text{H}_2\text{N}{-}\underset{\displaystyle \overset{|}{\text{COO}^-}}{\text{C}}{-}\text{R}^1 \xrightarrow{\text{N-Acylase}} \text{R}{-}\overset{\displaystyle O}{\overset{\|}{\text{C}}}{-}\underset{\displaystyle H}{\text{N}}{-}\underset{\displaystyle \overset{|}{\text{COO}^-}}{\text{C}}{-}\text{R}^1 \qquad (52c)$$

from mammalian tissues that catalyse the formation of N-benzoyl-
glycine[220], N-phenylacetylglutamine[221] and N-acetylaspartate[222] by
this pathway have been studied in detail.

Alternative mechanisms appear operative in fatty acid amide bond
formation.   Soluble rat liver preparations, which catalyse N-palmi-
toylphenylalanine biosynthesis, do not require the addition of ATP
or CoA[223].   Hydrolytic activity was associated with the enzyme and it
was concluded that synthesis was attributable to a reversible amino-
acylase (equation 53).   Diisopropyl fluorophosphate, a reagent which

$$\text{N-Acylamino acid} + \text{H}_2\text{O} \xrightleftharpoons{\text{Aminoacylase}} \text{Fatty acid} + \text{Amino acid} \qquad (53)$$

reacts with enzyme serine groups, inhibited activity suggesting a
mechanism analogous to esterases and peptidases, i.e. the formation
of an acyl–enzyme intermediate.

A microsomal system obtained from guinea pig and rat tissues has
been described[224] which catalysed the formation of fatty acid amides
of ethanolamine and a number of other amines.   Only aliphatic fatty
acids were found to be active as acyl donors.   The activity of the
microsomal preparation was not typical of the other amide bond
synthesizing systems in that there was no energy requirement and the
reversal of hydrolysis was ruled out.   It was known from preliminary
studies that fatty acids are bound to microsomes in a relatively stable
manner, and the authors suggested that free ^{14}C-labelled palmitoyl-
ethanolamide may be formed in a two-step reaction sequence.   It was
proposed that added ^{14}C-labelled palmitic acid exchanged with preformed microsomal-bound fatty acids and in a second step was transferred to the exogeneous amine.

X. FORMATION OF HYDROXAMATES

The oxidized amide bond, a hydroxamate, occurs in a substantial number of natural products. Within the last decade over two dozen such compounds have been isolated, mostly from fungi but also from bacteria and higher plants[225,226]. These compounds, for the most part, possess antibiotic properties. The ability to inhibit growth has been attributed to the hydroxamate function since it is the unique chemical feature of these otherwise diverse compounds. Moreover a number of chemically synthesized hydroxamates exhibit antibiotic activity. The physiological effects have most frequently been attributed to the special affinity of hydroxamate anions for ferric ions. These compounds are thought to complex and make inaccessable, ferric ions normally needed for growth. The ferrichromes, a group of trihydroxamate cyclic peptides, which are isolated as iron chelates, are able to reverse the toxic effects of other hydroxamates and, for some organisms, function as growth factors. It has been assumed the ferrichrome compounds act as iron transfer agents and that they are instrumental in providing iron for the protoporphyrin molecule.

The hydroxamate group occurs in aliphatic and cyclic structures. Representative structures among the aliphatic hydroxamates in-

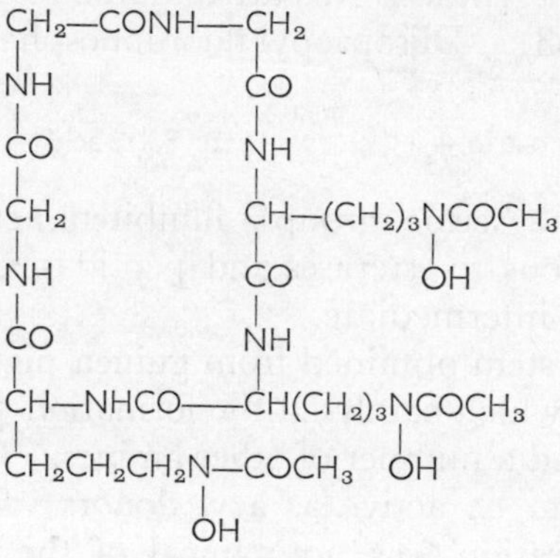

$$
\begin{array}{ll}
CH_2-CONH-CH_2 & \\
| & | \\
NH & CO \\
| & | \\
CO & NH \\
| & | \\
CH_2 & CH-(CH_2)_3NCOCH_3 \\
| & | \quad\quad\quad | \\
NH & CO \quad\quad\quad OH \\
| & | \\
CO & NH \\
| & | \\
CH-NHCO-CH(CH_2)_3NCOCH_3 \\
| \quad\quad\quad\quad\quad\quad\quad\quad\quad\quad | \\
CH_2CH_2CH_2N-COCH_3 \quad OH \\
| \\
OH
\end{array}
$$

ferrichrome

(**56**)

clude hadacidin (**57**), fusarinine (**58**), and the only primary hydroxamate, actinonin. Cyclic peptide structures which contain a hydroxamate group are cycloserine, mycobactin, which is a dihydroxamate, and the trihydroxamates, ferrioxamines and ferrichromes (e.g. **56**). A number of the cyclic hydroxamates are pyrazine derivatives and include the aspergillic acids (e.g. **59**) mycelianamide (**60**) and pul-

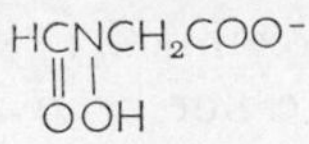

HCNCH$_2$COO$^-$
‖ |
O OH

hadacidin

(57)

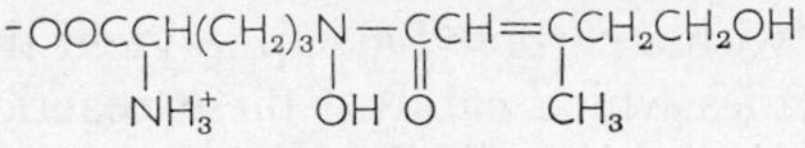

$^-$OOCCH(CH$_2$)$_3$N—CCH=CCH$_2$CH$_2$OH
| | ‖ |
NH$_3^+$ OH O CH$_3$

fusarinine

(58)

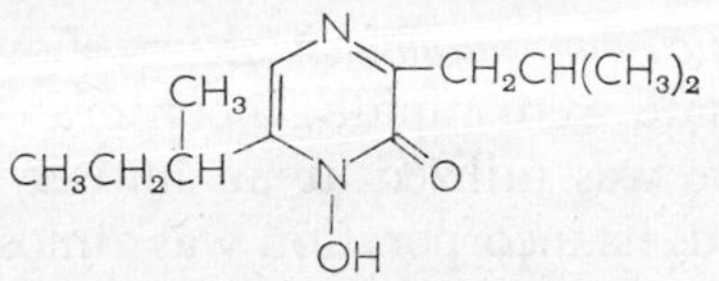

aspergillic acid

(59)

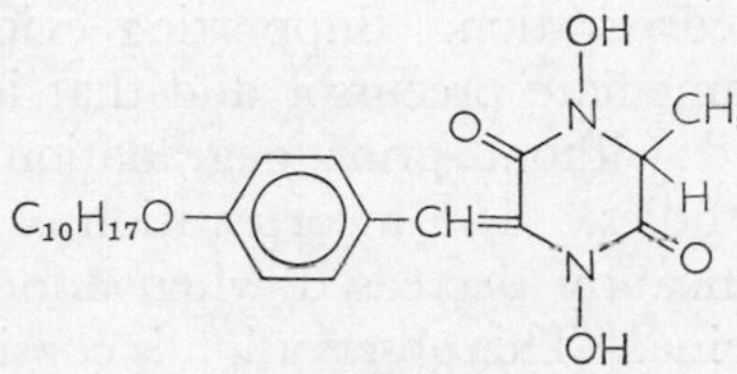

mycelianamide

(60)

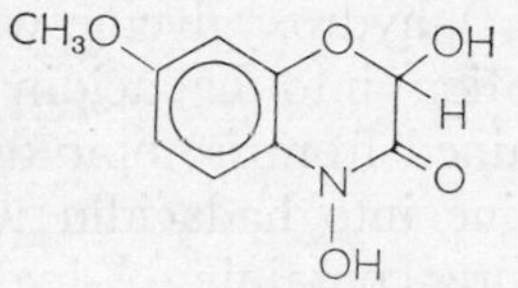

DIMBOA

(61)

cherrimin.   The cyclic hydroxamate DIMBOA (61) is a benzoxazine derivative.

The carbon skeletons of the aliphatic hydroxamates, in those compounds that have been subjected to biosynthetic studies, are formed from amino acids with corresponding carbon skeletons. The α-amino group for some of these compounds becomes the hydroxylamino group of the hydroxamate. For example hadacidin is biosynthesized from glycine and formate[227], aspergillic acid from a molecule each of leucine and isoleucine[228]; and mycelianamide from tyrosine and alanine[226]. Ornithine is incorporated into fusarinine and the ferrichromes[229] but in these cases the δ-amine is converted to a hydroxylamino group and the acyl substituent arises from a compound related to mevalonic acid or acetate. The cyclic hydroxamate DIMBOA, which occurs as the glycoside in certain plants, derives its benzene ring from the shikimic acid pathway for the biosynthesis of aromatic compounds and the carbon atoms of the heterocyclic ring arise from C$_{(1)}$ and C$_{(2)}$ of ribose[230].

The formation of the hydroxamate group can occur in one of two ways; by the oxidation of an amide bond or by hydroxylation of an amino group to form hydroxylamine followed by a condensation reaction with a carbonyl group to yield the hydroxamate. Although the enzymes which catalyse these reactions have not been isolated, evidence from biosynthetic studies suggests hydroxamate biosynthesis occurs by both pathways.

Precursor studies of hadacidin (N-formyl-N-hydroxyglycine) biosynthesis by *Penicillium aurantioviolaceum* showed that ^{14}C-labelled glycine and formate were rapidly incorporated into the compound. N-Hydroxyglycine was utilized at an initially slower rate, but over longer time periods its incorporation was almost twice that of glycine. Formylglycine was incorporated to a much lesser extent and double-label studies with ^{14}C in both the formyl and glycyl portions of the molecule indicated that the formylglycine was degraded, probably to formate and glycine, prior to incorporation. Supporting evidence that N-hydroxyglycine was an immediate precursor and that incorporation into hadacidin occurred without prior degradation was obtained from isotopic dilution studies. The incorporation of ^{14}C-glycine into hadacidin was significantly decreased when added to cultures containing N-hydroxyglycine. This observation is consistent with the hypothesis that the hydroxylamino compound is a direct intermediate in the biosynthetic pathway. Formylglycine and other compounds which were tested as possible precursors did not affect the incorporation of glycine. Stimulation of hadacidin synthesis by N-hydroxyglycine, which was not observed with other added compounds, was also suggestive of a precursor relationship. Stevens and Emery[227] concluded from these studies that the hydroxamate group of hadacidin is biosynthetically formed by N-formylation of the preformed N-hydroxyglycine.

The hydroxamate group of fusarinine and the ferrichromes is formed from the δ-amino group of ornithine, an oxygen atom and an acyl substituent. Experimental results employing isotopically labelled metabolites indicated that the biosynthetic pathway of ferrichrome formation is analogous to that of hadacidin[229]. Cultures of *Ustilage sphaerogena* utilized N-hydroxyornithine more readily than ornithine for synthesis of ferrichrome and during longer time periods, which probably compensated for permeability differences, δ-N-acetyl-δ-N-hydroxyornithine was the most efficient precursor. These results suggested a reaction sequence of N-hydroxyornithine synthesis, acetylation to yield the hydroxamate and finally, in the case of **56**, formation of the peptide bonds.

There is substantial evidence that hydroxamates can also be synthesized by oxidation of the amide bond. The formation of the aspergillic acid hydroxamates reportedly occurs by oxygenation of the nitrogen atom of the pyrazine ring, i.e. after amide bond formation (equation 54). Micetich and MacDonald[231] found that deoxyaspergillic acid and deoxyneoaspergillic acid (**62**) served as intermediates in the biosynthesis of these cyclic hydroxamates.

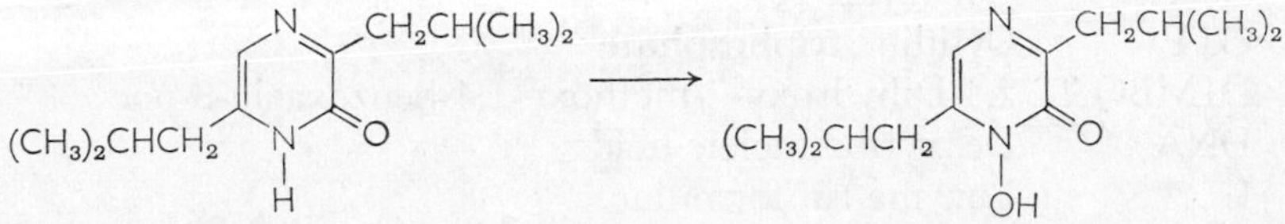

$$
\begin{array}{cc}
\text{flavacol (62)} & \text{neoaspergillic acid}
\end{array}
\tag{54}
$$

Although hydroxamates have not been found to occur normally in animal tissues, administered arylamines such as 2-aminofluorene and 2-naphthylamine are converted to their respective acetyl hydroxamates These compounds are excreted as conjugated glucuronides. From *in vivo* experiments it could not be determined whether the hydroxylation reaction preceded or followed amide bond formation. Irving[232], however, has demonstrated that liver microsomal preparations from several mammals are capable of hydroxylating 2-acetylaminofluorene (equation 55).

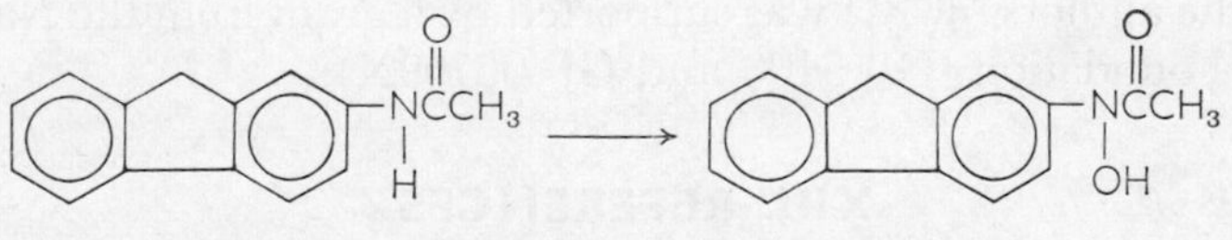

$$
\tag{55}
$$

The biosynthesis of the oxazine ring of DIMBOA from $C_{(1)}$ and $C_{(2)}$ of ribose suggested a biosynthetic pathway analogous to the formation of the imidazole of histidine, the pyrrole ring of tryptophan and the azine moiety of pteridines. The sequence of reactions for the formation of these heterocyclic compounds involves a condensation reaction between an amine and the aldehyde group of ribose followed by an Amadori-type rearrangement and ring closure to include $C_{(1)}$ and $C_{(2)}$ of ribose with elimination of the triose moiety. This reaction sequence infers hydroxylation of the nitrogen atom occurs after amide bond formation. Subsequently 2-(2-hydroxy-7-methoxy-1,4-benzoxazin-3-one)-β-D-glucopyranoside was isolated from corn roots[233] which suggests that the deoxy compound may serve as an intermediate in the biosynthesis of DIMBOA.

XI. ABBREVIATIONS

Ad	Adenine
ADP	Adenosine diphosphate
AMP	Adenosine monophosphate
ATP	Adenosine triphosphate
CoA	Coenzyme A
CTP	Cytidine triphosphate
DIMBOA	2,4-Dihydroxy-7-methoxy-1,4-benzoxazin-3-one
DNA	Deoxyribonucleic acid
E	Enzyme (in formulae)
EDTA	Ethylenediaminetetraacetic acid
GMP	Guanosine monophosphate
GTP	Guanosine triphosphate
IMP	Inosine monophosphate
NAD	Nicotinamide–adenine dinucleotide
P	Phosphoric acid residue (in formulae)
Pi	Phosphate
PPi	Pyrophosphate
RNA	Ribonucleic acid
XMP	Xanthosine monophosphate

XII. ACKNOWLEDGMENT

One of the authors (J. R.) was supported by a grant from the National Science Foundation (GB 3408 and GB 06709X).

XIII. REFERENCES

1. R. M. Allison and R. H. Burris, *J. Biol. Chem.*, **224**, 351 (1957).
2. A. P. Sims and B. F. Folkes, *Proc. Roy. Soc. (London), Ser. B*, **159**, 479 (1964).
3. K. W. Joy and B. F. Folkes, *J. Exptl. Botany*, **16**, 646 (1965).
4. D. P. Burma and R. H. Burris, *J. Biol. Chem.*, **225**, 287 (1957).
5. T. H. Benzinger and R. Hems, *Proc. Natl. Acad. Sci. U.S.*, **42**, 896 (1956).
6. L. Levintow and A. Meister, *J. Biol. Chem.*, **209**, 265 (1954).
7. A. Meister, in *The Enzymes*, Vol. 6 (Eds. P. D. Boyer, H. Lardy, and K. Myrbäck), Academic Press, New York, 1962, p. 443.
8. A. Meister, P. R. Krishnaswamy, and V. Pamiljans, *Federation Proc.*, **21**, 1013 (1962).
9. A. Meister, *Biochemistry of the Amino Acids*, Vol. 1, Academic Press, New York, 1965.
10. S. Black and N. G. Wright, *J. Biol. Chem.*, **213**, 27 (1955).
11. P. R. Krishnaswamy and A. Meister, *J. Biol. Chem.*, **235**, 408 (1960).
12. L. T. Webster, Jr. and E. W. Davie, *J. Biol. Chem.*, **236**, 479 (1961).
13. W. C. Rhodes and W. D. McElroy, *Science*, **128**, 253 (1958).

14. H. S. Kingdon, L. T. Webster, Jr., and E. W. Davie, *Proc. Natl. Acad. Sci. U.S.*, **44**, 757 (1958).
15. H. A. Krebs, *Biochem. J.*, **29**, 1951 (1935).
16. J. F. Speck, *J. Biol. Chem.*, **179**, 1405 (1949).
17. W. H. Elliott and E. F. Gale, *Nature*, **161**, 129 (1948).
18. W. H. Elliott, *Biochem. J.*, **49**, 106 (1951).
19. J. E. Varner and G. C. Webster, *Plant Physiol.*, **30**, 393 (1955).
20. P. D. Boyer, O. J. Koeppe, and W. W. Luchsinger, *J. Am. Chem. Soc.*, **78**, 356 (1956).
21. A. Kowalsky, C. Wyttenbach, L. Langer, and D. E. Koshland, Jr., *J. Biol. Chem.*, **219**, 719 (1956).
22. J. E. Varner, D. Slocum, and G. C. Webster, *Arch. Biochem. Biophys.*, **73**, 508 (1958).
23. L. Levintow and A. Meister, *Federation Proc.*, **15**, 299 (1956).
24. P. K. Stumpf and W. D. Loomis, *Arch. Biochem. Biophys.*, **25**, 451 (1950).
25. L. Levintow, A. Meister, G. H. Hogeboom, and E. I. Kuff, *J. Am. Chem. Soc.*, **77**, 5304 (1955).
26. P. R. Krishnaswamy, V. Pamiljans, and A. Meister, *J. Biol. Chem.*, **235**, PC 39 (1960).
27. P. R. Krishnaswamy, V. Pamiljans, and A. Meister, *J. Biol. Chem.*, **237**, 2932 (1962).
28. W. P. Jencks, in *The Enzymes*, Vol. 6 (Eds. P. D. Boyer, H. Lardy, and K. Myrbäck), Academic Press, New York, 1962, p. 339.
29. J. M. Buchanan and S. C. Hartman, *Advan. Enzymol.*, **21**, 199 (1959).
30. E. Khedouri, V. P. Wellner, and A. Meister, *Biochemistry*, **3**, 824 (1964).
31. R. A. Ronzio and A. Meister, *Proc. Natl. Acad. Sci. U.S.*, **59**, 164 (1968).
32. V. P. Wellner and A. Meister, *Biochemistry*, **5**, 872 (1966).
33. L. Levintow and A. Meister, *J. Am. Chem. Soc.*, **75**, 3039 (1953).
34. B. M. Braganca, J. H. Quastel, and R. Schucher, *Arch. Biochem. Biophys.*, **41**, 478 (1952).
35. N. Lichtenstein, H. E. Ross, and P. P. Cohen, *J. Biol. Chem.*, **201**, 117 (1953).
36. A. Goldstone and E. Adams, *Biochem. Biophys. Res. Commun.*, **16**, 71 (1964).
37. H. M. Kagan and A. Meister, *Biochemistry*, **5**, 725 (1966).
38. H. M. Kagan and A. Meister, *Biochemistry*, **5**, 2423 (1966).
39. A. Meister, *Federation Proc.*, **27**, 100 (1968).
40. I. B. Wilson and E. Cabib, *J. Am. Chem. Soc.*, **78**, 202 (1956).
41. C. Nicmann, *Science*, **143**, 1287 (1964).
42. B. Zerner and M. L. Bender, *J. Am. Chem. Soc.*, **86**, 3669 (1964).
43. H. Gutfreund and J. M. Sturtevant, *Biochem. J.*, **63**, 656 (1956).
44. V. Wellner, M. Zoukis, and A. Meister, *Biochemistry*, **5**, 3509 (1966).
45. C. A. Woolfolk and E. R. Stadtman, *Arch. Biochem. Biophys.*, **118**, 736 (1967).
46. D. Mecke, K. Wulff, K. Liess, and H. Holzer, *Biochem. Biophys. Res. Commun.*, **24**, 452 (1966).
47. B. M. Shapiro, H. S. Kingdon, and E. R. Stadtman, *Proc. Natl. Acad. Sci. U.S.*, **58**, 642 (1967).
48. H. S. Kingdon, B. M. Shapiro, and E. R. Stadtman, *Proc. Natl. Acad. Sci. U.S.*, **58**, 1703 (1967).

49. H. S. Kingdon, J. S. Hubbard, and E. R. Stadtman, *Biochemistry*, **7**, 2136 (1968).
50. B. M. Shapiro and A. Ginsburg, *Biochemistry*, **7**, 2153 (1968).
51. R. C. Valentine, B. M. Shapiro, and E. R. Stadtman, *Biochemistry*, **7**, 2143 (1968).
52. A. AlDawody and J. E. Varner, *Federation Proc.*, **20**, 10c (1961).
53. J. M. Ravel, S. J. Norton, J. S. Humphreys, and W. Shive, *J. Biol. Chem.*, **237**, 2845 (1962).
54. J. J. Burchall, E. C. Reichelt, and M. J. Wolin, *J. Biol. Chem.*, **239**, 1794 (1964).
55. S. Blumenthal-Goldschmidt, G. W. Butler, and E. E. Conn, *Nature*, **197**, 718 (1963).
56. C. Ressler, Y.-H. Giza, and S. N. Nigam, *J. Am. Chem. Soc.*, **85**, 2874 (1963).
57. L. Levintow, *Science*, **126**, 611 (1957).
58. J. Kidd, *J. Exptl. Med.*, **98**, 565, 583 (1953).
59. R. E. Neuman and T. A. McCoy, *Science*, **124**, 124 (1956).
60. J. D. Broome, *Nature*, **191**, 1114 (1961).
61. J. D. Broome, *J. Exptl. Med.*, **118**, 99 (1963).
62. E. A. Boyse, L. J. Old, H. A. Campbell, and L. T. Mashburn, *J. Exptl. Med.*, **125**, 17 (1967).
63. A. Schuegraf, S. Ratner, and R. C. Warner, *J. Biol. Chem.*, **235**, 3597 (1960).
64. J. E. Snoke and K. Bloch, *J. Biol. Chem.*, **213**, 825 (1955).
65. D. H. Stumeyer and K. Bloch, *J. Biol. Chem.*, **235**, PC27 (1960), and references therein.
66. J. S. Nishimura, E. A. Dodd, and A. Meister, *J. Biol. Chem.*, **238**, PC1179 (1963).
67. J. S. Nishimura, E. A. Dodd, and A. Meister, *J. Biol. Chem.*, **239**, 2553 (1964).
68. E. D. Mooz and A. Meister, *Biochemistry*, **6**, 1722 (1967).
69. J. M. Buchanan and S. C. Hartman, *Advan. Enzymol.*, **21**, 199 (1959).
70. J. C. Sonne, I. Lin, and J. M. Buchanan, *J. Biol. Chem.*, **220**, 369 (1956).
71. B. Levenberg, S. C. Hartman, and J. M. Buchanan, *J. Biol. Chem.*, **220**, 379 (1956).
72. S. C. Hartman and J. M. Buchanan, *J. Biol. Chem.*, **233**, 456 (1958).
73. L. N. Leukens and J. M. Buchanan, *J. Biol. Chem.*, **234**, 1791 (1959).
74. L. N. Leukens and J. M. Buchanan, *J. Biol. Chem.*, **234**, 1799 (1959).
75. R. W. Miller, L. N. Leukens, and J. M. Buchanan, *J. Biol. Chem.*, **234**, 1806 (1959).
76. B. Levenberg and J. M. Buchanan, *J. Biol. Chem.*, **224**, 1019 (1957).
77. I. Melnick and J. M. Buchanan, *J. Biol. Chem.*, **225**, 157 (1957).
78. S. C. Hartman, B. Levenberg, and J. M. Buchanan, *J. Am. Chem. Soc.*, **77**, 501 (1955).
79. B. Levenberg, I. Melnick, and J. M. Buchanan, *J. Biol. Chem.*, **225**, 163 (1957).
80. I. B. Dawid, T. C. French, and J. M. Buchanan, *J. Biol. Chem.*, **238**, 2178 (1963).
81. T. C. French, I. B. Dawid, R. A. Day, and J. M. Buchanan, *J. Biol. Chem.*, **238**, 2171 (1963).

82. T. C. French, I. B. Dawid, and J. M. Buchanan, *J. Biol. Chem.*, **238**, 2186 (1963).
83. C. E. Carter and L. H. Cohen, *J. Am. Chem. Soc.*, **77**, 499 (1955).
84. R. Abrams and M. Bentley, *J. Am. Chem. Soc.*, **77**, 4179 (1955).
85. C. E. Carter, *J. Biol. Chem.*, **223**, 139 (1956).
86. I. Lieberman, *J. Biol. Chem.*, **223**, 327 (1956).
87. I. Lieberman, *J. Biol. Chem.*, **222**, 765 (1956).
88. N. P. Salzman, H. Eagle, and E. Sebring, *J. Biol. Chem.*, **230**, 1001 (1958).
89. R. B. Hurlbert and H. O. Kammen, *J. Biol. Chem.*, **235**, 443 (1960).
90. K. P. Chakraborty and R. B. Hurlbert, *Biochim. Biophys. Acta*, **47**, 607 (1961).
91. J. Preiss and P. Handler, *J. Biol. Chem.*, **233**, 493 (1958).
92. R. L. Spencer and J. Preiss, *J. Biol. Chem.*, **242**, 385 (1967).
93. U. Lagerkvist, *J. Biol. Chem.*, **233**, 143 (1958).
94. R. Abrams and M. Bentley, *Arch. Biochem. Biophys.*, **79**, 91 (1959).
95. H. S. Moyed and B. Magasanik, *J. Biol. Chem.*, **226**, 351 (1957).
96. S. Udaka and H. S. Moyed, *J. Biol. Chem.*, **238**, 2797 (1963).
97. H. A. Krebs and K. Henseleit, *Z. Physiol. Chem.*, **210**, 33 (1932).
98. S. Ratner, in *The Enzymes*, Vol. 6, (Eds. P. D. Boyer, H. Lardy, and K. Myrbäck), Academic Press, New York, 1962, p. 495.
99. S. Ratner and A. Pappas, *J. Biol. Chem.*, **179**, 1199 (1949).
100. S. Ratner and B. Petrack, *J. Biol. Chem.*, **191**, 693 (1951).
101. B. Petrack and S. Ratner, *J. Biol. Chem.*, **233**, 1494 (1958).
102. J. B. Walker, *Proc. Natl. Acad. Sci. U.S.*, **38**, 561 (1952).
103. J. B. Walker and J. Myers, *J. Biol. Chem.*, **203**, 143 (1953).
104. D. C. Davison and W. H. Elliott, *Nature*, **169**, 313 (1952).
105. S. Ratner, B. Petrack, and O. Rochovansky, *J. Biol. Chem.*, **204**, 95 (1953).
106. O. Rochovansky and S. Ratner, *J. Biol. Chem.*, **236**, 2254 (1961).
107. A. Kornberg, I. Lieberman, and E. S. Simms, *J. Biol. Chem.*, **215**, 389 (1955).
108. H. G. Khorana, J. F. Fernandes, and A. Kornberg, *J. Biol. Chem.*, **230**, 941 (1958).
109. D. A. Goldthwait, *J. Biol. Chem.*, **222**, 1051 (1956).
110. A. Neidle and H. Waelsch, *J. Biol. Chem.*, **224**, 586 (1959).
111. H. S. Moyed and B. Magasank, *J. Biol. Chem.*, **235**, 149 (1960).
112. D. W. E. Smith and B. N. Ames, *J. Biol. Chem.*, **239**, 1848 (1964).
113. D. H. Meadows, J. L. Markley, J. S. Cohen, and O. Jardetzky, *Proc. Natl. Acad. Sci. U.S.*, **58**, 1307 (1967).
114. W. P. Jencks and J. Carriuolo, *J. Am. Chem. Soc.*, **82**, 1778 (1960).
115. M. Bender and F. J. Kézdy, *J. Am. Chem. Soc.*, **86**, 3704 (1964).
116. T. C. Bruice and S. J. Benkovic, *Bioorganic Mechanisms*, Vol. 1, W. A. Benjamin, Inc., New York, 1966.
117. B. M. Anderson, E. H. Cordes, and W. P. Jencks, *J. Biol. Chem.*, **236**, 455 (1961).
118. P. R. Srinivasan and A. Rivera, Jr., *Biochemistry*, **2**, 1059 (1963).
119. J. M. Edwards, F. Gibson, L. M. Jackman, and J. S. Shannon, *Biochim. Biophys. Acta*, **93**, 78 (1964).
120. F. Gibson, J. Pittard, and E. Reich, *Biochim. Biophys. Acta*, **136**, 573 (1967).
121. P. R. Srinivasan, *Biochemistry*, **4**, 2860 (1965).

122. M. I. Gibson and F. Gibson, *Biochem. J.*, **90**, 248 (1964).
123. T. I. Baker and I. P. Crawford, *J. Biol. Chem.*, **241**, 5577 (1966).
124. J. A. DeMoss, *J. Biol. Chem.*, **240**, 1231 (1965).
125. F. Gibson, *Biochem. J.*, **90**, 256 (1964).
126. A. F. Egan and F. Gibson, *Biochim. Biophys. Acta*, **130**, 276 (1966).
127. C. Ratledge, *Nature*, **203**, 428 (1964).
128. J. G. Levin and D. B. Sprinson, *J. Biol. Chem.*, **239**, 1142 (1964).
129. F. Lingen, B. Sprössller, and W. Goebel, *Biochim. Biophys. Acta*, **121**, 164 (1966).
130. J. R. D. McCormick, J. Reichenthal, U. Hirsch, and N. O. Sjolander, *J. Am. Chem. Soc.*, **84**, 3711 (1962).
131. V. F. Lingens, W. Lück, and W. Goebel, *Z. Naturforsch.*, Pt. b, **18**, 851 (1963).
132. B. Weiss and P. R. Srinivasan, *Proc. Natl. Acad. Sci. U.S.*, **45**, 1491 (1959).
133. P. R. Srinivasan and B. Weiss, *Biochim. Biophys. Acta*, **51**, 597 (1961).
134. S. Roseman, F. E. Moses, J. Ludowieg, and A. Dorfman, *J. Biol. Chem.*, **203**, 213 (1953).
135. Y. J. Topper and M. M. Lipton, *J. Biol. Chem.*, **203**, 135 (1953).
136. C. E. Becker and H. G. Day, *J. Biol. Chem.*, **201**, 795 (1953).
137. L. F. Leloir and C. E. Cardini, *Biochim. Biophys. Acta*, **12**, 15 (1953).
138. S. Ghosh, H. J. Blumenthal, E. Davidson, and S. Roseman, *J. Biol. Chem.*, **235**, 1265 (1960).
139. A. I. Virtanen, *Angew. Chem. Intern. Ed. Engl.*, **1**, 299 (1962).
140. J. F. Thompson, C. J. Morris, W. N. Arnold, and D. H. Turner in *Amino Acid Pools* (Ed. J. T. Holden), Elsevier Publishing Co., Amsterdam, 1962, p. 54.
141. L. Fowden, *Ann. Rev. Biochem.*, **33**, 173 (1964).
142. J. Jadot, J. Casimir, and E. Renard, *Biochim. Biophys. Acta*, **43**, 322 (1960).
143. B. Levenberg, *J. Biol. Chem.*, **239**, 2267 (1964).
144. S. G. Waley, *Biochem. J.*, **68**, 189 (1958).
145. S. G. Waley, *Biochem. J.*, **67**, 172 (1957).
146. D. L. Buchanan, E. E. Haley, and R. T. Markiw, *Biochemistry*, **1**, 612 (1962).
147. C. S. Hanes, F. R. S. Hird, F. J. R. Hird, and F. A. Isherwood, *Nature*, **166**, 288 (1950).
148. C. S. Hanes, F. J. R. Hird, and F. A. Isherwood, *Biochem. J.*, **51**, 25 (1952).
149. F. Binkley, *J. Biol. Chem.*, **236**, 1075 (1961).
150. A. Szewczuk and T. Baranowski, *Biochem. Z.*, **338**, 317 (1963).
151. M. Orlowski and A. Meister, *J. Biol. Chem.*, **240**, 338 (1965).
152. H. J. Gigliotti and B. Levenberg, *J. Biol. Chem.*, **239**, 2274 (1964).
153. W. E. Hanby and H. N. Rydon, *Biochem. J.*, **40**, 297 (1946).
154. C. G. Leonard and R. D. Housewright, *Biochim. Biophys. Acta*, **73**, 530 (1963).
155. M. Bovarnick, *J. Biol. Chem.*, **145**, 415 (1942).
156. W. Bruckner, J. Kovács, and H. Nagy, *J. Chem. Soc.*, **148** (1953).
157. W. J. Williams and C. B. Thorne, *J. Biol. Chem.*, **211**, 631 (1954).
158. W. J. Williams, J. Litwin, and C. B. Thorne, *J. Biol. Chem.*, **212**, 427 (1955).
159. J. P. Greenstein, *Advan. Enzymol.*, **8**, 117 (1948).

160. J. P. Greenstein and F. M. Leuthardt, *Arch. Biochem.*, **17**, 105 (1948).
161. M. Errera, *J. Biol. Chem.*, **178**, 483 (1949).
162. M. Errera and J. P. Greenstein, *J. Biol. Chem.*, **178**, 495 (1949).
163. A. Meister and S. V. Tice, *J. Biol. Chem.*, **187**, 173 (1950).
164. A. Meister, *Science*, **120**, 43 (1954).
165. J. D. Klingman and P. Handler, *J. Biol. Chem.*, **232**, 369 (1958).
166. F. W. Sayre and E. Roberts, *J. Biol. Chem.*, **233**, 1128 (1958).
167. D. R. Robinson and W. P. Jencks, *J. Am. Chem. Soc.*, **87**, 2462, 2470 (1965).
168. W. B. Jakoby and J. Fredericks, *J. Biol. Chem.*, **239**, 1978 (1964).
169. S. C. Hartman, *J. Biol. Chem.*, **243**, 853 (1968).
170. R. A. Hammer and S. C. Hartman, *J. Biol. Chem.*, **243**, 864 (1968).
171. S. C. Hartman, *J. Biol. Chem.*, **243**, 870 (1968).
172. H. A. Campbell, L. T. Mashburn, E. A. Boyse, and L. J. Old, *Biochemistry*, **6**, 721 (1967).
173. D. B. Tower, E. L. Peters, and W. C. Curtis, *J. Biol. Chem.*, **238**, 983 (1963).
174. M. E. Jones, L. Spector, and F. Lipmann, *J. Am. Chem. Soc.*, **77**, 819 (1955).
175. P. P. Cohen, in *The Enzymes*, Vol. 6 (Eds. P. D. Boyer, H. Lardy, and K. Myrbäck), Academic Press, New York, 1962, p. 477.
176. M. E. Jones, *Science*, **140**, 1373 (1963).
177. M. Marshall and P. P. Cohen, *J. Biol. Chem.*, **241**, 4197 (1966).
178. R. L. Metzenberg, M. Marshall, and P. P. Cohen, *J. Biol. Chem.*, **233**, 1560 (1958).
179. M. Marshall, R. L. Metzenberg, and P. P. Cohen, *J. Biol. Chem.*, **236**, 2229 (1961).
180. M. E. Jones and L. Spector, *J. Biol. Chem.*, **235**, 2897 (1960).
181. A. Meister, *Biochemistry of the Amino Acids*, Vol. 2, Academic Press, New York, 1965.
182. L. A. Fahien and P. P. Cohen, *J. Biol. Chem.*, **239**, 1925 (1964).
183. L. A. Fahien, J. M. Schooler, G. A. Gehred, and P. P. Cohen, *J. Biol. Chem.*, **239**, 1935 (1964).
184. F. Lacroute, *J. Gen. Microbiol.*, **40**, 127 (1965).
185. A. Pierard, N. Glansdorff, M. Mergeay, and J. M. Wiame, *J. Mol. Biol.*, **14**, 23 (1965).
186. S. E. Hager and M. E. Jones, *J. Biol. Chem.*, **242**, 5667 (1967).
187. S. E. Hager and M. E. Jones, *J. Biol. Chem.*, **242**, 5674 (1967).
188. P. M. Anderson and A. Meister, *Biochemistry*, **5**, 3157 (1966).
189. S. M. Kalman, P. H. Duffield, and T. Brzozowski, *J. Biol. Chem.*, **241**, 1871 (1966).
190. V. P. Wellner, J. I. Santos, and A. Meister, *Biochemistry*, **7**, 2848 (1968).
190a. R. B. Huston and P. P. Cohen, *Biochemistry*, **8**, 2658 (1969).
191. G. M. Hills, *Biochem. J.*, **34**, 1057 (1940).
192. H. D. Slade, *Arch. Biochem. Biophys.*, **42**, 204 (1953).
193. R. C. Valentine and R. S. Wolfe, *Biochim. Biophys. Acta*, **45**, 389 (1960).
194. D. P. Wallach and S. Grisolia, *J. Biol. Chem.*, **226**, 277 (1957).
195. S. Grisolia and D. P. Wallach, *Biochim. Biophys. Acta*, **18**, 449 (1955).
196. J. Szulmajster, *Biochim. Biophys. Acta*, **44**, 173 (1960).
197. P. P. Cohen and M. Marshall, in *The Enzymes*, Vol. 6 (Eds. P. D. Boyer, H. Lardy, and K. Myrbäck), Academic Press, New York, 1962, p. 327.

198. P. Reichard, *Acta Chem. Scand.*, **11**, 523 (1957).

199. G. H. Burnett and P. P. Cohen, *J. Biol. Chem.*, **229**, 337 (1957).

200. P. Reichard and G. Hanshoff, *Acta Chem. Scand.*, **10**, 548 (1956).

201. J. C. Gerhart and H. K. Schachman, *Biochemistry*, **4**, 1054 (1965).

202. J. Monod, J. Wyman, and J.-P. Changeux, *J. Mol. Biol.*, **12**, 88 (1965).

203. J.-P. Changeux, J. C. Gerhart, and H. K. Schachman, *Biochemistry*, **7**, 531 (1968).

204. J. C. Gerhart and H. K. Schachman, *Biochemistry*, **7**, 538 (1968).

205. J.-P. Changeux and M. M. Rubin, *Biochemistry*, **7**, 553 (1968).

206. J. D. Watson and F. H. C. Crick, *Nature*, **171**, 737 (1953).

207. A. Lezius, E. Ringelman, and F. Lynen, *Biochem. Z.*, **336**, 510 (1963).

208. S. Ratner, in *The Enzymes*, Vol. 6 (Eds. P. D. Boyer, H. Lardy, and K. Myrbäck), Academic Press, New York, 1962, p. 267.

209. J. B. Walker, in *Ciba Foundation Study Group No. 19* (Eds. G. E. W. Wolstenholme and M. P. Cameron), Little, Brown and Co., Boston, 1965, p. 43.

210. S. Ratner and O. Rochovansky, *Arch. Biochem. Biophys.*, **63**, 296 (1956).

211. J. B. Walker, *J. Biol. Chem.*, **231**, 1 (1958).

212. E. Grazi, F. Conconi, and V. Vigi, *J. Biol. Chem.*, **240**, 2465 (1965).

213. F. Conconi and E. Grazi, *J. Biol. Chem.*, **240**, 2461 (1965).

214. E. Grazi, G. Ronca, and V. Vigi, *J. Biol. Chem.*, **240**, 4267 (1965).

215. J. X. Khym, R. Shapira, and D. G. Doherty, *J. Am. Chem. Soc.*, **79**, 5663 (1957).

216. H. H. Tallan, *J. Biol. Chem.*, **224**, 41 (1957).

217. J. Lucas-Lenard and F. Lipmann, *Proc. Natl. Acad. Sci., U.S.*, **57**, 1050 (1967).

218. R. L. M. Synge, *Ann. Rev. Plant Physiol.*, **19**, 113 (1968).

219. P. Berg, *J. Biol. Chem.*, **222**, 991, 1015 (1956).

220. D. Schachter and J. V. Taggart, *J. Biol. Chem.*, **221**, 271 (1954).

221. K. Moldave and A. Meister, *J. Biol. Chem.*, **229**, 463 (1957).

222. F. B. Goldstein, *J. Biol. Chem.*, **234**, 2702 (1959).

223. T. Fukui and B. Axelrod, *J. Biol. Chem.*, **236**, 811 (1961).

224. N. R. Bachur and S. Udenfriend, *J. Biol. Chem.*, **241**, 1308 (1966).

225. J. B. Neilands, *Science*, **156**, 1443 (1967).

226. A. J. Birch and H. Smith in *Ciba Found. Symp. Amino Acids Peptides Antimetab. Activity* (Eds. G. E. W. Wolstenholme and C. M. O'Connor), Little, Brown and Co., Boston, 1958, p. 247.

227. R. L. Stevens and T. F. Emery, *Biochemistry*, **5**, 74 (1966).

228. J. C. MacDonald, *J. Biol. Chem.*, **236**, 512 (1961).

229. T. Emery, *Biochemistry*, **5**, 3694 (1966).

230. J. E. Reimann and R. U. Byerrum, *Biochemistry*, **3**, 847 (1964).

231. R. G. Micetich and J. C. MacDonald, *J. Biol. Chem.*, **240**, 1692 (1965).

232. C. C. Irving, *J. Biol. Chem.*, **239**, 1589 (1964).

233. H. E. Gahagan and R. O. Mumma, *Phytochem.*, **6**, 1441 (1967).

234. W. P. Jencks, *J. Am. Chem. Soc.*, **80**, 4581 (1958).

CHAPTER 12

Directing and activating effects of the amido group

J. A. SHAFER

The University of Michigan, Ann Arbor, Michigan, U.S.A.

I. SUBSTITUENT EFFECTS OF AMIDO GROUPS IN AROMATIC COMPOUNDS

Effects of amido substituents in reactions of aromatic compounds have not been extensively studied, since in many cases the possibility of destruction of amido substituents, for example, by solvolysis, must be considered. There is enough information available, however, to characterize effects of amido substituents on the reactivity of aromatic compounds. Acylamino substituents can donate electrons to aromatic systems through mesomeric interactions, whereas they can withdraw electrons through inductive effects. Inductive and mesomeric inter- actions of carboxamido substituents with aromatic systems result in withdrawal of electrons from aromatic nuclei.

A. Substituent Constants

The effect of a *meta* or *para* substituent, X, on the equilibrium con- stant, K_y^X, or the rate constant k_y^X, for a chemical reaction, y, can often be represented by the Hammett[1] equation (1). (See reference 2 for

$$\log (K_y^X/K_y^0) \quad \text{or} \quad \log (k_y^X/k_y^0) = \sigma_X \rho_y \qquad (1)$$

additional discussions of the scope and limitations of the Hammett equation.) The corresponding equilibrium and rate constants for the reaction of the unsubstituted derivatives are K_y^0 and k_y^0, respec- tively. The constant ρ_y depends on the reaction and the reaction conditions being studied, and ideally, the substituent constant, σ_X, is dependent only on the substituent, and is defined by equation (2)

$$\sigma_X = \log (K_a^X/K_a^0), \qquad (2)$$

where K_a^X/K_a^0 is the ratio of the acid dissociation constant of the sub- stituted benzoic acid to that of benzoic acid, both in water at 25°. Sometimes substituent constants are evaluated indirectly by determin- ing the value of ρ_y for a reaction resembling the ionization of benzoic acid in water from the effect of substituents with known σ's. Once the value of ρ_y is established, equation (1) may be solved for the un- known substituent constant from the effect of this substituent on the reaction. Hammett substituent constants for the acetylamino group are listed in Table 1. The substituent constant seems to be dependent on the solvent composition; however, additional data are required to firmly establish the solvent dependence of the substituent constant for the acetylamino group. Leffler and Grunwald[2c] have attributed the dependence of the substituent constant (determined from the acid dissociation constants of benzoic acids in ethanol–water mixtures) for the hydroxyl group on the solvent composition, to changes in the solvation of the hydroxyl group. For purposes of comparison, data from their tabulation of the change in the substituent constant for the hydroxyl group are also listed in Table 1 along with the corresponding substituent constants determined by Jaffé and coworkers[3]. The data

TABLE 1.   Hammett substituent constants for acetylamino and hydroxyl groups.

Reaction used to define σ^a	Acetylamino		Hydroxyl	
	σ_p	σ_m	σ_p	σ_m
Acid dissociation constants of benzoic acids				
in water	$-0\cdot06^b$	$0\cdot15^b$	$-0\cdot37^c$	$0\cdot12^c$
in 50% ethanol	$0\cdot053^b$	$0\cdot27^b$	$-0\cdot34^c$	$-0\cdot05^c$
First acid dissociation constants of phenylphosphoric acids^d				
in water			$-0\cdot21$	$0\cdot07$
in 50% ethanol	$-0\cdot01$		$-0\cdot30$	$0\cdot04$
Second acid dissociation constants of phenylphosphoric acids^d				
in water	$-0\cdot10$		$-0\cdot25$	$-0\cdot02$
in 50% ethanol	$-0\cdot02$		$-0\cdot31$	$-0\cdot03$

^{a} At 25°.
^{b} From substituent constants tabulated by D. M. McDaniel and H. C. Brown, *J. Org. Chem.*, **23**, 420 (1958).
^{c} From reference 2c, p. 174.
^{d} Reference 3.

in Table 1 indicate that the effects of ethanol on the substituent constants for the acetylamino group and the hydroxyl group are comparable in magnitude.

The tautomeric and inductive effects of a substituent on the reaction centre are represented by σ_X, whereas ρ_y represents the sensitivity of the reaction to these effects. Hammett's equation may be rewritten in terms of several parameters in order to separately represent the contributions of the various tautomeric and inductive effects to the molecule's reactivity:

$$\log (K_y^X/K_y^0) = \sigma_{X1}\rho_{y1} + \sigma_{X2}\rho_{y2} + \cdots + \sigma_{Xn}\rho_{yn}, \tag{3}$$

where

$$\sigma_X = \log (K_a^X/K_a^0) = \sigma_{X1}\rho_{a1} + \sigma_{X2}\rho_{a2} + \cdots + \sigma_{Xn}\rho_{an}, \tag{4}$$

$$\rho_y = \rho_{y1} + \rho_{y2} + \cdots + \rho_{yn}, \tag{5}$$

and

$$1 = \rho_{a1} + \rho_{a2} + \cdots + \rho_{an}. \tag{6}$$

When the values of ρ in the $\sigma\rho$ terms making significant contributions to the overall substituent effect have the same relative importance as they do in the ionization of benzoic acid, i.e.

$$\rho_{y1}/\rho_y = \rho_{a1}, \ \rho_{y2}/\rho_y = \rho_{a2}, \ \cdots \ \rho_{yn}/\rho_y = \rho_{an}, \qquad (7)$$

equation (3) simplifies to equation (1). Substituents in the *meta* position exert their influence on the reaction centre primarily through inductive effects. Therefore, the number of significant terms in equations (3–6) is markedly reduced for *meta* substituents (perhaps to one or two terms), thereby increasing the probability that equations (7) will hold. Thus, effects of *meta* substituents on almost all reactions of aromatic compounds can be correlated reasonably well by the two-parameter Hammett equation (1), where σ is defined by equation (2).

When the use of σ values as defined by equation (2), leads to pronounced deviations from the Hammett equation for certain reactions, special substituent constants are defined by determining the values of ρ_y from the effect of 'normal' substituents (usually *meta* substituents) on a given reaction. Equation (1) is then used to establish the value of σ_X for the 'abnormal' substituent. A special substituent constant (σ_X^-) for a substituent in the *para* position which withdraws electrons (through tautomeric effects) is recommended for use in reactions involving unshared electrons on an atom attached to the benzene ring. Another special substituent constant (σ_X^+) is recommended for use in aromatic electrophilic substitution reactions as well as in reactions involving the production of a positive charge on an atom attached to the benzene ring[4]. Use of σ^+ values (as opposed to σ values) in these reactions becomes important for *para* substituents which donate electrons (through mesomeric effects). Substituent constants for amido groups are listed in Table 2.

The values of σ^+ for the two acylanilides were defined from rates of bromination ($\sigma^+ = -0.75$) and chlorination ($\sigma^+ = -0.79$) at 25° in acetic acid given by de la Mare and Hassan[5,6] and the values of ρ for these reactions determined by Brown and Okamoto[4b]. It should be mentioned that the dissociation of triphenylcarbinol derivatives to triphenyl carbonium ions in aqueous sulphuric acid leads to σ^+ values of 0.47 and 0.42 for the acetylamino and benzoylamino groups[4b]. Special difficulties associated with determining the dissociation constant of acylamino-substituted triphenylcarbinols might be responsible for this discrepancy[7]. This difference in the values of σ^+ might also reflect changes in solvation of the acylamino substituents in the two reactions.

TABLE 2. Substituent constants for amido groups.

Substituent	σ	σ^+	σ^-
p-Acetylamino	-0.06^a	-0.77^b	
m-Acetylamino	0.15^a		
p-Benzoylamino	0.078^c	$-0.77^{b,d}$	
m-Benzoylamino	0.217^c		
p-Carboxamido	0.280^c		$0.627^{c,e}$

[a] Defined by the acid dissociation constant of benzoic acids in water at 25°. D. M. McDaniel and H. C. Brown, *J. Org. Chem.*, **23**, 420 (1958).
[b] Defined by rates of bromination and chlorination at 25°, see text.
[c] From the compilation in reference 2a.
[d] The rates of *para* chlorination of benzanilide and acetanilide were assumed equal at 25°, since they are essentially equal at 20°[10].
[e] Defined by the rate of reduction of nitrobenzenes by $TiCl_3$ in ethanol–HCl.

B. Chlorination and Bromination of Anilides in Acetic Acid

There is considerable evidence suggesting that molecular chlorine acts as the electrophile in chlorinations of benzene derivatives in acetic acid solution[8]. Partial rate factors (i.e. the rate of substitution at a particular position relative to the rate of substitution at one of the equivalent positions of benzene) for the monochlorination of some acetanilides are listed in Table 3. The large partial rate factor for *para* substitution of acetanilide, 25.2×10^5, reflects stabilization of the transition state through resonance.

When a 4-methyl substituent is introduced into acetanilide, the 2- and 6-positions become 5 times more reactive. The agreement between this factor and the partial rate factor of 5 for the chlorination of toluene in the *meta* position[9] has been pointed out by de la Mare and Hassan[6]. This result indicates that the contribution of the acetamido and methyl substituents to the free energy of activation are independent and additive. Significantly, the methyl group in 2-methylacetanilide does not enhance the rate of chlorination at the 4- and 6-positions. Instead of the expected partial rate factors of 125×10^5 and 31×10^5 for chlorination in the 4- and 6-positions, values of 6.2×10^5 and 1.6×10^5 are observed. This 20-fold decrease has been ascribed to the

23 + c.o.a.

TABLE 3.   Partial rate factors for chlorination of acetanilides[a]

Reactant	$10^{-5} \times$ Partial rate factor at position				
	2	3	4	5	6
Acetanilide	6·1		25·2		6·1
4-Methylacetanilide	31				31
2-Methylacetanilide			6·2		1·6
N-Methylacetanilide			0·02		
2,6-Dimethylacetanilide		0·23	0·012	0·23	
1,4-Diacetamidobenzene	2·3	2·3		2·3	2·3

[a] In acetic acid at 25°, from reference 6.

steric inhibition of resonance caused by the interactions between the methyl and acetylamino groups[6].   Interestingly, N-methylacetanilide, which should have similar steric restraints for resonance as 2-methyl-acetanilide is chlorinated in the 4-position about 1200 times slower than acetanilide.   This extra decrease caused by the N-methyl

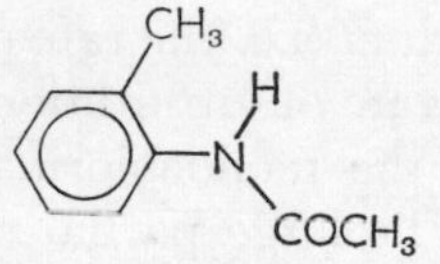

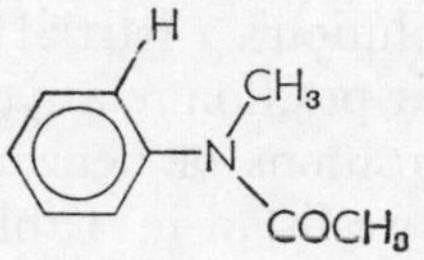

group (a factor of about 60) is in the opposite direction expected for polar effects, and has been taken by de la Mare and Hassan[6] as evidence for the contribution of N—H hyperconjugation to the reactivity of acetanilide.   It is not surprising, that a second methyl substituent in 2,6-dimethylacetanilide decreases the rate of 4-sub-stitution by a larger factor than the first methyl group (2600 vs. 20, i.e. $5 \times 6·2 \times 10^5/0·012 \times 10^5$), since an ortho-methyl group should interact more with an acetyl group than a hydrogen atom.   Compari-son of the partial rate factors for 2-chlorination of 1,4-diacetamido-benzene with acetanilide leads to a partial rate factor of 0·38 (i.e. $2·3 \times 10^5/6·1 \times 10^5$) for the 3-chlorination of acetanilide.   This result reflects the predominance of electron withdrawal from the meta position through direct inductive effects over any increase in electron density at the meta position through electrostatic interactions (second-order release) with the ortho and para positions, whose electron density is increased by mesomeric interactions with the acetylamino group. In 2,6-dimethylacetanilide these second-order mesomeric effects on the meta position are reduced, and the inductive effect of the acetamido group is more pronounced. Comparison of the partial rate factors for the 4-chlorination of 1,3-xylene and the 3-chlorination of 2,6-dimethylacetanilide leads to a partial rate factor of 0·05 for the 3-chlorination of acetanilide[6]. It should be noted, however, that resonance is probably not completely inhibited in 2,6-dimethylacetanilide, since the partial rate factor for the 4-chlorination of this compound is about 50 times higher than the partial rate factor for the 5-chlorination of 1,3-xylene[6].

Studies of Orton and Bradfield[10] indicate that the isomer distributions and rates of chlorination of anilides of organic acids are insensitive to changes in the organic acid (Table 4). Chlorination of

TABLE 4.　Products and rates of chlorination of anilides of organic acids[a].

Reactant	Product (%)		$k \, (sec^{-1} M^{-1})$
	ortho	para	
Formanilide	30	70	0·15
Acetanilide	33	67	1·0
Benzanilide	30	70	1·2
Benzenesulphonanilide	35	65	0·73

[a] In acetic acid containing 1% water at 20°, from reference 10.

acetanilide and benzanilide gives *ortho* isomers in 30% yield, but bromination* of these compounds under essentially identical conditions gives only the *para* isomer[5]. This change in selectivity might be reflecting differences in the mechanism of chlorination and bromination. Although Cl^+ and Br^+ have been ruled out as the active electrophiles, the rate of chlorination is strictly first order in Cl_2[6], whereas the rate law for bromination is primarily second order in Br_2 with an additional term first order in Br_2[11]. Moreover, specific rate constants for bromination have been reported to depend on the initial concentration of the aromatic reactant[12].

* Interestingly, acetanilide and benzanilide are also brominated at equal rates.

Bromination of 4-methoxy-2-nitroacetanilide can lead to displacement of its acetamido and nitro groups[13]. Harrison and McOmie[13] have suggested a concerted elimination reaction as a possible mechanism for this reaction (equation 8).

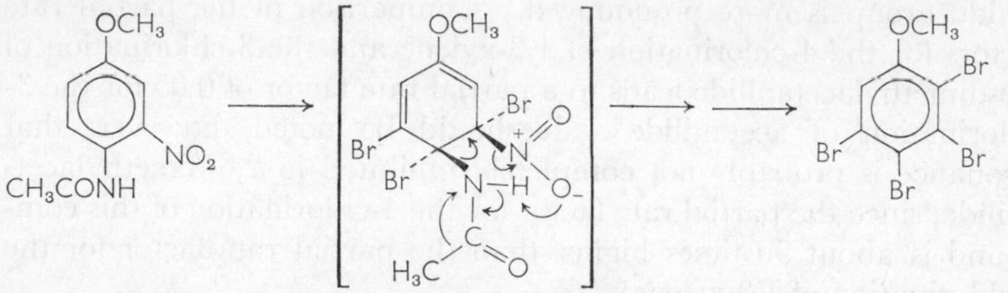

$$\tag{8}$$

C. Nitration of Acetanilide

As shown in Table 5 the directing effect of the acetylamino substituent in acetanilide is markedly dependent on the solvent. Paul[14] attributes the high yield of *ortho* isomer in acetic anhydride to the low dielectric constant of the medium, where the increased difficulty in separating charges would be expected to lead to an increased negative

TABLE 5. Product distribution in mononitrations of aromatic compounds.

Substrate and nitration medium	$T(°c)$	Relative yield[a] (0.5 ortho/para)
Acetanilide[b]		
HNO_3, H_2SO_4	20	0·12
HNO_3, acetic anhydride	20	1·1
90% aq. HNO_3	−20	0·15
80% aq. HNO_3	−20	0·34
Chlorobenzene[c]		
90% aq. HNO_3	0	0·21
HNO_3, acetic anhydride	0	0·056
Bromobenzene[c]		
90% aq. HNO_3	0	0·30
HNO_3, acetic anhydride	0	0·16
Anisole[b]		
HNO_3, H_2SO_4	45	0·23
HNO_3, acetic acid	65	0·34
HNO_3, acetic anhydride	10	1·3

[a] The yields of *meta* isomers were less than 3%.
[b] From reference 8, p. 53.
[c] From reference 14.

charge density in the *ortho* position relative to the *para* position. Paul's[14] rationale also explains the decreased yield of *ortho* isomer in the nitration of bromobenzene and chlorobenzene on going from an aqueous medium to acetic anhydride. With electron-withdrawing substituents the positive charge density at the *ortho* position would be expected to increase as the dielectric constant of the medium is decreased. The low yield of *o*-nitroanisole in acetic acid (Table 5) seems inconsistent with Paul's hypothesis. Perhaps changes in solvation of the methoxy substituent (and the difference in temperature) are responsible for the changes in the directive effects of the methoxy group listed in Table 5.

D. Nitration of Benzamide

Cooper and Ingold[15] determined the directing effects of the carboxamido group on the nitration of benzamide in fuming nitric acid. The yields of *ortho*, *meta*, and *para* isomers were 27%, 70%, and less than 3%, respectively. Obviously under these conditions, the *para* position is selectively deactivated. In light of Paul's arguments[14], it would be interesting to see if the yield of *para* isomer increases in acetic anhydride.

E. Hydroxylation of Acetanilide

The hydroxylation of acetanilide by several hydroxylating systems has recently been investigated[16]. Studies with p-^{2}H-acetanilide (using n.m.r.) indicate that 7.5% of the deuterium migrated to the 3-position on hydroxylation with trifluoroperacetic acid. Isotope studies with other hydroxylating systems showed in less than 2% retention of the hydrogen originally at the *para* position in the *p*-hydroxyacetanilide formed. A pathway suggested for hydroxylation of acetanilide in trifluoroperacetic acid is given by equation (9)[16].

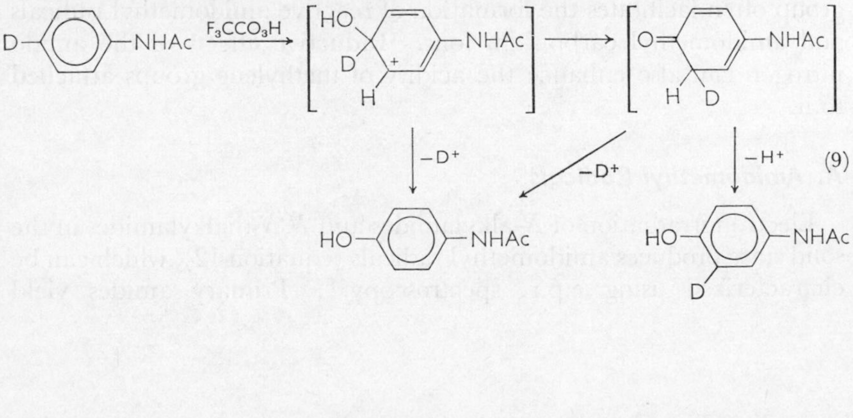

F. Ortho Metallation

In hexane–tetrahydrofuran, n-butyllithium reacts with N-methyl-benzamide forming o,N-dilithiobenzamide (1)[17]. This compound

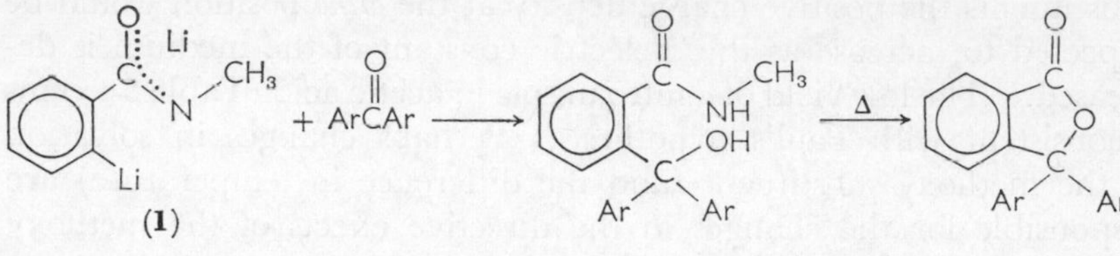

$$\text{(10)}$$

readily condenses with aldehydes and ketones, offering a convenient route to substituted phthalides and o-carbinols of N-methylbenzamide[17] (equation 10). Similarly, benzenesulphonamides are *ortho* metallated by excess n-butyllithium to o,N-dilithiobenzenesulphonamides, which may be condensed with aldehydes and ketones, to form substituted o-sulphamylbenzyl alcohols and substituted sultams (equation 11)[18].

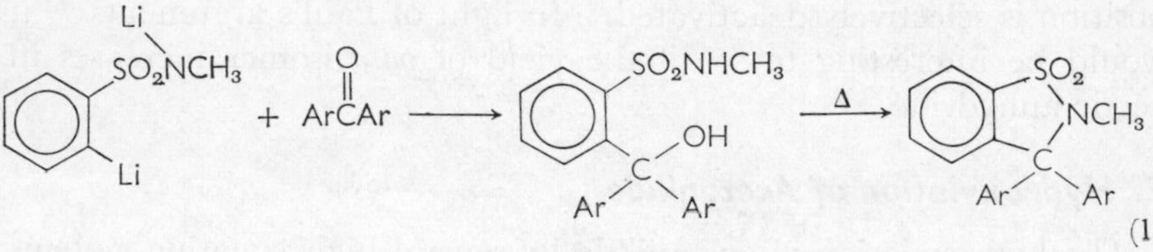

$$\text{(11)}$$

II. EFFECTS ON ALIPHATIC CARBON ATTACHED TO AMIDO NITROGEN

The enhanced reactivity of a methylene carbon attached to an amido nitrogen can be attributed to interactions between the unshared pair of electrons on the nitrogen atom and the methylene carbon atom. The unshared pair of electrons on the nitrogen atom of an amido group often facilitates the formation of reactive amidomethyl radicals and amidomethyl carbonium ions. Inductive effects of the amido nitrogen can also enhance the acidity of methylene groups attached to it.

A. Amidomethyl Radicals

Electron irradiation of N-alkylamides and N,N-dialkylamides in the solid state produces amidomethyl radicals (equation 12), which can be characterized using e.p.r. spectroscopy[19]. Primary amides yield radicals on the carbon atom adjacent to the amido carbonyl on ir-radiation in the solid state[19] (equation 13). Free radicals derived

$$CH_3CONHCH_3 \xrightarrow{\text{x-ray}} CH_3CONH\dot{C}H_2 \qquad (12)$$

$$CH_3CONH_2 \xrightarrow{\text{x-ray}} \dot{C}H_2CONH_2 \qquad (13)$$

from acetamide and formamide have been generated in solution[20] using the hydroxyl radical (from H_2O_2–Ti^{III}) for hydrogen atom abstraction. Surprisingly, in the radical from formamide, the un-paired electron appears to be located on nitrogen[20].

Amidomethyl radicals are probable intermediates in the per-sulphate-mediated dealkylation of N-substituted amides[21]. A possi-ble pathway for this reaction put forth by Needles and Whitfield[21] appears in equations (14a–14e). Amidomethyl radicals are also

$$S_2O_8^{2-} \longrightarrow 2SO_4^{\cdot-} \qquad (14a)$$

$$SO_4^{\cdot-} + HOH \longrightarrow HOSO_3^- + HO^\cdot \qquad (14b)$$

$$\overset{\overset{\displaystyle CH_3}{|}}{RCONCH_3} + SO_4^{\cdot-} \ (\text{or } HO^\cdot) \longrightarrow \overset{\overset{\displaystyle CH_3}{|}}{RCON\dot{C}H_2} + HOSO_3^- \ (\text{or } HOH) \quad (14c)$$

$$\overset{\overset{\displaystyle CH_3}{|}}{RCON\dot{C}H_2} + SO_4^{\cdot-} \longrightarrow \overset{\overset{\displaystyle CH_3}{|}}{RCONCH_2OSO_3^-} \qquad (14d)$$

$$\overset{\overset{\displaystyle CH_3}{|}}{RCONCH_2OSO_3^-} + H_2O \longrightarrow \overset{\overset{\displaystyle CH_3}{|}}{RCONH} + CH_2O + HOSO_3^- \qquad (14e)$$

likely intermediates in the formation of N-acetoxymethyl-N-methyl-acetamide from N,N-dimethylacetamide and peracetic acid[22a], and in the formation of N-benzoyloxymethyl-N-methylformamide from N,N-dimethylformamide and benzoyl peroxide[22b].

Formation of amidomethyl formates and acetates through the elec-trolysis of solutions containing a N,N-dimethylamide and an acid salt may also involve formation of amidomethyl radicals[23]. Stabilization of amidomethyl radicals through resonance may account for their facile formation.

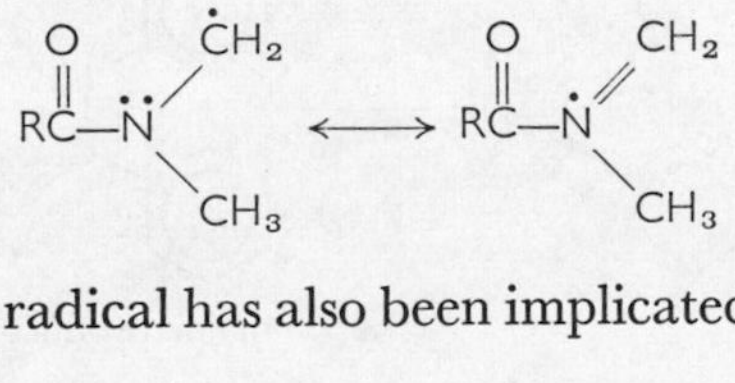

An amidomethyl radical has also been implicated as an intermediate in the decomposition of the diazonium ion derived from o-amino-N,N-dimethylbenzamide (equation 15)[24]. Cohen and coworkers[25] have shown that in this reaction, the rate of intramolecular hydrogen transfer to the benzene ring is faster than rotation about the carbon–nitrogen bond. Interestingly, in the decomposition of the diazonium

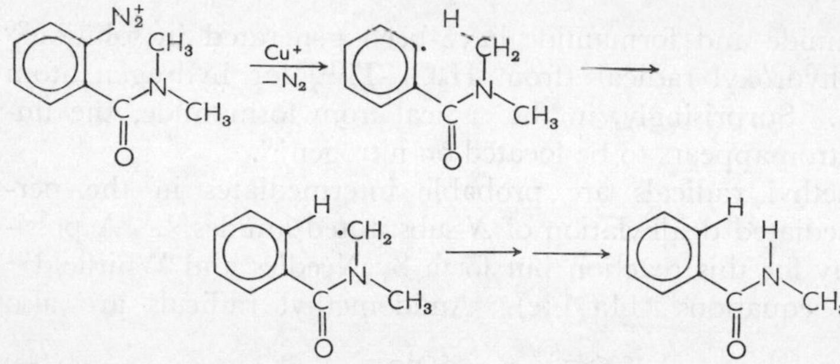

(15)

ion from substituted benzanilides, reactions between the phenyl radical and another benzene ring appears to be favoured over intramolecular hydrogen transfer (equation 16)[26]. Substituents (NO_2, and CH_3)

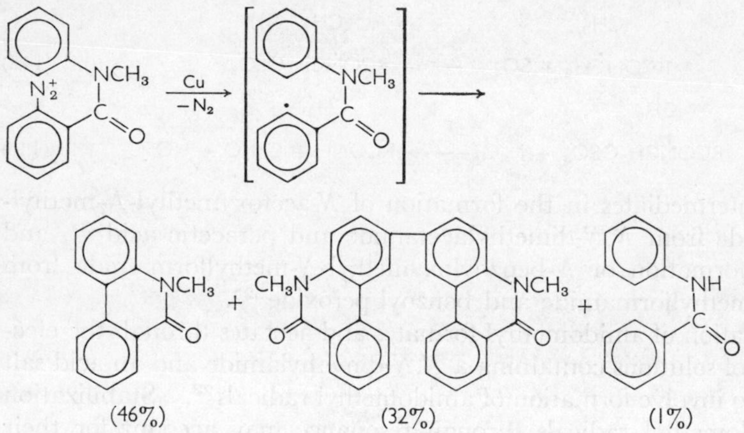

(16)

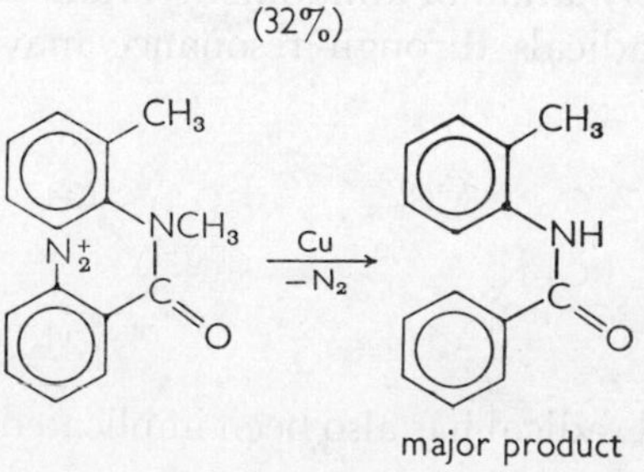

(46%)   (32%)   (1%)

(17)

major product

ortho to the nitrogen cause hydrogen transfer to the phenyl radical to become predominant (equation 17)[27]. Interactions between an *ortho* substituent and the *N*-methyl group probably force the two benzene rings out of coplanarity, thereby decreasing the susceptibility of the aniline ring to attack by the phenyl radical.

B. Amidomethyl Carbonium Ions

Kinetic studies of Firestone and coworkers[28] indicate that cyanide-catalysed racemization of α-acetamido-α-methyl nitriles in dimethyl sulphoxide proceeds via a carbonium ion rather than S_N2 attack by cyanide (equation 18).

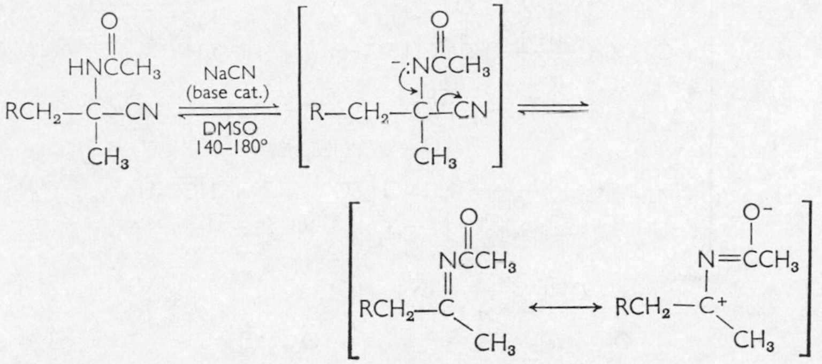

$$(18)$$

The acid-catalysed acyl interchange of *N*-formyloxymethyl-*N*-methylformamide (**2**) is also convincing evidence for the existence of amidomethyl carbonium ions (equation 19)[29]. The facile generation of carbonium ions from *N*-formyloxymethyl-*N*-methylformamide (**2**)

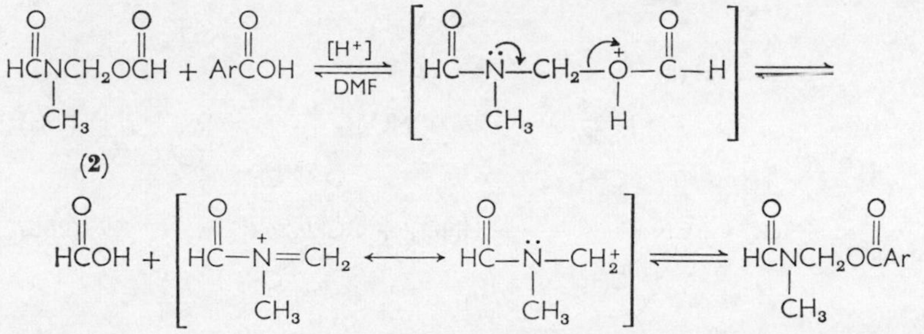

$$(19)$$

makes it a useful electrophile, as exemplified in equations (20)–(23)[29]. Similar examples of amidomethylation (and imidomethylations) using amidomethyl (and imidomethyl) -halogens, -alcohols and -amines

$$\underset{\substack{\mathrm{CH_3}\\(\mathbf{2})}}{\overset{\mathrm{O}\quad\quad\mathrm{O}}{\mathrm{HCNCH_2OCH}}} +$$

$$\mathrm{EtOH} \xrightarrow[\substack{\text{room}\\\text{temp.}}]{[\mathrm{H^+}]} \underset{\mathrm{CH_3}}{\overset{\mathrm{O}}{\mathrm{HCNCH_2OEt}}} + \mathrm{HCO_2H} \tag{20}$$

$$\mathrm{C_6H_5OH} \xrightarrow{[\mathrm{H^+}]} \underset{\mathrm{CH_3}}{\overset{\mathrm{O}}{\mathrm{HCNCH_2}}}\!\!-\!\!\bigcirc\!\!-\!\!\mathrm{OH} \;+ \tag{21}$$

$$\underset{\mathrm{CH_3}}{\overset{\mathrm{O}}{\mathrm{HCNCH_2}}}\!\!-\!\!\bigcirc\!\!-\!\!\mathrm{OH} + \mathrm{HCO_2H}$$

$$n\text{-}\mathrm{C_5H_{11}SH} \xrightarrow[\substack{\text{room}\\\text{temp.}}]{[\mathrm{H^+}]} \underset{\mathrm{CH_3}}{\overset{\mathrm{O}}{\mathrm{HCNCH_2SC_5H_{11}\text{-}n}}} + \mathrm{HCO_2H} \tag{22}$$

$$\underset{}{\overset{\mathrm{S}}{\mathrm{NH_2CNH_3^+Cl^-}}} \xrightarrow[\substack{\text{room}\\\text{temp.}}]{\mathrm{DMF}} \underset{\mathrm{CH_3}}{\overset{\mathrm{O}\quad\quad\mathrm{NH_2^+Cl^-}}{\mathrm{HCNCH_2SC}}}{}_{\mathrm{NH_2}} + \mathrm{HCO_2H} \tag{23}$$

$$\mathrm{NO_3^-} \longrightarrow e + \mathrm{NO_3^{\bullet}} \tag{24a}$$

$$\mathrm{NO_3^{\bullet}} + \underset{\mathrm{CH_3}}{\overset{\mathrm{O}\quad\ \mathrm{CH_3}}{\mathrm{RCN}}} \longrightarrow \mathrm{HNO_3} + \underset{\substack{\mathrm{CH_3}\\(\mathbf{3})}}{\overset{\mathrm{O}\quad\ \overset{\bullet}{\mathrm{C}}\mathrm{H_2}}{\mathrm{RCN}}} \tag{24b}$$

$$\underset{\substack{\mathrm{CH_3}\\(\mathbf{3})}}{\overset{\mathrm{O}\quad\ \overset{\bullet}{\mathrm{C}}\mathrm{H_2}}{\mathrm{RCN}}} \longrightarrow \underset{\substack{\mathrm{CH_3}\\(\mathbf{4})}}{\overset{\mathrm{O}\quad\ \mathrm{CH_2^+}}{\mathrm{RCN}}} + e \tag{24c}$$

$$\mathrm{NO_3^-} \longrightarrow 2e + \mathrm{NO_3^+} \tag{25a}$$

$$\mathrm{NO_3^+} + \underset{\mathrm{CH_3}}{\overset{\mathrm{O}\quad\ \mathrm{CH_3}}{\mathrm{RCN}}} \longrightarrow \mathrm{HNO_3} + \underset{\substack{\mathrm{CH_3}\\(\mathbf{4})}}{\overset{\mathrm{O}\quad\ \mathrm{CH_2^+}}{\mathrm{RCN}}} \tag{25b}$$

have been discussed by Hellmann[30]. However, the ease of preparation of amidomethyl esters should favour their use as amidomethylating agents. N-Methylamidomethyl esters or ethers can be conveniently produced by electrolysis of solutions containing ammonium nitrate, a dimethylamide and an organic acid or an alcohol[23b]. Amidomethyl carbonium ions are probable intermediates in these reactions. Two possible mechanisms proposed by Ross and coworkers[23b] are given in equations (24a–24c) and (25a,b).

Cohen and coworkers[31a–d] have shown that amidobenzyl carbonium ion **7** is an intermediate in the thermal decomposition of diazonium ion **5** to N-benzylbenzamide. Although in the absence of water, this carbonium ion slowly cyclizes to phthalimidine (**6**); in the presence of water, carbonium ion **7** is hydrolysed to N-benzylbenzamide without forming **6**[31d]. Thus, **7** is not an intermediate in the formation of **6** in the thermal decomposition of **5** in aqueous solution. Cohen

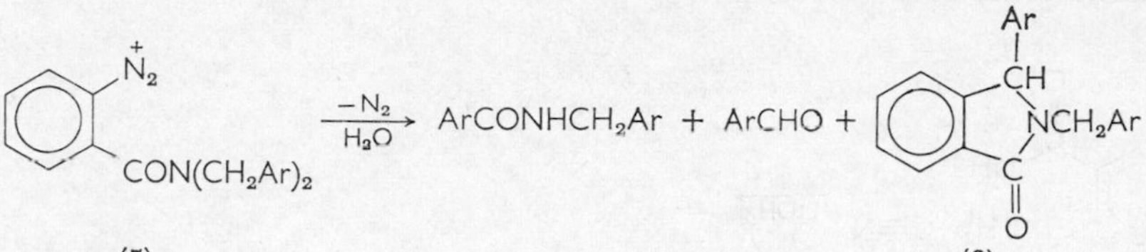

$$\xrightarrow[\text{H}_2\text{O}]{-\text{N}_2} \quad \text{ArCONHCH}_2\text{Ar} + \text{ArCHO} +$$

(5)   (6)

(26)

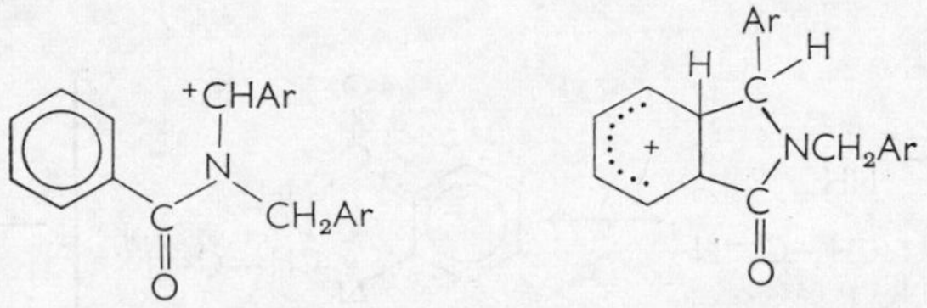

(7)   (8)

$$\xrightarrow[\Delta]{\text{Aq. sol.}}$$

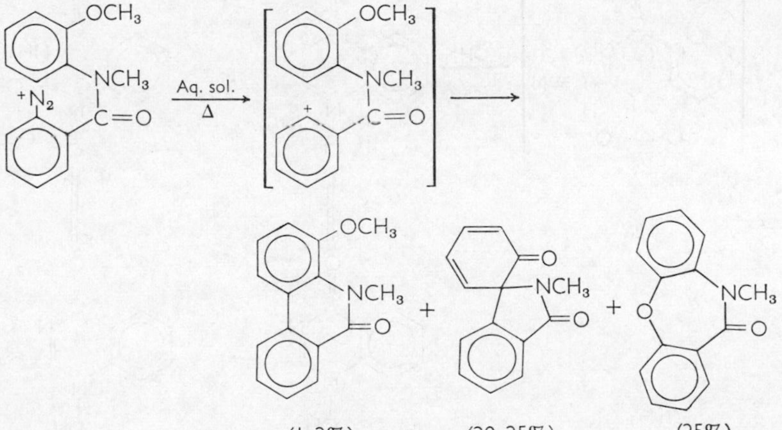

(27)

(1–2%)   (20–25%)   (25%)

and Lipowitz[31d] have suggested that the benzene carbonium ion formed from **5** inserts into a C—H bond of a benzyl residue forming **8** which then decomposes to **6**, and perhaps also to carbonium ion **7**. Other interesting examples of reactions of similar benzene carbonium ions have been reported by Hey and coworkers[32] (equation 27).

C. Amidomethyl Carbanions

Examples of reactions involving the development of a negative charge on a carbon atom attached to a nitrogen atom of an amido group have been reported by Tennant and Vaughan[33] (equations 28a,b). These authors assume that **10** arises from the reduction of intermediate **9**.

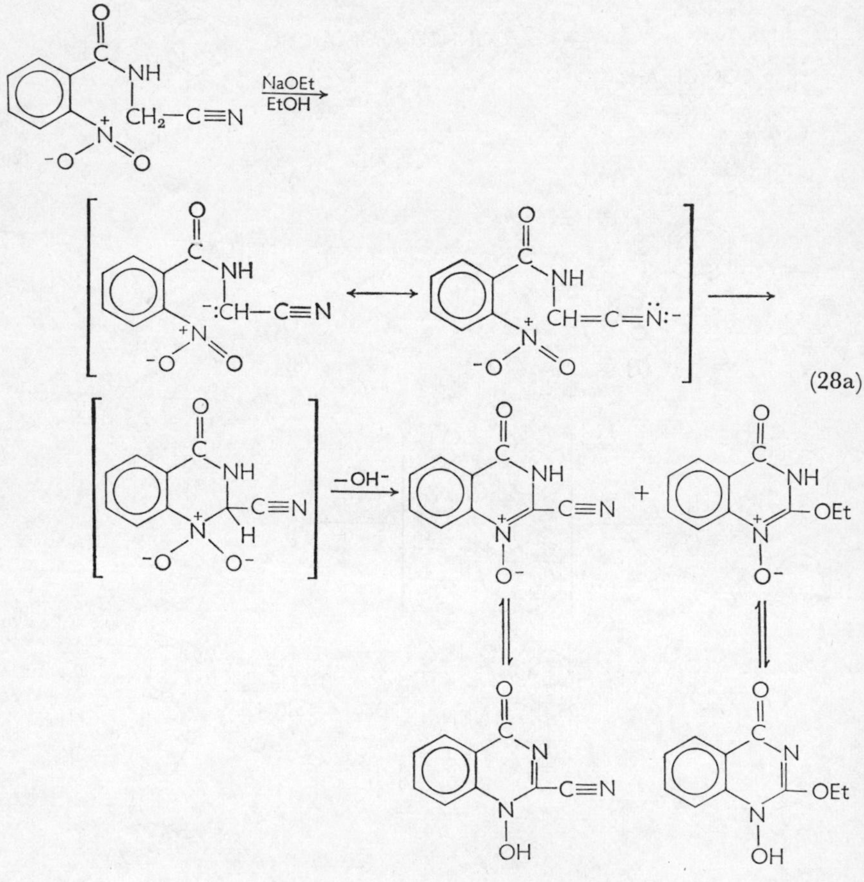

(28a)

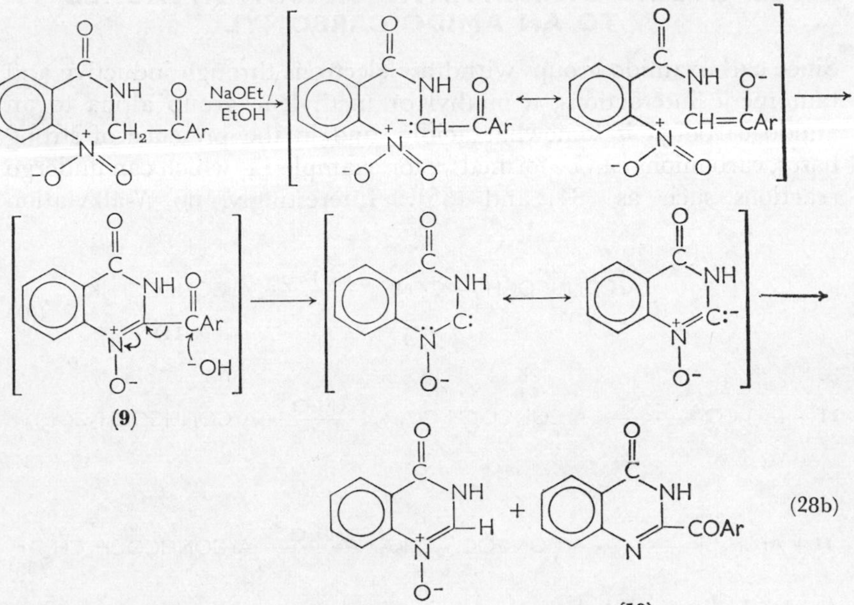

(9)

(28b)

(10)

Also, Chambers and Stirling[34] have observed promotion of β-elimination by a β-toluenesulphonamido substituent, as shown in equations (29) and (30). The effect of the β-toluenesulphonamido

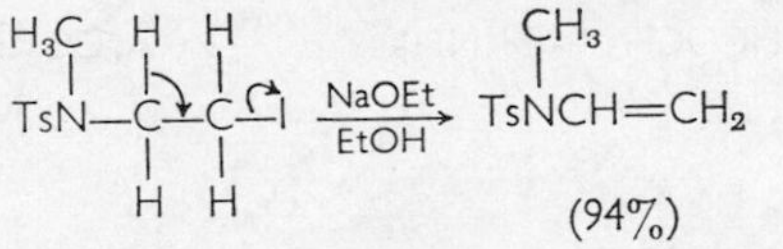

$$\text{TsN}\underset{\substack{|\\H}}{\overset{\substack{H_3C\\|}}{C}}\underset{\substack{|\\H}}{\overset{\substack{H\\|}}{C}}\text{Cl} \xrightarrow[\text{EtOH}]{\text{NaOEt}} \text{TsNCH}{=}\text{CH}_2 \tag{29}$$
$$(94\%)$$

$$\underset{\substack{|\\CH_3}}{\text{TsNCH}_2}{-}\text{CH}_2\text{OTs} \xrightarrow[\text{EtOH}]{\text{NaOEt}} \underset{\substack{|\\CH_3}}{\text{TsNCH}}{=}\text{CH}_2 + \underset{\substack{|\\CH_3}}{\text{TsNCH}_2}\text{CH}_2\text{OEt} \tag{30}$$
$$(31\%) \qquad\qquad (57\%)$$

group on the product distributions is similar to that reported by DePuy and Froemsdorf[35] for a β-phenyl substituent (equations 31 and 32).

$$\text{ArCH}_2\text{CH}_2\text{I} \xrightarrow[\text{EtOH}]{\text{NaOEt}} \text{ArCH}{=}\text{CH}_2 \tag{31}$$
$$(100\%)$$

$$\text{ArCH}_2\text{CH}_2\text{OTs} \xrightarrow[\text{EtOH}]{\text{NaOEt}} \text{ArCH}{=}\text{CH}_2 + \text{ArCH}_2\text{CH}_2\text{OEt} \tag{32}$$
$$(33\%) \qquad\qquad (67\%)$$

III. EFFECTS ON ALIPHATIC CARBON ATTACHED TO AN AMIDO CARBONYL

Since carboxamido groups withdraw electrons through inductive and tautomeric interactions, a methyl or methylene group alpha to an amido carbonyl is somewhat acidic, and in the presence of strong bases, carbanions can be formed[36], for example 11, which can undergo reactions such as (34) and (35). Interestingly, no N-alkylation

$$\text{ArCONHCOCH}_3 + 2\text{KNH}_2 \xrightarrow{\text{Liq. NH}_3} \overset{\overset{\text{K}}{|}}{\text{ArCONCOCH}_2\text{K}} \tag{33}$$
$$(\mathbf{11})$$

$$11 + (\text{Ar})_2\text{CO} \longrightarrow \overset{\overset{\text{K}}{|}}{\text{ArCONCOCH}_2}\underset{\underset{\text{OK}}{|}}{\text{C}(\text{Ar})_2} \xrightarrow{\text{NH}_4\text{Cl}} \text{ArCONHCOCH}_2\underset{\underset{\text{OH}}{|}}{\text{C}(\text{Ar})_2} \tag{34}$$

$$11 + \text{ArCH}_2\text{Cl} \longrightarrow \overset{\overset{\text{K}}{|}}{\text{ArCONCOCH}_2\text{CH}_2\text{Ar}} \xrightarrow{\text{NH}_4\text{Cl}} \text{ArCONHCOCH}_2\text{CH}_2\text{Ar} \tag{35}$$

occurs, indicating that the carbanion is a much more efficient nucleophile than the nitrogen anion. Wolfe and Mao[37] also showed that trialkali metal salts of imides (e.g. 12) could be selectively alkylated (equation 37). Here again, alkylation occurs mainly at the most basic

$$\text{ArCOCH}_2\text{CONHCOCH}_3 + 3\text{KNH}_2 \xrightarrow{\text{Liq. NH}_3} \overset{\overset{\text{K} \quad \text{K}}{| \quad |}}{\text{ArCOCHCONCOCH}_2\text{K}} \tag{36}$$
$$(\mathbf{12})$$

$$\overset{\overset{\text{K} \quad \text{K}}{| \quad |}}{\text{ArCOCHCONCOCH}_2\text{K}} + \text{ArCH}_2\text{Cl} \longrightarrow \longrightarrow$$
$$(\mathbf{12}) \qquad\qquad\qquad \text{ArCOCH}_2\text{CONHCOCH}_2\text{CH}_2\text{Ar} \tag{37}$$
$$(69\%)$$

anionic centre. Surprisingly, when the potassium in 12 is replaced by sodium, the yield of equation (37) is lowered to 8%. Wolfe and Mao[37] point out that the difference of reactivity between potassium and sodium salts towards alkylating agents seems to depend on the nature of the anion. For example, 1-phenyl-1,3,5-trihexanone trisodium salt is alkylated by certain halides which do not alkylate the tripotassium salt.

The trisodium analogue of 12 seems to be more effective than 12 itself in Claisen condensations with diphenyl ketones (equations 38 and 39)[37]. Wolfe and Mao[38] have used similar reactions to produce

$$\underset{(12)}{ArCOCHCONCOCH_2K} + (Ar)_2CO \longrightarrow \longrightarrow \underset{(16\%)}{ArCOCH_2CONHCOCH_2C(Ar)_2}$$

$$\begin{array}{cc} K & K \\ | & | \end{array}$$

$$\begin{array}{c} | \\ OH \end{array}$$

(38)

$$\underset{ArCOCHCONCOCH_2Na}{} + (Ar)_2CO \longrightarrow \longrightarrow \underset{(40\%)}{ArCOCH_2CONHCOCH_2C(Ar)_2}$$

$$\begin{array}{cc} Na & Na \\ | & | \end{array}$$

$$\begin{array}{c} | \\ OH \end{array}$$

(39)

derivatives of *N*-acetylsalicylamides, e.g. equation (40). Alkali metal salts of amides can also be aroylated with methyl benzoate (equation

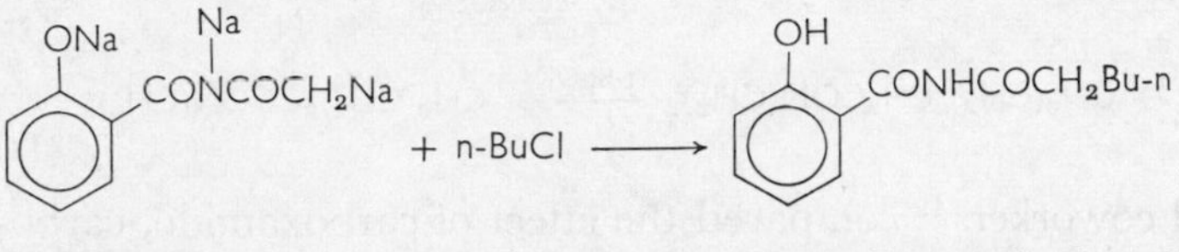

(40)

$$CH_3CONH_2 + 2ArCO_2CH_3 + 4NaH \longrightarrow \longrightarrow ArCOCH_2CONHCOAr$$

(41)

41)[39]. Aroylation and alkylation of alkali metal salts of amides and

imides should provide useful routes to carboxylic acids, since the resulting imides are readily hydrolysed to carboxylic acids by aqueous base.

Tennant[40] has demonstrated that methylene groups alpha to amido and keto carbonyl groups are acidic, by showing that anilides such as **13** are easily alkylated by n-alkyl iodides in the presence of potassium carbonate (equation 42). Furthermore, warming an alkylanilide

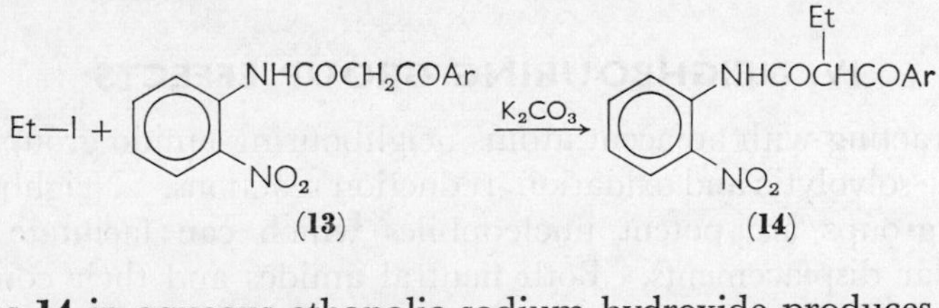

(42)

such as **14** in aqueous ethanolic sodium hydroxide produces a quinoxaline *N*-oxide (equation 43)[40].

Carbanions are also probable intermediates in the nitration of amides with amyl nitrate (equation 44)[41].

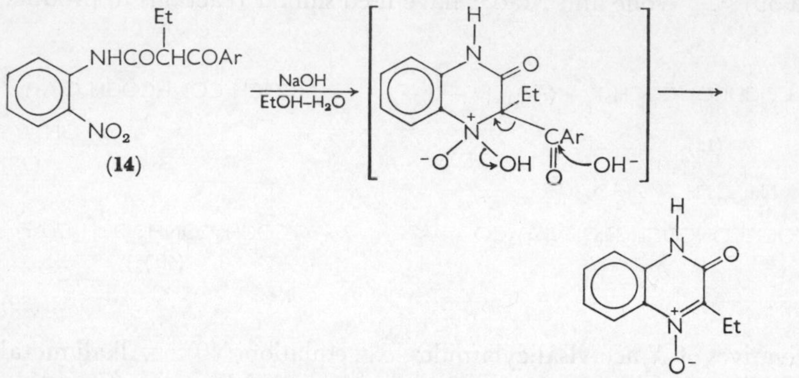

NHCOCHCOAr
Et
NO2
(14)
NaOH
EtOH–H2O
H
N
O
Et
N
CAr
-O
OH
O
OH-
H
N
O
N
Et
O-

(43)

n-C5H11ONO2 + CH3(CH2)2CH2CON(CH3)2
t-BuOK
CH3(CH2)2CCON(CH3)2
O-
O- K+
N

(44)

Patai and coworkers[42] compared the effect of carboxamido, carboethoxy and cyano substituents on the reactivity of active methylene compounds towards aromatic aldehydes (equations 45a,b). The

$$CH_2(CN)R \rightleftharpoons H^+ + {}^-CH(CN)R \qquad (45a)$$

$$ArCHO + {}^-CH(CN)R \longrightarrow ArCH(OH)CH(CN)R \longrightarrow ArCH{=}C(CN)R \qquad (45b)$$

relative order of reactivity of these active methylene compounds toward aromatic aldehydes is $R = CN > R = CO_2Et > R = CONH_2$. In the presence of excess aldehyde, the reactivity of these active methylene compounds appears to be dependent on their rate of ionization (equation 45a), rather than the nucleophilicity of their conjugate bases[42].

IV. NEIGHBOURING GROUP EFFECTS

By interacting with adjacent atoms neighbouring amido groups often facilitate solvolytic and oxidation–reduction reactions. Neighbouring amido groups are potent nucleophiles which can facilitate intramolecular displacements. Both neutral amides and their conjugate bases are effective as nucleophiles. Oxygen appears to be the nucleophilic centre of neutral amido groups, whereas both oxygen and nitrogen can serve as nucleophilic centres of conjugate bases of amido groups.

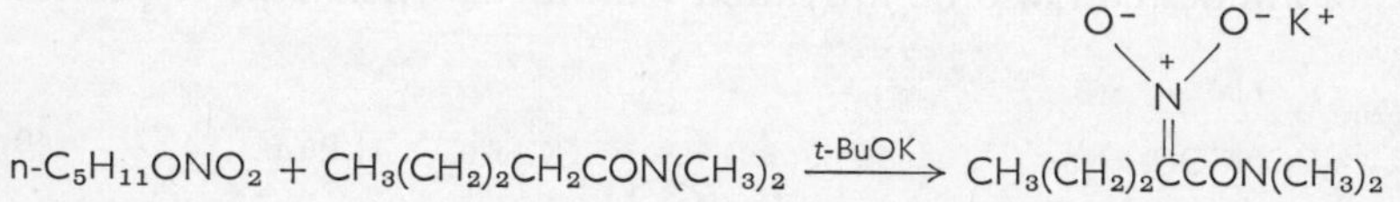

A. Intramolecular Displacements by Amido Anions

Methoxide ion facilitates displacement of bromide ion from N-aryl-4-bromobutyramides (equation 46a)[43a]. Failure to observe a

$$BrCH_2CH_2CH_2\overset{\overset{\displaystyle O}{\|}}{C}NHAr + CH_3O^- \underset{}{\overset{K}{\rightleftharpoons}}$$

$$\mathbf{(15)}$$

$$BrCH_2CH_2CH_2\overset{\overset{\displaystyle O}{\vdots^-}}{C} NAr + CH_3OH \xrightarrow{k_2} \quad \begin{array}{c} H_2C\text{——}CH_2 \\ | \qquad\quad | \\ H_2C\diagdown_{}\diagup C{=}O \\ N \\ | \\ Ar \end{array} \qquad (46a)$$

$$\mathbf{(16)}$$

product of oxygen attack (an iminolactone) is surprising, since **17** readily cyclizes to oxazaline **19** in the presence of methoxide ion (equation 46b)[43b]. Electron-withdrawing substituents on the aro-

$$Ar\overset{\overset{\displaystyle O}{\|}}{C}NHCH_2CH_2Br + CH_3O^- \underset{}{\overset{K}{\rightleftharpoons}}$$

$$\mathbf{(17)}$$

$$Ar\overset{\overset{\displaystyle O^-}{\vdots}}{C}NCH_2CH_2Br + CH_3OH \xrightarrow{k_2} \quad \begin{array}{c} H_2C\text{——}CH_2 \\ | \qquad\quad | \\ O\diagdown_{}\diagup N \\ C \\ | \\ Ar \end{array} \qquad (46b)$$

$$\mathbf{(18)} \qquad\qquad\qquad\qquad\qquad\qquad \mathbf{(19)}$$

matic residue of **15** and **17** facilitate cyclization, suggesting that the concentrations of intermediates **16** and **18** (rather than the nucleophilicity of the amido group) is limiting the reaction rate. The rate of the methoxide ion-catalysed cyclization of **15** is equal to $Kk_2[CH_3O^-][\mathbf{15}]/(1 + K[CH_3O^-])$. In calculating K, a value of one was assigned to the activity of the pure solvent. When $K[CH_3O^-]$ $\ll 1$ second-order kinetics are observed. The second-order rate constants (Kk_2) for the methoxide ion-catalysed cyclization of **15** ($3\cdot0 \times 10^{-3}$ sec^{-1}M^{-1} at $22\cdot9°$) and **17** ($2\cdot2 \times 10^{-3}$ sec^{-1}M^{-1} at $22\cdot9°$) are nearly equal. Since **15** is a stronger acid than **17**, cyclization of **18** via oxygen attack appears to be more facile than cyclization of **16** through nitrogen attack. It should be emphasized, however, that the reactivity of a neighbouring amido anion is a sensitive function of its local environment, and generalizations concerning the relative reactivity of oxygen versus nitrogen are difficult to deduce. For example, amide **20** cyclizes with attack by oxygen[44], whereas amides **21** and **25** cyclize primarily with attack by nitrogen[45,46].

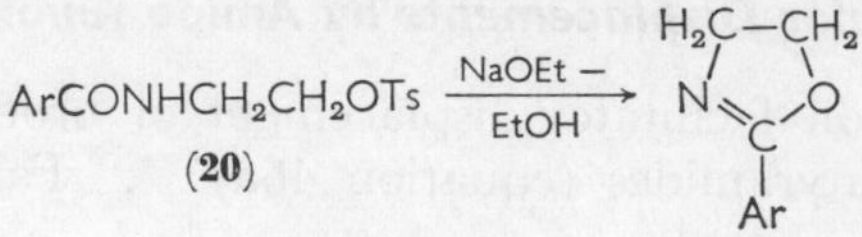

$$ArCONHCH_2CH_2OTs \xrightarrow[\text{EtOH}]{\text{NaOEt} -} \quad \underset{Ar}{\underset{C}{N=}} \overset{H_2C-CH_2}{\underset{}{\diagdown O}} \tag{47}$$

(20)

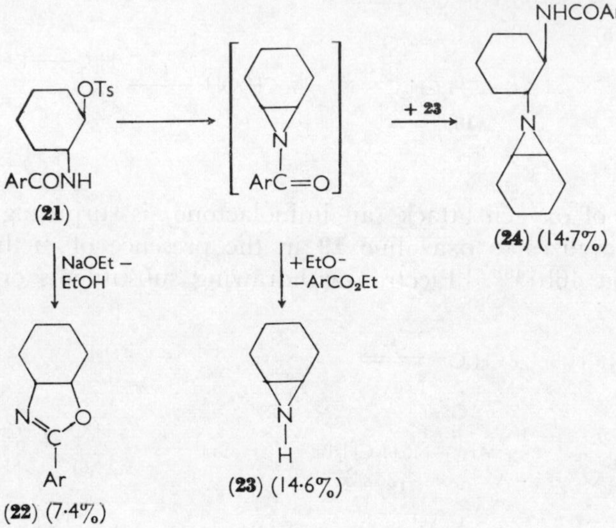

$$ \tag{48} $$

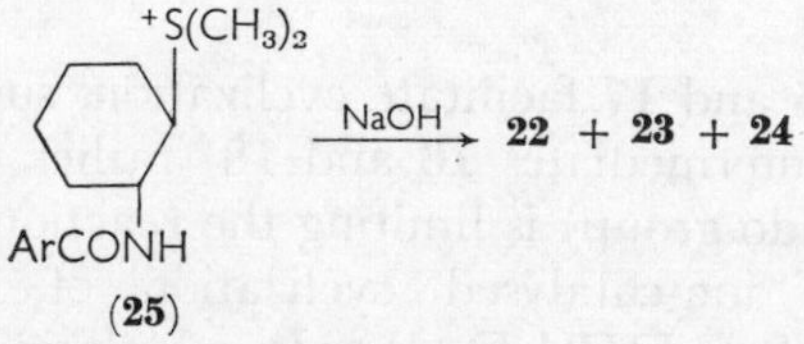

(21)

(24) (14·7%)

NaOEt −
EtOH

+EtO⁻
−ArCO₂Et

(22) (7·4%)

(23) (14·6%)

$$\xrightarrow{\text{NaOH}} \quad 22 + 23 + 24 \tag{49}$$

(25)

Zioudrou and Schmir[47] demonstrated that neighbouring amido anions facilitate displacements of phosphate esters from compounds like 26a. It is interesting that no evidence could be found for attack of anionic nitrogen on phosphorus, since other anionic nucleophiles such as oxide ions usually attack phosphate esters on phosphorus rather than carbon. These authors also studied the displacement of chloride and tosylate by the neighbouring amido group in 26b and 26c (Table 6). Assuming that the acidity of the amido group in 26c is not markedly different from the acidity of the amido groups in 26a or 26b, the data in Table 6 indicate that tosylate is most easily displaced by the neighbouring amido anion in 26.

Neighbouring benzenesulphonamido anions are also efficient nucleo-

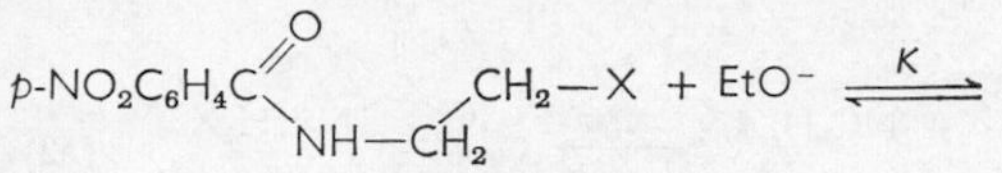

$$p\text{-}NO_2C_6H_4C \overset{\displaystyle O}{\underset{\displaystyle NH-CH_2}{\Big\backslash}} CH_2-X + EtO^- \underset{\displaystyle K}{\rightleftharpoons}$$

$$\mathbf{(26)}$$

(26a) $X = OP(OAr)_2$ (with O double-bonded)
(26b) $X = Cl$
(26c) $X = OTs$

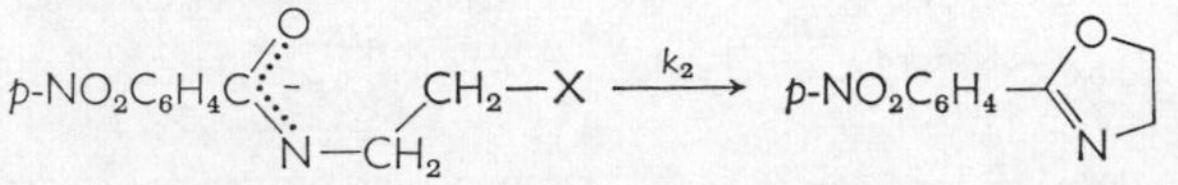

$$p\text{-}NO_2C_6H_4\overset{\displaystyle O}{C\underset{\displaystyle N-CH_2}{\overset{..}{\Big\backslash}}}CH_2-X \overset{k_2}{\longrightarrow} p\text{-}NO_2C_6H_4-\overset{\displaystyle O}{\underset{\displaystyle N}{\Big\langle}} \qquad (50)$$

philes. Scott and Flynn[48] studied effects of substituents (X) on the rate of cyclization of **27**. This reaction is facilitated by electron-

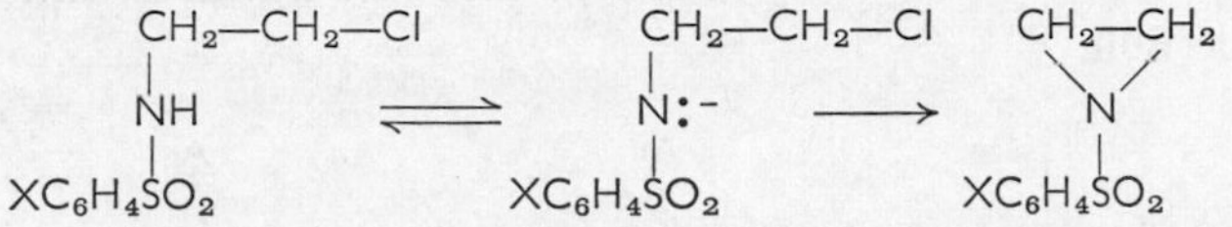

$$\begin{array}{ccc}
CH_2-CH_2-Cl & CH_2-CH_2-Cl & CH_2-CH_2 \\
| & | & \diagdown\diagup \\
NH & N:^- & N \\
| & | & | \\
XC_6H_4SO_2 & XC_6H_4SO_2 & XC_6H_4SO_2 \\
\mathbf{(27)}
\end{array} \qquad (51)$$

donating substituents ($\rho = -0.93$) indicating that the reaction velocity is limited by the nucleophilicity of the anion rather than the concentration of anion. On the other hand, electron-withdrawing substituents facilitate cyclization of **28** ($\rho = 1.72$) and **29** ($\rho = 0.8$), and like reactions (46a) and (46b) the velocity of reactions (52) and (53) also

TABLE 6. Rate constants for reaction (50)[a].

Substrate	k_2 (sec^{-1})	Kk_2 (sec^{-1}M^{-1})
26a	0·019	0·35
26b	0·016	0·095
26c		18

[a] Sodium ethoxide in ethanol at 30°. From reference 47.

appear to be limited by the concentration of amide anion[49,50]. The difference between substituent effects on sulphonamides and carboxamides is expected, since the more acidic sulphonamides are completely ionized in the basic medium. Scott, Glick, and Winstein[44] have determined relative rates of cyclization (equations 54 and 55) for **17, 30a** and **30b** in sodium ethoxide–ethanol at 25°. Urethane **30a**

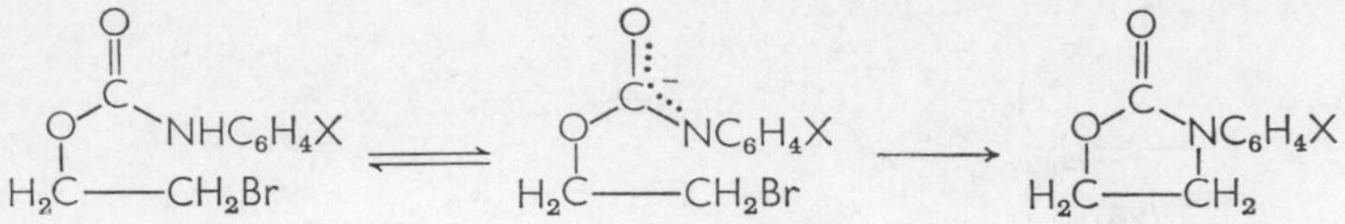

$$(52)$$

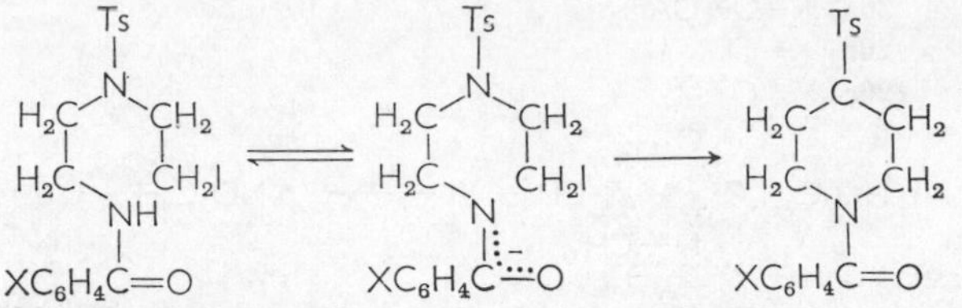

$$(53)$$

cyclized ten times faster than amide **17** and one hundred times faster than ureide **30b**.

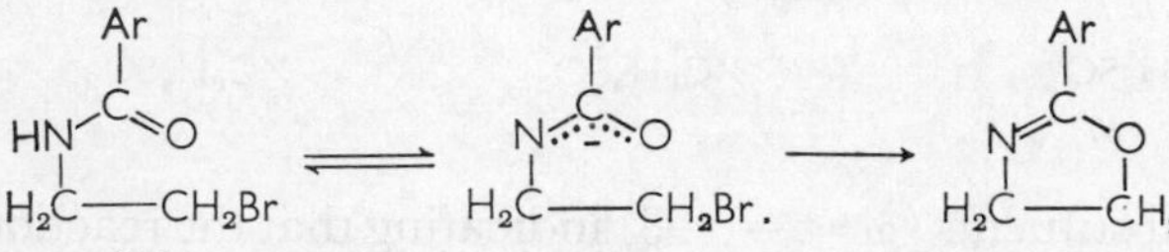

$$(54)$$

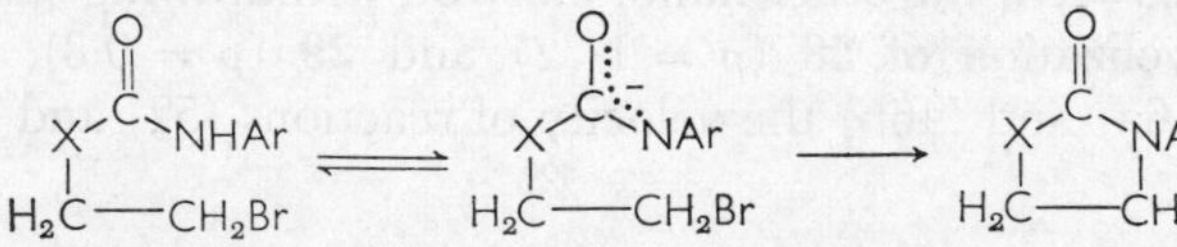

$$(55)$$

(30a) X = O

(30b) X = NH

Neighbouring amido anions can also carry out nucleophilic displacements on carbonyl carbon atoms. Hancock and Linstead[51] proposed the formation of an imide intermediate in the alkaline hydrolysis of esters of anilic acids of methylsuccinic acid (equation 56) in order to explain the migration of the amido group during alkaline hydrolysis. This pathway is also supported by the absence of re-

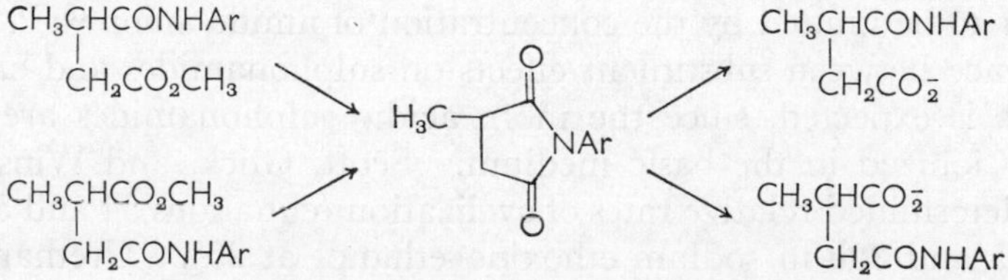

$$(56)$$

arrangement in the alkaline hydrolysis of the corresponding N-methylanilic esters which cannot form an imide. Alkaline hydrolysis of esters of asparagine and glutamine derivatives has been shown to proceed through imide intermediates[52,53]. In some instances, cyclic imide intermediates from asparagine derivatives were isolated from reaction mixtures. Esters of glutamine derivatives can be cyclized to the corresponding glutarimides in sodium methoxide–alcohol solutions (equation 57)[52,54].

$$
\begin{array}{ccc}
\underset{\substack{\displaystyle \\ CH_3O}}{\overset{\substack{CH_2 \\}}{\underset{\substack{NH_2}}{H_2C{\displaystyle \underset{O=C\quad C=O}{}} CHNHCO_2CH_2Ar}}} & \xrightarrow{\text{NaOCH}_3} & \underset{\underset{H}{N}}{\overset{CH_2}{H_2C \quad NHCO_2CH_2Ar}}
\end{array}
\qquad (57)
$$

Bernhard and coworkers[55] determined rate constants for the base-catalysed cyclization and hydrolysis of several β-benzyl esters of N-carbobenzoxyaspartyl amides and peptides. The hydroxyl group in β-benzyl-N-carbobenzoxy-L-aspartyl-L-serinamide (**31**) was found to further enhance (by a factor of 2 to 4) the rate of base-catalysed hydrolysis of the benzyl ester. Apparently, the effect of the hydroxyl

$$
\underset{(\mathbf{31})}{ArCH_2OCONHCH\underset{CH_2CO_2CH_2Ar}{\overset{CONHCH\overset{CONH_2}{\diagup}}{\diagdown}}}\xrightarrow{OH^-}
$$

$$
\underset{\substack{H_2C{-}C \\ O}}{ArCH_2OCONHCH\ \ NCH\overset{CONH_2}{\underset{CH_2OH}{\diagup\diagdown}}}\xrightarrow{OH^-}\ ArCH_2OCONHCH\underset{CH_2CO_2^-}{\overset{CONHCH\overset{CONH_2}{\diagup}}{\diagdown}}CH_2OH\ +
$$

$$
ArCH_2OCONHCH\underset{CH_2CONHCH}{\overset{CO_2^-}{\diagup}}{\diagdown}\underset{CH_2OH}{\overset{CONH_2}{\diagup}}\qquad (58)
$$

group in **32** is much more pronounced[56]. The base-catalysed hydrolysis of the corresponding p-hydroxyl isomer is reported to be 10^{-4} that of **32**. According to Shalitin and Bernhard[56], the pH dependence of the rate of hydrolysis of **32** is consistent with involvement of a phenolate anion (**33a,b**) as a general base or an unionized phenolic group (**34a,b**) as a general acid in the hydrolysis and cyclization of **32**.

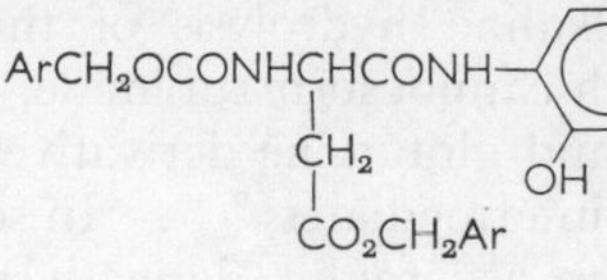

$$ArCH_2OCONHCHCONH{-}\text{(Ar-OH ring)}$$
$$|$$
$$CH_2$$
$$|$$
$$CO_2CH_2Ar$$

(32)

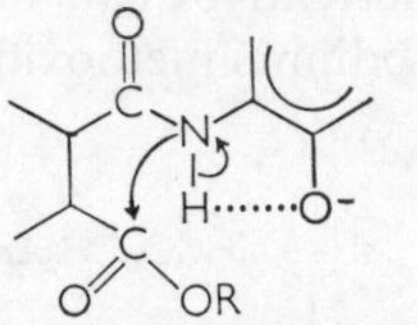

(33a) 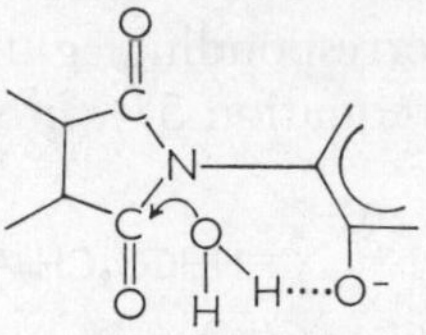   (33b)

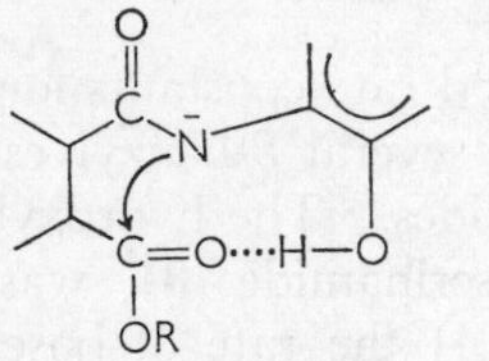

(34a) 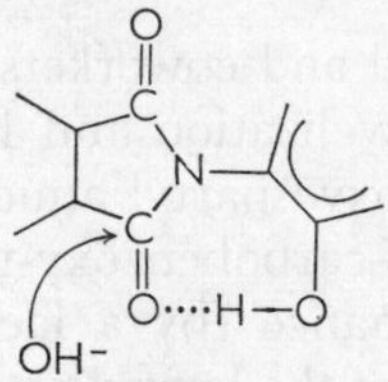   (34b)

Sodium methoxide catalyses cyclization of poly-(benzyl β-L-aspartate) to poly-L-succinimide (equation 59)[57]. This reaction probably proceeds via nucleophilic attack of an amido anion on a carbonyl carbon atom.   Under comparable conditions, poly-(benzyl γ-glutamate) does not react in the presence of catalytic quantities

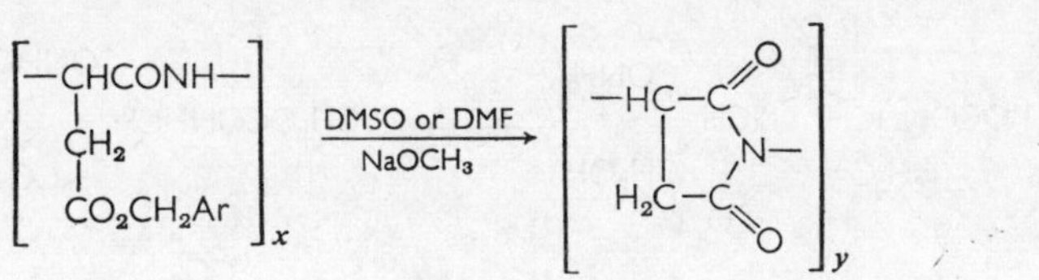

$$\left[\begin{array}{c}-CHCONH-\\ |\\ CH_2\\ |\\ CO_2CH_2Ar\end{array}\right]_x \xrightarrow[\text{NaOCH}_3]{\text{DMSO or DMF}} \left[\begin{array}{c}-HC-C{\diagup}^O\\ \quad\quad\backslash\\ \quad\quad N-\\ H_2C-C\diagdown_O\end{array}\right]_y \qquad (59)$$

$$\left[\begin{array}{c}-NHCHCO-\\ |\\ CH_2\\ |\\ CH_2\\ |\\ CO_2CH_2Ar\end{array}\right]_x \xrightarrow[\text{NaOCH}_3]{\text{DMSO or DMF}} \quad\text{HN}\diagup\diagdown\text{CO}_2\text{Na} \qquad (60)$$

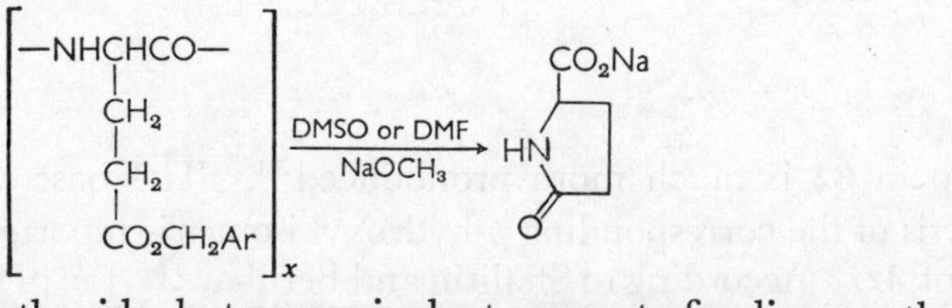

of sodium methoxide, but an equivalent amount of sodium methoxide converts the polymer to sodium D,L-2-pyrrolidone-5-carboxylate (equation 60)[57].   The mechanism for this reaction is unknown.

The base-catalysed cyclization of *N*-carbobenzoxyglycyl-L-proline *p*-nitrophenyl ester (**35**) probably involves attack by a neighbouring amido anion on a carbonyl carbon atom (equation 61)[58]. The ease

$$\text{ArCH}_2\text{OCONHCH}_2\text{CON} \quad \xrightarrow{\text{Base}} \quad \text{ArCH}_2\text{OCON} \qquad (61)$$

$$\text{CO}_2\text{C}_6\text{H}_4\text{NO}_2\text{-}p$$

(**35**)

with which this ester cyclizes was ascribed to the fact that the rigid proline ring holds the reacting groups in close proximity. No evidence could be found for base-catalysed intramolecular cyclization when phenylalanine was substituted for proline in **35**.

Cyclization of **36** was found to be insensitive to the nature of the aromatic substituent, X, ($\rho = -0\cdot1$ to $-0\cdot2$) indicating that a substituent effect which increases the nucleophilicity of an intermediate anion is almost completely counterbalanced by the accompanying decrease in the concentration of anionic intermediate caused by the decreased acidity of the amide[59].

$$\text{NHC}_6\text{H}_4\text{X} \longrightarrow \text{NC}_6\text{H}_4\text{X} \longrightarrow \text{NC}_6\text{H}_4\text{X} \qquad (62)$$

$$\text{CO}_2\text{CH}_3 \qquad\qquad \text{CO}_2\text{CH}_3$$

(**36**)

Several other examples of attack of a neighbouring amido anion on a carbonyl carbon atom have been studied (equations 63–67). Apparent second-order rate constants for the hydroxide ion-catalysed formation of an imide or oxazolinone intermediate (Kk_2) and hydrolysis of these intermediates (k_h) from amides **37–40** are compared in Table 7 with the second-order rate constants for the hydroxide ion-catalysed hydrolysis of esters and amides without neighbouring amido groups. An amido group in phthalamide (**38**) enhances the rate of hydrolysis of the other amido group by a factor of about $4\cdot5 \times 10^5$ ($4\cdot9/1\cdot1 \times 10^{-5}$)[59,60]. The amido group's ability to accelerate the hydrolysis of the adjacent ester group in methyl phthalamate (**37a**) is considerably reduced by the relatively slow rate of hydrolysis of the phthalimide intermediate, and the approximate rate enhancement which may be attributed to the neighbouring amido group is reduced to 870 ($20/2\cdot3 \times 10^{-2}$)[59,60]. The neighbouring amido group in *O*-acetylsalicylamide (**39**) efficiently displaces the *o*-phenoxy group[61].

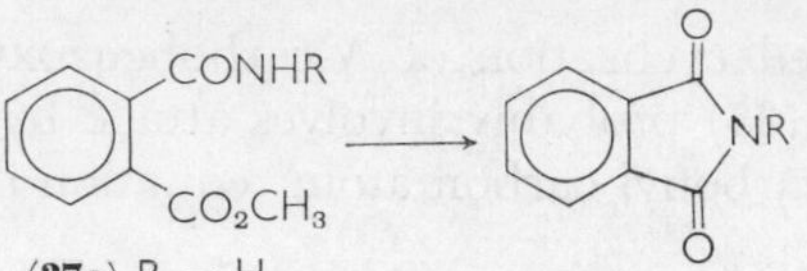

(37a) R = H
(37b) R = CH₃

refs. 59,60 (63)

(38)

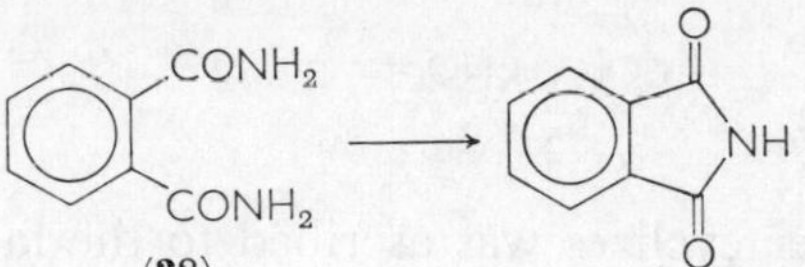

refs. 59,60 (63a)

(39)

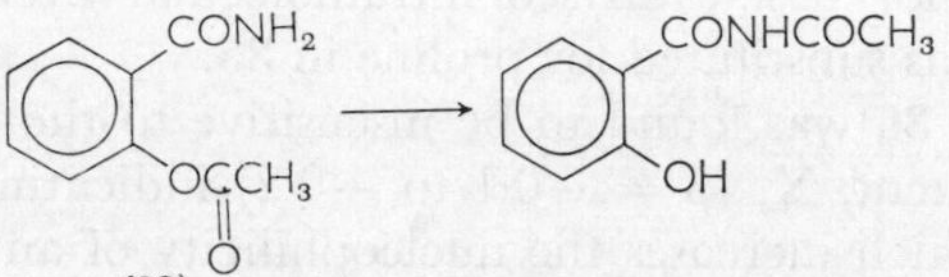

ref. 61 (64)

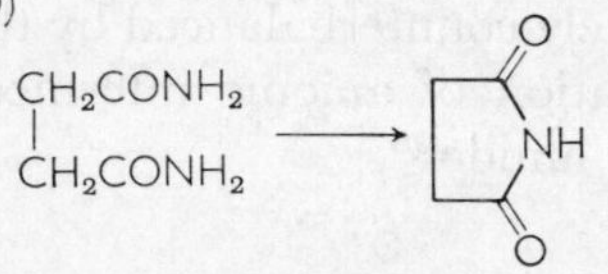

ref. 62 (65)

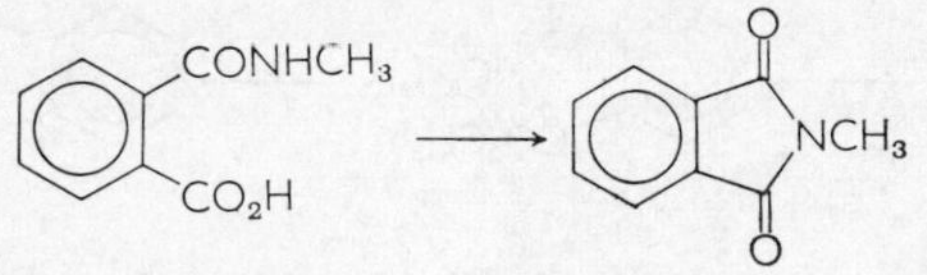

ref. 63 (66)

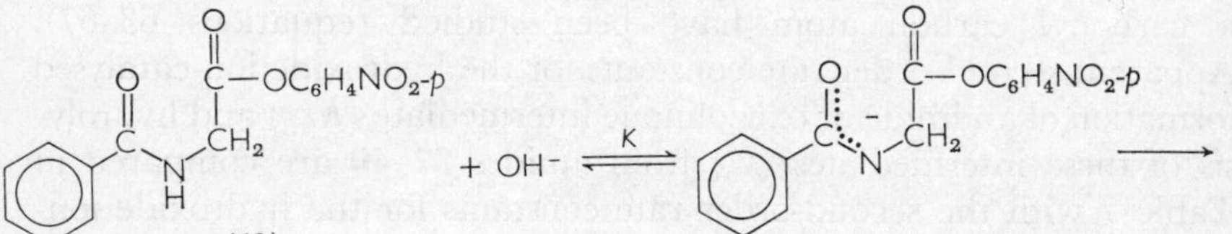

(40)

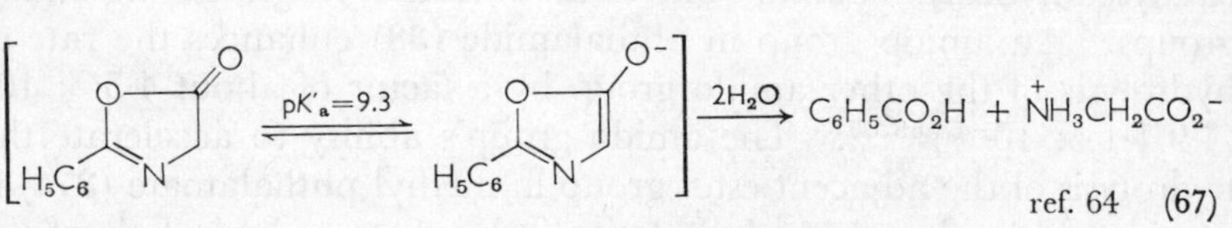

ref. 64 (67)

The inability of Behme and Cordes[61] to detect an acetamide-catalysed displacement of *p*-nitrophenol from *p*-nitrophenyl acetate led them to conclude that the rate of the intramolecular displacement by the neighbouring amido group in **39** is at least 6×10^4 faster than a comparable bimolecular process. Because of slow hydrolysis of the imide intermediate, the amido group in O-acetylsalicylamide (**39**) causes a decrease rather than an increase in the overall rate of hydrolysis of the adjacent phenyl ester group[61]. Neighbouring amido anions can increase the rate of hydrolysis of phenyl esters, if oxygen rather than nitrogen attacks the carbonyl carbon atom, as evidenced

TABLE 7.   Rate constants for cyclization and hydrolysis of some esters and amides[a].

Compound	$Kk_2{}^b$ (sec^{-1}M^{-1})	$k_\text{h}{}^c$ (sec^{-1}M^{-1})	Refs.
Phthalamide (**38**)	4·9	20	d,e
Benzamide		$1 \cdot 1 \times 10^{-5}$	f
Methyl phthalamate (**37a**)	$3 \cdot 1 \times 10^3$	20	d,e
Methyl benzoate		$2 \cdot 3 \times 10^{-2g}$	h
O-Acetylsalicylamide (**39**)	$2 \cdot 0 \times 10^4$	$1 \cdot 3 \times 10^{-2i}$	j
Phenyl acetate		$3 \cdot 7^k$	l
p-Nitrophenyl hippurate (**40**)	$1 \cdot 1 \times 10^{4k}$	m	n
p-Nitrophenyl acetate		24^k	l

^{a} In water at $25° \pm 1°$.

^{b} Apparent second-order rate constant for the hydroxide ion-catalysed cyclization to imide. K is the equilibrium constant for the reaction: Amide + OH$^-$ $\rightleftharpoons$ Amide anion + H$_2$O, and k_2 is the first-order rate constant for cyclization of an amide anion.

^{c} Second-order rate constant for the hydroxide ion-catalysed hydrolysis of an imide (to an amic acid) or for the hydroxide ion-catalysed hydrolysis of an amide or an ester without a neighbouring amido group.

^{d} Reference 59.

^{e} Reference 60.

^{f} Interpolated from the temperature dependence of rate constants listed by M. L. Bender, R. D. Ginger, and J. P. Unik, *J. Am. Chem. Soc.*, **80**, 1044 (1958).

^{g} In 1:3 dioxan–water.

^{h} M. L. Bender, H. Matsui, R. J. Thomas, and S. W. Tobey, *J. Am. Chem. Soc.*, **83**, 4193 (1961).

^{i} The rate expression also contains a term second order in the hydroxide ion concentration $(1 \cdot 3 \times 10^{-1}$ sec^{-1}M$^{-1})$.

^{j} Reference 61.

^{k} At 30°.

^{l} T. C. Bruice and M. F. Mayahi, *J. Am. Chem. Soc.*, **82**, 3067 (1960).

^{m} Above pH 7, the oxazolinone (pK_a' 9·3) appears to be hydrolysed through attack by water on its conjugate base $(k = 5 \cdot 6 \times 10^{-2}$ sec$^{-1})$.

^{n} Reference 64.

by the fact that the neighbouring amido group in p-nitrophenyl hippurate increases the rate of hydrolysis of this nitrophenyl ester by several-hundred-fold at pH 7[64]. Above pH 7, the rate of hydrolysis of oxazolinone begins to become rate determining, and the rate enhancement which can be assigned to the neighbouring amido group in **40** decreases with increasing pH*.

There is considerable evidence for the involvement of oxazolinones as intermediates in the racemization (through enolization of an oxazolinone) of nitrophenyl esters of peptides during hydrolytic and peptide coupling reactions in basic solutions[65–68]. Goodman and co-workers[65,67] have shown that oxazolinones racemize under conditions used for peptide coupling reactions. Williams and Young[68] have shown that N-benzoyl-L-leucine p-nitrophenyl ester is in equilibrium with 4-isobutyl-2-phenyloxazolin-5-one in a solution of N-methyl-piperidine and chloroform, and that this oxazolinone is an intermediate in the racemization of the nitrophenyl ester. The degree of racemization observed during a peptide coupling reaction would of course depend on the rate of oxazolinone formation relative to the rate of aminolysis of the ester as well as the relative rates of enolization and ring opening of the oxazolinone[65,67].

The cyclization of o-cyanobenzamide to iminophthalimide is another example of attack of an amido anion on an unsaturated carbon atom[69]. The second-order dependence of the rate of this reaction on the concentration of hydroxide ion suggests that this reaction may involve the addition of hydroxide ion to the cyano group (equation 68). Stabilization of this addition compound by the neighbouring amido group must be very effective, since the formation of phthal-amide (through tautomerization of the intermediate) is not observed[69].

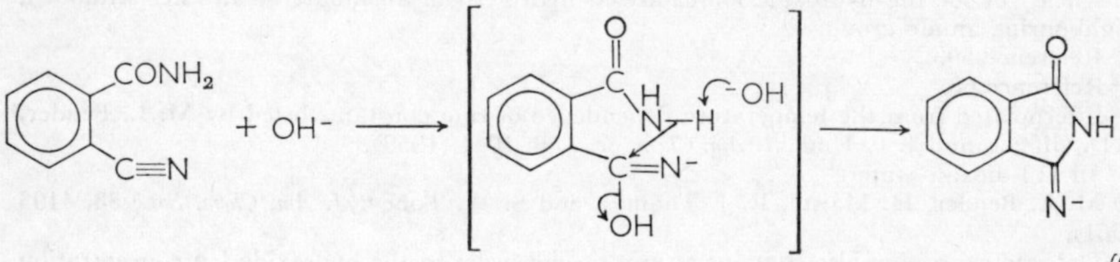

$$(68)$$

Attack of an amido anion on nitrogen is probably involved in the decomposition of o-nitrosobenzamide in ethanolic sodium hydroxide (equation 69)[70a]. This reaction supports Rosenblum's[70b] conclusion that o-nitrosobenzamides are intermediates in the von Richter reaction (equation 70).

* Above pH 7, the rate of hydrolysis of p-nitrophenyl acetate is essentially proportional to the hydroxide ion concentration, whereas the rate of hydrolysis of the oxazolinone is proportional to the fraction of oxazolinone (pK'_a 9·3) which is present as the conjugate base.

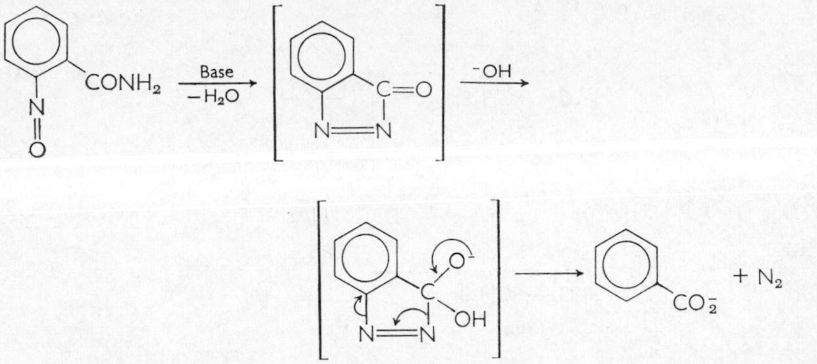

$$+ N_2 \quad (69)$$

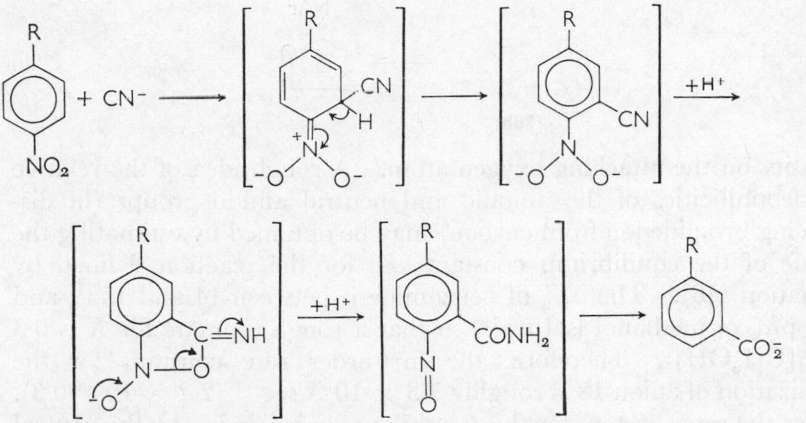

$$(70)$$

B. Intramolecular Displacements by Neutral Amido Groups

In the absence of added base, compounds **17**, **30a**, and **30b**, cyclize via a first-order process (equations 71 and 72)[44]. Cyclization probably involves attack by the neutral form of the neighbouring group. Interestingly, the centre of nucleophilicity of the urethano and ureido groups (in **30a**, **30b**) changes from nitrogen to oxygen on going from the anionic to the neutral form of the nucleophile (compare equations 55 and 72). As judged by the rates of cyclization in 80% aqueous ethanol at 50°, the amido group in **17** is 46 times more reactive as a nucleophile than the urethano group in **30a** and 2·5 times more reactive than the ureido group in **30b**[44]. In the absence of strong bases, the neutral form of the neighbouring amido group is the nucleophile, and electron-donating substituents increase the rate of cyclization of **17**[43], **30a**[49], and **30b**[44], probably by increasing the negative charge

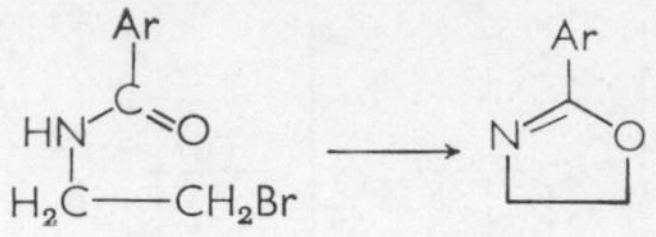

$$
\begin{array}{c}
\underset{(17)}{
\overset{\displaystyle Ar}{\underset{H_2C\!-\!\!-\!CH_2Br}{\overset{|}{\underset{|}{HN\diagdown C{=}O}}}}
}
\end{array}
\longrightarrow
\overset{\displaystyle Ar}{\underset{}{N{=}\!\diagup\!\!\diagdown O}}
\qquad (71)
$$

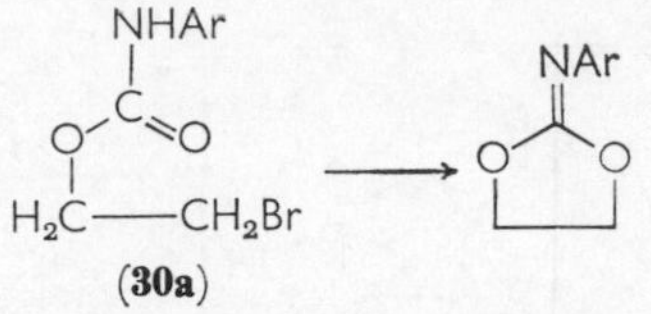

$$
\underset{(30a)}{
\overset{\displaystyle NHAr}{O\diagup\overset{|}{\underset{H_2C\!-\!\!-\!CH_2Br}{C{=}O}}}
}
\longrightarrow
\overset{\displaystyle NAr}{O\diagdown\diagup O}
\qquad (72a)
$$

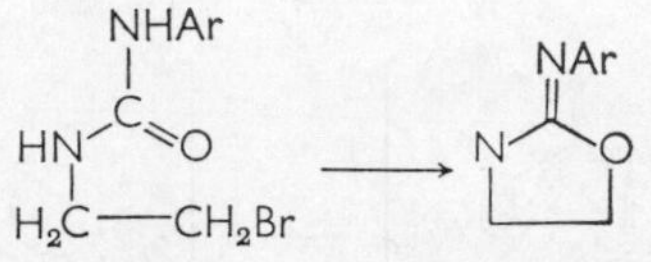

$$
\underset{(30b)}{
\overset{\displaystyle NHAr}{HN\diagdown\overset{|}{\underset{H_2C\!-\!\!-\!CH_2Br}{C{=}O}}}
}
\longrightarrow
\overset{\displaystyle NAr}{N\diagdown\diagup O}
\qquad (72b)
$$

density on the attacking oxygen atom. A rough idea of the relative nucleophilicities of the anionic and neutral amido groups (in displacing bromide ion from carbon) may be obtained by estimating the value of the equilibrium constant (K) for the reaction defined by equation (46). The pK'_a of benzamide is between 14 and 15[71], and the pK'_a of methanol is 15·5[72], so that a rough estimate for K is 0·3 $(10/[CH_3OH])$. Therefore, the first-order rate constant for the cyclization of anion **18** is roughly $7·3 \times 10^{-3}\,\mathrm{sec}^{-1}$ $(2·2 \times 10^{-3}/0·3)$. Since the rate constant for the expulsion of bromide ion by the neutral amido group in **17** is $2·4 \times 10^{-5}\,\mathrm{sec}^{-1}$ (in methanol at 22·9°)[43], the neutral amido group in **17** is about 1/300 as effective as its conjugate base (**18**) in displacing bromide ion from a neighbouring carbon atom.

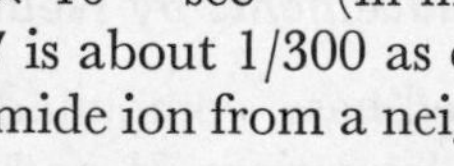

$$
\underset{(18)}{Ar\overset{O^-}{\overset{\|}{C}}\!\cdots\!NCH_2CH_2Br}
$$

Winstein and his coworkers[73,74] found that the amide-facilitated displacement of tosylate from **41** to form an oxazolinium ion was 200 times faster (at 75°) than the acetoxy-facilitated displacement of tosylate from **42**, and about 1,000 times faster than the displacement of

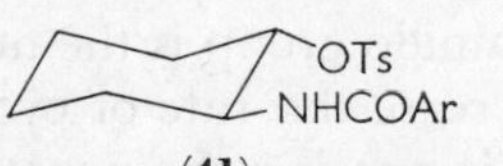

$$
\underset{(41)}{\text{cyclohexane–OTs, –NHCOAr}}
\qquad\qquad
\underset{(42)}{\text{cyclohexane–OTs, –OCOCH}_3}
$$

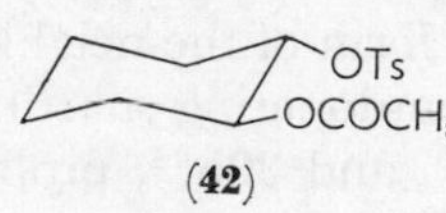

tosylate from the *cis* isomer of **41**.   Investigations of amide-facilitated displacements of 2-benzamidocyclohexyl methanesulphonates in which the conformation of cyclohexyl ring is fixed (by a *t*-butyl substituent in the 4-position), led to the conclusion that although a diaxial conformation is most favourable for intramolecular displacement ($k = 6{\cdot}28 \times 10^{-3}$ sec^{-1} for ethanolysis of **43**), intramolecular displacement is also possible when amido and ester groups are in the diequatorial conformation ($k = 0{\cdot}76 \times 10^{-4}$ sec^{-1} for ethanolysis of **44**)[75].   The similarity in the rate constants for the ethanolysis of **44** and **45** ($0{\cdot}76 \times 10^{-4}$ sec^{-1} vs. $2{\cdot}5 \times 10^{-4}$ sec^{-1}) indicates that displacement of methanesulphonate by the neighbouring amido group of **45** might proceed without inversion to the diaxial conformer of **45**[75].

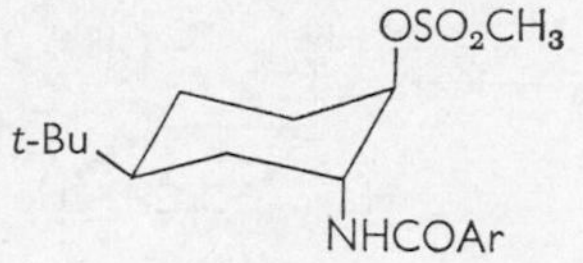

(43)    **(44)**

(45)

Displacements of sulphonate esters by neighbouring amido groups have been used in the synthesis of carbohydrates[76].   For example, Hanessian[76] used a neighbouring acetamido group to convert an arabinose derivative into a lyxose derivative (equation 73).

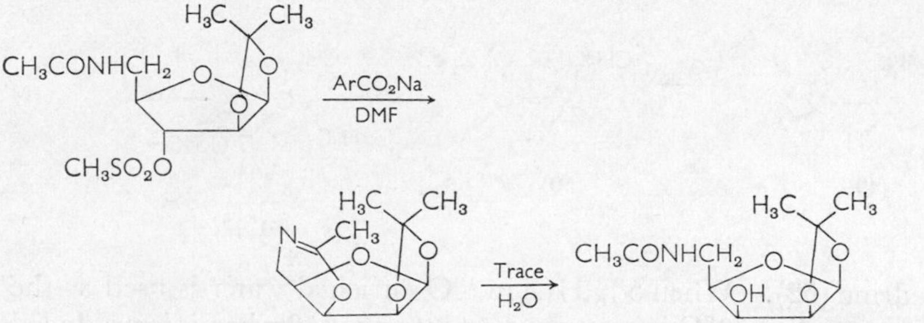

Treatment of β-hydroxy amides with thionyl chloride or phosphoryl chloride results in the formation of oxazoline salts[77].   Presumably, an ester formed on reaction with an acid chloride is displaced by a neighbouring amido group.   Acid hydrolysis of the oxazoline yields a β-amino ester, whereas hydrolysis in neutral or basic medium yields the original β-hydroxy amide with an inverted configuration about the β-carbon atom. Thus treatment of N-benzoylallothreonine methyl ester (46) with thionyl chloride yields the *trans*-oxazoline 47 which on treatment with water yields N-benzoylthreonine methyl ester (48)[78].

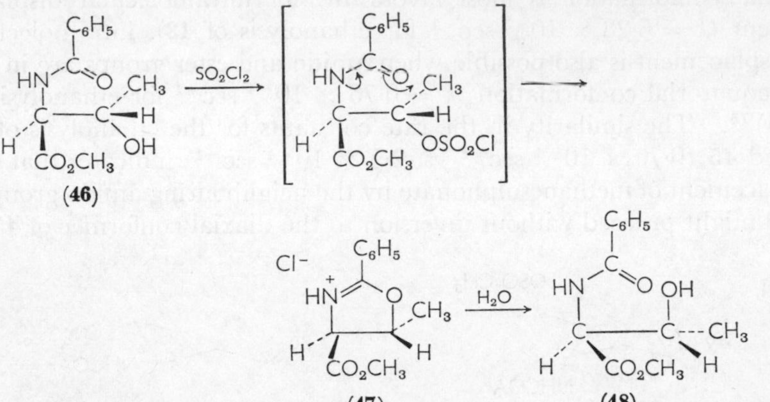

Welsh[79] has demonstrated that a neighbouring neutral amido group displaces a hydroxyl group from a carbon atom during the acid-catalysed $N \rightarrow O$ migration in N-benzoyl-$(-)$-Ψ-ephedrine (49). After ten minutes in 5% refluxing HCl 49 is converted to the benzoate esters 50 and 51. However, ester 50 with an inverted configuration about the β-carbon atom is obtained in 79% yield. No inversion of configuration accompanies the migration of N-benzoyl-$(+)$-Ψ-

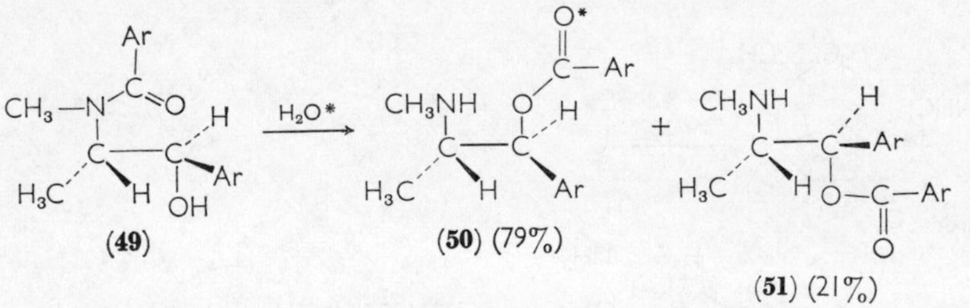

$$(74)$$

ephedrine (52). When 5% HCl in ^{18}O-enriched water is used as the reaction medium, ^{18}O is incorporated into ester 50 when it is produced via reaction (74), but no ^{18}O is incorporated into ester 51, nor into ester 50 when it is produced via reaction (75). Thus acyl migration with retention of configuration involves direct attack of the β-hydroxyl group on the amido carbonyl carbon atom, whereas acyl migration with inversion undoubtedly involves displacement of the β-hydroxyl

group by the neighbouring amido group.   Examination of the various
conformers of **49** and **52** reveals that a *trans* conformation of hydroxyl
and amido groups is more favourable in isomer **49**.   This result is

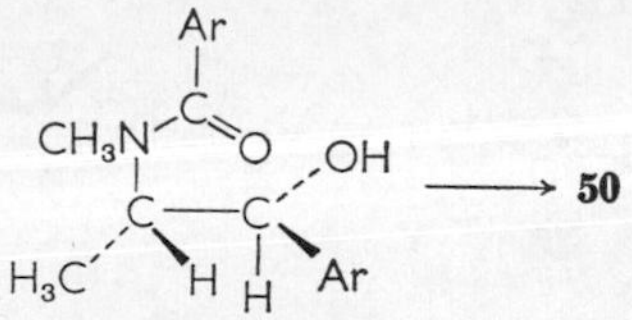

$$\text{(structure 52)} \longrightarrow \textbf{50} \tag{75}$$

(**52**)

consistent with the finding that acyl migration in **49** is accompanied
by inversion.

Effects of amido groups on the hydrolysis of glycosides have been
studied by Piszkiewicz and Bruice[80].   At neutrality, *o*- and *p*-nitro-
phenyl 2-acetamido-2-deoxy-β-D-glucopyranosides hydrolyse 10^5 times
faster than their α-anomers[80a].   This enhancement has been attributed to intramolecular nucleophilic attack on the anomeric carbon atom
by the neighbouring acetamido group (equation 76)[80a].   Although a
2-hydroxyl group also enhances the rate of hydrolysis (at neutrality) of

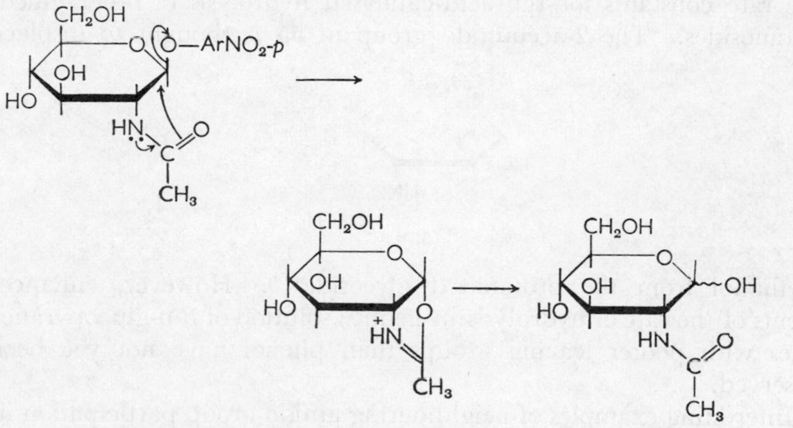

$$\text{(reaction scheme)} \tag{76}$$

nitrophenyl β-D-glucopyranosides, the 2-acetamido group is 218–344
times more effective than the 2-hydroxyl group in increasing the rate
of hydrolysis of *o*- and *p*-nitrophenyl β-D-glucopyranosides relative to
the α-anomers[80a].

The *o*-carboxyl group in *o*-carboxyphenyl β-D-glucopyranoside (**53**)
enhances the rate of hydrolysis of this glycoside by a factor of 6×10^3
above that which would be expected for specific acid-catalysed hydrolysis of this glucopyranoside[80b,81].   Surprisingly, replacement of

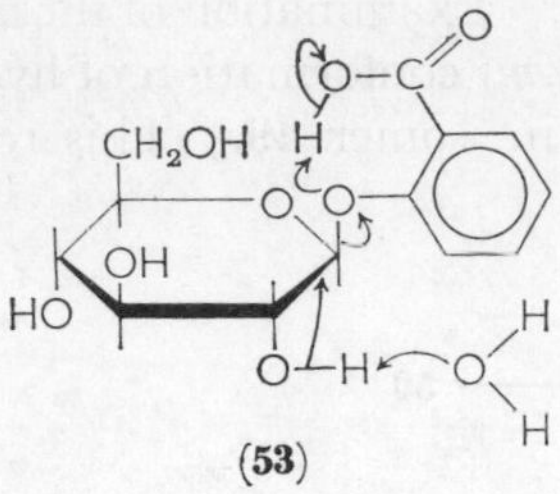

(53) 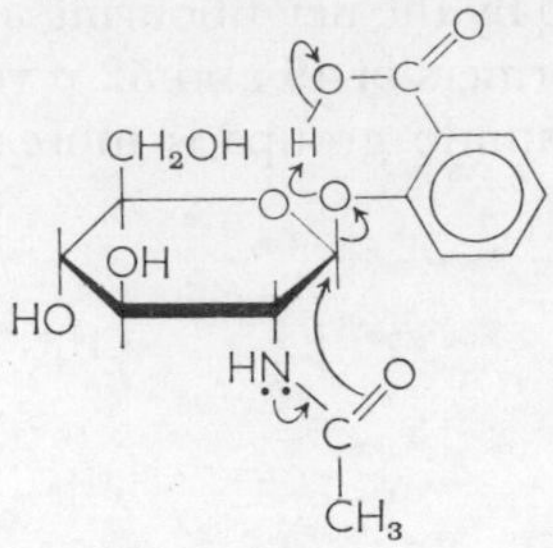   (54)

the 2-hydroxyl group in **53** with an acetamido group results in only a 7-fold enhancement in the rate of hydrolysis of **54**[80b]. The low efficiency observed for intramolecular bifunctional catalysis of the hydrolysis of glycoside **54** has been attributed to a decrease in the entropy of activation caused by the need to orient a second catalytic group[80b].

A 2-acetamido group also appears to increase the second-order rate constant for the specific acid-catalysed hydrolysis of methyl glycoside **55** by 50-fold over that which would be anticipated from the rate constants for the acid-catalysed hydrolysis of other glucopyranosides. The 2-acetamido group in **55** is thought to displace

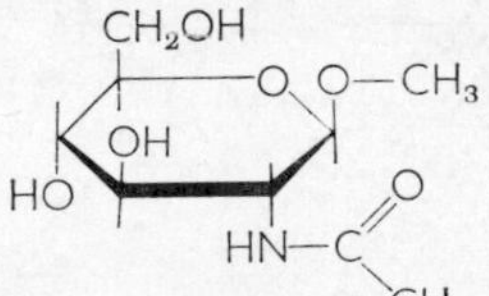

(55)

methanol from the protonated glycoside[80c]. However, enhancements of the rate of hydrolysis in neutral solution of β-D-glucopyranosides with poorer leaving groups than phenol have not yet been observed.

Interesting examples of neighbouring amido group participation in the bromination of olefins have been reported by Winstein and his coworkers (equations 77 and 78)[82]. Hydrolysis of oxazoline **56** completes a stereospecific synthesis of a trisubstituted cyclohexane derivative (equation 79).

Displacement of a substituent from a γ-substituted butyramide by a neighbouring amido group is often observed on fusion or heating in solution (equation 80)[83]. The resulting iminolactone is easily hydrolysed to a lactone and an amine or to a γ-hydroxybutyramide.

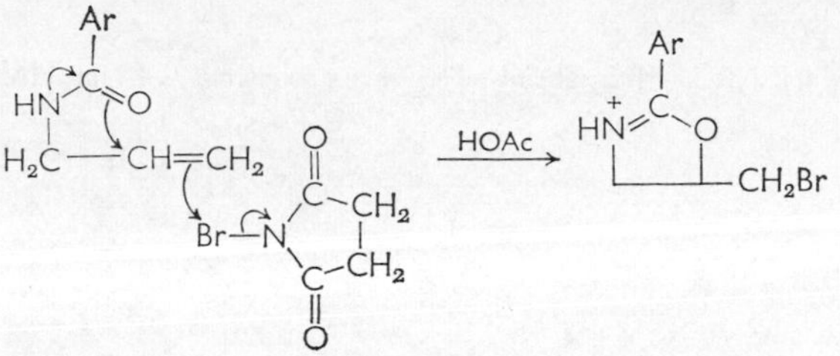

(77)

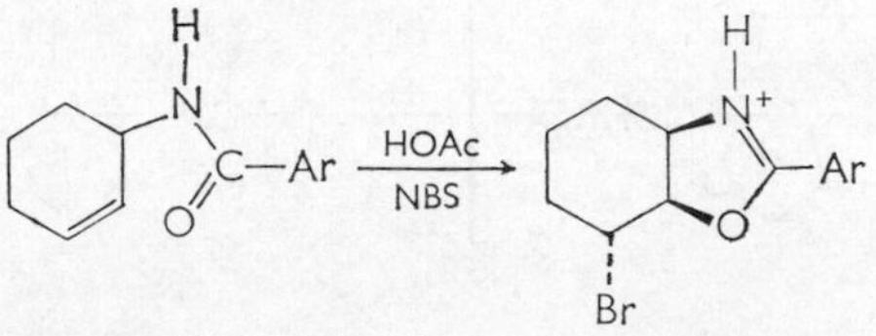

(78)

(56)

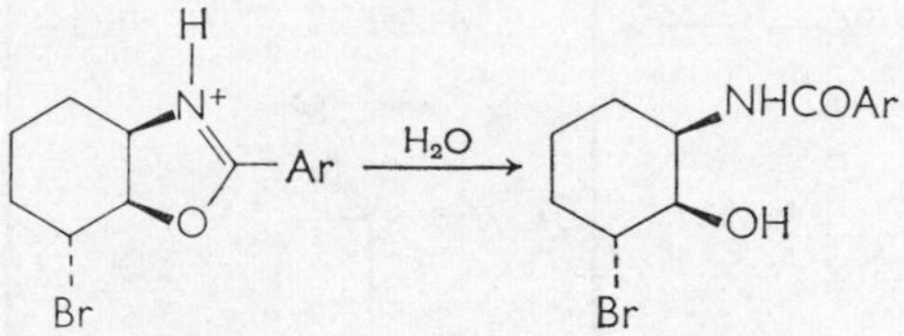

(79)

(56)   **(57)**

Thus, heating γ-bromobutyranilide (**15**) yields butyrolactone and aniline and not iminolactone **58**[83]. Hydrolysis of **58** could have led to butyrolactone and aniline[84]. Iminolactone **58** can be prepared by

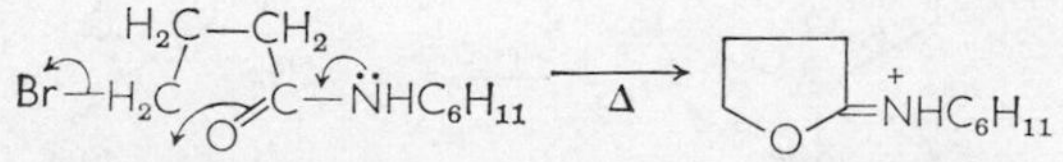

(80)

cyclizing **15** in benzene–methylene chloride solutions of silver tetra-fluoroborate (equation 81)[84]. Some of the important reactions used to cleave proteins selectively at specific amino acids residues probably

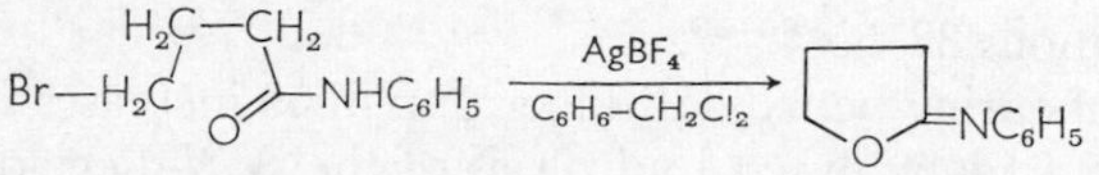

(81)

(15)   **(58)**

involve displacement of a substituent by a neighbouring neutral amido group to form an iminolactone intermediate (equations 82a–82c)[85].

Nucleophilic displacement by a neighbouring amido group on a

24+c.o.a.

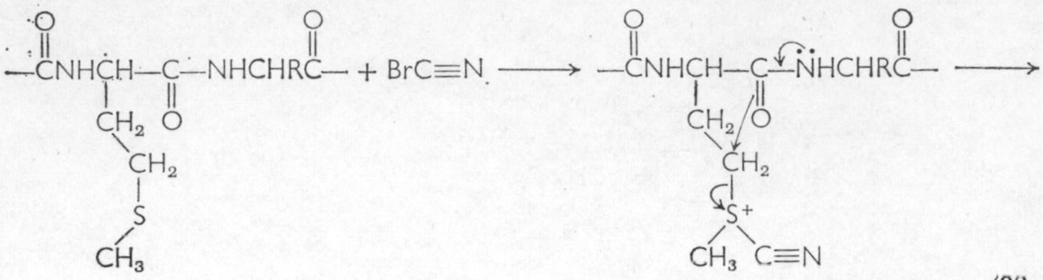

$$\text{—CNH\dot{C}H—C—NHCHRC— + BrC≡N} \longrightarrow \text{—CNHCH—C—NHCHRC—} \longrightarrow \tag{82a}$$

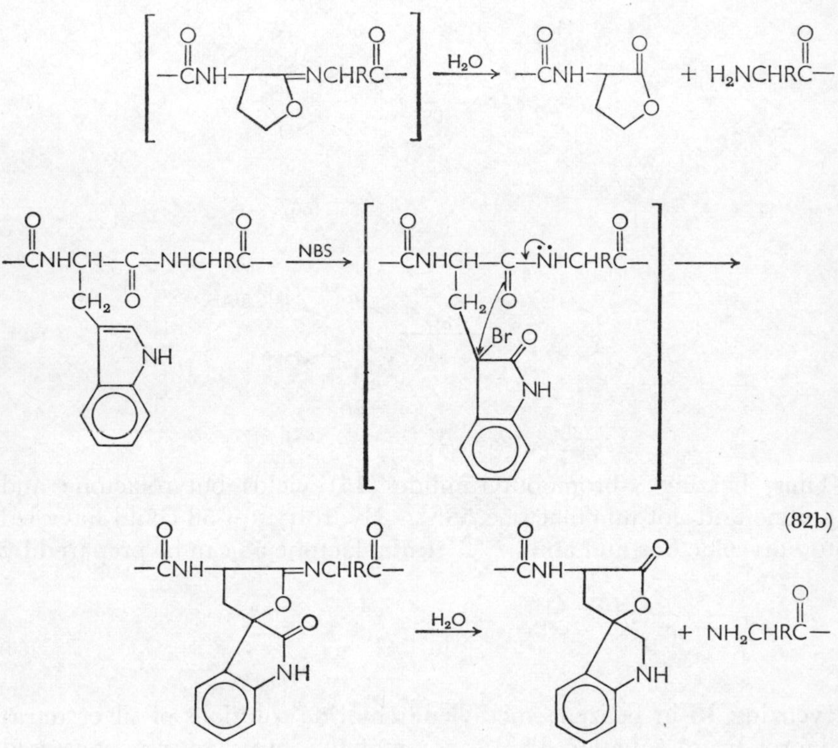

$$\left[\text{—CNH} \underset{O}{\bigcirc} \text{=NCHRC—} \right] \xrightarrow{H_2O} \text{—CNH} \underset{O}{\bigcirc} \text{ + } H_2NCHRC—$$

$$\text{—CNHCH—C—NHCHRC—} \xrightarrow{NBS} \left[\text{—CNHCH—C—NHCHRC—} \right] \longrightarrow \tag{82b}$$

$$\text{—CNH} \underset{O}{\bigcirc} \text{=NCHRC—} \xrightarrow{H_2O} \text{—CNH} \underset{O}{\bigcirc} \text{ + } NH_2CHRC—$$

carbonyl carbon atom is probably involved in the formation of iso-imides and nitriles from an amide with an adjacent activated carboxyl group (equations 83–85)[86–88].

The neighbouring benzoylamido group in **59** increases the rate constant for the acid-catalysed hydrolysis of the N,N-dicyclohexylamido group by more than $1·4 \times 10^4$[89]. This reaction probably involves attack by oxygen on the neighbouring carbonyl carbon atom, forming benzoylanthranil (**60**) as an intermediate (equation 86). Although benzoylanthranil (**60**) was not isolated from the reaction mixture, it

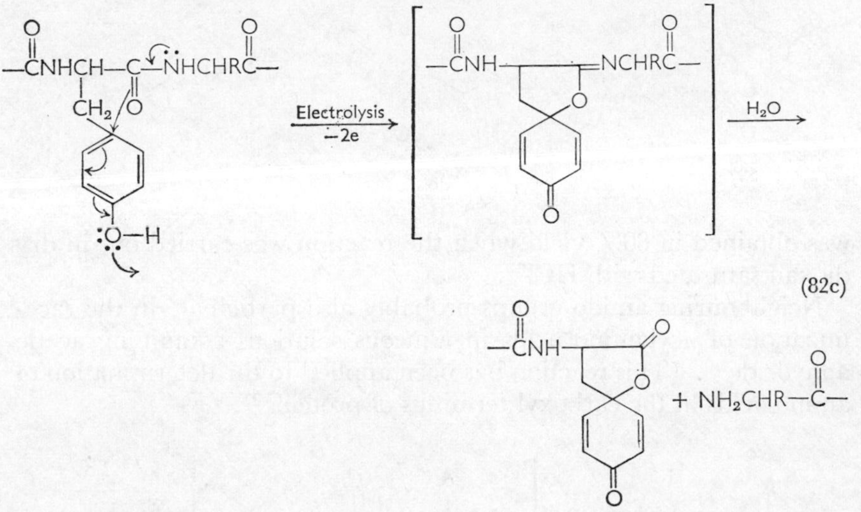

(82c)

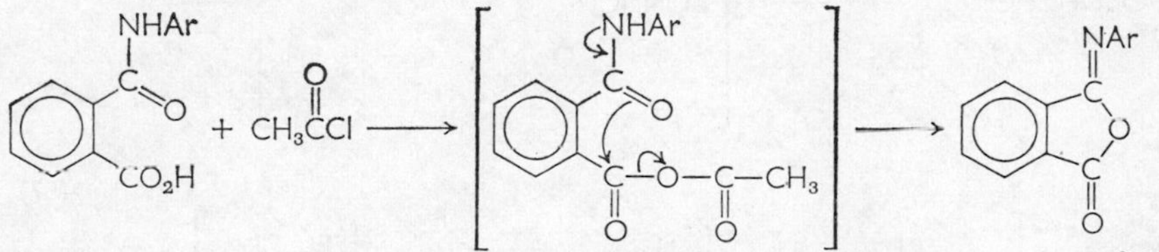

(83)

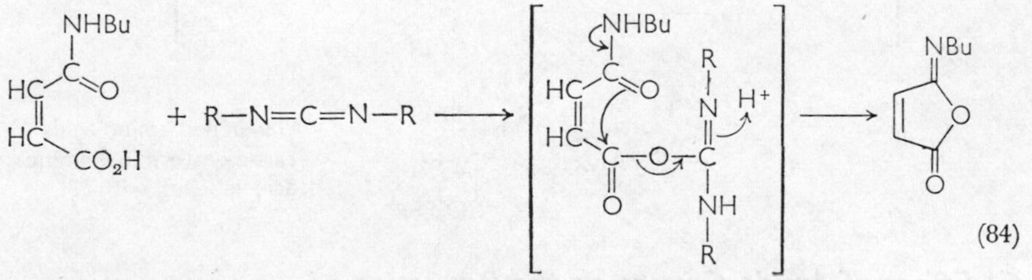

(84)

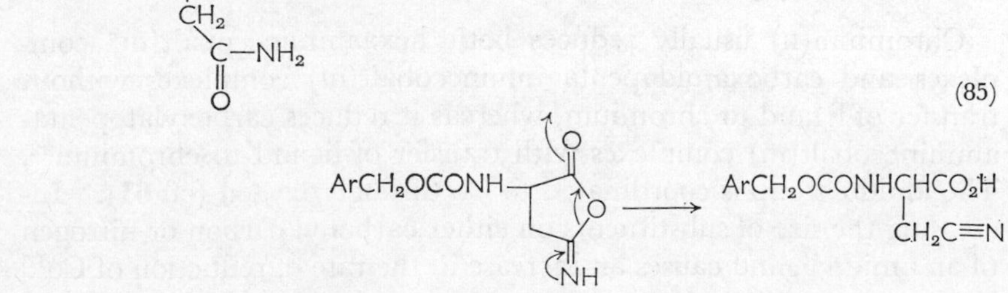

(85)

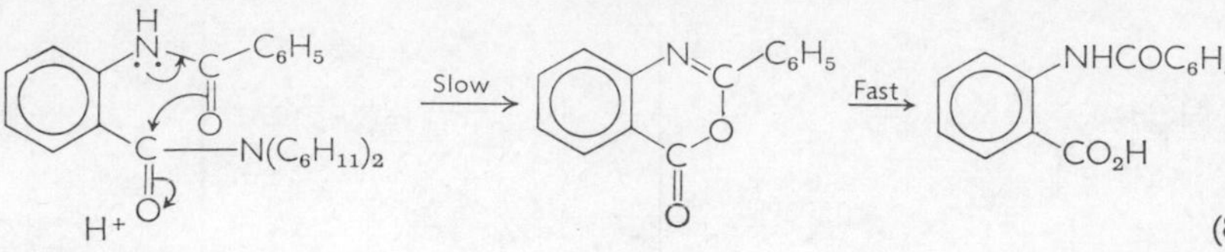

(86)

(59) (60)

was obtained in 80% yield when the reaction was carried out in dry dioxan saturated with HCl[89].

Neighbouring amido groups probably also participate in the racemization of acylamino acids in aqueous solutions containing acetic anhydride[90]. This reaction has been applied to the determination of amino acids at the carboxyl terminus of proteins[91].

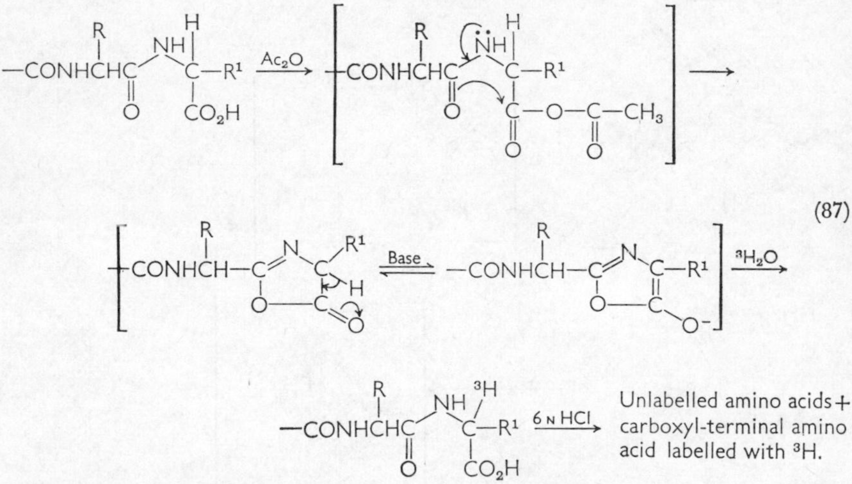

(87)

Unlabelled amino acids + carboxyl-terminal amino acid labelled with ^{3}H.

C. Effect of Amido Groups on the Reduction of Carboxamidopentaamminecobalt(III) Complexes by Chromium(II)

Chromium(II) usually reduces both hexaamminecobalt(III) complexes and carboxamidopentaamminecobalt(III) complexes without transfer of ligand to chromium, whereas it reduces carboxylatopentaamminecobalt(III) complexes with transfer of ligand to chromium[92]. The amido group is coordinated to Co through oxygen (cf. **61**). Increasing the size of substituents on either carbonyl carbon or nitrogen of an amido ligand causes an increase in the rate of reduction of Co^{III}

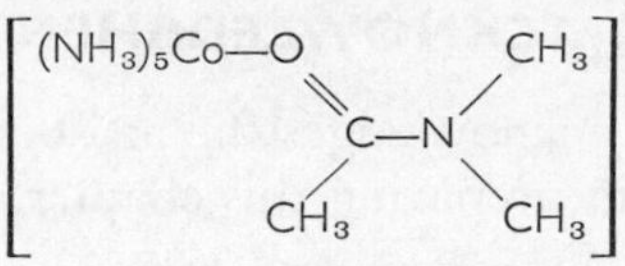

$$\left[(NH_3)_5Co-O \overset{}{\underset{CH_3}{\overset{}{C}}} - N \overset{CH_3}{\underset{CH_3}{}} \right]^{3+}$$

(61)

by Cr^{II}, suggesting that one or more Co-to-ligand bonds are stretched in the transition state[92]. Although pentaamminecobalt(III) complex **62** is reduced without transfer of ligand to Cr^{92}, ligand is transferred to Cr in the reduction of pentaamminecobalt(III) complexes **63** and **64**

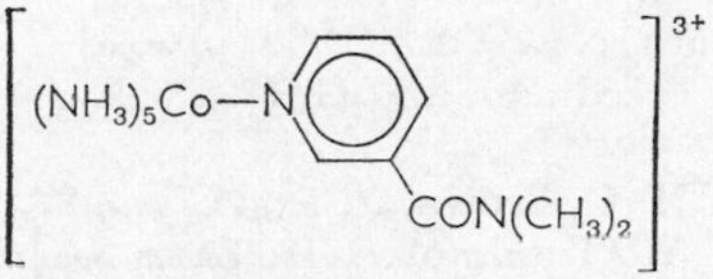

$$\left[(NH_3)_5Co-N \bigcirc \underset{CON(CH_3)_2}{} \right]^{3+}$$

(62)

forming respectively **65** and **66a** + **66b**[93]. In the reduction of **63** and **64** an electron is probably transferred from Cr^{II} to Co^{III} through an amido ligand[93].

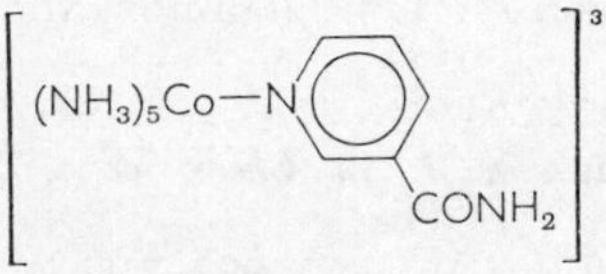

$$\left[(NH_3)_5Co-N \bigcirc \underset{CONH_2}{} \right]^{3+}$$

(63)

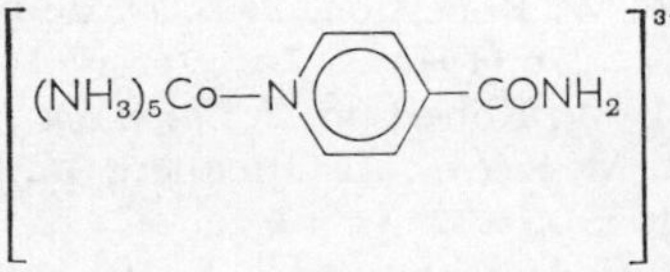

$$\left[(NH_3)_5Co-N \bigcirc -CONH_2 \right]^{3+}$$

(64)

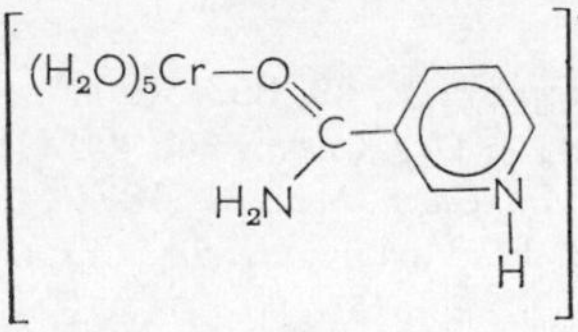

$$\left[(H_2O)_5Cr-O \underset{H_2N}{\overset{}{C}} \bigcirc \underset{\underset{H}{N}}{} \right]^{4+}$$

(65)

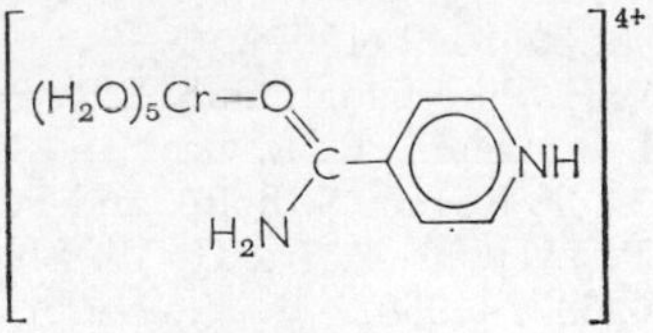

$$\left[(H_2O)_5Cr-O \underset{H_2N}{\overset{}{C}} \bigcirc NH \right]^{4+}$$

(66a)

$$\left[(H_2O)_5Cr-N \bigcirc -CONH_2 \right]^{3+}$$

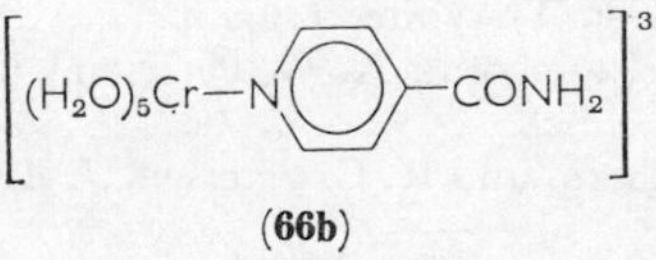

(66b)

V. ACKNOWLEDGMENT

The author gratefully acknowledges the assistance of Mr. C. J. Belke and Mr. S. C. K. Su in preparing this chapter.

VI. REFERENCES

1. L. P. Hammett, *Physical Organic Chemistry*, McGraw-Hill Book Co., New York, 1940.

2a. H. H. Jaffé, *Chem. Rev.*, **53**, 191 (1953).

2b. J. Hine, *Physical Organic Chemistry*, 2nd ed., McGraw-Hill Book Co., New York, 1962, pp. 85–95.

2c. J. E. Leffler and E. Grunwald, *Rates and Equilibria of Organic Reactions*, John Wiley and Sons, New York, 1963, pp. 171–216.

3. H. H. Jaffé, L. D. Freedman, and G. O. Doak, *J. Am. Chem. Soc.*, **75**, 2210 (1953).

4a. Y. Okamoto and H. C. Brown, *J. Org. Chem.*, **22**, 485 (1957).

4b. H. C. Brown and Y. Okamoto, *J. Am. Chem. Soc.*, **80**, 4979 (1958).

5. P. B. D. de la Mare, *J. Chem. Soc.*, 4450 (1954).

6. P. B. D. de la Mare and M. Hassan, *J. Chem. Soc.*, 1519 (1958).

7. N. C. Deno and W. L. Evans, *J. Am. Chem. Soc.*, **79**, 5804 (1957).

8. P. B. D. de la Mare and J. H. Ridd, *Aromatic Substitution—Nitration and Halogenation*, Butterworths, London, 1959, pp. 122–124.

9. H. C. Brown and L. M. Stock, *J. Am. Chem. Soc.*, **79**, 5175 (1957).

10. K. J. P. Orton and A. E. Bradfield, *J. Chem. Soc.*, 986 (1927).

11a. P. W. Robertson, P. B. D. de la Mare, and W. T. G. Johnston, *J. Chem. Soc.*, 276 (1943).

11b. P. W. Robertson, *J. Chem. Soc.*, 1267 (1954).

12. R. M. Keefer, A. Ottenberg, and L. J. Andrews, *J. Am. Chem. Soc.*, **78**, 255 (1956).

13. C. R. Harrison and J. F. W. McOmie, *J. Chem. Soc.*, (*C*), 997 (1966).

14. M. A. Paul, *J. Am. Chem. Soc.*, **80**, 5332 (1958).

15. K. E. Cooper and C. K. Ingold, *J. Chem. Soc.*, 836 (1927).

16. D. Jerina, J. Daly, W. Landis, B. Witkop, and S. Udenfriend, *J. Am. Chem. Soc.*, **89**, 3349 (1967).

17. W. H. Puterbaugh and C. R. Hauser, *J. Org. Chem.*, **29**, 853 (1964).

18. H. Watanabe, R. L. Gay, and C. R. Hauser, *J. Org. Chem.*, **33**, 900 (1968).

19a. M. T. Rogers, S. Bolte, and P. S. Rao, *J. Am. Chem. Soc.*, **87**, 1875 (1965).

19b. P. J. Hamrick, Jr., H. W. Shields, and S. H. Parkey, *J. Am. Chem. Soc.*, **90**, 5371 (1968).

20. P. Smith and P. B. Wood, *Can. J. Chem.*, **44**, 3085 (1966).

21. H. L. Needles and R. E. Whitfield, *J. Org. Chem.*, **29**, 3632 (1964).

22a. W. Walter, M. Steffen, and K. Heyns, *Chem. Ber.*, **94**, 2462 (1961).

22b. C. H. Bamford and E. F. T. White, *J. Chem. Soc.*, 1860 (1959).

23a. S. D. Ross, M. Finkelstein, and R. C. Petersen, *J. Org. Chem.*, **31**, 128 (1966).

23b. S. D. Ross, M. Finkelstein, and R. C. Petersen, *J. Am. Chem. Soc.*, **88**, 4657 (1966).

24. A. H. Lewin, A. H. Dinwoodie, and T. Cohen, *Tetrahedron*, **22**, 1527 (1966).
25. T. Cohen, C. H. McMullen, and K. Smith, *J. Am. Chem. Soc.*, **90**, 6866 (1968).
26. D. H. Hey, C. W. Rees, and A. R. Todd, *J. Chem. Soc.*, (*C*), 1518 (1967).
27. R. A. Heacock and D. H. Hey, *J. Chem. Soc.*, 1508 (1952).
28. R. A. Firestone, D. F. Reinhold, W. A. Gaines, J. M. Chemerda, and M. Sletzinger, *J. Org. Chem.*, **33**, 1213 (1968).
29. S. D. Ross, M. Finkelstein, and R. C. Petersen, *J. Org. Chem.*, **31**, 133 (1966).
30. H. Hellmann in *Newer Methods of Preparative Organic Chemistry*, Vol. II, (Ed. W. Foerst), Academic Press, New York, 1963, p. 277.
31a. T. Cohen, R. M. Moran, Jr., and G. Sowinski, *J. Org. Chem.*, **26**, 1 (1964).
31b. T. Cohen, A. H. Dinwoodie, and L. D. McKeever, *J. Org. Chem.*, **27**, 3385 (1962).
31c. T. Cohen and J. Lipowitz, *J. Am. Chem. Soc.*, **86**, 2514 (1964).
31d. T. Cohen and J. Lipowitz, *J. Am. Chem. Soc.*, **86**, 2515 (1964).
32. D. H. Hey, J. A. Leonard, C. W. Rees, and A. R. Todd, *J. Chem. Soc.*, (*C*), 1513 (1967).
33. G. Tennant and K. Vaughan, *J. Chem. Soc.*, (*C*), 2287 (1966).
34. A. Chambers and C. J. M. Stirling, *J. Chem. Soc.*, 4556 (1965).
35. C. H. DePuy and D. H. Froemsdorf, *J. Am. Chem. Soc.*, **79**, 3710 (1957).
36. S. D. Work, D. R. Bryant, and C. R. Hauser, *J. Am. Chem. Soc.*, **86**, 872 (1964).
37. J. F. Wolfe and C.-L. Mao, *J. Org. Chem.*, **32**, 1977 (1967).
38. J. F. Wolfe and C.-L. Mao, *J. Org. Chem.*, **32**, 3382 (1967).
39. J. F. Wolfe and G. B. Trimitsis, *J. Org. Chem.*, **33**, 894 (1968).
40. G. Tennant, *J. Chem. Soc.*, (*C*), 2285 (1966); and references therein.
41. H. Feuer and B. F. Vincent, Jr., *J. Org. Chem.*, **29**, 939 (1964).
42. S. Patai, J. Zabicky, and Y. Israeli, *J. Chem. Soc.*, 2038 (1960).
43a. H. W. Heine, P. Love, and J. L. Bove, *J. Am. Chem. Soc.*, **77**, 5420 (1955).
43b. H. W. Heine, *J. Am. Chem. Soc.*, **78**, 3708 (1956).
44. F. L. Scott, R. E. Glick, and S. Winstein, *Experientia*, **13**, 183 (1953).
45. T. Taguchi and M. Kojima, *J. Am. Chem. Soc.*, **81**, 4316 (1959).
46. T. Taguchi and M. Kojima, *J. Am. Chem. Soc.*, **81**, 4318 (1959).
47. C. Zioudrou and G. L. Schmir, *J. Am. Chem. Soc.*, **85**, 3258 (1963).
48. F. L. Scott and E. Flynn, *Tetrahedron Letters*, 1675 (1964).
49. F. L. Scott and D. F. Fenton, *Tetrahedron Letters*, 1681 (1964).
50. A. Chambers and C. J. M. Stirling, *J. Chem. Soc.*, 4558 (1965).
51. J. E. H. Hancock and R. P. Linstead, *J. Chem. Soc.*, 3490, (1953).
52. E. Sondheimer and R. W. Holley, *J. Am. Chem. Soc.*, **76**, 2467 (1953).
53. A. R. Battersby and J. C. Robinson, *J. Chem. Soc.*, 259 (1955).
54. E. Sondheimer and R. W. Holley, *J. Am. Chem. Soc.*, **79**, 3767 (1957).
55. S. A. Bernhard, A. Berger, J. H. Carter, E. Katchalski, M. Sela, and Y. Shalitin, *J. Am. Chem. Soc.*, **84**, 2421 (1962).
56. Y. Shalitin and S. A. Bernhard, *J. Am. Chem. Soc.*, **86**, 2292 (1964).
57. A. J. Adler, G. D. Fasman, and E. R. Blout, *J. Am. Chem. Soc.*, **85**, 90 (1963).
58. M. Goodman and K. C. Stueben, *J. Am. Chem. Soc.*, **84**, 1279 (1962),
59. J. A. Shafer and H. Morawetz, *J. Org. Chem.*, **28**, 1899 (1963).

60. J. A. Shafer, *Doctoral Dissertation*, Polytechnic Institute of Brooklyn, 1963.
61. M. T. Behme and E. H. Cordes, *J. Org. Chem.*, **29**, 1255 (1964).
62. M. B. Vigneron, P. Crooy, F. Kezdy, and A. Bruylants, *Bull. Soc. Chim. Belges*, **69**, 616 (1960).
63. J. Brown, S. C. K. Su, and J. A. Shafer, *J. Am. Chem. Soc.*, **88**, 4468 (1966).
64. J. de. Jersey, A. A. Kortt, and B. Zerner, *Biochem. Biophys. Res. Commun.*, **23**, 745 (1966).
65. M. Goodman and K. C. Stucben, *J. Org. Chem.*, **27**, 3409 (1962).
66. M. Goodman and W. J. McGahren, *J. Am. Chem. Soc.*, **87**, 3028 (1965).
67. M. Goodman and W. J. McGahren, *J. Am. Chem. Soc.*, **88**, 3887 (1966).
68. M. W. Williams and G. T. Young, *J. Chem. Soc.*, 3701 (1964).
69. J. Zabicky, *Chem. Ind. (London)*, 236 (1964).
70a. K. M. Ibne-Rasa and E. Koubek, *J. Org. Chem.*, **28**, 3240 (1963).
70b. M. Rosenblum, *J. Am. Chem. Soc.*, **82**, 3797 (1960).
71. G. E. K. Branch and J. O. Clayton, *J. Am. Chem. Soc.*, **50**, 1680 (1928).
72. P. Ballinger and F. A. Long, *J. Am. Chem. Soc.*, **82**, 795 (1960).
73. S. Winstein, L. Goodman, and R. Boschan, *J. Am. Chem. Soc.*, **72**, 2311 (1950).
74. S. Winstein and R. Boschan, *J. Am. Chem. Soc.*, **72**, 4669 (1950).
75. J. Sicher, M. Tichý, F. Šipoš, and M. Pánová, *Proc. Chem. Soc.*, 384 (1960); *Collection Czech. Chem. Commun.*, **26**, 2418 (1961).
76. S. Hanessian, *J. Org. Chem.*, **32**, 163 (1967); and references therein.
77a. M. Bergmann, E. Brand, and F. Dreyer, *Chem. Ber.*, **54**, 936 (1921).
77b. M. Bergmann and E. Brand, *Chem. Ber.*, **56**, 1280 (1923).
77c. M. Bergmann, E. Brand, and F. Weinmann, *Z. Physiol. Chem.*, **131**, 1 (1923).
77d. M. Bergmann and A. Miekeley, *Z. Physiol. Chem.*, **140**, 128 (1924).
77e. M. Bergmann, A. Miekeley, F. Weinmann, and E. Kann, *Z. Physiol. Chem.*, **143**, 108 (1925).
77f. M. Bergmann, A. Miekeley, and E. Kann, *Z. Physiol. Chem.*, **146**, 247 (1925).
78. J. Attenburrow, D. F. Elliott, and G. F. Penny, *J. Chem. Soc.*, 310 (1948).
79. L. H. Welsh, *J. Org. Chem.*, **32**, 119, (1967); and references therein.
80a. D. Piszkiewicz and T. C. Bruice, *J. Am. Chem. Soc.*, **89**, 6237 (1967).
80b. D. Piszkiewicz and T. C. Bruice, *J. Am. Chem. Soc.*, **90**, 2156 (1968).
80c. D. Piszkiewicz and T. C. Bruice, *J. Am. Chem. Soc.*, **90**, 5844 (1968).
81. B. Capon, *Tetrahedron Letters*, 911 (1963).
82a. L. Goodman and S. Winstein, *J. Am. Chem. Soc.*, **79**, 4788 (1957).
82b. L. Goodman, S. Winstein, and R. Boschan, *J. Am. Chem. Soc.*, **80**, 4312 (1958).
83. C. J. M. Stirling, *J. Chem. Soc.*, 255 (1960); and references therein.
84. G. L. Schmir and B. L. Cunningham, *J. Am. Chem. Soc.*, **87**, 5692 (1965); and references therein.
85a. E. Gross, in *Methods in Enzymology*, Vol. 11, (Ed. C. H. W. Hirs), Academic Press, New York, 1967, pp. 238–255.
85b. L. K. Ramachandran and B. Witkop in *Methods in Enzymology*, Vol. 11, (Ed. C. H. W. Hirs), Academic Press, New York, 1967, pp. 283–299.
85c. L. A. Cohen and L. Farber, in *Methods in Enzymology*, Vol. 11, (Ed. C. H.W. Hirs), Academic Press, New York, 1967, pp. 299–315.

86a. P. H. Van der Meulen, *Rec. Trav. Chim.*, **15,** 282 (1896).

86b. M. L. Sherrill, F. L. Schaeffer, and E. P. Shoyer, *J. Am. Chem. Soc.*, **50,** 474 (1928).

87. C. Ressler and H. Ratzkin, *J. Org. Chem.*, **26,** 3356 (1961).

88. R. Paul and A. S. Kende, *J. Am. Chem. Soc.*, **86,** 4162 (1964).

89. T. Cohen and J. Lipowitz, *J. Am. Chem. Soc.*, **86,** 5611 (1964).

90. V. Du Vigneaud and C. E. Meyer, *J. Biol. Chem.*, **98,** 295 (1932).

91. H. Matsuo, Y. Fujimoto, and T. Tatsuno, *Biochem. Biophys. Res. Commun.*, **22,** 69 (1966).

92. E. S. Gould, *J. Am. Chem. Soc.*, **90,** 1740 (1968).

93. F. R. Nordmeyer and H. Taube, *J. Am. Chem. Soc.*, **88,** 4295 (1966).

*

CHAPTER **13**

Reactions of the carboxamide group

BRIAN C. CHALLIS and JUDITH A. CHALLIS

Imperial College, London, England

I. INTRODUCTION

This chapter shall be concerned mainly with substitution and addition reactions of the amido group together with several of the important transformations that ensue from them. The situation is complicated by the reactivity of all three atoms in the O—C—N chain, arising mainly from delocalization of the π electrons along this chain. The consequence of such delocalization is to diminish both the inherent carbonyl reactivity and the nucleophilic properties of the amide group.

Nonetheless, the chemistry of amides is simplified considerably on recognizing that the great majority of their reactions proceed by one or other of two processes. The first involves nucleophilic attack by the oxygen or nitrogen atom on electrophilic centres in either positively charged or neutral species. The second, and less common process, involves nucleophilic addition to the carbonyl entity. Other transformations, such as elimination, dehydration, deamination, etc., are invariably parasitic to these two reactions.

An estimate of the nucleophilic properties of the amide function can be gathered from its behaviour towards the proton (i.e. conjugate acid formation). It is now well established (Chapter 3) that amides are relatively weak bases (the pK_a lower approximately by 10 units than that of a similar amine) and that protonation invariably takes place on the oxygen atom (equation 1). Thus amides should behave as feeble nucleophiles with oxygen as the most reactive site. This conclusion is borne out in practice. Amides in their neutral form react only with the more powerful electrophilic reagents and *initially* form an *O*-substituted derivative. Although *N*-substituted products are commonly isolated from reaction mixtures, these usually arise from rearrangement of the *O*-substituted precursor and therefore represent the thermodynamically stable product. Direct substitution at the nitrogen atom occurs only in special circumstances: either with powerful electrophilic species such as diazomethane (alkylation), and nitroso derivatives (and even here the evidence is far from conclusive) or under strongly alkaline or acidic conditions where the reactive species are

the anion $\left(\begin{array}{c} \quad\quad O \\ R-C \diagdown \\ \quad NH^- \end{array} \right)$ or the conjugate acid $\left(\begin{array}{c} \quad\quad O-H \\ R-C\overset{\cdot\cdot}{}+ \\ \quad NH_2 \end{array} \right)$

of the amide, respectively.

Because of π-electron delocalization along the O—C—N chain, the carbonyl reactivity is also subdued. Amides, in fact, are much less reactive than esters in this respect and bear a closer resemblance to their parent carboxylic acids. Thus carbonyl addition does not generally occur unless additional factors enhance the inherent polarization of the carbon–oxygen bond. These may be either protonation, as in hydrolysis, or O-complex formation with an electron-deficient species as in the reactions of inorganic acid halides and metallo-organic reagents.

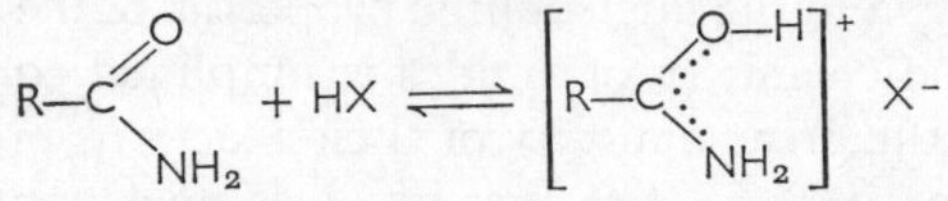

$$R-C\underset{NH_2}{\overset{O}{\big\langle}} + HX \rightleftharpoons \left[R-C\underset{NH_2}{\overset{O-H}{\big\langle}} \right]^+ X^- \tag{1}$$

The behaviour of amides towards many reagents often depends on the specific structure of either the acyl (RCO) or the amino part ($-NR^1R^2$) of the molecule, and it is necessary to identify these structural features clearly. For this reason we have adopted the convention of referring to the atom directly attached to the carbonyl as the $\alpha_{C=O}$ atom, and to those directly attached to the nitrogen as α_N atoms:

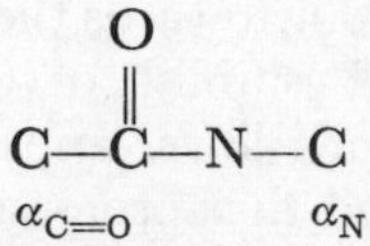

$$\underset{\alpha_{C=O} \qquad\quad \alpha_N}{C-\overset{\overset{\textstyle O}{\|}}{C}-N-C}$$

Our intention is to present a synopsis of the most important reactions of the amide group. Recent rather than historic developments are stressed from both synthetic and mechanistic standpoints. In many instances, however, sound quantitative information to establish the mechanism is entirely lacking and our conclusions are mostly speculative. Our approach can be justified, nonetheless, by the fresh research effort we hope to encourage.

II. ALKYLATION

Alkylation reactions have been widely studied and the results contribute significantly to our understanding of the nucleophilic properties of the amide moiety. Both substrate reactivity and the site of substitution are understandably related to the experimental conditions, because the molecular amide may be in equilibrium with both its anion and conjugate acid (equation 2). Alkylation does in fact occur at either

$$\left[RC\underset{NH_2}{\overset{OH}{\big\langle}} \right]^+ \underset{+H^+}{\overset{}{\rightleftharpoons}} RCONH_2 \overset{-H^+}{\underset{}{\rightleftharpoons}} RCONH^- \tag{2}$$

the nitrogen, the oxygen or the $\alpha_{C=O}$-carbon atom depending on the pH of the reaction medium. Accordingly, it is convenient to discuss alkylation from the standpoint of reaction in either neutral, alkaline or acidic media.

A. Alkylation under Neutral Conditions

Under these conditions, which include most reactions in aprotic solvents, the amide is present in its molecular (unionized) form. This species reacts sluggishly as expected from its feeble basicity and only those agents more active than alkyl halides, such as alkyl sulphates[1], oxonium salts[2] and diazoalkanes[3], are synthetically useful. The products are the *O*- and *N*-alkyl derivatives and the latter may arise either from direct substitution or by rearrangement (Scheme 1). With primary and secondary amides, the products are neutralized to

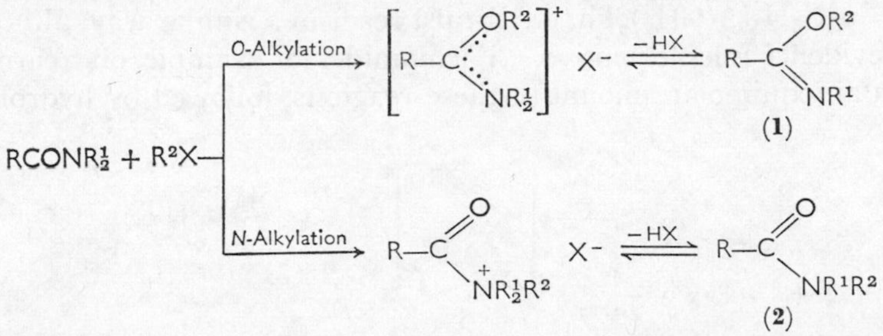
O-Alkylation
N-Alkylation
RCONR¹₂ + R²X
OR²
NR¹₂
X⁻
−HX
R—C
NR¹
(1)
O
NR¹₂R²
NR¹R²
(2)

SCHEME 1. *O*- and *N*-alkylation of amides under neutral conditions.

1 and **2** by proton loss from nitrogen. The proportion of *O*- and *N*-alkyl products depends on the reaction temperature and the reactivity of the alkylating agent.

1. Alkyl sulphates and alkyl sulphonates

Both reagents react readily with most amides at slightly elevated temperatures to form products arising predominantly from *O*-alkylation[1]. Dimethyl sulphate, for instance, reacts quantitatively with an equimolar proportion of either primary [1a,b], secondary[1a,c,d] or tertiary amide[1e], to give the corresponding *O*-alkylimidonium salt (**3**) but no *N*-alkyl products (equation 3). Reaction temperatures of

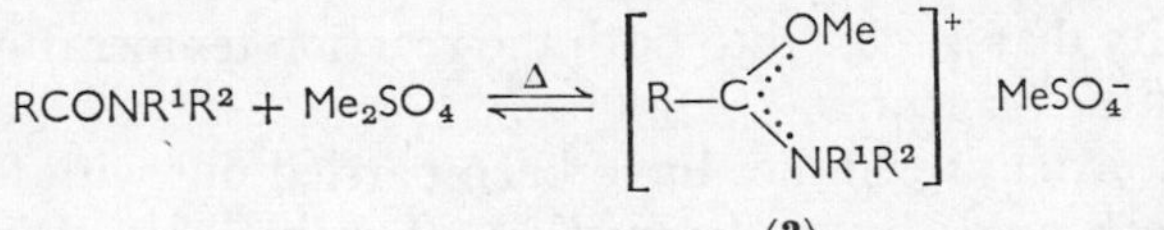
RCONR¹R² + Me₂SO₄
Δ
OMe
R—C
NR¹R²
MeSO₄⁻

(3)

(3)

20° to 60° are required. The salt **3** is readily attacked by nucleophiles including the amide itself, and this explains the need for equimolar reactant concentrations. These further reactions with excess amide have been studied in the case of formamide[1b]. With up to one molar excess, only the methoxy group is displaced to give the amidinium salt (**4**) and methyl formate (equation 4); but with an even larger excess of formamide, triformylaminomethane [$CH(NHCHO)_3$] is eventually formed via a series of successive substitutions.

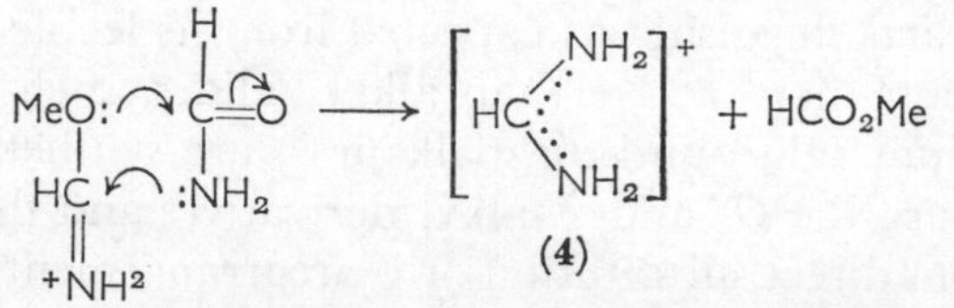

$$MeO: \overset{\overset{\displaystyle H}{|}}{C} = O \longrightarrow \left[HC \overset{NH_2}{\underset{NH_2}{\cdots}} \right]^+ + HCO_2Me \qquad (4)$$

$$\overset{|}{HC} \quad :NH_2$$
$$\overset{\parallel}{{}^+NH^2}$$

(4)

Alkyl benzene sulphonates ($PhSO_3Bu$-n, or p-$CH_3C_6H_4SO_3R$ where R = n-C_8H_{17}, $(CH_2)_nPh$, etc.) must react in a similar way, although the evidence is less extensive. Formamide, for example, on treatment with an equimolar amount of these reagents, followed by hydrolysis,

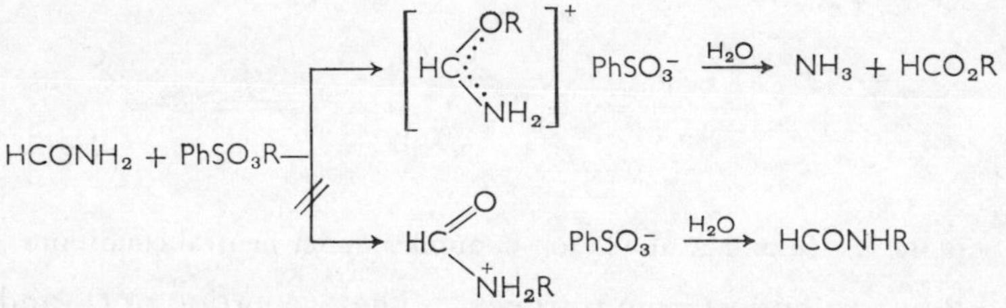

$$HCONH_2 + PhSO_3R \begin{cases} \longrightarrow \left[HC \overset{OR}{\underset{NH_2}{\cdots}} \right]^+ PhSO_3^- \xrightarrow{H_2O} NH_3 + HCO_2R \\ \\ \longrightarrow HC \overset{O}{\underset{{}^+NH_2R}{\Big\backslash}} \quad PhSO_3^- \xrightarrow{H_2O} HCONHR \end{cases}$$

Scheme 2. Alkylation of formamide with alkyl benzenesulphonates.

gives ammonia but apparently no *N*-alkylformamide[1b] (Scheme 2). Once again this is consistent with exclusive *O*-alkylation.

2. Alkyl halides

Generally higher temperatures ($\sim 150°$) or heavy metal catalysts such as silver salts are necessary to induce reaction with alkyl halides, and under these conditions the bifunctional nucleophilic behaviour of the amide moiety becomes important. Thus either *O*- or *N*-substituted products, or even a mixture of both, are commonly formed with a facility that is related to both the reaction temperature and the structure of the reagent.

The most extensive studies have been carried out with formamide by Bredereck and his colleagues[1b,4] (Scheme 3). With excess formamide, usually present as the solvent, reaction proceeds via two paths to give the N-alkylformamide (**5**) and the alkyl formate (**6**) respectively[1b,4]. Under anhydrous conditions, the alkyl formate is claimed to arise from solvolysis of the formamide by an alcohol molecule, itself produced through dehydration of the intermediate O-alkylimidonium salt[4a]. Obviously, direct hydrolysis of the O-alkylimidonium salt would give the same formate ester product.

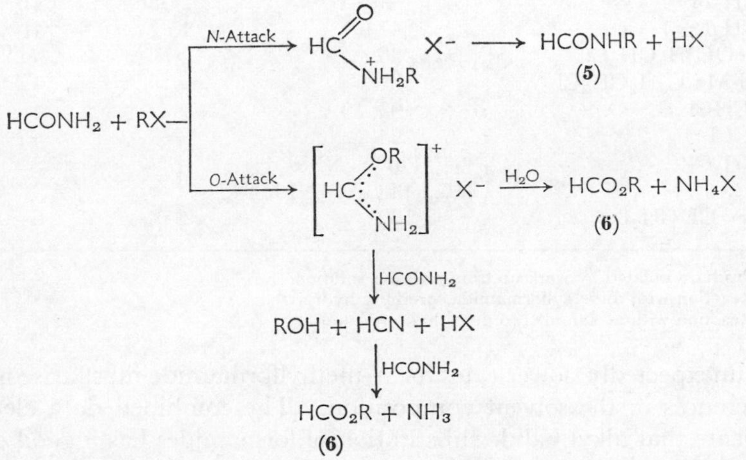

$$HCONH_2 + RX \longrightarrow$$

$$\text{N-Attack} \longrightarrow HC\!\!\begin{array}{c} O \\ \\ \overset{+}{NH_2R} \end{array}\!\! X^- \longrightarrow HCONHR + HX \quad (5)$$

$$\text{O-Attack} \longrightarrow \left[HC\!\!\begin{array}{c} OR \\ \\ NH_2 \end{array} \right]^+ X^- \xrightarrow{H_2O} HCO_2R + NH_4X \quad (6)$$

$$\downarrow HCONH_2$$

$$ROH + HCN + HX$$

$$\downarrow HCONH_2$$

$$HCO_2R + NH_3$$

$$(6)$$

SCHEME 3. Reaction of alkyl halides with formamide.

The ratio of O- to N-substituted products is of importance as far as synthetic considerations are concerned. The data of Table 1 show that usually N-alkylformamides (i.e. N-substitution) are favoured with alkyl halides forming relatively stable carbonium ions (e.g. Ph_3CCl, Ph_2CHCl), whereas less polarizable reagents, such as $PhCH_2Cl$ and $C_8H_{17}Br$, preferentially form the alkyl formate. Judging from the results for the benzyl halides, the nature of the halide ion may also have a bearing on product orientation; alternatively this may be an artifact stemming from important differences in the method of product isolation.

The overall reactivity of various alkyl halides towards formamide has also been investigated, both by conductance measurements of the extent of alkyl halide decomposition (Table 2) and by kinetic studies (Table 3) under pseudo-first-order conditions (with a ten-fold excess of formamide)[4b]. The kinetic experiments have been extended to N-methylformamide[4b] and these results, too, are listed in Table 3:

TABLE 1. O- and N-substituted products for the reaction of alkyl halides with formamide at $\sim 150°$.

| | Product (%) | | |
Alkyl halide	N-Alkylformamide	Alkyl formate	References
n-$C_8H_{17}Br$		92	4a
$PhCH_2Cl$	5	74	4a
$PhCH_2Br$ [a]	44·5	27·5	1b
$PhCH_2I$ [a]	47·6	15·2	1b
p-$MeOC_6H_4CH_2Cl$	36		4a
$2,4,6$-$Me_3C_6H_2CH_2Cl$	31		4a
Ph_2CHCl	95		4a
Ph_3CCl	94		4a
$PhCH_2Cl$ [b]	$\begin{cases} 36\% \ PhCH_2NMe_2 \\ 34\% \ (PhCH_2)_2NMe \end{cases}$		5
$CH_2{=}CHCH_2Br$ [c]	100		6

[a] Products isolated by work-up from aqueous solution.
[b] Reaction with dimethylformamide, products hydrolysed.
[c] Reaction with acetamide, products hydrolysed.

the unexpectedly slower rates for N-methylformamide must arise from differences in the solvent composition. The combined data clearly indicate that alkyl halide substitution of formamides has a good deal of S_N1 character, as factors which would stabilize the developing alkyl carbonium ion in the transition state also increase the reaction rate. For example, the observed reactivity of butyl chloride is $t > s > n$ and for *para*-substituted benzyl chloride is $MeO > Me > H > NO_2$. These findings led Gompper and his coworkers to account for the

TABLE 2. Decomposition of alkyl halides in excess formamide by conductivity measurements [4b].

Alkyl halide	Time (hr)	Temperature (°c)	Decomposition (%)
p-$NO_2C_6H_4CH_2Cl$	2	50	5
$C_6H_5CH_2Cl$	2	50	29
p-$MeOC_6H_4CH_2Cl$	2	50	78
n-BuBr	5	50	9
i-PrBr	5	50	21
t-BuBr	5	50	89
n-BuBr	5	70	42
n-$C_8H_{17}Br$	5	70	1·5

tendency towards O- and N-substitution by various reagents in terms of the transition state structure[4b], along the lines developed earlier by Kornblum for ambident nucleophilic anions[7]. The result is unsatisfactory, however, for it requires that the most S_N1-like transition state will be associated with substitution at the nitrogen atom, which is the atom of lower electronegativity. This is, of course, contrary to Kornblum's predictions.

TABLE 3. Pseudo-first-order rate coefficients for the reaction of alkyl halides with ten-fold excess formamide and N-methylformamide[4b].

	Formamide		N-Methylformamide[a]	
Alkyl halide	Temp. (°c)	k (hr^{-1})	Temp. (°c)	k (hr^{-1})
n-BuBr	80	0·040	80	0·036
s-BuBr	80	0·119	80	0·040
t-BuBr	20	0·147	80	0·056
p-MeC$_6$H$_4$CH$_2$Cl	55	0·104		
C$_6$H$_5$CH$_2$Cl	80	0·230		
p-NO$_2$C$_6$H$_4$CH$_2$Cl	90	0·110		
Ph$_2$CHCl	20	b		
Ph$_3$CCl	10	b		
CH$_2$=CHCH$_2$Br	45	0·087		

[a] Reaction in equimolar dioxan: N-methylformamide.
[b] Too fast to measure.

Other evidence suggests the tendency towards mixed O- and N-alkylation may arise from the high temperatures necessary to induce reaction. This comes from studies of both the highly reactive trityl chloride[4b] and the effect of silver salt catalysis[8]. Trityl chloride is sufficiently powerful to alkylate formamide at temperatures as low as 20°. Under these conditions the conductivity of the mixture composed of equimolar amounts of reactants reaches a maximum value almost instantaneously, and hydrolysis of this solution results in the isolation of triphenylcarbinol but not N-triphenylmethylformamide (Scheme 4)[4b]. When the same reaction is carried out at 110°, however, the latter is the sole hydrolysis product[4a]. Alkylation by methyl and ethyl iodide can also be effected under mild conditions in the presence of silver oxide catalysts[8]. The silver ion promotes polarization of the alkyl halide, thereby increasing its reactivity. In this way the same temperature-dependent substitution pattern

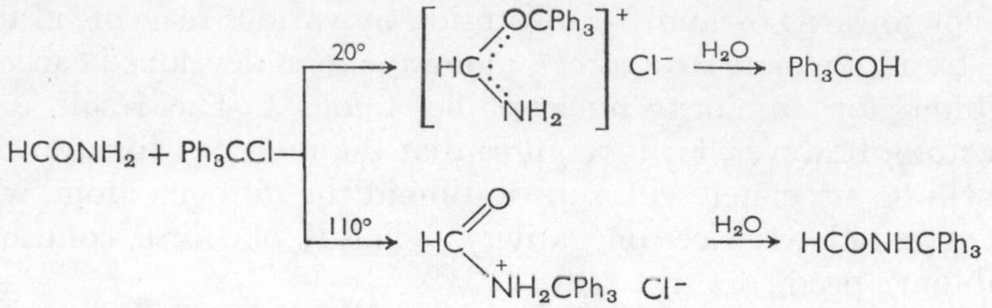

SCHEME 4. Reaction of formamide with trityl chloride.

emerges as that for trityl chloride. Thus the reaction of N-phenyl-formamide with ethyl iodide in the presence of silver oxide at 40° produces only the O-ethyl imidate (7), whereas mixed O- and N-alkylated products are obtained at 100°[8] (Scheme 5). Both these results point to a mechanism in which at least some of the N-alkylated product arises from a thermal rearrangement of the O-alkyl imidate salt. Since substantial support for this hypothesis is forthcoming from the reactions of other alkylating agents, further discussion of the mechanism is deferred until section II.A.6.

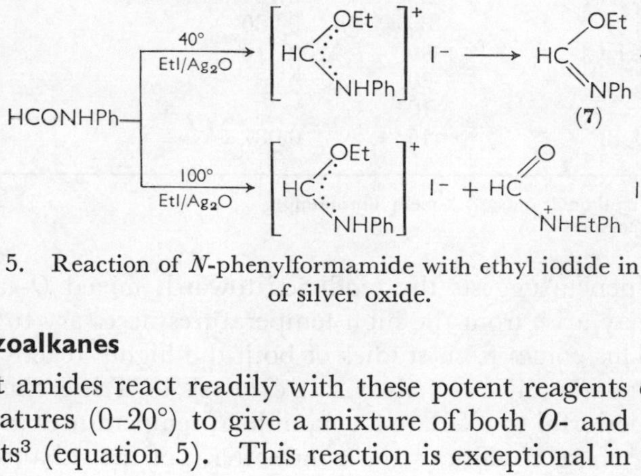

SCHEME 5. Reaction of N-phenylformamide with ethyl iodide in the presence of silver oxide.

3. Diazoalkanes

Most amides react readily with these potent reagents even at low temperatures (0–20°) to give a mixture of both O- and N-alkylated products[3] (equation 5). This reaction is exceptional in the sense of

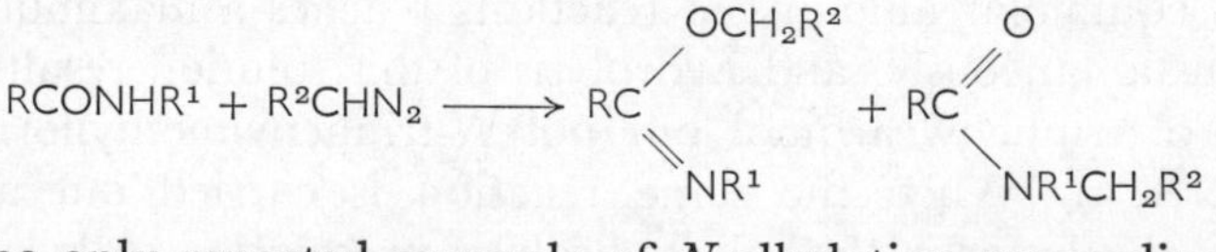

being the only reported example of N-alkylation proceeding at low temperatures, and the reasons for this behaviour are discussed later.

Various attempts have been made to develop a yardstick for predicting the site of substitution. The most reliable is due to Gompper[3a], based on the infrared stretching vibration frequency ($\nu_{C=O}$) of the

TABLE 4. Correlation between amide $\nu_{C=O}$ and the site of methylation by diazomethane[3a].

Amide $\nu_{C=O}$ (cm^{-1})[a]	Site of methylation
1620–1680	O
1680–1730	O, N
1730–1800	N

[a] $\nu_{C=O}$ measured by the KBr disc method.

carbonyl bond (Table 4). It is apparent that enhanced single-bond character of the carbonyl group (lower $\nu_{C=O}$) favours O-substitution and both phenomena are obviously related to increased delocalization of the nitrogen lone-pair electrons. The importance of this conjugative interaction to product orientation is also evident from studies with cyclic amides[3b] (equation 6). The results in Table 5 show that

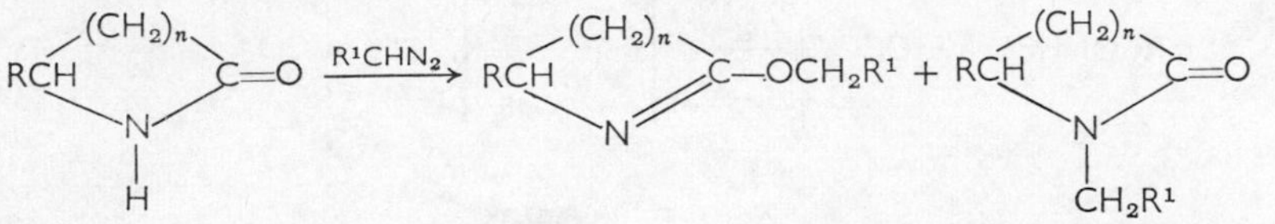

$$
\underset{\underset{\text{H}}{|}}{\overset{\text{(CH}_2)_n}{\underset{\text{N}}{\text{RCH}}}}\hspace{-0.5em}\text{C=O} \xrightarrow{\text{R}^1\text{CHN}_2} \underset{\underset{\text{N}}{}}{\overset{\text{(CH}_2)_n}{\text{RCH}}}\hspace{-0.5em}\text{C}-\text{OCH}_2\text{R}^1 + \underset{\underset{\text{CH}_2\text{R}^1}{|}}{\overset{\text{(CH}_2)_n}{\underset{\text{N}}{\text{RCH}}}}\hspace{-0.5em}\text{C=O} \qquad (6)
$$

five-membered ($n = 2$) cyclic amides undergo N-alkylation in contrast to the specific O-alkylation with six-($n = 3$) and seven-membered

TABLE 5. Alkylation of lactams of structure $\underset{\underset{\text{H}}{|}}{\overset{\text{(CH}_2)_n}{\underset{\text{N}}{\text{R—CH}}}}\hspace{-0.5em}\text{C=O}$ with diazoalkanes of structure R^1CHN_2.

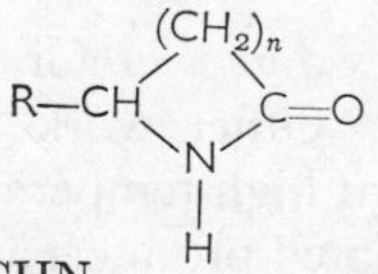

n	R	R^1	Amide $\nu_{C=O}$ (cm^{-1})[a]	Site of alkylation[b]
2	H	CH$_3$	1706[d], 1690[e]	N, N[c]
2	CO$_2$Et	CH$_3$		N
2	CO$_2$H	CH$_3$		N
3	H	H	1672[d], 1669[f]	O
4	H	H	1669[d], 1658[f]	O
4	H	CH$_3$	1669[d], 1658[f]	O

[a] From reference 9.
[b] From reference 3b.
[c] In the presence of HBF$_4$.
[d] Infrared spectrum taken in CCl$_4$ solution.
[e] Infrared spectrum taken in a liquid thin film.
[f] Infrared spectrum taken in a KBr disc.

($n = 4$) compounds. Clearly the size of the ring structure bears on the site of substitution, which can be explained at least partly by the steric requirements for conjugation between the oxygen and nitrogen atoms. Thus the specific N-alkylation of five-membered lactams arises from inhibition of this conjugation in the constrained small ring. It is interesting to note, too, that the specific O-alkylation of six- and seven-membered lactams is consistent with the infrared frequency correlation of Gompper.

4. Miscellaneous reagents

Innumerable other alkylating agents react with amides, but few have been subject to more than a cursory study. Two of the more useful ones are ethyl chloroformate[10] (equation 7) and triethyl-oxonium fluoroborate[2] (equation 8). Both react readily at room

$$RCONHR^1 + EtOCOCl \xrightarrow{20^\circ} \left[RC \begin{matrix} OEt \\ NHR^1 \end{matrix} \right]^+ Cl^- + CO_2 \qquad (7)$$

$$HCONR^1_2 + Et_3O^+BF_4^- \xrightarrow{20^\circ} \left[HC \begin{matrix} OEt \\ NR^1_2 \end{matrix} \right]^+ BF_4^- + Et_2O \qquad (8)$$

temperatures and therefore give only the O-alkylimidonium salts in good yield. Other work has shown that alcohols will react with formamide at high temperatures ($\sim 170^\circ$) to produce a mixture of O- and N-alkylated products as in the case of alkyl halides. Addition of catalytic amounts of mineral acid, however, leads to specific N-alkylation with yields of secondary amides in the region of 60–100%[4a].

5. Intramolecular alkylation

Examples of intramolecular rearrangements resulting in the alkylation of either the oxygen or the nitrogen atom of the amide moiety are known. An interesting point is that the experimental conditions seem to have an important bearing on the site of alkylation as with inter-molecular reactions. N-Substitution predominates in strongly alkaline conditions in which the anion of the amide is present (see section II.B), whereas in neutral solutions only the products of O-substitution are normally observed. For example, the thermal rearrangement of N-(bromoalkyl)amides (8) in water yields the oxazolines (9) via O-substitution rather than the corresponding aziridines (10) from cyclization at the nitrogen atom[11]. In this instance there is a possibility that steric factors favour formation of the larger (and therefore less strained) oxazoline product. This does not seem to be the overriding factor, however, for even when cyclic products of the same size

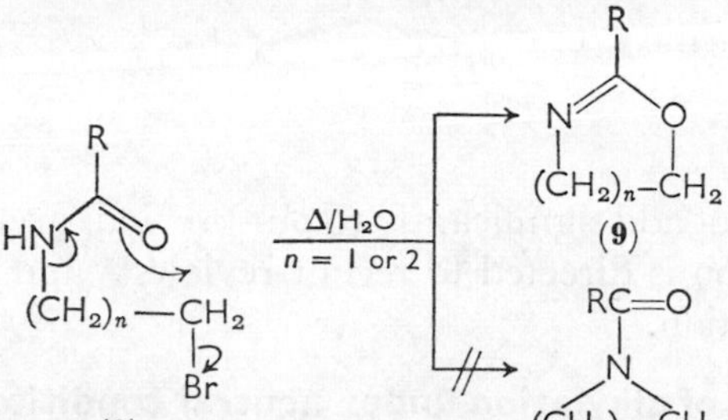

would result from either *O*- or *N*-substitution, the former pathway is still favoured under neutral conditions. Thus fusion of 4-bromo-*N*-cyclohexylbutyramide (**11**) by itself yields *only* the tetrahydrofuran **12**, whereas in the presence of KOH the pyrrolidone **13** is formed[12]. This result has a direct bearing on the mechanism of alkylation under neutral conditions (section II.A.6). We have noted earlier that *N*-alkylated products obtained from intermolecular reactions with alkyl halides at ∼170° probably arise from thermal rearrangement of an *O*-alkylimidonium precursor. For the intramolecular reaction with 4-bromo-*N*-cyclohexylbutyramide (**11**), a corresponding O to N rearrangement should be prohibited by the energetics of ring opening.

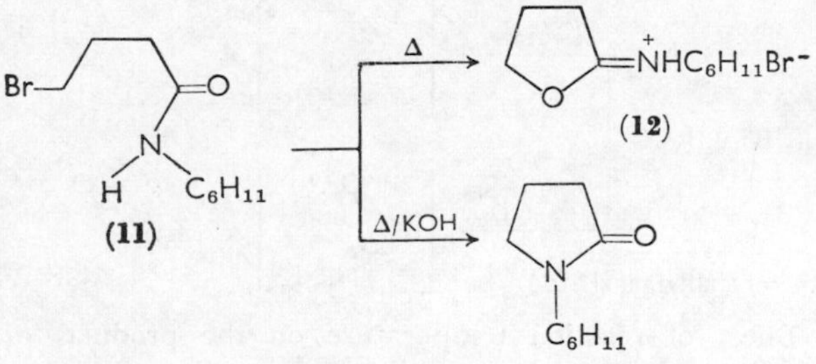

Other examples of amide-oxygen participation in solvolysis reactions (effectively *O*-alkylation) are numerous. An illustrative case is the enhanced ionization rate of *trans*-2-benzamidocyclohexyl *p*-toluene-sulphonate (**14**) relative to the *cis* isomer, and the oxazoline product (**14a**) is sufficiently stable to be isolated as the picrate derivative from feebly basic solutions[13]. Detailed discussion of these

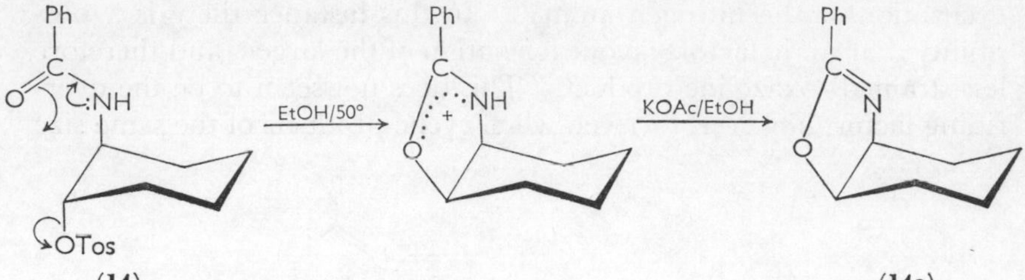

(14)   **(14a)**

reactions does not add significantly to our previous arguments, and the reader's attention is directed to recent reviews[14] and papers[13a,15] for further information.

6. Mechanisms of alkylation under neutral conditions

The ambident nucleophilic properties of amides are evident in alkylation reactions and it is not surprising that the most difficult mechanistic problem is associated with the site of substitution. We have already suggested that reaction conditions are of prime importance in this respect. This is a tentative hypothesis, however, that warrants further examination.

a. *Kinetic and thermodynamic products.* An overall appraisal of the experimental findings indicates that, with the exception of diazomethane, only O-alkylated products are associated with reactions conducted at 'low' temperatures, whereas both O- and N-substitution occur at 'high' temperatures ($> 60°$). The situation is summarized in Scheme 6. Thus all the highly reactive agents such as dimethyl

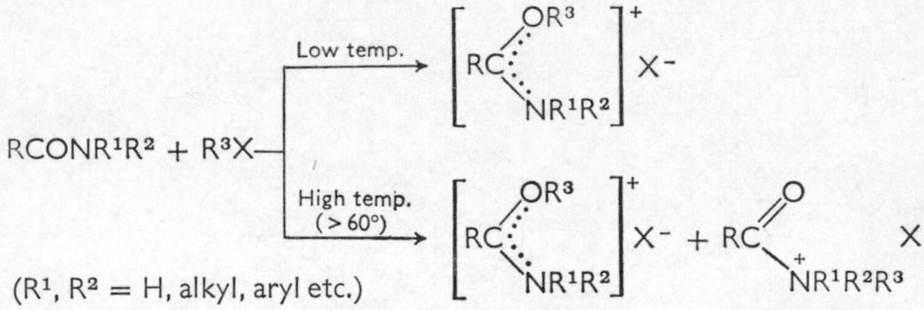

SCHEME 6. Effect of reaction temperature on the products of alkylation reactions.

sulphate[1], trityl chloride[4], ethyl chloroformate[10] and triethyloxonium fluoroborate[2] form only O-alkylated products at ambient temperatures, whereas trityl chloride at $100°$[4] and other alkyl halides[4-6] and alcohols[16] at even higher temperatures yield varying proportions of both O- and N-substituted derivatives.

This kind of temperature-dependent specificity clearly suggests that O-alkylimidonium salts arise from reactions carried out under kinetic control, but these may transform to the thermodynamically stable N-alkylamides at higher temperatures. Closer scrutiny of transition state and product structures (Scheme 7) qualitatively supports this assertion. Delocalization of the nitrogen lone-pair electrons should dissipate charge and lower the energy of the transition state (**15**) for

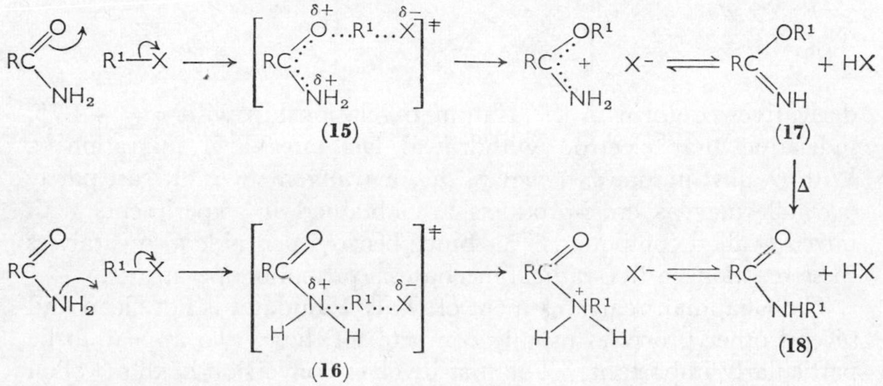

SCHEME 7. Transition states and products for O- and N-alkylation of primary amides.

O-substitution. No comparable effect is possible in the corresponding transition state for direct N-substitution (**16**); the induced positive charge is therefore localized on the nitrogen atom and the transition-state energy is accordingly higher. As far as the product stabilities are concerned, the structure of the amide is important. With both primary and secondary amides, the O-alkyl imidate (**17**) is more reactive (and presumably less stable) than the corresponding N-alkylamide (**18**). Furthermore, the rearrangement **17** → **18** is known to proceed readily on heating, as discussed in the next section. With tertiary amides, however, the corresponding O-alkylamidonium salt cannot be neutralized by proton loss, and a comparable O to N rearrangement will be favoured only if the displaced N-alkyl substituent forms a relatively stable carbonium ion.

b. *Rearrangement of O-alkyl imidates to N-alkylamides.* Either heating to about 180° or to lower temperatures in the presence of an alkylating reagent is known to induce alkyl migration in O-alkyl imidates. The purely thermal process, known as the Chapman rearrangement has been studied extensively[8,17], but even so the mechanism is not entirely understood. One important factor seems to be the structure of the migrating group, and reaction rates are usually faster when this is electron attracting*[8,17a]. With O-aryl imidates, the rearrangement is definitely *intra*molecular[17a] (equation 9) consistent with the formation of a tetrahedral intermediate stabilized by electron delocalization throughout the aromatic nucleus. The rates for *para*-substituted

$$
\begin{array}{ccc}
RC & \xrightarrow{\Delta} & \left[RC \right] \longrightarrow RC & \hspace{2cm} (9)
\end{array}
$$

derivatives conform to the Hammett relationship with $\rho = +1.75$, indicating that electron withdrawal facilitates aryl migration[17a]. With O-alkyl imidates, however, the rearrangement is at least partly *inter*molecular as cross-products are obtained in experiments with mixed O-alkyl compounds[17b]. Since benzoyl peroxide also catalyses these reactions, a free-radical mechanism probably operates[17b].

The Chapman rearrangement of O-alkyl imidates is not clean and several other processes usually compete: of these, two appear to be particularly important. The first involves conversion of the O-alkyl

$$
\begin{array}{ccc}
& \text{OMe} & & \text{NHPh} \\
HC & \xrightarrow{\Delta} & HC & \hspace{2cm} (10) \\
& \text{NPh} & & \text{NPh}
\end{array}
$$

formimidate to formamidine (equation 10)[17b]. The second is dehydration to the corresponding nitrile in the case of both O-alkyl and O-aryl imidates derived from primary amides[8] (equation 11). This is, of course, the reverse of the well-authenticated Pinner synthesis of

$$
\begin{array}{cc}
& \text{OR}^1 \\
RC & \xrightarrow[\text{HCl}]{\Delta} RCN + R^1OH \hspace{2cm} (11) \\
& \text{NH}
\end{array}
$$

imido esters[8]. We shall not engage in further discussion of these side-reactions, but clearly both influence the yield of N-alkylamides obtained with unreactive alkylating agents.

Excess alkylating reagent is known to catalyse the Chapman reaction. In the presence of alkyl iodide, for instance, rearrangement

* This is contrary, however, to the findings of Bredereck and his colleagues[1b,4a] for the reaction of alkyl halides with formamides (section II.A.2). On the limited evidence available, both sets of results are irreconcilable which implies the intervention of additional paths for the reactions with alkyl halides.

of the O-alkyl imidate occurs at temperatures as low as 100° [18]. The most convincing evidence, however, comes from studies by Benson and Cairns with dimethyl sulphate [1c]. They found that slow addition of an equimolar amount of dimethyl sulphate to caprolactam at 60° produces only the O-alkyl imidate (19), but with excess reagent a mixture of this and the N-methyl isomer (20) was obtained (Scheme 8). Furthermore, the conversion of 19 to 20, which normally requires temperatures in excess of 300°, occurs readily in the presence of dimethyl sulphate at 60°. Benson and Cairns suggested that dimethyl intermediates were involved in the rearrangement [1c], but a synchronous alkylation–dealkylation process as in 21 seems more likely.

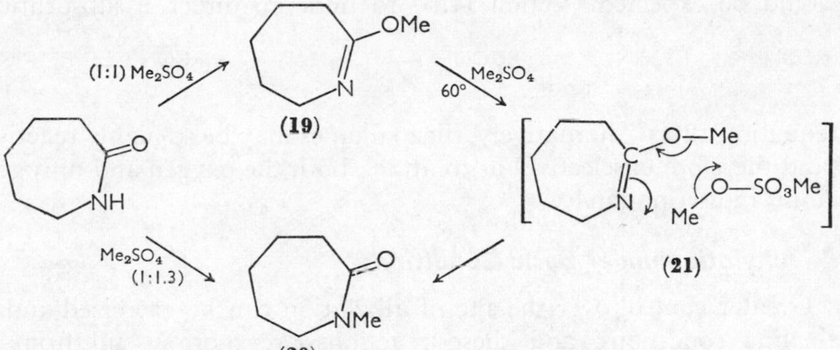

SCHEME 8. Reactions of methyl sulphate with caprolactam at 60°.

Both experimental evidence [19] and arguments against the formation of N-alkylamides via rearrangement of the O-alkyl imidates under the conditions of alkylation have been proffered from time to time, particularly in connexion with the influence of silver salts. Kornblum [7] has suggested that the tendency towards either O- or N-alkylation of ambident anions (including carboxamide anions) is related to the nature of the transition state. For reactions with 'S_N1-like' transition states, alkylation of the more electronegative oxygen atom is favoured, whereas greater S_N2 character in the transition state favours alkylation of the less electronegative nitrogen atom [7]. In this way silver salts are supposed to enhance the unimolecular nature of the reaction by polarization of the alkyl halide bond, thereby promoting alkylation at oxygen. The same arguments should also apply to reactions of the neutral amide. We do not discount this explanation, for it is probable that some of the N-alkylamide arises from direct substitution and supplements the yield obtained by rearrangement of the O-alkyl imidate. Deciphering the relative importance of each process, which must vary with the substrate, reagents and experimental conditions, awaits further investigation.

c. *Diazoalkanes.* We have already commented on the exceptional formation of N-alkylamides with diazoalkanes at low temperatures[3]. It seems unlikely that alkyl migration from oxygen to nitrogen occurs under these conditions, which implies that diazoalkanes react directly at the nitrogen atom. Two possible explanations may account for the phenomenon. One is that diazoalkanes may act as basic catalysts and abstract the N-proton. The resulting conjugate base of the amide would be expected (section II.B) to undergo direct N-substitution

$$RCONHR^1 + R^2CHN_2 \rightleftharpoons RCO\bar{N}R^1 + R^2CH_2N_2^+ \longrightarrow RCONR^1CH_2R^2 + N_2$$

$$(12)$$

(equation 12). Alternatively, diazoalkanes may be so highly reactive (and therefore unselective) as to attack both the oxygen and nitrogen atoms indiscriminately.

B. Alkylation under Basic Conditions

Greater control over the site of alkylation can be exercized under alkaline conditions, and these reactions are more useful from a synthetic standpoint. Primary and secondary amides in the presence of a strong base normally react at the nitrogen atom, with only small amounts of other products[20]. This selectivity can be associated with formation of the carboxamide anion (or at least an ion pair) in which alkylation of the nitrogen atom occurs directly (Scheme 9). O-

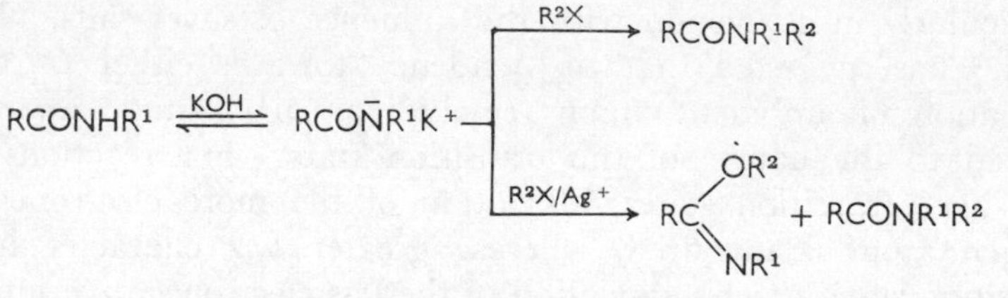

SCHEME 9. Alkylation of primary and secondary amides in alkaline media.

Alkylated products (O-alkyl imidates) are obtained, however, in the presence of silver salts[21], and in this respect the reactions are similar to those in neutral solutions. For compounds where additional structural features (e.g. $R = PhCH_2$) enhance the acidity of the $\alpha_{C=O}$ hydrogens, alkylation may also occur at this site via a carbanion intermediate[22].

For tertiary amides, of course, this is the only reaction of importance[23] (equation 13).

$$PhCH_2CONR_2^1 \underset{K^+}{\overset{KOH}{\rightleftharpoons}} Ph\overset{-}{C}HCONR_2^1 \xrightarrow{R^2X} Ph\overset{\overset{R^2}{|}}{C}HCONR_2^1 + KX \qquad (13)$$

I. N-Alkylation

Alkylation of primary and secondary amides with alkyl halides in the presence of such bases as sodium alkoxide, sodium hydride or sodamide is the most satisfactory way of synthesizing more highly substituted amides (Scheme 9). These reactions are well documented and details can be found elsewhere[12,20]. Clearly the carboxamide anions formed in basic solutions should be relatively powerful nucleophiles, but this property has not been exploited to any appreciable extent. Only recently has it been demonstrated that the amide nitrogen will readily attack a variety of electrophilic species. Several examples of Michael-type additions to activated olefins are now known[24] (equation 14), and for convenience the details are summarized

$$RCONHR^1 \overset{Base}{\rightleftharpoons} RCO\overset{-}{N}R^1 \xrightarrow{R^2CH=CR^3X} RCON\overset{\overset{R^1}{|}}{-}CH\overset{\overset{R^2}{|}}{-}\overset{\overset{R^3}{|}}{C}HX \qquad (14)$$

$$(X = COR, SO_2R, CN, \text{etc.})$$

in Table 6. Also the sodium salt of benzamide displaces both halogen and alkoxy groups from halogenated methyl ethers (equations 15a–c) to form[25] the bis-(benzamido)methane (**22**). Other work has shown that even epoxides suffer ring cleavage by secondary anilides in mild base to give several condensation products[26].

TABLE 6. Base-catalysed Michael addition to amide nitrogen (equation 14).

R	R^1	R^2	R^3	X	Base	Temp.	Solvent	Ref.
H	H	H	H	COR	Na	90°	$HCONH_2$	24a[a]
Pyrrolidone		Cl	H	CN	NaH	25°	C_6H_6/DMF	24b
RCH=CR	H	H	H	SO_2R	NaH/NaOR	~75°	Dioxan/THF	24c
Heterocyclic amides		H	Alkyl	CN $CONR_2$ CO_2R	NaOMe	~50°	MeOH	24d

[a] Only for this case were yields reported (20–50%).

One of the most interesting developments is the application of base-catalysed intramolecular alkylation to the synthesis of lactams and other heterocyclic nitrogen compounds. We have already referred to

$$PhCONH_2 \underset{}{\overset{NaOH}{\rightleftharpoons}} PhCON\bar{H} \xrightarrow{XCH_2OR} [PhCONHCH_2OR] + X^- \quad (15a)$$

$$PhCON\bar{H} + PhCONHCH_2OR \longrightarrow (PhCONH)_2CH_2 + OR^- \quad (15b)$$

$$(22)$$

this in connexion with the thermal cyclization of 4-bromo-N-cyclohexylbutyramide, where the tetrahydrofuran derivative **23a** is formed under neutral conditions in contrast to the pyrrolidone **23b** formed with fused KOH[12]. Related investigations confirm that nucleophilic participation by the nitrogen atom is normally favoured

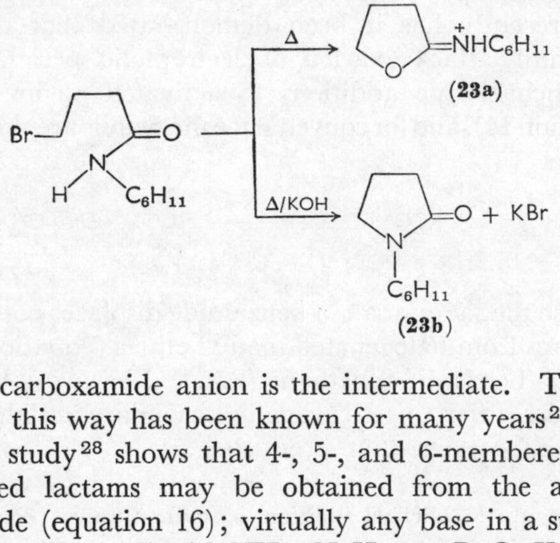

when the carboxamide anion is the intermediate. The synthesis of lactams in this way has been known for many years[27] and a recent systematic study[28] shows that 4-, 5-, and 6-membered, but not the 7-membered lactams may be obtained from the appropriate ω-bromoamide (equation 16); virtually any base in a suitable aprotic solvent (e.g. Na in liquid NH_3, NaH or t-BuO$^-$K$^+$ in dimethyl

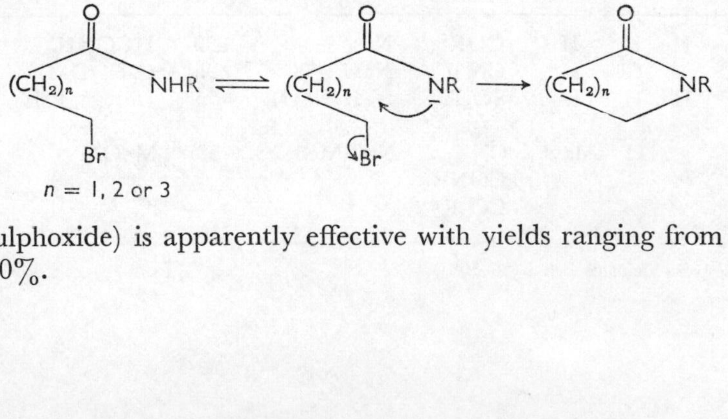

$$(16)$$

$$n = 1, 2 \text{ or } 3$$

sulphoxide) is apparently effective with yields ranging from 50% to 90%.

Synthesis of the corresponding 3-membered lactams (aziridones) by treating α-bromoamides with either NaH or t-BuO$^-$K$^+$ has also been reported by several workers[29]. These very reactive compounds can only be isolated at low temperatures and, although their structure has long been disputed, recent evidence confirms the lactam rather than the isomeric epoxide configuration.

Cyclization of the N-substituent itself can also be achieved in alkaline solutions. Stirling[30], for example, has studied this transformation for a series of N-substituted benzamides with various ω leaving groups and chain structures (Scheme 10). In dilute sodium

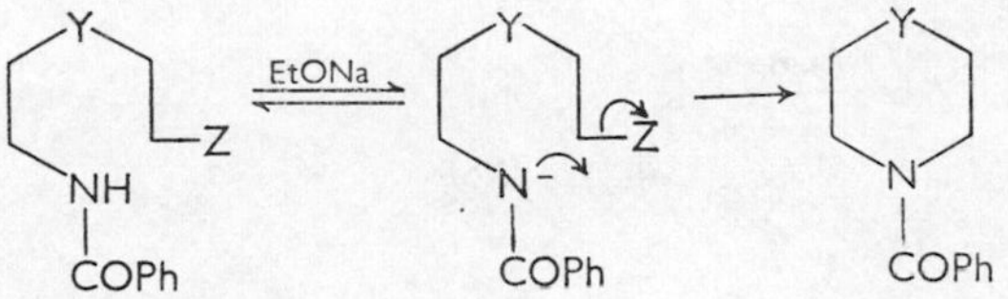

Y = O, CH$_2$N—tosyl, S, SO, SO$_2$
Z = O—tosyl, I, Cl

SCHEME 10. Cyclization of N-substituted benzamides.

ethoxide in ethanol cyclic products 24 arise only from N-substitution, although intermolecular substitution by ethoxide ion and elimination compete. The extent of cyclization, however, is insensitive to the leaving group but depends on the acidity of the N-proton: this indicates that the carboxamide anion is the reactive species.

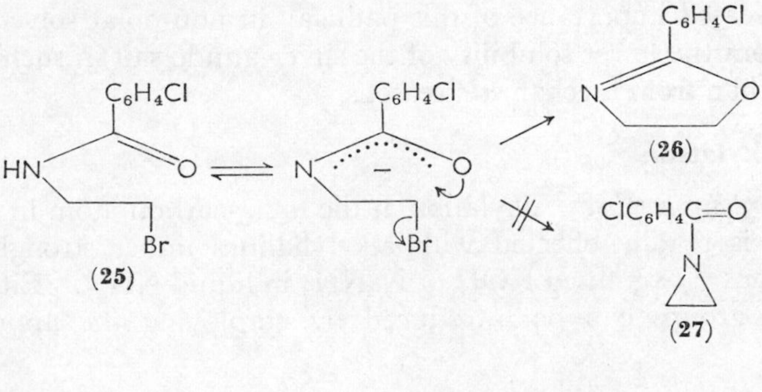

Intramolecular O-alkylation is rarely observed under alkaline conditions and then only when steric factors (i.e. much larger ring size) are particularly favourable. For example, N-(2-bromoethyl)-4-chlorobenzamide (25) on reaction with sodium methoxide in methanol yields the oxazoline 26 rather than the aziridine 27[31]. With trans-2-benzamidocyclohexyl p-toluenesulphonate under closely similar conditions, however, a mixture of the corresponding aziridine and oxazoline are obtained[32].

2. O-Alkylation

The preference for N-alkylamide formation in alkaline solutions is drastically altered by the addition of silver salts to the reaction mixture, and appreciable amounts of O-substituted products are then obtained[21]. This interesting phenomenon, which has been known for a long time, is often used to effect the synthesis of O-alkyl imidates. For example, treatment of the silver salt of N-phenylformamide (28)

$$\text{HCONHPh} \underset{}{\overset{\text{NaOH/AgNO}_3}{\rightleftharpoons}} \text{HCON}^-\text{Ph Ag}^+ \xrightarrow{\text{EtI}} \text{HC} \begin{matrix} \text{OEt} \\ \diagup \\ \diagdown \\ \text{NPh} \end{matrix}$$

$$(28) \hspace{5cm} (29)$$

with ethyl iodide in solvent petrol produces 74% of the corresponding O-ethyl imidate 29. Usually, however, a mixture of O- and N-alkylated products is formed unless the alkylation is carried out in non-polar solvents[21a].

The reason why silver salts promote O-alkylation has not been investigated thoroughly and no entirely satisfactory explanation is available. Since these reactions are conducted at ambient temperatures, it seems unlikely that O to N rearrangements of the kind encountered in neutral solutions are important. As discussed previously in section II.A.6.b, Kornblum[7] has suggested that in common with the alkylation of other ambident anions, silver salts may enhance the unimolecular character of the reaction with alkyl halides, thereby promoting alkylation at the more electronegative oxygen atom. Recent studies[21b] with 2-pyridones, however, suggest that the situation is more complicated. In particular, the heterogeneous nature of the silver salt reactions appears to favour O-alkylation, and the increased importance of this pathway in non-polar solvents may arise from the lower solubility of the silver amide salt in such media, rather than from mechanistic reasons.

3. C-Alkylation

a. *Tertiary amides.* Alkylation at the $\alpha_{C=O}$-carbon atom in tertiary amides is readily effected with alkyl halides under strongly basic conditions[23] (e.g. fused KOH or $NaNH_2$ in liquid NH_3). Either one or two groups can be introduced by employing the appropriate quantity of base and reagent (equation 17). Until recently it was generally assumed that only activated substrates such as phenyl acetamides ($PhCH_2CONR_2$) would react. This is now known to be

$$PhCH_2CONR_2 \xrightarrow{R^1X/KOH} PhCHR^1CONR_2 \xrightarrow{R^1X/KOH} PhCR^1_2CONR_2 \quad (17)$$

untrue and the reaction has wider applicability. By using sodamide in either benzene[23b] or liquid ammonia[23a] as the base, even tertiary acetamides, propionamides and butyramides undergo mono- and dialkylation at the $\alpha_{C=O}$-carbon atom in reasonable yields.

b. *Primary and secondary amides.* With compounds of this type that form relatively stable carbanions, alkylation of the $\alpha_{C=O}$-carbon atom competes with N-alkylation and may in some cases become the preferred reaction. The usual procedure is to use two molar equivalents of sodamide in liquid ammonia to produce the dianion **30**, which then undergoes either mono- or disubstitution according to the amount of alkyl halide added[22] (equations 17a, 17b). With just one equivalent

$$PhCH_2CONH_2 \underset{\text{Liq. }NH_3}{\overset{2\,NaNH_2/}{\rightleftharpoons}} Ph\bar{C}HCO\bar{N}H \begin{array}{l} \xrightarrow{RX} PhCHRCONH_2 \quad (17a) \\ \\ \xrightarrow{2\,RX} PhCHRCONHR \quad (17b) \end{array}$$

$$(30)$$

of alkylating reagent the C-substituted derivative is usually obtained indicating that the carbanion centre is the most nucleophilic site. It is therefore not surprising to find that Michael addition reactions take place with activated olefins[33]. In these cases, sodium hydride is used to generate the dianion intermediate, as in the reaction of phenyl-acetamides with ethyl cinnamate[33a] (equation 18).

$$PhCH_2CONHR \overset{2\,NaH}{\rightleftharpoons} Ph\bar{C}HCO\bar{N}R \xrightarrow{PhCH=CHCO_2Et} \begin{array}{l} PhCHCH_2CO_2Et \\ | \\ PhCHCONHR \end{array} \quad (18)$$

C. Alkylation under Acidic Conditions

Most alkylamides (and arylamides to a lesser extent) are sufficiently basic to exist mainly as their conjugate acids in even dilute (~ 0.5 N) acid solutions. Accordingly alkylation under these conditions is usually very difficult. One way, however, in which an acid may catalyse these reactions is by protonation of the alkylating agent and several examples of this effect are known. Thus alkylation by alcohols[1b], acetals[34], ethyl orthoformate[35], t-butyl acetate[36] and vinyl ethers[37] are all facilitated by trace amounts of mineral acid. An important observation is that only products arising from N-substitution are usually obtained, even with reactions at low temperature. One possible explanation, albeit speculative, is that hydrogen bonding

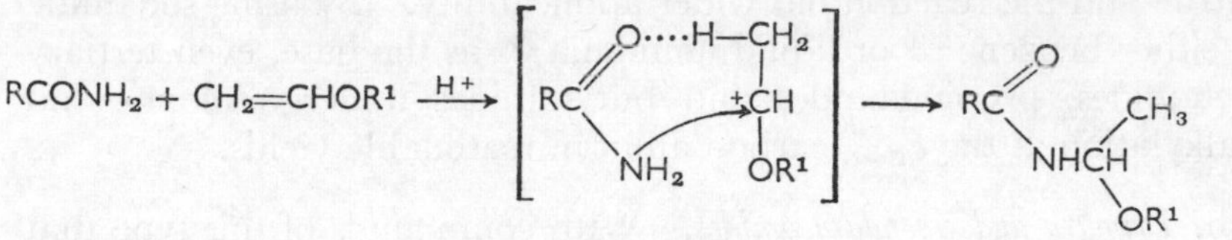

$$RCONH_2 + CH_2\!=\!CHOR^1 \xrightarrow{H^+} \left[\begin{matrix} O\cdots H\!-\!CH_2 \\ RC \qquad \overset{+}{\underset{OR^1}{C}H} \\ NH_2 \end{matrix} \right] \longrightarrow RC \overset{O}{\underset{\underset{OR^1}{NHCH}}{\diagdown}} CH_3 \quad (19)$$

between the protonated reagent and the carbonyl oxygen directs attack by the carbonium ion centre towards the nitrogen atom. This is illustrated for reaction by vinyl ether in equation (19).

D. Homolytic Alkylation

Direct O- or N-substitution of amides by alkyl radicals is not known, but alkylation can be effected by homolytic addition to olefinic compounds yielding only C-alkylated products (see also Chapter 5). Formamide, for example, reacts on initiation with ultraviolet[38a] and electron irradiation[38b], or in the presence of t-butyl peroxide[38c], to

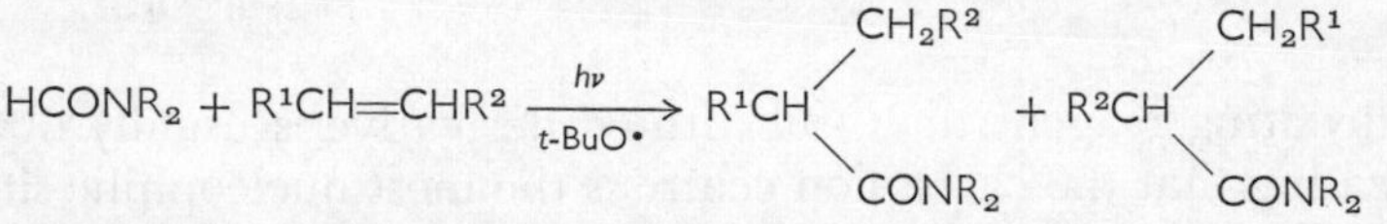

$$HCONR_2 + R^1CH\!=\!CHR^2 \xrightarrow[t\text{-BuO}\cdot]{h\nu} R^1CH \overset{CH_2R^2}{\underset{CONR_2}{\diagdown}} + R^2CH \overset{CH_2R^1}{\underset{CONR_2}{\diagdown}} \quad (20)$$

form mixed alkylamides (equation 20). With more highly substituted amides, such as N-methylacetamide, the products are consistent with the formation of $\dot{C}H_2CONHCH_3$ and $CH_3CONH\dot{C}H_2$ radical intermediates, which then add across the olefinic double bond (equation 21)[38d].

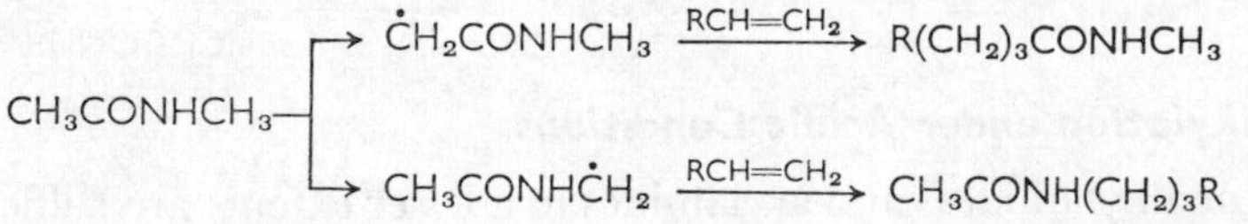

$$CH_3CONHCH_3 \left\{ \begin{matrix} \dot{C}H_2CONHCH_3 \xrightarrow{RCH=CH_2} R(CH_2)_3CONHCH_3 \\ \\ CH_3CONH\dot{C}H_2 \xrightarrow{RCH=CH_2} CH_3CONH(CH_2)_3R \end{matrix} \right. \quad (21)$$

III. REACTIONS WITH ALDEHYDES AND KETONES

In common with other nucleophiles, primary and secondary amides add to the carbonyl group of aldehydes and ketones. The initial product is usually the N-acylcarbinolamine $(RCONHC(R^1)_2OH)$

resulting from substitution by the amide nitrogen; unlike alkylation, no O-substituted products are obtained.   The carbinolamine derivative is stable in neutral and mildly basic solutions, but in the presence of acid, dehydration and further coupling occurs.   In this and many other respects, these reactions are similar to the carbonyl addition reactions of other nucleophilic species such as amines, alcohols, etc. Thus the addition of amides is both reversible and catalysed by acids and bases.   The catalysis is specially important because of the weak nucleophilic properties of amides.   For this reason, too, only activated carbonyl compounds react and even then relatively high temperatures are required.

Under neutral and mildly basic conditions, aldehydes, particularly formaldehyde[39] and those containing electron-withdrawing substituents such as chloral[40], combine with amides containing an N—H to produce the carbinolamine derivative **31** in good yield (equation 22).

$$RCONHR^1 + R^2CHO \xrightleftharpoons{\sim 150°} RCONR^1CHR^2OH \qquad (22)$$

$$(\mathbf{31})$$

High temperatures (100°–150°) are necessary to induce reaction[39,40], and this may explain the absence of any O-substituted products: O-alkyl imidates bearing electron-withdrawing substituents (such as OH) are known to rearrange rapidly at these temperatures to the corresponding N-alkylamide (section II.A.6.b).   Ketones, generally, are less reactive than aldehydes, but several examples of carbonyl addition are known[41,42].   For instance, both hexafluoroacetone and *sym*-dichlorotetrafluoroacetone condense with primary amides at 50°

$$RCONH_2 + (CF_2X)_2CO \xrightleftharpoons{50°} RCONHC(CF_2X)_2OH \qquad (23)$$

$$(X = F, Cl)$$

(equation 23)[41].   Other α-substituted ketones such as acyloins and α-aminoalkyl ketones react readily with formamide: the initial product is probably the corresponding carbinolamine derivative, but

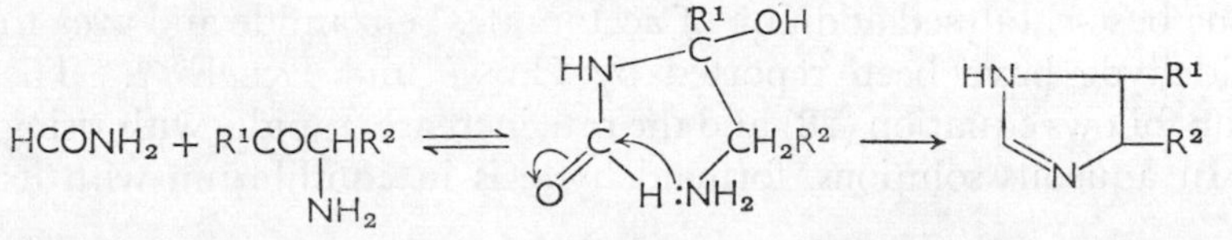

$$(24)$$

subsequent condensations result in the formation of imidazoles (equation 24)[42].

A. Hydroxymethylation

The most intensive studies have been concerned with the addition of formaldehyde[39] (and to a lesser extent of glyoxal)[43] to various amides. This process is usually referred to as 'hydroxymethylation'. The addition to formaldehyde is reversible (equation 25), but the N-

$$\text{RCONH}_2 + \text{HCHO} \rightleftharpoons \text{RCONHCH}_2\text{OH} \xrightarrow[\text{(MgO)}]{\text{HCHO}} \text{RCON(CH}_2\text{OH)}_2 \quad (25)$$
$$\textbf{(32)} \qquad\qquad\qquad \textbf{(33)}$$

hydroxymethylamide **32** is stable and can be isolated in good yield from neutral or mildly alkaline solutions. Primary amides react further to form the bis-hydroxymethylated derivative **33** on heating with excess formaldehyde in the presence of MgO catalyst (equation 25)[44].

N-Hydroxymethylamides (**32**) are useful intermediates that condense readily with compounds bearing labile hydrogen (equation 26;

$$\text{RCONH}_2 + \text{HCHO} \rightleftharpoons \text{RCONHCH}_2\text{OH} \xrightarrow{\text{XH}} \text{RCONHCH}_2\text{X} + \text{H}_2\text{O} \quad (26)$$

$\text{XH} = \text{HCR}_3, \text{H}_2\text{C(COR)}_2$, etc.)[45]. The best known example is the Einhorn reaction between amide, formaldehyde and amine ($\text{XH} = \text{R}^1\text{NH}_2$)[46], but in this case intermediacy of the N-hydroxymethylamide seems unlikely. The reaction is usually accomplished by heating the three reagents together at about 70°, which is much below that required for N-hydroxymethylamide formation[46]. This, and other evidence[46e], suggests the Einhorn reaction proceeds via the N-hydroxymethyl*amine* **34** instead, and is therefore a Mannich reaction (equation 27)[47]. For further information on the synthetic applica-

$$\text{R}^1\text{NH}_2 + \text{HCHO} \rightleftharpoons \text{R}^1\text{NHCH}_2\text{OH} \xrightarrow{\text{RCONH}_2} \text{R}^1\text{NHCH}_2\text{NHCOR} \quad (27)$$
$$\textbf{(34)}$$

tions of amidomethylation, the reader is referred to a recent review[45].

Nearly all the mechanistic information about the addition of amides to aldehydes and ketones comes from studies of hydroxymethylation. Like other carbonyl additions, this process is both reversible[39e] and catalysed by acids[48] and bases[39,43a]. Kinetic studies of the aqueous base-catalysed addition of acetamide, benzamide and urea to formaldehyde have been reported by Crowe and Lynch[39d]. The reaction follows equation (28) and the rate increases rapidly with rising pH. In aqueous solutions, formaldehyde is in equilibrium with its

$$\text{Rate} = k_2[\text{Amide}][\text{HCHO}] \quad (28)$$

hydrated form. In going from pH 8·6 to 12·7 the concentration of unhydrated formaldehyde increases by a factor of about 30, whereas the rate of hydroxymethylation of acetamide, for example, increases by a factor of 3000[39d]. This suggests, but does not prove, that at least part of the base catalysis is associated with formation of the carboxamide anion, which then reacts with the dehydrated formaldehyde (equation 29). This, in turn, may account for the incidence of N-substitution by these amides in alkaline solutions even at low temperatures. A

$$RCONH_2 \xrightleftharpoons{OH^-} RCO\bar{N}H$$
$$+ \rightleftharpoons RCONHCH_2OH \qquad (29)$$
$$HCHO(H_2O) \xrightleftharpoons{OH^-} HCHO$$

similar effect is observed in the alkylation of carboxamide anions (section II.B.1).

The acid-catalysed addition to formaldehyde has been investigated for a series of alkylamides and substituted benzamides[48]. These reaction rates also follow equation (28), and in addition there is an approximate first-order dependence on the hydronium ion concentration over a limited pH range[48]. Substituent effects for the benzamides and alkylamides correlate with Hammett ($\rho = -1\cdot1$) and Taft parameters ($\rho^* = -2\cdot16$), respectively[48], and the ρ values indicate that electron supply increases the reaction rate. The conclusion from these data is that the rate-controlling step involves attack by the protonated formaldehyde on the neutral amide (equations 30a,b)[48]. However, an alternative mechanism involving the protonated amide

$$H_2C{=}O + H^+ \xrightleftharpoons{Fast} H_2C{=}\overset{+}{O}{-}H \quad \text{(preequilibrium)} \qquad (30a)$$

$$H_2C{=}\overset{+}{O}{-}H + RCONH_2 \xrightarrow{Slow} \left[\begin{array}{c} O{\cdots}H{-}O \\ RC \qquad \qquad C^+ \\ N{:} \quad H \quad H \\ H_2 \end{array} \right]^{\ddagger} \longrightarrow RCONHCH_2OH \qquad (30b)$$

and the *neutral* formaldehyde is not excluded by these results. In either case, direct N-substitution would be expected from the findings for alkylation in acidic solutions (section II.C).

B. Strongly Acidic Conditions

Apart from increasing the rate of carbonyl addition, acids often facilitate dehydration of the carbinolamine derivative[35] which results in the formation of other products (Scheme 11). With primary amides, further coupling with the amide is promoted to give the alkylidene or arylidene bis-amide (36)[39e,49]. For aliphatic aldehydes carrying at least one α-hydrogen atom, elimination of water produces an N-vinylamide (or enamide) (37)[50]. In some cases, as with reaction between isobutyraldehyde and phenylacetamide or benzamide[51], and between acetaldehyde and lactams[52], N-vinylamide and bis-amide formation compete and mixed products are obtained. This suggests that dehydration of the initial carbinolamine derivative 35 produces a hybridized ionic intermediate (38), which can either lose a proton or react with amide.

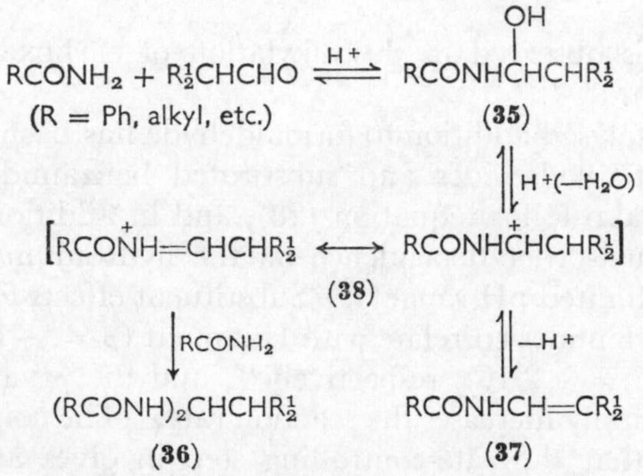

$$
\begin{array}{c}
\text{OH} \\
| \\
RCONH_2 + R_2^1CHCHO \underset{}{\overset{H^+}{\rightleftharpoons}} RCONHCHCHR_2^1 \\
(R = Ph, \text{ alkyl, etc.}) \qquad\qquad (35)
\end{array}
$$

$$ \Big\downarrow\Big\uparrow H^+(-H_2O) $$

$$ \left[RCO\overset{+}{N}H{=}CHCHR_2^1 \longleftrightarrow RCON\overset{+}{H}CHCHR_2^1 \right] $$

$$ (38) $$

$$ \downarrow RCONH_2 \qquad\qquad \Big\downarrow\Big\uparrow {-}H^+ $$

$$ (RCONH)_2CHCHR_2^1 \qquad RCONHCH{=}CR_2^1 $$

$$ (36) \qquad\qquad\qquad (37) $$

SCHEME 11. Reactions of amides and aldehydes under strongly acidic conditions.

The reaction of ketones with amides under acidic conditions is rare. The single example reported thus far, cyclohexanone with phenylacetamide, yields an N-vinylamide via an addition–elimination sequence similar to that of Scheme 11 (equation 31)[53].

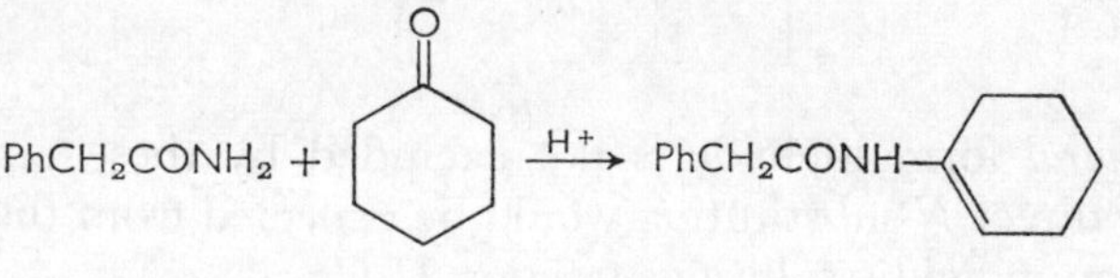

$$ PhCH_2CONH_2 + \text{(cyclohexanone)} \xrightarrow{H^+} PhCH_2CONH{-}\text{(cyclohexenyl)} \qquad (31) $$

C. Strongly Basic Conditions

Reaction in strongly basic conditions (e.g. with sodamide or sodium methoxide) has not been examined in detail, but the few known examples show clearly that substitution of the amide $\alpha_{C=O}$-carbon atom may compete with N-substitution. This can be related, of course, to carbanion formation, and in this sense the reactions with aldehydes and ketones are analogous to ordinary alkylation under similar conditions.

Phenylacetamides, for example, on treatment with an equimolar amount of either benzaldehyde[54] or benzophenone[22a] in the presence of sodamide in liquid ammonia react preferentially at the $\alpha_{C=O}$-carbon atom, presumably via a dianion intermediate (equations 32a,b).

$$PhCH_2CONHR \rightleftharpoons Ph\bar{C}HCO\bar{N}R \qquad (32a)$$

$$Ph\bar{C}HCO\bar{N}R + PhCHO \longrightarrow PhCH\begin{smallmatrix} Ph \\ | \\ CHO^- \\ \diagup \\ \diagdown \\ CO\bar{N}R \end{smallmatrix} \xrightarrow[\text{(NH}_4\text{Cl)}]{2H^+} PhCH\begin{smallmatrix} Ph \\ | \\ CHOH \\ \diagup \\ \diagdown \\ CONHR \end{smallmatrix} \qquad (32b)$$

Unactivated amides, however, such as acrylamides or isobutyramide, react only at the nitrogen atom with formaldehyde under similar conditions (equation 33)[49a,55]. The reaction is reversible, of course,

$$RCONH_2 + HCHO \underset{\text{(CCl}_4)}{\overset{\text{Na or NaOMe}}{\rightleftharpoons}} RCONHCH_2OH \qquad (33)$$

and it is necessary to precipitate the product (in this case by using CCl_4 solvent) in order to effect a high yield.

IV. ACYLATION

Primary and secondary amides undergo N-substitution by powerful acylating agents such as organic acyl chlorides and anhydrides; esters are less effective and carboxylic acids react in a different way resulting in addition to the amide carbonyl function (section X). With primary amides, dehydration to the nitrile invariably competes with substitution, and tertiary amides form only salt-like addition complexes. The usual products derived from each type of amide by esters, anhydrides and acyl halides are summarized in Scheme 12.

Under neutral conditions all these products seem to arise from either rearrangement or further substitution of a precursor that is a mixed anhydride of a carboximidic and a carboxylic acid (equation 34).

$$RCONHR^1 + R^2COX \rightleftharpoons RC\begin{smallmatrix} OCOR^2 \\ \diagup \\ \diagdown \\ NR^1 \end{smallmatrix} \begin{array}{l} \xrightarrow{(R^1=H)} RCN + R^2CO_2H \\ \xrightarrow{(R^1=H) + R^2COX} RCON(COR^2)_2 \qquad (34) \\ \longrightarrow RCONR^1COR^2 \end{array}$$

$(R^1 = H, \text{alkyl, etc.})$

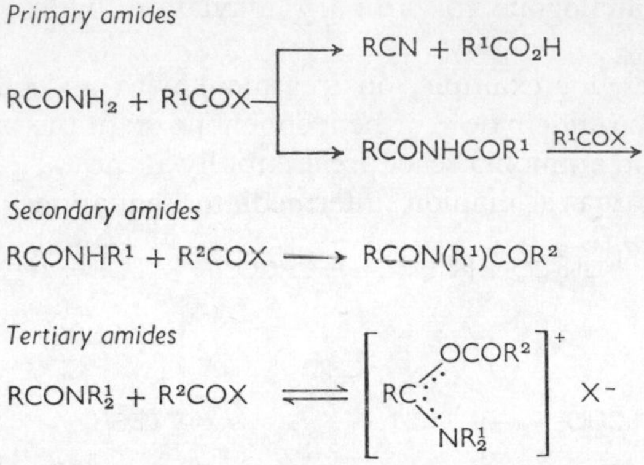

Primary amides

$$RCONH_2 + R^1COX \Biggl[\begin{array}{l} \longrightarrow RCN + R^1CO_2H \\ \\ \longrightarrow RCONHCOR^1 \xrightarrow{R^1COX} RCON(COR^1)_2 \end{array}$$

Secondary amides

$$RCONHR^1 + R^2COX \longrightarrow RCON(R^1)COR^2$$

Tertiary amides

$$RCONR_2^1 + R^2COX \;\rightleftharpoons\; \left[RC \begin{array}{c} OCOR^2 \\ \\ NR_2^1 \end{array} \right]^+ X^-$$

SCHEME 12. Normal products from the acylation of amides.

These reactions therefore resemble alkylation (section II), except that the intermediate mixed anhydrides are usually too unstable to be isolated. The analogy goes even further in that substitution by acyl halides and anhydrides under acidic or alkaline conditions probably involves direct substitution of the nitrogen atom.

Other, more specialized, reagents such as isocyanates, ketenes and oxalyl chloride also combine with amides, although the products are different on account of specific transformations subsequent to the initial acylation. However, circumstantial evidence points to *O*-acylation as the *initial* process under neutral conditions.

A. Acyl Halides

Reagents of this type react with most amides at ambient or even lower temperatures[56-58]. The products are those listed in Scheme 12. Substitution of the nitrogen atom (i.e. di- and triacylamine formation) is favoured in the presence of base and worthwhile yields may be obtained in this way. However, a number of other factors are also important, the relevance of which is best appreciated from considerations of the reaction mechanism.

Direct evidence for the formation of *O*-acylated intermediates comes from studies of tertiary amides[57]. These combine with acyl halides to form a salt-like 1:1 addition complex which can be isolated at low temperatures[57]. Various pieces of indirect evidence point to a mixed anhydride salt structure (**39**) for this adduct. For example, the complexes **40** obtained by condensation of the imidoyl chloride **41** with acetate ion, and of dimethylformamide with acetyl chloride, are identical (equation 35)[59], and it is also known that benzoyl chloride

$$RCONR^1_2 + R^2COCl \rightleftharpoons \left[RC\begin{subarray}{l} OCOR^2 \\ \\ NR^1_2 \end{subarray} \right]^+ Cl^-$$

(39)

does not complex with tertiary amino compounds lacking a carbonyl oxygen atom (e.g. $MeSO_2NMe_2$, $Et_2NC\equiv N$, Me_2NNO)[57]. More

$$HCON(CH_3)_2 \xrightarrow{CH_3COCl} \left[HC\begin{subarray}{l} OCOCH_3 \\ \\ N(CH_3)_2 \end{subarray} \right]^+ Cl^- \xleftarrow{CH_3CO_2^-} \left[HC\begin{subarray}{l} Cl \\ \\ N(CH_3)_2 \end{subarray} \right]^+ Cl^-$$

(40)   **(41)**   (35)

convincing evidence is that decomposition of the ionic complex derived from dimethylformamide and benzoyl bromide with water results in the production of benzoic but not formic acid; likewise, decomposition with aniline results in the formation of benzanilide but not formanilide (Scheme 13)[57]. This is only consistent with an *O*-acyl structure (**42**) for the complex and not with the corresponding *N*-acyl structure (**43**). The actual mechanism of complex formation

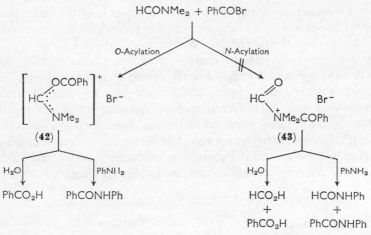

SCHEME 13. Decomposition products from a tertiary amide–acyl halide complex.

with tertiary amides has not been investigated, but it is known that the reactivity of acyl halides towards dimethylformamide decreases in the order $MeCOCl > PhCOCl > PhSO_2Cl > EtOCOCl$[57]: this

25*

suggests that steric factors are important and presumably the reaction is of the S_N2 type.

We have already noted that the reactions of primary amides are complicated by the formation of three products—both di- and tri-acylamines plus the nitrile from dehydration. The proportion of each product depends on the acyl halide and the experimental conditions, and the influence of these factors is summarized by Table 7. Bearing

TABLE 7. Influence of reagent, temperature and catalyst on products of acyl halides with primary amides[56].

Reaction variable	Major product		
	RCN	$(RCO)_3N$	$(RCO)_2NH$
Reagent	Strong (e.g. p-$NO_2C_6H_4COCl$)	Moderate (e.g. p-ClC_6H_4COCl)	Weak (e.g. C_3H_7COCl)
Temperature	$>0°$	$-60°$	$>0°$
Catalyst		Pyridine	Pyridine

in mind the capriciousness of generalizations, it is evident from Table 7 that N-acyl derivatives are best obtained in the presence of a base catalyst (usually pyridine) with increasing yields of the triacylamine at low temperatures[56]. The reason for these effects is most easily understood once the mechanism has been established.

By analogy with tertiary amides, it seems probable that mixed anhydrides are also formed intially from primary amides. This species has never been isolated, and it must rearrange rapidly even at low temperatures. Evidence for its participation, however, comes from studies of both nitrile and triacylamine formation. Titherly and Holden[60] showed many years ago that acetamide reacts with benzoyl chloride to form acetonitrile, whereas benzamide and acetyl chloride yield only N-acetylbenzamide (Scheme 14). This clearly rules out

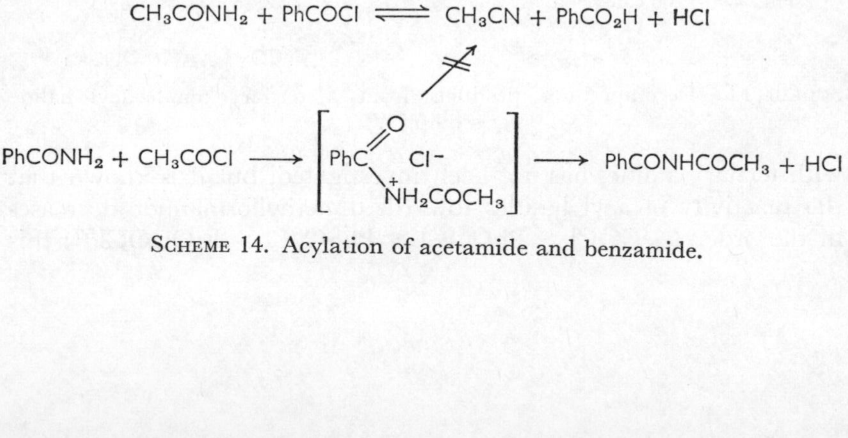

$$CH_3CONH_2 + PhCOCl \rightleftharpoons CH_3CN + PhCO_2H + HCl$$

$$PhCONH_2 + CH_3COCl \longrightarrow \left[PhC\underset{\overset{+}{NH_2COCH_3}}{\overset{O}{\diagup}} Cl^- \right] \longrightarrow PhCONHCOCH_3 + HCl$$

SCHEME 14. Acylation of acetamide and benzamide.

the *N*-acylamide as an intermediate in dehydration, but is consistent with a mixed anhydride intermediate.   Furthermore, it is known that the dehydration reaction is reversible—under acid catalysed conditions, nitriles react with carboxylic acids as shown in equation (36)[61].

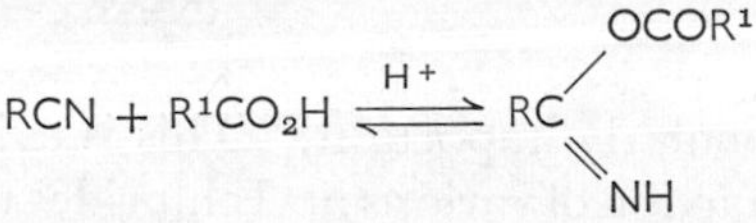

$$RCN + R^1CO_2H \underset{}{\overset{H^+}{\rightleftharpoons}} RC \overset{OCOR^1}{\underset{NH}{<}} \tag{36}$$

The importance of mixed anhydride intermediates in the substitution of primary amides is suggested by an interesting observation of Thompson's on the benzoylation of benzamides[56].   Under identical experimental conditions that favoured substitution at the expense of dehydration (excess benzoyl chloride with pyridine catalyst in chloroform at low temperature), he noticed that tribenzamide (**44**) was formed more slowly from dibenzamide (**45**) than from benzamide itself.   This seems to eliminate **45** as a viable intermediate in the formation of tribenzamide, but favours a mechanism involving a rapid second substitution of the mixed anhydride **46** followed by an O to N rearrangement.   This situation is represented by Scheme 15 where $k_1 > k_2 \gg k_3$.   Unfortunately, detailed kinetic studies to establish the relative magnitudes of the rate coefficients have not been undertaken.

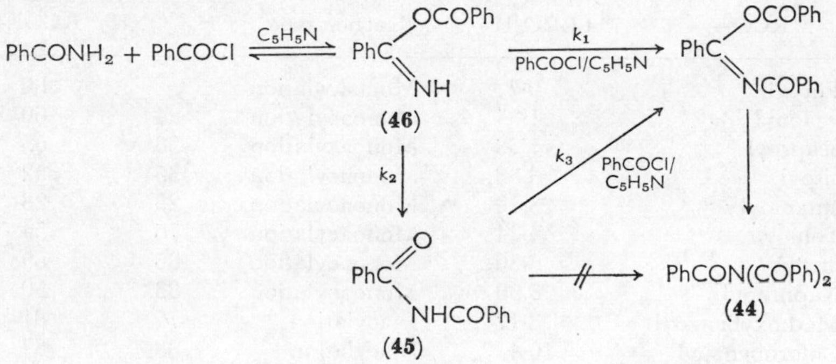

SCHEME 15.   Reaction of benzoyl chloride with benzamide.

From the available evidence, it seems likely that all three products with primary amides arise from either rearrangement or further substitution of a mixed anhydride formed in a rapid preequilibrium step.   The situation is summarized by equation (37).   In the light of this conclusion, we are in a better position to understand (and predict) the factors influencing product orientation.   Of these, reactivity of

$$\text{RCONH}_2 + \text{R}^1\text{COX} \rightleftharpoons \text{RC} \underset{\text{NH}}{\overset{\text{OCOR}^1}{<}} \longrightarrow \begin{cases} \longrightarrow \text{RCN} + \text{R}^1\text{CO}_2\text{H} \\ \longrightarrow \text{RCONH(COR}^1) \\ \xrightarrow{\text{R}^1\text{COX}} \text{RCON(COR}^1)_2 \end{cases} \quad (37)$$

the reagent is particularly important. This is evident from the data in Table 8 for the reaction of various acyl chlorides with both acetamide and the amide derivative of the acyl chloride: clearly, nitriles, triacylamines and diacylamines are favoured by strong, moderate and weak acyl chlorides, respectively, where the reactivity of the reagent is linked to the acidity of the parent carboxylic acid. The tendency towards nitrile formation with the most powerful reagents may arise from the enhanced stability of the expelled carboxylate ion, and disubstitution should be favoured with the more reactive agents whenever dehydration is not the dominant path. In addition, both temperature and pyridine catalyst concentration, affect the proportion

TABLE 8. Product variation with acyl chloride (RCOCl) reactivity in the acylation of primary amides[56].

RCO—	$10^5 K_a$ of RCO_2H^a	Reaction type	Product yield (%) with	
			CH_3CONH_2	$RCONH_2{}^b$
2-Furoyl	0·7	Monoacylation		80
Propionyl	1·34	Monoacylation	24	60
Isocaproyl	1.53	Monoacylation	38	65
Anisoyl	3.38	Monoacylation	35	52
Cinnamoyl	3·70	Monoacylation	25	23
p-Toluoyl	4·24	Monoacylation	76	54
Benzoyl	6.30	Monoacylation	86	89
2-Naphthoyl	6.90	Monoacylation	83	50
3-Methoxybenzoyl	8·17	Diacylation	75	81
4-Chlorobenzoyl	10·4	Diacylation	88	67
4-Bromobenzoyl	10·7	Diacylation	84	76
Diphenylacetyl	11·2	Diacylation	87	79
4-Iodobenzoyl		Diacylation	92	74
3-Bromobenzoyl	15·4	Diacylation	93	82
3-Nitrobenzoyl	32·1	Dehydration		84
4-Nitrobenzoyl	37·6	Dehydration		84
2-Nitrobenzoyl	671·0	Dehydration		57

^{a} The reactivity of the acyl chloride is expressed in terms of the acidity of the parent acid.
^{b} R is the same substituent as in the acyl chloride (column 1).

of each product. The first may be related to the stability of the mixed anhydride precursor. Triacylamine formation predominates at low temperatures ($-60°$) because the precursor has a sufficiently long half-life for further substitution to occur, whereas, at $0°$ and higher temperatures, the O to N rearrangement must occur so rapidly that mainly nitriles and diacylamines are produced. Pyridine is known to complex with acyl halides to form a very reactive acylating agent[56], and this, rather than carboxamide anion formation, accounts mainly for its catalytic effect. Since lower temperatures may then be used to effect reaction, substitution is favoured at the expense of dehydration. However, diacylamines are formed in greater yields with excess pyridine[56], and in this instance direct substitution at the nitrogen atom of the carboxamide anion may occur.

Transformations with secondary amides have not been examined thoroughly, although it is known that *N*-acylamides are normally obtained. For example, benzoyl chloride reacts with ε-caprolactam

$$\text{(structure) NH} + \text{PhCOCl} \xrightarrow{\text{PhNMe}_2} \text{(structure) NCOPh} \qquad (38)$$

(47)

in the presence of *N,N*-dimethylaniline to give **47**[58]. In view of our previous discussion, these reactions probably proceed via an *O*-acyl intermediate under neutral conditions.

B. Acyl Exchange Reactions

Exchange of acyl groups between the amide and the acyl chloride are known to occur. The simplest case is with tertiary amides[62,74d], where an equilibrium (equation 39) is set up, which may be displaced in either direction by removal of the more volatile acyl chloride. With dimethylformamide ($R = H$; $R^1 = Me$) this process is a useful synthetic route to tertiary amides since the equilibrium is displaced readily to the right-hand side by loss of carbon dioxide and hydrogen chloride[62]. Although no detailed investigation has been reported, it seems probable that the exchange process involves the formation of mixed anhydride salt, which then rearranges via a quaternary ammonium ion (equation 39).

Direct exchange of acyl groups does not occur with primary amides. Instead, these undergo substitution to the diacylamine,

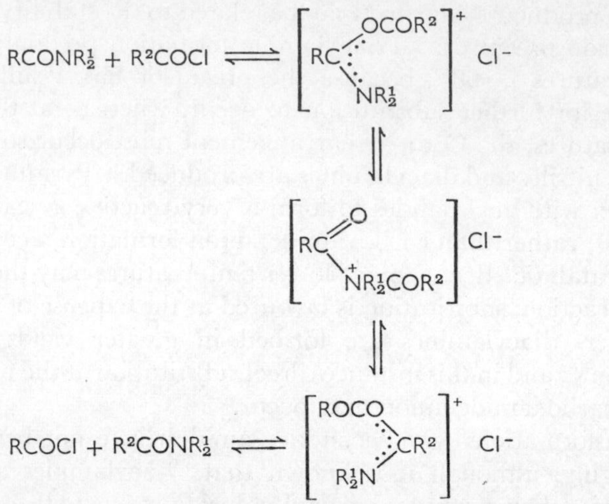

$$RCONR_2^1 + R^2COCl \rightleftharpoons \left[RC \substack{OCOR^2 \\ \\ NR_2^1} \right]^+ Cl^-$$

$$\left[RC \substack{O \\ \\ \overset{+}{N}R_2^1COR^2} \right] Cl^- \qquad (39)$$

$$RCOCl + R^2CONR_2^1 \rightleftharpoons \left[\substack{ROCO \\ \\ R_2^1N} CR^2 \right]^+ Cl^-$$

which then exchanges with the acyl halide[60,63,64]. An example is shown in equation (40)[63], but these transformations are not universal.

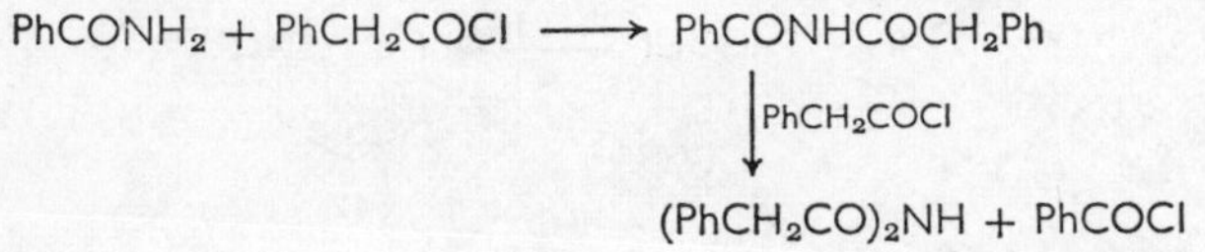

$$PhCONH_2 + PhCH_2COCl \longrightarrow PhCONHCOCH_2Ph$$

$$\downarrow PhCH_2COCl \qquad (40)$$

$$(PhCH_2CO)_2NH + PhCOCl$$

C. Anhydrides

Anhydrides react sluggishly with most amides unless a catalyst such as sulphuric acid, dry hydrogen chloride, an acyl chloride, or even ammonium chloride is used[65,66]. Both dehydration to the nitrile and mono-N-substitution is observed with primary derivatives[65], suggesting that mixed anhydride intermediates are involved as in the case of acyl halides (Scheme 16).

Most catalysts appear to speed both nitrile and diacylamine formation. Sulphuric acid is unusual, however, in that the proportion of each product depends on its concentration. With small quantities

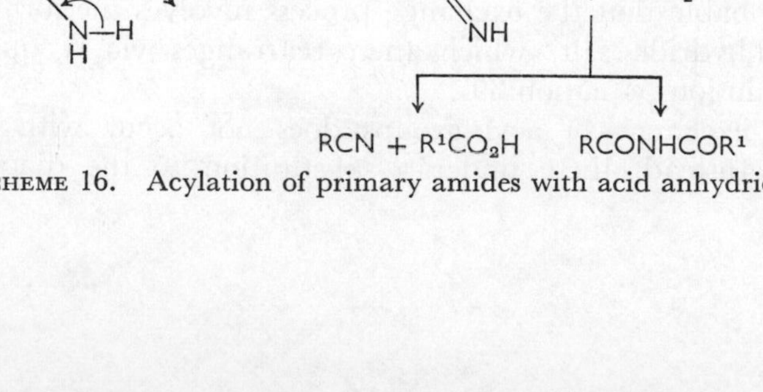

$$RC \substack{O \\ \\ N-H \\ H} \substack{COR^1 \\ \\ O-COR^2} \rightleftharpoons RC \substack{OCOR^1 \\ \\ NH} + R^2CO_2H$$

$$RCN + R^1CO_2H \qquad RCONHCOR^1$$

SCHEME 16. Acylation of primary amides with acid anhydride.

of the order of 0·01 equivalents, dehydration is favoured, whereas higher concentrations increase the amount of diacylamine[65]. A tentative explanation of this phenomenon is linked to the protonating power of the medium with respect to the amide. With low concentrations of sulphuric acid, the amide will exist mainly in its neutral form, but proton loss from the intermediate mixed anhydride ion **48** will be retarded. This would hinder rearrangement to the *N*-acylamine, but not the dehydration reaction, since the latter may still reasonably proceed via a six-centre transition state as depicted in **48**.

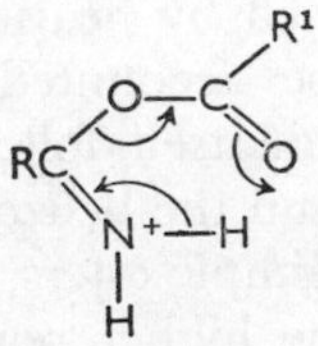

(48)

High concentrations of sulphuric acid will convert the amide to the *O*-protonated conjugate acid. This species, as in alkylation reactions, may undergo direct substitution at the nitrogen atom (Scheme 17). Only the diacylamine product would be expected from this pathway.

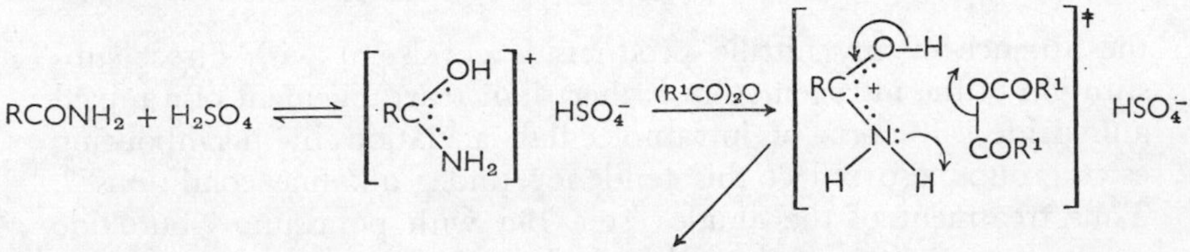

$$RCONHCOR^1 + R^1CO_2H + H_2SO_4$$

SCHEME 17. Sulphuric acid catalysed acylation of primary amides with anhydrides.

A similar situation may prevail with dry hydrogen chloride catalyst, as *N*-acylamides are also formed in preference to nitriles[66]. These studies are, however, much less complete. Little is known about the mode of catalytic action by acyl chlorides and ammonium chloride.

D. Amides and Esters

Disproportionation of primary amides occurs at elevated temperatures and the isolation of nitriles, diacylamines and ammonia from the reaction mixture is most consistent with the formation and decomposition of a mixed anhydride intermediate (equation 41)[67].

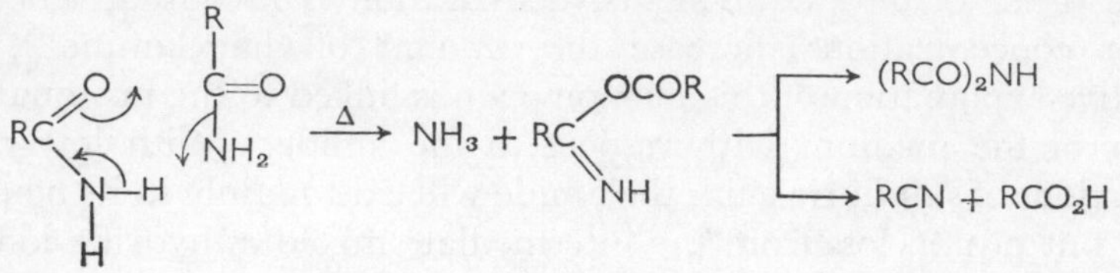

$$\text{(41)}$$

It is claimed, too, that unactivated esters react with amides on heating together at 200°[68], but the evidence is tenuous. The mixture of products (butyronitrile, p-nitrobenzonitrile, butyric acid, phenyl butyrate and phenol) obtained by heating butyramide with phenyl p-nitrobenzoate[68] can all be accounted for by unassisted disproportionation of the butyramide itself, followed by regular transformations between the products and the benzoate ester.

Reaction does occur with simple esters, however, in the presence of strong base[69], because of the higher reactivity of the carboxamide anion. Benzoate esters, for example, substitute both the $\alpha_{C=O}$-carbon and the nitrogen atom of various primary amides in the presence of excess sodium hydride (equation 42)[69]. This reaction is, of course, directly comparable with base-catalysed alkylation (section II.B) and

$$RCH_2CONH_2 + 2\,ArCO_2R^1 \xrightarrow{\text{NaH}} ArCOCHRCONHCOAr \qquad \text{(42)}$$

the absence of any nitrilic products is consistent with direct substitution of the nitrogen atom rather than rearrangement of a mixed anhydride. Reports of intramolecular acylation by neighbouring ester groups also reflect this tendency under alkaline conditions[70]. Thus treatment of the amido ester **49a** with potassium t-butoxide results in rearrangement to the alcohol **49b**[70a], most probably via nucleophilic attack by the carboxamide anion on the ester carbonyl

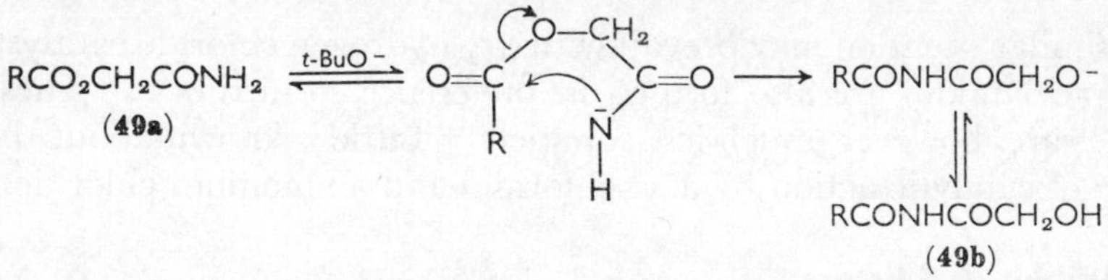

$$\text{(43)}$$

function (equation 43). More direct evidence comes from the isolation of the succinimide **50** alone, and not the anticipated hydrolysis product on treating the amido ester **51** with aqueous hydroxide (equation 44)[70b]. Conversion of phthalamic esters (**52**) to phthal-

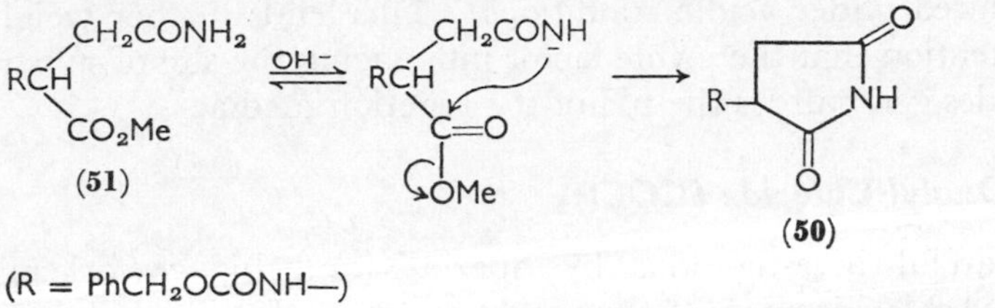

$$(44)$$

(R = PhCH$_2$OCONH—)

imides can also be effected readily at pH 7·8[71]. However, in view of the neutral conditions and the known stability of the anhydride derivative **53**[72], it seems possible that the normal O-attack and subsequent slow rearrangement mechanism for neutral amides is

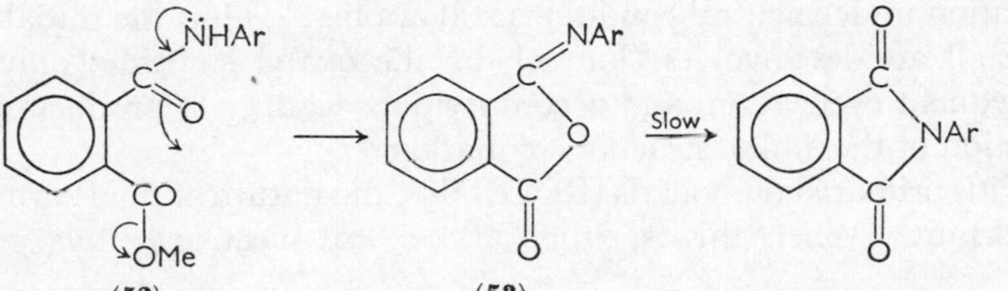

$$(45)$$

followed (equation 45). The low Hammett ρ value $(-0\cdot1)$ found for the various Ar substituents of **52** would not be inconsistent with this mechanism[71], but further studies are required to resolve this point.

Acidic species also appear to catalyse the reactions with esters. Isopropenyl esters, for instance, react with both primary and secondary amides in the presence of p-toluenesulphonic acid to form N-acyl-amides[73]. Here, too, the absence of any nitrilic products is suggestive of direct substitution of the nitrogen atom. It is not known whether the protonated amide (as shown in equation 46) or the protonated ester

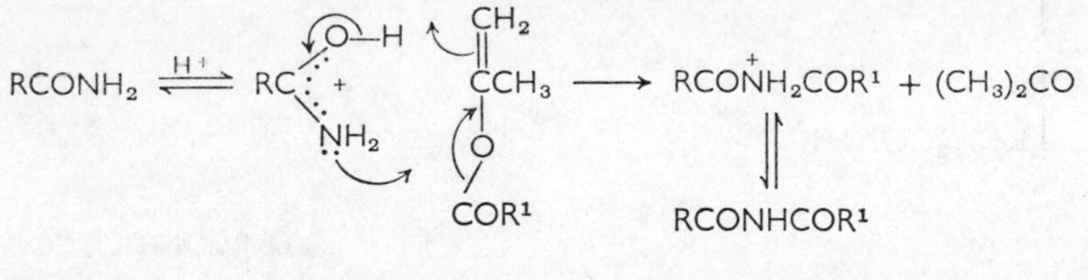

$$(46)$$

is involved, but in either case hydrogen bonding between the amide oxygen and the ester would direct attack by the acyl fragment towards the nitrogen atom. It is interesting to recall that a similar situation prevails in alkylation reactions, where N-substitution is also favoured under acidic conditions. This lends further weight to our contention that the prime factor influencing the site of substitution in amides generally is the pH of the reaction media.

E. Oxalyl Chloride (COCl)₂

Careful investigations by Speziale and his coworkers[74] make oxalyl chloride one of the few acylating agents for which the mechanism is reasonably well-established, and these results contribute significantly to our understanding of the acylation process with other reagents. The products depend on the amide structure as summarized in Table 9, and clearly reflect the bifunctional character of oxalyl chloride. It is important to note, however, that the general mechanism for acylation under neutral conditions still applies. Thus the initial step with all amides involves O-attack by the oxalyl chloride; only the subsequent cyclizations and decompositions leading to products are a function of the amide structure (equation 47).

With primary compounds ($RCONH_2$), the nature of the R group is important. When this is either aryl or an electronegative group

TABLE 9. Products from the reaction of various amides with oxalyl chloride[74].

Amide type	Products
Primary { $RCONH_2$ (R = t-alkyl, Ph, etc.)	$RCONCO + CO$
RCH_2CONH_2	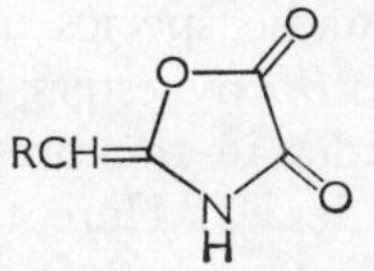
Secondary { $RCONHR^1$ (R = t-alkyl, Ph, etc.)	$RCONR^1COCOCl$
RCH_2CONHR^1	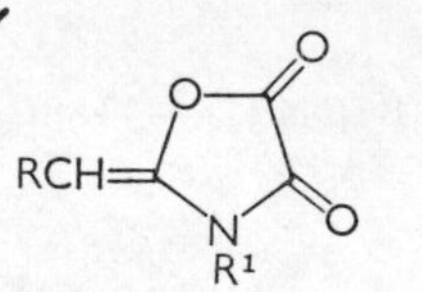 or $RCONR^1COCOCl$
Tertiary $RCH_2CONR_2^1$ {$+2(COCl)_2$}	

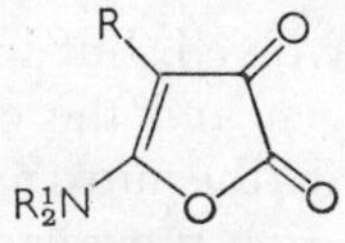

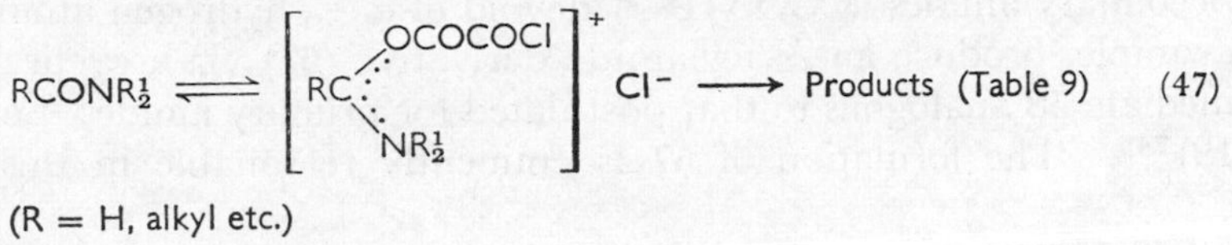

$$RCONR_2^1 \rightleftharpoons \left[RC \underset{NR_2^1}{\overset{OCOCOCl}{<}} \right]^+ Cl^- \longrightarrow \text{Products (Table 9)} \quad (47)$$

(R = H, alkyl etc.)

without $\alpha_{C=O}$-hydrogen atoms, the intermediate mixed anhydride decomposes on heating to give an isocyanate (**54**) in moderate to good yield together with hydrogen chloride and carbon monoxide[74a-c]. When the R substituent possesses $\alpha_{C=O}$-hydrogen atoms (e.g. R = PhCH$_2$), however, cyclization and elimination to an enamine **55** occurs instead[74b]. Both types of reaction are thought to involve an intramolecular cyclization of the mixed anhydride to an intermediate **56**, which then either eliminates the $\alpha_{C=O}$-hydrogen atom or decomposes to the isocyanate (Scheme 18).

Direct information on the initial formation of a mixed anhydride species actually comes from investigations with secondary amides. The mechanism of these reactions closely resembles those with primary amides although the ultimate products are different.

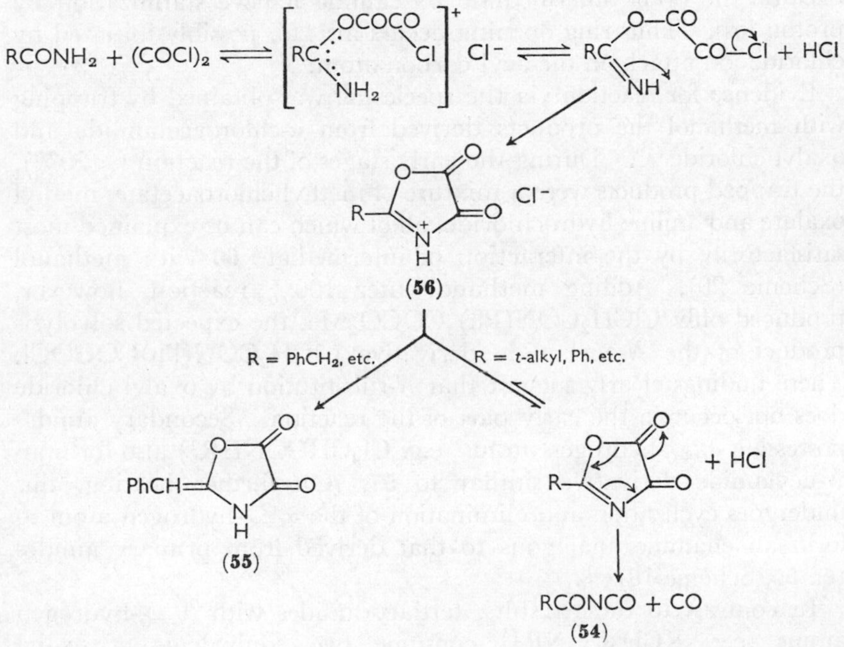

SCHEME 18. Reaction of primary amides with oxalyl chloride.

Secondary amides ($RCONHR^1$) devoid of $\alpha_{C=O}$-hydrogen atoms, for example, produce an *N*-acylamide derivative (**57**) via a cyclic intermediate **58** analogous to that postulated for primary amides (Scheme 19)[74b]. The formation of **57** is eminently reasonable in this case

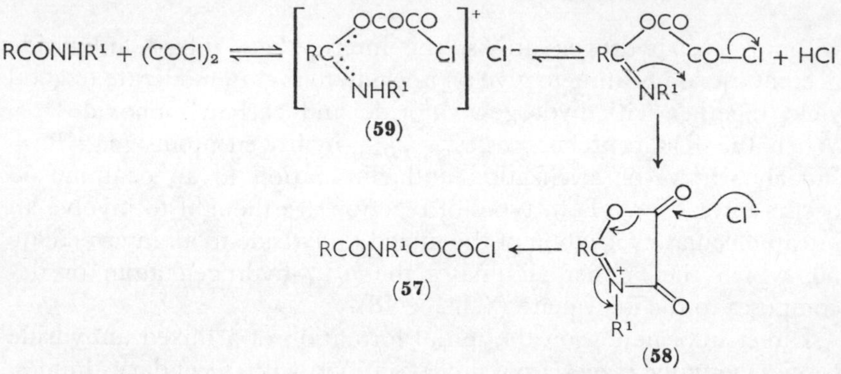

SCHEME 19.   Reaction of secondary amides with oxalyl chloride.

because the cyclic intermediate **58** cannot achieve stabilization by proton loss. Thus ring opening occurs instead, possibly initiated by chloride ion attack on the acyl carbon atom.

Evidence for reaction via the species **59** was obtained by trapping with methanol the products derived from α-chloroacetanilide and oxalyl chloride[74b]. During the early stages of the reaction (< 20%), the trapped products were a mixture of methyl chloroacetate, methyl oxalate and aniline hydrochloride, all of which can be explained most satisfactorily by the interaction of intermediate **60** with methanol (Scheme 20). Adding methanol after 100% reaction, however, produced only $ClCH_2CON(Ph) COCO_2Me$, the expected solvolysis product of the *N*-acylamide derivative $ClCH_2CON(Ph)COCOCl$. These findings clearly suggest that *N*-substitution by oxalyl chloride does not occur in the early part of the reaction. Secondary amides possessing $\alpha_{C=O}$-hydrogen atoms (e.g. $Cl_2CHCONHR^1$) also form an *N*-acylamide derivative similar to **57**. On further heating, this undergoes cyclization and elimination of the $\alpha_{C=O}$-hydrogen atom to form an enamine analogous to that derived from primary amides (cf. **55**, Scheme 18)[74b].

In contrast to these results, tertiary amides with $\alpha_{C=O}$-hydrogen atoms (e.g. $RCH_2CONR_2^1$) consume two equivalents of oxalyl chloride to produce an aminofuranone (**61**)[74d,e]. The proposed

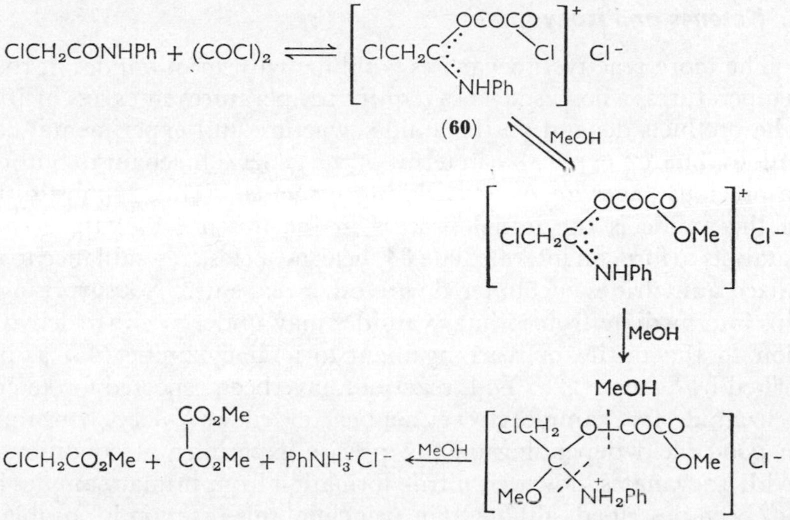

$ClCH_2CONHPh + (COCl)_2 \rightleftharpoons \left[ClCH_2C \underset{NHPh}{\overset{OCOCO}{\cdots}} Cl \right]^+ Cl^-$

(60)

$\searrow$ MeOH

$\left[ClCH_2C \underset{NHPh}{\overset{OCOCO}{\cdots}} OMe \right]^+ Cl^-$

$\downarrow$ MeOH

MeOH

$ClCH_2CO_2Me + \overset{CO_2Me}{\underset{CO_2Me}{|}} + PhNH_3^+Cl^- \overset{MeOH}{\longleftarrow} \left[\overset{ClCH_2}{\underset{MeO}{}} \underset{\overset{+}{NH_2Ph}}{\overset{O \vdash COCO}{\underset{\times}{C}}} OMe \right] Cl^-$

SCHEME 20. Products from the reaction of α-chloroacetanilide with oxalyl chloride in the presence of methanol (trapping of the intermediate **60**).

mechanism involves decomposition of the mixed anhydride salt **62** to a chloroenamine (**63**), which then reacts further with oxalyl chloride (Scheme 21). Tertiary amides lacking a pair of $\alpha_{C=O}$-hydrogen atoms (e.g. $Cl_2HCCONEt_2$) react only under forcing conditions[74d]. The products, as yet, have not been characterized.

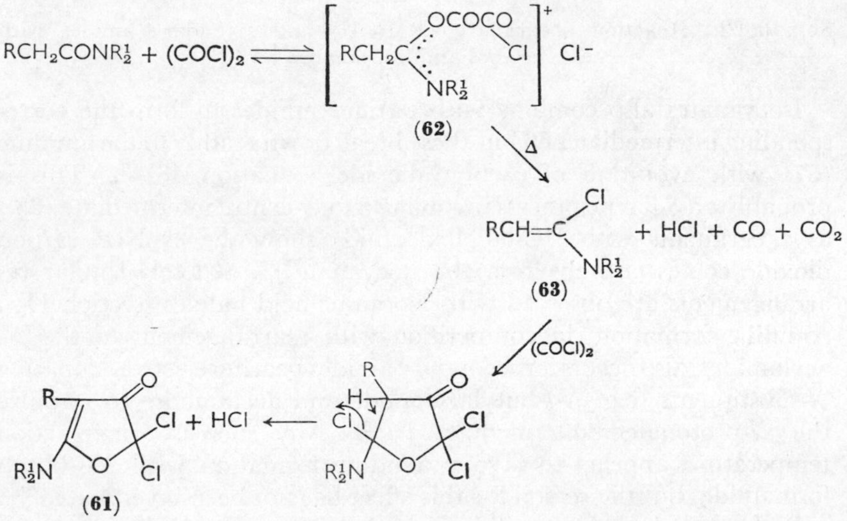

$RCH_2CONR_2^1 + (COCl)_2 \rightleftharpoons \left[RCH_2C \underset{NR_2^1}{\overset{OCOCO}{\cdots}} Cl \right]^+ Cl^-$

(62)

$\searrow \Delta$

$RCH=C \overset{Cl}{\underset{NR_2^1}{}} + HCl + CO + CO_2$

(63)

$\swarrow (COCl)_2$

$\overset{R-}{\underset{R_2^1N}{}} \overset{O}{\underset{O}{\parallel}} Cl + HCl \longleftarrow \overset{R}{\underset{R_2^1N}{}} \overset{H}{\underset{O}{}} \overset{O}{\underset{Cl}{\parallel}} Cl$

(61)

SCHEME 21. Reaction of oxalyl chloride with tertiary amides.

F. Ketenes and Isocyanates

The more reactive isocyanates combine with most amides at room temperatures, whereas ketenes require temperatures in excess of 100°. The products depend on the amide structure and experimental conditions, but are explicable in terms of the general mechanism outlined in previous pages for other acylating reagents. Thus O-substitution of the amide is the initial process in the absence of base or acid catalysts to form an intermediate **64**, bearing a close resemblance to the mixed anhydrides encountered with other reagents. Not surprisingly, this intermediate from primary amides may undergo either dehydration to the nitrile or rearrangement to a diacylamine (**65**) as described by Scheme 22. Both reactions have been reported for ketene: benzamide, for example, gives either benzonitrile or N-acetylbenzamide in good yield depending on the precise experimental conditions[75]. With isocyanates, however, nitrile formation from primary amides has not been described, although, in principle, this reaction is possible.

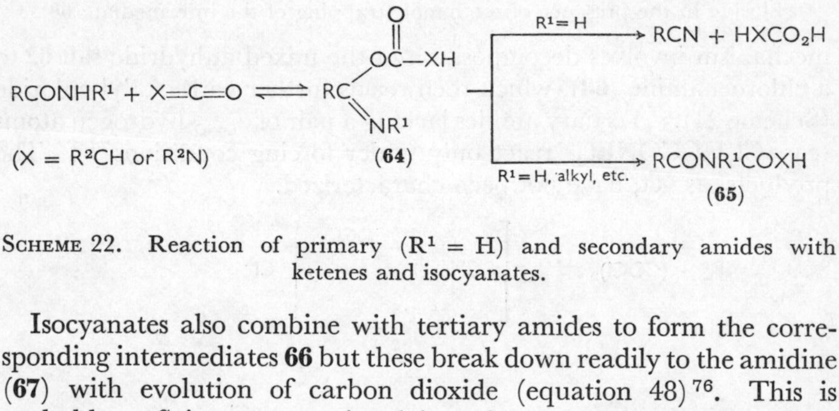

$$RCONHR^1 + X{=}C{=}O \rightleftharpoons RC\!\!\begin{array}{c} \overset{\displaystyle O}{\underset{\displaystyle }{\parallel}} \\ OC{-}XH \\ \diagdown NR^1 \end{array}$$

$$\xrightarrow{R^1 = H} RCN + HXCO_2H$$

$$\xrightarrow[R^1 = H, \text{ alkyl, etc.}]{} RCONR^1COXH$$

$(X = R^2CH \text{ or } R^2N)$ (**64**)   (**65**)

SCHEME 22. Reaction of primary ($R^1 = H$) and secondary amides with ketenes and isocyanates.

Isocyanates also combine with tertiary amides to form the corresponding intermediates **66** but these break down readily to the amidine (**67**) with evolution of carbon dioxide (equation 48)[76]. This is probably an $S_N i$ type process involving a four-centre intermediate (**68**), as experiments with ^{14}C-labelled amide show the evolved carbon dioxide comes entirely from the isocyanate[76a]. Closely similar rearrangements are observed with inorganic acid halides (section IX). Amidine formation, in competition with rearrangement to the N-acylamides, also occurs for secondary amides bearing electron-donating N-substituents (e.g. N-t-butylacetamide and acetanilide) to stabilize the N-protonated intermediate **66**[76]. Any increase in reaction temperature appears to favour amidine formation with N-s-butylformamide, but the reason for this effect has not been investigated[76a].

Both secondary (acetanilide)[77] and primary amides[78] react with

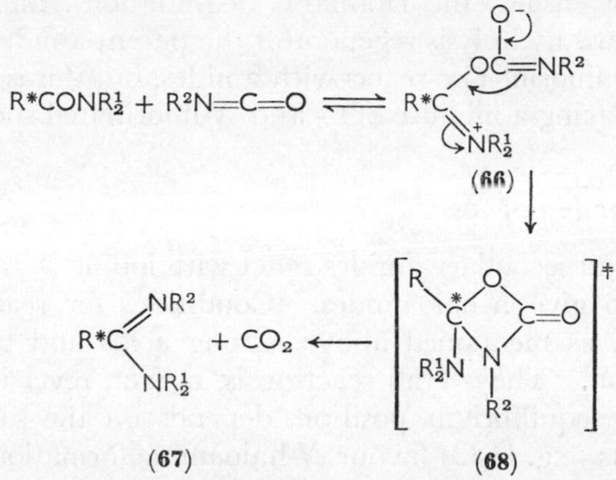

$$R^*CONR_2^1 + R^2N=C=O \rightleftharpoons R^*C \overset{\underset{\displaystyle +NR_2^1}{}}{\overset{\displaystyle OC=NR^2}{}}$$

(66)

(48)

$$R^*C\overset{NR^2}{\underset{NR_2^1}{}} + CO_2 \longleftarrow \left[\overset{R}{\underset{R_2^1N}{}} C \overset{O}{\underset{N}{}} C=O \right]^{\ddagger}$$

(67)   (68)

ketene in the presence of sulphuric acid to give nearly quantitative yields of the *N*-acylamide. No nitrilic products are formed. This is a familiar situation, of course, and, as with other unsaturated alkylating (e.g. vinyl ethers) and acylating (e.g. isopropenyl ester) agents, interaction between the amide and ketene must direct substitution to the nitrogen atom. This is represented in equation (49) as arising from preequilibrium protonation of the amide oxygen atom, although interaction between the neutral amide and the protonated ketene would lead to an identical transition state.

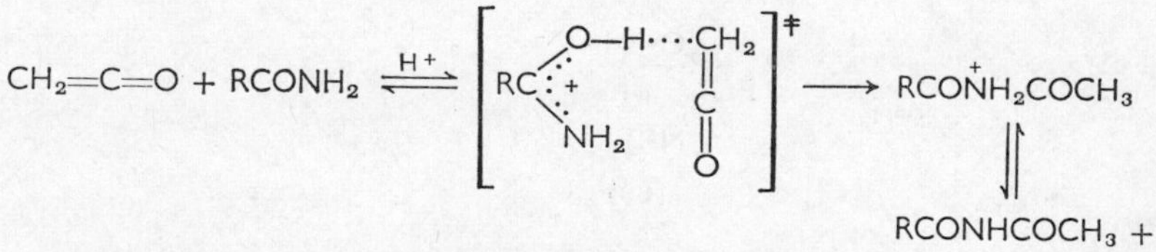

$$CH_2=C=O + RCONH_2 \overset{H^+}{\rightleftharpoons} \left[RC \overset{O-H\cdots CH_2}{\underset{NH_2}{}} \overset{\displaystyle C}{\underset{\displaystyle O}{}} \right]^{\ddagger} \longrightarrow RCO\overset{+}{N}H_2COCH_3$$

(49)

$$\Updownarrow$$

$$RCONHCOCH_3 + H^+$$

V. HALOGENATION

Both molecular halogens (other than fluorine) and hypohalites behave as ionic halogenating agents towards primary and secondary amides. The usual product is the *N*-haloamide, although in a few instances carbon-substituted derivatives are also obtained. This can be associated with the instability of the *N*-haloamides and their decomposition in acidic solutions to give 'positive' halogen, which may then attack other parts of the amide molecule. Of course, *N*-haloamides also decompose readily in basic solutions. With derivatives of primary amides, degradation to the amine with one less carbon atom ensues (the Hofmann degradation), and with others straightforward hydrolysis regenerates the parent amide.

Molecular fluorine also reacts with amides, but this is a free-radical process producing a mixture of C- and N-fluorinated species.

A. Molecular Halogens

Primary and secondary amides react with iodine[79], bromine[80] and chlorine[81] to give N-haloamides. Conditions for reaction are not critical, but, as mentioned above, strong acids and bases enhance decomposition. The overall reaction is, in fact, reversible (equation 50) and the equilibrium position depends on the solvent; highly polar solvents (e.g. H_2O) favour N-haloamide formation[81].

$$RCONHR^1 + Cl_2 \rightleftharpoons RCONClR^1 + HCl \qquad (50)$$
$$(R^1 = H, \text{ alkyl, aryl, etc.})$$

The mechanism has not been diligently investigated, although several findings (but not all) point to an initial substitution of the oxygen atom followed by an O to N rearrangement as with organic acylating reagents. It is known from infrared spectral measurements, for example, that tertiary amides form an O-complex (**69**) with molecular iodine[79]. On complex formation, the carbonyl stretching vibration shifts to a higher frequency and the enthalpy of formation

$$RC\underset{NR_2^1}{\overset{O\cdots I-I^{\delta-}}{\Big\langle}} {}_{\delta+}$$

(**69**)

(ΔH_f) correlates well with Taft σ^* parameters[79]. The ρ^* value of -0.60 indicates that complex stability, as expected, increases slightly with electron donation from the $\alpha_{C=O}$-substituent.

The reverse reaction, the hydrolysis of N-haloamides, has been studied kinetically in connexion with the Orton rearrangement of N-chloroacetanilides, which is catalysed by halogen acids with rates proportional to $[HX]^2$, i.e. to $[H^+][X^-]$[81]. This, together with other observations, has been interpreted as evidence for a rate-controlling nucleophilic attack by halide ion (X^-) on the N-protonated substrate (step a) followed by a rapid intermolecular halogenation (step c) of the aromatic nucleus (equation 51)[81]. The important point as far as we are concerned is that, by the principle of microscopic reversibility, the molecular halogenation of acetanilides (step *b*) must then proceed via direct substitution at the nitrogen atom (by XCl). This is, of course, countercurrent to our previous conclusions and therefore requires explanation. Two come to mind. Firstly,

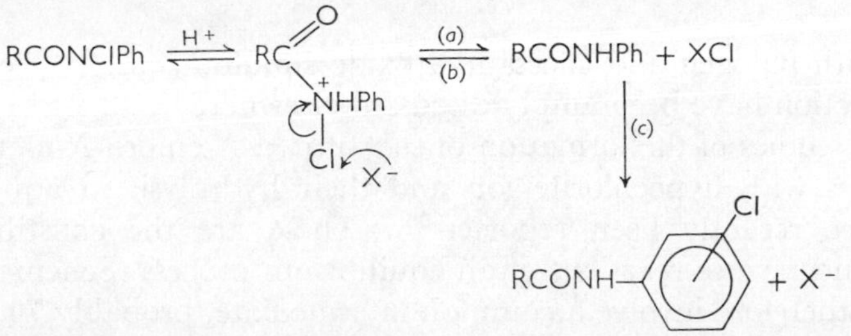

$$\text{RCONClPh} \underset{H^+}{\rightleftharpoons} \text{RC} \underset{(b)}{\overset{(a)}{\rightleftharpoons}} \text{RCONHPh} + \text{XCl} \tag{51}$$

it is possible that molecular halogens are sufficiently reactive to attack the nitrogen atom directly, as in the precedent set by diazomethane (section II.A.6.c). Secondly, and more likely, the Orton rearrangement may proceed via the *O*-protonated rather than the *N*-protonated conjugate acid, and the arguments outlined above are then invalid. Evidence from the base-catalysed hydrolysis discussed in the next section offers some support for the second explanation.

C-Halogenation competes with *N*-haloamide formation particularly when aromatic substituents are present, as in the Orton rearrangement discussed above (equation 51). This process is explored in detail in Chapter 4 and it is sufficient to mention here that, although the precise mechanism is contentious, the ionic intermolecular mechanism catalysed by halogen acids is supplemented by a free-radical pathway catalysed by either light or radical initiators[81]. Photocatalysed rearrangement of *N*-haloalkylcarboxamides to *C*-substituted products has also been discussed recently[82] (see also Chapter 5).

B. Hypohalites

These are preferable to molecular halogens for preparing *N*-haloamides because competing *C*-halogenation is less of a problem. Hypohalous acids are usually formed *in situ* by adding an equimolar amount of sodium hydroxide to a mixture of the molecular halogen and the amide (equation 52)[83]. Treatment of halogenated primary

$$X_2 + NaOH \rightleftharpoons HOX + NaX \xrightarrow{RCONHR^1} RCONXR^1 + H_2O \tag{52}$$
$$(R^1 = H, \text{alkyl, Ph, etc.})$$

amides (RCONHX) with further alkali induces the eliminating

rearrangement known as the Hofmann degradation (equation 53)[84].
In practice the N-haloamide is rarely isolated, instead the amide is

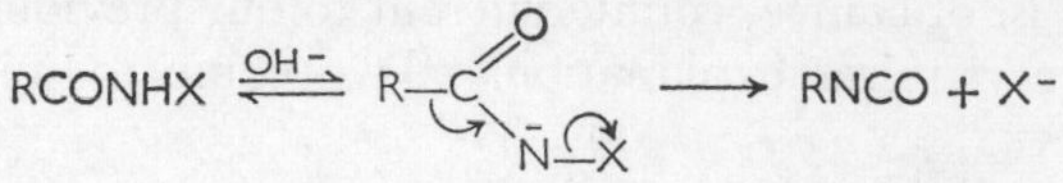

$$RCONHX \; \rightleftharpoons^{OH^-} \; R-C\begin{smallmatrix}O\\ \\N-X\end{smallmatrix} \; \longrightarrow \; RNCO + X^- \qquad (53)$$

treated with halogen and excess hydroxide simultaneously. Details
of this reaction have been amply discussed elsewhere[84].

Kinetic studies of the formation of substituted N-chloro-N-methyl-
benzamides with hypochlorite ion and their hydrolysis in aqueous
alkali have recently been reported[85]. These are the constituent
forward and reverse reactions of an equilibrium process (Scheme 23)
and must therefore involve a common intermediate, probably **70**. A
number of observations, but in particular the existence of an induction
period, suggest that the hydrolysis reaction involves an initial rearrangement (step k_{-3}) of the N-chloroamide to the O-chloro imidate
71, which then undergoes a rate-controlling hydrolysis by hydroxide
ion to the N-methylbenzamide (step k_{-2})[85]. Consequently, the
chlorination reaction must involve attack by the hypochlorite ion on
the amidic hydrogen (step k_1) to give the O-chlorinated intermediate
70, and studies of substituent effects on the rate of chlorination are in
accord with this being the rate-controlling step. Two important
points emerge from this investigation. The first is that halogenation
by hypochlorite ion initially involves substitution of the amide
oxygen atom with subsequent fast rearrangement to the stable N-

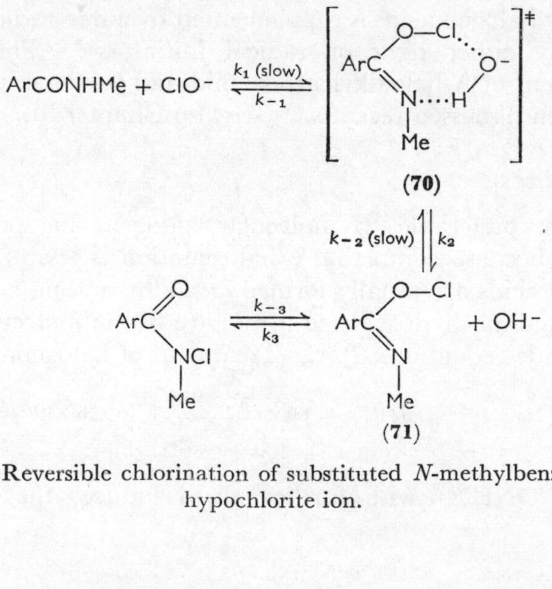

SCHEME 23.   Reversible chlorination of substituted N-methylbenzamides with
hypochlorite ion.

chlorinated product, and this agrees nicely with the general mechanism for electrophilic substitution of neutral amides. The second is that rearrangement of the *O*-chloro imidate to the *N*-chlorinated product is reversible (at least under alkaline conditions) and this may also be a factor in the Orton rearrangement.

The esters of hypohalous acids (e.g. *t*-BuOX) will also effect *N*-halogenation of primary and secondary amides[82,86]. The mixture usually consists of *t*-butyl hypochlorite and molecular halogen with the amide in carbon tetrachloride. Since the halogen monochloride (e.g. ICl, BrCl) is surprisingly ineffective under these conditions, it has been suggested, and proven[80], that the reagent is the *t*-butyl hypohalite formed *in situ* (equation 54). More recent work has shown that *t*-butyl hypochlorite itself is effective under similar conditions[82].

$$\text{t-BuOCl} + \text{I}_2 \rightleftharpoons \text{t-BuOI} + \text{ICl}$$

$$(R^1 = \text{H, alkyl, etc.}) \qquad\qquad \searrow^{\text{RCONHR}^1} \qquad\qquad (54)$$

$$\text{RCONIR}^1 + \text{t-BuOH}$$

C. Fluorination

A mixture of *C*- and *N*-fluorinated products is obtained by treating most amides with gaseous fluorine[87]. These are invariably free-radical processes. As an illustration, dimethylformamide gives a mixture of $(CF_3)_2NF$ and $(CF_3)_2NN(CF_3)_2$, possibly via the routes outlined in Scheme 24. Otherwise, fluoride ion is reported to react

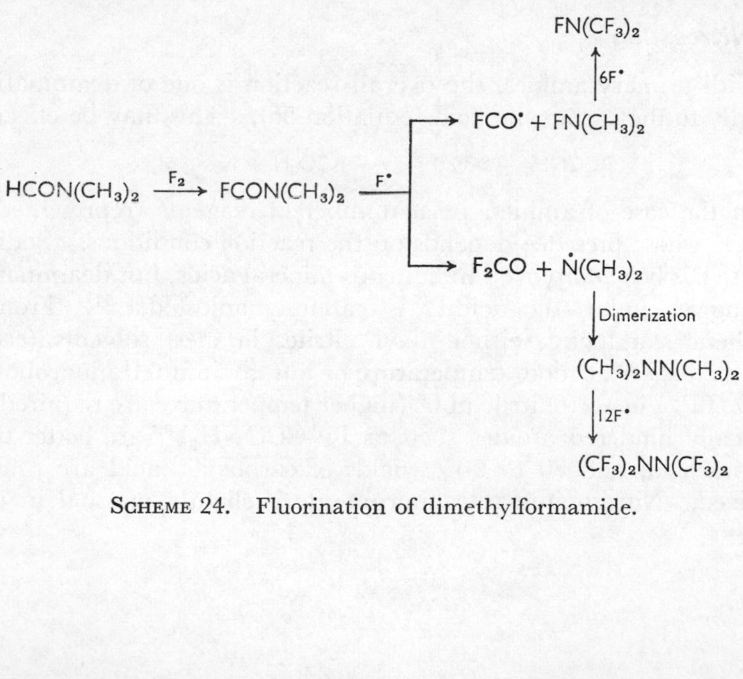

$$FN(CF_3)_2$$

$$\uparrow 6F^{\bullet}$$

$$\longrightarrow FCO^{\bullet} + FN(CH_3)_2$$

$$\text{HCON(CH}_3)_2 \xrightarrow{F_2} \text{FCON(CH}_3)_2 \xrightarrow{F^{\bullet}}$$

$$\longrightarrow F_2CO + \dot{N}(CH_3)_2$$

$$\downarrow \text{Dimerization}$$

$$(CH_3)_2NN(CH_3)_2$$

$$\downarrow 12F^{\bullet}$$

$$(CF_3)_2NN(CF_3)_2$$

SCHEME 24. Fluorination of dimethylformamide.

with tertiary acetamides at 160° to form only monofluorinated products (equation 55)[88]. None of these reactions, however, has been

$$CH_3CONPhMe \xrightarrow[160°]{F^-} FCH_2CONPhMe \tag{55}$$

investigated in detail or found appreciable synthetic application.

VI. NITROSATION AND NITRATION

Both nitrous and nitric acid combine with suitable weak bases to form powerful electrophilic reagents, which can be regarded as carriers of the nitrosonium (NO^+) and the nitronium (NO_2^+) ions, respectively. Many of these reagents react with primary and secondary amides, but not generally with tertiary compounds. All the reactions are much less facile than with amines, however, reflecting the lower nucleophilic strength of amides. Similar products result from both nitration and nitrosation: primary amides undergo deamination to the carboxylic acid, and secondary amides form their N-substituted derivatives. With aromatic substrates, substitution of this nucleus also occurs, but we shall not consider these transformations in detail. The mechanism of deamination and N-substitution has not been closely studied, particularly in regard to the site of initial attack by the reagent. Incidental evidence favours the nitrogen atom, but this is by no means proven. Nitrosation reactions have evoked the greater interest and these are considered first.

A. Nitrosation

With primary amides, the overall reaction is one of deamination, usually to the carboxylic acid (equation 56). This may be effected,

$$RCONH_2 + XNO \longrightarrow RCO_2H + N_2 + HX \tag{56}$$

as in the case of amines, by a number of reagents (represented as XNO) whose presence depends on the reaction conditions. Sodium nitrite has been employed in aqueous mineral acids, but deamination is sluggish unless the acidity is carefully adjusted[89-91]. From a synthetic standpoint, either alkyl nitrites in inert solvents (ether, dioxan, etc.)[92] at room temperature or nitrosonium tetrafluoroborate ($NO^+BF_4^-$) in acetonitrile at 0° (higher temperatures are required for sterically hindered amides such as Ph_3CCONH_2)[93] are better than nitrous acid, and 70 to 90% yields of carboxylic acid are usually realized. Nitrosonium tetrafluoroborate is slightly unusual in that with *ortho-t*-butylbenzamide, the corresponding aldehyde and not the carboxylic acid is formed[93].

Secondary amides are converted to their *N*-nitroso derivatives by similar reagents (equation 57). As with secondary amines, the reaction is reversible and to ensure high yields of products a suitable base is

$$RCONHR^1 + XNO \underset{}{\overset{\text{Base}}{\rightleftharpoons}} RCON(NO)R^1 + HX \tag{57}$$

added to remove the HX acid. The efficacy of several reagents has been tested by White[94], who concluded that nitrosyl chloride or nitrogen tetroxide in either carbon tetrachloride or acetic acid solvents are best: in both cases sodium acetate should be added to drive the equilibrium to the right. Other reagents such as sodium nitrite in mineral or acetic acid are both less productive and only successful when the $\alpha_{C=O}$ group is RCH_2[94].

N-Alkyl-*N*-nitrosoamides are highly reactive substances which undergo both thermal[95] and photolytic decomposition[96]. The thermal process (equation 58) produces both carboxylic acids and esters, probably via a diazo-ester intermediate (**72**)[95]. Photolysis,

$$
\begin{array}{c}
\overset{O}{\underset{R^1-N\overset{|}{\underset{}{\nwarrow}}N}{\overset{\|}{R-C\nwarrow O}}} \xrightarrow{\Delta} \left[\overset{O}{\underset{RCON=NR^1}{\overset{\|}{}}} \right]
\begin{array}{l}
\longrightarrow RCO_2R^1 + N_2 \\
\\
\longrightarrow RCO_2H + N_2 + \text{Olefins}
\end{array}
\tag{58}
\end{array}
$$

(**72**)

however, results in fission of the N—N bond to give a mixture of aldehyde and primary amide (equation 59)[96]. The thermal stability of *N*-nitrosoamides decreases with increasing branching at the $\alpha_{C=O}$-

$$
\underset{RCONCH_2R^1}{\overset{NO}{\overset{|}{}}} \xrightarrow{h\nu,\, H^+} RCON=CHR^1 + [NOH]
$$
$$
\searrow H_2O \tag{59}
$$
$$
RCONH_2 + R^1CHO
$$

carbon atom. Thus compounds carrying either an RCH_2 or R_2CH $\alpha_{C=O}$-group are stable up to 75° and 50°, respectively, whereas those where this group is tertiary decompose at temperatures below 0° and are therefore difficult to isolate[94]. Decomposition of *N*-aryl-*N*-nitroso-amides occurs similarly, but results in the formation of phenyl radicals[97] and possibly a benzyne intermediate[98].

Mechanistic studies have been concerned mainly with the deamination of primary amides in aqueous solvents. Under these conditions several reactive nitrous species are in equilibrium with nitrous acid.

A similar situation applies to the nitrosation of amines, for which the studies are more complete, and it is evident that the various species lie in the following order of increasing reactivity[99]:

N_2O_3	nitrous anhydride
HalNO	nitrosyl halides
H_2ONO^+	nitrous acidium ion
NO^+	nitrosonium ion

It is noteworthy that nitrous acid, itself, is quite ineffective. One expects the same schedule of reagents to be important for the nitrosation of amides in aqueous solutions, and most of the investigations thus far have been directed towards proving this point.

Studies by Bruylants and his colleagues[92a,100] have provided ample evidence that the nitrosation (deamination) of acetamide and related alkylamides in hydrochloric acid arises from reaction between nitrosyl chloride (formed in a rapid preequilibrium step) and the un-protonated amide (equations 60). Logarithmic rate coefficients can

$$HNO_2 + HCl \rightleftharpoons ClNO + H_2O \quad \text{(fast preequilibrium)}$$
$$RCONH_2 + ClNO \xrightarrow{\text{Slow}} RCO_2H + N_2 + HCl \tag{60}$$

be correlated with Taft σ^* parameters ($\rho^* = -3 \cdot 0$) providing one assumes that steric effects are the same as in ester hydrolysis[92a,100a]. It is surprising that steric interactions from the alkyl substituent are important at all in nitrosation, and this result may indicate a hitherto unrevealed subtlety in the mechanism.

In perchloric and sulphuric acids, the corresponding nitrosyl salts are fully ionized and this means that potential nitrosating agents will be limited to nitrous anhydride, the nitrous acidium ion and the nitrosonium ion in that order with increasing acidity[99]. There is no evidence that nitrous anhydride alone is capable of reacting with amides (although it does with amines) probably because of its low reactivity. There is good evidence, however, that the nitrosation of both benzamide[90] and acetamide[89] is strongly catalysed by mineral acids, although it is uncertain whether this corresponds to reaction of the nitrous acidium ion or the nitrosonium ion (or even both!). Thus the results for benzamide, originally regarded as evidence for a rate-controlling reaction between the nitrosonium ion and the neutral amide[90], have been reinterpreted[89] in the light of further data in favour of the nitrous acidium ion. The same conclusion has been reached for acetamide, but in both cases the results are not entirely consistent with this reagent and doubt remains[89].

All the mechanistic studies to date have assumed that the rate-controlling step involves attack by the nitrous species on the nitrogen atom of the neutral (unprotonated) amide (Scheme 25). There can be little doubt that the unprotonated amide is one reactant, because

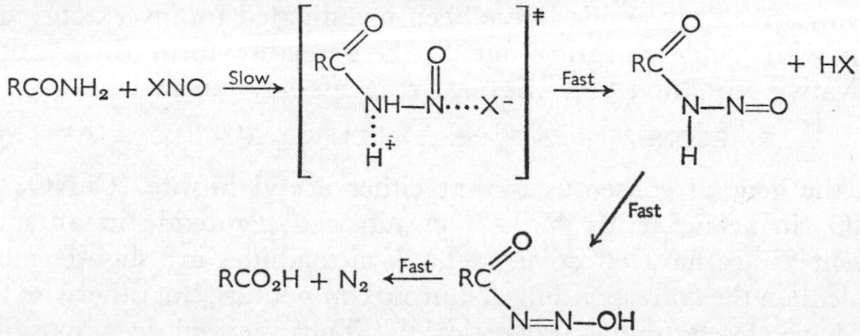

$$RCONH_2 + XNO \xrightarrow{Slow} \left[\begin{array}{c} RC{=}O \\ NH{-}N{\cdots}X^- \\ H^+ \end{array} \right]^{\ddagger} \xrightarrow{Fast} RC{=}O \quad N{-}N{=}O \quad + HX \\ H$$

$$RCO_2H + N_2 \xleftarrow{Fast} RC{=}O \quad N{=}N{-}OH \xleftarrow{Fast}$$

SCHEME 25. Deamination of primary amides with nitrous acid.

the rate of deamination slows down in concentrated mineral acid where the amide exists mainly in its conjugate acid form[89,90]. The site of initial substitution seems to be an unanswered question. Since other electrophilic reagents preferentially attack the amide oxygen atom, the same may be true for *neutral* nitrosating agents, with the product arising from a subsequent O to N rearrangement. With positively charged agents (e.g. H_2ONO^+ and NO^+) electrostatic interaction with the amidic oxygen would direct attack towards the nitrogen atom, as with alkylation and acylation under acidic conditions (equation 61).

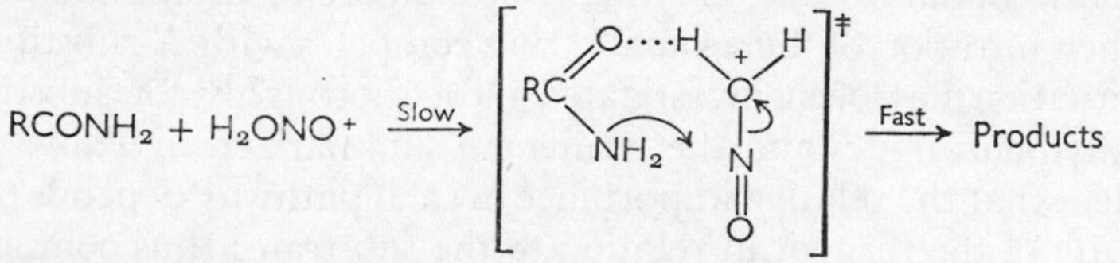

$$RCONH_2 + H_2ONO^+ \xrightarrow{Slow} \left[\begin{array}{c} RC{=}O{\cdots}H \quad H \\ NH_2 \quad O^+ \\ N \\ O \end{array} \right]^{\ddagger} \xrightarrow{Fast} Products \quad (61)$$

The only evidence bearing on this concerns the reversibility of *N*-nitroso formation with secondary amides. If the reverse reaction (decomposition) is acid catalysed, the simplest mechanism would involve *N*-protonation (equation 62). Then by the principle of

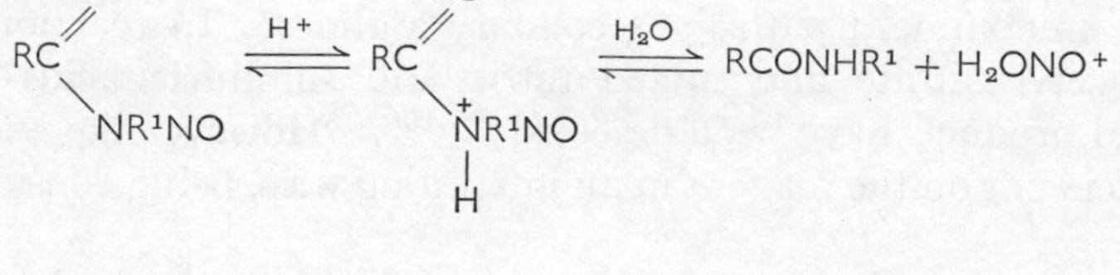

$$RC{\overset{O}{\underset{NR^1NO}{}}} \xrightleftharpoons{H^+} RC{\overset{O}{\underset{\overset{+}{N}R^1NO}{}}} \xrightleftharpoons{H_2O} RCONHR^1 + H_2ONO^+ \quad (62)$$

microscopic reversibility, the forward reaction, too, should involve a direct N-substitution by the nitrosating agent.

B. Nitration

Only secondary amides have been investigated to any extent, and even with feeble nitrating agents these readily form the N-nitro derivative (equation 63). Several reagents have in fact been used,

$$RCONHR^1 + XNO_2 \rightleftharpoons RCON(NO_2)R^1 + HX \qquad (63)$$

but the general consensus is that either acetyl nitrate ($CuNO_3$ or HNO_3 in acetic acid)[95a,101,102] or nitrogen pentoxide in an inert solvent[103] are most effective. The N-nitroamides are slightly more stable than the corresponding N-nitroso compounds, but otherwise the two have closely similar properties[104]. Thus thermal decomposition does occur (equation 64), but requires temperatures in the region of

$$(64)$$

25° to 75° depending on the structure of the acyl moiety[104]. The products, a mixture of the carboxylic acid and the ester are the same as from N-nitrosoamides, and decomposition probably proceeds via a similar pathway involving the diazoxy ester (**73**)—instead of the diazo-ester intermediate (cf. equation 58)[95a,104].

If the secondary amide is either a benzamide or an anilide derivative, then nitration of the aromatic ring competes with N-substitution. The situation has been investigated most thoroughly for substituted N-methylbenzamides and the results are summarized in Table 10. It is evident that the relative importance of each pathway depends on the reactivity of the reagent in relation to the substrate: thus compounds with strongly deactivated nuclei only form N-nitro products even with the powerful nitronium ion; mildly deactivated, monosubstituted N-methylbenzamides undergo N-substitution with feeble reagents, such as acetyl nitrate, but ring substitution with the nitronium ion; and activated compounds suffer ring substitution with both weak and powerful reagents[101]. Nitration of the aromatic ring also seems to be the predominant path for secondary anilides. In reactions with both acetyl nitrate and mixed nitric and sulphuric acids no N-nitrated products have been detected[105,106]. However the *ortho*:*para* ratio depends on the reagent in an interesting way, being considerably higher with acetyl nitrate (ca. 5·0) than with the mixed acids (ca. 0·05)[105]. A consistent rationale for this effect is that preferred *ortho* substitution with acetyl nitrate results from rearrangement of the *N*-nitro precursor, whereas the *para* substitution favoured in mixed

TABLE 10. Nitration of substituted *N*-methylbenzamides[101].

Reactivity of aromatic ring	Product orientation	
	$CH_3CO_2NO_2$[a]	NO_2^+[b]
Deactivated (e.g. 3,5-$(NO_2)_2$; 2,4,6-Cl_3; etc.)	N	N
Mildly deactivated (e.g. 4-NO_2; 4-Cl; etc.)	N	N + ring
Activated (e.g. 4-MeO; 4-Me; etc.)	ring	N + ring

[a] From either $CuNO_3$ or HNO_3 in acetic anhydride.
[b] From mixed HNO_3 and H_2SO_4.

acids arises from substitution of the conjugate acid of the anilide substrate[105].

Few studies with primary amides have been reported, although it appears that the *N*-nitro derivative is very unstable and deamination occurs almost as rapidly as with *N*-nitrosoamides. Benzamide, for example, reacts with acetyl nitrate (either $CuNO_3$ or HNO_3 in acetic anhydride) to form benzoic acid in high yield (equation 65)[101].

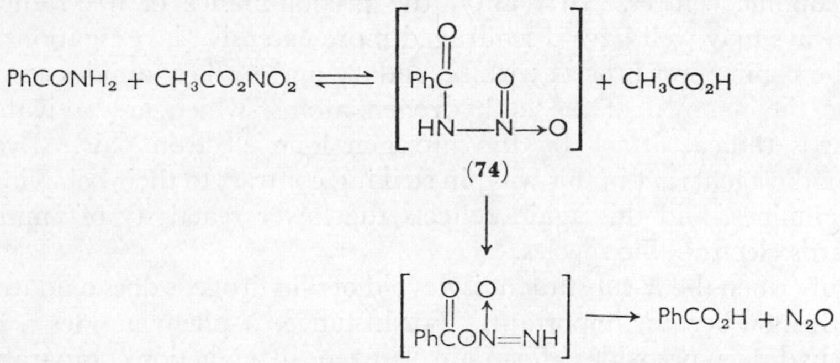

$$PhCONH_2 + CH_3CO_2NO_2 \rightleftharpoons \left[\begin{array}{c} O \\ \| \\ PhC \quad O \\ | \quad \| \\ HN\!-\!\!-\!N\!\rightarrow\!O \end{array} \right] + CH_3CO_2H$$

$$\textbf{(74)}$$

$$(65)$$

$$\downarrow$$

$$\left[\begin{array}{c} O \;\; O \\ \| \;\; \uparrow \\ PhCON\!\!=\!\!NH \end{array} \right] \longrightarrow PhCO_2H + N_2O$$

The half-life of the reaction is less than five minutes, which is considerably faster than conventional hydrolysis under these conditions (section X). Formation and decomposition of an *N*-nitro intermediate (**74**) is an obvious possibility by analogy with nitrosation. Support for this conclusion comes from the isolation of a crystalline

26+c.o.a.

solid in the low-temperature nitration of 3-hydroxy-4-pyridinecarbox-
amide with mixed nitric and sulphuric acids; this is believed to be the
N-nitro derivative and rapid decomposition occurs on melting[107].

Tertiary alkylamides appear to be inert towards all nitrating agents
unless the $\alpha_{C=O}$-carbon atom is unsubstituted ($RCH_2CONR_2^1$). In
this instance, nitration of the $\alpha_{C=O}$ site is reported with amyl nitrate in
the presence of potassium t-butoxide (equation 66)[108] presumably via
the carbanion intermediate **75**. The product is isolated as the
extremely hygroscopic potassium salt **76**, which decomposes violently

$$RCH_2CONMe_2 \xrightleftharpoons{t\text{-BuOK}} R\overset{-}{C}HCONMe_2 \xrightarrow{C_5H_{11}ONO_2} R\overset{\overset{NO_2^-\ K^+}{\|}}{C}CONMe_2 \quad (66)$$
$$\qquad\qquad\qquad\qquad (75) \qquad\qquad\qquad\qquad\qquad (76)$$

on exposure to the atmosphere on account of its high energy of hydration. Attempts to obtain the neutral α-nitroamide by acidification
failed without exception[108].

VII. OXIDATION

Little is known about the oxidation of amides—less than a dozen of
the many available oxidants have been investigated—and this subject
may be regarded as one of the underdeveloped areas of organic
chemistry. Even the limited information available, however, does
reveal a complex situation, with hydrogen abstraction from carbon
and nitrogen competing effectively with oxidative substitution at the
nucleophilic centres. In reality, the predominance of free-radical
pathways may well have discouraged more extensive investigations.

The commonest process with secondary and tertiary amides seems
to be the removal of the α_N-hydrogen atoms, which are activated
towards radical attack by the nitrogen lone electron pair. Even
peroxidic agents act in this way, in striking contrast to their behaviour
with amines, and this again reflects the lower reactivity of amides
towards electrophilic species.

Only when the N-substituent is devoid of α-hydrogens does oxidative
substitution become important. For instance, N-phenylamides react
with hydrogen peroxide to form nitrobenzene in a reaction comparable
to that of aromatic amines. The oxidation of primary amides
proceeds via different pathways for the same reason, with either
hydrogen abstraction from the nitrogen atom (with peroxydisulphates)
or oxidative substitution (with lead tetraacetate) being the more
important.

It is unwise to generalize further, because both the amide and the oxidant structure have an important bearing on the product-forming stages of the reactions. We shall therefore consider each case individually.

A. Autoxidation

On account of its interference to the commercial production of nylon, the autoxidation of N-alkyl- and N,N-dialkylamides has engendered considerable interest and attention. As expected, this is invariably a free-radical process which may be induced either by thermal $(>100°)$[109] or by photochemical means[109b,110-112] at lower temperatures both with and without suitable initiators. Three principal overall reactions have been identified (equations 67–71)[109b,110]:

(i) Formation of N-acylamides from N-n-alkylamides:

$$RCONHCH_2R^1 \longrightarrow RCONHCOR^1 \tag{67}$$

(ii) Formation of N-formylamides from N-n-alkylamides or N-acylamides from N-s-alkylamides, via carbon–carbon bond fission:

$$RCONHCH_2R^1 \longrightarrow RCONHCHO \tag{68}$$

$$RCONHCH\begin{smallmatrix}R^1\\ \\R^2\end{smallmatrix} \longrightarrow RCONHCOR^1 + RCONHCOR^2 \tag{69}$$

(iii) Oxidative dealkylation (carbon–nitrogen bond fission) to yield carbonyl derivatives:

$$RCONHCH_2R^1 \longrightarrow RCONH_2 + R^1CHO \tag{70}$$

$$RCONHCH\begin{smallmatrix}R^1\\ \\R^2\end{smallmatrix} \longrightarrow RCONH_2 + R^1COR^2 \tag{71}$$

What factors determine the relative importance of each path has not been clearly established, but recent investigations, particularly some elegant kinetic measurements by Sagar and his colleagues[109], have begun to unravel the complex mechanism of these reactions. There is little doubt that the initial steps common to all three pathways are removal of the α_N-hydrogen atom followed by oxygen addition to give the peroxy radical (77), which then decomposes to products by various routes (Scheme 26). Only the α_N-hydrogen atom seems sufficiently labile to suffer abstraction, which explains the resistance of both acetamide and *N-t*-alkylamides to autoxidation. This is clearly demonstrated by experiments with N-(1-[$^{14}C_1$]-pentyl)-

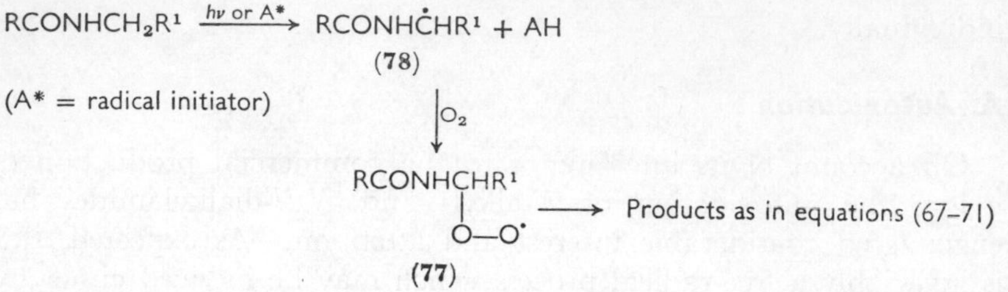

$$RCONHCH_2R^1 \xrightarrow{h\nu \text{ or } A^*} RCON H\overset{\cdot}{C}HR^1 + AH$$

(78)

(A* = radical initiator)

$\downarrow O_2$

RCONHCHR^1
|
O—O$^{\cdot}$ $\longrightarrow$ Products as in equations (67–71)

(77)

SCHEME 26.   Initial steps for the autoxidation of *N*-alkylamides.

hexanamide under conditions (uninitiated photooxidation at 50°) where oxidative dealkylation (cf. equation 70) is the principal product-forming route[113]. The isotopic label is found only in the valeraldehyde and valeric acid products, showing these originate solely from the *N*-pentyl fragment; the other products, n-hexanoic acid and n-hexanamide were inactive and must therefore come from the acyl fragment (equation 72). Various rate measurements have established

$$n\text{-}C_5H_{11}CONH^{14}CH_2Bu\text{-}n \xrightarrow{h\nu, O_2}$$
$$n\text{-}Bu^{14}CHO + n\text{-}Bu^{14}CO_2H + n\text{-}C_5H_{11}CO_2H + n\text{-}C_5H_{11}CONH_2 \quad (72)$$

oxygen uptake by the carboxamide radical **78**[109b,d,110]. In one instance, the photooxidation of ε-caprolactam (**79**), a hydroperoxide intermediate (**80**) has been isolated; on treatment with a cobalt(III) salt this is reduced to adipimide (**81**) (equation 73)[114].

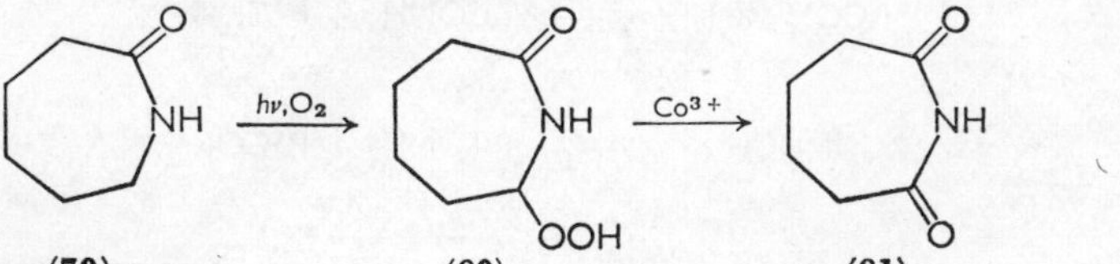

(79)   (80)   (81)   (73)

Other details of these reactions, in particular the nature of the product-forming steps, are not generally understood. Sagar[109d], however, has provided strong evidence for the operation of a radical-chain mechanism in the thermal oxidation (< 100°) of *N*-alkylamides, in which product formation is governed by four processes. The first is a chain reaction of the substrate with oxygen giving *N*-alkylamide hydroperoxide as the primary product[109c] in accord with our discussion above; the second is thermal decomposition of this hydro-

peroxide; the third is chain termination by condensation of two
N-alkylamide hydroperoxy radicals; the fourth, of importance only in
the later stages of the reaction, is interference with the oxidation chain
by products from the hydroperoxide. A much simplified version of the
first three processes is described by Scheme 27. However, the low-
temperature photooxidation using sodium anthraquinonesulphonate
initiator is claimed to proceed via a *non-chain* mechanism[110].

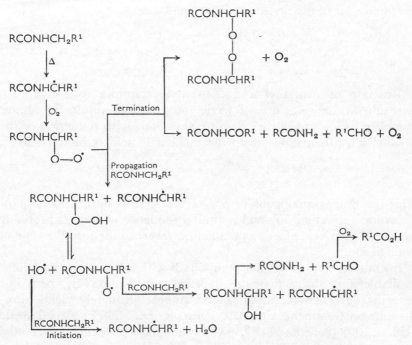

Scheme 27. Radical-chain thermal oxidation of N-alkylamides.

B. Hydrogen Peroxide and Diacyl Peroxides

The known electrophilic properties of peroxidic reagents are not
manifest in their reactions with amides. The products from N-
alkylamides resemble those from autoxidation and clearly similar
free-radical mechanisms operate. Oxidative substitution, reminiscent
of amine oxidation, is only important for N-arylamides. The
preference for hydrogen abstraction can be associated with the low
nucleophilic reactivity of the amide group. High reaction temperatures are therefore necessary, and this, in turn, promotes homolytic
fission of the weak peroxide linkage.

Oxidation of N,N-dialkylamides with diacyl peroxide results in the introduction of an acyloxy moiety on the α_N-carbon atom (equations 74)[115,116]. A similar process would be expected for N-alkylamides. Both photoinitiation[116] and inhibition by radical scavengers such as styrene[115] indicate a free-radical process, but no detailed mechanistic studies have been reported. A tentative mechanism involving removal of the α_N-hydrogen followed by radical coupling (equation 74) is reminiscent of the first stages of autoxidation.

$$(PhCO_2)_2 \xrightarrow{h\nu \text{ or } \Delta} 2\,PhCO_2^{\cdot}$$

$$HCONMe_2 + PhCO_2^{\cdot} \longrightarrow HCONMe\overset{\cdot}{C}H_2 + PhCO_2H \tag{74}$$

$$HCONMe\overset{\cdot}{C}H_2 + PhCO_2^{\cdot} \longrightarrow HCONMeCH_2OCOPh$$

Reaction of N-methyl and N-n-butylacetamides with hydrogen peroxide in the presence of ferric ion catalysts results in almost complete fission of the carbon–nitrogen bond, with oxidation of the N-alkyl substituent through to the carboxylic acid (equation 75)[117].

$$CH_3CONHCH_2R \xrightarrow{H_2O_2/Fe^{3+}} CH_3CONH_2 + RCO_2H \tag{75}$$
$$(R = H \text{ or n-Pr})$$

This reaction is analogous to one of the routes established for autoxidation (equation 70) and a similar sequence of steps is probably involved. No evidence is available, however, to identify the type of bond fission.

In contrast to these results, compounds without α_N-hydrogen atoms available for abstraction do undergo substitution by peroxidic reagents. N-Phenylacetamide, for instance, ultimately yields nitrobenzene on treatment with 30% hydrogen peroxide in glacial acetic acid at 100° (Scheme 28)[118]. The reaction is probably heterolytic, by analogy with peroxidic oxidations of N-arylamines, involving substitution of the nitrogen atom by electrophilic oxygen. Since acetylphenylamine N-oxide (82), but not phenylhydroxylamine (this couples with nitrosobenzene to form azoxybenzene in glacial acetic acid), is also oxidized to nitrobenzene under the reaction conditions[118], it seems probable that elimination of the acetoxy fragment results from an attack by a second hydrogen peroxide molecule on the hydroxamic acid intermediate (83). Nitrosobenzene (84) must be another intermediate since this is also isolated from the reaction mixture. A tentative mechanism embodying all these findings is reproduced in Scheme 28. Other studies[118] have shown that formanilide is more reactive than acetanilide and this may be attributed to steric differences. Substituents in the aromatic ring affect the yield of nitro product, but it is impossible to deduce a sensible explanation on the limited data available.

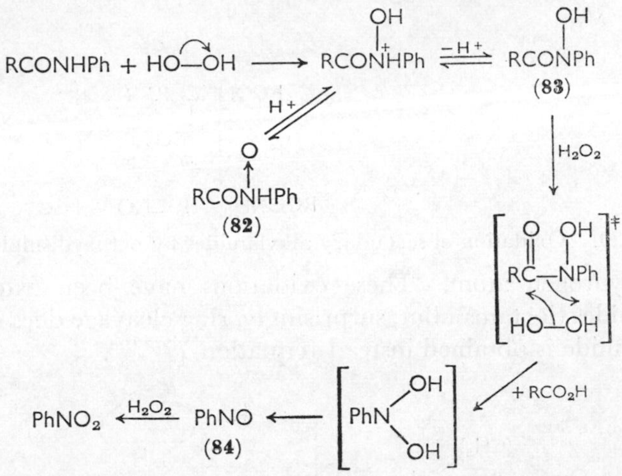

SCHEME 28. Oxidation of *N*-arylamides with hydrogen peroxide.

C. Peroxydisulphate

The lability of the α_N-hydrogen atoms towards radical abstraction is also evident with this reagent. Thus *N*-alkylamides (both secondary and tertiary) suffer dealkylation with peroxydisulphate salts at moderate temperatures (65°–90°) to yield a mixture of the less highly substituted amide and a carbonyl derivative (equation 76)[119a].

$$RCON(CH_2R^1)_2 \xrightarrow{S_2O_8^{2-}} RCONH(CH_2R^1) + R^1CHO \xrightarrow{S_2O_8^{2-}} RCONH_2 + R^1CHO \qquad (76)$$

Preliminary results of kinetic studies have been published and these show a $\frac{3}{2}$ order dependence in peroxydisulphate ion concentration, but a zero-order dependence in amide concentration[119b]. This fits nicely with the usual behaviour of the reagent, as the effective oxidant is thought to be a sulphate radical ion ($SO_4^{\cdot -}$) produced slowly from the peroxydisulphate ion[120]. It is also consistent with a radical-chain process (Scheme 29) involving induced decomposition of the peroxydisulphate ion by intermediate amidic radicals—hence the $\frac{3}{2}$ dependence in oxidant concentration. Both the products and related kinetic studies with *N*-acetyl-L-alanine[119b] are consistent with abstraction of

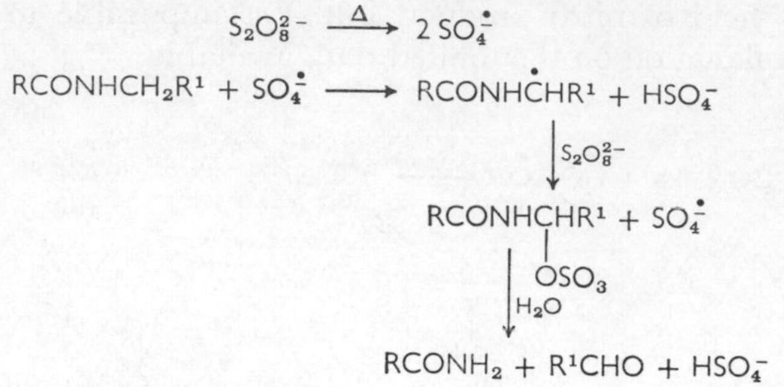

$$S_2O_8^{2-} \xrightarrow{\Delta} 2\,SO_4^{\overset{\cdot}{-}}$$

$$RCONHCH_2R^1 + SO_4^{\overset{\cdot}{-}} \longrightarrow RCON\overset{\cdot}{H}CHR^1 + HSO_4^{-}$$

$$\Big\downarrow S_2O_8^{2-}$$

$$\underset{\underset{\displaystyle OSO_3}{|}}{RCONH\overset{\cdot}{C}HR^1} + SO_4^{\overset{\cdot}{-}}$$

$$\Big\downarrow H_2O$$

$$RCONH_2 + R^1CHO + HSO_4^{-}$$

Scheme 29.   Oxidation of secondary alkylamides by peroxydisulphate salts.

the α_N-hydrogen atom. These oxidations have been extended to cyclic amides (lactams), but, surprisingly, ring cleavage does not occur and the imide is obtained instead (equation 77)[119c].

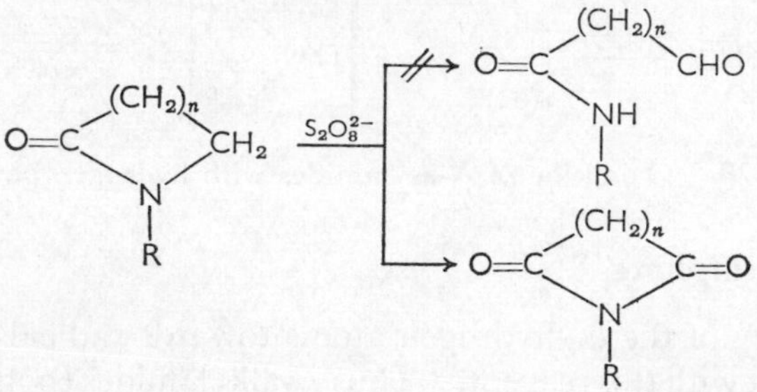

(77)

Primary amides may also be expected to react differently and this is indeed the case. Formamide decomposes to carbon dioxide and ammonia presumably via abstraction of the formyl hydrogen although this has not been proven[119a]. Smaller amounts of carbon dioxide and ammonia are also recovered in the reactions with N-alkylformamides and these probably arise from further decomposition of formamide obtained by dealkylation[119a]. Apparently acetamide does not react unless silver salts are added to the reaction mixture[119d]. Under these conditions, rapid deamination occurs even at 30° to give a quantitative yield of acetic acid. The rate of oxidation has a first-order dependence on both the peroxydisulphate and silver ion concentrations, but it is independent of the acetamide concentration[119d]. This again suggests that the effective oxidant is the sulphate radical-ion ($SO_4^{\overset{\cdot}{-}}$) produced slowly from the peroxydisulphate salt. Subsequent oxidation of the acetamide is not rate controlling (equation 78). The suggested mechanism for the oxidation (Scheme 30) involves hydrogen ab-

$$S_2O_8^{2-} \xrightarrow[\text{slow}]{Ag^+} 2\,SO_4^{\overset{\cdot}{-}} \xrightarrow[\text{fast}]{2\,CH_3CONH_2} 2\,CH_3CO_2H + 2\,N_2 \qquad (78)$$

straction from the nitrogen atom followed by dimerization of the resultant radical to 1,2-diacetylhydrazine (**85**)[119d]. This, on further oxidation, would form an azo intermediate (**86**) which is known to suffer homolytic fission of the carbon–nitrogen bonds giving nitrogen

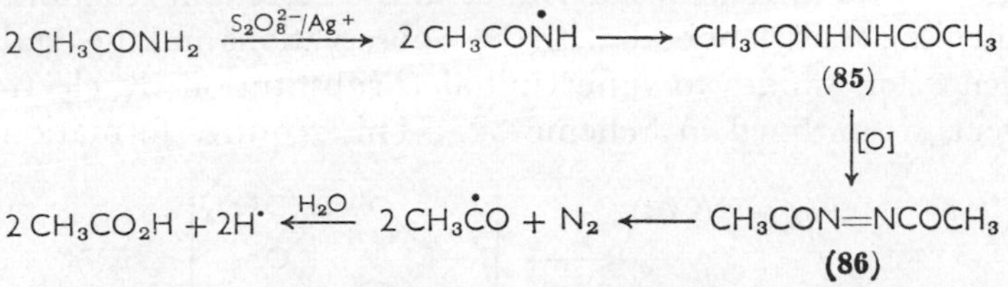

2 CH3CONH2
S2O8 2−/Ag+
2 CH3CONH
CH3CONHNHCOCH3
(85)
[O]
2 CH3CO2H + 2H·
H2O
2 CH3CO + N2
CH3CON=NCOCH3
(86)

SCHEME 30. Peroxydisulphate ion oxidation of acetamide.

and acetyl radicals. No one has yet investigated the effect of silver ion on the oxidation of secondary and tertiary amides, but this would appear to be a profitable exercise.

D. Lead Tetraacetate

Only primary amides react with lead tetraacetate via a peculiar oxidative rearrangement similar to the Hofmann reaction[121,122]. The initial product is an isocyanate (**87**) but this can only be isolated when an inert (aprotic), basic solvent such as dimethylformamide is the reaction medium[122]. Otherwise further reaction occurs rapidly with either the acetic acid coproduct to give acetylamines (**88**) and ureas (**89**), or with another proton source such as *t*-BuOH solvent to give urethanes (**90**) (Scheme 31)[121,122].

RCONH2 + Pb(OAc)4
RNCO + Pb(OAc)2 + 2 HOAc
(87)
HOAc
t-BuOH
RNHAc + RNHCONHR
RNHCO2Bu-t
(88)
(89)
(90)
40–80%
3–8%

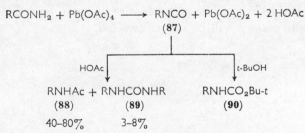

SCHEME 31. Reaction of primary amides with lead tetraacetate.

Some information about the mechanism of this interesting reaction is available. Rate measurements show a first-order dependence on both the amide and lead tetraacetate concentrations; also the oxidation is catalysed by base (pyridine, tertiary amines, etc.) and favoured by polar solvents (dimethylformamide is one of the better ones)[123]. By

analogy with the Curtius rearrangement in acetic acid (e.g. isocyanate formation from acyl azide) the reaction might be expected to proceed via a nitrene intermediate. Attempts to trap nitrenes have failed thus far[121], and this intimates that migration of the alkyl group is synchronous with nitrene formation, as in the Hofmann reaction. One mechanism, consistent both with these observations and the established tendency for amides to suffer initial *O*-substitution by electrophilic reagents, is outlined in Scheme 32. This requires formation of an

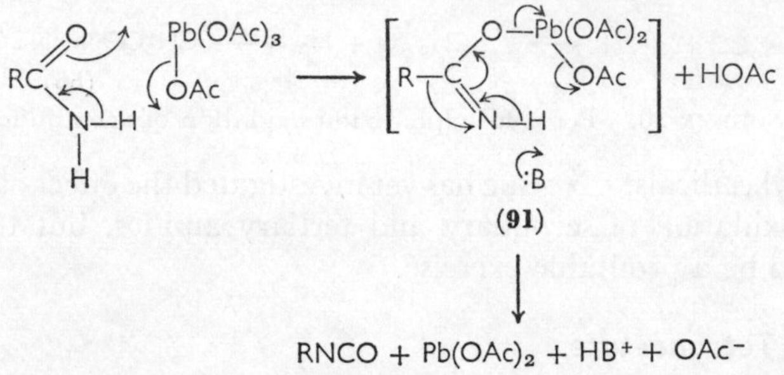

$$RNCO + Pb(OAc)_2 + HB^+ + OAc^-$$

SCHEME 32. Mechanism of the lead tetraacetate oxidation of primary amides.

O-imidate precursor (**91**) followed by base-induced rearrangement with elimination of $Pb(OAc)_2$ to the isocyanate. Other mechanisms have been suggested involving an initial substitution by lead tetraacetate on the amide nitrogen atom[121,122], but in the absence of definite evidence, further speculation is out of place.

An interesting application of this novel process is the preparation of various heterocyclic species through ring closure of *ortho*-substituted aromatic amides[123]. Two examples are given in equations (79a,b). Further applications will be keenly awaited.

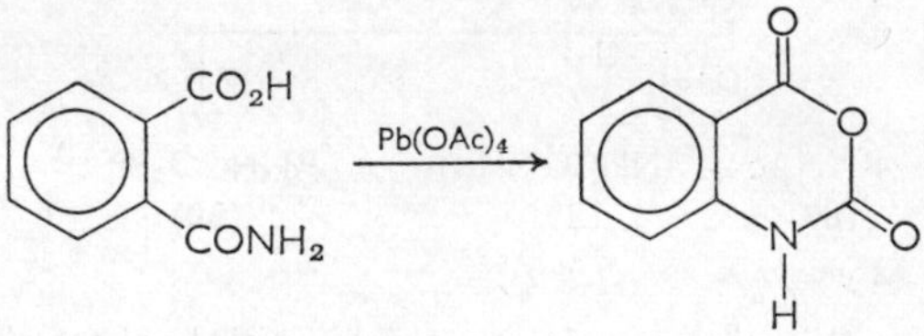

(79a)

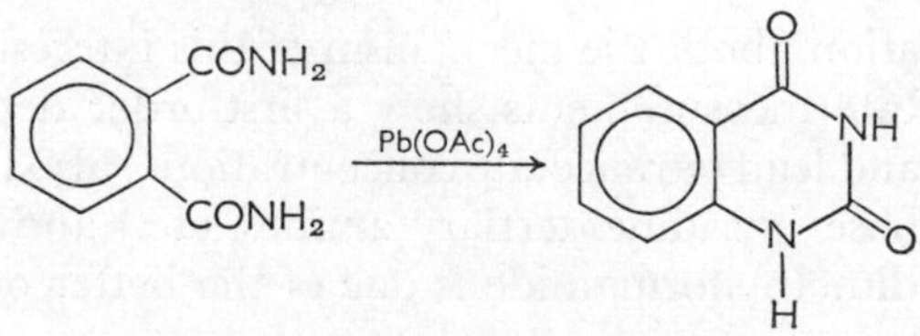

(79b)

VIII. REDUCTION

Amides, generally, are fairly resistant to reduction and only the more potent reagents such as alkali metal hydrides, sodium in liquid ammonia (Birch reduction) and electrolysis are effective; of these, the first method is foremost. Two distinct reaction paths have been recognized, whose importance depends primarily on the reagent and to a lesser extent on the type of amide as well. The first route (path a of Scheme 33) produces an amine with the same number of carbon

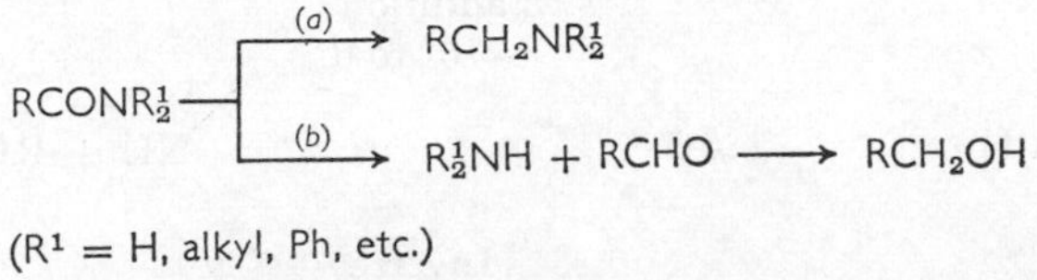

$$RCONR_2^1 \begin{cases} \xrightarrow{(a)} RCH_2NR_2^1 \\ \xrightarrow{(b)} R_2^1NH + RCHO \longrightarrow RCH_2OH \end{cases}$$

$(R^1 = H, alkyl, Ph, etc.)$

SCHEME 33.   General routes for the reduction of amides.

atoms as the original amide by reduction of the carbonyl group to methylene. The second route (path b) arises from fission of the carbon–nitrogen bond to give an aldehyde plus an amine (the aldehyde may then be reduced further to an alcohol). Both routes are synthetically useful and a high yield of a single product can usually be obtained by an appropriate choice of reagent and conditions.

A. Complex Metal Hydrides

Gaylord's excellent treatise[124] on these reagents has been outdated by recent developments. It is now known, for instance, that hydrides of both boron and aluminium are effective, although the latter are favoured on account of their higher reactivity. Products depend on several factors as discussed in the text below, but it is possible to make fairly reliable generalizations for the various amides and reagents, as summarized in Table 11. Exceptions are not unknown, however, and the cited references should also be consulted.

I. Primary and secondary amides

One can see from Table 11 that the majority of hydride reagents reduce these compounds to the corresponding amine (path a of Scheme 33) and not to the aldehyde (path b). There is not much to choose between the various hydrides in regard to their efficacy, and excellent yields may be expected throughout. It should be noted, however, that primary amides react sluggishly and periods of the

TABLE 11.   Normal products from complex metal hydride reductions of amides.

Amide[a]	Reagent	[Amide]/ [Reagent]	Special conditions[b]	Product	Refs.
$RCONH_2$	$LiAlH_4$	1:1		RCH_2NH_2	125,126
$RCONH_2$	$LiAlH_4$	1:0·5		RCN	125
$RCONHR^1$	$LiAlH_4$	1:0·75		RCH_2NHR^1	126
$RCONR_2^1$	$LiAlH_4$	1:0·5		$RCH_2NR_2^1$	126
$RCONR_2^{1\,c}$	$LiAlH_4$	1:0·25	Inverse addition $-70°$ to $0°$	RCHO	124
$RCON{\diagup}^d_{\diagdown}$	$LiAlH_4$	1:0·5		$^{\diagdown}_{\diagup}NH + RCH_2OH$	124
$RCON{\diagup}^d_{\diagdown}$	$LiAlH_4$	1:0·25	Inverse addition $-70°$ to $0°$	$^{\diagdown}_{\diagup}NH + RCHO$	124
$RCONR_2^1$	$LiAlH_4/AlCl_3$	1:0·5		$RCH_2NR_2^1$	127
$RCONH_2$	$LiAlH(OMe)_3$	1:4		RCH_2NH_2	128
$RCONR_2^1$	$LiAlH(OMe)_3$	1:2		$RCH_2NR_2^1$	128
$RCONR_2^1$	$LiAlH(OEt)_3$	1:1		RCHO	129
$RCONH_2$	AlH_3	1:1·33		RCH_2NH_2	128
$RCONR_2^1$	AlH_3	1:0·67		$RCH_2NR_2^1$	128
$RCONH_2$	$NaBH_4$	1:0·5	Reflux in diglyme	RCN	130
$RCONHR^1$	$NaBH_4$	1:0·75	Via O-ethyl imidate[e]	RCH_2NHR^1	2b
$RCONR_2^1$	$NaBH_4$	1:0·5	Via O-ethyl imidate[e]	$RCH_2NR_2^1$	2b
$RCONH_2$	BH_3	1:2·33		RCH_2NH_2	131,132
$RCONHR^1$	BH_3	1:2·0		RCH_2NHR^1	132
$RCONR_2^1$	BH_3	1:1·66		$RCH_2NR_2^1$	132

[a] R and R^1 = alkyl, aryl or heterocyclic.
[b] Usually the amide is added to the hydride solution, unless otherwise stated.
[c] Tendency to RCHO formation increases with bulky R^1 groups.
[d] Heteroaromatic compounds only.
[e] The O-ethyl imidate is prepared previous to the reduction (reaction 80).

order of 24 hr at 0° and 3 to 6 hr at 25° are required for completion[126,128]. Another important practical consideration to the attainment of high yields is the relative reactant concentrations. In addition to the expected two equivalents of hydride ion (H^-) for reduction of the carbonyl group, it is essential to add an *extra* one equivalent per

N-hydrogen atom present in the substrate, probably because the hydride also functions as a base for neutralization of the N-proton. The [amide]:[reagent] ratios listed in Table 11 take account of this requirement.

There is, in fact, fairly good evidence that reduction of primary amides proceeds via base-induced dehydration to a nitrile, which is then reduced to the amine (Scheme 34)[125]. It has been shown, for

$$RCONH_2 + 2 MH \longrightarrow RCN + M_2O + 2 H_2$$

$$RCN + 2 MH \longrightarrow RCH_2NM_2$$

$$RCH_2NM_2 + H_2O \longrightarrow RCH_2NH_2 + M_2O$$

$$(M = AlH_2, LiAlH_3, BH_2, etc.)$$

SCHEME 34. Overall stoicheiometry for the reduction of primary amides.

instance, that two molecules of hydrogen are evolved during reduction. Also, when a deficient quantity (i.e. only two equivalents of hydride ion) of $LiAlH_4$ is used, the nitrile derivative can be isolated from the reaction mixture as the major product[125]. Otherwise, not a great deal is known about the mechanism. Probably, an O-aluminate complex (**92a**) is formed first, but its exact structure has not been ascertained. It is known, however, that the second molecule of hydrogen is usually evolved more slowly than the first one, and that nitriles are reduced very rapidly to amines under the experimental conditions[126]. This tentatively fixes the rate-controlling step as hydrogen elimination from an O-imidate intermediate (**92b**). A

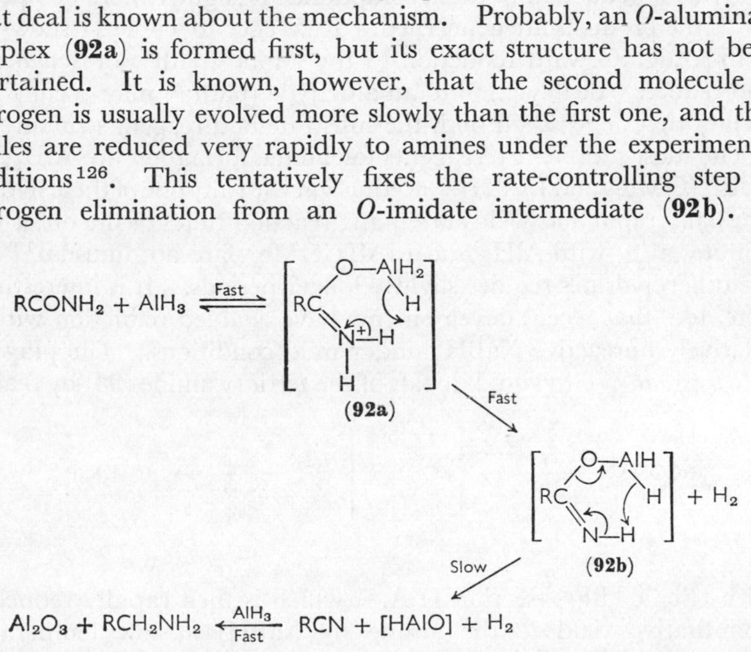

SCHEME 35. Tentative mechanism for the reduction of primary amides with aluminium hydride.

plausible mechanism based on these observations involving intramolecular elimination of hydrogen is illustrated for AlH_3 reduction in Scheme 35, but clearly some thorough investigations are required.

Reduction with borohydride reagents warrants special comment. For many years these were regarded as unreactive towards amides generally, but recent developments have invalidated that conclusion. Diborane, itself, is now known to reduce primary and secondary amides to their corresponding amines, with no C—N bond scission, in almost quantitative yields under relatively mild conditions[131,132]. With special techniques, $NaBH_4$ also reacts; primary amides, for example, are reduced to the nitrile on refluxing in solvent diglyme[130]. This latter reaction seems to be a useful alternative method for dehydrating primary amides in the absence of acidic reagents such as PCl_5 or $SOCl_2$ (section IX and section III of Chapter 4). Both reductions with $NaBH_4$ shown in Table 11 are obviously analogous to those with $LiAlH_4$.

2. Tertiary amides

Hydride reduction of these compounds is slightly more complex as far as the products are concerned. Reference to Table 11 shows that both straightforward reduction to the amine (path *a* of Scheme 33) and reductive dealkylation to an aldehyde (path *b*) may occur with a facility that depends on both the substrate and reagent structure.

The most satisfactory reagents for amine formation are AlH_3, BH_3, $LiAlH(OMe)_3$ and $LiAlH_4$. Reduction with any one of these hydrides is usually rapid and yields are high; reaction times of the order of 30 minutes at $0°$ with AlH_3 and $LiAlH(OMe)_3$ are not unusual[128], but the other hydrides require slightly longer periods. It is interesting to note, too, that recent developments have enabled reduction with the relatively unreactive $NaBH_4$ under mild conditions. The ploy here is to form an O-ethyl imidate salt of the tertiary amide (**93**) by reaction

$$RCONR_2^1 \xrightarrow[CH_2Cl_2]{(Et_3O)^+BF_4^-} \left[RC \underset{NR_2^1}{\overset{OEt}{\cdots}} \right]^+ BF_4^- \xrightarrow[10°/EtOH]{NaBH_4} RCH_2NR_2^1 \qquad (80)$$

$$(93)$$

with $(Et_3O)^+BF_4^-$ (section II.A.4) which is then rapidly reduced in quantitative yield to the amine by $NaBH_4$ at low temperatures (equation 80)[2b]. One notable advantage of this method is its high selectivity, and the amide moiety can be reduced in the presence of other functional groups. Secondary amides may be reduced in a similar way[2b].

Reductive deamination to an aldehyde (path *b* of Scheme 33) is less common, but often competes with amine formation (path *a*) in LiAlH$_4$ reductions. It is, in fact, the main reason for the low yields of amine sometimes obtained with LiAlH$_4$. Aldehyde formation is particularly prevalent when the *N*-substituents are bulky[124]. It will become the major pathway if deficient amounts (i.e. less than 0·5 equivalents) of LiAlH$_4$ are used at low temperature ($-70°$ to $0°$) and if the hydride is added to the amide (inverse addition) rather than vice versa as is more usual[124]. Also, reductive deamination to a mixture of aldehydes and alcohols is the usual reaction with *N*-heteroaromatic amides, irrespective of the experimental technique[124]. The most satisfactory synthetic method for aldehyde formation is, however, to use a substituted hydride such as LiAlH(OEt)$_3$ or LiAlH$_2$(OEt)$_2$: even unhindered tertiary alkylamides give yields of aldehyde to the extent of 60% to 90%, and the method seems to have general applicability[129].

The mechanism of tertiary amide reduction is not well understood, and there are almost as many theories as reagents employed. One of the earlier rationalizations, due to Weygand[133], invoking a common

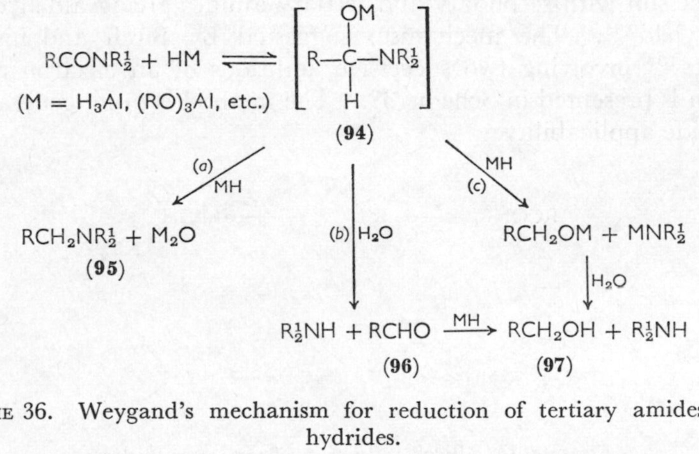

$$\text{RCONR}_2^1 + \text{HM} \rightleftharpoons \left[\begin{array}{c} \text{OM} \\ | \\ \text{R-C-NR}_2^1 \\ | \\ \text{H} \end{array} \right]$$

$$(\text{M} = \text{H}_3\text{Al}, (\text{RO})_3\text{Al}, \text{etc.})$$

(94)

(a) MH   (b) H$_2$O   (c) MH

$$\text{RCH}_2\text{NR}_2^1 + \text{M}_2\text{O} \quad\quad\quad\quad \text{RCH}_2\text{OM} + \text{MNR}_2^1$$

(95)

H$_2$O

$$\text{R}_2^1\text{NH} + \text{RCHO} \xrightarrow{\text{MH}} \text{RCH}_2\text{OH} + \text{R}_2^1\text{NH}$$

(96)   **(97)**

SCHEME 36. Weygand's mechanism for reduction of tertiary amides with hydrides.

tetrahedral intermediate (**94**) for all the products, seems to be the most satisfactory (Scheme 36). The intermediate **94**, some kind of *O*-aluminate complex, can react further by way of three routes; nucleophilic attack by the hydride reagent on the carbon–oxygen bond (path a), either intra- or intermolecularly, would form the tertiary amine **95**; hydrolysis would convert the intermediate **94** to the aldehyde **96**; and either further reduction of the aldehyde or nucleophilic attack by the hydride reagent on the carbon–nitrogen bond (path c) with subsequent hydrolysis would produce the alcohol **97**. However, any substantial proof for such a mechanism, and a suitable structure for the tetrahedral intermediate **94**, is entirely lacking.

It has been suggested that bulky N-substituents favour aldehyde formation by inhibiting lone-pair nitrogen electron delocalization and thereby increasing the susceptibility of the carbonyl group to nucleophilic attack by water[124]. However, the very same arguments should of course apply to substitution by the nucleophilic hydride reagent which leads to amine formation. Steric hindrance has also been cited to explain the prevalence of reductive deamination with substituted hydrides (e.g. $LiAlH(EtO)_3$, etc.)[129], but the absence of a second hydride ion (H^-) for intramolecular substitution of the carbon–oxygen bond must be another important factor.

B. Birch Reduction

This well known method of reduction using sodium in liquid ammonia, together with a proton source such as acetic acid or alcohol, is successful with secondary and tertiary amides giving aldehydes in high yield[134]. The mechanism suggested by Birch and his colleagues[134] involving two successive additions of an electron and a proton is presented in Scheme 37. The method is quite general and has wide applicability.

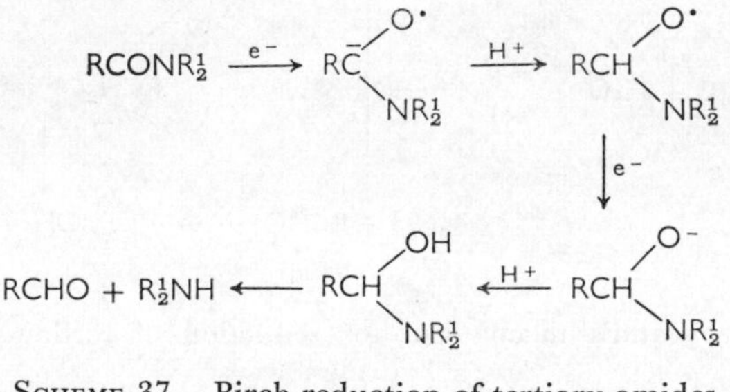

SCHEME 37. Birch reduction of tertiary amides.

C. Electrolytic Reduction

In contrast to the Birch method above, electrolytic reduction invariably produces an amine without C—N bond scission (path a of

Scheme 33)[135].   The process is usually carried out in acidic solutions (e.g. conc. H_2SO_4) using a lead cathode, and is most facile when the carbon or nitrogen atoms bear electron-donating substituents[135]. Presumably, the initial steps are similar to the Birch reaction, but further reduction of the amino alcohol **98** occurs (equation 81).

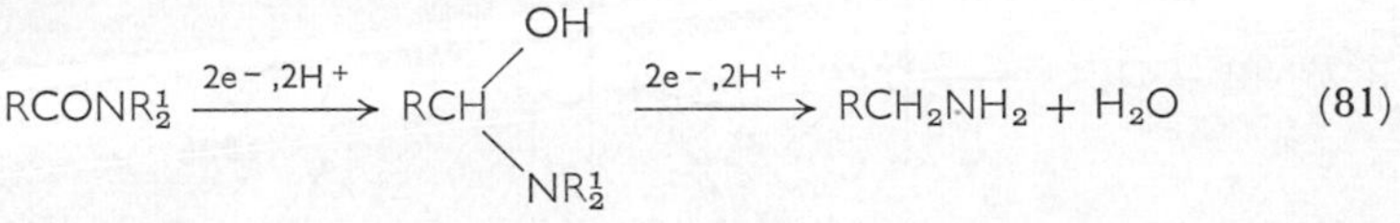

$$RCONR_2^1 \xrightarrow{2e^-,2H^+} RCH\begin{smallmatrix} OH \\ \diagup \\ \diagdown \\ NR_2^1 \end{smallmatrix} \xrightarrow{2e^-,2H^+} RCH_2NH_2 + H_2O \qquad (81)$$

$$(98)$$

D. Catalytic Reduction

These techniques have attracted little attention as far as amides are concerned.   It is known, however, that amines (path a of Scheme 33) are produced with copper chromite[136] or rhenium catalysts[137], but with substituted amides scrambling of the N-alkyl groups is an undesirable side-reaction. With common catalysts, such as Raney nickel or palladium on charcoal, high temperatures and pressures are usually required with the result that other active substituents are preferentially reduced or react. For example, with secondary or tertiary N-aminomethylamides, dealkylation is the major reaction (equation 82)[138].

$$RCONR^1CH_2NR_2^2 \xrightarrow{H_2/Ni \text{ or } Pd} RCONR^1H + CH_3NR_2^2 \qquad (82)$$

$$(R^1 = H, \text{alkyl, etc.})$$

IX. INORGANIC ACID HALIDES

Although the carbonyl function in neutral amides is normally un-reactive, nucleophilic addition will occur whenever external factors enhance the polarization of the carbon–oxygen bond.   In this way many reactive inorganic acid halides, such as phosgene, thionyl chloride and phosphorus pentachloride combine with amides usually to form an imidoyl chloride [$(RC(Cl)\!\!=\!\!NR_2^1)^+Cl^-$], although in some cases, and invariably with primary amides, further transformations in the reaction mixture result in dehydration to the nitrile. Acid fluorides interact similarly to produce the analogous covalent difluoro-alkylamine derivative $RCF_2NR_2^1$.

A reasoned explanation of these reactions is that formation of an O-complex (**99a**) between the amide and the acid halide polarizes the carbon–oxygen bond, which then suffers a rapid internal nucleophilic substitution by halide ion (an S_Ni process) with synchronous elimination of a neutral fragment such as CO_2, SO_2 or $POCl_3$ (equation 83). The imidoyl chloride (**99b**) may then react further, depending on its structure and the conditions, as is discussed in detail below.

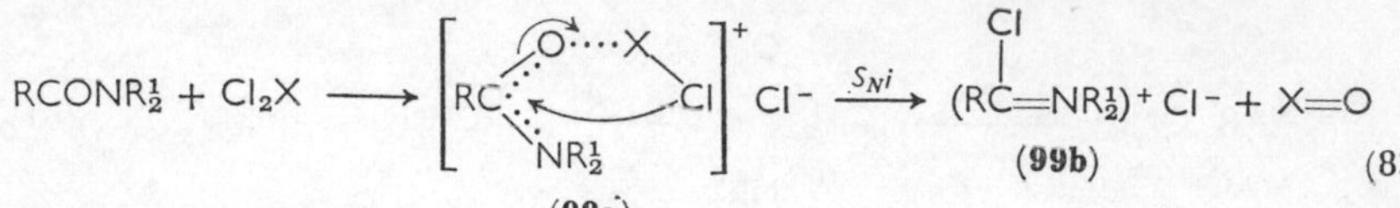

$$\text{RCONR}_2^1 + Cl_2X \longrightarrow \left[\text{RC} \begin{array}{c} \overset{+}{O}\cdots X \\ \diagdown \\ NR_2^1 \end{array} Cl \right]^+ Cl^- \xrightarrow{\,S_Ni\,} (RC{=}NR_2^1)^+ Cl^- + X{=}O \tag{83}$$

(**99a**)   (**99b**)

(X = CO, SO, PCl_3, etc.; R^1 = H, alkyl, Ar, etc.)

Not all acid halides react in this way. Two distinct structural features making the S_Ni process unfavourable can be envisaged and the initial O-complex may then be sufficiently stable to have more than a transient existence: the first is where the expelled fragment is unstable as, for example, in the Vilsmeier reactions with phosphorus oxychloride (equation 84)—the PO_2Cl fragment is known to be a hypo-

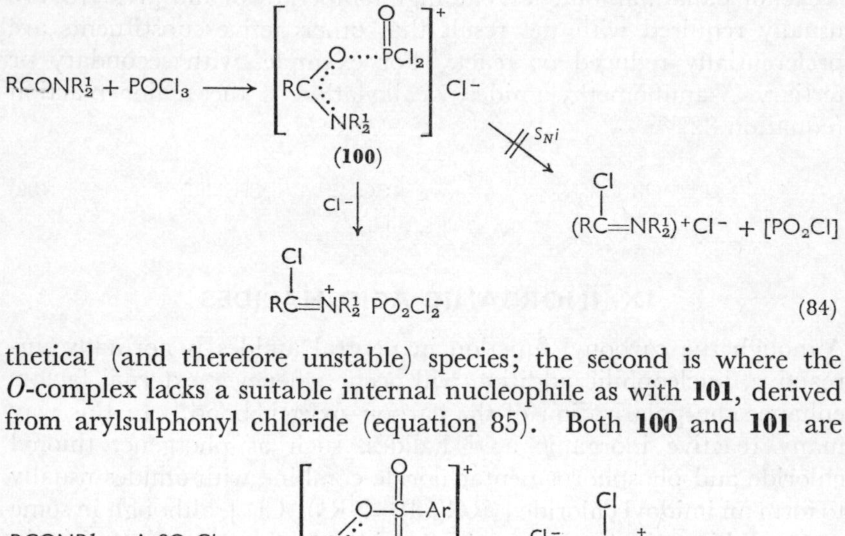

$$\text{RCONR}_2^1 + POCl_3 \longrightarrow \left[\text{RC} \begin{array}{c} \overset{+}{O}\cdots PCl_2 \\ \diagdown \\ NR_2^1 \end{array} \right]^+ Cl^- $$

(**100**) $\xrightarrow{\;\;\;\;\;\;S_Ni\;} (RC{=}NR_2^1)^+ Cl^- + [PO_2Cl]$

$$Cl^- \downarrow$$

$$\underset{RC{=}\overset{+}{N}R_2^1\ PO_2Cl_2^-}{\overset{Cl}{|}} \tag{84}$$

thetical (and therefore unstable) species; the second is where the O-complex lacks a suitable internal nucleophile as with **101**, derived from arylsulphonyl chloride (equation 85). Both **100** and **101** are

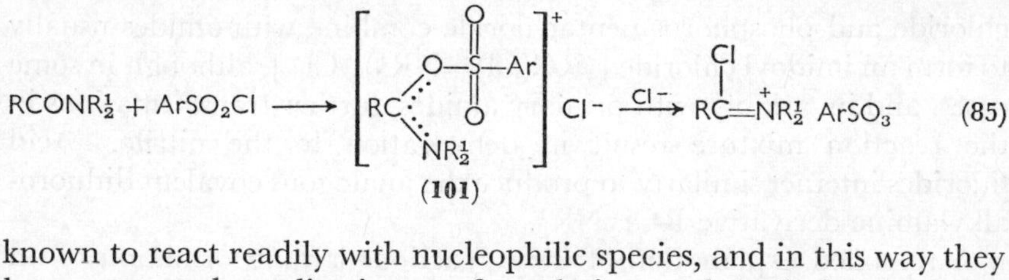

$$\text{RCONR}_2^1 + ArSO_2Cl \longrightarrow \left[\text{RC} \begin{array}{c} \overset{O}{\underset{O}{\overset{\|}{S}}}{-}Ar \\ \diagdown \\ NR_2^1 \end{array} \right]^+ Cl^- \xrightarrow{Cl^-} RC{=}\overset{+}{N}R_2^1\ ArSO_3^- \tag{85}$$

(**101**)

known to react readily with nucleophilic species, and in this way they have attracted application as formylating and acylating reagents

(Vilsmeier reactions). It therefore seems probable that imidoyl chloride formation in these instances results from an intermolecular attack by chloride ion.

A. Tertiary Amides

I. Phosgene, thionyl chloride and phosphorus pentachloride

The most stable imidoyl halides are derived from tertiary amides. Phosgene will react readily at room temperature[139] and is probably the best reagent for their preparation (equation 86). Recent work

$$RCONR_2^1 + COCl_2 \longrightarrow (R\overset{\overset{\displaystyle Cl}{|}}{C}=NR_2^1)^+Cl^- + CO_2 \qquad (86)$$

has shown that with excess phosgene (two equivalents) α-chloro-β-chlorocarbonyl enamines (**102a**) are also formed in up to 30% yield from tertiary amides bearing $\alpha_{C=O}$-hydrogen atoms (equation 87)[140]. These may be readily separated from the ionic imidoyl chloride (**102b**)

$$RCH_2CONR_2^1 + COCl_2 \longrightarrow RCH_2\overset{\overset{\displaystyle Cl}{|}}{C}\overset{+}{=}NR_2^1\ Cl^- + CO_2$$
$$\textbf{(102b)}$$
$$\Big\downarrow COCl_2$$
$$R\overset{\overset{\displaystyle COCl}{|}}{C}=CClNR_2^1 + 2\,HCl \qquad (87)$$
$$\textbf{(102a)}$$

by extraction with a non-polar solvent such as toluene. Thionyl chloride and phosphorus pentachloride also react to form an imidoyl chloride, but higher temperatures are required[139a]. The stronger conditions may account for some dealkylation to the nitrile with phosphorus pentachloride[141], as is commonly found for primary and secondary amides (cf. the Von Braun reaction in section IX.B.1).

Doubts existed for many years over the structure of these imidoyl chlorides. The ionic formulation $(R\overset{+}{CCl}=NR_2^1Cl^-)$ is now preferred over the isomeric covalent structure $(RCCl_2NR_2^1)$ on account of their typical 'salt-like' properties, their infrared spectrum and their ability to undergo nucleophilic substitution at carbon with a host of reagents such as alkoxides, amines, thiols and carboxylic acids[139]. A typical example is given in equation (88) for sodium alkoxide, in which subsequent hydrolysis yields the ester **103a** and the amine **103b**.

Further details of these useful transformations have been reviewed elsewhere [139a].

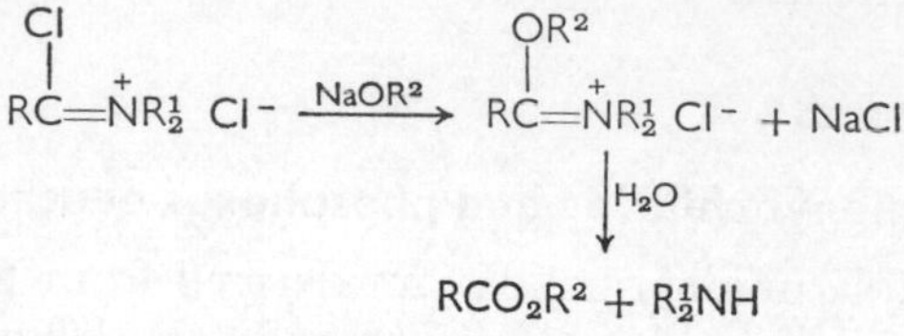

$$\underset{\substack{| \\ RC{=}\overset{+}{N}R_2^1 \ Cl^-}}{\overset{Cl}{}} \xrightarrow{\ NaOR^2\ } \underset{\substack{| \\ RC{=}\overset{+}{N}R_2^1 \ Cl^- \ + \ NaCl}}{\overset{OR^2}{}}$$

$$\Big\downarrow {}_{H_2O}$$

$$RCO_2R^2 \ + \ R_2^1NH \qquad\qquad (88)$$

$$\textbf{(103a)} \quad \textbf{(103b)}$$

2. Acid fluorides

Covalent products are obtained with acid fluorides. Carbonyl fluoride and sulphur tetrafluoride, for example, both react with tertiary amides to give the difluoroalkylamine **104**, usually a volatile liquid totally devoid of ionic properties (equation 89)[142,143].

$$RCONR_2^1 \xrightarrow{\ SF_4 \ or \ COF_2\ } RCF_2NR_2^1 \qquad\qquad (89)$$

$$\textbf{(104)}$$

Carbonyl fluoride appears to be the more effective because product purification is simpler, and the eliminated CO_2 has been shown by ^{14}C-radiotracer experiments to originate entirely from the reagent and not from the amide [142]. This suggests that the reaction also proceeds by way of an O-acyl intermediate (**105**) reminiscent of the transformations with acyl chlorides (equation 90).

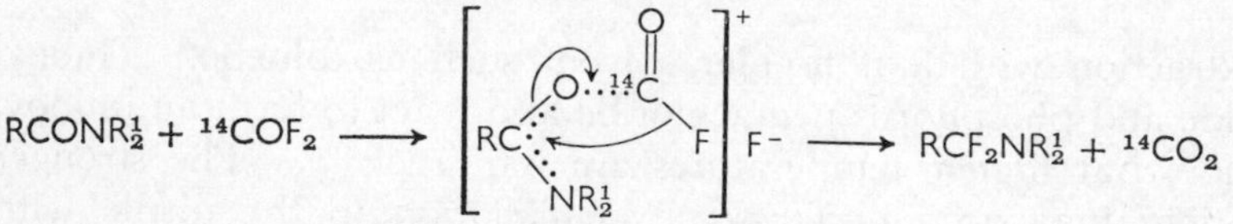

$$RCONR_2^1 + {}^{14}COF_2 \longrightarrow \left[\underset{\substack{| \\ NR_2^1}}{RC} \overset{O{\cdots}^{14}C}{\underset{}{\diagup}} \overset{\displaystyle\overset{O}{\|}}{} F \right]^{+} F^- \longrightarrow RCF_2NR_2^1 + {}^{14}CO_2 \quad (90)$$

$$\textbf{(105)}$$

The stability of the difluoroalkylamine product depends very much on the nature of the R substituent. When this is primary, hydrogen fluoride is readily eliminated (equation 91) and the resultant enamine

$$CH_3CF_2NR_2^1 \xrightarrow{\ -HF\ } RCH_2{=}CFNR_2^1 \xrightarrow{\ 2\,COF_2\ } (FCO)_2C{=}CFNR_2^1 \quad (91)$$

$$\textbf{(106)}$$

reacts further to yield **106** [142]. This is closely similar to the competing reaction observed with excess phosgene (cf. equation 87).

3. Phosphorus oxychloride and arenesulphonyl chlorides*

The reaction of phosphorus oxychloride with tertiary amides has been widely examined in connexion with the Vilsmeier–Haack method of aromatic formylation and the stable species isolated from these solutions is often referred to as the Vilsmeier complex[144]. The structure of this complex has been a contentious issue for several years and is not yet settled. Its typical salt-like properties are consistent, however, with an ionic formulation. Arguments based mainly on infrared spectral measurements[145,146] in favour of the O-acyl structure (107a) have been advanced, but more recent n.m.r. data (in particular the absence of hydrogen–phosphorus coupling in the dimethyl form-amide complex) are more consistent with the imidoyl chloride salt (107b)[147]. We have previously suggested, however, that structure

$$\left[\begin{array}{c} \overset{O}{\underset{\parallel}{}} \\ RC \overset{O-PCl_2}{\underset{NR_2^1}{\cdots}} \end{array} \right]^+ Cl^- \qquad RC \overset{Cl}{\underset{}{=}} \overset{+}{N}R_2^1 \; PO_2Cl_2^-$$

(107a)   (107b)

107a may be more stable for phosphorus oxychloride than for other acid halides and explained why imidoyl chloride formation may then be less facile. It therefore seems probable that *both* 107a and 107b can be isolated from solutions of phosphorus oxychloride and tertiary amides under the appropriate conditions. Vilsmeier complexes are readily attacked by nucleophilic species and in this way they fulfil an important role in synthetic organic chemistry. These reactions have been extensively reviewed[144] and we shall only discuss them briefly. One important point is that the low temperature conditions ($< 25°$) under which Vilsmeier complexes are normally prepared and allowed to react would not favour rearrangement of the O-acyl precursor 107a to the imidoyl salt 107b. Furthermore, the range of reactivity of these nucleophilic substitutions is much wider than with imidoyl chlorides prepared from other acid halides, and embraces aromatic and hetero-cyclic nuclei as well as amines, activated olefins and alkoxide ions. Both these observations suggest that the active species derived from phosphorus oxychloride and tertiary amides is therefore 107a.

The most important substitutions are those by aromatic and hetero-cyclic compounds, which are known as Vilsmeier–Haack formylation reactions. This process results in the introduction of the CHO

* See also section IV of Chapter 4.

substituent into the aromatic or heterocyclic nucleus.   It is usually accomplished by first complexing a disubstituted formamide with an equimolar proportion of phosphorus oxychloride, and then treating this mixture with the substrate; subsequent hydrolysis affords the aldehyde (e.g. **108**).   N-Methylformanilide is usually employed, but either dimethylformamide or N-formylpiperidine are suitable alternatives[144].   Simple aromatics such as benzene and naphthalene fail to react, but the more basic polynuclear aromatic hydrocarbons, heterocyclic compounds such as indole and pyrrole and benzene derivatives with electron-donating substituents (e.g. **OH, OCH$_3$, NMe$_2$, NHMe**) are successfully converted to their aldehyde derivatives[144].   This is illustrated in Scheme 38 for dimethylformamide and phenol: as expected, orientation is usually *ortho* and *para*.

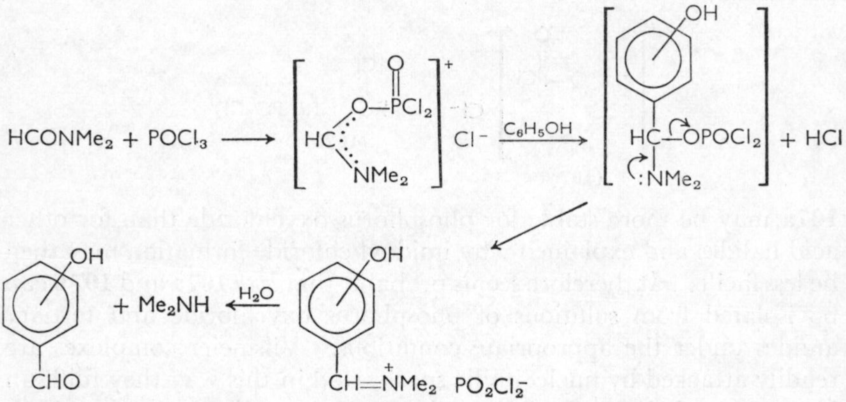

SCHEME 38.   Formylation of phenol by the Vilsmeier–Haack reaction.

Tertiary amides other than formamides also complex with phosphorus oxychloride and then react with suitable substrates to form the corresponding keto derivatives.   These transformations have not been widely exploited, but the early patent literature lists many examples of aromatic acylation[144] and recent investigations have demonstrated their viability with benzofuran[148] (Scheme 39) as well as with several indole[149,150] and pyrrole[149,151] derivatives.

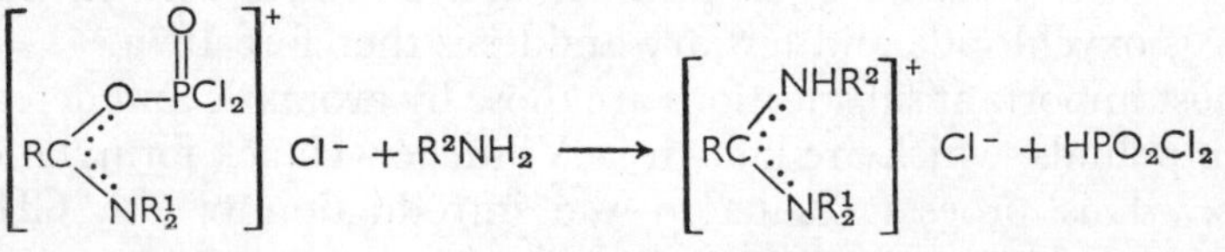

$$ (92) $$

(**109**)

(R^2 = alkyl, Ar, RNH)

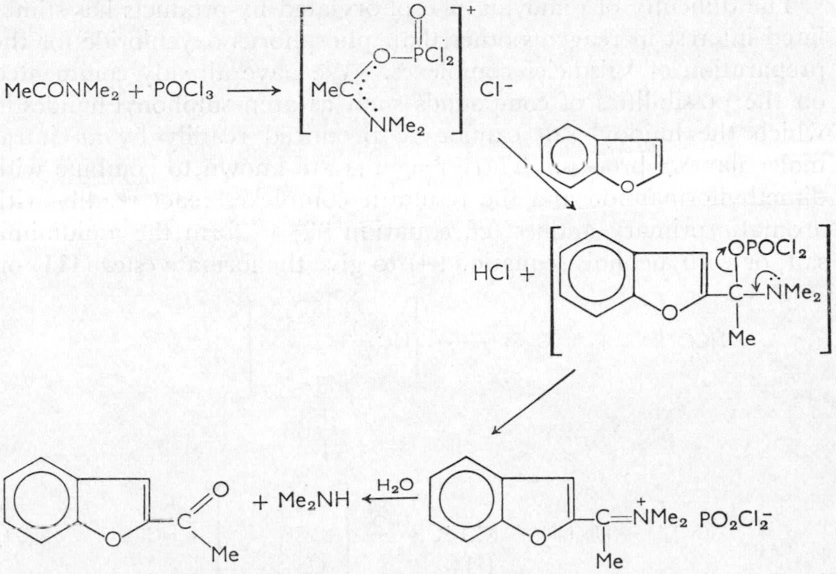

$$MeCONMe_2 + POCl_3 \longrightarrow \left[MeC \begin{array}{c} O-PCl_2 \\ \vdots \\ NMe_2 \end{array} \right]^+ Cl^-$$

HCl +

$$\left[\begin{array}{c} OPOCl_2 \\ C-NMe_2 \\ Me \end{array} \right]$$

$$\begin{array}{c} O \\ C \\ Me \end{array} + Me_2NH \xleftarrow{H_2O} \quad C=\overset{+}{N}Me_2 \ PO_2Cl_2^- \\ Me$$

SCHEME 39. Vilsmeier–Haack acetylation of benzofuran.

As with imiduyl halides (section IX.A.1), many other nucleophilic species readily attack the carbonyl carbon atom of the Vilsmeier complex. An amidinium salt (**109**), for example, is obtained from either aliphatic amines or hydrazines (equation 92)[146]. It is noteworthy that primary aromatic amines react in this way, too[152], in contrast to ring substitution observed with their secondary and tertiary counterparts.

Another illustrative case is the reaction with olefins (equation 93), which after hydrolysis yields the unsaturated aldehyde (**110**)[144].

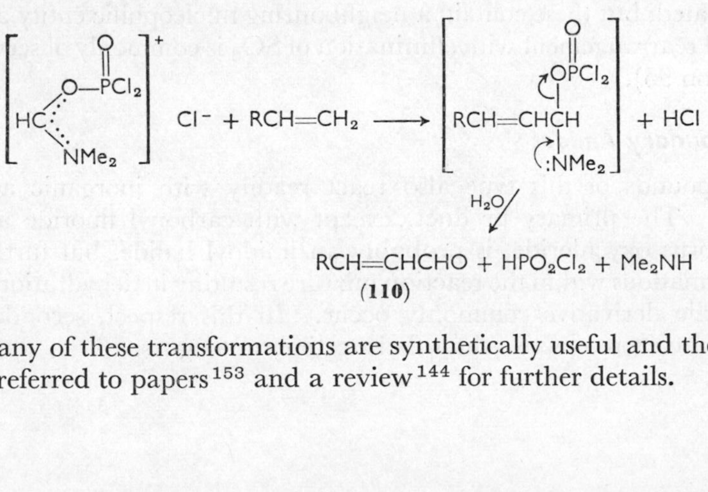

$$\left[HC \begin{array}{c} O-PCl_2 \\ \vdots \\ NMe_2 \end{array} \right]^+ Cl^- + RCH=CH_2 \longrightarrow \left[\begin{array}{c} O \\ OPCl_2 \\ RCH=CHCH \\ :NMe_2 \end{array} \right] + HCl$$

$$\xrightarrow{H_2O}$$

$$RCH=CHCHO + HPO_2Cl_2 + Me_2NH \quad (93)$$
$$(\mathbf{110})$$

Many of these transformations are synthetically useful and the reader is referred to papers[153] and a review[144] for further details.

The difficulty of removing phosphorylated by-products has stimulated interest in reagents other than phosphorus oxychloride for the preparation of Vilsmeier complexes. We have already commented on the possibilities of compounds such as arenesulphonyl halides in which the imidoyl salt cannot be produced readily by an intramolecular $S_N i$ process. These reagents are known to combine with dimethylformamide and the resultant complexes react readily with aromatic primary amines (cf. equation 92) to form the amidinium salt, or with alcohols (equation 94) to give the formate ester (111) on

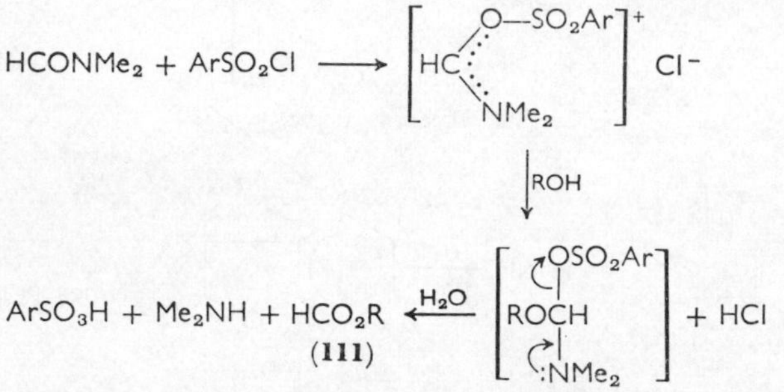

$$\text{HCONMe}_2 + \text{ArSO}_2\text{Cl} \longrightarrow \left[\text{HC} \underset{\text{NMe}_2}{\overset{\text{O—SO}_2\text{Ar}}{\cdots}} \right]^+ \text{Cl}^-$$

$$\downarrow \text{ROH}$$

$$\text{ArSO}_3\text{H} + \text{Me}_2\text{NH} + \text{HCO}_2\text{R} \xleftarrow{\text{H}_2\text{O}} \left[\begin{array}{c} \overset{\text{OSO}_2\text{Ar}}{\text{ROCH}} \\ \overset{|}{\text{:NMe}_2} \end{array} \right] + \text{HCl} \qquad (94)$$
$$\qquad\qquad (111)$$

hydrolysis[154]. Thus far, however, aromatic substitution with these complexes has not been reported. Other sulphonyl chloride derivatives, such as $\text{Me}_2\text{NSO}_2\text{Cl}$[155] and $\text{RO}_2\text{CNHSO}_2\text{Cl}$[156], have been

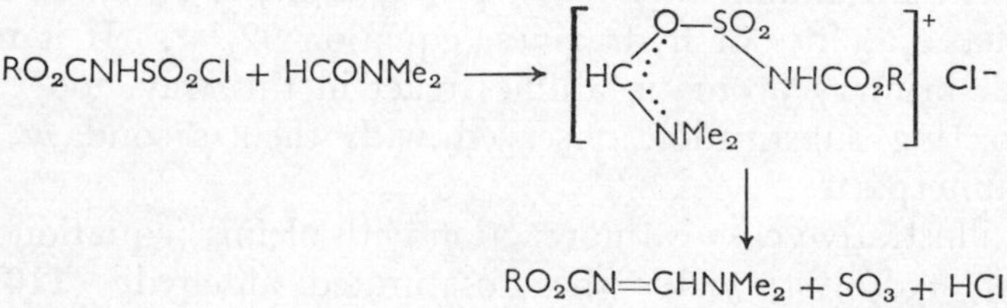

$$\text{RO}_2\text{CNHSO}_2\text{Cl} + \text{HCONMe}_2 \longrightarrow \left[\text{HC} \underset{\text{NMe}_2}{\overset{\text{O—SO}_2}{\cdots}} \text{NHCO}_2\text{R} \right]^+ \text{Cl}^-$$

$$\downarrow$$

$$\text{RO}_2\text{CN}{=}\text{CHNMe}_2 + \text{SO}_3 + \text{HCl} \qquad (95)$$

investigated, but these contain a neighbouring nucleophilic entity and internal rearrangement with elimination of SO_3 is commonly observed (equation 95).

B. Secondary Amides

Compounds of this type also react readily with inorganic acid halides. The primary product, except with carbonyl fluoride and phosphorus oxychloride, is probably an imidoyl halide, but further transformations within the reaction mixture resulting in dehydration to the nitrile derivative commonly occur. In this respect, secondary amides are more reactive than their tertiary analogues.

I. Thionyl chloride and phosphorus pentachloride*

Treatment of many (but not all!) N-alkylamides with either one of these reagents may induce N–alkyl bond fission on heating, with the formation of a nitrile and the alkyl halide (equation 96). This

$$\text{RCONHR}^1 + \text{PCl}_5 \xrightarrow{\Delta} \text{RCN} + \text{R}^1\text{Cl} + \text{POCl}_3 \qquad (96)$$

remarkable reaction was discovered in 1900 by von Pechmann using phosphorus pentachloride[157]; subsequently von Braun carried out extensive investigations and the reaction now bears his name[158]. More recently, thionyl chloride has proven to be a better reagent, since product isolation and purification is then simpler[159]. The reaction is by no means universal; formamides and some alkylamides are notably unreactive, but good yields may be obtained from benzamides. The reason for this selectivity is discussed later.

Although details of the mechanism await elucidation, there is evidence pointing to the intermediacy of imidoyl chlorides. It has been demonstrated independently, for example, that thermal decomposition of N-alkylbenzimidoyl chloride under von Braun conditions leads to benzonitrile and alkyl halide (equation 97)[158,159].

$$\underset{\underset{\displaystyle \text{PhC}=\text{NR}}{|}}{\overset{\displaystyle \text{Cl}}{}} \xrightarrow{\Delta} \text{PhCN} + \text{RCl} \qquad (97)$$

Also, it is possible to isolate the expected imidoyl chloride from the reaction of either benzanilide or N-benzylbenzamide with thionyl chloride[159] and of several N-alkylbenzamides with phosphorus pentachloride[160].

Suggestions have been made that decomposition of the imidoyl chloride to the nitrile may proceed via two limiting pathways (Scheme 40)[159]. The first involves heterolysis of the N–alkyl bond by an 'S_N1-like' mechanism, followed by rapid expulsion of the chloride ion (equation 98). The other is a reversal of this sequence, with chloride ion loss occurring first to give the intermediate **112**, which then interacts with the chloride ion in an 'S_N2-like' rate-determining step (equation 99). The necessity for two concurrent pathways comes from stereochemical studies with optically active substrates and from structural effects of both the R and R^1 groups. When R is aromatic, for instance, the S_N2 process should be favoured by stabilization of the intermediate carbonium ion **112**, and the inversion of optical rotation reported for benzamides with assymmetric R^1 substituents[161] is consistent with this argument. On the other hand, the S_N1 process

* See also section III.A.2 of Chapter 4.

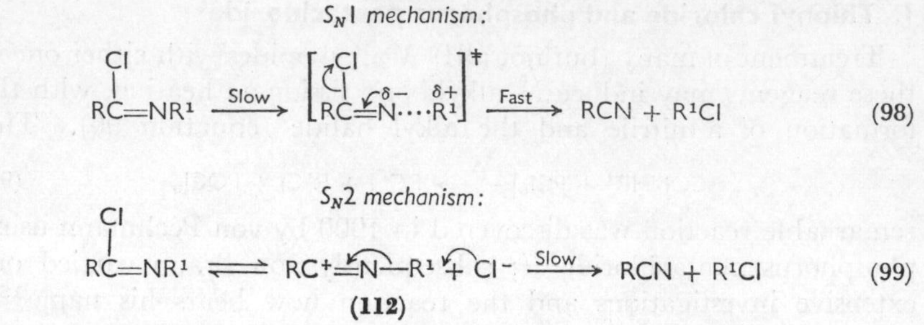

S_N1 *mechanism*:

$$RC\underset{\underset{Cl}{|}}{=}NR^1 \xrightarrow{\text{Slow}} \left[\underset{\underset{Cl}{|}}{RC} \xrightarrow{\delta-}{\underset{}{N}} \cdots \overset{\delta+}{R^1} \right]^{\ddagger} \xrightarrow{\text{Fast}} RCN + R^1Cl \qquad (98)$$

S_N2 *mechanism*:

$$RC\underset{\underset{Cl}{|}}{=}NR^1 \rightleftharpoons RC^+\underset{=}{=}N\!-\!R^1 + Cl^- \xrightarrow{\text{Slow}} RCN + R^1Cl \qquad (99)$$
$$\textbf{(112)}$$

SCHEME 40.　Thermal decomposition of imidoyl chlorides.

should be more facile for amides in which the displaced N-alkyl (R^1) group forms a stable carbonium ion. This is partially substantiated by the result for N-alkylacetamides; with optically active $(-)$-N-(α-methylbenzyl)acetamide, for example, racemization concurs with nitrile formation, and increasing yields of alkyl halide product along the series N-benzyl-, N-α-methylbenzyl- and N-benzhydrylacetamide suggest that steric factors are unimportant in their transition states[159]. Both observations are characteristic of an S_N1 mechanism.

In the light of these proposals, one can now account for the failure of the von Braun reaction with some secondary amides. The most reactive substrates should be those either in which the $\alpha_{C=O}$ substituent (R) stabilizes the intermediate carbonium ion (**112**) or in which the α_N substituent (R^1) exists as a relatively stable carbonium ion. These predictions are borne out in practice.

2. Phosgene

N-Alkylamides react with phosgene to form imidoyl chlorides (equation 100)[162]. There is no evidence that further reaction of the

$$RCONHR^1 + COCl_2 \xrightarrow{\Delta} RC\underset{\underset{Cl}{|}}{=}NR^1 + CO_2 + HCl \qquad (100)$$

$$RCH_2CONHR^1 + COCl_2 \xrightarrow{\Delta} RCH_2C\underset{\underset{Cl}{|}}{=}\overset{+}{N}HR^1 \; Cl^- + CO_2$$

$$\searrow$$

$$RCH\underset{\underset{Cl}{|}}{=}CNHR^1 + HCl \qquad (101)$$

von Braun type occurs, although in principle this must surely be possible. Enamine formation via elimination of the $\alpha_{C=O}$-hydrogen atom has been reported, however, in a few instances (equation 101)[163].

It will be recalled that similar reactions occur with tertiary amides on treatment with either phosgene or carbonyl fluoride. This kind of elimination may also compete with the von Braun reaction using thionyl chloride and phosphorus pentachloride, but any evidence is again lacking.

3. Phosphorus oxychloride

N-Alkylamides (other than formamides) usually suffer dealkylation to the corresponding nitrile with phosphorus oxychloride, and only in special circumstances does nucleophilic substitution, reminiscent of the Vilsmeier–Haack transformation with tertiary amides, occur instead. The dealkylation reactions have not been closely examined because thionyl chloride and phosphorus pentachloride do the same job more effectively. It is known, however, that high temperatures ($\sim 120°$) are required and that only compounds for which the α_N substituent will form a relatively stable carbonium ion (e.g. *N*-benzyl-, *N*-*t*-butyl-, or *N*-cyclohexylbenzamide) react readily[164]. This suggests that an S_N1 process operates, similar to one of the two limiting paths depicted in Scheme 39 for the von Braun reaction. Unlike the latter, however, alkyl halides are not the usual coproduct. Instead, the expelled carbonium ion (R^{1+}) either reacts with the aromatic solvent or eliminates a proton to give an olefin. This different behaviour is also consistent with an S_N1 process and further suggests that the initial *O*-acyl complex (**113**) rearranges to the imidoyl salt (**114**) before dealkylation occurs, probably because of the high temperatures employed. A mechanism consistent with these findings is outlined in Scheme 41.

An unusual reaction occurs with secondary formamides resulting in overall dehydration to the isonitrile (equation 102). Product

$$\text{HCONHR}^1 + \text{POCl}_3 \xrightarrow{\text{C}_5\text{H}_5\text{N}} \text{R}^1\text{—NC} + \text{HCl} + \text{HOPOCl}_2 \qquad (102)$$

formation is favoured by basic solvents such as pyridine or quinoline, but the addition of *t*-butoxide is recommended for formanilides[165]. It is interesting that arenesulphonyl chlorides are even more expeditious (and more convenient as far as work-up is concerned) than phosphorus oxychloride[166,167], with good yields of isonitrile obtained on standing at room temperature in solvent quinoline. The mechanism of dehydration has not been investigated, but the low-temperature conditions suggest that the relatively stable *O*-acyl complex **115**, only formed from phosphorus oxychloride or arenesulphonyl chlorides, is involved. A plausible explanation would then

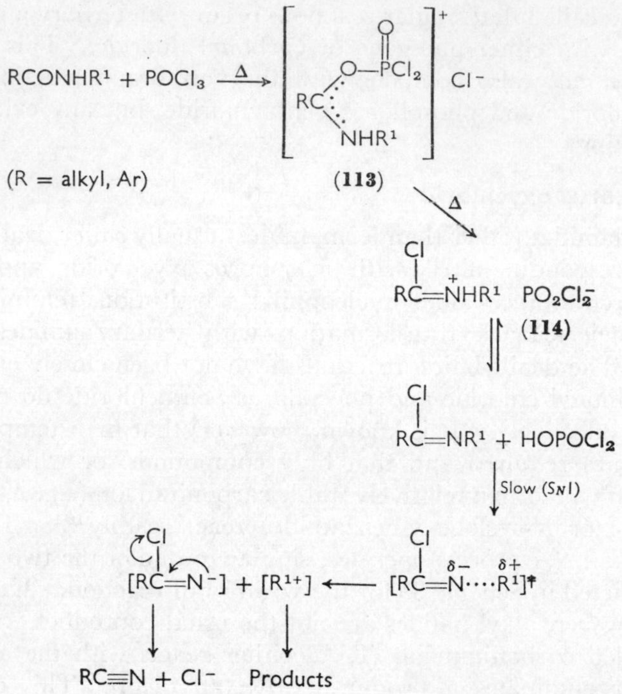

$$RCONHR^1 + POCl_3 \xrightarrow{\Delta} \left[\begin{array}{c} RC \underset{NHR^1}{\overset{O-PCl_2}{}} \end{array} \right]^+ Cl^-$$

(R = alkyl, Ar) **(113)**

$$\downarrow \Delta$$

$$RC \overset{Cl}{=} \overset{+}{N}HR^1 \quad PO_2Cl_2^- \quad \textbf{(114)}$$

$$\updownarrow$$

$$RC \overset{Cl}{=} NR^1 + HOPOCl_2$$

$$\downarrow \text{Slow (}S_N1\text{)}$$

$$[RC \overset{Cl}{=} N^-] + [R^{1+}] \longleftarrow [RC \overset{Cl}{=} N \cdots R^1]^{\ddagger}$$

$$\downarrow \qquad\qquad \downarrow$$

$$RC \equiv N + Cl^- \quad \text{Products}$$

SCHEME 41. Phosphorus oxychloride catalysed dealkylation of secondary amides.

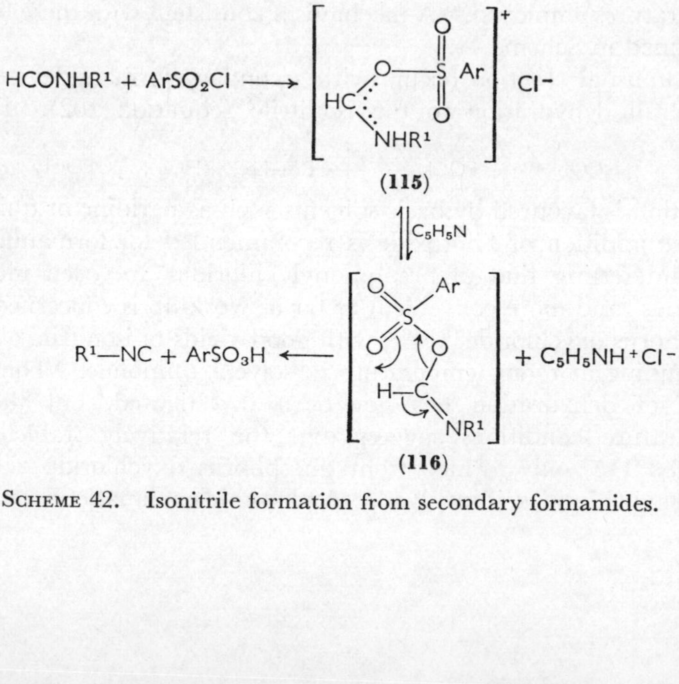

$$HCONHR^1 + ArSO_2Cl \longrightarrow \left[\begin{array}{c} HC \underset{NHR^1}{\overset{O-S-Ar}{}} \end{array} \right]^+ Cl^-$$

(115)

$$\updownarrow C_5H_5N$$

$$R^1{-}NC + ArSO_3H \longleftarrow \left[\begin{array}{c} O{=}S{-}Ar \\ H{-}C \\ NR^1 \end{array} \right] + C_5H_5NH^+Cl^-$$

(116)

SCHEME 42. Isonitrile formation from secondary formamides.

be an intramolecular proton abstraction from the neutral imidate (**116**) as shown in Scheme 42.

Vilsmeier–Haack-type aromatic formylations with secondary amides in the presence of phosphorus oxychloride are not common. The Bischler–Napieralski isoquinoline synthesis from acyl derivatives of (β-phenylethyl)amines (**117**) may involve a related sequence, however, in which the aromatic nucleus attacks the neutral *O*-acyl imidate (**118**) to effect cyclization (Scheme 43). No definitive evidence is available

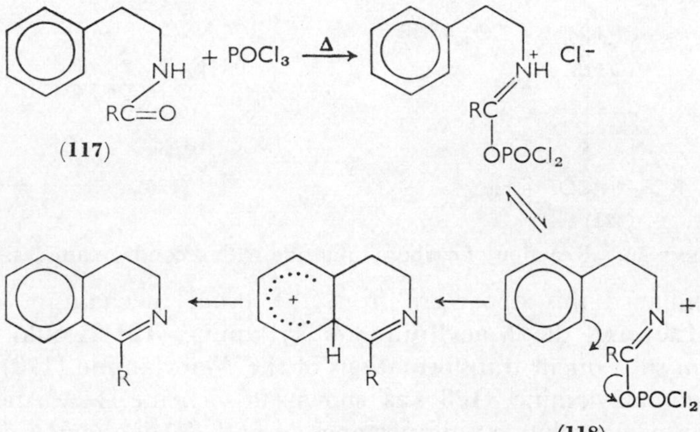

SCHEME 43. Bischler–Napieralski isoquinoline synthesis.

on this point, but it has been shown that nitrile formation competes with cyclization particularly when the substrate is substituted in the α_N position[164].

4. Carbonyl fluoride

Atypical reactions are observed with this reagent, which bear a much closer resemblance to acylation with organic rather than inorganic acid halides (section IV.A). The usual products are a mixture of the *N*-acylamide **119** and the ureide **120** together with a small amount of the *N*-acyl(difluoroalkyl)amine **121**[142]. With cyclic secondary amides (lactams), ring cleavage also occurs[142]. The most satisfactory and consistent way of accounting for these products is shown in Scheme 44. The usual *O*-acyl imidate salt (**122**) is formed initially, but O to N rearrangement (path *a*) must be faster than intramolecular ($S_N i$) attack by the fluoride ion (path *b*). Both the greater strength of the carbon–fluorine bond (relative to C—Cl, for example) and the poor nucleophilic properties of the fluoride ion

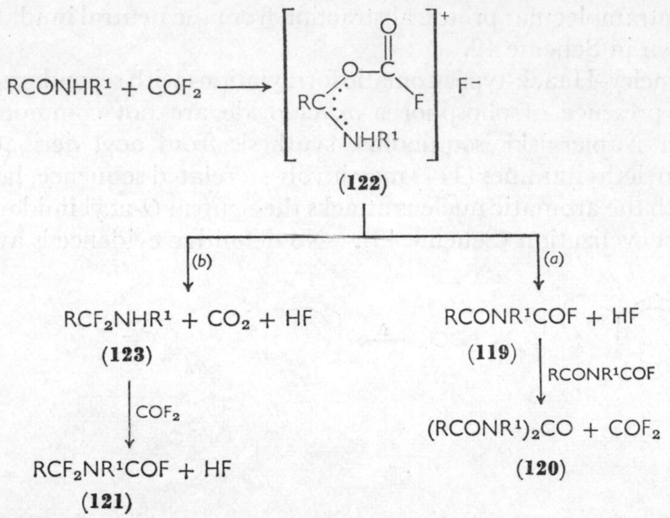

$$RCONHR^1 + COF_2 \longrightarrow \left[\begin{array}{c} O \\ \| \\ O-C \\ RC \diagup \quad \diagdown F \\ \vdots \\ NHR^1 \end{array} \right]^+ F^-$$

(122)

$$(b) \qquad\qquad\qquad\qquad (a)$$

$$RCF_2NHR^1 + CO_2 + HF \qquad\qquad RCONR^1COF + HF$$

(123)   (119)

$$\downarrow COF_2 \qquad\qquad\qquad\qquad \downarrow RCONR^1COF$$

$$\qquad\qquad\qquad\qquad\qquad (RCONR^1)_2CO + COF_2$$

$$RCF_2NR^1COF + HF \qquad\qquad\qquad (120)$$

(121)

SCHEME 44.   Reaction of carbonyl fluoride with secondary amides.

would enhance this departure from the usual mechanism. The ureide (120) and the N-acyl(difluoroalkyl)amine (121) would then arise from subsequent transformations of the N-acylamide (119) and the difluoroalkylamine (123) as shown in Scheme 44. Another indication of the relative unimportance of path (b) is the absence of enamine products from suitable substrates, unlike the reactions between phosgene and N-alkylamides discussed above (section IX.B.2).

C. Primary Amides*

The dehydration of primary amides to nitriles by almost any acid halide and phosphorus pentoxide is one of the better known aspects of amide chemistry. These reactions were last reviewed some time ago[168], but subsequent developments have been few. We have already discussed the application of organic acid halides in section IV.A. As far as the inorganic reagents are concerned, dehydration with phosphorus pentachloride, thionyl chloride, phosgene and phosphorus oxychloride are all of preparative value (equation 103). Both aliphatic and aromatic substrates react and the use of basic

$$RCONH_2 + XCl_2 \xrightarrow{\Delta} RCN + X{=}O + 2\,HCl \qquad\qquad (103)$$

$$(X = CO, SO, PCl_3, POCl)$$

* See also section III.A.1 of Chapter 4.

solvents (pyridine, dimethylaniline, N-alkylformanilides, etc.) is recommended to remove acid by-products[168]. The most convenient reagent is probably thionyl chloride: it is less toxic than phosgene and most of the by-products are volatile. Recent work has shown that in solvent dimethylformamide, dehydration with thionyl chloride can be effected at 0° with high yields of relatively pure nitrile[169]. With phosphorus oxychloride less than an equimolar amount of reagent is required (0·25 to 0·5 molar equivalents is typical)[170] and this has been attributed to regeneration of phosphorus oxychloride by dispro-portionation of dichlorophosphoric acid ($HOPOCl_2$) formed in the dehydration process (equation 104)[168].

$$3\ HOPOCl_2 \longrightarrow 2\ POCl_3 + H_3PO_4 \tag{104}$$

Although the synthetic applications of dehydration have been widely studied (reference 168 gives all the relevant details), sur-prisingly little is known about the mechanism. It seems possible that in some cases, at least, a sequence of O-acyl and imidoyl halide inter-mediates (Scheme 45) is involved, as in the dehydration of secondary and tertiary amides. No reliable evidence is available on this point.

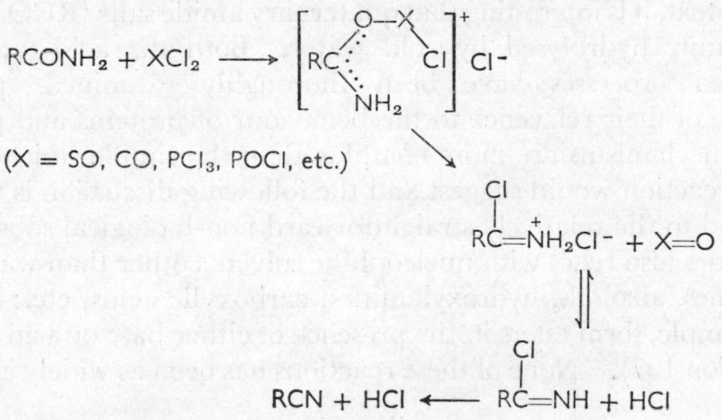

SCHEME 45. Dehydration of primary amides.

With phosphorus pentachloride the situation may be slightly different as the phosphoryl dichloride derivative of the iminochloride ($RCCl{=}NPOCl_2$) has been isolated from the reaction mixture for α-chloroacetamides[170]. Other more recent work has shown that conformation at the $\alpha_{C=O}$-carbon atom is not changed on dehydration with either thionyl chloride or phosphorus oxychloride[171,172]. This is illustrated for 4-t-butylcyclohexylcarboxamide in equation (105). This result is consistent with the mechanism outlined in Scheme 45.

$$t\text{-Bu} \diagdown \diagup \diagdown \diagup -\text{CONH}_2 \xrightarrow[\text{POCl}_3/106°]{\text{SOCl}_2/80°} t\text{-Bu} \diagdown \diagup \diagdown \diagup -\text{C}\equiv\text{N} \qquad (105)$$

(trans)   (trans)

X. HYDROLYSIS AND SOLVOLYSIS

Amides hydrolyse under suitable conditions to regenerate the parent carboxylic acid and an amine (equation 106). Since the initial step

$$\text{RCONR}_2^1 + \text{H}_2\text{O} \xrightarrow{\text{H}^+ \text{ or OH}^-} \text{RCO}_2\text{H} + \text{R}_2^1\text{NH} \qquad (106)$$

involves nucleophilic addition to the carbonyl function, it is not surprising that the reaction is usually sluggish. Water by itself, for instance, is virtually inert and many amides can be recrystallized successfully from this solvent. Hydrolysis can be effected, however, with the assistance of either base or acid catalysts. In basic conditions, the more powerful OH^- nucleophile is available, whereas protonation of the amide oxygen atom in acidic solutions renders the carbonyl carbon more susceptible to nucleophilic attack by water itself. In this context, it is interesting that quaternary amide salts ($\text{RCO}\overset{+}{\text{N}}\text{R}_3^1\text{X}^-$) are readily hydrolysed by cold water. Both the acid- and base-catalysed processes have been thoroughly examined, probably because of their relevence to the behaviour of proteins and peptides. Their mechanisms are more complex than the simple stoicheiometry of the reaction would suggest and the following discussion is therefore confined to the relatively straightforward non-biological substrates.

Amides also react with nucleophilic solvents other than water, such as amines, alcohols, hydroxylamines, carboxylic acids, etc.: alcohols, for example, form esters in the presence of either base or acid catalysts (equation 107). None of these reactions has been as widely examined

$$\text{RCONR}_2^1 + \text{R}^2\text{OH} \xrightarrow{\text{H}^+ \text{ or OH}^-} \text{RCO}_2\text{R}^2 + \text{R}_2^1\text{NH} \qquad (107)$$

as hydrolysis itself, and, with the notable exception of tertiary amides, they are of minor importance as synthetic procedures.

A. Alkaline Hydrolysis

Most amides hydrolyse in aqueous alkaline solutions. The reaction is usually less facile than with esters, for reasons discussed earlier, and unactivated compounds such as simple alkylamides require relatively high temperatures. It has been known for a long time[173] that, under these conditions, the hydrolysis rate usually follows equation (108), although other evidence to be discussed later

$$\text{Rate} = k_h[\text{Amide}][\text{OH}^-] \tag{108}$$

shows this to be one particular limiting form of a general, more complex kinetic expression. For the present we shall concentrate on reactions following equation (108), which is indicative of an attack by the hydroxide ion on the polarized carbonyl bond of the neutral amide. This interaction could occur in a couple of possible ways, and much of the early argument was devoted to this aspect of the mechanism. One would be a two-step sequence (equation 109) with the formation

$$\text{RCONR}_2^1 + \text{OH}^- \rightleftharpoons \underset{\substack{| \\ \text{OH}}}{\overset{\substack{\text{O}^- \\ |}}{\text{R—C—NR}_2^1}} \xrightarrow{\text{H}_2\text{O}} \text{RCO}_2\text{H} + \text{R}_2^1\text{NH} + \text{OH}^- \tag{109}$$

(124)

(R^1 = H, Ar, alkyl, etc.)

of a relatively stable tetrahedral (sp^3) intermediate (124); the other (equation 110) would be a direct, 'S_N2-like,' displacement via a square-planar transition state (125). Although steric factors (i.e. bond angles are only 90°) make transition state 125 seem unlikely,

$$\text{RCONR}_2^1 + \text{OH}^- \xrightarrow{\text{Slow}} \left[\underset{\substack{| \\ \text{R}}}{\overset{\substack{\text{O} \\ ||}}{\text{HO}\cdots\text{C}\cdots\text{NR}_2^1}} \right]^{-\ddagger} \xrightarrow{\text{H}_2\text{O}} \text{RCO}_2\text{H} + \text{R}_2^1\text{NH} + \text{OH}^- \tag{110}$$

(125)

Schowen and his colleagues[174] have pointed out that back-bonding by both the oxygen and nitrogen lone-pair electrons into the relatively low energy π^* orbital of the carbonyl group could well counteract the unfavourable geometry.

Extensive investigations within the last decade have established the importance of tetrahedral intermediates in nucleophilic displacements at carbonyl carbon[175], and in a few instances it has been possible to isolate such species: for example, the salt 126 has been obtained in

$$\left[\underset{\substack{| \\ \text{O}_-}}{\overset{\substack{\text{OEt} \\ |}}{\text{CF}_3\text{—C—NH}_2}} \right] \text{Na}^+$$

(126)

low yield from the reaction of trifluoroacetamide with sodium ethoxide[176]. Although similar intermediates have not been isolated from the alkaline hydrolysis of amides, there is now little doubt that the reaction does proceed via the mechanism outlined by equation (109). The evidence is largely indirect, but nonetheless sound, and comes from a careful analysis of kinetic data for oxygen exchange, buffer catalysis and structural effects in the hydrolysis reaction.

1. Oxygen-18 exchange

The observation of concurrent oxygen exchange between the carbonyl group and the solvent water during the alkaline hydrolysis of benzamides (Scheme 46) was the first real indication that tetrahedral

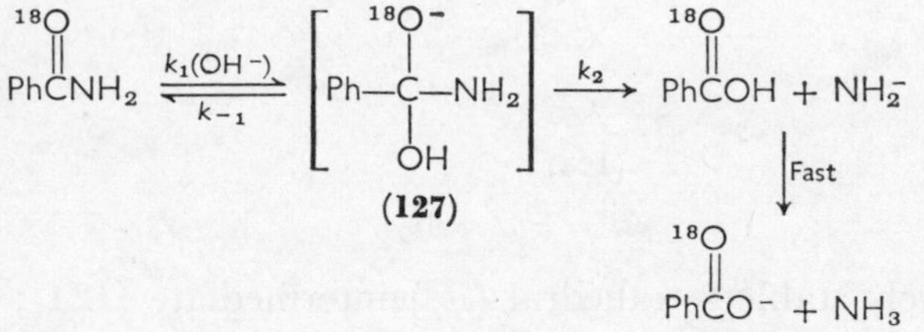

$$
\overset{^{18}O}{\underset{\|}{Ph\overset{}{C}NH_2}} \underset{k_{-1}}{\overset{k_1(OH^-)}{\rightleftharpoons}} \left[\begin{array}{c} ^{18}O^- \\ | \\ Ph-\overset{}{C}-NH_2 \\ | \\ OH \end{array} \right] \overset{k_2}{\longrightarrow} \overset{^{18}O}{\underset{\|}{Ph\overset{}{C}OH}} + NH_2^-
$$

$$(127)$$

$$\Big\downarrow \text{Fast}$$

$$
\overset{^{18}O}{\underset{\|}{Ph\overset{}{C}O^-}} + NH_3
$$

Scheme 46.   Oxygen-18 exchange during the alkaline hydrolysis of benzamide.

intermediates such as **127** might be involved, although it doesn't prove their existence[177]. For instance, it could be argued that ^{18}O exchange is irrelevant to the reaction path for hydrolysis, but recent studies (to be discussed in detail later) of buffer catalysis show this to be unlikely. However, if the tetrahedral intermediate **127** is assumed to partition either to give products or to regenerate reactants, the reverse step (k_{-1}) will result in ^{18}O exchange so long as the oxygen atoms in **127** equilibrate by means of rapid proton transfers. The rate coefficients for exchange (k_{ex}) and hydrolysis (k_h) are then related to k_{-1} and k_2 by equation (111).

$$k_{ex}/k_h = k_{-1}/2\,k_2 \tag{111}$$

Under conditions where the hydrolysis rate follows equation (108), ^{18}O exchange is faster than hydrolysis for benzamide and its N-methyl derivative (Table 12)[177]. This implies, of course, that decomposition of **127** rather than its formation is the rate-controlling step in alkaline hydrolysis. The lack of appreciable ^{18}O exchange for N,N-di-methylbenzamide $(k_{ex}/k_h = 0.05)$[177d] is surprising. It does not, however, signify any radical change in the hydrolysis mechanism, but probably arises from solvation of the tetrahedral intermediate. We

TABLE 12.   Ratio of ^{18}O exchange and alkaline hydrolysis rates for benzamides[a].

Substrate	k_{ex}/k_h
$C_6H_5CONH_2$	4·7
$C_6H_5CONHMe$	1·38
$C_6H_5CONMe_2$	0·05

[a] At about 100° in 0·1 M OH$^-$: this ratio decreases slightly with increasing [OH$^-$].   From reference 177.

shall return to this result later, once the properties of the tetrahedral intermediate have been more clearly defined.

Measurement of substituent effects for both ^{18}O exchange and alkaline hydrolysis rates of *para*-substituted acetanilides provides some information on this point[178].   From this data it is possible to assess by means of Hammett $\sigma\rho$ plots the electronic requirements for both the formation (k_1) of the tetrahedral intermediate and its partitioning (k_{-1}/k_2) as well as for the overall hydrolysis rate (k_h).   From the ρ values obtained for each process (Table 13), it is clear that substituent

TABLE 13.   Hammett ρ values for the alkaline hydrolysis of *para*-substituted acetanilides[178].

Process	ρ
Overall hydrolysis (ρ_h)	$+0·1$
Formation of **127** (ρ_1)	$+1·0$
Partitioning of **127** (ρ_{-1}/ρ_2)	$-1·0$

effects are negligible for the overall rate of hydrolysis, but electron-withdrawing groups, as expected, mildly facilitate hydroxide ion addition to the carbonyl group.   The result for partitioning of the tetrahedral intermediate is more interesting: the negative sign for ρ_{-1}/ρ_2 implies that electron-attracting substituents (e.g. NO_2) in the aniline fragment favour exchange over hydrolysis (since $k_{ex}/k_h = k_{-1}/2 k_2$).   Any simple hydrolysis mechanism (such as Scheme 46) in which the amine is expelled as an anionic species (i.e. as $ArNH^-$) is obviously inconsistent with this finding.   Perhaps this is not surprising as the low acidity of aniline ($pK \simeq 35$) would make the conjugate base a very poor leaving group.   Several other explanations come to mind, but the most satisfactory, particularly when considered in conjunction with solvent effects for the hydrolysis of benzamide[177c], is that outlined by Scheme 47 involving equilibrium formation of a

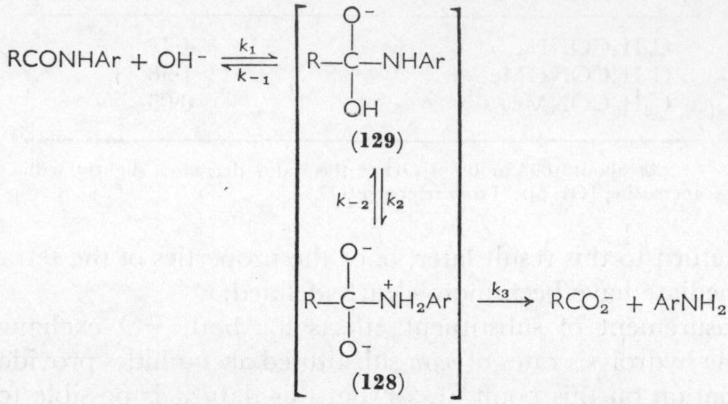

$$
RCONHAr + OH^- \underset{k_{-1}}{\overset{k_1}{\rightleftharpoons}}
\left[
\begin{array}{c}
\underset{\substack{| \\ OH}}{\overset{\substack{O^- \\ |}}{R-C-NHAr}} \\
(129) \\
\\
k_{-2} \big\updownarrow k_2 \\
\\
\underset{\substack{| \\ O^-}}{\overset{\substack{O^- \\ |}}{R-C-\overset{+}{N}H_2Ar}} \quad \xrightarrow{k_3} \quad RCO_2^- + ArNH_2 \\
(128)
\end{array}
\right]
$$

SCHEME 47.   Alkaline hydrolysis of acetanilides.

dipolar ion (128) from the initial addition intermediate (129), followed by rapid breakdown to products[178].   Electron-withdrawing substituents in the aniline fragment should lower the equilibrium concentration of 128 and thereby promote ^{18}O exchange relative to hydrolysis.   Subsequent kinetic studies of buffer-catalysed alkaline hydrolysis (discussed below in this section) strongly support this conclusion and show the general applicability of Scheme 47 to the alkaline hydrolysis of amides at low pH.   However, in more strongly alkaline conditions, the tetrahedral intermediate 129 can decompose to products via an alternative pathway.

Decomposition via a dipolar ion species such as 128 also explains the lack of ^{18}O exchange during the hydrolysis of N,N-dimethyl-benzamide noted previously (Table 12).   This odd result has never been well-understood and has fostered ideas that hydrolysis and ^{18}O exchange are unrelated processes.   It has also led to speculation that equilibration of the oxygen atoms in the tetrahedral intermediate results from rapid proton transfers involving high-energy structures such as 130.   A more plausible explanation comes directly from the

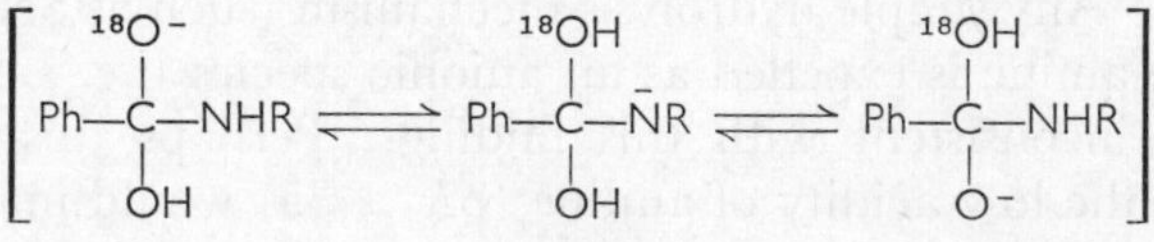

$$
\left[
\underset{\substack{| \\ OH}}{\overset{\substack{^{18}O^- \\ |}}{Ph-C-NHR}} \quad \rightleftharpoons \quad
\underset{\substack{| \\ OH}}{\overset{\substack{^{18}OH \\ |}}{Ph-C-\bar{N}R}} \quad \rightleftharpoons \quad
\underset{\substack{| \\ O^-}}{\overset{\substack{^{18}OH \\ |}}{Ph-C-NHR}}
\right]
$$

(130)

mechanism outlined in Scheme 47. It seems unlikely that conversion of the addition intermediate 129 to the dipolar ion 128 results from direct O to N proton transfer and a water molecule is probably the transfer agent. This water molecule must be strongly hydrogen bonded to the dipolar-ion structure so that free rotation about the C—N bond is inhibited. Thus equilibration of the oxygen atoms in the dipolar-ion intermediate for tertiary amides (131) is sluggish and negligible ^{18}O exchange is therefore observed during hydrolysis.

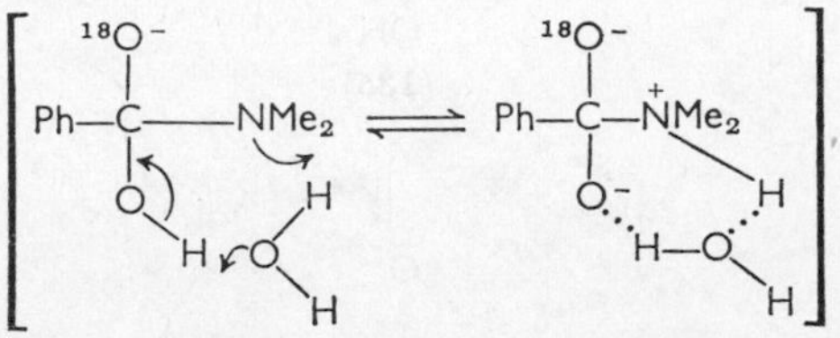

(132) (131)

2. General base- and general acid-catalysed alkaline hydrolysis

In relatively strong solutions of aqueous hydroxide ($> 0 \cdot 1$ M), the hydrolysis rate of acetanilides follows an expression containing both first- and second-order terms in hydroxide ion concentration (equation 112). This distinction was first noticed by Biechler and Taft[179], but

$$\text{Rate} = [\text{Amide}](k[\text{OH}^-] + k'[\text{OH}^-]^2) \tag{112}$$

similar behaviour has been reported subsequently for many other amides bearing electron-withdrawing substituents attached to the carbonyl group, as in 2,2,2-trifluoroacetanilides[174,180], chloroacetamide[181], urea[182] and N,N-diacylamines[183]. The function of the second hydroxide ion in the hydrolytic process has curried considerable speculation, but the most reasonable explanation is one (Scheme 48) where the products arise via two pathways involving intermediates 133 and 134[175c]. Structure 133 is familiar as the dipolar-ion intermediate suggested by the ^{18}O exchange experiments of Bender and Thomas[178]; structure 134 contains one proton less, and is obviously associated with the second-order hydroxide ion term.

The steady-state rate expression for hydrolysis in accordance with Scheme 48, making the simplifying assumption that the various intermediates are in equilibrium with each other, is given by equation (113), and the rate equations discussed previously are limiting forms

$$\text{Rate} = \left(\frac{k_1 k_3 K_y [\text{OH}^-] + k_1 k_2 K_x [\text{OH}^-]^2}{k_{-1} + k_2 K_x [\text{OH}^-] + k_3 K_y} \right) [\text{Amide}] \tag{113}$$

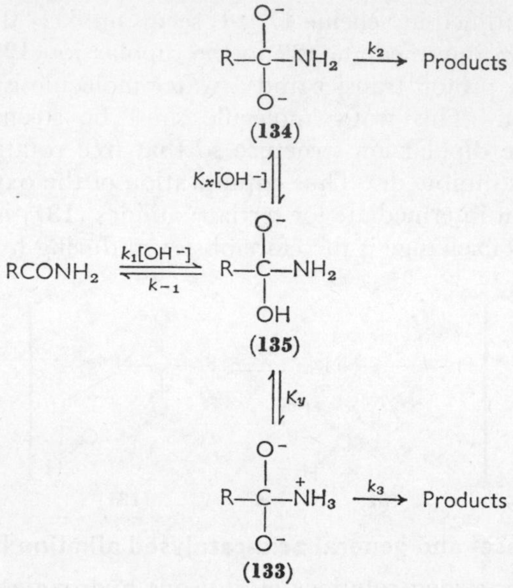

$$
\begin{array}{c}
\text{O}^- \\
| \\
\text{R—C—NH}_2 \xrightarrow{\ k_2\ } \text{Products} \\
| \\
\text{O}^-
\end{array}
$$

(134)

$$ K_x[\text{OH}^-] \updownarrow $$

$$
\begin{array}{c}
\text{O}^- \\
| \\
\text{RCONH}_2 \underset{k_{-1}}{\overset{k_1[\text{OH}^-]}{\rightleftharpoons}} \text{R—C—NH}_2 \\
| \\
\text{OH}
\end{array}
$$

(135)

$$ \updownarrow K_y $$

$$
\begin{array}{c}
\text{O}^- \\
| \\
\text{R—C—} \overset{+}{\text{N}}\text{H}_3 \xrightarrow{\ k_3\ } \text{Products} \\
| \\
\text{O}^-
\end{array}
$$

(133)

SCHEME 48.   Hydroxide ion catalysed alkaline hydrolysis of amides.

of this.   For instance, only the first-order term will be important at very low hydroxide ion concentrations: provided $(k_{-1} + k_3K_y) > k_2K_x[\text{OH}^-])$ in the denominator, the kinetic expression approximates to equation (108) and represents the hydrolysis rate via intermediate **133**.   Similar arguments may be used to derive equation (112).   The most interesting result, however, comes at very high pH where $k_2K_x[\text{OH}^-] > (k_{-1} + k_3K_y)$, and equation (113) then reduces to: Rate = $k_1[\text{Amide}][\text{OH}^-]$, i.e. the initial addition of hydroxide ion to the amide carbonyl is rate limiting.   The realization of this transition in the rate-limiting step with increasing hydroxide ion concentration for the alkaline hydrolysis of both 2,2,2-trifluoroacetanilide[180] and its N-methyl derivative[174] is convincing evidence that the addition intermediate **135** is kinetically important and actually lies on the reaction path.   In turn, this justifies the assumption made earlier for analysing the ^{18}O exchange data.

A difficult question to answer is the nature of the rate-limiting step at lower pH (i.e. when the hydrolysis rate follows equation 108 or 112). A tentative conclusion has already been drawn in respect of equation (108) from the ^{18}O exchange experiments with acetanilides[178], but further information is available from kinetic measurements in buffer solutions.

These show that in the alkaline hydrolysis of 2,2,2-trifluoroacet-anilides[184], for example, the second-order dependence on hydroxide ion concentration derives from a general base term superimposed on a first-order hydroxide ion term.   In other words, equation (112) is a special case (with $B_i = OH^-$) of the limiting rate expression defined by equation (114) in which B_i represents the general base species. For N-methyl-2,2,2-trifluoroacetanilide[184a], the rate coefficients for

$$\text{Rate} = [\text{Amide}]\left(k[\text{OH}^-] + [\text{OH}^-]\sum_i k_i'[\text{B}_i]\right) \qquad (114)$$

several catalysts fit a Brönsted relationship with $\beta \simeq 0.3$.   This establishes that proton transfer occurs in the rate-limiting process. The most likely explanation seems to be either general base-catalysed formation of the intermediate 134 or its general acid-catalysed de-composition, via transition states 136 and 137, respectively.   From the observation of appreciable solvent isotope effects for N-methyl-2,2,2-trifluoroacetanilide ($k_{H_2O}/k_{D_2O} = 2.2$), Schowen and his colleagues[185] have concluded that *only* proton transfer and no cleavage of the

(136)   (137)

carbon–nitrogen bond occurs in the transition state, but the generality of this conclusion remains to be tested.

Related investigations show that hydrolysis via the dipolar-ion intermediate 133 is also catalysed by buffer components.   Thus rate accelerations by HCO_3^-, $H_2PO_4^-$ and several amine cations for the alkaline hydrolysis of 2,2,2-trifluoroacetanilide have been interpreted[180b,184b] as general acid catalysis (i.e. Rate = $[\text{Amide}][\text{OH}^-]\sum_i k_i'[\text{HB}_i]$).   This catalysed pathway must involve the same addition intermediate 135 because an identical limiting rate (that of formation of 135) is observed with high concentrations of the general acid catalyst.   A plausible explanation is that conversion of the addition intermediate 135 to the dipolar-ion intermediate 133 is rate limiting via a transition state such as 138.   In the absence of buffer components, the first-order dependence on hydroxide ion concentration (equation 108) may then derive from a similar transition state with water acting as the proton donor $(HB_t = H_2O)$. This is in agreement with conclusions drawn both

$$\left[\begin{array}{c} O^- \\ | \\ F_3C-C-N \\ | \quad \diagdown \\ OH \quad {}^{\diagup}H-\overset{\frown}{B_t} \end{array} \right]^{\ddagger}$$

(138)

from the ^{18}O experiments of Bender and Thomas[178] discussed earlier and from studies of solvent isotope effects[185]. Undoubtedly, further detailed studies of buffer-catalysed hydrolysis are required to establish these promising conclusions.

3. Structure and reactivity

Generally the rate of alkaline hydrolysis of amides is mildly accelerated by a lowering of the electron density at the carbonyl carbon, but retarded by bulky groups. This is, of course, the expected structural influence for a multistep process involving the intermediacy of tetravalent structures in which the electronic demands of each step are of opposite sign. Thus rate coefficients for the first-order hydroxide ion catalysed hydrolysis (equation 108) of primary alkylcarboxamides fit a modified form of the Taft equation giving reasonable values of $\rho^* = +2\cdot08$ and a steric parameter $\delta = +0\cdot73$. Surprisingly, however, similar data for alkylcarboxanilides correlate poorly with Taft σ^* parameters, and this has been tentatively ascribed to an additional steric inhibition of resonance in the anilido fragment[179]. A more interesting consideration is the influence of structure on the relative importance of the first- (k) and second-order (k') hydroxide ion terms of equation (112). Electron-withdrawing substituents should clearly favour hydrolysis via the second-order process. This is confirmed nicely by the data of Biechler and Taft[179] for N-methyl-2,2,2-trifluoroacetanilide ($k'/k = 190$), N-methyl-2,2-difluoroacetanilide ($k'/k = 34$) and N-methyl-2-chloroacetanilide ($k'/k = 2$). It is therefore doubtful whether unsubstituted alkylcarboxamides undergo alkaline hydrolysis by anything other than the first-order process even at high pH.

B. Acid Hydrolysis

Strong mineral acids are effective reagents for the hydrolysis of amides at temperatures of about 100°. The products are usually a mixture of the amine and carboxylic acid, consistent with fission of the

$$RCONR_2^1 + H_3O^+ \longrightarrow RCO_2H + R_2^1NH_2^+ \tag{115}$$

$$(R^1 = H, \text{ alkyl, aryl, etc.})$$

N–acyl bond (equation 115), although examples of N–alkyl bond fission have also been reported[187].

I. N–acyl bond fission

The kinetics of these reactions have been widely examined and a good deal is therefore known about their mechanism. In dilute acid, reaction rates have a first-order dependence on the hydronium ion concentration and follow equation (116). The most significant

$$\text{Rate} = k[\text{Amide}][\text{H}_3\text{O}^+] \tag{116}$$

feature for most amides, however, is the existence of a rate maximum at some high acidity, dependent on both the solvent acid and amide structure (Table 14), but usually in the region of 2 M to 5 M for sulphuric acid[188]. This is where most amides are extensively protonated on the oxygen atom[189] and a logical deduction is that this

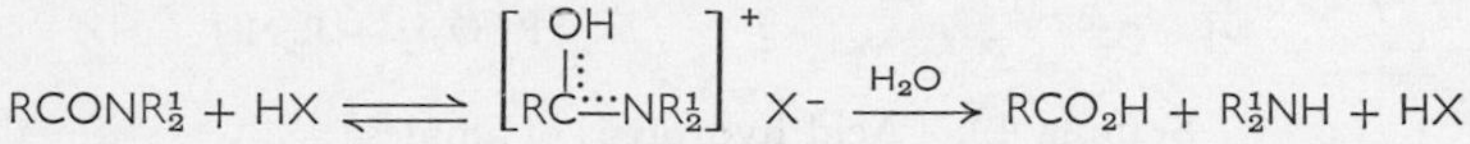

$$RCONR_2^1 + HX \rightleftharpoons \left[\begin{array}{c} OH \\ | \\ RC^{\cdots}NR_2^1 \end{array} \right]^+ X^- \xrightarrow{H_2O} RCO_2H + R_2^1NH + HX \tag{117}$$

species is the reactive intermediate (equation 117). Thus below the rate maximum increasing acidity raises the concentration of the

TABLE 14. Experimental conditions for rate maxima for the acid-catalysed hydrolysis of amides.

Amide	$-pK_a{}^a$	Acidity (M)	$-H_0$	Reference
Formamide		6 HCl	2·12	188b
		4·75 H$_2$SO$_4$	2·16	188b
Acetamide	0·6	3·25 HCl	1·14	188b
		2·5 H$_2$SO$_4$	1·12	188b
Propionamide	0·8	3·2 HCl	1·12	188b
		2·4 H$_2$SO$_4$	1·07	188b
Benzamide	1·74	4·5 HCl	1·58	188c
		3·5 H$_2$SO$_4$	1·62	188c
p-Methoxybenzamide	1·46	3·0 H$_2$SO$_4$	1·38	188c
p-Nitrobenzamide	2·70	4·5 H$_2$SO$_4$	2·05	188c
o-Nitroacetanilide	3·72	6·07 H$_2$SO$_4$	2·79	188d

^{a} From the compilation in reference 189a.

27*

reactive intermediate, whereas beyond the rate maximum the chief effect of increasing acidity is to decrease the concentration or activity of water.　Accordingly, these two effects would account for the rate maximum.　In the absence of definitive evidence to the contrary*, we shall assume the O-protonated amide is the reactive intermediate and this hypothesis fits with most of the experimental facts.

In contrast to alkaline hydrolysis, no measurable ^{18}O exchange between the carbonyl group and the solvent is observed during the acid-catalysed reaction with benzamides[177a,195] and N-acetyl-imidazole[196].　This implies, of course, that steps subsequent to the attack of water on the O-protonated amide are rapid, and the formation of the tetrahedral intermediate is the rate-controlling process (Scheme 49).　This change from rate-limiting decomposition to

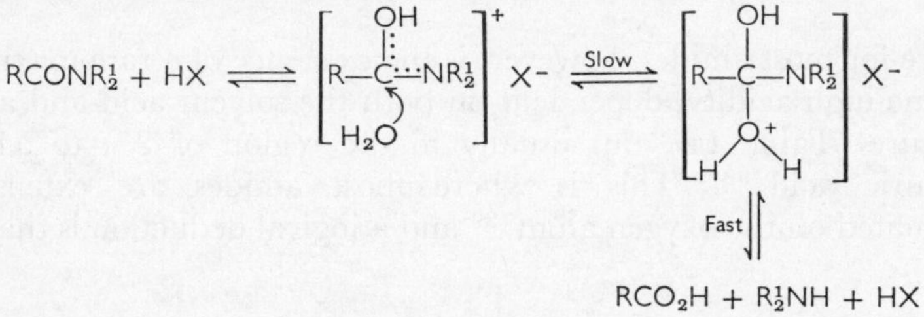

$$RCONR_2^1 + HX \rightleftharpoons \left[\begin{matrix} OH \\ \vdots \\ R-C\cdots NR_2^1 \\ H_2O \end{matrix} \right]^+ X^- \xrightleftharpoons{Slow} \left[\begin{matrix} OH \\ | \\ R-C-NR_2^1 \\ | \\ O^+ \\ H \quad H \end{matrix} \right] X^-$$

$$\Big\downarrow \text{Fast}$$

$$RCO_2H + R_2^1NH + HX$$

SCHEME 49.　Acid hydrolysis of amides.

rate-limiting formation of the tetrahedral intermediate in going from alkaline to acidic conditions has an interesting parallel in the hydrolysis of imidate esters[197].

* A similar rate–acidity profile would be observed if the reactive intermediate were the N-protonated amide (known from n.m.r. studies to be formed in low concentration[190]) in equilibrium with the more abundant O-conjugate acid. This alternative hypothesis is attractive because it is evident from investigations of alkaline hydrolysis that the amino fragment is difficult to expel as an anionic species.　Direct evidence on this point is not available.　A close correspondence between the hydrolytic rates for methyl benzimidate ($C_6H_5C(OMe)=NH$) and benzamide under identical experimental conditions suggests that both reactions involve rate-limiting hydration of a similar reactive intermediate (which must, of course, represent the O-protonated structure for benzamide)[191]; however, more recent data for benzimidate hydrolysis has been interpreted in terms of a rate-limiting decomposition of a tetrahedral intermediate[192].　Also Bunton and his colleagues[193] have tentatively suggested that acidity–rate profiles for benzamide hydrolysis are indicative of *some* reaction via the N-conjugate acid, but the theoretical basis of their argument has been severely criticized[194]. For a more ample discussion of the site of protonation of the amido group see Chapter 3.

Solvent isotope effects for benzamide[193] and acetamide[198] (Table 15) are also consistent with Scheme 49. Values of $k_{H_2O}/k_{D_2O} > 1$ in dilute acid reflect the higher concentration of protonated amide in heavy water. The inversion of this ratio at higher acidities $(k_{H_2O}/k_{D_2O} < 1)$ is consistent with complete protonation of the substrate, and only the lower nucleophilic strength of D_2O relative to

TABLE 15. Solvent isotope effects for the hydrolysis of amides catalysed by hydrochloric acid.

Substrate	Acidity (M)	k_{H_2O}/k_{D_2O} (100°)	Reference
CH_3CONH_2	0·1	1·45	198
	4·0	0·86	198
$C_6H_5CONH_2$	1	1·15	193
	6	0·90	193

H_2O is important. This factor also explains why the k_{H_2O}/k_{D_2O} ratio at lower acidities is only slightly larger than unity and not the usual value of 2 to 4 observed for the difference in preequilibrium protonation.

Attempts have been made recently to specify in a more exact way the role of water in the transition state of acid-catalysed reactions such as the hydrolysis of amides, by analysing the significance of linear free-energy relationships between observed rate coefficients and various combinations of water activity and acidity function data[189a,199]. A detailed account of these treatments is outside the scope of the present discussion, but the conclusions for amide hydrolysis are of interest. The most pertinent relationship is that derived by Yates and his colleagues[189a,199e] in which the experimental first-order rate coefficient (Rate = k_{obs}[Amide]) is correlated with the H_A acidity function (measured from the equilibrium protonation of amides) and the water activity (a_w) by equation (118), where r is the number of water molecules required to convert the protonated amide to the transition state structure. This equation is similar to one derived

$$\log k_{obs} + H_A = r \log a_w + \text{constant} \tag{118}$$

earlier by Bunnett[199a], but refers specifically to amide substrates. The r value obtained from the data for several amides is approximately 3, and this has led to the suggestion that the transition state for hydrolysis may be represented by **139**, where the water nucleophile on average is solvated by two additional solvent molecules. A similar conclusion for this aspect of the transition state has been reached by both Moodie[199d] using a more direct relationship involving the water activity but not acidity function data, and by Bunnett and Olsen[199b]

$$\left[\begin{array}{c} \text{H} \\ \diagdown \\ \text{O} \cdots \text{H} \\ \diagup \\ \text{H} \end{array} \quad \begin{array}{c} \text{R} \\ | \\ \text{O} \cdots \text{C} \cdots \text{OH(H}_2\text{O)}_x \\ | \\ \text{NR}_2^1 \end{array} \right]^+$$

(139)

using acidity function data but not the water activity. The unanimous agreement between these apparently independent treatments may, however, be misleading. Conceptually all are similar in that the hydration number (r) is derived, either directly or indirectly, from the dependence of the hydrolytic rate coefficient for the protonated amide on the water activity of the medium. A more serious criticism concerns the universal assumption that the ratio of the activity coefficient for the protonated amide to that of the transition state $(\gamma_{[RC(OH)NR_2^1]^+}/\gamma^\ddagger)$ is independent of the medium. Experiments with butyramide show this to be invalid, as the rates of hydrolysis in perchloric and sulphuric acid for the same water activity differ by an appreciable factor[194]. Thus the real significance of the hydration parameter (r) is questionable, although there seems little doubt that water is involved in a rate-limiting nucleophilic attack on the protonated amide, and the hydrolysis is therefore of the $A2$ type.

2. Substituent effects

Since the rate of acid hydrolysis for simple amides is not very dependent on their structure, the most important consideration from a practical standpoint is the rate maximum. This means it is not always expeditious to employ the most concentrated acid as the reaction medium. As a rough guide (see Table 14) the maximum rate prevails in acid solutions whose H_0 value equals the pK_a of the amide.

Systematic studies of substituent effects for alkylamides and benzamides have corroborated, by and large, previous conclusions in regard to the mechanism of hydrolysis. Data for alkylamides taken from Bolton's[186] compilation are listed in Table 16. These rate coefficients correlate well with a modified form of the Taft equation[200] containing terms for steric and hyperconjugative, but not polar (i.e. inductive), interactions (equation 119). The unimportance of polar

$$\log k/k_0 = 0.86\,E_S^C + 0.49\,(n - 3) \tag{119}$$

effects is rationalized by the ambiguous requirements of the mechanism in that increased electron density at the carbonyl carbon

TABLE 16.   Acid-catalysed hydrolysis of amides at $75°$[186].

Amide	$10^4 k^a$
Acetamide	10·3
Propionamide	12·0
2,2-Dimethylpropionamide	2·63
Butyramide	5·99
Isobutyramide	6·22
2-Methylbutyramide	2·08
3-Methylbutyramide	1·91
3,3-Dimethylbutyramide	0·395
Valeramide	5·15
Chloroacetamide	8·54
N-Methylacetamide	0·58
N,N-Dimethylacetamide	0·65

^{a} In units l/mole sec.   For equation (116).

assists protonation but inhibits nucleophilic attack by water. Steric retardation is consistent with the tetrahedral structure of the transition state: in many cases this is the overriding factor, and sterically congested amides are well known to be impassive to hydrolysis. The hyperconjugation is associated with stabilization of the conjugate acid by $\alpha_{C=O}$-hydrogen atoms, as reflected by the relative basicities of the compounds. One puzzling feature of these results is the anomalous reactivity of secondary amides, i.e. $NH_2 \gg NMe_2 > NHMe$, and the same trend is evident for derivatives of benzamide[193]. If only the steric factor were important, then the tertiary compounds should be the least reactive. Differences in solvation of the conjugate acids seem to be eliminated by the closely similar activation entropies for all three compounds[186] and one is therefore left with stabilization of the induced positive charge on the nitrogen atom of the conjugate acid to explain the anomaly.

Similar mechanistic inferences may be drawn from the examination of substituted benzamides[201]. Only small rate perturbations are found for *meta* and *para* substituents, but *ortho* groups inhibit hydrolysis through steric interaction (e.g. for nitrobenzamides, $k_{ortho}/k_{para} = 0\cdot03$). It has been possible to determine independently in terms of the Hammett equation the *meta* and *para* substituent effects on the pre-equilibrium protonation (ρ_1) and rate-limiting steps (ρ_2)[201] of Scheme 49. Their sum does represent, of course, the Hammett ρ factor (ρ_{ov}) for the overall hydrolysis reaction (equation 120). The close agreement between the calculated value of ρ_{ov} and that measured

$$\rho_{ov} = \rho_1 + \rho_2 \qquad\qquad (120)$$

directly constitutes additional evidence for the mechanism of acid hydrolysis.

3. N–alkyl bond fission

The introduction of bulky substituents on the nitrogen atom produces a more dramatic effect, with N–alkyl bond fission becoming the preferred process under suitable conditions (equation 121). This

$$\text{RCONHBu-}t + \text{H}_2\text{O} \xrightarrow{\text{H}_2\text{SO}_4} \text{RCONH}_2 + t\text{-BuOH} \qquad (121)$$

change of hydrolysis mechanism has been noted only for amides bearing *N-t*-alkyl groups (as well as ureas, thioureas and sulphonamides) in sulphuric acid[187a], and for *N,N*-dicyclohexylbenzamide in acetic acid[187b], although it must have a wider scope. The *N-t*-alkylamides have been examined thoroughly, and in 98% sulphuric acid N–alkyl bond fission is rapid $(t_{\frac{1}{2}} \simeq 5 \text{ min})$ giving high yields of primary amide (equation 121). At lower acidities, both rates and

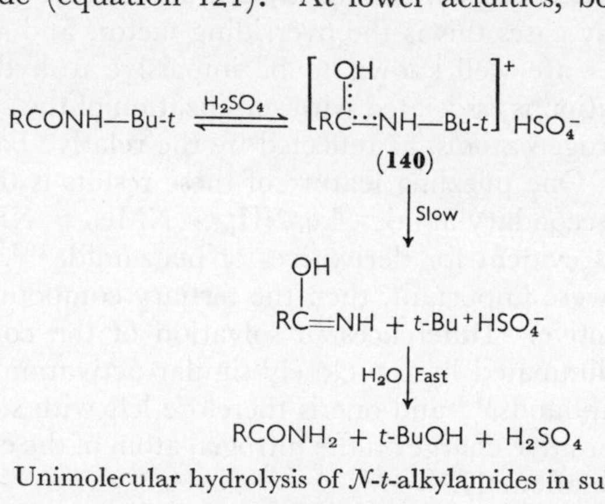

$$\text{RCONH—Bu-}t \underset{}{\overset{\text{H}_2\text{SO}_4}{\rightleftharpoons}} \left[\begin{array}{c} \text{OH} \\ \overset{\shortparallel}{\text{RC}}\text{-}\text{-}\text{NH—Bu-}t \end{array} \right]^{+} \text{HSO}_4^{-}$$

(**140**)

$$\downarrow \text{Slow}$$

$$\overset{\text{OH}}{\underset{|}{\text{RC}}}\text{=NH} + t\text{-Bu}^{+}\text{HSO}_4^{-}$$

$$\text{H}_2\text{O} \bigg\downarrow \text{Fast}$$

$$\text{RCONH}_2 + t\text{-BuOH} + \text{H}_2\text{SO}_4$$

SCHEME 50.   Unimolecular hydrolysis of *N-t*-alkylamides in sulphuric acid.

yields are less, the latter probably because of competition from the normal hydrolysis reaction. There is little doubt that N–alkyl bond fission results from unimolecular decomposition of the O-protonated amide (**140**), and substituent effects for a series of N-t-alkylbenzamides accord with this hypothesis[187a]. The unimolecular pathway should be promoted by the combination of low water activity in concentrated acids, the stability of the carbonium ion fragment and steric congestion about the carbonyl group. In this context, it is interesting that N-t-alkylformamides react by the normal bimolecular process with N–acyl bond fission even in 98% sulphuric acid[187a].

C. General Acid- and General Base-catalysed Hydrolysis

In addition to the mechanisms discussed above, hydrolysis may also occur by other pathways arising purely from general acid or general base catalysis. These reactions are not usually observed with ordinary amides, but they are important for reactive substrates, such as N-acetylimidazoles, and for intramolecularly catalysed hydrolysis.

I. Intermolecular reactions

The hydrolysis of N-acetyl-N-methylimidazolium ion (AcImMe$^+$) in buffer solutions is accelerated by basic species such as acetate or phosphate ions, and N-methylimidazole: the reaction rates follow equation (122) which is consistent with general base catalysis[202].

$$\text{Rate} = [\text{AcImMe}^+] \sum_i k'_{\text{B}_i}[\text{B}_i] \qquad (122)$$

The rate of disappearence of N-acetylimidazole (AcIm) under similar conditions, however, depends on the concentration of both the acidic and basic buffer components (equation 123), apparently indicating the existence of a general acid- as well as a general base-catalysed path[203]. If these two sets of results are compared, it is apparent right away that

$$\text{Rate} = [\text{AcIm}] \sum_i k_{\text{HB}_i}[\text{HB}_i] + [\text{AcIm}] \sum_i k_{\text{B}_i}[\text{B}_i] \qquad (123)$$

the first term of equation (123) actually refers to a general base-catalysed hydrolysis of the N-acetylimidazolium ion; i.e. the kinetic expression is better represented by equation (124).

$$\text{Rate} = [\text{AcImH}^+] \sum_i k'_{\text{B}_i}[\text{B}_i] + [\text{AcIm}] \sum_i k_{\text{B}_i}[\text{B}_i] \qquad (124)$$

The rate of hydrolysis of N-acetyl-N-methylimidazolium ion is exactly the same as that of fully protonated imidazole and the k'_{B_i}

coefficients for both compounds over a wide range of catalytic activity are remarkably similar (Table 17)[202]. Clearly the hydrolysis mechanism for these two ions must be identical, with substitution of N-methyl for N-H making little difference in reactivity. Although nucleophilic catalysis is rarely observed for simple amides, this process is important with N-acetylimidazoles and related compounds (see section X.D): it is known, for example, that phenolates interact with

TABLE 17.   Solvolysis rates for the N-acetyl-N-methylimidazolium ion and N-acetylimidazolium ion at $25°$[202].

B_t	k'_{B_t} [a]	
	$(AcImMe^+)$	$(AcImH^+)$
H_2O [b]	0·051	0·051
$CH_3CO_2^-$	17	19
$(CH_2CO_2^-)_2$	42	78
HPO_4^{2-}	2230	2100
NH_3	62000	81000

[a] In units l/mole min.   For equations (122) and (124) respectively.
[b] Assuming $[H_2O] = 55\cdot5$ M.

the N-acetylimidazolium ion exclusively by a nucleophilic path[204]. However, Jencks and his colleagues have clearly demonstrated that catalysis by acetate ion results to the extent of 78% from a general base path and 22% from direct nucleophilic interaction (Scheme 51)[203c]. Thus the most plausible explanation for the catalytic mechanism with N-acetylimidazolium ions is a combination of nucleophilic attack via

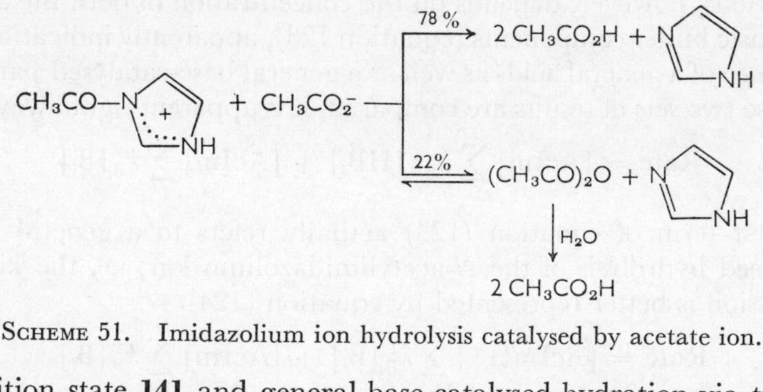

SCHEME 51.   Imidazolium ion hydrolysis catalysed by acetate ion.

transition state **141** and general base-catalysed hydration via transition state **142**, with the relative importance of each path dependent on the catalyst structure.

(141)   **(142)**

The second term in equation (124) represents general base-catalysed hydrolysis of the neutral N-acetylimidazole. It has been observed only for catalysis by imidazole[203a,b] and acetate ion[203c], although other bases must also be effective. The mechanistic implications of this term are ill-defined at present. Arguments in favour of a base-assisted hydration of the neutral amide (analogous to **142**) have been presented[203a] and it may be significant that steric effects in this reaction are remarkably similar to those for the base-catalysed hydration of the N-acetylimidazolium ion[203b].

Acetate ions also catalyse the hydrolysis of the acetylpyridinium ion[205]. Here, too, the mechanism has not been examined, but it should be the same as for the N-acetylimidazolium ion. Acetamide is the only unactivated compound for which general acid-catalysed hydrolysis has been reported, in this case by Wyness for acetic acid buffers[206]. Although this finding was originally interpreted in terms of nucleophilic catalysis by acetate ion on the conjugate acid of the amide[206], a base-catalysed hydration seems more in line with the other conclusions.

2. Intramolecular catalysis

Some of the best examples of general acid-catalysed hydrolysis come from studies of intramolecular catalysis. An unusual feature of these reactions is that, in addition to requiring a favourable configuration of the interacting groups with the formation of a relatively unstable intermediate, the most powerful catalysis is obtained from neighbouring groups, such as carboxylic acid and imidazolium ion, that may act as a proton donor as well as a powerful nucleophile. The proton requirement may be associated with either the low reactivity of the neutral amide or the poor leaving ability of the amino fragment. The overall effect, of course, is one of intramolecular general acid catalysis, in striking contrast to the intramolecular nucleophilic catalysis commonly observed in the hydrolysis of esters and other carboxylic acid derivatives. These reactions also have a wide significance in peptide and protein chemistry, and this particular aspect has been reviewed elsewhere[175b].

From a mechanistic standpoint, one of the best examples is phthalamic acid[207], which hydrolyses rapidly in dilute acid: at pH = 3, for example, its hydrolysis rate is nearly 10^5 times faster than that of benzamide. The shape of the pH–rate profile (Figure 1) shows

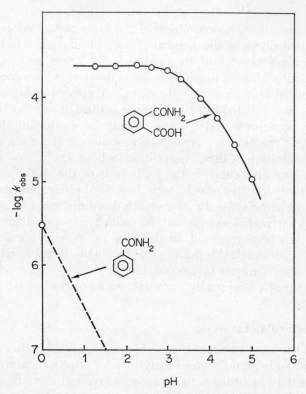

FIGURE 1. pH–rate profiles for the hydrolysis of phthalamic acid and benzamide in aqueous solutions at about 48°. From reference 207. [Reproduced with permission of the American Chemical Society.]

unambiguously that the un-ionized carboxylic group rather than its anion is involved. Intermediate formation of phthalic anhydride has been demonstrated indirectly by reacting phthalamic acid, labelled with ^{13}C on the amide carbonyl, with $H_2^{18}O$—the ^{18}O-enriched carbon dioxide obtained from the resulting phthalic acid comes from both the ^{13}C-enriched and isotopically normal carboxylic groups (equation 125). The mechanism favoured by Bender and his colleagues[207] (Scheme 52) involves simultaneous nucleophilic attack and proton

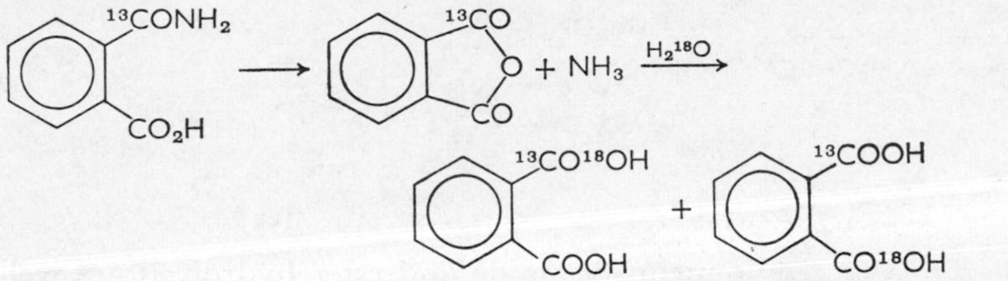

$$(125)$$

transfer to the amide nitrogen, termed 'nucleophilic–electrophilic catalysis'. The experimental data, however, are equally consistent with intramolecular displacement by the carboxylate anion on the protonated amide (Scheme 53). Participation by neighbouring

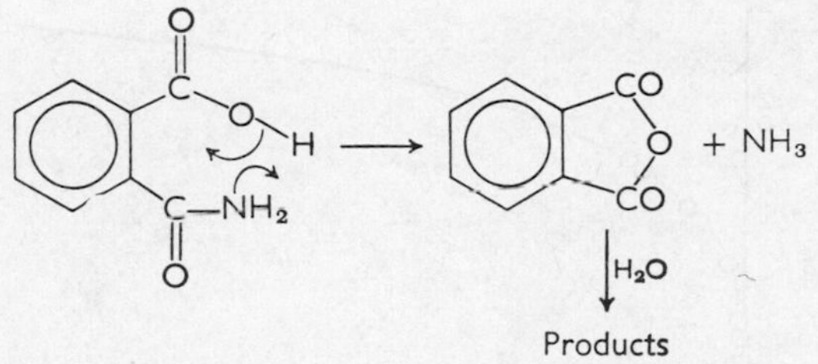

SCHEME 52. 'Nucleophilic–electrophilic catalysis' for the hydrolysis of phthalamic acid.

carboxylic acid groups has also been cited in the hydrolysis of both glycyl-L-asparagine (**143**, R = H) and L-leucyl-L-asparagine (**143**, R = i-Bu)[208] and of the aromatic amide **144**[209]. In these reactions, however, the evidence is less complete.

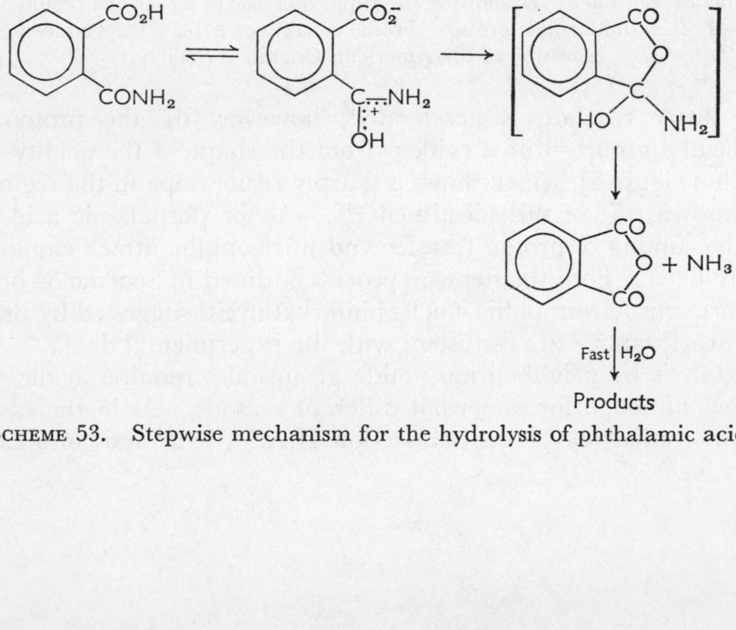

SCHEME 53. Stepwise mechanism for the hydrolysis of phthalamic acid.

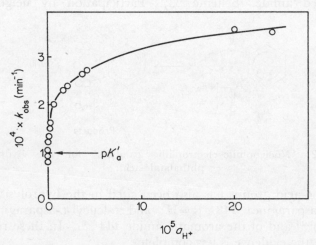

(143)   (144)

The different requirements of amide and ester hydrolysis are well illustrated by γ-(4-imidazolyl)butyric acid derivatives (145). The neutral imidazolyl function strongly catalyses hydrolysis of the phenyl ester (R = OPh), but not of the corresponding amide (R = NH$_2$).

FIGURE 2.   Acidity–rate profile for the hydrolysis of γ-(4-imidazolyl)butyramide in aqueous solution at 78° showing the rapid increase of k_{obs} in the region of the pK_a of the imidazolyl group.   From reference 210. [Reproduced with permission of the American Chemical Society.]

The latter reaction is accelerated, however, by the protonated imidazolyl group—this is evident from the shape of the acidity–rate profile (Figure 2), which shows a sharply rising slope in the region of the known pK_a of this substituent[210]. As for phthalamic acid, the relative timing of proton transfer and nucleophilic attack cannot be firmly fixed.   Both the stepwise process outlined in Scheme 54 or the synchronous 'electrophilic–nucleophilic catalysis' suggested by Bruice and Sturtevant[210] are consistent with the experimental data.

Catalysis by neighbouring amide groups also requires acidic conditions, although for somewhat different reasons.   As in the case of intramolecular alkylation (see section II.A.5), hydrolysis in alkaline

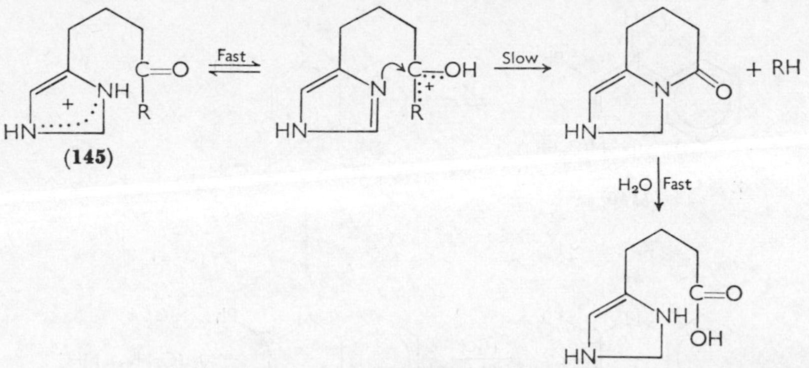

SCHEME 54. Hydrolysis of γ-(4-imidazolyl)butyramide (R = NH$_2$).

solutions usually involves nucleophilic attack by the amide nitrogen to form an imide, which is resistant to subsequent hydrolysis (equation 126)[211]. Catalysed hydrolysis may occur, however, under acidic

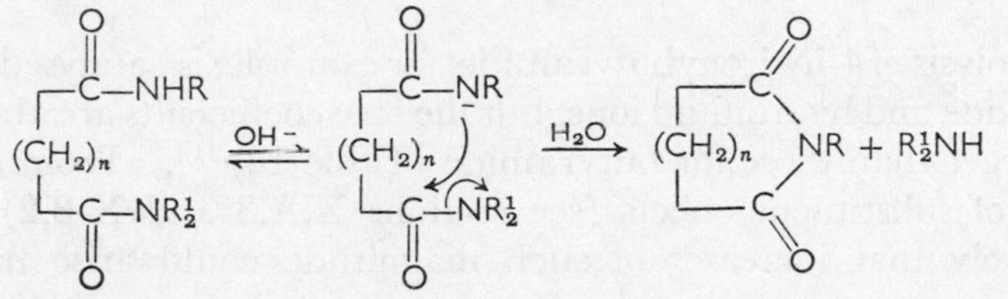

$$(126)$$

conditions where the carbonyl oxygen of the neutral amide acts as the nucleophilic reagent. This situation has been realized for o-benzamido-N,N-dicyclohexylbenzamide (**146**) in acetic–sulphuric acid mixtures, as the hydrolysis rate is about 10^4 times faster than N,N-dicyclohexylbenzamide under similar conditions[187b]. Although a portion of the increased rate may arise from steric acceleration, the isolation of benzoylanthranil intermediate **147** in 80% yield under milder conditions strongly supports the claim for anchimeric assistance[187b]. A plausible mechanism involving preequilibrium protonation of the more basic amide oxygen is outlined in Scheme 55.

Intramolecular catalysis by aliphatic hydroxyl groups has been widely recognized and enhanced solvolysis rates for aldonamides[212], 4-hydroxybutyramide[213] (as well as the corresponding anilide[214]), 5-hydroxyvaleramide[213a] and tertiary N-(2-chloroethyl)carboxamides[215] (these first hydrolyse to the 2-hydroxy derivative) have all been attributed to this cause. Reaction via lactone intermediates is almost certainly involved in each case.

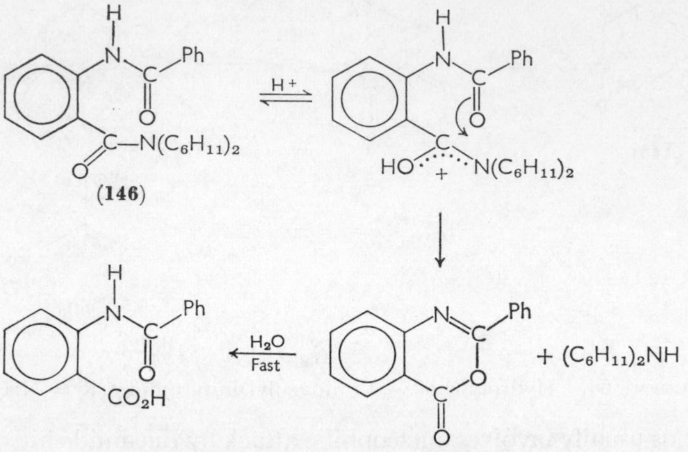

SCHEME 55.   Hydrolysis of o-benzamido-N,N-dicyclohexylbenzamide in acetic–
sulphuric acid mixtures.

The hydrolysis of 4-hydroxybutyramide, for example, is catalysed by
both hydroxide and hydronium ions, but the rate coefficients are about
18 times larger than those for butyramide (Table 18)[213]. From our
knowledge of substituent effects (see sections X.A.3 and X.B.2), it
seems unlikely that increases of such magnitude could arise from

TABLE 18.   Hydrolysis rates for butyramide and
4-hydroxybutyramide[213b].

	$10^4 k_{H^+} (30°)$ [a]	$10^4 k_{OH^-} (100°)$ [a]
$CH_3(CH_2)_2CONH_2$	3·7	0·44
$HOCH_2(CH_2)_2CONH_2$	65	9·5

[a] In units of l/mole min.

inductive electron withdrawal by the hydroxide group. A more
reasonable explanation is that both the neutral OH and the anionic
O^- forms of the substituent react intramolecularly to produce the
lactone (148), which then decomposes rapidly to products (Scheme 56).
For the acid-catalysed path, it is possible that proton transfer to
nitrogen and attack by O^- on the carbonyl carbon are synchronous,
and this represents another example of 'electrophilic–nucleophilic
catalysis'[213b]. The unassisted participation by the O^- entity for the
base-catalysed path is in direct contrast to the carboxylate ion and

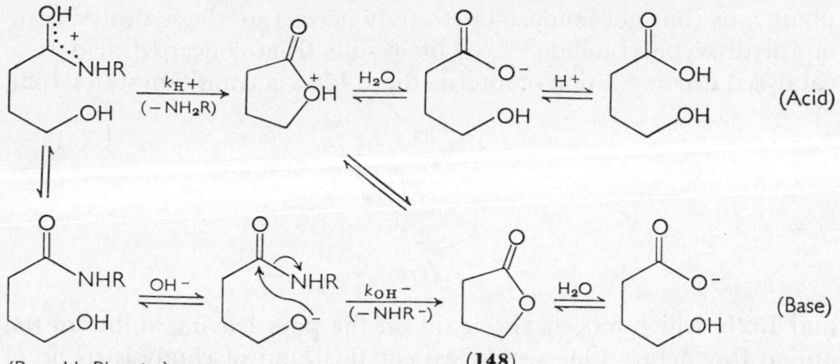

Scheme 56. Hydrolysis of 4-hydroxybutyramides.

neutral imidazole groups, but this merely reflects the greater nucleo-philicity of alkoxide ions. A closer examination of the base-catalysed pathway for the anilide derivative (R = Ph) has produced some convincing kinetic evidence for the formation of tetrahedral pre-cursors (**149** and **150**) to the lactone intermediate (Scheme 57), as would be expected from our knowledge of intermolecular alkaline hydrolysis[214]. An interesting point is that decomposition via the neutral intermediate **150** produces a plateau in the pH–rate profile at neutral pH. The occurrence of a similar plateau in the hydrolysis of 4-hydroxybutyramide was originally attributed[213b] to a reaction of the zwitterion **151**, but in the light of the anilide data this conclusion needs to be reexamined. Another finding, with far-reaching im-plications, is that bifunctional species such as bicarbonate and phos-

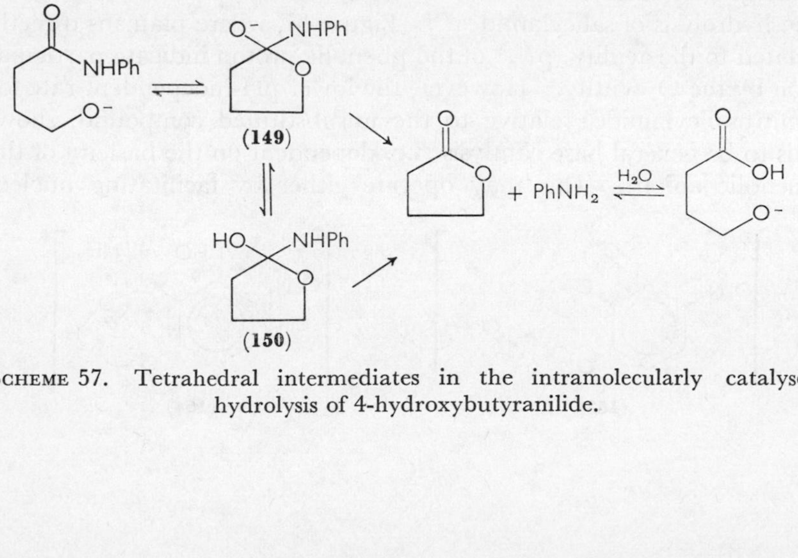

Scheme 57. Tetrahedral intermediates in the intramolecularly catalysed hydrolysis of 4-hydroxybutyranilide.

phate ions (but not imidazole) strongly accelerate the hydrolysis rate of 4-hydroxybutyranilide[214]. This results from concerted acid–base catalysed decomposition of intermediate **150** via transition states **152a**

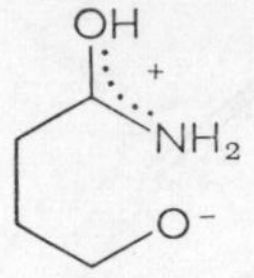

(151)

and **152b**, which reflects yet again on the poor leaving ability of the amino fragment. One would expect this kind of catalysis to be of major importance to intermolecular hydrolysis, although only tentative indications have been found thus far.

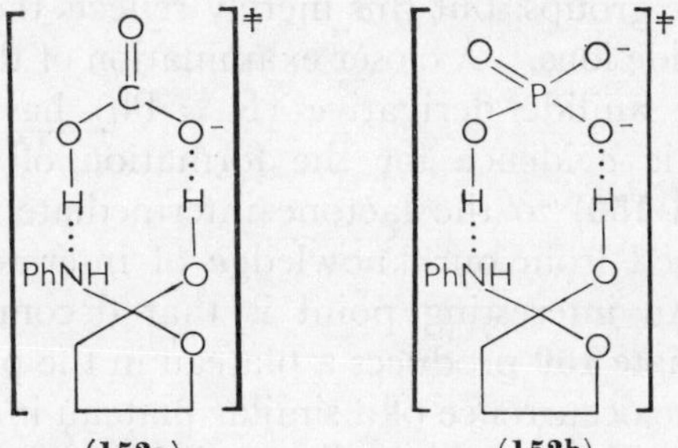

(152a)   (152b)

When direct nucleophilic catalysis through intermediate lactone formation is prohibited by the molecular geometry, neighbouring hydroxide groups may still assist amide hydrolysis by functioning as general base catalysts. This is evident from the pH–rate profiles for the hydrolysis of salicylamides[216] (Figure 3), where plateaus directly related to the acidity (pK_a) of the phenolic proton indicate participation by the O$^-$ entity. However, the lower pH-independent rate for 5-nitrosalicylamide, relative to the unsubstituted compound, shows this to be general base catalysis (i.e. dependent on the basicity of the phenolic anion). This may operate either by facilitating nucleo-

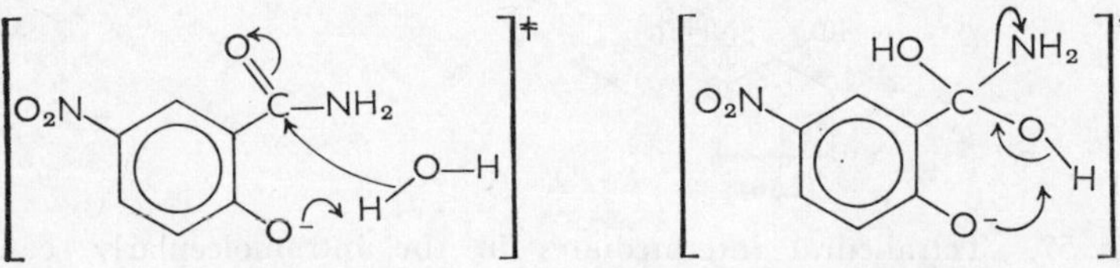

(153)   (154)

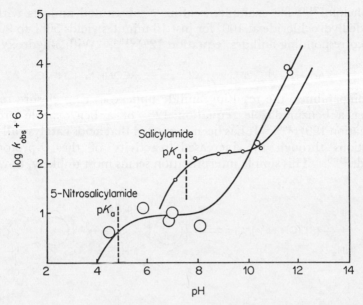

FIGURE 3. pH–rate profiles for the hydrolysis of salicylamides in aqueous solution at 100° showing plateaus in the vicinity of the pK_a of the phenolic proton. From reference 216. [Reproduced with permission of the American Chemical Society.]

philic attack by water (**153**) as suggested by Bruice and Tanner[216], or by assisting decomposition of the tetrahedral intermediate **154**. The second explanation is more consistent with the maxim that 'breakdown of the tetrahedral intermediate is usually rate limiting for the alkaline hydrolysis of amides'.

D. Solvolysis Reactions

Nucleophilic solvents other than water, such as amines, alcohols and carboxylic acids, will also interact with amides under suitable conditions. Few of these reactions have been investigated thoroughly, but it would be reasonable to suppose that many of the mechanistic intricacies of hydrolysis apply to solvolysis reactions in general. Most of the available information supports this assertion.

Amino-group exchange occurs on heating with primary amines, often without additional catalysts, but usually the reactions only proceed to completion when the expelled amine is relatively volatile[217]. Aromatic amines react sluggishly, however, and mild acid catalysts are helpful[218]; for example, heating p-substituted anilines with the amide hydrochloride at $100°$ for just 10 minutes yields 50% to 90% of the corresponding anilides (equation 127)[218d]. With o-hydroxy- and

$$ArNH_2 + [RC(OH)NH_2]^+Cl^- \xrightarrow{\Delta} RCONHAr + NH_4Cl \qquad (127)$$

o-aminoanilines, the resulting anilide undergoes ring closure to give either a benzimidazole (equation 128) or a benzoxazole product (equation 129)[218d]. It has been suggested that acids catalyse all these reactions through the increased reactivity of the O-protonated amide[218b]. This simple interpretation seems most unlikely, however,

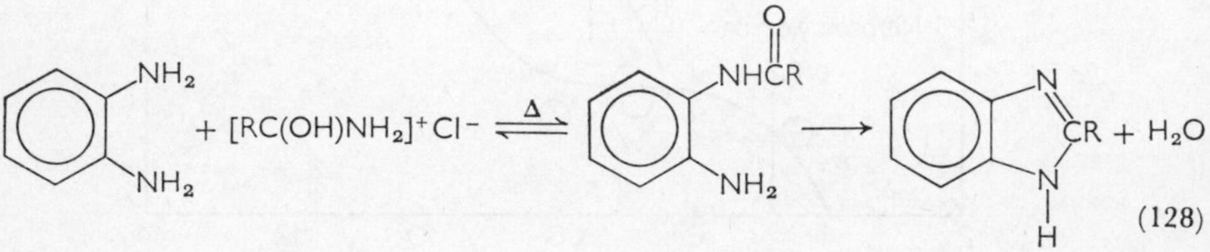

$$ (128) $$

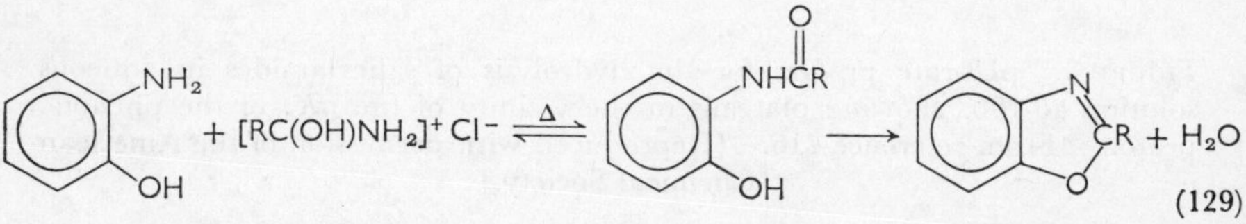

$$ (129) $$

because the *amine* is by far the more basic species. Accordingly, a more plausible explanation is that acids catalyse the decomposition of a tetrahedral intermediate.

Nucleophilic displacement is particularly facile with amino derivatives containing electron-donating substituents, as in hydrazine and hydroxylamine. Kinetic studies have been reported for the reaction of hydroxylamine with alkylamides in which displacement of the amino fragment yields a hydroxamic acid (equation 130)[219]. The

$$RCONH_2 + HONH_2 \xrightarrow{25°} RCONHOH + NH_3 \qquad (130)$$

pH–rate profile is characteristically bell-shaped, with a maximum rate at about $pH = 6.5$, because of general acid catalysis by the hydroxyl-ammonium ion ($pK_a = 6.04$). A more telling observation is a steady decrease in the catalytic coefficient for the general acid (both hydroxylammonium ion and imidazolium ion) as its concentration is increased (Figure 4)[219]. This sort of fall-off in the rate-coefficient has already been discussed in connexion with alkaline hydrolysis, and is indicative of a change in the rate-limiting step and, therefore, of the

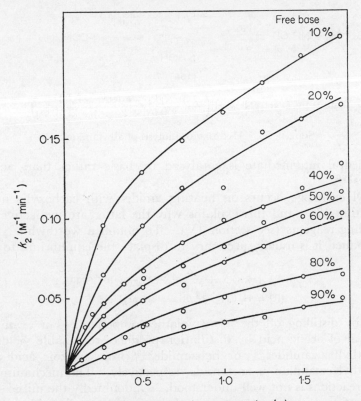

FIGURE 4. Dependence of the second-order rate coefficient for formohydrox-amic acid formation ($k_2' = k_{obs}/[NH_2OH]$) on hydroxylamine concentration at different fractions of hydroxylamine neutralization in aqueous solution at 39°. From reference 219. [Reproduced with permission of the American Chemical Society.]

formation of a tetrahedral intermediate. One possible mechanism, involving general acid-catalysed formation together with both spontaneous and acid-catalysed decomposition of a tetrahedral intermediate **155**, is outlined by Scheme 58. The present data do not distinguish, however, between this and other permutations of similar rate coefficients for a two-step reaction. Other kinetic investigations suggest that the multiplicity of pathways discussed earlier for hydrolysis will also be found in solvolysis reactions. Thus preliminary results for displacements from 2,2,2-trifluoroacetanilide by both hydrazine and hydroxylamine show that breakdown of the

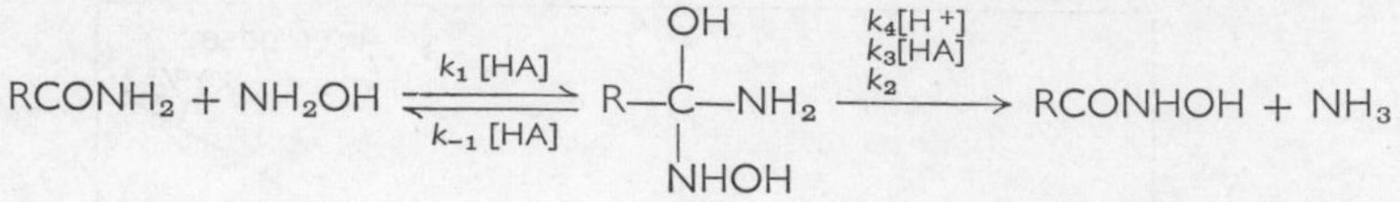

$$RCONH_2 + NH_2OH \xrightleftharpoons[k_{-1}\,[HA]]{k_1\,[HA]} R-\underset{\underset{NHOH}{|}}{\overset{\overset{OH}{|}}{C}}-NH_2 \xrightarrow{\overset{k_4[H^+]}{\overset{k_3[HA]}{k_2}}} RCONHOH + NH_3$$

(155)

$$\text{Rate} = [RCONH_2][NH_2OH]\left(\frac{k_1[HA](k_2 + k_3[HA] + k_4[H^+])}{k_2 + k_3[HA] + k_4[H^+] + k_{-1}[HA]}\right)$$

Scheme 58. Hydroxyaminolysis of alkylamides.

tetrahedral intermediate is catalysed by basic rather than acidic species[220].

Acyl exchange occurs on heating amides with carboxylic acids (equation 131), and this explains why the latter are ineffective N-acylating reagents (see section IV). To obtain a worthwhile yield of product, it is usually necessary to displace the equilibrium to the

$$RCONR^1_2 + R^2CO_2H \overset{\Delta}{\rightleftharpoons} RCO_2H + R^2CONR^1_2 \tag{131}$$

(R^1 = H, CH_3; R^2, R = alkyl, aryl, etc.)

right by distilling off the more volatile component. For example, removal of acetic acid by distillation produces reasonable yields of N-methylbenzanilides[221] or benzamide[218c] from benzoic acid and N-methylacetanilide or acetamide, respectively. The mechanism of these reactions is not well-understood. Undoubtedly the initial step is one of nucleophilic substitution of the O-protonated amide by the carboxylate ion to give a tetrahedral intermediate 156, which might then be expected to form an anhydride via elimination of the amino fragment (as in the acetolysis of N-acetylimidazole[203c]). The anhydride could then react with the displaced amine to regenerate an amide. However, tests for both anhydride and amine have proved negative (for example, NH_3 is not evolved in the transformation between acetamide and benzoic acid[218c]) and this suggests that 156 decomposes to products directly via a four-centre transition state (equation 132)[221].

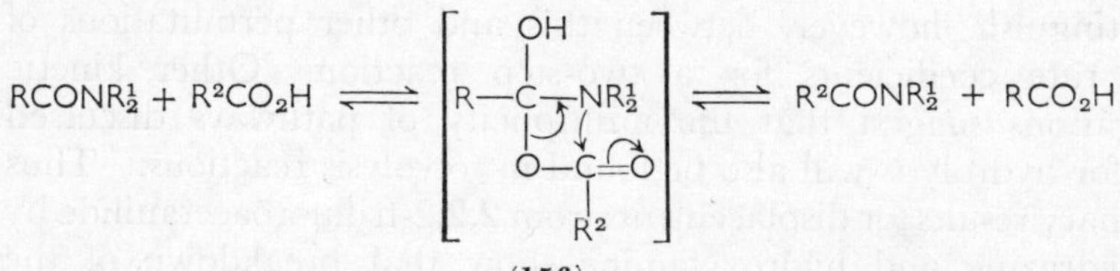

$$RCONR^1_2 + R^2CO_2H \rightleftharpoons \left[R-\underset{\underset{R^2}{\underset{|}{O-C-O}}}{\overset{\overset{OH}{|}}{C}}-NR^1_2 \right] \rightleftharpoons R^2CONR^1_2 + RCO_2H \tag{132}$$

(156)

The most important and extensive studies of solvolysis have concerned the acyl derivatives of tertiary amines such as pyridine and imidazole. We have already commented on some of these reactions in connexion with general acid- and general base-catalysed hydrolysis (section X.C). Current interest in these compounds stems from their utility as acylating agents for a wide range of nucleophilic species, and their performance in this respect has earned them the title of 'energy-rich compounds'.

The advantages of pyridine as a solvent for the acylation of amines, alcohols and phenols with either acid chloride or anhydrides has been widely recognized. Incidental evidence suggests the intermediacy of the N-acetylpyridinium ion (**157**), which then reacts rapidly with any available nucleophilic species (equation 133)[222]. The acetyl-pyridinium ion has not been detected directly in the reaction solutions,

$$(CH_3CO)_2O + N\bigcirc \rightleftharpoons CH_3\overset{O}{\overset{\|}{C}}-N\langle + \rangle + CH_3CO_2^-$$

Nu = a nucleophile

(157)

$$\downarrow Nu$$

$$CH_3CONu + N\bigcirc \qquad (133)$$

but it has been isolated from acetyl chloride and pyridine mixtures under anhydrous conditions[222b].

Acyl derivatives of imidazole and related azoles are sufficiently stable to be obtained as crystalline compounds[223]. They, too, have found wide applicability as acylating reagents, and this aspect of their chemistry has been reviewed by Staab[223]. It is evident that their solvolysis by water, alcohols, phenols, thiols, amines, carboxylic acids and various other nucleophiles is both clean and rapid. N-Acetyl-imidazole, for example, undergoes rapid hydrolysis even in neutral solutions, in striking contrast to other amides. Some information about the mechanism of these reactions is available from kinetic studies, but much more remains to be done. The number of terms in the rate expression depends on the experimental conditions (solvent nucleophile, pH, buffer components, etc.) as in the case of hydrolysis, but Jencks and Carruiolo[203a] have established the general importance of one term with a first-order dependence on the N-acetylimidazole and nucleophile concentrations. The pH–rate profile, as well as comparison with the data for N-acetyl-N-methylimidazolium ion[202], leave no doubt, however, that this refers to attack by the conjugate base of the nucleophile on the N-acetylimidazolium ion; i.e. the kinetics follow equation (134) where $HX = RNH_3^+$, RCO_2H, ROH, etc. Since most of the measurements have been made in aqueous solutions, some contribution to the overall rate may come from general base-catalysed hydrolysis (which would also follow equation 134). Product studies have not been extensive for these reactions, but it is

$$\text{Rate} = k_1[\text{AcImH}^+][\text{X}] \qquad (134)$$

known that hydroxylamines[203a] and phenolate ions[204] interact exclusively by the nucleophilic path, whereas the acetate reaction occurs 78% via the general base-catalysed hydrolytic path and 22% via the nucleophilic reaction[203c] (see section X.C.1). It has been estimated that the point of change of hydrolytic to nucleophilic mechanism for the N-acetylimidazolium ion with changing basicity of the nucleophile occurs in the region of pH = 4 to 5[224].

There is good evidence, too, that under the appropriate conditions the nucleophilic substitution is catalysed by general bases[203a]. In this sense, these reactions parallel hydrolysis. For example, the reaction with amines in imidazole buffers follows equation (135),

$$\text{Rate} = k_1[\text{AcImH}^+][\text{RNH}_2] + k_2[\text{AcImH}^+][\text{RNH}_2][\text{Im}] +$$
$$k_3[\text{AcIm}][\text{RNH}_2][\text{Im}] + k_4[\text{AcIm}][\text{RNH}_2]^2 \qquad (135)$$

which contains terms for imidazole catalysis of the N-acetylimidazolium ion reaction (k_2) and for imidazole- (k_3) and amine- (k_4) catalysed substitution of neutral N-acetylimidazole.

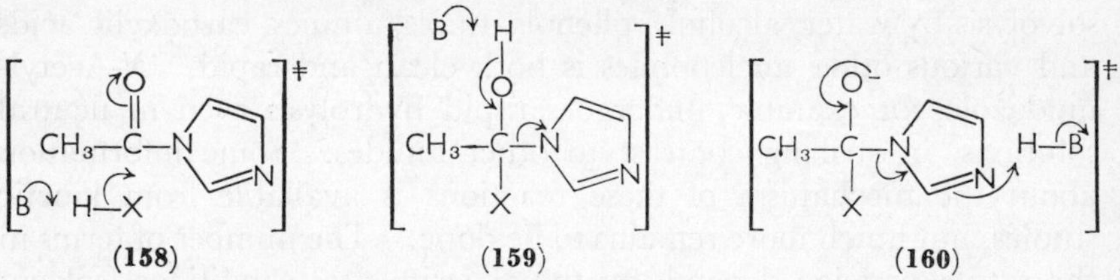

(158)   (159)   (160)

The exact mechanism of the catalysis has not been ascertained and the kinetic terms for the neutral N-acetylimidazole are equally consistent with general base-catalysed attack by the nucleophile (158), general base-assisted breakdown of the tetrahedral intermediate (159) or general acid-catalysed decomposition of the tetrahedral intermediate (**160**). Arguments in favour of **158** have been tentatively advanced[203a], but it is by no means unlikely that the catalytic mechanism changes with both substrate (*N*-acetylimidazolium ion versus neutral *N*-acetylimidazole) and the experimental conditions.

XI. ADDITION OF ORGANOMETALLIC REAGENTS

Nucleophilic addition to the amide carbonyl function also takes place with organometallic substances such as Grignard reagents and organolithium compounds. This process probably involves a concerted attack by the metal atom and the carbanion fragment on the amide oxygen and carbon atoms respectively.

In practice, the reactions with primary and secondary amides are not very useful, as much of the organometallic reagent is converted to the corresponding hydrocarbon through interaction with the feebly acidic *N*-proton[225]. However, phenylacetamide and diphenylacetamide dehydrate readily to the corresponding nitriles with three mole equivalents of n-butyllithium in inert solvents[226].

Tertiary amides react cleanly but slowly with Grignard reagents to form a stable addition complex (**161**), which on treatment with aqueous acid produces a ketone (equation 136)[227]. Dimethylformamide gives the corresponding aldehyde[228]. The addition of excess Grignard reagent to tertiary amides such as dialkylformamides[228], *N*-methylpyridone[229] or *N*-methylcaprolactam[229],

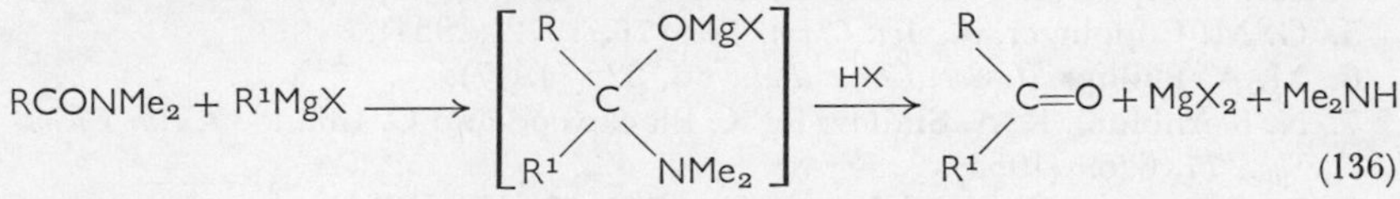

$$RCONMe_2 + R^1MgX \longrightarrow \begin{bmatrix} R & OMgX \\ & C \\ R^1 & NMe_2 \end{bmatrix} \xrightarrow{HX} \begin{array}{c} R \\ C{=}O + MgX_2 + Me_2NH \\ R^1 \end{array} \quad (136)$$

$$(\mathbf{161})$$

however, results in reductive alkylation, i.e. replacement of the amidic oxygen by two alkyl groups (equation 137).

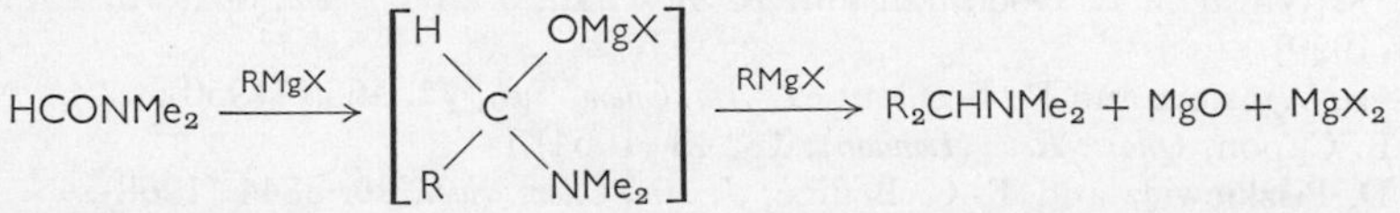

$$HCONMe_2 \xrightarrow{RMgX} \begin{bmatrix} H & OMgX \\ & C \\ R & NMe_2 \end{bmatrix} \xrightarrow{RMgX} R_2CHNMe_2 + MgO + MgX_2 \quad (137)$$

With organolithium compounds, tertiary amides also form addition complexes which on hydrolysis yield the ketone[230], as for the Grignard reagent. When the $\alpha_{C{=}O}$ proton is acidic, however, the strongly

$$+ \text{OH}^- + \text{Li}^+ \quad (138)$$

basic conditions induce elimination from the addition complex with the formation of enamine products[231] (equation 138). Few of these reactions have generated much interest as synthetic procedures.

XII. REFERENCES

1a. M. Matsui, *Mem. Coll. Sci. Eng. Kyoto*, **2**, 37 (1909); *Chem. Abstr.*, **5**, 475 (1911); A. Bühner, *Ann. Chem.*, **333**, 289 (1904).

1b. H. Bredereck, R. Gompper, H. Rempfer, K. Klemm and H. Keck, *Chem. Ber.*, **92**, 329 (1959).

1c. R. E. Benson and T. L. Cairns, *J. Am. Chem. Soc.*, **70**, 2115 (1948).

1d. H. Bredereck, F. Effenberger and E. Hanseleit, *Chem. Ber.*, **98**, 2754 (1965).

1e. H. Bredereck, F. Effenberger and G. Simchen, *Angew. Chem.*, **73**, 493 (1961); H. Bredereck, *Ger. Pat.*, 1,156,779 (1963); H. Bredereck, F. Effenberger and H. P. Bayerlin, *Chem. Ber.*, **97**, 3076 (1964).

2a. H. Meerwein, W. Florian, N. Schön and G. Stopp, *Ann. Chem.*, **641**, 1 (1961); H. Meerwein, P. Borner, O. Fuchs, H. J. Sasse, H. Schrodt and J. Spille, *Chem. Ber.*, **89**, 2060 (1956).

2b. R. F. Borch, *Tetrahedron Letters*, 61 (1968).

3a. R. Gompper, *Chem. Ber.*, **93**, 187, 198 (1960).

3b. J. W. Ralls, *J. Org. Chem.*, **26**, 66 (1961).

4a. H. Bredereck, R. Gompper and G. Theilig, *Chem. Ber.*, **87**, 537 (1954).

4b. R. Gompper and O. Christmann, *Chem. Ber.*, **92**, 1935 (1959).

5. G. M. Coppinger, *J. Am. Chem. Soc.*, **76**, 1372 (1954).

6. M. A. Phillips, *J. Soc. Chem. Ind.*, **66**, 325 (1947).

7. N. Kornblum, R. A. Smiley, R. K. Blackwood and C. Iffland, *J. Am. Chem. Soc.*, **77**, 6269 (1955).

8. R. Roger and D. G. Nielson, *Chem. Rev.*, **61**, 179 (1961).

9. R. Mecke and R. Mecke, *Chem. Ber.*, **89**, 343 (1956).

10. W. Hechelhammer, *Ger. Pat.*, 948,973 (1956).

11. P. Rehländer, *Chem. Ber.*, **27**, 2157 (1894).

12. C. J. M. Stirling, *J. Chem. Soc.*, 255 (1960).

13a. S. Winstein, L. Goodman and R. Boschan, *J. Am. Chem. Soc.*, **72**, 2311 (1950).

13b. S. Winstein and R. Boschan, *J. Am. Chem. Soc.*, **72**, 4669 (1950).

14. B. Capon, *Quart. Rev. (London)*, **18**, 45 (1964).

15. D. Piszkiewicz and T. C. Bruice, *J. Am. Chem. Soc.*, **90**, 5844 (1968).

16. S. Dayagi, *Israel J. Chem.*, **6**, 477 (1968).

17a. K. B. Wiberg and B. I. Rowland, *J. Am. Chem. Soc.*, **77**, 2205 (1955).

17b. K. B. Wiberg, T. M. Shryne and R. R. Kintner, *J. Am. Chem. Soc.*, **79**, 3160 (1957).

18. G. D. Lander, *J. Chem. Soc.*, 406 (1903); A. E. Arbuzov and V. E. Shiskin, *J. Gen. Chem. USSR (Eng. Transl.)*, **34**, 3628 (1964); *Dokl. Akad. Nauk SSSR*, **141**, 81, 349, 611 (1961); *Chem. Abstr.*, **56**, 10038, 15424, 11491 (1962).
19. G. D. Lander, *J. Chem. Soc.*, 1414 (1903).
20a. W. J. Hickinbottom, *Reactions of Organic Compounds*, 3rd ed., Longmans, Green and Co., New York, 1957, pp. 409–411.
20b. R. Levine and W. C. Fernelius, *Chem. Rev.*, **54**, 467 (1954).
21a. L. E. Smith and J. Nichols, *J. Org. Chem.*, **6**, 502 (1941).
21b. G. C. Hopkins, J. P. Jonak, H. J. Minnemeyer and H. Tieckelmann, *J. Org. Chem.*, **32**, 4040 (1967).
22a. S. D. Work, D. R. Bryant and C. R. Hauser, *J. Org. Chem.*, **29**, 722 (1964).
22b. R. B. Meyer and C. R. Hauser, *J. Org. Chem.*, **26**, 3187 (1961); K. Shimo, S. Wakamatsu and T. Inoue, *J. Org. Chem.*, **26**, 4868 (1961); K. Shimo, S. Wakamatsu and F. Haruyama, *Kogyo Kagaku Zasshi*, **62**, 201 (1959); *Chem. Abstr.*, **55**, 27334 (1961).
23a. P. G. Gassman and B. L. Fox, *J. Org. Chem.*, **31**, 982 (1966).
23b. H. L. Needles and R. E. Whitfield, *J. Org. Chem.*, **31**, 989 (1966).
23c. O. Martensson and E. Nilsson, *Acta Chem. Scand.*, **14**, 1129 (1960); **15**, 1026 (1961).
24a. H. Bredereck, R. Gompper, H. Herlinger and E. Woitun, *Chem. Ber.*, **93**, 2423 (1960).
24b. F. Scotti and E. J. Frazza, *U.S. Pat.*, 3,138,606 (1964).
24c. U. Bahr, E. Siegal and G. E. Nischk, *Fr. Pat.*, 1,392,760 (1965).
24d. N. M. Bortnick, *Brit. Pat.*, 875,134 (1961); 875,135 (1961); *U.S. Pat.*, 3,021,338 (1962).
25. A. E. Arbuzov and O. M. Shapshinskaya, *Tr. Kazansk. Khim.-Tekhnol. Inst.*, **30**, 22 (1962); *Chem. Abstr.*, **60**, 7951 (1964).
26. I. Yoshio and I. Shimichi, *Bull. Chem. Soc. Japan*, **39**, 2490 (1966).
27. I. L. Knunyants and N. P. Gambaryan, *Izv. Akad. Nauk SSSR Otd. Khim. Nauk*, 1037 (1955); *Chem. Abstr.*, **50**, 11277 (1956); H. W. Heine, P. Love and J. L. Bove, *J. Am. Chem. Soc.*, **77**, 5420 (1955).
28. M. S. Manhas and S. J. Jeng, *J. Org. Chem.*, **32**, 1246 (1967).
29. S. Sarel and H. Leader, *J. Am. Chem. Soc.*, **82**, 4752 (1960); H. E. Baumgarten, R. L. Zey and U. Krolls, *J. Am. Chem. Soc.*, **83**, 4469 (1961); H. E. Baumgarten, *J. Am. Chem. Soc.*, **84**, 4975 (1962); H. E. Baumgarten, J. F. Fuerholzer, R. D. Clark and R. D. Thompson, *J. Am. Chem. Soc.*, **85**, 3303 (1963); J. C. Sheehan and I. Lengyel, *J. Am. Chem. Soc.*, **86**, 746, 1356 (1964).
30. C. J. M. Stirling, *J. Chem. Soc.*, 3676 (1962).
31. H. W. Heine, *J. Am. Chem. Soc.*, **78**, 3708 (1956).
32. T. Taguchi and M. J. Kojima, *J. Am. Chem. Soc.*, **81**, 4316 (1959).
33a. R. B. Meyer and C. H. Hauser, *J. Org. Chem.*, **26**, 3187 (1961).
33b. S. Portnoy, *J. Heterocyclic Chem.*, **3**, 363 (1966).
34. H. E. Johnson and D. G. Crosby, *J. Org. Chem.*, **27**, 2205 (1962); H. E. Winberg, *U.S. Pat.*, 3,121,084 (1964).
35. H. Bredereck, F. Effenberger and H. J. Treiber, *Chem. Ber.*, **96**, 1505 (1963).

28+c.o.a.

36. E. Schnabel and H. Schuelssler, *Ann. Chem.*, **686,** 229 (1965).

37. A. J. Speziale, K. W. Ratts and G. J. Marco, *J. Org. Chem.*, **26,** 4311 (1961); C. Glacet and G. Troude, *Bull. Soc. Chim. France*, 292 (1964); J. Furukawa, A. Onishi and T. Tsuruta, *J. Org. Chem.*, **23,** 672 (1958); S. Adomaitene, A. M. Sladkov and V. P. Shishkov, *J. Gen. Chem. USSR* (*Eng. Transl.*), **34,** 434 (1964); **34,** 2992 (1964); M. G. Voronokov, *J. Gen. Chem. USSR* (*Eng. Transl.*), **21,** 1631 (1951).

38a. G. Lauermann, *Ger.* (*East*) *Pat.*, 45,079 (1966).

38b. D. P. Gush, N. S. Marans, F. Wessells, W. D. Addy and S. J. Olfky, *J. Org. Chem.*, **31,** 3829 (1966).

38c. D. Elad and J. Rokach, *J. Org. Chem.*, **30,** 3361 (1965).

38d. J. C. Allen, J. I. G. Cadogan and D. H. Hey, *J. Chem. Soc.*, 1918 (1965).

39a. H. Böhme, A. Dick and G. Driesen, *Chem. Ber.*, **94,** 1879 (1961); H. Böhme, R. Broese, A. Dick and F. Eiden, *Chem. Ber.*, **92,** 1599 (1959).

39b. J. P. Chupp and A. J. Speziale, *J. Org. Chem.*, **28,** 2592 (1963).

39c. J. F. Walker, *Formaldehyde*, 2nd ed., Reinhold Publishing Corp., New York, 1953, p. 290.

39d. G. A. Crowe and C. C. Lynch, *J. Am. Chem. Soc.*, **72,** 3622 (1950).

39e. R. D. Haworth, D. H. Peacock, W. R. Smith and R. MacGillivray, *J. Chem. Soc.*, 2972 (1952).

39f. S. L. Vail, C. M. Moran and H. B. Moore, *J. Org. Chem.*, **27,** 2067 (1962).

40. J. P. La Rocca, J. M. Leonard and W. E. Weaver, *J. Org. Chem.*, **16,** 47 (1951); A. Boucherle, G. Carraz and J. Vigier, *Bull. Trav. Soc. Pharm. Lyon*, **10,** 3 (1966); C. Broquet-Borgel, *Compt. Rend.*, **250,** 1075 (1960); J. Kaupp and E. Hambsch, *Ger. Pat.*, 1,126,404 (1962); F. Weygand, W. Steglich, I. Lengyel, F. Fraunberger, A. Maierhofer and W. Oettmeier, *Chem. Ber.*, **99,** 1944 (1966).

41. P. E. Newallis and E. J. Rumanowski, *J. Org. Chem.*, **29,** 3114 (1964).

42. H. Bredereck and G. Theilig, *Chem. Ber.*, **86,** 88 (1953); *Angew. Chem.*, **71,** 753 (1959).

43a. S. L. Vail, R. H. Barker and C. M. Moran, *J. Org. Chem.*, **31,** 1642 (1966); S. L. Vail, C. M. Moran, H. B. Moore and R. M. H. Kullman, *J. Org. Chem.*, **27,** 2071 (1962).

43b. Nobel-Bozel, *Neth. Appl.*, 6,507,348 (1965).

44. Farbwerke-Hoechst, *Belg. Pat.*, 647,478 (1964).

45. H. Hellmann, *Angew. Chem.*, **69,** 463 (1957).

46a. A. Einhorn, *Ann. Chem.*, **343,** 207 (1905); **361,** 113 (1908).

46b. W. Seidel, A. Soeder and F. Lindner, *Med. Chem.*, **6,** 316 (1958); W. J. Gotstein, W. I. Minor and L. C. Cheney, *J. Am. Chem. Soc.*, **81,** 1198 (1959).

46c. E. Mueller, *Ger. Pat.*, 1,102,157 (1954); G. Wenner and H. G. Trieschmann, *Ger. Pat.*, 1,021,575 (1957); S. Foldeak, B. Matkovics and J. Porszasz, *Hung. Pat.*, 150,858 (1962).

46d. V. I. Stavrovskaya, S. K. Drusvyatskaya and M. O. Kolosova, *Zh. Organ. Khim.*, **4,** 488 (1968).

46e. H. Fraenkel-Conrat and H. S. Olcott, *J. Am. Chem. Soc.*, **70,** 2673 (1948).

47. B. C. Challis and A. R. Butler, *The Chemistry of the Amino Group*, (Ed. S. Patai), Interscience Publishers, London, 1968, p. 301.

48. M. Imoto and M. Kobayashi, *Bull. Chem. Soc. Japan*, **33,** 1651 (1960).

49a. H. Feuer and V. E. Lynch, *J. Am. Chem. Soc.*, **75,** 5027 (1953).

49b. N. Yanaihara and M. Saito, *Chem. Pharm. Bull. Japan*, **15**, 128 (1967).

50. H. Bestian, W. Schwiersch and R. Hartwimmer, *Ger. Pat.*, 1,170,936 (1964); F. Eiden and B. S. Nagar, *Naturwissenschaften*, **48**, 599 (1961); *Arch. Pharm.*, **296**, 445 (1963).

51. R. Giger and D. Ben-Ishai, *Israel J. Chem.*, **5**, 253 (1967).

52. T. Falbe and H.-J. Schulze-Steinen, *Brennstoff-Chem.*, **48**, 136 (1967).

53. U. Zehavi and D. Ben-Ishai, *J. Org. Chem.*, **26**, 1097 (1961); W. A. Noyes and D. B. Forman, *J. Am. Chem. Soc.*, **55**, 3493 (1933).

54. D. M. von Schriltz, E. M. Kaiser and C. R. Hauser, *J. Org. Chem.*, **32**, 2610 (1967).

55. I. K. Mosevich, I. A. Arbuzova and A. I. Kol'tsov, *J. Gen. Chem. USSR (Eng. Transl.)*, **38**, 1180 (1968).

56. Q. E. Thompson, *J. Am. Chem. Soc.*, **73**, 5841 (1951).

57. H. K. Hall, *J. Am. Chem. Soc.*, **78**, 2717 (1956).

58. R. Tull, R. C. O'Neill, E. P. McCarthy, J. J. Pappas and J. M. Chemerda, *J. Org. Chem.*, **29**, 2425 (1964).

59. D. E. Horning and J. M. Muchowski, *Can. J. Chem.*, **45**, 1247 (1967).

60. A. W. Titherly and T. H. Holden, *J. Chem. Soc.*, **101**, 1871 (1912).

61. W. G. Toland and L. L. Ferstandig, *J. Org. Chem.*, **23**, 1350 (1958); R. H. Wiley and W. B. Geurrant, *J. Am. Chem. Soc.*, **71**, 981 (1949).

62. A. E. Kulikova, E. N. Zil'berman and F. A. Ekstrin, *Zh. Vses. Khim. Obshchestva im. D. I. Mendeleeva*, **11**, 704 (1966); *Chem. Abstr.*, **66**, 115426 (1967).

63. C. D. Hurd and A. G. Propas, *J. Org. Chem.*, **24**, 388 (1959).

64. A. W. Titherly and T. H. Holden, *J. Chem. Soc.*, **85**, 1673 (1904).

65. D. Davidson and H. Skovronek, *J. Am. Chem. Soc.*, **80**, 376 (1958).

66. J. B. Polya and P. L. Tardrew, *J. Chem. Soc.*, 1081 (1948); *Rec. Trav. Chim.*, **67**, 927 (1948); **71**, 676 (1952).

67. D. Davidson and M. Karten, *J. Am. Chem. Soc.*, **78**, 1066 (1956).

68. F. Suzuki and M. Hasegawa, *Nippon Kagaku Zasshi*, **86**, 256 (1965); *Chem. Abstr.*, **63**, 13021 (1965).

69. J. F. Wolfe and G. B. Trimitsis, *J. Org. Chem.*, **33**, 894 (1968).

70a. M. Brenner, *Ger. Pat.*, 1,068,721 (1959).

70b. E. Sondheimer and R. W. Holley, *J. Am. Chem. Soc.*, **76**, 2467 (1954).

70c. R. M. Topping and D. E. Tutt, *J. Chem. Soc.* (B), 1346 (1967).

71. J. A. Schafer and H. Morawetz, *J. Org. Chem.*, **28**, 1899 (1963).

72. M. L. Sherrill, F. L. Schaeffer and E. P. Shoyer, *J. Am. Chem. Soc.*, **50**, 474 (1928).

73. E. S. Rothman, S. Serota and D. Swern, *J. Org. Chem.*, **29**, 646 (1964).

74a. A. J. Speziale and L. R. Smith, *J. Org. Chem.*, **27**, 3742 (1962).

74b. A. J. Speziale and L. R. Smith, *J. Org. Chem.*, **28**, 1805 (1963).

74c. A. J. Speziale, L. R. Smith and J. E. Fedder, *J. Org. Chem.*, **30**, 4306 (1965).

74d. A. J. Speziale and L. R. Smith, *J. Org. Chem.*, **27**, 4361 (1962).

74e. A. J. Speziale, L. R. Smith and J. E. Fedder, *J. Org. Chem.*, **30**, 4303 (1965).

75. F. O. Rice, J. Greenberg, C. E. Waters and R. E. Volrath, *J. Am. Chem. Soc.*, **56**, 1760 (1934); D. N. Padgham and J. B. Polya, *Australian J. Sci., Res., Ser. A*, **13**, 113 (1951).

76a. W. Logemann and D. Artini, *Chem. Ber.*, **90**, 2527 (1957); **91**, 2574 (1958); W. Logemann, D. Artini and G. Tosolini, *Chem. Ber.*, **91**, 2566 (1958); W. Logemann, D. Artini, G. Tosolini and F. Piccinini, *Chem. Ber.*, **91**, 951 (1958).

76b. C. King, *J. Org. Chem.*, **25**, 352 (1960); P. F. Wiley, *J. Am. Chem. Soc.*, **71**, 1310, 3746 (1949).

77. N. M. Smirnova, A. P. Skoldinov and K. A. Kocheshkov, *Dokl. Akad. Nauk SSSR*, **84**, 737 (1952).

78. R. E. Dunbar and G. C. White, *J. Org. Chem.*, **23**, 915 (1958).

79. C. D. Schmulback and R. S. Drago, *J. Am. Chem. Soc.*, **82**, 4484 (1960); R. S. Drago, R. L. Carlson, N. J. Rose and D. A. Wenz, *J. Am. Chem. Soc.*, **83**, 3572 (1961); R. S. Drago, D. A. Wenz and R. L. Carlson, *J. Am. Chem. Soc.*, **84**, 1106 (1962).

80. J. D. Park, H. J. Gerjovich, W. R. Lycan and J. R. Lacher, *J. Am. Chem. Soc.*, **74**, 2189 (1952).

81. C. K. Ingold, *Structure and Mechanism in Organic Chemistry*, G. Bell and Sons, London, 1953, pp. 604–610; D. V. Banthorpe, *The Chemistry of the Amino Group*, (Ed. S. Patai), Interscience Publishers, London, 1968, p. 637.

82. R. S. Neale, N. L. Marcus and R. G. Schepers, *J. Am. Chem. Soc.*, **88**, 3051 (1966).

83. E. P. Olivetto and C. Gerold, *Organic Syntheses*, Vol. 25, (Ed. R. S. Schreider), John Wiley and Sons, New York, 1945, p. 17; ref. 20a, p. 337.

84. E. S. Wallis and J. F. Lane, *Organic Reactions*, Vol. 3, (Ed. A. C. Cope), John Wiley and Sons, New York, 1946, p. 267; P. A. S. Smith, in *Molecular Rearrangements* (Ed. P. DeMayo), Interscience Publishers, New York, 1963, Chap. 8.

85. F. E. Hardy and P. Robson, *J. Chem. Soc. (B)*, 1151 (1967).

86. D. H. R. Barton, A. L. J. Beckwith and A. Goosen, *J. Chem. Soc.*, 181 (1965).

87. J. A. Attaway, R. H. Groth and L. A. Bigelow, *J. Am. Chem. Soc.*, **81**, 3599 (1959); F. P. Avonda, J. A. Gervasi and L. A. Bigelow, *J. Am. Chem. Soc.*, **78**, 2798 (1956).

88. Japan Soda Co., *Neih. Appl.*, 6,500,874 (1965).

89. M. N. Hughes and G. Stedman, *J. Chem. Soc. Suppl.*, **1**, 5840 (1964).

90. H. Ladenheim and M. L. Bender, *J. Am. Chem. Soc.*, **82**, 1895 (1960).

91. A. E. Kulikova, E. N. Zil'berman and T. K. Golubeva, *Zh. Obshch. Khim.*, **34**, 4080 (1964).

92a. Z. Kricsfalussy and A. Bruylants, *Bull. Soc. Chim. Belges*, **73**, 96 (1964).

92b. N. Sperber, D. Papa and E. Schwenck, *J. Am. Chem. Soc.*, **70**, 3091 (1948).

93. G. A. Olah and J. A. Olah, *J. Org. Chem.*, **30**, 2386 (1965).

94. E. H. White, *J. Am. Chem. Soc.*, **77**, 6008 (1955).

95a. E. H. White, *J. Am. Chem. Soc.*, **77**, 6011 (1955).

95b. E. H. White, *J. Am. Chem. Soc.*, **77**, 6014 (1955).

96. Y. L. Chow and A. C. H. Lee, *Can. J. Chem.*, **45**, 311 (1967); *Chem. Ind. (London)*, 827 (1967).

97. D. R. Augood and G. H. Williams, *Chem. Rev.*, **57**, 129 (1957); C. Rüchardt, C.-C. Tan and B. Freudenberg, *Tetrahedron Letters*, 4019 (1968); P. Miles and H. Suschitzky, *Tetrahedron*, **18**, 1369 (1962).

98. D. L. Brydon, J. I. G. Cadogan, D. M. Smith and J. B. Thomson, *Chem. Commun.*, 727 (1967).

99. Reference 47, p. 305.

100a. J. Jaz and A. Bruylants, *Bull. Soc. Chim. Belges*, **70**, 99 (1961).

100b. F. J. Kezdy, J. Jaz and A. Bruylants, *Bull. Soc. Chim. Belges*, **67**, 687 (1958).

101. R. Campbell and C. J. Peterson, *J. Org. Chem.*, **28**, 2294 (1963).

102. W. E. Bachmann, W. J. Horton, E. L. Jenner, N. W. MacNaughton and C. E. Maxwell, *J. Am. Chem. Soc.*, **72**, 3132 (1950).

103. J. Runge and W. Triebs, *J. Prakt. Chem.*, **15**, 223 (1962).

104. E. H. White and D. W. Grisley, *J. Am. Chem. Soc.*, **83**, 1191 (1961).

105. B. M. Lynch, C. M. Chen and Y.-Y. Wigfield, *Can. J. Chem.*, **46**, 1141 (1967).

106. S. Bogdal, *Zeszyty Nauk Politech. Wroclaw Chem.*, **7**, 44 (1961); *Chem. Abstr.*, **58**, 2348 (1963).

107. E. F. Reuk and N. Clauson-Kaas, *U.S. Pat.*, 3,228,950 (1966).

108. H. Feuer and B. F. Vincent, *J. Org. Chem.*, **29**, 939 (1964).

109a. M. V. Lock and B. F. Sagar, *Proc. Chem. Soc.*, 358 (1960).

109b. M. V. Lock and B. F. Sagar, *J. Chem. Soc. (B)*, 690 (1966).

109c. B. F. Sagar, *J. Chem. Soc. (B)*, 428 (1967).

109d. B. F. Sagar, *J. Chem. Soc. (B)*, 1047 (1967).

110. G. M. Burnett and K. M. Riches, *J. Chem. Soc.(B)*, 1229 (1966).

111. R. F. Moore, *Polymer*, **4**, 493 (1963).

112. A. T. Betts and N. Uri, *Chem. Ind. (London)*, 512 (1967).

113. W. H. Sharkey and W. E. Mochel, *J. Am. Chem. Soc.*, **81**, 3000 (1959).

114. A. Rieche and W. Schoen, *Chem. Ber.*, **99**, 3238 (1966).

115. C. H. Bamford and E. F. T. White, *J. Chem. Soc.*, 1860 (1959); *J. Chem. Soc.*, 4490 (1960).

116. W. Walter, M. Steffen and C. Heyns, *Chem. Ber.*, **99**, 3204 (1966).

117. E. Mikolajewski, J. E. Swallow and M. W. Webb, *J. Appl. Polymer Sci.*, **8**, 2067 (1964).

118. H. Ito, *Chem. Pharm. Bull. (Japan)*, **12**, 326, 329 (1964); T. Kosuge and M. Miyashita, *Chem. Pharm. Bull. (Japan)*, **2**, 397 (1954).

119a. H. L. Needles and R. E. Whitfield, *J. Org. Chem.*, **29**, 3632 (1964).

119b. D. E. Remy, R. E. Whitfield and H. L. Needles, *Chem. Commun.*, 681 (1967).

119c. H. L. Needles and R. E. Whitfield, *J. Org. Chem.*, **31**, 341 (1966).

119d. M. C. Agrawal and S. P. Mushran, *J. Indian Chem. Soc.*, **43**, 343 (1966).

120. D. A. House, *Chem. Rev.*, **62**, 185 (1962).

121. B. Acott and A. L. J. Beckwith, *Chem. Commun.*, 161 (1965); B. Acott, A. L. J. Beckwith and A. Hassanali, *Australian J. Chem.*, **21**, 185, 197 (1968).

122. H. E. Baumgarten and A. Staklis, *J. Am. Chem. Soc.*, **87**, 1141 (1965).

123. A. L. J. Beckwith, unpublished results.

124. N. G. Gaylord, *Reduction with Complex Metal Hydrides*, Interscience Publishers, New York, 1956, pp. 544–594.

125. M. S. Newman and T. Fukunaga, *J. Am. Chem. Soc.*, **82**, 693 (1960).

126. H. C. Brown, P. M. Weissman and N. M. Yoon, *J. Am. Chem. Soc.*, **88**, 1458 (1966).

127. R. F. Nystrom and C. R. A. Berger, *J. Am. Chem. Soc.*, **80**, 2896 (1958).

128. H. C. Brown and N. M. Yoon, *J. Am. Chem. Soc.*, **88**, 1464 (1966).

129. H. C. Brown and A. Tsukamoto, *J. Am. Chem. Soc.*, **86**, 1089 (1964).

130. S. E. Ellzey, C. H. Mack and W. J. Connick, *J. Org. Chem.*, **32,** 846 (1967).
131. Z. B. Papanastassiou and R. J. Bruni, *J. Org. Chem.*, **29,** 2870 (1964).
132. H. C. Brown and P. Heim, *J. Am. Chem. Soc.*, **86,** 3566 (1964).
133. F. Weygand and G. Eberhardt, *Angew. Chem.*, **64,** 458 (1952); F. Weygand, G. Eberhardt, H. Linden, F. Schäfer and I. Eigen, *Angew. Chem.*, **65,** 525 (1953); F. Weygand and H. Linden, *Angew. Chem.*, **66,** 174 (1954).
134. A. J. Birch, J. Cymerman-Craig and M. Slaytor, *Australian J. Chem.*, **8,** 512 (1955); A. J. Birch and H. Smith, *Quart. Rev. (London)*, **12,** 17 (1958).
135. M. J. Allen, *Organic Electrode Processes*, Chapman and Hall Ltd., London, 1958, p. 73; F. D. Popp and H. P. Schultz, *Chem. Rev.*, **62,** 19 (1962).
136. N. V. Sidgewick, *The Organic Chemistry of Nitrogen*, 3rd ed., Clarendon Press, Oxford, 1966, pp. 234–235.
137. H. S. Broadbent, G. C. Campbell, W. J. Bartley and J. H. Johnson, *J. Org. Chem.*, **24,** 1847 (1959).
138. M. Sekiya and K. Ito, *Chem. Pharm. Bull. (Japan)*, **14,** 996 (1966); M. Sekiya and M. Tomie, *Chem. Pharm. Bull. (Japan)*, **15,** 238 (1967).
139a. H. Eilingsfeld, M. Seefelder and H. Weidinger, *Angew. Chem.*, **72,** 836 (1960).
139b. H. Eilingsfeld, M. Seefelder and H. Weidinger, *Chem. Ber.*, **96,** 2671 (1963); Z. Arnold, *Collection Czech. Chem. Commun.*, **28,** 2047 (1963).
140. R. Buyle and H. G. Viehe, *Tetrahedron*, **24,** 4217 (1968).
141. J. von Braun, F. Jostes and A. Heymons, *Chem. Ber.*, **60,** 92 (1927); A. Heymons, *Chem. Ber.*, **65,** 321 (1932).
142. F. S. Fawcett, C. W. Tullock and D. D. Coffman, *J. Am. Chem. Soc.*, **84,** 4275 (1962).
143. M. Brown, *U.S. Pat.*, 3,092,637 (1963).
144. V. I. Minkin and G. N. Dorofeenko, *Russ. Chem. Rev.*, **29,** 599 (1960).
145. H. H. Bosshard and H. Zollinger, *Helv. Chim. Acta*, **42,** 1659 (1959).
146. H. Bredereck, R. Gompper, K. Klemm and H. Rempfer, *Chem. Ber.*, **92,** 837 (1959).
147. G. J. Martin and M. Martin, *Bull. Soc. Chim. France*, 1637 (1963); M. L. Filleux-Blanchard, M. T. Quemeneur and G. J. Martin, *Chem. Commun.*, 837 (1968).
148. D. S. Deorha and P. Gupta, *Chem. Ber.*, **97,** 616 (1964).
149. W. C. Anthony, *J. Org. Chem.*, **25,** 2049 (1960).
150. Upjohn Co., *Brit. Pat.*, 869,775 (1961).
151. A. Ermili, A. J. Castro and P. A. Westfall, *J. Org. Chem.*, **30,** 339 (1965); G. G. Kleinspehn and A. E. Briod, *J. Org. Chem.*, **26,** 1652 (1961).
152. H. Bredereck, R. Gompper, K. Klemm and B. Foehlisch, *Chem. Ber.*, **94,** 3119 (1961).
153. H. Bredereck and K. Bredereck, *Chem. Ber.*, **94,** 2278 (1961); H. Bredereck, F. Effenberger and G. Simchen, *Chem. Ber.*, **97,** 1403 (1964).
154. J. D. Albright, E. Benz, A. E. Lanzilotti and L. Goldman, *Chem. Commun.*, 413 (1965).
155. F. L. Scott and J. A. Barry, *Tetrahedron Letters*, 2457 (1968).
156. G. Cohaus, *Chem. Ber.*, **100,** 2719 (1967).
157. H. von Pechmann, *Chem. Ber.*, **33,** 611 (1900).
158. J. von Braun, *Angew, Chem.*, **47,** 611 (1934).
159. W. R. Vaughan and R. D. Carlson, *J. Am. Chem. Soc.*, **84,** 769 (1962).

160. G. D. Lander, *J. Chem. Soc.*, **83**, 320 (1903).
161. N. J. Leonard and E. W. Nommensen, *J. Am. Chem. Soc.*, **71**, 2808 (1949).
162. Stamicarbon N. V., *Belg. Pat.*, 609,822 (1962).
163. J. H. Ottenheym and J. W. Garritsen, *Ger. Pat.*, 1,157,210 (1963).
164. B. Prajsnar and C. Troszkiewicz, *Roczniki Chem.*, **36**, 265, 843, 853, 1029 (1962); B. Prajsnar, *Zeszyty Nauk Politech. Slask. Chem.*, **24**, 241 (1964); *Chem. Abstr.*, **63**, 11289 (1965).
165. I. Ugi and R. Meyr, *Chem. Ber.*, **93**, 239 (1960); I. Ugi, R. Meyr, M. Lupinski, F. Bodesheim and F. Rosendahl, *Organic Syntheses*, Vol. 41, (Ed. J. D. Roberts), John Wiley and Sons, New York, 1961, p. 13.
166. W. R. Hertler and E. J. Corey, *J. Org. Chem.*, **23**, 1221 (1958).
167. J. Casanova, R. E. Schuster and N. D. Werner, *J. Chem. Soc.*, 4280 (1963).
168. D. T. Mowry, *Chem. Rev.*, **42**, 257 (1948).
169. J. C. Thurman, *Chem. Ind. (London)*, 752 (1964).
170. J. von Braun and W. Rudolph, *Chem. Ber.*, **67**, 1769 (1934).
171. D. J. Cram and P. Haberfield, *J. Am. Chem. Soc.*, **83**, 2354, 2363 (1961).
172. B. Rickborn and F. R. Jensen, *J. Org. Chem.*, **27**, 4608 (1962).
173. E. Reid, *J. Am. Chem. Soc.*, **24**, 397 (1900).
174. R. L. Schowen, H. Jayaraman and L. Kershner, *J. Am. Chem. Soc.*, **88**, 3373 (1966).
175a. M. L. Bender, *Chem. Rev.*, **60**, 53 (1960).
175b. T. C. Bruice and S. J. Benkovic, *Bio-organic Mechanisms*, Vol. 1, W. A. Benjamin Inc., New York, 1966.
175c. S. L. Johnson, 'General Base and Nucleophilic Catalysis' in *Advances in Physical Organic Chemistry*, Vol. 5, (Ed. V. Gold), Academic Press, London, 1967, p. 237.
176. M. L. Bender, *J. Am. Chem. Soc.*, **75**, 5986 (1953).
177a. M. L. Bender and R. D. Ginger, *J. Am. Chem. Soc.*, **77**, 348 (1955).
177b. M. L. Bender, R. D. Ginger and J. P. Unik, *J. Am. Chem. Soc.*, **80**, 1044 (1958).
177c. M. L. Bender and R. D. Ginger, *Suomen Kemi*, **B33**, 25 (1960); C. A. Bunton, B. Nayak and C. O'Connor, *J. Org. Chem.*, **33**, 572 (1968).
178. M. L. Bender and R. J. Thomas, *J. Am. Chem. Soc.*, **83**, 4183 (1961).
179. S. S. Biechler and R. W. Taft, *J. Am. Chem. Soc.*, **79**, 4927 (1957).
180a. P. M. Mader, *J. Am. Chem. Soc.*, **87**, 3191 (1965).
180b. S. O. Eriksson and C. Holst, *Acta Chem. Scand.*, **20**, 1892 (1966).
181. A. Bruylants and F. Kezdy, *Record Chem. Progr.*, **21**, 213 (1960).
182. K. R. Lynn, *J. Phys. Chem.*, **69**, 687 (1965).
183. M. T. Behme and E. H. Cordes, *J. Org. Chem.*, **29**, 1255 (1964).
184a. R. L. Schowen and G. W. Zuorick, *J. Am. Chem. Soc.*, **88**, 1223 (1966).
184b. S. O. Eriksson and C. Bratt, *Acta Chem. Scand.*, **21**, 1812 (1967).
185. R. L. Schowen, H. Jayaraman, L. Kershner and G. W. Zuorick, *J. Am. Chem. Soc.*, **88**, 4008 (1966).
186. P. D. Bolton, *Australian J. Chem.*, **19**, 1013 (1966).
187a. R. N. Lacey, *J. Chem. Soc.*, 1633 (1960).
187b. T. Cohen and J. Lipowitz, *J. Am. Chem. Soc.*, **86**, 5611 (1964).
188a. T. W. J. Taylor, *J. Chem. Soc.*, 2741 (1930).
188b. V. K. Krieble and K. A. Holst, *J. Am. Chem. Soc.*, **60**, 2976 (1938).
188c. J. T. Edward and S. C. R. Meacock, *J. Chem. Soc.*, 2000 (1957).

188d. M. I. Vinnick, I. M. Medvetskaya, L. R. Andreeva and A. E. Tiger, *J. Phys. Chem. USSR* (*Eng. Transl.*), **41**, 128 (1967).

189a. K. Yates and J. B. Stevens, *Can. J. Chem.*, **43**, 529 (1965).

189b. A. R. Katritzky, A. J. Waring and K. Yates, *Tetrahedron*, **19**, 465 (1963).

190a. A. Berger, A. Loewenstein and S. Meiboom, *J. Am. Chem. Soc.*, **81**, 62 (1959).

190b. R. J. Gillespie and T. Birchall, *Can. J. Chem.*, **41**, 148 (1963).

190c. A. R. Katritzky and R. A. Y. Jones, *Chem. Ind.* (*London*), 722 (1961).

191. J. T. Edward and S. C. R. Meacock, *J. Chem. Soc.*, 2009 (1957).

192. R. H. DeWolfe and F. B. Augustine, *J. Org. Chem.*, **30**, 699 (1965).

193. C. A. Bunton, C. O'Connor and T. A. Turney, *Chem. Ind.* (*London*), 1835 (1967).

194. V. C. Armstrong, D. W. Farlow and R. B. Moodie, *J. Chem. Soc.* (*B*), 1099 (1968).

195. C. A. Bunton, T. A. Lewis and D. R. Llewellyn, *Chem. Ind.* (*London*), 1154 (1954).

196. C. A. Bunton, *J. Chem. Soc.*, 6045 (1963).

197. M. Kandel and E. H. Cordes, *J. Org. Chem.*, **32**, 3061 (1967); G. L. Schmir, *J. Am. Chem. Soc.*, **90**, 3478 (1968); R. K. Chaturvedi and G. L. Schmir, *J. Am. Chem. Soc.*, **90**, 4413 (1968).

198. K. Wiberg, *Chem. Rev.*, **55**, 719 (1955).

199a. J. F. Bunnett, *J. Am. Chem. Soc.*, **83**, 4956, 4968, 4973, 4978 (1963).

199b. J. F. Bunnett and F. P. Olsen, *Can. J. Chem.*, **44**, 1899, 1917 (1966).

199c. R. B. Martin, *J. Am. Chem. Soc.*, **84**, 4130 (1962).

199d. R. B. Moodie, P. D. Wale and T. J. Whaite, *J. Chem. Soc.*, 4273 (1963); R. B. Homer and R. B. Moodie, *J. Chem. Soc.*, 4377 (1963).

199e. K. Yates and J. C. Riordan, *Can. J. Chem.*, **43**, 2328 (1965).

200. C. K. Hancock, E. A. Myers and B. J. Yager, *J. Am. Chem. Soc.*, **83**, 4211 (1961).

201. J. A. Leisten, *J. Chem. Soc.*, 765 (1959); cf. J. T. Edward, H. S. Chang, K. Yates and R. Stewart, *Can. J. Chem.*, **38**, 1518, 2271 (1960).

202. R. Wolfenden and W. P. Jencks, *J. Am. Chem. Soc.*, **83**, 4390 (1961).

203a. W. P. Jencks and J. Carruiolo, *J. Biol. Chem.*, **234**, 1272, 1280 (1959); W. P. Jencks and J. Carruiolo, *J. Am. Chem. Soc.*, **82**, 1778 (1960).

203b. T. H. Fife, *J. Am. Chem. Soc.*, **87**, 4597 (1965).

203c. W. P. Jencks, F. Barley, R. Barnett and M. Gilchrist, *J. Am. Chem. Soc.*, **88**, 4464 (1966).

204. J. Gerstein and W. P. Jencks, *J. Am. Chem. Soc.*, **86**, 4655 (1964).

205. A. R. Butler and V. Gold, *J. Chem. Soc.*, 4362 (1961); C. A. Bunton, N. A. Fuller, S. G. Perry and V. J. Shiner, *Tetrahedron Letters*, 458 (1961).

206. K. G. Wyness, *J. Chem. Soc.*, 2934 (1958).

207. M. L. Bender, Y.-L. Chow and F. Chloupek, *J. Am. Chem. Soc.*, **80**, 5380 (1958).

208. S. J. Leach and H. Lindley, *Trans. Faraday Soc.*, **49**, 921 (1953).

209. H. Morawetz and J. Schafer, *J. Am. Chem. Soc.*, **84**, 3783 (1962).

210. T. C. Bruice and J. M. Sturtevant, *J. Am. Chem. Soc.*, **81**, 2860 (1959).

211. J. A. Schafer and H. Morawetz, *J. Org. Chem.*, **28**, 1899 (1963).

212. M. L. Wolfrom, R. B. Bennett and J. D. Cram, *J. Am. Chem. Soc.*, **80**, 944 (1958).

213a. L. Zürn, *Ann. Chem.*, **631,** 56 (1960).
213b. T. C. Bruice and F.-H. Marquardt, *J. Am. Chem. Soc.*, 84,
213c. R. B. Martin, R. Hendrick and A. Parcell, *J. Org. Chem.*, 2
214. B. A. Cunningham and G. L. Schmir, *J. Am. Chem. Soc.*, 8
215. F. Nerdel, P. Weyerstahl and R. Dahl, *Ann. Chem.*, **716,** 12
216. T. C. Bruice and D. W. Tanner, *J. Org. Chem.*, **30,** 1668 (1
217. C. D. Hurd, M. F. Dull and K. E. Martin, *J. Am. Chem. Soc.*, 54, 1971
      (1932).
218a. R. Juavin, *Helv. Chim. Acta*, **35,** 1414 (1952).
218b. R. Juavin, M. B. Piccoli and T. Charlambous, *Helv. Chim. Acta*, **37,** 216
      (1954).
218c. D. Davidson and M. Karten, *J. Am. Chem. Soc.*, **78,** 1066 (1956).
218d. E. N. Zil'berman, A. E. Kulikova, N. M. Teplyakov and A. A. Rushin-
      skaya, *J. Gen. Chem. USSR* (*Eng. Transl.*), **32,** 2989 (1962).
219. W. P. Jencks and M. Gilchrist, *J. Am. Chem. Soc.*, **86,** 5616 (1964).
220. S. O. Eriksson, *Acta Chem. Scand.*, **22,** 892 (1968).
221. R. N. Ring, J. G. Sharefkin and D. Davidson, *J. Org. Chem.*, **27,** 2428
      (1962).
222a. V. Gold and E. G. Jefferson, *J. Chem. Soc.*, 1409 (1953).
222b. D. E. Koshland, *J. Am. Chem. Soc.*, **74,** 2286 (1952).
223. H. A. Staab, *Angew. Chem. Intern. Ed. Engl.*, **1,** 351 (1962).
224. S. L. Johnson, 'General Base and Nucleophilic Catalysis' in *Advances in
      Physical Organic Chemistry*, Vol. 5 (Ed. V. Gold), Academic Press, London,
      1967, p. 273.
225. P. T. Izzo and S. R. Safir, *J. Org. Chem.*, **24,** 701 (1959); F. Kuffner and
      W. Koechlin, *Monatsh. Chem.*, **93,** 476 (1962); H. E. Zaugg, R. W. Denet
      and M. Freifelder, *J. Am. Chem. Soc.*, **80,** 2773 (1958).
226. E. M. Kaiser, R. L. Vaux and C. R. Hauser, *J. Org. Chem.*, **32,** 3640
      (1967).
227a. E. E. Royals, *Advanced Organic Chemistry*, Prentice-Hall, New York,
      1954, p. 582.
227b. B. Castro, *Bull. Soc. Chim. France*, 1540 (1967).
228. J. G. Sharefkin, *Anal. Chem.*, **35,** 1616 (1963).
229. R. Lukeš, V. Dudek, O. Sedlakova and J. Koran, *Collection Czech. Chem.
      Commun.*, **26,** 1105 (1961); R. Lukes and A. Fabriova, *Collection Czech.
      Chem. Commun.*, **25,** 1618 (1960).
230. E. A. Braude and E. A. Evans, *J. Chem. Soc.*, 3334 (1955); E. A. Evans,
      J. Chem. Soc., 4691 (1956); E. A. Evans, *Chem. Ind.* (*London*), 1596 (1957);
      A. Zaitseva, E. M. Panov and K. A. Kocheshkov, *Izv. Akad. Nauk SSSR,
      Otd. Khim. Nauk*, 831 (1961); *Chem. Abstr.*, **55,** 23300 (1961).
231. R. Lukes and O. Cervinka, *Collection Czech. Chem. Commun.*, **26,** 1893
      (1961).

Author index

This author index is designed to enable the reader to locate an author's name and work with the aid of the reference numbers appearing in the text. The page numbers are printed in normal type in ascending numerical order, followed by the reference numbers in brackets. The numbers in *italics* refer to the pages on which the references are actually listed.

If reference is made to the work of the same author in different chapters, the above arrangement is repeated separately for each chapter.

859

29—c.o.a.

30—C.O.A.

Subject index